BASIC LAWS OF ARITHMETIC

Basic Laws of Arithmetic.

Derived using concept-script

BY

Gottlob Frege

VOLUMES I & II

TRANSLATED AND EDITED BY

Philip A. Ebert & Marcus Rossberg

WITH

Crispin Wright

AND THE ADVICE OF

Michael Beaney
Roy T. Cook
Gottfried Gabriel
Michael Hallett
Richard G. Heck, Jr.
Robert May
Eva Picardi
William Stirton
Christian Thiel
Kai F. Wehmeier

APPENDIX BY

Roy T. Cook

OXFORD
UNIVERSITY PRESS

OXFORD
UNIVERSITY PRESS

Great Clarendon Street, Oxford, OX2 6DP,
United Kingdom

Oxford University Press is a department of the University of Oxford.
It furthers the Universitys objective of excellence in research, scholarship,
and education by publishing worldwide. Oxford is a registered trade mark of
Oxford University Press in the UK and in certain other countries

© Philip A. Ebert, Marcus Rossberg, Crispin Wright, and Roy T. Cook 2013

The moral rights of the authors have been asserted

First Edition published in 2013

Impression: 2

All rights reserved. No part of this publication may be reproduced, stored in a retrieval system, or transmitted, in any form or by any means, without the prior permission in writing of Oxford University Press, or as expressly permitted by law, by licence or under terms agreed with the appropriate reprographics rights organization. Enquiries concerning reproduction outside the scope of the above should be sent to the Rights Department, Oxford University Press, at the address above

You must not circulate this work in any other form
and you must impose this same condition on any acquirer

British Library Cataloguing in Publication Data
Data available

Library of Congress Cataloging in Publication Data
Data available

ISBN 978–0–19–928174–9

As printed and bound by
CPI Group (UK) Ltd, Croydon, CR0 4YY

Links to third party websites are provided by Oxford in good faith
and for information only. Oxford disclaims any responsibility for the materials
contained in any third party website referenced in this work.

Contents

Foreword
Crispin Wright vii

Translators' Introduction
Philip A. Ebert and Marcus Rossberg xiii

Basic Laws of Arithmetic
Gottlob Frege

First volume

Foreword V
Table of contents XXVII
Introduction 1
Part I: Exposition of the concept-script 5
Part II: Proofs of the basic laws of cardinal number 70
Appendices 239

Second volume

Table of contents V
Part II: Proofs of the basic laws of cardinal number (continued) 1
Part III: The real numbers 69
Appendices 244
Afterword 253

Translators' Notes 269
Corrections 275
Bibliography 281

Appendix: How to Read *Grundgesetze*
Roy T. Cook A-1

Index I-1

Foreword

Crispin Wright

The importance of Frege's ideas within contemporary English-speaking analytical philosophy would be hard to exaggerate. Yet while two of Frege's three principal books—*Begriffsschrift* of 1879 and *Die Grundlagen der Arithmetik* of 1884—have been available in good quality English translation for many years, his most major work, *Grundgesetze der Arithmetik*, published in two volumes in 1893 and 1903, has never previously been completely translated into English.

Part of the explanation of the omission has no doubt been the long-standing impression that Frege's project in that book had completely failed. *Grundgesetze* was to have been the climax of his life's researches—a rigorous demonstration within the system of concept-script how the fundamental laws of the classical pure mathematics of the natural and real numbers could be derived from principles that were purely logical. Bertrand Russell's famous letter, received shortly before the publication of the second volume, disclosed that Basic Law V of Frege's system, governing identity for value-ranges, led to contradiction. A hastily proposed revision was later independently shown by a number of writers to be satisfiable only within singleton domains but Frege would himself have rapidly realised that it would not underwrite the proofs his project required, including crucially, of course, the derivation of the infinity of the series of natural numbers. Within a short space of time, Frege had apparently convinced himself that the problem was not one of infelicitous formulation but undermined his whole approach. Volume III of *Grundgesetze*, which was to contain the formal derivations of the theories of the real and (presumably) the complex numbers, never appeared.

Frege's own despairing assessment of his achievement prevailed for the first four-fifths of the twentieth century. Writing in 1973, no less an authority than Sir Michael Dummett remarked that Frege's philosophy of mathematics remained of "merely historical interest". Yet quite soon afterwards that assessment began to seem completely wrong. A sea change began with developments effected in work by the late George Boolos and, independently, by the present writer in the 1980s. The broad upshot of these researches was that the replacement of Basic Law V in the system of *Grundgesetze* by what came to be known as *Hume's Principle*, associating bijectable concepts with the same cardinal number, provided for, first, a consistent theory in which classical number theory could be developed and, second, a theory of considerable philosophical interest. Principles of the form shared by Basic Law V and Hume's Principle—principles licensing the transition from an equivalence relation on entities of a certain kind to identity statements configuring terms standing for objects of a new kind (the abstracts)—came to be known as *abstraction principles.* These have provided the central focus of study in a neo-Fregean (sometimes called neo-logicist, or "abstractionist") programme for the foundations for classical mathematics aiming at a broad subservience to Frege's own original philosophical objective: a demonstration that arithmetic and analysis flow from logic and definitions—with abstraction principles, so it is claimed, playing the definitional, or quasi-definitional role.

The last 30 years have seen considerable philosophical and technical advances in this programme, with much of the work being done by the core group and wider network of scholars associated with the AHRC-funded project on the Logical and Metaphysical Foundations of Classical Mathematics, which I had the good fortune to lead at the St Andrews philosophical research centre, Arché, from 2000 to 2005. On the technical side, it emerged that (presumptively) consistent abstractionist systems can be developed for theories which are at least interpretable as real and complex analysis, but that the development of a reasonably strong pure abstractionist set theory—strong enough, for instance, to recover the sets in the standard iterative hierarchy up to but excluding the first inaccessible cardinal—seems to present serious obstacles of principle. Philosophically, much interesting work has been done on a wide range of central questions, including the claim of at least some abstraction principles to constitute good implicit definitions of the concepts they introduce; the nature of their distinction from "bad" cases where no such claim can be upheld; the connection in general between implicit definition and basic a priori knowledge; the ramifications of the issues here for the inferentialist conception of the meanings of the logical constants developed in work of Gentzen and Prawitz; and, most generally, the metaphysics of abstract entities that belongs with abstractionism, and the resources it offers to address traditional empiricist (nominalist) scruples. All this has provided a vibrant research context, provoking major contributions from some of the world's most distinguished contemporary philosophers of mathematics, and with many of the best contributions—as so often in philosophy—being made, of course, from a critical standpoint.

By the early years of the millennium, a point had thus been reached when Frege's approach to foundations, and modern descendants of his ideas, had become a central preoccupation of much of the best work on the philosophy of mathematics and on associated metaphysical and epistemological issues. There was now, for the first time, a compelling argument for an authoritative complete English translation of *Grundgesetze* based purely on grounds of contemporary scholarly interests and need. Moreover with the advent of the LaTeX typesetting system, the formidable technical barriers confronting any attempt to republish a book containing large tracts of Frege's idiosyncratic logical notation had become surmountable.

Still, there remained another, potentially insuperable barrier: the sheer amount of labour, and sheer range of linguistic, philosophical and technical expertise, necessary to do justice to more than 500 pages of intricate formal proofs and conceptually challenging nineteenth century German prose. There are vanishingly few single scholars in the world with the range of competences demanded by such a project and, among that small number, still fewer who would be willing and able to take it on, in a scholarly milieu on both sides of the Atlantic that places a large and ever-increasing premium on personal research publication on top of the normal pressures exerted by teaching and administration on a typical professional scholar's working time.

However it occurred to me, about mid-way through the five years of the Logical and Metaphysical Foundations of Classical Mathematics project, that a perhaps unique opportunity had now ripened to accomplish the task by teamwork. In Philip Ebert and Marcus Rossberg, both then approaching completion of doctoral theses written in association with the AHRC project, the project team included two native German speakers who were fluent in English, who had trained intensively in Frege's philosophy over a period of years and were completely familiar with his philosophy of mathematics in particular. My own experience, before I took up philosophy, in the translation of Latin and Greek encouraged me to think that I could help with the accurate rendition of Frege's prose into decent philosophical English and with the development and maintenance of a suitable authorial "voice". Furthermore I knew we could count on the support of a group who believed absolutely in the importance of making Frege's legacy more fully available to the mostly non-German-speaking community of analytical philosophers worldwide with interests in the field. A small foreseeable surplus in the AHRC funding suggested that it might be possible to dedicate a couple of half-time research fellowships to the translation over the final year of the award—surely ample time to accomplish at least a complete basic translation...

It remained to persuade Philip and Marcus to take the project on. In fact, they needed no persuading. In the context, then as now, of an extremely bearish job market in philosophy and the pressing need to complete their doctorates and get some publications out if they were to compete effectively in it, their decision to do so impressed and continues to impress me as a remarkably courageous one.

So we launched the project in April 2003 using the tried and tested Arché seminar method, a small group receiving weekly small translation bulletins—no more than two to three pages of text—from Marcus and Philip and then working over them together, struggling with alternative interpretations, debating suggestions and agreeing revisions. The aim at this stage was merely to produce a passably accurate text in passably intelligible English. Initially our practice in the seminars was simply to come up with a list of suggestions, leaving the translators to collate them later and work them into the translation text. This had the drawbacks both that progress was excruciatingly slow and that we had little opportunity to get a sense of the collective effect of the changes being proposed or check for consistency of renditions of particular phrases. But things improved exponentially when Walter Pedriali assumed the role of court stenographer and armed with personal laser pointers, whose use somehow occasioned much hilarity, we took to editing the text of the translation online on the Arché seminar room smartboard. Even so, it rapidly became clear that there was no hope whatever of completing anything but a fragment of the task within the nine months or so for which we had AHRC funding. In February 2005 we therefore decided to submit an outline research grant application for a full two-years of further sponsorship, including fully funded 24-month postdoctoral fellowships for the translators, to the Leverhulme Trust. We subsequently received the go-ahead to submit a full application and learned of its success in December 2005. At that time, as now, the Trust's policy was to reward applications that prioritised the research vision of a principal investigator. Since I was not offering to lead any research, as usually conceived, and the research vision involved was entirely that of Frege, we had cause to be extremely grateful to the Trust for the flexibility it showed in responding positively to our application. It was the resulting crucial two-year period of Leverhulme support, providing the opportunity to devote several days a week to the translation, that gave us the opportunity to get real momentum into the project and hone the skills it required.

The backing of the Leverhulme Trust was one of two crucial factors that have enabled us to bring this work to completion. The other was the unstinting support of our group of international expert *Grundgesetze* enthusiasts who have given us continual aid, comfort, and advice throughout the process and to whom a great proportion of the credit is due if what we now publish measures up to the scholarly standards which Frege's flawed masterpiece deserves. Many of this group attended at three delightful Grundgesetze workshops, at St. Andrews in 2006, when we worked over the entire first draft translation of volume I, and again in 2008, where the entire draft translation of volume II was discussed, and then with the support of the NYU Philosophy Department, in New York in the spring of 2010, where selected passages from both volumes were subjected to yet another collaborative overhaul. I would like to express my thanks here to all who attended these workshops or sent us written comments at various stages of the translation: to Michael Beaney, Roy Cook, Gottfried Gabriel, Michael Hallett, Richard Heck, Michael Kremer, Wolfgang Künne, Robert May, Eva Picardi, Peter Simons, William Stirton, Peter Sullivan, Jamie Tappenden, Christian Thiel, and Kai Wehmeier. A collaboration on this scale takes a lot of logistical management and I would also like to record my gratitude to Sharon Coull for administrating the liaison with these colleagues and working so hard and ably to ensure that the workshops ran smoothly and enjoyably.

Three members of the wider team played especially important roles. William Stirton regularly travelled from Edinburgh to attend project seminars in the early days and subsequently provided an immense amount of help with the setting of the formulas and proof-reading the final files. Roy Cook offered expert advice on Frege's logic and proofs throughout the process, and has contributed the invaluable guide to Frege's logic and symbolism included here as an Appendix. Richard Heck enlisted his unsurpassed scholarship and philosophical understanding of Frege's writings to the aid of the project from the start, providing both intellectual and moral support in equal munificent measure.

At the end of the Leverhulme funding, the core group broke up, Marcus and Philip leaving Arché for posts at the Universities of Connecticut and Stirling, and I to work at New York University and to the launch of the Northern Institute of Philosophy at Aberdeen. The translation continued by Skype, with only occasional face-to-face meetings when the three of us were on the same continent. This was the period of "rants and frustrations" and "ruined weekends" that Marcus and Philip refer to below.

Now, however, at last, it is done, in the best Odyssean tradition almost ten years after we started. Two things are worth emphasising about how it has been done. First, the work has been, despite the uncertainties and delays, extraordinarily enjoyable. I have never previously, and don't expect again, to experience such a warmth of camaraderie around an academic project, and it has been a delight to work closely, over an extended period, with so many talented scholars whom I now number among my friends. Second, I believe that the whole process stands as a tribute to the mode of interactive, collaborative philosophical working which the long-term members of the Logical and Metaphysical Foundations of Classical Mathematics project—Marcus, Philip, Roy Cook, Nikolaj Jang Pedersen, Graham Priest, Agustín Rayo, Stewart Shapiro, Robbie Williams, and Elia Zardini—pioneered at Arché in the early years of the millennium and which has provided a model for a whole new wave of similarly inspired research centres across the UK and Europe. The *Grundgesetze* translation project was conceived and entered upon in the sprit of that paradigm, and conducted and completed with the energy, dedication, and good humour that it characteristically fosters. I think the result shines like a beacon as evidence of what can be accomplished by collaborative processes in philosophical research and scholarship, and will be a lasting resource to generations of students of Frege, and of the foundations of mathematics.

New York University and
The Northern Institute of Philosophy, Aberdeen *December 2012*

Translators' Introduction

The importance of *Grundgesetze der Arithmetik*

Gottlob Frege's *Grundgesetze der Arithmetik* was originally published in two volumes: the first in 1893, the second in 1903. It was to be the pinnacle of Frege's life's work. The aim was to demonstrate that arithmetic and analysis are reducible to logic—a position later called "Logicism". Frege's project began with the publication of *Begriffsschrift* in 1879, which contains the first version of his logical system, also named '*Begriffsschrift*' (concept-script), the mature formulation of which Frege would present in *Grundgesetze*. His work was groundbreaking in many ways: it contained the first occurrence of the quantifier in formal logic, including a treatment of second-order quantification; the first formal treatment of multiple and embedded generality; and it also offered the first formulation of a logical system containing relations. *Begriffsschrift* is widely acknowledged to constitute one of the greatest advances in logic since Aristotle. As W.V. Quine put it: "Logic is an old subject, and since 1879 it has been a great one" (Quine 1950, p. vii).

Begriffsschrift was followed by *Die Grundlagen der Arithmetik* in 1884. Having previously completed a manuscript of a more formal treatment of Logicism around 1882—a lost ancestor of *Grundgesetze*—Frege developed a philosophical foundation for his position in *Grundlagen*. It was likely on the advice of Carl Stumpf[1] that Frege decided to write a less technical and more accessible introduction to Logicism. *Grundlagen* is regarded by many as a philosophical masterpiece and by some, most notably Sir Michael Dummett, as "the most brilliant piece of philosophical writing of its length ever penned" (Auxier and Hahn, 2007, p. 9). The first part contains devastating criticisms of well-known approaches to the philosophy of arithmetic, including those of John Stuart Mill and Immanuel Kant, as well as those of Ernst Schröder, Hermann Hankel, and others. In the second part, Frege develops his Logicism. In *Grundlagen*, Frege eschews the use of a formal system, merely offering non-formal sketches of how a version of the Peano–Dedekind axioms for arithmetic can be derived from pure logic. He closes *Grundlagen* by suggesting that the Logicist approach is not restricted to arithmetic: it also has the potential to account for higher mathematics, in particular,

[1] Stumpf was a German philosopher and psychologist, a student of Franz Brentano and Hermann Lotze, who at the time was professor of philosophy in Prague. In a letter by Stumpf (September 9, 1882)—in response to a letter that Frege wrote to him (but, it seems, wrongly filed as a letter to Anton Marty who was Stumpf's colleague in Prague)—Frege is encouraged to spell out his Logicist programme in a more accessible manner, without recourse to his formal language. See Gabriel et al. (1976), p. 257, and the editor's note on p. 162.

real and complex analysis. At the end of *Grundlagen*, Frege was thus left with the monumental task properly to establish Logicism: he needed to identify a small number of basic laws of logic; offer a small number of indisputably sound rules of inference; and, finally, provide gapless proofs in his formal system of the basic laws of arithmetic, using only the identified basic laws and rules of logic together with suitable explicit definitions. This was to be the task of his *magnum opus*: *Grundgesetze der Arithmetik*.

The first volume of *Grundgesetze* is structured to reflect these three main tasks and includes a substantial philosophical foreword. In this foreword, Frege provides extensive criticism of psychological approaches to logic—in particular, against Benno Erdmann's *Logik* (1892)—and also offers an explanation of how his logical system has changed since 1879 and why the publication of *Grundgesetze* was delayed. Frege notes that a nearly completed manuscript was discarded in light of "a deep-reaching development" (vol. I, p. X) in his logical views, in particular his adoption of the now infamous Basic Law V. The remainder of the volume is divided into two parts. In part I, Frege introduces the language of concept-script, his modes of inference, his basic laws, and a number of important explicit definitions. Part II of *Grundgesetze* contains the proofs of the basic laws of cardinal number—i.e., the most important theorems of arithmetic proven within Frege's logical system—as well as Frege's treatment of recursion and countable infinity (see Heck (2012) for details). The proofs are contained in sections labelled "Construction". Each such section is preceded by a section called "Analysis" in which Frege outlines in non-technical prose the general proof strategy in order to facilitate understanding of the subsequent formal proofs. Volume I finishes rather abruptly. One may speculate that Frege was given a page limit by the publisher.

It was to be another ten years until the publication of volume II. It seamlessly continues where volume I left off: with no introduction, but merely a brief reminder of two theorems of volume I that had not previously been indexed for further use but which are employed in the subsequent proofs. Frege thus finishes part II of *Grundgesetze*, providing sixty-eight further pages of proofs of arithmetical theorems. Part III occupies the rest of the second volume. Following a strategy analogous to the one he uses in *Grundlagen*, Frege first provides strong, and in parts polemical, criticisms of alternative approaches to real analysis of many of his contemporaries. One particular focus is his criticism of (game) formalism, as advocated by Hermann Hankel and Frege's colleague in Jena, Johannes Thomae. The critical sections of part III are followed by a brief non-formal description of the approximate shape of Frege's approach to a logicist foundation of real analysis as based on the notion of "magnitude". (Frege points out that he will not "follow this path in every detail" (vol. II, p. 162).) Part III.2 contains the beginnings of a formal treatment of real analysis, and finishes with an outline of what remains to be accomplished at the close of volume II. Evidently Frege had plans for, or may even had already written parts of, a third volume in order to finish part III. Perhaps it was also to contain a part IV dedicated—as already noted in *Grundlagen*—to a logicist treatment of the complex numbers.[2]

---

[2] Whether and how much of the continuation of part III was written is unfortunately unclear. We may again speculate that Frege was, in turn, pushed to the publication of volume II by having exceeded a certain page limit. It seems natural to assume that he had some of the proofs that volume III was meant to contain. Indeed there is evidence that Frege worked on further aspects of his theory of irrational numbers—see Dummett (1991), p. 242, fn. 3. According to the inventory of Frege's *Nachlaß*, published in Veraart (1976), there were manuscripts that might have contained such material. Frege's *Nachlaß*, except for a small part that was transcribed by Heinrich Scholz

However, it was not to be. As is well known, in 1902 Frege received a letter from Bertrand Russell, when the second volume was already in press. In this letter, Russell proposes his famous antinomy, which made Frege realise that his Basic Law V, governing the identity of value-ranges, leads into inconsistency.[3] Frege discusses a revision to Basic Law V—which he labels: V′—in the afterword to volume II of *Grundgesetze*. However, it can be surmised that Frege himself realised (probably sometime after 1906) that V′ was unsuitable for his project—it is inconsistent with the assumption that there are at least two distinct objects.[4] Frege did not publish a third volume. Given the inconsistency of his Basic Law V and Frege's inability to find a suitable substitute that could be regarded as a basic law of logic, Frege gave up on Logicism late in his life.[5]

A recent resurgence of Logicism, and so-called Neo-Fregeanism, has again sparked interest in Frege's original writings and, in particular, in *Grundgesetze*. Despite the inconsistency of the formal system of *Grundgesetze*, Frege's proof strategy and technical accomplishment have come under renewed investigation. There is, for example, the recent "discovery" of so-called *Frege's Theorem*:[6] the proof that the axioms of arithmetic can be derived in second-order logic using *Hume's Principle*—a principle governing the identity of cardinal numbers: the number of Fs equals the number of Gs if, and only if, the Fs and the Gs are in one-to-one correspondence. Frege first introduced Hume's Principle in *Grundlagen* (see §§63 and 72) but rejected it as a foundation for arithmetic because of the infamous Julius Caesar Problem (*Grundlagen*, §§56 and 66): it cannot be decided on the basis of Hume's Principle alone, whether Julius Caesar (or any other object that is not given as a number) is identical to the number Two, say. In *Grundgesetze*, Frege proves both directions of Hume's Principle (vol. I, propositions (32), §65, and (49), §69; see also §38 where Frege mentions the principle). Once the two directions of Hume's Principle are proven Frege makes no further essential use of Basic Law V in the development of arithmetic.[7] There have also been attempts to offer revisions to Basic Law V, e.g., in Boolos (1998); lastly, there has been recent work on identifying consistent fragments of the logic of *Grundgesetze*.[8] Yet it is not just these formal aspects of *Grundgesetze* that called for a new engagement with Frege's *magnum opus*. The philosophical arguments in part III, in particular, Frege's account of definitions, his conception of Basic Law V, his critical assessment of contemporary accounts of real analysis, and Frege's own approach to real analysis have again become relevant to current debates in the philosophy of logic and mathematics. Given that only parts of *Grundgesetze* had previously been translated, the need for a complete English translation became more and more pressing.

(published as Frege (1983)), is presumed lost due to an airstrike towards the end of the second world war—however, see Wehmeier and Schmidt am Busch (2005).

[3] The antinomy Russell suggests is not well formed in Frege's system—in his response to Russell, Frege provides the correct formulation. See the Frege–Russell correspondence from 1902, in Gabriel et al. (1976), and, in particular, Russell (1902) and Frege (1902).

[4] See Quine (1955) and Cook (2013).

[5] See his posthumously published note written around 1924/25 in Frege (1924/25).

[6] It was published in Wright (1983). More recent presentations of the proof can be found in Boolos (1987) (discursive), Boolos (1990) (rigourous), Boolos (1995), Boolos (1996), and Zalta (2012). A very insightful discussion of Frege's Theorem, including its origins and recent (re-)discovery, can be found in Heck (2011).

[7] For an excellent discussion see Heck (2012).

[8] See e.g. Wehmeier (1999), Heck (2000), Fine (2002), and Burgess (2005).

Translating Frege's *Grundgesetze*

The first translation of parts of *Grundgesetze* was published in 1915 by Johann Stachelroth and Philip E. B. Jordain. Between 1915 and 1917, they published three articles in the journal *The Monist*, translating Frege's philosophical foreword of *Grundgesetze* (Frege, 1915, 1916), the introduction, and §§1–7 of the main text (Frege, 1917). In 1952, Peter Geach and Max Black published their edition *Translations from the Philosophical Writings of Gottlob Frege*, reprinting Stachelroth and Jourdain's translation of the foreword, as well as producing English translations of parts of volume II. More specifically, they offer the first translation of "Frege on Definitions I" (vol. II, §§56–67), "Frege on Definitions II" (vol. II, §§139–144, 146–147), "Frege against the Formalists" (vol. II, §§86–137),[9] and "Frege on Russell's Paradox" (vol. II, Afterword). Their book was substantially revised for its third edition, published in 1980, with changes made to the translation of '*Bedeutung*' and its cognates (see below). In 1964, Montgomery Furth provided the first full translation of part I of volume I of *Grundgesetze* with a detailed translator's introduction (Frege, 1964). After these translations went out of print, Michael Beaney published the *Frege Reader* (Beaney, 1997) which contains, amongst other essential works by Frege, selected passages from *Grundgesetze*. Perhaps not surprisingly, there is no uniformity in the translation of Frege's technical terms across the various translations. Controversy about how to translate, for example, the term '*Bedeutung*' led Beaney to leave the term untranslated.

Similar difficulties naturally also affected the present translation. Deciding how to translate technical terms was not always easy. We had the good fortune, however, to be able to draw on the advice of a team of experts who assisted us extensively throughout the process and provided invaluable feedback on important translation decisions. What follows are elucidations regarding our choices for the translation of some important technical terms and a more extensive glossary.

Anzahl

We translate the German '*Anzahl*' by 'cardinal number', and '*Zahl*' by 'number'. This is in contrast to Furth's choice to translate '*Anzahl*' using the capitalised 'Number' while translating '*Zahl*' as 'number'. 'Cardinal number' exactly covers Frege's intended technical use of '*Anzahl*'; by '*Zahl*', on the other hand, Frege intends a wider class of numbers, at least including the reals (compare his characterisation of the contrast between '*Anzahl*' and '*Zahl*' in vol. II, §157, p. 155–156). There are, however, two special cases worth mentioning. Firstly, Frege once uses '*Anzahl*' in a non-technical context where the English 'cardinal number' seems inappropriate. In that case only we use 'number' instead (vol. I, p. VI): "One must strive to reduce the number ['*Anzahl*'] of these fundamental laws as far as possible by proving everything that is provable." Secondly, throughout *Grundgesetze* Frege only once uses the German expression '*Nummern*' (vol. I, p. 70, §53) which we translate as 'number' as well. These occurrences are marked by translators' notes—all other occurrences of 'number' correspond to the German '*Zahl*'; all occurrences of 'cardinal number' correspond to the German '*Anzahl*'.

[9] This was previously published in Black (1950).

Aussage

The German '*Aussage*' is to be translated as 'predication'. For example, Frege's insight of §46 of *Grundlagen*, which he repeats in vol. I, p. IX, thus reads 'a statement of number contains a predication about a concept' in our translation. Austin uses 'assertion' in his translation of *Grundlagen*, but his choice is not quite correct. An '*Aussage*' in the sense relevant here can be made even when no assertion is involved, for example in questions, hypotheses, or in the antecedent of conditionals. There are a number of occurrences of '*Aussage*' that clearly indicate that Frege's use of the term is best captured by 'predication' and not 'assertion' (see for example vol. I, p. XXI).[10] We made an exception on p. XX, where we use 'Erdmann's statements' to capture the phrase '*Erdmanns Aussagen*'.

When it comes to the verb '*aussagen*', however, we did not always use 'to predicate'. For example, we translate '*das Prädicat* algebraisch *von einer Curve ausgesagt*' (vol. I, p. XIX) as 'the predicate *algebraic* as applied to a curve'—in addition to the obvious problem, 'the predicate *algebraic* as predicated' gives the false impression that cognates are used in the original. Moreover, in vol. II, §65, p. 76, and on some other occasions, we use the verb 'to say' for '*aussagen*' since 'to predicate' does not quite capture the intended meaning in these passages; to wit: "Consequently, one could truthfully say [*aussagen*] neither that it coincides with the reference of 'One' nor that it does not coincide with it."

Bedeutung, bedeuten, andeuten

Translators of Frege's writings face the difficulty of translating '*Bedeutung*' and its cognates '*bedeuten*', '*gleichbedeutend*', '*bedeutungsvoll*', etc. Furth uses 'denotation' (and cognates) in his translation of part I of *Grundgesetze*. Geach and Black's volume uses 'meaning' (and cognates) from the third edition onwards; in the first two editions, 'reference' is used for '*Bedeutung*' and 'stand for' or 'designate' for '*bedeuten*'. The changes to Geach and Black's third edition, however, were not implemented consistently which led to some confusion.[11] Beaney discusses at length the problems of the various options and ultimately decides to leave the German '*Bedeutung*' untranslated in his *Frege Reader*.[12] The main difficulty that Beaney, as well as Geach and Black, face is that they translate works that span Frege's entire career. That means, in particular, that they translate writings both before and after his 1892 article "*Über Sinn und Bedeutung*", in which he draws his famous, eponymous distinction. Given that after 1892 Frege takes the *Bedeutung* of a term to be the object referred to/denoted by that term, we decided to make this aspect clear and use 'reference' instead of Geach and Black's 'meaning'. The reason we decided to adopt 'reference' and not to follow Furth in using 'denotation' is three-fold: firstly, 'reference' as a translation of '*Bedeutung*' in Frege's writings after 1892 is better entrenched in the literature than 'denotation' (it is Frege's famous "sense/reference distinction"); secondly, 'denotation' has the

[10] See Dummett (1991), p. 88, and Künne (2009), p. 422, for further criticism of Austin's choice of translation.

[11] See in particular Beaney (1997), p. 46, fn. 106, for a discussion of the "unsystematic nature" of these changes.

[12] See his extensive discussion on the difficulty of translating '*Bedeutung*' on pp. 36–46 in Beaney (1997).

ring of an artificial technical term that both 'reference' and '*Bedeutung*' lack; thirdly, some of the cognates of '*Bedeutung*' are more easily translated using cognates of 'reference'. '*Gleichbedeutend*', for example, can easily be translated as 'co-referential', while 'co-denotational' seems somewhat unnatural;[13] the same holds of the triple '*bedeutungsvoll*', 'referential', 'denotational'.

There are two further important decisions we made in our translation that relate to '*Bedeutung*'. Firstly, when the term '*Bedeutung*' occurs in works of other authors such as Peano (vol. II, §58, fn. 1) or Thomae ('*formale Bedeutung*', vol. II, §97), we decided to opt for uniformity and use 'reference' ('formal reference' for Thomae) in our translation. The main reason is that Frege discusses these quotations in his own text and here too uses the term '*Bedeutung*'. It would seem awkward to have Frege use 'meaning', or whatever alternative translation may be used for '*Bedeutung*', in these passages. Moreover, it would obscure the fact that Frege takes the quoted author's use of '*Bedeutung*' to be in line with his specific use of '*Bedeutung*'. As a result, whenever the term 'reference' (and cognates) occurs in our translation it will correspond to the German '*Bedeutung*' (and cognates), whether it is Frege's writings or the writings of other authors' Frege is quoting. Indeed, our translation of '*Bedeutung*' is single-valued in both directions: no other English term is used as translation of '*Bedeutung*' and *vice versa.* Thus, although 'mean' occurs in our translation, it never occurs as a translation of '*bedeuten*'; instead, it translates '*heissen*' or '*meinen*'. Lastly, Frege uses '*andeuten*' to describe the function of Roman letters in order to draw a distinction between a name that *refers to* (*bedeutet*) an object, and a Roman letter (i.e., one of his devices for generality) that merely *indicates* (*andeutet*) an object. Unfortunately, the fact that 'be*deuten*' and 'an*deuten*' have the same stem is lost in translation.

Begriffsschrift

We translate the German '*Begriffsschrift*' as 'concept-script' when Frege is referring to his system of logic or the formal language it is formulated in but we leave the term untranslated when Frege refers to his 1879 book *Begriffsschrift.* Throughout our translation, we use original titles of Frege's works in italics, e.g., '*Die Grundlagen der Arithmetik*'; the same applies to titles of works by other writers that Frege is citing.

Begründung, Grundlage

The German word '*Begründung*' presents a difficulty. It can be translated either as 'foundation', or even 'basis' (see vol. II, p. 154), to indicate that Frege is offering a foundation for a theory, for example, for arithmetic; or it can be used to state that a justification is provided for a certain principle without necessarily also providing a foundation. As a result, we translate '*Begründung*' as 'foundation'/'basis' or 'justification', depending on context. A further complication arises, given that we also use 'foundation' as a translation of '*Grundlage*': to provide a *Grundlage* is to provide a *Begründung* in the foundational sense noted above.

[13] Furth translates '*gleichbedeutend*' using either 'has the same denotation' or 'becomes the same in meaning' (see e.g. vol. I, §27, p. 45) suggesting that Frege is not using the term in the technical sense in the latter passage. We disagree with Furth's assessment and use 'co-referential' here as elsewhere.

Bestimmung, bestimmen

We translate '*Bestimmung*' (and cognates) either as 'determination' (and cognates) or as 'specification' (and cognates). In particular in passages where confusion is possible—e.g., when 'determination' could be misunderstood as 'resolution' or 'willpower'—we use 'specification' (and cognates), but we also do so where 'specify' is significantly more idiomatic than 'determine'.

Definition, Erklärung, Erläuterung

Most translators have not distinguished between Frege's uses of the terms '*Erklärung*' and '*Definition*', but use the English word 'definition' as a translation for both. Admittedly, Frege sometimes uses the verb '*erklären*' or the noun '*Erklärung*' to describe his definitions. However, there are other occurences of these terms where he does not intend to give a definition (cf. vol. I, §8, p. 12). There is no satisfactory account of what precisely distinguishes a Fregean '*Erklärung*' from a '*Definition*' and how they relate—nevertheless, Frege uses different words in the original text, and we decided to respect the distinction in our translation. Thus, in our translation, 'definition', 'define', always stand for '*Definition*', '*definieren*', respectively (and *vice versa*), while nearly all occurrences of '*Erklärung*', '*erklären*' are translated as 'explanation', 'explain'. The only exceptions appear in two passages (on p. XIII and in §6, vol. I) where we use 'to declare' as a translation of '*erklären*', as Frege here uses the verb in this sense (compare '*Erklärung der Menschenrechte*': 'declaration of human rights').

It is worth noting that both *Definition* and *Erklärung* are distinct from a Fregean *Erläuterung*, for which we choose the term 'elucidation'. An elucidation may be provided to assist the understanding of a definition or explanation: it is a mere heuristics that is, strictly speaking, inessential and can be imprecise without thereby undermining the correctness and precision of a proof that uses the technical notion (compare vol. I, §§34–35.) The purpose of an elucidations is simply to aid the reader and point in the right direction until the definition proper is given or when the technical notion in question is primitive (and thus governed by a basic law).

Lastly, we should note that both '*Definition*' as well as '*Definieren*'—the latter being a substantivised verb referring to the act of giving definitions—are translated as 'definition'. Only in cases in which the context does not clearly disambiguate between those two uses of 'definition', we use the gerund 'defining' as a translation of '*Definieren*' (see, for instance, vol. I, p. XXV: 'mathematician's defining').

Eindeutigkeit

We translate '*eindeutige Beziehung*' as 'single-valued relation', and likewise '*eindeutige Zuordnung*' as 'single-valued correlation'. In contrast, Furth uses 'many-one' (as in 'many-one relation' and 'many-one correspondence'). The phrase '*eindeutig bestimmt*', however, is translated as 'uniquely determined', as we wish to avoid the impossible phrase 'single-valuedly determined'. This should not lead to any misunderstandings, as 'unique relation' instead of 'single-valued relation' would. '*Beiderseits eindeutig*' is translated as 'single-valued in both directions' instead of the phrase 'one to one' that Austin uses in his translation of *Grundlagen*. We avoid introducing a new technical term where the same technical term is reused in the original.

Erkenntnis

'*Erkenntnis*' is standardly translated as 'knowledge' or 'cognition'. There are well-known difficulties with the translation of '*Erkenntnis*' in Kant (think, in particular, of the difference between '*Wissen*' and '*Erkenntnis*') as well as in other writings, such as Carnap's. One problem is that '*Erkenntnis*' can be used to indicate a certain process of coming to know—as in '*Erkenntnisthat*' ('act of cognition', vol. I, p. VII)—or the result of this process—'*Als Ziel muss die Erkenntnis dastehen*' ('Knowledge must stand as the goal', vol. II, p. 101). We thus did not translate '*Erkenntnis*' uniformly: we use 'knowledge' or 'cognition', and once 'insight' (vol. II, p. 85), depending on context. To allow the reader to track Frege's use of '*Erkenntnis*' (and to distinguish it from occurrences of 'knowledge' as a translation of '*Wissen*', as, e.g., in vol. II, §56), we mark each occurrence of a translation of '*Erkenntnis*' with a translators' note.

Festsetzung, festsetzen

There are two exceptions to our translation of '*Festsetzung*' and '*festsetzen*' as 'stipulation' and 'stipulate'. In the foreword, p. XV, we use 'legislate' to reflect a legal connotation of '*festsetzen*'. In this passage, Frege is discussing the normative aspect of laws of thought and the extent to which they 'legislate' ('*festsetzen*') how one ought to think. Given that laws do not, strictly speaking, 'stipulate' anything in the sense in which 'stipulate' is used elsewhere in *Grundgesetze*, Frege's use of '*festsetzen*' here is better captured by 'legislate'. In vol. II, p. 94, §83, we use 'fix' instead of 'stipulate' as a translation of '*festsetzen*': 'These definitions can fix the references of the new words and signs with at least the same right as [...]'. Using 'stipulate' here could give the impression that Frege intends to stipulate objects into existence, while the German original clearly does not invite this reading.

Ganze Zahl

'Integer' is usually chosen as a translation of '*ganze Zahl*', as, e.g., in Geach and Black's translation. In vol. II, §101, however, Frege quotes Thomae, who uses '*ganze Zahl*' intending the positive integers only. In other passages, Frege uses '*ganze Zahlen*' to encompass all integers (compare vol. II, §57, where Frege speaks of positive as well as negative *ganze Zahlen*). We chose 'whole number' as a translation since it has a similar ambiguity in English, despite some preference amongst mathematicians to use the term to designate only the positive integers. A further advantage is that 'number' is retained as a constituent of the expression 'whole number', as it is the case with '*Zahl*' and '*ganze Zahl*'.

Gleichheit, Gleichung, gleich, Identität

All occurrences of '*Identität*', '*Gleichheit*', and '*Gleichung*' are translated as 'identity', 'equality', and 'equation', respectively. The adjective '*gleich*' is usually translated as 'equal', with some exceptions: for example, in vol. I, p. 62b, we use 'common', vol. I, p. 63b, 'the same as', and vol. II, p. 73, 'same'.

Relation

We translate '*Beziehung*' as 'relation' and the German term '*Relation*' using the capitalised English term 'Relation'. '*Relation*' is defined as '*Umfang einer Beziehung*' (extension of a relation). There simply is no other suitable English word apart from 'relation' that we could use as a translation of the German '*Relation*'—or indeed of '*Beziehung*'. We had no choice but to resort to the technique of capitalising an already used expression to capture an important difference in the original. The alternative, employed by Furth, as well as Geach and Black, in their translations of the afterword, is to use 'extension of a relation' as a translation of the German '*Relation*'. This might be acceptable given that the phrase only occurs twice in the afterword, but it is not a viable option for part III, where Frege introduces '*Relation*' as a technical term for 'extension of a relation' and uses it abundantly.

Satz

The translation of '*Satz*' was subject to much discussion with our advisors (see *Acknowledgments* below). Stachelroth and Jordain do not opt for a consistent translation and use 'theorem' as well as 'proposition'. Furth uses 'proposition' throughout; Geach and Black sometimes use 'sentence' and sometimes 'proposition'. We initially translated '*Satz*' as 'sentence', partly motivated by Frege's introduction of the term '*Satz*' in vol. I, §5, as short for '*Begriffsschriftsatz*'. He introduces the term to refer to the concept-script *representation* of a judgement—i.e., the judgement-stroke followed by a formula in concept-script notation—and in doing so gives '*Satz*' a decidedly syntactic flavour. Moreover, Frege's own index to the first volume of *Grundgesetze* contains '*Satz*' and refers to pages 9 and 44, and thus to passages where Frege describes *Sätze* as signs (*Zeichen*).

However, after much discussion, we decided to follow Furth and use 'proposition' rather than 'sentence' throughout volume I and II: it better captures the numerous ways in which Frege uses '*Satz*' while allowing for a uniform translation of the German term. Certain phrases—'sense of a proposition', or 'the proposition "Scylla had six dragon gullets"'—sound somewhat unusual to a modern ear, but readers should quickly get used to this type of phrasing. Obviously, Frege does not intend the modern sense of 'proposition' (i.e., what is expressed by a sentence), as he has his own terminology for objects of this sort: what a sentence expresses for Frege is its sense—that is, a *thought*.

In general, it seems that Frege's conception of '*Satz*' is something that essentially possesses both syntactic and semantic properties; nowadays, we might call this an 'interpreted sentence'. Yet, for Frege, a *Satz* is not something that could be characterised as a sentence that is interpreted, because this would suggest the possibility of an uninterpreted sentence or the reinterpretation of a sentence, which he dismisses.[14]

As Furth eloquently writes in his translator's introduction (Frege (1964), pp. lv–lvi),

> [t]he rendering of '*Satz*' presents great difficulties. In various contexts it can mean sentence, theorem, proposition, clause. In some of Frege's other writings (of 1891 and later) he uses it for that variety of expression (name) which, for him, denotes a truth-value; in such cases 'sentence'

[14] See, in particular, the so-called *Frege–Hilbert debate* in Gabriel et al. (1976), XV/3–9.

> would be appropriate. In this work however, '*Satz*' is almost without exception applied to expressions *with a judgment-stroke prefixed* and [...] such expressions, for Frege, are *not* names. 'Assertion' might then be considered, yet it seems that '*Satz*' ought to be rendered differently from '*Behauptung*'. 'Theorem' is ruled out as both too wide and too narrow: too narrow because Frege applies '*Satz*' quite generally, and not merely to theorems of his logical theory; too wide because, for example, in his discussion of the Russell paradox where he shows that a self-contradictory statement can in fact be derived from the axioms, Frege gives the derivation informally and does *not* prefix the suspect expressions with the judgment-stroke, apparently on the ground that although they are indeed (unhappily) theorems, he does not believe that they are *true*. Thus we are forced onto 'proposition'. Some later writers have used this word for the *sense* expressed by a sentence, Frege's *Gedanke*. Therefore the reader must take care here to understand "proposition" in something nearer to its vague English meaning of a 'propounding'. The situation is unsatisfactory, but Frege has left the translator little choice.

We should remark, however, that '*Satz*' also occurs in compound nouns such as '*Behauptungssatz*', '*Lehrsatz*', and '*Bedingungssatz*' (e.g., vol. II, §§65, 140). We translate these phrases as 'declarative sentence', 'theorem', and 'conditional clause', respectively.

Selbstverständlich, einleuchten

We translate '*selbstverständlich*' as 'self-evident' when it is used in a technical epistemic sense, and as 'evidently' when it is used colloquially. Further problems are raised by the translation of '*einleuchten*'. In contrast to other scholars,[15] we do not think the term is best rendered using 'self-evident' as well. Frege uses the term to indicate that something, such as a proof, is easily understood or accepted. For example, in vol. I, p. VIII, Frege writes: "*Man begnügt sich ja meistens damit, dass jeder Schritt im Beweise als richtig einleuchte*", which we render as "Mostly, no doubt, one contents oneself with the obvious correctness of each step in a proof" (similarly, on p. 1, vol. I., which is the only other occurrence of the term in volume I). In volume II, the phrase is used most prominently in §156, p. 154: "*Dass kein Widerspruch bestehe—meinen nun wohl Manche—leuchte unmittelbar ein, da* [...]", which we translate as "That there is no contradiction—some may now claim—is immediately obvious since [...]". Clearly, using 'self-evident' here would not express what Frege intends. Perhaps most importantly, Frege uses the term in the afterword when he writes about Basic Law V: "*Ich habe mir nie verhehlt, dass es nicht so einleuchtend ist, wie die andern, und wie es eigentlich von einem logischen Gesetze verlangt werden muss*", which we translate as "I have never concealed from myself that it is not as obvious as the others nor as obvious as must properly be required of a logical law". The only exception to our choice of translation occurs in §140, p. 142: "*die Gesetze in einleuchtender Weise zu entwickeln*" is translated as "a lucid development of the laws"—"obvious development" would be misleading.

---

[15] See for example Jeshion (2001).

vertreten

Frege uses '*vertreten*' in a formal context in relation to small Greek letters, i.e., such a letter *vertritt* an argument place. In such contexts, we use 'proxy for' as our translation. Frege also writes that functions can be "*vertreten durch ihre Werthverläufe*", a phrase we translate as 'represented by their value-range' (compare vol. I, §25). We here follow Furth in making this distinction. When Frege uses '*vertreten*' in a less formal context we use either 'stand in for' or, where appropriate, 'represent' (in particular, in the discussion of Thomae in vol. II, §§131–132). It is worth noting that we also use 'represent' as a translation of '*darstellen*'. In context, this conflation will not lead to any misunderstandings.

Vorhanden sein, Bestand haben, existieren

There is a temptation to use 'exist' to translate '*vorhanden sein*' or the phrase '*Bestand haben*'. However, since 'exists' is a second-level predicate according to Frege, expressions like 'a thing exists' are, strictly speaking, ill-formed. Moreover, the fact that Frege uses the German '*bestehen*' or '*vorhanden sein*' rather than '*existieren*' on many occasions is something we wanted our translation to reflect. We thus treat '*existieren*' (and cognates) as a technical term to be translated as 'exist' (and cognates). For the German '*vorhanden*' we sometimes use 'is present' or simply the verb 'to be', depending on context. In the case of '*Bestand haben*' (e.g., vol. I, p. XXIV) we use the phrase 'has being'. We also note that the German phrase '*Bestand ausmachen*' is translated as 'constitute'.

In vol. II, §155, p. 153, fn. 2, Frege refers to Kant's criticism of the ontological argument for the existence of God. Here, the German phrase we translate as 'existence of God' is not '*Existenz Gottes*', but rather '*Dasein Gottes*'.

Vorstellung

'*Vorstellung*' presents another notorious difficulty when translating German texts into English. A natural choice is 'idea' but often a '*Vorstellung*' is a '*Vorstellung*' of something, which has led some translators to use 'representation' instead. The latter has an additional advantage: the verb '*vorstellen*' can be translated as 'represent', allowing cognates to be retained in the translation.

But this immediately gives rise to a difficulty: the German '*vertreten*' too is naturally translated as 'represent' (see above). Moreover, a '*Vorstellung*' is always subjective and requires a personal bearer; it is not clear that 'representation' carries this connotation as strongly as 'idea' does, if at all. An option may be 'mental representation', if it was not for the somewhat technical feel. Also, the noun '*das Vorstellen*' cannot easily be rendered using cognates of 'represent' (only the cumbersome 'the act of representing' could be entertained).

We opt for 'idea' and use the (admittedly uncommon) verb 'to ideate' as the translation of '*vorstellen*'.[16] Also, we translate '*Vorstellen*' as 'ideation' and '*das Vorgestelle*' as 'what is ideated', thereby respecting in our translation the fact that

[16] It is worth noting that 'ideate' was not "created" by Furth for his translation, as sometimes suggested. In fact, the verb 'to ideate' has been in use since the late 17th century and can be found in the Oxford English Dictionary.

the terms are cognates in German. These terms occur in the introduction to volume I, where Frege is criticising Benno Erdmann—the Erdmann quotes proved to be some of the most difficult passages to translate in all of *Grundgesetze.* Erdmann's arguments for strongly idealist conclusions provide further reason to use 'idea': it carries the right connotation.

der Wahrheitswerth davon, dass

The phrase '*der Wahrheitswerth davon, dass*' recurs on numerous occasions. For instance, in vol. I, §5, p. 10, Frege writes:

> *Wir können also sagen, dass*
>
> $$\Delta = (-\!\!\!-\!\!\!-\, \Delta)$$
>
> *der Wahrheitswerth davon ist, dass* Δ *ein Wahrheitswerth sei.*

Furth does not translate the phrase uniformly. Often, he adopts a gerundial construction. For instance, he translates the sentence above as:

> We can therefore say that
>
> $$\Delta = (-\!\!\!-\!\!\!-\, \Delta)$$
>
> is the truth-value of Δ's *being a truth-value.*

This, however, generates a number of difficulties. Firstly, there are stylistic reasons why this construction should be avoided. For example, vol. I, §8, p. 13, Furth translates:

> But if we want to designate the truth-value of the function
>
> $$(\xi + \xi = 2\,.\,\xi) = (\overset{\mathfrak{a}}{\smile}\, \xi = \mathfrak{a})\text{'s}$$
>
> *having the True as value for every argument,* then [...]

The possessive 's' appended to a formula is probably best avoided in any case; but here the gerund does not aid comprehension either. Secondly, this option requires the introduction of italics where there are none in the original. Finally, a gerund construction seems to refer to a concept, rather than a proposition, and so it would not normally be the bearer of a truth-value. As a result, we experimented with a number of alternatives: among them, 'is the truth-value of the proposition that ...'—which would have required the addition of 'proposition'—and Alonzo Church's 'is the truth-value thereof that ...' (as used in Church (1951), p. 108) which barely sounds English. In the end, we opted for the construction 'the truth-value of: that ...'. This adds a colon where there is none in the original, but it ensures making clear what the truth-value is a truth-value of, namely the ensuing proposition. Hence, the above two occurrences are now translated:

> We can accordingly say that
>
> $$\Delta = (-\!\!\!-\!\!\!-\, \Delta)$$
>
> is the truth-value of: that Δ is a truth value.

and

> If, however, one wants to designate the truth-value of: that the function
>
> $$(\xi + \xi = 2\,.\,\xi) = (\text{-}\!\overset{\mathfrak{a}}{\smile}\!\text{-}\ \xi = \mathfrak{a})$$
>
> has the True as value for every argument, then [. . .]

Werthverlauf

We decided to translate '*Werthverlauf*' and '*Werthverläufe*' using the English 'value-range' and 'value-ranges'. Furth uses the unwieldy 'course(s)-of-values'. Recent Frege scholarship also seems to prefer 'value-range'.[17] Moreover, we translate '*rechter Werthverlaufsname*' (vol. I, §31, p. 49) as 'regular value-range name' instead of 'fair course-of-values-name' (Furth).

Wortsprache

Literally, '*Wortsprache*' would be 'word-language', in contrast to an "artificial" symbolic language, and, in particular, to Frege's concept-script. Accordingly, 'natural language' might seem to be an option, but Esperanto, for instance, would count as a *Wortsprache*, while it is not a natural language. We thus translate '*Wortsprache*' as 'ordinary language'. While one may hesitate to call Esperanto an "ordinary" language, it is less jarring than calling it a "natural" language. Note that we also translate '*gewöhnliche Sprache*' as 'ordinary language'. Our translation is less than ideal; however, the text gains in readability. Nevertheless, note that some aspects of the sense of '*Wortsprache*' may be lost in its translation as 'ordinary language'.

Zeichen, bezeichnen

We generally use 'sign' rather than 'symbol' as a translation of '*Zeichen*'. This choice is in part motivated by our preference for using English cognates where cognates are used in the German original. We translate '*Bezeichnen*' as 'designate' and '*Bezeichnung*' as 'designation' (although sometimes the use of 'notation' was unavoidable), and hence 'sign' is the preferred translation for most occurrences. However, Frege also uses the plural '*Zeichen*' in phrases like '*in Zeichen*' or '*mit meinen/unseren Zeichen*' to indicate in what follows the thought in question is expressed in concept-script. We translate 'in signs' or 'in my/our symbolism', depending on the context (see for example vol. I, p. V).

Zuordnung, zuordnen

In part III, Frege discusses the theories of the real numbers by Cantor and by the formalists. According to these approaches, a certain real number is *assigned* to a series or sequence of numbers. The German word we translate as 'assign' here is '*zuordnen*'; indeed, we translate all occurrences of '*zuordnen*' as 'assign', and all occurrences of '*Zuordnung*' as 'assignment' in part III. Using 'correlate' and 'correlation' as a translation in these contexts would be inappropriate because '*Zuordnung*' here points

[17] See, for example, the recent Potter and Ricketts (2010).

to a priority—numbers are assigned to certain series or sequences and are thereby defined. This sense is better captured by 'assignment' than 'correlation'. However, in parts I and II, Frege uses '*zuordnen*' and '*Zuordnung*' in the context of his own theory, and we use 'correlate' and 'correlation' as translations. No priority is suggested here. Rather, objects falling under a certain concept are correlated (*zugeordnet*) with objects falling under a different concept. 'Correlation' rather than 'assignment' is here the appropriate translation of '*Zuordnung*'.

Note that 'assign' occurs on a number of occasions in vol. I, but only as a translation of German words other than '*zuordnen*'—for example, in §14, p. 25: "assign a label [to a proposition]" ("[*einem Satz*] *ein Abzeichen geben*"). This should not lead to any serious misunderstandings.

Glossary of technical terms

Allgemeingültigkeit	general validity
Allgemeingewissheit	general certainty
andeuten (*unbestimmt*)	indicate (indeterminately)
Anzahl	cardinal number
Anzahlreihe	cardinal number series
Aussagc	prcdication
Bedeutung	reference
bedeutungslos	without reference
bedeutungsvoll	referential
Bedingungsstrich	conditional stroke
Begriff	concept
Begriffsschrift [book]	*Begriffsschrift*
Begriffsschrift [system, language]	concept-script
bestimmen	determine, specify
bezeichnen	designate
Bezeichnung	designation, notation
Beziehung	relation
Definition	definition
Definitionsdoppelstrich	double-stroke of definition
Definitionsstrich	definition-stroke
deutscher Buchstabe	German letter
Doppelwerthverlauf	double value-range
eindeutig(*e Beziehung*)	single-valued (relation)
eindeutig bestimmt	uniquely determined
Eigenname	proper name
endlich	finite
Endlos	Endlos
endlos fortlaufen	proceed endlessly
erfüllen	fill in, instantiate (for concepts)
ergänzen	complete (for argument-places)
Erkenntnis	knowledge, cognition, insight
Erklärung	explanation
Erläuterung	elucidation

festsetzen	stipulate, fix
Festsetzung	stipulation
Folge	sequence
folgen (in einer Reihe)	following (in a series)
Forderungssatz	postulate
formale Arithmetik	formal arithmetic
Function	function
Fürwahrhalten	taking to be true
ganze Zahl	whole number
Gegenstand	object
gekoppelt(e Beziehung)	coupled (relation)
gesättigt	saturated
Gestalt, gleichgestaltet	shape, equal-shaped
gleichbedeutend	co-referential
Gleichheit	equality
Gleichung	equation
gleichstufig(e Beziehung)	equal-levelled (relation)
gleichzahlig	equinumerous
griechischer Vokalbuchstabe	Greek vowel
Grösse	magnitude
Grössengebiet	domain of magnitudes
Grössenverhältnis	magnitude-ratio
Höhlung	concavity
Inhalt	content
inhaltliche Arithmetik	contentual arithmetic
Identität	identity
Kennzeichen (zur Wiedererkennung)	criterion (for recognition)
Klasse	class
lateinischer Buchstabe	Roman letter
Lehre	theory
Marke (Functions-, Gegenstands-)	marker (function-, object-)
Maasszahl	measuring number
Menge	collection, set
Merkmal	characteristic mark
Null	Zero
Nullrelation	null Relation
Oberglied	supercomponent
Positivalklasse	positival class
Positivklasse	positive class
Quantität	quantity
Relation	Relation [capitalised]
Reihe	series
reihende Beziehung	series-forming relation
Satz	proposition
Sinn	sense
Spiritus lenis	smooth breathing
Stufe	level
Theorie	theory

übergeordnet	superordinate
Umfang	extension
Umkehrung (einer Beziehung)	converse (of a relation)
unendlich	infinite
ungesättigt	unsaturated
ungleichstufig(e Beziehung)	unequal-levelled (relation)
unscharf begrenzt	without sharp boundaries
untergeordnet	subordinate
Unterglied	subcomponent
Urtheil	judgement
Urtheilsstrich	judgement-stroke
vertreten	represent, stand in for
Verneinungsstrich	negation-stroke
Verschmelzung	fusion
Vorstellung	idea
Wendung	contraposition
Werthverlauf	value-range
Wortsprache	ordinary language
Zahl	number
Zahlangabe	statement of number
Zahlgrösse, Zahlengrösse	numerical magnitude
Zahlzeichen	number-sign
Zeichen	sign
zugehörige Function	corresponding function
zuordnen	correlate, assign
zusammengesetzte Beziehung	composite relation
zusammengesetzter Name	complex name
zusammensetzen (Beziehungen)	compose (relations)
Zwischenzeichen	transition-sign

General remarks on the translation

We follow the original pagination of Frege's *Grundgesetze*, and we also respect Frege's use of columns. Given that our translation usually differs in length from the original, some pages contain more text than others, and sometimes there is a blank space between main text and footnotes. This minor cosmetic oddity is outweighed by the benefits of respecting the original pagination. We also follow Frege's original numbering of footnotes. As in the original, Frege's notes are listed at the bottom of the page and are numbered using arabic numerals, relative to page or indeed column. Translators' notes, in contrast, are indicated by small Roman letters and appear as endnotes. We corrected obvious and minor typos in the original without explicitly acknowledging it. Substantial typographical errors are corrected and listed in a section labelled "Corrections" which is appended to Frege's text. We made use of Frege's own *corrigenda* to both volumes, and also those suggested by Christian Thiel in a recent edition of *Grundgesetze* (Frege, 1998); moreover, we added a number of corrections. Two corrections suggested by Scholz and Bachmann which Thiel mentions have been omitted: they are incorrect.

We do not translate the titles of books or articles that Frege cites—with only one exception. In a footnote on p. 106, vol. II, Frege quotes the title of an article, and we translate the quotation into English. Frege is not merely referencing the article; the title contains the phrase "a function of given *letters*" ("*eine Function gegebener* Buchstaben"), which Frege uses as evidence to support his claim that the confusion of signs and their reference is rampant among the mathematicians of his time. The point would be lost if the title remained untranslated.

We added a bibliography containing all publications Frege cites, providing complete bibliographical information and, where available, references to English translations of these works. The translations of passages that Frege quotes from other writers are our own, regardless of whether English translations of these works exist.

We followed a principle of exegetical neutrality. As a result, our translation is usually close to the original. Technical terms are translated uniformly and not translated away. Passages that are ambiguous or otherwise unclear are often purposefully translated so as to retain the unclarities. We did not attempt to "improve" on the original; our goal was to translate the text so that it is suitable for scholarly work.

We made use of a number of dictionaries for our translation, in particular: Langenscheidt's *Fachwörterbuch Mathematik: Englisch–Deutsch–Französisch–Russisch*, fourth edition, 1996, for mathematical terminology. We drew on *The New and Complete Dictionary of the German and English Languages*, by Johann Ebers, published in 1796, H. E. Lloyd and G. H. Nöhden's *New Dictionary of the English and German Languages*, 1836, Georg W. Mentz's *New English–German and German–English Dictionary Containing All the Words in General Use*, 1841, and *Cassell's German and English Dictionary* by Karl Breul, 1909, to provide a better idea of 19th-century usage of German and English words. We consulted the *Oxford English Dictionary*, the *Oxford Thesaurus for English*, the *Oxford German Dictionary*, the *Langenscheidt English–German Dictionary*, the Macintosh OS X dictionary, and various online resources such as the *LEO German–English Dictionary* (`http://dict.leo.org`).

Typesetting *Grundgesetze*

The question might naturally arise: since we translated *Grundgesetze* prose into English, why did we not also "translate" Frege's formulae into modern notation? Early on in the project, we briefly considered such a change but decided against it. There are several reasons that decisively speak against such an endeavour.

Firstly, Frege elucidates his formalism in the first forty-six sections of volume I, and in doing so mentions (rather than uses) his notation and describes the formulae and their components. A rendering of his formalism in modern notation would have made these passages nonsensical. Keeping Frege's formalism up to §46 of the first volume and changing over to modern notation thereafter was obviously not an option. Rewriting the prose in which Frege mentions and describes his formalism—e.g., exchanging 'subcomponent' with 'antecedent'—would not have led to a translation suitable for scholarly purposes, and thus was not an option either.[18]

---

[18] This last strategy was adopted in a new German edition of *Grundgesetze*, see Frege (2009), whose purpose is different from ours.

Moreover, transforming Frege's notation into a more familiar formalism would generate the need for numerous parentheses which would hinder readability. For instance, the fairly easily readable proposition (25) of vol. I, p. 83:

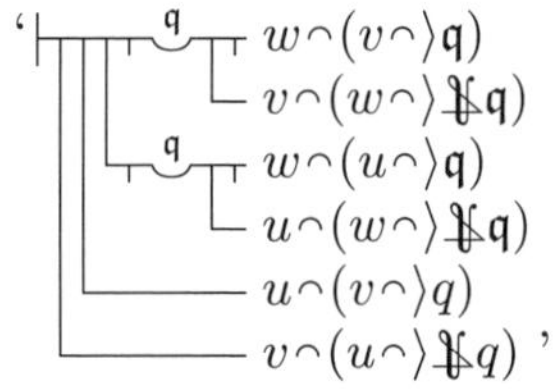

turns into an unsurveyable forest of parentheses in modern notation:

$$\text{`}\vdash (v^\frown(u^\frown)\not\rangle q) \supset (u^\frown(v^\frown)q) \supset (\neg\forall\mathfrak{q}(u^\frown(w^\frown)\not\rangle\mathfrak{q}) \supset \neg w^\frown(u^\frown)\mathfrak{q})) \supset \neg\forall\mathfrak{q}(v^\frown(w^\frown)\not\rangle\mathfrak{q}) \supset \neg w^\frown(v^\frown)\mathfrak{q})))))\text{'}.$$

Adopting the left-association convention for embedded conditionals in order to reduce the number of brackets provides little improvement:

$$\text{`}\vdash v^\frown(u^\frown)\not\rangle q) \supset u^\frown(v^\frown)q) \supset \neg\forall\mathfrak{q}(u^\frown(w^\frown)\not\rangle\mathfrak{q}) \supset \neg w^\frown(u^\frown)\mathfrak{q})) \supset \neg\forall\mathfrak{q}(v^\frown(w^\frown)\not\rangle\mathfrak{q}) \supset \neg w^\frown(v^\frown)\mathfrak{q}))\text{'}.$$

Proposition (25) is far from being the longest or most complicated formula when it comes to the structure of embedded conditionals. There is little value in rendering Frege's formulae in this way as far as readability is concerned. Attempting to make the modern rendering more surveyable by choosing equivalent formulae that utilise signs for conjunction and disjunction, as well as those for the conditional and negation, would clash with Frege's rules of inference. In fact, Frege's rules may in any case seem bewildering, albeit valid, when applied to modern formulae, while they are natural and, dare we say, elegant in the concept-script system.

Finally, despite all similarities, concept-script differs in significant respects from modern logic—compare Roy Cook's appendix to this volume for details. Presenting Frege's logic in the formalism of modern second-order predicate logic would obscure this fact and fail to do justice to his system.

Thus, the only responsible way to render Frege's notation from a scholarly perspective—namely, rendering it in exactly the way Frege did—turned out to be the only sensible solution on the whole. Frege's concept-script is unfamiliar to most, and perhaps somewhat more difficult to learn than modern notation. But learning to understand Frege's formalism is certainly not an insurmountable task, and once some familiarity is achieved, Frege's system is surprisingly easy to read. Working through Frege's own introduction to the concept-script, the reader will soon appreciate this. Roy Cook's appendix will further aid the reader in this process and offer additional help in understanding the details of some of the more arcane features of Frege's system.

Typesetting concept-script could only take one form: using TEX. When we started our project, there was, however, no straightforward way to do so. TEX-expert Josh Parsons (post-doc at Arché, St Andrews, at the time) wrote the first LATEX-style for concept-script (the *begriff* package, which is now included in most TEX-distributions) that allowed us to typeset the formulae in an elegant way.[19] Josh's style-file renders

[19] LATEX is the widely used macro-package for the TEX typesetting system.

concept-script formulae as they appear in Frege's first book, *Begriffsschrift.* With the help of Richard Heck, J. J. Green, and Agustín Rayo, we wrote a style file based on Josh's that allowed us to render concept-script formulae in the way they appear in *Grundgesetze.* It required many changes and additions, but being able to build on Josh's work made the task significantly less demanding.

This type of free collaboration is of course characteristic of anything to do with TEX. TEX is open-source freeware, which is continually developed, improved, and supplemented by thousands of enthusiasts. For our translation, we used more TEX resources than we knew existed before we started. We are indebted to this great community without whom we could not have produced this edition. We would also like to thank the creators and community surrounding TEXShop and TEXLive.

All of Frege's formulae had to be rendered in LATEX-code, and despite the convenience of LATEX, doing so by typing up each formula using only a keyboard would have been even more daunting a task than it turned out to be. Robert MacInnis (then a computer-science student at St Andrews), under the supervision of Roy Dyckhoff, wrote a graphical user interface (GUI) that enables one to create concept-script formulae by mouse-click and outputs a choice of LATEX- or XML-code (see MacInnis et al. (2004)). Some bug-fixes and adjustments of the software to suit our specific purposes were implemented by Guðmundur Andri Hjálmarsson (then a philosophy Ph.D. student at Arché, St Andrews).

In addition to the unfamiliar representation of the logic, Frege employs symbols for defined functions, such as '℘', 'ℳ', 'ƺ', '⅄̲', and individual constants, 'Ɵ', 'Ɨ', '∽'. Frege chose these unfamiliar signs to follow his own advice regarding definitions. Referring to the abhorred practice of 'piecemeal definition', he writes: "Indeed, it would have been possible to replace the old signs and notations by new ones, and actually, this is what logic requires; but this is a decision that is hard to make. And this reluctance to introduce new signs or words is the cause of many unclarities in mathematics." (vol. II, §58, pp. 70–71).

Frege chose his signs from whatever stock of metal types his publisher, Hermann Pohle, had in his printshop (the *Frommannsche Buchdruckerei*), but appears to have picked the signs, where possible, to be suggestive of the respective function: a sign for pound ('℔', an 'lb'-ligature) is turned over to resemble a cursive 'A', or perhaps an 'An'-ligature: '℘', and so serves as the sign for *Anzahl* (cardinal number); 'ℳ', an old currency sign for Mark is used for *Umkehrung* (converse); '∽', apparently constructed from metrical signs (for the annotation of classical poetry) and a lying bracket, is importantly distinct from, but still suggestive of, John Wallis's '∞' and Georg Cantor's 'ω'—it is Frege's sign for *Endlos* ("Endless"), the transfinite cardinal number of the natural numbers.[20]

Frege found these symbols in his publisher's stock—we were not so lucky to find all of them in the stock of symbols LATEX was able to provide at the time. Creating symbols, using Metafont, was beyond the abilities of everyone involved at that stage, but we could once more rely on the TEX community. Richard Heck sent out a plea for help online and found TEX-wizard J. J. Green who enthusiastically contributed his time and skills in TEX and Metafont. Jim created the *fge* package, a LATEX-package that contains all of Frege's function symbols that were missing from the common stock.

[20] See Green et al. (2012) for a more detailed discussion of the typography of *Grundgesetze.*

All above mentioned LaTeX style-files (`begriff.sty`, `grundgesetze.sty`, `fge.sty`) are available on `www.ctan.org` and also on our website `www.frege.info`. The *Begriffsschrift* GUI mentioned above is also available on the latter website.

While we stuck closely to Frege's page-breaks as well as his formalism in all respects, we took more liberty with other typesetting features. The German original uses two different means of emphasis: italics (as in '*Begriffsschrift*') and letter-spacing (as in 'A n z a h l'). Letter-spacing is used for personal names, in the introduction of technical terms, and sometimes (vol. I, pp. 30–34) for the statement of rules; italics is used for Latin phrases and as a means to refer to concepts. We do not track the distinction between italics and letter-spacing, but set both as italics. No more confusion should arise from this than does from using italics for both Latin phrases and concepts. Moreover, the original is not consistent in applying the distinction: in vol. II, the introduction of technical terms sometimes uses italics instead of letter-spaced (see e.g., p. 171); the titles of works Frege cites (both articles and books) are often letter-spaced (e.g., vol. I, p. 5 fn. 1), but also sometimes italics (e.g., vol. I, p. IX–X; vol. 2, p. 152 fn. 1), and sometimes not emphasised at all (e.g., vol. I, p. XI fn. 1, p. 1 fn. 2 and 3, p. 3 fn. 4).

Another divergence from the original lies in the use of quotation marks. Frege uses single German-style quotation marks for logical and mathematical symbols, as in: ‚— $2^2 = 4$'. German-style double quotation marks are used for quoting prose: „Unser Denken". It was an obvious decision to change these to English-style single and double quotation marks, respectively: '— $2^2 = 4$', "our thinking". Larger formulae present a difficulty. Where Frege quotes concept-script propositions with one or more subcomponents, the opening quotation mark is vertically aligned with the lowest subcomponent, and the closing quotation mark with the supercomponent:

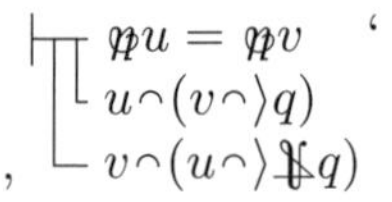

Upon reflection, the most consistent rendering using English-style quotation was to reverse Frege's alignment of the quotation marks for displayed formulae:

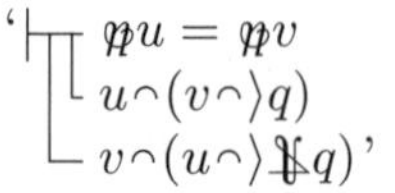

and to use both opening and closing quotation marks aligned with the supercomponent where the proposition is in line with the prose: 

Matters get more confusing in the second volume. The end of part II (the first fifty-four sections of vol. II) follows the quotation conventions described above. However, for almost all of part III.1 (§§55–164, pp. 69–162) French quotation marks replace the single quotation marks for logical and mathematical signs: »$(2 - 1) + 2$«, »$A \smile\!\rangle\!\!\!\!\times B$«. The only exception in part III.1 is a long footnote on pp. 70–71, where single quotation marks are used, as in the first volume. Part III.2 uses single quotation marks as parts I and II do—up until the last four pages: §241, p. 240, uses French quotation marks again, and their use is continued in the afterword.

We follow the quotation conventions of the first volume throughout the whole text and therefore replace all French quotation marks by single English-style quotation marks. Frege draws no distinction in the changing from single German to French quotation marks. It is merely a quirk.

We close this section with a few final oddities involving quotations within quotations in *Grundgesetze*. In vol. II, §85, p. 95, Frege quotes Cantor who, in turn, has a quoted expression in his sentence: Frege's original uses the French-style quotation marks for the inner quotation marks here; we replace them by single English quotation marks as per the convention above. In §127, p. 131, Frege quotes Thomae and again needs quotation within a quotation. The solution in the original is to use quadruple quotation marks:

> „Da alle Terme nicht angeschrieben werden können, so ist unter „„alle“ “ hier wie in ähnlichen Fällen zu verstehen [...].“

We follow this solution. Lastly, in vol. I, §54, p. 71b, Frege uses single quotation marks to quote his own, semi-formal expressions which, in turn, contain the quotation of logical symbols, *viz.* Roman letters. Idiosyncratic semi-circles are employed in lieu of quotation marks:

> Des bequemern Ausdrucks halber sage ich nun statt ‚Begriff, dessen Umfang durch ˓*u*˒ angedeutet wird‘ ‚*u*-Begriff‘ [...]

This use of semi-circles occurs thrice in this section (pp. 71b and 72b). We use single quotation marks for both inner and outer quotation marks.

Acknowledgments

It is difficult properly to acknowledge, or even express in words, the amazing support and guidance we received in the many years since this project started. Philosophers around the world—many of whom we are now grateful to call our friends—have generously supported us by selflessly spending many hours giving us feedback on the translation. In describing how the translation came together, we hope to acknowledge all those who participated in the project.

From its very beginning, the translation project was a team effort. It was initiated in 2003 by Crispin Wright when he approached us with the proposal to work together on a full translation of Frege's *Grundgesetze*. We were enthusiastic and immediately agreed to it—after all, we thought, it couldn't take much longer than two years to translate this book. The project was inspired by the strongly collaborative working methods that prevailed at the Arché research centre in St Andrews. The work was initially supported and hosted under the auspices of the AHRC-funded project on The Logical and Metaphysical Foundations of Classical Mathematics, which was running at Arché from 2000 to 2005 under Crispin's leadership. Both of us participated in the project as graduate students.

The very first *Grundgesetze* translation meeting took place in April 2003. It was attended by Peter Clark (then Head of the Philosophy Departments at St Andrews), Roy Cook, Walter Pedriali, Stephen Read, Crispin Wright, Elia Zardini, and the two of us. The task was to discuss drafts of our translation, which were usually very close to the original. During the very first meeting, the challenge lying ahead was brutally laid bare: we scarcely managed to discuss half a page of prose in a two hour session. The *Grundgesetze* translation project meetings continued to take place once or twice a week between 2003 and 2008, for two- to four-hour sessions. The work was gruelling, and the group shrank quickly to the core members Crispin, Walter, and the two of us. Walter was part of the group from 2003–5 and then rejoined us in the academic year 2006–7. Throughout these years, Walter provided feedback on the translation and helped by editing and collating corrections made during the meetings. We are very grateful for his help.

Given the difficulties in putting together a translation, and our lack of experience, Crispin suggested asking well-known Frege scholars to act as advisors to our project. The most we hoped for was that a few of the people we asked would help by providing feedback on specific questions regarding our translation and the choice of technical terms, or that they might assist with difficult passages. We were amazed by the enthusiasm with which our invitation was met, and by how much more of their time our advisors were willing to invest than we had expected. To organise the timing of feedback and advice more effectively, we held three workshops that brought many of our consultants together. We discussed a draft of volume I at the first *Grundgesetze* workshop in St Andrews in 2006, which was attended by Michael Beaney, Roy Cook, Gottfried Gabriel, Michael Hallett, Robert May, Eva Picardi, Stephen Read, Stewart Shapiro, William Stirton, Kai Wehmeier, and Elia Zardini. Volume II was discussed at the second workshop in December 2008 in St Andrews, where we received substantial feedback from Michael Beaney, Roy Cook, Gottfried Gabriel, Michael Hallett, Richard Heck, Robert May, Eva Picardi, William Stirton, Christian Thiel, and Kai Wehmeier. The last translation workshop, sponsored by the Northern Institute of Philosophy, Aberdeen, and New York University, took place in May 2010 in New York. We discussed a full draft of our translation (already completely revised several times) and received detailed, final comments from Michael Beaney, Roy Cook, Michael Hallett, Richard Heck, Robert May, William Stirton, and Kai Wehmeier. Many of our advisors not only invested a significant amount of their time in reading our translation and attending the workshops, but also gave us line-by-line comments on hundreds of pages of our translation. Richard Heck visited St Andrews several times in the early phases of the project and participated in the translation meetings. "Beyond the call of duty" scarcely captures the work our advisors invested.

Of course we did not always all agree on how various passages should best be translated, but this did not curtail the extremely fruitful, friendly, and collaborative working experience. The workshops were the most productive and collaborative events any of us has ever experienced—they were invaluable to improving the translation, but they also assured us that we would have strong support from our advisors during this long and difficult project. We would also like to acknowledge the feedback we received in written correspondence from Patricia Blanchette, Werner Holly, Michael Kremer, Wolfgang Künne, Matthias Schirn, Lionel Shapiro, Peter Simons, Peter Sullivan, Jamie Tappenden, and Joan Weiner.

We would like to thank all our advisors for their work, commitment, and great spirit in helping putting this translation together. We doubt it would have seen the light of day had it not been for your advice and support throughout these year: thank you all very much!

Special thanks go to Roy T. Cook. Roy was not only part of the translation project from its inception but also put together a detailed and superbly helpful technical appendix to this translation. The aim of the appendix is to help the reader grasp some of the formal aspects of Frege's work by providing a modern introduction to the symbolic language, logic, and defined notions.

The *Grundgesetze*-project would have never run as smoothly and as successfully had it not been for another core team-member: Sharon Coull. Her support and enthusiasm for the "*Grrrrrundgesetze*" was simply amazing. Many thanks, Sharon, for all your help.

Typesetting *Grundgesetze* proved to be another challenge. In the previous section we have already mentioned the many people who helped us representing Frege's formulae using LaTeX. First and foremost we owe a great debt to William Stirton, who rendered all of Frege's over four thousand formulae in LaTeX code and who also proof-read all of part II—formulae and prose. We would like to thank Charlie Siu who helped us extensively with the final typesetting of Frege's proofs (i.e., the sections entitled 'Construction'), and Nilanjan Bhowmick, Nora Hanson, and Daniel Massey, whose laborious task it was to proof-read the formulae of both volumes and who caught a number of mistakes we missed. Another arduous task was the compilation of the index. This was achieved by Thomas Hodgson, assisted by Michael Hughes and Thomas Cunningham—many thanks! And many thanks also to Kathleen Hanson for editing Frege's original title pages.

We would like to give special thanks to Richard Heck. Richard's enthusiasm for the project, his unwavering moral support, and his thoughtful advice throughout the years have been extremely valuable—not to mention his extensive comments on a close-to-final manuscript of the translation in summer 2012.

Thanks also go to Peter Momtchiloff at Oxford University Press for his patience and continued support throughout the years.

This project would not have been possible had it not been for our colleagues' appreciation of the value of spending many years working on the translation. In a time when single-authored publications in top journals are a must for a successful tenure case; when the quantity of original articles published in top journals are used as the main guide to assessing the quality of one's research; when funding for British philosophy departments depends greatly on individual scores in the so-called *Research Excellence Framework*; when politicians force "buzz words" like *knowledge transfer* or *impact* high on the agenda of funding bodies and thereby on the agenda of vice-chancellors, deans, and heads of departments, we were extremely lucky to have had the continuing support from our heads of department who shared the long-term vision of this project. Thus, we would like to thank our heads of department: Peter Clark at the University of St Andrews, Antony Duff and Peter Sullivan at the University of Stirling, and Crawford Elder and Donald Baxter at the University of Connecticut. They not only supported our project financially but did so knowing that much of our individual research time would be spent on a translation of a 110 year-old book—work few will regard as *original research*, with no prospect of having *impact* outside of academia, with seemingly little material for *knowledge transfer*, and with an "output" that might not be *REF*-able. Thank you for thinking and acting "outside the box".

Also we would like to thank the funding bodies who supported the project, especially given the aforementioned constraints under which these organisations operate. In particular, we would like to acknowledge the Arts and Humanities Research Council, who provided funding through the Arché centre and who awarded Philip with a six-month early-career research fellowship in 2012; the Philip Leverhulme Trust, for supporting the project with a two-year research grant from 2005 to 2007; the British Academy, whose support enabled Marcus to work in Stirling for three months as a visiting scholar in the summer of 2010; and the University of Connecticut Humanities Institute, who granted Marcus a fellowship for the academic year 2010–11 and through whom we received the Felberbaum Family Award to support the project.

Many thanks to Nora Hanson and Charlotte Geniez, for their continued support, understanding, and encouragement through many years, and not least for putting up with our absences, our rants and frustrations, and our never-ending video-chat sessions. We are grateful and also amazed that you believed in us.

Lastly, there can be only one person to whom we are most deeply indebted: Crispin Wright. Crispin not only initiated the project, guided it from its beginning to its very end, provided moral and financial support over ten years, and helped us receive the required grants—most importantly, he spent many years with us in weekly seminars working on the translation. Crispin, many thanks for all your work, and the many memorable moments we had at our *Grundgesetze* sessions. It's been quite a journey, and we finally made it—thanks boss!

Philip Ebert, Stirling, Scotland
Marcus Rossberg, Storrs, Conn., USA *November 2012*

References

Auxier, Randall E. and Lewis Edwin Hahn, eds. (2007). *The Philosophy of Michael Dummett*, vol. XXXI of *The Library of Living Philosophers*. Chicago and La Salle, Ill.: Open Court.

Beaney, Michael, ed. (1997). *The Frege Reader*. Oxford: Blackwell.

Black, Max (1950). 'Frege Against the Formalists', *The Philosophical Review* 59:77–93, 202–220, 332–345.

Boolos, George (1987). 'The Consistency of Frege's Foundations of Arithmetic', in J. J. Thomson (ed.), *On Being and Saying: Essays for Richard Cartwright*. Cambridge, Mass.: The MIT Press, pages 3–20. Reprinted in Boolos (1998), pages 183–201.

——— (1990). 'The Standard of Equality of Numbers', in George Boolos (ed.), *Meaning and Method: Essays in Honor of Hilary Putnam*. Cambridge: Cambridge University Press, pages 261–278. Reprinted in Boolos (1998), pages 202–219.

——— (1995). 'Frege's Theorem and the Peano Postulates', *The Bulletin of Symbolic Logic* 1:317–26. Reprinted in Boolos (1998), pages 291–300.

——— (1996). 'On the proof of Frege's Theorem', in A. Morton and S. Stich (eds.), *Benacerraf and his Critics*. Oxford: Blackwell Publishers. Reprinted in Boolos (1998), pages 275–290.

Boolos, George (1998). *Logic, Logic, and Logic.* Cambridge, Mass.: Harvard University Press.

Burgess, John P. (2005). *Fixing Frege.* Princeton: Princeton University Press.

Church, Alonzo (1951). 'The Need for Abstract Entities in Semantic Analysis', *Proceedings of the American Academy of Arts and Sciences* 80(1):100–112.

Cook, Roy T. (2013). 'Frege's Little Theorem and Frege's Way Out', in Philip A. Ebert and Marcus Rossberg (eds.), *Essays on Frege's Basic Laws of Arithmetic.* Oxford: Oxford University Press. Forthcoming.

Dummett, Michael (1991). *Frege: Philosophy of Mathematics.* London: Duckworth.

Erdmann, Benno (1892). *Logik.* Halle a. S.: Max Niemeyer.

Fine, Kit (2002). *Limits of Abstraction.* Oxford: Blackwell Publisher.

Frege, Gottlob (1879). *Begriffsschrift: Eine der arithmetischen nachgebildete Formelsprache des reinen Denkens.* Halle a. d. Saale: Verlag L. Nebert. English translation by S. Bauer-Mengelberg in van Heijenoort (1967), pages 1–82; and by T. W. Bynum in Frege (1972).

——— (1884). *Die Grundlagen der Arithmetik. Eine logisch mathematische Untersuchung über den Begriff der Zahl.* Breslau: Wilhelm Koebner. English translation: Frege (1950).

——— (1893). *Grundgesetze der Arithmetik. Begriffsschriftlich abgeleitet.* I. Band. Jena: Verlag H. Pohle.

——— (1902). 'Letter to Russell, June 22, 1902', in Gabriel et al. (1976), pages 212–213. English translation: Gabriel et al. (1980), pages 131–133; first published in English translation in van Heijenoort (1967), pages 126–128.

——— (1903). *Grundgesetze der Arithmetik. Begriffsschriftlich abgeleitet.* II. Band. Jena: Verlag H. Pohle.

——— (1915). 'The Fundamental Laws Of Arithmetic', translated by Johann Stachelroth and Philip E. B. Jourdain, *The Monist* 25(4):481–494.

——— (1916). 'The Fundamental Laws Of Arithmetic: Psychological Logic', translated by Johann Stachelroth and Philip E. B. Jourdain, *The Monist* 26(2):182–199.

——— (1917). 'Class, Function, Concept, Relation', translated by Johann Stachelroth and Philip E. B. Jourdain, *The Monist* 27(1):114–127.

——— (1924/25). 'Neuer Versuch der Grundlegung der Arithmetik', in Frege (1983), pages 298–302. English translation: Frege (1979a).

——— (1950). *The Foundations of Arithmetic.* Translated by J. L. Austin. Oxford: Blackwell.

——— (1962). *Grundgesetze der Arithmetik I/II.* Facsimile reprint. Hildesheim: Olms.

Frege, Gottlob (1964). *The Basic Laws of Arithmetic: exposition of the system.* Edited and translated, with an introduction, by Montgomery Furth. Berkeley and Los Angeles: University of California Press.

——— (1972). *Conceptual Notation and Related Articles.* Translated and edited by T. W. Bynum. Oxford: Clarendon Press.

——— (1979a). 'A new Attempt at a Foundation of Arithmetic', in Frege (1979b), pages 278–281.

——— (1979b). *Posthumous Writings.* Edited by Hans Hermes, Friedrich Kambartel, and Friedrich Kaulbach, translated by Peter Long and Roger White. Oxford: Basil Blackwell.

——— (1983). *Nachgelassene Schriften.* (1st ed. 1969) 2nd rev. ed., by Hans Hermes, Friedrich Kambartel, and Friedrich Kaulbach. Hamburg: Meiner. English translation of the first edition: Frege (1979b).

——— (1998). *Grundgesetze der Arithmetik I/II.* Facsimile reprint, with critical additions by Christian Thiel. Hildesheim: Olms.

——— (2009). *Grundgesetze der Arithmetik: Begriffsschriftlich abgeleitet.* Band I und II. In moderne Formelnotation transkribiert und mit einem ausführlichen Sachregister versehen von Thomas Müller, Bernhard Schröder, und Rainer Stuhlmann-Laeisz. Paderborn: mentis Verlag.

Gabriel, Gottfried, H. Hermes, F. Kambartel, Christian Thiel, A. Veraart, and Brian McGuinness, eds. (1980). *Gottlob Frege: Philosophical and Mathematical Correspondence.* Translated by H. Kaal. Chicago: University of Chicago Press.

Gabriel, Gottfried, H. Hermes, F. Kambartel, Christian Thiel, and A. Veraart, eds. (1976). *Gottlob Frege: Wissenschaftlicher Briefwechsel.* Hamburg: Meiner. Partial English Translation: Gabriel et al. (1980).

Geach, Peter and Max Black (1952). *Translations from the Philosophical Writings of Gottlob Frege*, 1st edition. Oxford: Basil Blackwell.

Green, J. J., Marcus Rossberg, and Philip Ebert (2012). 'The convenience of the typesetter: notation in Frege's *Grundgesetze der Arithmetik*'. Unpublished manuscript.

Heck, Jr, Richard G. (2000). 'Cardinality, counting, and equinumerosity', *Notre Dame Journal of Formal Logic* 41:187–209. Reprinted in Heck (2011), pages 156–179.

——— (2011). *Frege's Theorem.* Oxford: Clarendon Press.

——— (2012). *Reading Frege's* Grundgesetze. Oxford: Clarendon Press.

Jeshion, Robin (2001). 'Frege's Notion of Self-Evidence', *Mind* 110(440):937–976.

Künne, Wolfgang (2009). *Die Philosophische Logik Gottlob Freges. Ein Kommentar.* Frankfurt: Klostermann.

MacInnis, Rob, James McKinna, Josh Parsons, and Roy Dyckhoff (2004). 'A mechanised environment for Frege's Begriffsschrift notation', in *Proceedings of Third Mathematical Knowledge Management Conference, 18–21 September 2004: The Mathematical User-Interfaces Workshop*. Białowieża, Poland.

Potter, Michael and Tom Ricketts, eds. (2010). *The Cambridge Companion to Frege*. Cambridge: Cambridge University Press.

Quine, W. V. (1950). *Methods of Logic*. New York: Henry Holt.

——— (1955). 'On Frege's Way Out', *Mind* 64(254):145–159.

Russell, Bertrand (1902). 'Letter to Frege, June 16, 1902', in Gabriel et al. (1976), pages 211–212. English translation: Gabriel et al. (1980), pages 130–131; first published in English translation in van Heijenoort (1967), pages 124–125.

Schirn, Matthias, ed. (1976). *Studien zu Frege*. 3 vols. Stuttgart–Bad Cannstatt: Friedrich Frommann Verlag.

van Heijenoort, Jean, ed. (1967). *From Frege to Gödel: A Source Book in Mathematical Logic, 1879–1931*. Cambridge, Mass.: Harvard University Press.

Veraart, Albert (1976). 'Geschichte des wissenschaftlichen Nachlasses Gottlob Freges und seiner Edition. Mit einem Katalog des ursprünglichen Bestands der nachgelassenen Schriften Freges', in Schirn (1976), pages 49–106.

Wehmeier, Kai F. (1999). 'Consistent Fragments of *Grundgesetze* and the Existence of Non-Logical Objects', *Synthese* 121:309–328.

Wehmeier, Kai F. and Hans-Christoph Schmidt am Busch (2005). 'The Quest for Frege's *Nachlass*', in Michael Beaney and Erich H. Reck (eds.), *Critical Assessments of Leading Philosophers: Gottlob Frege*, vol. I. London: Routledge, pages 54–68. English translation of Wehmeier and Schmidt am Busch (2000). 'Auf der Suche nach Freges Nachlaß', in Gottfried Gabriel and Uwe Dathe (eds.), *Gottlob Frege – Werk und Wirkung*. Paderborn: mentis Verlag, pages 267–281.

Wright, Crispin (1983). *Frege's Conception of Numbers as Objects*. Aberdeen: Aberdeen University Press.

Zalta, Edward N. (2012). 'Frege's Logic, Theorem, and Foundations for Arithmetic', in E. N. Zalta (ed.), *The Stanford Encyclopedia of Philosophy* (Winter 2012 Edition). `http://plato.stanford.edu/archives/win2012/entries/frege-logic/`.

GRUNDGESETZE DER ARITHMETIK.

Begriffsschriftlich abgeleitet

von

Dr. G. FREGE

A. O. PROFESSOR AN DER UNIVERSITÄT JENA.

I. Band.

JENA

Verlag von Hermann Pohle

1893.

BASIC LAWS

OF ARITHMETIC.

Derived using concept-script

by

DR. G. FREGE
ASSOCIATE PROFESSOR AT THE UNIVERSITY OF JENA

Volume I

JENA
Verlag von Hermann Pohle
1893

Foreword

In this book one finds theorems on which arithmetic is based, proven using signs that collectively I call concept-script. The most important of these propositions, some with an accompanying translation appended, are listed at the end. As may be seen, the investigation does not yet include the negative, rational, irrational and complex numbers, nor addition, multiplication, etc. Moreover, propositions about the cardinal numbers are not yet present with the completeness initially planned. Missing, in particular, is the proposition that the cardinal number of objects falling under a concept is finite, if the cardinal number of objects falling under a superordinate concept is finite. External reasons have made me postpone both this and the treatment of other numbers, and mathematical operations, to a sequel whose publication will depend on the reception of this first volume. What I have offered here may suffice to give an idea of my method. It might be thought that the propositions concerning the cardinal number Endlos[1] could have been omitted. To be sure, they are not needed for the foundation of arithmetic in its traditional extent; but their derivation is often easier than those of the corresponding propositions concerning finite cardinal numbers and can serve as preparation for the latter. Propositions also occur which are not about cardinal numbers but which are needed in proofs. They treat, for example, of following in a series, of single-valuedness of relations, of composite and coupled relations, of mapping by means of relations, and such like. These propositions could perhaps be allocated to an extended theory of combinations.

The proofs are contained solely in the sections entitled "Construction", while those headed "Analysis" are meant to facilitate understanding by providing a preliminary and rough sketch of the proof. The proofs themselves contain no words but are carried out solely in my symbolism. They are presented as a series of formulae separated by

[1] Cardinal number of a countably infinite set.

continuous or broken lines or other signs. Each of these formulae is a complete proposition displaying all the conditions on which its validity depends. This completeness, which does not tolerate any tacit addition of assumptions in thought, seems to me indispensable for the rigorous conduct of proof.

The progression from one proposition to the next proceeds by the rules which are listed in §48, and no transition is made that does not accord with these rules. How, and according to which rule, an inference is drawn is indicated by the sign standing between the formulae, while ——•—— marks the termination of a chain of inferences. For this purpose there have to be propositions which are not derived from others. Some of these are the basic laws listed in §47; others are definitions which are collected in a table at the end, together with a reference to their first occurrence. Time and again, the pursuit of this project will generate a need for definitions. Their governing principles are listed in §33. Definitions themselves are not creative, and in my view must not be; they merely introduce abbreviative notations (names), which could be dispensed with were it not for the insurmountable external difficulties that the resulting prolixity would cause.

The ideal of a rigorous scientific method for mathematics that I have striven to realise here, and which could be named after Euclid, can be characterised as follows. It cannot be required that everything be proven, as this is impossible; but it can be demanded that all propositions appealed to without proof are explicitly declared as such, so that it can be clearly recognised on what the whole structure rests. One must strive to reduce the number[a] of these fundamental laws as far as possible by proving everything that is provable. Furthermore, and in this I go beyond Euclid, I demand that all modes of inference and consequence which are used be listed in advance. Otherwise compliance with the first demand cannot be secured. This ideal I believe I have now essentially achieved. Only in a few points could one impose even more rigorous demands. In order to attain more flexibility and to avoid excessive length, I have allowed myself tacit use of permutation of subcomponents (conditions) and fusion of equal subcomponents, and have not reduced the modes of inference and consequence to a minimum. Anyone acquainted with my little book *Begriffsschrift* will gather from it how here too one could satisfy the strictest demands, but also that this would result in a considerable increase in extent.

Furthermore, I believe that the criticisms that can justifiably be made of this book will pertain not to rigour but rather only to the choice of the course of proof and of the intermediate steps. Often several ways of conducting a proof are available; I have not tried to pursue them all and it is possible, indeed likely, that I have not always chosen the shortest. Let whoever has complaints on this score try to do better. Other matters will be disputable. Some might have preferred to increase the circle of permissible modes of inference and consequence, in order to achieve greater flexibility and brevity. However, one has to draw a line somewhere if one approves of my stated ideal at all; and wherever one does so, people could always say: it would have been better to allow even more modes of inference.

The gaplessness of the chains of inferences contrives to bring to light each axiom, each presupposition, hypothesis, or whatever one may want to call that on which a proof rests; and thus we gain a basis for an assessment of the epistemological nature of the proven law. Although it has already been announced many times that arithmetic is merely logic further developed, still this remains disputable as long as there occur transitions in the proofs which do not conform to acknowledged logical laws but rather seem to rest on intuitive knowledge.[b] Only when these transitions are analysed into simple logical steps can one be convinced that nothing but logic forms the basis. I have listed everything that can facilitate an assessment whether the chains of inferences are properly connected and the buttresses are solid. If anyone should believe that there is some fault, then he must be able to state precisely where, in his view, the error lies: with the basic laws, with the definitions, or with the rules or a specific application of them. If everything is considered to be in good order, one thereby knows precisely the grounds on which each individual theorem rests. As far as I can see, a dispute can arise only concerning my basic law of value-ranges (V), which perhaps has not yet been explicitly formulated by logicians although one thinks in accordance with it if, e.g., one speaks of extensions of concepts. I take it to be purely logical. At any rate, the place is hereby marked where there has to be a decision.

My purpose demands some divergences from what is common in mathematics. Rigour of proof requires, as an inescapable consequence, an increase in length. Whoever fails to keep an eye on this will indeed be surprised how cumbersome our proofs often are of propositions into which he would suppose he had an immediate insight, through a single act of cognition.[c] This will be especially striking if one compares Mr Dedekind's essay, *Was sind und was sollen die Zahlen?*, the most thorough study I have seen in recent times concerning the foundations of arithmetic. It pursues, in much less space, the laws of arithmetic to a much higher level than here. This concision is achieved, of course, only because much is not in fact proven at all. Often, Mr Dedekind merely states that a proof follows from such and such propositions; he uses dots, as in "$\mathfrak{M}(A, B, C \ldots)$"; nowhere in his essay do we find a list of the logical or other laws he takes as basic; and even if it were there, one would have no chance to verify whether in fact no other laws were used, since, for this, the proofs would have to be not merely indicated but carried out gaplessly. Mr Dedekind too is of the opinion that the theory of numbers is a part of logic; but his essay barely contributes to the confirmation of this opinion since his use of the expressions "system", "a thing belongs to a thing" are neither customary in logic nor reducible to something acknowledged as logical. I do not say this as a complaint; his procedure may have been the most appropriate for his purpose; I say this only to cast a brighter light upon my own intentions by contrast. The length of a proof should not be measured by the ell. It is easy to make a proof appear short on paper, by missing out many intermediate steps in the chain of inferences or by merely gesturing at them. Mostly, no doubt, one contents oneself with the obvious correctness of each step in a proof; and permissibly so, if the aim is merely to persuade of the truth of the proposition to be proven. However, if the aim is to convey insight into the nature of this obviousness, this procedure does not suffice; rather, one must write out all intermediate steps, so that the full light of awareness may fall upon them. Usually, mathematicians are merely concerned with the content of a proposition and that it be proven. Here the novelty is not the content of the proposition, but how its proof is conducted, on what foundations it rests. That this essentially different perspective also requires another kind of treatment must not put us off. When one of our propositions is proven in the usual manner, then a proposition that appears to be unnecessary for the proof will easily be overlooked. In a thorough examination of my proof given here, I believe, one will indeed realise its indispensability, unless an entirely different path is taken. Here and there one will perhaps also encounter conditions in our propositions that strike one as redundant at first, but which will prove to be necessary after all, or at least eliminable only by using a proposition to be proven for this specific purpose.

I here carry out a project that I already had in mind at the time of my *Begriffsschrift* of the year 1879 and which I announced in my *Grundlagen der Arithmetik* of the year 1884.[1] By this act I aim to confirm the conception of cardinal number

[1] Compare the introduction and §§90 and 91 in my *Grundlagen der Arithmetik*, Breslau, Verlag von Wilhelm Koebner, 1884.

which I set forth in the latter book. The basis for my results is articulated there in §46, namely that a statement of number contains a predication about a concept; and the exposition here rests upon it. If someone takes a different view, he should try to develop a sound and usable symbolic exposition on that basis; he will find that it will not work. No doubt in language the point is not so transparent; but if one pays close attention, one finds that even here there is mention of a concept, rather than of a group, an aggregate or suchlike, whenever a statement of number is made; and even if exceptions sometimes occur, the group or the aggregate is always determined by a concept, i.e., by the properties an object must have in order to belong to the group, while what unites the group into a group, or makes the system into a system, the relations of the members to each other, has absolutely no bearing on the cardinal number.

The reason why the implementation appears so late after the announcement is owing in part to internal changes within the concept-script which forced me to jettison a nearly completed handwritten work. This progress might be mentioned here briefly. The primitive signs used in my *Begriffsschrift* occur again here with one exception. Instead of the three parallel lines, I have chosen the usual equality-sign, for I have convinced myself that in arithmetic it possesses just that reference that I too want to designate. Thus, I use the word "equal" with the same reference as "coinciding with" or "identical with", and this is also how the equality-sign is actually used in arithmetic. The objection to this which might be raised would rest on insufficiently distinguishing between sign and what is designated. No doubt, in the equation '$2^2 = 2 + 2$' the sign on the left is different from the one on the right; but both designate or refer to the same number.[1] To the original primitive signs two have now been added: the smooth breathing, designating the value-range of a function, and a sign to play the role of the definite article in language. The introduction of value-ranges of functions is an essential step forward, thanks to which we achieve far greater flexibility. What previously had been derived signs can now be replaced by other, and indeed simpler, ones, although the definitions of single-valuedness of a relation, of following in a series, of mapping are essentially the same as those given partly in my *Begriffsschrift*, partly in my *Grundlagen der Arithmetik*. Value-ranges, however, have a much more

[1] To be sure, I also say: the sense of the sign on the right is different from the one on the left; but the reference is the same. Compare my essay "Über Sinn und Bedeutung" in the *Zeitschrift f. Philos. u. philos. Kritik*, vol. 100, p. 25.

fundamental importance; for I define cardinal numbers themselves as extensions of concepts, and extensions of concepts are value-ranges, according to my specification. So without the latter one would never be able to get by. The old primitive signs that re-occur outwardly unaltered, and whose algorithm has hardly changed, have however been provided with different explanations. What was formerly the content-stroke reappears as the horizontal. These are consequences of a deep-reaching development in my logical views. Previously I distinguished two components in that whose external form is a declarative sentence: 1) acknowledgement of truth, 2) the content, which is acknowledged as true. The content I called judgeable content. This now splits for me into what I call thought and what I call truth-value. This is a consequence of the distinction between the sense and the reference of a sign. In this instance, the thought is the sense of a proposition and the truth-value is its reference. In addition, there is the acknowledgment that the truth-value is the True. For I distinguish two truth-values: the True and the False. I have justified this in more detail in my above mentioned essay *Über Sinn und Bedeutung*. Here, it might merely be mentioned that only in this way can indirect speech be accounted for correctly. For in indirect speech, the thought, which is normally the sense of the proposition, becomes its reference. Only a thorough engagement with the present work can teach how much simpler and more precise everything is made by the introduction of the truth-values. These advantages alone already weigh heavily in favour of my conception, which at first sight might admittedly seem strange. Moreover, the nature of functions, in contrast to objects, is characterised more precisely than in my *Begriffsschrift*. Further, from this the distinction between functions of first and second level results. As elaborated in my lecture *Function und Begriff*,[1] concepts and relations are functions as I extend the reference of the term, and so we also must distinguish concepts of first and second level and relations of equal and unequal level.

As one can see, the years since the publication of my *Begriffsschrift* and *Grundlagen* have not passed in vain: they have seen the work mature. But the very thing which I regard as essential progress serves, as I cannot conceal from myself, as a major obstruction to the dissemination and influence of this book. Moreover, what I regard as not the least of its virtues, strict gaplessness of the chains of inferences, will earn it, I fear, scant appreciation. I have departed further from traditional conceptions

[1] Jena, Verlag von Hermann Pohle.

and thereby impressed on my views a paradoxical character. An expression, cropping up here and there as one leafs through the pages, will all too easily seem strange and provoke negative prejudice. I can myself gauge somewhat the resistance which my innovations will encounter, as I too had first to overcome something similar in order to make them. To be sure, I have arrived at them not arbitrarily and out of a craze for novelty, but was forced by the very subject matter itself.

With this, I arrive at a second reason for the delay: the despondency that at times overcame me as a result of the cool reception, or rather, the lack of reception, by mathematicians[1] of the writings mentioned above, and the unfavourable scientific currents against which my book will have to struggle. The first impression alone can only be off-putting: strange signs, pages of nothing but alien formulae. Thus sometimes I concerned myself with other subjects. Yet as time passed, I simply could not contain these results of my thinking, which seemed to me valuable, locked up in my desk; and work expended always called for further work if it was not to be in vain. Thus the subject matter kept me captive. In such a case, when the value of a book cannot be appreciated on a swift reading, the reviewer should step in to assist. But in general the remuneration will be too poor. The critic can never hope to be compensated in money for the effort that a thoroughgoing study of this book will demand. All that is left for me is to hope that someone may from the outset have sufficient confidence in the work to anticipate that his inner reward will be repayment enough, and will then publicise the results of a thorough examination. It is not that only a complimentary review could satisfy me; quite the contrary! I would always prefer a critical assault based on a thorough study to praise that indulges in generalities without engaging the heart of the matter. Now I would like to offer some hints to assist the work of a reader approaching the book with these intentions.

In order to gain an initial rough idea of how I express thoughts with my signs, it will be helpful to look at some of the easier cases in the table of the more important theorems, to which a translation is appended. It will then be possible to surmise what is intended in further, similar examples which are not followed by a translation. Next, one should begin with the introduction and start to tackle the exposition of the concept-script. However, I advise first to make merely a summary overview of it

[1] One searches in vain for my *Grundlagen der Arithmetik* in the *Jahrbuch über die Fortschritte der Mathematik*. Researchers in the same area, Mr Dedekind, Mr Otto Stolz, Mr von Helmholtz seem not to be acquainted with my works. Kronecker does not mention them in his essay on the concept of number either.

and not to dwell on particular concerns. In order to meet all objections, some issues have had to be taken up which are not required for understanding concept-script propositions. I include in this the second half of §8 which starts on p. 12 with "If we now give the following explanation", and also the second half of §9, which starts on p. 15 with the words "If I say in general", together with the whole of §10. These passages should be omitted on a first reading. The same applies to §26 and §§28–32. By contrast, I wish to lay stress on the first half of §8, as well as §§12 and 13, as particularly important for understanding. A more detailed reading should start with §34 and continue to the end. Occasionally, one will have to revisit §§ merely fleetingly read. The index at the end and the table of contents will facilitate this. The derivations in §§49–52 can be used as preparation for an understanding of the proofs themselves. Here, all modes of inference and nearly all of the applications of our basic laws already occur. When one has reached the end, one should reread the entire exposition of the concept-script with this as background, keeping in mind that those stipulations that will not be used later, and therefore appear unnecessary, serve to implement the principle that all correctly formed signs ought to refer to something—a principle that is essential for full rigour. In this way, I believe, the mistrust that my innovations may initially provoke will gradually disappear. The reader will recognise that my principles will in no case lead to consequences other than ones he must acknowledge as correct himself. Perhaps he will then admit that he had overestimated the labour, that, in fact, my gapless approach facilitates understanding, once the barrier presented by the novelty of the signs is overcome. May I be fortunate enough to find such a reader or reviewer! For a review based on a superficial reading might easily do more harm than good.

Otherwise, of course, the prospects for my book are dim. In any case, we must give up on those mathematicians who, encountering logical expressions like "concept", "relation", "judgement", think: *metaphysica sunt, non leguntur!*[d] and also on those philosophers who, sighting a formula, cry out: *mathematica sunt, non leguntur!* and the exceptions will be very few. Perhaps the number of mathematicians who care about the foundation of their science is not large in any case, and even these often seem to be in a great hurry until they leave the fundamentals behind them. Moreover, I hardly dare hope that many of them will be convinced by my reasons for the painstaking rigour, and the lengthiness connected with it. Custom exerts great power over the mind. If I compare arithmetic with a tree that high up unfolds in a multiplicity of methods and theorems, while the root stretches into the depths, then it seems to me that the growth of the root, at least in Germany, is weak. Even in the *Algebra der Logik* of Mr E. Schröder, a work one would want to count as pursuing this direction, upper growth soon dominates before any greater depth is attained, causing an upward bent and a ramification into methods and theorems.

Of further disadvantage for my book is a widespread tendency to accept only what can be sensed as being. What cannot be perceived with the senses one tries to disown, or at least to ignore. Now the objects of arithmetic, the numbers, are imperceptible; how to come to terms with this? Very simple! Declare the number-signs to be the numbers. In the signs, one then has something visible; and this, of course, is the main thing. To be sure, the signs have properties completely different from the numbers; but so what? Just credit them with the desired properties by so-called definitions. To be sure, it is a puzzle how there can be a definition where there is no question of a connection between sign and what is designated. One kneads together sign and what is designated as indistinguishably as possible; depending on what is required, one can assert existence by appeal to their tangibility[1] or bring the true properties of the numbers to the foreground. On occasion, it seems that the number-signs are regarded like chess pieces, and the so-called definitions like rules of the game. In that case the sign designates nothing, but is rather the thing itself. One small detail is overlooked in all this, of course; namely that a thought is expressed by means of '$3^2 + 4^2 = 5^2$', whereas a configuration of chess pieces says nothing. When one is content with such superficialities, there is surely no basis for a deeper understanding.

Here it is crucial to get clear about what definition is and what it can achieve. Often one seems to credit it with a creative power, although in truth nothing takes place except to make something prominent by demarcation and designate it with a name. Just as the geographer does not create a sea when he draws borderlines and says: the part of the water surface bordered by these lines I will call Yellow Sea, so too the mathematician cannot properly create anything by his definitions. Moreover, a property which a thing just does not have cannot be magically attached to it by mere definition, except for the property of now being called by the name that one has given to it. That, however, an egg-shaped figure, produced with ink on paper, may

[1] Compare E. Heine, *Die Elemente der Functionslehre*, in Crelle's *Journal*, vol. 74, p. 173: "Concerning definitions, I take the purely formal standpoint in calling certain tangible signs numbers, so that the existence of these numbers is thus not in question."

be endowed by definition with the property of resulting in One if added to One, I can only regard as scientific superstition. A lazy student could just as well be turned into a diligent one by means of definition alone. Unclarity develops easily here for want of the distinction between concept and object. If one says: "A square is a rectangle in which adjacent sides are equal", then one defines the concept *square* by stating what properties something must have in order to fall under it. I call these properties characteristic marks of the concept. Yet note that these characteristic marks of the concept are not its properties. The concept *square* is not a rectangle, it is only the objects that fall under this concept that are rectangles, just as the concept *black cloth* is neither black nor a cloth. Whether there are such objects is not immediately known on the basis of the definition. One wants to define the number Zero, for example, by saying: it is something which when added to One, results in One. Thus a concept is defined by stating what property an object must have in order to fall under it. This property, however, is not a property of the defined concept. Yet, as it seems, it is often imagined that something which added to One results in One is created by definition. What a great illusion! The defined concept does not possess this property, nor does the definition guarantee that the concept is instantiated. This first requires an investigation. Only when one has shown that there is one and only one object with the requisite property is one in a position to give this object the proper name "Zero". To create Zero is hence impossible. I have repeatedly spelt these things out but, seemingly, without success.[1]

A proper appreciation of the distinction I draw, between a characteristic mark of a concept and a property of an object, can scarcely be hoped for from the prevailing logic either,[2] for that seems to be contaminated with psychology through and through. If instead of the things themselves, one considers only their subjective images, their ideas, then naturally all finer-grained, objective distinctions are lost and others appear in their place that are logically completely worthless. Thus I come to speak about the obstacle to the influence of my book on the logicians. It is the ruinous incursion of psychology into logic. Decisive for the treatment of this science is how the logical laws are conceived, and this in turn connects with how one understands

[1] Mathematicians who prefer not to enter into the mazes of philosophy are requested to stop reading the foreword here.

[2] In the logic of Mr B. Erdmann I find no trace of this important distinction.

the word "true". It is commonly granted that the logical laws are guidelines which thought should follow to arrive at the truth; but it is too easily forgotten. The ambiguity of the word "law" here is fatal. In one sense it says what is, in the other it prescribes what ought to be. Only in the latter sense can the logical laws be called laws of thought, in so far as they legislate[e] how one ought to think. Every law stating what is the case can be conceived as prescriptive, one should think in accordance with it, and in that sense it is accordingly a law of thought. This holds for geometrical and physical laws no less than for the logical. The latter better deserve the title "laws of thought" only if thereby it is supposed to be said that they are the most general laws, prescribing how to think wherever there is thinking at all. But the phrase "laws of thought" seduces one to form the opinion that these laws govern thinking in the same way that the laws of nature govern events in the external world. In that case they can be nothing other than psychological laws; for thinking is a mental process. And if logic had to do with psychological laws, it would be a part of psychology. And thus it is in fact conceived. These laws of thought may then be conceived as guidelines merely in the manner of stating a mean, similar to the way one can say how healthy digestion proceeds in humans, or how grammatically correct speech goes, or how one dresses fashionably. Then one can merely say: humans' taking to be true conforms on average to these laws, both at present and wherever human beings are found; so, if one wants to stay in harmony with the mean, one had better follow suit. However, what is fashionable today will be out of fashion sometime, and is at present not fashionable amongst the Chinese; so, likewise, one can present psychological laws of thought as setting a standard only with restrictions. Indeed so, if logic deals with being taken to be true and not, rather, with being true! And that is what the psychological logicians conflate. Thus in the first volume of his *Logik*,[1] pp. 272 to 275, Mr B. Erdmann equates truth with general validity, grounding the latter on general certainty regarding the object judged, and this in turn on general consensus amongst those judging. And so, in the end, truth is reduced to being taken to be true by individuals. In opposition to this, I can only say: being true is different from being taken to be true, be it by one, be it by many, be it by all, and is in no way reducible to it. It is no contradiction

[1] Halle a. S., Max Niemeyer, 1892.

that something is true that is universally held to be false. By logical laws I do not understand psychological laws of taking to be true, but laws of being true. If it is true that I write this in my room on 13th July, 1893, while the wind is howling outside, then it remains true even if all humans should later hold it to be false. If being true is thus independent of anyone's acknowledgement, then the laws of being true are not psychological laws either but boundary stones which are anchored in an eternal ground, which our thinking may wash over but yet cannot displace. And because of this they set the standards for our thinking if it wants to attain the truth. Their relation to thinking is not like that of the grammatical laws to language, as if they were to give expression to the nature of our human thinking and vary with it. The conception of the logical laws according to Mr Erdmann is, of course, entirely different. He doubts their unconditional, eternal validity and wants to restrict them to our thinking as it is now (pp. 375ff). But "our thinking" can surely only mean the thinking of humanity up until now. Accordingly, the possibility remains open that human or other beings might be discovered who could execute judgements contradicting our logical laws. What if this were to happen? Mr Erdmann would say: so we see that those principles are not valid everywhere. Certainly! if they are to be psychological laws, they ought to be formulated in a way that makes explicit the genus of beings whose thinking is empirically governed by them. I would say: there are therefore beings who do not recognise certain truths immediately in the manner we do but are reliant, perhaps, on the more protracted way of induction. What, however, if beings were even found whose laws of thought directly contradicted ours, so that their application often led to opposite results? The psychological logician could only accept this and say: for them, those laws hold, for us these. I would say: here we have a hitherto unknown kind of madness. He who thinks of logical laws as prescriptive of what ought to be thought, or as laws of what is true, rather than as natural laws concerning humans' taking to be true, will ask: Who is right? Whose laws of taking to be true are in accord with the laws of being true? The psychological logician cannot admit this question; for by so doing he would acknowledge laws of being true that were not psychological. Can the sense of the word "true" be subjected to a more damaging corruption than by the attempt to incorporate a relation to the judging subject! Surely no-one will here object that the proposition "I am hungry" could be true for one but false for another? The proposition, no doubt, but not the thought; for the word "I" in the mouth of the other refers to a different person, and the proposition, accordingly, expresses a different thought when it is uttered by him. All determinations of place, time, and so on, belong to the thought whose truth is at issue; being true itself is place- and timeless. How, then, is the principle of identity to be read? Is it like this: "It is impossible for humans in the year 1893 to acknowledge an object as being different from itself"? Or like this: "Every object is identical to itself"? The former law is about humans and contains a determination of time; in the latter, there is mention neither of humans nor of time. The latter is a law of being true; the former one of human taking to be true. Their content is entirely different, and they are independent of each other so that neither can be inferred from the other. This is why it is very confusing to designate both by the same name of the basic law of identity. Such confusions of fundamentally different things are to blame for the appalling unclarity which we find in the psychological logicians.

As to the question, why and with what right we acknowledge a logical law to be true, logic can respond only by reducing it to other logical laws. Where this is not possible, it can give no answer. Stepping outside logic, one can say: our nature and external circumstances force us to judge, and when we judge we cannot discard this law—of identity, for example—but have to acknowledge it if we do not want to lead our thinking into confusion and in the end abandon judgement altogether. I neither want to dispute nor to endorse this opinion, but merely note that what we have here is not a logical conclusion. What is offered here is not a ground of being true but of our taking to be true. And further: this impossibility, to which we are subject, of rejecting the law does not prevent us from supposing beings who do so; but it does prevent us from supposing that such beings do so rightly; and it prevents us, moreover, from doubting whether it is we or they who are right. At least this is true of myself. If others dare in the same breath to both acknowledge a law and doubt it, then that seems to me to be an attempt to jump out of one's own skin against which I can only urgently warn. Whoever has once acknowledged a law of being true has thereby also acknowledged a law that prescribes what ought to be judged, wherever, whenever and by whomsoever the judgement may be made.

Surveying the whole matter, it seems to me that different conceptions of truth lie at the source of the dispute. For me, truth is something objective, independent of the judging subject, for psychological logicians, it is not. What Mr B. Erdmann calls "objective certainty" is only a general acknowledgement by those who judge and cannot, accordingly, be independent of them but is liable to change with their mental nature.

We can capture this more generally still: I acknowledge a realm of the objective, non-actual, while the psychological logicians take the non-actual to be subjective without further ado. Yet it is utterly incomprehensible why something that has being independently of the judging subject has to be actual, i.e., has to be capable of acting, directly or indirectly, upon the senses. No such connection between the concepts is to be found. One can even give examples to show the opposite. The number One, e.g., is not easily regarded as actual, unless one is a follower of J. S. Mill. On the other hand, it is impossible to credit each human with his own number One; for in that case we should first have to investigate to what extent the properties of these Ones agreed. And if someone said, "One times One is One", and another, "One times One is Two", then we could only register the difference and say: your One has that property, mine this. There could be no talk of a dispute about who is right or of an attempt to instruct; for there is no common object. Obviously this runs entirely contrary to the sense of the word "One" and the sense of the proposition "One times One is One". Since One, as the same for everybody, confronts everyone in the same way, it can no more be investigated by means of psychological observation than the Moon. Should there after all be ideas of the number One in individual minds, then these are still to be distinguished from the number One, just as ideas of the Moon are to be distinguished from the Moon itself. Since the psychological logicians fail to appreciate the possibility of the objective non-actual, they take concepts to be ideas and thereby assign them to psychology. But the true state of affairs asserts itself too forcefully for this to be accomplished easily. And hence a vacillation afflicts the use of the word "idea", so that sometimes it seems to refer to something which belongs to the mental life of the individual and which, in accordance with the psychological laws, amalgamates with other ideas, associates with them; while at other times, to something that confronts everyone in the same way, so that no bearer of ideas[f] is either mentioned or even presupposed. These two uses are incompatible; for the former, associations, amalgamations merely occur within the individual bearer of ideas and merely occur at something that is as private to the bearer of ideas as his joy or pain. It must never be forgotten that the ideas of different people, however similar they may be, which, by the way, we cannot ascertain precisely, nevertheless do not coincide but are to be distinguished. Everyone has his own ideas which cannot also belong to another. Here, of course, I understand "idea" in the psychological sense. The vacillating use of the word causes unclarity and helps the psychological logicians conceal their weakness. When will this finally be put to an end! This way everything will eventually be dragged down into the realm of psychology; the boundary between the objective and the subjective is eroded further and further, and even actual objects are treated psychologically as ideas. For what is *actual* other than a predicate? And what are logical predicates other than ideas? Everything leads thus into idealism and therefore, as an unavoidable consequence, into solipsism. If everyone designated something different by the name "Moon", namely one of his ideas, much like he voices his pain with the exclamation "ouch!", then of course a psychological viewpoint would be justified; but a dispute concerning the properties of the Moon would be pointless: one could perfectly well assert of his moon the opposite of what another says of his with the same right. If we could apprehend nothing but what is internal to ourselves, then a conflict of opinion, a mutual understanding would be impossible since a common ground would be lacking, and such a common ground cannot be an idea in the sense of psychology. There would be no logic appointed to be arbiter in a conflict of opinions.

But lest I give the impression that I am tilting at windmills, let me illustrate this inescapable sinking into idealism with reference to a particular book. For this, I choose Mr B. Erdmann's above mentioned *Logik* as one of the most recent works of the psychological trend, one which might not be denied all significance. First, let us observe the following proposition (I, p. 85):

> "Thus psychology teaches with certainty that the objects of memory and imagination, just as those of deranged hallucinatory and illusionary ideation,[g] are of an ideal nature. ... Ideal, moreover, is the whole range of properly mathematical ideas, from the number-series down to the objects of mechanics."

What a motley! So, the number Ten should stand on the same level as hallucinations! Here obviously the objective non-actual is being conflated with the subjective. Some objective things are actual, others not. *Actual* is only one of many predicates and is of no more concern to logic than, for instance, the predicate *algebraic* as applied to a curve. Naturally, this conflation ensnares Mr Erdmann in metaphysics, however much he strives to distance himself from it. I take it to be a sure sign of error should logic have to rely on metaphysics and psychology, sciences which themselves require logical principles. Where in that case is the real basic ground on which everything rests? Or is the situation like that of Münchhausen who pulled himself out of the bog by his own hair? I strongly doubt that this is possible and surmise that Mr Erdmann remains enmired in the psychologico-metaphysical bog.

There is no real objectivity for Mr Erdmann; for everything is idea. Let us convince ourselves of this on the basis of his own statements. We read on p. 187 of the first volume:

> "As a relation between what is ideated,[h] a judgement presupposes at least two relata between which the relation holds. As a *predication* about what is ideated, it demands that one of these relata be determined as the object of which is predicated, the subject, ... the other as the object that is predicated, the predicate ...".[i]

To begin with, we see here that both the subject of the predication and the predicate are designated as object or what is ideated. Here "what is ideated" could have been written instead of "object", for we read (I, p. 81): "For objects are what is ideated." And also conversely, everything ideated is meant to be object. On p. 38 one finds:

> "According to its origin, the ideated divides into objects of sense perception and self-consciousness on the one hand, and into primitive and derived on the other."

But what has its source in sense perception or self-consciousness is of course mental in nature. The objects, what is ideated, and hence also subject and predicate, are thereby assigned to psychology. This is confirmed by the following passage (I, pp. 147 and 148):

> "It is the ideated or the idea in general. For both are one and the same: the ideated is the idea, the idea what is ideated."

The word "idea" is indeed usually taken in a psychological sense; that this is also Mr Erdmann's use can be seen from the passages:

> "Consciousness therefore is the genus of feeling, ideation, wanting" (p. 35)

and

> "Ideation is composed of the ideas ... and the passages of ideas" (p. 36).[j]

After this we should not be surprised that an object comes into being in a psychological manner:

> "Insofar as a perception-mass ... presents the same as earlier stimuli and the excitations triggered by them, it *reproduces* the memory traces that originated from this same of the earlier stimuli and *amalgamates* with them into an object of the apperceived idea" (I, p. 42).[k]

On p. 43 it is then shown by way of an example how a steel engraving of Raphael's Sistine Madonna comes into being in a purely psychological way, without steel press, ink and paper. After all this, no doubt can remain that the object about which a predication is made, the subject, is in Mr Erdmann's opinion taken to be an idea in the psychological sense of the word, as is the predicate, the object that is predicated. If this were right, then it could not be truthfully predicated of any subject that it was green, since there are no green ideas. Moreover, I could not predicate of any subject that it was independent of its being ideated or of myself, the bearer of ideas, any more than my decisions can be independent of my wanting and of myself, the wanting subject; rather they would be destroyed with me, if I were destroyed. So there is no real objectivity for Mr Erdmann, which follows also from his taking the ideated or ideas in general, objects in the most general sense of the word, as highest genus (*γενικώτατον*, *genus summum*) (p. 147). He is thus an idealist. If the idealists were consistent, they would regard the proposition "Charlemagne conquered the Saxons" neither as true nor as false but as fiction, just as we are accustomed to understand, for example, the proposition "Nessus carried Deïanira across the river Euenus"; for the proposition "Nessus did not carry Deïanira across the river Euenus" could likewise only be true if the name "Nessus" had a bearer. It would probably not be straightforward to drive the idealists out of this point of view. But one does not have to tolerate that they corrupt the sense of the proposition in this way, as if I wanted to predicate something of my idea when I speak of Charlemagne; what I want is to designate a man who is independent of myself and my ideation and to predicate something of him. One can grant the idealists that the achievement of this intention is not entirely certain, that without wanting to, I perhaps lapse from truth into fiction. But this has no bearing on the sense. With the proposition "This blade of grass is green", I predicate nothing of any idea of mine; I am not designating any of my ideas by means of the words "this blade of grass"; and were I doing so, the proposition would be false. At this point a second falsification intrudes, namely, that my idea of the green is being predicated of my idea of this blade of grass. I repeat: there is in no way any mention of my ideas in this proposition; an entirely different sense is being smuggled in here. Incidentally, I do not understand at all how an idea can be predicated of something. It would equally be a falsification if one were to say that in the proposition "The Moon is independent of me and my ideation", my idea of independence of myself and my ideation is predicated of my idea of the Moon. This would be to surrender objectivity in the proper sense of the word, and to put something entirely different in its place. No doubt it is possible that, in making a judgement, such a play of ideas should occur; but that is not the sense of the proposition. It may also be observed that for one and the same proposition, and one and the same sense of the proposition, the play of ideas can be entirely different. Yet it is this logically irrelevant side-show which our logicians take as the proper object of their research.

How understandable it is that the nature of the subject matter recoils against sinking into idealism, and that Mr Erdmann does not want to admit that, for him, there is no real objectivity; but equally understandable is the futility of his endeavor. For if all subjects and predicates are ideas, and if all thinking is nothing but production, connection, change of ideas, then it is impossible to see how anything objective can ever be achieved. An indication of this futile resistance is the very use of the words "what is ideated" and "object" which at first apparently designate something objective, rather than an idea, but only apparently; for it becomes manifest that they refer to the same. To what purpose, then, this superfluity of expressions? This is not hard to guess. One may notice in addition that there is mention of the object of an idea, although the object is taken to be itself an idea. That would then be an idea of an idea. What relation between ideas might be designated by this? Unclear as this is, it is intelligible enough how, in the clash between the nature of the subject matter and idealism, such maelstroms can arise. Everywhere, we find the object of which I form an idea confused with this idea itself, only for their differences to come into prominence later. This conflict is manifest in the following proposition:

> "For an idea whose object is general is thus, as such, as an event of consciousness, no more general than an idea itself is real because its object is posited as real, or than an object that we experience as sweet ... is presented by ideas which themselves are sweet" (I, p. 86).[1]

Here, the true state of affairs asserts itself with force. I could almost agree; but note that according to Erdmann's principles the object of an idea and the object which is presented by ideas are themselves ideas; and so we can see that all struggle is futile. Further, I ask to keep in mind the words "as such" that are similarly used on p. 83 in the following passage:

> "When actuality is predicated of an object, the real subject of a judgement is not the object or the ideated as such but is rather *the transcendent*, which is presupposed as the ground of being of the ideated, through which the ideated presents itself. Here, the transcendent should not be regarded

> as the unknowable ... rather its transcendence is only to consist in its independence from being ideated."[m]

Again, a vain attempt to haul oneself out of the bog! If we take these words seriously, then it is claimed that in this case the subject is not an idea. Yet if that is possible, then it cannot be seen why with other predicates, which express specific kinds of efficacy or actuality, the real subject must surely be an idea, e.g., as in the judgement "the earth is magnetic". So we would then arrive at the view that the real subject will be an idea in only a few judgements. However, once it is granted that it is not essential for either the subject or the predicate to be an idea, the rug is pulled out from under the whole psychological logic. All psychological considerations, which now swell our logic texts, thus prove to be pointless.

In fact, however, we probably should not take Mr Erdmann's notion of transcendence too seriously. I merely have to remind him of one of his statements (I, p. 148):

> "Also subordinate to the highest genus is the *metaphysical* limit of our ideation, the transcendent",

and he is sunk; for the highest genus (*γενικώτατον*, *genus summum*) is, according to him, just the ideated, or the idea in general. Or might the word "transcendent" be used above in a different sense from here? In any case, one would suppose, the transcendent should be subordinate to the highest genus.

Let us dwell a moment longer on the expression "as such". I take the case where someone wants me to think that all objects are nothing but images on the retina of my eyes. Very well! I make no comment yet. But now he maintains that the tower is bigger than the window through which I take myself to be seeing it. To this, I would then say: either not both the tower and the window are retinal images in my eye, in which case the tower may be bigger than the window; or the tower and the window are, as you say, images on my retina, in which case the tower is not bigger but, rather, smaller than the window. At this point, he tries to relieve his embarrassment by resort to "as such", and says: the retinal image of the tower as such is, admittedly, not bigger than that of the window. Here I almost want to jump out of my skin and shout at him: well, in that case the retinal image of the tower is not at all bigger than that of the window; and if the tower were the retinal image of the tower and the window were the retinal image of the window, then the tower simply would not be bigger than the window, and if your logic teaches you otherwise, then it is good for nothing. This "as such" is an excellent invention of unclear writers who want to say neither yes nor no. However I do not tolerate such wavering between the two, but rather ask: if actuality is predicated of an object, is the real subject of the judgement the idea, yes or no? If not, then it arguably is the transcendent, which is presupposed as the ground of being of such an idea. But the transcendent is itself what is ideated or an idea. Thus we are driven to assume that the subject of the judgement is not the ideated transcendent, but rather the transcendent which is presupposed as the ground of being of this ideated transcendent. So we would have to go on forever; and no matter how far we were to go, we could never get past the subjective. Incidentally, the same game could also be initiated with the predicate, and not only with the predicate *actual* but just as well with, for example, *sweet*. We should then first say: if one predicates actuality or sweetness of an object, then the real predicate is not the ideated actuality or sweetness, but rather the transcendent which is presupposed as ground of the ideated. Yet we would not be able to come to rest with this, but would always be driven further. What can we learn from this? That psychological logic is on the wrong track when it conceives of the subject and predicate of judgements as ideas in the psychological sense, that psychological considerations are no more appropriate in logic than in astronomy and geology. If we ever want to get past the subjective, then we have to think of cognition[n] as an activity that does not create what is cognised, but grasps what is already there. The image of grasping is well suited to elucidate the issue. When I grasp a pencil, many things take place in my body: stimulation of the nerves, changes in the tension and the pressure of muscles, tendons and bones, changes in the circulation of the blood. The sum of these processes, however, is not the pencil, nor do they create it. The latter has being independently of these processes. It is essential to grasping that there is something which is grasped; the inner changes alone are not the grasping. Similarly, what we mentally apprehend has being independently of this activity, of the ideas and their changes that are part of or accompany the apprehension; it is neither the sum of these processes nor is it created as part of our mental life.

Let us see further how subtler differences in the subject matter are smudged over by the psychological logicians. The point was already made in the case of characteristic mark and property. This is connected with the distinction I have emphasised between object and concept, as well as that between concepts of first and second level. Naturally, these differences are indiscernible by psychological logicians; for them everything is idea. For this reason, the proper conception of those judgements which we express in English by "there is"[o] also eludes them. This existence is mixed up by Mr B. Erdmann (*Logik* I, p. 311) with actuality, which, as we saw, is also not clearly distinguished from objectivity. Of what are we in fact asserting that it is actual when we say, there are square roots of Four? Is it Two or −2? But neither the one nor the other is in any way named. And if I wanted to say that the number Two acted or was active or actual, then this would be false and quite different from what I want to say with the proposition "There are square roots of Four". The confusion here before us is almost as bad as can be; since it does not involve concepts of the same level, but rather collapses a concept of the first level with a concept of the second. This is a hallmark of the obtuseness of psychological logic. Someone who has, generally, attained a more open point of view may wonder how such a mistake could be made by a professional logician; but before one can gauge the scale of such an error, one obviously has to recognise the distinction between concepts of first and second level in the first place, and psychological logic will presumably be incapable of that. The greatest barrier to this will be that the proponents are so exceedingly in awe of the psychological profundity, which however is nothing but psychological corruption of logic. And thus our thick logic books come about, bloated with unhealthy psychological lard, concealing all finer details. A fruitful cooperation between mathematicians and logicians is thereby rendered impossible. While the mathematician defines objects, concepts and relations, the psychological logician is listening in on the coming and going of ideas, and in the end the mathematician's defining can only appear foolish to him, since it does not convey the nature of ideas. He looks into his psychological peep box[p] and says to the mathematician: I see nothing at all of what you are defining. And the latter can merely answer: no wonder! For it is not where you are looking for it.

This may suffice to put my logical standpoint into a clearer light by the contrast. The distance from psychological logic seems to me to be as wide as the sky, so much so that there is no prospect that my book will have an effect on it immediately. My impression is that the tree that I have planted has to heave an incredible load of stone to make space and light for itself. Still, I will not give up all hope that my book will eventually aid the overthrow of psychological logic. To make the proponents of the latter come to terms with my book, some acknowledgement from the mathematicians will not come amiss. And indeed, I believe that I can expect some support from this quarter, since the mathematicians have in the end to make common cause against the psychological logicians. As soon as the latter deign to engage with my book seriously, even if only in order to refute it, I shall take myself to have won. For the whole of part II is really a test of my logical convictions. It is from the outset unlikely that such a construction could be built on an insecure, defective basis. But if anyone has different convictions, let him try to build a similar construction on them and he will find, I believe, that it does not work, or at least that it does not work so well. And I could only acknowledge it as a refutation if someone indeed showed that a better, more enduring building can be erected on different basic convictions, or if someone proved to me that my basic principles lead to manifestly false conclusions. But no one will succeed in doing so. And so may this book, even if belatedly, contribute to a renaissance of logic.

Jena in July, 1893.

G. Frege

Table of contents

Introduction
Task, demands on the conduct of proof, Dedekind's system, Schröder's class .... page 1

I. Exposition of the concept-script

1. The primitive signs

Introduction to function, concept, relation

§ 1. The function is unsaturated .... page 5
§ 2. Truth-values, reference and sense, thought, object .... " 6
§ 3. Value-range of a function, concept, extension of a concept .... " 7
§ 4. Functions with two arguments .... " 8

Signs for functions

§ 5. Judgement and thought, judgement-stroke and horizontal .... page 9
§ 6. Negation-stroke, fusion of horizontals .... " 10
§ 7. Equality-sign .... " 11
§ 8. Generality, German letter, its scope, fusion of horizontals .... " 11
§ 9. Designation of the value-range, small Greek vowel, its scope .... " 14
§10. More precise determination of what the value-range of a function is supposed to be .... " 16
§11. Replacement of the definite article, the function $\backslash\xi$ .... " 18
§12. Conditional-stroke, and, neither–nor, or, subcomponents, supercomponent .... " 20
§13. If, all, each, subordination, particular affirmative proposition, some ... " 23

Inferences and consequences

§14. First mode of inference .... page 25
§15. Second mode of inference, contraposition .... " 26
§16. Third mode of inference .... " 30
§17. Roman letters, transition from Roman to German letters .... " 31
§18. Laws in concept-script notation (I, IV, VI) .... " 34

Extension of the notation for generality

§19. Generality with respect to functions; function-, object-letters .... page 34
§20. Laws in concept-script notation (IIa, III, V) .... " 35
§21. Functions and concepts of first- and second-level .... " 36
§22. Examples of second-level functions, unequal-levelled functions and relations .... " 38
§23. Kinds of arguments and argument-places, second-level functions with arguments of the second and third kind .... " 39
§24. General explanation of the use of function-letters .... " 41
§25. Generality with respect to second-level functions, law IIb .... " 42

2. Definitions

General remarks

§26. Classification of signs, names, markers, concept-script proposition, transition sign .... page 43
§27. The double-stroke of definition .... " 44
§28. Correct formation of names .... " 45
§29. When does a name refer to something? .... " 45
§30. Two ways to form a name .... " 46
§31. Our primitive names refer to something .... " 48
§32. Every concept-script proposition expresses a thought .... " 50
§33. Principles of definition .... " 51

Special definitions

§34. Definition of the function $\xi \frown \zeta$ .... page 52
§35. Representation of second-level functions by first-level ones .... " 54
§36. The double value-range. The extension of a relation .... " 54
§37. Definition of the function Iξ .... " 55
§38. Definition of the function $\rangle\xi$ .... " 56
§39. Definition of the function ꝁξ .... " 57
§40. Definition of the function ℘ξ .... " 57
§41. Definition of 0 .... " 57
§42. Definition of 1, concept of cardinal number .... " 58
§43. Definition of f .... " 58
§44. Some concept-script propositions as examples .... " 58
§45. Definition of the function $\preceq\xi$, following and preceding in a series ...... " 59
§46. Definition of the function $\smile\xi$ .... " 60

3. Derived laws

§47. Summary of the basic laws .... page 60
§48. Summary of the rules .... " 61
§49. Derivation of some propositions from (I) .... " 64
§50. Derivation of the main propositions of the function $\xi = \zeta$ .... " 65

§51. Derivation of some propositions from (IV) .... page 68
§52. Derivation of some propositions from (V) and (VI) .... " 69

II. Proofs of the basic laws of cardinal number

§53. Preliminaries .... page 70

A. *Proof of the proposition*

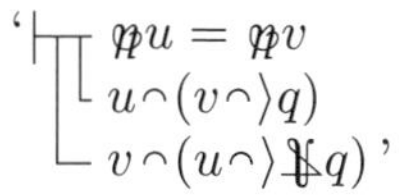

a) Proof of the proposition

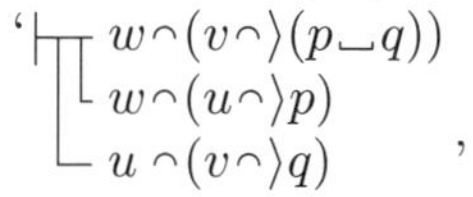

§54 to §59. Definition of the function $\xi \smile \zeta$, propositions (1) to (19) .... page 70

b) Proof of the proposition

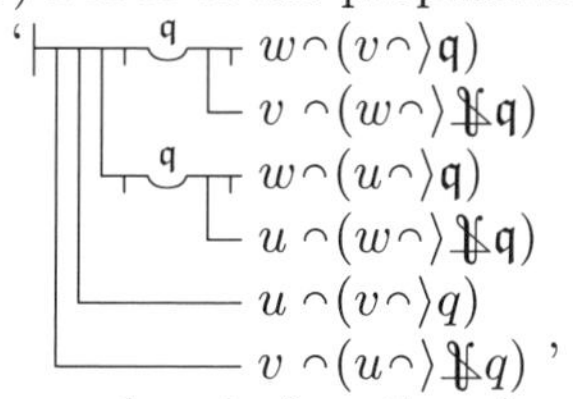

and end of section A

§60 to §65. Propositions up to (32) .... page 80

B. *Proof of the proposition* ‘$\vdash$ If’

a) Proof of the proposition

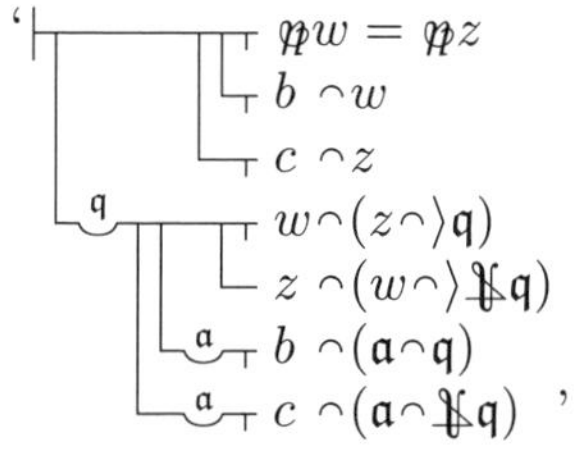

§66 to §77. Propositions up to (56) .... page 86

b) Proof of the proposition

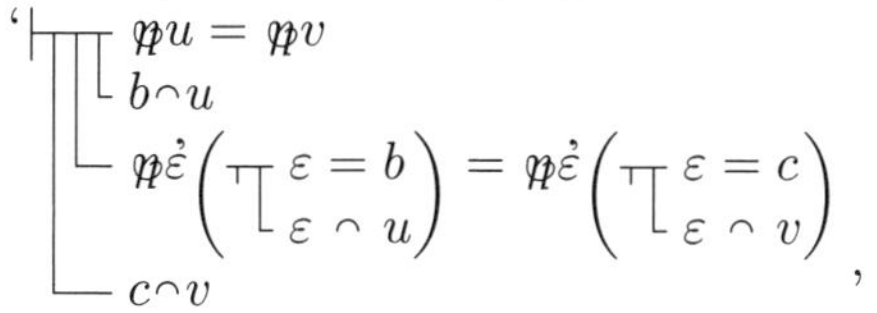

and end of section B

§78 to §87. Propositions up to (71) .... page 101

Γ. *Proof of the proposition* ‘⊢ I∦f’

a) Proof of the proposition

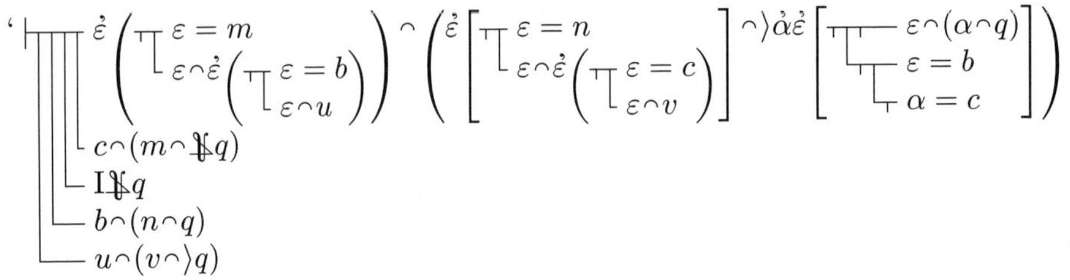

§88 to §91. Propositions up to (84) .... page 113

b) Proof of the proposition

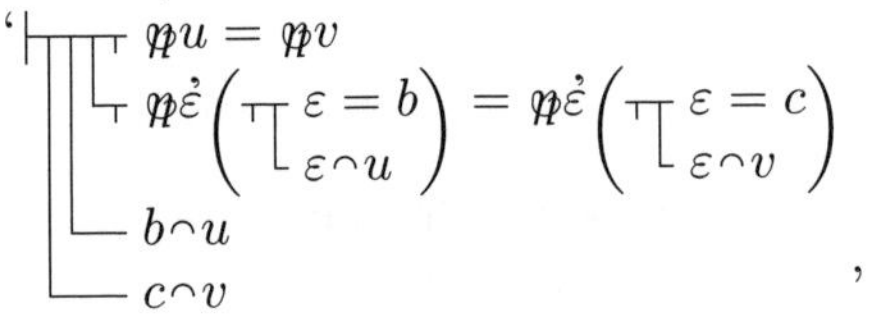

and end of section Γ

§92 to §95. Propositions up to (90) .... page 121

Δ. *Proofs of some propositions concerning the cardinal number Zero*

a) Proof of the proposition

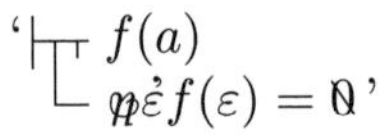

§96 to §97. Propositions up to (95) .... page 127

b) Proof of the proposition

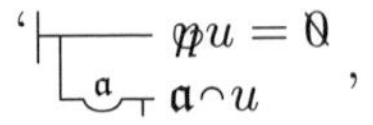

and of some corollaries

§98 to §101. Propositions up to (107) .... page 128

E. *Proofs of some propositions of the cardinal number One*

§102 to §107. Propositions up to (122) .... page 131

Z. *Proof of the proposition*

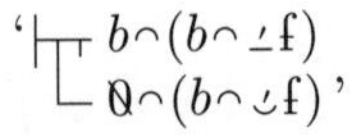

a) Proof of the proposition

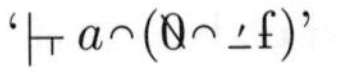

§108 and §109. Propositions up to (126) .... page 137

b) Proof of the proposition

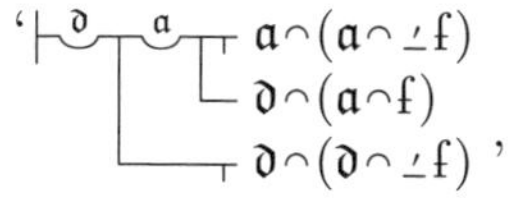

and end of section Z

§110 to §113. Propositions up to (145) .... page 139

H. *Proof of the proposition*

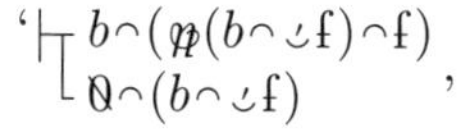

§114 to §119. Propositions up to (155) .... page 144

Θ. *Some corollaries*

§120 and §121. Propositions up to (161) .... page 149

I. *Proof of some propositions of the cardinal number Endlos*

a) Proof of the proposition

§122 to §125. Definition of ∾, propositions up to (167) .... page 150

b) Proof of the proposition

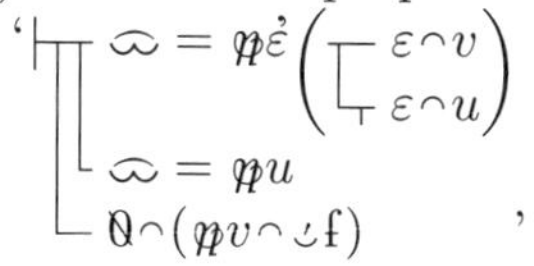

§126 to §127. Propositions up to (172) .... page 154

c) Proof of the proposition

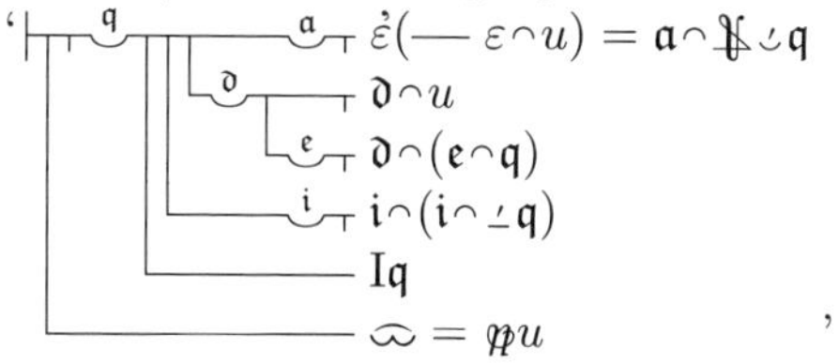

§128 to §143. Propositions up to (207) .... page 160

d) Proof of the proposition

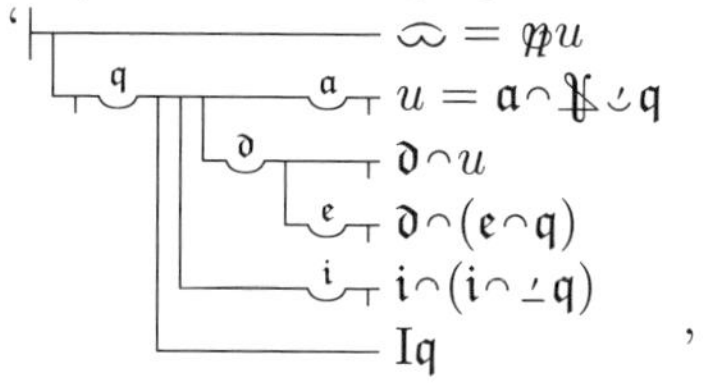

§144 to §157. Definitions of the functions $\xi;\zeta$, $\xi \simeq \zeta$, $\xi \nwarrow \zeta$. Propositions up to (263) .... page 178

K. *Proof of the proposition*

$$`\vdash \neg \big(\forall \mathfrak{A}, \mathfrak{q}\ \neg (u = \mathfrak{A} \mathbin{\underline{\measuredangle}} \mathfrak{q}) \big) \rightarrow \mathfrak{d} \frown (\mathfrak{p}u \frown \smile \mathfrak{f})\text{'}$$

a) Proof of the proposition

$$`\vdash x; \mathfrak{1} \frown (y; n \frown \smile (q \succeq \mathfrak{f})) \rightarrow \big(\mathrm{I}q \rightarrow \big(\neg (y \frown (y \frown \underline{\smile} q)) \rightarrow \mathfrak{p}(x; y \mathbin{\underline{\measuredangle}} q) = \mathfrak{p}(\mathfrak{1}; n \mathbin{\underline{\measuredangle}} \mathfrak{f}) \big) \big)\text{'}$$

§158 to §165. Definition of the function $\xi \mathbin{\underline{\measuredangle}} \zeta$. Propositions up to (298) ... page 201

b) Proof of the proposition

$$`\vdash \mathfrak{d} \frown (n \frown \smile \mathfrak{f}) \rightarrow n = \mathfrak{p}(\mathfrak{1}; n \mathbin{\underline{\measuredangle}} \mathfrak{f})\text{'}$$

and end of section K

§166 to §171. Propositions up to (327) .... page 217

Λ. *Proof of the proposition*

$$`\vdash \mathfrak{d} \frown (\mathfrak{p}u \frown \smile \mathfrak{f}) \rightarrow \neg \forall \mathfrak{A}, \mathfrak{q}\ \neg \big(\grave{\varepsilon}(-\ \varepsilon \frown u) = \mathfrak{A} \mathbin{\underline{\measuredangle}} \mathfrak{q} \big)\text{'}$$

§172 to §179. Propositions up to (348) .... page 224

Appendices

1. Table of the basic laws and propositions immediately following from them page 239
2. Table of definitions .... " 240
3. Table of important theorems .... " 242
Index " 252

Introduction

In my *Grundlagen der Arithmetik*[1] I aimed to make it plausible that arithmetic is a branch of logic and needs to rely neither on experience nor intuition as a basis for its proofs. In the present book this is now to be established by deduction of the simplest laws of cardinal number by logical means alone. In order for this to be done convincingly, significantly higher demands have to be imposed on the conduct of proof than is usual in arithmetic.[2] We have to mark out in advance a few modes of inference and consequence, and no step is allowed to occur which is not in accordance with one of them. Thus, in the transition to a new judgement, one is no longer to be satisfied, as mathematicians up until now nearly always have been, with its obvious correctness, rather it must be analysed into the simple logical steps of which it consists, and these are often not particularly few. Herein, no presupposition can remain unnoticed; every required axiom must be uncovered. For it is precisely the presuppositions that are made tacitly or without clear awareness that bar insight into the epistemological nature of a law.

For such an undertaking to succeed, the required concepts must, obviously, be sharply characterised. This applies especially to what mathematicians intend to designate by the word 'set'. *Dedekind*[3] seems to use the word 'system' with the same intention. But, the considerations which appeared four years earlier in my *Grundlagen* notwithstanding, no clear insight into the nature of the subject matter can be found in Dedekind, although he sometimes comes close to the core of things, as in this passage (p. 2): "Such a system S ... is completely determined if it is determined of every thing whether it is an element of S or not. Therefore the system S is the same as the system T, in signs $S = T$, if every element of S is also an element of T

[1] Breslau 1884.

[2] Compare my *Grundlagen* §90.

[3] *Was sind und was sollen die Zahlen?* Braunschweig 1888.

and every element of T is also an element of S." In contrast, other passages go astray, e.g., the following (pp. 1 and 2): "It very often happens that different things a, b, c ... considered for whatever reason under a common aspect, are joined together in the mind, so that they are said to form a *system* S." Here the common aspect admittedly provides an inkling of the truth; but this considering, this joining together in the mind, is no objective characteristic. I ask: in whose mind? If they are joined together in one mind and not in another, do they still make up a system? What is supposed to be joined together in my mind surely must be in my mind. Do the things outside of me not compose systems? Is a system a subjective construction within a single mind? Is then the constellation Orion a system? And what are its elements? The stars, the molecules or the atoms? The following passage (p. 2) is noteworthy: "It is advantageous for the uniformity of expression to allow for the special case where a system S consists of a *single* (one and only one) element a; i.e., where the thing a is an element of S, but each thing distinct from a is not an element of S." This is later (p. 3) so understood that every element s of a system S may itself be viewed as a system. Since in this case element and system coincide, it here becomes very perspicuous that according to *Dedekind*, it is in fact the elements that constitute the system. In his *Vorlesungen über die Algebra der Logik*,[1] E. *Schröder* takes a step further than *Dedekind* by drawing attention to a connection which the latter has seemingly overlooked between his systems and concepts. In fact, what *Dedekind* actually means when he calls a system part of a system (p. 2) is the subordination of a concept under a concept, or the falling of an object under a concept, cases that he, like *Schröder*, fails to distinguish owing to a common error in their views; for *Schröder* too, at bottom, takes the elements to be what his *class* consists in. Thus, on his view, an empty class should not really occur any more than an empty system should on *Dedekind*'s view; yet the demand arising from the nature of the subject matter makes itself felt on both authors in different ways. *Dedekind* continues the passage quoted above: "In contrast, for specific reasons we here want wholly to exclude the empty system which contains no element, although it may be convenient for other investigations to invent such a fiction." Such a fiction would thus be admissible; it is dispensed with only for certain reasons. *Schröder* ventures the fiction of an empty class. So, it seems that both agree with

[1] Leipzig 1890, p. 253.

many mathematicians that one may freely invent something that is not there, even something unthinkable; for if the system consists in its elements then the system will be abolished together with its elements. As concerns where the limits of such capricious fiction are, indeed whether there are any at all, little clarity and agreement may be found; yet the correctness of a proof may hinge on it. I believe I have settled this issue for all reasonable people, in my *Grundlagen der Arithmetik* (§92 and ff.) and in my lecture *Ueber formale Theorien der Arithmetik.*[1] *Schröder* invents his Zero and thereby entangles himself in great difficulties.[2] Accordingly, although there is lack of insight in both *Schröder* and *Dedekind*, the true situation makes itself felt every time a system has to be specified. *Dedekind* then cites the properties a thing must have in order to belong to the system, i.e., he defines a concept in virtue of its characteristic marks.[3] Now, if it is the characteristic marks that constitute the concept, rather than the objects falling under it, then there are no problems and worries concerning an empty concept. Then, certainly, an object can never simultaneously be a concept; and a concept under which only one object falls must not be confused with it. So, in the end, the point will stand that a statement of number contains a predication about a concept.[4] I have reduced cardinal number to the relation of equinumerosity and this to single-valued correlation. Much the same applies to the word 'correlation' as to the word 'set'. Both are not uncommonly used in contemporary mathematics, usually without any deeper understanding of what one really wants to designate by them. If my thought, that arithmetic is a branch of pure logic, is correct then a purely logical expression for 'correlation' must be selected. I use 'relation' for this purpose. Concept and relation are the foundation stones on which I build my construction.

Yet even after the concepts are sharply circumscribed, it would be hard, almost impossible, to satisfy the demands necessarily imposed here on the conduct of proof without special auxiliary means. Such an auxiliary means is my concept-script, whose exposition will be my first task. The following may be noted in advance. It will not

[1] *Sitzungsberichte der Jenaischen Gesellschaft für Medicin und Naturwissenschaft, Jahrg. 1885*, session of 17th July.

[2] Compare E. G. *Husserl* in *Göttinger gel. Anzeigen*, 1891, no. 7, p. 272, who however does not seem to untie the knot.

[3] On *concept, object, property, characteristic marks* compare my *Grundlagen* §§38, 47, 53, and my essay "Ueber Begriff und Gegenstand" in *Vierteljahresschrift für wissenschaftliche Philosophie*, XVI, 2.

[4] §46 of my *Grundlagen*.

always be possible to give a regular definition of everything, simply because our ambition has to be to reduce matters to what is logically simple, and this as such allows of no proper definition. In such a case, I have to make do with gesturing at what I mean. The important thing is that I be understood and therefore I will aim to unfold the subject matter gradually, rather than strive at the outset for full generality and final expression. Someone may perhaps wonder about the frequent use of quotation marks. It is by this means that I distinguish cases in which I speak of the sign itself from cases in which I speak of its reference. Pedantic though it may seem, I nevertheless take this to be necessary. It is remarkable how an imprecise manner of speaking and writing, perhaps originally used only for ease and brevity yet with full awareness of its imprecision, can finally addle the thinking after this awareness has disappeared. Thus people have managed to regard number-signs as numbers, the name as what is named, the mere auxiliary means as the object of arithmetic itself. Such experiences teach how necessary it is to place the highest demands on the accuracy of our manner of speaking and writing. And I have made every effort to meet them, at least in all cases where it seemed to me that something depends upon it.

I. Exposition of the concept-script

1. The primitive signs

Introduction to function, concept, relation[1]

§1. If the task is to give the original reference of the word 'function' as used in mathematics, then it is easy to slip into calling a function of x any expression that is formed from 'x' and certain specific numbers by means of the notations for sum, product, power, difference, etc. This is incorrect, since it presents the function as an *expression*, as a combination of signs, and not as what is designated by these. One will therefore be tempted to say 'reference of an expression' instead of 'expression'. Now, in this expression the letter 'x' occurs, which does not refer to a number like the sign '2' does but only indicates one indeterminately. For different number-signs put in place of 'x' we generally obtain different references. E.g., if we insert into the expression '$(2 + 3 . x^2) . x$' the number-signs '0', '1', '2', '3' one after the other for x, then we obtain as reference the numbers 0, 5, 28, 87, respectively. None of these references can claim to be our function. The nature of the function reveals itself, rather, in the bond that it establishes between the numbers whose signs we put for 'x' and the numbers that then result as the reference of our expression; a bond which is manifested by the course of the curve whose equation in rectangular coordinates is

$$`y = (2 + 3 . x^2) . x'$$

The nature of the *function* lies therefore in that part of the expression that is present without the 'x'. The expression of a *function* is *in need of completion, unsaturated.*

[1] Compare my lecture on *Function und Begriff* (Jena 1891) and my essay on "Begriff und Gegenstand" in the *Vierteljahrszeitschrift für wissensch. Phil.* XVI, 2. My *Begriffsschrift* (Halle a. S. 1879) no longer corresponds entirely to my present standpoint; it is therefore to be consulted as an elucidation of what is presented here only with caution.

The letter 'x' serves only to hold open places for a number-sign that is to complete the expression, and so marks the special kind of need for completion that constitutes the peculiar nature of the function just designated. In the sequel, instead of 'x' the letter 'ξ' will be used for this purpose.[1] This place-holding is to be understood in such a way that all places occupied by 'ξ' must always be filled just by the same, and never by different signs. I call these places *argument places*, that whose sign (name) takes this place in a particular case I call the *argument* of the function for this case. The function is completed by the argument; and that which results from this completion I call the *value* of the function for the argument. Thus we obtain a name of the value of a function for an argument if we fill in the argument places of the name of the function with the name of the argument. So, e.g., '$(2 + 3 . 1^2) . 1$' is a name of the number 5, composed of the function-name '$(2 + 3 . \xi^2) . \xi$' and '1'. So the argument is thus not to be considered part of the *function*, but serves rather to complete the, in itself, *unsaturated function*. When in what follows an expression like 'the function $\Phi(\xi)$' is used, it is always to be borne in mind that 'ξ' contributes to the designation of the function only insofar as it marks its argument places, and that the nature of the function would be unchanged if any other sign were put for 'ξ'.

§2. As means of generating functions, one has added to the fundamental arithmetical operations the taking of limits in its various forms as infinite series, differential quotient, integral; and the word 'function' has come finally to be understood in such a general way that the connection between the value of a function and its argument may no longer be expressible by the signs of analysis, but only by words. A further extension has consisted in admitting complex numbers as arguments and hence also as function-values. In both these directions I have gone further. For on the one hand the signs of analysis were not always sufficient, and on the other hand not all of them were being used for the formation of function-names in that, e.g., '$\xi^2 = 4$' and '$\xi > 2$' were not accepted as names of functions, as I accepted them to be. But with this it is acknowledged at the same time that the range of function-values cannot remain restricted to numbers; for if I take the numbers 0, 1, 2, 3, one after the other, as the argument of the function, $\xi^2 = 4$, then I do not obtain numbers.

$$\text{`}0^2 = 4\text{'},\ \text{`}1^2 = 4\text{'},\ \text{`}2^2 = 4\text{'},\ \text{`}3^2 = 4\text{'}$$

are expressions of thoughts, some true, some false. I express it like this: the value of

[1] With this, however, nothing is meant to be stipulated for the concept-script. Rather, 'ξ' itself will never occur in the concept-script developments; I will only use it in the exposition of the concept-script and in elucidation.

the function $\xi^2 = 4$ is either the *truth-value* of the true, or that of the false.[1] It is already clear from this that I do not want to assert anything yet when I simply write down an equation, but that I merely *designate* a truth-value; just as I assert nothing when I simply write down '2^2', but merely *designate* a number. I say: the names '$2^2 = 4$' and '$3 > 2$' *refer to* the same truth-value, which I call for short *the True.* Likewise, for me, '$3^2 = 4$' and '$1 > 2$' *refer to* the same truth-value, which I call for short *the False*, exactly as the name '2^2' *refers to* the number Four. Accordingly, I call the number Four the *reference* of '4' and '2^2', and I call the True the reference of '$3 > 2$'. I distinguish, however, the *reference* of a name from its *sense.* '2^2' and '$2 + 2$' do not have the same *sense*, and nor do '$2^2 = 4$' and '$2 + 2 = 4$' have the same *sense.* The sense of a name of a truth-value I call a *thought.* I say further that a name *expresses* its sense and *refers to* its reference. I *designate* with a name that which it refers to.

The function $\xi^2 = 4$ can therefore only have two values, namely the True for the arguments 2 and -2 and the False for every other argument.

Moreover, the domain of what is admissible as an argument has to be expanded and extended to objects in general. *Objects* stand opposed to functions. Accordingly, I count as an *object* everything that is not a function, e.g., numbers, truth-values and the value-ranges introduced below. Thus, names of objects, the *proper names*, do not in themselves carry argument places; like the objects themselves, they are saturated.

§3. I use the words

> "the function $\Phi(\xi)$ has the same *value-range* as the function $\Psi(\xi)$"

always as co-referential with the words

> "the functions $\Phi(\xi)$ and $\Psi(\xi)$ always have the same value for the same argument."

This is the case with the functions $\xi^2 = 4$ and $3 . \xi^2 = 12$, at least if numbers are taken as arguments. We can, however, also think of the signs for square and multiplication as defined in such a way that the function

$$(\xi^2 = 4) = (3 . \xi^2 = 12)$$

has the True as value for every arbitrary argument. Here an expression of logic can also be used: "the concept *square root of 4* has the same extension as the concept

[1] I argued for this in more detail in my essay "Über Sinn und Bedeutung" in the *Zeitschrift für Philosophie und philosophische Kritik*, vol. 100.

something such that the triple of its square is 12". For such functions, whose value is always a truth-value, we can hence say 'extension of the concept' instead of 'value-range of the function'; and it seems appropriate simply to call a *concept* any function whose value is always a truth-value.

§4. So far only functions with a single argument have been talked about; but we can easily pass on to *functions with two arguments.* These stand *in need of double completion* insofar as a function with one argument is obtained after their completion by one argument has been effected. Only after yet another completion do we arrive at an object, and this object is then called the *value* of the function for the two arguments. Just as the letter 'ξ' served us in the case of functions with one argument, so here we use the letters 'ξ' and 'ζ' to indicate the double unsaturatedness of functions with two arguments, as in

$$`(\xi + \zeta)^2 + \zeta'.$$

By inserting e.g., '1' for 'ζ', we saturate the function in such a way that we have in $(\xi + 1)^2 + 1$ only a function with one argument. This use of the letters 'ξ' and 'ζ' must always be kept in view, whenever an expression like 'the function $\Psi(\xi, \zeta)$' occurs (cf. 2nd fn. in §1). I call *ξ-argument places* the places in which 'ξ' stands and *ζ-argument places* those in which 'ζ' stands. I say that the ξ-argument places are *related* to one another, and likewise the ζ-argument places, while I do not describe a ξ-argument place as *related* to a ζ-argument place.

The functions with two arguments, $\xi = \zeta$ and $\xi > \zeta$, always have a truth-value as value (at least if the signs '=' and '>' are explained in the appropriate way). Such functions we will suitably call relations. In the first relation, for example, 1 stands to 1, in general every object to itself; in the second, for example, 2 stands to 1. We say that an object Γ *stands in the relation* $\Psi(\xi, \zeta)$ *to* the object Δ if $\Psi(\Gamma, \Delta)$ is the True. Likewise we say that the object Δ *falls under* the concept $\Phi(\xi)$ if $\Phi(\Delta)$ is the True. It is here presupposed, of course, that the function $\Phi(\xi)$, and equally $\Psi(\xi, \zeta)$, always has a truth-value as value.[1]

[1] There is a difficulty here which can easily obscure the true state of affairs and thereby arouse suspicion concerning the correctness of my conception. If we compare the expression 'the truth-value of: that Δ falls under the concept $\Phi(\xi)$' with '$\Phi(\Delta)$', then we see that the '$\Phi(\ )$' really corresponds to 'the truth-value of: that ( ) falls under the concept $\Phi(\xi)$', and not to 'the concept $\Phi(\xi)$'. So the latter words do not really designate a concept (in our sense), even though the linguistic form makes it look as if they do. On the predicament in which language here finds itself, cf. my essay "Ueber Begriff und Gegenstand".

Signs for functions

§5. Above it is already stated that within a mere equation no assertion is yet to be found; with '$2+3=5$' only a truth-value is designated, without its being said which one of the two it is. Moreover, if I wrote '$(2+3=5)=(2=2)$' and presupposed that one knows that $2=2$ is the True, even then I would not thereby have asserted that the sum of 2 and 3 is 5; rather I would only have designated the truth-value of: that '$2+3=5$' refers to the same as '$2=2$'. We are therefore in need of another special sign in order to be able to assert something as true. To this end, I let the sign '$\vdash$' precede the name of the truth-value, in such a way that, e.g., in

$$\text{`}\vdash 2^2 = 4\text{'}$$ [1]

it is asserted that the square of 2 is 4. I distinguish the *judgement* from the *thought* in such a way that I understand by a *judgement* the acknowledgement of the truth of a *thought*. The concept-script representation of a judgement by means of the sign '$\vdash$' I call a *concept-script proposition* or *proposition* for short. I regard '$\vdash$' as composed of the vertical stroke, which I call the *judgement-stroke*, and the horizontal stroke, which I now propose to label simply the *horizontal*.[2] The horizontal will mostly occur conjoined with other signs, as it does here with the judgement-stroke, and will thereby be protected from confusion with the minus-sign. Where it does occur separately, it has to be made somewhat longer than the *minus-sign*. I regard it as a function-name such that

$$—\Delta$$

is the True when Δ is the True, and is the False when Δ is not the True.[3] Accordingly,

$$—\xi$$

is a function whose value is always a truth-value, or a concept according to our stipu-

[1] Here, I often use notations for sum, product, power, in a provisional way, although they are not yet defined, in order to be able to form convenient examples and to facilitate the understanding through hints. It should be kept in mind, though, that nothing rests on the references of these notations.

[2] Earlier I called it the *content-stroke*, when I combined under the expression 'judgeable content' that which I now have learnt to distinguish as truth-value and thought. Cf. my essay "Über Sinn und Bedeutung".

[3] Evidently, the sign 'Δ' must not be without reference, but it has to refer to an object. Names without reference must not occur in concept-script. The stipulation is made such that under all circumstances '$—\Delta$' refers to something, provided only that 'Δ' refers to something. Otherwise $—\xi$ would not be a concept with sharp boundaries, thus in our sense not be a concept at all. I here use the *capital Greek letters* as if they were names referring to something, without stating their reference. Proceeding within concept-script itself, they, just as 'ξ' and 'ζ', will not occur.

lation. Under this concept falls the True and only the True. Thus

$$\text{`}{—}\ 2^2 = 4\text{'}$$

refers to the same as '$2^2 = 4$', namely the True. For, in order to dispense with brackets, I specify that everything standing right of the horizontal, occupying the argument place of the function $—\ \xi$, should be conceived of as a whole unless brackets prohibit this.

$$\text{`}{—}\ 2^2 = 5\text{'}$$

refers to the False, and hence the same as '$2^2 = 5$', whereas

$$\text{`}{—}\ 2\text{'}$$

refers to the False, and hence something different from the number 2. If Δ is a truth-value, then $—\ \Delta$ is the same truth-value, with the result that

$$\Delta = (—\ \Delta)$$

is the True. The latter, however, is the False if Δ is not a truth-value. We can accordingly say that

$$\Delta = (—\ \Delta)$$

is the truth-value of: that Δ is a truth-value.

Accordingly, the function $—\ \Phi(\xi)$ is a concept, and the function $—\ \Psi(\xi, \zeta)$ is a relation, irrespective of whether or not $\Phi(\xi)$ is a concept or $\Psi(\xi, \zeta)$ is a relation.

Of the two signs of which '$\vdash$' is composed, only the judgment-stroke contains the assertion.

§6. We do not need a specific sign to declare a truth-value to be the False provided we have a sign by means of which every truth-value is transformed into its opposite, which is in any case indispensable. I now stipulate:

The value of the function

$$\top\ \xi$$

is to be the False for every argument for which the value of the function

$$—\ \xi$$

is the True; and it is to be the True for all other arguments.

We thus have in

$$\top\ \xi$$

a function whose value is always a truth-value: it is a concept under which all objects fall with the sole exception of the True. From this it follows that '$\top\ \Delta$' always refers to the same as '$\top\ (—\ \Delta)$', as '$—\ \top\ \Delta$', and as '$—\ \top\ (—\ \Delta)$'. We therefore regard '$\top$' as composed of the small vertical stroke, the *negation-stroke*, and the two parts of the horizontal stroke each of which can be regarded as a *horizontal* in our sense. The transition from '$\top\ (—\ \Delta)$' or from '$—\ \top\ \Delta$' to '$\top\ \Delta$', as well as that from '$—\ —\ \Delta$' to '$—\ \Delta$', I will call the *fusion* of horizontals.

According to our stipulation, $\top\, 2^2 = 5$ is the True; thus

$$\vdash\!\top\, 2^2 = 5,$$

in words: $2^2 = 5$ is not the True; or: the square of 2 is not 5.

Thus also: $\vdash\!\top\, 2$.

§7. We have already used the equality-sign rather casually to form examples but it is necessary to stipulate something more precise regarding it.

$$\text{‘}\Gamma = \Delta\text{’}$$

refers to the True, if Γ is the same as Δ; in all other cases it is to refer to the False.

In order to dispense with brackets, I specify that everything standing to the left of the equality-sign up to the nearest horizontal, as a whole refers to the ξ-argument of the function $\xi = \zeta$, insofar as *brackets* do not prevent this; that everything standing to the right of the equality-sign up to the nearest equality-sign collectively refers to the ζ-argument of this function, insofar as *brackets* do not prevent this (compare p. 10).

§8. We considered in §3 the case where an equation such as

$$\text{‘}\Phi(x) = \Psi(x)\text{’}$$

always yields a name for the True, whatever proper name we might insert for ‘x’, provided only that this really refers to an object. We then have the generality of an equality, while in ‘$2^2 = 4$’ we merely have an equality. This difference manifests itself thus: in the former case we have a letter ‘x’ that only indicates indeterminately, while in ‘$2^2 = 4$’ every sign has a determinate reference.[a] In order to obtain an expression for generality, one might have the idea of defining: “Let us understand ‘$\Phi(x)$’ as the True, if the value of the function $\Phi(\xi)$ is the True for every argument; otherwise it shall refer to the False.” Here, it would be presupposed, as in all our considerations of this kind, that ‘$\Phi(\xi)$’ always acquires a reference, if we replace ‘ξ’ by a name that refers to an object. Otherwise, I would not call $\Phi(\xi)$ a *function*. Accordingly, ‘$x\,.\,(x-1) = x^2 - x$’ would refer to the True, at least if the notations for multiplication, subtraction and squaring were defined to apply also to objects that are not numbers, so as to allow the equation to hold generally. In contrast, ‘$x\,.\,(x-1) = x^2$’ would refer to the False, because we obtain the False as reference, if we insert ‘1’ for ‘x’, although we obtain the True if we insert ‘0’. But in this stipulation the scope of generality is not sufficiently demarcated. One would, e.g., be in doubt whether ‘$\top\, 2 + 3\,.\,x = 5\,.\,x$’ would have to be understood as the negation of a generality or as the generality of a negation; more precisely, whether this should refer to the truth-value of: that not for every argument the value of the function $2 + 3 . \xi = 5 . \xi$ is the True, or whether it should refer to the truth-value of: that for every argument the value of the function $-\!\top 2 + 3 . \xi = 5 . \xi$ is the True. In the first case '$-\!\top 2 + 3 . x = 5 . x$' would refer to the True, in the other the False. It must, however, be possible to express the generality of a negation, as well as the negation of a generality. I will express the former as follows:

$$\text{‘}-\overset{\mathfrak{a}}{\smile}\!\top\, 2 + 3 . \mathfrak{a} = 5 . \mathfrak{a}\text{’}$$

and the negation of a generality thus:

$$\text{‘}-\!\top\!\overset{\mathfrak{a}}{\smile}-\, 2 + 3 . \mathfrak{a} = 5 . \mathfrak{a}\text{’}$$

and the generality itself thus:

$$\text{‘}-\overset{\mathfrak{a}}{\smile}-\, 2 + 3 . \mathfrak{a} = 5 . \mathfrak{a}\text{’}.$$

The latter would refer to the True, if for every argument the value of the function $2 + 3 . \xi = 5 . \xi$ were the True. Because this is not the case,

$$-\overset{\mathfrak{a}}{\smile}-\, 2 + 3 . \mathfrak{a} = 5 . \mathfrak{a}$$

is the False, and therefore

$$-\!\top\!\overset{\mathfrak{a}}{\smile}-\, 2 + 3 . \mathfrak{a} = 5 . \mathfrak{a}$$

is the True.

$$-\overset{\mathfrak{a}}{\smile}\!\top\, 2 + 3 . \mathfrak{a} = 5 . \mathfrak{a}$$

is the False, since it is not the case that the value of the function $-\!\top 2 + 3 . \xi = 5 . \xi$ is the True for every argument; because for the argument 1 it is the False. Accordingly,

$$-\!\top\!\overset{\mathfrak{a}}{\smile}\!\top\, 2 + 3 . \mathfrak{a} = 5 . \mathfrak{a}$$

is the True and

$$\text{‘}\vdash\!\top\!\overset{\mathfrak{a}}{\smile}\!\top\, 2 + 3 . \mathfrak{a} = 5 . \mathfrak{a}\text{’}$$

says: *there is* at least one solution for the equation '$2 + 3 . x = 5 . x$'. Likewise:

$$\vdash\!\top\!\overset{\mathfrak{a}}{\smile}\!\top\, \mathfrak{a}^2 = 1;$$

in words: *there is* at least one square root of 1. One sees from this how 'there is' is rendered in concept-script.

If we now give the following explanation:

> let
>
> $$\text{‘}-\overset{\mathfrak{a}}{\smile}-\, \Phi(\mathfrak{a})\text{’}$$
>
> refer to the True if the value of the function $\Phi(\xi)$ is the True for every argument, and otherwise the False;

then this requires supplementation in that one needs to state more precisely which function $\Phi(\xi)$ is in each case. We will call it the *corresponding* function. For there could be doubts. $\Delta = \Delta$ is the value both of the function $\Delta = \xi$ and the value of the function $\xi = \xi$, in both cases for the argument Δ. So one might, starting from $-\overset{\mathfrak{a}}{\smile}-\, \mathfrak{a} = \mathfrak{a}$, want to take as the corresponding function $\xi = \mathfrak{a}$, $\mathfrak{a} = \xi$, or $\xi = \xi$. With our use of the German letters, however, we would in the first two cases not even have a *function* because '$\xi = \mathfrak{a}$' and '$\mathfrak{a} = \xi$' always remain without reference, whatever one may insert for 'ξ'; because the German letter '$\mathfrak{a}$' ought not occur without '$-\overset{\mathfrak{a}}{\smile}-$' prefixed, except in '$-\overset{\mathfrak{a}}{\smile}-$' itself. Here only $\xi = \xi$ can thus be considered as the corresponding function. It is not so easy in the case of an expression like

$$`-\overset{\mathfrak{a}}{\smile}-((\mathfrak{a} + \mathfrak{a} = 2 \,.\, \mathfrak{a}) = (-\overset{\mathfrak{a}}{\smile}- \mathfrak{a} = \mathfrak{a}))'$$

If one were to proceed blindly one might think to have the corresponding function in

$$`(\xi + \xi = 2 \,.\, \xi) = (-\overset{\xi}{\smile}- \xi = \xi)'$$

We now want to say that '$\mathfrak{a}$' stands above a *concavity* in '$-\overset{\mathfrak{a}}{\smile}-$'. The place above the concavity is never an *argument place*; thus at least the '$\mathfrak{a}$' standing above the second concavity has to be preserved. But since '$-\overset{\mathfrak{a}}{\smile}-$' must always be followed by a combination of signs that contain '$\mathfrak{a}$', '$\mathfrak{a}$' must be preserved in at least one of the two places in '$\mathfrak{a} = \mathfrak{a}$'. Accordingly one could surmise that the following functions were the corresponding ones

$$\begin{aligned}
&(\xi + \xi = 2 \,.\, \xi) = (-\overset{\mathfrak{a}}{\smile}- \xi = \mathfrak{a}),\\
&(\xi + \xi = 2 \,.\, \xi) = (-\overset{\mathfrak{a}}{\smile}- \mathfrak{a} = \xi),\\
&(\xi + \xi = 2 \,.\, \xi) = (-\overset{\mathfrak{a}}{\smile}- \mathfrak{a} = \mathfrak{a});
\end{aligned}$$

but the first two conceptions contradict the fact that the reference of the '$-\overset{\mathfrak{a}}{\smile}- \mathfrak{a} = \mathfrak{a}$' occurring in

$$`-\overset{\mathfrak{a}}{\smile}-((\mathfrak{a} + \mathfrak{a} = 2 \,.\, \mathfrak{a}) = (-\overset{\mathfrak{a}}{\smile}- \mathfrak{a} = \mathfrak{a}))'$$

is already established and must not be called into question again.

We now call that which follows a concavity with a *German letter*, which together with this same concavity forms the name of the truth-value of: that, for every argument, the value of the corresponding function is the True, the *scope* of the German letter standing over the concavity. Now, the *corresponding* function is determined by the rule:

1. All places, in which a German letter occurs in its own scope, but not within a subordinate scope of the same letter nor above a concavity, are related argument places, namely those of the corresponding function.

If, however, one wants to designate the truth-value of: that the function

$$(\xi + \xi = 2 \,.\, \xi) = (-\overset{\mathfrak{a}}{\smile}- \xi = \mathfrak{a})$$

has the True as value for every argument, then one will choose a different German letter:

$$-\overset{\mathfrak{e}}{\smile}-(\mathfrak{e} + \mathfrak{e} = 2 \,.\, \mathfrak{e}) = (-\overset{\mathfrak{a}}{\smile}- \mathfrak{e} = \mathfrak{a}).$$

I capture this in the following rule:

2. If in the name of a function German letters already occur, within whose scopes lie argument places of this function, then a German letter distinct from these is to be chosen in order to form the corresponding expression of generality.

According to our specifications, one German letter is in general as good as any other, with the restriction, however, that the distinctness of these letters can be essential. For some German letters we will stipulate later a slightly different kind of use.

$$\text{`}\!-\!\overset{\mathfrak{a}}{\smile}\!-\, \Phi(\mathfrak{a})\text{'}$$

refers to the same as

$$\text{`}\!-\!\overset{\mathfrak{a}}{\smile}\!-(\!-\!\!-\, \Phi(\mathfrak{a}))\text{'}$$

and as

$$\text{`}\!-\!\!-(-\!\overset{\mathfrak{a}}{\smile}\!-\, \Phi(\mathfrak{a}))\text{'}$$

I therefore consider the horizontal stroke left and right of the concavity in '$-\!\overset{\mathfrak{a}}{\smile}\!-$' as *horizontals* in our special sense of the word, so that by the *fusion* of horizontals we can immediately pass from the forms '$-\!\!-(-\!\overset{\mathfrak{a}}{\smile}\!-\, \Phi(\mathfrak{a}))$' and '$-\!\overset{\mathfrak{a}}{\smile}\!-(-\!\!-\, \Phi(\mathfrak{a}))$' to '$-\!\overset{\mathfrak{a}}{\smile}\!-\, \Phi(\mathfrak{a})$'.

§9. If $-\!\overset{\mathfrak{a}}{\smile}\!-\, \Phi(\mathfrak{a}) = \Psi(\mathfrak{a})$ is the True, we can, according to our previous specification (§3), also say that the function $\Phi(\xi)$ has the same value-range as the function $\Psi(\xi)$; that is: we can convert the generality of an equality into a value-range equality and *vice versa*. This possibility must be regarded as a logical law of which, incidentally, use has always been made, even if tacitly, whenever extensions of concepts were mentioned. The entire calculating logic of Leibniz and Boole rests upon it. One could perhaps regard this conversion as unimportant or even dispensable. Against this, I remind the reader, that in my *Grundlagen der Arithmetik*, I defined a cardinal number as the extension of a concept, and I had already then pointed out that the negative, irrational, in brief, all numbers are also to be defined as extensions of concepts. We can fix a simple sign for a value-range, and this is, e.g., how the name of the cardinal number Zero will be introduced. In contrast, in '$-\!\overset{\mathfrak{a}}{\smile}\!-\, \Phi(\mathfrak{a}) = \Psi(\mathfrak{a})$' we cannot put a simple sign for '$\Phi(\mathfrak{a})$', because the letter '$\mathfrak{a}$' always has to occur in what can be put for '$\Phi(\mathfrak{a})$', for example.

The transformation of the generality of an equality into a value-range equality must also be possible in our signs. Thus I write, e.g., for

$$\text{`}\!-\!\overset{\mathfrak{a}}{\smile}\!-\, \mathfrak{a}^2 - \mathfrak{a} = \mathfrak{a}\,.\,(\mathfrak{a} - 1)\text{'}$$

$$\text{`}\dot{\varepsilon}(\varepsilon^2 - \varepsilon) = \dot{\alpha}(\alpha\,.\,(\alpha - 1))\text{'}$$

by understanding '$\dot{\varepsilon}(\varepsilon^2 - \varepsilon)$' as the value-range of the function $\xi^2 - \xi$ and '$\dot{\alpha}(\alpha\,.\,(\alpha - 1))$'

as the value-range of the function $\xi \,.\, (\xi - 1)$. Equally, $\grave{\varepsilon}(\varepsilon^2 = 4)$ is the value-range of the function $\xi^2 = 4$, or, as we can also say, the extension of the concept *square root of 4*.

If I say in general:

> let
>
> $$\text{‘}\grave{\varepsilon}\Phi(\varepsilon)\text{’}$$
>
> refer to the value-range of the function $\Phi(\xi)$,

then this too requires supplementation, just like our explanation of ‘$\overset{\mathfrak{a}}{\smile}\, \Phi(\mathfrak{a})$’ above. Specifically, the question is which function is to be regarded as the *corresponding* function $\Phi(\xi)$ in each case. That $\grave{\varepsilon}(\varepsilon^2 - \varepsilon)$ is the value-range of the function $\xi^2 - \xi$ and not of $\xi^2 - \varepsilon$ nor of $\varepsilon^2 - \xi$ is readily understood because in our usage of the *small Greek vowel* neither ‘$\xi^2 - \varepsilon$’ nor ‘$\varepsilon^2 - \xi$’ would acquire a reference for any object whose name were inserted for ‘ξ’, or, as we can also put it, because those combinations of signs do not refer to functions, but lack reference if detached from ‘$\grave{\varepsilon}$’. A combination of signs like ‘$\grave{\varepsilon}\Psi(\varepsilon, \grave{\varepsilon}\mathrm{X}(\varepsilon))$’ has to be judged similar to ‘$\overset{\mathfrak{a}}{\smile}\, \Psi(\mathfrak{a}, \overset{\mathfrak{a}}{\smile}\, \mathrm{X}(\mathfrak{a}))$’ in §8. The place under the smooth breathing is no more an *argument place* than the one above the concavity. Let us call the *scope* of a *small Greek vowel* that which follows this Greek letter with a smooth breathing, and together with it forms the name of the value-range of the *corresponding* function, so we can lay down the rule:

1. All places in which a small Greek vowel occurs in its own scope but not within a subordinate scope of the same letter nor with the smooth breathing, are related argument places, namely those of the corresponding function.

This function is hereby determined. Accordingly, $\grave{\varepsilon}(\varepsilon = \grave{\varepsilon}(\varepsilon^2 - \varepsilon))$ is the value-range of the function $\xi = \grave{\varepsilon}(\varepsilon^2 - \varepsilon)$, and $\grave{\alpha}(\alpha = \grave{\varepsilon}(\varepsilon = \alpha))$ is the value-range of the function $\xi = \grave{\varepsilon}(\varepsilon = \xi)$. The following rule thus applies to the formation of a name for a value-range:

2. If small Greek vowels already occur in the name of a function, in whose scope argument places of this function lie, then one is to choose one that is different from those in order to form the name of the value-range of this function.

According to our specifications, one *small Greek vowel* is in general as good as any other, with the restriction, however, that the distinctness of these letters can be essential.

The introduction of the notation for value-ranges seems to me one of the most consequential additions to my concept-script that I made since my first publication on this subject matter. Thereby, also the domain of that which can occur as an argument of a function is extended. For example, $\grave{\varepsilon}(\varepsilon^2 - \varepsilon) = \grave{\alpha}(\alpha\,.\,(\alpha - 1))$ is the value of the function $\xi = \grave{\alpha}(\alpha\,.\,(\alpha - 1))$ for the argument $\grave{\varepsilon}(\varepsilon^2 - \varepsilon)$.

§10. By presenting the combination of signs '$\grave{\varepsilon}\Phi(\varepsilon) = \grave{\alpha}\Psi(\alpha)$' as co-referential with '$\smile^{\mathfrak{a}}\ \Phi(\mathfrak{a}) = \Psi(\mathfrak{a})$', we have admittedly by no means yet completely fixed the reference of a name such as '$\grave{\varepsilon}\Phi(\varepsilon)$'. We have a way always to recognise a value-range as the same if it is designated by a name such as '$\grave{\varepsilon}\Phi(\varepsilon)$', whereby it is already recognisable as a value-range. However, we cannot decide yet whether an object that is not given to us as a value-range is a value-range or which function it may belong to; nor can we decide in general whether a given value-range has a given property if we do not know that this property is connected with a property of the corresponding function. If we assume that

$$\mathrm{X}(\xi)$$

is a function that never receives the same value for different arguments, then exactly the same criterion for recognition holds for the objects whose names have the form '$\mathrm{X}(\grave{\varepsilon}\Phi(\varepsilon))$' as for the objects whose signs have the form '$\grave{\varepsilon}\Phi(\varepsilon)$'. For then '$\mathrm{X}(\grave{\varepsilon}\Phi(\varepsilon)) = \mathrm{X}(\grave{\alpha}\Psi(\alpha))$' too is co-referential with '$\smile^{\mathfrak{a}}\ \Phi(\mathfrak{a}) = \Psi(\mathfrak{a})$'.[1] From this it follows that by equating the reference of '$\grave{\varepsilon}\Phi(\varepsilon) = \grave{\alpha}\Psi(\alpha)$' with that of '$\smile^{\mathfrak{a}}\ \Phi(\mathfrak{a}) = \Psi(\mathfrak{a})$', the reference of a name such as '$\grave{\varepsilon}\Phi(\varepsilon)$' is by no means completely determined; at least if there is such a function $\mathrm{X}(\xi)$ whose value for a value-range as argument is not always equal to the value-range itself. Now, how is this indeterminacy resolved? By determining for every function, when introducing it, which value it receives for value-ranges as arguments, just as for all other arguments. Let us do this for the functions hitherto considered. These are the following:

$$\xi = \zeta,\ \text{—}\,\xi,\ \top\,\xi$$

The last one can be left out of consideration, since its argument may always be taken to be a truth-value. It makes no difference whether one takes as argument an object or the value that the function —ξ has for this object as argument. In addition, we can now reduce the function —ξ to the function $\xi = \zeta$. For based on our stipulations the function $\xi = (\xi = \xi)$ has the same value as the function —ξ for every argument; for the value of the function $\xi = \xi$ is the True for every argument. It follows from

[1] Thereby it is not said that the sense is the same.

this that the value of the function $\xi = (\xi = \xi)$ is the True only for the True as argument, and that it is the False for all other arguments, just as for the function $-\!\!-\, \xi$. After having thus reduced everything to the consideration of the function $\xi = \zeta$, we ask which values it has when a value-range appears as argument. Since so far we have only introduced the truth-values and value-ranges as objects, the question can only be whether one of the truth-values might be a value-range. If that is not the case, then it is thereby also decided that the value of the function $\xi = \zeta$ is always the False when a truth-value is taken as one of its arguments and a value-range as the other. If, on the other hand, the True is at the same time the value-range of a function $\Phi(\xi)$, then it is thereby also decided what the value of the function $\xi = \zeta$ is in all cases where the True is taken as one of the arguments; and matters are similar if the False is at the same time the value-range of a certain function. Now, the question whether one of the truth-values is a value-range cannot possibly be decided on the basis of '$\acute{\varepsilon}\Phi(\varepsilon) = \acute{\alpha}\Psi(\alpha)$' having the same reference as '$-\!\overset{\mathfrak{a}}{\smile}\!-\, \Phi(\mathfrak{a}) = \Psi(\mathfrak{a})$'. It is possible to stipulate generally that '$\tilde{\eta}\Phi(\eta) = \tilde{\alpha}\Psi(\alpha)$' is to refer to the same as '$-\!\overset{\mathfrak{a}}{\smile}\!-\, \Phi(\mathfrak{a}) = \Psi(\mathfrak{a})$', without it being possible to infer from that to the equality of $\acute{\varepsilon}\Phi(\varepsilon)$ and $\tilde{\eta}\Phi(\eta)$. We would then have, for example, a class of objects with names of the form '$\tilde{\eta}\Phi(\eta)$' for whose differentiation and recognition the same criterion would hold as for the value-ranges. We could now determine the function $X(\xi)$ by saying that its value is to be the True for $\tilde{\eta}\Lambda(\eta)$ as argument, and it is to be $\tilde{\eta}\Lambda(\eta)$ for the True as argument; further, the value of the function, $X(\xi)$, is to be the False for the argument $\tilde{\eta}M(\eta)$, and it is to be $\tilde{\eta}M(\eta)$ for the False as argument; for every other argument, the value of the function $X(\xi)$ is to coincide with the argument itself. So, provided the functions $\Lambda(\xi)$ and $M(\xi)$ do not always have the same value for the same argument, our function $X(\xi)$ never has the same value for different arguments, and therefore '$X(\tilde{\eta}\Phi(\eta)) = X(\tilde{\alpha}\Psi(\alpha))$' is then also always co-referential with '$-\!\overset{\mathfrak{a}}{\smile}\!-\, \Phi(\mathfrak{a}) = \Psi(\mathfrak{a})$'. The objects whose names would be of the form '$X(\tilde{\eta}\Phi(\eta))$' would then also be recognised by the same means as the value-ranges, and $X(\tilde{\eta}\Lambda(\eta))$ would be the True and $X(\tilde{\eta}M(\eta))$ would be the False. Thus, without contradicting our equating '$\acute{\varepsilon}\Phi(\varepsilon) = \acute{\varepsilon}\Psi(\varepsilon)$' with '$-\!\overset{\mathfrak{a}}{\smile}\!-\, \Phi(\mathfrak{a}) = \Psi(\mathfrak{a})$', it is always possible to determine that an arbitrary value-range be the True and another arbitrary value-range be the False. Let us therefore stipulate that $\acute{\varepsilon}(-\!\!-\, \varepsilon)$ be the True and that $\acute{\varepsilon}(\varepsilon = (\,\top\!\overset{\mathfrak{a}}{\smile}\!-\, \mathfrak{a} = \mathfrak{a}))$ be the False. $\acute{\varepsilon}(-\!\!-\, \varepsilon)$ is the value-range of the function $-\!\!-\, \xi$, whose value is the True only if the argument is the True, and whose value is the False for all other arguments. All functions of which this holds have the same value-range and, according to our stipulation, this is the True. Thus — $\grave{\varepsilon}\Phi(\varepsilon)$ is the True only if the function $\Phi(\xi)$ is a concept under which only the True falls; in all other cases — $\grave{\varepsilon}\Phi(\varepsilon)$ is the False. Further, $\grave{\varepsilon}(\varepsilon = (\top\!\overset{\mathfrak{a}}{\smile}\, \mathfrak{a} = \mathfrak{a}))$ is the value-range of the function, $\xi = (\top\!\overset{\mathfrak{a}}{\smile}\, \mathfrak{a} = \mathfrak{a})$, whose value is the True only if the argument is the False, and whose value is the False for all other arguments. All functions of which this holds have the same value-range and, according to our stipulation, this is the False. Every concept, therefore, under which the False and only it falls, has as its extension the False.[1]

We have hereby determined the *value-ranges* as far as is possible here. Only when the further issue arises of introducing a function that is not completely reducible to the functions already known will we be able to stipulate what values it should have for value-ranges as arguments; and this can then be viewed as a determination of the value-ranges as well as of that function.

§11. Indeed we do still require such functions. If the equating of '$\grave{\varepsilon}(\Delta = \varepsilon)$' with '$\Delta$' could be maintained generally,[2] then we would have a substitute for the definite

[1] It suggests itself to generalise our stipulation so that every object is conceived as a value-range, namely, as the extension of a concept under which it falls as the only object. A concept under which only the object Δ falls is $\Delta = \xi$. We attempt the stipulation: let $\grave{\varepsilon}(\Delta = \varepsilon)$ be the same as Δ. Such a stipulation is possible for every object that is given to us independently of value-ranges, for the same reason that we have seen for truth-values. But before we may generalise this stipulation, the question arises whether it is not in contradiction with our criterion for recognising value-ranges if we take an object for Δ which is already given to us as a value-range. It is out of the question to allow it to hold only for such objects which are not given to us as value-ranges, because the way an object is given must not be regarded as its immutable property, since the same object can be given in different ways. Thus, if we insert '$\grave{\alpha}\Phi(\alpha)$' for '$\Delta$' we obtain

$$\text{`}\grave{\varepsilon}(\grave{\alpha}\Phi(\alpha) = \varepsilon) = \grave{\alpha}\Phi(\alpha)\text{'}$$

and this would be co-referential with

$$\text{`}\overset{\mathfrak{a}}{\smile}\, (\grave{\alpha}\Phi(\alpha) = \mathfrak{a}) = \Phi(\mathfrak{a})\text{'},$$

which, however, only refers to the True, if $\Phi(\xi)$ is a concept under which only a single object falls, namely $\grave{\alpha}\Phi(\alpha)$. Since this is not necessary, our stipulation cannot be upheld in its generality.

The equation '$\grave{\varepsilon}(\Delta = \varepsilon) = \Delta$' with which we attempted this stipulation, is a special case of '$\grave{\varepsilon}\Omega(\varepsilon, \Delta) = \Delta$', and one can ask how the function $\Omega(\xi, \zeta)$ would have to be constituted, so that it could generally be specified that Δ be the same as $\grave{\varepsilon}\Omega(\varepsilon, \Delta)$. Then

$$\grave{\varepsilon}\Omega(\varepsilon, \grave{\alpha}\Phi(\alpha)) = \grave{\alpha}\Phi(\alpha)$$

also has to be the True, and thus also

$$\overset{\mathfrak{a}}{\smile}\, \Omega(\mathfrak{a}, \grave{\alpha}\Phi(\alpha)) = \Phi(\mathfrak{a}),$$

no matter what function $\Phi(\xi)$ might be. We shall later be acquainted with a function having this property in $\xi \frown \zeta$; however we shall define it with the aid of the value-range, so that it cannot be of use for us here.

[2] Compare note 1.

article in language in the form '$\dot{\varepsilon}\Phi(\varepsilon)$'. For, if we assumed that $\Phi(\xi)$ were a concept under which the object Δ and only this fell, then $\overset{\mathfrak{a}}{\smile}\, \Phi(\mathfrak{a}) = (\Delta = \mathfrak{a})$ would be the True and hence also $\dot{\varepsilon}\Phi(\varepsilon) = \dot{\varepsilon}(\Delta = \varepsilon)$ would be the True, and following our equating of '$\dot{\varepsilon}(\Delta = \varepsilon)$' and '$\Delta$', $\dot{\varepsilon}\Phi(\varepsilon)$ would be the same as Δ; i.e., in case $\Phi(\xi)$ were a concept under which one and only one object fell, '$\dot{\varepsilon}\Phi(\varepsilon)$' would designate this object. This is admittedly not possible, because the former equation had to be abandoned in its full generality; nevertheless we can help ourselves by introducing the function

$$\backslash\xi$$

with the specification to distinguish two cases:

1) if, for the argument, there is an object Δ such that $\dot{\varepsilon}(\Delta = \varepsilon)$ is the argument, then the value of the function $\backslash\xi$ is to be Δ itself;

2) if, for the argument, there is no object Δ such that $\dot{\varepsilon}(\Delta = \varepsilon)$ is the argument, then the argument itself is to be the value of the function $\backslash\xi$.

Accordingly, $\backslash\dot{\varepsilon}(\Delta = \varepsilon) = \Delta$ is the True, and then '$\backslash\dot{\varepsilon}\Phi(\varepsilon)$' refers to the object which falls under the concept $\Phi(\xi)$, if $\Phi(\xi)$ is a concept under which one and only one object falls; in all other cases '$\backslash\dot{\varepsilon}\Phi(\varepsilon)$' refers to the same as '$\dot{\varepsilon}\Phi(\varepsilon)$'. So, e.g., $2 = \backslash\dot{\varepsilon}(\varepsilon + 3 = 5)$ is the True, because 2 is the only object that falls under the concept,

that which increased by 3 yields 5

presupposing here a suitable definition of the plus-sign. $\dot{\varepsilon}(\varepsilon^2 = 1) = \backslash\dot{\varepsilon}(\varepsilon^2 = 1)$ is the True, because not just one object falls under the concept, *square-root of 1*. $\dot{\varepsilon}(\top\, \varepsilon = \varepsilon) = \backslash\dot{\varepsilon}(\top\, \varepsilon = \varepsilon)$ is the True because no object falls under the concept *not equal to itself*. $\dot{\varepsilon}(\varepsilon + 3) = \backslash\dot{\varepsilon}(\varepsilon + 3)$ is the True because the function $\xi + 3$ is not a concept.

Here, then, we have a substitute for the definite article of language, which serves to form proper names out of concept-words. For example, out of the words

'positive square-root of 2',

that refer to a concept, we form the proper name

'the positive square-root of 2'.

Here is a logical risk. For if we were to form out of the words 'square-root of 2' the proper name 'the square-root of 2', we would commit a logical error, since this proper name would be, without further stipulation, ambiguous[1] and just for that reason without reference. If there were no irrational numbers, as has indeed been asserted, then the proper name 'the positive square-root of 2' would also be without reference,

[1] I am here taking for granted that there are negative and irrational numbers.

at least according to the immediate sense of the word, without special stipulation. And if we were specifically to assign a reference to this proper name, then this would have no connection with its formation and it would not be permissible to infer that it was a positive square-root of 2, and yet we would be all too inclined to conclude that. This risk carried by the definite article is now avoided altogether, since '$\backslash\grave{\varepsilon}\Phi(\varepsilon)$' always has a reference, whether the function $\Phi(\xi)$ is not a concept, or a concept under which more than one or no object falls, or whether it is a concept under which one and only one object falls.

§12. Next, in order to be able to designate the subordination of concepts and other important relations, I introduce the function with two arguments

$$\begin{array}{l} \top\!\!\!-\, \xi \\ \llcorner\, \zeta \end{array}$$

by means of the specification that its value shall be the False if the True is taken as the ζ-argument, while any object that is not the True is taken as ξ-argument; that in all other cases the value of the function shall be the True. According to this and the previous stipulations, the value of this function is also determined for value-ranges as arguments. It follows that

$$\begin{array}{l} \top\!\!\!-\, \Gamma \\ \llcorner\, \Delta \end{array}$$

is the same as

$$-\left(\begin{array}{l} \top\!\!\!-\, (-\Gamma) \\ \llcorner\, (-\Delta) \end{array}\right)$$

and therefore that in

$$\text{`}\begin{array}{l} \top\!\!\!-\, \Gamma \\ \llcorner\, \Delta \end{array}\text{'}$$

we can regard the horizontal stroke before 'Δ', as well as each of the two parts of the upper horizontal stroke partitioned by the vertical, as *horizontals* in our particular sense. We speak here, just as previously, of the *fusion of horizontals*. I call the vertical stroke the *conditional-stroke*. It may be lengthened as required.

The following propositions hold:

$$\text{`}\vdash\!\!\!\begin{array}{l} \top\!\!\!-\, 3^2 > 2 \\ \llcorner\, 3\ \ > 2 \end{array}\text{'};\quad \text{`}\vdash\!\!\!\begin{array}{l} \top\!\!\!-\, 2^2 > 2 \\ \llcorner\, 2\ \ > 2 \end{array}\text{'};\quad \text{`}\vdash\!\!\!\begin{array}{l} \top\!\!\!-\, 1^2 > 2 \\ \llcorner\, 1\ \ > 2 \end{array}\text{'}.$$

The function $-\!\!\begin{array}{l} \top\!\!\!-\, \xi \\ \llcorner\, \zeta \end{array}$, or $-\!\!\top\!\!-\!\begin{array}{l} \top\!\!\!-\, \xi \\ \llcorner\, \zeta \end{array}$, always has the True as value, when the function $\begin{array}{l} \top\!\!\!-\, \xi \\ \llcorner\, \zeta \end{array}$ has the False as value, and conversely. Hence $-\!\!\begin{array}{l} \top\!\!\!-\, \Gamma \\ \llcorner\, \Delta \end{array}$ is the True if and only if Δ is the True and

Γ is not the True. Accordingly,

$$\begin{array}{l} \vdash\!\top\!\!\top\; 2 > 3 \\ \quad\;\; \llcorner\; 2 + 3 = 5 \end{array}$$

in words: 2 is not greater than 3 *and* the sum of 2 and 3 is 5.

$$\begin{array}{l} \vdash\!\top\!\!\top\; 3 > 2 \\ \quad\;\; \llcorner\!\!-\; 2 + 3 = 5 \end{array}$$

in words: 3 is greater than 2 *and* the sum of 2 and 3 is 5. For $\begin{array}{l} \top\!\!\top\; 3 > 2 \\ \;\llcorner\; 2 + 3 = 5 \end{array}$ is the value of the function, $\begin{array}{l} \top\!\!\top\; \xi \\ \;\llcorner\; \zeta \end{array}$, when $\top\; 3 > 2$ is the ξ-argument, and 2+3=5 is the ζ-argument.

$$\begin{array}{l} \vdash\!\top\!\!-\; 2^3 = 3^2 \\ \quad\;\; \llcorner\!\top\; 1^2 = 2^1 \end{array}$$

in words: *neither* is the third power of 2 the second power of 3, *nor* is the second power of 1 the first power of 2.

By way of the propositions

$$\text{‘}\begin{array}{l} \vdash\!\!-\; 3^2 > 3 \\ \;\llcorner\!\top\; 3 \;\; < 3 \end{array}\text{’};\quad \text{‘}\begin{array}{l} \vdash\!\!-\; 2^2 > 3 \\ \;\llcorner\!\top\; 2 \;\; < 3 \end{array}\text{’};\quad \text{‘}\begin{array}{l} \vdash\!\!-\; 1^2 > 3 \\ \;\llcorner\!\top\; 1 \;\; < 3 \end{array}\text{’}$$

one has the following

$$\text{‘}\begin{array}{l} \vdash\!\top\!\!\top\!\!-\; 3^2 > 3 \\ \quad\;\; \llcorner\!\top\; 3 \;\; < 3 \end{array}\text{’};\quad \text{‘}\begin{array}{l} \vdash\!\top\!\!\top\!\!-\; 2^2 > 3 \\ \quad\;\; \llcorner\!\top\; 2 \;\; < 3 \end{array}\text{’};\quad \text{‘}\begin{array}{l} \vdash\!\top\!\!\top\!\!-\; 1^2 > 3 \\ \quad\;\; \llcorner\!\top\; 1 \;\; < 3 \end{array}\text{’}.$$

Now, since $\begin{array}{l} \top\!\!\top\!\!-\; 1^2 > 3 \\ \;\llcorner\!\top\; 1 \;\; < 3 \end{array}$ is the truth-value of: that neither is the square of 1 greater than 3, nor is 1 less than 3, this is negated by our last proposition, so it asserts at least one of the two is true, that the square of 1 is greater than 3 *or* that 1 is less than 3. One can see from these examples how the ‘*and*’ of language, when it connects propositions, the ‘*neither* — *nor*’, and the ‘*or*’ between propositions, are to be rendered.

One can insert into ‘$\begin{array}{l} \top\; \xi \\ \llcorner\; \Delta \end{array}$’ any proper name for ‘ξ’, even for example ‘$\begin{array}{l} \top\; \Theta \\ \llcorner\; \Lambda \end{array}$’. Thus we obtain

$$\text{‘}\begin{array}{l} \top\!\left(\begin{array}{l} \top\; \Theta \\ \llcorner\; \Lambda \end{array}\right) \\ \llcorner\!\!-\!\!-\; \Delta \end{array}\text{’},$$

wherein we can now *fuse* the horizontals:

$$\text{‘}\begin{array}{l} \top\!\!\top\; \Theta \\ \;\;\llcorner\; \Lambda \\ \llcorner\!\!-\; \Delta \end{array}\text{’},$$

This refers to the False if Δ is the True and $\displaystyle\mathop{\vphantom{\Big|}}_{\substack{\top\\\bot}}{\Theta\atop\Lambda}$ is not the True; i.e., in this case, if $\displaystyle\mathop{\vphantom{\Big|}}_{\substack{\top\\\bot}}{\Theta\atop\Lambda}$ is the False. The latter, however, is the case if and only if Λ is the True and Θ is not the True. Thus,

$$\begin{array}{c}\rule{0pt}{0pt}\\[-2pt]\mathop{\hbox{\vrule height 20pt width 0.5pt}}\limits^{\displaystyle-\Theta}_{\displaystyle\substack{-\Lambda\\-\Delta}}\end{array}$$

is the False if Δ and Λ are the True while Θ is not the True; in all other cases it is the True. From this follows the permutability of Λ and Δ:

$$\mathop{\hbox{\vrule height 20pt width 0.5pt}}\limits^{\displaystyle-\Theta}_{\displaystyle\substack{-\Lambda\\-\Delta}}$$

is the same truth-value as

$$\mathop{\hbox{\vrule height 20pt width 0.5pt}}\limits^{\displaystyle-\Theta}_{\displaystyle\substack{-\Delta\\-\Lambda}}\,.$$

In

'$\mathop{\hbox{\vrule height 20pt width 0.5pt}}\limits^{\displaystyle-\Theta}_{\displaystyle\substack{-\Delta\\-\Lambda}}$'

we shall term '— Θ' *supercomponent*, '— Δ' and '— Λ' *subcomponents*. We can, however, also regard '$\mathop{\vphantom{\Big|}}\limits^{\displaystyle-\Theta}_{\displaystyle-\Delta}$' as *supercomponent* and '— Λ' alone as *subcomponent*. The subcomponents are therefore *permutable*. Likewise, we can see that

$$\mathop{\hbox{\vrule height 28pt width 0.5pt}}\limits^{\displaystyle-\Theta}_{\displaystyle\substack{-\Lambda\\-\Delta\\-\Xi}}$$

is the False if and only if Λ, Δ, and Ξ are all the True, while Θ is not the True. In all other cases it is the True. Once more we thus have the *permutability of the subcomponents*, '— Λ', '— Δ', '— Ξ'. Strictly speaking, this permutability must be proven for each occurring case, and I have done this for some cases in my little book *Begriffsschrift*, in such a way that it will be straightforward to treat all cases accordingly. In order not to be tied up in excessive prolixity, I will here assume this permutability to be generally granted and in what follows make use of it without further reminder.

is the True, if and only if Λ, Δ, and Ξ are all the True, while Θ is not the True. Thus,

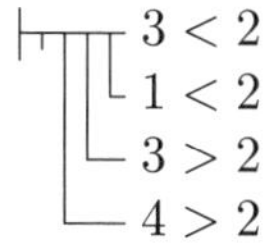

in words: 3 is not less than 2 and 1 is less than 2 and 3 is greater than 2 and 4 is greater than 2;

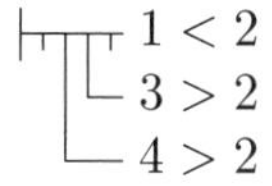

in words: 1 is less than 2 and 3 is greater than 2 and 4 is greater than 2. One can think of the latter as dissected thus:

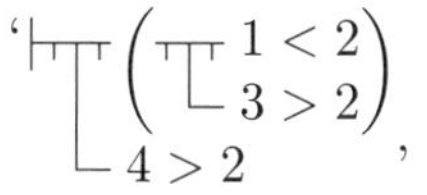

The negation-strokes between the conditional-strokes cancel each other and the horizontals may be fused. We have in

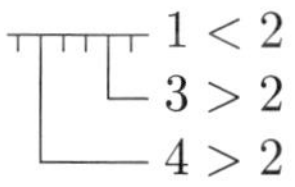

the value of the function 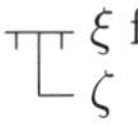 for 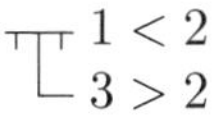 as ξ-argument and $4 > 2$ as the ζ-argument, wherein 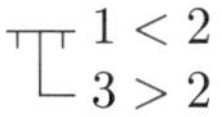 is, in turn, the value of the same function for $1 < 2$ as ξ-argument and $3 > 2$ as ζ-argument.

§13. To justify the name, 'conditional-stroke', I point out that the names '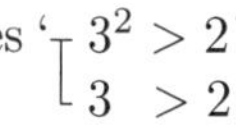', '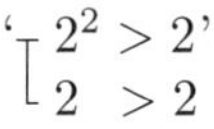', '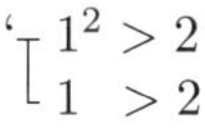' result from '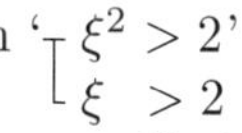' by replacing 'ξ' with '3', '2' and '1'. If we now use the sign, '>', in such a way that '$\Gamma > \Delta$' refers to the True if Γ and Δ are real numbers and Γ is greater than Δ, and that in all other cases '$\Gamma > \Delta$' refers to the False; if we assume further that the notation 'Γ^2' is explained so that it has a reference whenever Γ is an object, then the value of the function

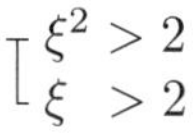

is the True for every argument; hence

$$\vdash\!\overset{\mathfrak{a}}{\smile}\!\!\begin{array}{l} \top\; \mathfrak{a}^2 > 2 \\ \llcorner\; \mathfrak{a} \;\; > 2 \end{array}$$

in words: *if* something is greater than 2 then its square is also greater than 2. So also

$$\vdash\!\overset{\mathfrak{a}}{\smile}\!\!\begin{array}{l} \top\; \mathfrak{a}^4 = 1 \\ \llcorner\; \mathfrak{a}^2 = 1 \end{array}$$

in words: *if* the square of something is 1 then its fourth power is also 1. One can, however, also say: *every* square root of 1 is also a fourth root of 1; or: *all* square roots of 1 are also fourth roots of 1.[1] Here we have the *subordination* of a concept under a concept, a *universal* affirmative proposition. We have called any function with one argument whose value is always a truth-value a concept. Such functions are here $\xi^4 = 1$ and $\xi^2 = 1$; the latter is the *subordinate*, the former the *superordinate* concept. $\begin{array}{l} \top\!\!\top\; \xi^4 = 1 \\ \;\llcorner\; \xi^2 = 1 \end{array}$ is composed from these concepts as characteristic marks. Under this concept falls, e.g., the number -1:

$$\vdash\!\!\top\!\!\top\!\begin{array}{l} (-1)^4 = 1 \\ \llcorner\; (-1)^2 = 1 \end{array}$$

in words: -1 is square root of 1 and fourth root of 1. We have seen in §8 how the 'there is' of ordinary language is rendered. We apply this to say that there is something that is square root of 1 and fourth root of 1: $\vdash\!\!\top\!\overset{\mathfrak{a}}{\smile}\!\top\!\!\top\!\!\top\!\begin{array}{l} \mathfrak{a}^4 = 1 \\ \llcorner\; \mathfrak{a}^2 = 1 \end{array}$. Obviously the two negation-strokes cancel each other out here: $\vdash\!\!\top\!\overset{\mathfrak{a}}{\smile}\!\top\!\!\top\!\begin{array}{l} \mathfrak{a}^4 = 1 \\ \llcorner\; \mathfrak{a}^2 = 1 \end{array}$. Let us have a look at this from another side. $\overset{\mathfrak{a}}{\smile}\!\top\!\!\top\!\begin{array}{l} \mathfrak{a}^4 = 1 \\ \llcorner\; \mathfrak{a}^2 = 1 \end{array}$ is the truth-value of: that, if anything is a square root of 1, then it is not a fourth root of 1; or, as we can also say, *no* square root of 1 is a fourth root of 1. This truth-value is the False, and hence: $\vdash\!\!\top\!\overset{\mathfrak{a}}{\smile}\!\top\!\!\top\!\begin{array}{l} \mathfrak{a}^4 = 1 \\ \llcorner\; \mathfrak{a}^2 = 1 \end{array}$. We have here the negation of a *universal* negative proposition, i.e., a *particular* affirmative proposition,[2] which we can also say: '*some* square roots of 1 are fourth roots of 1',

[1] One easily connects this with the accompanying thought that there is something that is a square root of 1. This must be kept entirely at a distance. Likewise, the accompanying thought that there is more than one square root of 1 is likewise to be fended off here.

[2] The *particular* affirmative proposition, on the one hand, says less than the *universal* affirmative one, but, on the other hand, as is easily overlooked, also says more, since it asserts the instantiation of the concept, while subordination also, and indeed always, occurs with empty concepts. Some logicians seem to assume concepts to be instantiated without further ado and to overlook completely the very important case of the empty concept, perhaps because they, quite wrongly, do not acknowledge

where, however, the form of the plural is not to be understood as requiring that there must be more than one.

$$\vdash\!\!\overset{\mathfrak{a}}{\smile}\!\!\top\!\!\top\ \mathfrak{a}^4 = 1 \quad\text{[conditional stroke]}\quad \mathfrak{a}^3 = 1$$

in words: there is at least one cube root of 1 which is also a fourth root of 1; or: *some*—or at least one—cube root of 1 is a fourth root of 1.

In our symbolism, the 'and' that connects propositions appears less simple than the function-name '$\top\ \xi$ / $\llcorner\ \zeta$', for which a simple ordinary language expression is wanting. The relationship present in ordinary language easily seems more natural and appropriate, because it is familiar. However, which is simpler from a logical point of view is not easy to say: using 'and' and negation one can explain our '$\top\ \xi$ / $\llcorner\ \zeta$', but also conversely using the function-name '$\top\ \xi$ / $\llcorner\ \zeta$' and the negation-stroke one can explain 'and'. Obviously, for example, '$\vdash\!\top\ 2+3=5$ / $\llcorner\ 2+2=4$' says less than '$\vdash\!\!\top\!\!\top\ 2+3=5$ / $\llcorner\ 2+2=4$' and could therefore be considered simpler. The ultimate reason for the introduction of '$\top\ \xi$ / $\llcorner\ \zeta$' is the ease and perspicuity with which one can thereby represent inference, to which we now proceed.

Inferences and consequences

§14. From the propositions '$\vdash\!\top\ \Gamma$ / $\llcorner\ \Delta$' and '$\vdash \Delta$' one can infer: '$\vdash \Gamma$'; for if Γ were not the True, then, since Δ is the True, $\top\ \Gamma$ / $\llcorner\ \Delta$ would be the False. To each proposition put forward in concept-script symbolism, if it is to be used later in a further conduct of proof, I will assign a *label* for the purpose of citation. Accordingly, if the proposition '$\vdash\!\top\ \Gamma$ / $\llcorner\ \Delta$' has received the label 'α' and '$\vdash \Delta$' the label 'β', then I will write the inference either as

'
$$\begin{array}{l} \vdash\!\top\ \Gamma \\ \quad\llcorner\ \Delta \\ (\beta){::}\ \overline{\qquad\qquad} \\ \vdash \Gamma \end{array}$$
'

with double colon

or as

'
$$\begin{array}{l} \vdash \Delta \\ (\alpha){:}\ \overline{\qquad\qquad} \\ \vdash \Gamma \end{array}$$
'

with single colon.

---

empty concepts as legitimate. It is because of this that I do not use the expressions 'subordination', 'universal affirmative', 'particular affirmative' in exactly the same sense as these logicians, and so arrive at pronouncements which they will be wrongly inclined to regard as false.

This is the sole mode of inference that I used in my *Begriffsschrift*,[b] and one can even manage with it alone. The demand of scientific parsimony would now usually require this; but considerations of practicality pull in the opposite direction, and here, where I will have to form long chains of inferences, I will have to make some concessions. For an inordinate lengthiness would result if I were not to allow some other modes of inference, as already anticipated in the preface of my little work.

If we are given the propositions

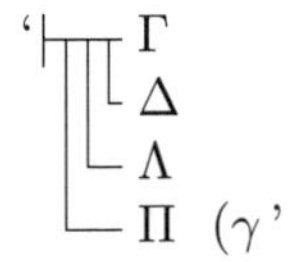

and ‘⊢ Δ (β’

then we cannot immediately make the inference as above but only after having transformed (γ) by making use of the permutability of the subcomponents thus:

However, in order to avoid excessive elaboration, I will not write this out explicitly but rather write immediately

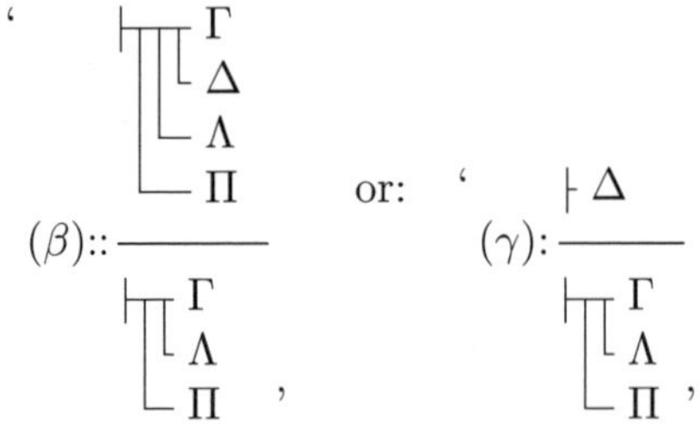

in which the subcomponents of the conclusion could also be ordered differently.

If a subcomponent of a proposition differs from a second proposition only in lacking the judgement-stroke, then one may infer a proposition which results from the first by suppressing that subcomponent.

We also combine two inferences of this kind as can be seen from what follows. Let there be given the further propostion ‘⊢ Λ (ϱ’. Then we write the double inference in this way:

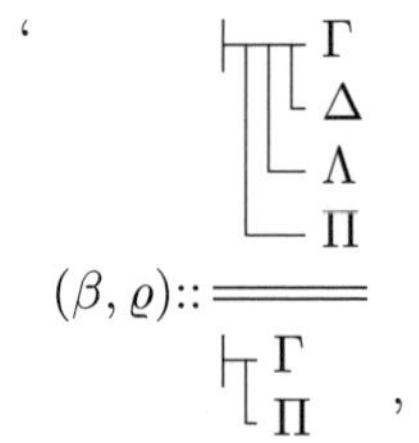

§15. The following mode of inference is a little less simple. From the two propositions

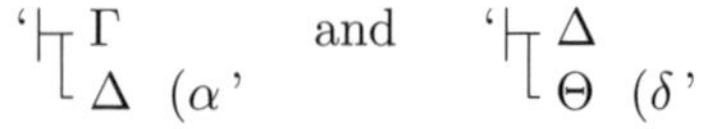

we can infer the proposition ‘⊢ Γ Θ’. For ⊤ Γ Θ is only the False if Θ is the True and Γ is not the True. However if Θ is the True then also Δ must be the True because otherwise ⊤ Δ Θ would be the False. If, however, Δ is the True and Γ were not the True, then Γ Δ would be the False. The case in which Γ Θ is the False cannot, therefore, occur, and Γ Θ is the True.

This inference I write either like this:

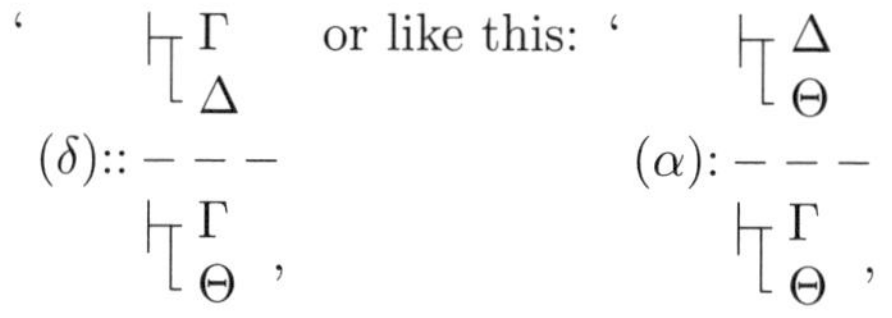

If instead of the proposition (α) we have as premiss the proposition labeled 'γ' in §14, then we should have to carry out a transformation before making the inference, as we did there. Yet, for the sake of brevity, we perform this, as above, mentally and write:

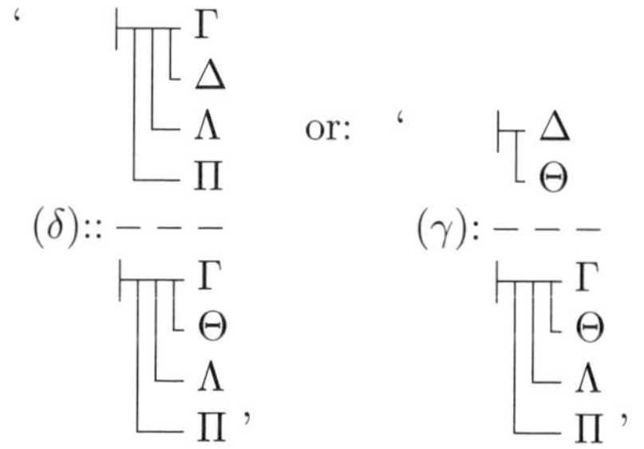

Δ Γ is the False if Γ is the True and Δ is not the True; i.e. if — Γ is the False and Δ is the True. In all other cases Δ Γ is the True. The same holds, however, also for Γ Δ, so that the functions ζ ξ and ξ ζ always have the same value for the same arguments. Equally, the functions ζ ξ and ξ ζ always have the same value for the same arguments. One may reduce this case to the previous one by putting 'ζ' for 'ζ' and canceling adjacent negation-strokes. Further, the functions ξ ζ and ζ ξ always have the same value for the same arguments. We can therefore pass from the proposition 'Γ Δ' to the proposition 'Δ Γ' and conversely from the latter to the former. We write these transitions as follows:

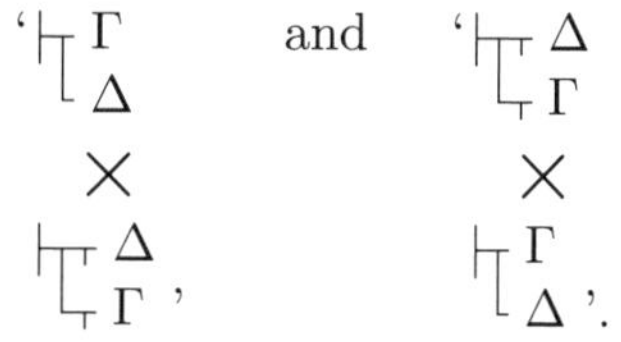

Likewise also:

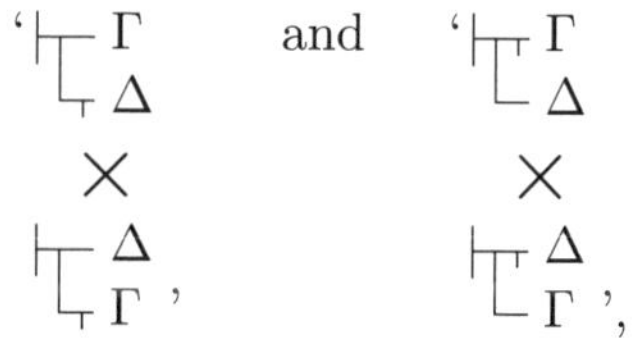

cases which are reducible to the first of the above by the elimination of negation-strokes. We can capture this in a rule thus:

> One may permute a subcomponent with a supercomponent provided one simultaneously *reverses* the truth-value of each.

We call this transition *contraposition*. However, there may also be several subcomponents present. In that case we have the transition

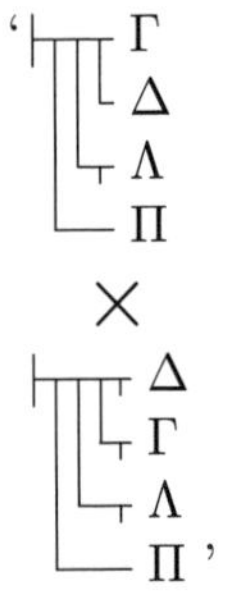

However, by tacit appeal to the permutability of the subcomponents, we may also write:

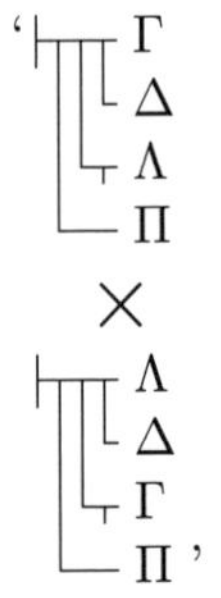

By two steps of contraposition all subcomponents may be combined in one, as follows:

For in the second contraposition we regard

as supercomponent and '⊤ Γ' as subcomponent. Let 'Θ' be an abbreviation for the truth-value

The penultimate proposition then becomes '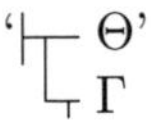' from which '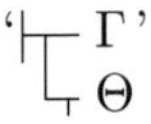' follows. If we then again insert for 'Θ' the unabbreviated expression, we obtain the conclusion. As can be seen from §12, we have in

the truth-value of: that Δ is the True, Λ is not the True and Π is the True.

If we take the propositions

 and 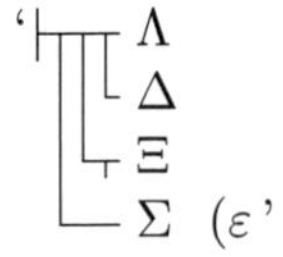

as given, we may make the inference as follows: we first combine the subcomponents of (ε):

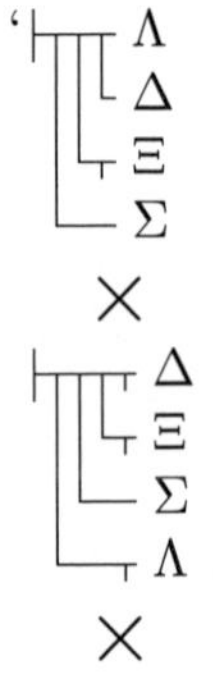

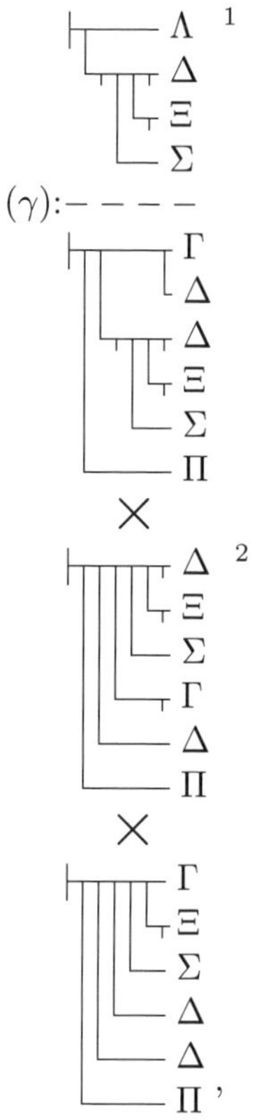

This can be simplified by writing 'Δ' only once:

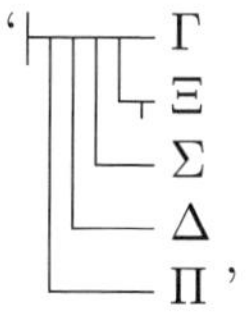

for

is always the same truth-value as .

A subcomponent occurring twice need only be written once.

We call this the *fusion* of equal subcomponents.

I will write this transition as:

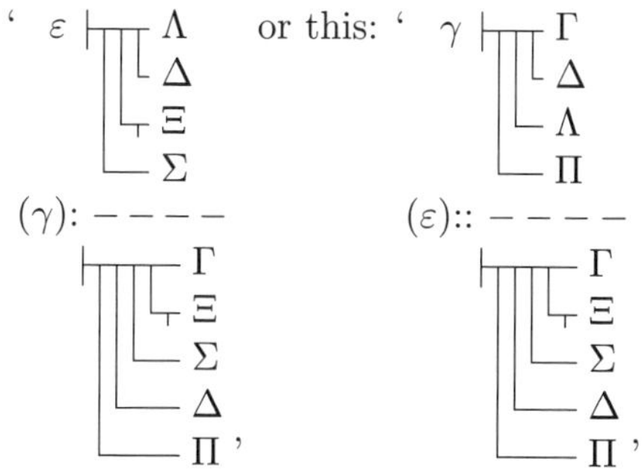

and lay down the following rule for it:

If the same combination of signs occurs in one proposition as supercomponent and in another as subcomponent, then a proposition may be inferred in which the supercomponent of the latter features as supercomponent and all subcomponents of both, save that mentioned, feature as subcomponents. However, subcomponents that occur in both need only be written once.

In a manner similar to that of §14, we can here combine two inferences. For example, if in addition to (ε) we are given the propositions

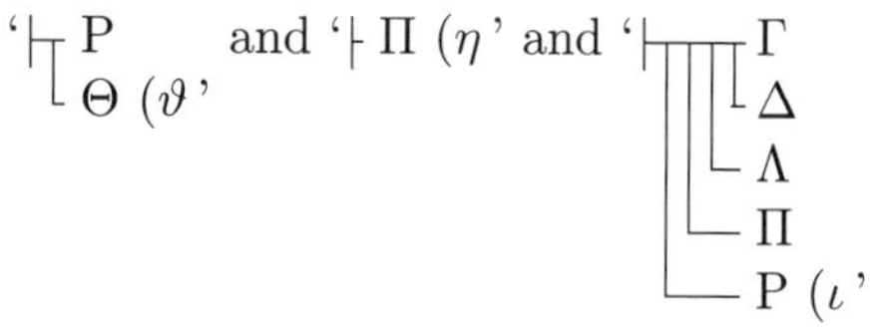

then we can write

[1] Here we can infer in the same way as at the start of this paragraph since this proposition has the same form as (δ) there.

[2] We now resolve the complex subcomponent again.

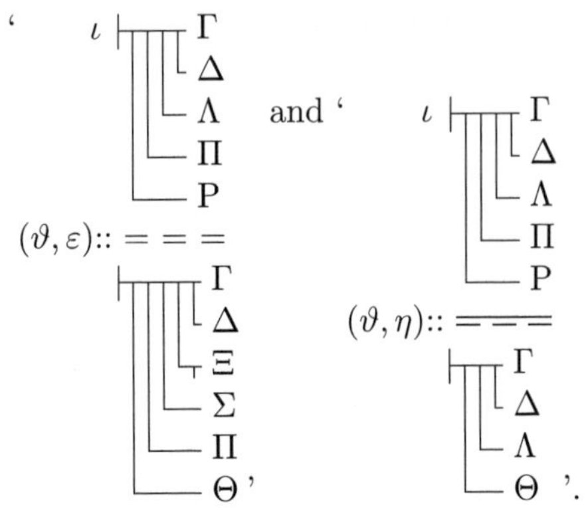

§**16.** If we assume the propositions

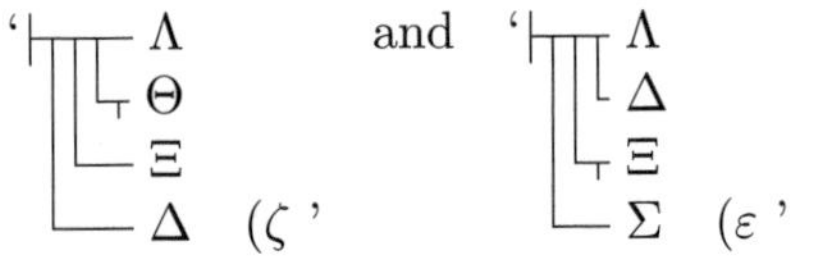

are given, then we can reduce this case to the one just treated as follows:

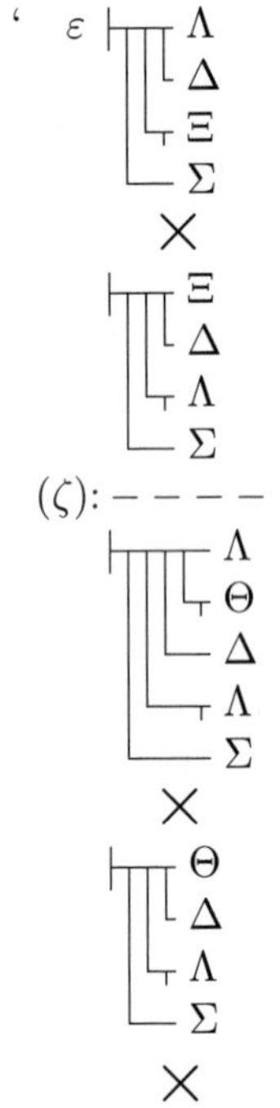

The purpose of these two contrapositions is to do away with one occurrence of '┬ Λ' by fusion of equal subcomponents. That

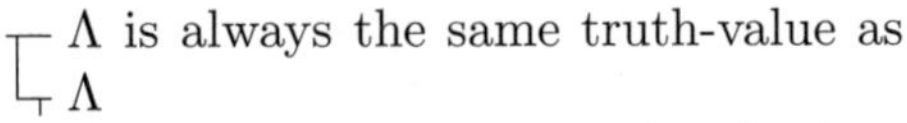

— Λ can also be seen immediately since

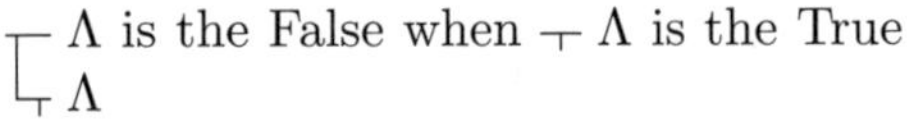

and Λ is not the True, and is otherwise the True, for the latter condition contains the first. However, — Λ is also the False when Λ is not the True, and is otherwise the True. We now abbreviate this transition as:

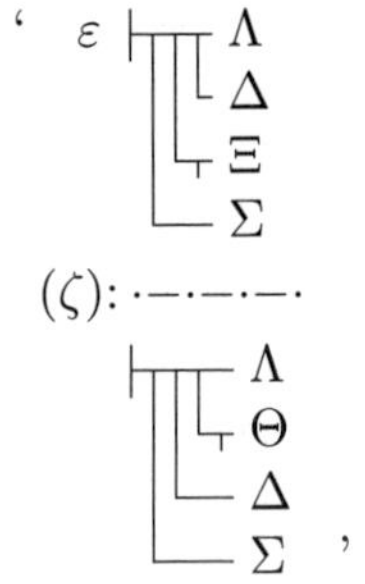

and articulate the rule like this:

> *If two propositions agree in their supercomponents while a subcomponent of the one differs from a subcomponent of the other only by a prefixed negation-stroke, then we can infer a proposition in which the common*

supercomponent features as supercomponent, and all subcomponents of both propositions with the exception of the two mentioned feature as subcomponents. In this, subcomponents which occur in both propositions need only be written down once (fusion of equal subcomponents).

§17. Let us now see how the inference called 'Barbara' in logic fits in here. From the two propositions:

'All square roots of 1 are fourth roots of 1'

and

'All fourth roots of 1 are eighth roots of 1'

we can infer:

'All square roots of 1 are eighth roots of 1'

If we now write the premisses thus:

$$`\vdash\!\overset{\mathfrak{a}}{\smile}\!\!\top \begin{array}{l} \mathfrak{a}^4 = 1 \\ \mathfrak{a}^2 = 1 \end{array}\text{'} \quad \text{and} \quad `\vdash\!\overset{\mathfrak{a}}{\smile}\!\!\top \begin{array}{l} \mathfrak{a}^8 = 1 \\ \mathfrak{a}^4 = 1 \end{array}\text{'}$$

then we cannot apply our modes of inference; however we can if we write the premisses as follows:

$$`\vdash\!\!\top \begin{array}{l} x^4 = 1 \\ x^2 = 1 \end{array}\text{'} \quad \text{and} \quad `\vdash\!\!\top \begin{array}{l} x^8 = 1 \\ x^4 = 1 \end{array}\text{'}$$

Here we have the case of §15. Above we attempted to express generality in this way using a *Roman letter*, but abandoned it because we observed that the scope of generality would not be adequately demarcated. We now address this concern by stipulating that the *scope* of a *Roman letter* is to include everything that occurs in the proposition apart from the judgement-stroke.[1] Accordingly, one can never express the negation of a generality by means of a Roman letter, although we can express the generality of a negation. An ambiguity is thus no longer present. Nevertheless, it is clear that the expression of generality with German letters and concavity is not rendered superfluous. Our stipulation regarding the *scope* of a *Roman letter* is only to demarcate its narrowest extent and not its widest. It thus remains permissible to let the scope extend to multiple propositions so that the Roman letters are suitable to serve in inferences in which the German letters, with their strict demarcation of scope, cannot serve. So, when our premisses are '$\vdash\!\!\top \begin{array}{l} x^8 = 1 \\ x^4 = 1 \end{array}$' and '$\vdash\!\!\top \begin{array}{l} x^4 = 1 \\ x^2 = 1 \end{array}$', then in order to make the inference to the conclusion '$\vdash\!\!\top \begin{array}{l} x^8 = 1 \\ x^2 = 1 \end{array}$', we temporarily expand the scope of 'x' to include both premisses and conclusion, although each of these propositions holds even without this extension.

We do not say of a Roman letter that it *refers to* an object but that it *indicates* an object.[c]

[1] The use of Roman letters is hereby explained only for the case in which a judgement-stroke occurs. This, however, is always the case in a pure concept-development; since then we always progress from proposition to proposition.

Likewise, we say that a German letter *indicates* an object where it does not stand over a concavity.

A proposition with a Roman letter can always be transformed into one with a German letter whose concavity is separated from the judgement-stroke only by a horizontal. We write such a transition thus:

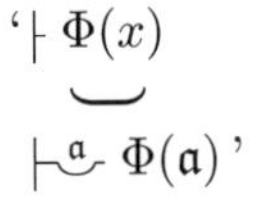

In doing so, the second rule of §8 must be observed, as in the following example where '$\mathfrak{e}$' may not be chosen as the German letter newly to be introduced.

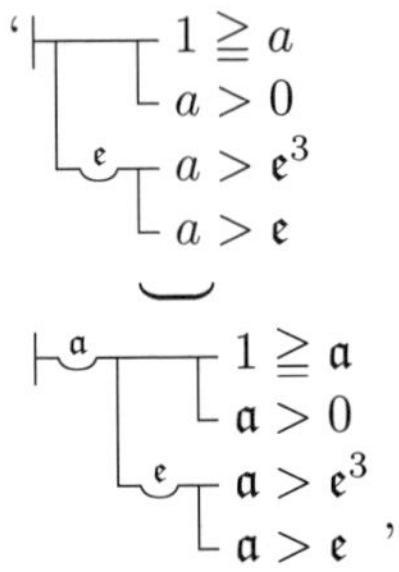

In the transition from a Roman to a German letter the following case also must be mentioned. Consider the proposition '$\vdash\overset{\mathfrak{a}}{\smile}\!\top\begin{array}{l}\Phi(\mathfrak{a})\\ \Gamma\end{array}$'; where 'Γ' is a proper name and '$\Phi(\xi)$' a function-name. $-\overset{\mathfrak{a}}{\smile}\!\top\begin{array}{l}\Phi(\mathfrak{a})\\ \Gamma\end{array}$ is the False if the function $\top\begin{array}{l}\Phi(\xi)\\ \Gamma\end{array}$ has the False as value for any particular argument. This case obtains if Γ is the True and the value of the function $-\Phi(\xi)$ is the False for some argument. In all other cases $-\overset{\mathfrak{a}}{\smile}\!\top\begin{array}{l}\Phi(\mathfrak{a})\\ \Gamma\end{array}$ is the True. '$-\overset{\mathfrak{a}}{\smile}\!\top\begin{array}{l}\Phi(\mathfrak{a})\\ \Gamma\end{array}$' thus says that either Γ is not the True or that the value of the function $\Phi(\xi)$ is the True for every argument. Compare this with '$\top\begin{array}{l}\overset{\mathfrak{a}}{\smile}\,\Phi(\mathfrak{a})\\ \Gamma\end{array}$'. The latter refers to the False if Γ is the True and $-\overset{\mathfrak{a}}{\smile}-\Phi(\mathfrak{a})$ is the False. But this is the case if for some argument the value of the function $-\Phi(\xi)$ is the False. In all other cases $\top\begin{array}{l}\overset{\mathfrak{a}}{\smile}\,\Phi(\mathfrak{a})\\ \Gamma\end{array}$ is the True. The proposition '$\vdash\!\top\begin{array}{l}\overset{\mathfrak{a}}{\smile}\,\Phi(\mathfrak{a})\\ \Gamma\end{array}$' thus says the same as '$\vdash\overset{\mathfrak{a}}{\smile}\!\top\begin{array}{l}\Phi(\mathfrak{a})\\ \Gamma\end{array}$'. If for 'Γ' and '$\Phi(\xi)$' combinations of signs are put which do not refer to an object and a function but merely indicate by containing Roman letters, then the above still holds if for each Roman letter a name is put, whichever name it may be, and thus it holds generally.

In order to be able to express myself more precisely, I will introduce the following terminology. I shall call *names* only those signs or combinations of signs that refer to something. Roman letters, and combinations of signs in which those occur, are thus not *names* as they merely *indicate*. A combination of signs which contains Roman letters, and which always results in a proper name when every Roman letter is replaced by a name, I will call a *Roman object-marker*. In addition, a combination of signs which contains Roman letters and which always results in a function-name when every Roman letter is replaced by a name, I will call a *Roman function-marker* or *Roman marker* of a function.

We can now say: the proposition ‘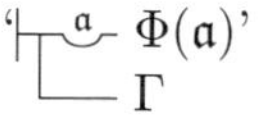’ always says the same as the proposition ‘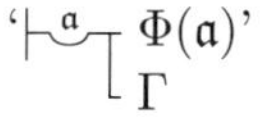’ not only when ‘$\Phi(\xi)$’ is a function-name and ‘Γ’ is a proper name, but also when ‘$\Phi(\xi)$’ is a Roman function-marker and ‘Γ’ is a Roman object-marker.

Let us apply this to the following example.

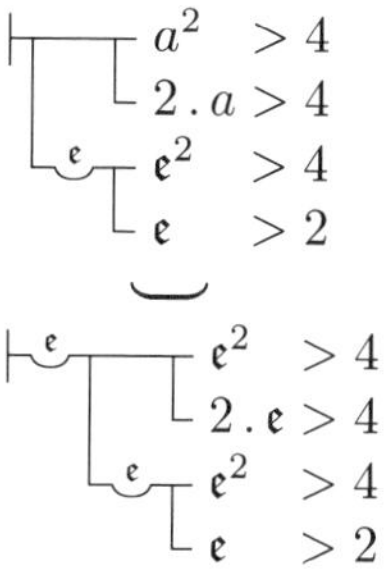
[1]

According to what has just been said, for the last proposition we can write instead

‘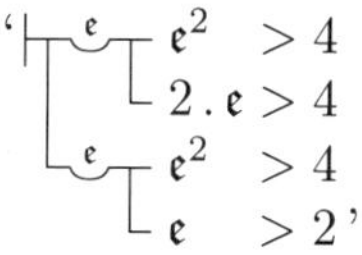’

It is clear that only those subcomponents that do not contain the Roman letter being replaced may be excluded from the scope of the German letter newly to be introduced. I will write such transitions like this:

‘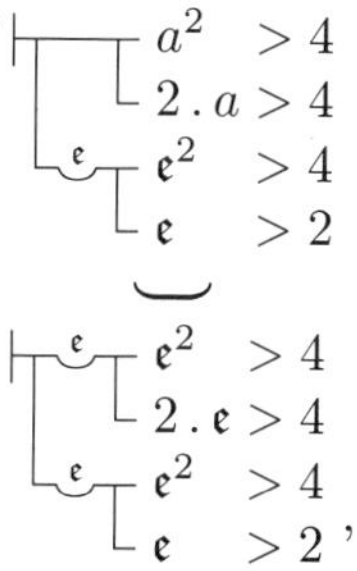’

Instead of introducing several German letters one after the other, we will write the final result immediately under the sign ‘⌣’.

We will summarise this in the following rule:

> *A Roman letter may be replaced wherever it occurs in a proposition by one and the same German letter. At the same time, the latter has to be placed above a concavity in front of one such supercomponent outside of which the Roman letter does not occur.*[1] *If in this supercomponent the scope of a German letter is contained*

[1]The second rule of §8 does not prohibit here the repeated usage of ‘𝔢’ since ‘a’ does not occur in the first proposition within the scope of the ‘𝔢’.

[1]So, if the Roman letter occurs in every subcomponent, then the whole proposition excluding the judgement-stroke has to be regarded as the supercomponent, and the concavity with the German letter may be separated from the judgement-stroke only by a horizontal.

and the Roman letter occurs within this scope, then the German letter that is to be introduced for the latter must be distinct from the former (second rule of §8).

§18. We will now set up some general laws for Roman letters which we will have to make use of later. According to §12

$$\begin{array}{l}\top\!\!\top\,\Gamma\\ \;\;\llcorner\,\Delta\\ \llcorner\,\Gamma\end{array}$$

would be the False only if Γ and Δ were the True while Γ was not the True. This is impossible; accordingly

$$\begin{array}{l}\vdash\!\!\top\!\!\top\,a\\ \quad\;\;\llcorner\,b\\ \quad\llcorner\,a\end{array}\qquad\qquad (\mathrm{I}$$

'I' is given to this proposition as label (§14) and indices will be assigned in this manner to propositions in what follows. If we write 'a' instead of 'b' we can fuse equal subcomponents so as to obtain in '$\begin{array}{l}\vdash\!\!\top\,a\\ \;\llcorner\,a\end{array}$' a special case of (I), which is, without reminder, also to be understood as an instance of (I).

$—\,\Delta$ and $\top\,\Delta$ are always different and are truth-values. Since $—\,\Gamma$ is likewise always a truth-value, it must coincide with either $—\,\Delta$ or $\top\,\Delta$. From this it follows that $\begin{array}{l}\top(—\,\Gamma)=(—\,\Delta)\\ \llcorner\!\!\top(—\,\Gamma)=(\top\,\Delta)\end{array}$ is always the True; for it would only be the False if $\top(—\,\Gamma)=(\top\,\Delta)$ were the True, i.e., if $(—\,\Gamma)=(\top\,\Delta)$ were the False, and $(—\,\Gamma)=(—\,\Delta)$ were not the True, i.e., the False. In other words: $\begin{array}{l}\top(—\,\Gamma)=(—\,\Delta)\\ \llcorner\!\!\top(—\,\Gamma)=(\top\,\Delta)\end{array}$ would only be the False if both $(—\,\Gamma)=(—\,\Delta)$ and $(—\,\Gamma)=(\top\,\Delta)$ were the False, which as we have just seen is not possible. Therefore

$$\begin{array}{l}\vdash\!\!\top(—\,a)=(—\,b)\\ \quad\llcorner\!\!\top(—\,a)=(\top\,b)\end{array}\qquad\qquad (\mathrm{IV}$$

The brackets on the right side of the equality-sign can be dispensed with if one wishes.

From the reference of the function-name $\backslash\xi$ (§11)

$$\vdash a=\backslash\grave{\varepsilon}(a=\varepsilon)\qquad\qquad (\mathrm{VI}$$

follows.

Extension of the notation for generality

§19. So far, generality has been expressed only with respect to objects. In order to be able to do the same for functions, we distinguish as *function-letters* the letters 'f', 'g','h','F','G','H' and the corresponding German letters, in contrast to the others that we call *object-letters*,[1] so that they indicate only functions and never, like the latter, objects. We also count the small Greek vowels

[1] With the exception of 'M' which is reserved for a special purpose.

amongst the *object-letters*, since their only occurrences without the smooth breathing are in places where a proper name may also stand. Within its scope, a function-letter is always followed by a *bracket*, whose interior contains either one place, or two places separated by a comma, depending on whether the letter is to indicate a function with one or with two arguments. Such a place serves to receive a simple or complex sign which either refers to or indicates an argument or, like a small Greek vowel, occupies the argument place. It is clear that, within its scope, a function-letter must always occur with one argument place or always with two argument places. In the case of a Roman function-letter, the *scope* comprises everything occurring in the proposition except the judgement-stroke; in the case of a German function-letter, it is demarcated by a concavity together with the German letter standing alone. In this regard, the use of function-letters conforms entirely with that of object-letters. To begin with, this may be illustrated by examples.

§20. $\smile^{\mathfrak{a}}\, \Phi(\mathfrak{a})$ is the True only if the value of the corresponding function $\Phi(\xi)$ is the True for every argument. So, $\Phi(\Gamma)$ likewise has to be the True. From this it follows that $\left[\smile^{\mathfrak{a}}\, \Phi(\mathfrak{a}) \multimap \Phi(\Gamma)\right]$ is always the True, whatever function with one argument $\Phi(\xi)$ may be. Here, in order to identify the corresponding function $\Phi(\xi)$, the first rule of §8 is to be observed. If, e.g., one were to write '$\smile^{\mathfrak{a}}\, \Psi(\mathfrak{a}, \smile^{\mathfrak{a}}\, \mathrm{X}(\mathfrak{a}, \mathfrak{a})) \multimap \Psi(\Gamma, \smile^{\mathfrak{a}}\, \mathrm{X}(\Gamma, \mathfrak{a}))$' one would only appear to have names of the same function in both the super- and the subcomponent; in fact, the subcomponent would have been formed with the function-name '$\Psi(\xi, \smile^{\mathfrak{a}}\, \mathrm{X}(\mathfrak{a}, \mathfrak{a}))$' and the supercomponent with the function-name '$\Psi(\xi, \smile^{\mathfrak{a}}\, \mathrm{X}(\xi, \mathfrak{a}))$'. Now, '$\smile^{\mathfrak{f}}\left[\smile^{\mathfrak{a}}\, \mathfrak{f}(\mathfrak{a}) \multimap \mathfrak{f}(\Gamma)\right]$' is to be understood as the truth-value of: that a name of the True is always obtained no matter what function-name is inserted at the place of '$\mathfrak{f}$' into '$\smile^{\mathfrak{a}}\, \mathfrak{f}(\mathfrak{a}) \multimap \mathfrak{f}(\Gamma)$'. This truth-value is the True whatever object is referred to by 'Γ': $\vdash \smile^{\mathfrak{f}}\left[\smile^{\mathfrak{a}}\, \mathfrak{f}(\mathfrak{a}) \multimap \mathfrak{f}(a)\right]$. Since the concavity with the '$\mathfrak{f}$' is only separated from the judgement-stroke by a horizontal, we can omit the concavity and write a Roman letter instead of the German letter:

$$\vdash \smile^{\mathfrak{a}}\, f(\mathfrak{a}) \multimap f(a) \qquad \text{(IIa}$$

This law could be expressed in words like this: what holds of all objects, also holds of any.

According to §7, the function with two arguments $\xi = \zeta$ always has a truth-value as its value, namely the True if and only if the ζ-argument coincides with the ξ-argument. If $\Gamma = \Delta$ is the True, then so is $\smile^{\mathfrak{f}} \top\!\!\!\!\!\!\!\!\!\!\quad \mathfrak{f}(\Gamma)$ / $\llcorner\, \mathfrak{f}(\Delta)$; i.e., if Γ is the same as Δ, then Γ falls under every concept under which Δ falls, or, as one may also say, every predication that holds of Δ also holds of Γ. But also conversely: if $\Gamma = \Delta$ is the False, then not every predication that holds of Δ holds of Γ; i.e., in that case $\smile^{\mathfrak{f}} \top\, \mathfrak{f}(\Gamma)$ / $\llcorner\, \mathfrak{f}(\Delta)$ is the False. For example, Γ will not fall under the concept $\xi = \Delta$, under which Δ falls. Thus, $\Gamma = \Delta$ is always the same truth-value as $\smile^{\mathfrak{f}} \top\, \mathfrak{f}(\Gamma)$ / $\llcorner\, \mathfrak{f}(\Delta)$.

Consequently, $\smile^{\mathfrak{f}} \top\, \mathfrak{f}(\Gamma)$ / $\llcorner\, \mathfrak{f}(\Delta)$ falls under every concept under which $\Gamma = \Delta$ falls; thus

$$\vdash\!\top\; g\left(\smile^{\mathfrak{f}} \top\, \mathfrak{f}(a) \;/\; \llcorner\, \mathfrak{f}(b)\right) \;/\; \llcorner\; g(a = b) \qquad \text{(III}$$

We saw (§3, §9) that a value-range equality can always be converted into the generality of an equality, and *vice versa*:

$$\vdash (\grave{\varepsilon} f(\varepsilon) = \grave{\alpha} g(\alpha)) = (\smile^{\mathfrak{a}}\, f(\mathfrak{a}) = g(\mathfrak{a})) \qquad \text{(V}$$

In this, the first rules of §§8 and 9 are to be observed.

§21. In order to explain the use of function-letters in general, we require a further notational device that will now be explained.

Considering the names

$$\text{`}\!\smile^{\mathfrak{a}}\, \mathfrak{a}^2 = 4\text{'} \qquad \text{`}\!\smile^{\mathfrak{a}}\, \mathfrak{a} > 0\text{'}$$

$$\text{`}\!\smile^{\mathfrak{a}} \top\top\; \mathfrak{a}^2 = 1 \;/\; \llcorner\; \mathfrak{a} > 0\text{'}$$

we can easily see that we obtain them from '$\smile^{\mathfrak{a}}\, \Phi(\mathfrak{a})$'[1], by replacing the function-name '$\Phi(\xi)$' with, respectively, the names of the functions, $\xi^2 = 4$, $\xi > 0$, $\top\top\; \xi^2 = 1$ / $\llcorner\; \xi > 0$. It is clear that only names of functions with one argument, not proper names or names of functions with two arguments, can be inserted; for the combination of signs that is to be inserted always has to have open argument places to receive the letter '$\mathfrak{a}$'[2] and if we wanted to insert a name of a function with two arguments then the ζ-argument place would remain unfilled. In order, e.g., to insert the name of the function $\Psi(\xi, \zeta)$, one might be tempted to write '$\smile^{\mathfrak{a}}\, \Psi(\mathfrak{a}, \mathfrak{a})$'; but, in truth, one would have inserted not the name of the function $\Psi(\xi, \zeta)$ but that of the function with one argument, $\Psi(\xi, \xi)$ (first rule of §8). If one were to write '$\smile^{\mathfrak{a}}\, \Psi(\mathfrak{a}, 2)$' then one would again only insert the name of a function with one argument $\Psi(\xi, 2)$. One could, for example, leave the 'ζ' in place: '$\smile^{\mathfrak{a}}\, \Psi(\mathfrak{a}, \zeta)$', and one would then have a function whose argument is indicated by 'ζ'. We take the foregoing into consideration along with the case where the argument-sign in '$X(\xi)$' is replaced by '$\Phi(\xi)$': '$X(\Phi(\xi))$'. It is common, yet inaccurate, to speak here

[1] Compare §13.

[2] That functions such as $\xi = \xi$ or $\xi^2 - \xi \cdot \xi$ which have the same value for every argument—one may call them constant—must nevertheless be distinguished from this value (object) is shown in my lecture *Function und Begriff* (p. 8).

of a function of a function; for if we remind ourselves that functions are fundamentally different from objects and that the value of a function for an argument is to be distinguished from the function itself, then we can see that a function-name can never take the place of a proper name, because it will involve empty places corresponding to the unsaturatedness of the function. If we say 'the function $\Phi(\xi)$', then we should never forget that 'ξ' belongs with the function-name only by way of making the unsaturatedness recognisable. Another function thus can never occur as argument of the function $\mathrm{X}(\xi)$, although the value of a function for an argument can do so, for instance $\Phi(2)$, in which case the value is $\mathrm{X}(\Phi(2))$. If we write '$\mathrm{X}(\Phi(\xi))$', then by '$\Phi(\xi)$' we merely indicate an argument, just as 'ξ' in '$\mathrm{X}(\xi)$' merely indicates. The function-name is really just a part of '$\Phi(\xi)$' and so the function does not here occur as the argument of $\mathrm{X}(\xi)$, since the function-name only fills part of the argument place. Likewise, one cannot say that in '$-\!\overset{\mathfrak{a}}{\smile}\!-\, \Psi(\mathfrak{a}, \zeta)$' the function-name $\Psi(\xi, \zeta)$ occupies the place of the function-name '$\Phi(\xi)$' in '$-\!\overset{\mathfrak{a}}{\smile}\!-\, \Phi(\mathfrak{a})$'; since it fills only one part while the other, namely the place of 'ζ', remains open for a proper name. *Functions with two arguments* are just as fundamentally distinct from *functions with one argument* as the latter are from *objects*. For, while the latter are fully *saturated*, functions with two arguments are less saturated than those with one argument, which are already *unsaturated*.

Thus, in '$-\!\overset{\mathfrak{a}}{\smile}\!-\, \Phi(\mathfrak{a})$' we have an expression in which we can replace the name of the function $\Phi(\xi)$ by names of functions with one argument, but not by names of objects nor by names of functions with two arguments. This leads us to regard $-\!\overset{\mathfrak{a}}{\smile}\!-\, \mathfrak{a}^2 = 4$, $-\!\overset{\mathfrak{a}}{\smile}\!-\, \mathfrak{a} > 0$, $-\!\overset{\mathfrak{a}}{\smile}\!\!\begin{array}{l}\!\top\, \mathfrak{a}^2 = 1 \\ \!\llcorner\, \mathfrak{a} > 0\end{array}$ as values of the same function $-\!\overset{\mathfrak{a}}{\smile}\!-\, \varphi(\mathfrak{a})$ with different *arguments*. However, here these arguments are again themselves functions, namely functions with one argument $\xi^2 = 4$, $\xi > 0$, $\begin{array}{l}\!\top\, \xi^2 = 1 \\ \!\llcorner\, \xi > 0\end{array}$; and only functions with one argument can be arguments of our function $-\!\overset{\mathfrak{a}}{\smile}\!-\, \varphi(\mathfrak{a})$. If we say, 'the function $-\!\overset{\mathfrak{a}}{\smile}\!-\, \varphi(\mathfrak{a})$', then '$\varphi$' is proxy for the sign of an argument, just as 'ξ' in the expression 'the function $\xi^2 = 4$' is proxy for a proper name which could feature as argument-sign. In this case 'φ' no more belongs to the function than 'ξ' in the latter case. Functions whose arguments are objects we now call *first-level functions*; in contrast those functions whose arguments are first-level functions will be called *second-level functions.* The value of our function $-\!\overset{\mathfrak{a}}{\smile}\!-\, \varphi(\mathfrak{a})$ is always a truth-value whatever first-level function we take as argument. In agreement with our earlier terminology, we will thus call it a concept, more precisely a *second-level concept*, in contrast to the *first-level concepts* which are first-level functions.

Our function $\unicode{x2510}^{a}\unicode{x2510} \varphi(\mathfrak{a})$ had, for the arguments taken above, the True as its value. If we now take as its argument the function $\unicode{x2510}\unicode{x2510} \xi^3 = -1$, then we obtain in $\unicode{x2510}\xi > 0$ $\unicode{x2510}^{a}\unicode{x2510} \mathfrak{a}^3 = -1$ the False, since there is $\unicode{x2510}\mathfrak{a} > 0$ no positive cubic root of -1. Likewise, the value of our function for the argument $\xi + 3$ is the False; for we can always replace $\unicode{x2510}^{a}\unicode{x2510} \mathfrak{a} + 3$ by $\unicode{x2510}^{a}\unicode{x2510} (-\mathfrak{a} + 3)$, and this is the False, since the value of the function $-\xi + 3$ is always the False, that is, if we assume that the plus-sign is explained in such a way that for no argument the value of the function $\xi + 3$ is the True.

§22.
We have a different second-level function in

$$\unicode{x2510}^{a}\unicode{x2510}^{e}\unicode{x2510} \begin{array}{l} \mathfrak{a} = \mathfrak{e} \\ \varphi(\mathfrak{e}) \\ \varphi(\mathfrak{a}) \end{array}$$

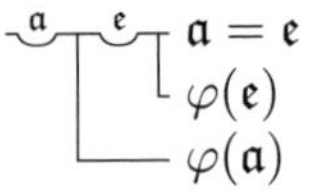

where 'φ' again is a proxy for the sign of the argument. Its value is the True for every first-level concept as argument under which at most a single object falls. Accordingly,

$$\vdash \unicode{x2510}^{a}\unicode{x2510}^{e}\unicode{x2510} \begin{array}{l} \mathfrak{a} = \mathfrak{e} \\ \mathfrak{e} + 1 = 3 \\ \mathfrak{a} + 1 = 3 \end{array} \qquad \vdash \unicode{x2510}^{a}\unicode{x2510}^{e}\unicode{x2510} \begin{array}{l} \mathfrak{a} = \mathfrak{e} \\ \mathfrak{e} = \mathfrak{e} \\ \mathfrak{a} = \mathfrak{a} \end{array}$$

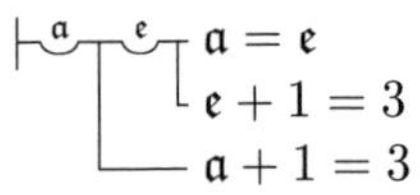

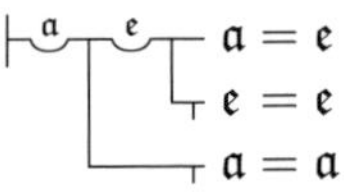

In contrast:

$$\vdash \unicode{x2510}^{a}\unicode{x2510}^{e}\unicode{x2510} \begin{array}{l} \mathfrak{a} = \mathfrak{e} \\ \mathfrak{e}^2 = 1 \\ \mathfrak{a}^2 = 1 \end{array}$$

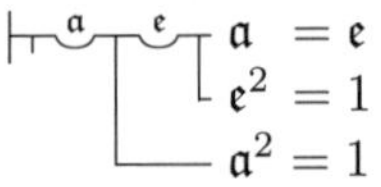

We also have a second-level function in $\varphi(2)$. Some of the values of this function are truth-values, as for example when the arguments are $\xi + \xi = \xi \cdot \xi$, $\xi + 1 = 4$ to which correspond the values $2 + 2 = 2 \cdot 2$ and $2 + 1 = 4$, while others are other objects, for instance the number 3 when the argument is $\xi + 1$. This second-level function is distinct from the mere number 2 since, like all functions, it is unsaturated.

The second-level function $\unicode{x2510}\varphi(2)$ is distinguished from the above by its value always being a truth-value. It is thus a second-level concept which may be called property of the number 2; for every concept falls under this second-level concept under which the number 2 falls, while all other first-level functions with one argument do not fall under this second-level concept.[1]

Also in

$$\unicode{x2510}\unicode{x2510}^{a}\unicode{x2510} \begin{array}{l} \varphi(2) \\ \mathfrak{a} = 2 \\ \varphi(\mathfrak{a}) \end{array}$$

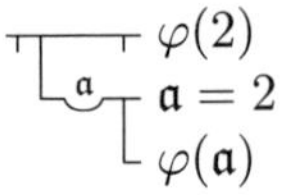

we have a second-level concept which we might call: property exclusively of the number 2.

A further second-level concept is $\unicode{x2510}^{a}\unicode{x2510} \varphi(\mathfrak{a})$. In $\grave\varepsilon\varphi(\varepsilon)$ we have an example of a second-level function which is not a concept.

In order to take an example from

---

[1] Compare note p. 8.

analysis we will consider the differential quotient of a function. We regard the latter as argument. Let us take a specific function, e.g., ξ^2, as argument; then we first obtain another first-level function $2 \,.\, \xi$ and only when we take as its argument an object, e.g., the number 3, do we obtain an object as value: the number 6. The differential quotient is therefore to be regarded as a function with two arguments, of which the one has to be a first-level function with one argument, and the other an object. We can therefore call it an *unequal-levelled* function with two arguments. From this we obtain a second-level function with one argument by saturating it with one object-argument—e.g., the number 3—i.e., by determining that the differential quotient is to be formed for the argument 3.[1]

A further example of an unequal-levelled function with two arguments is provided by $— \varphi(\xi)$, where 'ξ' occupies the place of an object-argument and '$\varphi(\ )$' that of the function-argument and makes them salient. Since the value of this function is always a truth-value, we may call it an unequal-levelled relation. It is the relation of an object to a concept under which it falls. Examples of equal-levelled relations of second-level are $\smile^{\mathfrak{a}}\!\top\, \varphi(\mathfrak{a}) \;\llcorner\, \psi(\mathfrak{a})$ and $\top\!\smile^{\mathfrak{a}}\!\top\!\top\, \varphi(\mathfrak{a}) \;\llcorner\, \psi(\mathfrak{a})$, where '$\varphi$' and '$\psi$' mark the argument places. The latter relation holds between, e.g., the concepts $\xi^3 = 1$ and $\xi^2 = 1$, because we have $\vdash\!\top\!\smile^{\mathfrak{a}}\!\top\!\top\, \mathfrak{a}^3 = 1 \;\llcorner\, \mathfrak{a}^2 = 1$. In words: at least one square root of 1 is also a cube root of 1.

§23. In the examples given so far, functions with one argument featured as arguments; $\top\!\smile^{\mathfrak{a}}\!\smile^{\mathfrak{e}}\!\top\, \varphi(\mathfrak{a}, \mathfrak{e})$ is a second-level concept, whose argument has to be a function with two arguments. Under this concept falls every relation in which some objects stand. For one can also give relations—one might call them empty— in which no objects stand to one another; e.g., $\top\!\top\, 2 \,.\, \xi = 2 \,.\, \zeta \;\llcorner\, \xi = \zeta$, since

$$\vdash\!\top\!\top\!\smile^{\mathfrak{a}}\!\smile^{\mathfrak{e}}\!\top\!\top\, 2 \,.\, \mathfrak{a} = 2 \,.\, \mathfrak{e} \;\llcorner\, \mathfrak{a} = \mathfrak{e}.$$

In order to have a further example of this kind, let us try to express the *single-valuedness* of a relation. By this we understand that for every ξ-argument there is no more than one ζ-argument such that the value of our function (relation) $\mathrm{X}(\xi, \zeta)$ is the True. We can also say: if, whenever a stands to b in this relation and a stands to c in this relation,

[1] Here, as with all examples taken from arithmetic, it must be presupposed that the signs of addition, multiplication, and so on, as well as that of the differential quotient, are so defined that a name correctly formed from these and proper names always has a reference. The usual definitions certainly do not achieve this, for they invariably take into consideration only numbers, for the most part without saying what numbers are.

it follows that b and c coincide, then we say that the relation is single-valued. Or, if, whenever $X(a,b)$ is the True and $X(a,c)$ is the True, it follows that $c = b$ is the True, then we call the function $X(\xi, \zeta)$ a single-valued relation, as long as it is a relation.

$$\begin{array}{c}\underset{\epsilon\ \mathfrak{d}}{\smile}\ \overset{\mathfrak{a}}{\frown}\begin{array}{l}\mathfrak{d}=\mathfrak{a}\\ X(\epsilon,\mathfrak{a})\\ X(\epsilon,\mathfrak{d})\end{array}\end{array}$$

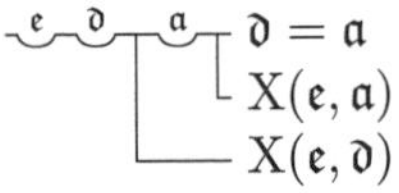

must be the True, if the relation — $X(\xi, \zeta)$ is to be single-valued. If we introduce 'φ' to mark the argument place for 'X', then we obtain in

$$\begin{array}{c}`\underset{\epsilon\ \mathfrak{d}}{\smile}\ \overset{\mathfrak{a}}{\frown}\begin{array}{l}\mathfrak{d}=\mathfrak{a}\\ \varphi(\epsilon,\mathfrak{a})\\ \varphi(\epsilon,\mathfrak{d})\end{array}\text{,}\end{array}$$

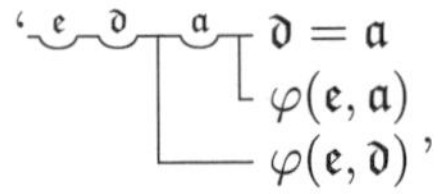

the name of a second-level function which requires as argument a function with two arguments. This second-level function is a second-level concept under which all single-valued relations fall, but also those functions $X(\xi, \zeta)$ for which — $X(\xi, \zeta)$ is a single-valued relation. The single-valuedness is always to be understood as running in the direction from the ξ to the ζ-argument. If we take the function $\xi^2 = \zeta$ as the argument of our second-level function, then we obtain the function-value

$$\begin{array}{c}\underset{\epsilon\ \mathfrak{d}}{\smile}\ \overset{\mathfrak{a}}{\frown}\begin{array}{l}\mathfrak{d}=\mathfrak{a}\\ \epsilon^2=\mathfrak{a}\\ \epsilon^2=\mathfrak{d}\end{array}\end{array}$$

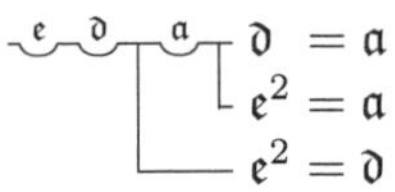

that is, the True; while the False is obtained as the value of the function if we take as the argument the function $\xi = \zeta^2$:

$$\begin{array}{c}\underset{\epsilon\ \mathfrak{d}}{\smile}\ \overset{\mathfrak{a}}{\frown}\begin{array}{l}\mathfrak{d}=\mathfrak{a}\quad {}^{1}\\ \epsilon=\mathfrak{a}^2\\ \epsilon=\mathfrak{d}^2\end{array}\end{array}$$

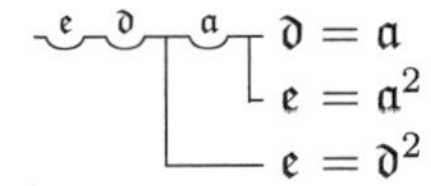

We recognise in these examples the great multiplicity of functions. We can also see that there are fundamentally different functions, since the argument places are fundamentally different. In particular, those which are suited to take proper names cannot receive names of functions, and conversely. Moreover, argument places that can take names of first-level functions with one argument are unable to take names of first-level functions with two arguments. Accordingly, we distinguish:

arguments of the first kind: objects;

arguments of the second kind: first-level functions with one argument;

arguments of the third kind: first-level functions with two arguments.

Likewise, we distinguish:

argument places of the first kind, that are suitable to take proper names;

argument places of the second kind, that are suitable to take names of first-level functions with one argument;

argument places of the third kind, that are suitable to take names of

---

[1] Subject to a suitable definition of ξ^2 for arguments that are not numbers.

first-level functions with two arguments.

Proper names and object-letters *fit* the argument places of the first kind; names of first-level functions with one argument *fit* the argument places of the second kind; names of first-level functions with two arguments *fit* argument places of the third kind. The objects and functions whose names fit the argument places of names of functions are *fitting* arguments for these functions. Functions with one argument, for which arguments of the second kind are fitting, we call *second-level functions with an argument of the second kind*; functions with one argument, for which arguments of the third kind are fitting, we call *second-level functions with an argument of the third kind.*

Just as in $-\overset{\mathfrak{a}}{\smile}-\ \mathfrak{a} = \mathfrak{a}$ we have the value of the second-level function $-\overset{\mathfrak{a}}{\smile}-\ \varphi(\mathfrak{a})$ for the argument $\xi = \xi$, so too we can regard $-\overset{\mathfrak{f}}{\smile}\!\!\begin{array}{l} \top\ \mathfrak{f}(1+1) \\ \llcorner\ \mathfrak{f}(2) \end{array}$ as the value of a *third-level* function for the argument $\begin{array}{l} \top\ \varphi(1+1) \\ \llcorner\ \varphi(2) \end{array}$ which is itself a second-level function with one argument of the second kind.

§24. It is now possible to give a general explanation of the use of function-letters.

If a concavity with a German function-letter is followed by a combination of signs consisting of the name of a second-level function with one argument together with the function-letter in question in its argument places, then this whole refers to the True provided the value of that second-level function is the True for every fitting argument; in all other cases it refers to the False. Which places are argument places of the *corresponding* function is to be judged in accordance with the first rule of §8. The second rule of §8 also applies to function-letters, just as it applies to object-letters.

We have hereby introduced two third-level functions, whose names can be displayed like this:

$$\text{`}-\overset{\mathfrak{f}}{\smile}-\ \mu_\beta(\mathfrak{f}(\beta))\text{'} \quad \text{and} \quad \text{`}-\overset{\mathfrak{f}}{\smile}-\ \mu_{\beta\gamma}(\mathfrak{f}(\beta,\gamma))\text{'},$$

where we mark the argument places by 'μ_β' and '$\mu_{\beta\gamma}$', just as we mark the argument places of the second and third kind by 'φ' and 'ψ', and those of the first kind by 'ξ' and 'ζ'. Like the former letters, incidentally, 'μ_β' and '$\mu_{\beta\gamma}$' are not to be regarded as signs of concept-script, but serve us only provisionally. If we take as arguments for the first of these functions the second-level functions with one argument of the second kind, $-\overset{\mathfrak{a}}{\smile}-\ \varphi(\mathfrak{a})$, $\varphi(2)$, $-\overset{\mathfrak{a}}{\smile}\!\!-\!\!\overset{\mathfrak{e}}{\smile}\!\!\begin{array}{l} \top\ \mathfrak{a} = \mathfrak{e} \\ \llcorner\ \varphi(\mathfrak{e}) \\ \llcorner\ \varphi(\mathfrak{a}) \end{array}$, then we respectively obtain as value: $-\overset{\mathfrak{f}}{\smile}\!\!-\!\!\overset{\mathfrak{a}}{\smile}-\ \mathfrak{f}(\mathfrak{a})$, $-\overset{\mathfrak{f}}{\smile}-\ \mathfrak{f}(2)$, and

$$-\overset{\mathfrak{f}}{\smile}\!\!-\!\!\overset{\mathfrak{a}}{\smile}\!\!-\!\!\overset{\mathfrak{e}}{\smile}\!\!\begin{array}{l} \top\ \mathfrak{a} = \mathfrak{e}\,. \\ \llcorner\ \mathfrak{f}(\mathfrak{e}) \\ \llcorner\ \mathfrak{f}(\mathfrak{a}) \end{array}$$

§25. We still need a way to express generality with respect to second-level functions with one argument of the second kind. It might be thought that this would not nearly suffice; however we will see that we can get by with this expression and indeed that it is required only in one single proposition. Here it may merely be briefly remarked that this economy is made possible by the fact that second-level functions are representable, in a way, by first-level functions where the functions that appear as arguments of the former are represented by their the value-ranges. However the notational device that is necessary for this is not included in the primitives of concept-script; we will later introduce it by means of our primitive signs. Since this expressive device is only used in one single proposition, it is unnecessary to explain it in full generality.

We indicate a second-level function with one argument of the second kind by using the *Roman function-letter* 'M'[1] in this way:

$$\text{`}M_\beta(\varphi(\beta))\text{'}$$

just as by '$f(\xi)$' we indicate a first-level function with one argument. Here, '$\varphi(\ )$' marks the argument place, just as 'ξ' in '$f(\xi)$'. The bracketed letter 'β' here fills the argument place of the function that occurs as argument. The use of '$M_\beta(\varphi(\beta))$' for second-level functions is completely analogous to that of $f(\xi)$ for first-level functions. We avail ourselves of this expression of generality in the following law:

$$\begin{array}{l} \vdash\!\!\top\!\!-\, M_\beta(f(\beta)) \\ \;\;\;\;\llcorner\!\smile^{\mathfrak{f}}\!\!-\, M_\beta(\mathfrak{f}(\beta)) \end{array} \qquad \text{(IIb}$$

in words: what holds of all first-level functions with one argument also holds of any. Obviously this law is to our second-level functions what (IIa) is to first-level functions. Here the letter 'f' in (IIa) corresponds to 'M_β', the 'a' in (IIa) to 'f', and the '$\mathfrak{a}$' to '$\mathfrak{f}$'. Let $\Omega_\beta(\varphi(\beta))$ be a second-level function with one argument of the second kind whose places are marked by 'φ'. Then $-\!\smile^{\mathfrak{f}}\!\!-\, \Omega_\beta(\mathfrak{f}(\beta))$ is the True only if the value of our second-level function is the True for every fitting argument. Then $\Omega_\beta(\Phi(\beta))$ also has to be the True. Thus,

$$\begin{array}{l} \top\!\!-\, \Omega_\beta(\Phi(\beta)) \\ \llcorner\!\smile^{\mathfrak{f}}\!\!-\, \Omega_\beta(\mathfrak{f}(\beta)) \end{array}$$

is always the True, no matter what first-level function with one argument $\Phi(\xi)$ may be, regardless of whether $-\!\smile^{\mathfrak{f}}\!\!-\, \Omega_\beta(\mathfrak{f}(\beta))$ is the True or the False; and our law (IIb) states this in general, for every second-level function with one argument of the second kind.

[1] This letter is thus not an *object-letter*.

2. Definitions

General remarks

§26. The signs explained so far will now be used to introduce new names. But, before I discuss the rules that have to be followed here, it will promote comprehension to classify the signs and combinations of signs into kinds and to label them accordingly.[1]

I will not call the German, Roman and Greek letters occurring in concept-script *names* since they are not to refer to anything. In contrast, I do call, for example, ‘$\overset{\mathfrak{a}}{\smile}\, \mathfrak{a} = \mathfrak{a}$’ a *name* since it refers to the True; it is a *proper name.* Thus, I call a *proper name* or *name* of an object any sign, be it simple or complex, that is to refer to an object, but not such a sign which merely indicates an object.

If we remove from a proper name some or all occurrences of another proper name that forms part of or coincides with it, but in such a way that these places remain marked as fillable by one and the same arbitrary proper name (as *argument places of the first kind*), then I call that which we obtain in this way a *name* of a first-level function with one argument. Such a name forms together with a proper name which fills the argument places a proper name. Accordingly, we also have in ‘ξ’ itself a function-name if the letter ‘ξ’ is merely to mark the argument place. The function so named has the property that its value for any argument coincides with the argument itself.

If we remove from a name of a first-level function with one argument all or some occurrences of a proper name that forms part of it, but in such a way that these places remain marked as fillable by one and the same arbitrary proper name (as argument places of the first kind), then I call that which we obtain in this way a *name* of a first-level function with two arguments.

If we remove from a proper name all or some occurrences of a name of a first-level function that forms part of it, but in such a way that these places remain marked as fillable by one and the same arbitrary name of a first-level function (as argument places of the second or third kind), then I call that which we obtain in this way a

[1] Compare §17.

name of a second-level function with one argument, and specifically, of the second or third kind, depending on whether the argument places are of second or third kind.

Names of functions I call *function-names* for short.

It is not necessary to continue further this explanation of kinds of names.

If, in a proper name, we replace proper names that form part of or coincide with it by object-letters, function-names by function-letters, then I call that which we obtain in this way an *object-marker*, or *marker* of an object. If this replacement involves only Roman letters, then I call the marker obtained a *Roman object-marker*. Thus, object-letters are also object-markers and Roman object-letters are Roman object-markers.

A sign (proper name or object-marker) that only consists of the function-name '$\xi = \zeta$' and proper names or object-markers occurring at the two argument places, I call an *equation*.

If, in a function-name, we replace proper names by object-letters, function-names by function-letters, then I call that which we obtain in this way a *function-marker*, specifically, a *marker* of a function of the same kind as the one from whose name it originates. If the replacement involves only Roman letters, then I call the marker obtained a *Roman* function-marker. Function-letters are also function-markers and Roman function-letters are Roman function-markers.

I reckon the judgement-stroke to belong neither with the *names* nor with the *markers*; it is a sign of its own kind. A sign which consists of a judgement-stroke and a name of a truth-value with a prefixed horizontal, I call a *concept-script proposition*, or *proposition*, where there can be no doubt. Likewise, I call a *concept-script proposition* (or *proposition*) a sign which consists of a judgement-stroke and a Roman marker of a truth-value with a prefixed horizontal.

Signs like

$$\text{`}(\alpha) : \overline{\qquad\qquad}\text{'}, \quad \text{`}(\alpha,\beta) :: = = = = =\text{'}, \quad \text{`}(\alpha) :: - - - - -\text{'}, \quad \text{`}\times\text{'}$$

that stand between propositions to indicate how the lower results from the one above, I call *transition-signs*.

§27. In order to introduce new signs by means of those already known, we now require the *double-stroke of definition* which appears as a double judgement-stroke combined with a horizontal:

$$\text{`}\Vdash\text{'}$$

and which is used instead of the judgement-stroke where something is to be defined, rather than judged. By means of a *definition* we introduce a new name by determining that it is to have the same sense and the same reference as a name composed of already known signs. The new sign thereby becomes co-referential with the explaining sign; the definition thus immediately turns into a proposition. Accordingly, we are allowed to cite a definition just like a proposition replacing the definition-stroke by a judgement-stroke.

Here a definition is always presented in the form of an equation with a prefixed ‘╟’. We will always write the explaining sign on the left side of the equality-sign, and the explained on the right side. The former will be composed of known signs.

§28. I now lay down the following governing principle for definitions:

Correctly formed names must always refer to something.

I call a name *correctly* formed if it consists only of such signs that are primitive or introduced by definition, and if these signs are only used as they were introduced to be used, that is, proper names as proper names, names of functions of first-level with one argument as names of such functions, and so on, so that the argument places are always filled with fitting names or markers. For *correct* formation it is further required, that German and small Greek letters are always used in accordance with their purpose. Thus a German letter may only stand above a concavity if the concavity is immediately followed by a marker of a truth-value composed of the name, or marker, of a function with one argument whose argument places are filled by the same German letter. Throughout its scope, a function-letter must occur everywhere either with one or with two argument places. A German letter may occur in an argument place only if a concavity with the same letter stands to the left of it, demarcating its scope. Only a German letter may stand above a concavity. A small Greek vowel may stand beneath a smooth breathing only when immediately followed by an object-marker consisting of a name or the marker of a first-level function with one argument with the same Greek vowel filling its argument places. A small Greek vowel may stand in an argument place only when preceded by the same vowel with a smooth breathing demarcating its scope. Under the smooth breathing only a small Greek vowel may occur.

§29. We now answer the question: when does a name refer to something? We confine ourselves to the following cases.

A name of a first-level function with one argument has a *reference* (*refers to* something, is *referential*) if the proper name which results from this function-name when the argument places are filled by a proper name always has a reference provided the inserted name refers to something.

A proper name has a *reference* if, whenever it fills the argument places of a referential name of a first-level function with one argument, the resulting proper name has a reference, and if the name of a first-level function with one argument which results from the relevant proper name's filling the ξ-argument places of a referential name of a first-level function with two arguments, always has a reference, and if the same also holds for the ζ-argument places.

A name of a first-level function with two arguments has a *reference* if the proper name which results from this function-name by filling both the ξ-argument places with a referential proper name and the ζ-argument places with a referential proper name always has a reference.

A name of a second-level function with one argument of the second kind has a *reference* if, whenever a name of a first-level function with one argument refers to something, it follows that the proper name that results by its insertion into the argument places of our second-level function has a reference.

Accordingly, every name of a first-level function with one argument which forms a referential proper name with every referential proper name, also forms a referential name with every referential name of a second-level function with one argument of the second kind.

The name '$\smile^{\mathfrak{f}}\ \mu_\beta(\mathfrak{f}(\beta))$' of a third-level function is referential if, whenever a name of a second-level function with one argument of the second kind refers to something, it follows that also the proper name that results by its insertion into the argument place of '$\smile^{\mathfrak{f}}\ \mu_\beta(\mathfrak{f}(\beta))$' has a reference.

§30. These propositions are not to be regarded as explanations of the expressions 'to have a reference' or 'to refer to something', since their application always presupposes that one has already recognised some names as referential; but they can serve to widen the circle of such names gradually. It follows from them that every name formed out of referential names refers to something. This formation takes place in such a way that a name fills argument places of another name that are fitting for it. Thus a proper name results from a proper name and a name of a first-level function with one argument, or from a name of a first-level function and a name of a second-level function with one argument, or from a name of a second-level function with one argument of the second kind and the name '$\overset{\mathfrak{f}}{\smile}\,\mu_\beta(\mathfrak{f}(\beta))$' of a third-level function. Thus a name of a first-level function with one argument results from a proper name and a name of a first-level function with two arguments. The names so formed can in turn be used in the same way for the formation of names, and all names resulting in this way are referential provided the primitive simple ones are.

A proper name can only come to be employed in this formation insofar as it fills the argument places of one of the simple or complex first-level functions. Complex names of first-level functions result in the manner described only from simple names of first-level functions with two arguments by a proper name's filling the ξ- or the ζ-argument place. The remaining argument places of the complex function-name are thus always also those of a simple name of a function with two argument places. From this it follows that a proper name that is part of a name so formed always stands, wherever it occurs, in an argument place of one of the simple names of first-level functions. If we now replace this proper name by another at some or all places, then the resulting proper name[d] is also formed in the manner described, so it too has a reference provided all simple names used therein are referential. To be sure, it is here presupposed that all simple names of first-level functions with one argument have only one argument place and that the simple names of first-level functions with two arguments have only one ξ- and one ζ-argument place. If this were not the case, it could happen in the replacement just mentioned that related argument places of simple function-names are filled with different names, and in such a case an explanation of the reference would be lacking. However, this can always be avoided, and must be avoided in order to prevent the occurrence of names without reference. Indeed, it would serve no purpose to introduce several ξ-argument places and several ζ-argument places for the simple function-names. If we assume this, we can recognise the possibility of a second way to form names of first-level functions. For we first form a name in the first way and then omit a proper name that forms a part of it (or coincides with it entirely) at some or all places, but in such a manner that the latter remain recognisable as argument places of the first kind. The function-name thereby resulting also always has a reference provided the simple names from which it is formed refer to something, and it can in turn be used to form referential names in the first or second way.

Thus we can, e.g., form the function-name '$\Delta = \zeta$' in the first way from the proper name 'Δ' and the function-name '$\xi = \zeta$'; and then the proper name '$\Delta = \Delta$' from this and 'Δ'. From this we can form in the second way the function-name '$\xi = \xi$', and from the latter and the function-name '$\overset{\mathfrak{a}}{\smile}\, \varphi(\mathfrak{a})$', in the first way, the proper name '$\overset{\mathfrak{a}}{\smile}\, \mathfrak{a} = \mathfrak{a}$'.

All correctly formed names are so formed.

§31. We apply this to show that proper names and names of first-level functions that we can form in this manner from our previously introduced simple names, always have a reference. According to what has been said, it is required merely to demonstrate of our primitive names that they refer to something. These are

1. names of first-level functions with one argument:

$$\text{`}{-\!\!\!-}\,\xi\text{'},\quad \text{`}{\top}\,\xi\text{'},\quad \text{`}\backslash\xi\text{'};$$

2. names of first-level functions with two arguments:

$$\text{`}\begin{array}{l}\top\,\xi\\ \llcorner\,\zeta\end{array}\text{'},\quad \text{`}\xi = \zeta\text{'};$$

3. names of second-level functions with one argument of the second kind:

$$\text{`}\overset{\mathfrak{a}}{\smile}\, \varphi(\mathfrak{a})\text{'},\quad \text{`}\dot{\varepsilon}\varphi(\varepsilon)\text{'};$$

4. names of third-level functions:

$$\text{`}\overset{\mathfrak{f}}{\smile}\, \mu_{\beta}(\mathfrak{f}(\beta))\text{'},\quad \text{`}\overset{\mathfrak{f}}{\smile}\, \mu_{\beta\gamma}(\mathfrak{f}(\beta,\gamma))\text{'}$$

   of which the last may remain out of consideration since it will not be made use of.

First it may be noted that always only one ξ- and only one ζ-argument place occurs. We assume that the names of truth-values refer to something, namely either the True or the False. We will then gradually widen the circle of names to be recognised as referential, by demonstrating that the names that are to be added form referential names with those already added, by way of one occupying fitting argument places of the other.

Now, in order first to show that the function-names '${-\!\!\!-}\,\xi$' and '$\top\,\xi$' refer to something, we have to show that the names that result if we put a name of a truth-value for 'ξ' are referential (here, we are not yet recognising any other objects). This follows immediately from our explanations. The names obtained are again names of truth-values.

If we put names of truth-values for 'ξ' and 'ζ' in the function-names '$\begin{array}{l}\top\,\xi\\ \llcorner\,\zeta\end{array}$' and '$\xi = \zeta$', then we obtain names that refer to truth-values. Consequently, our names of first-level functions with two arguments have references.

In order to examine whether the name of a second-level function, '$\overset{\mathfrak{a}}{\smile}\,\varphi(\mathfrak{a})$', refers to something, we ask whether, whenever a function-name '$\Phi(\xi)$' refers to something, it follows that '$\overset{\mathfrak{a}}{\smile}\,\Phi(\mathfrak{a})$' is referential. Now, '$\Phi(\xi)$' has a reference if, for every referential proper name 'Δ', '$\Phi(\Delta)$' refers to something. If so, this reference is either always the True (whatever 'Δ' may refer to), or not always. In the first case, '$\overset{\mathfrak{a}}{\smile}\,\Phi(\mathfrak{a})$' refers to the True, in the second, the False. Thus, whenever an inserted function-name '$\Phi(\xi)$' refers to something, it follows that '$\overset{\mathfrak{a}}{\smile}\,\Phi(\mathfrak{a})$' refers to something. Therefore, the function-name '$\overset{\mathfrak{a}}{\smile}\,\varphi(\mathfrak{a})$' is to be added to the circle of referential names. The same follows in a similar way for '$\overset{\mathfrak{f}}{\smile}\,\mu_{\beta}(\mathfrak{f}(\beta))$'.

The matter is less simple with '$\dot{\varepsilon}\varphi(\varepsilon)$'; for with this we introduce not only a new function-name but at the same time a new proper name (value-range name) for every name of a first-level function with one argument, and indeed not only for the ones already known, but also, in advance, for any that may yet be introduced. In investigating whether a value-range name refers to something, we need only consider those which are formed from referential names of first-level functions with one argument. For short, we will call them *regular* value-range names. We must examine whether a regular value-range name that is put into the argument places of '$—\,\xi$' and '$\top\,\xi$' results in a referential proper name, and further whether, put in the ξ- or ζ-argument places of '$\begin{array}{l}\top\,\xi\\ \llcorner\,\zeta\end{array}$' and '$\xi = \zeta$', it in each case forms a referential name of a first-level function with one argument. If we put the value-range name '$\dot{\varepsilon}\Phi(\varepsilon)$' for ζ in '$\xi = \zeta$', then the question becomes whether '$\xi = \dot{\varepsilon}\Phi(\varepsilon)$' is a referential name of a first-level function with one argument, and for this we need in turn to ask whether all proper names refer to something that result from putting either a name of a truth-value or a regular value-range name in the argument place. Owing to our stipulations that '$\dot{\varepsilon}\Psi(\varepsilon) = \dot{\varepsilon}\Phi(\varepsilon)$' is always to be co-referential with '$\overset{\mathfrak{a}}{\smile}\,\Psi(\mathfrak{a}) = \Phi(\mathfrak{a})$', that '$\dot{\varepsilon}(—\,\varepsilon)$' is to refer to the True and that '$\dot{\varepsilon}(\varepsilon = \top\overset{\mathfrak{a}}{\smile}\,\mathfrak{a} = \mathfrak{a})$' is to refer to the False, every proper name of the form '$\Gamma = \Delta$' is guaranteed a reference if 'Γ' and 'Δ' are regular value-range names or names of truth-values. Thereby it is also known that we always obtain a referential proper name from the function-name '$\xi = (\xi = \xi)$', if we put a regular value-range name in the argument places. Now, since according to our specifications the function $—\,\xi$ always has the same value for the same argument as the function $\xi = (\xi = \xi)$, then it is also known of the function-name '$—\,\xi$' that it always results in a proper name of a truth-value by insertion of a regular value-range name.

According to our specifications the names '$-\!\!\!\top\!\!\!-\, \Delta$' and '$\begin{array}{l}\top\, \Gamma\\ \llcorner\, \Delta\end{array}$' always have references if the names '$—\, \Delta$' and '$—\, \Gamma$' refer to something. Now, since this is the case when 'Γ' and 'Δ' are regular value-range names, we always obtain from the function-names '$-\!\!\!\top\!\!\!-\, \xi$' and '$\begin{array}{l}\top\, \xi\\ \llcorner\, \zeta\end{array}$' referential proper names whenever we put regular value-range names or names of truth-values in the argument places. We have seen that each of our names of simple first-level functions hitherto recognised as referential, '$—\, \xi$', '$-\!\!\!\top\!\!\!-\, \xi$', '$\begin{array}{l}\top\, \xi\\ \llcorner\, \zeta\end{array}$', '$\xi = \zeta$', results in a referential name when we put regular value-range names in the argument places. The regular value-range names may thus be added to the circle of referential names. Thereby, however, the same is settled for our function-name '$\dot{\varepsilon}\varphi(\varepsilon)$', since whenever a name of a first-level function with one argument refers to something, it follows that the proper name that results by its insertion into '$\dot{\varepsilon}\varphi(\varepsilon)$' refers to something.

Of our primitive names only '$\backslash\xi$' now remains. We have specified that '$\backslash\Delta$' is to refer to Γ if 'Δ' is a name of a value-range $\dot{\varepsilon}(\varepsilon = \Gamma)$, that on the other hand '$\backslash\Delta$' is to refer to Δ if there is no object Γ such that 'Δ' is a name of the value-range $\dot{\varepsilon}(\varepsilon = \Gamma)$. In this way, a reference is secured in all cases for proper names of the form '$\backslash\Delta$' and therefore for the function-name '$\backslash\xi$'.

§32. Thus it is shown that our eight primitive names have a reference and thereby that the same holds of all names correctly formed out of them. However, not only a reference but also a sense belongs to all names correctly formed from our signs. Every such name of a truth-value *expresses* a sense, a *thought.* For owing to our stipulations, it is determined under which conditions it refers to the True. The sense of this name, the *thought*, is: that these conditions are fulfilled. Now, a concept-script proposition consists of a judgement-stroke with a name, or a Roman marker, of a truth-value. However, such a marker is transformed into the name of a truth-value when German letters are introduced for the Roman ones with concavities put in front, in accordance with §17. If we suppose this has been carried out, then there is only the case where the proposition is composed of the judgement-stroke and a name of a truth-value. By means of such a proposition it is then asserted that this name refers to the True. Now, since it at the same time expresses a thought, we have in every correctly formed concept-script proposition a judgement that a thought is true; and so a thought cannot be missing. It will be for the reader to clarify for himself the thought of each concept-script proposition that occurs and I will strive to facilitate this as much as possible at the outset.

Now, the simple or complex names of which the name of a truth-value consists contribute to expressing the thought, and this contribution of the individual name is its *sense.* If a name is part of the name of a truth-value, then the sense of the former name is part of the thought expressed by the latter.

§33. The following principles govern the use of definitions.

1. Every name correctly formed from the defined names must have a reference. Thus, for each case it must be possible to supply a name, composed of our eight primitive names, that is co-referential with it, and the latter must be unambiguously determined by the definitions, except for inessential choices of German and Greek letters.

2. From this it follows that the same must never be defined twice, since it would remain in doubt whether these definitions were in harmony with one another.

3. A defined name must be simple; i.e., it must not be composed of names known or still to be explained; for otherwise it would remain in doubt whether the explanations of the names are in accord with one another.

4. If on the left-hand side of a definitional equation we have a proper name that is correctly formed from our primitive names or defined names, then this will always have a reference and we can put on the right a simple, hitherto unused sign which is now introduced by our definition as a co-referential proper name, so that this sign may in future be replaced wherever it occurs by the name standing on the left. Evidently, it must never be used as a function-name, since the path back to the primitive names would then be cut off.

5. A name that is introduced for a first-level function with one argument may only contain a single argument place. In the case of several argument places, it would be possible to fill these with different names, and then the defined name would be used as one of a function with several arguments, although not defined as such. In defining a name of a first-level function with one argument, the argument places on the left side of the definitional equation must be filled with the same Roman object-letter that marks the argument place of the new function-name on the right. The definition then states that the proper name resulting from insertion of a referential

proper name into the argument place on the right, is always to be co-referential with the one which results from insertion of the same proper name into all argument places on the left. The single argument place of the explained name thus represents all those of the explaining name. Whenever the defined function-name occurs subsequently, its argument place must always be filled by a proper name or an object-marker.

6. A name that is introduced for a first-level function with two arguments must contain two and no more than two argument places. The mutually related argument places on the left must be occupied by one and the same Roman object-letter, which also marks one of the two argument places on the right; the non-related argument places must contain different Roman letters. The definition then states that the proper name resulting from insertion of referential proper names into the argument places on the right, is always to be co-referential with the one which results from insertion of the same proper names into the corresponding argument places on the left. The one argument place on the right represents all ξ-argument places on the left, the other, all ζ-argument places.

7. A Roman letter must accordingly never occur on one side of such a definitional equation which does not also occur on the other. If the object-marker on the left-hand side turns into a correctly formed proper name when Roman letters are replaced by proper names, then by our stipulations the explained function-name always has a reference.

Cases other than the foregoing will not appear in the sequel.

Special definitions

§34. It has already been observed in §25 that first-level functions can be used instead of second-level functions in what follows. This will now be shown. As was indicated, this is made possible by the fact that the functions appearing as arguments of second-level functions are represented by their value-ranges, although of course not in such a way that they simply concede their places to them, for that is impossible. In the first instance, our concern is only to designate the value of the function $\Phi(\xi)$ for the argument Δ, that is, $\Phi(\Delta)$, using 'Δ' and '$\grave{\varepsilon}\Phi(\varepsilon)$'. I do so in this way:

$$`\Delta \frown \grave{\varepsilon}\Phi(\varepsilon)'$$

which is to be co-referential with '$\Phi(\Delta)$'. The object $\Phi(\Delta)$ appears as the value of the function $\xi \frown \zeta$ with two arguments, Δ as the ξ-argument, and $\grave{\varepsilon}\Phi\varepsilon$ as the ζ-argument.

Next, however, we have to explain $\xi\frown\zeta$ for all possible objects as arguments. This can be done thus:

$$\Vdash \backslash\dot{\alpha}\left(\smile^{\mathfrak{g}}\begin{array}{l}\mathfrak{g}(a)=\alpha\\ u=\dot{\varepsilon}\mathfrak{g}(\varepsilon)\end{array}\right)=a\frown u \qquad \text{(A}$$

Here, since a function with two arguments is being defined, two Roman letters occur both on the right and on the left. Although the explaining expression contains only known signs, some elucidation may not be superfluous. On the left, we have a Roman marker which results from the proper name '$\backslash\dot{\alpha}\left(\smile^{\mathfrak{g}}\begin{array}{l}\mathfrak{g}(\Theta)=\alpha\\ \Gamma=\dot{\varepsilon}\mathfrak{g}(\varepsilon)\end{array}\right)$' by replacing '$\Theta$' by '$a$' and '$\Gamma$' by '$u$'. This proper name has the form of $\backslash\dot{\alpha}\Phi(\alpha)$. In accordance with §11, two cases are to be distinguished here, depending on whether or not an object Δ can be supplied that is the only one falling under the concept $—\Phi(\xi)$. If so, Δ itself is $\backslash\dot{\alpha}\Phi(\alpha)$. Applied to the case at hand, this means that if there is an object Δ such that $\smile^{\mathfrak{g}}\begin{array}{l}\mathfrak{g}(\Theta)=\Delta\\ \Gamma=\dot{\varepsilon}\mathfrak{g}(\varepsilon)\end{array}$ is the True, while for every argument distinct from Δ the function $\smile^{\mathfrak{g}}\begin{array}{l}\mathfrak{g}(\Theta)=\xi\\ \Gamma=\dot{\varepsilon}\mathfrak{g}(\varepsilon)\end{array}$ has the False as value, then Δ itself is $\backslash\dot{\alpha}\left(\smile^{\mathfrak{g}}\begin{array}{l}\mathfrak{g}(\Theta)=\alpha\\ \Gamma=\dot{\varepsilon}\mathfrak{g}(\varepsilon)\end{array}\right)$. Now, $\smile^{\mathfrak{g}}\begin{array}{l}\mathfrak{g}(\Theta)=\Delta\\ \Gamma=\dot{\varepsilon}\mathfrak{g}(\varepsilon)\end{array}$ is the True provided that there is a first-level function with one argument whose value is Δ for Θ as argument and whose value-range is Γ. Otherwise $\smile^{\mathfrak{g}}\begin{array}{l}\mathfrak{g}(\Theta)=\Delta\\ \Gamma=\dot{\varepsilon}\mathfrak{g}(\varepsilon)\end{array}$ is the False. If we assume that Γ is a value-range, then for any function whose value-range is Γ, it is determined by Γ what value the function has for the argument Θ. Then there is always one and only one such value, and this value is $\backslash\dot{\alpha}\left(\smile^{\mathfrak{g}}\begin{array}{l}\mathfrak{g}(\Theta)=\alpha\\ \Gamma=\dot{\varepsilon}\mathfrak{g}(\varepsilon)\end{array}\right)$ or $\Theta\frown\Gamma$. If, however, Γ is not a value-range at all, then the function $\smile^{\mathfrak{g}}\begin{array}{l}\mathfrak{g}(\Theta)=\xi\\ \Gamma=\dot{\varepsilon}\mathfrak{g}(\varepsilon)\end{array}$ has the False as value for every argument, and then our stipulation is to be drawn upon that '$\backslash\Lambda$' is to refer to Λ itself if there is no object Δ such that Λ is the value-range $\dot{\varepsilon}(\Delta=\varepsilon)$. Thus, when Γ is not a value-range, '$\Theta\frown\Gamma$' refers to the value-range of the function whose value is the False for every argument, namely $\dot{\varepsilon}(\top\, \varepsilon=\varepsilon)$.

To summarise: two cases have to be distinguished in order for the value of the function $\xi\frown\zeta$ to be determined. When the ζ-argument is a value-range, then the value of the function $\xi\frown\zeta$ is the value of that function whose value-range is the ζ-argument for the ξ-argument as argument. If on the other hand the ζ-argument is not a value-range, then the value of the function $\xi\frown\zeta$ is $\dot{\varepsilon}(\top\, \varepsilon=\varepsilon)$ for every ξ-argument.

§35. Here we see confirmed what we could gather from our earlier considerations, namely, that the function-name '$\xi \frown \zeta$' has a reference. This alone is fundamental for the conduct of the proofs to come; in other respects, our elucidation could be wrong without thereby calling into question the correctness of these proofs; for only the definition itself is the foundation for this construction. As mentioned at the outset, the goal was to enable us to use a first-level function instead of a second-level one. Let us now see how this goal is achieved. In §22 we introduced the second-level function $\varphi(2)$. Now, we can write '$2 \frown \grave{\varepsilon}\varphi(\varepsilon)$' for '$\varphi(2)$'. This is still the name of a second-level function; but if we write 'ξ' for '$\grave{\varepsilon}\varphi(\varepsilon)$', then we have in '$2 \frown \xi$' the name of a first-level function. The function $\varphi(2)$ has for the function $\Phi(\xi)$ as argument the same value, $\Phi(2)$, as has the function $2 \frown \xi$ for $\grave{\varepsilon}\Phi(\varepsilon)$ as argument. If an object is taken as argument of the function $2 \frown \xi$ that is not a value-range, then we have no corresponding argument for the second-level function $\varphi(2)$ and so the mutual representability of functions of the first and second levels lapses.

The second-level functions

$$\top\!\overset{\mathfrak{a}}{\smile}\!\top\ \varphi(\mathfrak{a}) \quad \text{and} \quad \smile\!\overset{\mathfrak{a}}{\smile}\!\top\!\overset{\mathfrak{e}}{\smile}\!\top \begin{array}{l} \mathfrak{a} = \mathfrak{e} \\ \varphi(\mathfrak{e}) \\ \varphi(\mathfrak{a}) \end{array}$$

correspond in the same manner to the first-level functions

$$\top\!\overset{\mathfrak{a}}{\smile}\!\top\ \mathfrak{a} \frown \xi \quad \text{and} \quad \smile\!\overset{\mathfrak{a}}{\smile}\!\top\!\overset{\mathfrak{e}}{\smile}\!\top \begin{array}{l} \mathfrak{a} = \mathfrak{e} \\ \mathfrak{e} \frown \xi \\ \mathfrak{a} \frown \xi \end{array}$$

§36. To find different examples, we seek to represent functions with two arguments by objects in a way similar to that followed for functions with one argument. A simple value-range cannot be used here, but only a double value-range, which is for a function with two arguments what the former is for a function with one argument.

For example, let us start with the function with two arguments $\xi + \zeta$. If we take, e.g., the number 3 as ζ-argument, then we have in $\xi + 3$ a function with just one argument, whose value-range is $\grave{\varepsilon}(\varepsilon + 3)$. The same holds for every ζ-argument, and we have in $\grave{\varepsilon}(\varepsilon + \zeta)$ a function with one argument, whose value is always a value-range. If we take the ξ- and the ζ-argument together with the value of the function $\xi + \zeta$ to be represented as rectangular co-ordinates in space, then we can display the value-range $\grave{\varepsilon}(\varepsilon + 3)$ as a straight line. If we allow the ζ-argument to vary continuously, then the straight line moves accordingly and thereby describes a plane. In each of its positions it displays a value-range, the value of the function $\grave{\varepsilon}(\varepsilon+\zeta)$ for a given ζ-argument. The value-range of the function $\grave{\varepsilon}(\varepsilon+\zeta)$ is now $\grave{\alpha}\grave{\varepsilon}(\varepsilon+\alpha)$, and this is what I call a *double value-range*. Thus

$$\Delta\frown\grave{\alpha}\dot{\grave{\varepsilon}}(\varepsilon+\alpha) = \dot{\grave{\varepsilon}}(\varepsilon+\Delta)$$

is the True and so is

$$\Gamma\frown(\Delta\frown\grave{\alpha}\dot{\grave{\varepsilon}}(\varepsilon+\alpha)) = \Gamma\frown\dot{\grave{\varepsilon}}(\varepsilon+\Delta),$$

and since

$$\Gamma\frown\dot{\grave{\varepsilon}}(\varepsilon+\Delta) = \Gamma+\Delta$$

is the True,

$$\Gamma\frown(\Delta\frown\grave{\alpha}\dot{\grave{\varepsilon}}(\varepsilon+\alpha)) = \Gamma+\Delta$$

is also the True. Here on the left we see a double value-range representing a function with two arguments on the right, not however in such a way that the representing simply takes the place of the represented, which is impossible, but only in such a way that the double value-range on the left captures what differentiates the function on the right from other first-level functions with two-arguments. If a function with two arguments is a relation, then we may say '*extension* of the relation' as an alternative to 'double value-range'.

One can still ask what $\Gamma\frown(\Delta\frown\Theta)$ is when Θ is not a double value-range but merely a simple value-range or not a value-range at all. In the first case, $\Delta\frown\Theta$ is not a value-range and consequently $\Gamma\frown(\Delta\frown\Theta)$ is the same as $\dot{\grave{\varepsilon}}(\top\,\varepsilon=\varepsilon)$. In the other case, $\Delta\frown\Theta$ coincides with $\dot{\grave{\varepsilon}}(\top\,\varepsilon=\varepsilon)$, and

$$\Gamma\frown(\Delta\frown\Theta) = \Gamma\frown\dot{\grave{\varepsilon}}(\top\,\varepsilon=\varepsilon)$$

is the True; therefore, $\Gamma\frown(\Delta\frown\Theta) = (\top\,\Gamma = \Gamma)$ is also the True; i.e., $\Gamma\frown(\Delta\frown\Theta)$ is then the False.

§37. Instead of the second-level function

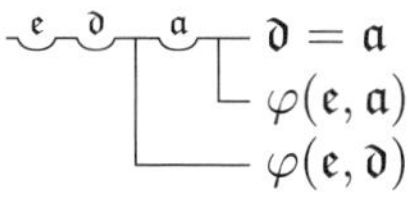

(§23) we can now consider the first-level function

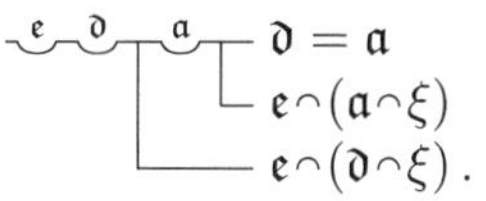

We will introduce a simple notation for this by the following definition:

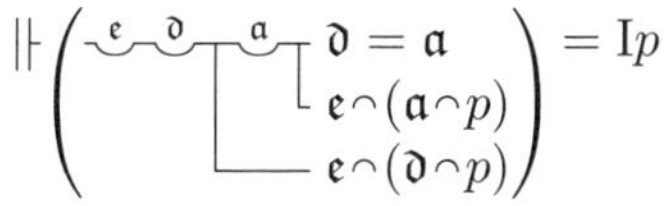

(Γ

According to §23,

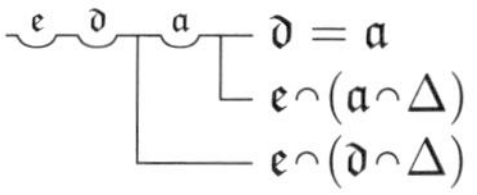

is the truth-value of: that the relation $—\,\xi\frown(\zeta\frown\Delta)$ is single-valued, i.e., that for every ξ-argument there is no or only one ζ-argument for which the value of the function is the True, or as we can also say, that for every object there is at most one object to which it stands in the relation $—\,\xi\frown(\zeta\frown\Delta)$. If Δ is not a double value-range, then, according to §36, the value of the function $—\,\xi\frown(\zeta\frown\Delta)$ is either the False or $\dot{\grave{\varepsilon}}(\top\,\varepsilon=\varepsilon)$. Since the latter is not the True, the value of the function $—\,\xi\frown(\zeta\frown\Delta)$ is thus always the False when Δ is not a double value-range; i.e., $—\,\xi\frown(\zeta\frown\Delta)$ is then a relation in which no object stands to any object. In that case, $\mathrm{I}\Delta$ is the True. The function-name '$\mathrm{I}\xi$' is introduced particularly in view of the cases in which an extension of a relation occurs as argument. If this relation is $\mathrm{X}(\xi,\zeta)$ then $\mathrm{I}\grave{\alpha}\dot{\grave{\varepsilon}}\mathrm{X}(\varepsilon,\alpha)$ is the True when the relation $\mathrm{X}(\xi,\zeta)$ is single-valued (going from the ξ- to the

ζ-argument). So, e.g., $\vdash \mathrm{I}\grave{\alpha}\grave{\varepsilon}(\varepsilon^2 = \alpha)$. According to our definitions, 'I' may be used only as a function-sign that precedes an argument-sign or its proxy.

§38. Now we can approach our objective of the definition of number. In my *Grundlagen der Arithmetik* I based this on a relation that I called equinumerosity. In §72 (p. 85) of my *Grundlagen*, I define:

The expression 'the concept F is equinumerous with the concept G' is co-referential with the expression 'there is a relation φ that is single-valued in both directions and correlates^e the objects falling under the concept F with the objects falling under the concept G'.

What is it for a relation φ to correlate the objects falling under F with the objects falling under G? It is (§71 of *Grundlagen*) for every object that falls under F to stand in the relation φ with some object falling under G, or, more precisely, the two propositions 'a falls under F' and 'a stands to no object falling under G in the relation φ' cannot both hold for any a.

We now take $-\xi \frown \Gamma$ as concept F, $-\xi \frown \Delta$ as concept G, and $-\xi \frown (\zeta \frown \Upsilon)$ as the relation φ. Then we can express what is said in the symbolism of concept-script like this:

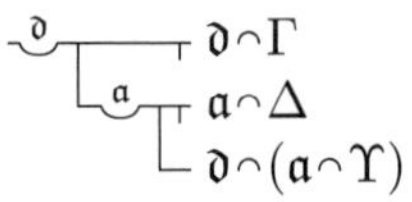

$$\underset{\mathfrak{a}}{\vdash}\!\!\begin{array}{l}\partial\frown\Gamma\\ \mathfrak{a}\frown\Delta\\ \partial\frown(\mathfrak{a}\frown\Upsilon)\end{array} \qquad 1$$

---
[1] The 'a' of the informal characterisation corresponds here to the '∂'.

The relation has to be single-valued. If we now add this, then we obtain

$$\underset{\mathfrak{a}}{\vdash}\!\!\begin{array}{l}\partial\frown\Gamma\\ \mathfrak{a}\frown\Delta\\ \partial\frown(\mathfrak{a}\frown\Upsilon)\\ \mathrm{I}\Upsilon\end{array}$$

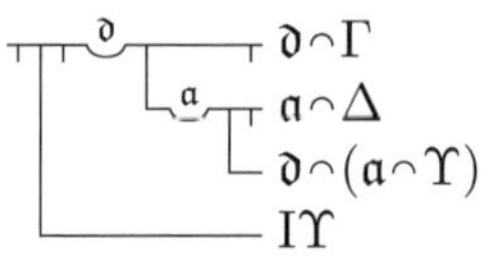

(Concerning 'and' compare §12). We consider this as the value of the function with two arguments

$$\underset{\mathfrak{a}}{\vdash}\!\!\begin{array}{l}\partial\frown\xi\\ \mathfrak{a}\frown\zeta\\ \partial\frown(\mathfrak{a}\frown\Upsilon)\\ \mathrm{I}\Upsilon\end{array}$$

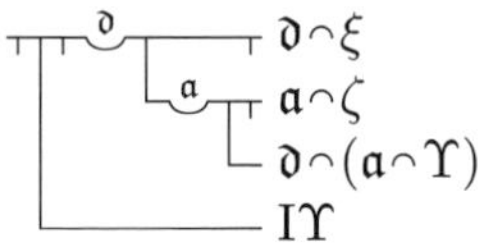

for the arguments Γ and Δ. This function is a relation. Its double-value-range is:

$$\grave{\alpha}\grave{\varepsilon}\left[\underset{\mathfrak{a}}{}\!\!\begin{array}{l}\partial\frown\varepsilon\\ \mathfrak{a}\frown\alpha\\ \partial\frown(\mathfrak{a}\frown\Upsilon)\\ \mathrm{I}\Upsilon\end{array}\right]$$

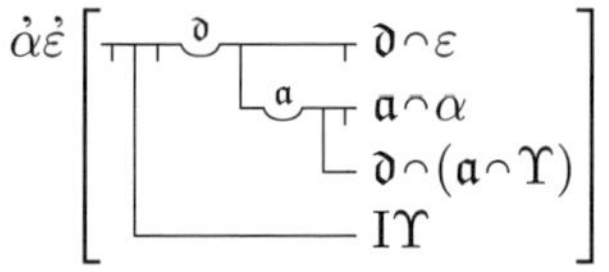

We consider the latter as the value of the function

$$\grave{\alpha}\grave{\varepsilon}\left[\underset{\mathfrak{a}}{}\!\!\begin{array}{l}\partial\frown\varepsilon\\ \mathfrak{a}\frown\alpha\\ \partial\frown(\mathfrak{a}\frown\xi)\\ \mathrm{I}\xi\end{array}\right]$$

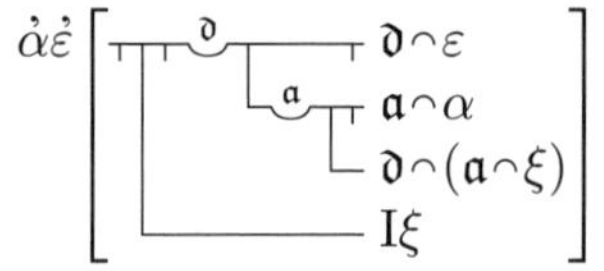

for the argument Υ. The following definition introduces a shortened name for this function:

$$\Vdash \grave{\alpha}\grave{\varepsilon}\left[\underset{\mathfrak{a}}{}\!\!\begin{array}{l}\partial\frown\varepsilon\\ \mathfrak{a}\frown\alpha\\ \partial\frown(\mathfrak{a}\frown p)\\ \mathrm{I}p\end{array}\right] = \mathop{\rangle}p \qquad (\Delta$$

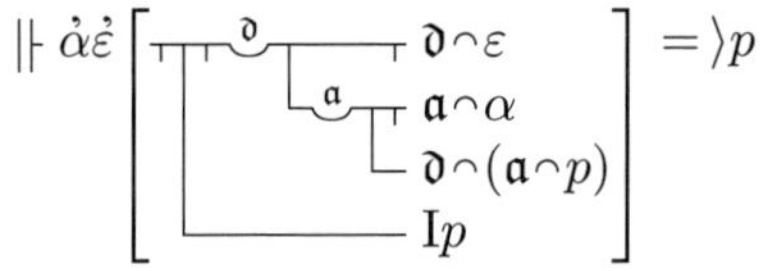

The value of this function is always the extension of a relation. What accordingly is $\Gamma \frown (\Delta \frown) \Upsilon)$? According to the definition we obtain the following:

$$\Gamma \frown \left[\Delta \frown \grave{\alpha}\grave{\varepsilon}\left[\underset{\mathfrak{a}}{}\!\!\begin{array}{l}\partial\frown\varepsilon\\ \mathfrak{a}\frown\alpha\\ \partial\frown(\mathfrak{a}\frown\Upsilon)\\ \mathrm{I}\Upsilon\end{array}\right]\right]$$

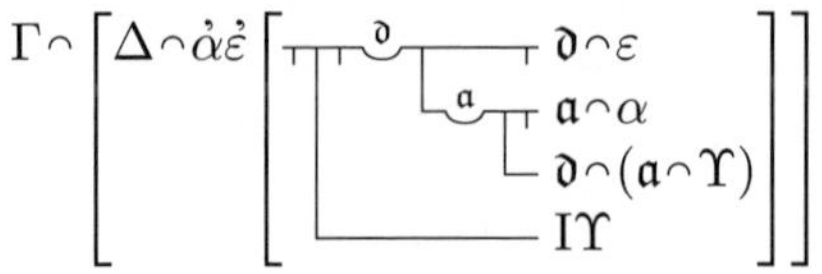

or

$$\begin{array}{l} \smile^{\mathfrak{d}}\!\!-\!\!\smile^{\mathfrak{a}}\!\!\top\; \mathfrak{a}\frown\Delta \\ \qquad\qquad\;\; \llcorner\; \mathfrak{d}\frown(\mathfrak{a}\frown\Upsilon) \\ \qquad\quad \mathfrak{d}\frown\Gamma \\ \quad \llcorner\; \mathrm{I}\Upsilon \end{array}$$

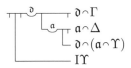

This is the truth-value of: that the relation, $-\xi\frown(\zeta\frown\Upsilon)$, is single-valued and correlates the objects falling under the concept $-\xi\frown\Gamma$ with the objects falling under the concept $-\xi\frown\Delta$. For this, we will introduce the abbreviation, 'the Υ-relation *maps* the Γ-concept into the Δ-concept', calling, in general, a concept whose extension is Γ the Γ-*concept*, and a relation whose extension is Υ the Υ-*relation.*

§39. Now, if equinumerosity is to obtain between concepts, then there must be a relation of which holds not only what we just said of the Υ-relation but of whose converse also holds the same, when the roles of Γ and Δ are exchanged, so that it maps the Δ-concept into the Γ-concept. For this purpose, it is desirable to introduce a function-name '$\mathfrak{U}\xi$' in such a way that if Υ is the extension of a relation, then $\mathfrak{U}\Upsilon$ is the extension of its converse. For this purpose we define

$$\Vdash \dot{\alpha}\dot{\varepsilon}(\alpha\frown(\varepsilon\frown p)) = \mathfrak{U}p \qquad \text{(E}$$

The relation

$$-\xi\frown(\zeta\frown\mathfrak{U}\Upsilon) \quad \text{or}$$
$$-\xi\frown(\zeta\frown\dot{\alpha}\dot{\varepsilon}(\alpha\frown(\varepsilon\frown\Upsilon)))$$

is then the same as $-\zeta\frown(\xi\frown\Upsilon)$.

§40. Thus, in order to say the same of the converse of the relation $-\xi\frown(\zeta\frown\Upsilon)$ as we have said of the relation itself, we only need to replace 'Υ' by '$\mathfrak{U}\Upsilon$'. Accordingly, $\begin{array}{l}\top\!\!\top\; \Gamma\frown(\Delta\frown\rangle\Upsilon) \\ \;\llcorner\; \Delta\frown(\Gamma\frown\rangle\mathfrak{U}\Upsilon)\end{array}$ is the truth-value of: that the Υ-relation maps the Γ-concept into the Δ-concept and that its converse maps the latter into the former, assuming of course that Γ and Δ are extensions of concepts and Υ an extension of a relation. In order for these concepts to be equinumerous there has to be such a relation. $-\xi\frown(\zeta\frown\Upsilon)$ is always a relation, whatever object 'Υ' may refer to, and every relation may be designated in this form, '$-\xi\frown(\zeta\frown\Upsilon)$', by taking its extension for Υ. Accordingly, $\begin{array}{l}\smile^{\mathfrak{q}}\!\top\!\!\top\; \Gamma\frown(\Delta\frown\rangle\mathfrak{q}) \\ \quad\;\llcorner\; \Delta\frown(\Gamma\frown\rangle\mathfrak{U}\mathfrak{q})\end{array}$ is the truth-value of: that the concepts $-\xi\frown\Gamma$ and $-\xi\frown\Delta$ are equinumerous. We can regard this as the value of the function $\begin{array}{l}\smile^{\mathfrak{q}}\!\top\!\!\top\; \xi\frown(\Delta\frown\rangle\mathfrak{q}) \\ \quad\;\llcorner\; \Delta\frown(\xi\frown\rangle\mathfrak{U}\mathfrak{q})\end{array}$ for the argument Γ. This function is a concept whose extension is $\dot{\varepsilon}\left(\begin{array}{l}\smile^{\mathfrak{q}}\!\top\!\!\top\; \varepsilon\frown(\Delta\frown\rangle\mathfrak{q}) \\ \quad\;\llcorner\; \Delta\frown(\varepsilon\frown\rangle\mathfrak{U}\mathfrak{q})\end{array}\right)$. And, in accordance with my definition (*Grundlagen* §68), this extension of a concept is the *cardinal number* that belongs to the concept $-\xi\frown\Delta$. Instead of 'cardinal number that belongs to the Δ-concept', I also say for short 'cardinal number of the Δ-concept'. I now define:

$$\Vdash \dot{\varepsilon}\left(\begin{array}{l}\smile^{\mathfrak{q}}\!\top\!\!\top\; \varepsilon\frown(u\frown\rangle\mathfrak{q}) \\ \quad\;\llcorner\; u\frown(\varepsilon\frown\rangle\mathfrak{U}\mathfrak{q})\end{array}\right) = \natural u \qquad \text{(Z}$$

§41. Accordingly, $\natural\dot{\varepsilon}(\top\,\varepsilon = \varepsilon)$ is the cardinal number that belongs to the $\dot{\varepsilon}(\top\,\varepsilon = \varepsilon)$-concept, or the cardinal number that belongs to the concept,

$\top\, \xi = \xi$, and this is the cardinal number Zero (*Grundlagen* §74). Later, it will prove necessary to distinguish the cardinal number Zero from the number Zero, and so I will mark the former with a slanting stroke. I define

$$\Vdash \wp\grave{\varepsilon}(\top\, \varepsilon = \varepsilon) = \not0 \qquad (\Theta$$

§42. Likewise, I also define (*Grundlagen* §77)

$$\Vdash \wp\grave{\varepsilon}(\varepsilon = \not0) = \not1 \qquad (\mathrm{I}$$

The slanting stroke in '$\not1$' is meant to distinguish the cardinal number One from the number One. Accordingly, $\not1$ is the cardinal number that belongs to the concept $\xi = \not0$.

$\top\overset{\mathfrak{u}}{\smile}\top\, \wp\mathfrak{u} = \Gamma$ is the truth-value of: that there is a concept to which the cardinal number Γ belongs or, as we can also say, that Γ is a cardinal number. Therefore, we call the function $\top\overset{\mathfrak{u}}{\smile}\top\, \wp\mathfrak{u} = \xi$ the concept *cardinal number*.

§43. The relation in which one member of the cardinal number series stands to that immediately following it still remains to be explained. I will here give my definition (*Grundlagen* §76) in a slightly modified formulation:

If there is a concept, $—\, \xi\frown\Gamma$, and an object Δ falling under it, such that the cardinal number belonging to the concept, $—\, \xi\frown\Gamma$, is Λ and the cardinal number belonging to the concept, $\begin{array}{l}\top\, \xi = \Delta,\\ \llcorner\, \xi\frown\Gamma\end{array}$ is Θ, then I say: Λ follows immediately after Θ in the cardinal number series.

We now have in

$$\begin{array}{l} \wp\Gamma = \Lambda \\ \Delta\frown\Gamma \\ \wp\grave{\varepsilon}\left(\begin{array}{l}\top\, \varepsilon = \Delta \\ \llcorner\, \varepsilon\frown\Gamma\end{array}\right) = \Theta \end{array}$$

the truth-value of: that Λ is the cardinal number that belongs to the concept, $—\, \xi\frown\Gamma$, that Δ falls under this concept, and that Θ is the cardinal number of the $\grave{\varepsilon}\left(\begin{array}{l}\top\, \varepsilon = \Delta \\ \llcorner\, \varepsilon\frown\Gamma\end{array}\right)$-concept. Accordingly, we have in

$$\top\overset{\mathfrak{u}}{\smile}\overset{\mathfrak{a}}{\smile} \begin{array}{l} \wp\mathfrak{u} = \Lambda \\ \mathfrak{a}\frown\mathfrak{u} \\ \wp\grave{\varepsilon}\left(\begin{array}{l}\top\, \varepsilon = \mathfrak{a} \\ \llcorner\, \varepsilon\frown\mathfrak{u}\end{array}\right) = \Theta \end{array}$$

the truth-value of: that Λ follows immediately after Θ in the *cardinal number series*. We regard this as the value of the function

$$\top\overset{\mathfrak{u}}{\smile}\overset{\mathfrak{a}}{\smile} \begin{array}{l} \wp\mathfrak{u} = \zeta \\ \mathfrak{a}\frown\mathfrak{u} \\ \wp\grave{\varepsilon}\left(\begin{array}{l}\top\, \varepsilon = \mathfrak{a} \\ \llcorner\, \varepsilon\frown\mathfrak{u}\end{array}\right) = \xi \end{array}$$

for the arguments Θ and Λ. The extension of this relation is

$$\grave{\alpha}\grave{\varepsilon}\left[\top\overset{\mathfrak{u}}{\smile}\overset{\mathfrak{a}}{\smile} \begin{array}{l} \wp\mathfrak{u} = \alpha \\ \mathfrak{a}\frown\mathfrak{u} \\ \wp\grave{\varepsilon}\left(\begin{array}{l}\top\, \varepsilon = \mathfrak{a} \\ \llcorner\, \varepsilon\frown\mathfrak{u}\end{array}\right) = \varepsilon \end{array}\right]$$

and for this a simple name will be introduced:

$$\Vdash \grave{\alpha}\grave{\varepsilon}\left[\top\overset{\mathfrak{u}}{\smile}\overset{\mathfrak{a}}{\smile} \begin{array}{l} \wp\mathfrak{u} = \alpha \\ \mathfrak{a}\frown\mathfrak{u} \\ \wp\grave{\varepsilon}\left(\begin{array}{l}\top\, \varepsilon = \mathfrak{a} \\ \llcorner\, \varepsilon\frown\mathfrak{u}\end{array}\right) = \varepsilon \end{array}\right] = \mathfrak{f} \qquad (\mathrm{H}$$

Accordingly, '$\not0\frown(\not1\frown\mathfrak{f})$' expresses that $\not1$ follows immediately after $\not0$ in the cardinal number series.

§44. The six propositions listed in §78 of my *Grundlagen* may now be expressed in our symbolism as follows:

'$\vdash \begin{array}{l} a = \not1 \\ \llcorner\, \not0\frown(a\frown\mathfrak{f}) \end{array}$'   '$\vdash\overset{\mathfrak{a}}{\smile} \begin{array}{l} \mathfrak{a}\frown u \\ \wp u = \not1 \end{array}$'

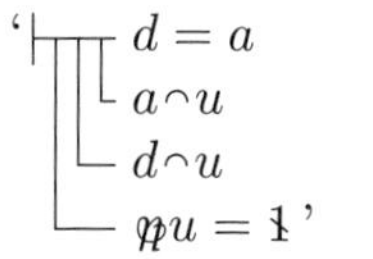

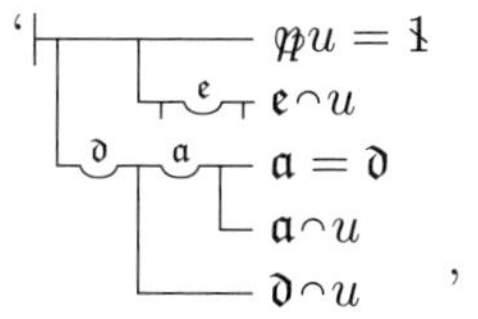

I leave it to the reader to make the sense clear to himself. 'I𝔣' expresses that the 𝔣-relation is single-valued, in other words: that for every cardinal number there is no more than a single one which follows it immediately in the cardinal number series. And 'I⟍𝔣' expresses that for every cardinal number there is no more than a single one which it immediately follows. By 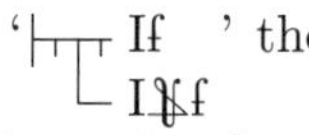 the fifth of the propositions is rendered.

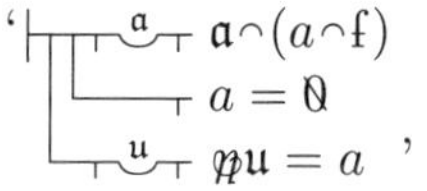

says that for every cardinal number, with the exception of 𝔒, there is one that immediately precedes it in the cardinal number series.

§45. The 𝔣-relation orders the cardinal numbers in such a way that a series results. We now have to explain in general what is stated by 'an object follows after an object in a series', where the type of this series is determined by the relation in which a member of the series always stands to the one following. I repeat in slightly different words the explanation that I gave in §79 in my *Grundlagen* and in the *Begriffsschrift.*

If the proposition

'If every object to which Δ stands in the ϒ-relation falls under the concept — $F(\xi)$ and if, whenever an object falls under this concept, it follows that every object to which it stands in the ϒ-relation also falls under the concept — $F(\xi)$ then Θ falls under this concept'

holds generally for every concept — $F(\xi)$, then we say: 'Θ *follows in the* ϒ-*series after* Δ'. Accordingly,

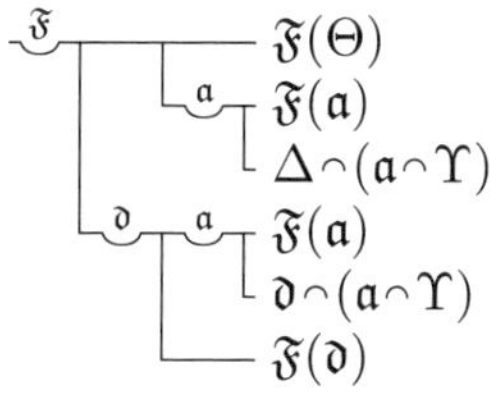

is the truth-value of: that Θ follows in the ϒ-series after Δ. We can also regard this as the value of the function

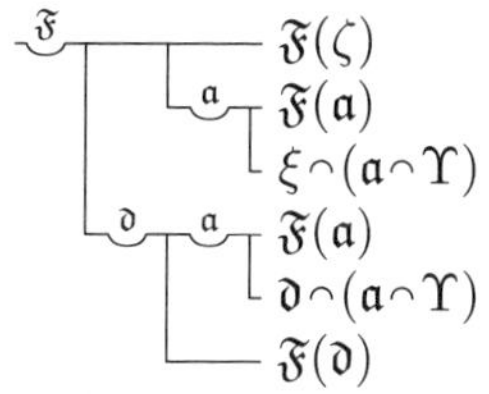

for Δ and Θ as arguments. The extension of this relation is

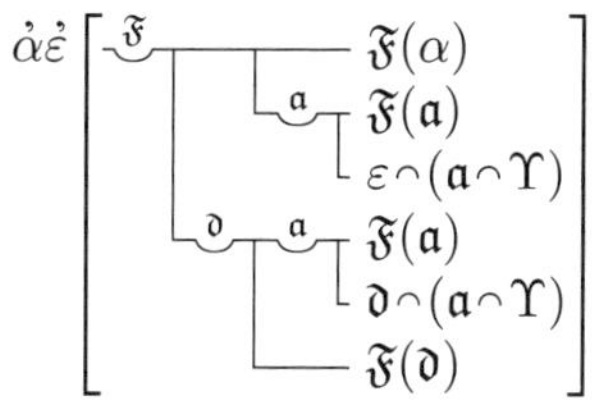

We can regard it as the value of the function

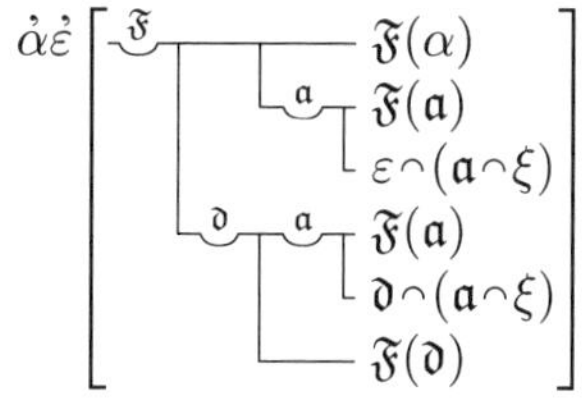

for Υ as argument. For this function I introduce a simple name by defining:

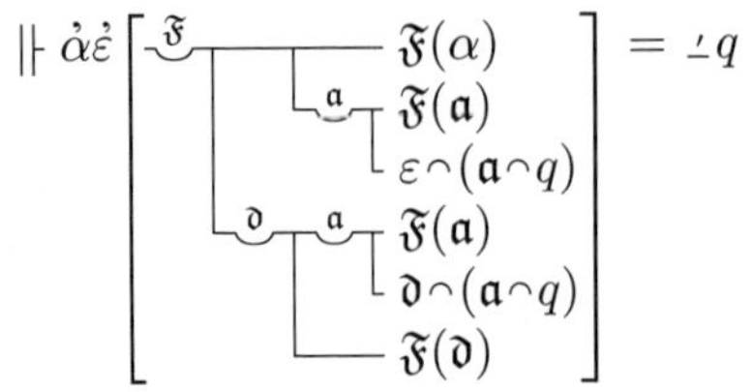

(K

Accordingly, '$\Delta \frown (\Theta \frown \lrcorner \Upsilon)$' expresses that Θ follows after Δ in the Υ-series. And '$\Delta \frown (\Theta \frown \lrcorner \mathfrak{f})$' expresses that Θ follows after Δ in the cardinal number series. Instead of 'Θ follows after Δ in the Υ-series' I also say 'Δ *precedes* Θ in the Υ-*series*'.

§46. $\begin{array}{l}\Theta = \Delta \\ \Delta \frown (\Theta \frown \lrcorner \Upsilon)\end{array}$ is the truth-value of: that Θ either follows after Δ in the Υ-series or coincides with Δ. For short, I say that Θ *belongs to the* Υ-*series starting with* Δ, or that Δ *belongs to the* Υ-*series ending with* Θ. I regard this as the value of the function $\begin{array}{l}\zeta = \xi \\ \xi \frown (\zeta \frown \lrcorner \Upsilon)\end{array}$ for Δ and Θ as arguments. The extension of this relation is $\dot{\alpha}\dot{\varepsilon}\left(\begin{array}{l}\alpha = \varepsilon \\ \varepsilon \frown (\alpha \frown \lrcorner \Upsilon)\end{array}\right)$. I regard it as the value of the function $\dot{\alpha}\dot{\varepsilon}\left(\begin{array}{l}\alpha = \varepsilon \\ \varepsilon \frown (\alpha \frown \lrcorner \xi)\end{array}\right)$ for Υ as argument, and I introduce a simple name by defining:

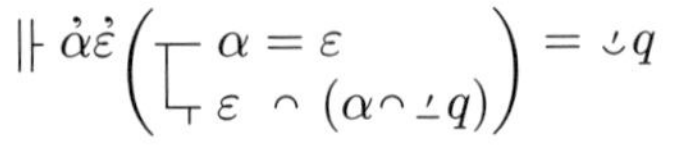

(Λ

Accordingly, $\Delta \frown (\Theta \frown \llcorner \Upsilon)$ is the truth-value of: that Θ belongs to the Υ-series starting with Δ. Thus, $\emptyset \frown (\Theta \frown \llcorner \Upsilon)$ is the truth-value of: that Θ belongs to the cardinal number series starting with $\emptyset$, for which I can also say that Θ is a *finite* cardinal number.

In §82 of my *Grundlagen*, I mentioned the proposition that the cardinal number that belongs to the concept

> *belonging to the cardinal number series ending with n*

follows n immediately in the cardinal number series if n is a finite cardinal number. This can now be represented thus:

'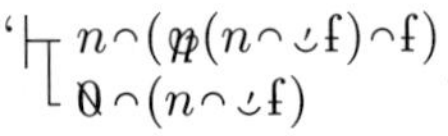';

since $(\Theta \frown \llcorner \mathfrak{f})$ is the extension of the concept, *belonging to the cardinal number series ending with* Θ.

3. Derived laws

§47. We have been seeing how concepts and objects with which we will be occupied later can be designated with our signs. This would be of little significance, however, if one could not also calculate with them, if, without mixing in words, series of inferences could not be represented, proofs not be conducted. We have now become familiar with the basic laws and modes of inference that are going to be employed for this purpose. We will now derive laws from them for later use, in such a way that the method of calculating is illustrated at the same time. I will first summarise the basic laws and rules, and add some supplementary points.

Summary of the basic laws

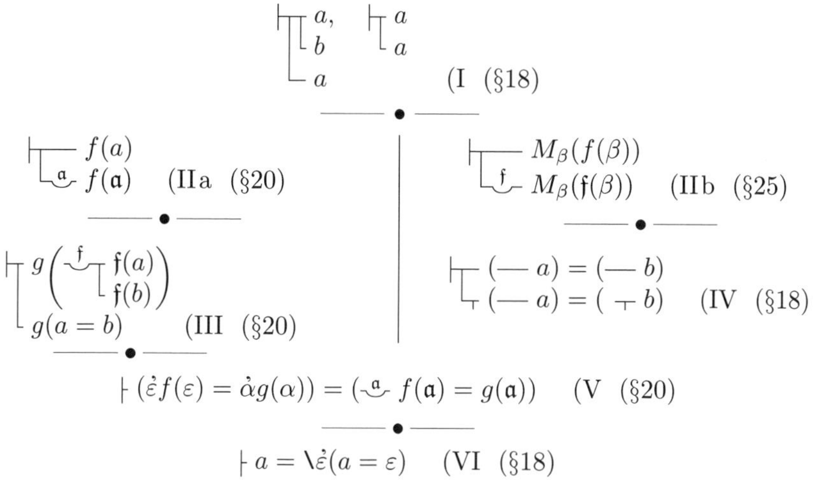

$$\vdash \begin{matrix} a, \\ b \\ a \end{matrix} \qquad \vdash \begin{matrix} a \\ a \end{matrix} \qquad \text{(I (§18))}$$

$$\vdash \begin{matrix} f(a) \\ \stackrel{a}{\smile} f(\mathfrak{a}) \end{matrix} \quad \text{(IIa (§20))} \qquad \vdash \begin{matrix} M_\beta(f(\beta)) \\ \stackrel{\mathfrak{f}}{\smile} M_\beta(\mathfrak{f}(\beta)) \end{matrix} \quad \text{(IIb (§25))}$$

$$\vdash \begin{matrix} g\left(\stackrel{\mathfrak{f}}{\smile} \begin{matrix} \mathfrak{f}(a) \\ \mathfrak{f}(b) \end{matrix} \right) \\ g(a = b) \end{matrix} \quad \text{(III (§20))} \qquad \vdash \begin{matrix} (-a) = (-b) \\ (-a) = (\top b) \end{matrix} \quad \text{(IV (§18))}$$

$$\vdash (\dot{\varepsilon}f(\varepsilon) = \dot{\alpha}g(\alpha)) = (\stackrel{\mathfrak{a}}{\smile} f(\mathfrak{a}) = g(\mathfrak{a})) \quad \text{(V (§20))}$$

$$\vdash a = \backslash \dot{\varepsilon}(a = \varepsilon) \quad \text{(VI (§18))}$$

§48. Summary of the rules

1. Fusion of horizontals

If as argument of the function $-\xi$ there occurs the value of this same function for some argument, then the horizontals may be fused.

The two parts of the horizontal line that are separated by the negation-stroke in '$\top \xi$' are horizontals in our sense.

The lower and the two parts of the upper horizontal stroke in '$\top\!\!\begin{smallmatrix}\xi\\\zeta\end{smallmatrix}$' are horizontals in our sense.

Finally, the two straight strokes which are adjoined to the concavity in '$\stackrel{\mathfrak{a}}{\smile} \varphi(\mathfrak{a})$' are horizontals in our sense.

2. Permutation of subcomponents

Subcomponents of the same proposition may be permuted with one another arbitrarily.

3. Contraposition

A subcomponent in a proposition may be permuted with a supercomponent provided one also inverts their truth-values.[f]

Transition-sign: ✕

4. Fusion of equal subcomponents

A subcomponent that occurs repeatedly in the same proposition only needs to be written once.

5. *Transformation of a Roman letter into a German letter*

A Roman letter may be replaced wherever it occurs in a proposition by one and the same German letter, namely, an object-letter by an object-letter and a function-letter by a function-letter. At the same time, the latter has to be placed above a concavity in front of a supercomponent outside of which the Roman letter does not occur. If in this supercomponent the scope of a German letter is wholly contained and the Roman letter occurs within this scope, then the German letter that is to be introduced for the latter must be distinct from the former.

Transition-sign: $\smile$

This sign is also used if several German letters are to be introduced in this way. Although one may write down the final product straightaway, one must think of them as introduced one after another.

6. *Inferring* (*a*)

If a subcomponent of a proposition differs from another proposition only in lacking the judgement-stroke, then one may infer a proposition which results from the first by suppressing that subcomponent.

Transition-signs: ( ) : ———

and ( ) :: ———

Combined inferences thus:

(,) :: ═══

7. *Inferring* (*b*)

If the same combination of signs (either a proper name or a Roman object-marker) occurs in one proposition as supercomponent and in another as subcomponent, then a proposition may be inferred in which the supercomponent of the second proposition appears as supercomponent and all subcomponents of both, save that mentioned, appear as subcomponents. Equal subcomponents may here be fused according to rule (4).

Transition-signs: ( ) : – – –

and ( ) :: – – –

Combined inferences thus:

(,) :: = = = = and (,) :: ═ ═ ═ ═

8. *Inferring* (*c*)

If two propositions agree in their supercomponents while a subcomponent of the one differs from a subcomponent of the other only by a prefixed negation-stroke, then we can infer a proposition in which the common supercomponent features as supercomponent, and all subcomponents of both propositions with the exception of the two mentioned feature as subcomponents.

Transition-sign: ( ) : ·—·—·—·—·

9. *Citing propositions: replacement of Roman letters*

When citing a proposition by its label, we may effect a simple inference by uniformly replacing a Roman letter within the proposition by the same proper name or the same Roman object-marker.

Likewise, we may replace all occurrences in a proposition of a Roman function-letter, 'f', 'g', 'h', 'F', 'G', 'H', by the same name or Roman marker of a first-level function with one or two arguments, depending on whether the Roman letter indicates a function with one or two arguments.

When we cite law (IIb) we may replace both occurrences of 'M_β' by the same name or Roman marker of a second-level function with one argument of the second kind.

Concerning the word 'same'[g] in the second and third paragraphs of this rule, it is to be observed that the argument is not a part of the function; that is, that a change of the argument-sign is not a change of the function-name. In order for the same function-name to occur in different locations, it is required that the related argument places appropriately match. Concerning the question which argument places are to be regarded as related, these rules are to be observed:

All occurrences of a German letter within its scope, save those within an enclosed scope of the same letter or above a concavity, are related argument places of the corresponding function;

All occurrences of a small Greek vowel within its scope, save those within an enclosed scope of the same letter or together with the smooth breathing, are related argument places of the corresponding function.

10. *Citing propositions: replacement of German letters*

When citing a proposition by its label, we may uniformly replace a German letter above a concavity and at all argument places of the corresponding function by one and the same distinct letter, that is, an object-letter by an object-letter and a function-letter by a function-letter, just provided no German letter occurring in a scope enclosed within its own thereby becomes the same as the one whose scope is enclosed.

11. *Citing propositions: replacement of Greek vowels*

When citing a proposition by its label, we may uniformly replace a Greek vowel under a smooth breathing and at all argument places of the corresponding function by one and the same distinct letter, just provided no Greek letter occurring in a scope enclosed within its own thereby becomes the same as the one whose scope is enclosed.

12. *Citing definitions*

When citing a definition by its label, we may replace the definition-stroke by the judgement-stroke and make those modifications that are allowed according to (9), (10), (11) when citing a proposition.

Stipulations concerning the use of brackets

13. Everything standing together to the right of a horizontal is to be considered as a whole that stands in place of 'ξ' in '$—\,\xi$', provided brackets do not forbid it.

14. Everything standing together to the left of an equality-sign up to but excluding the nearest horizontal, is to be considered as a whole that stands in place of 'ξ' in '$\xi = \zeta$', provided brackets do not forbid it.

Accordingly, '$a = b = c$', e.g., is to be understood as '$(a = b) = c$'. Since, however, '$a = b = c$' is commonly used in a different sense, I will in such a case write down the brackets.

15. Everything standing to the right of an equality-sign up to but excluding the nearest equality-sign, is to be considered as a whole that stands in place of 'ζ' in '$\xi = \zeta$', provided brackets do not forbid it.

16. We have names of functions with two arguments, e.g., '$\xi = \zeta$', '$\xi \frown \zeta$', which have argument places both to the left and to the right. I will call such function-signs *two-sided*. For two-sided function-signs, except the equality-sign, the following is specified.

Everything standing together to the left of such a sign up to the nearest equality-sign or horizontal, is to be considered as a whole that stands at the left argument place, provided brackets do not forbid it, and everything standing together to the right of such a sign up to the nearest two-sided function-sign, is to be considered as a whole that stands at the right argument place, provided brackets do not forbid it.

17. So far, simple names of first-level functions with one argument have been, and will continue to be, formed so that the argument place stands to the right of the function-sign proper, as in '$\mathrm{I}\xi$', '$\rangle\xi$', '$\mathfrak{J}\xi$', '$\not{\mathfrak{p}}\xi$', '$\dot{-}\xi$', '$\dot{\smile}\xi$'. For such *one-sided* function-signs, except the horizontal, I specify the following.

Everything standing together to the right of a one-sided function-sign up to the nearest two-sided function-sign, is to be considered as a whole that stands at the argument place.

18. If a horizontal ends free on the left, then we enclose it together with its argument-sign in brackets.

§49. Let us first derive some propositions from (I).

I will now cite (I) in such a way that I will write '$\top b$' for 'b' by rule (9) of §48, and fuse the horizontals by rule (1). What follows will illustrate how a proposition is cited.

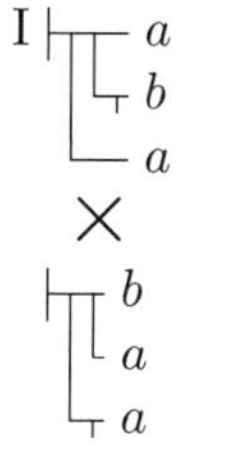

(Ia

(Ia) is hereby made the label of the new proposition. Concerning the transition, compare rule (3). Here we also made use of the permutability of the subcomponents according to rule (2).

In the following derivation, I cite (I) in a way that involves writing '$\top$ *a*' for '*a*'.

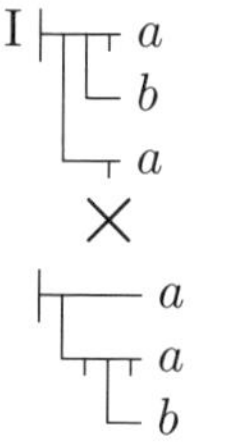

(Ib

In the application of rule (3), ' 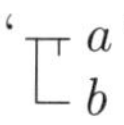 ' is to be regarded as the supercomponent.

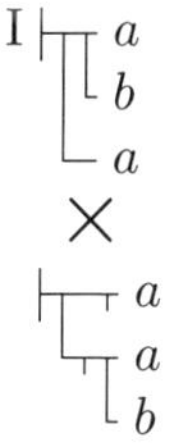

(Ic

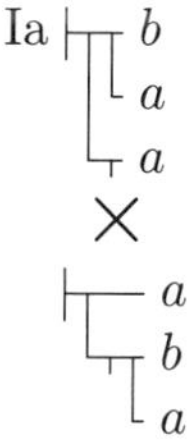

(Id

In the following derivation, (I) is to be thought of in the form, ' 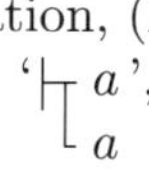 ', where ' 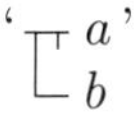 ' is now written in place of '*a*'. If we assume (I) in its original form and write ' 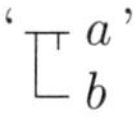 ' for '*a*', then initially we obtain:

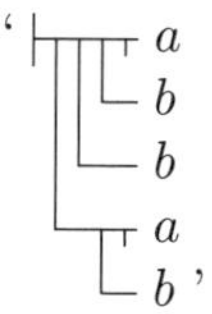

where the equal subcomponents may be fused according to rule (4). The following can also be understood in this way.

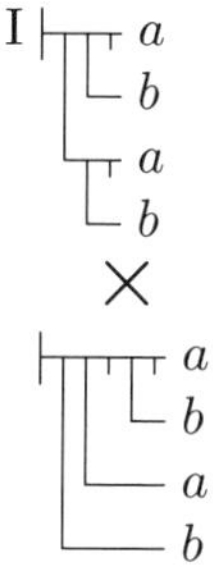

(Ie

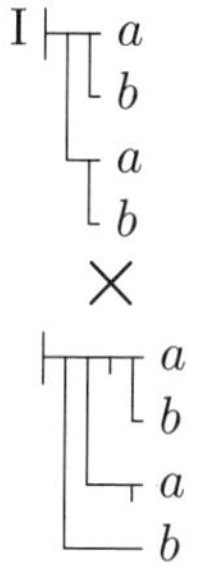

(If

Compare here what was said about 'and' in §12.

In the following (Ie) is cited in a way that involves writing '*a*' for '*b*' and fusing the equal subcomponents:

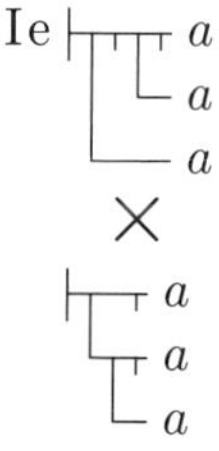

(Ig

§50. Now the principal laws for the function $\xi = \zeta$ will be derived. First, by rule (9) §48, we replace the function-letter '*g*' in (III) by the name of the function — ξ and fuse the horizontals.

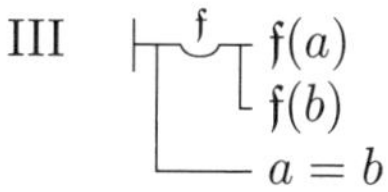

(IIb): — — — — — —

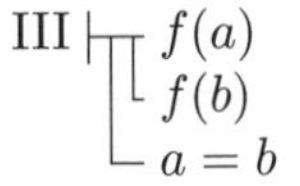

$$\text{III} \;\begin{array}{l}\vdash\!\!\top\!\!\top\; f(a)\\ \!\!\!\!\lfloor\; f(b)\\ \!\!\!\!\llcorner\; a=b\end{array} \qquad\text{(IIIa)}$$

The transition here is made by rule (7), and (IIb) is cited in the form

$$`\;\begin{array}{l}\vdash\!\!\top\!\!\top\; f(a)\\ \!\!\!\!\lfloor\; f(b)\\ \!\!\!\!\llcorner_{\mathfrak{f}}\!\!\top\; \mathfrak{f}(a)\\ \phantom{\mathfrak{f}}\!\!\!\!\lfloor\; \mathfrak{f}(b)\end{array}\;{}'$$

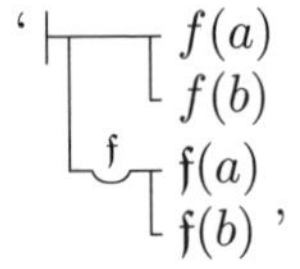

in which, by rule (9), '$M_\beta(\varphi(\beta))$' is replaced by the Roman marker of a second-level function '$\top\!\!\begin{array}{l}\varphi(a)\\ \varphi(b)\end{array}$'.

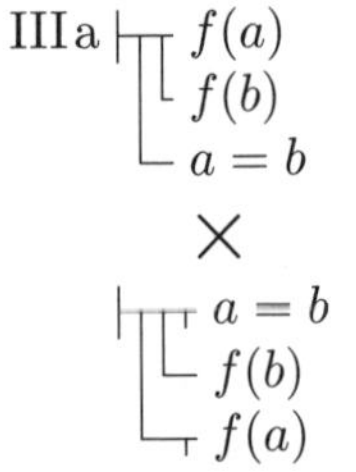

$$\text{IIIa} \;\begin{array}{l}\vdash\!\!\top\!\!\top\; f(a)\\ \!\!\!\!\lfloor\; f(b)\\ \!\!\!\!\llcorner\; a=b\end{array}$$
$$\times$$
$$\vdash\!\!\top\!\!\top\; a=b\\ \!\!\!\!\lfloor\; f(b)\\ \!\!\!\!\llcorner_{\top}\; f(a) \qquad\text{(IIIb)}$$

In the following derivation the function-letter 'f' in (IIIa) is replaced by the Roman function-marker '$\top f(\xi)$', and then the horizontals are fused.

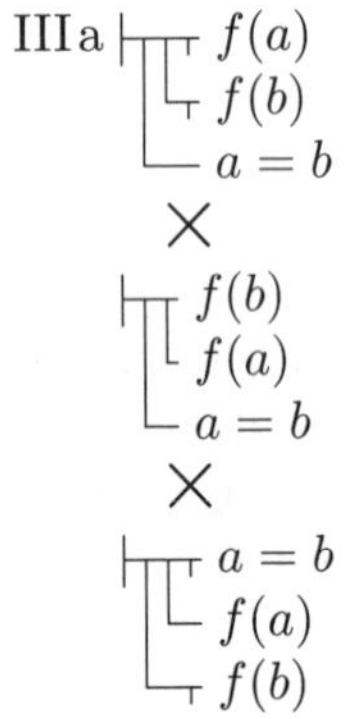

$$\text{IIIa} \;\begin{array}{l}\vdash\!\!\top\!\!\top_\top\; f(a)\\ \!\!\!\!\lfloor_\top\; f(b)\\ \!\!\!\!\llcorner\; a=b\end{array}$$
$$\times$$
$$\vdash\!\!\top\!\!\top\; f(b)\\ \!\!\!\!\lfloor\; f(a)\\ \!\!\!\!\llcorner\; a=b \qquad\text{(IIIc)}$$
$$\times$$
$$\vdash\!\!\top\!\!\top_\top\; a=b\\ \!\!\!\!\lfloor\; f(a)\\ \!\!\!\!\llcorner_\top\; f(b) \qquad\text{(IIId)}$$

We can render (IIIa) in words roughly like this: if a and b coincide, then everything that holds of b holds of a. Similarly, (IIIc). (IIId) may be paraphrased like this: if a predication holds of a that does not hold of b, then a and b do not coincide.

In the following derivation the function-letter 'g' in (III) is replaced by the function-name '$\top\xi$', and 'b' by 'a'.

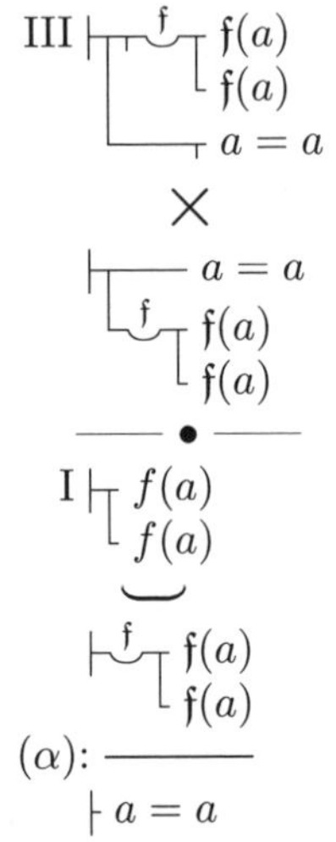

$$\text{III} \;\begin{array}{l}\vdash\!\!\top\!\!\overset{\mathfrak{f}}{\frown}\!\top\; \mathfrak{f}(a)\\ \phantom{\mathfrak{f}}\!\!\!\!\lfloor\; \mathfrak{f}(a)\\ \!\!\!\!\llcorner_\top\; a=a\end{array}$$
$$\times$$
$$\vdash\!\!\top\!\!\top\; a=a\\ \!\!\!\!\llcorner_{\mathfrak{f}}\!\!\top\; \mathfrak{f}(a)\\ \phantom{\mathfrak{f}}\!\!\!\!\lfloor\; \mathfrak{f}(a) \qquad(\alpha$$
$$\underline{\qquad\bullet\qquad}$$
$$\text{I} \;\vdash\!\!\top\; f(a)\\ \!\!\!\lfloor\; f(a)$$
$$\vdash\!\!\overset{\mathfrak{f}}{\frown}\!\top\; \mathfrak{f}(a)\\ \phantom{\mathfrak{f}}\!\!\!\!\lfloor\; \mathfrak{f}(a) \qquad(\beta$$
$$(\alpha):\;\underline{\qquad\qquad}$$
$$\vdash a=a \qquad\text{(IIIe)}$$

In the light of our explanation of the equality-sign, this proposition is indeed self-evident, but it is worth the effort to see how it can be established from (III). Further, doing so provides an opportunity to note some points that are also to hold for later derivations. This second proposition received the label 'α'. When so used, a small Greek letter is only required to remain fixed as a label within the same derivation, and so may be used to label a different proposition in a different derivation. A derivation terminates in a proposition to which a label other than a small Greek letter is first assigned. In our derivation the proposition (α) is followed by the sign

'—— • ——'

to note that we break a series of inferences here and start a new one which is only connected to the former where we cite (α). The transition to (β) is made by rule (5), that from (β) to (IIIe) by rule (6). Let us now replace the function-letter 'f' in (IIIa) by the Roman function-marker '$b = \xi$':

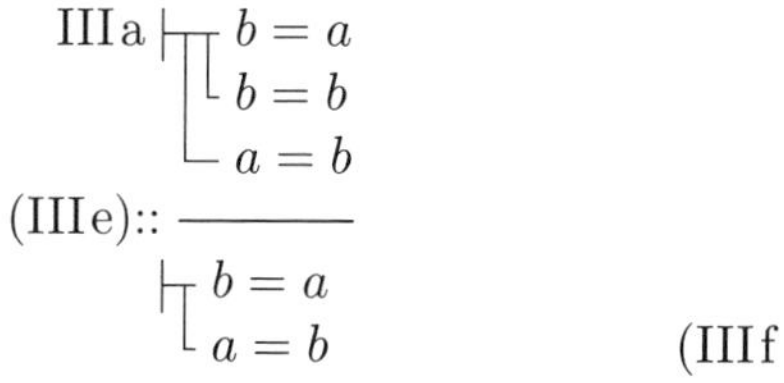
IIIa
b = a
b = b
a = b
(IIIe)::
b = a
a = b
(IIIf

This inference is made by rule (6).

In the following derivation, 'a' is replaced by '— a', 'b' by '⊤ a', and the function-letter 'f' by the function-name '— ξ' in (IIIc) and (IIIa), and the horizontals are fused where possible.

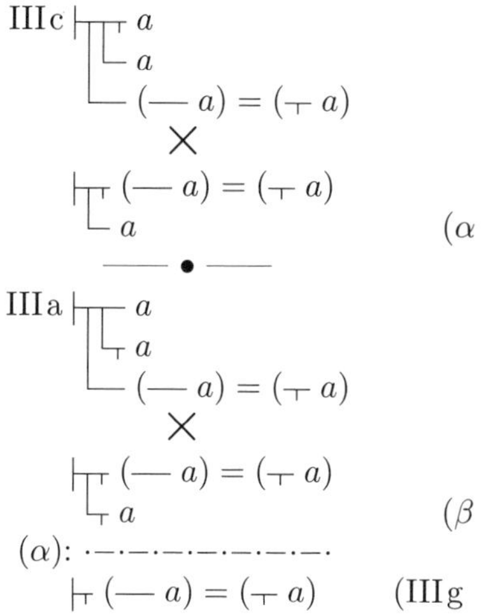
IIIc
a
a
(— a) = (⊤ a)
(— a) = (⊤ a)
a
(α
IIIa
a
a
(— a) = (⊤ a)
(— a) = (⊤ a)
a
(β
(α):
(— a) = (⊤ a)
(IIIg

In the transitions to (α) and (β) equal subcomponents are fused in accordance with rule (4). The final inference is made by rule (8).

In the following derivation we replace the function-letter 'f' in (IIIc) by the Roman function-marker '$f(a) = f(\xi)$'.

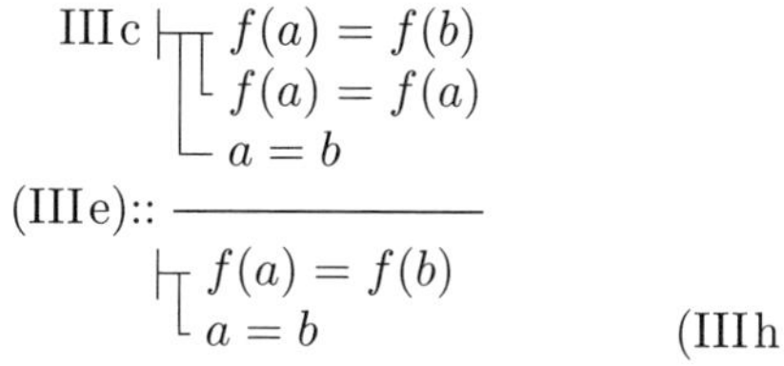
IIIc
f(a) = f(b)
f(a) = f(a)
a = b
(IIIe)::
f(a) = f(b)
a = b
(IIIh

In the following derivation the function-letter 'g' in (III) is replaced by '⊤ F(— ξ)' and the horizontals are fused.

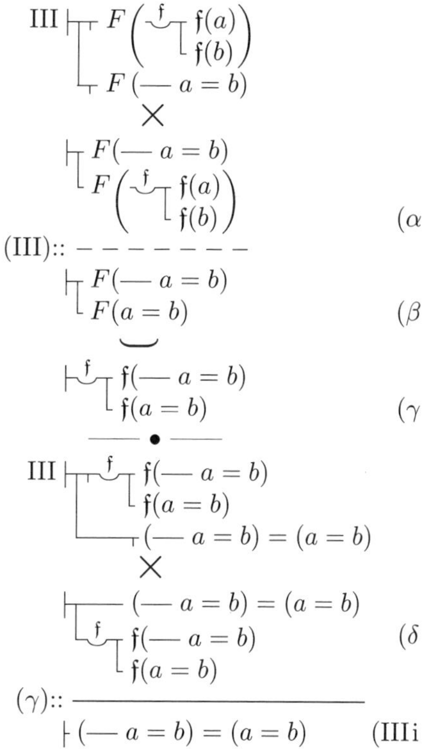
III
F(— a = b)
F(— a = b)
F(a = b)
(α
(III)::
F(— a = b)
F(a = b)
(β
f(— a = b)
f(a = b)
(γ
III
f(— a = b)
f(a = b)
(— a = b) = (a = b)
(— a = b) = (a = b)
f(— a = b)
f(a = b)
(δ
(γ)::
(— a = b) = (a = b)
(IIIi

In the second citation of (III), 'g' is replaced by 'F'. In the final citation of

(III) '$g(\xi)$' is replaced by '$\top\, \xi$', 'a' by '— $a = b$', 'b' by '$a = b$'.

§51. Some propositions are now to be derived from (IV).

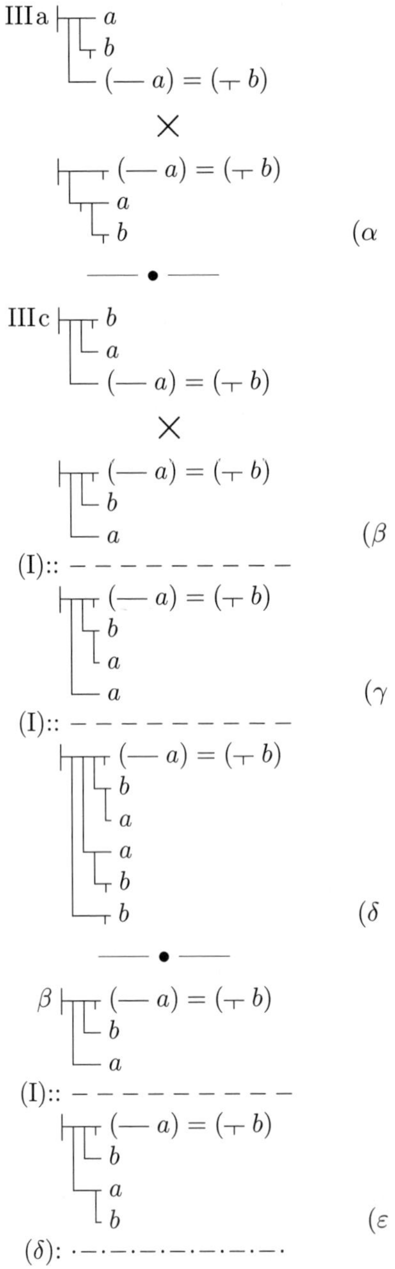

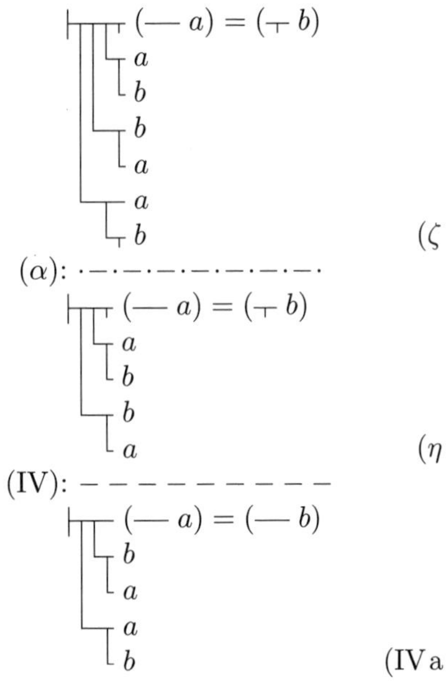

In its first application, (I) is here to be thought of in the form

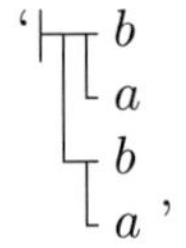

in its second in the form

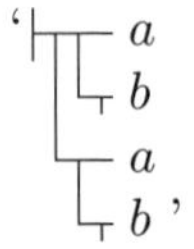

and in its third in the form

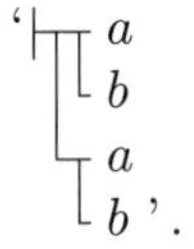

Note the effect of applying (I) in the transitions to (γ), (δ) and (ε). (I) will often be used in this way in what follows. Compare here the derivation of (Ie) in §49. Proposition (IV a) will often be used to prove equality of truth-values.

$$\text{IV} \vdash\!\!\!\top (-\, a) = (\top\, a)$$
$$\quad \llcorner_{\top} (-\, a) = (\top\!\top\, a)$$
$$\times$$
$$\vdash\!\!\!\top (-\, a) = (\top\!\top\, a)$$
$$\quad \llcorner_{\top} (-\, a) = (\top\, a)$$
$$\text{(IIIg)::} \;\rule{8em}{0.4pt}$$
$$\vdash (-\, a) = (\top\!\top\, a) \qquad \text{(IVb}$$
$$\text{(IIIa):} \;\rule{8em}{0.4pt}$$
$$\vdash\!\!\!\top f(-\, a)$$
$$\quad \llcorner f(\top\!\top\, a) \qquad \text{(IVc}$$

——— • ———

$$\text{IVb} \vdash (-\, a) = (\top\!\top\, a)$$
$$\text{(IIIc):} \;\rule{8em}{0.4pt}$$
$$\vdash\!\!\!\top f(\top\!\top\, a)$$
$$\quad \llcorner f(-\, a) \qquad \text{(IVd}$$

An example of an application of (IVa) is the following:

$$\text{IIIf} \vdash\!\!\!\top a = b$$
$$\quad \llcorner b = a$$
$$\text{(IVa):} \;\rule{4em}{0.4pt}$$
$$\vdash\!\!\!\top (-\, a = b) = (-\, b = a)$$
$$\quad \llcorner\!\!\top b = a$$
$$\quad\quad \llcorner a = b \qquad (\alpha$$
$$\text{(IIIf)::} \;\rule{12em}{0.4pt}$$
$$\vdash (-\, a = b) = (-\, b = a) \qquad (\beta$$
$$\text{(IIIc):} \;\rule{12em}{0.4pt}$$
$$\vdash\!\!\!\top (a = b) = (-\, b = a)$$
$$\quad \llcorner (-\, a = b) = (a = b) \qquad (\gamma$$
$$\text{(IIIi)::} \;\rule{10em}{0.4pt}$$
$$\vdash (a = b) = (-\, b = a) \qquad (\delta$$
$$\text{(IIIc):} \;\rule{10em}{0.4pt}$$
$$\vdash\!\!\!\top (a = b) = (b = a)$$
$$\quad \llcorner (-\, b = a) = (b = a) \qquad (\varepsilon$$
$$\text{(IIIi)::} \;\rule{10em}{0.4pt}$$
$$\vdash (a = b) = (b = a) \qquad \text{(IVe}$$

In the transition to (γ), (IIIc) is here to be thought of in the form

$$\text{‘}\vdash\!\!\!\top\!\!\top (a = b) = (-\, b = a)$$
$$\quad \llcorner (-\, a = b) = (-\, b = a)$$
$$\quad \llcorner (-\, a = b) = (a = b)\text{’},$$

where ‘$f(\xi)$’ is replaced by

$$\text{‘}\xi = (-\, b = a)\text{’},$$

‘a’ by ‘$(-\, a = b)$’, ‘b’ by ‘$(a = b)$’. For the transition to (ε), we have to think of ‘$f(\xi)$’ in (IIIc) as replaced by ‘$(a = b) = \xi$’, ‘a’ by ‘$(-\, b = a)$’, ‘b’ by ‘$(b = a)$’.

§52. Finally, some propositions may be derived from (V) and (VI).

$$\text{V} \vdash (\grave{\varepsilon} f(\varepsilon) = \grave{\alpha} g(\alpha)) = (\smile^{\mathfrak{a}}\, f(\mathfrak{a}) = g(\mathfrak{a}))$$
$$\text{(IIIa):} \;\rule{20em}{0.4pt}$$
$$\vdash\!\!\!\top \grave{\varepsilon} f(\varepsilon) = \grave{\alpha} g(\alpha)$$
$$\quad \llcorner\!\smile^{\mathfrak{a}} f(\mathfrak{a}) = g(\mathfrak{a})$$
$$\text{(IIIh):} \;- - - - - - - - -$$
$$\vdash\!\!\!\top F(\grave{\varepsilon} f(\varepsilon)) = F(\grave{\alpha} g(\alpha))$$
$$\quad \llcorner\!\smile^{\mathfrak{a}} f(\mathfrak{a}) = g(\mathfrak{a}) \qquad \text{(Va}$$

——— • ———

$$\text{V} \vdash (\grave{\varepsilon} f(\varepsilon) = \grave{\alpha} g(\alpha)) = (\smile^{\mathfrak{a}}\, f(\mathfrak{a}) = g(\mathfrak{a}))$$
$$\text{(IIIc):} \;\rule{20em}{0.4pt}$$
$$\vdash\!\!\!\top\!\smile^{\mathfrak{a}} f(\mathfrak{a}) = g(\mathfrak{a})$$
$$\quad \llcorner \grave{\varepsilon} f(\varepsilon) = \grave{\alpha} g(\alpha) \qquad (\alpha$$
$$\text{(IIa):} \;- - - - - - - - -$$
$$\vdash\!\!\!\top f(a) = g(a)$$
$$\quad \llcorner \grave{\varepsilon} f(\varepsilon) = \grave{\alpha} g(\alpha) \qquad \text{(Vb}$$

In (IIa), ‘$f(\xi)$’ is here to be thought of as replaced by ‘$f(\xi) = g(\xi)$’.

In the following derivation ‘$g(\xi)$’ in (Va) is replaced by ‘$a = \xi$’, while ‘ε’ is written for ‘α’ by rule (11) §48.

$$\text{Va} \vdash\!\!\!\top \grave{\varepsilon} f(\varepsilon) = \grave{\varepsilon}(a = \varepsilon)$$
$$\quad \llcorner\!\smile^{\mathfrak{a}} f(\mathfrak{a}) = (a = \mathfrak{a})$$
$$\text{(IIIa):} \;- - - - - - - - -$$
$$\vdash\!\!\!\top\!\!\top a = \backslash \grave{\varepsilon} f(\varepsilon)$$
$$\quad\quad \llcorner a = \backslash \grave{\varepsilon}(a = \varepsilon)$$
$$\quad \llcorner\!\smile^{\mathfrak{a}} f(\mathfrak{a}) = (a = \mathfrak{a}) \qquad (\alpha$$
$$\text{(VI)::} \;\rule{10em}{0.4pt}$$
$$\vdash\!\!\!\top a = \backslash \grave{\varepsilon} f(\varepsilon)$$
$$\quad \llcorner\!\smile^{\mathfrak{a}} f(\mathfrak{a}) = (a = \mathfrak{a}) \qquad \text{(VIa}$$

II. Proofs of the basic laws of cardinal number

Preliminaries

§53. Concerning the proofs to follow I would emphasize that the commentaries that I regularly give in advance under the heading 'analysis' are merely intended to serve the convenience of the reader; they could be omitted without compromising the force of the proof, which is to be sought under the heading 'construction' only.

The rules that I cite in the analysis are listed above in §48 under their respective numbers.[a] The laws derived above are collected in a special table at the end of the volume, together with the basic laws summarised in §47. In addition, the definitions of section I.2, and others, are also collected at the end of the volume.

First we prove the proposition:

The cardinal number of a concept is equal to the cardinal number of a second concept, if a relation maps the first into the second, and if the converse of this relation maps the second into the first.

A. Proof of the proposition

$$`\vdash\begin{array}{l} \mathfrak{P}u = \mathfrak{P}v \\ u\frown(v\frown\rangle q) \\ v\frown(u\frown\rangle \curlyvee q) \end{array}'$$

a) Proof of the proposition

$$`\vdash\begin{array}{l} w\frown(v\frown\rangle(p\smile q)) \\ w\frown(u\frown\rangle p) \\ u\frown(v\frown\rangle q) \end{array}'$$

§54. *Analysis*
According to definition (Z) the proposition

$$`\vdash\begin{array}{l} \mathfrak{P}u = \mathfrak{P}v \\ u\frown(v\frown\rangle q) \\ v\frown(u\frown\rangle \curlyvee q) \end{array}' \qquad (\alpha$$

is a consequence of

$$`\vdash\begin{array}{l} \grave{\varepsilon}\left(\overset{\mathfrak{q}}{\smile}\begin{array}{l}\varepsilon\frown(u\frown\rangle\mathfrak{q})\\ u\frown(\varepsilon\frown\rangle\curlyvee\mathfrak{q})\end{array}\right) = \grave{\varepsilon}\left(\overset{\mathfrak{q}}{\smile}\begin{array}{l}\varepsilon\frown(v\frown\rangle\mathfrak{q})\\ v\frown(\varepsilon\frown\rangle\curlyvee\mathfrak{q})\end{array}\right) \\ u\frown(v\frown\rangle q) \\ v\frown(u\frown\rangle \curlyvee q) \end{array}' \qquad (\beta$$

This proposition is to be derived using (V a) and rule (5) from the proposition

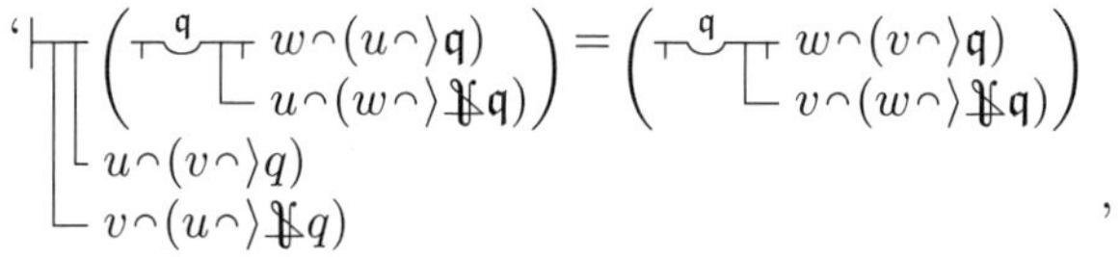

(γ

which is to be proven using (IV a). For this, we require the propositions

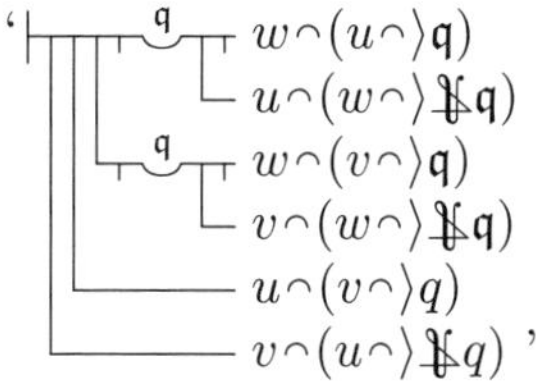

(δ

and

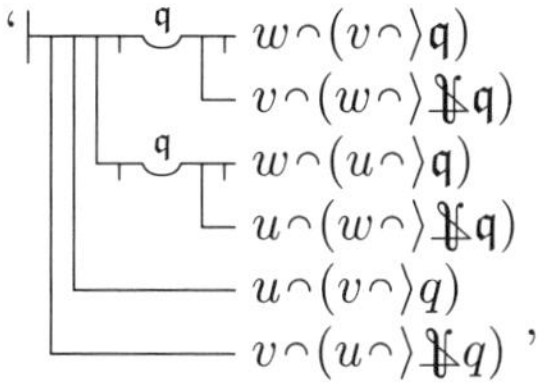

(ε

If we interchange 'u' with 'v' in (ε) and write '$\rangle q$' for 'q', we obtain:

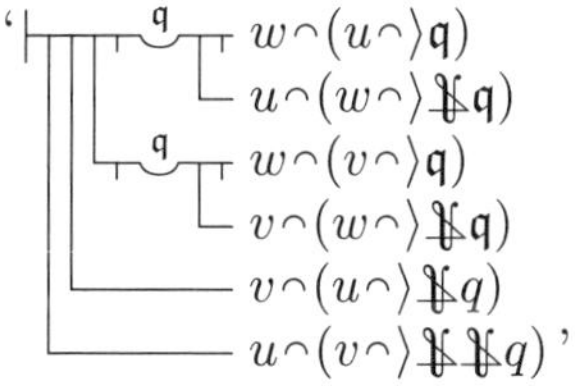

(ζ

This proposition almost coincides with (δ). To derive (δ) from (ζ) by rule (7) we require the proposition

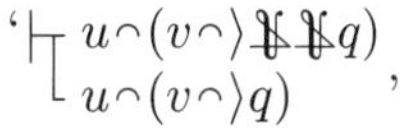

(η

So we first try to prove the proposition (ε). This results by contraposition (rule 3) from the proposition

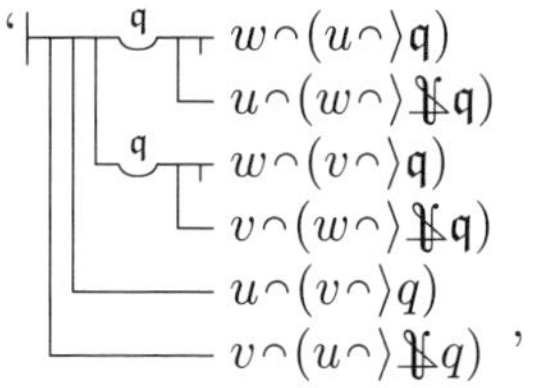

(ϑ

which follows by rule (5) from

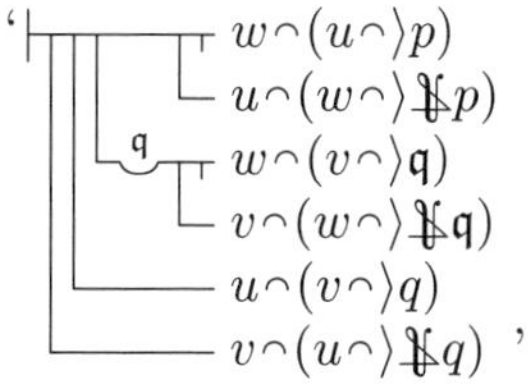

(ι

In order to grasp the sense of this better, we transform it by contraposition into

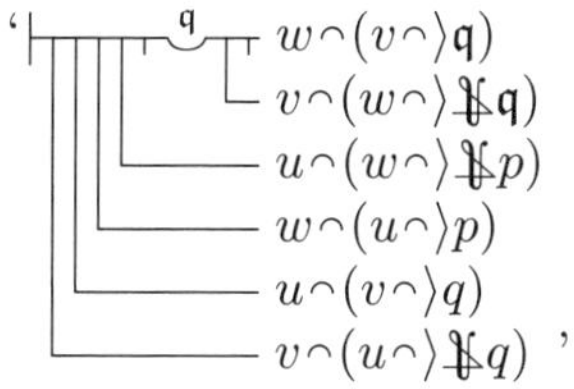

(κ

For ease of expression, I will now say 'u-concept' instead of 'concept whose extension is indicated by 'u'',[b] 'p-relation' instead of 'relation whose extension is indicated by 'p'', 'the p-relation maps the w-concept into the u-concept' instead of 'the objects falling under the w-concept are correlated single-valuedly with the objects falling under the u-concept by the p-relation'. We may now rephrase (κ) in words like this:

'If the converse of the p-relation maps the u-concept into the w-concept and the p-relation maps the w-concept into the u-concept, if further the q-relation maps the u-concept into the v-concept and the $\rangle q$-relation maps the v-concept into the u-concept, then there is a relation that maps the w-concept into the v-concept and whose converse maps the v-concept into the w-concept.'

Such a relation is evidently one that is composed from the p-relation and the q-relation,[1] as the following figure illustrates:

$$w \underset{p}{\twoheadrightarrow} u \underset{q}{\twoheadrightarrow} v$$

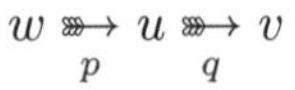

I now introduce the abbreviation '$p \smile q$' for the extension of a relation which is composed from the p-relation and the q-relation by defining:

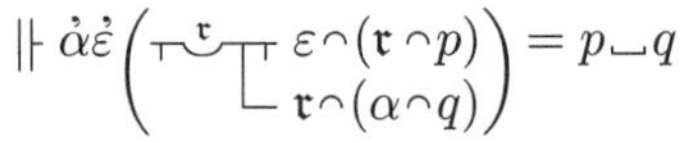

$$\Vdash \grave{\alpha}\grave{\varepsilon}\left(\stackrel{\mathfrak{r}}{\smile} \begin{array}{l} \top\; \varepsilon \frown (\mathfrak{r} \frown p) \\ \llcorner\; \mathfrak{r} \frown (\alpha \frown q) \end{array} \right) = p \smile q \qquad (\mathrm{B}$$

It now turns on the proposition

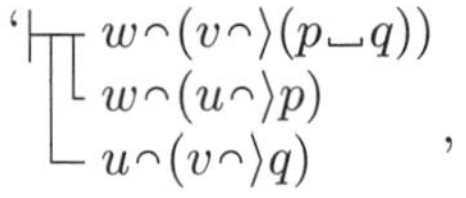

$$`\vdash \begin{array}{l} \top\; w \frown (v \frown \rangle (p \smile q)) \\ \llcorner\; w \frown (u \frown \rangle p) \\ \llcorner\; u \frown (v \frown \rangle q) \end{array} \text{'} \qquad (\lambda$$

in words:

'If the p-relation maps the w-concept into the u-concept and if the q-relation maps the u-concept into the v-concept, then the $(p \smile q)$-relation that is composed from the two maps the w-concept into the v-concept.'

In addition, we require the proposition

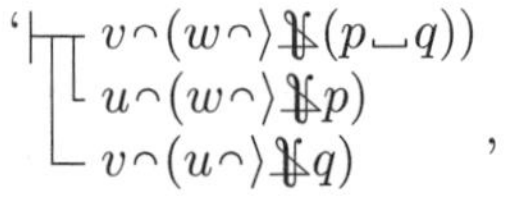

$$`\vdash \begin{array}{l} \top\; v \frown (w \frown \rangle \mathfrak{X}(p \smile q)) \\ \llcorner\; u \frown (w \frown \rangle \mathfrak{X}p) \\ \llcorner\; v \frown (u \frown \rangle \mathfrak{X}q) \end{array} \text{'} \qquad (\mu$$

which can be reduced to (λ) using the proposition

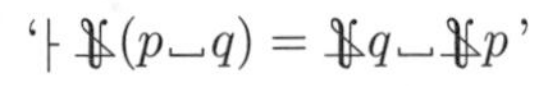

$$`\vdash \mathfrak{X}(p \smile q) = \mathfrak{X}q \smile \mathfrak{X}p\text{'} \qquad (\nu$$

We first attempt to prove the proposition (λ). From definition (Δ) it can be gathered that two things must be proven, namely first

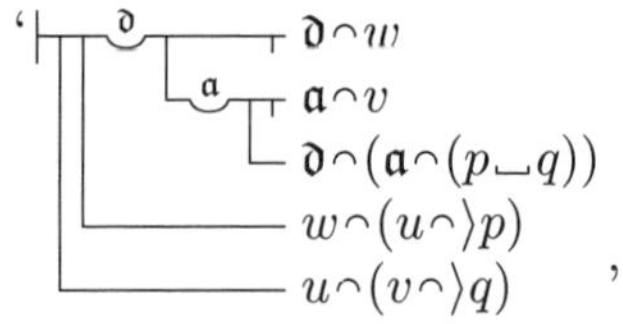

$$`\vdash \begin{array}{l} \stackrel{\mathfrak{d}}{\smile}\top\; \mathfrak{d} \frown w \\ \quad \llcorner \stackrel{\mathfrak{a}}{\smile} \top\; \mathfrak{a} \frown v \\ \qquad\quad \llcorner\; \mathfrak{d} \frown (\mathfrak{a} \frown (p \smile q)) \\ \llcorner\; w \frown (u \frown \rangle p) \\ \llcorner\; u \frown (v \frown \rangle q) \end{array} \text{'} \qquad (\xi$$

and second

$$`\vdash \begin{array}{l} \top\; \mathrm{I}(p \smile q) \\ \llcorner\; \mathrm{I}q \\ \llcorner\; \mathrm{I}p \end{array} \text{'} \qquad (o$$

(ξ) results from

$$`\vdash \begin{array}{l} \top\; d \frown w \\ \quad \llcorner \stackrel{\mathfrak{a}}{\smile} \top\; \mathfrak{a} \frown v \\ \qquad\quad \llcorner\; d \frown (\mathfrak{a} \frown (p \smile q)) \\ \llcorner\; w \frown (u \frown \rangle p) \\ \llcorner\; u \frown (v \frown \rangle q) \end{array} \text{'} \qquad (\pi$$

according to rule (5). In order to express (π) in words, it is convenient first to transform it by contraposition into

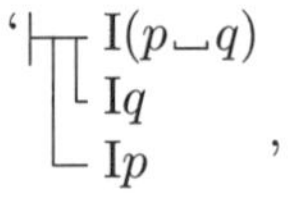

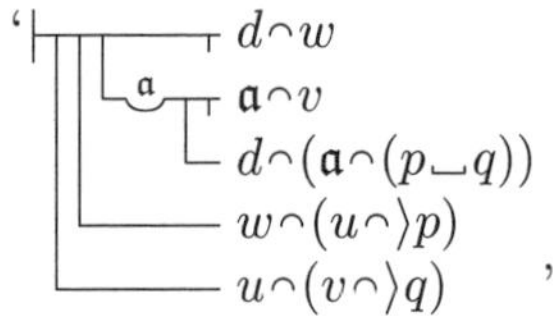

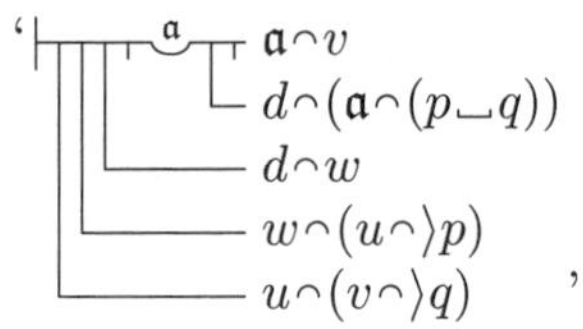

$$`\vdash \begin{array}{l} \stackrel{\mathfrak{a}}{\smile} \top\; \mathfrak{a} \frown v \\ \quad \llcorner\; d \frown (\mathfrak{a} \frown (p \smile q)) \\ \llcorner\; d \frown w \\ \llcorner\; w \frown (u \frown \rangle p) \\ \llcorner\; u \frown (v \frown \rangle q) \end{array} \text{'} \qquad (\varrho$$

Letting '*d*' now abbreviate 'object which is indicated by '*d*'',[c] our proposition may be rendered in words like this:

'If *d* falls under the w-concept, and if the w-concept is mapped into the u-concept by the p-relation, and if the u-concept is mapped into the v-concept by the q-relation, then there is an object that falls under the v-concept and to which *d* stands in the $(p \smile q)$-relation.'

[1] Compare *Grundlagen*, p. 86.

The proof will need to rely on the proposition

$$\text{‘}\vdash\begin{array}{l} d\frown(m\frown(p\smile q)) \\ e\frown(m\frown q) \\ d\frown(e\frown p) \end{array}\text{’} \qquad (\sigma$$

In words:

‘If d stands in the p-relation to e, and if e stands in the q-relation to m, then d stands in the $(p\smile q)$-relation to m.’

This will need to be derived from the proposition

$$\text{‘}\vdash\left(\neg\overset{\mathfrak{r}}{\smile}\neg\begin{array}{l} d\frown(\mathfrak{r}\frown p) \\ \mathfrak{r}\frown(m\frown q)\end{array}\right)=d\frown(m\frown(p\smile q))\text{’} \qquad (\tau$$

which follows from definition (B). In order to prove it, we require the proposition

$$\text{‘}\vdash f(a,b)=a\frown(b\frown\dot{\alpha}\dot{\varepsilon}f(\varepsilon,\alpha))\text{’} \qquad (\upsilon$$

since the left-hand side of the definitional equation (B) is a double value-range. (υ) is to be reduced to the proposition

$$\text{‘}\vdash f(a)=a\frown\dot{\varepsilon}f(\varepsilon)\text{’} \qquad (\varphi$$

which is to be derived from definition (A). Accordingly, we have to prove

$$\text{‘}\vdash f(a)=\backslash\dot{\alpha}\left(\neg\overset{\mathfrak{g}}{\smile}\neg\begin{array}{l}\mathfrak{g}(a)=\alpha \\ \dot{\varepsilon}f(\varepsilon)=\dot{\varepsilon}\mathfrak{g}(\varepsilon)\end{array}\right)\text{’} \qquad (\chi$$

This has to be done by appeal to (VIa) and the proposition

$$\text{‘}\vdash\overset{\mathfrak{a}}{\smile}\left(\neg\overset{\mathfrak{g}}{\smile}\neg\begin{array}{l}\mathfrak{g}(a)=\mathfrak{a} \\ \dot{\varepsilon}f(\varepsilon)=\dot{\varepsilon}\mathfrak{g}(\varepsilon)\end{array}\right)=(f(a)=\mathfrak{a})\text{’} \qquad (\psi$$

taking

$$\text{‘}\neg\overset{\mathfrak{g}}{\smile}\neg\begin{array}{l}\mathfrak{g}(a)=\xi \\ \dot{\varepsilon}f(\varepsilon)=\dot{\varepsilon}\mathfrak{g}(\varepsilon)\end{array}\text{’}$$

for ‘$f(\xi)$’ in (VIa) and replacing ‘a’ by ‘$f(a)$’. (ψ) is obtained by rule (5) from

$$\text{‘}\vdash\left(\neg\overset{\mathfrak{g}}{\smile}\neg\begin{array}{l}\mathfrak{g}(a)=b \\ \dot{\varepsilon}f(\varepsilon)=\dot{\varepsilon}\mathfrak{g}(\varepsilon)\end{array}\right)=(f(a)=b)\text{’} \qquad (\omega$$

which is to be proven by appeal to (IVa). For this, we require the propositions

$$\text{‘}\vdash\begin{array}{l}\neg\overset{\mathfrak{g}}{\smile}\neg\begin{array}{l}\mathfrak{g}(a)=b \\ \dot{\varepsilon}f(\varepsilon)=\dot{\varepsilon}\mathfrak{g}(\varepsilon)\end{array} \\ f(a)=b\end{array}\text{’} \qquad (\alpha'$$

and

$$\text{‘}\vdash\begin{array}{l} f(a)=b \\ \neg\overset{\mathfrak{g}}{\smile}\neg\begin{array}{l}\mathfrak{g}(a)=b \\ \dot{\varepsilon}f(\varepsilon)=\dot{\varepsilon}\mathfrak{g}(\varepsilon)\end{array}\end{array}\text{’} \qquad (\beta'$$

the first of which follows by contraposition according to rule (3) from

$$\text{‘}\vdash\begin{array}{l} \neg\, f(a)=b \\ \overset{\mathfrak{g}}{\smile}\neg\begin{array}{l}\mathfrak{g}(a)=b \\ \dot{\varepsilon}f(\varepsilon)=\dot{\varepsilon}\mathfrak{g}(\varepsilon)\end{array}\end{array}\text{’} \qquad (\gamma'$$

If we now write (IIb) in the form

$$\text{‘}\vdash\begin{array}{l}\neg\begin{array}{l} f(a)=b \\ \dot{\varepsilon}f(\varepsilon)=\dot{\varepsilon}f(\varepsilon)\end{array} \\ \overset{\mathfrak{g}}{\smile}\neg\begin{array}{l}\mathfrak{g}(a)=b \\ \dot{\varepsilon}f(\varepsilon)=\dot{\varepsilon}\mathfrak{g}(\varepsilon)\end{array}\end{array}\text{’}$$

then we see that (γ') follows from it and (IIIe). The proposition (β') follows by contraposition from

$$\text{‘}\vdash\begin{array}{l}\overset{\mathfrak{g}}{\smile}\neg\begin{array}{l}\mathfrak{g}(a)=b \\ \dot{\varepsilon}f(\varepsilon)=\dot{\varepsilon}\mathfrak{g}(\varepsilon)\end{array} \\ \neg\, f(a)=b\end{array}\text{’} \qquad (\delta'$$

and this in turn by rule (5) from

$$\text{‘}\vdash\begin{array}{l}\neg\begin{array}{l} g(a)=b \\ \dot{\varepsilon}f(\varepsilon)=\dot{\varepsilon}g(\varepsilon)\end{array} \\ \neg\, f(a)=b\end{array}\text{’} \qquad (\varepsilon'$$

This proposition is obtained by rule (7) and (Vb) from

$$\vdash\begin{array}{l}\neg\, g(a)=b \\ f(a)=g(a) \\ \neg\, f(a)=b\end{array},$$

which, by permutation of subcomponents, is just a special case of (IIIc). We now construct the proof accordingly. Concerning the derivation of (2), it is to be remarked in addition that in the first citation of (1) by rule (9), '$f(\xi)$' is replaced by the function-marker '$f(\xi, b)$'. Next, (IIIc) is to be thought of in the form

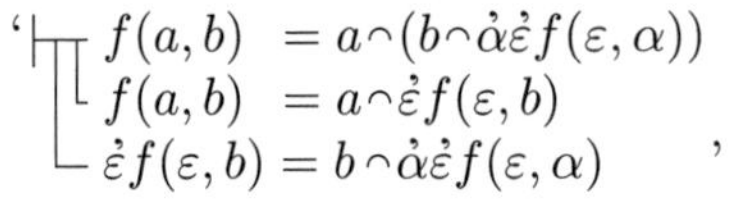

In the second citation of (1), it is to be thought of in the form

$$\text{'}\vdash \grave{\varepsilon} f(\varepsilon, b) = b \frown \grave{\alpha}\grave{\varepsilon} f(\varepsilon, \alpha)\text{'}$$

by putting '$\grave{\varepsilon} f(\varepsilon, \xi)$' for '$f(\xi)$', '$b$' for '$a$', and '$\alpha$' for '$\varepsilon$', according to the rules (9) and (11).

§55. *Construction*

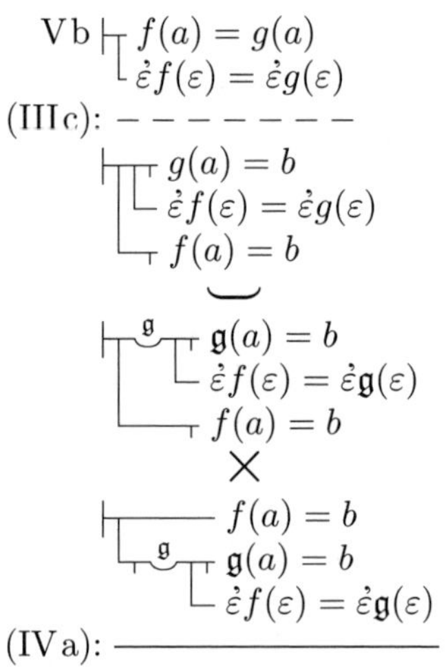

(α

(β

(γ

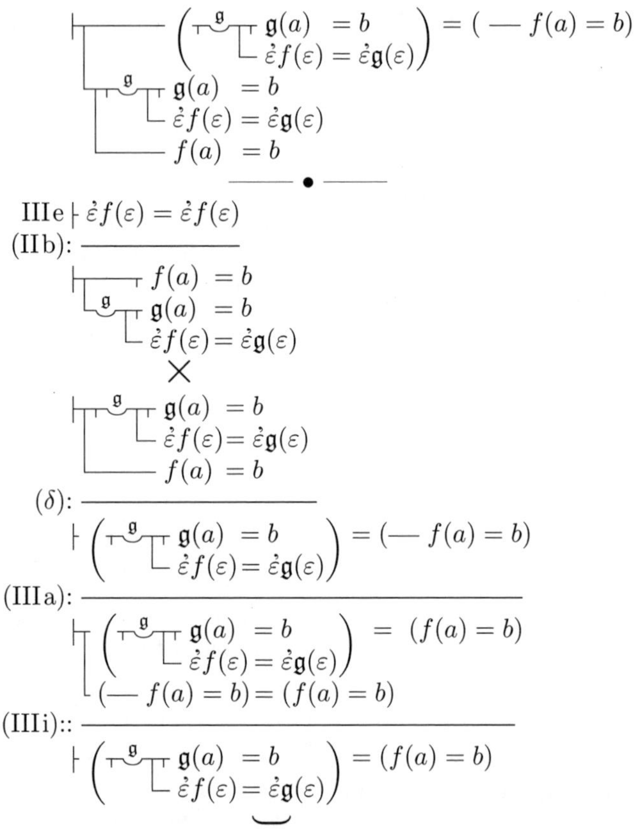

(δ

(ε

(ζ

(η

(ϑ

(ι

$$\vdash^{\mathfrak{a}}\!\left(\,\overset{\mathfrak{g}}{\smile}\; \mathfrak{g}(a) = \mathfrak{a} \;\Big|\; \grave{\varepsilon}f(\varepsilon) = \grave{\varepsilon}\mathfrak{g}(\varepsilon)\right) = (f(a) = \mathfrak{a}) \qquad (\kappa$$

(VIa): ———————————————

$$\vdash f(a) = \backslash\grave{\alpha}\left(\,\overset{\mathfrak{g}}{\smile}\; \mathfrak{g}(a) = \alpha \;\Big|\; \grave{\varepsilon}f(\varepsilon) = \grave{\varepsilon}\mathfrak{g}(\varepsilon)\right) \qquad (\lambda$$

(IIIa): ———————————————

$$\vdash f(a) = a\frown\grave{\varepsilon}f(\varepsilon) \;\Big|\; \backslash\grave{\alpha}\left(\,\overset{\mathfrak{g}}{\smile}\; \mathfrak{g}(a) = \alpha \;\Big|\; \grave{\varepsilon}f(\varepsilon) = \grave{\varepsilon}\mathfrak{g}(\varepsilon)\right) = a\frown\grave{\varepsilon}f(\varepsilon) \qquad (\mu$$

(A):: ———————————————

$$\vdash f(a) = a\frown\grave{\varepsilon}f(\varepsilon) \qquad (1$$

——— • ———

$$1 \vdash f(a,b) = a\frown\grave{\varepsilon}f(\varepsilon,b)$$

(IIIc): ———————————————

$$\vdash f(a,b) = a\frown(b\frown\grave{\alpha}\grave{\varepsilon}f(\varepsilon,\alpha)) \;\Big|\; \grave{\varepsilon}f(\varepsilon,b) = b\frown\grave{\alpha}\grave{\varepsilon}f(\varepsilon,\alpha) \qquad (\alpha$$

(1):: ———————————————

$$\vdash f(a,b) = a\frown(b\frown\grave{\alpha}\grave{\varepsilon}f(\varepsilon,\alpha)) \qquad (2$$

(IIIc): ———————————————

$$\vdash f(a,b) = a\frown(b\frown q) \;\Big|\; \grave{\alpha}\grave{\varepsilon}f(\varepsilon,\alpha) = q \qquad (3$$

——— • ———

$$\vdash \grave{\alpha}\grave{\varepsilon}\left(\,\overset{\mathfrak{r}}{\smile}\; \varepsilon\frown(\mathfrak{r}\frown p) \;\Big|\; \mathfrak{r}\frown(\alpha\frown q)\right) = p\_q \qquad (B$$

(3): ———————————————

$$\vdash \left(\,\overset{\mathfrak{r}}{\smile}\; d\frown(\mathfrak{r}\frown p) \;\Big|\; \mathfrak{r}\frown(m\frown q)\right) = d\frown(m\frown(p\_q)) \qquad (4$$

(IIIc): ———————————————

$$\vdash d\frown(m\frown(p\_q)) \;\Big|\; \overset{\mathfrak{r}}{\smile}\; d\frown(\mathfrak{r}\frown p) \;\Big|\; \mathfrak{r}\frown(m\frown q) \qquad (\alpha$$

×

$$\vdash^{\mathfrak{r}}\; d\frown(\mathfrak{r}\frown p) \;\Big|\; \mathfrak{r}\frown(m\frown q) \;\Big|\; \top d\frown(m\frown(p\_q)) \qquad (\beta$$

(IIa): – – – – – – – – – – –

$$\vdash d\frown(e\frown p) \;\Big|\; e\frown(m\frown q) \;\Big|\; \top d\frown(m\frown(p\_q)) \qquad (\gamma$$

×

$$\vdash d\frown(m\frown(p\_q)) \;\Big|\; e\frown(m\frown q) \;\Big|\; d\frown(e\frown p) \qquad (5$$

§56. *Analysis*

Now, in order to prove the proposition

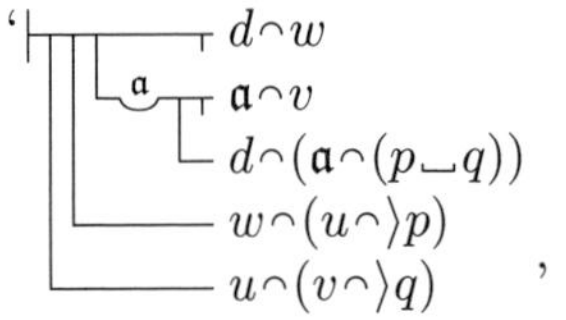

(α

(§54, π) we have to go back to (Δ). From this we derive the proposition

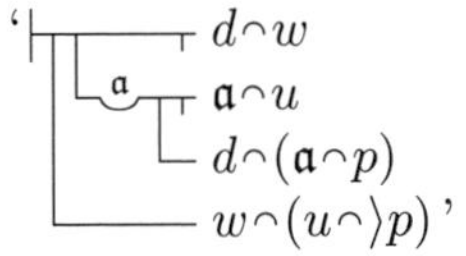

(β

In order to reach (α) from this, we need to have the proposition

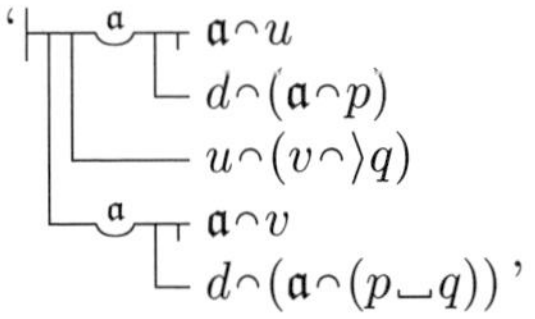

(γ

which we obtain by rule (5) from

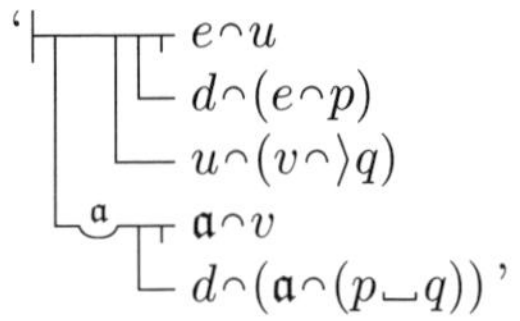

(δ

We now can also write the proposition (β) like this:

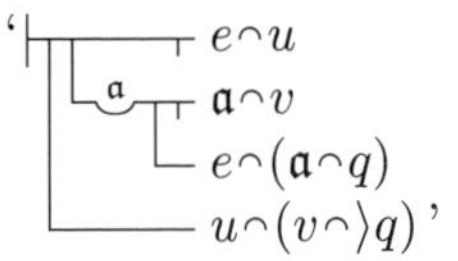

(β

To attain (δ) from this, we require the proposition

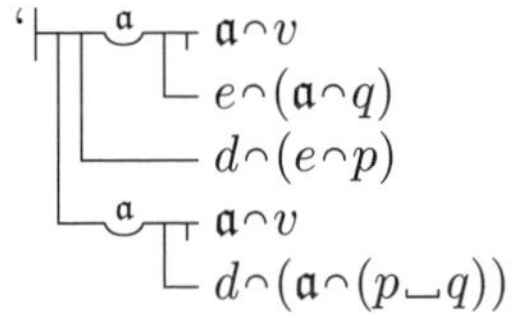

(ε

which follows by rule (5) from

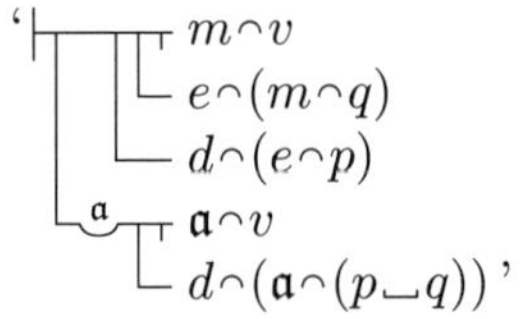

(ζ

This proposition is easily proven by means of (IIa) and (5). It thus turns on deriving the proposition (β) from (Δ). This is accomplished by means of

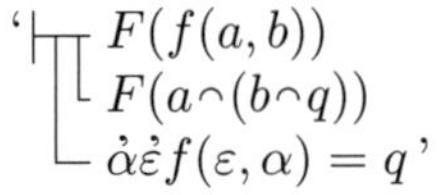

(η

which follows from (3).

§57. *Construction*

3

(IIIa):

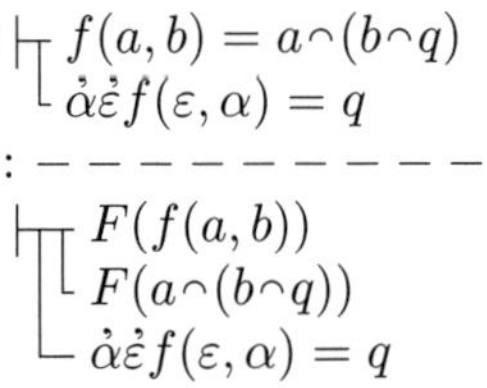

(6

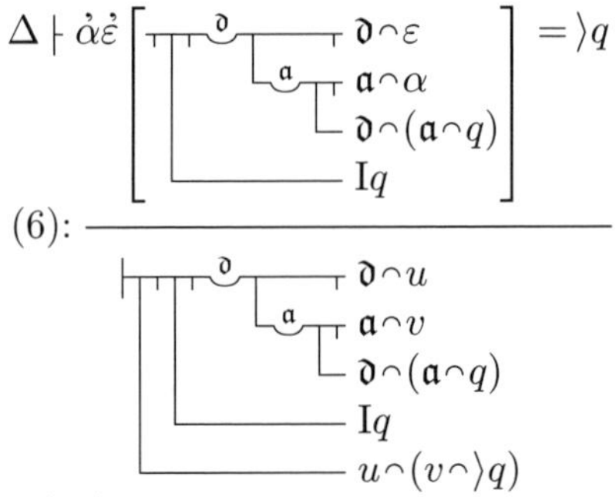

(7

(Ib):

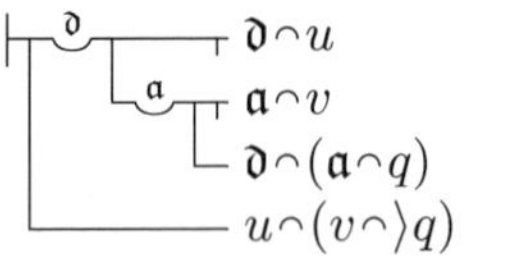

(α

(IIa):

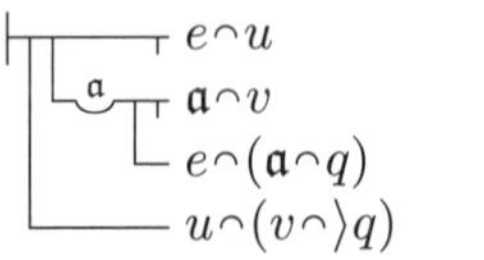

(8

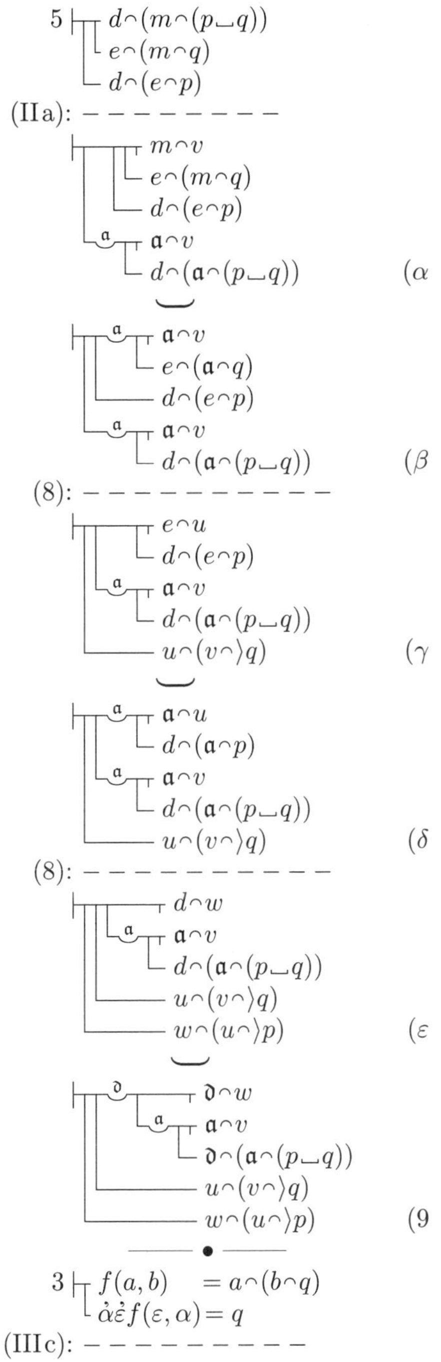

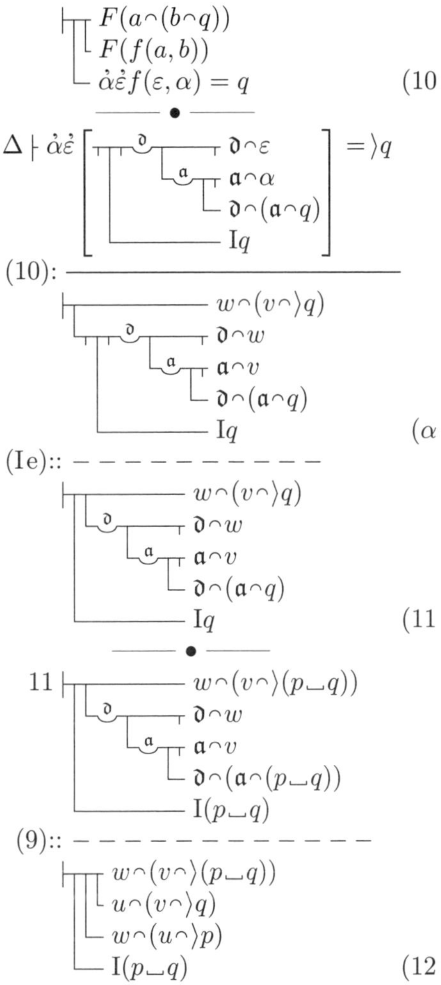

§58. *Analysis*

We now have to prove the proposition (§54, o)

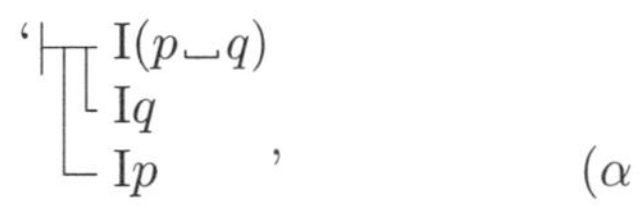

i.e.: 'the relation that is composed from the *p*-relation and the *q*-relation is single-valued if both the *p*-relation and the

q-relation are single-valued.' By definition (Γ), what has to be proven is

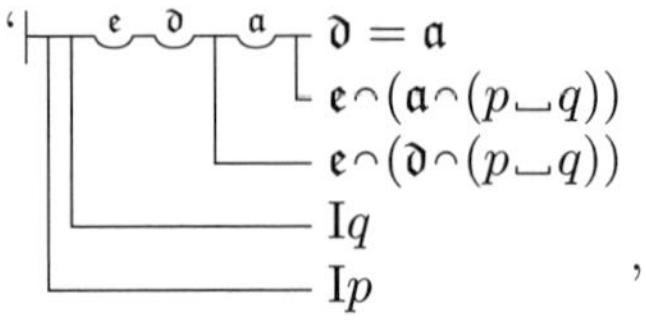
𝔡 = 𝔞
𝔢⌒(𝔞⌒(p⌣q))
𝔢⌒(𝔡⌒(p⌣q))
Iq
Ip

, $(\beta$

which is obtained by rule (5) from

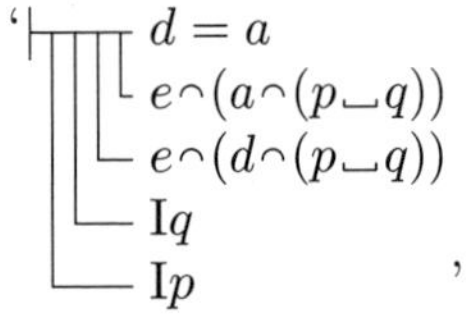
$d = a$
e⌒(a⌒(p⌣q))
e⌒(d⌒(p⌣q))
Iq
Ip

, $(\gamma$

From definition (B) it is easy to deduce

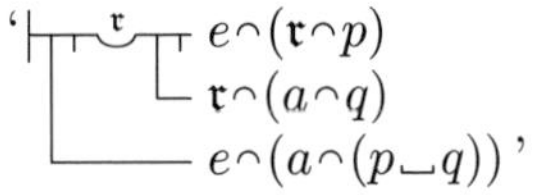
e⌒(𝔯⌒p)
𝔯⌒(a⌒q)
e⌒(a⌒(p⌣q))

' $(\delta$

or

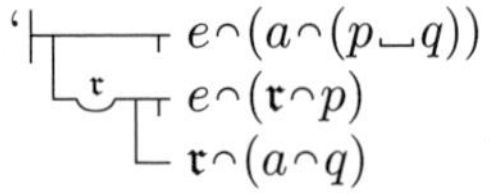
e⌒(a⌒(p⌣q))
e⌒(𝔯⌒p)
𝔯⌒(a⌒q)

, $(\varepsilon$

With the latter, one easily passes from

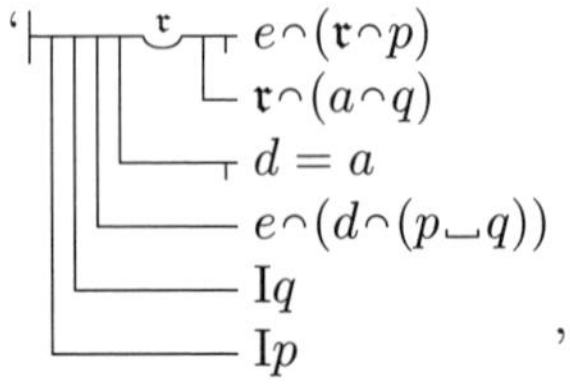
e⌒(𝔯⌒p)
𝔯⌒(a⌒q)
$d = a$
e⌒(d⌒(p⌣q))
Iq
Ip

, $(\zeta$

to the proposition

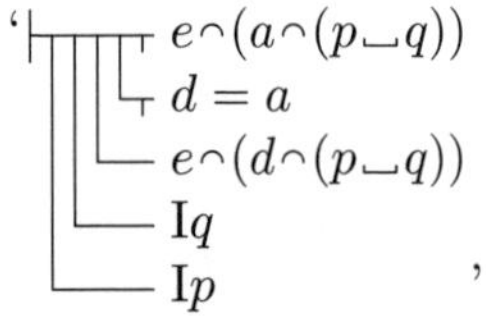
e⌒(a⌒(p⌣q))
$d = a$
e⌒(d⌒(p⌣q))
Iq
Ip

, $(\eta$

from which (γ) follows by contraposition. The proposition (ζ) is obtained by rule (5) from

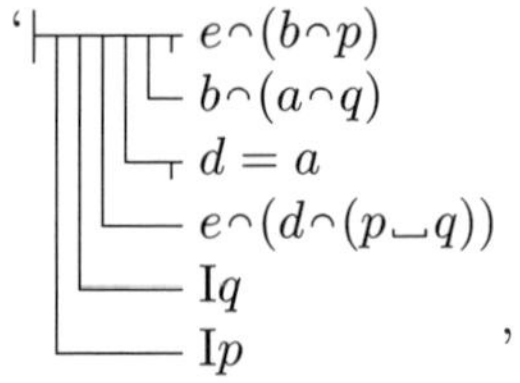
e⌒(b⌒p)
b⌒(a⌒q)
$d = a$
e⌒(d⌒(p⌣q))
Iq
Ip

, $(\vartheta$

This proposition is to be derived by means of

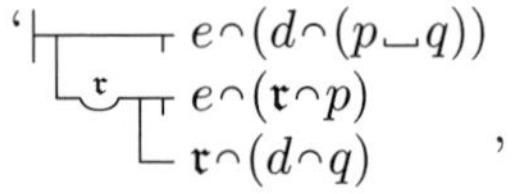
e⌒(d⌒(p⌣q))
e⌒(𝔯⌒p)
𝔯⌒(d⌒q)

, $(\varepsilon$

from

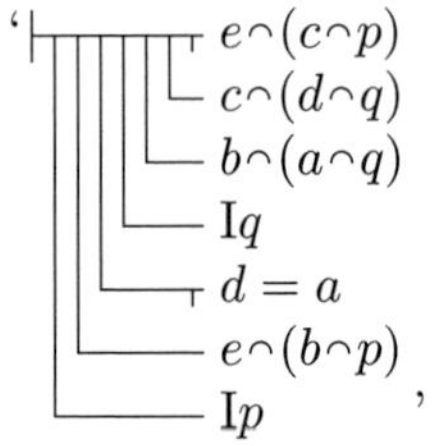
e⌒(c⌒p)
c⌒(d⌒q)
b⌒(a⌒q)
Iq
$d = a$
e⌒(b⌒p)
Ip

, $(\iota$

in a manner similar to how (η) is derived from (ϑ). Here 'c' will need to be replaced by '$\mathfrak{r}$'. Therefore the letters 'b' and 'c' must be distinct; otherwise, by rule (5), '$\mathfrak{r}$' would be introduced not only at places now occupied by 'c', but also at those occupied by 'b'. Now,

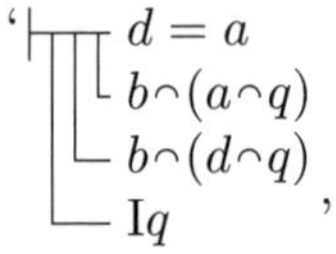
$d = a$
b⌒(a⌒q)
b⌒(d⌒q)
Iq

, $(\kappa$

follows from our definition (Γ), and from this by means of (IIIc)

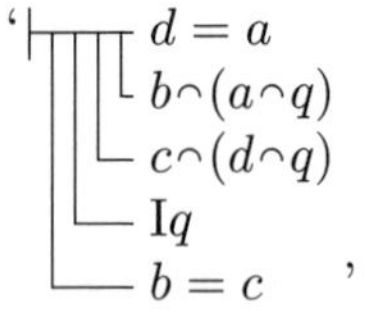
$d = a$
b⌒(a⌒q)
c⌒(d⌒q)
Iq
$b = c$

, $(\lambda$

If the proposition (κ) is applied to this in the form

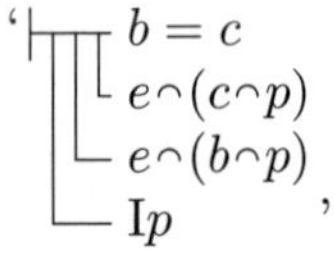
$b = c$
e⌒(c⌒p)
e⌒(b⌒p)
Ip

, $(\kappa$

then the proposition (ι) is proven, on the basis of which we can reach our proposition (α), as we saw.

§59. Construction

$$\Gamma \vdash \left(\vphantom{\begin{array}{c}a\\a\\a\end{array}} \begin{array}{l} \mathfrak{d} = \mathfrak{a} \\ \mathfrak{e}\frown(\mathfrak{a}\frown q) \\ \mathfrak{e}\frown(\mathfrak{d}\frown q) \end{array} \right) = Iq$$

(IIIa): ───────────────

$$\begin{array}{l} \mathfrak{d} = \mathfrak{a} \\ \mathfrak{e}\frown(\mathfrak{a}\frown q) \\ \mathfrak{e}\frown(\mathfrak{d}\frown q) \\ Iq \end{array} \qquad (\alpha$$

(IIa): ─ ─ ─ ─ ─ ─ ─

$$\begin{array}{l} \mathfrak{d} = \mathfrak{a} \\ b\frown(\mathfrak{a}\frown q) \\ b\frown(\mathfrak{d}\frown q) \\ Iq \end{array} \qquad (\beta$$

(IIa): ─ ─ ─ ─ ─ ─ ─

$$\begin{array}{l} d = a \\ b\frown(\mathfrak{a}\frown q) \\ b\frown(d\frown q) \\ Iq \end{array} \qquad (\gamma$$

(IIa): ─ ─ ─ ─ ─ ─ ─

$$\begin{array}{l} d = a \\ b\frown(a\frown q) \\ b\frown(d\frown q) \\ Iq \end{array} \qquad (13$$

───●───

$$6 \; \begin{array}{l} f(a,b) \\ a\frown(b\frown q) \\ \dot\alpha\dot\varepsilon(\mathbin{\dot{\mathbin{-}}} f(\varepsilon,\alpha)) = q \end{array}$$

×

$$\begin{array}{l} a\frown(b\frown q) \\ f(a,b) \\ \dot\alpha\dot\varepsilon(\mathbin{\dot{\mathbin{-}}} f(\varepsilon,\alpha)) = q \end{array} \qquad (14$$

───●───

$$B \vdash \dot\alpha\dot\varepsilon\left(\mathfrak{r} \begin{array}{l} \varepsilon\frown(\mathfrak{r}\frown p) \\ \mathfrak{r}\frown(\alpha\frown q) \end{array} \right) = p\smile q$$

(14): ───────────────

$$\begin{array}{l} e\frown(d\frown(p\smile q)) \\ e\frown(\mathfrak{r}\frown p) \\ \mathfrak{r}\frown(d\frown q) \end{array} \qquad (15$$

───●───

$$\Gamma \vdash \left(\vphantom{\begin{array}{c}a\\a\\a\end{array}} \begin{array}{l} \mathfrak{d} = \mathfrak{a} \\ \mathfrak{e}\frown(\mathfrak{a}\frown q) \\ \mathfrak{e}\frown(\mathfrak{d}\frown q) \end{array} \right) = Iq$$

(IIIc): ───────────────

$$\begin{array}{l} \mathfrak{d} = \mathfrak{a} \\ \mathfrak{e}\frown(\mathfrak{a}\frown q) \\ \mathfrak{e}\frown(\mathfrak{d}\frown q) \\ Iq \end{array} \qquad (16$$

───●───

$$13 \; \begin{array}{l} d = a \\ b\frown(a\frown q) \\ b\frown(d\frown q) \\ Iq \end{array}$$

(IIIc): ───────────

$$\begin{array}{l} d = a \\ b\frown(a\frown q) \\ c\frown(d\frown q) \\ Iq \\ b = c \end{array} \qquad (\alpha$$

(13):: ─ ─ ─ ─ ─ ─

$$\begin{array}{l} d = a \\ b\frown(a\frown q) \\ c\frown(d\frown q) \\ Iq \\ e\frown(c\frown p) \\ e\frown(b\frown p) \\ Ip \end{array} \qquad (\beta$$

×

$$\begin{array}{l} e\frown(c\frown p) \\ c\frown(d\frown q) \\ b\frown(a\frown q) \\ Iq \\ d = a \\ e\frown(b\frown p) \\ Ip \end{array} \qquad (\gamma$$

$$\begin{array}{l} e\frown(\mathfrak{r}\frown p) \\ \mathfrak{r}\frown(d\frown q) \\ b\frown(a\frown q) \\ Iq \\ d = a \\ e\frown(b\frown p) \\ Ip \end{array} \qquad (\delta$$

(15): ─ ─ ─ ─ ─ ─ ─ ─

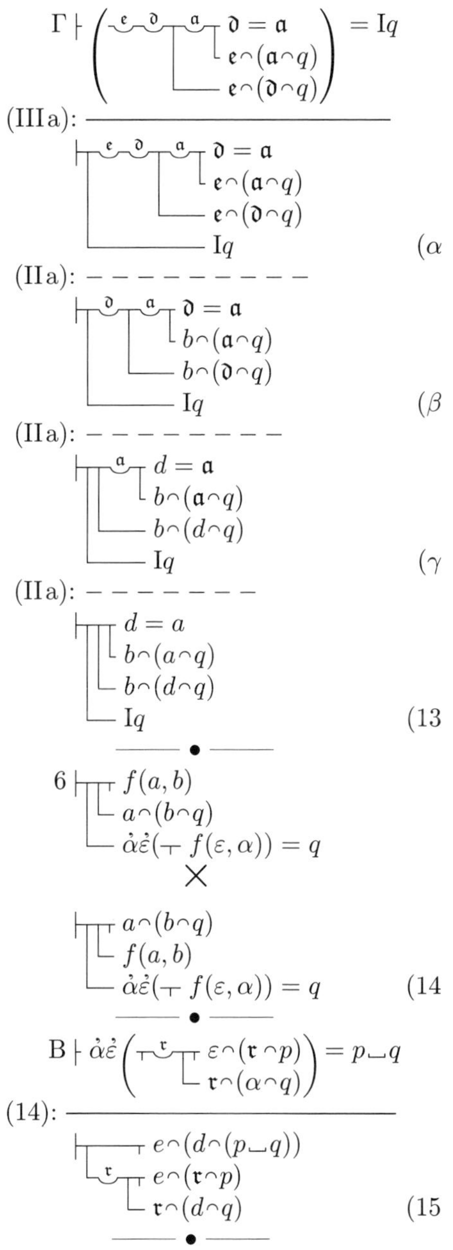

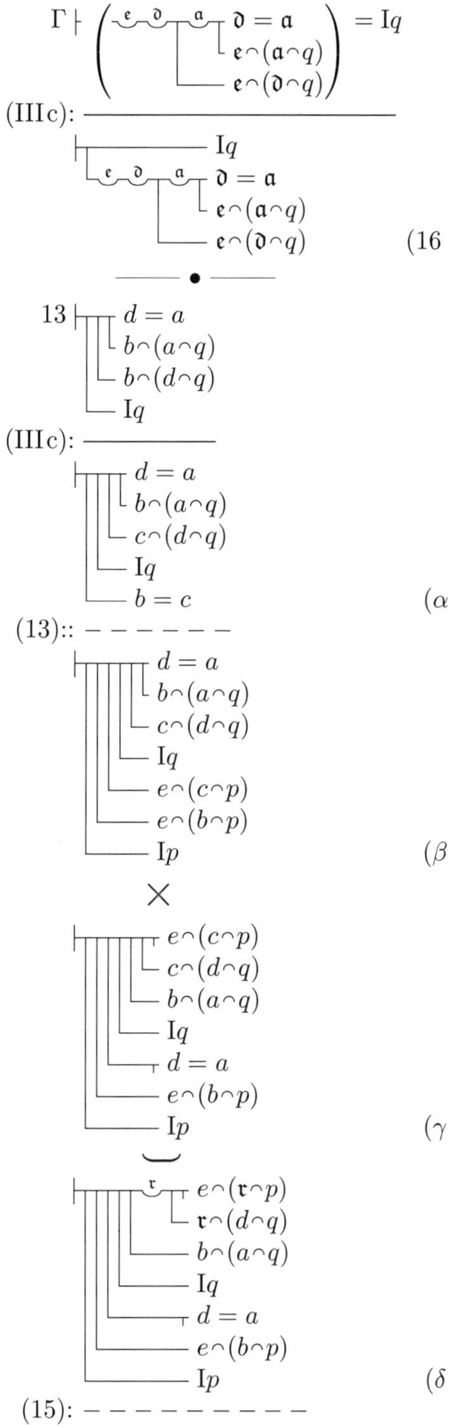

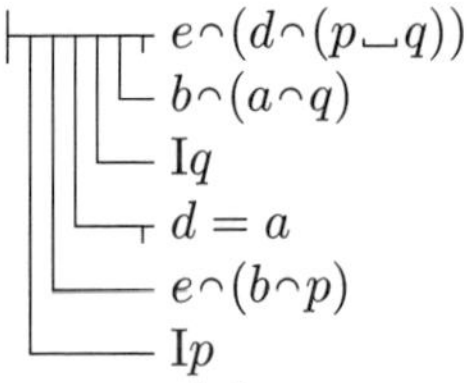

(ε

×

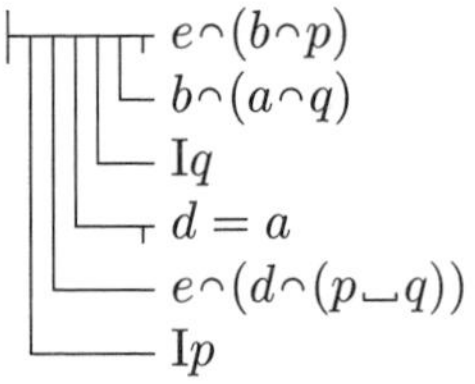

(ζ

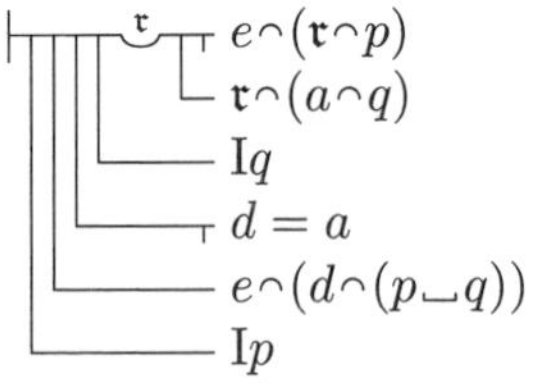

(η

(15): – – – – – – – – – – –

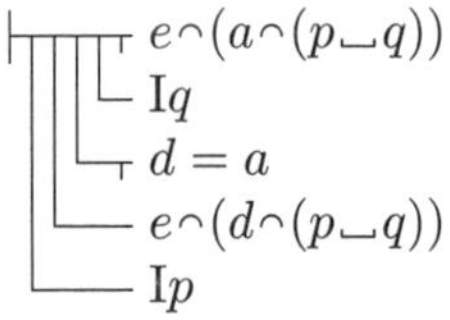

(ϑ

×

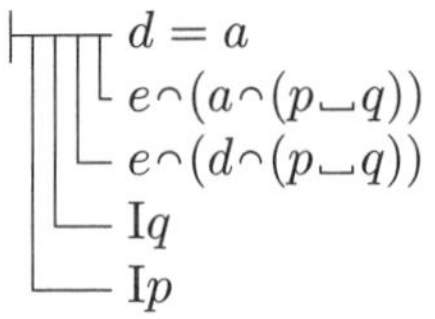

(ι

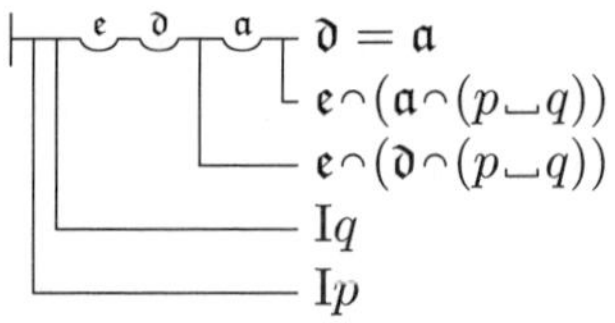

(κ

(16): – – – – – – – – – – – –

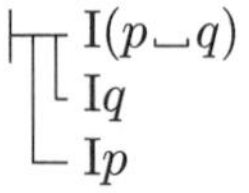

(17

——— • ———

7

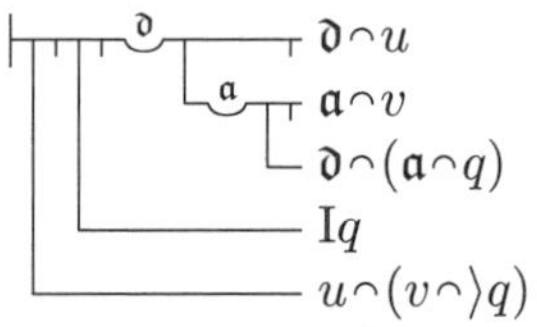

(Id): – – – – – – – – – – –

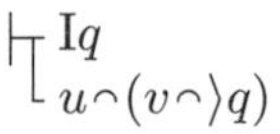

(18

(17): – – – – –

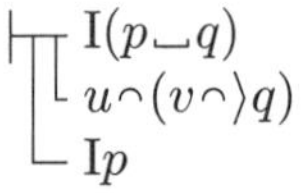

(α

(18):: – – – – – –

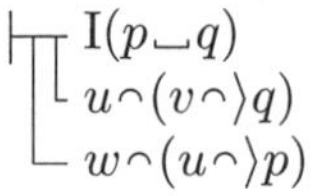

(β

(12): – – – – – –

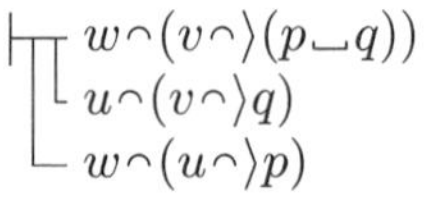

(19

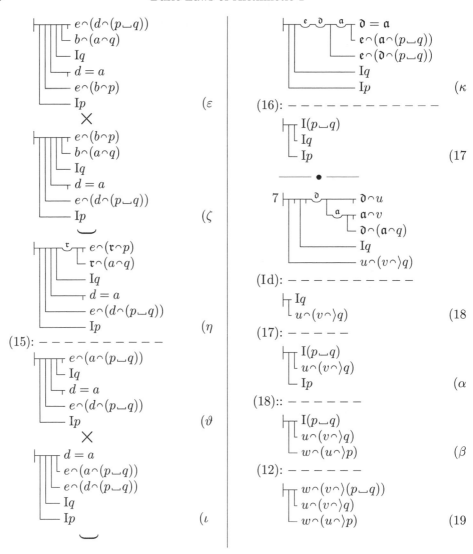

b) Proof of the proposition

‘

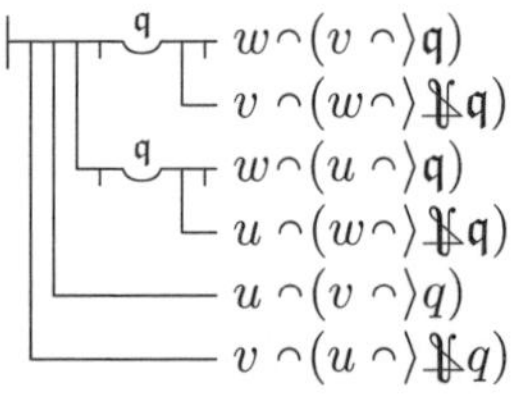

’

and end of section A

§60. *Analysis*

We now have to prove the proposition (§54, ν)

$$\text{‘}\vdash \natural(p\smile q) = \natural q \smile \natural p\text{’} \qquad (\alpha$$

According to definitions (E) and (B), this comes down to deriving the proposition

$$\text{‘}\vdash \grave{\alpha}\grave{\varepsilon}(\alpha\frown(\varepsilon\frown(p\smile q))) = \grave{\alpha}\grave{\varepsilon}\left(\begin{array}{l}\overset{\mathfrak{r}}{\smile}\ \varepsilon\frown(\mathfrak{r}\frown\natural q)\\ \quad \mathfrak{r}\frown(\alpha\frown\natural p)\end{array}\right)\text{’} \qquad (\beta$$

For this we may draw upon the proposition

$$\text{‘}\vdash\begin{array}{l}\grave{\alpha}\grave{\varepsilon}f(\varepsilon,\alpha) = \grave{\alpha}\grave{\varepsilon}g(\varepsilon,\alpha)\\ \underset{\mathfrak{d}\ \mathfrak{a}}{\smile\smile}\ f(\mathfrak{a},\mathfrak{d}) = g(\mathfrak{a},\mathfrak{d})\end{array}\text{’} \qquad (\gamma$$

which we will also use on other occasions and which can be proven by a double application of the proposition (V a). In order to apply this proposition here, we have to obtain the proposition

$$\text{‘}\vdash b\frown(a\frown(p\smile q)) = \left(\begin{array}{l}\overset{\mathfrak{r}}{\smile}\ a\frown(\mathfrak{r}\frown\natural q)\\ \quad \mathfrak{r}\frown(b\frown\natural p)\end{array}\right)\text{’} \qquad (\delta$$

which follows by (4) from

$$\text{‘}\vdash\left(\begin{array}{l}\overset{\mathfrak{r}}{\smile}\ b\frown(\mathfrak{r}\frown p)\\ \quad \mathfrak{r}\frown(a\frown q)\end{array}\right) = \left(\begin{array}{l}\overset{\mathfrak{r}}{\smile}\ a\frown(\mathfrak{r}\frown\natural q)\\ \quad \mathfrak{r}\frown(b\frown\natural p)\end{array}\right)\text{’} \qquad (\varepsilon$$

(ε) is to be proven by means of (IV a). For this we need the propositions

$$\text{‘}\vdash\begin{array}{l}\overset{\mathfrak{r}}{\smile}\ \begin{array}{l} b\frown(\mathfrak{r}\frown p)\\ \mathfrak{r}\frown(a\frown q)\end{array}\\ \lnot\overset{\mathfrak{r}}{\smile}\ \begin{array}{l} a\frown(\mathfrak{r}\frown\natural q)\\ \mathfrak{r}\frown(b\frown\natural p)\end{array}\end{array}\text{’} \qquad (\zeta$$

and

$$\text{‘}\vdash\begin{array}{l}\overset{\mathfrak{r}}{\smile}\ \begin{array}{l} a\frown(\mathfrak{r}\frown\natural q)\\ \mathfrak{r}\frown(b\frown\natural p)\end{array}\\ \lnot\overset{\mathfrak{r}}{\smile}\ \begin{array}{l} b\frown(\mathfrak{r}\frown p)\\ \mathfrak{r}\frown(a\frown q)\end{array}\end{array}\text{’} \qquad (\eta$$

We derive them from the proposition

$$\text{‘}\vdash r\frown(a\frown q) = a\frown(r\frown\natural q)\text{’} \qquad (\vartheta$$

that straightforwardly follows from (E). The proposition (α), thus proven, will be deployed as indicated in §54 in the proof of the proposition (§54, μ)

$$\text{‘}\vdash\begin{array}{l} v\frown(w\frown)\natural(p\smile q))\\ u\frown(w\frown)\natural p)\\ v\frown(u\frown)\natural q)\end{array}\text{’} \qquad (\iota$$

and from this and (19) we derive the proposition (§54, ε).

§61. *Construction*

$$\text{V a}\ \vdash\begin{array}{l}\grave{\varepsilon}f(\varepsilon,d) = \grave{\varepsilon}g(\varepsilon,d)\\ \underset{\mathfrak{a}}{\smile}\ f(\mathfrak{a},d) = g(\mathfrak{a},d)\end{array}$$

$$\text{(II a)::}\ \text{- - - - - - - - - - - -}$$

$$\vdash\begin{array}{l}\grave{\varepsilon}f(\varepsilon,d) = \grave{\varepsilon}g(\varepsilon,d)\\ \underset{\mathfrak{d}\ \mathfrak{a}}{\smile\smile}\ f(\mathfrak{a},\mathfrak{d}) = g(\mathfrak{a},\mathfrak{d})\end{array} \qquad (\alpha$$

$$\vdash\begin{array}{l}\underset{\mathfrak{a}}{\smile}\ \grave{\varepsilon}f(\varepsilon,\mathfrak{a}) = \grave{\varepsilon}g(\varepsilon,\mathfrak{a})\\ \underset{\mathfrak{d}\ \mathfrak{a}}{\smile\smile}\ f(\mathfrak{a},\mathfrak{d}) = g(\mathfrak{a},\mathfrak{d})\end{array} \qquad (\beta$$

$$\text{(V a):}\ \text{- - - - - - - - - - - -}$$

$$\vdash\begin{array}{l}\grave{\alpha}\grave{\varepsilon}f(\varepsilon,\alpha) = \grave{\alpha}\grave{\varepsilon}g(\varepsilon,\alpha)\\ \underset{\mathfrak{d}\ \mathfrak{a}}{\smile\smile}\ f(\mathfrak{a},\mathfrak{d}) = g(\mathfrak{a},\mathfrak{d})\end{array} \qquad (20$$

—— • ——

$$\text{E}\vdash\grave{\alpha}\grave{\varepsilon}(\alpha\frown(\varepsilon\frown q)) = \natural q$$

$$\frac{}{\vdash r\frown(a\frown q) = a\frown(r\frown\natural q)}\ \text{(3)} \qquad (21$$

$$\frac{}{\vdash\begin{array}{l}F(a\frown(r\frown\natural q))\\ F(r\frown(a\frown q))\end{array}}\ \text{(III c)} \qquad (22$$

—— • ——

$$21\vdash r\frown(a\frown q) = a\frown(r\frown\natural q)$$

$$\frac{}{\vdash\begin{array}{l}F(r\frown(a\frown q))\\ F(a\frown(r\frown\natural q))\end{array}}\ \text{(III a)} \qquad (23$$

—— • ——

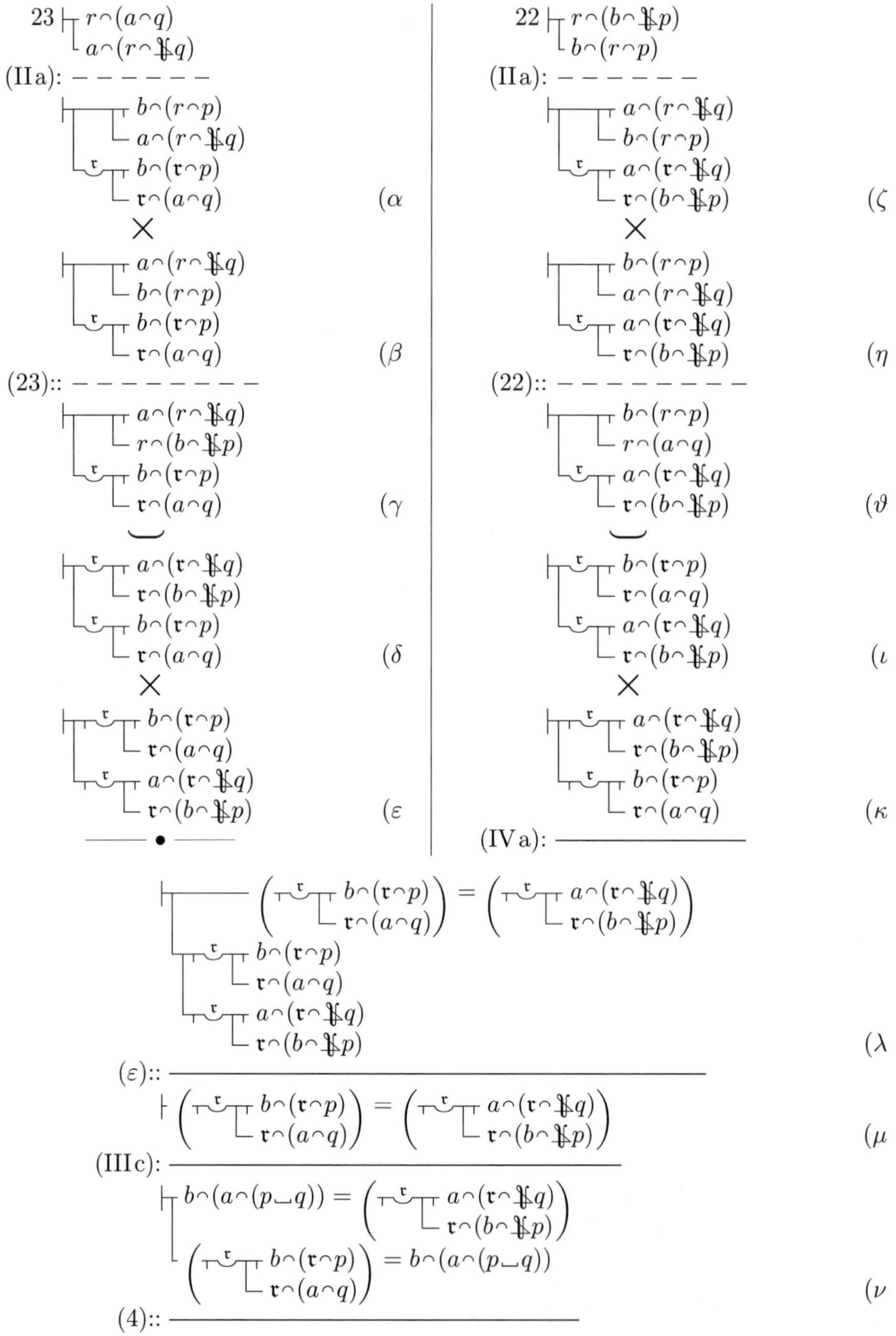
23
(IIa):
(α
(β
(23)::
(γ
(δ
(ε
22
(IIa):
(ζ
(η
(22)::
(ϑ
(ι
(κ
(IVa):
(λ
(ε)::
(μ
(IIIc):
(ν
(4)::

$$\vdash b\smallfrown(a\smallfrown(p\llcorner q)) = \left(\overset{\mathfrak{r}}{\smile}\!\!\top\!\!\begin{array}{l} a\smallfrown(\mathfrak{r}\smallfrown\natural q) \\ \mathfrak{r}\smallfrown(b\smallfrown\natural p)\end{array}\right) \qquad (\xi$$

$$\vdash\overset{\mathfrak{d}}{\smile}\overset{\mathfrak{a}}{\smile}\, \mathfrak{d}\smallfrown(\mathfrak{a}\smallfrown(p\llcorner q)) = \left(\overset{\mathfrak{r}}{\smile}\!\!\top\!\!\begin{array}{l} \mathfrak{a}\smallfrown(\mathfrak{r}\smallfrown\natural q) \\ \mathfrak{r}\smallfrown(\mathfrak{d}\smallfrown\natural p)\end{array}\right) \qquad (o$$

(20): ———————————————

$$\vdash \dot{\alpha}\dot{\varepsilon}(\alpha\smallfrown(\varepsilon\smallfrown(p\llcorner q))) = \dot{\alpha}\dot{\varepsilon}\left(\overset{\mathfrak{r}}{\smile}\!\!\top\!\!\begin{array}{l} \varepsilon\smallfrown(\mathfrak{r}\smallfrown\natural q) \\ \mathfrak{r}\smallfrown(\alpha\smallfrown\natural p)\end{array}\right) \qquad (\pi$$

(IIIc): ———————————————

$$\vdash\!\!\top\begin{array}{l} \dot{\alpha}\dot{\varepsilon}(\alpha\smallfrown(\varepsilon\smallfrown(p\llcorner q))) = \natural q\llcorner\natural p \\ \dot{\alpha}\dot{\varepsilon}\left(\overset{\mathfrak{r}}{\smile}\!\!\top\!\!\begin{array}{l} \varepsilon\smallfrown(\mathfrak{r}\smallfrown\natural q) \\ \mathfrak{r}\smallfrown(\alpha\smallfrown\natural p)\end{array}\right) = \natural q\llcorner\natural p\end{array} \qquad (\rho$$

(B):: ———————————————

$$\vdash \dot{\alpha}\dot{\varepsilon}(\alpha\smallfrown(\varepsilon\smallfrown(p\llcorner q))) = \natural q\llcorner\natural p \qquad (\sigma$$

(IIIc): ———————————————

$$\vdash\!\!\top\begin{array}{l} \natural(p\llcorner q) = \natural q\llcorner\natural p \\ \dot{\alpha}\dot{\varepsilon}(\alpha\smallfrown(\varepsilon\smallfrown(p\llcorner q))) = \natural(p\llcorner q)\end{array} \qquad (\tau$$

(E):: ———————————————

$$\vdash \natural(p\llcorner q) = \natural q\llcorner\natural p \qquad (24$$

(IIIa): ———————————————

$$\vdash\!\!\top\begin{array}{l} v\smallfrown(w\smallfrown\rangle\natural(p\llcorner q)) \\ v\smallfrown(w\smallfrown\rangle(\natural q\llcorner\natural p))\end{array} \qquad (\alpha$$

(19):: — — — — — — — — — —

$$\vdash\!\!\top\begin{array}{l} v\smallfrown(w\smallfrown\rangle\natural(p\llcorner q)) \\ u\smallfrown(w\smallfrown\rangle\natural p) \\ v\smallfrown(u\smallfrown\rangle\natural q)\end{array} \qquad (\beta$$

(IIa): — — — — — — — — —

$$\vdash\!\!\top\begin{array}{l} w\smallfrown(v\smallfrown\rangle(p\llcorner q)) \\ u\smallfrown(w\smallfrown\rangle\natural p) \\ v\smallfrown(u\smallfrown\rangle\natural q) \\ \overset{\mathfrak{q}}{\smile}\!\!\top\!\begin{array}{l} w\smallfrown(v\smallfrown\rangle\mathfrak{q}) \\ v\smallfrown(w\smallfrown\rangle\natural\mathfrak{q})\end{array}\end{array} \qquad (\gamma$$

——— • ———

$$19 \vdash\!\!\top\begin{array}{l} w\smallfrown(v\smallfrown\rangle(p\llcorner q)) \\ u\smallfrown(v\smallfrown\rangle q) \\ w\smallfrown(u\smallfrown\rangle p)\end{array}$$

×

$$\vdash\!\!\top\begin{array}{l} w\smallfrown(u\smallfrown\rangle p) \\ u\smallfrown(v\smallfrown\rangle q) \\ \top w\smallfrown(v\smallfrown\rangle(p\llcorner q))\end{array} \qquad (\delta$$

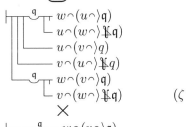

(γ):: — — — — — — — — —

$$\vdash\!\!\top\begin{array}{l} w\smallfrown(u\smallfrown\rangle p) \\ u\smallfrown(w\smallfrown\rangle\natural p) \\ u\smallfrown(v\smallfrown\rangle q) \\ v\smallfrown(u\smallfrown\rangle\natural q) \\ \overset{\mathfrak{q}}{\smile}\!\!\top\!\begin{array}{l} w\smallfrown(v\smallfrown\rangle\mathfrak{q}) \\ v\smallfrown(w\smallfrown\rangle\natural\mathfrak{q})\end{array}\end{array} \qquad (\varepsilon$$

$$\vdash\!\!\top\begin{array}{l} \overset{\mathfrak{q}}{\smile}\!\!\top\!\begin{array}{l} w\smallfrown(u\smallfrown\rangle\mathfrak{q}) \\ u\smallfrown(w\smallfrown\rangle\natural\mathfrak{q})\end{array} \\ u\smallfrown(v\smallfrown\rangle q) \\ v\smallfrown(u\smallfrown\rangle\natural q) \\ \overset{\mathfrak{q}}{\smile}\!\!\top\!\begin{array}{l} w\smallfrown(v\smallfrown\rangle\mathfrak{q}) \\ v\smallfrown(w\smallfrown\rangle\natural\mathfrak{q})\end{array}\end{array} \qquad (\zeta$$

×

$$\vdash\!\!\top\begin{array}{l} \top\overset{\mathfrak{q}}{\smile}\!\!\top\!\begin{array}{l} w\smallfrown(v\smallfrown\rangle\mathfrak{q}) \\ v\smallfrown(w\smallfrown\rangle\natural\mathfrak{q})\end{array} \\ \top\overset{\mathfrak{q}}{\smile}\!\!\top\!\begin{array}{l} w\smallfrown(u\smallfrown\rangle\mathfrak{q}) \\ u\smallfrown(w\smallfrown\rangle\natural\mathfrak{q})\end{array} \\ u\smallfrown(v\smallfrown\rangle q) \\ v\smallfrown(u\smallfrown\rangle\natural q)\end{array} \qquad (25$$

§62. *Analysis*

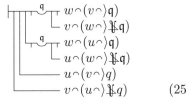

In order to derive the proposition (§54, δ) from (25), the proposition (§54, η) is needed

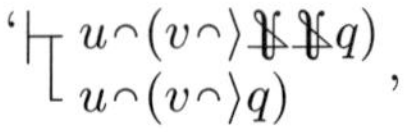

(α

For this, according to (11), we have to prove the propositions

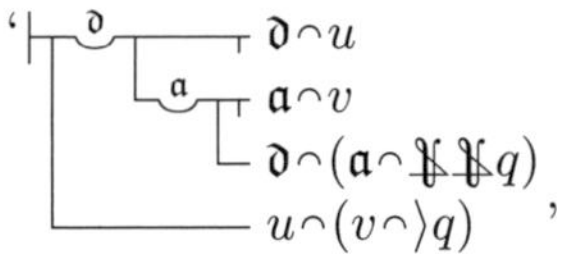

(β

and

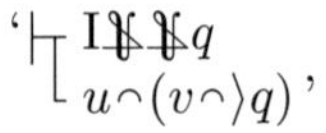

(γ

(β) is obtained by rule (5) from

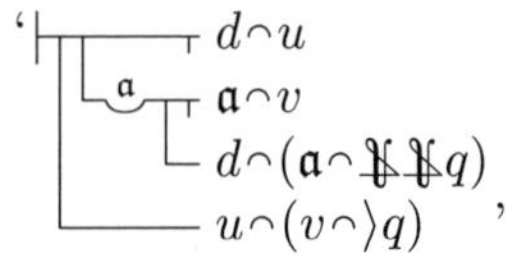

(δ

By (8) we now have

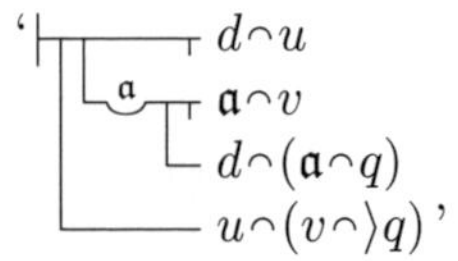

(ε

What thus remains to be proven is

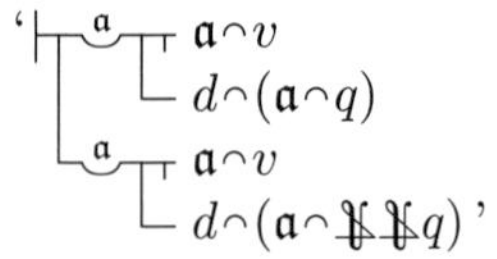

(ζ

which, by rule (5), follows from

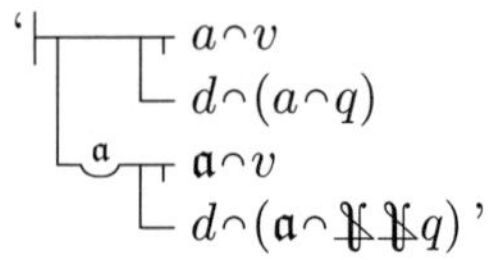

(η

If we write (IIa) in this way

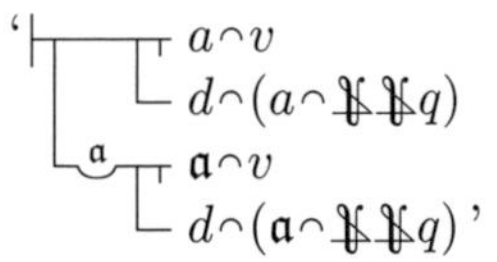

(ϑ

then we can see that the proposition

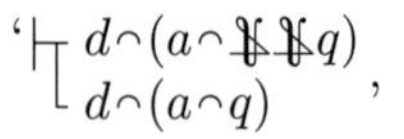

(ι

is to be derived, which is easily done by (22).

§63. *Construction*

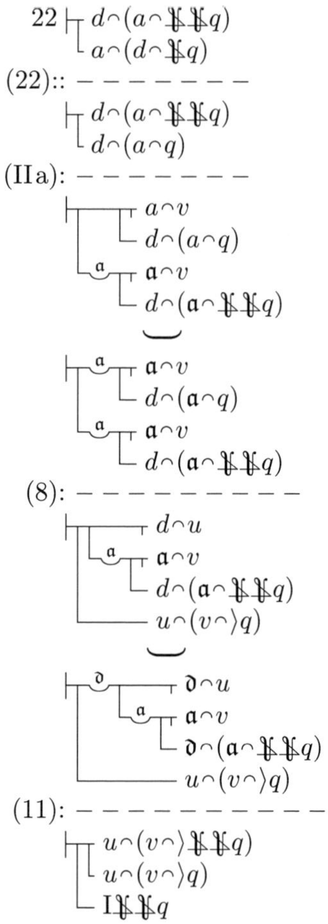

(26, (α, (β, (γ, (δ, (27

§64. *Analysis*

We are still lacking the proposition (§62, γ). We first prove

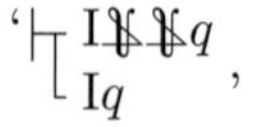

(α

from which the former follows together with (18). According to (16) we have to derive

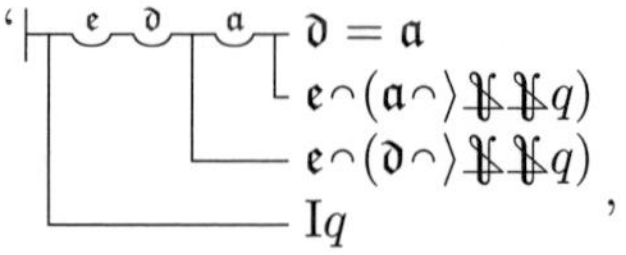

(β

or

‘⊢ $d = a$
$e⌢(a⌢‡‡q)$
$e⌢(d⌢‡‡q)$
Iq ’ (γ

By (13) we now have

‘⊢ $d = a$
$e⌢(a⌢q)$
$e⌢(d⌢q)$
Iq ’ (δ

(γ) follows from this and the proposition

‘⊢ $e⌢(a⌢q)$
$e⌢(a⌢‡‡q)$ ’ (ε

which follows from (23) in a manner similar to that in which (26) follows from (22). After we have thus proven proposition (§62, γ), we will employ it to eliminate the subcomponent ‘I‡‡q’ in (27) by fusion of subcomponents. Thus we arrive at the goal of our section A, as stated in §54.

§65. *Construction*

23 ⊢ $a⌢(e⌢‡q)$
$e⌢(a⌢‡‡q)$

(23): – – – – – – –

⊢ $e⌢(a⌢q)$
$e⌢(a⌢‡‡q)$ (28

(13): – – – – – –

⊢ $d = a$
$e⌢(a⌢‡‡q)$
$e⌢(d⌢q)$
Iq (α

(28):: – – – – – – – – –

⊢ $d = a$
$e⌢(a⌢‡‡q)$
$e⌢(d⌢‡‡q)$
Iq (β

⊢ ⌣𝔢 ⌣𝔡 ⌣𝔞 𝔡 = 𝔞
𝔢⌢(𝔞⌢‡‡q)
𝔢⌢(𝔡⌢‡‡q)
Iq (γ

(16): – – – – – – – – – – – – –

⊢ $I‡‡q$
Iq (29

(18):: – – – –

⊢ $I‡‡q$
$u⌢(v⌢)q)$ (30

(27): – – – – –

⊢ $u⌢(v⌢)‡‡q)$
$u⌢(v⌢)q)$ (31

(25): – – – – – – – –

⊢ ⌣𝔮 $w⌢(u⌢)$𝔮)
$u⌢(w⌢)‡$𝔮)
⌣𝔮 $w⌢(v⌢)$𝔮)
$v⌢(w⌢)‡$𝔮)
$v⌢(u⌢)‡q)$
$u⌢(v⌢)q)$ (α

(IV a): – – – – – – – – – – – –

⊢ (⌣𝔮 $w⌢(u⌢)$𝔮) / $u⌢(w⌢)‡$𝔮)) = (⌣𝔮 $v⌢(w⌢)$𝔮) / $w⌢(v⌢)‡$𝔮))
⌣𝔮 $w⌢(v⌢)$𝔮)
$v⌢(w⌢)‡$𝔮)
⌣𝔮 $w⌢(u⌢)$𝔮)
$u⌢(w⌢)‡$𝔮)
$v⌢(u⌢)‡q)$
$u⌢(v⌢)q)$ (β

(25):: –

⊢ (⌣𝔮 $w⌢(u⌢)$𝔮) / $u⌢(w⌢)‡$𝔮)) = (⌣𝔮 $w⌢(v⌢)$𝔮) / $v⌢(w⌢)‡$𝔮))
$u⌢(v⌢)q)$
$v⌢(u⌢)‡q)$ (γ

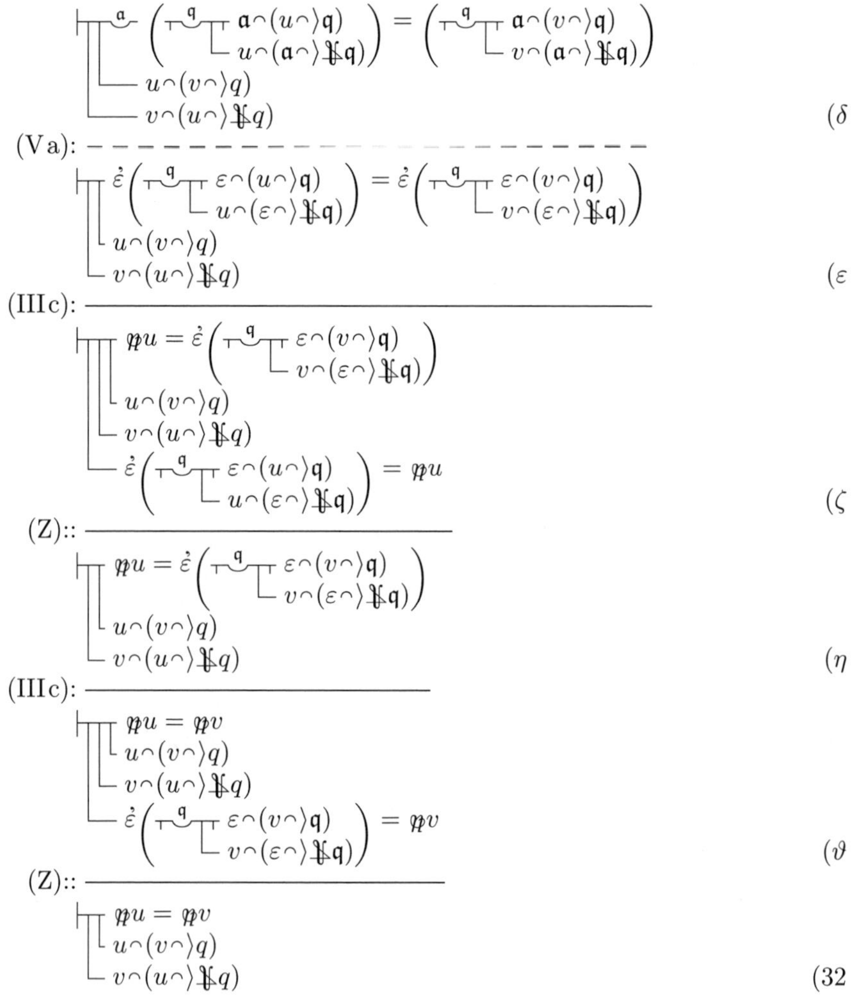

B. Proof of the proposition

⊦ If

a) Proof of the proposition

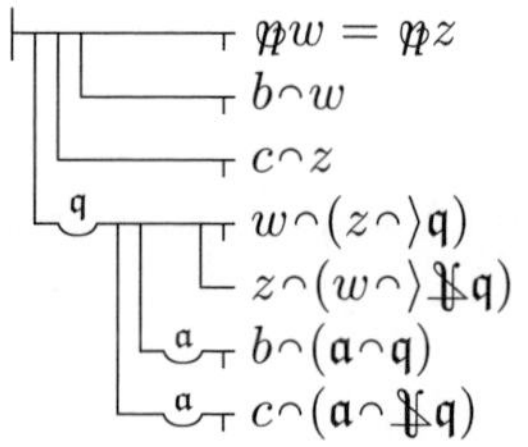

§66. Analysis

In order to prove the proposition that the relation of a cardinal number to the one immediately following is single-valued, or as one can also say, that for every cardinal number there is no more than one which immediately follows it in the cardinal number series,[1] we have to use proposition (16) and so we have to derive

$$\vdash \begin{array}{l} \mathfrak{d} = \mathfrak{a} \\ e \frown (\mathfrak{a} \frown f) \\ e \frown (\mathfrak{d} \frown f) \end{array} \quad (\alpha$$

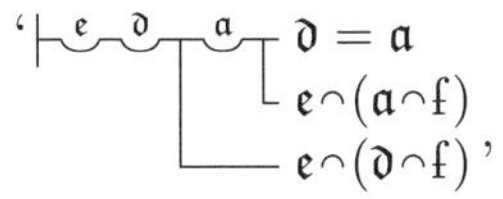

which follows from

$$\vdash \begin{array}{l} d = a \\ e \frown (a \frown f) \\ e \frown (d \frown f) \end{array} \quad (\beta$$

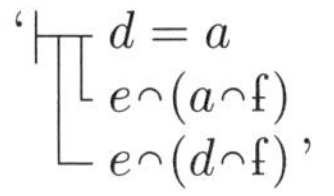

From definition (H) the following is easily derived

$$\vdash \begin{array}{l} \mathfrak{p}u = a \\ \mathfrak{a} \frown u \\ \mathfrak{p}\grave{\varepsilon}\left(\prod_{\varepsilon \frown u} \varepsilon = \mathfrak{a}\right) = e \\ e \frown (\mathfrak{a} \frown f) \end{array} \quad (\gamma$$

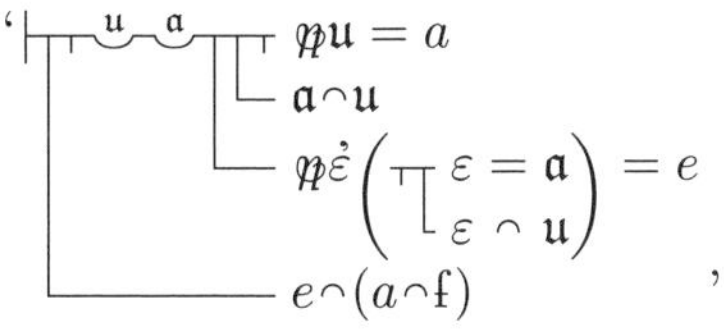

Accordingly, what needs to be proven is

$$\vdash \begin{array}{l} d = a \\ \mathfrak{p}u = a \\ \mathfrak{a} \frown u \\ \mathfrak{p}\grave{\varepsilon}\left(\prod_{\varepsilon \frown u} \varepsilon = \mathfrak{a}\right) = e \\ \mathfrak{p}u = d \\ \mathfrak{a} \frown u \\ \mathfrak{p}\grave{\varepsilon}\left(\prod_{\varepsilon \frown u} \varepsilon = \mathfrak{a}\right) = e' \end{array} \quad (\delta$$

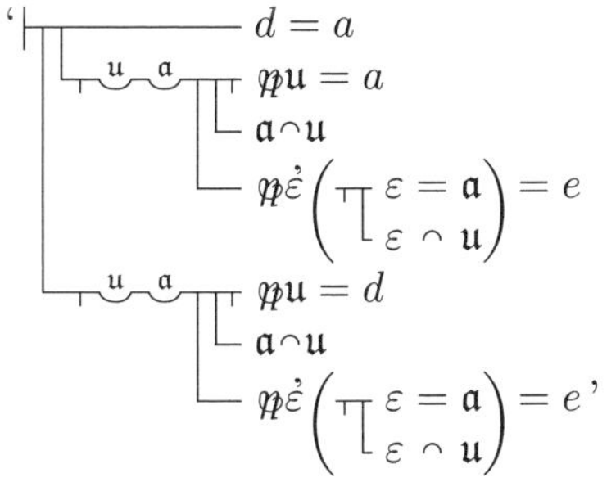

a proposition which, by multiple applications of contraposition and the introduction of German letters, is obtained from

$$\vdash \begin{array}{l} d = a \\ \mathfrak{p}u = a \\ b \frown u \\ \mathfrak{p}\grave{\varepsilon}\left(\prod_{\varepsilon \frown u} \varepsilon = b\right) = e \\ \mathfrak{p}v = d \\ c \frown v \\ \mathfrak{p}\grave{\varepsilon}\left(\prod_{\varepsilon \frown v} \varepsilon = c\right) = e' \end{array} \quad (\varepsilon$$

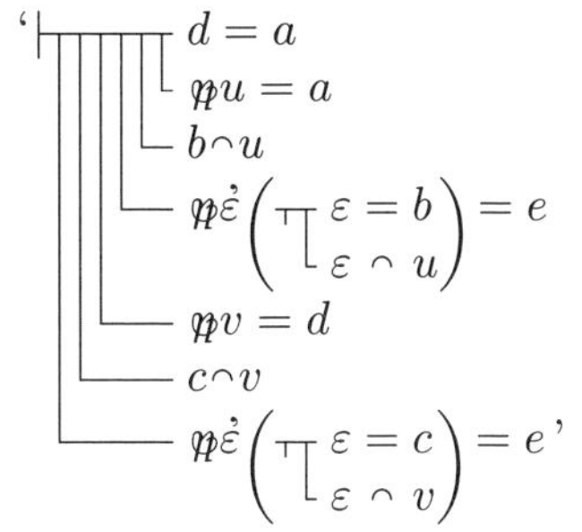

The latter proposition can be derived from

$$\vdash \begin{array}{l} \mathfrak{p}u = \mathfrak{p}v \\ b \frown u \\ \mathfrak{p}\grave{\varepsilon}\left(\prod_{\varepsilon \frown u} \varepsilon = b\right) = \mathfrak{p}\grave{\varepsilon}\left(\prod_{\varepsilon \frown v} \varepsilon = c\right) \\ c \frown v \end{array} \quad (\zeta$$

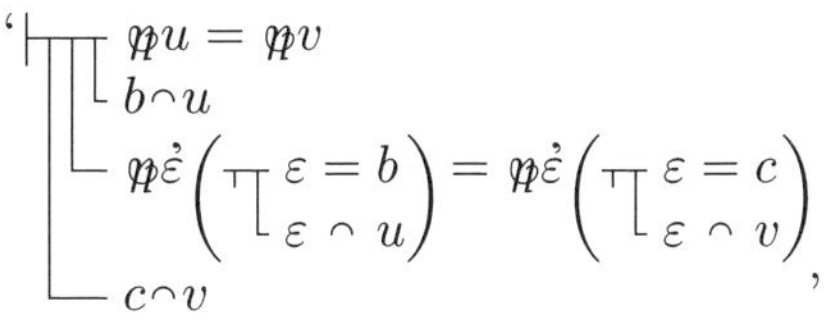

According to the recently proven proposition (32), we now merely need to show that there is a relation which maps the u-concept into the v-concept and whose converse maps the v-concept into the u-concept. That there is a relation which maps the $\grave{\varepsilon}\left(\prod_{\varepsilon \frown u} \varepsilon = b\right)$-concept into the $\grave{\varepsilon}\left(\prod_{\varepsilon \frown v} \varepsilon = c\right)$-concept follows from the equality of the cardinal numbers belonging to these concepts, which of course still has to be proven. Now, the v-concept differs in its extension from the $\grave{\varepsilon}\left(\prod_{\varepsilon \frown v} \varepsilon = c\right)$-concept only in that the object c falls under it but not under the latter; and the u-concept differs in its extension from the $\grave{\varepsilon}\left(\prod_{\varepsilon \frown u} \varepsilon = b\right)$-concept only in that the object b falls under it but not under the latter. From this it must now

---

[1] Compare §43.

be possible to conclude that there is also a relation that maps the u-concept into the v-concept. If one were to follow the usual practice of mathematicians, one might say something like this: we correlate the objects, other than b, falling under the u-concept with the objects, other than c, falling under the v-concept by means of the known relation, and we correlate b with c. In this way, we have mapped the u-concept into the v-concept and, conversely, the latter into the former. So, according to the proposition just proven, the cardinal numbers that belong to them are equal. This is indeed much briefer than the proof to follow which some, misunderstanding my project, will deplore on account of its length. What is it that we are doing when we correlate objects for the purpose of a proof? Seemingly something similar to drawing an auxiliary line in geometry. Euclid, whose method can still often serve as a model of rigour, has his postulates for this purpose, stating that certain lines may be drawn. However, the drawing of a line should no more be regarded as a creation, than the specification of a point of intersection. Rather, in both cases we merely bring to attention, apprehend, what is already there. What is essential to a proof is only that there be such a thing. In proofs, Euclid's postulates thus have the force of axioms that assert that there are certain lines, certain points. Since here we are aiming to reach down to the deepest foundations in every case, we ask on what the possibility of such correlation is based. If one wanted to propose a postulate, in the style of Euclid, it might be phrased like this: 'It is postulated that any object is correlated with any object' or 'It is possible to correlate any object with any object'. This should no more be regarded as a psychological proposition, than should Euclid's postulates be regarded as asserting an ability of our minds; for, so understood, it would indeed be false, since not all objects are known to us and the same ones are not known to everyone. In this way, a subjective element would intrude, which is completely alien to the subject-matter. Correlations also have to be possible of infinitely many with infinitely many objects, but only a few of these infinitely many correlations could actually be carried out if correlating were a creative activity of the mind. Rather, the postulates would have to be understood in this way: 'Any object is correlated with any object' or 'There is a correlation between any object and any object'. What then is such a correlation if it is nothing subjective, created only by our making? However, a particular correlation of an object to an object is not what can be at issue here, and what corresponds to an auxiliary line in geometry; rather we require a genus of correlations, so to speak, something that we have so-far called, and will continue to call, a relation. The desired correlation is thus achieved if we have found[1] a relation in which the object b stands to the object c, and which maps the $\dot{\varepsilon}\left(\begin{array}{l}\top\!\!\top \varepsilon = b\\ \;\llcorner\, \varepsilon \frown u\end{array}\right)$-concept into the $\dot{\varepsilon}\left(\begin{array}{l}\top\!\!\top \varepsilon = c\\ \;\llcorner\, \varepsilon \frown v\end{array}\right)$-concept, and whose converse maps the latter into the former. Here a q-relation which maps the $\dot{\varepsilon}\left(\begin{array}{l}\top\!\!\top \varepsilon = b\\ \;\llcorner\, \varepsilon \frown u\end{array}\right)$-concept into the $\dot{\varepsilon}\left(\begin{array}{l}\top\!\!\top \varepsilon = c\\ \;\llcorner\, \varepsilon \frown v\end{array}\right)$-concept, and whose converse maps the latter into the former, may be presupposed as known. What is not known, however, is whether b stands in this relation to any object, nor whether any object stands in this relation to c. We can now give a relation in which every pair of q-related objects stand, and by which b is related to c. This is the $\dot{\alpha}\dot{\varepsilon}\left(\begin{array}{l}\top\!\!\!-\!\!- \varepsilon \frown (\alpha \frown q)\\ \;\llcorner\!\top\, \varepsilon = b\\ \;\llcorner\, \alpha = c\end{array}\right)$-relation.[2] Although this has the other desired properties, it cannot be said whether it and its converse are single-valued as long as nothing more specific is known of the q-relation. E.g., it might be that b stands in the q-relation to an object d distinct from c. Then, b would stand in the given relation to two objects, namely c and d, and so the relation would not be single-valued, even though the q-relation is single-valued by assumption. In order to avoid this, we will seek a relation which shares the properties of the q-relation that are desirable for our purposes, but in which b stands to no object and in which no object stands to c. The $\dot{\alpha}\dot{\varepsilon}\left(\begin{array}{l}\top\!\!\top\!\!- \varepsilon \frown (\alpha \frown q)\\ \;\llcorner\!\top\, \varepsilon = b\\ \;\;\llcorner\!\top\, \alpha = c\end{array}\right)$-relation is such a relation. If we first abbreviate '$\dot{\varepsilon}\left(\begin{array}{l}\top\!\!\top \varepsilon = b\\ \;\llcorner\, \varepsilon \frown u\end{array}\right)$' by '$w$' and '$\dot{\varepsilon}\left(\begin{array}{l}\top\!\!\top \varepsilon = c\\ \;\llcorner\, \varepsilon \frown v\end{array}\right)$' by '$z$', what we have to prove is the proposition: 'If there is a q-relation which maps the w-concept into the z-concept and whose converse maps the latter into the former, then there is also a relation which does the same, but in which b stands to no object and in which no object stands to c, provided b does not fall under the w-concept and c does not fall under the z-concept.'

For the derivation in concept-script it is more convenient to prove the proposition which results from the latter by contraposition:

---

[1] To pursue the model of geometry, one could say 'constructed'; however, one has to remember that this is not a creation.

[2] Without essential changes, we may write

$$\text{`}\begin{array}{l}\top\!\!\!-\!\!-\!\!- x \frown (y \frown q)\text{'}\\ \;\llcorner\!\top\!\!\top\!\!\top\, x = b\\ \;\;\;\;\llcorner\, y = c\end{array}$$

instead of

$$\text{`}\begin{array}{l}\top\!\!\!-\!\!- x \frown (y \frown q)\text{'.}\\ \;\llcorner\!\top\, x = b\\ \;\llcorner\, y = c\end{array}$$

Compare here what is said about 'or' and 'and' in §12.

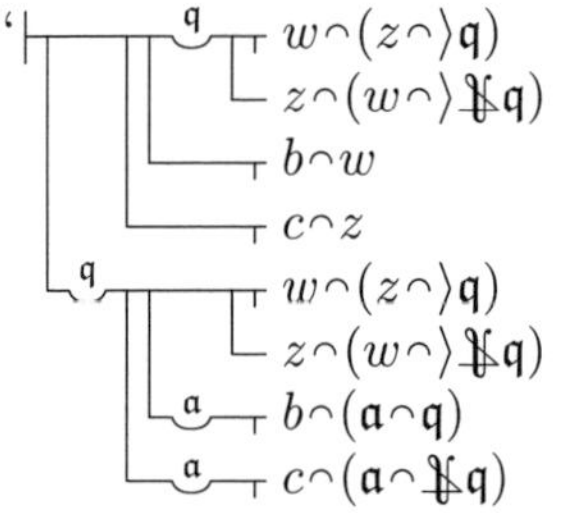

$$\begin{array}{l}w\frown(z\frown)\mathfrak{q}\\z\frown(w\frown)\mathbf{\mathfrak{k}}\mathfrak{q}\\b\frown w\\c\frown z\\w\frown(z\frown)\mathfrak{q}\\z\frown(w\frown)\mathbf{\mathfrak{k}}\mathfrak{q}\\b\frown(a\frown\mathfrak{q})\\c\frown(a\frown\mathbf{\mathfrak{k}}\mathfrak{q})\end{array}\quad(\eta)$$

Next, we will derive the proposition: 'If there is a relation which maps the $\grave{\varepsilon}\!\left(\underset{\varepsilon\frown u}{\prod}\varepsilon=b\right)$-concept into the $\grave{\varepsilon}\!\left(\underset{\varepsilon\frown v}{\prod}\varepsilon=c\right)$-concept and whose con-

verse maps the latter into the former, and which is so constituted that b stands in it to no object and that no object stands in it to c, then there is a relation which maps the u-concept into the v-concept and whose converse maps the latter into the former, provided b falls under the u-concept and c falls under the v-concept.'

The consequent can also be expressed as: 'then the cardinal number of the u-concept is equal to the cardinal number of the v-concept'. Contraposing, the proposition looks like this:

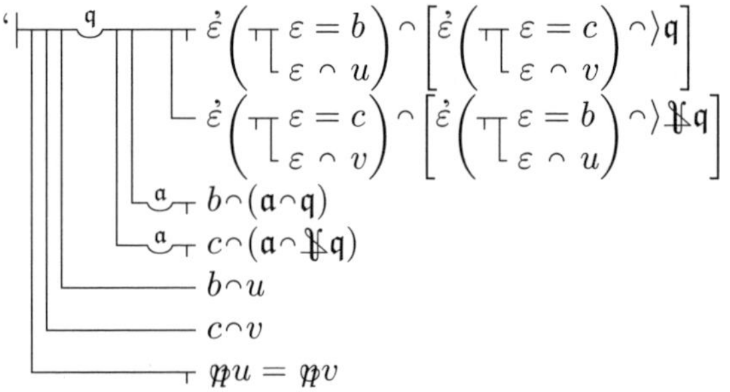

$$\quad(\vartheta)$$

From the propositions (η) and (ϑ) together with the above proposition

$$\begin{array}{l}w\frown(z\frown)\mathfrak{q}\\z\frown(w\frown)\mathbf{\mathfrak{k}}\mathfrak{q}\\\mathfrak{p}w=\mathfrak{p}z\end{array}\quad(\iota)$$

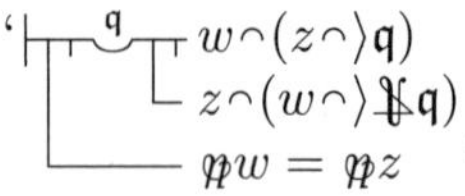

or

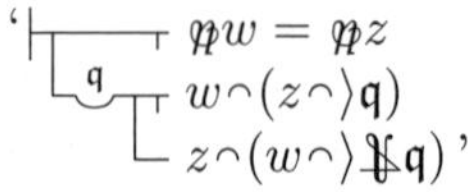

$$\begin{array}{l}\mathfrak{p}w=\mathfrak{p}z\\w\frown(z\frown)\mathfrak{q}\\z\frown(w\frown)\mathbf{\mathfrak{k}}\mathfrak{q}\end{array}\quad(\kappa)$$

our proposition (ζ) follows. The proposition (κ) follows straightforwardly from (Z) and

$$\begin{array}{l}w\frown\mathfrak{p}z\\\mathfrak{p}w=\mathfrak{p}z\end{array}\quad(\lambda)$$

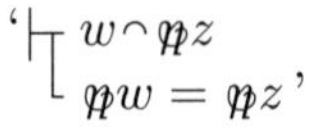

and the latter from (IIIc) in the form

$$\begin{array}{l}w\frown\mathfrak{p}z\\w\frown\mathfrak{p}w\\\mathfrak{p}w=\mathfrak{p}z\end{array}$$

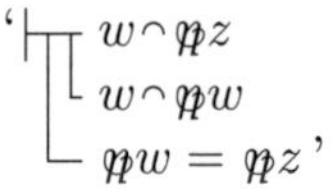

and the proposition

$$\vdash w\frown\mathfrak{p}w \quad(\mu)$$

This proposition is straightforwardly proven by showing that equality is a relation that maps every concept into itself and whose converse does the same. Accordingly, these propositions are to be derived:

$$\vdash w\frown(w\frown)\grave{a}\grave{\varepsilon}(\varepsilon=\alpha))' \quad(\nu)$$

$$\vdash w\frown(w\frown)\mathbf{\mathfrak{k}}\grave{a}\grave{\varepsilon}(\varepsilon=\alpha))' \quad(\xi)$$

Instead of (ν), we first prove the somewhat more comprehensive proposition,

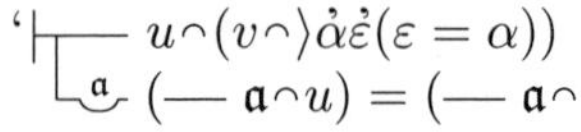

$$\begin{array}{l}u\frown(v\frown)\grave{a}\grave{\varepsilon}(\varepsilon=\alpha))\\(-a\frown u)=(-a\frown v)\end{array}\quad(o)$$

which we will also use later. For this we need the propositions

$$\begin{array}{l}d\frown u\\a\frown v\\d\frown(a\frown\grave{a}\grave{\varepsilon}(\varepsilon=\alpha))\\(-a\frown u)=(-a\frown v)\end{array}\quad(\pi)$$

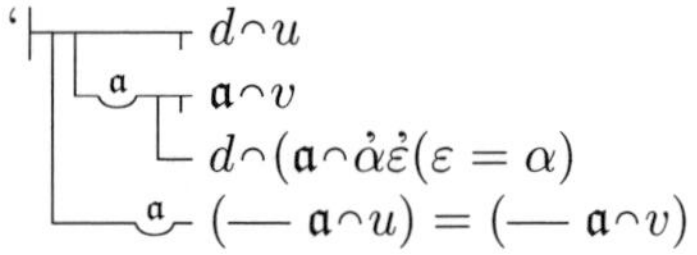

and

‘$\vdash \mathrm{I}\dot{\alpha}\dot{\varepsilon}(\varepsilon=\alpha)$’ (ϱ

The latter follows from (IIIc) and (16) using (2). In order to prove (π), we write (IIa) thus:

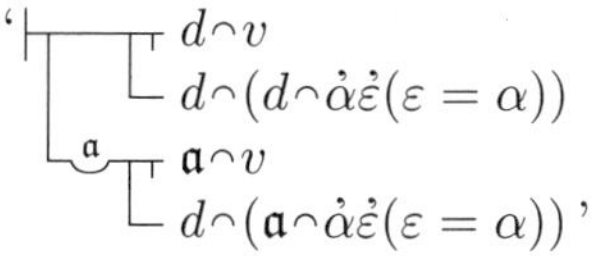

d⌢v
d⌢(d⌢ὰέ(ε = α))
𝔞
𝔞⌢v
d⌢(𝔞⌢ὰέ(ε = α))’

It remains to be shown

‘$\vdash d\frown(d\frown\dot{\alpha}\dot{\varepsilon}(\varepsilon=\alpha))$’ (σ

which can be inferred from (IIIe) and (2). The supercomponent ‘⊤ $d\frown v$’ may, by means of the subcomponent

‘ 𝔞 (— $\mathfrak{a}\frown u$) = (— $\mathfrak{a}\frown v$) ’

easily be transformed into ‘⊤ $d\frown u$’.

§67. *Construction*

2 $\vdash f(a,b) = a\frown(b\frown\dot{\alpha}\dot{\varepsilon}f(\varepsilon,\alpha))$

(IIIa): ———————

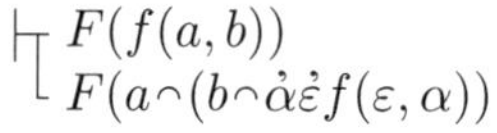

F(f(a, b))
F(a⌢(b⌢ὰέf(ε, α))

(33

—— • ——

IIa
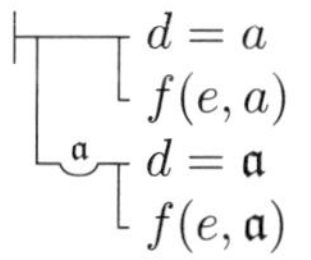

d = a
f(e, a)
𝔞
d = 𝔞
f(e, 𝔞)

(33):: – – – – – –

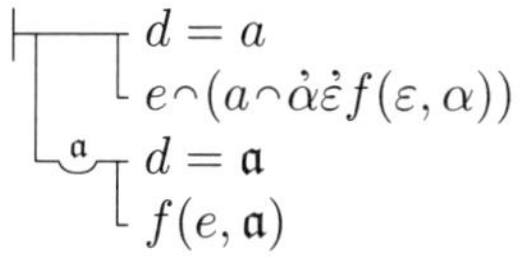

d = a
e⌢(a⌢ὰέf(ε, α))
𝔞
d = 𝔞
f(e, 𝔞)

(α

(IIa):: – – – – – – – – – – –

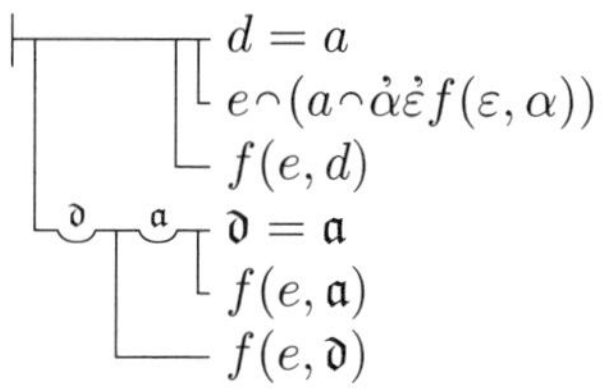

d = a
e⌢(a⌢ὰέf(ε, α))
f(e, d)
𝔡 𝔞
𝔡 = 𝔞
f(e, 𝔞)
f(e, 𝔡)

(β

(33):: – – – – – – – – – – – – –

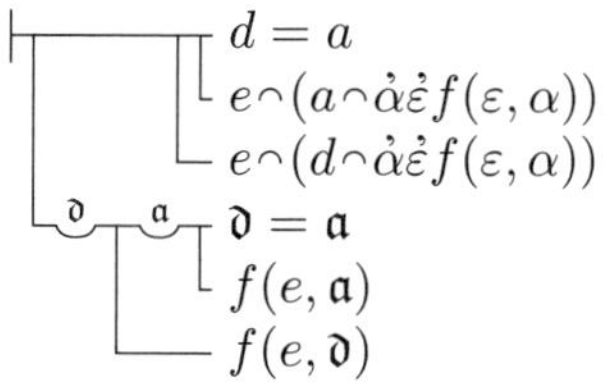

d = a
e⌢(a⌢ὰέf(ε, α))
e⌢(d⌢ὰέf(ε, α))
𝔡 𝔞
𝔡 = 𝔞
f(e, 𝔞)
f(e, 𝔡)

(γ

(IIa):: – – – – – – – – – – – – –

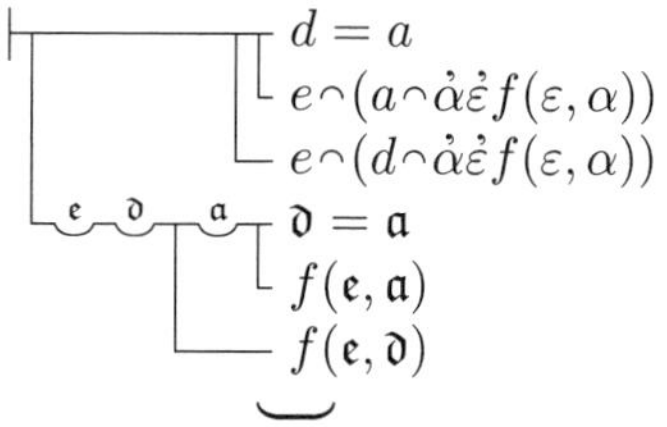

d = a
e⌢(a⌢ὰέf(ε, α))
e⌢(d⌢ὰέf(ε, α))
𝔢 𝔡 𝔞
𝔡 = 𝔞
f(𝔢, 𝔞)
f(𝔢, 𝔡)

(δ

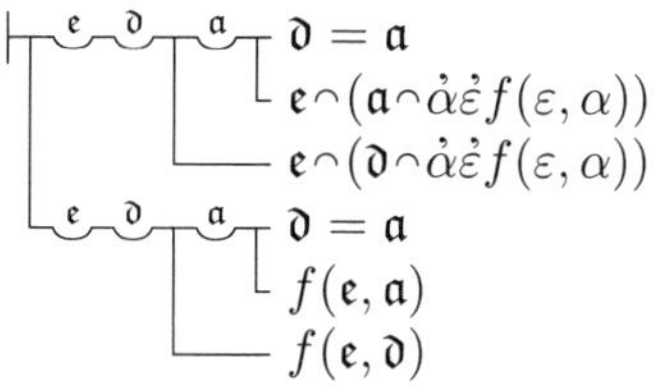

𝔢 𝔡 𝔞
𝔡 = 𝔞
𝔢⌢(𝔞⌢ὰέf(ε, α))
𝔢⌢(𝔡⌢ὰέf(ε, α))
𝔢 𝔡 𝔞
𝔡 = 𝔞
f(𝔢, 𝔞)
f(𝔢, 𝔡)

(ε

(16): – – – – – – – – – – – – – –

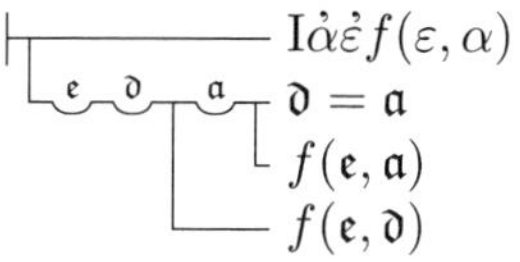

Iὰέf(ε, α)
𝔢 𝔡 𝔞
𝔡 = 𝔞
f(𝔢, 𝔞)
f(𝔢, 𝔡)

(34

—— • ——

IIIc
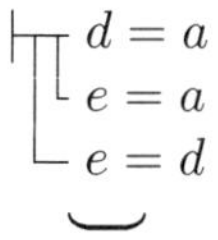

d = a
e = a
e = d

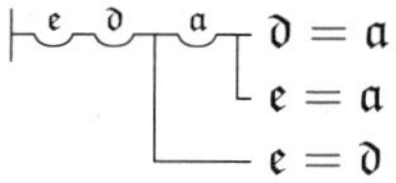

𝔢 𝔡 𝔞
𝔡 = 𝔞
𝔢 = 𝔞
𝔢 = 𝔡

(α

(34): ———————

$\vdash \mathrm{I}\dot{\alpha}\dot{\varepsilon}(\varepsilon=\alpha)$ (35

—— • ——

2 $\vdash f(a,b) = a\frown(b\frown\dot{\alpha}\dot{\varepsilon}f(\varepsilon,\alpha))$

(IIIc): ———————

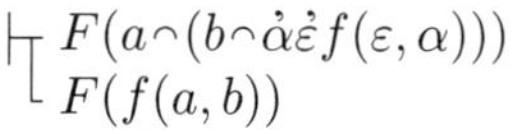

F(a⌢(b⌢ὰέf(ε, α)))
F(f(a, b))

(36

—— • ——

IIIe $\vdash d = d$

(36): ———

$\vdash d\frown(d\frown\grave{\alpha}\grave{\varepsilon}(\varepsilon=\alpha))$ (37

(IIa): ——————

$d\frown v$

$\mathfrak{a}\frown v$

$d\frown(\mathfrak{a}\frown\grave{\alpha}\grave{\varepsilon}(\varepsilon=\alpha))$ (α

(IIIa): ——————

$d\frown u$

$\mathfrak{a}\frown v$

$d\frown(\mathfrak{a}\frown\grave{\alpha}\grave{\varepsilon}(\varepsilon=\alpha))$

$(-\, d\frown u)=(-\, d\frown v)$ (β

(IIa):: – – – – – – – – – – – – – –

$d\frown u$

$\mathfrak{a}\frown v$

$d\frown(\mathfrak{a}\frown\grave{\alpha}\grave{\varepsilon}(\varepsilon=\alpha))$

$\mathfrak{a}$ $(-\, \mathfrak{a}\frown u)=(-\, \mathfrak{a}\frown v)$ (γ

$\mathfrak{d}\frown u$

$\mathfrak{a}\frown v$

$\mathfrak{d}\frown(\mathfrak{a}\frown\grave{\alpha}\grave{\varepsilon}(\varepsilon=\alpha))$

$\mathfrak{a}$ $(-\,\mathfrak{a}\frown u)=(-\,\mathfrak{a}\frown v)$ (δ

(11):– – – – – – – – – – – – – – –

$u\frown(v\frown)\grave{\alpha}\grave{\varepsilon}(\varepsilon=\alpha))$

$\mathfrak{a}$ $(-\, \mathfrak{a}\frown u)=(-\, \mathfrak{a}\frown v)$

$\mathrm{I}\grave{\alpha}\grave{\varepsilon}(\varepsilon=\alpha)$ (ε

(35):: ——————

$u\frown(v\frown)\grave{\alpha}\grave{\varepsilon}(\varepsilon=\alpha))$

$\mathfrak{a}$ $(-\, \mathfrak{a}\frown u)=(-\, \mathfrak{a}\frown v)$ (38

———•———

IIIe $\vdash(-\, a\frown w)=(-\, a\frown w)$

$\vdash$ $\mathfrak{a}$ $(-\, \mathfrak{a}\frown w)=(-\, \mathfrak{a}\frown w)$ (α

(38): ——————

$\vdash w\frown(w\frown)\grave{\alpha}\grave{\varepsilon}(\varepsilon=\alpha))$ (39

§68. *Analysis*

In order to prove the proposition

'$\vdash w\frown(w\frown)\mathrm{\maltese}\grave{\alpha}\grave{\varepsilon}(\varepsilon=\alpha))$'

we make use of the proposition

'$\vdash \grave{\alpha}\grave{\varepsilon}(\varepsilon=\alpha)=\mathrm{\maltese}\grave{\alpha}\grave{\varepsilon}(\varepsilon=\alpha)$'

which we can derive from the propositions

'$\vdash \grave{\alpha}\grave{\varepsilon}f(\alpha,\varepsilon)=\mathrm{\maltese}\grave{\alpha}\grave{\varepsilon}f(\varepsilon,\alpha)$'
'$\vdash \grave{\alpha}\grave{\varepsilon}(\varepsilon=\alpha)=\grave{\alpha}\grave{\varepsilon}(\alpha=\varepsilon)$'

The former may be derived from (2) and definition (E), the latter from (IVe), in both cases by use of (20).

§69. *Construction*

2 $\vdash f(a,b)=a\frown(b\frown\grave{\alpha}\grave{\varepsilon}f(\varepsilon,\alpha))$

$\vdash$ $\mathfrak{d}$ $\mathfrak{a}$ $f(\mathfrak{d},\mathfrak{a})=\mathfrak{d}\frown(\mathfrak{a}\frown\grave{\alpha}\grave{\varepsilon}f(\varepsilon,\alpha))$ (α

(20): ——————

$\vdash \grave{\alpha}\grave{\varepsilon}f(\alpha,\varepsilon)=\grave{\alpha}\grave{\varepsilon}(\alpha\frown(\varepsilon\frown\grave{\alpha}\grave{\varepsilon}f(\varepsilon,\alpha)))$ (β

(IIIa): ——————

$\grave{\alpha}\grave{\varepsilon}f(\alpha,\varepsilon)=\mathrm{\maltese}\grave{\alpha}\grave{\varepsilon}f(\varepsilon,\alpha)$

$\grave{\alpha}\grave{\varepsilon}(\alpha\frown(\varepsilon\frown\grave{\alpha}\grave{\varepsilon}f(\varepsilon,\alpha)))=\mathrm{\maltese}\grave{\alpha}\grave{\varepsilon}f(\varepsilon,\alpha)$ (γ

(E):: ——————

$\vdash \grave{\alpha}\grave{\varepsilon}f(\alpha,\varepsilon)=\mathrm{\maltese}\grave{\alpha}\grave{\varepsilon}f(\varepsilon,\alpha)$ (40

———•———

IVe $\vdash(a=b)=(b=a)$

$\vdash$ $\mathfrak{d}$ $\mathfrak{a}$ $(\mathfrak{a}=\mathfrak{d})=(\mathfrak{d}=\mathfrak{a})$ (α

(20): ——————

$\vdash \grave{\alpha}\grave{\varepsilon}(\varepsilon=\alpha)=\grave{\alpha}\grave{\varepsilon}(\alpha=\varepsilon)$ (β

(IIIa): ——————

$\grave{\alpha}\grave{\varepsilon}(\varepsilon=\alpha)=\mathrm{\maltese}\grave{\alpha}\grave{\varepsilon}(\varepsilon=\alpha)$

$\grave{\alpha}\grave{\varepsilon}(\alpha=\varepsilon)=\mathrm{\maltese}\grave{\alpha}\grave{\varepsilon}(\varepsilon=\alpha)$ (γ

(40):: ——————

$\vdash \grave{\alpha}\grave{\varepsilon}(\varepsilon=\alpha)=\mathrm{\maltese}\grave{\alpha}\grave{\varepsilon}(\varepsilon=\alpha)$ (41

(IIIc): ——————

$w\frown(w\frown)\mathrm{\maltese}\grave{\alpha}\grave{\varepsilon}(\varepsilon=\alpha))$

$w\frown(w\frown)\grave{\alpha}\grave{\varepsilon}(\varepsilon=\alpha))$ (α

(39):: ——————

$\vdash w\frown(w\frown)\mathrm{\maltese}\grave{\alpha}\grave{\varepsilon}(\varepsilon=\alpha))$ (42

———•———

1 $\vdash f(a)=a\frown\grave{\varepsilon}f(\varepsilon)$

(IIIc): ——————

$f(a)=a\frown v$

$\grave{\varepsilon}f(\varepsilon)=v$ (43

(IIIc): – – – – – –

$F(a\frown v)$

$F(f(a))$

$\grave{\varepsilon}f(\varepsilon)=v$ (44

———•———

$Z \vdash \dot{\varepsilon}\left([\mathfrak{q}]\ \varepsilon \frown (u \frown \rangle \mathfrak{q}) ;\ u \frown (\varepsilon \frown \rangle \mathfrak{q}) \right) = \wp u$

(44): ————————

$\vdash$ $v \frown \wp u$
$[\mathfrak{q}]$ $v \frown (u \frown \rangle \mathfrak{q})$
$u \frown (v \frown \rangle \mathfrak{q})$ (α

×

$\vdash$ $[\mathfrak{q}]$ $v \frown (u \frown \rangle \mathfrak{q})$
$u \frown (v \frown \rangle \mathfrak{q})$
$v \frown \wp u$ (β

(IIa): - - - - - - - - -

$\vdash$ $v \frown (u \frown \rangle q)$
$u \frown (v \frown \rangle q)$
$v \frown \wp u$ (γ

×

$\vdash$ $v \frown \wp u$
$u \frown (v \frown \rangle q)$
$v \frown (u \frown \rangle q)$ (45

——— • ———

43 $\vdash$ $f(a) = a \frown v$
$\dot{\varepsilon} f(\varepsilon) = v$

(IIIa): - - - - - - -

$\vdash$ $F(f(a))$
$F(a \frown v)$
$\dot{\varepsilon} f(\varepsilon) = v$ (46

——— • ———

$Z \vdash \dot{\varepsilon}\left([\mathfrak{q}]\ \varepsilon \frown (u \frown \rangle \mathfrak{q}) ;\ u \frown (\varepsilon \frown \rangle \mathfrak{q}) \right) = \wp u$

(46): ————————

$\vdash$ $[\mathfrak{q}]$ $v \frown (u \frown \rangle \mathfrak{q})$
$u \frown (v \frown \rangle \mathfrak{q})$
$v \frown \wp u$ (α

×

$\vdash$ $v \frown \wp u$
$[\mathfrak{q}]$ $v \frown (u \frown \rangle \mathfrak{q})$
$u \frown (v \frown \rangle \mathfrak{q})$ (47

——— • ———

45 $\vdash$ $w \frown \wp w$
$w \frown (w \frown \rangle \grave{\alpha}\dot{\varepsilon}(\varepsilon = \alpha))$
$w \frown (w \frown \rangle \grave{\alpha}\dot{\varepsilon}(\varepsilon = \alpha))$

(42, 39):: ═══════════

$\vdash w \frown \wp w$ (48

(IIIc): ————

$\vdash$ $w \frown \wp z$
$\wp w = \wp z$ (α

×

$\vdash$ $\wp w = \wp z$
$w \frown \wp z$ (β

(47):: - - - - - - -

$\vdash$ $\wp w = \wp z$
$[\mathfrak{q}]$ $w \frown (z \frown \rangle \mathfrak{q})$
$z \frown (w \frown \rangle \mathfrak{q})$ (49

§70. *Analysis*

We now prove the proposition

'$\vdash$ $[\mathfrak{q}]$ $w \frown (z \frown \rangle \mathfrak{q})$
$z \frown (w \frown \rangle \mathfrak{q})$
$b \frown w$
$c \frown z$
$[\mathfrak{q}]$ $w \frown (z \frown \rangle \mathfrak{q})$
$z \frown (w \frown \rangle \mathfrak{q})$
$[\mathfrak{a}]$ $b \frown (\mathfrak{a} \frown \mathfrak{q})$
$[\mathfrak{a}]$ $c \frown (\mathfrak{a} \frown \mathfrak{q})$ '

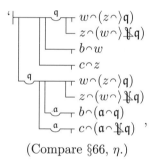

(Compare §66, η.)

We saw in §66 that

'$\xi \frown (\zeta \frown q)$
$\xi = b$
$\zeta = c$ ',

indicates a relation that agrees with the q-relation in the relevant properties, but in which b stands to no object, and in which no object stands to c. We thus write (IIa) in the form

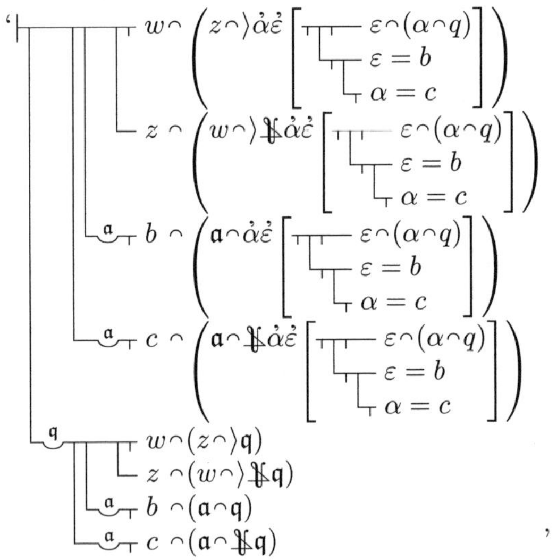

and now have to prove, amongst others, the proposition:

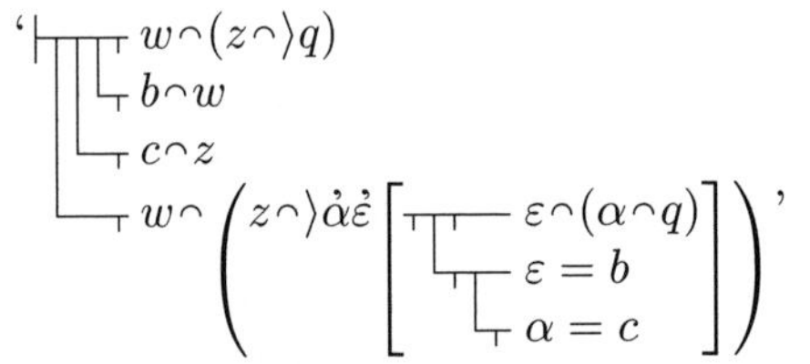

which follows by contraposition from

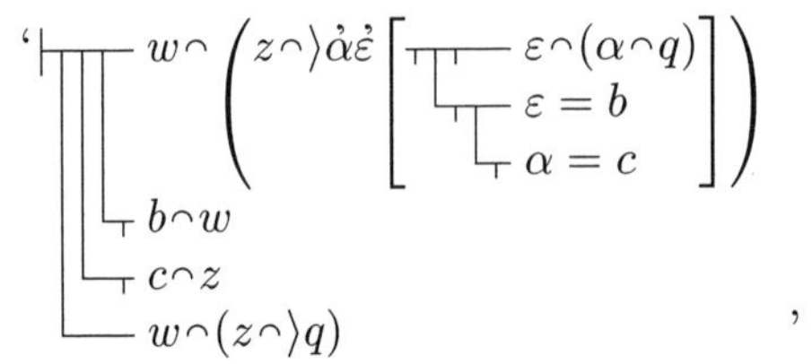

For this, we require the proposition

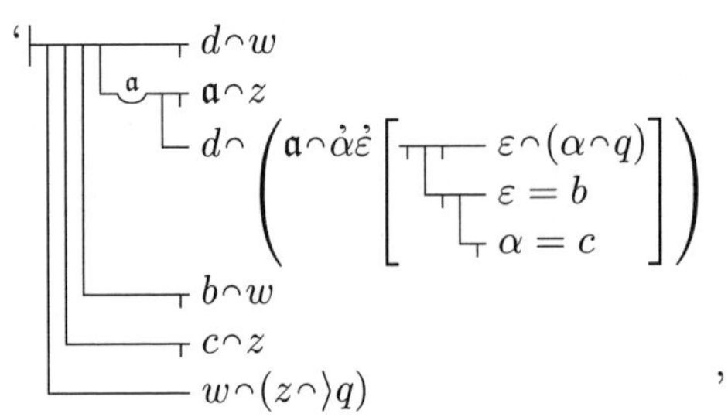

By (8) we now have

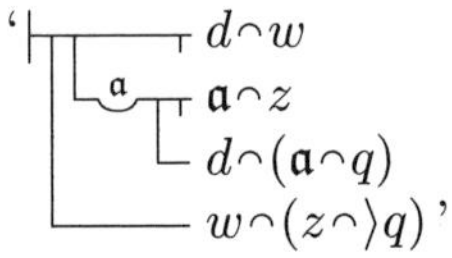

Accordingly, what needs to be proven is something like

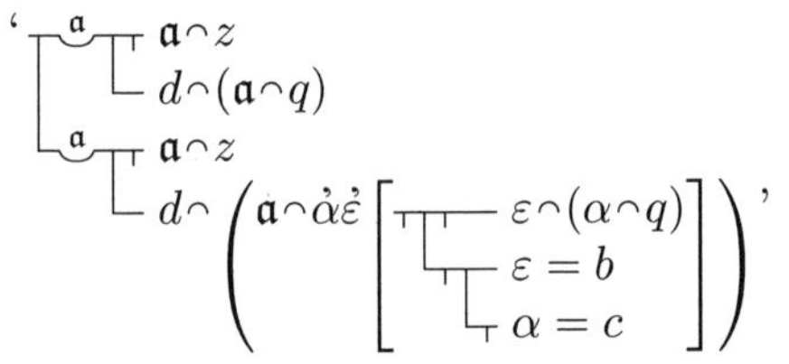

in which I have not yet put a judgement-stroke, because of conditions (subcomponents) that may have to be added. In the proper proof, expressions containing Roman letters may not occur without a judgement-stroke; here, where we concerned with a preliminary exploration, they may be admissible. The latter can be obtained by rule (5) from an expression like

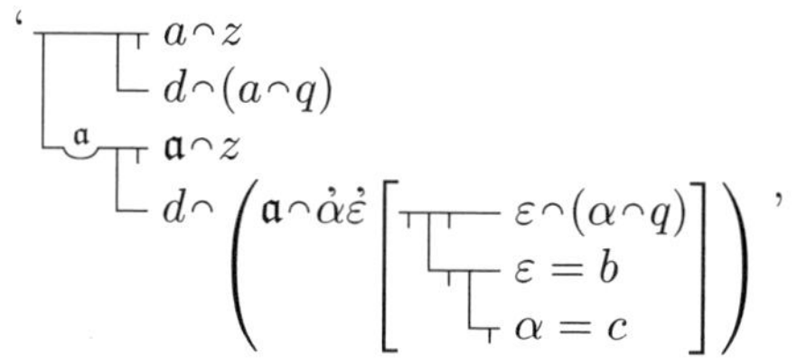

According to (II a), we now have

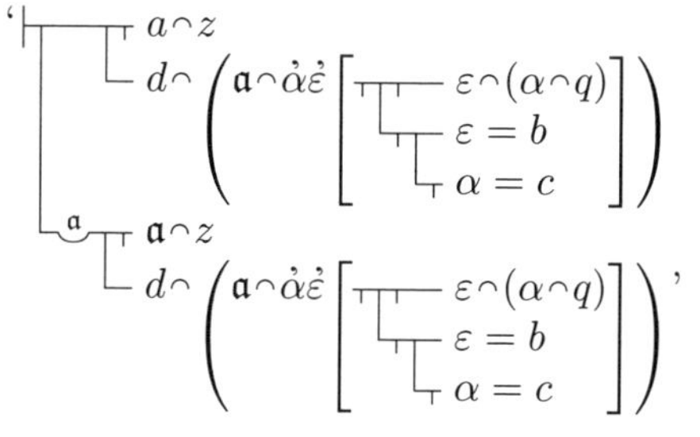

and it would remain to prove, using (36),

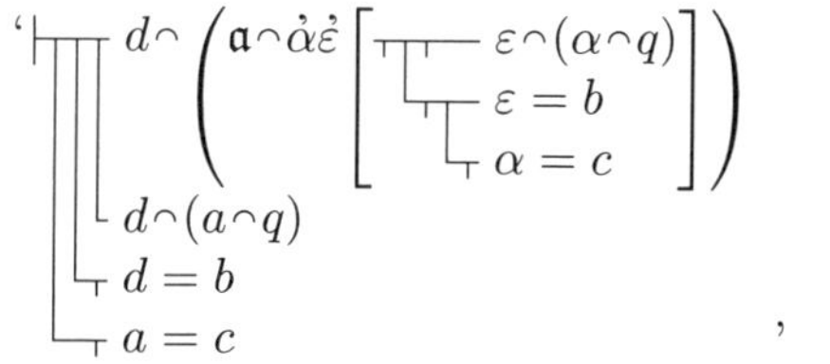

in which the subcomponents '$\top\, d = b$' and '$\top\, a = c$' occur. From the latter two propositions we infer by rule (7)

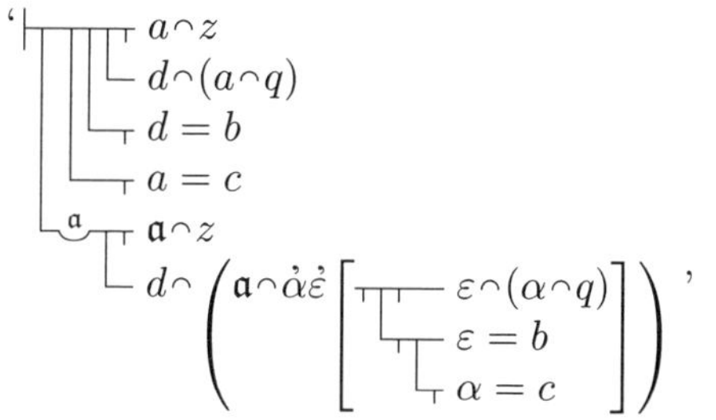

If we now attempted to introduce the German '$\mathfrak{a}$' instead of the Roman 'a' here, by rule (5), we would not attain the desired goal because the subcomponent, '$\top\, a = c$', would have to be included in the scope of '$\mathfrak{a}$'. But now, using (III a), we have

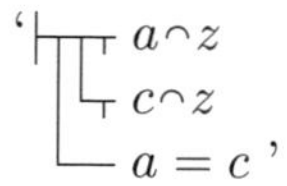

and by rule (8), the subcomponent '$\top\, a = c$' can be replaced by '$\top\, c \frown z$'. Later, the subcomponent '$\top\, d = b$' is likewise to be replaced by '$\top\, b \frown w$'.

§71. *Construction*

Ie

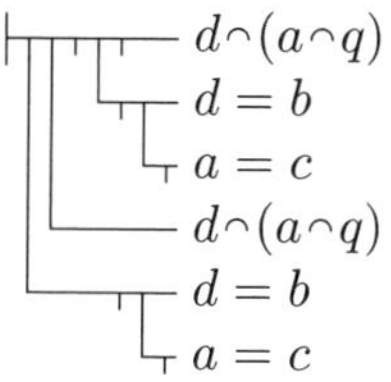

(I f)::

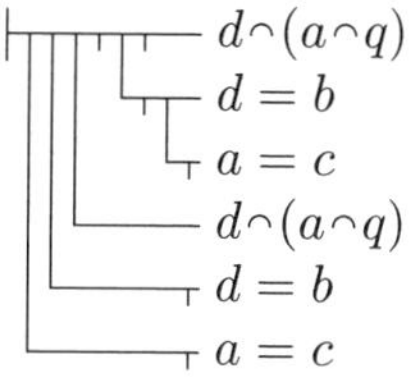

(α

(36):

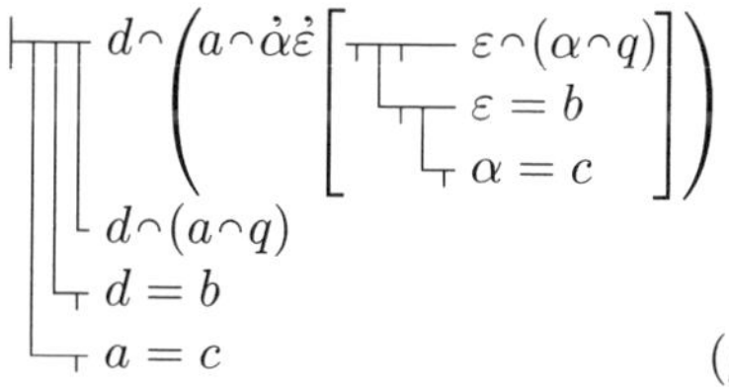

(β

(II a):

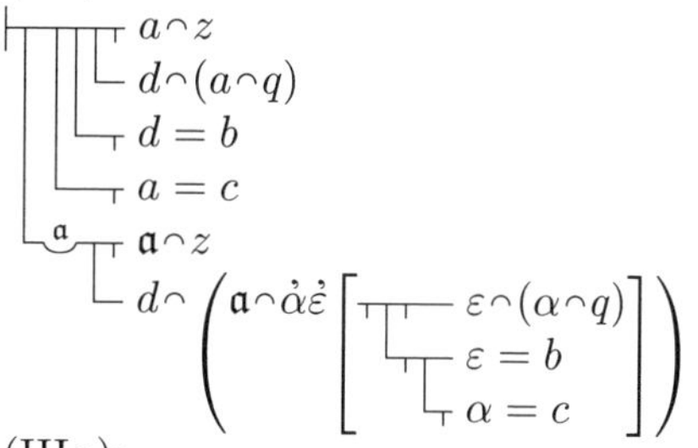

(γ

(III a): ·—·—·—·—·—·—·—·—·—·—·

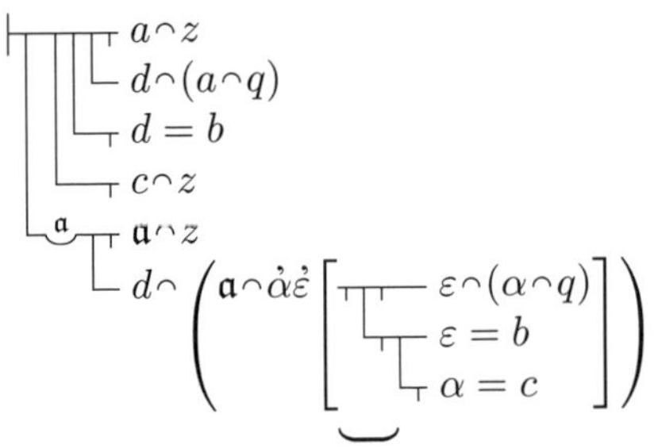

(δ

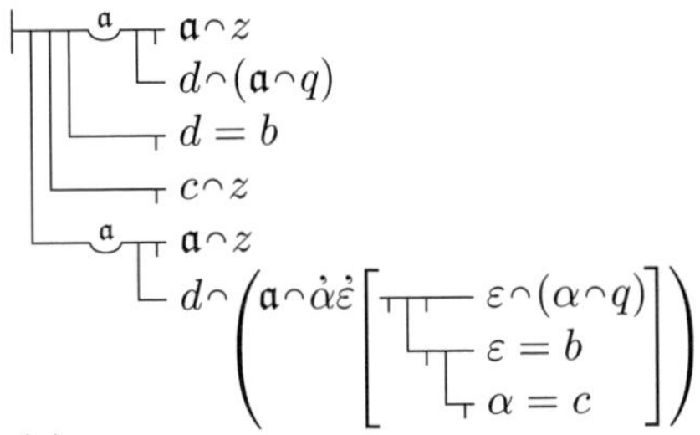

(ε

(8): - - - - - - - - - - - - - - - -

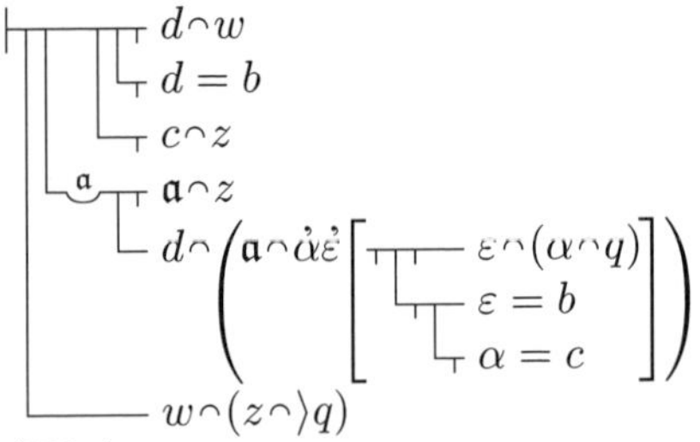

(ζ

(IIIa): ·—·—·—·—·—·—·—·—·—·—·—·—·

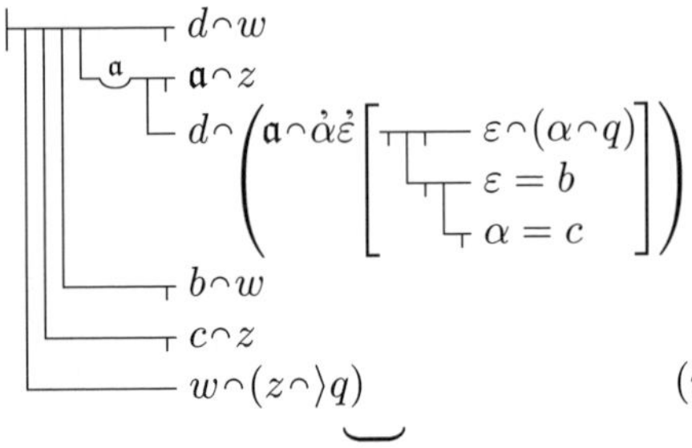

(η

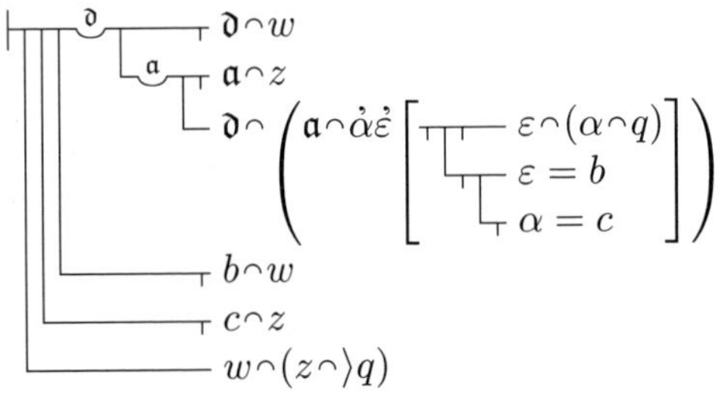

(ϑ

(11): -

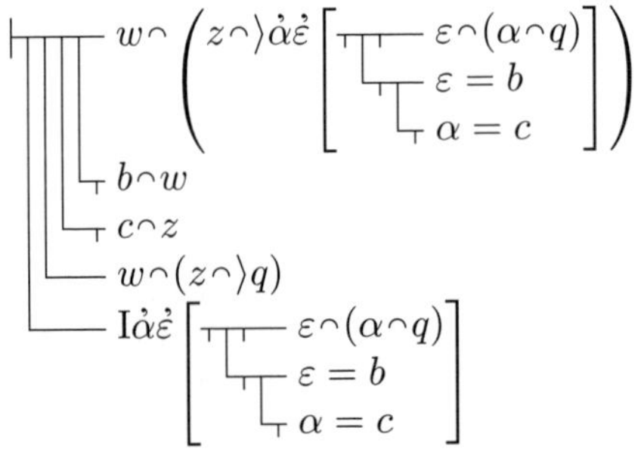

(50

§72. *Analysis*

In the proposition (50), we have to remove the subcomponent

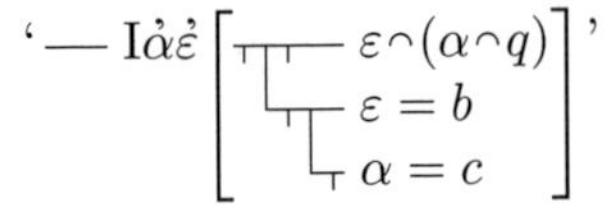

This is achieved by using the proposition

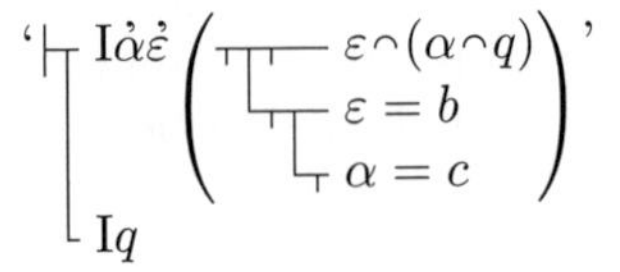

for whose proof we use (34). For this we require the proposition

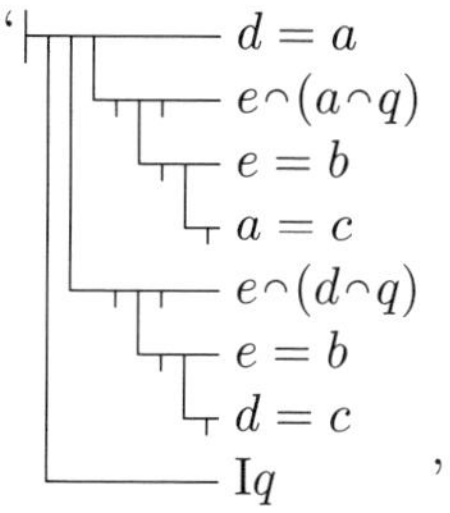

which easily follows from (13).

§73. *Construction*

13

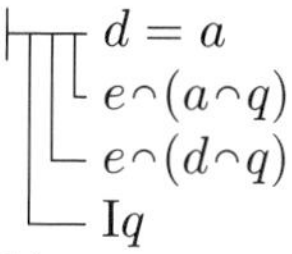

(Ib, Ib):: = = = =

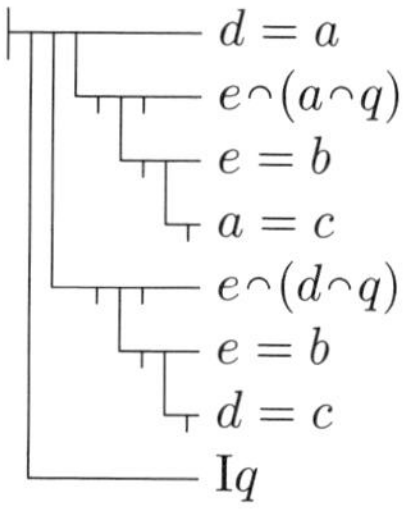

(α

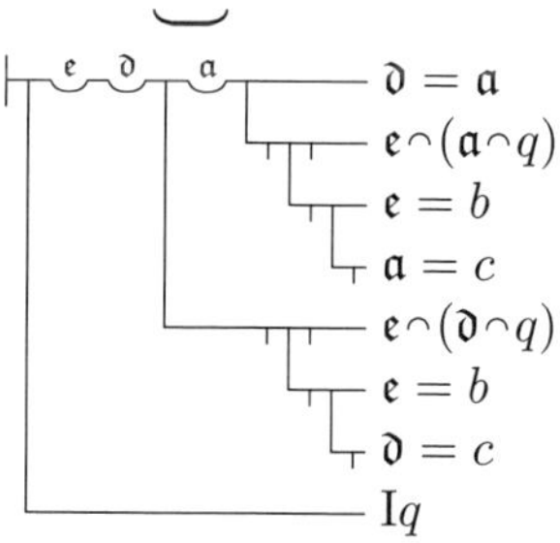

(β

(34): – – – – – – – – – – –

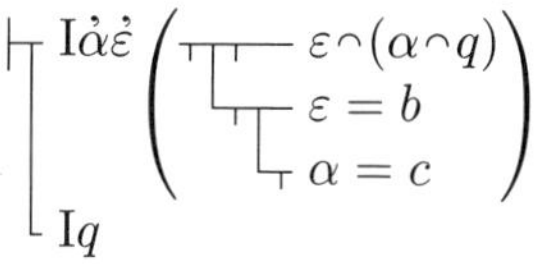

(γ

(18):: – – – – – – – – – – –

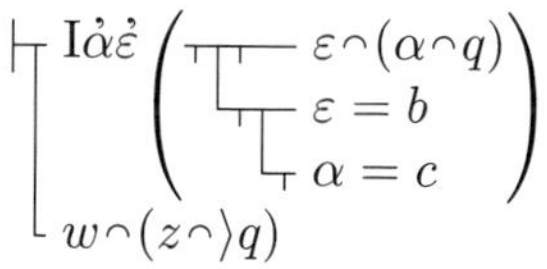

(δ

(50): – – – – – – – – – – –

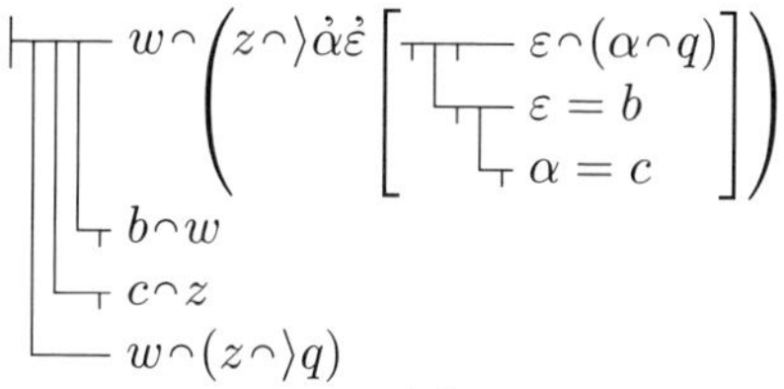

(51

×

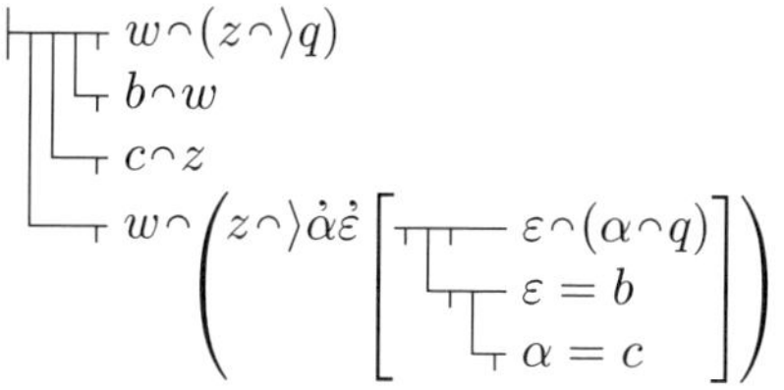

(52

§74. *Analysis*

In order now to prove in addition the proposition

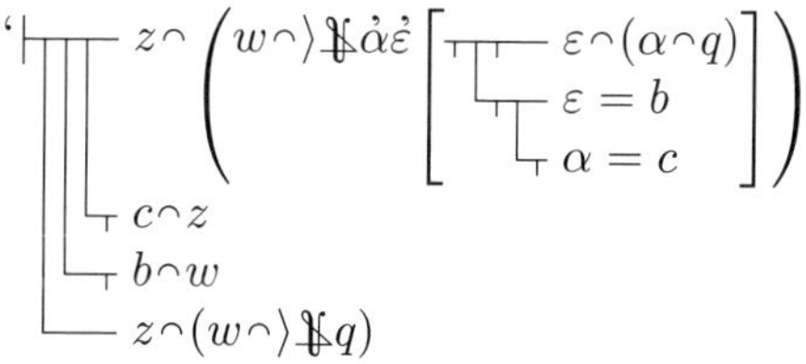

we first write (51) thus:

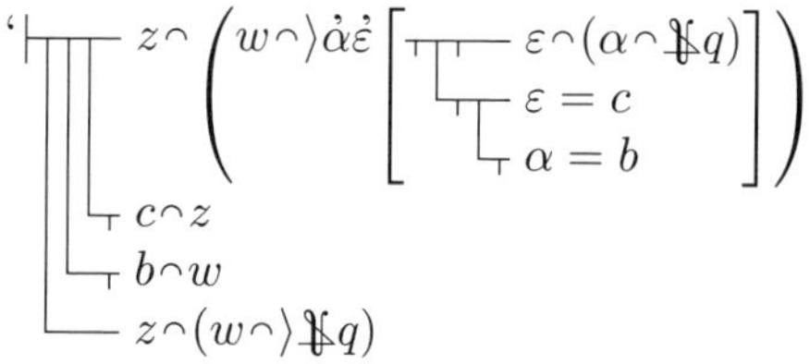

interchanging ‘*q*’ with ‘Ʉ*q*’, ‘*c*’ with ‘*b*’, ‘*z*’ with ‘*w*’. We now have to prove

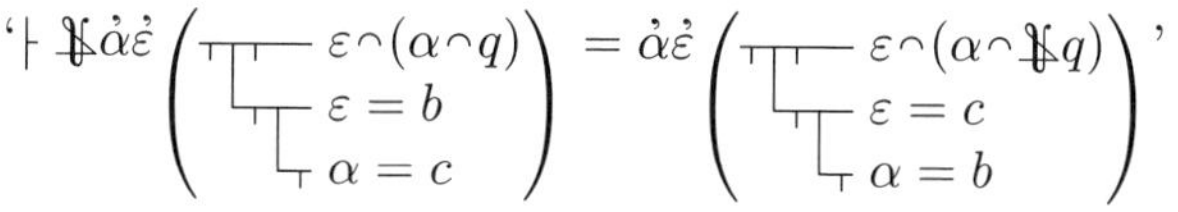

which follows from

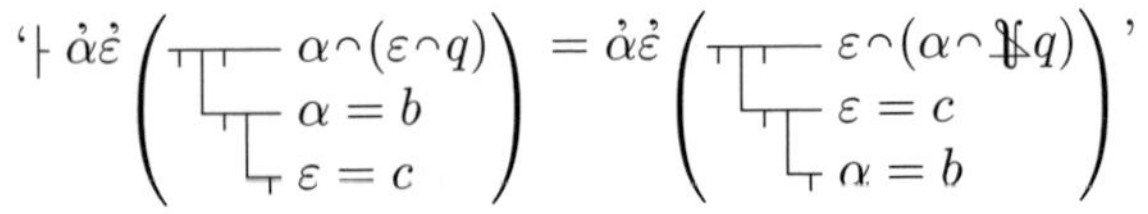

using (40). This proposition is to be proven using (20). For this, we require the proposition

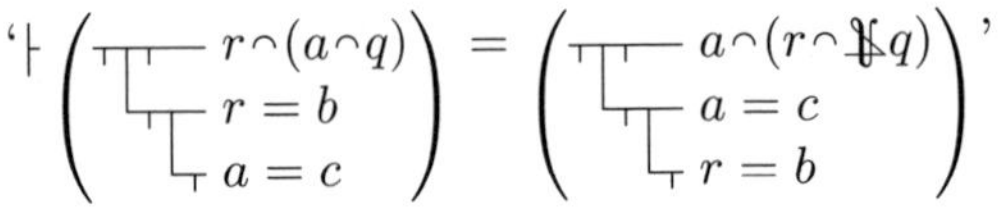

which follows from (21) and the proposition

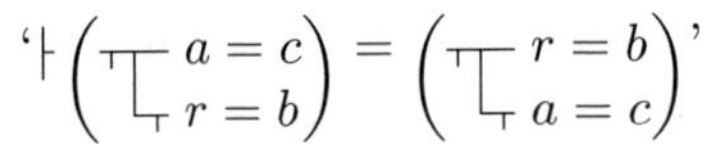

This is to be proven using (IV a).

§75. *Construction*

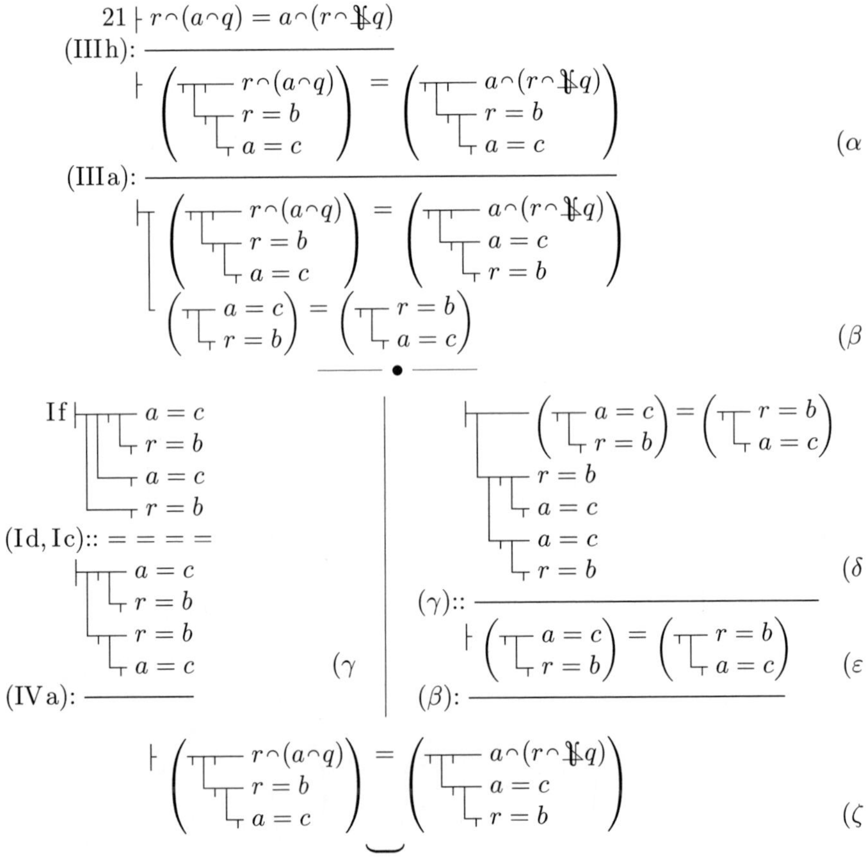

$$\vdash\overset{\mathfrak{a}}{\smile}\overset{\mathfrak{e}}{\smile}\left(\begin{array}{l}\mathfrak{a}\frown(\mathfrak{e}\frown q)\\ \mathfrak{a}=b\\ \mathfrak{e}=c\end{array}\right)=\left(\begin{array}{l}\mathfrak{e}\frown(\mathfrak{a}\frown\nparallel q)\\ \mathfrak{e}=c\\ \mathfrak{a}=b\end{array}\right) \qquad (\eta$$

(20): ———————————————

$$\vdash\dot{\alpha}\dot{\varepsilon}\left(\begin{array}{l}\alpha\frown(\varepsilon\frown q)\\ \alpha=b\\ \varepsilon=c\end{array}\right)=\dot{\alpha}\dot{\varepsilon}\left(\begin{array}{l}\varepsilon\frown(\alpha\frown\nparallel q)\\ \varepsilon=c\\ \alpha=b\end{array}\right) \qquad (\vartheta$$

(IIIc): ———————————————

$$\vdash\begin{array}{l}\nparallel\dot{\alpha}\dot{\varepsilon}\left(\begin{array}{l}\varepsilon\frown(\alpha\frown q)\\ \varepsilon=b\\ \alpha=c\end{array}\right)=\dot{\alpha}\dot{\varepsilon}\left(\begin{array}{l}\varepsilon\frown(\alpha\frown\nparallel q)\\ \varepsilon=c\\ \alpha=b\end{array}\right)\\ \dot{\alpha}\dot{\varepsilon}\left(\begin{array}{l}\alpha\frown(\varepsilon\frown q)\\ \alpha=b\\ \varepsilon=c\end{array}\right)=\nparallel\dot{\alpha}\dot{\varepsilon}\left(\begin{array}{l}\varepsilon\frown(\alpha\frown q)\\ \varepsilon=b\\ \alpha=c\end{array}\right)\end{array} \qquad (\iota$$

(40):: ———————————————

$$\vdash\nparallel\dot{\alpha}\dot{\varepsilon}\left(\begin{array}{l}\varepsilon\frown(\alpha\frown q)\\ \varepsilon=b\\ \alpha=c\end{array}\right)=\dot{\alpha}\dot{\varepsilon}\left(\begin{array}{l}\varepsilon\frown(\alpha\frown\nparallel q)\\ \varepsilon=c\\ \alpha=b\end{array}\right) \qquad (\kappa$$

(IIIa): ———————————————

$$\vdash\begin{array}{l}F\left(\nparallel\dot{\alpha}\dot{\varepsilon}\left[\begin{array}{l}\varepsilon\frown(\alpha\frown q)\\ \varepsilon=b\\ \alpha=c\end{array}\right]\right)\\ F\left(\dot{\alpha}\dot{\varepsilon}\left[\begin{array}{l}\varepsilon\frown(\alpha\frown\nparallel q)\\ \varepsilon=c\\ \alpha=b\end{array}\right]\right)\end{array} \qquad (53$$

———— • ————

$$51\;\vdash\begin{array}{l}z\frown\left(w\frown\rangle\dot{\alpha}\dot{\varepsilon}\left[\begin{array}{l}\varepsilon\frown(\alpha\frown\nparallel q)\\ \varepsilon=c\\ \alpha=b\end{array}\right]\right)\\ c\frown z\\ b\frown w\\ z\frown(w\frown\rangle\nparallel q)\end{array}$$

(53): ———————————————

$$\vdash\begin{array}{l}z\frown\left(w\frown\rangle\nparallel\dot{\alpha}\dot{\varepsilon}\left[\begin{array}{l}\varepsilon\frown(\alpha\frown q)\\ \varepsilon=b\\ \alpha=c\end{array}\right]\right)\\ c\frown z\\ b\frown w\\ z\frown(w\frown\rangle\nparallel q)\end{array} \qquad (54$$

§76. *Analysis*

If we write the proposition (IIa) as in §70, then we see that the propositions

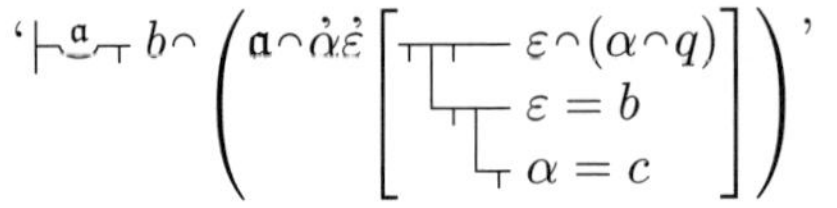

and

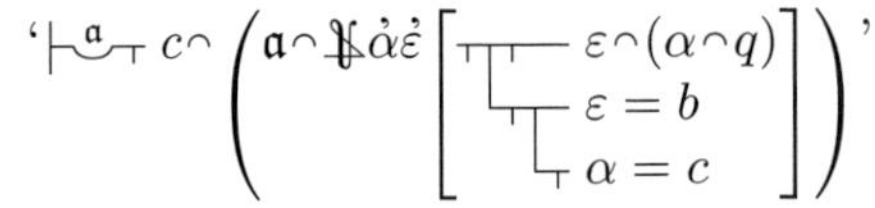

are still missing, of which the latter can, by means of (53), be reduced to the former, which in turn can be derived using (33).

§77. *Construction*

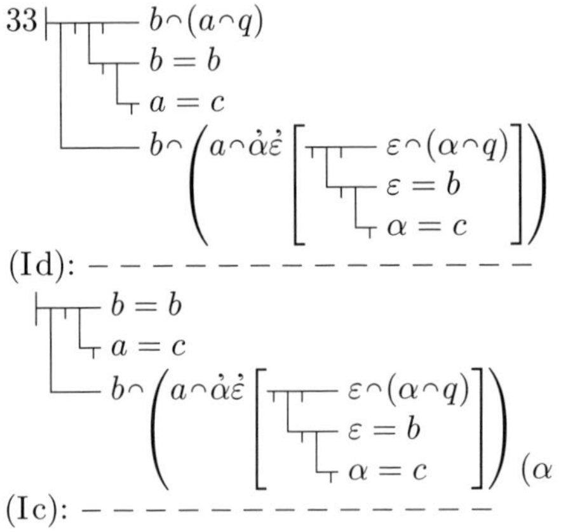

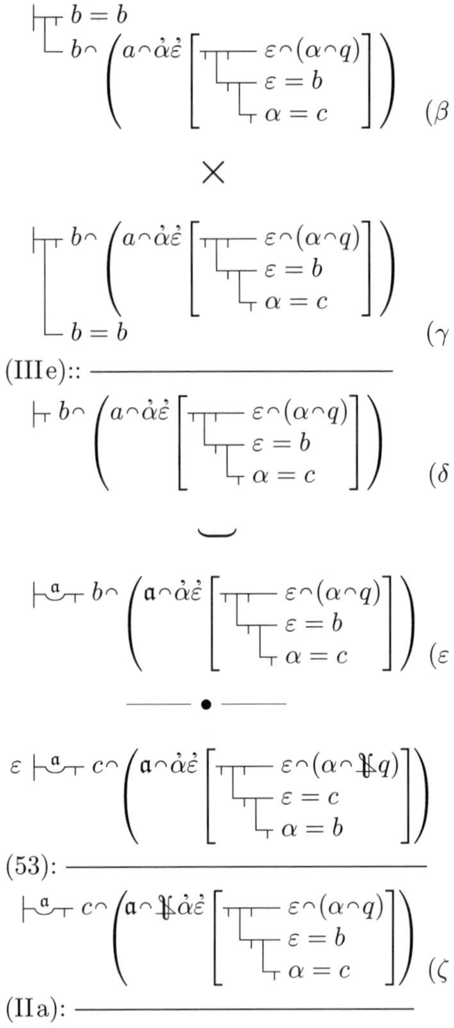

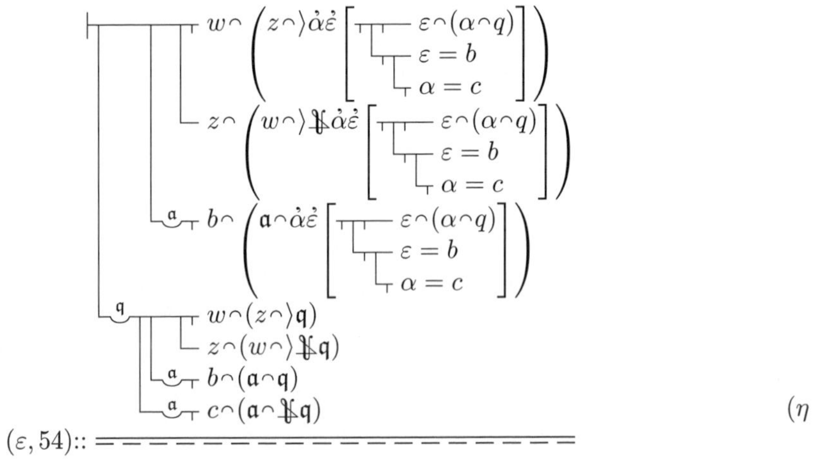

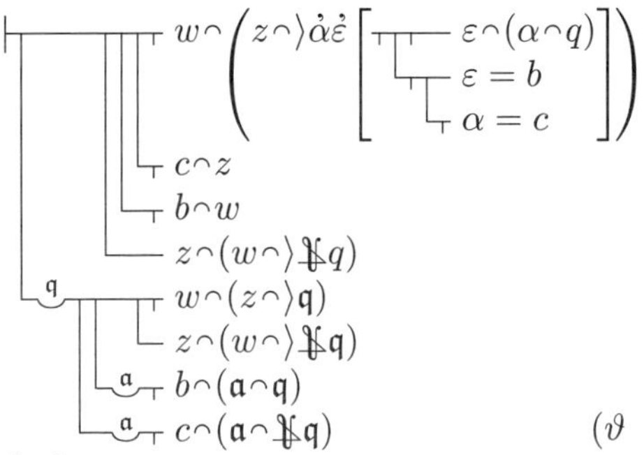

(ϑ

(52): - - - - - - - - - - - - - - - -

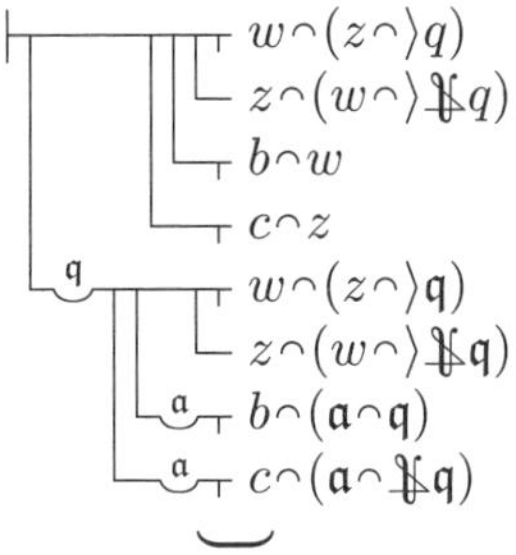

(ι

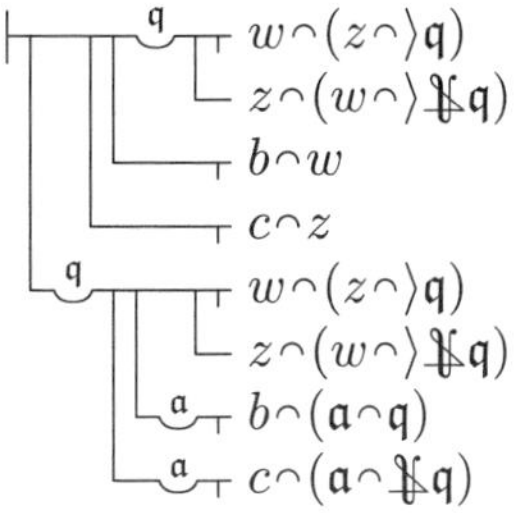

(55

(49): - - - - - - - - - - -

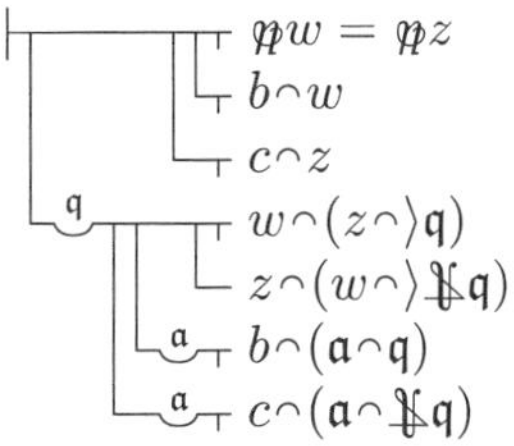

(56

b) Proof of the proposition

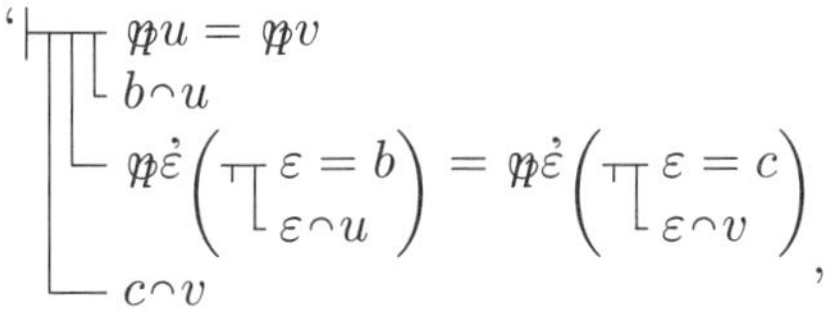

and end of section B

§78. *Analysis*

We now prove the proposition (ϑ) of §66 which, by rule (5), is obtained from

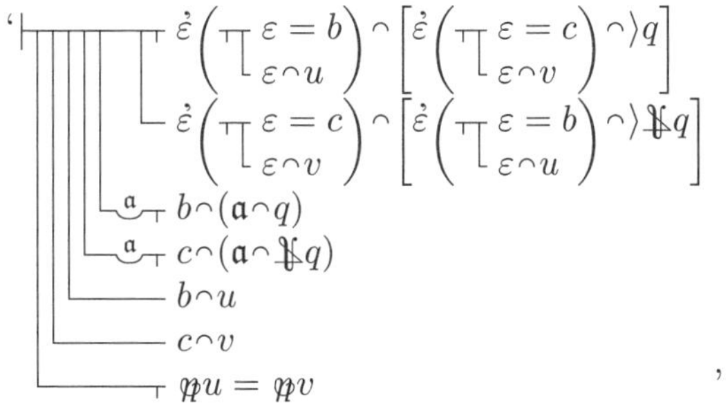

(α

For ease of understanding it is more convenient to consider the proposition resulting from the latter by contraposition:

$$\vdash \begin{array}{l} \mathfrak{p}u = \mathfrak{p}v \\ c \frown v \\ \dot{\varepsilon}\left(\begin{array}{l}\varepsilon = b \\ \varepsilon \frown u\end{array}\right) \frown \left[\dot{\varepsilon}\left(\begin{array}{l}\varepsilon = c \\ \varepsilon \frown v\end{array}\right) \frown \rangle q\right] \\ \smile^{\mathfrak{a}}\; b \frown (\mathfrak{a} \frown q) \\ b \frown u \\ \dot{\varepsilon}\left(\begin{array}{l}\varepsilon = c \\ \varepsilon \frown v\end{array}\right) \frown \left[\dot{\varepsilon}\left(\begin{array}{l}\varepsilon = b \\ \varepsilon \frown u\end{array}\right) \frown \rangle \rlap{/}{\mathfrak{X}} q\right] \\ \smile^{\mathfrak{a}}\; c \frown (\mathfrak{a} \frown \rlap{/}{\mathfrak{X}} q) \end{array} , \qquad (\beta$$

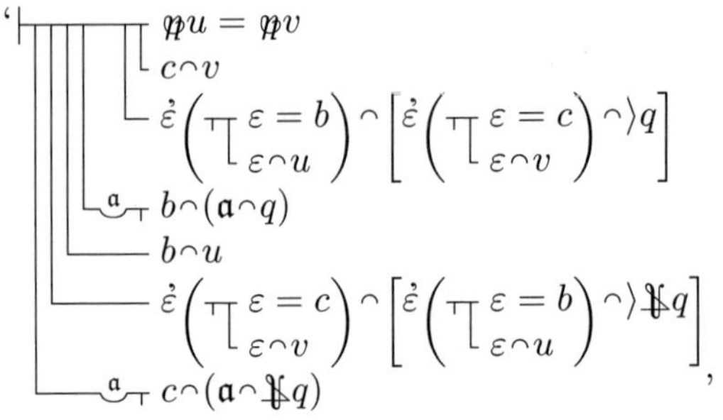

In order to prove that the cardinal number of the u-concept is equal to the cardinal number of the v-concept, it will be sufficient, according to (32), to supply any relation which maps the u-concept into the v-concept and whose converse maps the latter into the former. We have already encountered such a relation in §66. Accordingly, we will first derive the proposition

$$\vdash \begin{array}{l} u \frown \left(v \frown \rangle \dot{\alpha}\dot{\varepsilon}\left[\begin{array}{l} \varepsilon \frown (\alpha \frown q) \\ \varepsilon = b \\ \alpha = c \end{array}\right]\right) \\ c \frown v \\ \dot{\varepsilon}\left(\begin{array}{l}\varepsilon = b \\ \varepsilon \frown u\end{array}\right) \frown \left[\dot{\varepsilon}\left(\begin{array}{l}\varepsilon = c \\ \varepsilon \frown v\end{array}\right) \frown \rangle q\right] \\ \smile^{\mathfrak{a}}\; b \frown (\mathfrak{a} \frown q) \end{array} , \qquad (\gamma$$

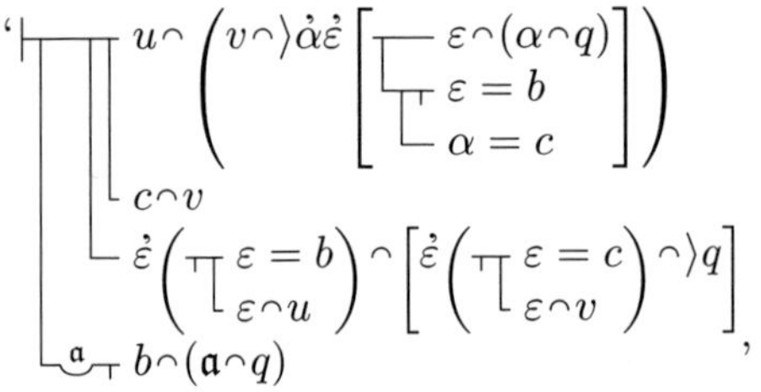

and then the proposition

$$\vdash \begin{array}{l} v \frown \left(u \frown \rangle \rlap{/}{\mathfrak{X}} \dot{\alpha}\dot{\varepsilon}\left[\begin{array}{l} \varepsilon \frown (\alpha \frown q) \\ \varepsilon = b \\ \alpha = c \end{array}\right]\right) \\ b \frown u \\ \dot{\varepsilon}\left(\begin{array}{l}\varepsilon = c \\ \varepsilon \frown v\end{array}\right) \frown \left[\dot{\varepsilon}\left(\begin{array}{l}\varepsilon = b \\ \varepsilon \frown u\end{array}\right) \frown \rangle \rlap{/}{\mathfrak{X}} q\right] \\ \smile^{\mathfrak{a}}\; c \frown (\mathfrak{a} \frown \rlap{/}{\mathfrak{X}} q) \end{array} , \qquad (\delta$$

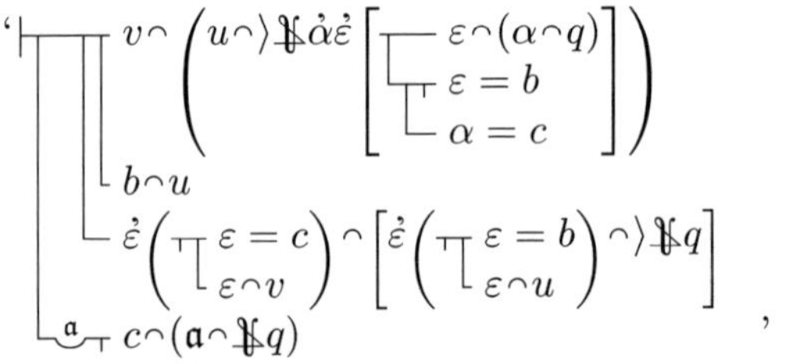

To prove the former using (11), we first require the proposition

$$\vdash \begin{array}{l} d \frown u \\ \smile^{\mathfrak{a}} \begin{array}{l} \mathfrak{a} \frown v \\ d \frown \left(\mathfrak{a} \frown \dot{\alpha}\dot{\varepsilon}\left[\begin{array}{l} \varepsilon \frown (\alpha \frown q) \\ \varepsilon = b \\ \alpha = c \end{array}\right]\right) \end{array} \\ c \frown v \\ \dot{\varepsilon}\left(\begin{array}{l}\varepsilon = b \\ \varepsilon \frown u\end{array}\right) \frown \left[\dot{\varepsilon}\left(\begin{array}{l}\varepsilon = c \\ \varepsilon \frown v\end{array}\right) \frown \rangle q\right] \end{array} , \qquad (\varepsilon$$

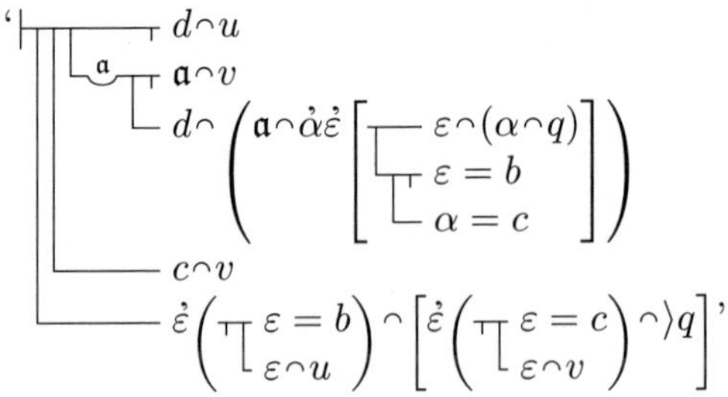

The cases $d = b$ and $\top\, d = b$ are to be distinguished. We write (8) in the form

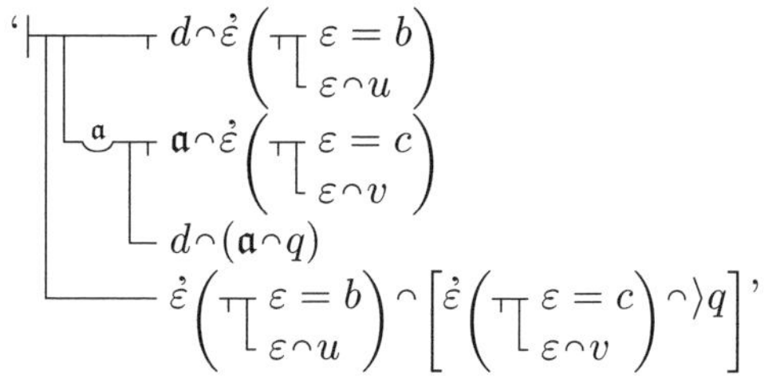

(ζ

from which straightforwardly follows

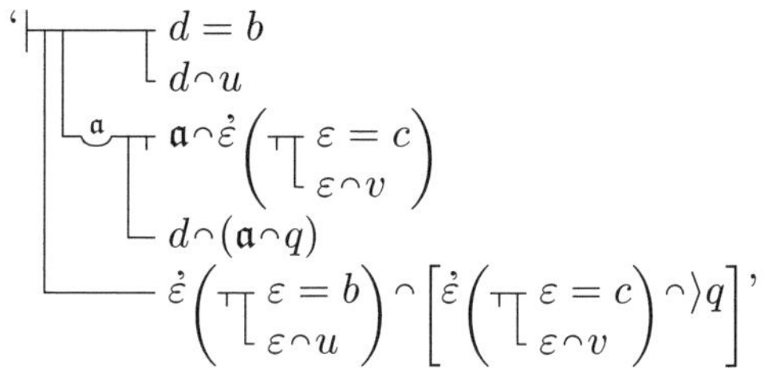

(η

From this we easily arrive at our proposition for the case $\top\, d = b$ by contraposition, once we have proven

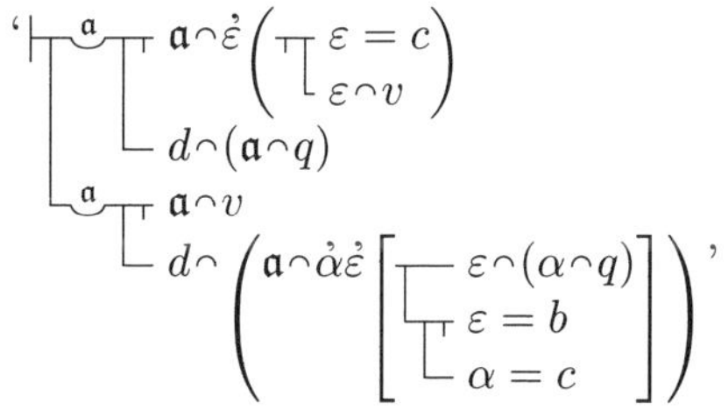

(ϑ

If, for this purpose, we write (IIa) in the form

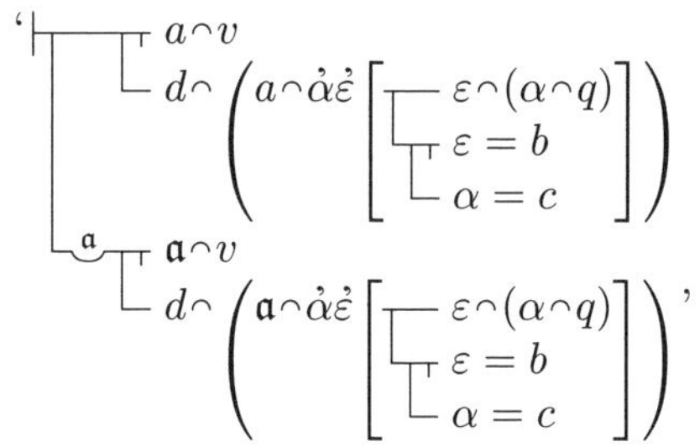

(ι

then we have to prove

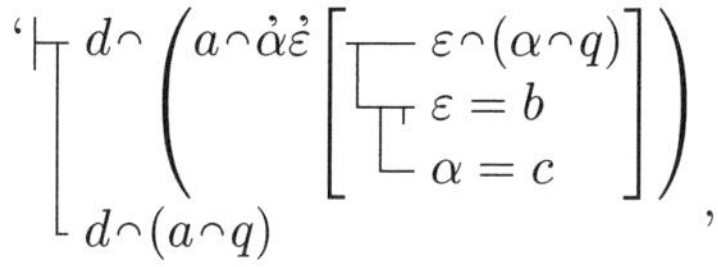

(κ

which is straightforward using (I) and (36). Now

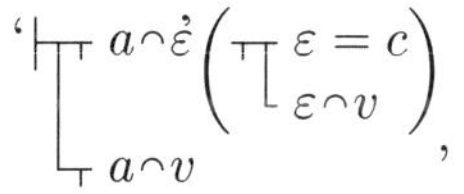

(λ

still has to be derived from (Ia) in the form

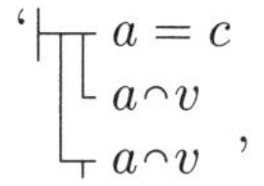

(μ

and the proposition

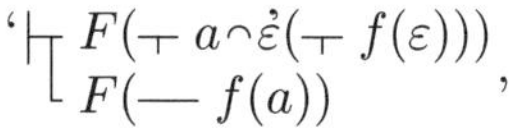

(ν

which follows from

$$‘\vdash (—\, f(a)) = (\top\, a\frown\dot{\varepsilon}(\top\, f(\varepsilon)))’ \qquad (\xi$$

This last proposition is straightforwardly proven from (1) using (IVb).

§79. *Construction*

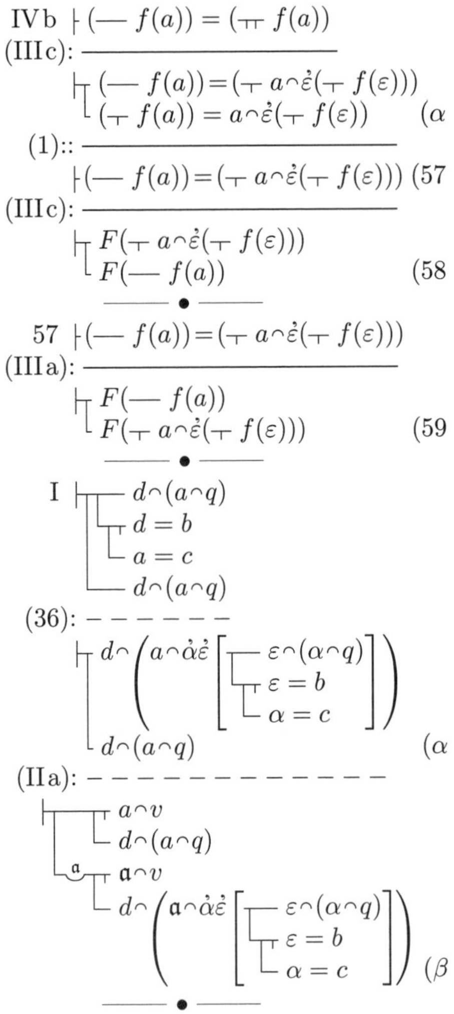

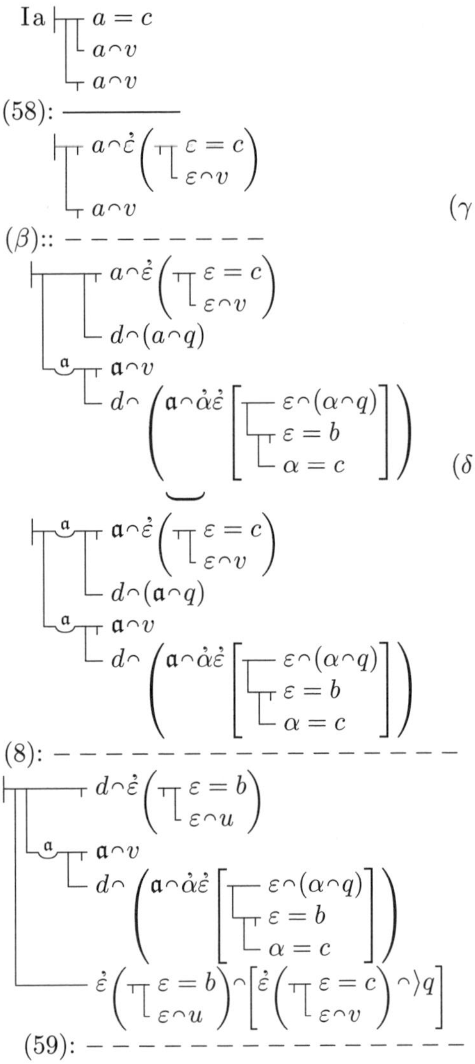

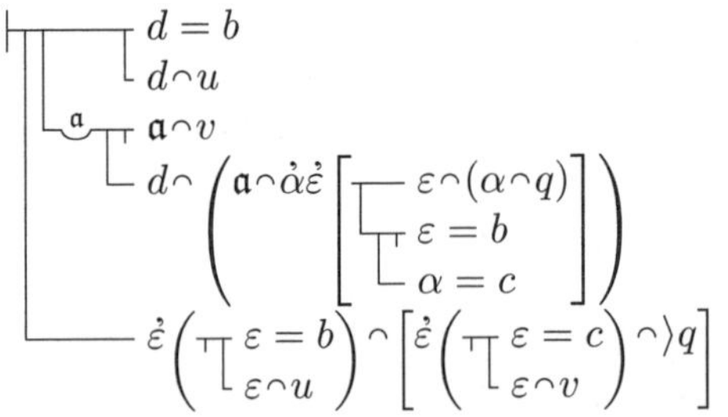

(60

§80. *Analysis*

We now prove the proposition

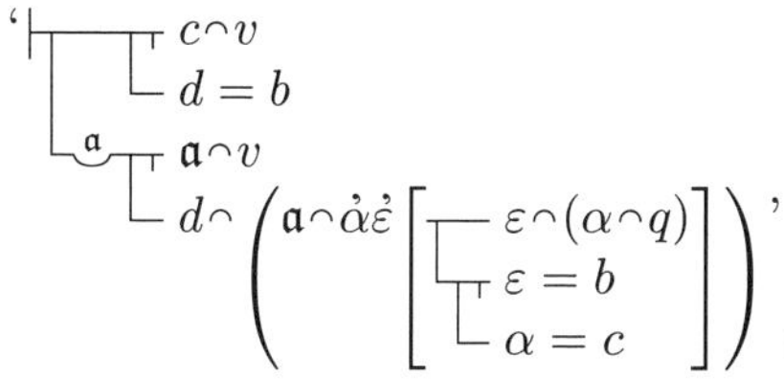

(α

which in conjunction with (60) leads to the proposition (ε) §78. If we write (IIa) in the form

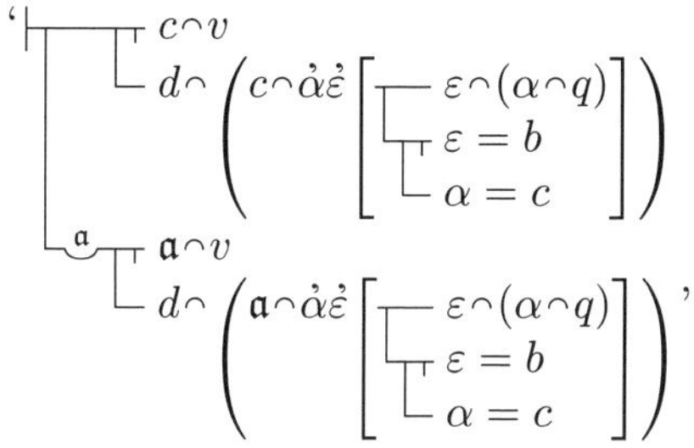

(β

then

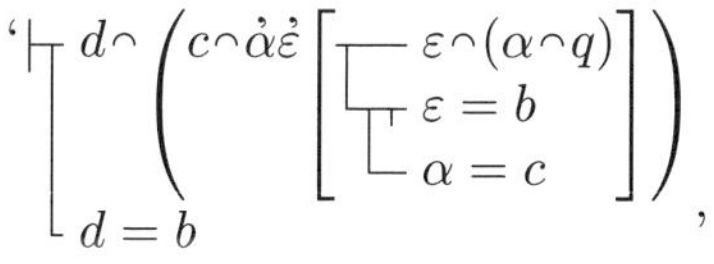

(γ

is still to be proven, which follows, using (36), from

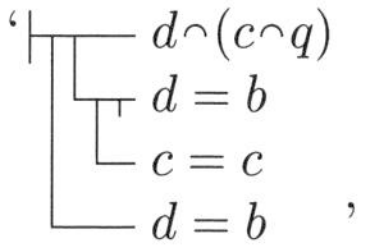

(δ

This proposition results from (Ia) in the form

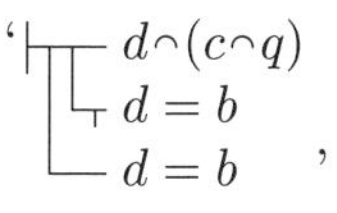

and (I) and (IIIe).

§81. *Construction*

IIIe $\vdash c = c$

(I): ——

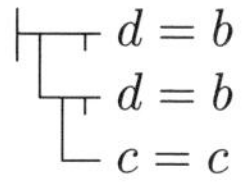

(Ia): – – – –

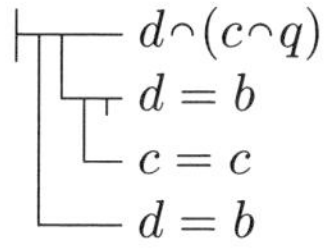

(36): – – – – – –

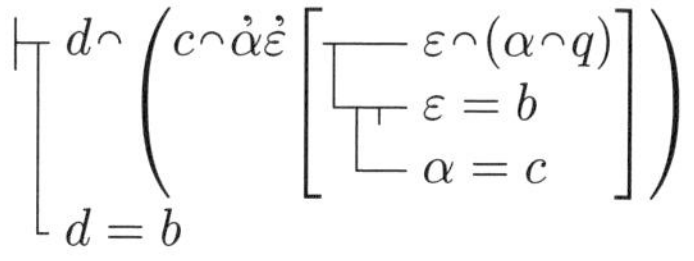

(IIa): – – – – – – – – – – – – – –

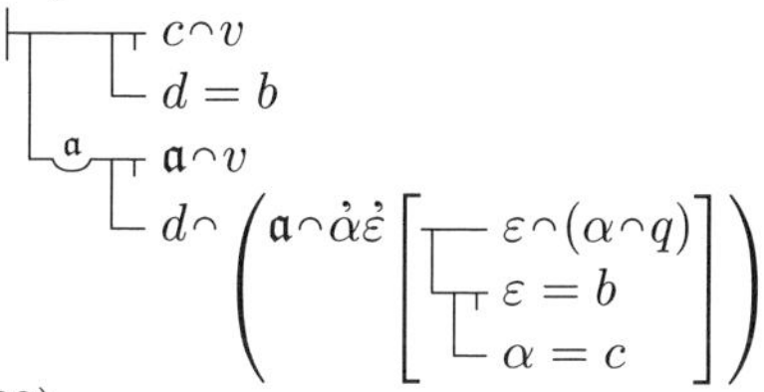

(60):: – – – – – – – – – – – – – –

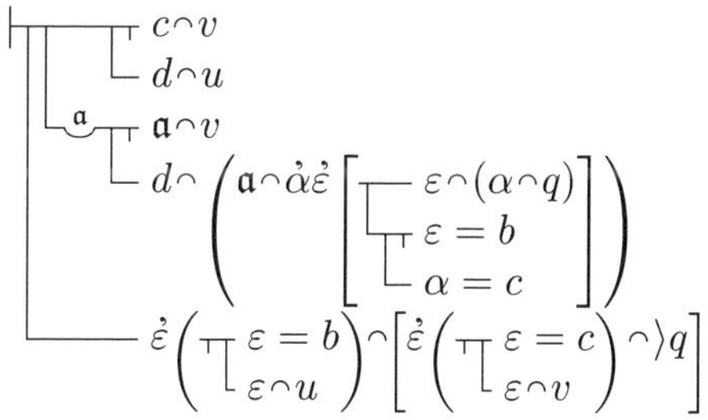

×

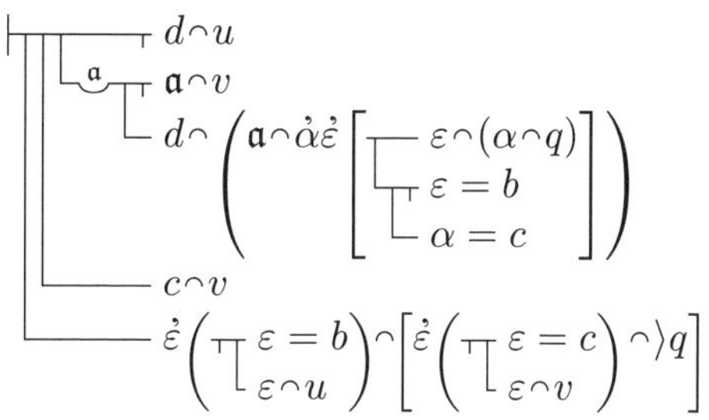

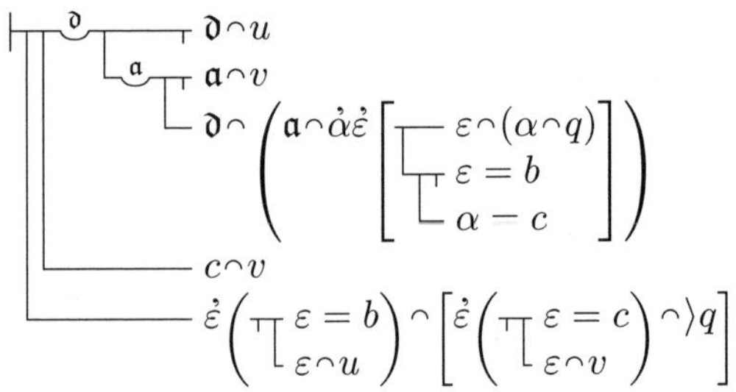

$$(11): \quad \begin{array}{l} \vdash\!\!\!\!\!\begin{array}{l} \partial\frown u \\ \mathfrak{a}\frown v \\ \partial\frown \left(\mathfrak{a}\frown\dot{\alpha}\dot{\varepsilon}\left[\begin{array}{l} \varepsilon\frown(\alpha\frown q) \\ \varepsilon = b \\ \alpha = c \end{array}\right]\right) \\ c\frown v \\ \dot{\varepsilon}\left(\prod_{\varepsilon\frown u}\varepsilon = b\right)\frown\left[\dot{\varepsilon}\left(\prod_{\varepsilon\frown v}\varepsilon = c\right)\frown q\right] \end{array} \\ \text{\rule{6cm}{0.4pt}} \\ \vdash\!\!\!\!\!\begin{array}{l} u\frown \left(v\frown\right)\dot{\alpha}\dot{\varepsilon}\left[\begin{array}{l} \varepsilon\frown(\alpha\frown q) \\ \varepsilon = b \\ \alpha = c \end{array}\right] \\ c\frown v \\ \dot{\varepsilon}\left(\prod_{\varepsilon\frown u}\varepsilon = b\right)\frown\left[\dot{\varepsilon}\left(\prod_{\varepsilon\frown v}\varepsilon = c\right)\frown q\right] \\ I\dot{\alpha}\dot{\varepsilon}\left[\begin{array}{l} \varepsilon\frown(\alpha\frown q) \\ \varepsilon = b \\ \alpha = c \end{array}\right] \end{array} \end{array} \qquad (61$$

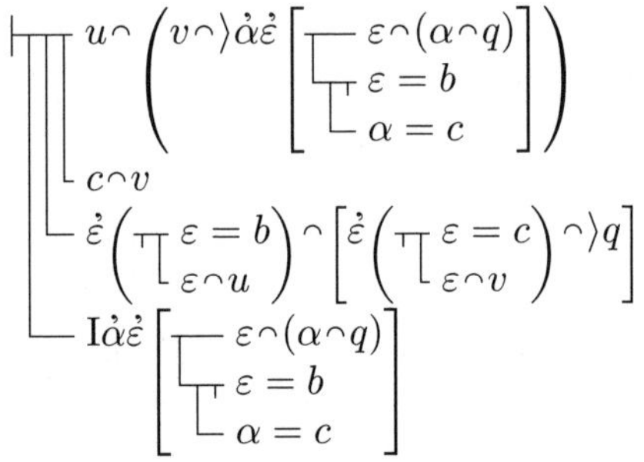

§82. Analysis

We lack a proof of the proposition

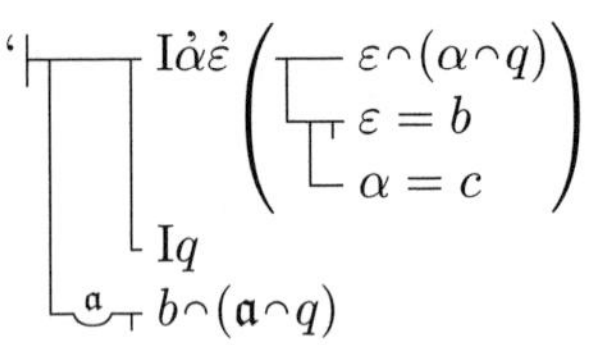

$$\text{`} \vdash\!\!\!\!\!\begin{array}{l} I\dot{\alpha}\dot{\varepsilon}\left(\begin{array}{l} \varepsilon\frown(\alpha\frown q) \\ \varepsilon = b \\ \alpha = c \end{array}\right) \\ Iq \\ \mathfrak{a}\!\!\!\!-\!\!\!\!\top b\frown(\mathfrak{a}\frown q) \end{array} \text{'} \qquad (\alpha$$

(compare §78, γ). In order to conduct it using (34) we need the proposition

$$\text{`} \vdash\!\!\!\!\!\begin{array}{l} d = a \\ e\frown(a\frown q) \\ e = b \\ a = c \\ e\frown(d\frown q) \\ e = b \\ d = c \\ Iq \\ \mathfrak{a}\!\!\!\!-\!\!\!\!\top b\frown(\mathfrak{a}\frown q) \end{array} \text{'} \qquad (\beta$$

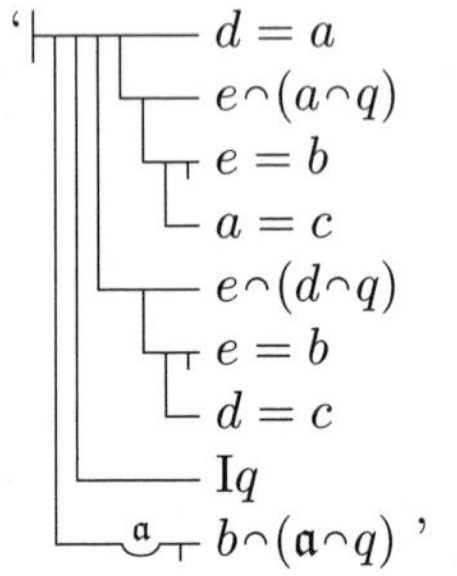

which we may derive from the propositions

$$\text{`} \vdash\!\!\!\!\!\begin{array}{l} d = a \\ e = b \\ e\frown(a\frown q) \\ e = b \\ a = c \\ e\frown(d\frown q) \\ e = b \\ d = c \\ Iq \end{array} \text{'} \qquad (\gamma$$

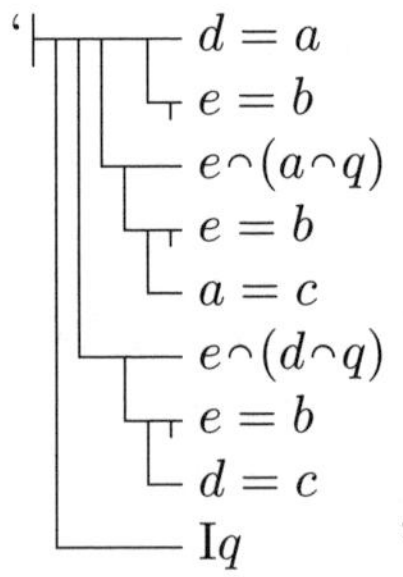

and

$$\text{`} \vdash\!\!\!\!\!\begin{array}{l} d = a \\ \mathfrak{a}\!\!\!\!-\!\!\!\!\top b\frown(\mathfrak{a}\frown q) \\ e\frown(a\frown q) \\ e = b \\ a = c \\ e\frown(d\frown q) \\ e = b \\ d = c \\ e = b \end{array} \text{'} \qquad (\delta$$

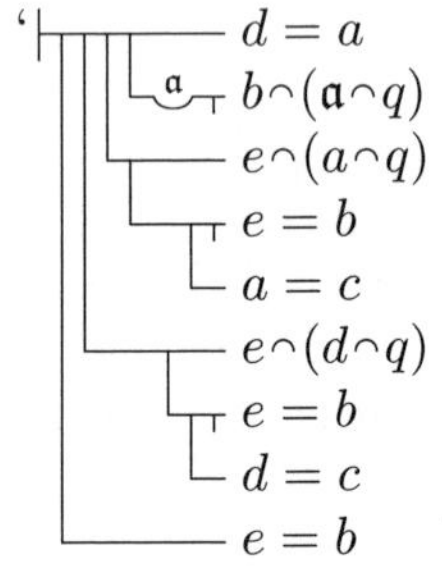

by rule (8). (γ) follows from (13) using (I) in the forms

‘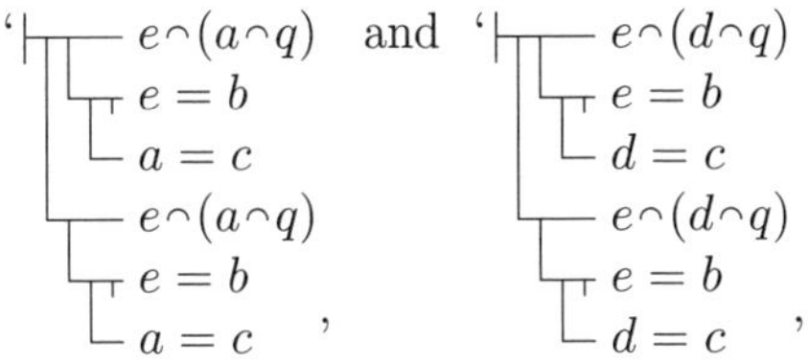’
e⌒(a⌒q)
e = b
a = c
e⌒(a⌒q)
e = b
a = c
and
e⌒(d⌒q)
e = b
d = c
e⌒(d⌒q)
e = b
d = c

The subcomponents ‘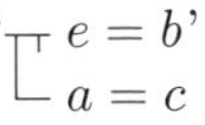’ and ‘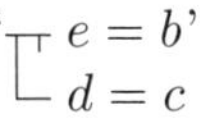’ initially occur here. These can be replaced by ‘⊤ $e = b$’, using (I) in the forms
e = b
a = c
e = b
d = c

‘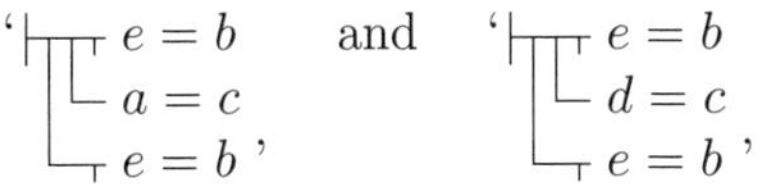’
e = b
a = c
e = b
and
e = b
d = c
e = b

§83. *Construction*

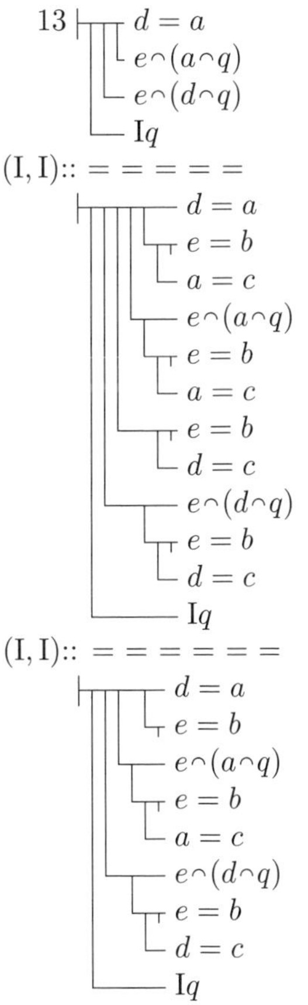
13
d = a
e⌒(a⌒q)
e⌒(d⌒q)
Iq
(I, I)::
d = a
e = b
a = c
e⌒(a⌒q)
e = b
a = c
e = b
d = c
e⌒(d⌒q)
e = b
d = c
Iq
(I, I)::
d = a
e = b
e⌒(a⌒q)
e = b
a = c
e⌒(d⌒q)
e = b
d = c
Iq

(α appears beside the first derived formula; (62 beside the last.

§84. *Analysis*

In order to prove the proposition (δ) of §82, we now note that

‘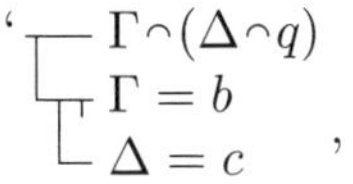’
Γ⌒(Δ⌒q)
Γ = b
Δ = c

indicates the truth-value of: that Γ stands in the q-relation to Δ, or that Γ coincides with b and Δ with c. If we now take Γ for b, then only the latter of the two cases can occur if there is no object to which b stands in the q-relation; i.e., Δ must then coincide with c. Accordingly, it will be possible to prove the proposition,

‘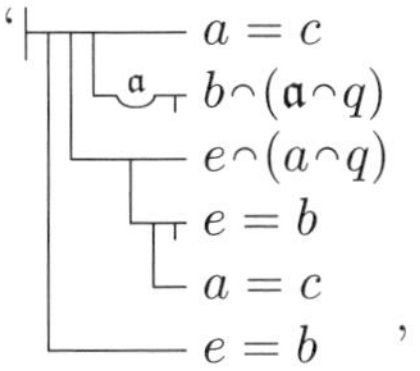’ (α
a = c
𝔞
b⌒(𝔞⌒q)
e⌒(a⌒q)
e = b
a = c
e = b

which, in a first step, follows from

‘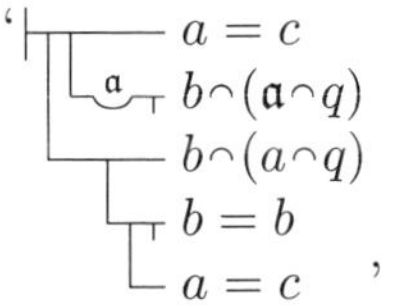’ (β
a = c
𝔞
b⌒(𝔞⌒q)
b⌒(a⌒q)
b = b
a = c

If we now write (I) in the form

‘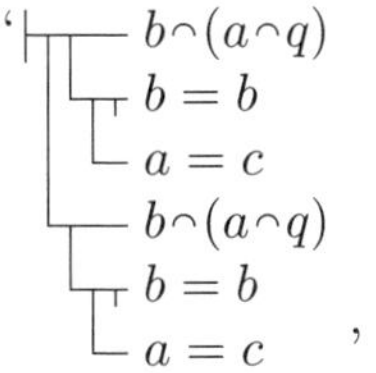’ (γ
b⌒(a⌒q)
b = b
a = c
b⌒(a⌒q)
b = b
a = c

we can then apply (Ia) to it in the form

‘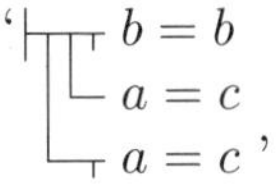’
b = b
a = c
a = c

and then, by contraposition and (IIa), easily arrive at our proposition (β). We then replace 'a' by 'd' in the proposition (α) and, using (IIIa) in the form,

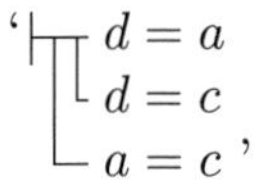

we obtain our goal.

§85. *Construction*

Ia 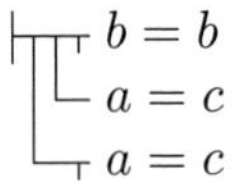

(I): – – – –

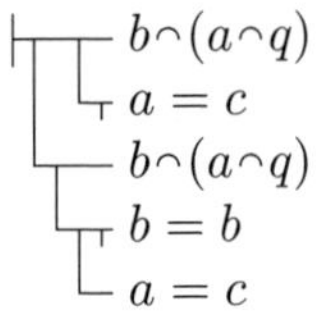

(α

×

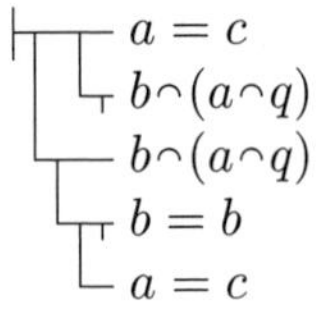

(β

(IIa):: – – – – – –

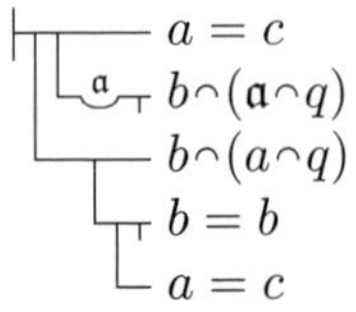

(γ

(IIIa): ————————

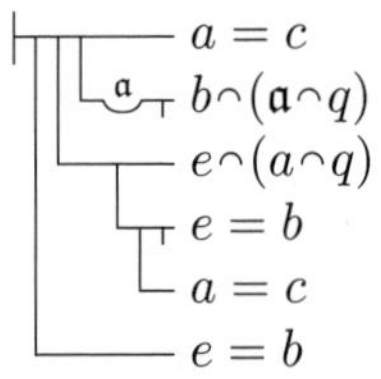

(δ

(IIIa): – – – – – – –

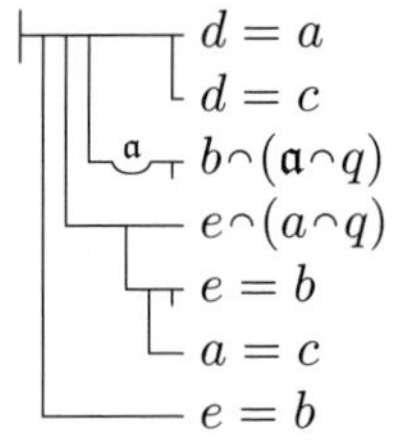

(ε

(δ):: – – – – – – –

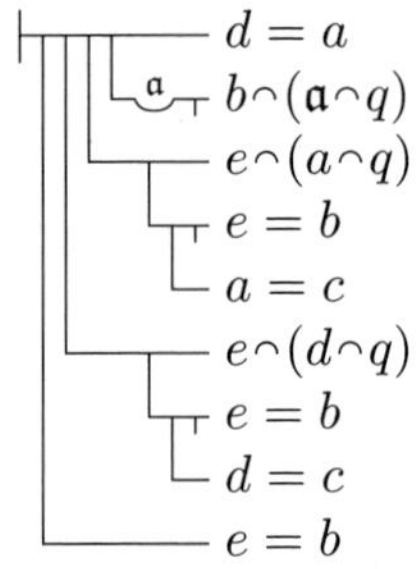

(ζ

(62): ·–·–·–·–·–·–·–·

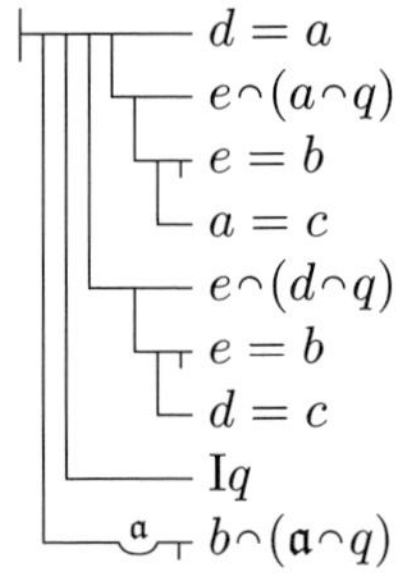

(η

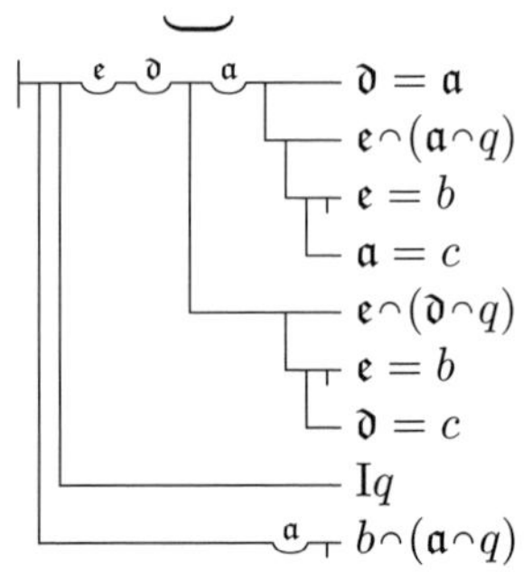

(ϑ

(34): – – – – – – – – – – – –

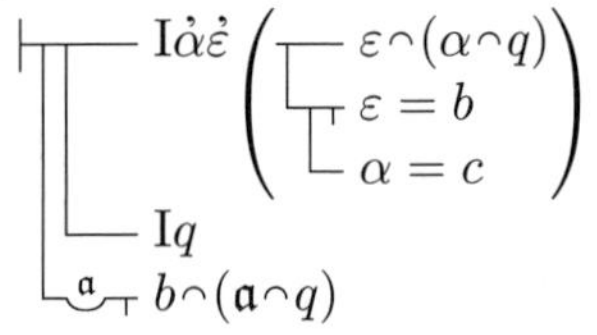

(ι

(61): – – – – – – – – – – – –

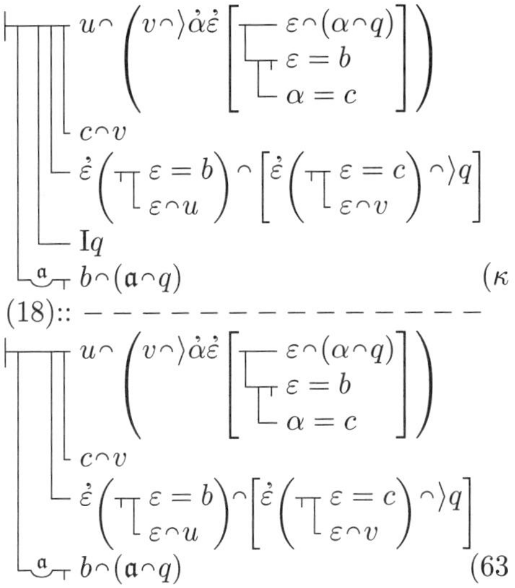

§86. *Analysis*

We have thus proven the proposition (γ) of §78. In order to derive (δ), we interchange 'q' with '$\rotatebox{90}{$\curlyvee$}q$', 'b' with 'c', 'u' with 'v' in (63). We thus get (63) in the form

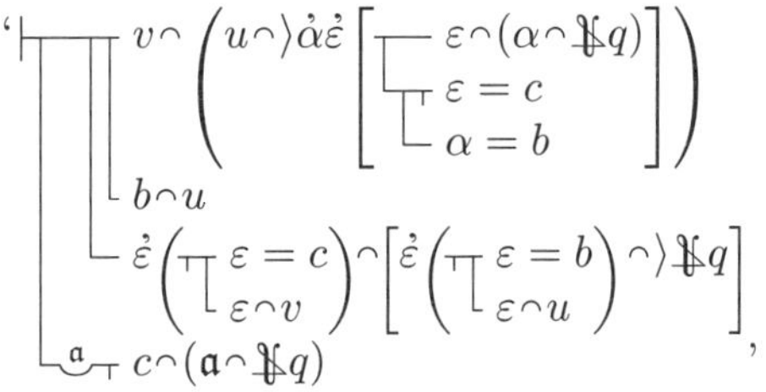

and merely have to supply the proof of the proposition

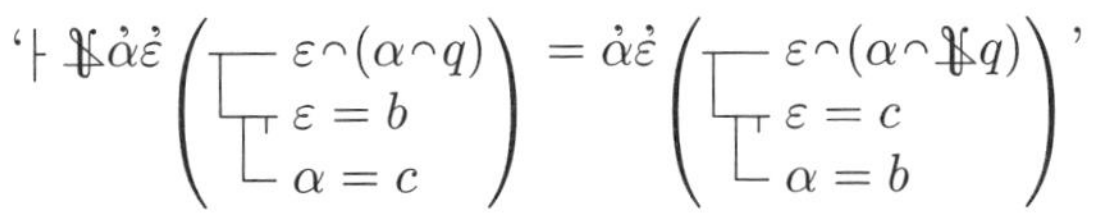

which is similar to that of §75 (κ).

§87. *Construction*

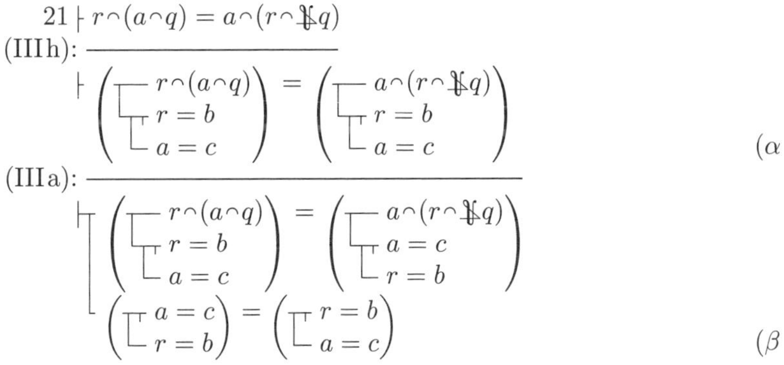

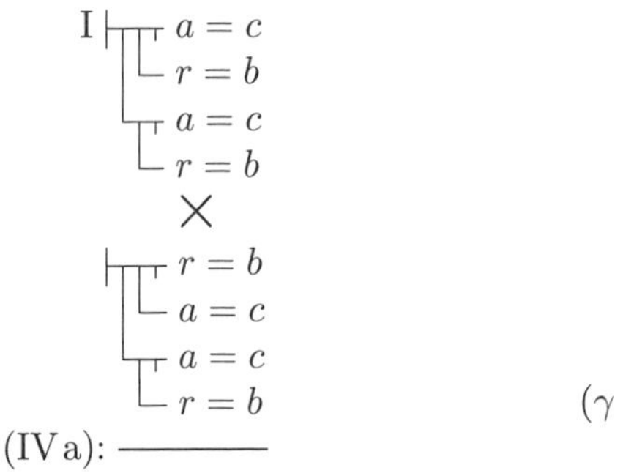

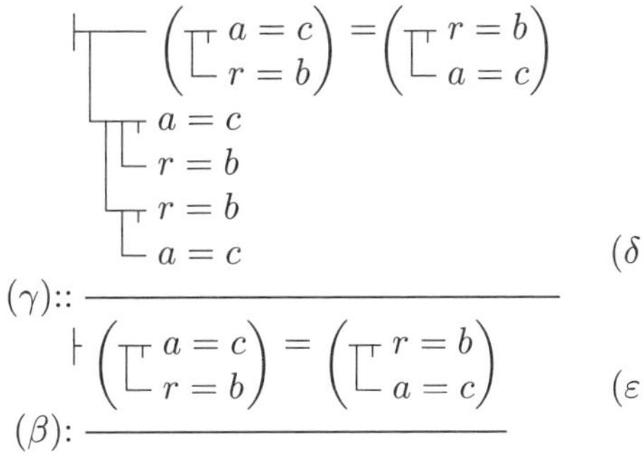

$$\vdash \left(\begin{array}{l} \top\!\!-\, r\smile(a\smile q) \\ \;\;\vdash r = b \\ \;\;\llcorner\, a = c \end{array}\right) = \left(\begin{array}{l} \top\!\!-\, a\smile(r\smile \natural q) \\ \;\;\vdash a = c \\ \;\;\llcorner\, r = b \end{array}\right)$$

$$\vdash^{\mathfrak{a}}\!\!\smile\!\!{}^{\mathfrak{e}}\!\!\smile \left(\begin{array}{l} \top\!\!-\, \mathfrak{a}\smile(\mathfrak{e}\smile q) \\ \;\;\vdash \mathfrak{a} = b \\ \;\;\llcorner\, \mathfrak{e} = c \end{array}\right) = \left(\begin{array}{l} \top\!\!-\, \mathfrak{e}\smile(\mathfrak{a}\smile \natural q) \\ \;\;\vdash \mathfrak{e} = c \\ \;\;\llcorner\, \mathfrak{a} = b \end{array}\right) \qquad (\zeta$$

(20): ——————————————————————

$$\vdash \dot{\alpha}\dot{\varepsilon}\left(\begin{array}{l} \top\!\!-\, \alpha\smile(\varepsilon\smile q) \\ \;\;\vdash \alpha = b \\ \;\;\llcorner\, \varepsilon = c \end{array}\right) = \dot{\alpha}\dot{\varepsilon}\left(\begin{array}{l} \top\!\!-\, \varepsilon\smile(\alpha\smile \natural q) \\ \;\;\vdash \varepsilon = c \\ \;\;\llcorner\, \alpha = b \end{array}\right) \qquad (\eta$$

(IIIc): ——————————————————————

$$\begin{array}{l} \vdash\!\top\, \natural\dot{\alpha}\dot{\varepsilon}\left(\begin{array}{l} \top\!\!-\, \varepsilon\smile(\alpha\smile q) \\ \;\;\vdash \varepsilon = b \\ \;\;\llcorner\, \alpha = c \end{array}\right) = \dot{\alpha}\dot{\varepsilon}\left(\begin{array}{l} \top\!\!-\, \varepsilon\smile(\alpha\smile \natural q) \\ \;\;\vdash \varepsilon = c \\ \;\;\llcorner\, \alpha = b \end{array}\right) \\ \;\llcorner\, \dot{\alpha}\dot{\varepsilon}\left(\begin{array}{l} \top\!\!-\, \alpha\smile(\varepsilon\smile q) \\ \;\;\vdash \alpha = b \\ \;\;\llcorner\, \varepsilon = c \end{array}\right) = \natural\dot{\alpha}\dot{\varepsilon}\left(\begin{array}{l} \top\!\!-\, \varepsilon\smile(\alpha\smile q) \\ \;\;\vdash \varepsilon = b \\ \;\;\llcorner\, \alpha = c \end{array}\right) \end{array} \qquad (\vartheta$$

(40):: ——————————————————————

$$\vdash \natural\dot{\alpha}\dot{\varepsilon}\left(\begin{array}{l} \top\!\!-\, \varepsilon\smile(\alpha\smile q) \\ \;\;\vdash \varepsilon = b \\ \;\;\llcorner\, \alpha = c \end{array}\right) = \dot{\alpha}\dot{\varepsilon}\left(\begin{array}{l} \top\!\!-\, \varepsilon\smile(\alpha\smile \natural q) \\ \;\;\vdash \varepsilon = c \\ \;\;\llcorner\, \alpha = b \end{array}\right) \qquad (\iota$$

(IIIa): ——————————————————————

$$\begin{array}{l} \vdash\!\top\, v\smile\left(u\smile\rangle\natural\dot{\alpha}\dot{\varepsilon}\left[\begin{array}{l} \top\!\!-\, \varepsilon\smile(\alpha\smile q) \\ \;\;\vdash \varepsilon = b \\ \;\;\llcorner\, \alpha = c \end{array}\right]\right) \\ \;\llcorner\, v\smile\left(u\smile\rangle\dot{\alpha}\dot{\varepsilon}\left[\begin{array}{l} \top\!\!-\, \varepsilon\smile(\alpha\smile \natural q) \\ \;\;\vdash \varepsilon = c \\ \;\;\llcorner\, \alpha = b \end{array}\right]\right) \end{array} \qquad (\kappa$$

(63):: − − − − − − − − − − − − − − − −

$$\begin{array}{l} \vdash\!\top\!\top\!\top\, v\smile\left(u\smile\rangle\natural\dot{\alpha}\dot{\varepsilon}\left[\begin{array}{l} \top\!\!-\, \varepsilon\smile(\alpha\smile q) \\ \;\;\vdash \varepsilon = b \\ \;\;\llcorner\, \alpha = c \end{array}\right]\right) \\ \;\;\;\;\llcorner\, b\smile u \\ \;\;\;\llcorner\, \dot{\varepsilon}\left(\begin{array}{l} \top\, \varepsilon = c \\ \llcorner\, \varepsilon\smile v \end{array}\right)\smile\left[\dot{\varepsilon}\left(\begin{array}{l} \top\, \varepsilon = b \\ \llcorner\, \varepsilon\smile u \end{array}\right)\smile\rangle\natural q\right] \\ \;\llcorner\!\overset{\mathfrak{a}}{\smile}\!\top\, c\smile(\mathfrak{a}\smile q) \end{array} \qquad (\lambda$$

(32): −

$$\vdash \; \mathfrak{p}u = \mathfrak{p}v$$
$$u \frown \left(v \frown \rangle \dot{\alpha}\dot{\varepsilon} \left[\begin{array}{l} \varepsilon \frown (\alpha \frown q) \\ \varepsilon = b \\ \alpha = c \end{array} \right] \right)$$
$$b \frown u$$
$$\dot{\varepsilon}\left(\begin{array}{l} \varepsilon = c \\ \varepsilon \frown v \end{array} \right) \frown \left[\dot{\varepsilon}\left(\begin{array}{l} \varepsilon = b \\ \varepsilon \frown u \end{array} \right) \frown \rangle \rotatebox{180}{$\rangle$}q \right]$$
$$\mathfrak{a} \; c \frown (\mathfrak{a} \frown \rotatebox{180}{$\rangle$}q) \qquad (\mu$$

(63):: -

$$\vdash \; \mathfrak{p}u = \mathfrak{p}v$$
$$c \frown v$$
$$\dot{\varepsilon}\left(\begin{array}{l} \varepsilon = b \\ \varepsilon \frown u \end{array} \right) \frown \left[\dot{\varepsilon}\left(\begin{array}{l} \varepsilon = c \\ \varepsilon \frown v \end{array} \right) \frown \rangle q \right]$$
$$\mathfrak{a} \; b \frown (\mathfrak{a} \frown q)$$
$$b \frown u$$
$$\dot{\varepsilon}\left(\begin{array}{l} \varepsilon = c \\ \varepsilon \frown v \end{array} \right) \frown \left[\dot{\varepsilon}\left(\begin{array}{l} \varepsilon = b \\ \varepsilon \frown u \end{array} \right) \frown \rangle \rotatebox{180}{$\rangle$}q \right]$$
$$\mathfrak{a} \; c \frown (\mathfrak{a} \frown \rotatebox{180}{$\rangle$}q) \qquad (\nu$$

×

$$\vdash \; \dot{\varepsilon}\left(\begin{array}{l} \varepsilon = b \\ \varepsilon \frown u \end{array} \right) \frown \left[\dot{\varepsilon}\left(\begin{array}{l} \varepsilon = c \\ \varepsilon \frown v \end{array} \right) \frown \rangle q \right]$$
$$\dot{\varepsilon}\left(\begin{array}{l} \varepsilon = c \\ \varepsilon \frown v \end{array} \right) \frown \left[\dot{\varepsilon}\left(\begin{array}{l} \varepsilon = b \\ \varepsilon \frown u \end{array} \right) \frown \rangle \rotatebox{180}{$\rangle$}q \right]$$
$$\mathfrak{a} \; b \frown (\mathfrak{a} \frown q)$$
$$\mathfrak{a} \; c \frown (\mathfrak{a} \frown \rotatebox{180}{$\rangle$}q)$$
$$b \frown u$$
$$c \frown v$$
$$\mathfrak{p}u = \mathfrak{p}v \qquad (\xi$$

$$\vdash \; \mathfrak{q} \; \dot{\varepsilon}\left(\begin{array}{l} \varepsilon = b \\ \varepsilon \frown u \end{array} \right) \frown \left[\dot{\varepsilon}\left(\begin{array}{l} \varepsilon = c \\ \varepsilon \frown v \end{array} \right) \frown \rangle \mathfrak{q} \right]$$
$$\dot{\varepsilon}\left(\begin{array}{l} \varepsilon = c \\ \varepsilon \frown v \end{array} \right) \frown \left[\dot{\varepsilon}\left(\begin{array}{l} \varepsilon = b \\ \varepsilon \frown u \end{array} \right) \frown \rangle \rotatebox{180}{$\rangle$}\mathfrak{q} \right]$$
$$\mathfrak{a} \; b \frown (\mathfrak{a} \frown \mathfrak{q})$$
$$\mathfrak{a} \; c \frown (\mathfrak{a} \frown \rotatebox{180}{$\rangle$}\mathfrak{q})$$
$$b \frown u$$
$$c \frown v$$
$$\mathfrak{p}u = \mathfrak{p}v \qquad (o$$

(56): -

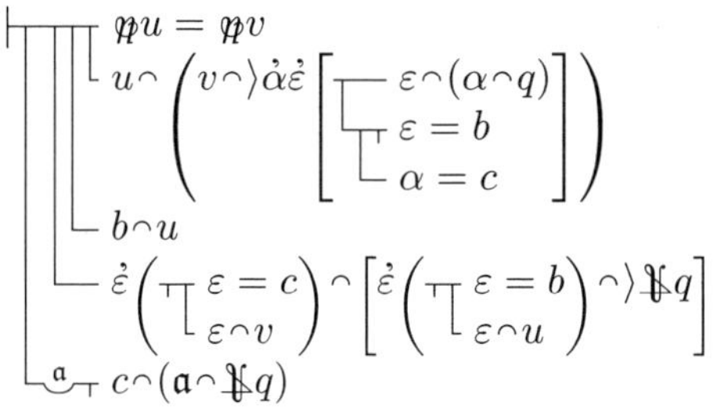

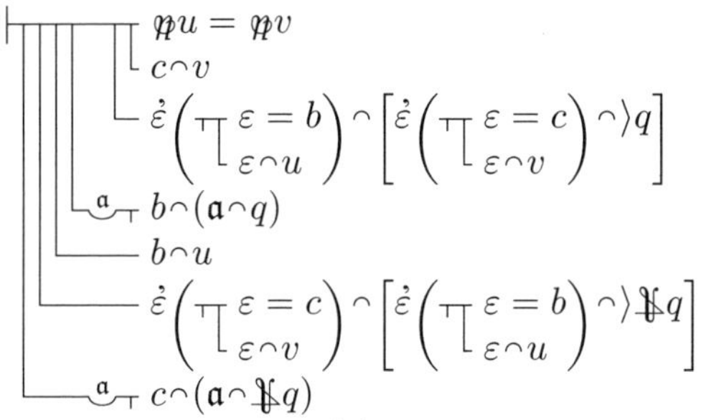

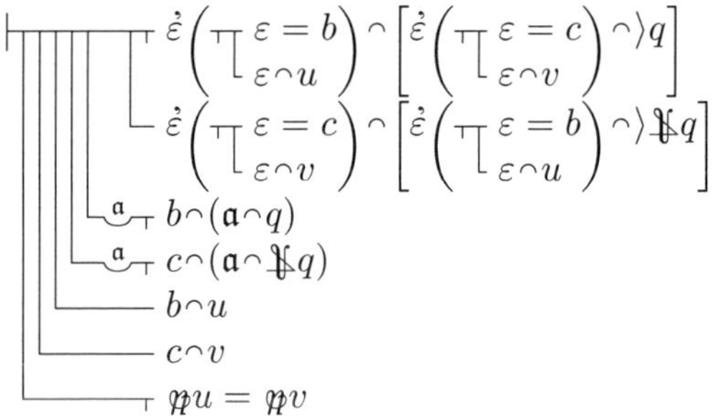

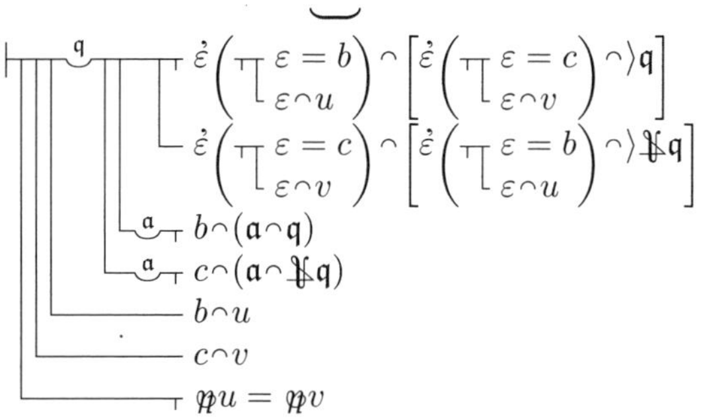

$$\vdash \mathfrak{p}\dot{\varepsilon}\left(\top\!\!\left[\begin{matrix}\varepsilon = b\\ \varepsilon\frown u\end{matrix}\right]\right) = \mathfrak{p}\dot{\varepsilon}\left(\top\!\!\left[\begin{matrix}\varepsilon = c\\ \varepsilon\frown v\end{matrix}\right]\right)$$
$$\quad \top\, b\frown\dot{\varepsilon}\left(\top\!\!\left[\begin{matrix}\varepsilon = b\\ \varepsilon\frown u\end{matrix}\right]\right)$$
$$\quad \top\, c\frown\dot{\varepsilon}\left(\top\!\!\left[\begin{matrix}\varepsilon = c\\ \varepsilon\frown v\end{matrix}\right]\right)$$
$$\quad b\frown u$$
$$\quad c\frown v$$
$$\quad \top\, \mathfrak{p}u = \mathfrak{p}v \qquad (64$$

——— • ———

IIIe $\vdash b = b$

(I): ———

$$\vdash b = b$$
$$\quad b\frown u \qquad (\alpha$$

(58): ———

$$\vdash b\frown\dot{\varepsilon}\left(\top\!\!\left[\begin{matrix}\varepsilon = b\\ \varepsilon\frown u\end{matrix}\right]\right) \qquad (65$$

(64): ———

$$\vdash \mathfrak{p}\dot{\varepsilon}\left(\top\!\!\left[\begin{matrix}\varepsilon = b\\ \varepsilon\frown u\end{matrix}\right]\right) = \mathfrak{p}\dot{\varepsilon}\left(\top\!\!\left[\begin{matrix}\varepsilon = c\\ \varepsilon\frown v\end{matrix}\right]\right)$$
$$\quad \top\, c\frown\dot{\varepsilon}\left(\top\!\!\left[\begin{matrix}\varepsilon = c\\ \varepsilon\frown v\end{matrix}\right]\right)$$
$$\quad b\frown u$$
$$\quad c\frown v$$
$$\quad \top\, \mathfrak{p}u = \mathfrak{p}v \qquad (\alpha$$

(65):: ———

$$\vdash \mathfrak{p}\dot{\varepsilon}\left(\top\!\!\left[\begin{matrix}\varepsilon = b\\ \varepsilon\frown u\end{matrix}\right]\right) = \mathfrak{p}\dot{\varepsilon}\left(\top\!\!\left[\begin{matrix}\varepsilon = c\\ \varepsilon\frown v\end{matrix}\right]\right)$$
$$\quad b\frown u$$
$$\quad c\frown v$$
$$\quad \top\, \mathfrak{p}u = \mathfrak{p}v \qquad (\beta$$

×

$$\vdash \mathfrak{p}u = \mathfrak{p}v$$
$$\quad b\frown u$$
$$\quad \mathfrak{p}\dot{\varepsilon}\left(\top\!\!\left[\begin{matrix}\varepsilon = b\\ \varepsilon\frown u\end{matrix}\right]\right) = \mathfrak{p}\dot{\varepsilon}\left(\top\!\!\left[\begin{matrix}\varepsilon = c\\ \varepsilon\frown v\end{matrix}\right]\right)$$
$$\quad c\frown v \qquad (66$$

(IIIa): - - - - - - - - - - - - - -

$$\vdash \top\, \mathfrak{p}u = a$$
$$\quad \top\, \mathfrak{p}v = a$$
$$\quad b\frown u$$
$$\quad \mathfrak{p}\dot{\varepsilon}\left(\top\!\!\left[\begin{matrix}\varepsilon = b\\ \varepsilon\frown u\end{matrix}\right]\right) = \mathfrak{p}\dot{\varepsilon}\left(\top\!\!\left[\begin{matrix}\varepsilon = c\\ \varepsilon\frown v\end{matrix}\right]\right)$$
$$\quad c\frown v \qquad (\alpha$$

(IIIc): - - - - - - - - - - - - - -

$$\vdash \top\, \mathfrak{p}u = a$$
$$\quad b\frown u$$
$$\quad \mathfrak{p}\dot{\varepsilon}\left(\top\!\!\left[\begin{matrix}\varepsilon = b\\ \varepsilon\frown u\end{matrix}\right]\right) = e$$
$$\quad \top\, \mathfrak{p}v = a$$
$$\quad c\frown v$$
$$\quad \mathfrak{p}\dot{\varepsilon}\left(\top\!\!\left[\begin{matrix}\varepsilon = c\\ \varepsilon\frown v\end{matrix}\right]\right) = e \qquad (\beta$$

⏟

$$\vdash \overset{\mathfrak{u}\;\;\mathfrak{a}}{\smile\smile}\top\, \mathfrak{p}\mathfrak{u} = a$$
$$\quad \mathfrak{a}\frown\mathfrak{u}$$
$$\quad \mathfrak{p}\dot{\varepsilon}\left(\top\!\!\left[\begin{matrix}\varepsilon = \mathfrak{a}\\ \varepsilon\frown \mathfrak{u}\end{matrix}\right]\right) = e$$
$$\quad \top\, \mathfrak{p}v = a$$
$$\quad c\frown v$$
$$\quad \mathfrak{p}\dot{\varepsilon}\left(\top\!\!\left[\begin{matrix}\varepsilon = c\\ \varepsilon\frown v\end{matrix}\right]\right) = e \qquad (67$$

——— • ———

$$\text{H} \vdash \dot{\alpha}\dot{\varepsilon}\left[\overset{\mathfrak{u}\;\;\mathfrak{a}}{\smile\smile}\top\, \mathfrak{p}\mathfrak{u} = \alpha \quad \mathfrak{a}\frown\mathfrak{u} \quad \mathfrak{p}\dot{\varepsilon}\left(\top\!\!\left[\begin{matrix}\varepsilon = \mathfrak{a}\\ \varepsilon\frown \mathfrak{u}\end{matrix}\right]\right) = \varepsilon\right] = \mathfrak{f}$$

(14): ———

$$\vdash \top\, e\frown(a\frown\mathfrak{f})$$
$$\quad \overset{\mathfrak{u}\;\;\mathfrak{a}}{\smile\smile}\top\, \mathfrak{p}\mathfrak{u} = a$$
$$\quad \mathfrak{a}\frown\mathfrak{u}$$
$$\quad \mathfrak{p}\dot{\varepsilon}\left(\top\!\!\left[\begin{matrix}\varepsilon = \mathfrak{a}\\ \varepsilon\frown \mathfrak{u}\end{matrix}\right]\right) = e \qquad (68$$

(67):: - - - - - - - - - - - - -

$$\vdash \top\, e\frown(a\frown\mathfrak{f})$$
$$\quad \top\, \mathfrak{p}v = a$$
$$\quad c\frown v$$
$$\quad \mathfrak{p}\dot{\varepsilon}\left(\top\!\!\left[\begin{matrix}\varepsilon = c\\ \varepsilon\frown v\end{matrix}\right]\right) = e \qquad (69$$

(IIIc): ———

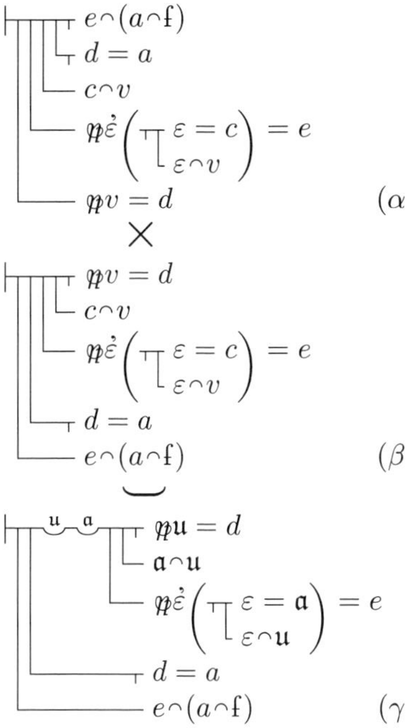

(68): – – – – – – – – – – – – –

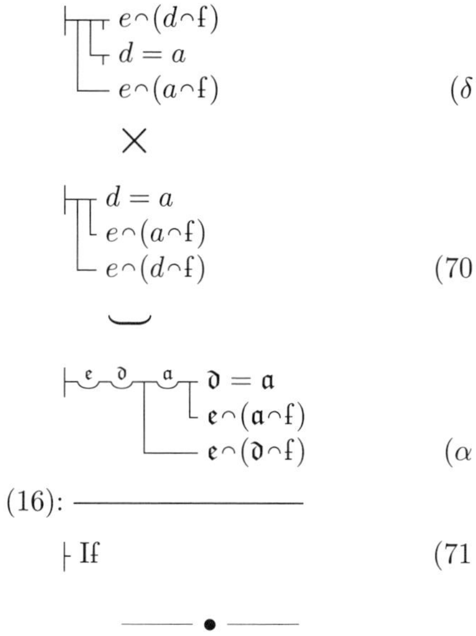

Γ. Proof of the proposition

‘⊦ I𝔉f’

a) Proof of the proposition

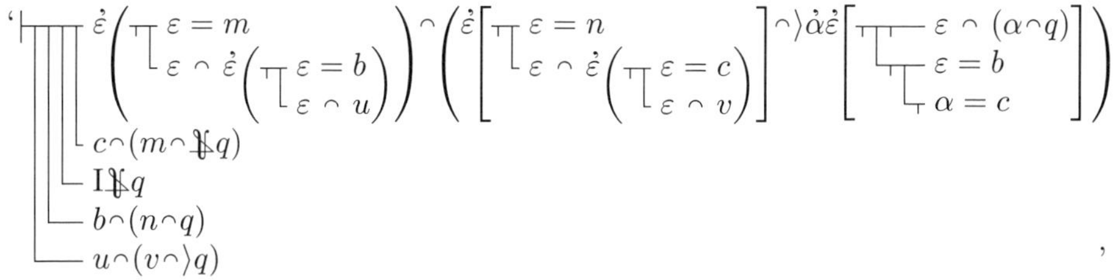

§88. *Analysis*

We now want to prove the proposition that, for every cardinal number, there is no more than one that immediately precedes it in the cardinal number series. This takes us back to the proposition

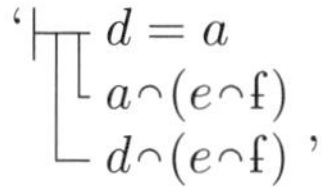

(α

If we here introduce the expressions resulting according to definition (H), then we have

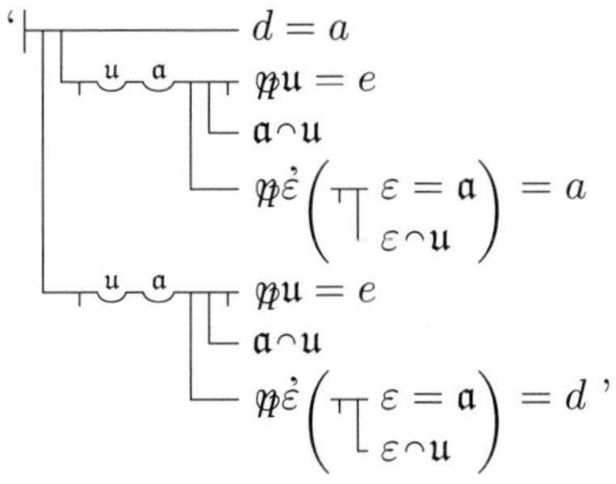

(β

a proposition which is obtained by repeated applications of contraposition and introduction of German letters from

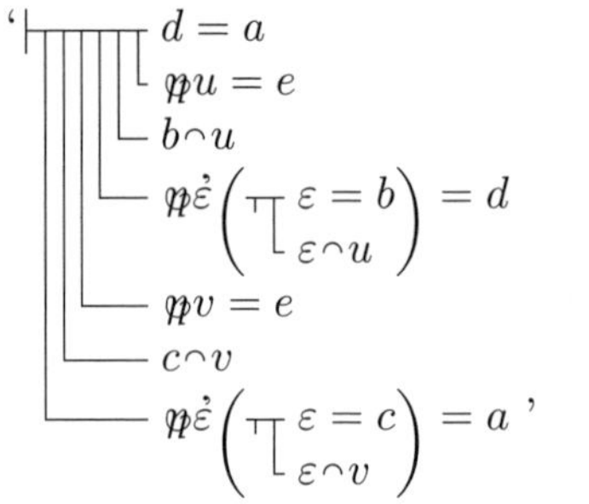

(γ

This proposition can be derived from

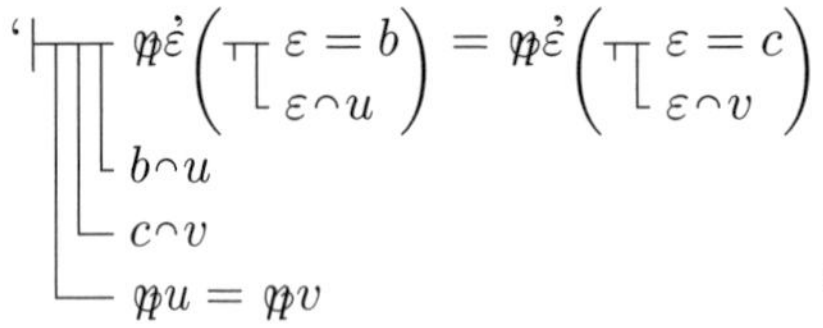

’ (δ

According to proposition (32), we now have only to disclose a relation that maps the $\dot{\varepsilon}\left(\begin{smallmatrix}\varepsilon = b\\ \varepsilon \frown u\end{smallmatrix}\right)$-concept into the $\dot{\varepsilon}\left(\begin{smallmatrix}\varepsilon = c\\ \varepsilon \frown v\end{smallmatrix}\right)$-concept and whose converse maps the latter into the former. The subcomponent ‘$\mathfrak{p}u = \mathfrak{p}v$’ tells us that there is a relation which maps the u-concept into the v-concept and whose converse maps the latter into the former. Let the q-relation be such a relation. We now know of the $\dot{\alpha}\dot{\varepsilon}\left(\begin{smallmatrix}\varepsilon \frown (\alpha \frown q)\\ \varepsilon = b\\ \alpha = c\end{smallmatrix}\right)$-relation that no object stands to c, and b stands to no object in this relation.[1] Moreover, no object stands to n and m stands to no object in this relation if m stands to c and b to n in the q-relation, since the latter, like its converse, is single-valued. The former relation maps the $\dot{\varepsilon}\left(\begin{smallmatrix}\varepsilon = m\\ \varepsilon \frown \dot{\varepsilon}\left(\begin{smallmatrix}\varepsilon = b\\ \varepsilon \frown u\end{smallmatrix}\right)\end{smallmatrix}\right)$-concept into the $\dot{\varepsilon}\left(\begin{smallmatrix}\varepsilon = n\\ \varepsilon \frown \dot{\varepsilon}\left(\begin{smallmatrix}\varepsilon = c\\ \varepsilon \frown v\end{smallmatrix}\right)\end{smallmatrix}\right)$-concept, and its converse maps the latter into the former. By (32) the cardinal number of the latter concept is equal to the cardinal number of the former. Using (66) we can then hope to reach our goal.

First, we turn to the proof of the proposition

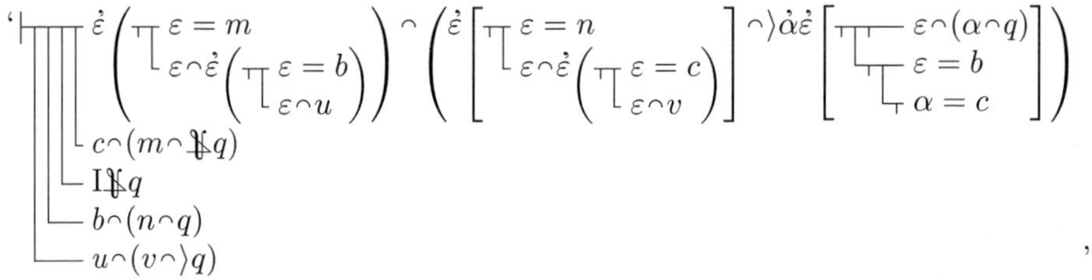

(ε

[1] Compare §66.

If we write (51) in the form

$$`\vdash\begin{array}{l}
\dot{\varepsilon}\left(\begin{array}{l}\top\,\varepsilon = m\\ \llcorner\,\varepsilon\frown\dot{\varepsilon}\left(\begin{array}{l}\top\,\varepsilon=b\\ \llcorner\,\varepsilon\frown u\end{array}\right)\end{array}\right)\frown\left(\dot{\varepsilon}\left[\begin{array}{l}\top\,\varepsilon=n\\ \llcorner\,\varepsilon\frown\dot{\varepsilon}\left(\begin{array}{l}\top\,\varepsilon=c\\ \llcorner\,\varepsilon\frown v\end{array}\right)\end{array}\right]\frown\rangle\dot{\alpha}\dot{\varepsilon}\left[\begin{array}{l}\varepsilon\frown(\alpha\frown q)\\ \varepsilon=b\\ \top\,\alpha=c\end{array}\right]\right)\\
\top\, b\frown\dot{\varepsilon}\left(\begin{array}{l}\top\,\varepsilon = m\\ \llcorner\,\varepsilon\frown\dot{\varepsilon}\left(\begin{array}{l}\top\,\varepsilon=b\\ \llcorner\,\varepsilon\frown u\end{array}\right)\end{array}\right)\\
\top\, c\frown\dot{\varepsilon}\left(\begin{array}{l}\top\,\varepsilon = n\\ \llcorner\,\varepsilon\frown\dot{\varepsilon}\left(\begin{array}{l}\top\,\varepsilon=c\\ \llcorner\,\varepsilon\frown v\end{array}\right)\end{array}\right)\\
\dot{\varepsilon}\left(\begin{array}{l}\top\,\varepsilon = m\\ \llcorner\,\varepsilon\frown\dot{\varepsilon}\left(\begin{array}{l}\top\,\varepsilon=b\\ \llcorner\,\varepsilon\frown u\end{array}\right)\end{array}\right)\frown\left(\dot{\varepsilon}\left[\begin{array}{l}\top\,\varepsilon=n\\ \llcorner\,\varepsilon\frown\dot{\varepsilon}\left(\begin{array}{l}\top\,\varepsilon=c\\ \llcorner\,\varepsilon\frown v\end{array}\right)\end{array}\right]\frown\rangle q\right)
\end{array}\quad ,$$

then we see that what primarily remains to be proven is

$$`\vdash\begin{array}{l}
\dot{\varepsilon}\left(\begin{array}{l}\top\,\varepsilon = m\\ \llcorner\,\varepsilon\frown\dot{\varepsilon}\left(\begin{array}{l}\top\,\varepsilon=b\\ \llcorner\,\varepsilon\frown u\end{array}\right)\end{array}\right)\frown\left(\dot{\varepsilon}\left[\begin{array}{l}\top\,\varepsilon=n\\ \llcorner\,\varepsilon\frown\dot{\varepsilon}\left(\begin{array}{l}\top\,\varepsilon=c\\ \llcorner\,\varepsilon\frown v\end{array}\right)\end{array}\right]\frown\rangle q\right)\\
m\frown(c\frown q)\\
\mathrm{I}\nsucc q\\
b\frown(n\frown q)\\
u\frown(v\frown\rangle q)
\end{array}\quad ,\qquad (\zeta$$

because the proposition

$$`\vdash b\frown\dot{\varepsilon}\left(\begin{array}{l}\top\,\varepsilon = m\\ \llcorner\,\varepsilon\frown\dot{\varepsilon}\left(\begin{array}{l}\top\,\varepsilon=b\\ \llcorner\,\varepsilon\frown u\end{array}\right)\end{array}\right)'$$

presents no difficulty. Once we have proven the proposition

$$`\vdash\begin{array}{l}
\dot{\varepsilon}\left(\begin{array}{l}\top\,\varepsilon=b\\ \llcorner\,\varepsilon\frown u\end{array}\right)\frown\left[\dot{\varepsilon}\left(\begin{array}{l}\top\,\varepsilon=n\\ \llcorner\,\varepsilon\frown v\end{array}\right)\frown\rangle q\right]\\
b\frown(n\frown q)\\
\mathrm{I}\nsucc q\\
u\frown(v\frown\rangle q)
\end{array}\quad ,\qquad (\eta$$

then we can apply it twice and thereby arrive at the proposition

$$`\vdash\begin{array}{l}
\dot{\varepsilon}\left(\begin{array}{l}\top\,\varepsilon = m\\ \llcorner\,\varepsilon\frown\dot{\varepsilon}\left(\begin{array}{l}\top\,\varepsilon=b\\ \llcorner\,\varepsilon\frown u\end{array}\right)\end{array}\right)\frown\left(\dot{\varepsilon}\left[\begin{array}{l}\top\,\varepsilon=c\\ \llcorner\,\varepsilon\frown\dot{\varepsilon}\left(\begin{array}{l}\top\,\varepsilon=n\\ \llcorner\,\varepsilon\frown v\end{array}\right)\end{array}\right]\frown\rangle q\right)\\
m\frown(c\frown q)\\
\mathrm{I}\nsucc q\\
b\frown(n\frown q)\\
u\frown(v\frown\rangle q)
\end{array}\quad ,\qquad (\vartheta$$

We now still require the propositions

$$`\vdash \dot{\varepsilon}g(\varepsilon, f(\varepsilon)) = \dot{\varepsilon}g(\varepsilon, \varepsilon\frown\dot{\varepsilon}f(\varepsilon))' \qquad (\iota$$

and

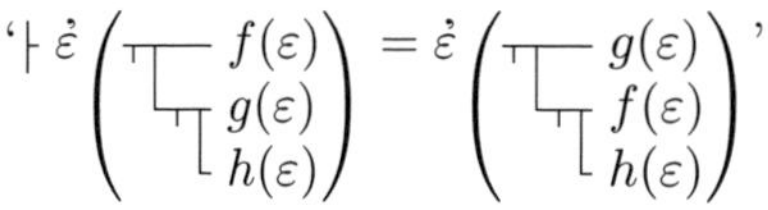

(κ

in order to be able to replace

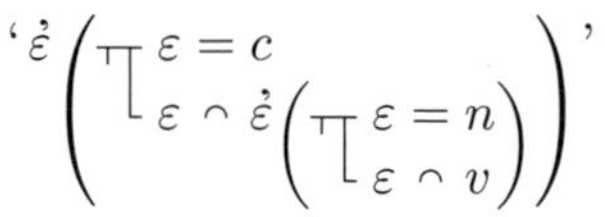

by

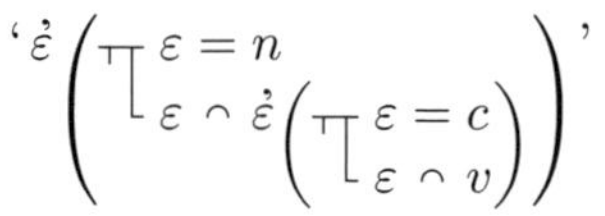

These propositions will be derived first.

§89. *Construction*

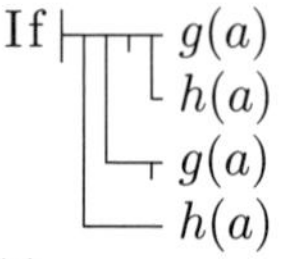

(I): – – – –

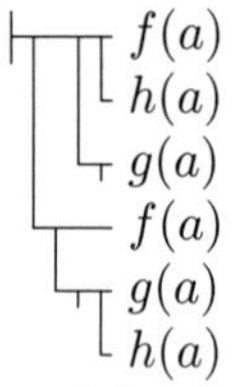

(α

×

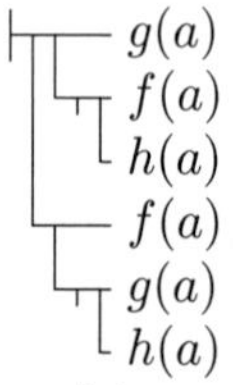

(β

×

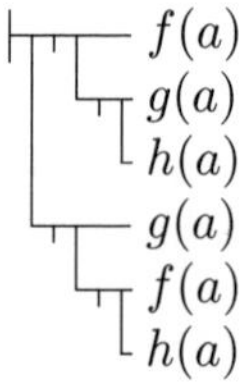

(γ

(IV a): ———

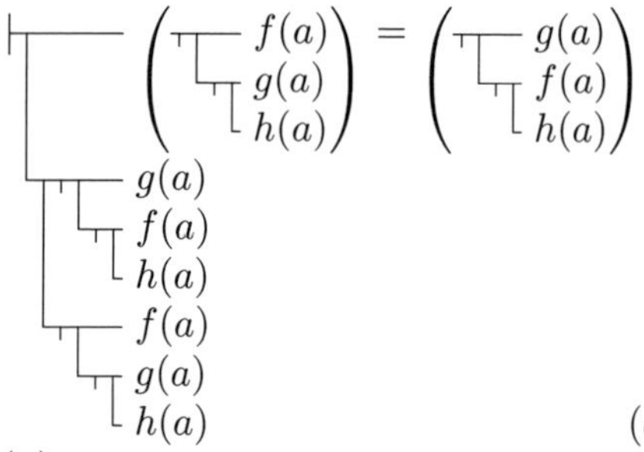

(δ

(γ):: ———

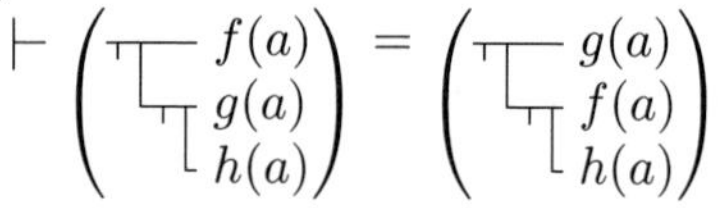

(ε

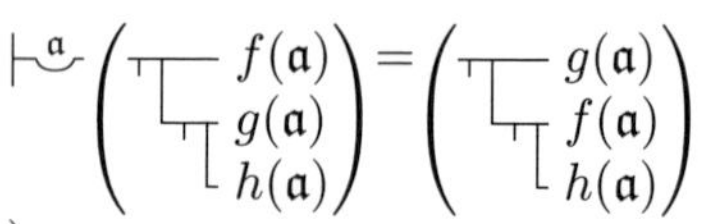

(ζ

(V a): ———

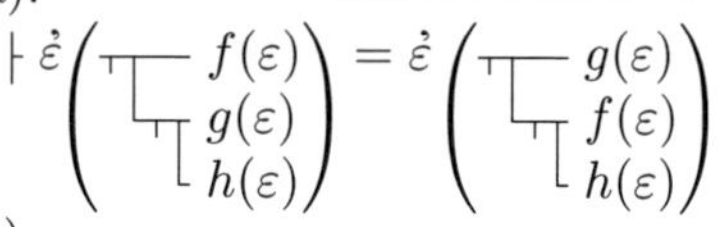

(η

(III a): ———

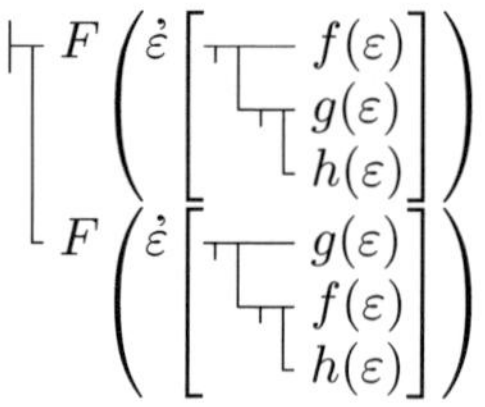

(72

——— • ———

$1 \vdash f(a) = a \frown \dot{\varepsilon} f(\varepsilon)$

(III h): ———

$\vdash g(a, f(a)) = g(a, a \frown \dot{\varepsilon} f(\varepsilon))$ (α

$\vdash\!\!\smile^{\mathfrak{a}}\; g(\mathfrak{a}, f(\mathfrak{a})) = g(\mathfrak{a}, \mathfrak{a} \frown \dot{\varepsilon} f(\varepsilon))$ (β

(V a): ———

$\vdash \dot{\varepsilon} g(\varepsilon, f(\varepsilon)) = \dot{\varepsilon} g(\varepsilon, \varepsilon \frown \dot{\varepsilon} f(\varepsilon))$ (73

(III a): ———

$\vdash \begin{array}{l} F(\dot{\varepsilon} g(\varepsilon, f(\varepsilon))) \\ F(\dot{\varepsilon} g(\varepsilon, \varepsilon \frown \dot{\varepsilon} f(\varepsilon))) \end{array}$ (74

——— • ———

$73 \vdash \dot{\varepsilon} g(\varepsilon, f(\varepsilon)) = \dot{\varepsilon} g(\varepsilon, \varepsilon \frown \dot{\varepsilon} f(\varepsilon))$

(III c): ———

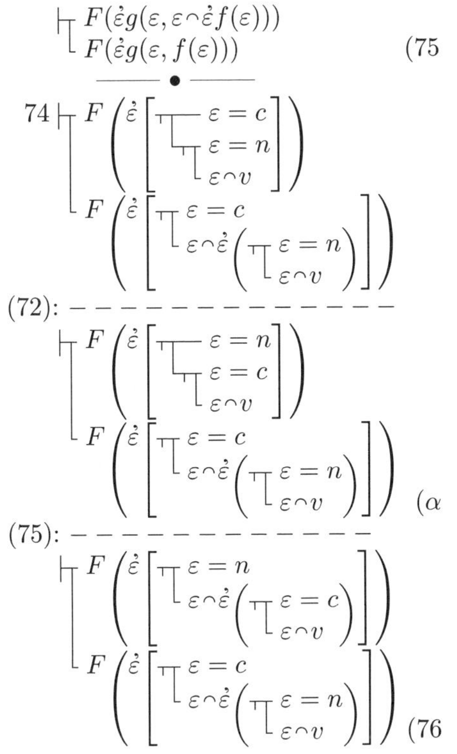

§90. *Analysis*

We now prove proposition (η) of §88. If the q-relation maps the u-concept into the v-concept, then there is, for every object falling under the u-concept, one falling under the v-concept to which it stands in the q-relation. Now, every object falling under the $\dot{\varepsilon}\left(\begin{smallmatrix}\varepsilon = b\\ \varepsilon \frown u\end{smallmatrix}\right)$-concept falls under the u-concept, and thus, given our condition, there is, for every object falling under the $\dot{\varepsilon}\left(\begin{smallmatrix}\varepsilon = b\\ \varepsilon \frown u\end{smallmatrix}\right)$-concept, one falling under the v-concept to which the former stands in the q-relation; yet n falls under the v-concept but does not fall under the $\dot{\varepsilon}\left(\begin{smallmatrix}\varepsilon = n\\ \varepsilon \frown v\end{smallmatrix}\right)$-concept. Now, if an object which stood to n in the q-relation fell under the $\dot{\varepsilon}\left(\begin{smallmatrix}\varepsilon = b\\ \varepsilon \frown u\end{smallmatrix}\right)$-concept, then there would not have to be an object to which it stood in the q-relation and which fell under the $\dot{\varepsilon}\left(\begin{smallmatrix}\varepsilon = n\\ \varepsilon \frown v\end{smallmatrix}\right)$-concept. This case is excluded, however, by the subcomponents '— $b \frown (n \frown q)$' and 'I$\rangle q$'.

§91. *Construction*

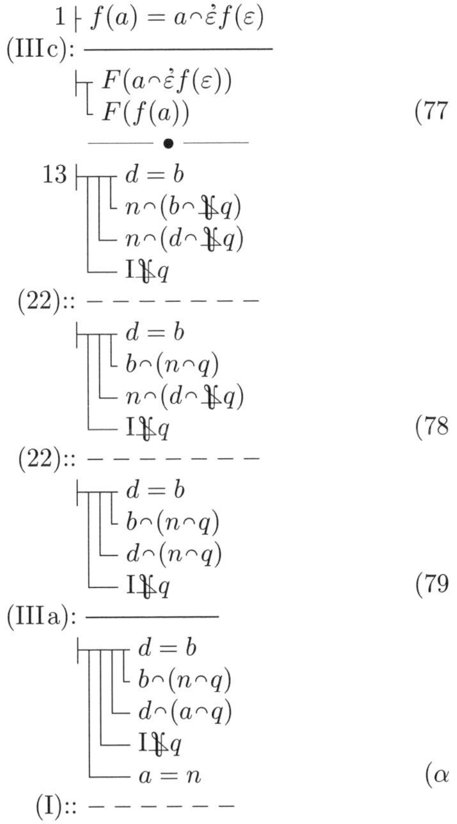

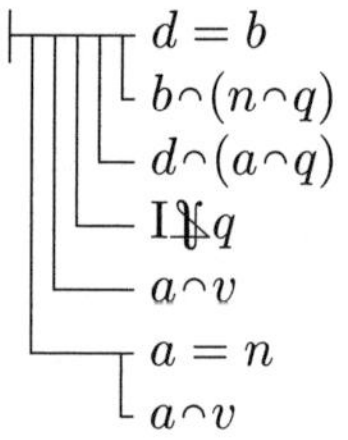

(β

×

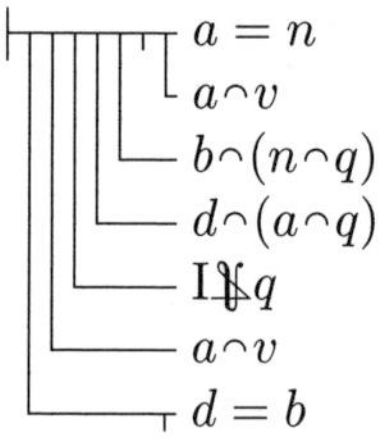

(γ

(77): - - - - - - - - -

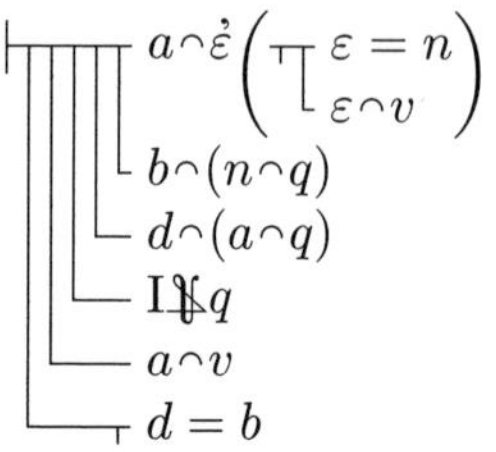

(δ

×

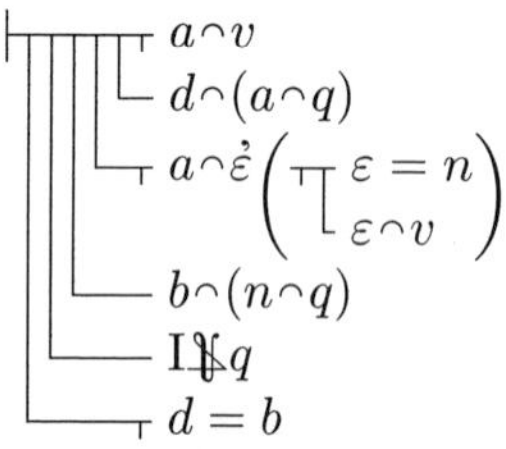

(ε

(IIa):: - - - - - - - - - - -

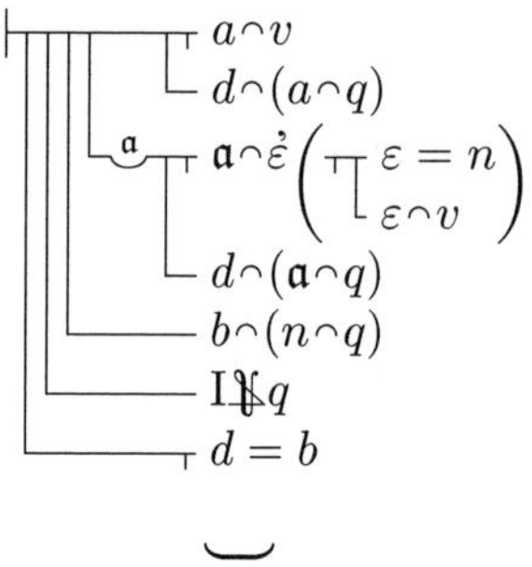

(ζ

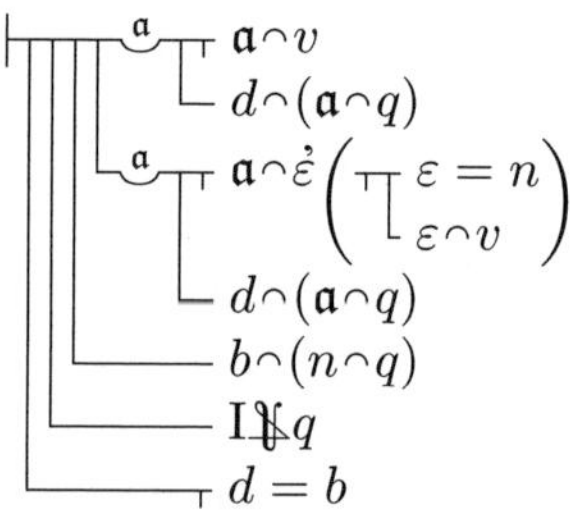

(η

(8): - - - - - - - - - - - -

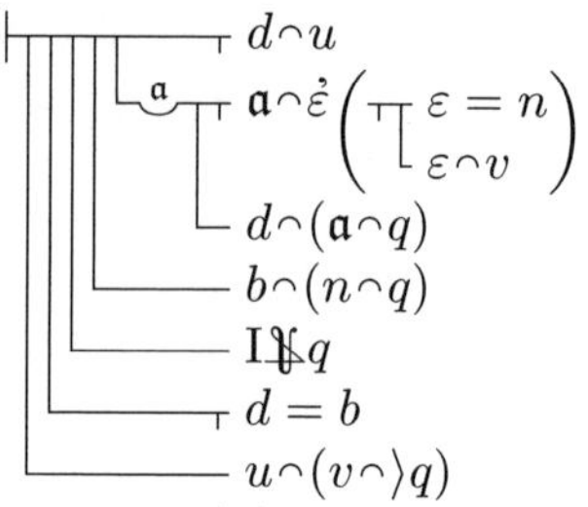

(ϑ

×

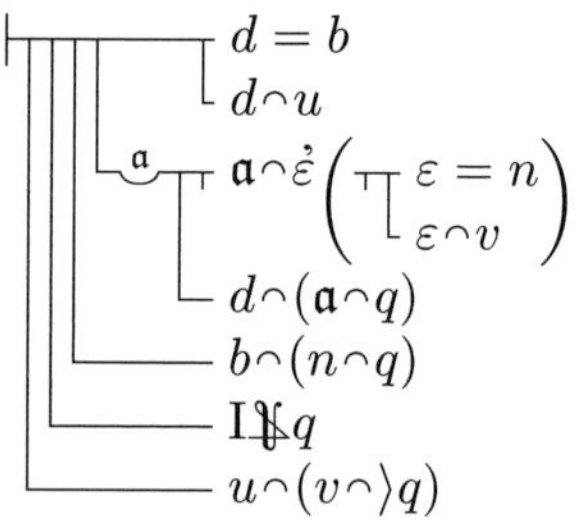

(ι

(58): - - - - - - - - - - - -

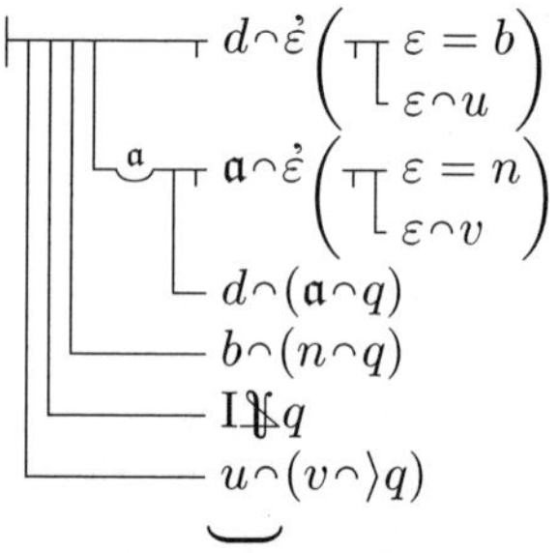

(κ

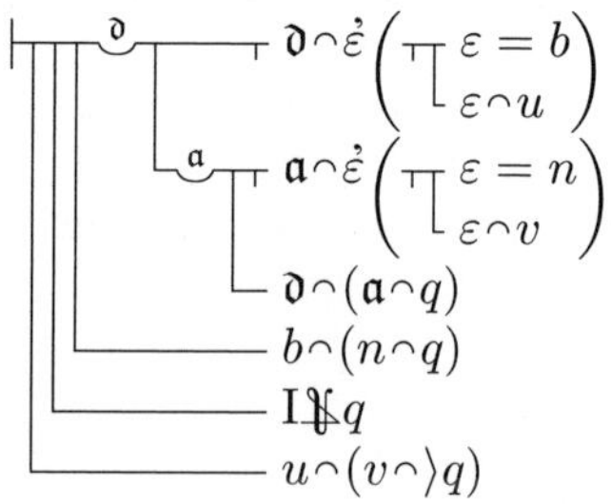

(λ

(11): - - - - - - - - - - - - -

$$\vdash\begin{array}{l}\dot\varepsilon\left(\top\!\begin{array}{l}\varepsilon=b\\ \llcorner\,\varepsilon\frown u\end{array}\right)\frown\left[\dot\varepsilon\left(\top\!\begin{array}{l}\varepsilon=n\\ \llcorner\,\varepsilon\frown v\end{array}\right)\frown\rangle q\right]\\ \llcorner\, b\frown(n\frown q)\\ \llcorner\, \mathrm{I}\rangle\!\!\!\mid q\\ \llcorner\, u\frown(v\frown\rangle q)\\ \llcorner\, \mathrm{I}q\end{array}\qquad(\mu$$

(18):: - - - - - - - - - - - - - - - - - -

$$\vdash\begin{array}{l}\dot\varepsilon\left(\top\!\begin{array}{l}\varepsilon=b\\ \llcorner\,\varepsilon\frown u\end{array}\right)\frown\left[\dot\varepsilon\left(\top\!\begin{array}{l}\varepsilon=n\\ \llcorner\,\varepsilon\frown v\end{array}\right)\frown\rangle q\right]\\ \llcorner\, b\frown(n\frown q)\\ \llcorner\, \mathrm{I}\rangle\!\!\!\mid q\\ \llcorner\, u\frown(v\frown\rangle q)\end{array}\qquad(80$$

(80): - - - - - - - - - - - - - - - - - -

$$\vdash\begin{array}{l}\dot\varepsilon\left(\top\!\begin{array}{l}\varepsilon=m\\ \llcorner\,\varepsilon\frown\dot\varepsilon\left(\top\!\begin{array}{l}\varepsilon=b\\ \llcorner\,\varepsilon\frown u\end{array}\right)\end{array}\right)\frown\left(\dot\varepsilon\left[\top\!\begin{array}{l}\varepsilon=c\\ \llcorner\,\varepsilon\frown\dot\varepsilon\left(\top\!\begin{array}{l}\varepsilon=n\\ \llcorner\,\varepsilon\frown v\end{array}\right)\end{array}\right]\frown\rangle q\right)\\ \llcorner\, m\frown(c\frown q)\\ \llcorner\, \mathrm{I}\rangle\!\!\!\mid q\\ \llcorner\, b\frown(n\frown q)\\ \llcorner\, u\frown(v\frown\rangle q)\end{array}\qquad(\alpha$$

(76): -

$$\vdash\begin{array}{l}\dot\varepsilon\left(\top\!\begin{array}{l}\varepsilon=m\\ \llcorner\,\varepsilon\frown\dot\varepsilon\left(\top\!\begin{array}{l}\varepsilon=b\\ \llcorner\,\varepsilon\frown u\end{array}\right)\end{array}\right)\frown\left(\dot\varepsilon\left[\top\!\begin{array}{l}\varepsilon=n\\ \llcorner\,\varepsilon\frown\dot\varepsilon\left(\top\!\begin{array}{l}\varepsilon=c\\ \llcorner\,\varepsilon\frown v\end{array}\right)\end{array}\right]\frown\rangle q\right)\\ \llcorner\, m\frown(c\frown q)\\ \llcorner\, \mathrm{I}\rangle\!\!\!\mid q\\ \llcorner\, b\frown(n\frown q)\\ \llcorner\, u\frown(v\frown\rangle q)\end{array}\qquad(\beta$$

(51): -

$$\vdash\begin{array}{l}\dot\varepsilon\left(\top\!\begin{array}{l}\varepsilon=m\\ \llcorner\,\varepsilon\frown\dot\varepsilon\left(\top\!\begin{array}{l}\varepsilon=b\\ \llcorner\,\varepsilon\frown u\end{array}\right)\end{array}\right)\frown\left(\dot\varepsilon\left[\top\!\begin{array}{l}\varepsilon=n\\ \llcorner\,\varepsilon\frown\dot\varepsilon\left(\top\!\begin{array}{l}\varepsilon=c\\ \llcorner\,\varepsilon\frown v\end{array}\right)\end{array}\right]\frown\rangle\dot\alpha\dot\varepsilon\left[\top\!\begin{array}{l}\varepsilon\frown(\alpha\frown q)\\ \llcorner\top\!\begin{array}{l}\varepsilon=b\\ \llcorner_{\top}\,\alpha=c\end{array}\end{array}\right]\right)\\ \llcorner_{\top}\, b\frown\dot\varepsilon\left(\top\!\begin{array}{l}\varepsilon=m\\ \llcorner\,\varepsilon\frown\dot\varepsilon\left(\top\!\begin{array}{l}\varepsilon=b\\ \llcorner\,\varepsilon\frown u\end{array}\right)\end{array}\right)\\ \llcorner_{\top}\, c\frown\dot\varepsilon\left(\top\!\begin{array}{l}\varepsilon=n\\ \llcorner\,\varepsilon\frown\dot\varepsilon\left(\top\!\begin{array}{l}\varepsilon=c\\ \llcorner\,\varepsilon\frown v\end{array}\right)\end{array}\right)\\ \llcorner\, m\frown(c\frown q)\\ \llcorner\, \mathrm{I}\rangle\!\!\!\mid q\\ \llcorner\, b\frown(n\frown q)\\ \llcorner\, u\frown(v\frown\rangle q)\end{array}\qquad(81$$

$$1 \vdash f(a) = a \frown \grave{\varepsilon} f(\varepsilon)$$

(IIIa): ————

$$\vdash F(f(a)) \;\text{—}\; F(a \frown \grave{\varepsilon} f(\varepsilon))$$ (82

—— • ——

$$\text{Id} \vdash b \frown \grave{\varepsilon}\left(\varepsilon = b \;\text{—}\; \varepsilon \frown u\right) \;\text{—}\; b = m \;\text{—}\; b \frown \grave{\varepsilon}\left(\varepsilon = b \;\text{—}\; \varepsilon \frown u\right)$$ (α

(82):: - - - - - - - - - -

$$\vdash b \frown \grave{\varepsilon}\left(\varepsilon = b \;\text{—}\; \varepsilon \frown u\right) \;\text{—}\; b \frown \grave{\varepsilon}\left(\varepsilon = m \;\text{—}\; \varepsilon \frown \grave{\varepsilon}\left(\varepsilon = b \;\text{—}\; \varepsilon \frown u\right)\right)$$ (β

×

$$\vdash b \frown \grave{\varepsilon}\left(\varepsilon = m \;\text{—}\; \varepsilon \frown \grave{\varepsilon}\left(\varepsilon = b \;\text{—}\; \varepsilon \frown u\right)\right) \;\text{—}\; b \frown \grave{\varepsilon}\left(\varepsilon = b \;\text{—}\; \varepsilon \frown u\right)$$ (γ

(65):: ————

$$\vdash b \frown \grave{\varepsilon}\left(\varepsilon = m \;\text{—}\; \varepsilon \frown \grave{\varepsilon}\left(\varepsilon = b \;\text{—}\; \varepsilon \frown u\right)\right)$$ (δ

(81): ————

$$\vdash \grave{\varepsilon}\left(\varepsilon = m \;\text{—}\; \varepsilon \frown \grave{\varepsilon}\left(\varepsilon = b \;\text{—}\; \varepsilon \frown u\right)\right) \frown \left(\grave{\varepsilon}\left[\varepsilon = n \;\text{—}\; \varepsilon \frown \grave{\varepsilon}\left(\varepsilon = c \;\text{—}\; \varepsilon \frown v\right)\right] \frown\rangle \grave{\alpha}\grave{\varepsilon}\left[\varepsilon \frown (\alpha \frown q) \;\text{—}\; \varepsilon = b \;\text{—}\; \alpha = c\right]\right)$$
$$\text{—}\; c \frown \grave{\varepsilon}\left(\varepsilon = n \;\text{—}\; \varepsilon \frown \grave{\varepsilon}\left(\varepsilon = c \;\text{—}\; \varepsilon \frown v\right)\right)$$
$$\text{—}\; m \frown (c \frown q)$$
$$\text{—}\; \mathrm{I} \rangle\!\langle q$$
$$\text{—}\; b \frown (n \frown q)$$
$$\text{—}\; u \frown (v \frown\rangle q)$$ (ε

(δ):: ————

$$\vdash \grave{\varepsilon}\left(\varepsilon = m \;\text{—}\; \varepsilon \frown \grave{\varepsilon}\left(\varepsilon = b \;\text{—}\; \varepsilon \frown u\right)\right) \frown \left(\grave{\varepsilon}\left[\varepsilon = n \;\text{—}\; \varepsilon \frown \grave{\varepsilon}\left(\varepsilon = c \;\text{—}\; \varepsilon \frown v\right)\right] \frown\rangle \grave{\alpha}\grave{\varepsilon}\left[\varepsilon \frown (\alpha \frown q) \;\text{—}\; \varepsilon = b \;\text{—}\; \alpha = c\right]\right)$$
$$\text{—}\; m \frown (c \frown q)$$
$$\text{—}\; \mathrm{I} \rangle\!\langle q$$
$$\text{—}\; b \frown (n \frown q)$$
$$\text{—}\; u \frown (v \frown\rangle q)$$ (83

(23):: -

$$\vdash \grave{\varepsilon}\left(\varepsilon = m \;\text{—}\; \varepsilon \frown \grave{\varepsilon}\left(\varepsilon = b \;\text{—}\; \varepsilon \frown u\right)\right) \frown \left(\grave{\varepsilon}\left[\varepsilon = n \;\text{—}\; \varepsilon \frown \grave{\varepsilon}\left(\varepsilon = c \;\text{—}\; \varepsilon \frown v\right)\right] \frown\rangle \grave{\alpha}\grave{\varepsilon}\left[\varepsilon \frown (\alpha \frown q) \;\text{—}\; \varepsilon = b \;\text{—}\; \alpha = c\right]\right)$$
$$\text{—}\; c \frown (m \frown \rangle\!\langle q)$$
$$\text{—}\; \mathrm{I} \rangle\!\langle q$$
$$\text{—}\; b \frown (n \frown q)$$
$$\text{—}\; u \frown (v \frown\rangle q)$$ (84

b) Proof of the proposition

‘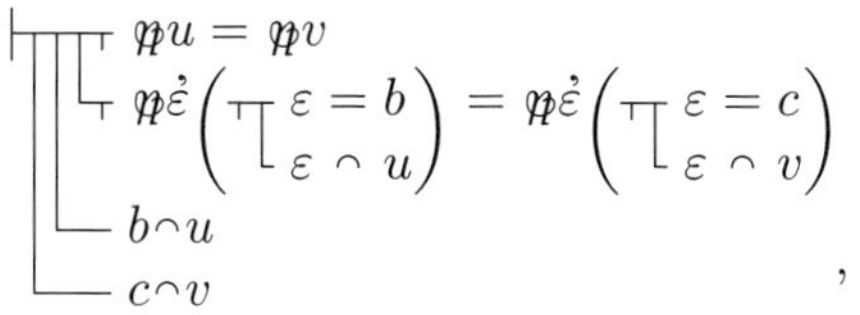’,

and end of section Γ

§92. *Analysis*

From (83) we can derive, using (53) and (22), the proposition

‘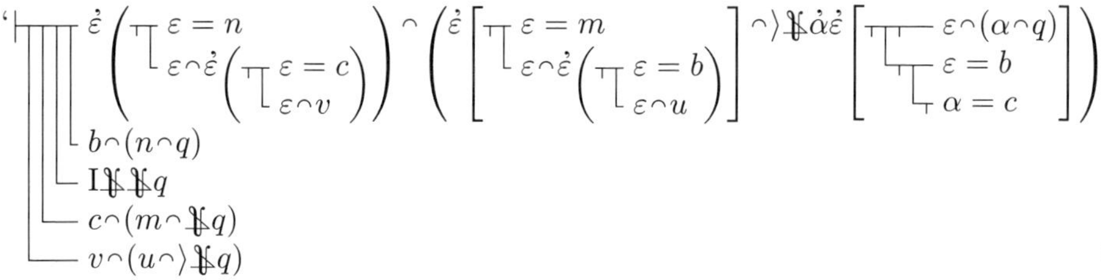’ (α

From this proposition and (84), we arrive, using (66), at a proposition with the supercomponent

‘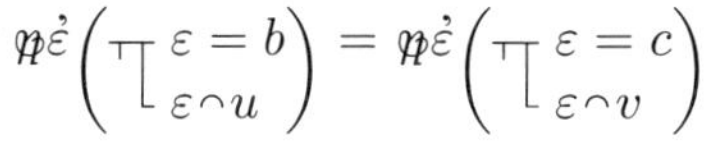’

§93. *Construction*

83 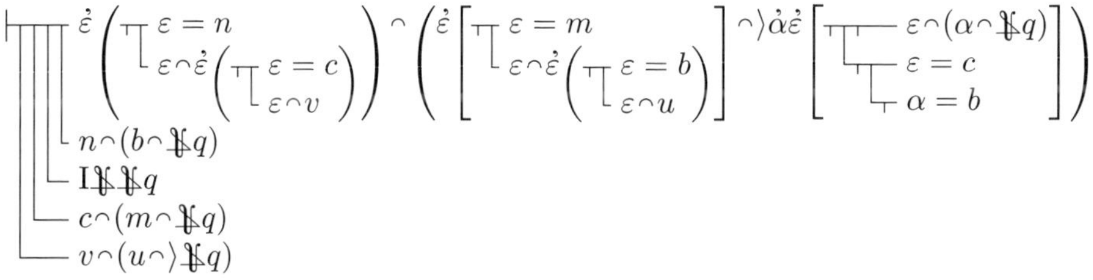

(53): –

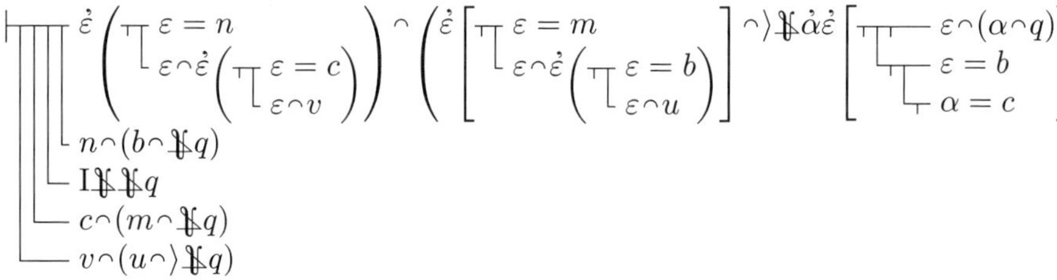 (α

(22):: –

$$\vdash \dot{\varepsilon}\left(\begin{array}{l}\varepsilon = n\\ \varepsilon \smallfrown \dot{\varepsilon}\left(\begin{array}{l}\varepsilon = c\\ \varepsilon \smallfrown v\end{array}\right)\end{array}\right) \smallfrown \left(\dot{\varepsilon}\left[\begin{array}{l}\varepsilon = m\\ \varepsilon \smallfrown \dot{\varepsilon}\left(\begin{array}{l}\varepsilon = b\\ \varepsilon \smallfrown u\end{array}\right)\end{array}\right] \smallfrown \rangle \nparallel \dot{\alpha}\dot{\varepsilon}\left[\begin{array}{l}\varepsilon \smallfrown (\alpha \smallfrown q)\\ \varepsilon = b\\ \alpha = c\end{array}\right]\right)$$

$$\begin{array}{l} b \smallfrown (n \smallfrown q)\\ \mathrm{I}\nparallel\nparallel q\\ c \smallfrown (m \smallfrown \nparallel q)\\ v \smallfrown (u \smallfrown \rangle \nparallel q)\end{array} \qquad (\beta$$

(32): -

$$\vdash \wp\dot{\varepsilon}\left(\begin{array}{l}\varepsilon = m\\ \varepsilon \smallfrown \dot{\varepsilon}\left(\begin{array}{l}\varepsilon = b\\ \varepsilon \smallfrown u\end{array}\right)\end{array}\right) = \wp\dot{\varepsilon}\left(\begin{array}{l}\varepsilon = n\\ \varepsilon \smallfrown \dot{\varepsilon}\left(\begin{array}{l}\varepsilon = c\\ \varepsilon \smallfrown v\end{array}\right)\end{array}\right)$$

$$\dot{\varepsilon}\left(\begin{array}{l}\varepsilon = m\\ \varepsilon \smallfrown \dot{\varepsilon}\left(\begin{array}{l}\varepsilon = b\\ \varepsilon \smallfrown u\end{array}\right)\end{array}\right) \smallfrown \left(\dot{\varepsilon}\left[\begin{array}{l}\varepsilon = n\\ \varepsilon \smallfrown \dot{\varepsilon}\left(\begin{array}{l}\varepsilon = c\\ \varepsilon \smallfrown v\end{array}\right)\end{array}\right] \smallfrown \rangle \dot{\alpha}\dot{\varepsilon}\left[\begin{array}{l}\varepsilon \smallfrown (\alpha \smallfrown q)\\ \varepsilon = b\\ \alpha = c\end{array}\right]\right)$$

$$\begin{array}{l} b \smallfrown (n \smallfrown q)\\ \mathrm{I}\nparallel\nparallel q\\ c \smallfrown (m \smallfrown \nparallel q)\\ v \smallfrown (u \smallfrown \rangle \nparallel q)\end{array} \qquad (\gamma$$

(84):: -

$$\vdash \wp\dot{\varepsilon}\left(\begin{array}{l}\varepsilon = m\\ \varepsilon \smallfrown \dot{\varepsilon}\left(\begin{array}{l}\varepsilon = b\\ \varepsilon \smallfrown u\end{array}\right)\end{array}\right) = \wp\dot{\varepsilon}\left(\begin{array}{l}\varepsilon = n\\ \varepsilon \smallfrown \dot{\varepsilon}\left(\begin{array}{l}\varepsilon = c\\ \varepsilon \smallfrown v\end{array}\right)\end{array}\right)$$

$$\begin{array}{l} c \smallfrown (m \smallfrown \nparallel q)\\ \mathrm{I}\nparallel q\\ b \smallfrown (n \smallfrown q)\\ u \smallfrown (v \smallfrown \rangle q)\\ \mathrm{I}\nparallel\nparallel q\\ v \smallfrown (u \smallfrown \rangle \nparallel q)\end{array} \qquad (\delta$$

(66): -

$$\vdash \wp\dot{\varepsilon}\left(\begin{array}{l}\varepsilon = b\\ \varepsilon \smallfrown u\end{array}\right) = \wp\dot{\varepsilon}\left(\begin{array}{l}\varepsilon = c\\ \varepsilon \smallfrown v\end{array}\right)$$

$$\begin{array}{l} m \smallfrown \dot{\varepsilon}\left(\begin{array}{l}\varepsilon = b\\ \varepsilon \smallfrown u\end{array}\right)\\ c \smallfrown (m \smallfrown \nparallel q)\\ \mathrm{I}\nparallel q\\ b \smallfrown (n \smallfrown q)\\ u \smallfrown (v \smallfrown \rangle q)\\ \mathrm{I}\nparallel\nparallel q\\ v \smallfrown (u \smallfrown \rangle \nparallel q)\\ n \smallfrown \dot{\varepsilon}\left(\begin{array}{l}\varepsilon = c\\ \varepsilon \smallfrown v\end{array}\right)\end{array} \qquad (\varepsilon$$

×

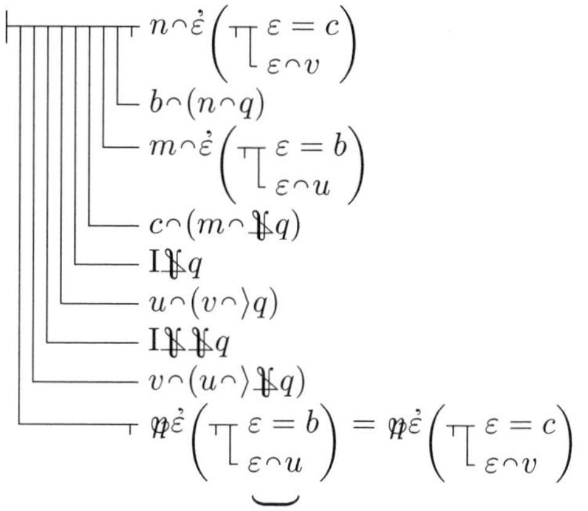
n⌢ἐ(ε = c, ε⌢v)
b⌢(n⌢q)
m⌢ἐ(ε = b, ε⌢u)
c⌢(m⌢ʮq)
Iʮq
u⌢(v⌢)q)
Iʮʮq
v⌢(u⌢)ʮq)
𝔭ἐ(ε = b, ε⌢u) = 𝔭ἐ(ε = c, ε⌢v)

(ζ

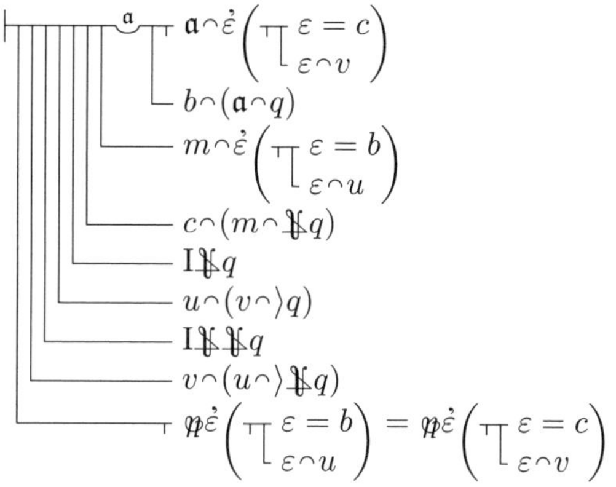
𝔞
𝔞⌢ἐ(ε = c, ε⌢v)
b⌢(𝔞⌢q)
m⌢ἐ(ε = b, ε⌢u)
c⌢(m⌢ʮq)
Iʮq
u⌢(v⌢)q)
Iʮʮq
v⌢(u⌢)ʮq)
𝔭ἐ(ε = b, ε⌢u) = 𝔭ἐ(ε = c, ε⌢v)

(85

§94. *Analysis*

The preceding two transitions already point to the path that is to be followed from here on. In (ε) we had n and m as auxiliary objects similar to auxiliary lines in geometry. These should not appear in our proposition and so must be removed. This is achieved, as always, by showing that there is something with the required constitution, or, more conveniently in concept-script, that if there were nothing of this kind, then one of the assumptions we make would not hold.

Our task is now to prove the proposition

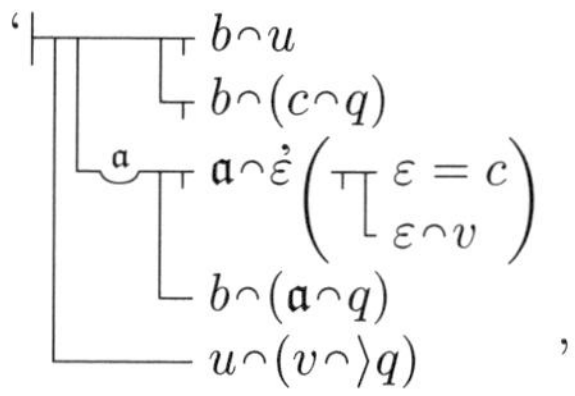
b⌢u
b⌢(c⌢q)
𝔞
𝔞⌢ἐ(ε = c, ε⌢v)
b⌢(𝔞⌢q)
u⌢(v⌢)q)

(α

Hereby we add the subcomponent '┬ b⌢(c⌢q)'; but we can also prove the otherwise unaltered proposition with the contrary subcomponent '— $b \frown (c \frown q)$'. In order to derive (α)

from (8) we require the proposition

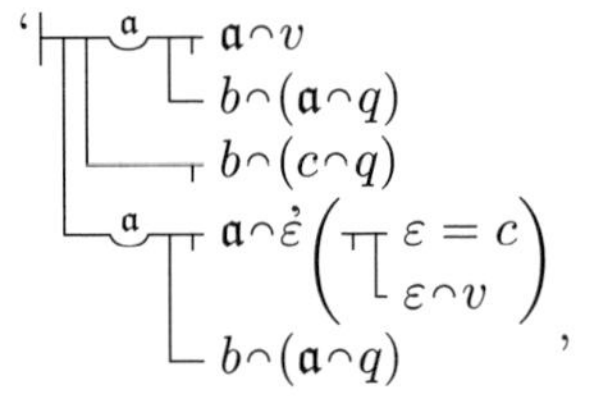

which follows from (77). Like '*n*', we remove '*m*' from our proposition and then, as was just indicated, prove the proposition

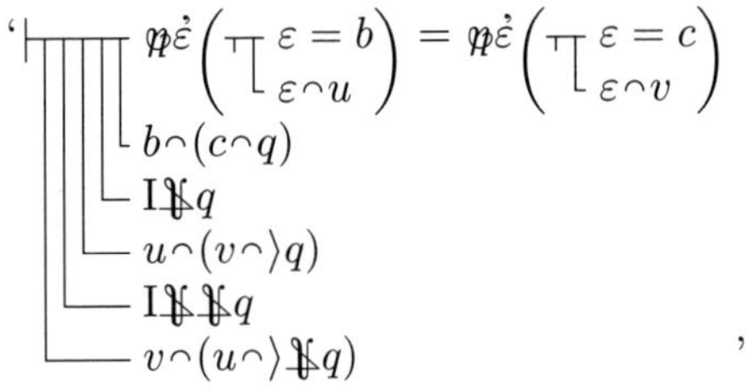

using (80) and (32), and eliminate the subcomponents '— $b \frown (c \frown q)$' and '┬ $b \frown (c \frown q)$' by rule (8). After removing the subcomponents 'I∦q' and 'I∦∦q' using (30) and (18), we employ (49) to attain the goal of section (b), and thereafter, as was indicated in §88, arrive at the proposition '⊢ I∦f' by means of (68).

§95. *Construction*

77

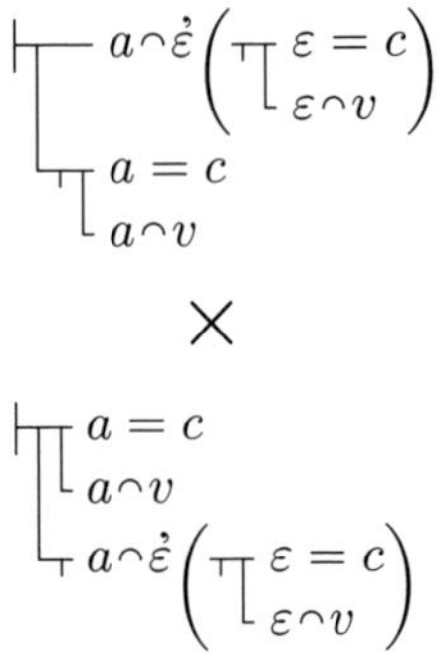

(α

(IIIa): - - - - - - - -

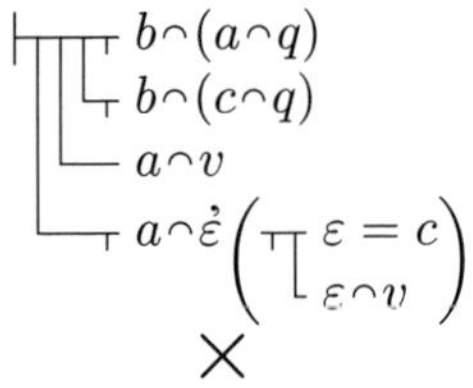

(β

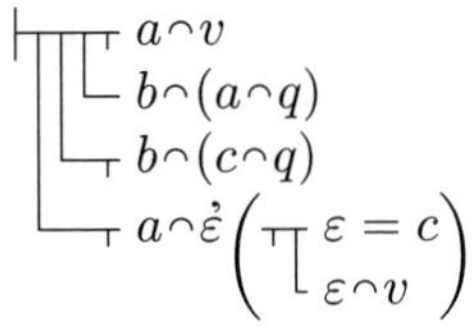

(γ

(IIa):: - - - - - - - - - -

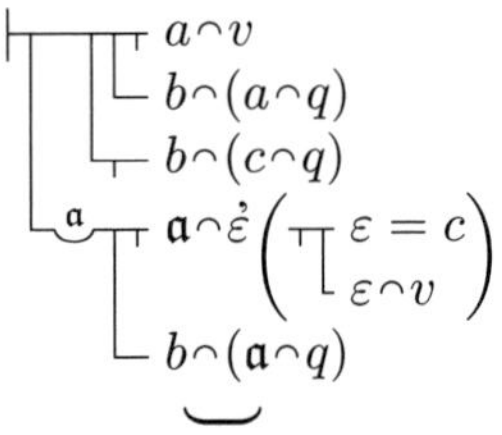

(δ

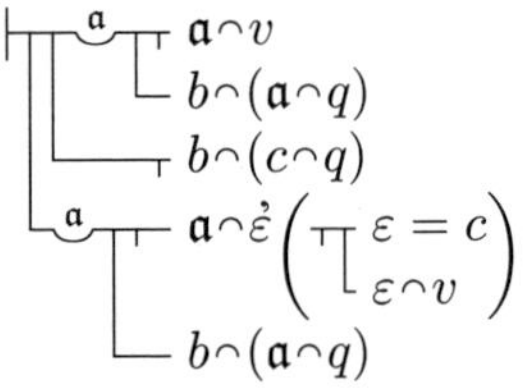

(ε

(8): - - - - - - - - - - -

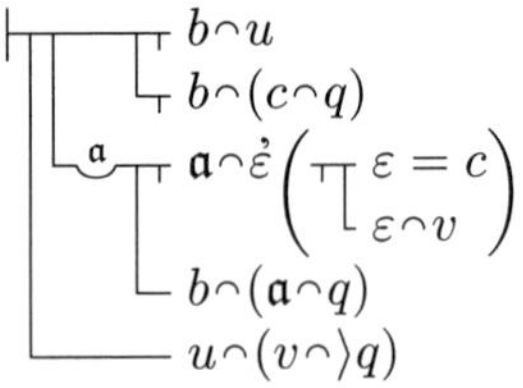

(ζ

(85):: - - - - - - - - - - -

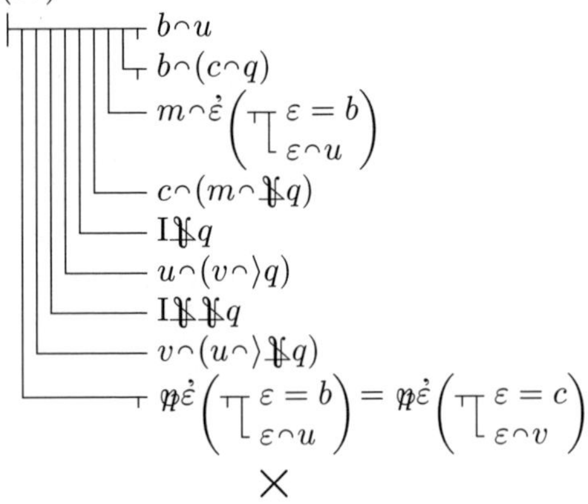

(η

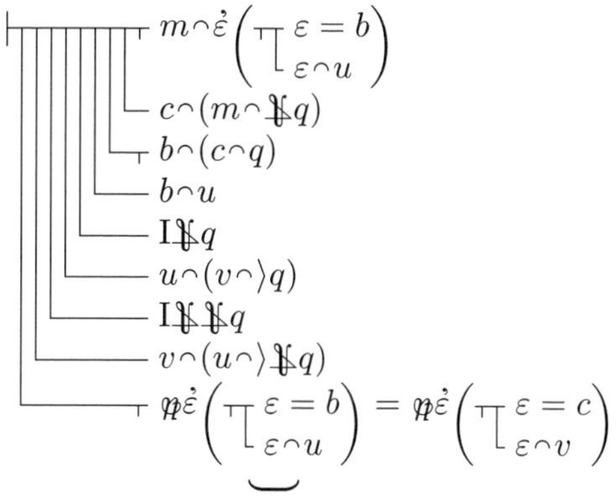

(ϑ

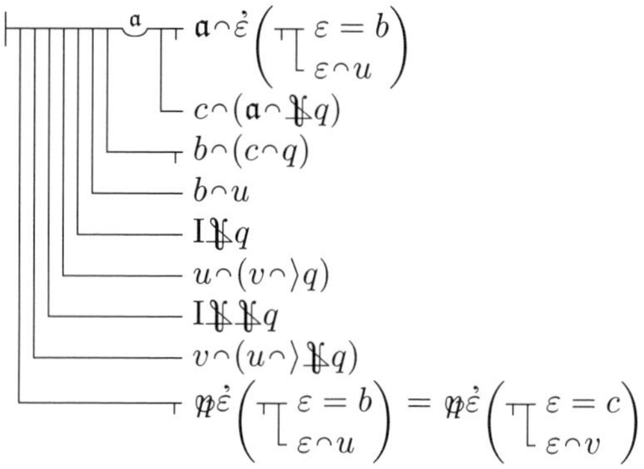

(ι

(ζ): –

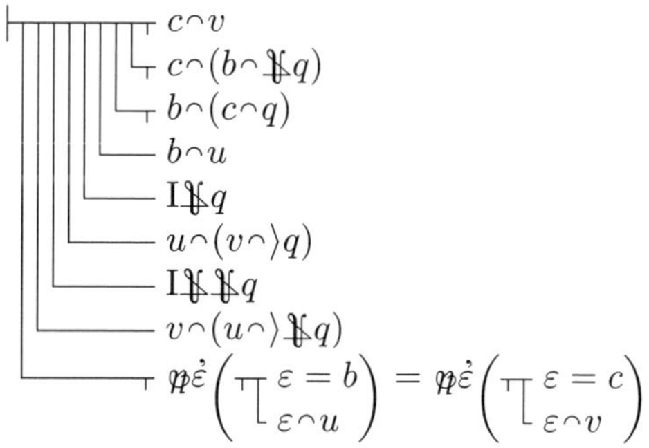

(κ

(22):: –

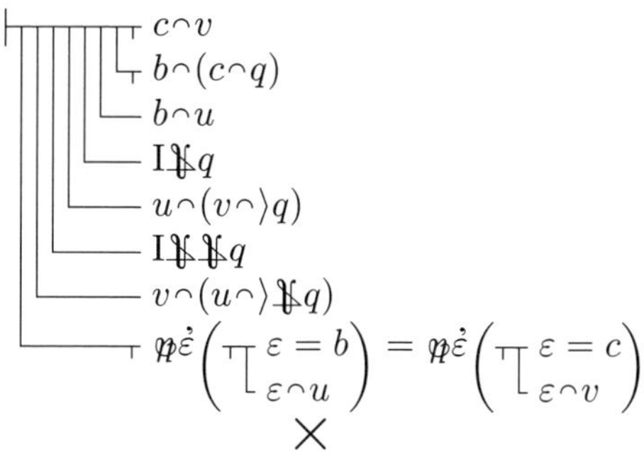

(λ

$$
\begin{array}{l}
\vdash\ \mathfrak{p}\dot{\varepsilon}\left(\begin{smallmatrix}\top\ \varepsilon = b\\ \llcorner\ \varepsilon\frown u\end{smallmatrix}\right) = \mathfrak{p}\dot{\varepsilon}\left(\begin{smallmatrix}\top\ \varepsilon = c\\ \llcorner\ \varepsilon\frown v\end{smallmatrix}\right)\\
\quad \neg\ b\frown(c\frown q)\\
\quad b\frown u\\
\quad \mathrm{I}\not\rangle q\\
\quad u\frown(v\frown\rangle q)\\
\quad \mathrm{I}\not\rangle\not\rangle q\\
\quad v\frown(u\frown\rangle\not\rangle q)\\
\quad c\frown v
\end{array}
\qquad (86
$$

——•——

$$
\begin{array}{l}
32\ \vdash\ \mathfrak{p}\dot{\varepsilon}\left(\begin{smallmatrix}\top\ \varepsilon = b\\ \llcorner\ \varepsilon\frown u\end{smallmatrix}\right) = \mathfrak{p}\dot{\varepsilon}\left(\begin{smallmatrix}\top\ \varepsilon = c\\ \llcorner\ \varepsilon\frown v\end{smallmatrix}\right)\\
\quad \dot{\varepsilon}\left(\begin{smallmatrix}\top\ \varepsilon = b\\ \llcorner\ \varepsilon\frown u\end{smallmatrix}\right)\frown\left[\dot{\varepsilon}\left(\begin{smallmatrix}\top\ \varepsilon = c\\ \llcorner\ \varepsilon\frown v\end{smallmatrix}\right)\frown\rangle q\right]\\
\quad \dot{\varepsilon}\left(\begin{smallmatrix}\top\ \varepsilon = c\\ \llcorner\ \varepsilon\frown v\end{smallmatrix}\right)\frown\left[\dot{\varepsilon}\left(\begin{smallmatrix}\top\ \varepsilon = b\\ \llcorner\ \varepsilon\frown u\end{smallmatrix}\right)\frown\rangle\not\rangle q\right]
\end{array}
$$

(80):: – – – – – – – – – – – – – – – –

$$
\begin{array}{l}
\vdash\ \mathfrak{p}\dot{\varepsilon}\left(\begin{smallmatrix}\top\ \varepsilon = b\\ \llcorner\ \varepsilon\frown u\end{smallmatrix}\right) = \mathfrak{p}\dot{\varepsilon}\left(\begin{smallmatrix}\top\ \varepsilon = c\\ \llcorner\ \varepsilon\frown v\end{smallmatrix}\right)\\
\quad \dot{\varepsilon}\left(\begin{smallmatrix}\top\ \varepsilon = b\\ \llcorner\ \varepsilon\frown u\end{smallmatrix}\right)\frown\left[\dot{\varepsilon}\left(\begin{smallmatrix}\top\ \varepsilon = c\\ \llcorner\ \varepsilon\frown v\end{smallmatrix}\right)\frown\rangle q\right]\\
\quad c\frown(b\frown\not\rangle q)\\
\quad \mathrm{I}\not\rangle\not\rangle q\\
\quad v\frown(u\frown\rangle\not\rangle q)
\end{array}
\qquad (\alpha
$$

(22):: – – – – – – – – – – – – – –

$$
\begin{array}{l}
\vdash\ \mathfrak{p}\dot{\varepsilon}\left(\begin{smallmatrix}\top\ \varepsilon = b\\ \llcorner\ \varepsilon\frown u\end{smallmatrix}\right) = \mathfrak{p}\dot{\varepsilon}\left(\begin{smallmatrix}\top\ \varepsilon = c\\ \llcorner\ \varepsilon\frown v\end{smallmatrix}\right)\\
\quad \dot{\varepsilon}\left(\begin{smallmatrix}\top\ \varepsilon = b\\ \llcorner\ \varepsilon\frown u\end{smallmatrix}\right)\frown\left[\dot{\varepsilon}\left(\begin{smallmatrix}\top\ \varepsilon = c\\ \llcorner\ \varepsilon\frown v\end{smallmatrix}\right)\frown\rangle q\right]\\
\quad b\frown(c\frown q)\\
\quad \mathrm{I}\not\rangle\not\rangle q\\
\quad v\frown(u\frown\rangle\not\rangle q)
\end{array}
\qquad (\beta
$$

(80):: – – – – – – – – – – – – – – –

$$
\begin{array}{l}
\vdash\ \mathfrak{p}\dot{\varepsilon}\left(\begin{smallmatrix}\top\ \varepsilon = b\\ \llcorner\ \varepsilon\frown u\end{smallmatrix}\right) = \mathfrak{p}\dot{\varepsilon}\left(\begin{smallmatrix}\top\ \varepsilon = c\\ \llcorner\ \varepsilon\frown v\end{smallmatrix}\right)\\
\quad b\frown(c\frown q)\\
\quad \mathrm{I}\not\rangle q\\
\quad u\frown(v\frown\rangle q)\\
\quad \mathrm{I}\not\rangle\not\rangle q\\
\quad v\frown(u\frown\rangle\not\rangle q)
\end{array}
\qquad (\gamma
$$

(86): ·–·–·–·–·–·–·–·–·–·–·–·–·

$$
\begin{array}{l}
\vdash\ \mathfrak{p}\dot{\varepsilon}\left(\begin{smallmatrix}\top\ \varepsilon = b\\ \llcorner\ \varepsilon\frown u\end{smallmatrix}\right) = \mathfrak{p}\dot{\varepsilon}\left(\begin{smallmatrix}\top\ \varepsilon = c\\ \llcorner\ \varepsilon\frown v\end{smallmatrix}\right)\\
\quad \mathrm{I}\not\rangle q\\
\quad u\frown(v\frown\rangle q)\\
\quad \mathrm{I}\not\rangle\not\rangle q\\
\quad v\frown(u\frown\rangle\not\rangle q)\\
\quad b\frown u\\
\quad c\frown v
\end{array}
\qquad (\delta
$$

(30, 18):: = = = = = = = = = = = = = =

$$
\begin{array}{l}
\vdash\ \mathfrak{p}\dot{\varepsilon}\left(\begin{smallmatrix}\top\ \varepsilon = b\\ \llcorner\ \varepsilon\frown u\end{smallmatrix}\right) = \mathfrak{p}\dot{\varepsilon}\left(\begin{smallmatrix}\top\ \varepsilon = c\\ \llcorner\ \varepsilon\frown v\end{smallmatrix}\right)\\
\quad u\frown(v\frown\rangle q)\\
\quad v\frown(u\frown\rangle\not\rangle q)\\
\quad b\frown u\\
\quad c\frown v
\end{array}
\qquad (\varepsilon
$$

×

$$
\begin{array}{l}
\vdash\ \neg\ \left(\begin{smallmatrix}\top\ u\frown(v\frown\rangle q)\\ \llcorner\ v\frown(u\frown\rangle\not\rangle q)\end{smallmatrix}\right)\\
\quad \neg\ \mathfrak{p}\dot{\varepsilon}\left(\begin{smallmatrix}\top\ \varepsilon = b\\ \llcorner\ \varepsilon\frown u\end{smallmatrix}\right) = \mathfrak{p}\dot{\varepsilon}\left(\begin{smallmatrix}\top\ \varepsilon = c\\ \llcorner\ \varepsilon\frown v\end{smallmatrix}\right)\\
\quad b\frown u\\
\quad c\frown v
\end{array}
\qquad (\zeta
$$

$\smile$

$$
\begin{array}{l}
\vdash\ \overset{\mathfrak{q}}{\smile}\ \left(\begin{smallmatrix}\top\ u\frown(v\frown\rangle \mathfrak{q})\\ \llcorner\ v\frown(u\frown\rangle\not\rangle \mathfrak{q})\end{smallmatrix}\right)\\
\quad \neg\ \mathfrak{p}\dot{\varepsilon}\left(\begin{smallmatrix}\top\ \varepsilon = b\\ \llcorner\ \varepsilon\frown u\end{smallmatrix}\right) = \mathfrak{p}\dot{\varepsilon}\left(\begin{smallmatrix}\top\ \varepsilon = c\\ \llcorner\ \varepsilon\frown v\end{smallmatrix}\right)\\
\quad b\frown u\\
\quad c\frown v
\end{array}
\qquad (\eta
$$

(49): – – – – – – – – – – – – – – –

$$
\begin{array}{l}
\vdash\ \neg\ \mathfrak{p}u = \mathfrak{p}v\\
\quad \neg\ \mathfrak{p}\dot{\varepsilon}\left(\begin{smallmatrix}\top\ \varepsilon = b\\ \llcorner\ \varepsilon\frown u\end{smallmatrix}\right) = \mathfrak{p}\dot{\varepsilon}\left(\begin{smallmatrix}\top\ \varepsilon = c\\ \llcorner\ \varepsilon\frown v\end{smallmatrix}\right)\\
\quad b\frown u\\
\quad c\frown v
\end{array}
\qquad (\vartheta
$$

(IIIc): ————————————————

$$
\begin{array}{l}
\vdash\ \neg\ \mathfrak{p}u = \mathfrak{p}v\\
\quad b\frown u\\
\quad \mathfrak{p}\dot{\varepsilon}\left(\begin{smallmatrix}\top\ \varepsilon = b\\ \llcorner\ \varepsilon\frown u\end{smallmatrix}\right) = d\\
\quad c\frown v\\
\quad \neg\ d = \mathfrak{p}\dot{\varepsilon}\left(\begin{smallmatrix}\top\ \varepsilon = c\\ \llcorner\ \varepsilon\frown v\end{smallmatrix}\right)
\end{array}
\qquad (\iota
$$

$\smile$

$\vdash\!\smile^{\mathfrak{u}}\!\smile^{\mathfrak{a}}$ $\wp\mathfrak{u} = \wp v$
$\mathfrak{a} \frown \mathfrak{u}$
$\wp\dot{\varepsilon}\left(\top\, \varepsilon = \mathfrak{a} ;\ \varepsilon \frown \mathfrak{u}\right) = d$
$c \frown v$
$d = \wp\dot{\varepsilon}\left(\top\, \varepsilon = c ;\ \varepsilon \frown v\right)$ (κ

(68): – – – – – – – – – – – – – –

$\vdash$ $d \frown (\wp v \frown \mathfrak{f})$
$c \frown v$
$d = \wp\dot{\varepsilon}\left(\top\, \varepsilon = c ;\ \varepsilon \frown v\right)$ (87

(IIIc): ————————

$\vdash$ $d \frown (e \frown \mathfrak{f})$
$c \frown v$
$d = \wp\dot{\varepsilon}\left(\top\, \varepsilon = c ;\ \varepsilon \frown v\right)$
$\wp v = e$ (α

×

$\vdash$ $\wp v = e$
$c \frown v$
$d = \wp\dot{\varepsilon}\left(\top\, \varepsilon = c ;\ \varepsilon \frown v\right)$
$d \frown (e \frown \mathfrak{f})$ (β

(IIIc): ————————

$\vdash$ $\wp v = e$
$c \frown v$
$\wp\dot{\varepsilon}\left(\top\, \varepsilon = c ;\ \varepsilon \frown v\right) = a$
$d = a$
$d \frown (e \frown \mathfrak{f})$ (γ

$\smile$

$\vdash\!\smile^{\mathfrak{u}}\!\smile^{\mathfrak{a}}$ $\wp\mathfrak{u} = e$
$\mathfrak{a} \frown \mathfrak{u}$
$\wp\dot{\varepsilon}\left(\top\, \varepsilon = \mathfrak{a} ;\ \varepsilon \frown \mathfrak{u}\right) = a$
$d = a$
$d \frown (e \frown \mathfrak{f})$ (δ

(68): – – – – – – – – – – – – – –

$\vdash$ $a \frown (e \frown \mathfrak{f})$
$d = a$
$d \frown (e \frown \mathfrak{f})$ (ε

×

$\vdash$ $d = a$
$a \frown (e \frown \mathfrak{f})$
$d \frown (e \frown \mathfrak{f})$ (88

(23, 23):: = = = =

$\vdash$ $d = a$
$e \frown (a \frown \rangle\!\!\!\!\!\!\!\!\mathfrak{f})$
$e \frown (d \frown \rangle\!\!\!\!\!\!\!\!\mathfrak{f})$ (α

$\smile$

$\vdash\!\smile^{\mathfrak{e}}\!\smile^{\mathfrak{d}}\!\smile^{\mathfrak{a}}$ $\mathfrak{d} = \mathfrak{a}$
$\mathfrak{e} \frown (\mathfrak{a} \frown \rangle\!\!\!\!\!\!\!\!\mathfrak{f})$
$\mathfrak{e} \frown (\mathfrak{d} \frown \rangle\!\!\!\!\!\!\!\!\mathfrak{f})$ (β

(16): ————————

$\vdash \mathrm{I}\rangle\!\!\!\!\!\!\!\!\mathfrak{f}$ (89

(Ie): —

$\vdash$ $\mathrm{I}\mathfrak{f}$
$\mathrm{I}\rangle\!\!\!\!\!\!\!\!\mathfrak{f}$
$\mathrm{I}\mathfrak{f}$ (α

(71):: ————

$\vdash$ $\mathrm{I}\mathfrak{f}$
$\mathrm{I}\rangle\!\!\!\!\!\!\!\!\mathfrak{f}$ (90

Δ. Proofs of some propositions concerning the cardinal number Zero

a) *Proof of the proposition*

‘$\vdash$ $f(a)$
$\wp\dot{\varepsilon}f(\varepsilon) = \mathbb{0}$’

§96. *Analysis*

We now prove the proposition that no object falls under a concept whose associated cardinal number is $\mathbb{0}$. The proposition mentioned in the heading is actually somewhat more general, because the function-letter ‘*f*’ does not only indicate concepts. Our proposition straightforwardly follows from the proposition

$$`\begin{array}{l}\vdash\!\top\, \wp u = \emptyset \\ \;\llcorner\, a \frown u\end{array}\text{'},$$

According to definition (Θ) we need to prove

$$`\begin{array}{l}\vdash\!\top\, \wp u = \wp\dot{\varepsilon}(\top\, \varepsilon = \varepsilon) \\ \;\llcorner\, a \frown u\end{array}\text{'},$$

which can be achieved by means of (49), by proving $`\begin{array}{l}\vdash\!\top\, a \frown u \\ \;\llcorner\, u \frown (\dot{\varepsilon}(\top\, \varepsilon = \varepsilon) \frown \rangle q)\end{array}$'. This requires use of (8) and the proposition $`\begin{array}{l}\vdash\!\top\, b \frown \dot{\varepsilon}(\top\, \varepsilon = \varepsilon) \\ \;\llcorner\, a \frown (b \frown q)\end{array}$' which is straightforwardly obtained from (58).

§97. *Construction*

$$\Theta \vdash \wp\dot{\varepsilon}(\top\, \varepsilon = \varepsilon) = \emptyset$$

(IIIc): ———————

$$\begin{array}{l}\vdash\!\top\, F(\emptyset) \\ \;\llcorner\, F(\wp\dot{\varepsilon}(\top\, \varepsilon = \varepsilon))\end{array} \qquad (91$$

——— • ———

$$\text{IIIe} \vdash b = b$$

(58): ———

$$\vdash b \frown \dot{\varepsilon}(\top\, \varepsilon = \varepsilon) \qquad (92$$

(I): ———————

$$\begin{array}{l}\vdash\!\top\, b \frown \dot{\varepsilon}(\top\, \varepsilon = \varepsilon) \\ \;\llcorner\, a \frown (b \frown q)\end{array} \qquad (\alpha$$

$$\smile$$

$$\begin{array}{l}\vdash\!\overset{\mathfrak{a}}{\smile}\!\top\, \mathfrak{a} \frown \dot{\varepsilon}(\top\, \varepsilon = \varepsilon) \\ \;\;\;\llcorner\, a \frown (\mathfrak{a} \frown q)\end{array} \qquad (\beta$$

(8): ———————

$$\begin{array}{l}\vdash\!\top\, a \frown u \\ \;\llcorner\, u \frown (\dot{\varepsilon}(\top\, \varepsilon = \varepsilon) \frown \rangle q)\end{array} \qquad (\gamma$$

$$\times$$

$$\begin{array}{l}\vdash\!\top\, u \frown (\dot{\varepsilon}(\top\, \varepsilon = \varepsilon) \frown \rangle q) \\ \;\llcorner\, a \frown u\end{array} \qquad (\delta$$

(I): ———————

$$\begin{array}{l}\vdash\!\top\!\top\, u \frown (\dot{\varepsilon}(\top\, \varepsilon = \varepsilon) \frown \rangle q) \\ \;\;\llcorner\, \dot{\varepsilon}(\top\, \varepsilon = \varepsilon) \frown (u \frown \rangle \not\models q) \\ \;\llcorner\, a \frown u\end{array} \qquad (\varepsilon$$

$$\smile$$

$$\begin{array}{l}\vdash\!\overset{\mathfrak{q}}{\smile}\!\top\!\top\, u \frown (\dot{\varepsilon}(\top\, \varepsilon = \varepsilon) \frown \rangle \mathfrak{q}) \\ \;\;\;\;\llcorner\, \dot{\varepsilon}(\top\, \varepsilon = \varepsilon) \frown (u \frown \rangle \not\models \mathfrak{q}) \\ \;\llcorner\, a \frown u\end{array} \qquad (\zeta$$

(49): – – – – – – – – – – – –

$$\begin{array}{l}\vdash\!\top\, \wp u = \wp\dot{\varepsilon}(\top\, \varepsilon = \varepsilon) \\ \;\llcorner\, a \frown u\end{array} \qquad (\eta$$

(91): ———————

$$\begin{array}{l}\vdash\!\top\, \wp u = \emptyset \\ \;\llcorner\, a \frown u\end{array} \qquad (93$$

$$\times$$

$$\begin{array}{l}\vdash\!\top\, a \frown u \\ \;\llcorner\, \wp u = \emptyset\end{array} \qquad (94$$

——— • ———

$$94 \begin{array}{l}\vdash\!\top\, a \frown \dot{\varepsilon} f(\varepsilon) \\ \;\llcorner\, \wp\dot{\varepsilon} f(\varepsilon) = \emptyset\end{array}$$

(82): – – – – – – –

$$\begin{array}{l}\vdash\!\top\, f(a) \\ \;\llcorner\, \wp\dot{\varepsilon} f(\varepsilon) = \emptyset\end{array} \qquad (95$$

b) Proof of the proposition

$$`\begin{array}{l}\vdash\!\top\!\!-\!\!-\, \wp u = \emptyset \\ \;\llcorner\!\overset{\mathfrak{a}}{\smile}\!\top\, \mathfrak{a} \frown u\end{array}\text{'}$$

and of some corollaries

§98. *Analysis*

The proposition mentioned in the heading is somewhat more general than the one we express in words like this: "If no object falls under a concept, then Zero is the cardinal number that belongs to this concept."

We first prove, using (32) and (38), the proposition

$$`\begin{array}{l}\vdash\!\top\!\!-\!\!-\, \wp u = \wp v \\ \;\llcorner\!\overset{\mathfrak{a}}{\smile}\, (\!-\!\!-\, \mathfrak{a} \frown u) = (\!-\!\!-\, \mathfrak{a} \frown v)\end{array}\text{'}$$

and then

$$`\begin{array}{l}\vdash\!\top\!\!-\!\!-\, (\!-\!\!-\, a \frown u) = (\!-\!\!-\, a \frown \dot{\varepsilon}(\top\, \varepsilon = \varepsilon)) \\ \;\llcorner\!\overset{\mathfrak{a}}{\smile}\!\top\, \mathfrak{a} \frown u\end{array}\text{'},$$

§99. *Construction*

IIIf $\vdash (-\, a\frown v) = (-\, a\frown u)$
$(-\, a\frown u) = (-\, a\frown v)$

(IIa):: - - - - - - - - - - -

$\vdash (-\, a\frown v) = (-\, a\frown u)$
$\mathfrak{a}\ (-\, \mathfrak{a}\frown u) = (-\, \mathfrak{a}\frown v)$ (α

$\vdash \mathfrak{a}\ (-\, \mathfrak{a}\frown v) = (-\, \mathfrak{a}\frown u)$
$\mathfrak{a}\ (-\, \mathfrak{a}\frown u) = (-\, \mathfrak{a}\frown v)$ (β

(38): - - - - - - - - - - - - -

$\vdash v\frown(u\frown\rangle\dot{\alpha}\dot{\varepsilon}(\varepsilon = \alpha))$
$\mathfrak{a}\ (-\, \mathfrak{a}\frown u) = (-\, \mathfrak{a}\frown v)$ (γ

(IIIc): ———————————

$\vdash v\frown(u\frown\rangle\rlap{\,/}\rlap{\,/}\dot{\alpha}\dot{\varepsilon}(\varepsilon = \alpha))$
$\mathfrak{a}\ (-\, \mathfrak{a}\frown u) = (-\, \mathfrak{a}\frown v)$
$\dot{\alpha}\dot{\varepsilon}(\varepsilon = \alpha) = \rlap{\,/}\rlap{\,/}\dot{\alpha}\dot{\varepsilon}(\varepsilon = \alpha)$ (δ

(41):: ———————————

$\vdash v\frown(u\frown\rangle\rlap{\,/}\rlap{\,/}\dot{\alpha}\dot{\varepsilon}(\varepsilon = \alpha))$
$\mathfrak{a}\ (-\, \mathfrak{a}\frown u) = (-\, \mathfrak{a}\frown v)$ (ε

(32): - - - - - - - - - - - - -

$\vdash \mathfrak{P}u = \mathfrak{P}v$
$u\frown(v\frown\rangle\dot{\alpha}\dot{\varepsilon}(\varepsilon = \alpha))$
$\mathfrak{a}\ (-\, \mathfrak{a}\frown u) = (-\, \mathfrak{a}\frown v)$ (ζ

(38):: - - - - - - - - - - - - -

$\vdash \mathfrak{P}u = \mathfrak{P}v$
$\mathfrak{a}\ (-\, \mathfrak{a}\frown u) = (-\, \mathfrak{a}\frown v)$ (96

——— • ———

92 $\vdash a\frown\dot{\varepsilon}(\top\ \varepsilon = \varepsilon)$

(Ia): ———————

$\vdash a\frown u$
$a\frown\dot{\varepsilon}(\top\ \varepsilon = \varepsilon)$ (α

(IVa): ———————

$\vdash (-\, a\frown u) = (-\, a\frown\dot{\varepsilon}(\top\ \varepsilon = \varepsilon))$
$a\frown\dot{\varepsilon}(\top\ \varepsilon = \varepsilon)$
$a\frown u$ (β

——— • ———

IIa $\vdash a\frown u$
$\mathfrak{a}\top\ \mathfrak{a}\frown u$

(Ia): - - - - -

$\vdash a\frown\dot{\varepsilon}(\top\ \varepsilon = \varepsilon)$
$a\frown u$
$\mathfrak{a}\top\ \mathfrak{a}\frown u$ (γ

(β): - - - - - - - -

$\vdash (-\, a\frown u) = (-\, a\frown\dot{\varepsilon}(\top\ \varepsilon = \varepsilon))$
$\mathfrak{a}\top\ \mathfrak{a}\frown u$ (δ

$\vdash \mathfrak{a}\ (-\, \mathfrak{a}\frown u) = (-\, \mathfrak{a}\frown\dot{\varepsilon}(\top\ \varepsilon = \varepsilon))$
$\mathfrak{a}\top\ \mathfrak{a}\frown u$ (ε

(96): - - - - - - - - - - - - - - -

$\vdash \mathfrak{P}u = \mathfrak{P}\dot{\varepsilon}(\top\ \varepsilon = \varepsilon)$
$\mathfrak{a}\top\ \mathfrak{a}\frown u$ (ζ

(91): ———————————

$\vdash \mathfrak{P}u = \mathbb{0}$
$\mathfrak{a}\top\ \mathfrak{a}\frown u$ (97

§100. *Analysis*

We begin by deriving some straightforward consequences from (97), and then turn our attention to the proposition: "If a cardinal number is not Zero, then there is one that immediately precedes it in the cardinal number series", in signs:

'$\vdash \mathfrak{a}\top\ \mathfrak{a}\frown(a\frown\mathfrak{f})$
$\top\ a = \mathbb{0}$
$\top\ \mathfrak{u}\top\ \mathfrak{P}\mathfrak{u} = a$',

We first derive the simpler proposition

'$\vdash\top\ c\frown u$
$\mathfrak{a}\top\ \mathfrak{a}\frown(\mathfrak{P}u\frown\mathfrak{f})$'

For this we require the proposition

'$\vdash \mathfrak{P}\dot{\varepsilon}\left(\top \begin{array}{l}\varepsilon = c\\ \varepsilon\frown u\end{array}\right)\frown(\mathfrak{P}u\frown\mathfrak{f})$
$c\frown u$',

which follows from definition (H).

§101. *Construction*

94 $\vdash\top\ a\frown u$
$\mathfrak{P}u = \mathbb{0}$

(I): ———————

$\vdash\top\ a\frown u$
$a\frown v$
$\mathfrak{P}u = \mathbb{0}$ (α

(58): - - - - -

$\vdash\!\!\top\; a \frown \dot{\varepsilon}\left(\top\!\top\, \varepsilon \frown u \;\; \llcorner\, \varepsilon \frown v\right)$
$\quad\llcorner\; \wp u = \emptyset$ (β

$\vdash\!\smile^{\mathfrak{a}}\!\top\; \mathfrak{a} \frown \dot{\varepsilon}\left(\top\!\top\, \varepsilon \frown u \;\; \llcorner\, \varepsilon \frown v\right)$
$\quad\llcorner\; \wp u = \emptyset$ (γ

(97): - - - - - - - - - - -

$\vdash\!\top\; \wp\dot{\varepsilon}\left(\top\!\top\, \varepsilon \frown u \;\; \llcorner\, \varepsilon \frown v\right) = \emptyset$
$\quad\llcorner\; \wp u = \emptyset$ (98

——— • ———

93 $\vdash\!\top\; \wp u = \emptyset$
$\quad\llcorner\; a \frown u$

(IIa):: - - - -

$\vdash\!\top\!\top\; \wp u = \emptyset$
$\quad\llcorner\; a \frown v$
$\quad\llcorner\!\smile^{\mathfrak{a}}\!\top\; \mathfrak{a} \frown u$
$\qquad\llcorner\; \mathfrak{a} \frown v$ (α

×

$\vdash\!\top\!\top\; a \frown v$
$\quad\llcorner\; \wp u = \emptyset$
$\quad\llcorner\!\smile^{\mathfrak{a}}\!\top\; \mathfrak{a} \frown u$
$\qquad\llcorner\; \mathfrak{a} \frown v$ (β

$\vdash\!\top\!\smile^{\mathfrak{a}}\!\top\; \mathfrak{a} \frown v$
$\quad\llcorner\; \wp u = \emptyset$
$\quad\llcorner\!\smile^{\mathfrak{a}}\!\top\; \mathfrak{a} \frown u$
$\qquad\llcorner\; \mathfrak{a} \frown v$ (γ

(97): - - - - - - - -

$\vdash\!\top\!\top\; \wp v = \emptyset$
$\quad\llcorner\; \wp u = \emptyset$
$\quad\llcorner\!\smile^{\mathfrak{a}}\!\top\; \mathfrak{a} \frown u$
$\qquad\llcorner\; a \frown v$ (99

——— • ———

10 $\vdash\!\top\!\top\; a \frown (b \frown q)$
$\quad\llcorner\!\top\; f(a, b)$
$\quad\llcorner\; \dot{\alpha}\dot{\varepsilon}(\top f(\varepsilon, \alpha)) = q$

×

$\vdash\!\top\!\top\; f(a, b)$
$\quad\llcorner\!\top\; a \frown (b \frown q)$
$\quad\llcorner\; \dot{\alpha}\dot{\varepsilon}(\top f(\varepsilon, \alpha)) = q$ (100

——— • ———

H $\vdash \dot{\alpha}\dot{\varepsilon}\left[\top\!\smile^{\mathfrak{u}}\!\smile^{\mathfrak{a}}\!\top\!\top\; \wp\mathfrak{u} = \alpha \;\; \llcorner\, \mathfrak{a} \frown \mathfrak{u} \;\; \llcorner\, \wp\dot{\varepsilon}\left(\top\, \varepsilon = \mathfrak{a} \;\; \llcorner\, \varepsilon \frown \mathfrak{u}\right) = \varepsilon\right] = \mathfrak{f}$

(100): ————————————

$\vdash\!\smile^{\mathfrak{u}}\!\smile^{\mathfrak{a}}\!\top\!\top\; \wp\mathfrak{u} = n$
$\quad\llcorner\; \mathfrak{a} \frown \mathfrak{u}$
$\quad\llcorner\; \wp\dot{\varepsilon}\left(\top\, \varepsilon = \mathfrak{a} \;\; \llcorner\, \varepsilon \frown \mathfrak{u}\right) = m$
$\quad\llcorner\!\top\; m \frown (n \frown \mathfrak{f})$ (α

(IIa): - - - - - - - - - - - - - -

$\vdash\!\smile^{\mathfrak{a}}\!\top\!\top\; \wp u = n$
$\quad\llcorner\; \mathfrak{a} \frown u$
$\quad\llcorner\; \wp\dot{\varepsilon}\left(\top\, \varepsilon = \mathfrak{a} \;\; \llcorner\, \varepsilon \frown u\right) = m$
$\quad\llcorner\!\top\; m \frown (n \frown \mathfrak{f})$ (β

(IIa): - - - - - - - - - - - - - -

$\vdash\!\top\!\top\!\top\; \wp u = n$
$\quad\llcorner\; c \frown u$
$\quad\llcorner\; \wp\dot{\varepsilon}\left(\top\, \varepsilon = c \;\; \llcorner\, \varepsilon \frown u\right) = m$
$\quad\llcorner\!\top\; m \frown (n \frown \mathfrak{f})$ (γ

×

$\vdash\!\top\!\top\!\top\; m \frown (n \frown \mathfrak{f})$
$\quad\llcorner\; c \frown u$
$\quad\llcorner\; \wp\dot{\varepsilon}\left(\top\, \varepsilon = c \;\; \llcorner\, \varepsilon \frown u\right) = m$
$\quad\llcorner\; \wp u = n$ (101

——— • ———

IIIe $\vdash \wp u = \wp u$

(101): ————————

$\vdash\!\top\!\top\; m \frown (\wp u \frown \mathfrak{f})$
$\quad\llcorner\; c \frown u$
$\quad\llcorner\; \wp\dot{\varepsilon}\left(\top\, \varepsilon = c \;\; \llcorner\, \varepsilon \frown u\right) = m$ (102

——— • ———

IIIe $\vdash \wp\dot{\varepsilon}\left(\top\, \varepsilon = c \;\; \llcorner\, \varepsilon \frown u\right) = \wp\dot{\varepsilon}\left(\top\, \varepsilon = c \;\; \llcorner\, \varepsilon \frown u\right)$

(102): ————————————

$\vdash\!\top\; \wp\dot{\varepsilon}\left(\top\, \varepsilon = c \;\; \llcorner\, \varepsilon \frown u\right) \frown (\wp u \frown \mathfrak{f})$
$\quad\llcorner\; c \frown u$ (103

×

$$\vdash\!\!\top\ c \frown u$$
$$\quad\llcorner\ \mathfrak{p}\dot{\varepsilon}\left(\begin{array}{l}\top\ \varepsilon = c\\ \llcorner\ \varepsilon \frown u\end{array}\right) \frown (\mathfrak{p}u \frown \mathfrak{f}) \qquad (\alpha$$

(IIa):: - - - - - - - - - - - -

$$\vdash\!\!\top\ c \frown u$$
$$\quad\llcorner\!\smile^{\mathfrak{a}}\!\top\ \mathfrak{a} \frown (\mathfrak{p}u \frown \mathfrak{f}) \qquad (104$$

$$\smile$$

$$\vdash\!\smile^{\mathfrak{a}}\!\top\ \mathfrak{a} \frown u$$
$$\quad\llcorner\!\smile^{\mathfrak{a}}\!\top\ \mathfrak{a} \frown (\mathfrak{p}u \frown \mathfrak{f}) \qquad (\alpha$$

(97): - - - - - - - -

$$\vdash\!\!\top\ \mathfrak{p}u = 0$$
$$\quad\llcorner\!\smile^{\mathfrak{a}}\!\top\ \mathfrak{a} \frown (\mathfrak{p}u \frown \mathfrak{f}) \qquad (105$$

$$\times$$

$$\vdash\!\!\top\!\smile^{\mathfrak{a}}\!\top\ \mathfrak{a} \frown (\mathfrak{p}u \frown \mathfrak{f})$$
$$\quad\llcorner\ \mathfrak{p}u = 0 \qquad (106$$

(IIId): ————————

$$\vdash\!\!\top\ \mathfrak{p}u = a$$
$$\quad\llcorner\!\top\!\smile^{\mathfrak{a}}\!\top\ \mathfrak{a} \frown (a \frown \mathfrak{f})$$
$$\qquad\llcorner\ a = 0 \qquad (\alpha$$

$$\smile$$

$$\vdash\!\!\top\!\smile^{\mathfrak{u}}\!\top\ \mathfrak{p}\mathfrak{u} = a$$
$$\quad\llcorner\!\top\!\smile^{\mathfrak{a}}\!\top\ \mathfrak{a} \frown (a \frown \mathfrak{f})$$
$$\qquad\llcorner\ a = 0 \qquad (\beta$$

$$\times$$

$$\vdash\!\!\top\!\smile^{\mathfrak{a}}\!\top\ \mathfrak{a} \frown (a \frown \mathfrak{f})$$
$$\quad\llcorner\ a = 0$$
$$\quad\llcorner\!\smile^{\mathfrak{u}}\!\top\ \mathfrak{p}\mathfrak{u} = a \qquad (107$$

E. Proofs of some propositions concerning the cardinal number One

§102. *Analysis*

We prove the proposition

$$\text{‘}\vdash\!\!\top\!\smile^{\mathfrak{a}}\!\top\ \mathfrak{a} \frown u$$
$$\quad\llcorner\ \mathfrak{p}u = 1\text{’} \qquad (\alpha$$

which may be expressed in words like this:

"There is an object which falls under a concept, if One is the cardinal number of this concept."

If this were not correct, then according to proposition (97) the cardinal number One would coincide with the cardinal number Zero. What needs to be shown is that this cannot be. To this end we prove the propositions

‘$\vdash 0 \frown (1 \frown \mathfrak{f})$’ $(\beta$, ‘$\vdash\!\!\top\ 0 \frown (0 \frown \mathfrak{f})$’ $(\gamma$

Of these, (β) follows from (101) by definition (I), (γ) from (68) using proposition (93).

§103. *Construction*

$$93\ \vdash\!\!\top\ \mathfrak{p}u = 0$$
$$\quad\llcorner\ a \frown u$$

(I): ————

$$\vdash\!\!\top\ \mathfrak{p}u = 0$$
$$\quad\llcorner\ a \frown u$$
$$\quad\llcorner\ \mathfrak{p}\dot{\varepsilon}\left(\begin{array}{l}\top\ \varepsilon = a\\ \llcorner\ \varepsilon \frown u\end{array}\right) = c \qquad (\alpha$$

$$\smile$$

$$\vdash\!\smile^{\mathfrak{u}}\!\smile^{\mathfrak{a}}\!\top\ \mathfrak{p}\mathfrak{u} = 0$$
$$\quad\llcorner\ \mathfrak{a} \frown \mathfrak{u}$$
$$\quad\llcorner\ \mathfrak{p}\dot{\varepsilon}\left(\begin{array}{l}\top\ \varepsilon = \mathfrak{a}\\ \llcorner\ \varepsilon \frown \mathfrak{u}\end{array}\right) = c \qquad (\beta$$

(68): ————————

$$\vdash\!\!\top\ c \frown (0 \frown \mathfrak{f}) \qquad (108$$

—— • ——

$$\text{IIIe}\ \vdash c = c$$

(77): ——

$$\vdash c \frown \dot{\varepsilon}(\varepsilon = c) \qquad (109$$

—— • ——

$$82\ \vdash\!\!\top\ a = 0$$
$$\quad\llcorner\ a \frown \dot{\varepsilon}(\varepsilon = 0)$$

(58): ————————

$$\vdash\!\!\top\ a \frown \dot{\varepsilon}\left(\begin{array}{l}\top\ \varepsilon = 0\\ \llcorner\ \varepsilon \frown \dot{\varepsilon}(\varepsilon = 0)\end{array}\right) \qquad (\alpha$$

$$\smile$$

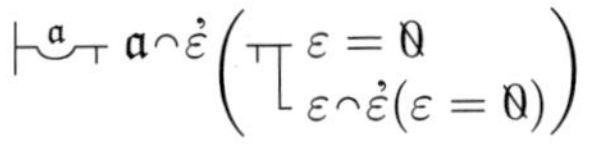 (β

(97):

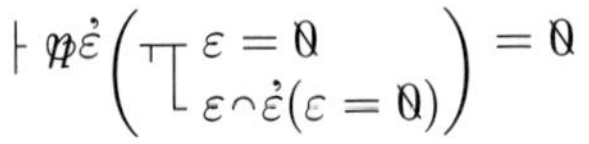 (γ

(101):

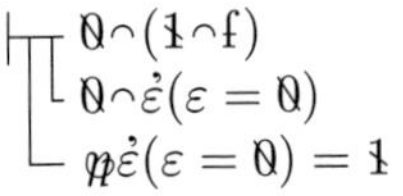 (δ

(109, I)::

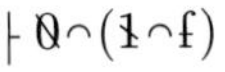 (110

(IIIb):

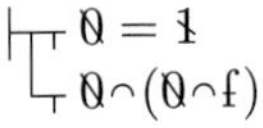 (α

(108)::

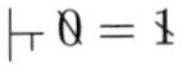 (111

•

97 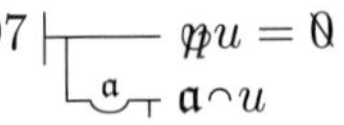

×

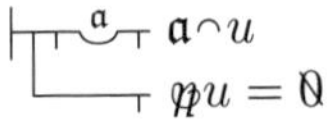 (112

(IIId)::

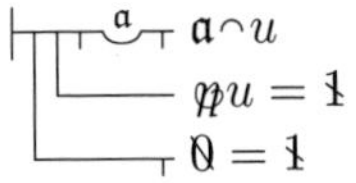

(111)::

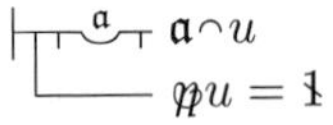 (113

•

§104. *Analysis*

Using (110) and (71) it is straightforward to prove the proposition that a cardinal number is One if it immediately follows Zero in the cardinal number series.

In order to prove the proposition

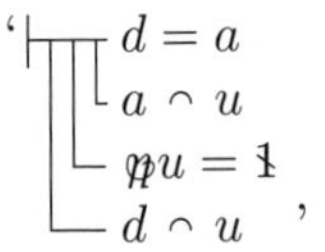 (α

we apply (49) in the form

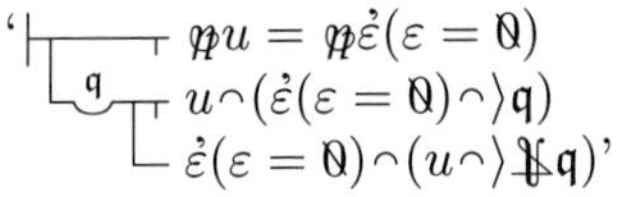

and now require the proposition

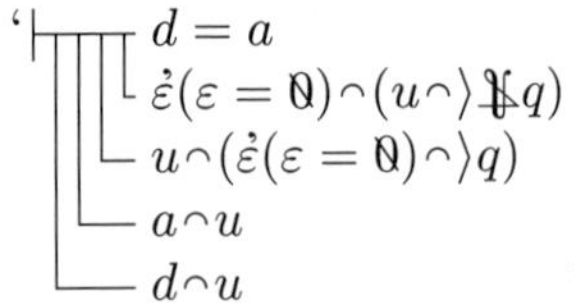 (β

From (79) and (18) we have the proposition

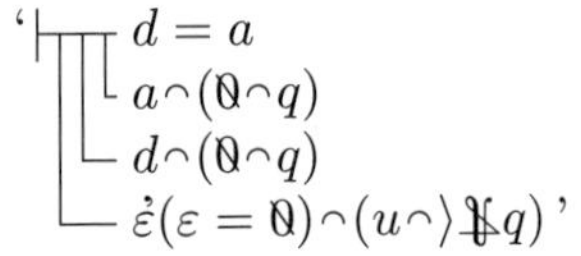 (γ

and now apply the proposition

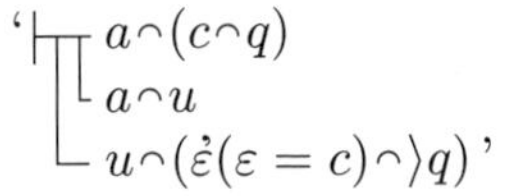

which is straightforwardly derived using (77) and (8).

§105. *Construction*

13 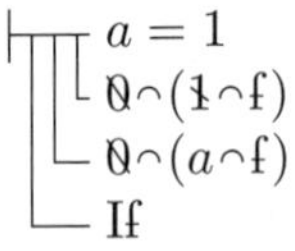

(110, 71)::

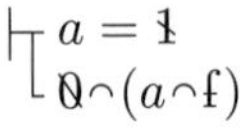 (114

•

IIId 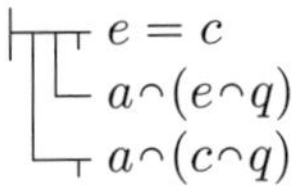

(77):

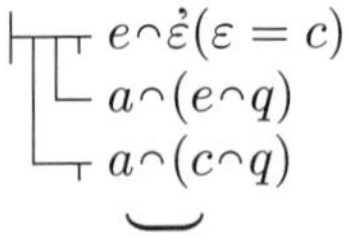 (α

$\vdash$ ($\mathfrak{a}$) $\mathfrak{a} \frown \dot{\varepsilon}(\varepsilon = c)$
$\quad a \frown (\mathfrak{a} \frown q)$
$\quad a \frown (c \frown q)$ $\quad$ (β

(8): - - - - - - - - -

$\vdash a \frown u$
$\quad a \frown (c \frown q)$
$\quad u \frown (\dot{\varepsilon}(\varepsilon = c) \frown \rangle q)$ $\quad$ (γ

$\times$

$\vdash a \frown (c \frown q)$
$\quad a \frown u$
$\quad u \frown (\dot{\varepsilon}(\varepsilon = c) \frown \rangle q)$ $\quad$ (115

——— • ———

I $\vdash \mathfrak{p}\dot{\varepsilon}(\varepsilon = \mathbb{0}) = \mathbb{1}$

(IIIc): ————————

$\vdash F(\mathbb{1})$
$\quad F(\mathfrak{p}\dot{\varepsilon}(\varepsilon = \mathbb{0}))$ $\quad$ (116

——— • ———

18 $\vdash \mathrm{I}\rangle\!\!\!\!\not\,q$
$\quad \dot{\varepsilon}(\varepsilon = \mathbb{0}) \frown (u \frown \rangle\!\!\!\!\not\,q)$

(79): - - - - - - - - - -

$\vdash d = a$
$\quad a \frown (\mathbb{0} \frown q)$
$\quad d \frown (\mathbb{0} \frown q)$
$\quad \dot{\varepsilon}(\varepsilon = \mathbb{0}) \frown (u \frown \rangle\!\!\!\!\not\,q)$ $\quad$ (α

(115, 115):: = = = = = = = =

$\vdash d = a$
$\quad a \frown u$
$\quad u \frown (\dot{\varepsilon}(\varepsilon = \mathbb{0}) \frown \rangle q)$
$\quad d \frown u$
$\quad \dot{\varepsilon}(\varepsilon = \mathbb{0}) \frown (u \frown \rangle\!\!\!\!\not\,q)$ $\quad$ (β

$\times$

$\vdash u \frown (\dot{\varepsilon}(\varepsilon = \mathbb{0}) \frown \rangle q)$
$\quad \dot{\varepsilon}(\varepsilon = \mathbb{0}) \frown (u \frown \rangle\!\!\!\!\not\,q)$
$\quad a \frown u$
$\quad d = a$
$\quad d \frown u$ $\quad$ (γ

$\smile$

$\vdash$ ($\mathfrak{q}$) $u \frown (\dot{\varepsilon}(\varepsilon = \mathbb{0}) \frown \rangle \mathfrak{q})$
$\quad \dot{\varepsilon}(\varepsilon = \mathbb{0}) \frown (u \frown \rangle\!\!\!\!\not\,\mathfrak{q})$
$\quad a \frown u$
$\quad d = a$
$\quad d \frown u$ $\quad$ (δ

(49): - - - - - - - - - - - - - -

$\vdash \mathfrak{p}u = \mathfrak{p}\dot{\varepsilon}(\varepsilon = \mathbb{0})$
$\quad a \frown u$
$\quad d = a$
$\quad d \frown u$ $\quad$ (ε

(116): ————————————

$\vdash \mathfrak{p}u = \mathbb{1}$
$\quad a \frown u$
$\quad d = a$
$\quad d \frown u$ $\quad$ (ζ

$\times$

$\vdash d = a$
$\quad a \frown u$
$\quad \mathfrak{p}u = \mathbb{1}$
$\quad d \frown u$ $\quad$ (117

——— • ———

§106. *Analysis*

We now prove the proposition

'$\vdash \mathfrak{p}u = \mathbb{1}$
$\quad$ ($\mathfrak{e}$) $\mathfrak{e} \frown u$
$\quad$ ($\mathfrak{d}$) ($\mathfrak{a}$) $\mathfrak{a} = \mathfrak{d}$
$\quad\quad \mathfrak{a} \frown u$
$\quad\quad \mathfrak{d} \frown u$ ', $\quad$ (α

i.e., "One is the cardinal number of a concept under which an object falls if, whenever an object a and an object d fall under that concept, it follows that a is the same as d."

This proposition is a consequence of the proposition

'$\vdash \mathfrak{p}u = \mathbb{1}$
$\quad$ ($\mathfrak{a}$) $\mathfrak{a} = c$
$\quad\quad \mathfrak{a} \frown u$
$\quad c \frown u$ ', $\quad$ (β

which we reduce to the following proposition

'$\vdash \mathbb{1} = \mathfrak{p}\dot{\varepsilon}(\varepsilon = c)$' or
'$\vdash \mathfrak{p}\dot{\varepsilon}(\varepsilon = n) = \mathfrak{p}\dot{\varepsilon}(\varepsilon = c)$' $\quad$ (γ

The $\dot{\alpha}\dot{\varepsilon}\left(\begin{smallmatrix} \alpha = c \\ \varepsilon = n \end{smallmatrix}\right)$-relation suggests itself as a mapping relation. Its single-valuedness, and that of its converse, has to be proven.

§107. Construction

IIIa $\vdash\vdash\begin{array}{l} d = a \\ a = c \\ d = c \end{array}$

(Ib, Ib):: = = = =

$\vdash\begin{bmatrix} d = a \\ \vdash\vdash\begin{array}{l} a = c \\ e = n \end{array} \\ \vdash\vdash\begin{array}{l} d = c \\ e = n \end{array} \end{bmatrix}$ (α

(33, 33):: = = = = =

$\vdash\begin{bmatrix} d = a \\ e\frown\left[a\frown\dot\alpha\dot\varepsilon\left(\vdash\vdash\begin{array}{l}\alpha = c \\ \varepsilon = n\end{array}\right)\right] \\ e\frown\left[d\frown\dot\alpha\dot\varepsilon\left(\vdash\vdash\begin{array}{l}\alpha = c \\ \varepsilon = n\end{array}\right)\right] \end{bmatrix}$ (β

$\vdash\overset{e\ \mathfrak{d}\ \mathfrak{a}}{\smile\smile\smile}\begin{bmatrix} \mathfrak{d} = \mathfrak{a} \\ e\frown\left[\mathfrak{a}\frown\dot\alpha\dot\varepsilon\left(\vdash\vdash\begin{array}{l}\alpha = c \\ \varepsilon = n\end{array}\right)\right] \\ e\frown\left[\mathfrak{d}\frown\dot\alpha\dot\varepsilon\left(\vdash\vdash\begin{array}{l}\alpha = c \\ \varepsilon = n\end{array}\right)\right] \end{bmatrix}$ (γ

(16): ─────────

$\vdash I\dot\alpha\dot\varepsilon\left(\vdash\vdash\begin{array}{l}\alpha = c \\ \varepsilon = n\end{array}\right)$ (δ

─── • ───

Ie $\vdash\vdash\begin{array}{l} a \\ b \\ a \\ b \end{array}$

(Ib, Id):: = = = =

$\vdash\vdash\begin{array}{l} a \\ b \\ b \\ a \end{array}$ (ε

(IVa): ───

$\vdash\vdash\begin{array}{l} \left(\vdash\vdash\begin{array}{l}a\\b\end{array}\right) = \left(\vdash\vdash\begin{array}{l}b\\a\end{array}\right) \\ b \\ a \\ a \\ b \end{array}$ (ζ

(ε):: ────────────────

$\vdash\left(\vdash\vdash\begin{array}{l}a\\b\end{array}\right) = \left(\vdash\vdash\begin{array}{l}b\\a\end{array}\right)$ (η

─── • ───

$\eta\vdash\left(\vdash\vdash\begin{array}{l}y = n \\ x = c\end{array}\right) = \left(\vdash\vdash\begin{array}{l}x = c \\ y = n\end{array}\right)$

$\vdash^{\mathfrak{a}}\left(\vdash\vdash\begin{array}{l}y = n \\ \mathfrak{a} = c\end{array}\right) = \left(\vdash\vdash\begin{array}{l}\mathfrak{a} = c \\ y = n\end{array}\right)$ (ϑ

(Va): ────────────────────

$\vdash\dot\varepsilon\left(\vdash\vdash\begin{array}{l}y = n \\ \varepsilon = c\end{array}\right) = \dot\varepsilon\left(\vdash\vdash\begin{array}{l}\varepsilon = c \\ y = n\end{array}\right)$ (ι

$\vdash^{\mathfrak{a}}\dot\varepsilon\left(\vdash\vdash\begin{array}{l}\mathfrak{a} = n \\ \varepsilon = c\end{array}\right) = \dot\varepsilon\left(\vdash\vdash\begin{array}{l}\varepsilon = c \\ \mathfrak{a} = n\end{array}\right)$ (κ

(Va): ────────────────────

$\vdash\dot\alpha\dot\varepsilon\left(\vdash\vdash\begin{array}{l}\alpha = n \\ \varepsilon = c\end{array}\right) = \dot\alpha\dot\varepsilon\left(\vdash\vdash\begin{array}{l}\varepsilon = c \\ \alpha = n\end{array}\right)$ (λ

(IIIc): ────────────────────

$\vdash\begin{bmatrix} F\left[\dot\alpha\dot\varepsilon\left(\vdash\vdash\begin{array}{l}\varepsilon = c \\ \alpha = n\end{array}\right)\right] \\ F\left[\dot\alpha\dot\varepsilon\left(\vdash\vdash\begin{array}{l}\alpha = n \\ \varepsilon = c\end{array}\right)\right] \end{bmatrix}$ (μ

(IIIc): ────────────────────

$\vdash\begin{bmatrix} F\left[\mathfrak{X}\dot\alpha\dot\varepsilon\left(\vdash\vdash\begin{array}{l}\alpha = c \\ \varepsilon = n\end{array}\right)\right] \\ F\left[\dot\alpha\dot\varepsilon\left(\vdash\vdash\begin{array}{l}\alpha = n \\ \varepsilon = c\end{array}\right)\right] \end{bmatrix}$

$\vdash\dot\alpha\dot\varepsilon\left(\vdash\vdash\begin{array}{l}\varepsilon = c \\ \alpha = n\end{array}\right) = \mathfrak{X}\dot\alpha\dot\varepsilon\left(\vdash\vdash\begin{array}{l}\alpha = c \\ \varepsilon = n\end{array}\right)$ (ν

(40):: ────────────────────

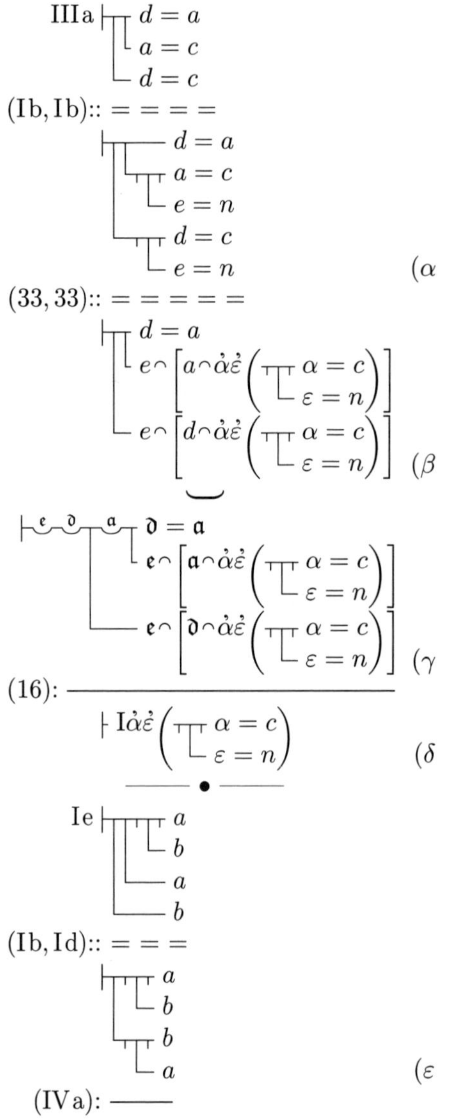

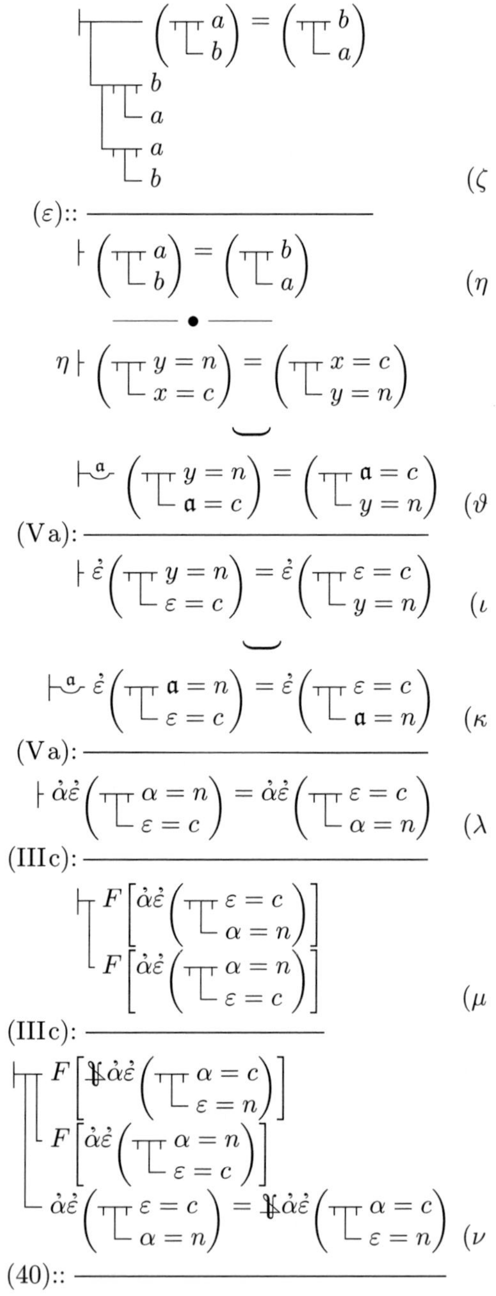

$$\begin{array}{l}\vdash\!\top\ F\left[\text{⅄}\dot\alpha\dot\varepsilon\left(\begin{array}{l}\top\!\top\ \alpha = c\\ \quad\llcorner\ \varepsilon = n\end{array}\right)\right]\\ \llcorner\ F\left[\dot\alpha\dot\varepsilon\left(\begin{array}{l}\top\!\top\ \alpha = n\\ \quad\llcorner\ \varepsilon = c\end{array}\right)\right]\end{array} \qquad (\xi$$

———•———

$$\begin{array}{l}36 \vdash\!\!-\ d\frown\left[c\frown\dot\alpha\dot\varepsilon\left(\begin{array}{l}\top\!\top\ \alpha = c\\ \quad\llcorner\ \varepsilon = n\end{array}\right)\right]\\ \quad\llcorner\!\top\!\top\ c = c\\ \qquad\llcorner\ d = n\end{array}$$

(Ie):: - - - - - - - - - - - - -

$$\begin{array}{l}\vdash\!\top\!\top\ d\frown\left[c\frown\dot\alpha\dot\varepsilon\left(\begin{array}{l}\top\!\top\ \alpha = c\\ \quad\llcorner\ \varepsilon = n\end{array}\right)\right]\\ \quad\llcorner\ c = c\\ \llcorner\ d = n\end{array} \qquad (o$$

(IIIe):: ————————————

$$\begin{array}{l}\vdash\!\top\ d\frown\left[c\frown\dot\alpha\dot\varepsilon\left(\begin{array}{l}\top\!\top\ \alpha = c\\ \quad\llcorner\ \varepsilon = n\end{array}\right)\right]\\ \llcorner\ d = n\end{array} \qquad (\pi$$

(82):: - - - - - - - - - - - -

$$\begin{array}{l}\vdash\!\top\ d\frown\left[c\frown\dot\alpha\dot\varepsilon\left(\begin{array}{l}\top\!\top\ \alpha = c\\ \quad\llcorner\ \varepsilon = n\end{array}\right)\right]\\ \llcorner\ d\frown\dot\varepsilon(\varepsilon = n)\end{array} \qquad (\rho$$

×

$$\begin{array}{l}\vdash\!\top\ d\frown\dot\varepsilon(\varepsilon = n)\\ \llcorner\!\top\ d\frown\left[c\frown\dot\alpha\dot\varepsilon\left(\begin{array}{l}\top\!\top\ \alpha = c\\ \quad\llcorner\ \varepsilon = n\end{array}\right)\right]\end{array} \qquad (\sigma$$

———•———

$$\begin{array}{l}\text{IIa} \vdash\!\!-\!\!\top\ c\frown\dot\varepsilon(\varepsilon = c)\\ \qquad\quad\llcorner\ d\frown\left[c\frown\dot\alpha\dot\varepsilon\left(\begin{array}{l}\top\!\top\ \alpha = c\\ \quad\llcorner\ \varepsilon = n\end{array}\right)\right]\\ \quad\llcorner\!\overset{\mathfrak{a}}{\smile}\!\top\ \mathfrak{a}\frown\dot\varepsilon(\varepsilon = c)\\ \qquad\quad\llcorner\ d\frown\left[\mathfrak{a}\frown\dot\alpha\dot\varepsilon\left(\begin{array}{l}\top\!\top\ \alpha = c\\ \quad\llcorner\ \varepsilon = n\end{array}\right)\right]\end{array}$$

×

$$\begin{array}{l}\vdash\!\!-\!\!\top\ d\frown\left[c\frown\dot\alpha\dot\varepsilon\left(\begin{array}{l}\top\!\top\ \alpha = c\\ \quad\llcorner\ \varepsilon = n\end{array}\right)\right]\\ \qquad\llcorner\ c\frown\dot\varepsilon(\varepsilon = c)\\ \quad\llcorner\!\overset{\mathfrak{a}}{\smile}\!\top\ \mathfrak{a}\frown\dot\varepsilon(\varepsilon = c)\\ \qquad\quad\llcorner\ d\frown\left[\mathfrak{a}\frown\dot\alpha\dot\varepsilon\left(\begin{array}{l}\top\!\top\ \alpha = c\\ \quad\llcorner\ \varepsilon = n\end{array}\right)\right]\end{array} \qquad (\tau$$

(109):: ————————————

$$\begin{array}{l}\vdash\!\!-\!\!\top\ d\frown\left[c\frown\dot\alpha\dot\varepsilon\left(\begin{array}{l}\top\!\top\ \alpha = c\\ \quad\llcorner\ \varepsilon = n\end{array}\right)\right]\\ \quad\llcorner\!\overset{\mathfrak{a}}{\smile}\!\top\ \mathfrak{a}\frown\dot\varepsilon(\varepsilon = c)\\ \qquad\quad\llcorner\ d\frown\left[\mathfrak{a}\frown\dot\alpha\dot\varepsilon\left(\begin{array}{l}\top\!\top\ \alpha = c\\ \quad\llcorner\ \varepsilon = n\end{array}\right)\right]\end{array} \qquad (\upsilon$$

(σ): - - - - - - - - - - - - - -

$$\begin{array}{l}\vdash\!\!-\!\!\top\ d\frown\dot\varepsilon(\varepsilon = n)\\ \quad\llcorner\!\overset{\mathfrak{a}}{\smile}\!\top\ \mathfrak{a}\frown\dot\varepsilon(\varepsilon = c)\\ \qquad\quad\llcorner\ d\frown\left[\mathfrak{a}\frown\dot\alpha\dot\varepsilon\left(\begin{array}{l}\top\!\top\ \alpha = c\\ \quad\llcorner\ \varepsilon = n\end{array}\right)\right]\end{array} \qquad (\varphi$$

$$\begin{array}{l}\vdash\!\overset{\mathfrak{d}}{\smile}\!\!-\!\!\top\ \mathfrak{d}\frown\dot\varepsilon(\varepsilon = n)\\ \quad\llcorner\!\overset{\mathfrak{a}}{\smile}\!\top\ \mathfrak{a}\frown\dot\varepsilon(\varepsilon = c)\\ \qquad\quad\llcorner\ \mathfrak{d}\frown\left[\mathfrak{a}\frown\dot\alpha\dot\varepsilon\left(\begin{array}{l}\top\!\top\ \alpha = c\\ \quad\llcorner\ \varepsilon = n\end{array}\right)\right]\end{array} \qquad (\chi$$

(11): ————————————

$$\begin{array}{l}\vdash\!\top\ \dot\varepsilon(\varepsilon = n)\frown\left[\dot\varepsilon(\varepsilon = c)\frown\rangle\dot\alpha\dot\varepsilon\left(\begin{array}{l}\top\!\top\ \alpha = c\\ \quad\llcorner\ \varepsilon = n\end{array}\right)\right]\\ \llcorner\ \mathrm{I}\dot\alpha\dot\varepsilon\left(\begin{array}{l}\top\!\top\ \alpha = c\\ \quad\llcorner\ \varepsilon = n\end{array}\right)\end{array} \qquad (\psi$$

(δ):: ————————————

$$\vdash \dot\varepsilon(\varepsilon = n)\frown\left[\dot\varepsilon(\varepsilon = c)\frown\rangle\dot\alpha\dot\varepsilon\left(\begin{array}{l}\top\!\top\ \alpha = c\\ \quad\llcorner\ \varepsilon = n\end{array}\right)\right] \qquad (\omega$$

(32): ————————————

$$\begin{array}{l}\vdash\!\top\ \not{p}\dot\varepsilon(\varepsilon = n) = \not{p}\dot\varepsilon(\varepsilon = c)\\ \llcorner\ \dot\varepsilon(\varepsilon = c)\frown\left[\dot\varepsilon(\varepsilon = n)\frown\rangle\text{⅄}\dot\alpha\dot\varepsilon\left(\begin{array}{l}\top\!\top\ \alpha = c\\ \quad\llcorner\ \varepsilon = n\end{array}\right)\right]\end{array} \qquad (\alpha'$$

(ξ):: - - - - - - - - - - - - - - - - - -

$$\vdash\!\!\top\ \wp\dot{\varepsilon}(\varepsilon = n) = \wp\dot{\varepsilon}(\varepsilon = c)$$
$$\quad\llcorner\ \dot{\varepsilon}(\varepsilon = c)\frown\left[\dot{\varepsilon}(\varepsilon = n)\frown\rangle\dot{\alpha}\dot{\varepsilon}\left(\top\!\!\top\ \begin{matrix}\alpha = n\\ \varepsilon = c\end{matrix}\right)\right]$$ (β'

(ω)::
____

$\vdash \wp\dot{\varepsilon}(\varepsilon = n) = \wp\dot{\varepsilon}(\varepsilon = c)$ (118

———•———

118 $\vdash \wp\dot{\varepsilon}(\varepsilon = 0) = \wp\dot{\varepsilon}(\varepsilon = c)$

(116): ______________________________

$\vdash 1 = \wp\dot{\varepsilon}(\varepsilon = c)$ (119

(IIIa): ______________________________

$\vdash\!\top\ F(1)$
$\quad\llcorner\ F(\wp\dot{\varepsilon}(\varepsilon = c))$ (120

———•———

IIIa $\vdash\!\top\!\top\ a\frown u$
$\quad\llcorner\ a = c$
$\quad\llcorner\ c\frown u$

(IVa):: – – – –

$\vdash\!\top\!\top\ (\!-\ a\frown u) = (\!-\ a = c)$
$\quad\top\ a = c$
$\quad\llcorner\ a\frown u$
$\quad\llcorner\ c\frown u$ (α

(IIa):: – – – – – – – – – – – – – –

$\vdash\!\top\!\top\ (\!-\ a\frown u) = (\!-\ a = c)$
$\quad\overset{\mathfrak{a}}{\smile}\top\ \mathfrak{a} = c$
$\quad\quad\llcorner\ \mathfrak{a}\frown u$
$\quad\llcorner\ c\frown u$ (β

(77): ______________________________

$\vdash\!\top\!\top\ (\!-\ a\frown u) = (\!-\ a\frown\dot{\varepsilon}(\varepsilon = c))$
$\quad\overset{\mathfrak{a}}{\smile}\top\ \mathfrak{a} = c$
$\quad\quad\llcorner\ \mathfrak{a}\frown u$
$\quad\llcorner\ c\frown u$ (γ

$\smile$

$\vdash\!\top\!\top\overset{\mathfrak{a}}{\smile}\ (\!-\ \mathfrak{a}\frown u) = (\!-\ \mathfrak{a}\frown\dot{\varepsilon}(\varepsilon = c))$
$\quad\overset{\mathfrak{a}}{\smile}\top\ \mathfrak{a} = c$
$\quad\quad\llcorner\ \mathfrak{a}\frown u$
$\quad\llcorner\ c\frown u$ (δ

(96): – – – – – – – – – – – – – –

$\vdash\!\top\!\top\ \wp u = \wp\dot{\varepsilon}(\varepsilon = c)$
$\quad\overset{\mathfrak{a}}{\smile}\top\ \mathfrak{a} = c$
$\quad\quad\llcorner\ \mathfrak{a}\frown u$
$\quad\llcorner\ c\frown u$ (ε

(120): ______________________________

$\vdash\!\top\!\top\ \wp u = 1$
$\quad\overset{\mathfrak{a}}{\smile}\top\ \mathfrak{a} = c$
$\quad\quad\llcorner\ \mathfrak{a}\frown u$
$\quad\llcorner\ c\frown u$ (121

(IIa):: – – – – – –

$\vdash\!\top\!\top\ \wp u = 1$
$\quad\overset{\mathfrak{d}}{\smile}\top\overset{\mathfrak{a}}{\smile}\top\ \mathfrak{a} = \mathfrak{d}$
$\quad\quad\quad\llcorner\ \mathfrak{a}\frown u$
$\quad\quad\llcorner\ \mathfrak{d}\frown u$
$\quad\llcorner\ c\frown u$ (α

$\times$

$\vdash\!\top\!\top\ c\frown u$
$\quad\quad\llcorner\top\ \wp u = 1$
$\quad\overset{\mathfrak{d}}{\smile}\top\overset{\mathfrak{a}}{\smile}\top\ \mathfrak{a} = \mathfrak{d}$
$\quad\quad\quad\llcorner\ \mathfrak{a}\frown u$
$\quad\quad\llcorner\ \mathfrak{d}\frown u$ (β

$\smile$

$\vdash\!\top\!\top\overset{\mathfrak{e}}{\smile}\top\ \mathfrak{e}\frown u$
$\quad\quad\llcorner\top\ \wp u = 1$
$\quad\overset{\mathfrak{d}}{\smile}\top\overset{\mathfrak{a}}{\smile}\top\ \mathfrak{a} = \mathfrak{d}$
$\quad\quad\quad\llcorner\ \mathfrak{a}\frown u$
$\quad\quad\llcorner\ \mathfrak{d}\frown u$ (γ

$\times$

$\vdash\!\top\!\top\ \wp u = 1$
$\quad\quad\llcorner\top\overset{\mathfrak{e}}{\smile}\top\ \mathfrak{e}\frown u$
$\quad\overset{\mathfrak{d}}{\smile}\top\overset{\mathfrak{a}}{\smile}\top\ \mathfrak{a} = \mathfrak{d}$
$\quad\quad\quad\llcorner\ \mathfrak{a}\frown u$
$\quad\quad\llcorner\ \mathfrak{d}\frown u$ (122

Z. Proof of the proposition

$$\vdash \begin{array}{l} b \frown (b \frown \_f) \\ 0 \frown (b \frown \smile f) \end{array}$$

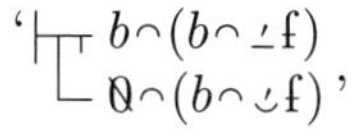

a) Proof of the proposition

'$\vdash a \frown (0 \frown \_f)$'

§108. *Analysis*

The proposition mentioned in the main heading states that no object belonging to the cardinal number series starting with Zero follows after itself in the cardinal number series. Instead, we could also say: "No *finite* cardinal number follows after itself in the cardinal number series." The importance of this proposition may be brought out more clearly through the following consideration. When we determine the cardinal number belonging to a concept $\Phi(\xi)$, or, as one normally says, when we count the objects falling under a concept $\Phi(\xi)$, we correlate these with the number-signs, one after the other, beginning with 'One' up to that number-sign 'N' which is determined by the correlating relation mapping the concept $\Phi(\xi)$ into the concept "member of the series of number-signs from 'One' to 'N'" and the converse relation mapping the latter into the former. 'N' then designates the desired cardinal number; i.e., N is that cardinal number. This process allows of manifold execution since the correlating relation is not completely specified. The question arises whether a different number-sign 'M' could have been reached under a different choice of relation. According to our specifications, M would then be the same cardinal number as N; at the same time, however, one of the number-signs would follow the other, e.g., 'N' would follow 'M'. In that case, N would also follow M in the cardinal number series, that is, it would follow itself. This is excluded for finite cardinal numbers by our proposition. We prove it using the propositions

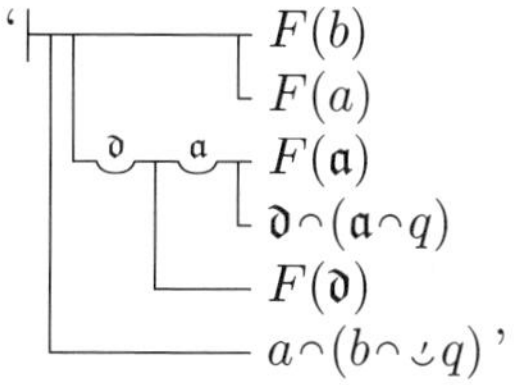

$$\vdash \begin{array}{l} F(b) \\ F(a) \\ F(\mathfrak{a}) \\ \mathfrak{d} \frown (\mathfrak{a} \frown q) \\ F(\mathfrak{d}) \\ a \frown (b \smile q) \end{array} \quad (\alpha$$

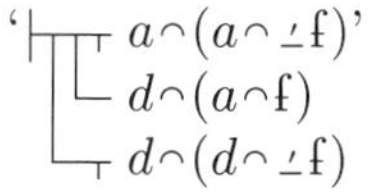

'$\vdash \begin{array}{l} a \frown (a \frown \_f) \\ d \frown (a \frown f) \\ d \frown (d \frown \_f) \end{array}$' and '$\vdash 0 \frown (0 \frown \_f)$'

The last one is a special case of

'$\vdash a \frown (0 \frown \_f)$'

which says that the cardinal number 0 does not follow any object in the cardinal number series. We prove this proposition first. For this we need the proposition

$$\vdash \begin{array}{l} a \frown (b \frown q) \\ e \frown (b \frown q) \end{array},$$

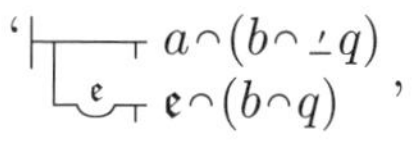

and (108). The former says that an object follows no object in the q-series if no object stands to it in the q-relation. In order to prove this we require the proposition

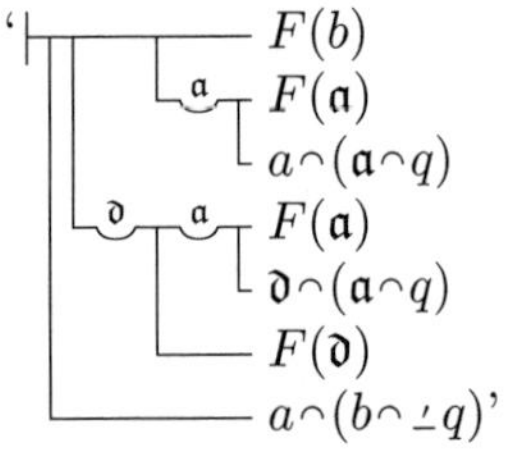

which follows from (K) and (6). We then replace the function-marker '$F(\xi)$' by '⊤⌣e⊤ e⌢(ξ⌢q)' and then have to prove the propositions

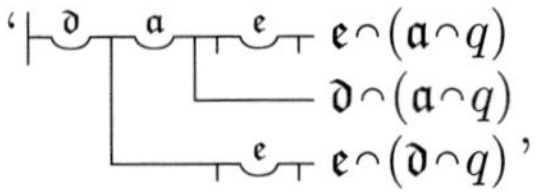

and

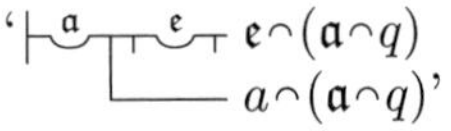

both of which follow from

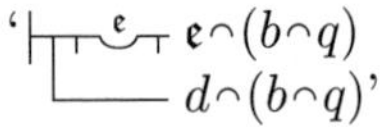

§109. *Construction*

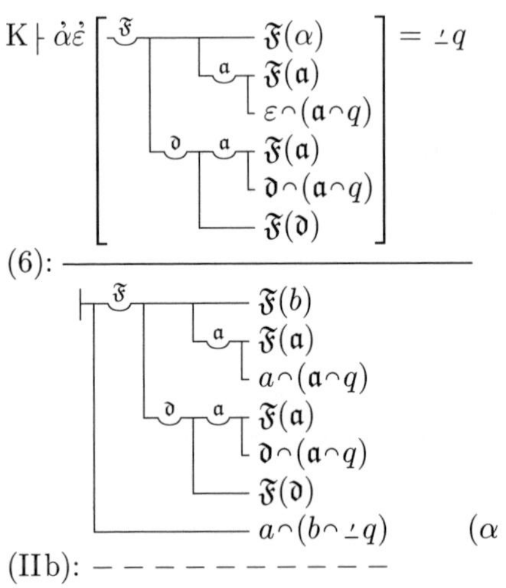

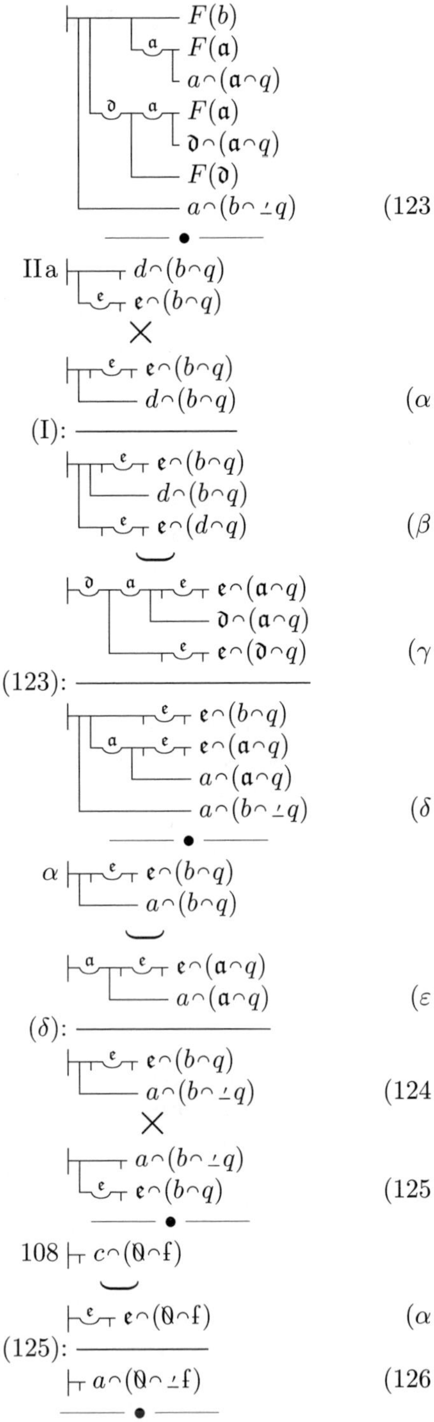

b) Proof of the proposition

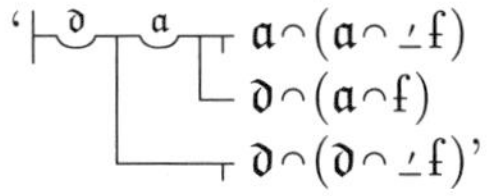

and end of section Z

§110. *Analysis*
The proposition

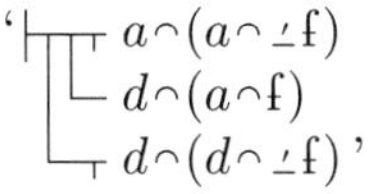
(α

is obtained by contraposition from

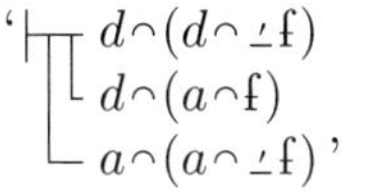
(β

This proposition can be concluded from the propositions

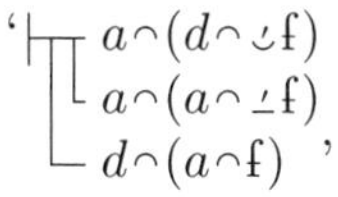
(γ

and

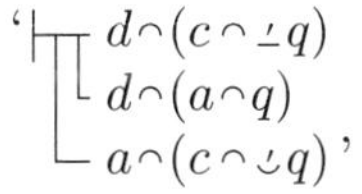
(δ

by replacing 'c' by 'd' and 'q' by 'f' in the latter. We prove (δ) from the propositions

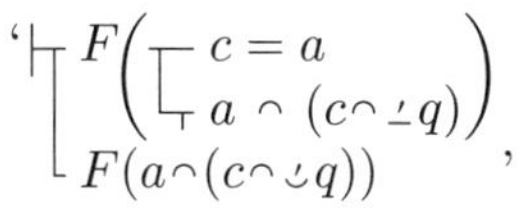
(ε

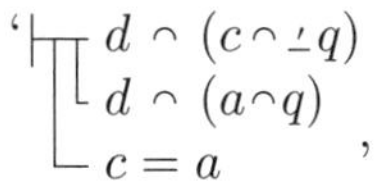
(ζ

and

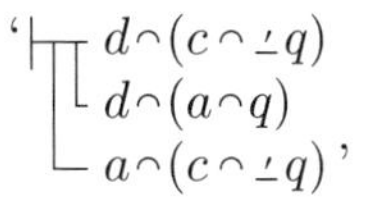
(η

which follow straightforwardly from (Λ), (K), and (123).

§111. *Construction*

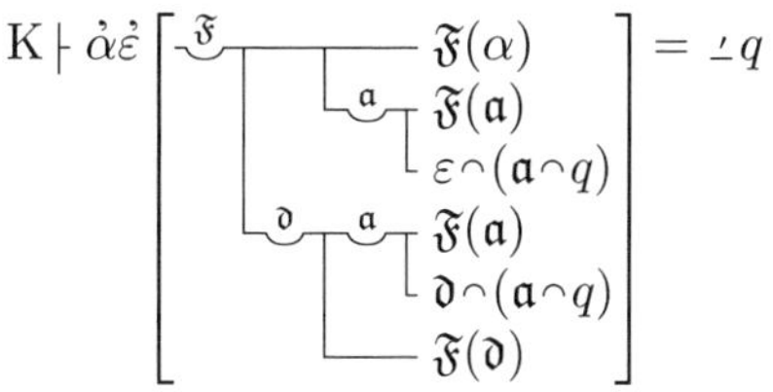

(10): ——————————————

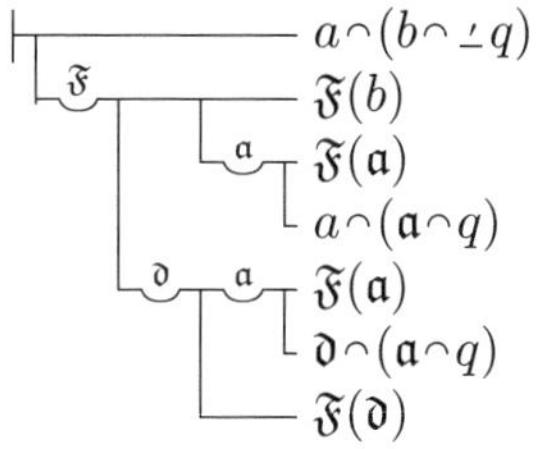
(127

——— • ———

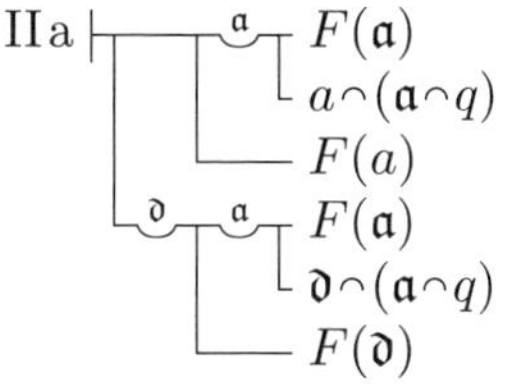

(123): - - - - - - - -

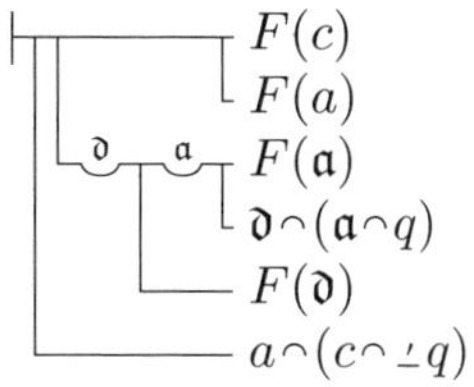
(128

(IIa):: - - - - - - - - -

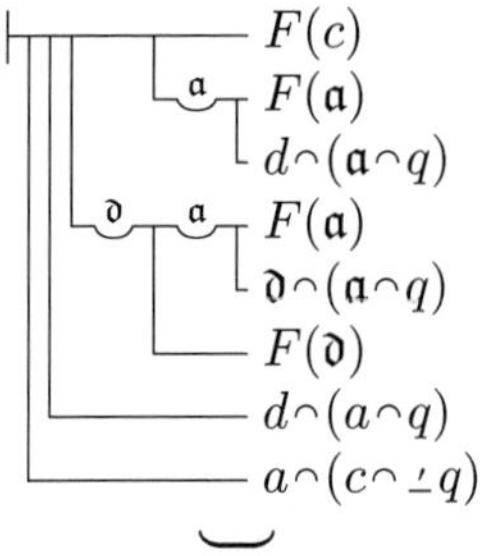
F(c)
F(𝔞)
d⌢(𝔞⌢q)
𝔡 𝔞
F(𝔞)
𝔡⌢(𝔞⌢q)
F(𝔡)
d⌢(a⌢q)
a⌢(c⌢⊥q)
(α

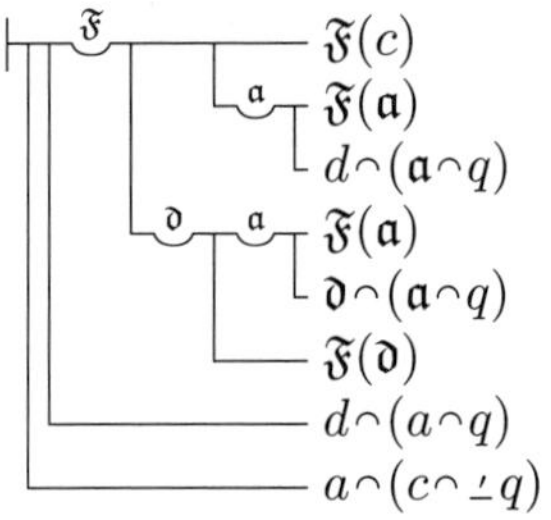
𝔉
𝔉(c)
𝔉(𝔞)
d⌢(𝔞⌢q)
𝔡 𝔞
𝔉(𝔞)
𝔡⌢(𝔞⌢q)
𝔉(𝔡)
d⌢(a⌢q)
a⌢(c⌢⊥q)
(β

(127): - - - - - - - - - - - -

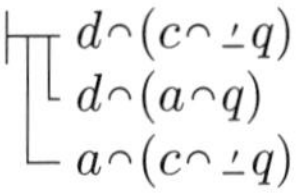
d⌢(c⌢⊥q)
d⌢(a⌢q)
a⌢(c⌢⊥q)
(129

———•———

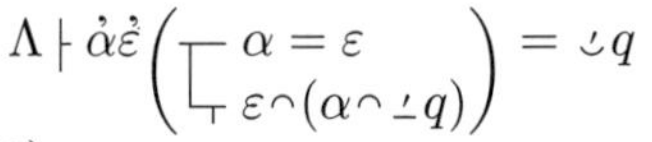
Λ
α = ε
ε⌢(α⌢⊥q)
= ⌣q

(6): ————————

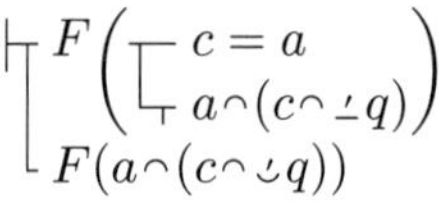
F
c = a
a⌢(c⌢⊥q)
F(a⌢(c⌢⌣q))
(130

———•———

IIa

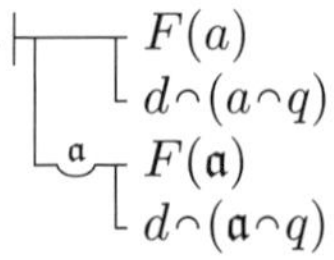
F(a)
d⌢(a⌢q)
𝔞
F(𝔞)
d⌢(𝔞⌢q)

(I): ————————

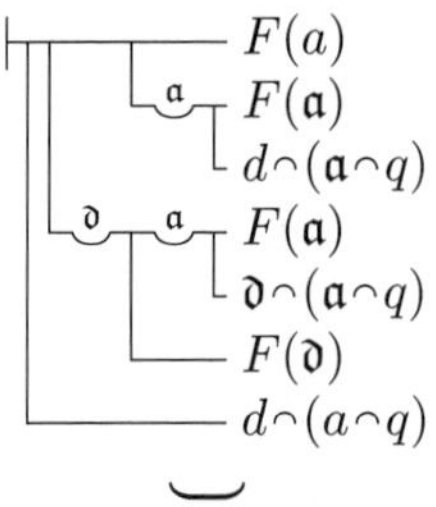
F(a)
F(𝔞)
d⌢(𝔞⌢q)
𝔡 𝔞
F(𝔞)
𝔡⌢(𝔞⌢q)
F(𝔡)
d⌢(a⌢q)
(α

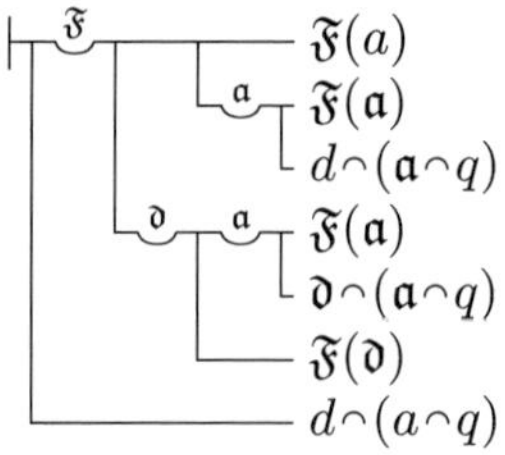
𝔉
𝔉(a)
𝔉(𝔞)
d⌢(𝔞⌢q)
𝔡 𝔞
𝔉(𝔞)
𝔡⌢(𝔞⌢q)
𝔉(𝔡)
d⌢(a⌢q)
(β

(127): - - - - - - - - - - - -

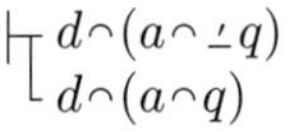
d⌢(a⌢⊥q)
d⌢(a⌢q)
(131

(IIIa): ————————

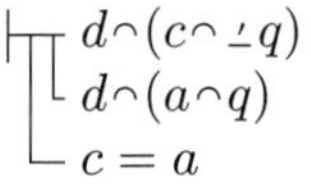
d⌢(c⌢⊥q)
d⌢(a⌢q)
c = a
(α

(130):: - - - - - -

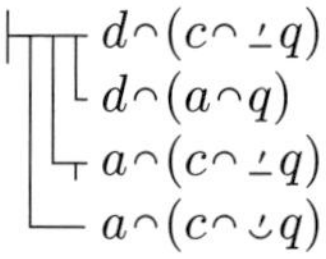
d⌢(c⌢⊥q)
d⌢(a⌢q)
a⌢(c⌢⊥q)
a⌢(c⌢⌣q)
(β

(129): -·-·-·-·-·-·-

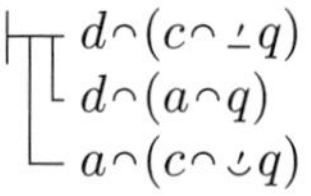
d⌢(c⌢⊥q)
d⌢(a⌢q)
a⌢(c⌢⌣q)
(132

§112. *Analysis*

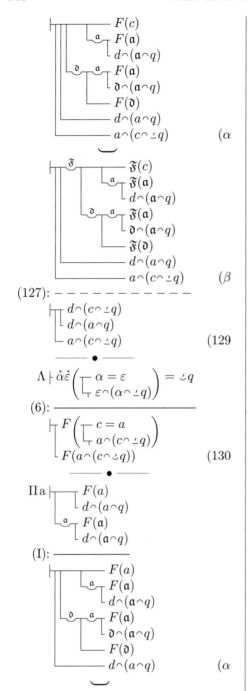

We now have to prove proposition (γ) of §110. It is a special case of

‘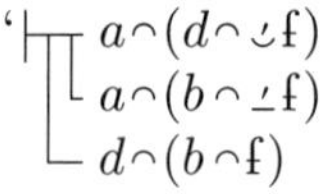’
a⌢(d⌢⌣𝔣)
a⌢(b⌢⊥𝔣)
d⌢(b⌢𝔣)

in words:

“If a cardinal number (b) follows a second cardinal number (a) in the cardinal number series, and immediately follows a third (d) in the cardinal number series, then the third (d) belongs to the cardinal number series starting with the second (a).”

Clearly, the analogue would not hold in general for an arbitrary series. It is essential here that predecession in the cardinal number series is single-valued (88). We rely on the proposition that if an object (b) follows a second (a) in an arbitrary (q-)series then there is an object that belongs to the (q-)series which starts with the second (a) and which stands to the first in the series-forming (q-)relation; in signs:

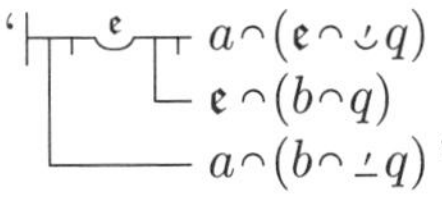
a⌒(𝔢⌒⌣q)
𝔢⌒(b⌒q)
a⌒(b⌒⊥q)
(α

If one now knows that there is no more than one object that stands in the (q-)relation to the first (b), then this object must also belong to the (q-)series starting with the second (a). For the proof of this proposition we use (123) in which we replace the function-marker ‘$F(\xi)$’ by ‘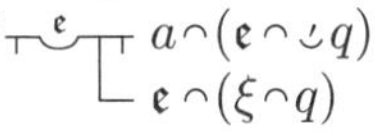’. We thus require the two propositions

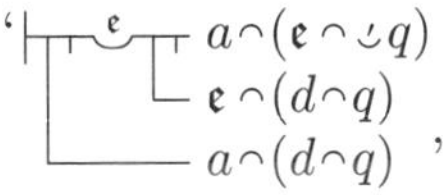
a⌒(𝔢⌒⌣q)
𝔢⌒(d⌒q)
a⌒(d⌒q)
(β

and

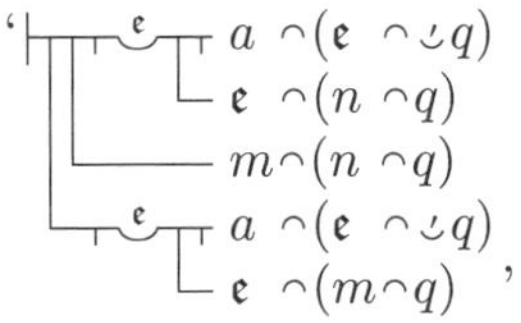
a⌒(𝔢⌒⌣q)
𝔢⌒(n⌒q)
m⌒(n⌒q)
a⌒(𝔢⌒⌣q)
𝔢⌒(m⌒q)
(γ

The former follows, using (IIa), from the proposition

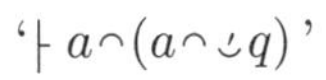

‘⊢ $a \frown (a \frown \smile q)$’

which is a consequence of definition (Λ). The latter is obtained by introduction of the German ‘𝔢’ and contraposition from

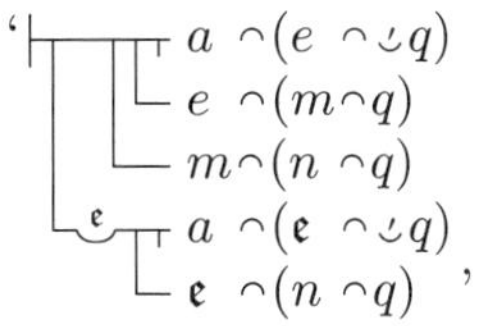
a⌒(e⌒⌣q)
e⌒(m⌒q)
m⌒(n⌒q)
a⌒(𝔢⌒⌣q)
𝔢⌒(n⌒q)
(δ

By (IIa) we now have

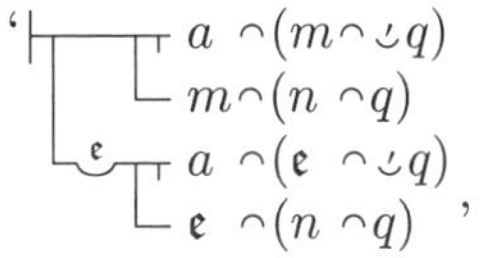
a⌒(m⌒⌣q)
m⌒(n⌒q)
a⌒(𝔢⌒⌣q)
𝔢⌒(n⌒q)

What remains to be shown is

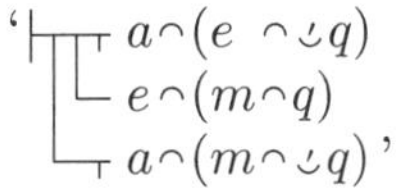
a⌒(e⌒⌣q)
e⌒(m⌒q)
a⌒(m⌒⌣q)

or

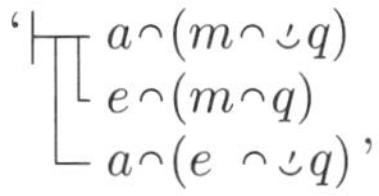
a⌒(m⌒⌣q)
e⌒(m⌒q)
a⌒(e⌒⌣q)
(ε

(ε) is a consequence of

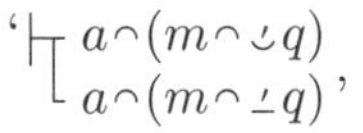
a⌒(m⌒⌣q)
a⌒(m⌒⊥q)
(ζ

and

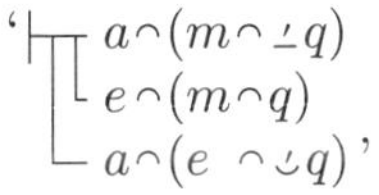
a⌒(m⌒⊥q)
e⌒(m⌒q)
a⌒(e⌒⌣q)
(η

The latter proposition is to be proven in a manner similar to (132).

§113. *Construction*

IIa 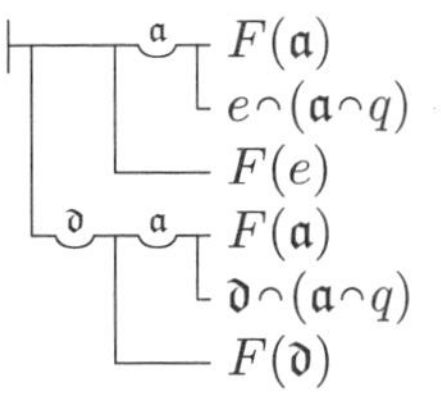
F(𝔞)
e⌒(𝔞⌒q)
F(e)
F(𝔞)
𝔡⌒(𝔞⌒q)
F(𝔡)

(IIa): – – – – – – – –

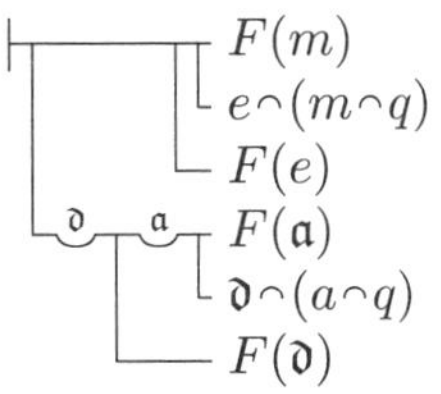
F(m)
e⌒(m⌒q)
F(e)
F(𝔞)
𝔡⌒(𝔞⌒q)
F(𝔡)
(α

(123):: – – – – – – – –

$F(m)$
$\mathfrak{a}$ $F(\mathfrak{a})$
$a\frown(\mathfrak{a}\frown q)$
$\mathfrak{d}$ $\mathfrak{a}$ $F(\mathfrak{a})$
$\mathfrak{d}\frown(\mathfrak{a}\frown q)$
$F(\mathfrak{d})$
$e\frown(m\frown q)$
$a\frown(e\frown\dot{\bot}q)$ (β

$\smile$

$\mathfrak{F}$ $\mathfrak{F}(m)$
$\mathfrak{a}$ $\mathfrak{F}(\mathfrak{a})$
$a\frown(\mathfrak{a}\frown q)$
$\mathfrak{d}$ $\mathfrak{a}$ $\mathfrak{F}(\mathfrak{a})$
$\mathfrak{d}\frown(\mathfrak{a}\frown q)$
$\mathfrak{F}(\mathfrak{d})$
$e\frown(m\frown q)$
$a\frown(e\frown\dot{\bot}q)$ (γ

(127): – – – – – – – – – – – –

$a\frown(m\frown\dot{\bot}q)$
$e\frown(m\frown q)$
$a\frown(e\frown\dot{\bot}q)$ (133

——— • ———

131 $a\frown(m\frown\dot{\bot}q)$
$a\frown(m\frown q)$

(IIIa): ————————

$a\frown(m\frown\dot{\bot}q)$
$e\frown(m\frown q)$
$e=a$ (α

(130):: – – – – – – –

$a\frown(m\frown\dot{\bot}q)$
$e\frown(m\frown q)$
$a\frown(e\frown\dot{\bot}q)$
$a\frown(e\frown\dot{\smile}q)$ (β

(133): ·–·–·–·–·–·–·

$a\frown(m\frown\dot{\bot}q)$
$e\frown(m\frown q)$
$a\frown(e\frown\dot{\smile}q)$ (134

——— • ———

$\Lambda \vdash \dot{\alpha}\dot{\varepsilon}\left(\begin{array}{l}\alpha=\varepsilon\\ \varepsilon\frown(\alpha\frown\dot{\bot}q)\end{array}\right)=\dot{\smile}q$

(10): ————————————————

$F(a\frown(m\frown\dot{\smile}q))$
$F\left(\begin{array}{l}m=a\\ a\frown(m\frown\dot{\bot}q)\end{array}\right)$ (135

——— • ———

135 $a\frown(m\frown\dot{\smile}q)$
$m=a$
$a\frown(m\frown\dot{\bot}q)$

(Ia):: – – – – – – – –

$a\frown(m\frown\dot{\smile}q)$
$a\frown(m\frown\dot{\bot}q)$ (136

(134):: – – – – – – –

$a\frown(m\frown\dot{\smile}q)$
$e\frown(m\frown q)$
$a\frown(e\frown\dot{\smile}q)$ (137

×

$a\frown(e\frown\dot{\smile}q)$
$e\frown(m\frown q)$
$a\frown(m\frown\dot{\smile}q)$ (α

(IIa):: – – – – – – – –

$a\frown(e\frown\dot{\smile}q)$
$e\frown(m\frown q)$
$m\frown(n\frown q)$
$\mathfrak{e}$ $a\frown(\mathfrak{e}\frown\dot{\smile}q)$
$\mathfrak{e}\frown(n\frown q)$ (β

$\smile$

$\mathfrak{e}$ $a\frown(\mathfrak{e}\frown\dot{\smile}q)$
$\mathfrak{e}\frown(m\frown q)$
$m\frown(n\frown q)$
$\mathfrak{e}$ $a\frown(\mathfrak{e}\frown\dot{\smile}q)$
$\mathfrak{e}\frown(n\frown q)$ (γ

×

$\mathfrak{e}$ $a\frown(\mathfrak{e}\frown\dot{\smile}q)$
$\mathfrak{e}\frown(n\frown q)$
$m\frown(n\frown q)$
$\mathfrak{e}$ $a\frown(\mathfrak{e}\frown\dot{\smile}q)$
$\mathfrak{e}\frown(m\frown q)$ (δ

$\smile$

$\mathfrak{d}$ $\mathfrak{a}$ $\mathfrak{e}$ $a\frown(\mathfrak{e}\frown\dot{\smile}q)$
$\mathfrak{e}\frown(\mathfrak{a}\frown q)$
$\mathfrak{d}\frown(\mathfrak{a}\frown q)$
$\mathfrak{e}$ $a\frown(\mathfrak{e}\frown\dot{\smile}q)$
$\mathfrak{e}\frown(\mathfrak{d}\frown q)$ (ε

(123): ————————————————

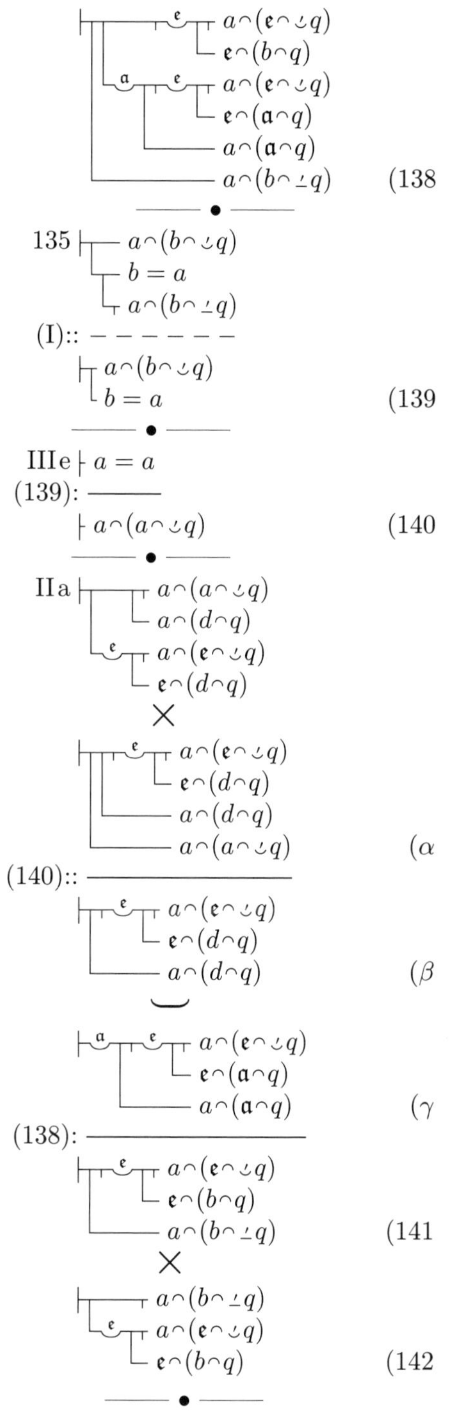
a⌢(𝔢⌢⌣q)
𝔢⌢(b⌢q)
a⌢(𝔢⌢⌣q)
𝔢⌢(𝔞⌢q)
a⌢(𝔞⌢q)
a⌢(b⌢⌣q) (138
135 a⌢(b⌢⌣q)
b = a
a⌢(b⌢⌣q)
(I)::
a⌢(b⌢⌣q)
b = a (139
IIIe a = a
(139):
a⌢(a⌢⌣q) (140
IIa a⌢(a⌢⌣q)
a⌢(d⌢q)
a⌢(𝔢⌢⌣q)
𝔢⌢(d⌢q)
a⌢(𝔢⌢⌣q)
𝔢⌢(d⌢q)
a⌢(d⌢q)
a⌢(a⌢⌣q) (α
(140)::
a⌢(𝔢⌢⌣q)
𝔢⌢(d⌢q)
a⌢(d⌢q) (β
a⌢(𝔢⌢⌣q)
𝔢⌢(𝔞⌢q)
a⌢(𝔞⌢q) (γ
(138):
a⌢(𝔢⌢⌣q)
𝔢⌢(b⌢q)
a⌢(b⌢⌣q) (141
a⌢(b⌢⌣q)
a⌢(𝔢⌢⌣q)
𝔢⌢(b⌢q) (142

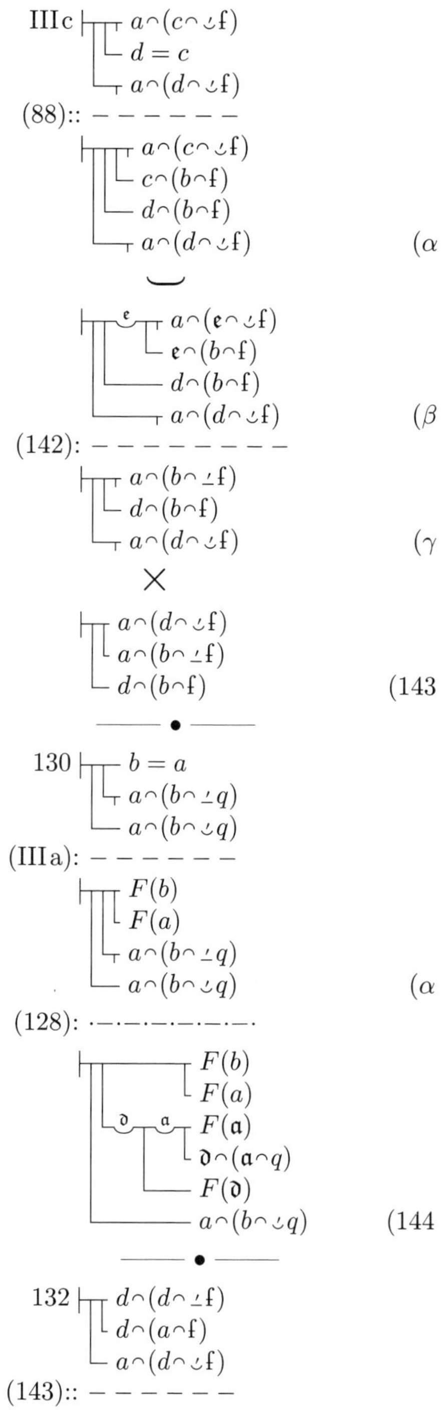
IIIc a⌢(c⌢⌣𝔣)
d = c
a⌢(d⌢⌣𝔣)
(88)::
a⌢(c⌢⌣𝔣)
c⌢(b⌢𝔣)
d⌢(b⌢𝔣)
a⌢(d⌢⌣𝔣) (α
a⌢(𝔢⌢⌣𝔣)
𝔢⌢(b⌢𝔣)
d⌢(b⌢𝔣)
a⌢(d⌢⌣𝔣) (β
(142):
a⌢(b⌢⌣𝔣)
d⌢(b⌢𝔣)
a⌢(d⌢⌣𝔣) (γ
a⌢(d⌢⌣𝔣)
a⌢(b⌢⌣𝔣)
d⌢(b⌢𝔣) (143
130 b = a
a⌢(b⌢⌣q)
a⌢(b⌢⌣q)
(IIIa):
F(b)
F(a)
a⌢(b⌢⌣q)
a⌢(b⌢⌣q) (α
(128):
F(b)
F(a)
F(𝔞)
𝔡⌢(𝔞⌢q)
F(𝔡)
a⌢(b⌢⌣q) (144
132 d⌢(d⌢⌣𝔣)
d⌢(a⌢𝔣)
a⌢(d⌢⌣𝔣)
(143)::

$$\vdash\begin{array}{l} d\frown(d\frown\underline{\prime}\mathfrak{f})\\ d\frown(a\frown\mathfrak{f})\\ a\frown(a\frown\underline{\prime}\mathfrak{f})\end{array}\qquad(\alpha$$

$$\times$$

$$\vdash\begin{array}{l} \neg\, a\frown(a\frown\underline{\prime}\mathfrak{f})\\ d\frown(a\frown\mathfrak{f})\\ \neg\, d\frown(d\frown\underline{\prime}\mathfrak{f})\end{array}\qquad(\beta$$

$$\smile$$

$$\vdash\overset{\mathfrak{d}}{\smile}\;\overset{\mathfrak{a}}{\smile}\begin{array}{l} \neg\,\mathfrak{a}\frown(\mathfrak{a}\frown\underline{\prime}\mathfrak{f})\\ \mathfrak{d}\frown(\mathfrak{a}\frown\mathfrak{f})\\ \neg\,\mathfrak{d}\frown(\mathfrak{d}\frown\underline{\prime}\mathfrak{f})\end{array}\qquad(\gamma$$

(144): ———

$$\vdash\begin{array}{l} \neg\, b\frown(b\frown\underline{\prime}\mathfrak{f})\\ \neg\,\emptyset\frown(\emptyset\frown\underline{\prime}\mathfrak{f})\\ \emptyset\frown(b\frown\dot{\smile}\mathfrak{f})\end{array}\qquad(\delta$$

(126):: ———

$$\vdash\begin{array}{l} \neg\, b\frown(b\frown\underline{\prime}\mathfrak{f})\\ \emptyset\frown(b\frown\dot{\smile}\mathfrak{f})\end{array}\qquad(145$$

H. Proof of the proposition

$$‘\vdash\begin{array}{l} b\frown(\wp(b\frown\dot{\smile}\mathfrak{f})\frown\mathfrak{f})\\ \emptyset\frown(b\frown\dot{\smile}\mathfrak{f})\end{array}’$$

§114. *Analysis*

We want to prove the proposition that the cardinal number that belongs to the concept

belonging to the cardinal number series ending with b

immediately follows *b* in the cardinal number series, provided *b* is a finite cardinal number. This has the immediate corollary that the cardinal number series is infinite; i.e., that for every finite cardinal number, there is one that immediately follows it.

We first attempt the proof with the proposition (144) by replacing the function-marker ‘$F(\xi)$’ by ‘$\xi\frown(\wp(\xi\frown\dot{\smile}\mathfrak{f})\frown\mathfrak{f})$’. For this we require the proposition

$$‘\vdash\begin{array}{l} a\frown(\wp(a\frown\dot{\smile}\mathfrak{f})\frown\mathfrak{f})\\ d\frown(a\frown\mathfrak{f})\\ d\frown(\wp(d\frown\dot{\smile}\mathfrak{f})\frown\mathfrak{f})\end{array}’\;[1]\qquad(\alpha$$

If we put ‘$(a\frown\dot{\smile}\mathfrak{f})$’ for ‘$u$’, ‘$a$’ for ‘$m$’ and for ‘$c$’ in (102), then we obtain

$$‘\vdash\begin{array}{l} a\frown(\wp(a\frown\dot{\smile}\mathfrak{f})\frown\mathfrak{f})\\ a\frown(a\frown\dot{\smile}\mathfrak{f})\\ \wp\grave{\varepsilon}\left(\begin{array}{l}\varepsilon = a\\ \varepsilon\frown(a\frown\dot{\smile}\mathfrak{f})\end{array}\right) = a\end{array}’$$

from which we can remove the subcomponent

$$‘—\; a\frown(a\frown\dot{\smile}\mathfrak{f})’$$

using (140). The question is whether the subcomponent

$$‘—\;\wp\grave{\varepsilon}\left(\begin{array}{l}\varepsilon = a\\ \varepsilon\frown(a\frown\dot{\smile}\mathfrak{f})\end{array}\right) = a’$$

can be established as a consequence of

‘$d\frown(a\frown\mathfrak{f})$’ and ‘$d\frown(\wp(d\frown\dot{\smile}\mathfrak{f})\frown\mathfrak{f})$’

Because of the single-valuedness of progression in the cardinal number series (70) we have

$$‘\vdash\begin{array}{l} \wp(d\frown\dot{\smile}\mathfrak{f}) = a\\ d\frown(a\frown\mathfrak{f})\\ d\frown(\wp(d\frown\dot{\smile}\mathfrak{f})\frown\mathfrak{f})\end{array}’\qquad(\beta$$

We thus attempt to establish whether

$$‘\,\wp\grave{\varepsilon}\left(\begin{array}{l}\varepsilon = a\\ \varepsilon\frown(a\frown\dot{\smile}\mathfrak{f})\end{array}\right) = \wp(d\frown\dot{\smile}\mathfrak{f})’$$

[1] This proposition is, it seems, unprovable, but is not asserted as true here either, since it stands in quotation marks.

is a consequence of '$d\frown(a\frown\mathfrak{f})$'. This will require (96). For this

$$
`\left(\begin{array}{l}\top\!\!\top\; b=a\\ \quad\llcorner\; b\frown(a\frown\acute{\smile}\mathfrak{f})\end{array}\right)=(\!-\!\!-\; b\frown(d\frown\acute{\smile}\mathfrak{f}))'
$$

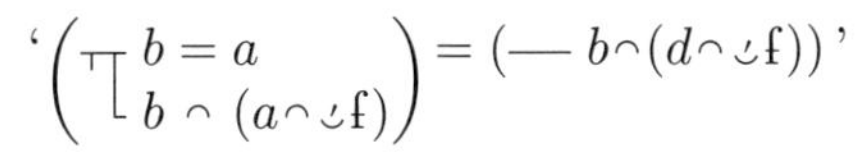

has to be established as a consequence of '$d\frown(a\frown\mathfrak{f})$', for which (IV a) will have to be used. It would thus have to be shown that the cardinal numbers belonging to the cardinal number series ending with a first cardinal number (a), except the latter itself, are the same as those that belong to the cardinal number series ending with a second cardinal number (d) if the first cardinal number (a) immediately follows the second cardinal number (d) in the cardinal number series. For this it is necessary to establish

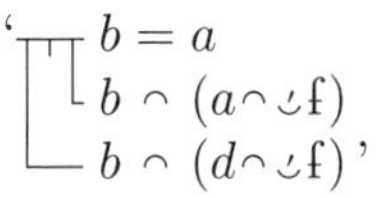

$$
`\begin{array}{l}\top\!\!\top\; b=a\\ \;\;\llcorner\; b\frown(a\frown\acute{\smile}\mathfrak{f})\\ \llcorner\; b\frown(d\frown\acute{\smile}\mathfrak{f})'\end{array}
\quad\text{and}\quad
`\begin{array}{l}\top\; b\frown(d\frown\acute{\smile}\mathfrak{f})\\ \llcorner\!\top\; b=a\\ \;\;\llcorner\; b\frown(a\frown\acute{\smile}\mathfrak{f})'\end{array}
$$

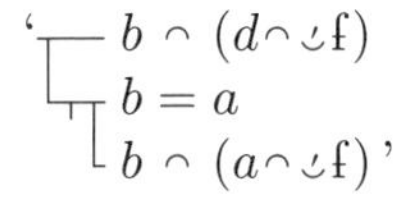

as consequences of '$d\frown(a\frown\mathfrak{f})$'. It turns out, however, that a further condition has to be added. Namely, '$\top\, b=a$' would have to be shown to be a consequence of '$b\frown(d\frown\acute{\smile}\mathfrak{f})$' and '$d\frown(a\frown\mathfrak{f})$'. Now, according to (134) we have

$$
`\begin{array}{l}\vdash\!\top\!\!\top\; b\frown(a\frown\acute{\underline{\ }}\mathfrak{f})\\ \quad\;\;\llcorner\; d\frown(a\frown\mathfrak{f})\\ \quad\llcorner\; b\frown(d\frown\acute{\smile}\mathfrak{f})'\end{array}
$$

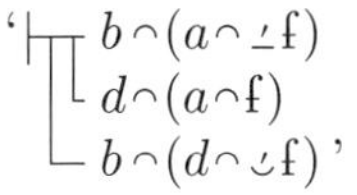

If b and a were to coincide, then the supercomponent would turn into '$-\!\!-\; a\frown(a\frown\acute{\underline{\ }}\mathfrak{f})$'. According to (145), this is excluded when a is a finite cardinal number. Thus the subcomponent

$$
`-\!\!-\; \emptyset\frown(a\frown\acute{\smile}\mathfrak{f})'
$$

is added. Thereby, applying (144) in the way we had intended becomes of course impossible; however, using (137) we can replace this subcomponent by '$-\!\!-\; \emptyset\frown(d\frown\acute{\smile}\mathfrak{f})$' and derive from (144) the proposition

$$
`\begin{array}{l}\vdash\!\!\top\!\!\top\!\!-\!\!-\!\!-\!\!-\!\!\top\; F(b)\\ \qquad\qquad\quad\;\;\llcorner\; F(a)\\ \quad\;\llcorner\!\smile^{\mathfrak{d}}\!\!\top\!\!\top\!\!\smile^{\mathfrak{a}}\!\!\top\; F(\mathfrak{a})\\ \qquad\qquad\quad\;\;\llcorner\; \mathfrak{d}\frown(\mathfrak{a}\frown q)\\ \qquad\quad\;\;\llcorner\; a\frown(\mathfrak{d}\frown\acute{\smile}q)\\ \qquad\quad\llcorner\; F(\mathfrak{d})\\ \quad\llcorner\; a\frown(b\frown\acute{\smile}q)'\end{array}
\qquad (\gamma
$$

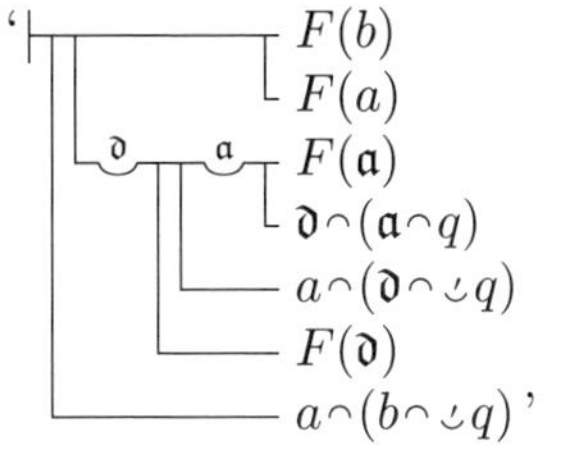

which then brings us to our goal. First, in order to have the proposition

$$
`\begin{array}{l}\vdash\!\!\top\!\!\top\!\!\top\!\!\top\; b=a\\ \qquad\quad\llcorner\; b\frown(a\frown\acute{\smile}\mathfrak{f})\\ \qquad\llcorner\; b\frown(d\frown\acute{\smile}\mathfrak{f})\\ \quad\;\llcorner\; \emptyset\frown(a\frown\acute{\smile}\mathfrak{f})\\ \quad\llcorner\; d\frown(a\frown\mathfrak{f})\;'\end{array}
\qquad (\delta
$$

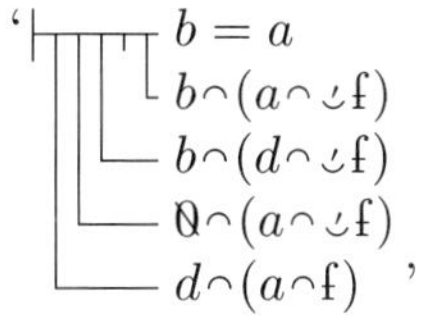

in full, we need to draw upon proposition (137) in the form

$$
`\begin{array}{l}\vdash\!\top\!\!\top\; b\frown(a\frown\acute{\smile}\mathfrak{f})\\ \quad\;\;\llcorner\; d\frown(a\frown\mathfrak{f})\\ \quad\llcorner\; b\frown(d\frown\acute{\smile}\mathfrak{f})'\end{array}
$$

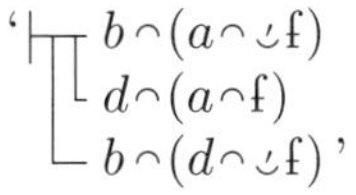

Then the proposition

$$
`\begin{array}{l}\vdash\!\top\!\!\top\; b\frown(d\frown\acute{\smile}\mathfrak{f})\\ \quad\;\;\llcorner\!\top\; b=a\\ \qquad\llcorner\; b\frown(a\frown\acute{\smile}\mathfrak{f})\\ \quad\llcorner\; d\frown(a\frown\mathfrak{f})\;'\end{array}
\qquad (\varepsilon
$$

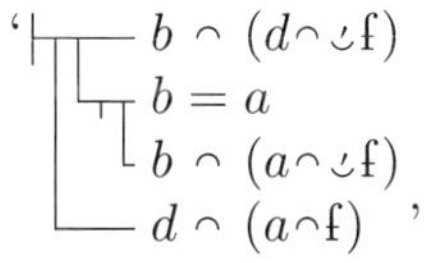

remains to be proven. According to (143) we have

$$
`\begin{array}{l}\vdash\!\top\!\!\top\; b\frown(d\frown\acute{\smile}\mathfrak{f})\\ \quad\;\;\llcorner\; b\frown(a\frown\acute{\underline{\ }}\mathfrak{f})\\ \quad\llcorner\; d\frown(a\frown\mathfrak{f})\;'\end{array}
$$

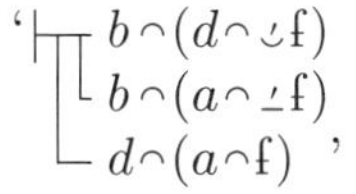

Additionally, we now also require the proposition

$$
`\begin{array}{l}\vdash\!\top\; b\frown(a\frown\acute{\underline{\ }}q)\\ \quad\llcorner\!\top\; b=a\\ \qquad\llcorner\; b\frown(a\frown\acute{\smile}q)'\end{array}
\qquad (\zeta
$$

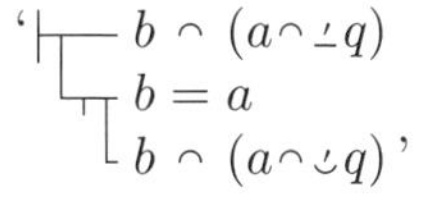

which follows straightforwardly from (130).

§115. *Construction*

130 $\vdash$
$$a = b$$
$$b \frown (a \frown \dot{\bot} q)$$
$$b \frown (a \frown \dot{\smile} q)$$

(III f): ------

$$b = a$$
$$b \frown (a \frown \dot{\smile} q)$$
$$b \frown (a \frown \dot{\bot} q) \qquad (146$$

×

$$b \frown (a \frown \dot{\bot} q)$$
$$b = a$$
$$b \frown (a \frown \dot{\smile} q) \qquad (147$$

——— • ———

147 $\vdash$
$$b \frown (a \frown \dot{\bot} \mathfrak{f})$$
$$b = a$$
$$b \frown (a \frown \dot{\smile} \mathfrak{f})$$

(143): ------

$$b \frown (d \frown \dot{\smile} \mathfrak{f})$$
$$b = a$$
$$b \frown (a \frown \dot{\smile} \mathfrak{f})$$
$$d \frown (a \frown \mathfrak{f}) \qquad (\alpha$$

(IV a): -------

$$\left(\begin{array}{l} b = a \\ b \frown (a \frown \dot{\smile} \mathfrak{f}) \end{array} \right) = (\text{—}\, b \frown (d \frown \dot{\smile} \mathfrak{f}))$$
$$d \frown (a \frown \mathfrak{f})$$
$$b = a$$
$$b \frown (a \frown \dot{\smile} \mathfrak{f})$$
$$b \frown (d \frown \dot{\smile} \mathfrak{f}) \qquad (\beta$$

——— • ———

134 $\vdash$
$$b \frown (a \frown \dot{\bot} \mathfrak{f})$$
$$d \frown (a \frown \mathfrak{f})$$
$$b \frown (d \frown \dot{\smile} \mathfrak{f})$$

(III d): ------

$$b = a$$
$$d \frown (a \frown \mathfrak{f})$$
$$b \frown (d \frown \dot{\smile} \mathfrak{f})$$
$$a \frown (a \frown \dot{\bot} \mathfrak{f}) \qquad (\gamma$$

(145):: -------

$$b = a$$
$$d \frown (a \frown \mathfrak{f})$$
$$b \frown (d \frown \dot{\smile} \mathfrak{f})$$
$$\emptyset \frown (a \frown \dot{\smile} \mathfrak{f}) \qquad (\delta$$

(I f): -------

$$b = a$$
$$b \frown (a \frown \dot{\smile} \mathfrak{f})$$
$$d \frown (a \frown \mathfrak{f})$$
$$b \frown (d \frown \dot{\smile} \mathfrak{f})$$
$$\emptyset \frown (a \frown \dot{\smile} \mathfrak{f})$$
$$b \frown (a \frown \dot{\smile} \mathfrak{f}) \qquad (\varepsilon$$

(137):: ---------

$$b = a$$
$$b \frown (a \frown \dot{\smile} \mathfrak{f})$$
$$b \frown (d \frown \dot{\smile} \mathfrak{f})$$
$$\emptyset \frown (a \frown \dot{\smile} \mathfrak{f})$$
$$d \frown (a \frown \mathfrak{f}) \qquad (\zeta$$

(β): -------

$$\left(\begin{array}{l} b = a \\ b \frown (a \frown \dot{\smile} \mathfrak{f}) \end{array} \right) = (\text{—}\, b \frown (d \frown \dot{\smile} \mathfrak{f}))$$
$$\emptyset \frown (a \frown \dot{\smile} \mathfrak{f})$$
$$d \frown (a \frown \mathfrak{f}) \qquad (\eta$$

(77): ---------------

$$\vdash \left[\text{—}\, b \frown \dot{\varepsilon} \left(\begin{array}{l} \varepsilon = a \\ \varepsilon \frown (a \frown \dot{\smile} \mathfrak{f}) \end{array} \right) \right] = (\text{—}\, b \frown (d \frown \dot{\smile} \mathfrak{f}))$$
$$\emptyset \frown (a \frown \dot{\smile} \mathfrak{f})$$
$$d \frown (a \frown \mathfrak{f}) \qquad (148$$

$$\smile$$

$$\vdash \overset{\mathfrak{a}}{\smile} \left[\text{—}\, \mathfrak{a} \frown \dot{\varepsilon} \left(\begin{array}{l} \varepsilon = a \\ \varepsilon \frown (a \frown \dot{\smile} \mathfrak{f}) \end{array} \right) \right] = (\text{—}\, \mathfrak{a} \frown (d \frown \dot{\smile} \mathfrak{f}))$$
$$\emptyset \frown (a \frown \dot{\smile} \mathfrak{f})$$
$$d \frown (a \frown \mathfrak{f}) \qquad (\alpha$$

(96): -------------------------

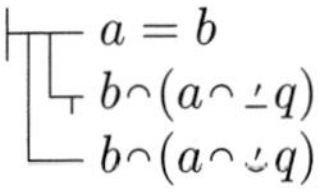

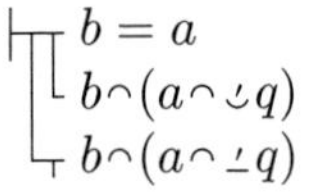

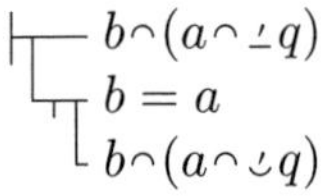

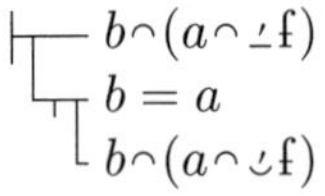

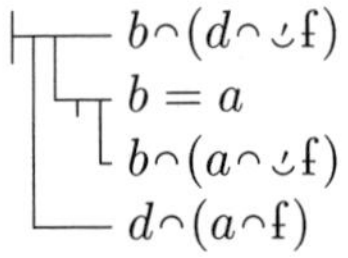

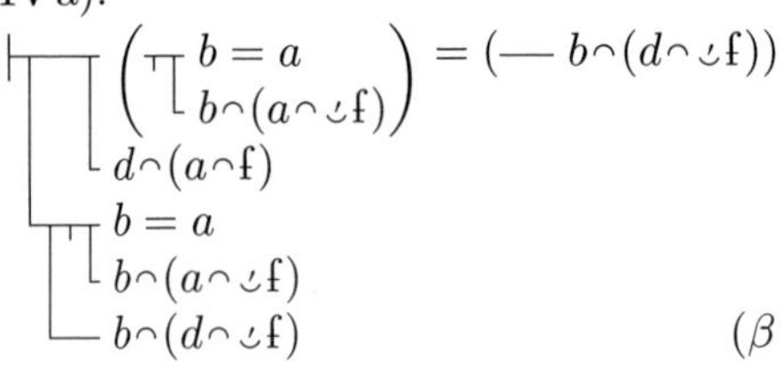

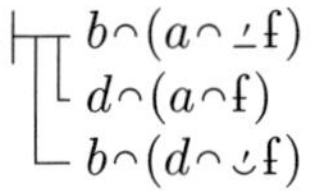

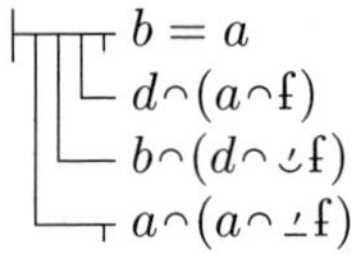

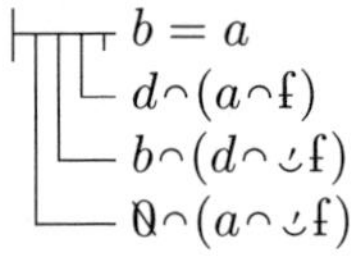

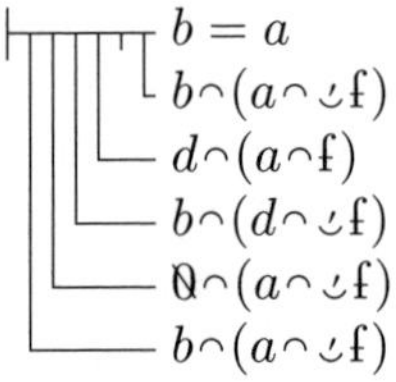

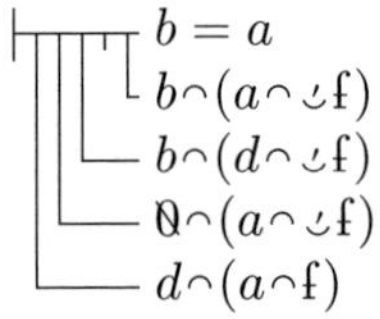

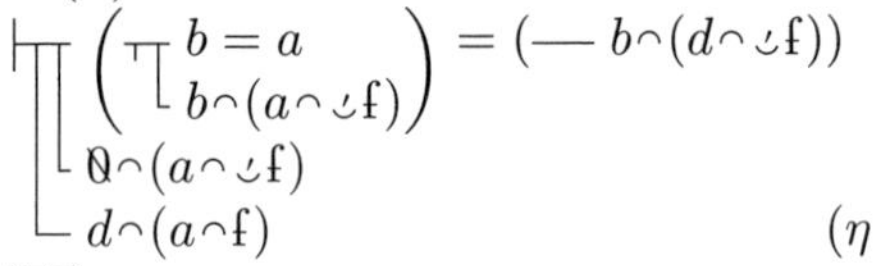

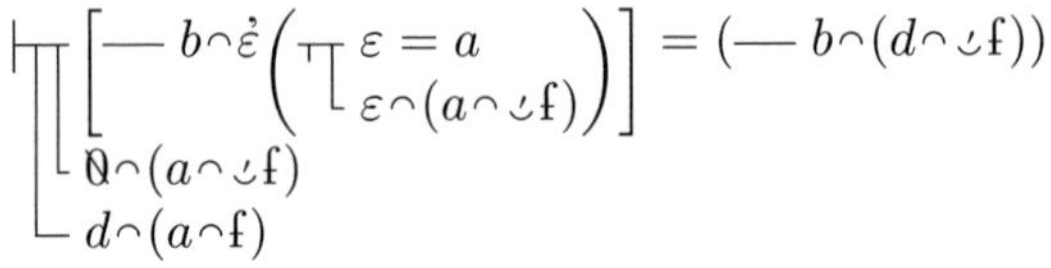

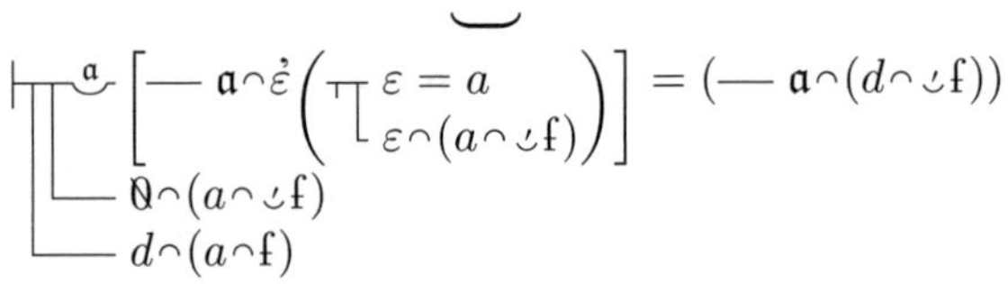

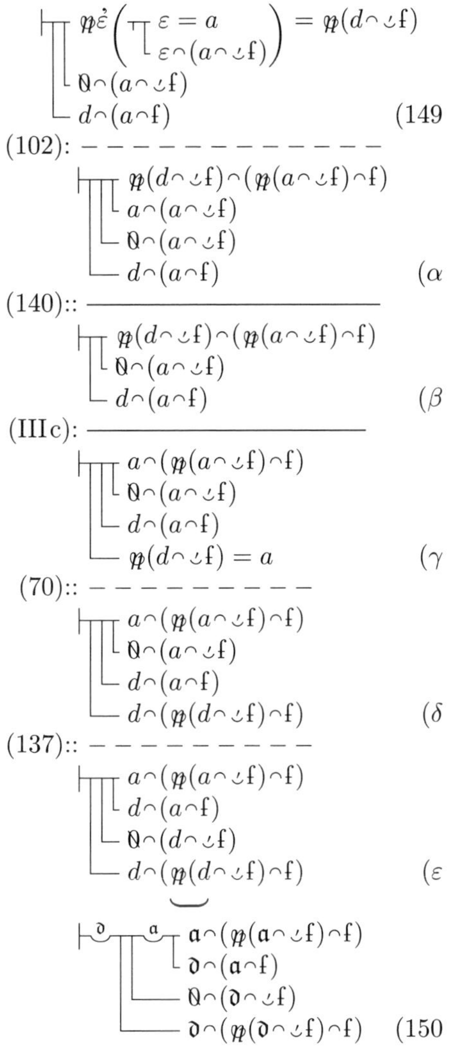

§116. *Analysis*

In order to prove the proposition (γ) of §114, we put ‘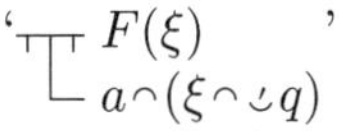’ in place of the function-marker ‘$F(\xi)$’ in (144). We then have to prove

‘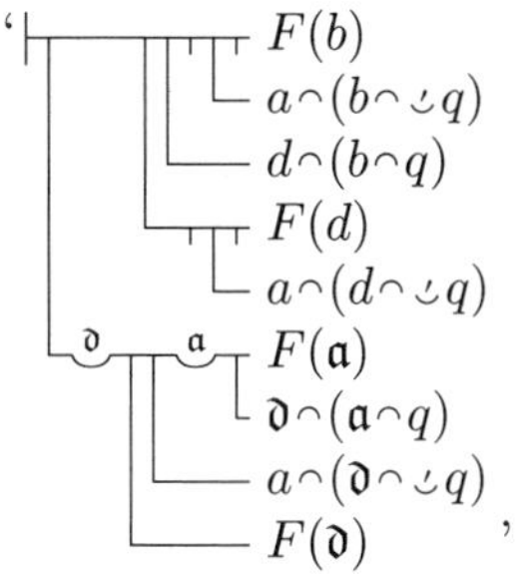’,

which can easily be done using (137). For the transition to (γ) compare p. 68.

§117. *Construction*

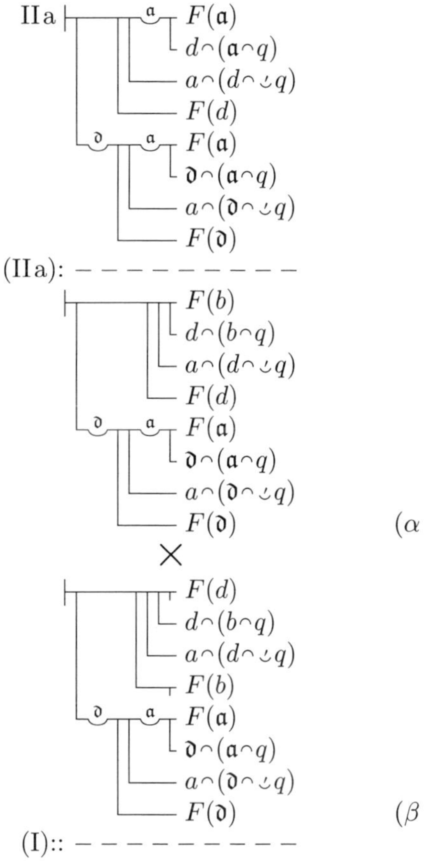

$\vdash$ $F(d)$
$d \frown (b \frown q)$
$a \frown (d \frown \smile q)$
$a \frown (b \frown \smile q)$
$F(b)$
$a \frown (b \frown \smile q)$
$\mathfrak{d}$ $\mathfrak{a}$ $F(\mathfrak{a})$
$\mathfrak{d} \frown (a \frown q)$
$a \frown (\mathfrak{d} \frown \smile q)$
$F(\mathfrak{d})$ (γ

(137):: - - - - - - - - - -

$\vdash$ $F(d)$
$a \frown (d \frown \smile q)$
$d \frown (b \frown q)$
$F(b)$
$a \frown (b \frown \smile q)$
$\mathfrak{d}$ $\mathfrak{a}$ $F(\mathfrak{a})$
$\mathfrak{d} \frown (a \frown q)$
$a \frown (\mathfrak{d} \frown \smile q)$
$F(\mathfrak{d})$ (δ

$\times$

$\vdash$ $F(b)$
$a \frown (b \frown \smile q)$
$d \frown (b \frown q)$
$F(d)$
$a \frown (d \frown \smile q)$
$\mathfrak{d}$ $\mathfrak{a}$ $F(\mathfrak{a})$
$\mathfrak{d} \frown (\mathfrak{a} \frown q)$
$a \frown (\mathfrak{d} \frown \smile q)$
$F(\mathfrak{d})$ (ε

$\smile$

$\vdash$ $\mathfrak{d}$ $\mathfrak{a}$ $F(\mathfrak{a})$
$a \frown (\mathfrak{a} \frown \smile q)$
$\mathfrak{d} \frown (\mathfrak{a} \frown q)$
$F(\mathfrak{d})$
$a \frown (\mathfrak{d} \frown \smile q)$
$\mathfrak{d}$ $\mathfrak{a}$ $F(\mathfrak{a})$
$\mathfrak{d} \frown (\mathfrak{a} \frown q)$
$a \frown (\mathfrak{d} \frown \smile q)$
$F(\mathfrak{d})$ (151

——— • ———

140 $\vdash a \frown (a \frown \smile q)$

(Ie): ————————

$\vdash$ $F(a)$
$a \frown (a \frown \smile q)$
$F(a)$ (α

(144): - - - - - - - -

$\vdash$ $F(b)$
$a \frown (b \frown \smile q)$
$F(a)$
$\mathfrak{d}$ $\mathfrak{a}$ $F(\mathfrak{a})$
$a \frown (\mathfrak{a} \frown \smile q)$
$\mathfrak{d} \frown (\mathfrak{a} \frown q)$
$F(\mathfrak{d})$
$a \frown (\mathfrak{d} \frown \smile q)$
$a \frown (b \frown \smile q)$ (β

(151):: - - - - - - - - - - - -

$\vdash$ $F(b)$
$a \frown (b \frown \smile q)$
$F(a)$
$\mathfrak{d}$ $\mathfrak{a}$ $F(\mathfrak{a})$
$\mathfrak{d} \frown (\mathfrak{a} \frown q)$
$a \frown (\mathfrak{d} \frown \smile q)$
$F(\mathfrak{d})$
$a \frown (b \frown \smile q)$ (γ

(Ib): - - - - - - - - - - -

$\vdash$ $F(b)$
$F(a)$
$\mathfrak{d}$ $\mathfrak{a}$ $F(\mathfrak{a})$
$\mathfrak{d} \frown (\mathfrak{a} \frown q)$
$a \frown (\mathfrak{d} \frown \smile q)$
$F(\mathfrak{d})$
$a \frown (b \frown \smile q)$ (152

——— • ———

150 $\vdash$ $\mathfrak{d}$ $\mathfrak{a}$ $\mathfrak{a} \frown (\not{p}(\mathfrak{a} \frown \smile \mathfrak{f}) \frown \mathfrak{f})$
$\mathfrak{d} \frown (\mathfrak{a} \frown \mathfrak{f})$
$\mathfrak{O} \frown (\mathfrak{d} \frown \smile \mathfrak{f})$
$\mathfrak{d} \frown (\not{p}(\mathfrak{d} \frown \smile \mathfrak{f}) \frown \mathfrak{f})$

(152): ————————————————

$\vdash$ $b \frown (\not{p}(b \frown \smile \mathfrak{f}) \frown \mathfrak{f})$
$\mathfrak{O} \frown (\not{p}(\mathfrak{O} \frown \smile \mathfrak{f}) \frown \mathfrak{f})$
$\mathfrak{O} \frown (b \frown \smile \mathfrak{f})$ (153

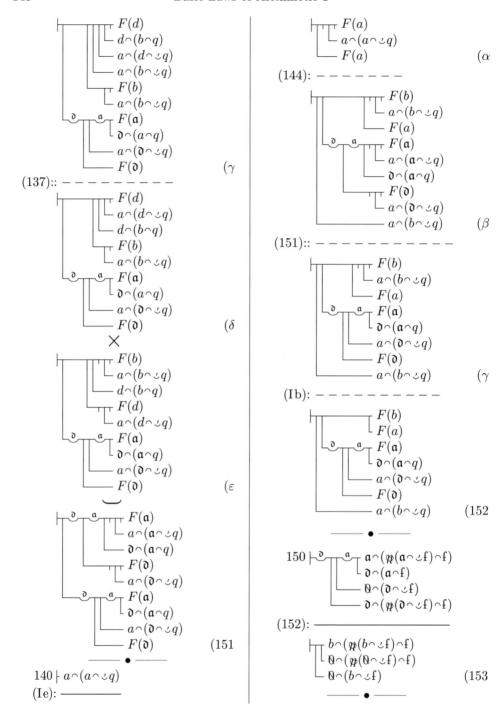

——— • ———

§118. *Analysis*

It now remains to prove the proposition

$$`\vdash \emptyset \frown (\mathfrak{p}(\emptyset \frown \smile\!\mathfrak{f}) \frown \mathfrak{f})'$$

By (102) we have

$$`\vdash\!\!\begin{array}{l} \top\ \emptyset \frown (\mathfrak{p}(\emptyset \frown \smile\!\mathfrak{f}) \frown \mathfrak{f}) \\ \mid\ \llcorner\ \emptyset \frown (\emptyset \frown \smile\!\mathfrak{f}) \\ \llcorner\ \mathfrak{p}\dot{\varepsilon}\left(\begin{array}{l} \top\ \varepsilon = \emptyset \\ \llcorner\ \varepsilon \frown (\emptyset \frown \smile\!\mathfrak{f}) \end{array}\right) = \emptyset' \end{array}$$

Here we can use (140). We then still have to prove the proposition

$$`\vdash \mathfrak{p}\dot{\varepsilon}\left(\begin{array}{l} \top\ \varepsilon = \emptyset \\ \llcorner\ \varepsilon \frown (\emptyset \frown \smile\!\mathfrak{f}) \end{array}\right) = \emptyset'$$

We apply the proposition (97) by showing that no object falls under the $\dot{\varepsilon}\left(\begin{array}{l} \top\ \varepsilon = \emptyset \\ \llcorner\ \varepsilon \frown (\emptyset \frown \smile\!\mathfrak{f}) \end{array}\right)$-concept. This straightforwardly follows from

$$`\vdash\!\!\begin{array}{l} \top\ a = \emptyset \\ \llcorner\ a \frown (\emptyset \frown \smile\!\mathfrak{f})' \end{array}$$

i.e., to the cardinal number series ending with $\emptyset$ only $\emptyset$ itself belongs. This proposition follows from (126) and (130).

§119. *Construction*

$$126 \vdash\!\top\ a \frown (\emptyset \frown \smile\!\mathfrak{f})$$

(130): ————

$$\vdash\!\!\begin{array}{l} \top\ \emptyset = a \\ \llcorner\ a \frown (\emptyset \frown \smile\!\mathfrak{f}) \end{array} \qquad (\alpha$$

(IIIf): − − − − −

$$\vdash\!\!\begin{array}{l} \top\ a = \emptyset \\ \llcorner\ a \frown (\emptyset \frown \smile\!\mathfrak{f}) \end{array} \qquad (\beta$$

(58): ————

$$\vdash\!\top\ a \frown \dot{\varepsilon}\left(\begin{array}{l} \top\ \varepsilon = \emptyset \\ \llcorner\ \varepsilon \frown (\emptyset \frown \smile\!\mathfrak{f}) \end{array}\right) \qquad (\gamma$$

$$\vdash\!\overset{\mathfrak{a}}{\smile}\!\top\ \mathfrak{a} \frown \dot{\varepsilon}\left(\begin{array}{l} \top\ \varepsilon = \emptyset \\ \llcorner\ \varepsilon \frown (\emptyset \frown \smile\!\mathfrak{f}) \end{array}\right) \qquad (\delta$$

(97): ————————

$$\vdash \mathfrak{p}\dot{\varepsilon}\left(\begin{array}{l} \top\ \varepsilon = \emptyset \\ \llcorner\ \varepsilon \frown (\emptyset \frown \smile\!\mathfrak{f}) \end{array}\right) = \emptyset \qquad (\varepsilon$$

(102): ————————

$$\vdash\!\!\begin{array}{l} \top\ \emptyset \frown (\mathfrak{p}(\emptyset \frown \smile\!\mathfrak{f}) \frown \mathfrak{f}) \\ \llcorner\ \emptyset \frown (\emptyset \frown \smile\!\mathfrak{f}) \end{array} \qquad (\zeta$$

(140):: ————————

$$\vdash \emptyset \frown (\mathfrak{p}(\emptyset \frown \smile\!\mathfrak{f}) \frown \mathfrak{f}) \qquad (154$$

(153): ————————

$$\vdash\!\!\begin{array}{l} \top\ b \frown (\mathfrak{p}(b \frown \smile\!\mathfrak{f}) \frown \mathfrak{f}) \\ \llcorner\ \emptyset \frown (b \frown \smile\!\mathfrak{f}) \end{array} \qquad (155$$

Θ. Some corollaries

§120. *Analysis*

First, we can easily conclude from (155) that for every finite cardinal number there is one immediately following it. Thereby it is stated that the cardinal number series starting with $\emptyset$ proceeds without end.

Moreover, we prove the proposition that provides a foundation for our counting by stating that n is the cardinal number belonging to a concept if a relation maps this concept into the cardinal number series up to and including n and excluding $\emptyset$, and if the converse of this relation maps that cardinal number series into the concept, provided n is a finite cardinal number.

This proposition follows straightforwardly from the proposition

$$`\vdash\!\!\begin{array}{l} \top\ n = \mathfrak{p}\dot{\varepsilon}\left(\begin{array}{l} \top\ \varepsilon = \emptyset \\ \llcorner\ \varepsilon \frown (n \frown \smile\!\mathfrak{f}) \end{array}\right) \\ \llcorner\ \emptyset \frown (n \frown \smile\!\mathfrak{f}) \end{array}',$$

which we prove using (87) and (155).

§121. *Construction*

155 ⊢ $b \frown (\mathfrak{p}(b \frown \smile\mathfrak{f}) \frown \mathfrak{f})$
└ $0 \frown (b \frown \smile\mathfrak{f})$

×

⊢ $0 \frown (b \frown \smile\mathfrak{f})$
└ $b \frown (\mathfrak{p}(b \frown \smile\mathfrak{f}) \frown \mathfrak{f})$ (α

(IIa):: − − − − − − − − −

⊢ $0 \frown (b \frown \smile\mathfrak{f})$
└ $\mathfrak{a}$ $b \frown (\mathfrak{a} \frown \mathfrak{f})$ (156

×

⊢ $\mathfrak{a}$ $b \frown (\mathfrak{a} \frown \mathfrak{f})$
└ $0 \frown (b \frown \smile\mathfrak{f})$ (157

——— • ———

87 ⊢ $n \frown (\mathfrak{p}v \frown \mathfrak{f})$
└ $c \frown v$
└ $n = \mathfrak{p}\grave{\varepsilon}\left(\begin{array}{l}\varepsilon = c \\ \varepsilon \frown v\end{array}\right)$

×

⊢ $n = \mathfrak{p}\grave{\varepsilon}\left(\begin{array}{l}\varepsilon = c \\ \varepsilon \frown v\end{array}\right)$
└ $c \frown v$
└ $n \frown (\mathfrak{p}v \frown \mathfrak{f})$ (158

(IIIa): ————————————

⊢ $n = \mathfrak{p}\grave{\varepsilon}\left(\begin{array}{l}\varepsilon = c \\ \varepsilon \frown v\end{array}\right)$
└ $c \frown v$
└ $n \frown (a \frown \mathfrak{f})$
└ $a = \mathfrak{p}v$ (159

——— • ———

155 ⊢ $n \frown (\mathfrak{p}(n \frown \smile\mathfrak{f}) \frown \mathfrak{f})$
└ $0 \frown (n \frown \smile\mathfrak{f})$

(158): − − − − − − − − −

⊢ $n = \mathfrak{p}\grave{\varepsilon}\left(\begin{array}{l}\varepsilon = 0 \\ \varepsilon \frown (n \frown \smile\mathfrak{f})\end{array}\right)$
└ $0 \frown (n \frown \smile\mathfrak{f})$ (160

(IIIa): − − − − − − − − − − − −

⊢ $\mathfrak{p}u = n$
└ $\mathfrak{p}u = \mathfrak{p}\grave{\varepsilon}\left(\begin{array}{l}\varepsilon = 0 \\ \varepsilon \frown (n \frown \smile\mathfrak{f})\end{array}\right)$
└ $0 \frown (n \frown \smile\mathfrak{f})$ (α

(32):: − − − − − − − − − − − − − −

⊢ $\mathfrak{p}u = n$
└ $u \frown \left[\grave{\varepsilon}\left(\begin{array}{l}\varepsilon = 0 \\ \varepsilon \frown (n \frown \smile\mathfrak{f})\end{array}\right) \frown \rangle q\right]$
└ $\grave{\varepsilon}\left(\begin{array}{l}\varepsilon = 0 \\ \varepsilon \frown (n \frown \smile\mathfrak{f})\end{array}\right) \frown (u \frown \rangle \rangle\!\!\!\!\mid q)$
└ $0 \frown (n \frown \smile\mathfrak{f})$ (161

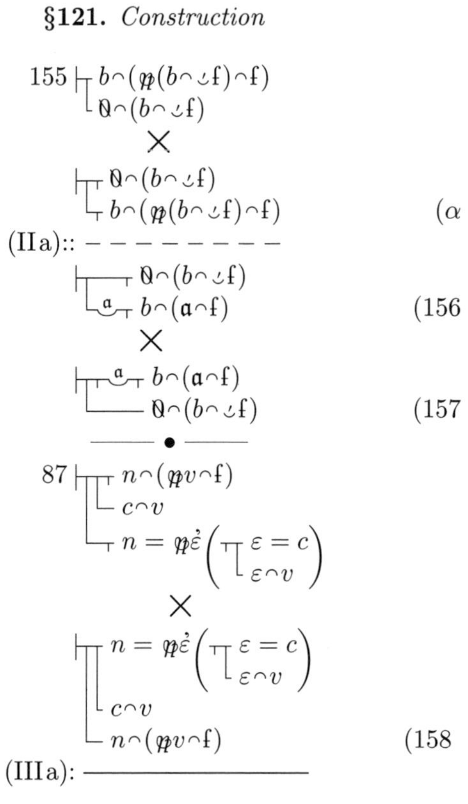

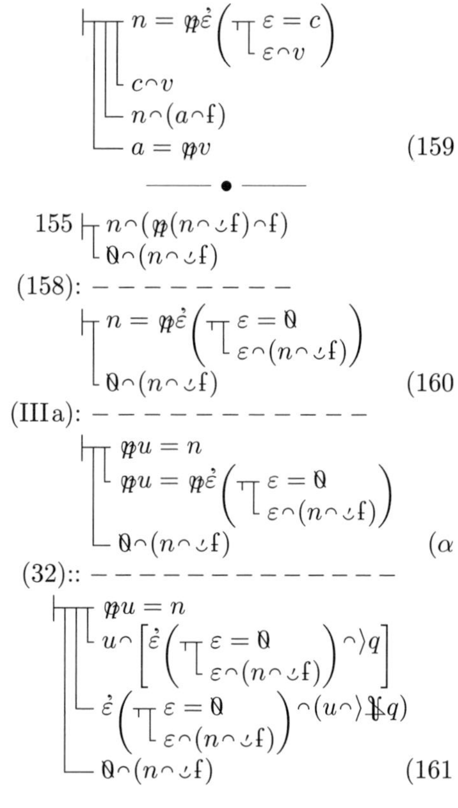

I. Proof of some propositions of the cardinal number Endlos

a) Proof of the proposition

'⊢ $0 \frown (\varpi \frown \smile\mathfrak{f})$'

§122. *Analysis*

There are cardinal numbers that do not belong to the cardinal number series beginning with 0, or, as we also say, that are not finite, that are infinite. One such cardinal number is that of the concept *finite cardinal number*; I propose to call it *Endlos* and designate it with 'ϖ'. I define it thus:

$$\Vdash \mathfrak{p}(0 \frown \rangle\!\!\!\mid\!\smile\mathfrak{f}) = \varpi \quad \text{(M}$$

For $0 \frown \rangle\!\!\!\mid\!\smile\mathfrak{f}$ is the extension of the concept *finite cardinal number*. The proposition mentioned in the heading says that the cardinal number Endlos is not a finite cardinal number. We prove it, as is indicated in §84 of my *Grundlagen*, by showing that the cardinal number Endlos follows after itself in the cardinal number series, which according to (145) no finite cardinal number does. First, it is to be shown that Endlos stands in the f-relation to itself:

$$\text{'}\vdash \varpi \frown (\varpi \frown \mathfrak{f})\text{'} \quad (\alpha$$

We reduce this proposition to

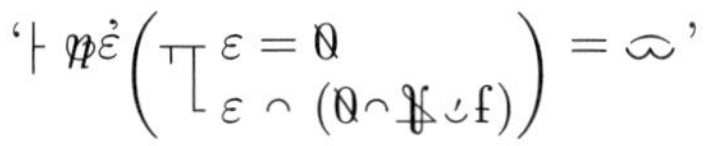

(β

which follows from the propositions

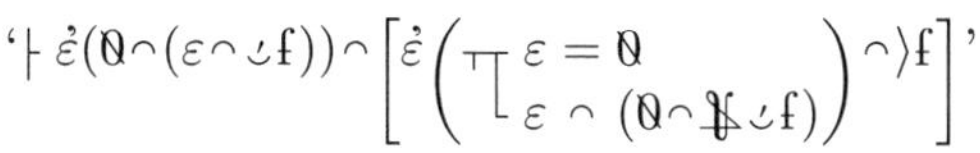

(γ

and

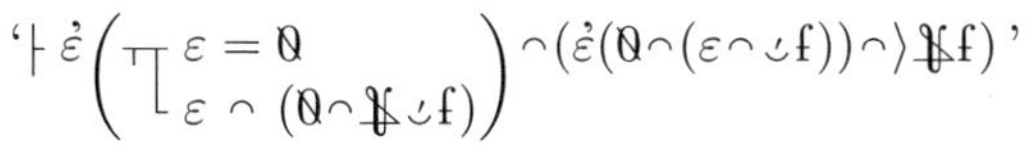

(δ

According to (11), in order to derive (γ) we have to show

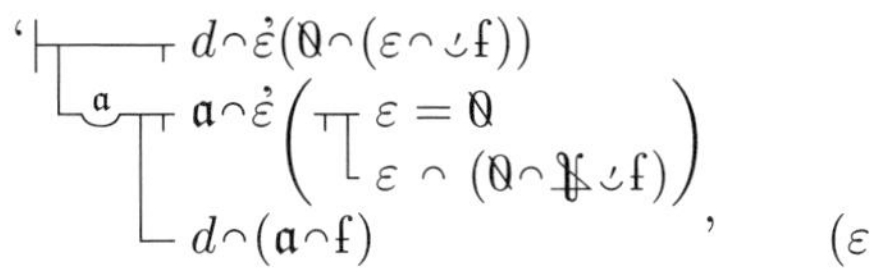

which is easily reduced to the proposition

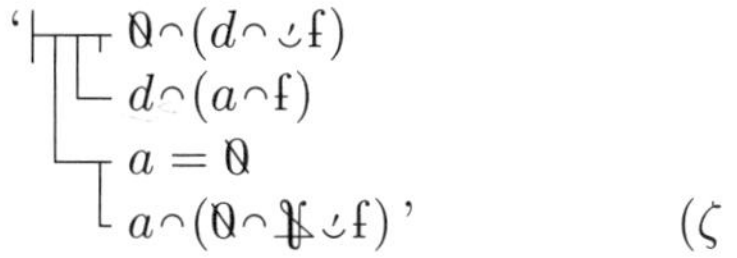

which breaks down into the propositions

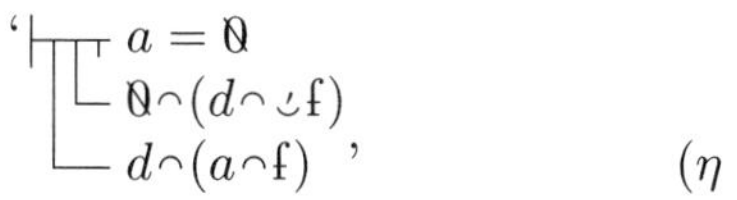

and (137).

§123. *Construction*

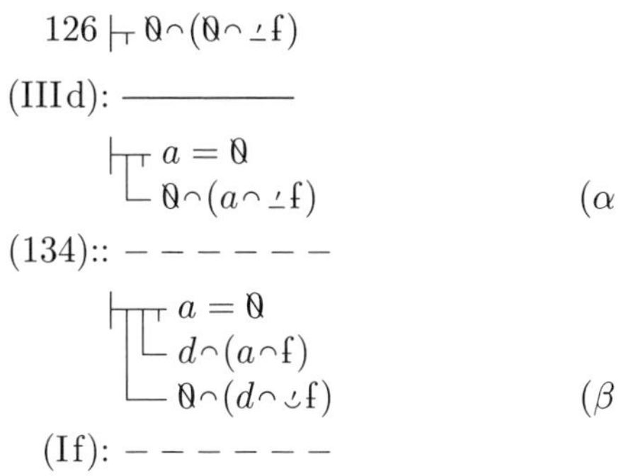

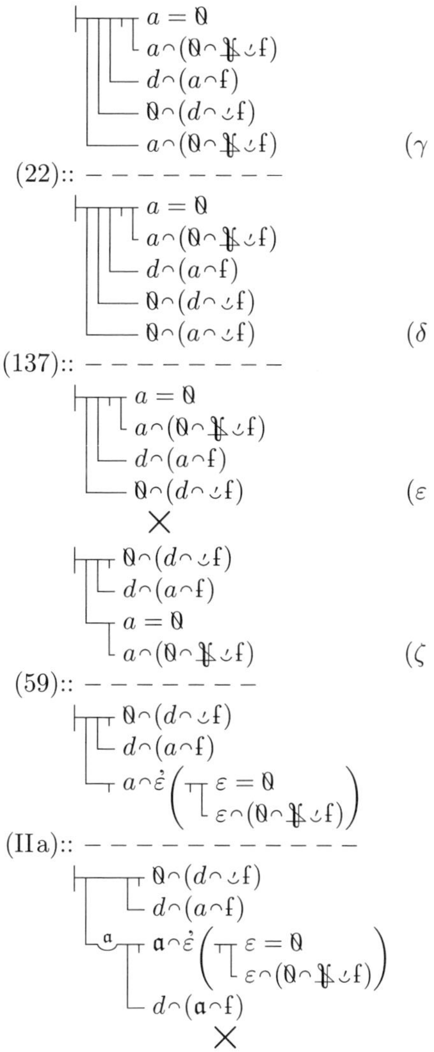

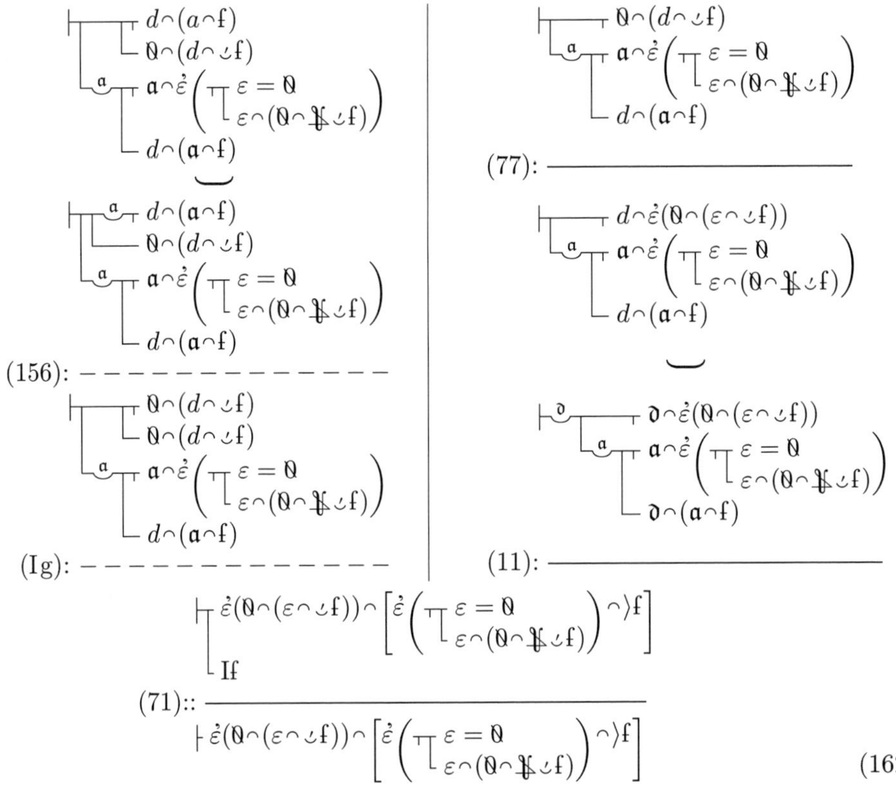

(162

§124. *Analysis*

Instead of proving proposition (δ) of §122, we first prove the following:

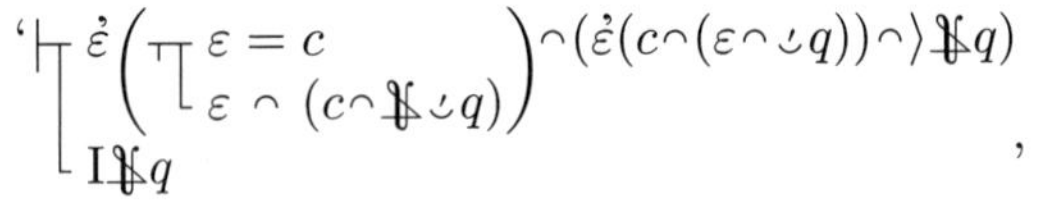

For this we require the proposition

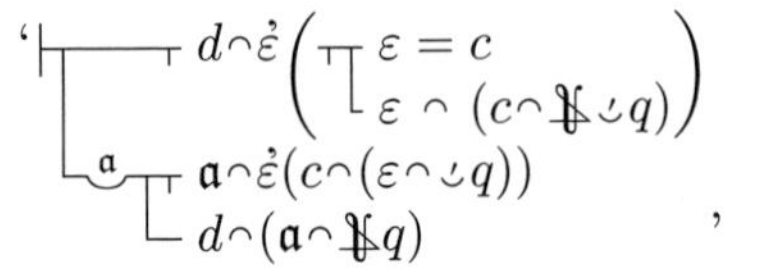

which is to be reduced to the proposition

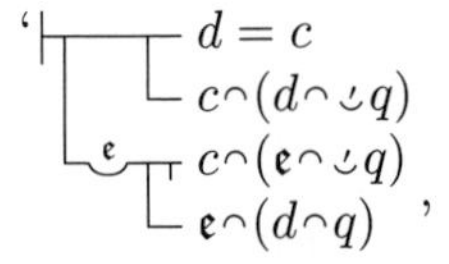

This follows easily from (142).

§125. *Construction*

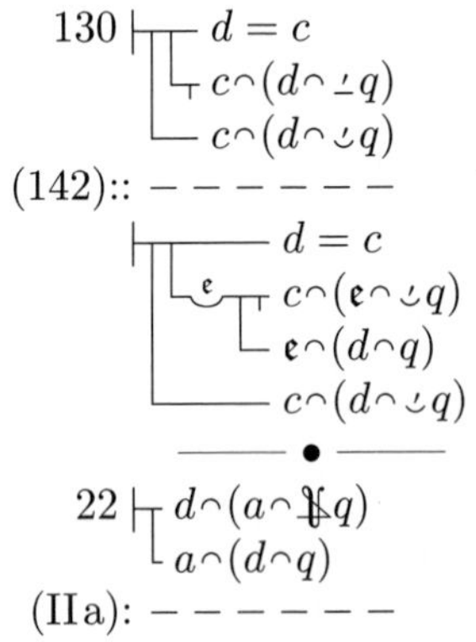

(α

$$\vdash \left[\begin{array}{l} a\frown\dot{\varepsilon}(c\frown(\varepsilon\frown\smile q)) \\ a\frown(d\frown q) \\ \overset{\mathfrak{a}}{\smile}\left[\begin{array}{l}\mathfrak{a}\frown\dot{\varepsilon}(c\frown(\varepsilon\frown\smile q)) \\ d\frown(\mathfrak{a}\frown\nparallel q)\end{array}\right]\end{array}\right] \qquad (\beta$$

(82): ——————

$$\vdash \left[\begin{array}{l} c\frown(a\frown\smile q) \\ a\frown(d\frown q) \\ \overset{\mathfrak{a}}{\smile}\left[\begin{array}{l}\mathfrak{a}\frown\dot{\varepsilon}(c\frown(\varepsilon\frown\smile q)) \\ d\frown(\mathfrak{a}\frown\nparallel q)\end{array}\right]\end{array}\right] \qquad (\gamma$$

$$\vdash \left[\begin{array}{l} \overset{\mathfrak{e}}{\smile}\left[\begin{array}{l} c\frown(\mathfrak{e}\frown\smile q) \\ \mathfrak{e}\frown(d\frown q)\end{array}\right] \\ \overset{\mathfrak{a}}{\smile}\left[\begin{array}{l}\mathfrak{a}\frown\dot{\varepsilon}(c\frown(\varepsilon\frown\smile q)) \\ d\frown(\mathfrak{a}\frown\nparallel q)\end{array}\right]\end{array}\right] \qquad (\delta$$

(α): – – – – – – – – – – –

$$\vdash \left[\begin{array}{l} d = c \\ c\frown(d\frown\smile q) \\ \overset{\mathfrak{a}}{\smile}\left[\begin{array}{l}\mathfrak{a}\frown\dot{\varepsilon}(c\frown(\varepsilon\frown\smile q)) \\ d\frown(\mathfrak{a}\frown\nparallel q)\end{array}\right]\end{array}\right] \qquad (\varepsilon$$

(23):: – – – – – – – – – – –

$$\vdash \left[\begin{array}{l} d = c \\ d\frown(c\frown\nparallel\smile q) \\ \overset{\mathfrak{a}}{\smile}\left[\begin{array}{l}\mathfrak{a}\frown\dot{\varepsilon}(c\frown(\varepsilon\frown\smile q)) \\ d\frown(\mathfrak{a}\frown\nparallel q)\end{array}\right]\end{array}\right] \qquad (\zeta$$

(58): ——————

$$\vdash \left[\begin{array}{l} d\frown\dot{\varepsilon}\left(\begin{array}{l}\varepsilon = c \\ \varepsilon\frown(c\frown\nparallel\smile q)\end{array}\right) \\ \overset{\mathfrak{a}}{\smile}\left[\begin{array}{l}\mathfrak{a}\frown\dot{\varepsilon}(c\frown(\varepsilon\frown\smile q)) \\ d\frown(\mathfrak{a}\frown\nparallel q)\end{array}\right]\end{array}\right] \qquad (\eta$$

$$\vdash \overset{\mathfrak{d}}{\smile}\left[\begin{array}{l} \mathfrak{d}\frown\dot{\varepsilon}\left(\begin{array}{l}\varepsilon = c \\ \varepsilon\frown(c\frown\nparallel\smile q)\end{array}\right) \\ \overset{\mathfrak{a}}{\smile}\left[\begin{array}{l}\mathfrak{a}\frown\dot{\varepsilon}(c\frown(\varepsilon\frown\smile q)) \\ \mathfrak{d}\frown(\mathfrak{a}\frown\nparallel q)\end{array}\right]\end{array}\right] \qquad (\vartheta$$

(11): ——————

$$\vdash \left[\begin{array}{l} \dot{\varepsilon}\left(\begin{array}{l}\varepsilon = c \\ \varepsilon\frown(c\frown\nparallel\smile q)\end{array}\right)\frown(\dot{\varepsilon}(c\frown(\varepsilon\frown\smile q))\frown\rangle\nparallel q) \\ \mathrm{I}\nparallel q\end{array}\right] \qquad (163$$

——— • ———

$$\mathrm{E} \vdash \dot{\alpha}\dot{\varepsilon}(\alpha\frown(\varepsilon\frown q)) = \nparallel q$$

(44): ——————

$$\vdash \left[\begin{array}{l} F(a\frown\nparallel q) \\ F(\dot{\varepsilon}(a\frown(\varepsilon\frown q)))\end{array}\right] \qquad (164$$

——— • ———

$$89 \vdash \mathrm{I}\nparallel\mathfrak{f}$$

(163): ——

$$\vdash \dot{\varepsilon}\left(\begin{array}{l}\varepsilon = 0 \\ \varepsilon\frown(0\frown\nparallel\smile\mathfrak{f})\end{array}\right)\frown(\dot{\varepsilon}(0\frown(\varepsilon\frown\smile\mathfrak{f}))\frown\rangle\nparallel\mathfrak{f}) \qquad (\alpha$$

(32): ——————

$$\vdash \left[\begin{array}{l} \mathfrak{p}\dot{\varepsilon}(0\frown(\varepsilon\frown\smile\mathfrak{f})) = \mathfrak{p}\dot{\varepsilon}\left(\begin{array}{l}\varepsilon = 0 \\ \varepsilon\frown(0\frown\nparallel\smile\mathfrak{f})\end{array}\right) \\ \dot{\varepsilon}(0\frown(\varepsilon\frown\smile\mathfrak{f}))\frown\left[\dot{\varepsilon}\left(\begin{array}{l}\varepsilon = 0 \\ \varepsilon\frown(0\frown\nparallel\smile\mathfrak{f})\end{array}\right)\frown\rangle\mathfrak{f}\right]\end{array}\right] \qquad (\beta$$

(162):: ——————

$$\vdash \mathfrak{p}\dot{\varepsilon}(0\frown(\varepsilon\frown\smile\mathfrak{f})) = \mathfrak{p}\dot{\varepsilon}\left(\begin{array}{l}\varepsilon = 0 \\ \varepsilon\frown(0\frown\nparallel\smile\mathfrak{f})\end{array}\right) \qquad (\gamma$$

(164): ——————

$$\vdash \mathfrak{p}(0\frown\nparallel\smile\mathfrak{f}) = \mathfrak{p}\dot{\varepsilon}\left(\begin{array}{l}\varepsilon = 0 \\ \varepsilon\frown(0\frown\nparallel\smile\mathfrak{f})\end{array}\right) \qquad (\delta$$

(IIIc): ——————

$$\vdash \wp\dot{\varepsilon}\left(\begin{array}{l} \varepsilon = 0 \\ \varepsilon \frown (0 \frown \not\Vdash \smile \mathfrak{f}) \end{array}\right) = \infty$$
$$\quad \wp(0 \frown \not\Vdash \smile \mathfrak{f}) = \infty \qquad (\varepsilon$$

(101): - - - - - - - - - - - - - -

$$\vdash \infty \frown (\infty \frown \mathfrak{f})$$
$$\quad \wp(0 \frown \not\Vdash \smile \mathfrak{f}) = \infty$$
$$\quad 0 \frown (0 \frown \not\Vdash \smile \mathfrak{f}) \qquad (\zeta$$

(22, M):: ═══════

$$\vdash \infty \frown (\infty \frown \mathfrak{f})$$
$$\quad 0 \frown (0 \frown \smile \mathfrak{f}) \qquad (\eta$$

(140):: ————

$$\vdash \infty \frown (\infty \frown \mathfrak{f}) \qquad (165$$

(131): ————

$$\vdash \infty \frown (\infty \frown \underline{\smile}\,\mathfrak{f}) \qquad (166$$

——— • ———

$$145 \vdash \infty \frown (\infty \frown \underline{\smile}\,\mathfrak{f})$$
$$\quad 0 \frown (\infty \frown \smile \mathfrak{f})$$

×

$$\vdash 0 \frown (\infty \frown \smile \mathfrak{f})$$
$$\quad \infty \frown (\infty \frown \underline{\smile}\,\mathfrak{f}) \qquad (\alpha$$

(166):: ————

$$\vdash 0 \frown (\infty \frown \smile \mathfrak{f}) \qquad (167$$

b) *Proof of the proposition*

$$\text{`}\vdash \infty = \wp\dot{\varepsilon}\left(\begin{array}{l} \varepsilon \frown v \\ \varepsilon \frown u \end{array}\right)$$
$$\quad \infty = \wp u$$
$$\quad 0 \frown (\wp v \frown \smile \mathfrak{f})\text{'}$$

§126. *Analysis*

We now prove the proposition

"If Endlos is the cardinal number of a concept and if the cardinal number of another concept is finite, then Endlos is the cardinal number of the concept *falling under either the first or the second concept.*"

using (144), by taking

$$\text{`}\smile\!\mathfrak{v}\!\!\top \infty = \wp\dot{\varepsilon}\left(\begin{array}{l} \varepsilon \frown \mathfrak{v} \\ \varepsilon \frown u \end{array}\right)$$
$$\quad \xi = \wp\mathfrak{v}\text{'},$$

in place of the function-marker '$F(\xi)$', and first have to derive the proposition

$$\text{`}\vdash \infty = \wp\dot{\varepsilon}\left(\begin{array}{l} \varepsilon \frown v \\ \varepsilon \frown u \end{array}\right)$$
$$\quad a = \wp v$$
$$\quad d \frown (a \frown \mathfrak{f})$$
$$\quad \smile\!\mathfrak{v}\!\!\top \infty = \wp\dot{\varepsilon}\left(\begin{array}{l} \varepsilon \frown \mathfrak{v} \\ \varepsilon \frown u \end{array}\right)$$
$$\qquad d = \wp\mathfrak{v}\text{'} \qquad (\alpha$$

According to (IIa) we have

$$\text{`}\vdash \infty = \wp\dot{\varepsilon}\left(\begin{array}{l} \varepsilon \frown \dot{\varepsilon}\left(\begin{array}{l} \varepsilon = c \\ \varepsilon \frown v \end{array}\right) \\ \varepsilon \frown u \end{array}\right)$$
$$\quad d = \wp\dot{\varepsilon}\left(\begin{array}{l} \varepsilon = c \\ \varepsilon \frown v \end{array}\right)$$
$$\quad \smile\!\mathfrak{v}\!\!\top \infty = \wp\dot{\varepsilon}\left(\begin{array}{l} \varepsilon \frown \mathfrak{v} \\ \varepsilon \frown u \end{array}\right)$$
$$\qquad d = \wp\mathfrak{v}\text{'},$$

To this we can now apply (159). In order to get the desired supercomponent, we must prove the proposition

$$\text{`}\vdash \infty = \wp\dot{\varepsilon}\left(\begin{array}{l} \varepsilon \frown v \\ \varepsilon \frown u \end{array}\right)$$
$$\quad \infty = \wp\dot{\varepsilon}\left(\begin{array}{l} \varepsilon \frown \dot{\varepsilon}\left(\begin{array}{l} \varepsilon = c \\ \varepsilon \frown v \end{array}\right) \\ \varepsilon \frown u \end{array}\right)\text{'} \qquad (\beta$$

To this end we distinguish the cases where c falls under the u-concept and its contrary. Thus, we have the propositions

$$‘\vdash \begin{array}{l} \dot{\varepsilon}\left(\begin{array}{l}\varepsilon\frown\dot{\varepsilon}\left(\begin{array}{l}\varepsilon = c\\ \varepsilon\frown v\end{array}\right)\\ \varepsilon\frown u\end{array}\right) = \dot{\varepsilon}\left(\begin{array}{l}\varepsilon\frown v\\ \varepsilon\frown u\end{array}\right)\\ c\frown u\end{array}\text{'}\qquad (\gamma$$

$$‘\vdash \begin{array}{l} \dot{\varepsilon}\left(\begin{array}{l}\varepsilon\frown\dot{\varepsilon}\left(\begin{array}{l}\varepsilon = c\\ \varepsilon\frown v\end{array}\right)\\ \varepsilon\frown u\end{array}\right) = \dot{\varepsilon}\left(\begin{array}{l}\varepsilon = c\\ \varepsilon\frown\dot{\varepsilon}\left(\begin{array}{l}\varepsilon\frown v\\ \varepsilon\frown u\end{array}\right)\end{array}\right)\\ c\frown u\end{array}\text{'}\qquad (\delta$$

In the second case we also require the proposition

$$‘\vdash \begin{array}{l} \varpi = \mathfrak{p}w\\ \varpi = \mathfrak{p}\dot{\varepsilon}\left(\begin{array}{l}\varepsilon = c\\ \varepsilon\frown w\end{array}\right)\end{array}\text{'}$$

which straightforwardly follows from (165) and (69).

§127. *Construction*

$$\text{IIId}\vdash \begin{array}{l} a = c\\ a\frown w\\ c\frown w\end{array}$$

(If): — — — — —

$$\vdash \begin{array}{l} a = c\\ a\frown w\\ a\frown w\\ c\frown w\end{array}\qquad (\alpha$$

(IVa): — — — — —

$$\vdash \begin{array}{l} (—\, a\frown w) = \left(\begin{array}{l}a = c\\ a\frown w\end{array}\right)\\ c\frown w\\ a\frown w\\ a = c\\ a\frown w\end{array}\qquad (\beta$$

(Id):: ————————————

$$\vdash \begin{array}{l} (—\, a\frown w) = \left(\begin{array}{l}a = c\\ a\frown w\end{array}\right)\\ c\frown w\end{array}\qquad (\gamma$$

(77): ————————————

$$\vdash \begin{array}{l} (—\, a\frown w) = \left[—\, a\frown\dot{\varepsilon}\left(\begin{array}{l}\varepsilon = c\\ \varepsilon\frown w\end{array}\right)\right]\\ c\frown w\end{array}\qquad (\delta$$

$$\vdash \begin{array}{l} \overset{\mathfrak{a}}{\smile}(—\, \mathfrak{a}\frown w) = \left[—\, \mathfrak{a}\frown\dot{\varepsilon}\left(\begin{array}{l}\varepsilon = c\\ \varepsilon\frown w\end{array}\right)\right]\\ c\frown w\end{array}\qquad (\varepsilon$$

(96): — — — — — — — — — — — — — — —

$$\vdash \begin{array}{l} \mathfrak{p}w = \mathfrak{p}\dot{\varepsilon}\left(\begin{array}{l}\varepsilon = c\\ \varepsilon\frown w\end{array}\right)\\ c\frown w\end{array}\qquad (\zeta$$

(IIIa): — — — — — — — — — —

$$\vdash \begin{array}{l} n = \mathfrak{p}w\\ n = \mathfrak{p}\dot{\varepsilon}\left(\begin{array}{l}\varepsilon = c\\ \varepsilon\frown w\end{array}\right)\\ c\frown w\end{array}\qquad (168$$

—— • ——

$$69\vdash \begin{array}{l} m\frown(n\frown\mathfrak{f})\\ \mathfrak{p}w = n\\ c\frown w\\ \mathfrak{p}\dot{\varepsilon}\left(\begin{array}{l}\varepsilon = c\\ \varepsilon\frown w\end{array}\right) = m\end{array}$$

$$\times$$

$$\vdash \begin{array}{l} \mathfrak{p}w = n\\ m\frown(n\frown\mathfrak{f})\\ c\frown w\\ \mathfrak{p}\dot{\varepsilon}\left(\begin{array}{l}\varepsilon = c\\ \varepsilon\frown w\end{array}\right) = m\end{array}\qquad (169$$

(IIIf):: — — — — — — — — — —

$$\vdash \begin{array}{l} \mathfrak{p}w = n\\ m\frown(n\frown\mathfrak{f})\\ c\frown w\\ m = \mathfrak{p}\dot{\varepsilon}\left(\begin{array}{l}\varepsilon = c\\ \varepsilon\frown w\end{array}\right)\end{array}\qquad (\alpha$$

(IIIf): — — — — — — — — — —

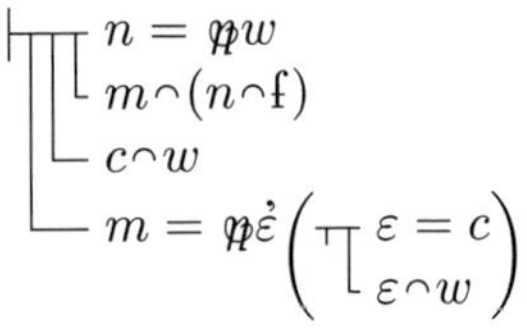
(170

———— • ————

(170): ————————

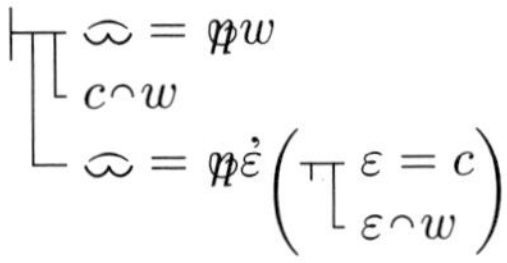
(α

(168): ·—·—·—·—·—·—·—·—·

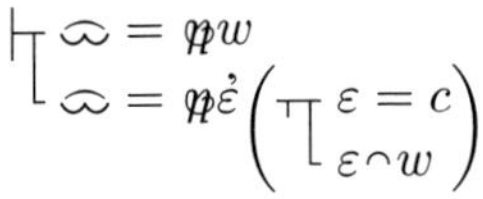
(171

———— • ————

If 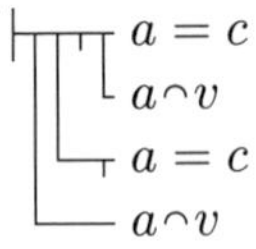

(Ic, Id):: = = = =

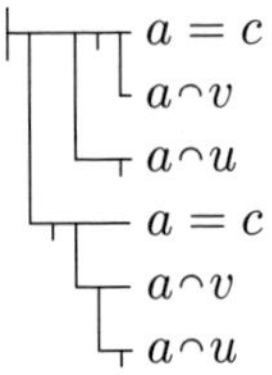
(α

(IVa): ————————

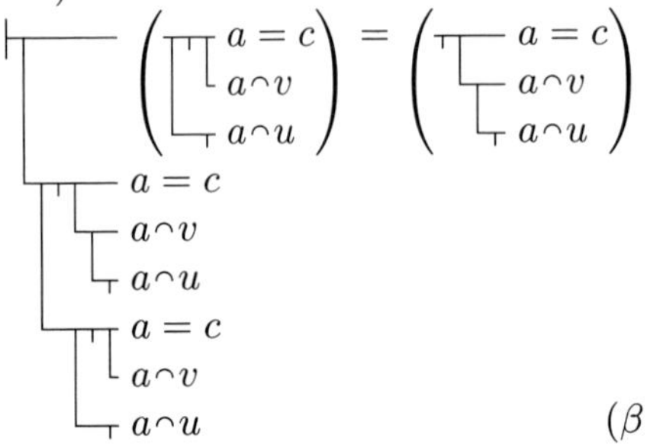
(β

———— • ————

Ic 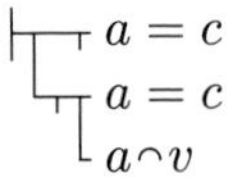

(I):: — — — — —

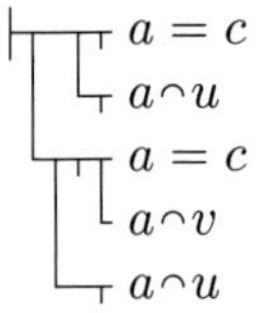
(γ

(IIId): ·—·—·—·—·

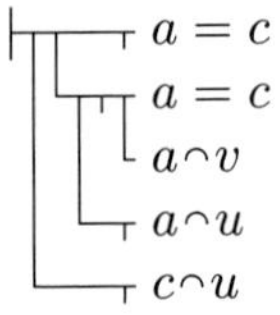
(δ

(If): — — — — —

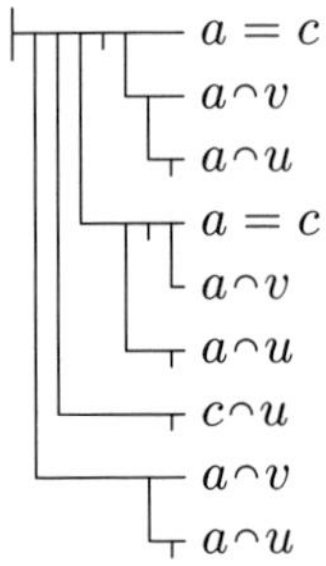
(ε

———— • ————

Id 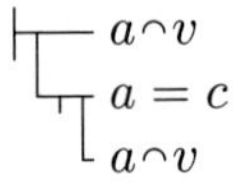

(I):: — — — — —

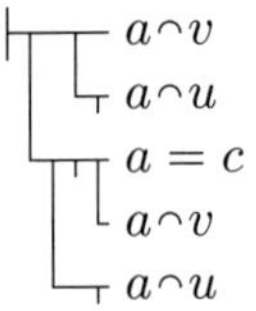
(ζ

(ε): — — — — —

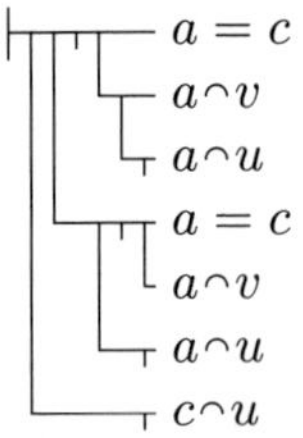
(η

(β): — — — — — —

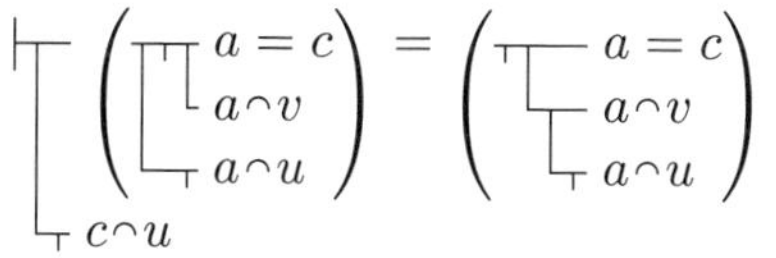

(ϑ

(77): ————————————

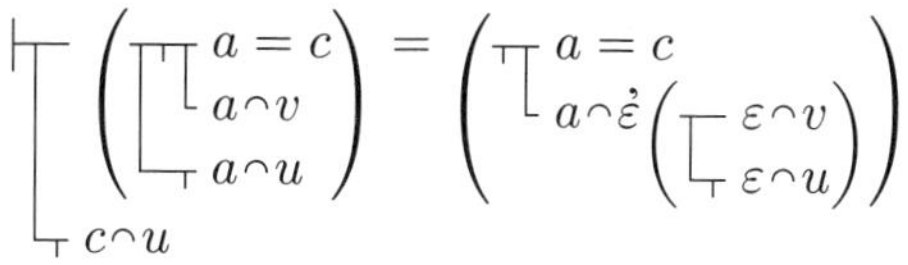

(ι

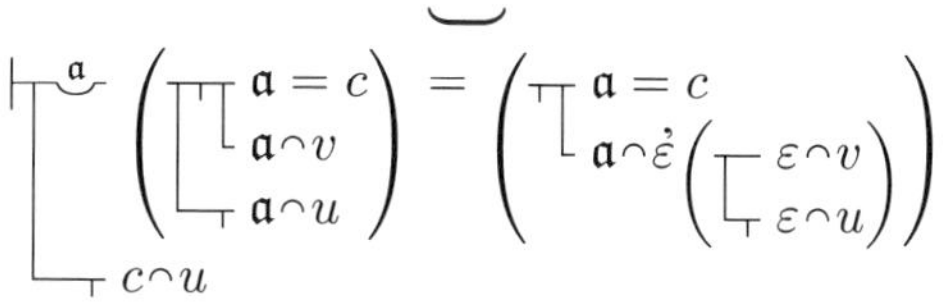

(κ

(Va): - - - - - - - - - - - - - - - - - - -

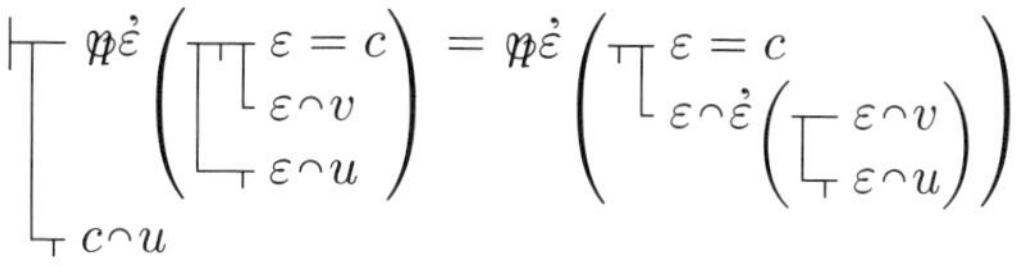

(λ

(IIIc): -

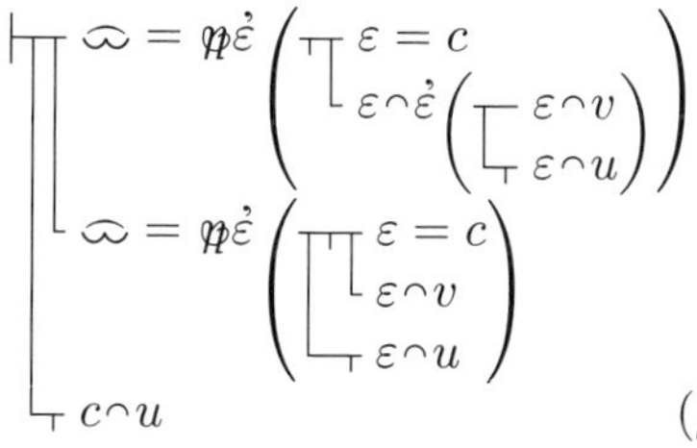

(μ

(171): - - - - - - - - - - - - - - - -

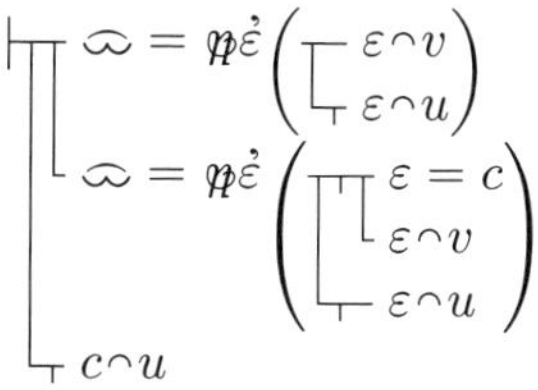

(ν

——— • ———

Ia 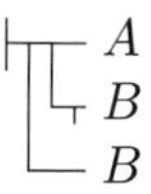

(I): - - -

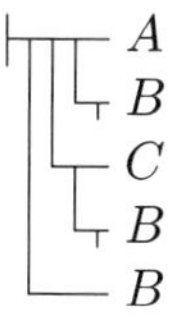

(ξ

(IVa): - - -

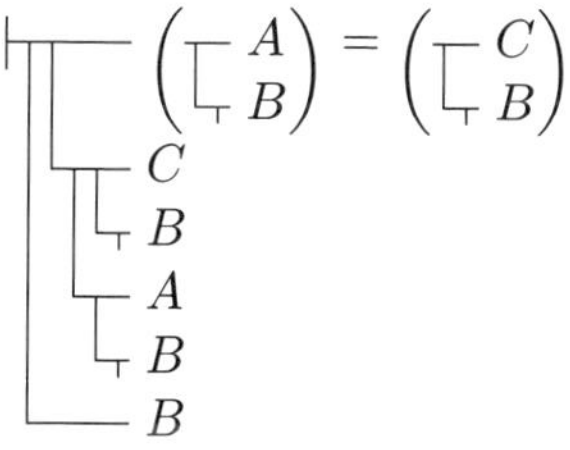

(o

(ξ):: - - - - - - - - - - - - - -

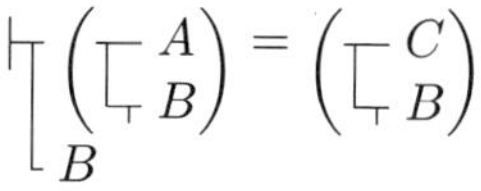

(π

——— • ———

Id 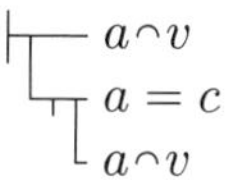

(IVa): ————

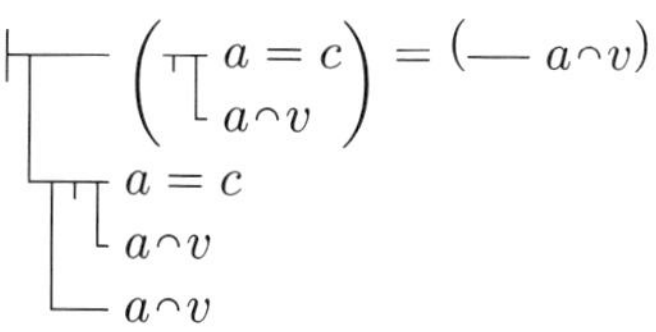

(ρ

(If):: - - - - - - - - - - - - - - -

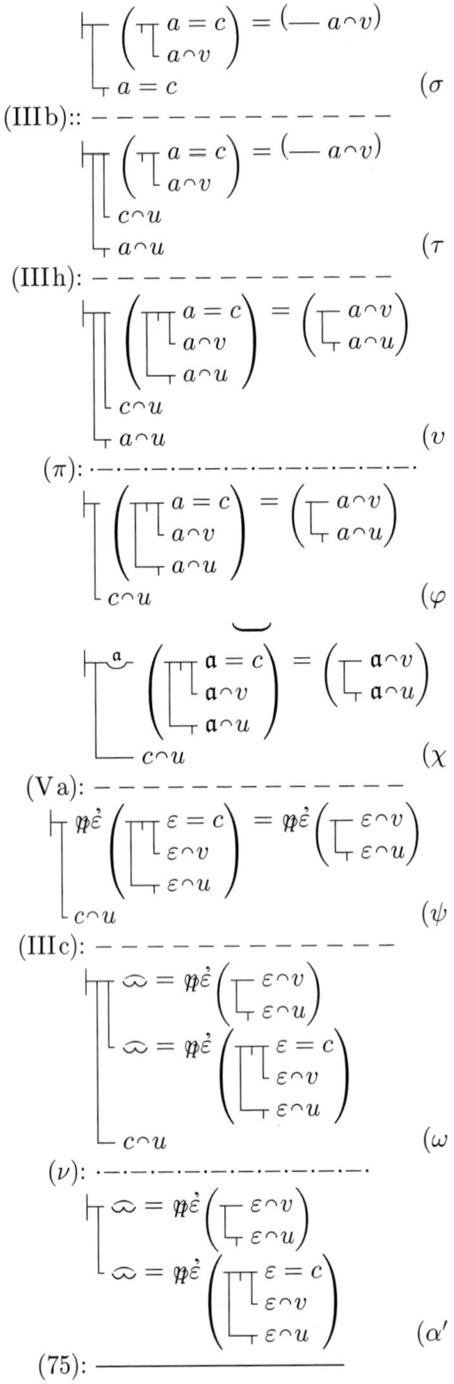
(σ
(IIIb)::
(τ
(IIIh):
(υ
(π):
(φ
(χ
(Va):
(ψ
(IIIc):
(ω
(ν):
(α′
(75):

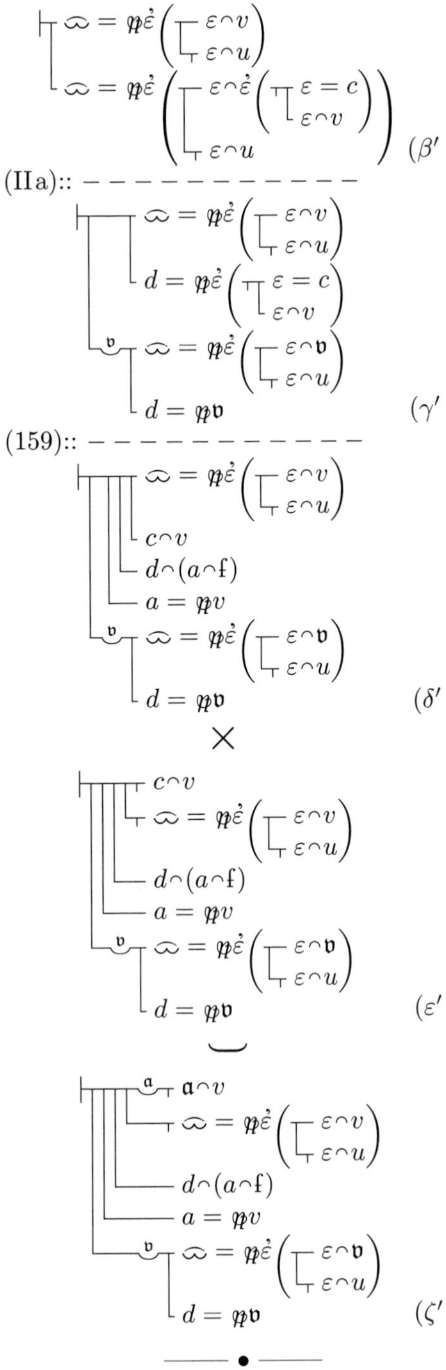
(β′
(IIa)::
(γ′
(159)::
(δ′
(ε′
(ζ′

108 ⊢ d⌢(0⌢f)

(IIIa): ————

⊢ d⌢(a⌢f)
  a = 0 (η′

(IIIa): ————

⊢ d⌢(a⌢f)
  a = ℘v
  ℘v = 0 (ϑ′

(97):: – – – – – –

⊢ d⌢(a⌢f)
  a = ℘v
  𝔞 𝔞⌢v (ι′

(ζ′):: – – – – – –

⊢ d⌢(a⌢f)
  a = ℘v
  $\frown = \wp\dot{\varepsilon}$(ε⌢v, ε⌢u)
  d⌢(a⌢f)
  𝔳 $\frown = \wp\dot{\varepsilon}$(ε⌢𝔳, ε⌢u)
  d = ℘𝔳 (κ′

×

⊢ $\frown = \wp\dot{\varepsilon}$(ε⌢v, ε⌢u)
  a = ℘v
  d⌢(a⌢f)
  𝔳 $\frown = \wp\dot{\varepsilon}$(ε⌢𝔳, ε⌢u)
  d = ℘𝔳 (λ′

‿

⊢ 𝔡 𝔞 𝔳 $\frown = \wp\dot{\varepsilon}$(ε⌢𝔳, ε⌢u)
  𝔞 = ℘𝔳
  𝔡⌢(𝔞⌢f)
  𝔳 $\frown = \wp\dot{\varepsilon}$(ε⌢𝔳, ε⌢u)
  𝔡 = ℘𝔳 (μ′

(144): ————

⊢ 𝔳 $\frown = \wp\dot{\varepsilon}$(ε⌢𝔳, ε⌢u)
  ℘v = ℘𝔳
  𝔳 $\frown = \wp\dot{\varepsilon}$(ε⌢𝔳, ε⌢u)
  0 = ℘𝔳
  0⌢(℘v⌢⌣f) (ν′

———— • ————

Ia ⊢ a⌢v
  a⌢u
  a⌢u

(IVa): ————

⊢ (a⌢v, a⌢u) = (— a⌢u)
  a⌢u
  a⌢v
  a⌢u (ξ′

———— • ————

I ⊢ a⌢v
  a⌢u
  a⌢v
  a⌢u

×

⊢ a⌢u
  a⌢v
  a⌢u
  a⌢v (ο′

(ξ′): – – – –

⊢ (a⌢v, a⌢u) = (— a⌢u)
  a⌢v (π′

(94):: – – – – – – – – – – – – –

⊢ (a⌢v, a⌢u) = (— a⌢u)
  ℘v = 0 (ρ′

(77): ————

⊢ [— $a\frown\dot{\varepsilon}$(ε⌢v, ε⌢u)] = (— a⌢u)
  ℘v = 0 (σ′

‿

⊢ 𝔞 [— $\mathfrak{a}\frown\dot{\varepsilon}$(ε⌢v, ε⌢u)] = (— 𝔞⌢u)
  ℘v = 0 (τ′

(96): – – – – – – – – – – – – – – –

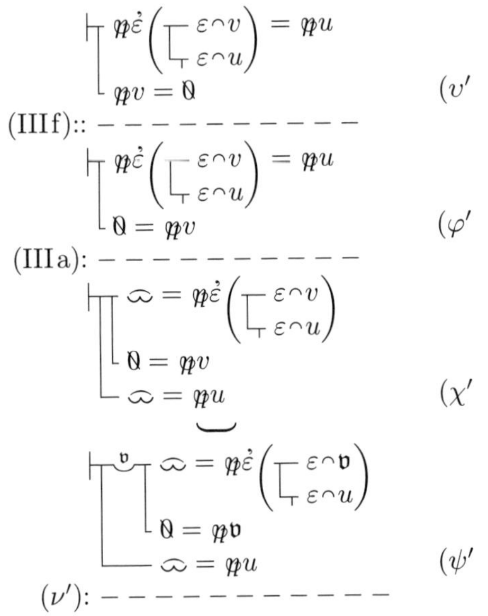

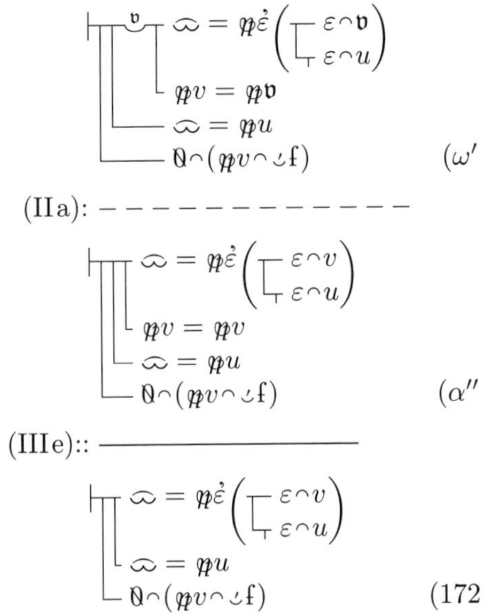

c) *Proof of the proposition*

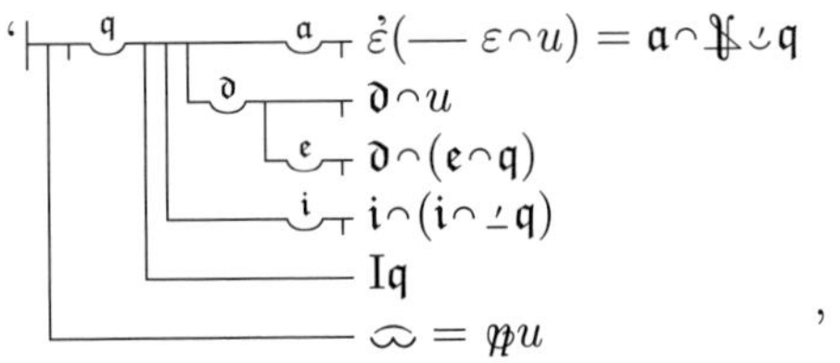

§128. *Analysis*

The proposition now to be proven can be expressed in words thus:

> "If Endlos is the cardinal number of a concept, then the objects falling under it can be ordered in a non-branching series which starts with a specific object and, without returning into itself, proceeds endlessly."

If ∞ is the cardinal number of the *u*-concept, then there must be a relation which maps the concept *finite cardinal number* into the *u*-concept and whose converse maps the latter into the former. Let the *p*-relation be of this kind; we now ask whether then the ($\nmid p \smile \mathfrak{f} \smile p$)-relation satisfies our requirements as series-forming if we take as initial member that to which 0 stands in the *p*-relation. Using (17), (18) and (71), we easily prove the single-valuedness of our series-forming relation. That the series proceeds without end we can derive from (156) and (8).

§129. *Construction*

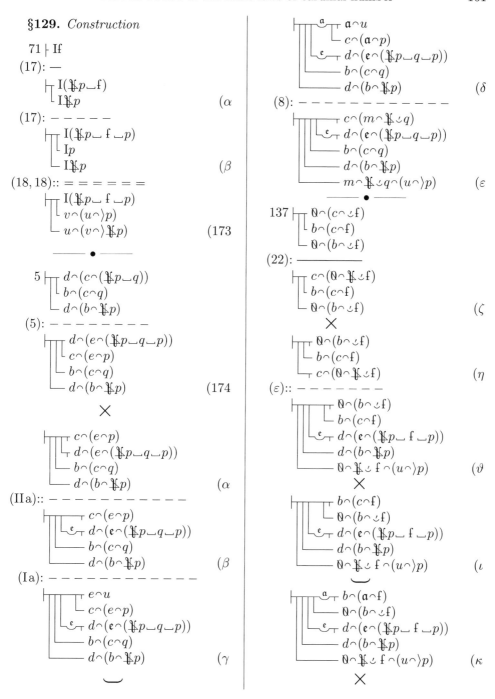

71 ⊢ If

(17): —

⊢ I(∦p⌣f)
  I∦p (α

(17): – – – – –

⊢ I(∦p⌣f⌣p)
  Ip
  I∦p (β

(18, 18):: = = = = = =

⊢ I(∦p⌣f⌣p)
  v⌢(u⌢⟩p)
  u⌢(v⌢⟩∦p) (173

——— • ———

5 ⊢ d⌢(c⌢(∦p⌣q))
  b⌢(c⌢q)
  d⌢(b⌢∦p)

(5): – – – – – – – – –

⊢ d⌢(e⌢(∦p⌣q⌣p))
  c⌢(e⌢p)
  b⌢(c⌢q)
  d⌢(b⌢∦p) (174

×

⊢ c⌢(e⌢p)
  d⌢(e⌢(∦p⌣q⌣p))
  b⌢(c⌢q)
  d⌢(b⌢∦p) (α

(IIa):: – – – – – – – – – – – –

⊢ c⌢(e⌢p)
  ⌣ e d⌢(e⌢(∦p⌣q⌣p))
  b⌢(c⌢q)
  d⌢(b⌢∦p) (β

(Ia): – – – – – – – – – – – – –

⊢ e⌢u
  c⌢(e⌢p)
  ⌣ e d⌢(e⌢(∦p⌣q⌣p))
  b⌢(c⌢q)
  d⌢(b⌢∦p) (γ

⏟

⊢ ⌣ a a⌢u
  c⌢(a⌢p)
  ⌣ e d⌢(e⌢(∦p⌣q⌣p))
  b⌢(c⌢q)
  d⌢(b⌢∦p) (δ

(8): – – – – – – – – – – – – –

⊢ c⌢(m⌢∦⌣q)
  ⌣ e d⌢(e⌢(∦p⌣q⌣p))
  b⌢(c⌢q)
  d⌢(b⌢∦p)
  m⌢∦⌣q⌢(u⌢⟩p) (ε

——— • ———

137 ⊢ Ø⌢(c⌢⌣f)
  b⌢(c⌢f)
  Ø⌢(b⌢⌣f)

(22): ————

⊢ c⌢(Ø⌢∦⌣f)
  b⌢(c⌢f)
  Ø⌢(b⌢⌣f) (ζ

×

⊢ Ø⌢(b⌢⌣f)
  b⌢(c⌢f)
  c⌢(Ø⌢∦⌣f) (η

(ε):: – – – – – – – –

⊢ Ø⌢(b⌢⌣f)
  b⌢(c⌢f)
  ⌣ e d⌢(e⌢(∦p⌣f⌣p))
  d⌢(b⌢∦p)
  Ø⌢∦⌣f⌢(u⌢⟩p) (ϑ

×

⊢ b⌢(c⌢f)
  Ø⌢(b⌢⌣f)
  ⌣ e d⌢(e⌢(∦p⌣f⌣p))
  d⌢(b⌢∦p)
  Ø⌢∦⌣f⌢(u⌢⟩p) (ι

⏟

⊢ ⌣ a b⌢(a⌢f)
  Ø⌢(b⌢⌣f)
  ⌣ e d⌢(e⌢(∦p⌣f⌣p))
  d⌢(b⌢∦p)
  Ø⌢∦⌣f⌢(u⌢⟩p) (κ

×

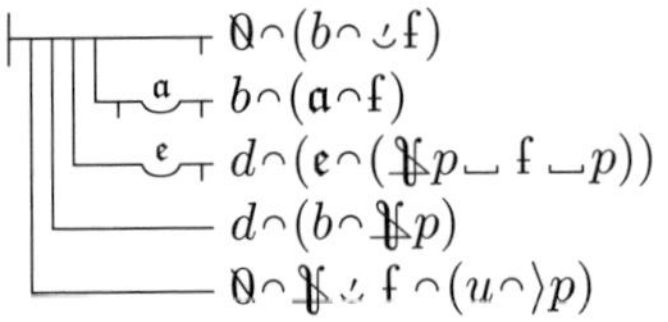

(λ

(156): ·—·—·—·—·—·—·—·—·—·—·—·—·

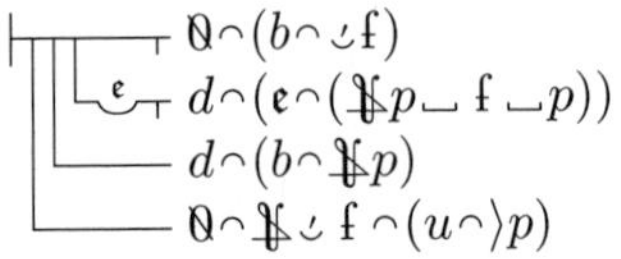

(μ

(22): ————————————————

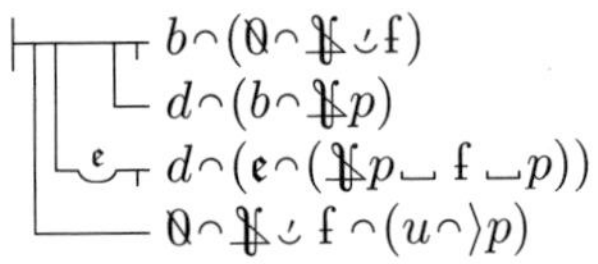

(ν

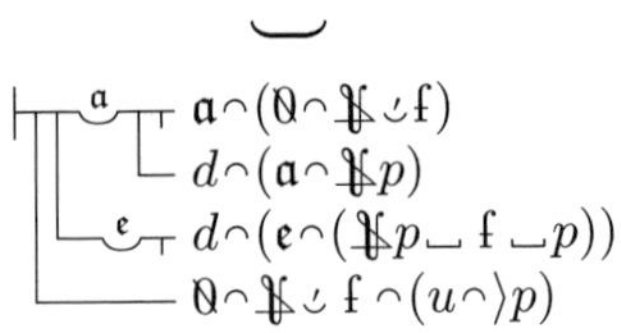

(ξ

(8): — — — — — — — — — — — — —

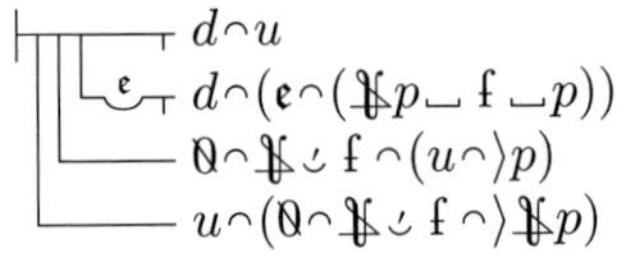

(175

§130. *Analysis*

It is not provable that no object follows after itself in the ($\not\!\!\!\!\, p$— f —p)-series but only that no object falling under the u-concept follows after itself in this series, if the u-concept can be mapped by means of the $\not\!\!\!\!\, p$-relation into the concept *finite cardinal number.* For now, we make do with such a series, so that, using our ($\not\!\!\!\!\, p$—f—p)-relation, we can then define another one which will share the properties here relevant with the former but will have, in addition, the property that no object follows after itself in its series.

We prove the proposition

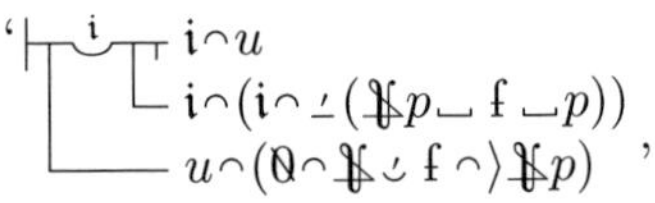

from the propositions

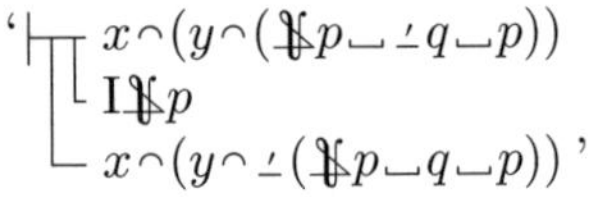

(α

and

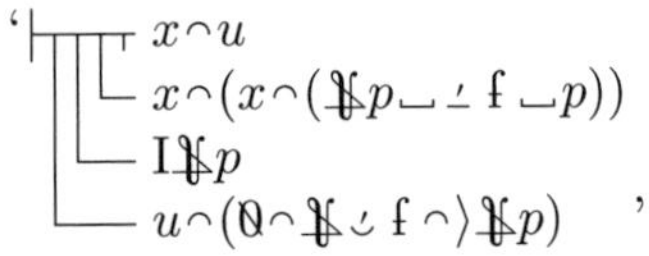

(β

The former we prove with (123), and for this we require the proposition

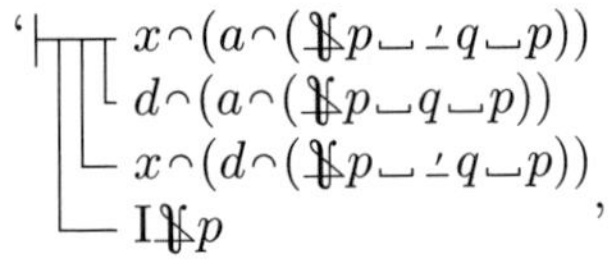

(γ

which can be derived from the more general proposition

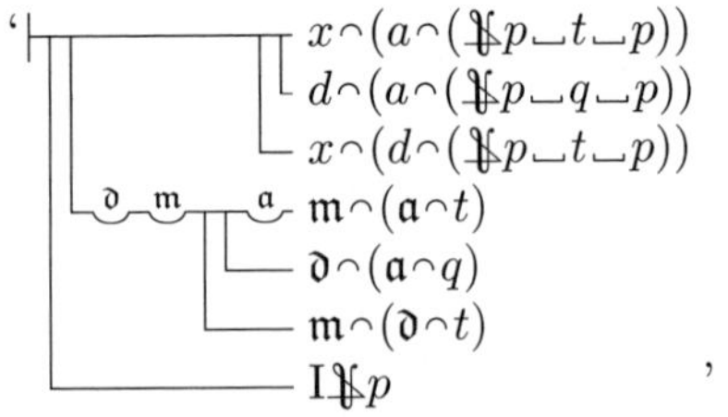

(δ

For the proof of (δ) we will draw on the objects—say, m, b, c—which stand to x, d, a in the p-relation. That there are such objects is to be shown by means of (15). Here, b occurs twice: first, by m's standing to it in the t-relation, and second, by standing in the q-relation to c. The following diagram may help make things more surveyable:

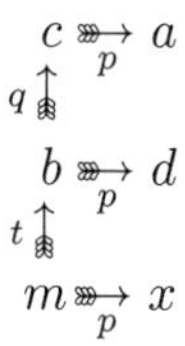

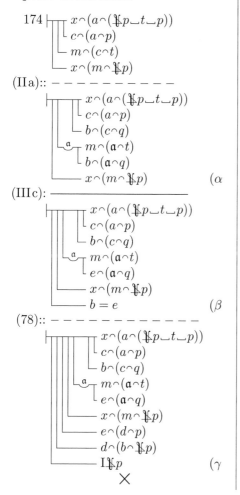

From the single-valuedness of the converse of the p-relation we must conclude that there is only a single object of this kind that can come into consideration.

§131. *Construction*

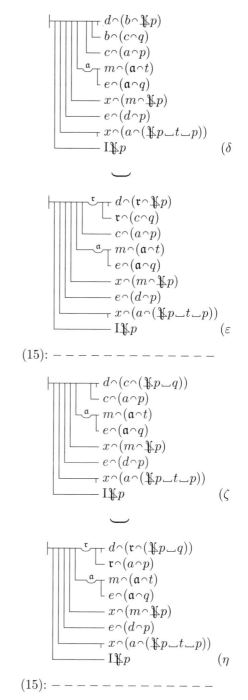

$$174 \vdash \begin{array}{l} x\frown(a\frown(\between p\smile t\smile p)) \\ c\frown(a\frown p) \\ m\frown(c\frown t) \\ x\frown(m\frown\between p) \end{array}$$

(IIa):: - - - - - - - - - - -

$$\vdash \begin{array}{l} x\frown(a\frown(\between p\smile t\smile p)) \\ c\frown(a\frown p) \\ b\frown(c\frown q) \\ \overset{\mathfrak{a}}{\smile}\; m\frown(\mathfrak{a}\frown t) \\ \quad b\frown(\mathfrak{a}\frown q) \\ x\frown(m\frown\between p) \end{array} \qquad (\alpha$$

(IIIc): ————————

$$\vdash \begin{array}{l} x\frown(a\frown(\between p\smile t\smile p)) \\ c\frown(a\frown p) \\ b\frown(c\frown q) \\ \overset{\mathfrak{a}}{\smile}\; m\frown(\mathfrak{a}\frown t) \\ \quad e\frown(\mathfrak{a}\frown q) \\ x\frown(m\frown\between p) \\ b = e \end{array} \qquad (\beta$$

(78):: - - - - - - - - - - - - -

$$\vdash \begin{array}{l} x\frown(a\frown(\between p\smile t\smile p)) \\ c\frown(a\frown p) \\ b\frown(c\frown q) \\ \overset{\mathfrak{a}}{\smile}\; m\frown(\mathfrak{a}\frown t) \\ \quad e\frown(\mathfrak{a}\frown q) \\ x\frown(m\frown\between p) \\ e\frown(d\frown p) \\ d\frown(b\frown\between p) \\ \mathrm{I}\between p \end{array} \qquad (\gamma$$

×

$$\vdash \begin{array}{l} d\frown(b\frown\between p) \\ b\frown(c\frown q) \\ c\frown(a\frown p) \\ \overset{\mathfrak{a}}{\smile}\; m\frown(\mathfrak{a}\frown t) \\ \quad e\frown(\mathfrak{a}\frown q) \\ x\frown(m\frown\between p) \\ e\frown(d\frown p) \\ x\frown(a\frown(\between p\smile t\smile p)) \\ \mathrm{I}\between p \end{array} \qquad (\delta$$

$\smile$

$$\vdash \begin{array}{l} \overset{\mathfrak{r}}{\smile}\; d\frown(\mathfrak{r}\frown\between p) \\ \quad \mathfrak{r}\frown(c\frown q) \\ c\frown(a\frown p) \\ \overset{\mathfrak{a}}{\smile}\; m\frown(\mathfrak{a}\frown t) \\ \quad e\frown(\mathfrak{a}\frown q) \\ x\frown(m\frown\between p) \\ e\frown(d\frown p) \\ x\frown(a\frown(\between p\smile t\smile p)) \\ \mathrm{I}\between p \end{array} \qquad (\varepsilon$$

(15): - - - - - - - - - - - - - -

$$\vdash \begin{array}{l} d\frown(c\frown(\between p\smile q)) \\ c\frown(a\frown p) \\ \overset{\mathfrak{a}}{\smile}\; m\frown(\mathfrak{a}\frown t) \\ \quad e\frown(\mathfrak{a}\frown q) \\ x\frown(m\frown\between p) \\ e\frown(d\frown p) \\ x\frown(a\frown(\between p\smile t\smile p)) \\ \mathrm{I}\between p \end{array} \qquad (\zeta$$

$\smile$

$$\vdash \begin{array}{l} \overset{\mathfrak{r}}{\smile}\; d\frown(\mathfrak{r}\frown(\between p\smile q)) \\ \quad \mathfrak{r}\frown(a\frown p) \\ \overset{\mathfrak{a}}{\smile}\; m\frown(\mathfrak{a}\frown t) \\ \quad e\frown(\mathfrak{a}\frown q) \\ x\frown(m\frown\between p) \\ e\frown(d\frown p) \\ x\frown(a\frown(\between p\smile t\smile p)) \\ \mathrm{I}\between p \end{array} \qquad (\eta$$

(15): - - - - - - - - - - - - - -

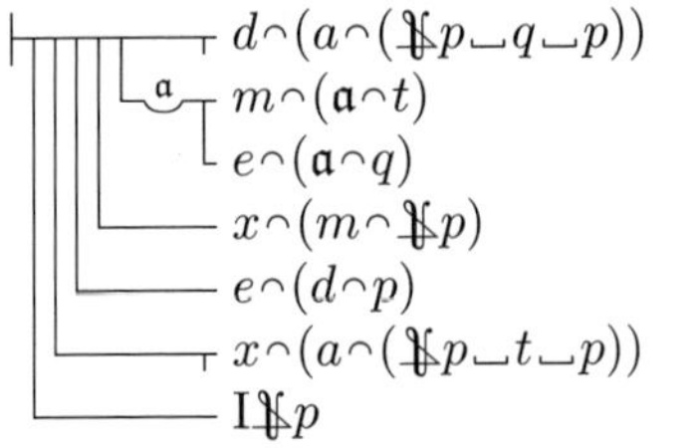

(ϑ

×

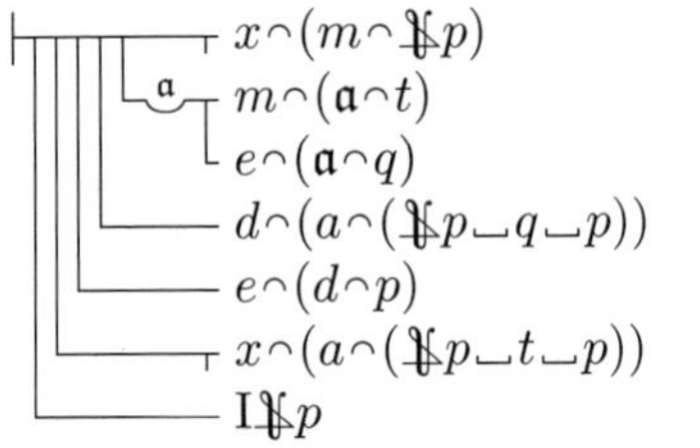

(ι

(IIa):: - - - - - - - - - - - - - -

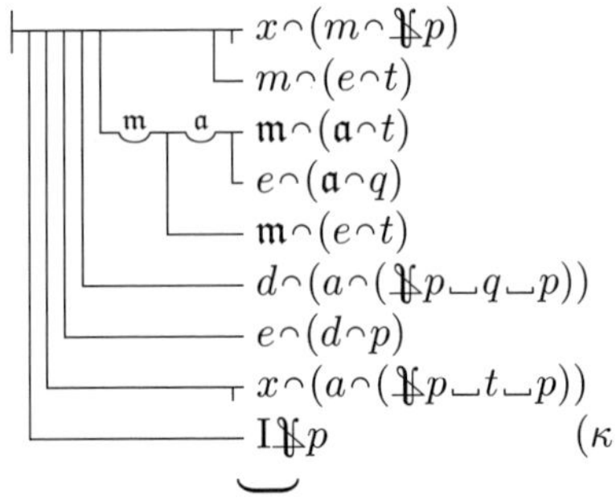

(κ

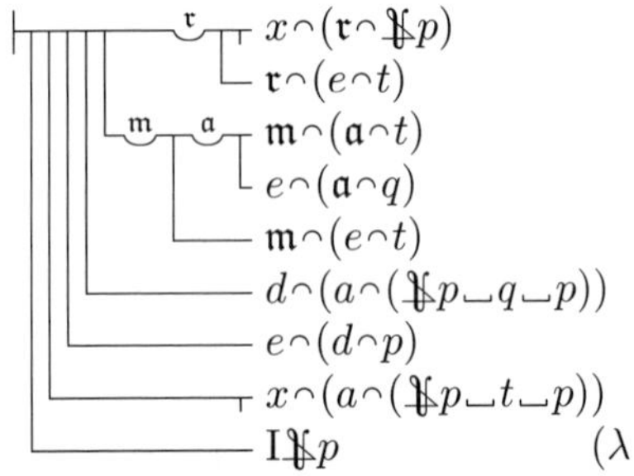

(λ

(15): - - - - - - - - - - - - - - -

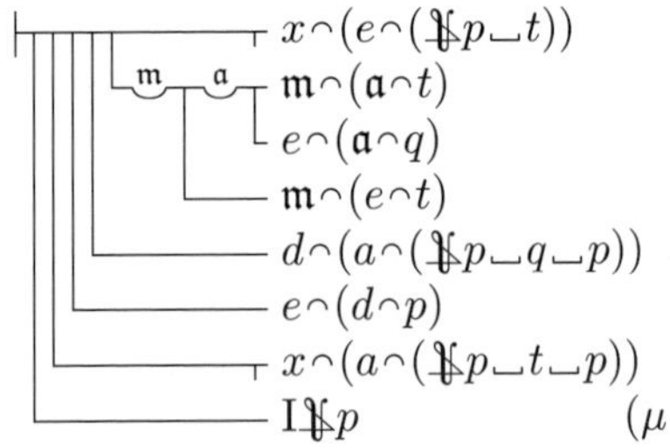

(μ

(IIa):: - - - - - - - - - - - - - - -

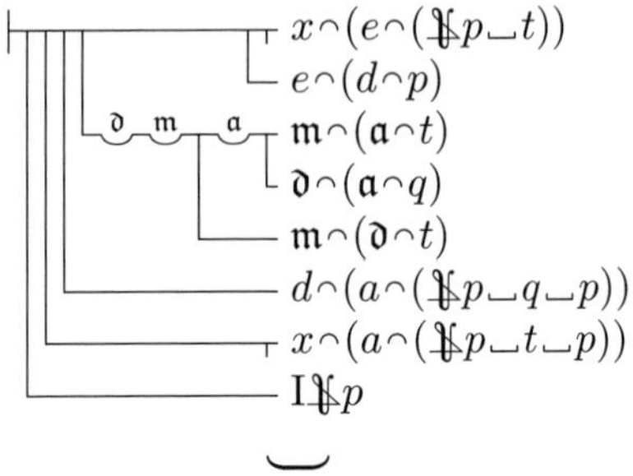

(ν

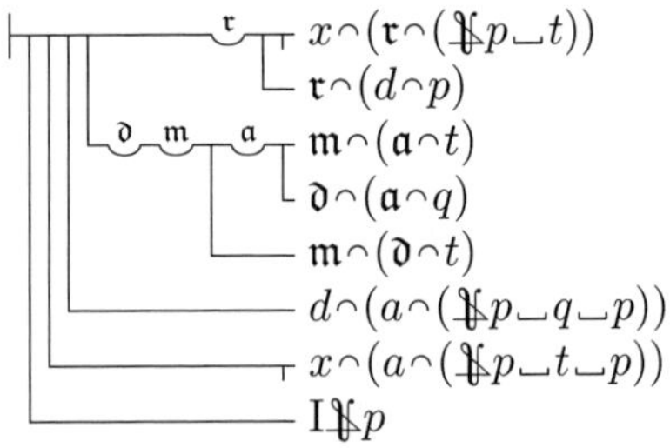

(ξ

(15): - - - - - - - - - - - - - -

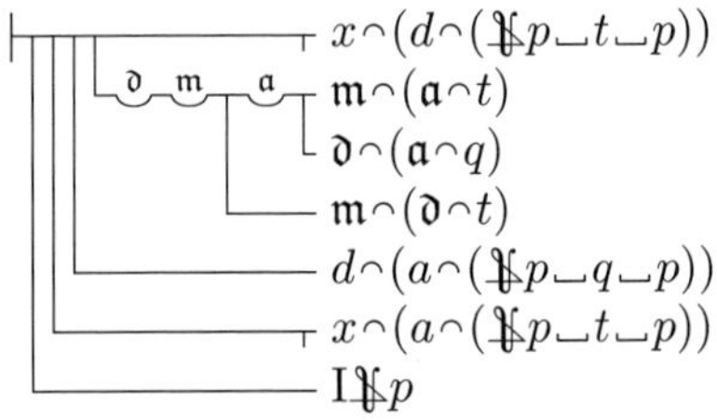

(o

×

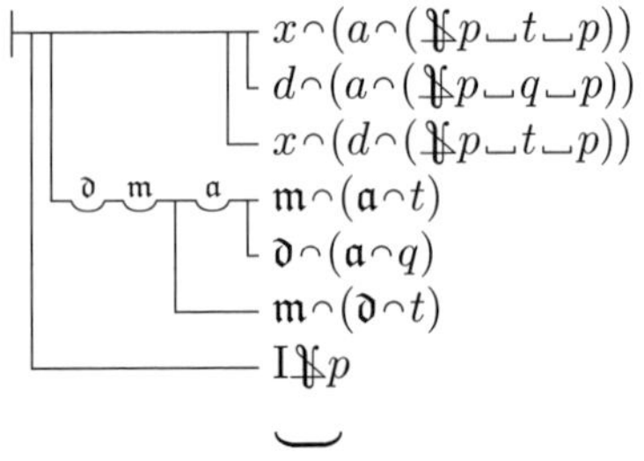

(π

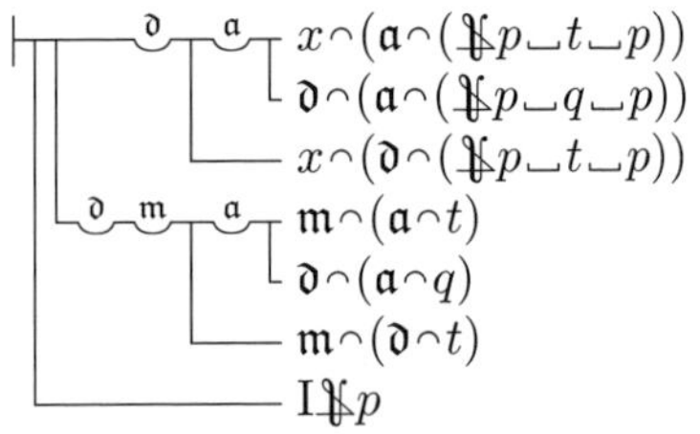

(176

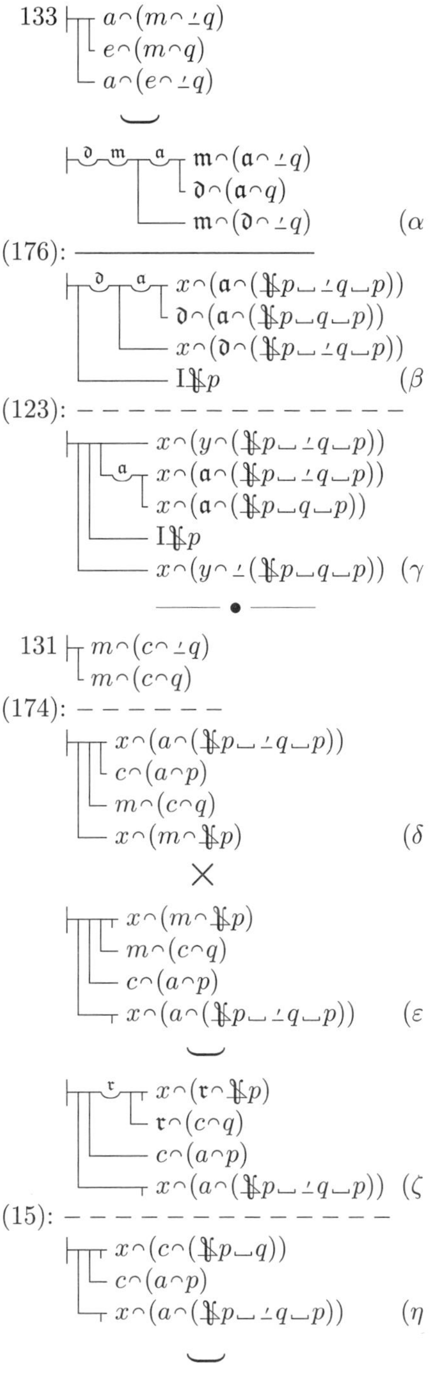

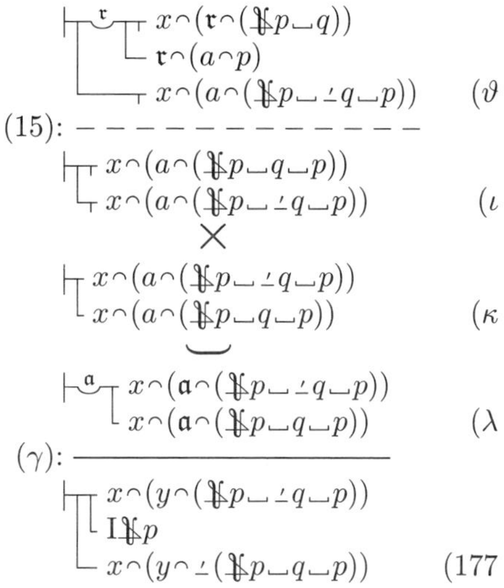

§132. *Analysis*

We will now prove the proposition (β) of §130 by inferring from the single-valuedness of the $\nparallel p$-relation that there is only one object which stands in this relation to x, while if x were to stand to itself in the ($\nparallel p$ ⌣ ⊥ f ⌣ p)-relation, there would have to be, according to (15), at least one such object which followed after itself in the cardinal number series and which thus, according to (145), could not be a finite cardinal number. From this it follows, in accordance with (8), that x could not fall under the u-concept if the u-concept is mapped by the $\nparallel p$-relation into the concept *finite cardinal number.*

§133. *Construction*

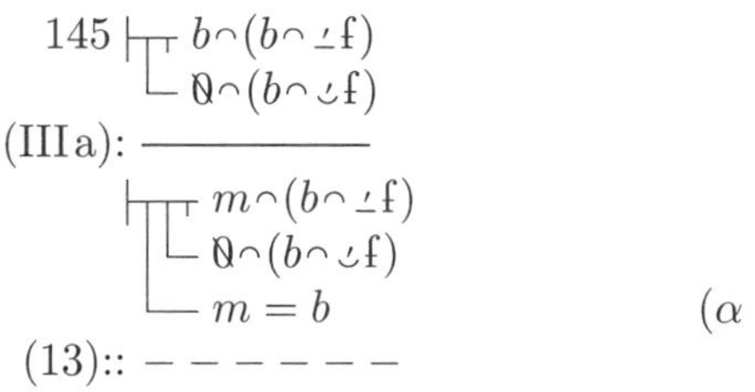

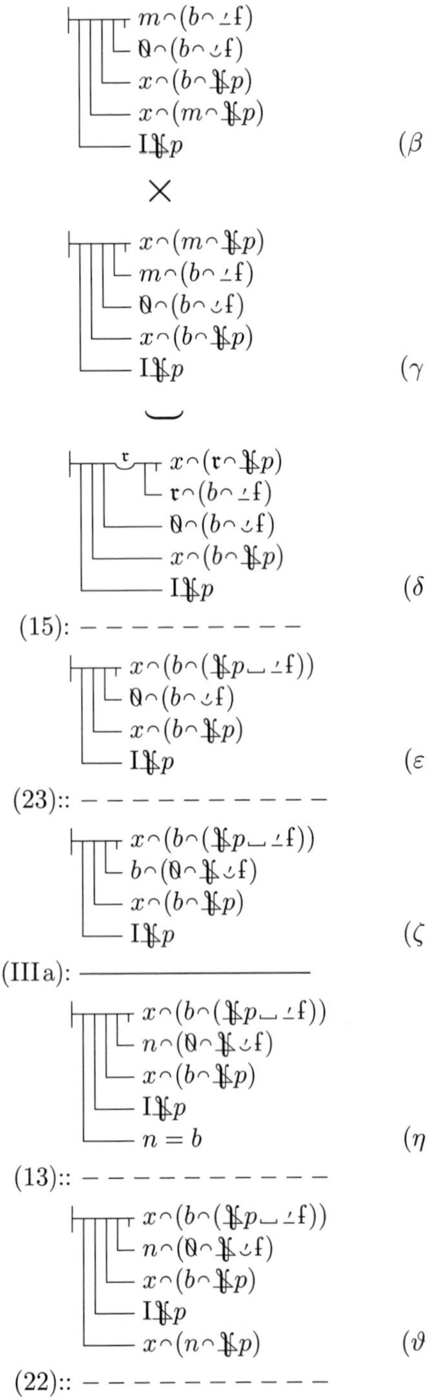
(β
×
(γ
(δ
(15):
(ε
(23)::
(ζ
(IIIa):
(η
(13)::
(ϑ
(22)::

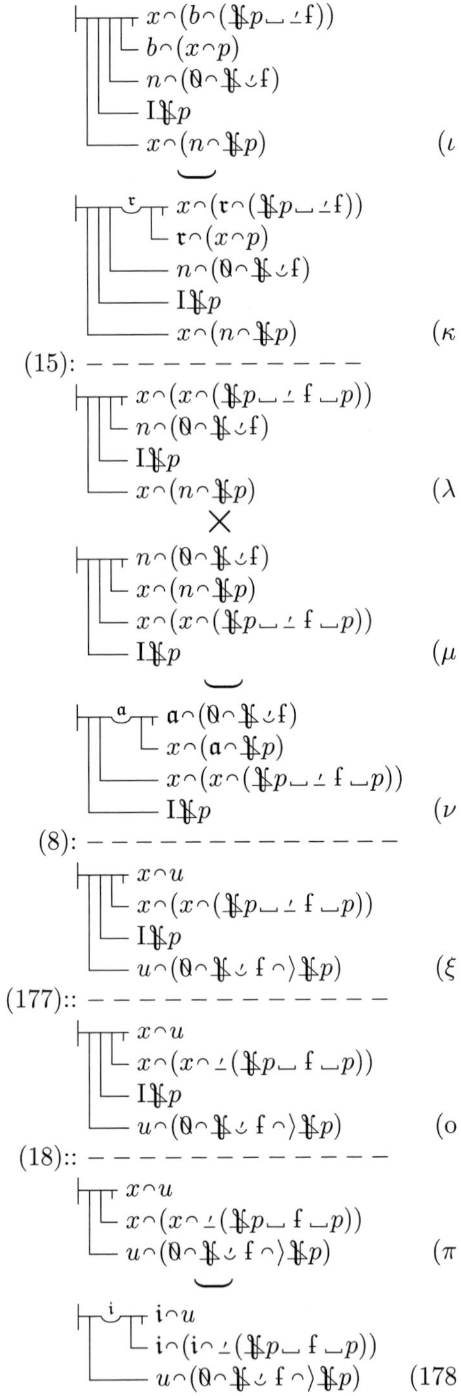
(ι
(κ
(15):
(λ
×
(μ
(ν
(8):
(ξ
(177)::
(o
(18)::
(π
(178

§134. *Analysis*

It now only remains to show that all the members of our series fall under the u-concept and conversely that all objects falling under the u-concept are members of our series. These are the two propositions

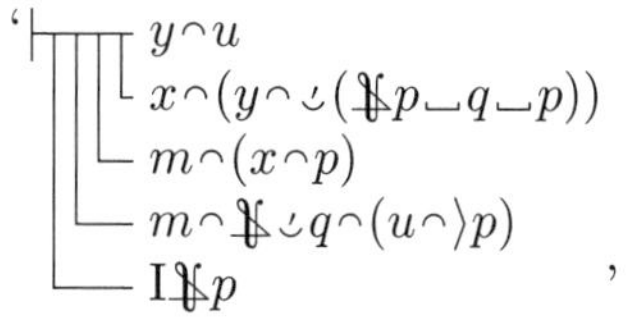
(α

and

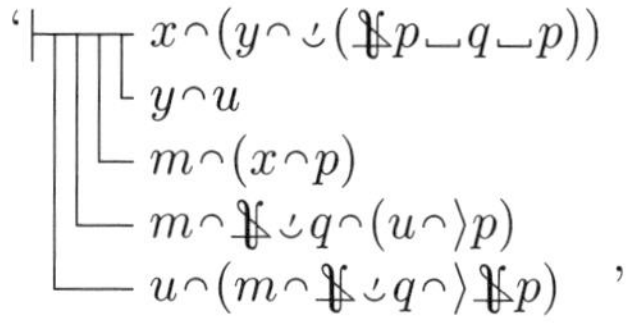
(β

where the q-series starting with m is taken for the cardinal number series more generally. We prove (α) from the propositions

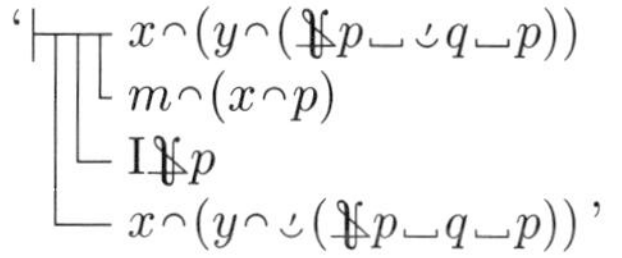
(γ

and

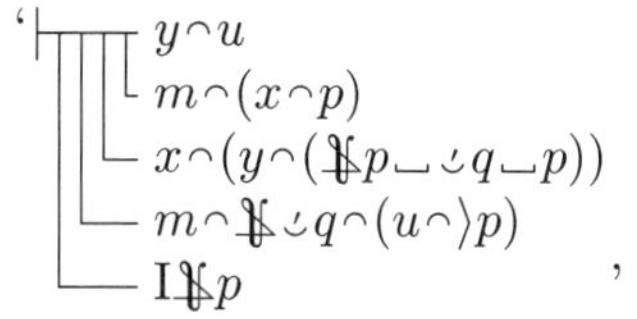
(δ

of which (γ) is easily deduced similarly to (177). In order to prove (δ) we infer from the single-valuedness of the $\not\!\!\!\!\smallsmile p$-relation that there is only one object which stands to x in the $\not\!\!\!\!\smallsmile p$-relation, and we infer from x's standing to y in the ($\not\!\!\!\!\smallsmile p \_ \smallsmile q \_ p$)-relation that there is such an object that belongs to a q-series ending with an object n which stands to y in the p-relation. Thus if an object m stands in the $\not\!\!\!\!\smallsmile p$-relation to x, it will also belong to the q-series ending with n. We will further prove the proposition

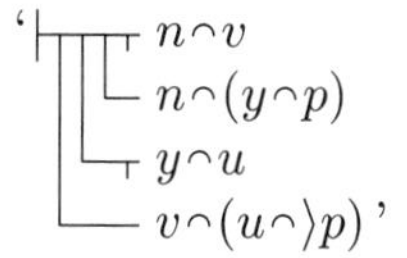

and arrive at our goal by taking the ($m \frown \not\!\!\!\!\smallsmile \smallsmile q$)-concept here as the v-concept.

§135. *Construction*

IIIc 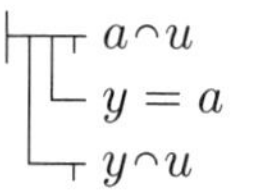

(13)::

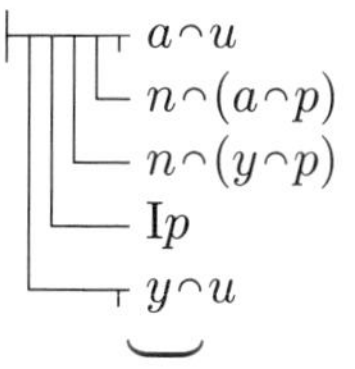
(α

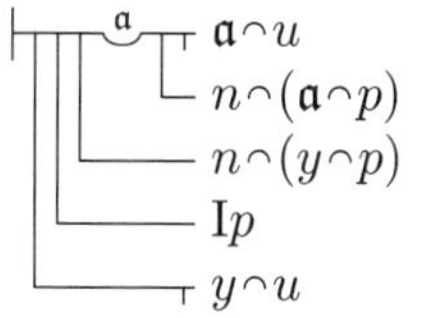
(β

(8):

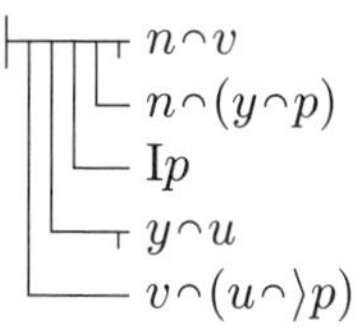
(γ

(18)::

⊢ n⌢v
  n⌢(y⌢p)
  y⌢u
  v⌢(u⌢⟩p) (179

———•———

22 ⊢ n⌢(m⌢⅌⌣q)
  m⌢(n⌢⌣q)

(IIIa): ————————

⊢ n⌢(m⌢⅌⌣q)
  a⌢(n⌢⌣q)
  a = m (α

(13):: - - - - - - -

⊢ n⌢(m⌢⅌⌣q)
  a⌢(n⌢⌣q)
  x⌢(m⌢⅌p)
  x⌢(a⌢⅌p)
  I⅌p (β

×

⊢ x⌢(a⌢⅌p)
  a⌢(n⌢⌣q)
  x⌢(m⌢⅌p)
  n⌢(m⌢⅌⌣q)
  I⅌p (γ

⊢ 𝔯: x⌢(𝔯⌢⅌p)
  𝔯⌢(n⌢⌣q)
  x⌢(m⌢⅌p)
  n⌢(m⌢⅌⌣q)
  I⅌p (δ

(15): - - - - - - - - - - -

⊢ x⌢(n⌢(⅌p⌴⌣q))
  x⌢(m⌢⅌p)
  n⌢(m⌢⅌⌣q)
  I⅌p (ε

(179):: - - - - - - - - - - -

⊢ x⌢(n⌢(⅌p⌴⌣q))
  n⌢(y⌢p)
  x⌢(m⌢⅌p)
  y⌢u
  m⌢⅌⌣q⌢(u⌢⟩p)
  I⅌p (ζ

⊢ 𝔯: x⌢(𝔯⌢(⅌p⌴⌣q))
  𝔯⌢(y⌢p)
  x⌢(m⌢⅌p)
  y⌢u
  m⌢⅌⌣q⌢(u⌢⟩p)
  I⅌p (η

(15): - - - - - - - - - - - - -

⊢ x⌢(y⌢(⅌p⌴⌣q⌴p))
  x⌢(m⌢⅌p)
  y⌢u
  m⌢⅌⌣q⌢(u⌢⟩p)
  I⅌p (ϑ

×

⊢ y⌢u
  x⌢(m⌢⅌p)
  x⌢(y⌢(⅌p⌴⌣q⌴p))
  m⌢⅌⌣q⌢(u⌢⟩p)
  I⅌p (ι

(22):: - - - - - - - - - - - -

⊢ y⌢u
  m⌢(x⌢p)
  x⌢(y⌢(⅌p⌴⌣q⌴p))
  m⌢⅌⌣q⌢(u⌢⟩p)
  I⅌p (179a^{d}

———•———

137 ⊢ a⌢(m⌢⌣q)
  e⌢(m⌢q)
  a⌢(e⌢⌣q)

⊢ 𝔡 𝔪 𝔞: 𝔪⌢(𝔞⌢⌣q)
  𝔡⌢(𝔞⌢q)
  𝔪⌢(𝔡⌢⌣q) (λ

(176): ————————————

⊢ 𝔡 𝔞: x⌢(𝔞⌢(⅌p⌴⌣q⌴p))
  𝔡⌢(𝔞⌢(⅌p⌴q⌴p))
  x⌢(𝔡⌢(⅌p⌴⌣q⌴p))
  I⅌p (μ

(144): - - - - - - - - - - - -

⊢ x⌢(y⌢(⅌p⌴⌣q⌴p))
  x⌢(x⌢(⅌p⌴⌣q⌴p))
  I⅌p
  x⌢(y⌢⌣(⅌p⌴q⌴p)) (ν

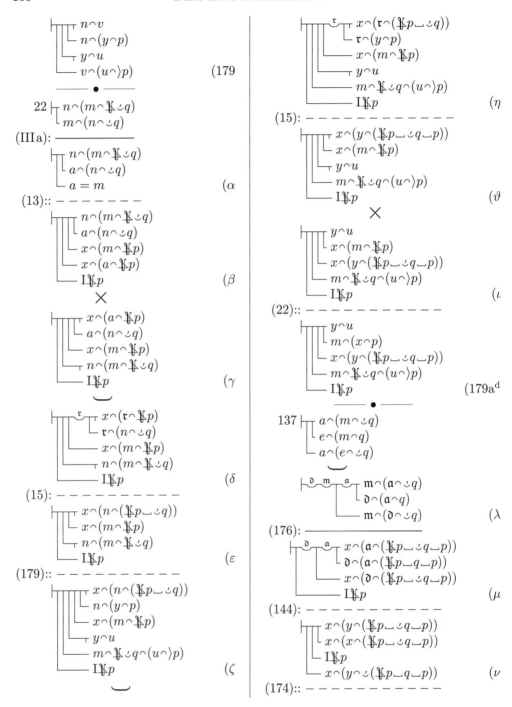

(174):: - - - - - - - - - - - -

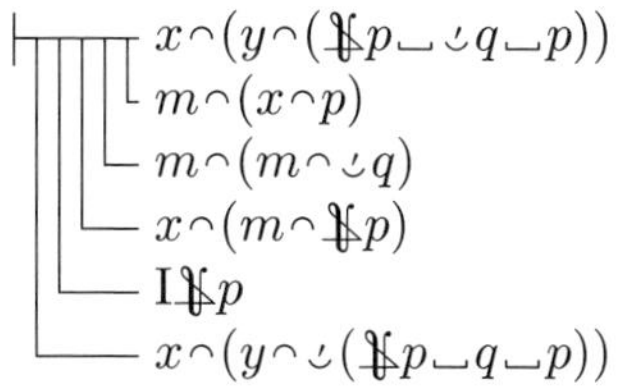

$$\begin{array}{l}
\vdash x\frown(y\frown(\between p\smile\underset{\cdot}{\smile}q\smile p)) \\
\quad m\frown(x\frown p) \\
\quad m\frown(m\frown\underset{\cdot}{\smile}q) \\
\quad x\frown(m\frown\between p) \\
\quad \mathrm{I}\between p \\
\quad x\frown(y\frown\underset{\cdot}{\smile}(\between p\smile q\smile p))
\end{array} \qquad (\xi$$

(22, 140):: ═══════════

$$\begin{array}{l}
\vdash x\frown(y\frown(\between p\smile\underset{\cdot}{\smile}q\smile p)) \\
\quad m\frown(x\frown p) \\
\quad \mathrm{I}\between p \\
\quad x\frown(y\frown\underset{\cdot}{\smile}(\between p\smile q\smile p))
\end{array} \qquad (180$$

(179a)[d]: – – – – – – – – – –

$$\begin{array}{l}
\vdash y\frown u \\
\quad x\frown(y\frown\underset{\cdot}{\smile}(\between p\smile q\smile p)) \\
\quad m\frown(x\frown p) \\
\quad m\frown\between\underset{\cdot}{\smile}q\frown(u\frown\rangle p) \\
\quad \mathrm{I}\between p
\end{array} \qquad (\alpha$$

(18):: – – – – – – – – – – – –

$$\begin{array}{l}
\vdash y\frown u \\
\quad x\frown(y\frown\underset{\cdot}{\smile}(\between p\smile q\smile p)) \\
\quad m\frown(x\frown p) \\
\quad m\frown\between\underset{\cdot}{\smile}q\frown(u\frown\rangle p) \\
\quad u\frown(m\frown\between\underset{\cdot}{\smile}q\frown\rangle\between p)
\end{array} \qquad (181$$

§136. *Analysis*

We now have to prove the proposition (β) of §134. From the u-concept's being mapped into the ($m\frown\between\underset{\cdot}{\smile}q$)-concept by means of the $\between p$-relation, and y's falling under the u-concept, we can infer that there is an object (n) to which y stands in the $\between p$-relation and which falls under the ($m\frown\between\underset{\cdot}{\smile}q$)-concept, i.e., it belongs to a q-series beginning with m. We now prove the proposition

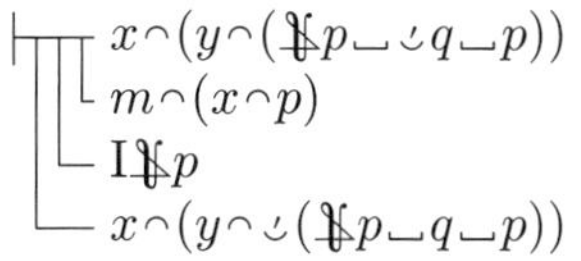

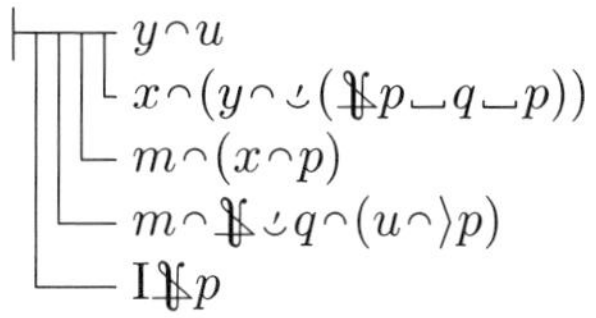

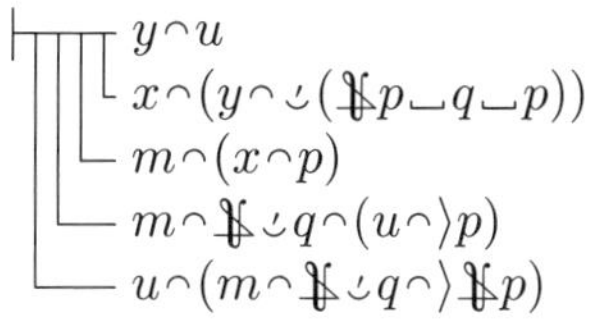

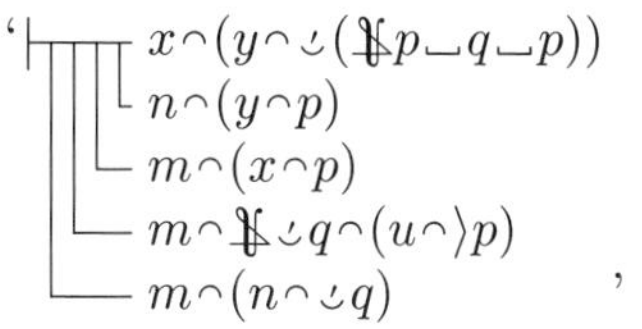

$$\begin{array}{l}
\text{‘}\vdash x\frown(y\frown\underset{\cdot}{\smile}(\between p\smile q\smile p)) \\
\quad n\frown(y\frown p) \\
\quad m\frown(x\frown p) \\
\quad m\frown\between\underset{\cdot}{\smile}q\frown(u\frown\rangle p) \\
\quad m\frown(n\frown\underset{\cdot}{\smile}q)\text{’}
\end{array} \qquad (\alpha$$

with (152). We require for this the proposition

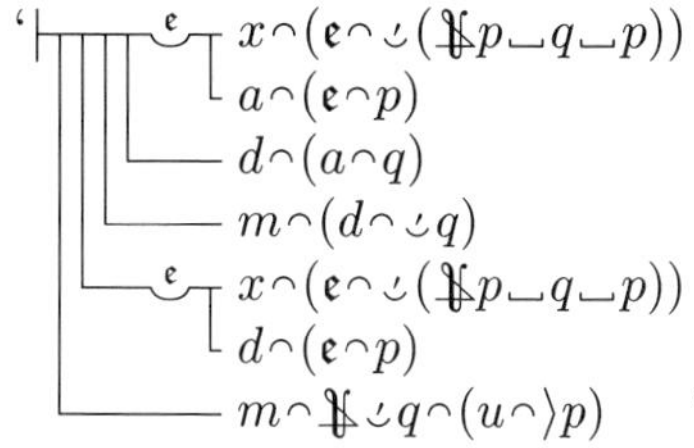

$$\begin{array}{l}
\text{‘}\vdash \forall\mathfrak{e}\; x\frown(\mathfrak{e}\frown\underset{\cdot}{\smile}(\between p\smile q\smile p)) \\
\qquad a\frown(\mathfrak{e}\frown p) \\
\quad d\frown(a\frown q) \\
\quad m\frown(d\frown\underset{\cdot}{\smile}q) \\
\quad \forall\mathfrak{e}\; x\frown(\mathfrak{e}\frown\underset{\cdot}{\smile}(\between p\smile q\smile p)) \\
\qquad d\frown(\mathfrak{e}\frown p) \\
\quad m\frown\between\underset{\cdot}{\smile}q\frown(u\frown\rangle p)\text{’}
\end{array} \qquad (\beta$$

From the ($m\frown\between\underset{\cdot}{\smile}q$)-concept's being mapped into the u-concept by the p-relation, we infer that there is an object (e) to which d stands in the p-relation if d belongs to the q-series starting with m. From this and the proposition

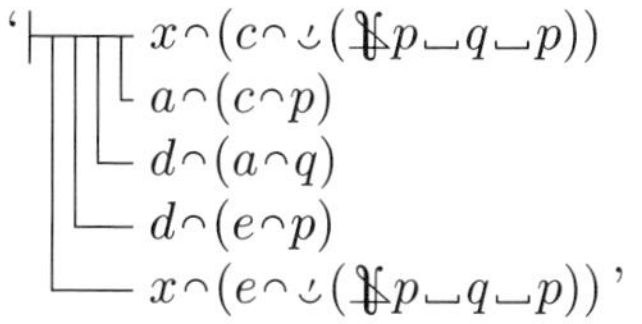

$$\begin{array}{l}
\text{‘}\vdash x\frown(c\frown\underset{\cdot}{\smile}(\between p\smile q\smile p)) \\
\quad a\frown(c\frown p) \\
\quad d\frown(a\frown q) \\
\quad d\frown(e\frown p) \\
\quad x\frown(e\frown\underset{\cdot}{\smile}(\between p\smile q\smile p))\text{’}
\end{array} \qquad (\gamma$$

(β) easily follows.

§137. *Construction*

$$\begin{array}{l}
139 \vdash x\frown(e\frown\underset{\cdot}{\smile}(\between p\smile q\smile p)) \\
\quad e = x
\end{array}$$

(13):: – – – – – – – – – – –

$$\begin{array}{l}
\vdash x\frown(e\frown\underset{\cdot}{\smile}(\between p\smile q\smile p)) \\
\quad m\frown(e\frown p) \\
\quad m\frown(x\frown p) \\
\quad \mathrm{I}p
\end{array} \qquad (\alpha$$

$\smile$

$$\begin{array}{l}
\vdash \forall\mathfrak{e}\; x\frown(\mathfrak{e}\frown\underset{\cdot}{\smile}(\between p\smile q\smile p)) \\
\qquad m\frown(\mathfrak{e}\frown p) \\
\quad m\frown(x\frown p) \\
\quad \mathrm{I}p
\end{array} \qquad (\beta$$

(18):: – – – – – – – – – – – –

$$\begin{array}{l}
\vdash \forall\mathfrak{e}\; x\frown(\mathfrak{e}\frown\underset{\cdot}{\smile}(\between p\smile q\smile p)) \\
\qquad m\frown(\mathfrak{e}\frown p) \\
\quad m\frown(x\frown p) \\
\quad v\frown(u\frown\rangle p)
\end{array} \qquad (182$$

———— • ————

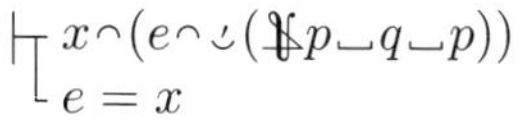

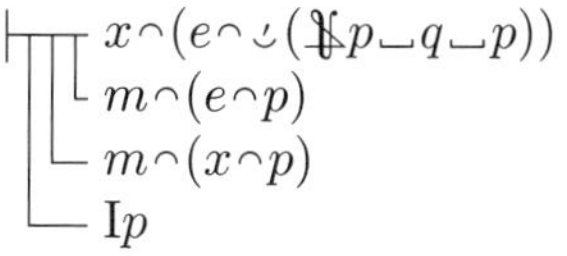

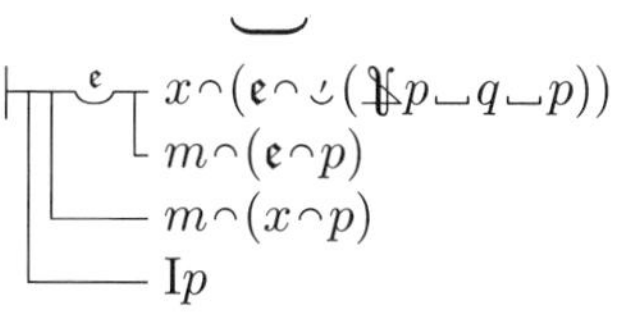

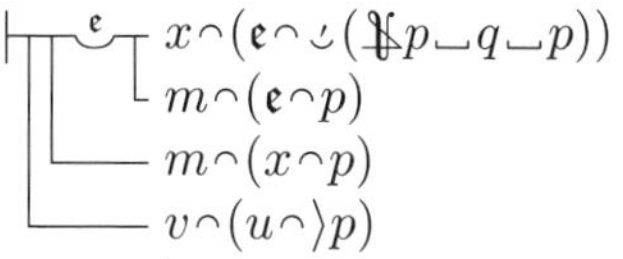

IIa 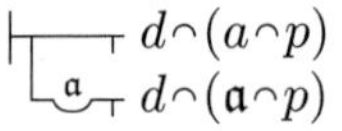

(Ia): - - - - - - - -

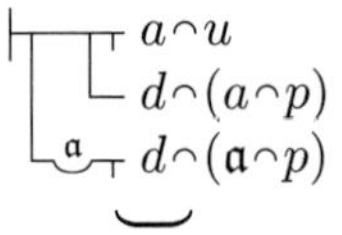 (α

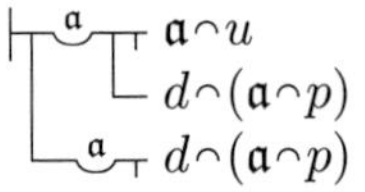 (β

(8): - - - - - - - -

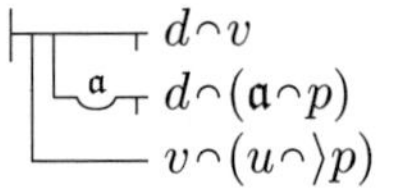 (183

•

174 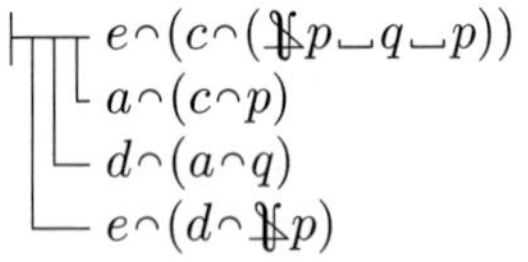

(22):: - - - - - - - - - - -

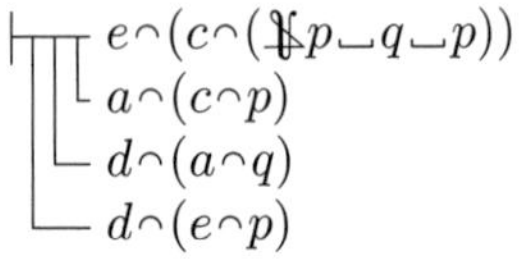 (184

(137): - - - - - - - - - - -

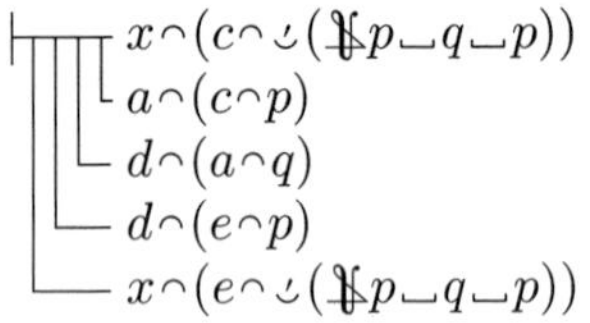 (α

(IIa):: - - - - - - - - - - - -

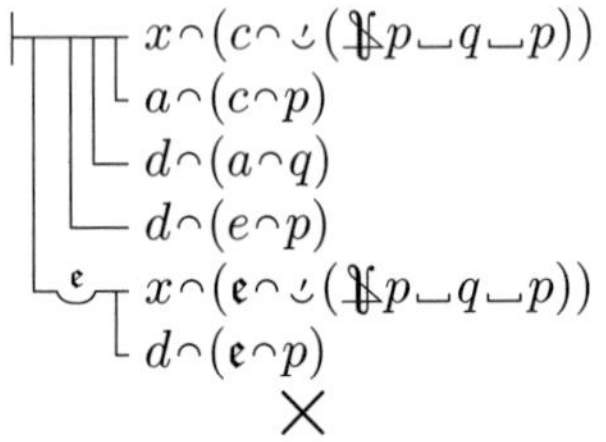 (β

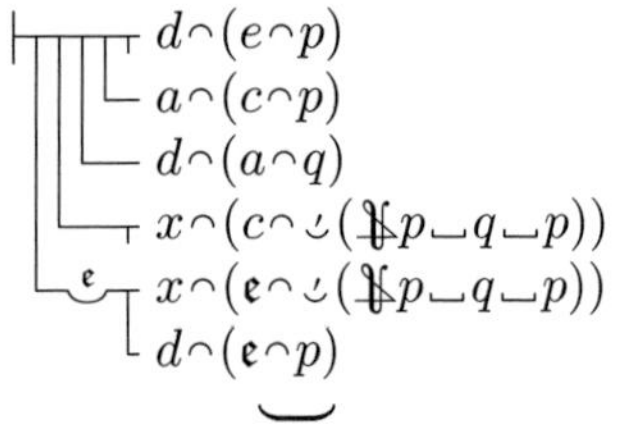 (γ

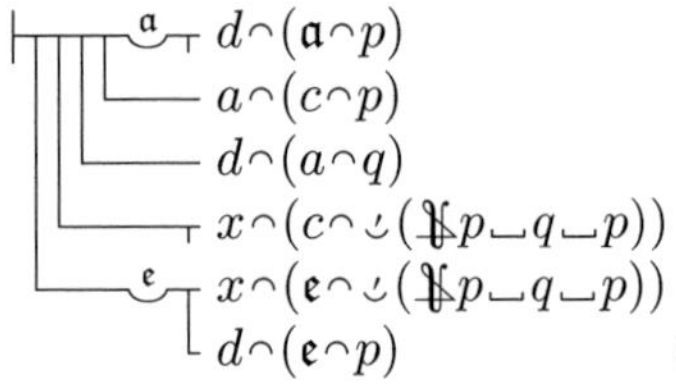 (185

(183): - - - - - - - - - - - - - -

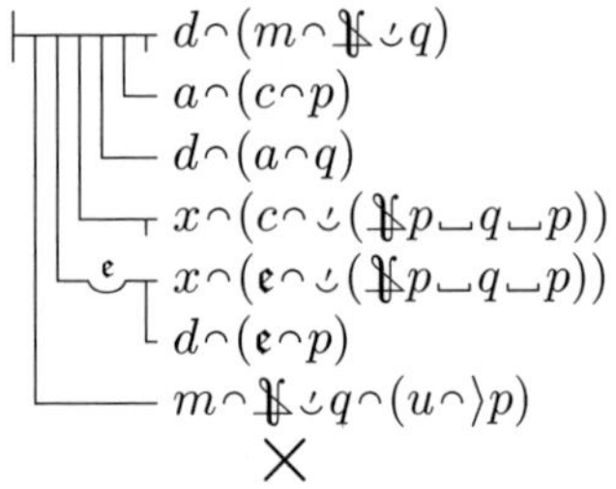 (α

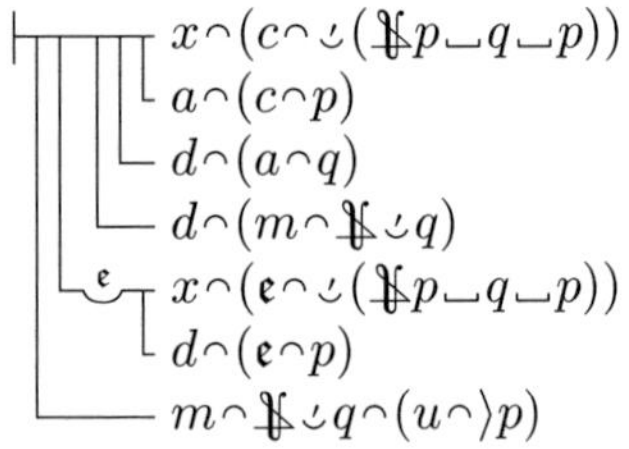 (β

(22):: - - - - - - - - - - - - - -

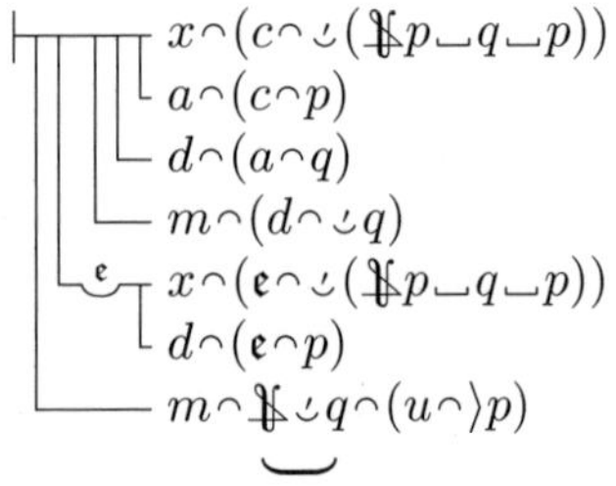 (γ

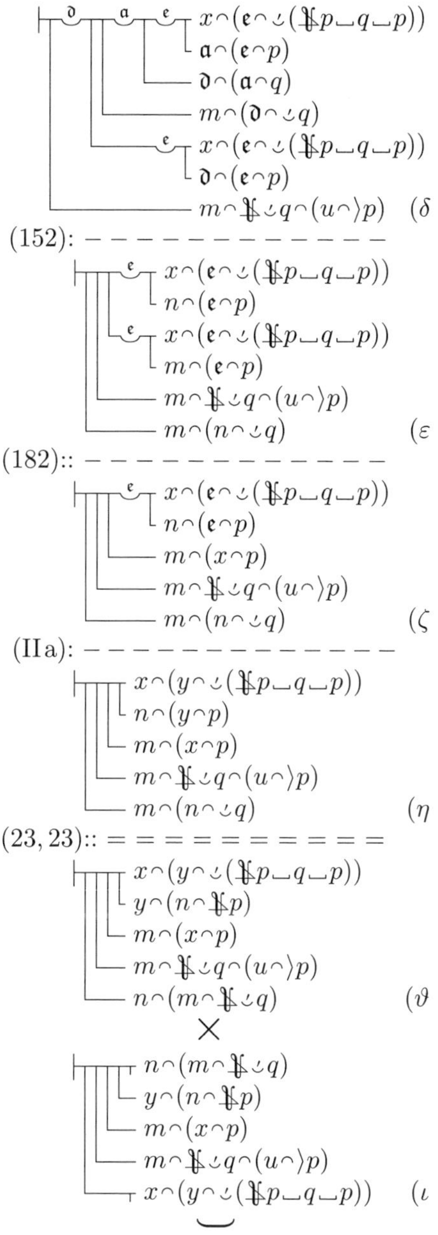

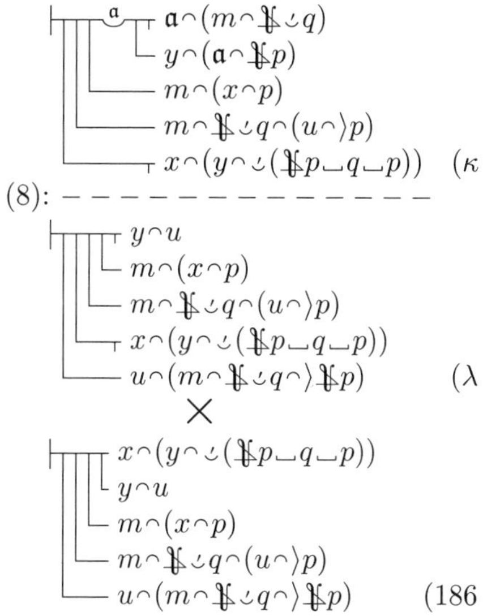

§138. *Analysis*

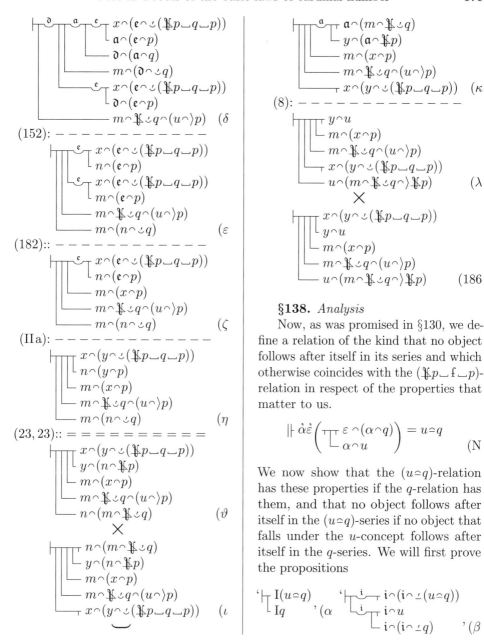

Now, as was promised in §130, we define a relation of the kind that no object follows after itself in its series and which otherwise coincides with the ($\nparallel p \smile \mathfrak{f} \smile p$)-relation in respect of the properties that matter to us.

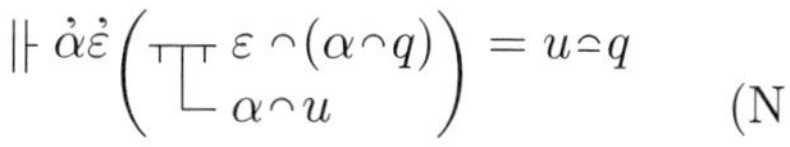

We now show that the ($u \frown q$)-relation has these properties if the q-relation has them, and that no object follows after itself in the ($u \frown q$)-series if no object that falls under the u-concept follows after itself in the q-series. We will first prove the propositions

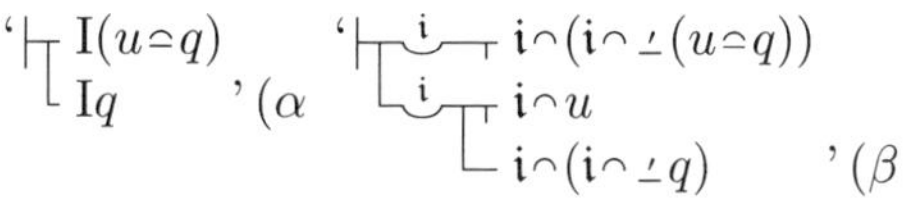

The first does not present any difficulty; (β) can be broken up into the propositions

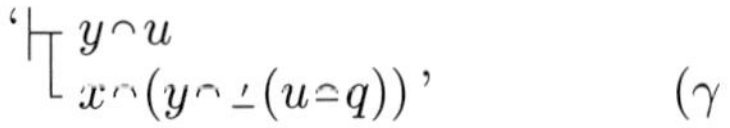

'$\vdash$ $y \frown u$
$x \frown (y \frown \dot{-}(u \frown q))$' ($\gamma$

'$\vdash$ $x \frown (y \frown \dot{-} q)$
$x \frown (y \frown \dot{-}(u \frown q))$' ($\delta$

§139. *Construction*

$\mathrm{N} \vdash \dot{\alpha}\dot{\varepsilon}\left(\vdash \varepsilon \frown (\alpha \frown q) ;\ \alpha \frown u\right) = u \frown q$

(6): ———————

$\vdash$ $e \frown (a \frown q)$
$a \frown u$
$e \frown (a \frown (u \frown q))$ (187

(Ib): – – – – – – – –

$\vdash$ $e \frown (a \frown q)$
$e \frown (a \frown (u \frown q))$ (188

(13): – – – – – – –

$\vdash$ $d = a$
$e \frown (a \frown (u \frown q))$
$e \frown (d \frown q)$
$\mathrm{I}q$ (α

(188):: – – – – – – – –

$\vdash$ $d = a$
$e \frown (a \frown (u \frown q))$
$e \frown (d \frown (u \frown q))$
$\mathrm{I}q$ (β

$\vdash^{\mathfrak{e}\ \mathfrak{d}\ \mathfrak{a}}$ $\mathfrak{d} = \mathfrak{a}$
$\mathfrak{e} \frown (\mathfrak{a} \frown (u \frown q))$
$\mathfrak{e} \frown (\mathfrak{d} \frown (u \frown q))$
$\mathrm{I}q$ (γ

(16): – – – – – – – – – – –

$\vdash$ $\mathrm{I}(u \frown q)$
$\mathrm{I}q$ (189

——— • ———

189 $\vdash$ $\mathrm{I}(u \frown (\nparallel p \smile \mathfrak{f} \smile p))$
$\mathrm{I}(\nparallel p \smile \mathfrak{f} \smile p)$

(173):: – – – – – – – – – –

$\vdash$ $\mathrm{I}(u \frown (\nparallel p \smile \mathfrak{f} \smile p))$
$v \frown (u \frown \rangle p)$
$u \frown (v \frown \rangle \nparallel p)$ (190

——— • ———

187 $\vdash$ $d \frown (y \frown q)$
$y \frown u$
$d \frown (y \frown (u \frown q))$

(Id): – – – – – – – –

$\vdash$ $y \frown u$
$d \frown (y \frown (u \frown q))$ (191

×

$\vdash$ $d \frown (y \frown (u \frown q))$
$\top\ y \frown u$ (192

$\vdash^{\mathfrak{e}}$ $\mathfrak{e} \frown (y \frown (u \frown q))$
$\top\ y \frown u$ (α

(125): – – – – – – – – –

$\vdash$ $x \frown (y \frown \dot{-}(u \frown q))$
$\top\ y \frown u$ (193

——— • ———

188 $\vdash$ $d \frown (a \frown q)$
$d \frown (a \frown (u \frown q))$

(133): – – – – – – –

$\vdash$ $x \frown (a \frown \dot{-} q)$
$d \frown (a \frown (u \frown q))$
$x \frown (d \frown \dot{-} q)$ (α

$\vdash^{\mathfrak{d}\ \mathfrak{a}}$ $x \frown (\mathfrak{a} \frown \dot{-} q)$
$\mathfrak{d} \frown (\mathfrak{a} \frown (u \frown q))$
$x \frown (\mathfrak{d} \frown \dot{-} q)$ (β

(123): ———————

$\vdash$ $x \frown (y \frown \dot{-} q)$
$^{\mathfrak{a}}$ $x \frown (\mathfrak{a} \frown \dot{-} q)$
$x \frown (\mathfrak{a} \frown (u \frown q))$
$x \frown (y \frown \dot{-}(u \frown q))$ (γ

——— • ———

188 $\vdash$ $x \frown (a \frown q)$
$x \frown (a \frown (u \frown q))$

(131): – – – – – – –

$\vdash$ $x \frown (a \frown \dot{-} q)$
$x \frown (a \frown (u \frown q))$ (δ

$\vdash^{\mathfrak{a}}$ $x \frown (\mathfrak{a} \frown \dot{-} q)$
$x \frown (\mathfrak{a} \frown (u \frown q))$ (ε

(γ): ———————

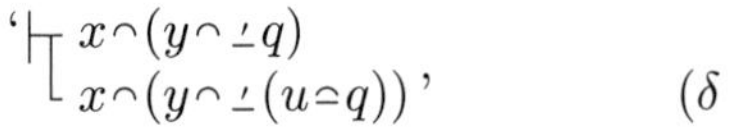

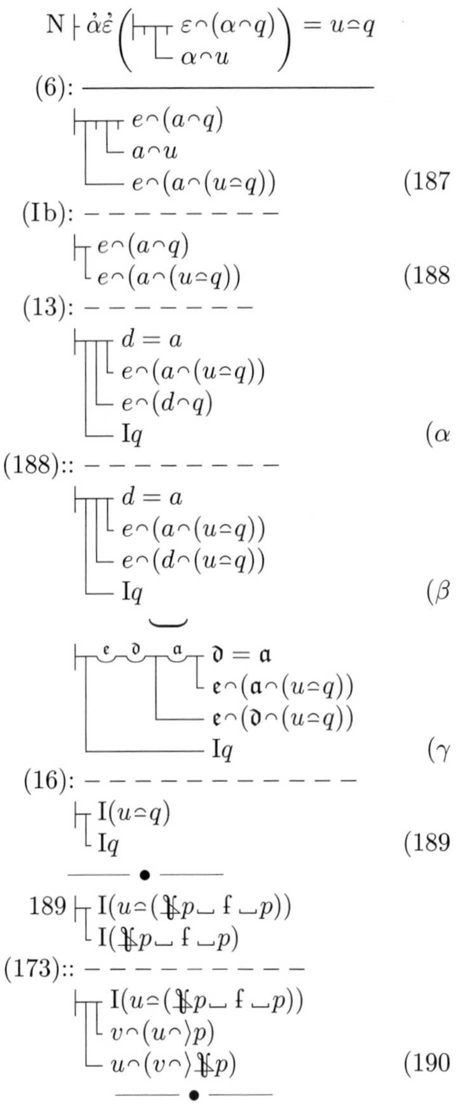

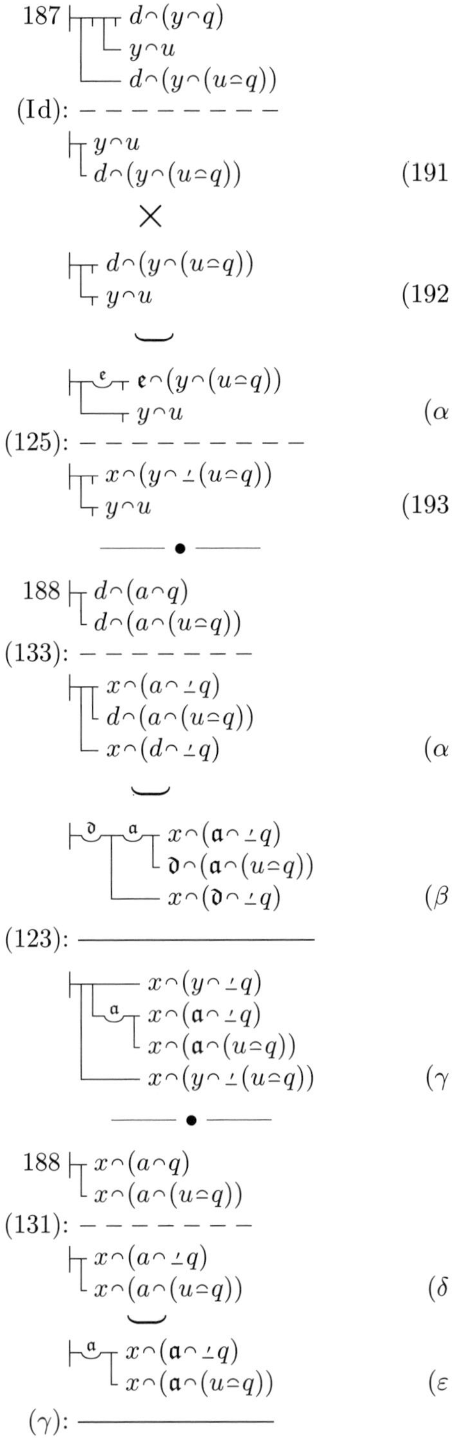

$$\begin{array}{l} \vdash x \frown (y \frown \dot{-} q) \\ \quad x \frown (y \frown \dot{-} (u \bumpeq q)) \end{array} \qquad (194$$

———— • ————

$$194 \vdash \begin{array}{l} y \frown (y \frown \dot{-} q) \\ y \frown (y \frown \dot{-} (u \bumpeq q)) \end{array}$$

(IIa): – – – – – – – – –

$$\begin{array}{l} \vdash y \frown u \\ \quad y \frown (y \frown \dot{-} (u \bumpeq q)) \\ \quad \overset{\mathfrak{i}}{\smile}\ \mathfrak{i} \frown u \\ \quad\quad \mathfrak{i} \frown (\mathfrak{i} \frown \dot{-} q) \end{array} \qquad (\alpha$$

×

$$\begin{array}{l} \vdash y \frown (y \frown \dot{-} (u \bumpeq q)) \\ \quad y \frown u \\ \quad \overset{\mathfrak{i}}{\smile}\ \mathfrak{i} \frown u \\ \quad\quad \mathfrak{i} \frown (\mathfrak{i} \frown \dot{-} q) \end{array} \qquad (\beta$$

(193): ·–·–·–·–·–·–·–·–·

$$\begin{array}{l} \vdash y \frown (y \frown \dot{-} (u \bumpeq q)) \\ \quad \overset{\mathfrak{i}}{\smile}\ \mathfrak{i} \frown u \\ \quad\quad \mathfrak{i} \frown (\mathfrak{i} \frown \dot{-} q) \end{array} \qquad (\gamma$$

$\smile$

$$\begin{array}{l} \vdash \overset{\mathfrak{i}}{\smile}\ \mathfrak{i} \frown (\mathfrak{i} \frown \dot{-} (u \bumpeq q)) \\ \quad \overset{\mathfrak{i}}{\smile}\ \mathfrak{i} \frown u \\ \quad\quad \mathfrak{i} \frown (\mathfrak{i} \frown \dot{-} q) \end{array} \qquad (195$$

———— • ————

$$178 \vdash \begin{array}{l} \overset{\mathfrak{i}}{\smile}\ \mathfrak{i} \frown u \\ \quad \mathfrak{i} \frown (\mathfrak{i} \frown \dot{-} (\nparallel p \smile \mathfrak{f} \smile p)) \\ u \frown (\Theta \frown \nparallel \smile \mathfrak{f} \frown \rangle \nparallel p) \end{array}$$

(195): – – – – – – – – – – – – –

$$\begin{array}{l} \vdash \overset{\mathfrak{i}}{\smile}\ \mathfrak{i} \frown (\mathfrak{i} \frown \dot{-} (u \bumpeq (\nparallel p \smile \mathfrak{f} \smile p))) \\ \quad u \frown (\Theta \frown \nparallel \smile \mathfrak{f} \frown \rangle \nparallel p) \end{array} \qquad (196$$

§140. *Analysis*

We now have to prove that under these assumptions the $(u \bumpeq (\nparallel p \smile \mathfrak{f} \smile p))$-series proceeds endlessly. This turns on the proposition

$$\text{‘}\vdash \begin{array}{l} d \frown (a \frown (\nparallel p \smile \mathfrak{f} \smile p)) \\ \quad \overset{\mathfrak{e}}{\smile}\ d \frown (\mathfrak{e} \frown (u \bumpeq (\nparallel p \smile \mathfrak{f} \smile p))) \\ \quad d \frown u \\ \quad \Theta \frown (x \frown p) \\ \quad \Theta \frown \nparallel \smile \mathfrak{f} \frown (u \frown \rangle p) \\ \quad u \frown (\Theta \frown \nparallel \smile \mathfrak{f} \frown \rangle \nparallel p) \end{array} \text{’} \qquad (\alpha$$

which is to be proven using the proposition

$$\text{‘}\vdash \begin{array}{l} d \frown (a \frown (u \bumpeq q)) \\ \quad d \frown (a \frown q) \\ \quad a \frown u \end{array} \text{’},$$

which follows from (N).

§141. *Construction*

$$\text{N} \vdash \dot{\alpha}\dot{\varepsilon}\left(\vdash \begin{array}{l} \varepsilon \frown (\alpha \frown q) \\ \alpha \frown u \end{array} \right) = u \bumpeq q$$

(10): ————————————

$$\begin{array}{l} \vdash d \frown (a \frown (u \bumpeq q)) \\ \quad d \frown (a \frown q) \\ \quad\quad a \frown u \end{array} \qquad (\alpha$$

(Ie):: – – – – – – – – –

$$\begin{array}{l} \vdash d \frown (a \frown (u \bumpeq q)) \\ \quad d \frown (a \frown q) \\ \quad a \frown u \end{array} \qquad (197$$

———— • ————

$$137 \vdash \begin{array}{l} x \frown (a \frown \smile (\nparallel p \smile \mathfrak{f} \smile p)) \\ \quad d \frown (a \frown (\nparallel p \smile \mathfrak{f} \smile p)) \\ \quad x \frown (d \frown \smile (\nparallel p \smile \mathfrak{f} \smile p)) \end{array}$$

(181): – – – – – – – – – – – –

$$\begin{array}{l} \vdash a \frown u \\ \quad d \frown (a \frown (\nparallel p \smile \mathfrak{f} \smile p)) \\ \quad x \frown (d \frown \smile (\nparallel p \smile \mathfrak{f} \smile p)) \\ \quad \Theta \frown (x \frown p) \\ \quad \Theta \frown \nparallel \smile \mathfrak{f} \frown (u \frown \rangle p) \\ \quad u \frown (\Theta \frown \nparallel \smile \mathfrak{f} \frown \rangle \nparallel p) \end{array} \qquad (\alpha$$

(197): – – – – – – – – – – – – –

$$\begin{array}{l} \vdash d \frown (a \frown (u \bumpeq (\nparallel p \smile \mathfrak{f} \smile p))) \\ \quad d \frown (a \frown (\nparallel p \smile \mathfrak{f} \smile p)) \\ \quad x \frown (d \frown \smile (\nparallel p \smile \mathfrak{f} \smile p)) \\ \quad \Theta \frown (x \frown p) \\ \quad \Theta \frown \nparallel \smile \mathfrak{f} \frown (u \frown \rangle p) \\ \quad u \frown (\Theta \frown \nparallel \smile \mathfrak{f} \frown \rangle \nparallel p) \end{array} \qquad (198$$

(186):: – – – – – – – – – – – – – –

$$\begin{array}{l} \vdash d \frown (a \frown (u \bumpeq (\nparallel p \smile \mathfrak{f} \smile p))) \\ \quad d \frown (a \frown (\nparallel p \smile \mathfrak{f} \smile p)) \\ \quad d \frown u \\ \quad \Theta \frown (x \frown p) \\ \quad \Theta \frown \nparallel \smile \mathfrak{f} \frown (u \frown \rangle p) \\ \quad u \frown (\Theta \frown \nparallel \smile \mathfrak{f} \frown \rangle \nparallel p) \end{array} \qquad (\alpha$$

×

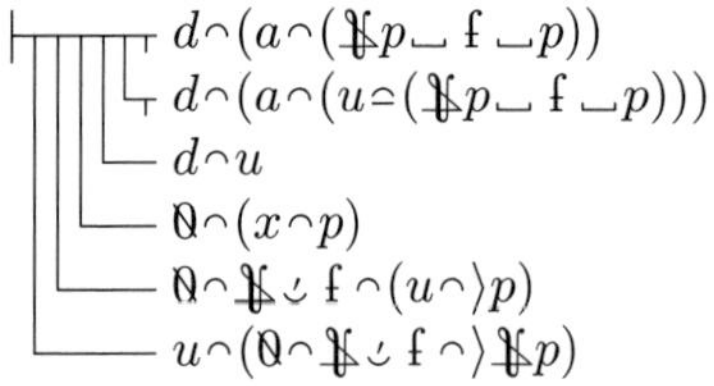

(β

(IIa):: — — — — — — — — — — — —

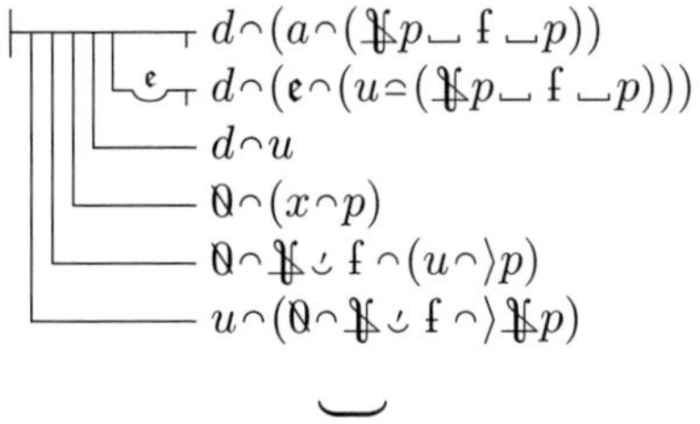

(γ

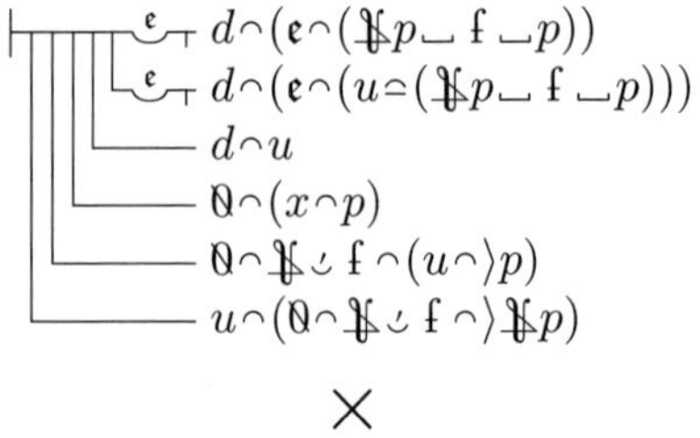

(δ

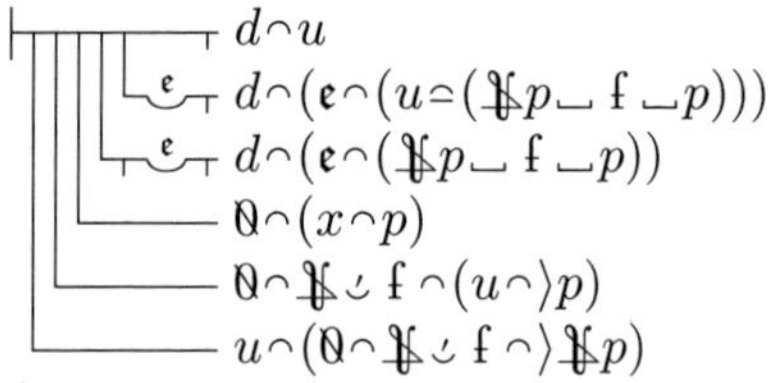

(ε

(175): ·—·—·—·—·—·—·—·—·—·—·—·

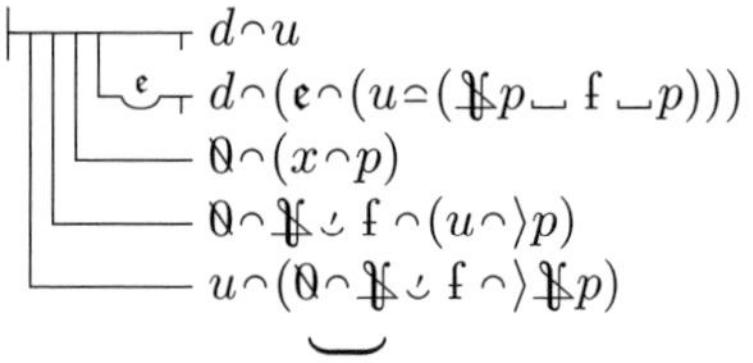

(ζ

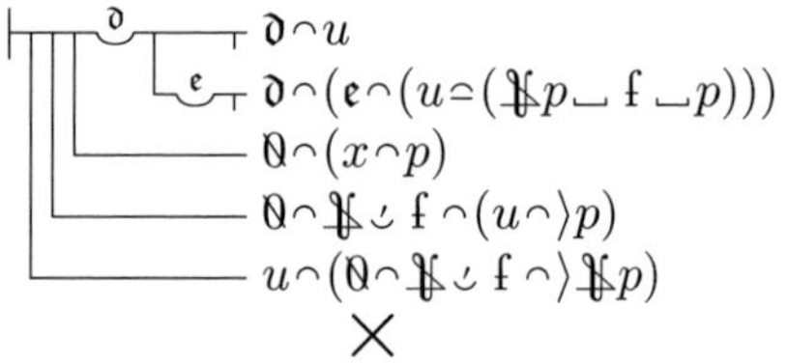

(η

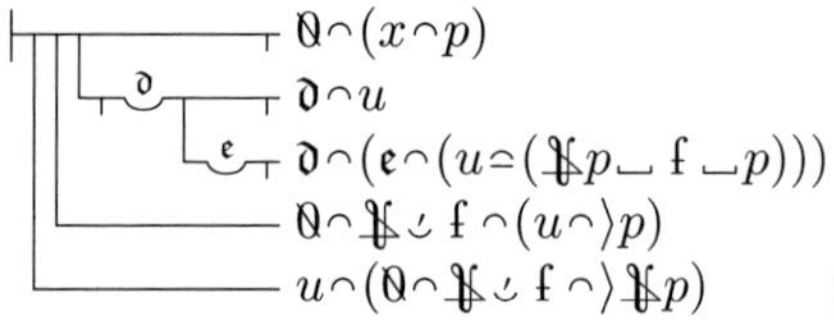

(199

§142. *Analysis*

It now remains to prove the proposition

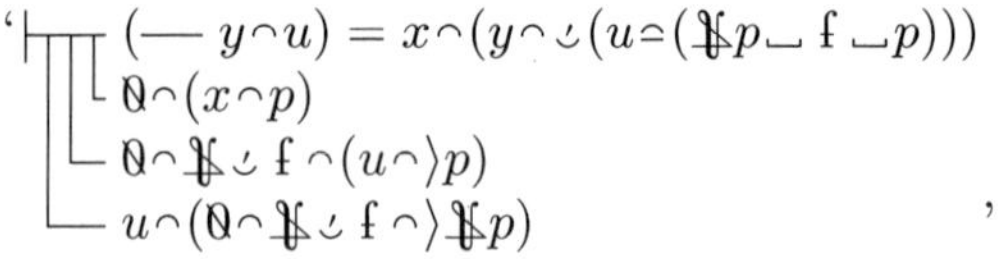

(α

Using the propositions (181) and (186) this is to be reduced to the propositions

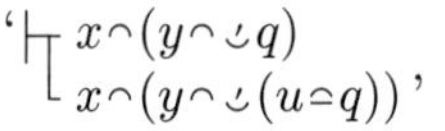

(β

and

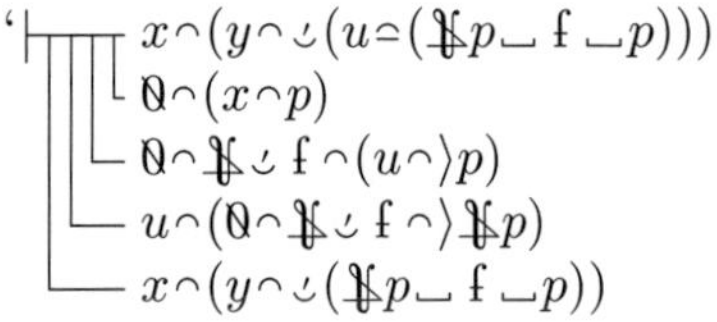

(γ

Of these (β) can be derived by means of (194), whereas (γ) can be proven using (152). For this we require the proposition

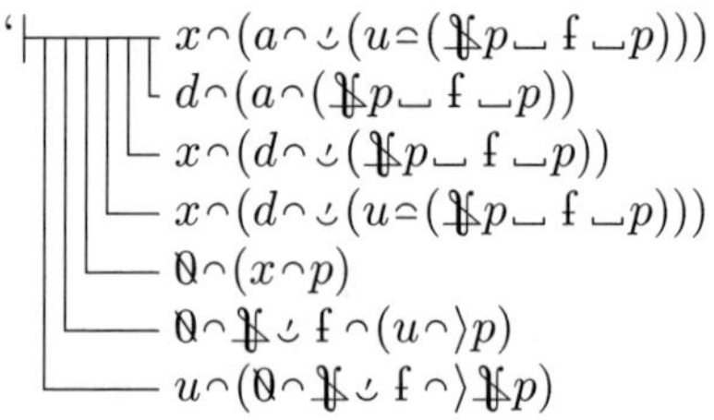

(δ

which follows from (198) and (137).

We then easily arrive at the end of our section (*c*).

§143. *Construction*

130 ⊢ y = x
  x⌢(y⌢⍊q)
  x⌢(y⌢⍙q)

×

⊢ x⌢(y⌢⍊q)
  y = x
  x⌢(y⌢⍙q) (200

———•———

136 ⊢ x⌢(y⌢⍙q)
  x⌢(y⌢⍊q)

(194):: – – – – – –

⊢ x⌢(y⌢⍙q)
  x⌢(y⌢⍊(u⍧q)) (α

(200):: – – – – – – – –

⊢ x⌢(y⌢⍙q)
  y = x
  x⌢(y⌢⍙(u⍧q)) (β

(139): ·–·–·–·–·–·–·

⊢ x⌢(y⌢⍙q)
  x⌢(y⌢⍙(u⍧q)) (201

———•———

201 ⊢ x⌢(y⌢⍙(∦p⌴ f ⌴p))
  x⌢(y⌢⍙(u⍧(∦p⌴ f ⌴p)))

(181): – – – – – – – – – – – – –

⊢ y⌢u
  x⌢(y⌢⍙(u⍧(∦p⌴ f ⌴p)))
  0⌢(x⌢p)
  0⌢∦⍙ f ⌢(u⌢⟩p)
  u⌢(0⌢∦⍙ f ⌢⟩∦p) (202

———•———

130 ⊢ F(⊢ c = a ; a⌢(c⌢⍊q))
  F(— a⌢(c⌢⍙q))

(135): – – – – – – – – –

⊢ F(a⌢(c⌢⍙q))
  F(— a⌢(c⌢⍙q)) (203

———•———

140 ⊢ a⌢(a⌢⍙q)

(22): ————————

⊢ a⌢(a⌢∦⍙q) (204

———•———

M ⊢ 𝔭(0⌢∦⍙f) = ∽

(IIIc): ——————————

⊢ F(∽)
  F(𝔭(0⌢∦⍙f)) (205

———•———

137 ⊢ x⌢(a⌢⍙(u⍧(∦p⌴ f ⌴p)))
  d⌢(a⌢(u⍧(∦p⌴ f ⌴p)))
  x⌢(d⌢⍙(u⍧(∦p⌴ f ⌴p)))

(198):: – – – – – – – – – – – – – –

⊢ x⌢(a⌢⍙(u⍧(∦p⌴ f ⌴p)))
  d⌢(a⌢(∦p⌴ f ⌴p))
  x⌢(d⌢⍙(∦p⌴ f ⌴p))
  x⌢(d⌢⍙(u⍧(∦p⌴ f ⌴p)))
  0⌢(x⌢p)
  0⌢∦⍙ f ⌢(u⌢⟩p)
  u⌢(0⌢∦⍙ f ⌢⟩∦p) (α

⊢ 𝔡 𝔞 x⌢(𝔞⌢⍙(u⍧(∦p⌴ f ⌴p)))
  𝔡⌢(𝔞⌢(∦p⌴ f ⌴p))
  x⌢(𝔡⌢⍙(∦p⌴ f ⌴p))
  x⌢(𝔡⌢⍙(u⍧(∦p⌴ f ⌴p)))
  0⌢(x⌢p)
  0⌢∦⍙ f ⌢(u⌢⟩p)
  u⌢(0⌢∦⍙ f ⌢⟩∦p) (β

(152): – – – – – – – – – – – – – – –

⊢ x⌢(y⌢⍙(u⍧(∦p⌴ f ⌴p)))
  x⌢(x⌢⍙(u⍧(∦p⌴ f ⌴p)))
  0⌢(x⌢p)
  0⌢∦⍙ f ⌢(u⌢⟩p)
  u⌢(0⌢∦⍙ f ⌢⟩∦p)
  x⌢(y⌢⍙(∦p⌴ f ⌴p)) (γ

(140):: ————————————————

⊢ x⌢(y⌢⍙(u⍧(∦p⌴ f ⌴p)))
  0⌢(x⌢p)
  0⌢∦⍙ f ⌢(u⌢⟩p)
  u⌢(0⌢∦⍙ f ⌢⟩∦p)
  x⌢(y⌢⍙(∦p⌴ f ⌴p)) (δ

(186):: – – – – – – – – – – – – – –

```
├┬┬┬┬ x⌢(y⌢˙(u≏(ℌp⌣ f ⌣p)))
│││└ y⌢u
││└── 𝔔⌢(x⌢p)
│└─── 𝔔⌢ℌ˙ f ⌢(u⌢⟩p)
└──── u⌢(𝔔⌢ℌ˙ f ⌢⟩ℌp)                                        (ε
(IVa): - - - - - - - - - - - - - - - -
├┬┬┬┬ (— y⌢u) = [— x⌢(y⌢˙(u≏(ℌp⌣ f ⌣p)))]
││││└ 𝔔⌢(x⌢p)
│││└─ 𝔔⌢ℌ˙ f ⌢(u⌢⟩p)
││└── u⌢(𝔔⌢ℌ˙ f ⌢⟩ℌp)
│└──┬ y⌢u
│   └ x⌢(y⌢˙(u≏(ℌp⌣ f ⌣p)))                                   (ζ
(202):: - - - - - - - - - - - - - - - - - - - - - - -
├┬┬┬ (— y⌢u) = [— x⌢(y⌢˙(u≏(ℌp⌣ f ⌣p)))]
│││└ 𝔔⌢(x⌢p)
││└─ 𝔔⌢ℌ˙ f ⌢(u⌢⟩p)
│└── u⌢(𝔔⌢ℌ˙ f ⌢⟩ℌp)                                          (η
(203): ————————————————————————————————————
├┬┬┬ (— y⌢u) = x⌢(y⌢˙(u≏(ℌp⌣ f ⌣p)))
│││└ 𝔔⌢(x⌢p)
││└─ 𝔔⌢ℌ˙ f ⌢(u⌢⟩p)
│└── u⌢(𝔔⌢ℌ˙ f ⌢⟩ℌp)                                          (ϑ
              ‿
├┬┬┬─𝔞─ (— 𝔞⌢u) = x⌢(𝔞⌢˙(u≏(ℌp⌣ f ⌣p)))
││└──── 𝔔⌢(x⌢p)
│└───── 𝔔⌢ℌ˙ f ⌢(u⌢⟩p)
└────── u⌢(𝔔⌢ℌ˙ f ⌢⟩ℌp)                                        (ι
(Va): - - - - - - - - - - - - - - - - - - - - - -
├┬┬┬ έ(— ε⌢u) = έ[x⌢(ε⌢˙(u≏(ℌp⌣ f ⌣p)))]
│││└ 𝔔⌢(x⌢p)
││└─ 𝔔⌢ℌ˙ f ⌢(u⌢⟩p)
│└── u⌢(𝔔⌢ℌ˙ f ⌢⟩ℌp)                                          (κ
(164): ————————————————————————————————————
├┬┬┬ έ(— ε⌢u) = x⌢ℌ˙(u≏(ℌp⌣ f ⌣p))
│││└ 𝔔⌢(x⌢p)
││└─ 𝔔⌢ℌ˙ f ⌢(u⌢⟩p)
│└── u⌢(𝔔⌢ℌ˙ f ⌢⟩ℌp)                                          (λ
                 ×
├┬┬┬┬ 𝔔⌢(x⌢p)
│││└┬ έ(— ε⌢u) = x⌢ℌ˙(u≏(ℌp⌣ f ⌣p))
││└── 𝔔⌢ℌ˙ f ⌢(u⌢⟩p)
│└─── u⌢(𝔔⌢ℌ˙ f ⌢⟩ℌp)                                         (μ
(IIa):: - - - - - - - - - - - - - - - - - - - -
```

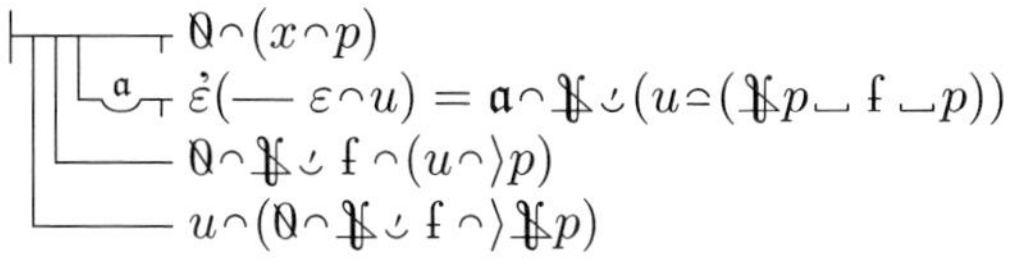

(ν

(IIa):: -

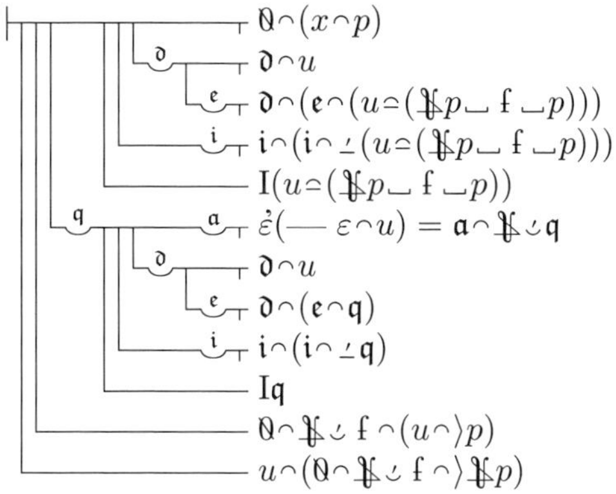

(ξ

(196, 190):: = = = = = = = = = = = = = = = = = = =

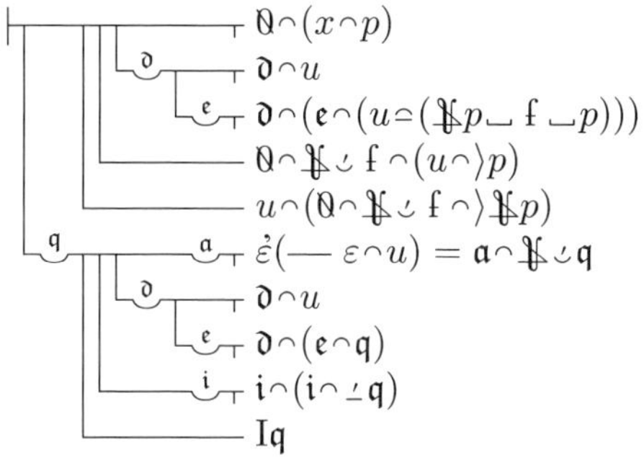

(ο

(199): ·–·–·–·–·–·–·–·–·–·–·–·–·–·–·–·

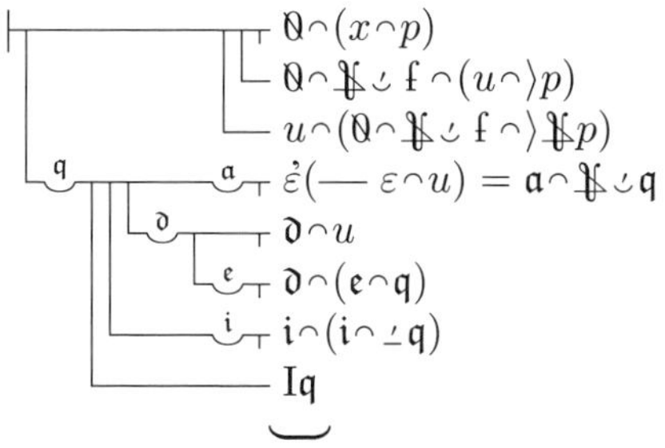

(π

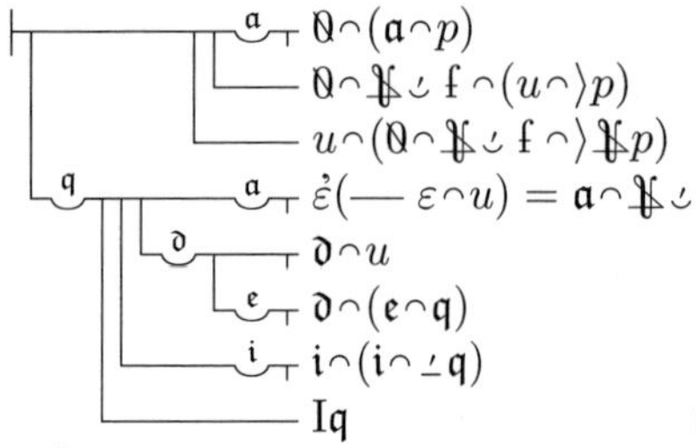

(ρ

(183): – – – – – – – – – – – – – –

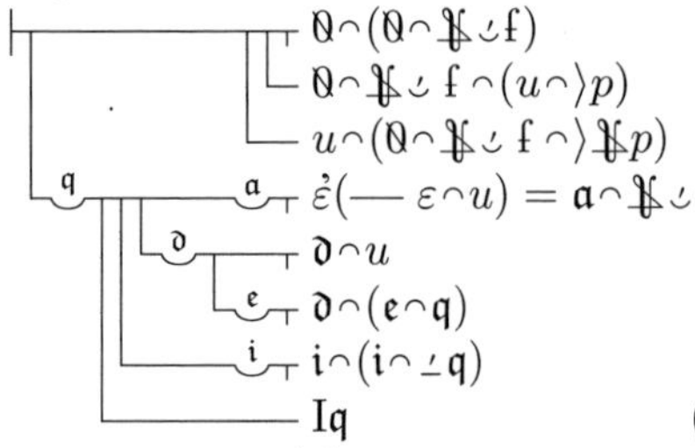

(σ

×

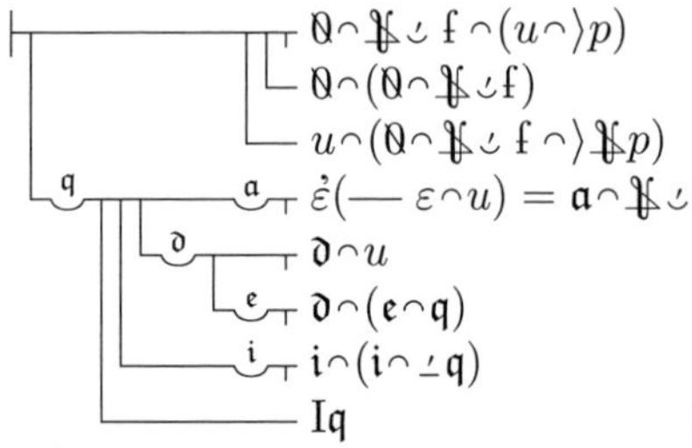

(τ

(204):: ————————————

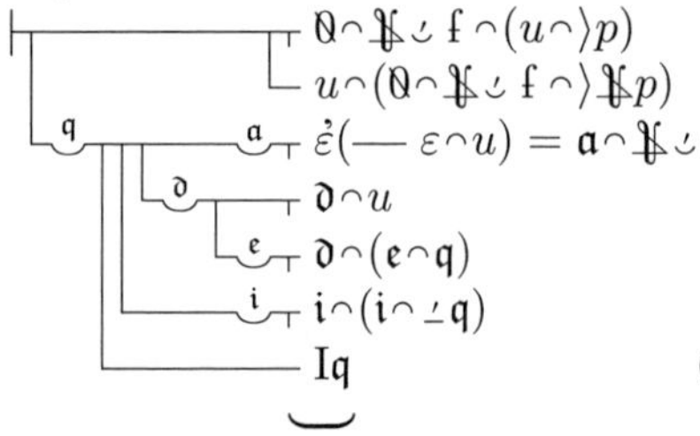

(υ

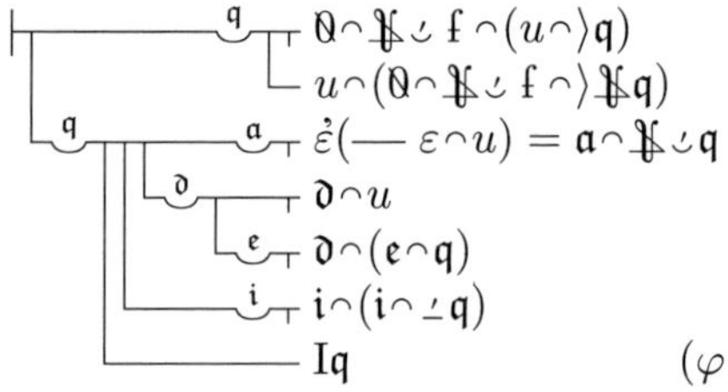

(φ

(49): – – – – – – – – – – – – – – –

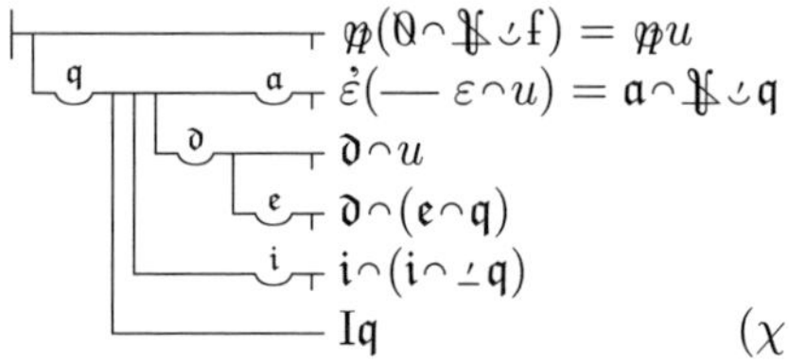

(χ

(205): ————————————

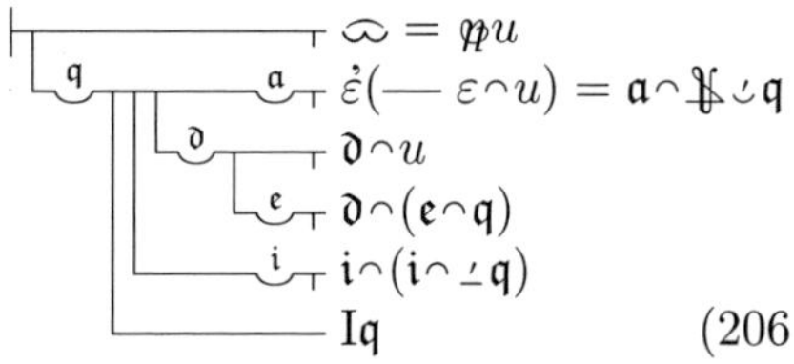

(206

×

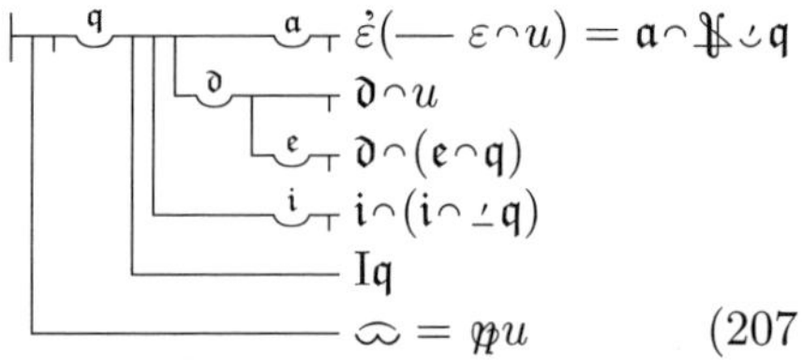

(207

d) Proof of the proposition

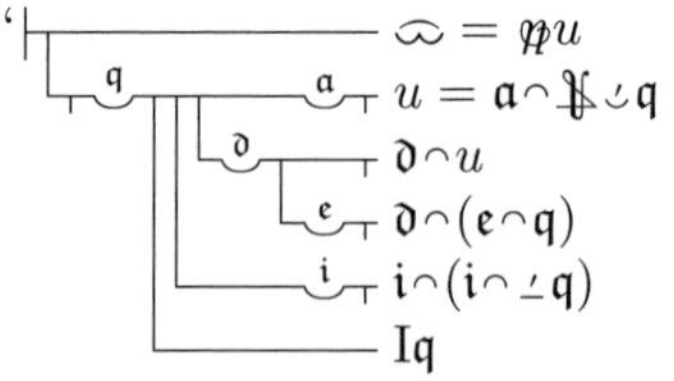

,

§144. *Analysis*

We now prove the converse of proposition (207), namely that Endlos is the cardinal number which belongs to a concept, if the objects falling under that concept can be ordered into a series that starts with a certain object and proceeds endlessly without looping back into itself and without branching. It turns on showing that Endlos is the cardinal number that belongs to the concept *member of such a series*; in signs

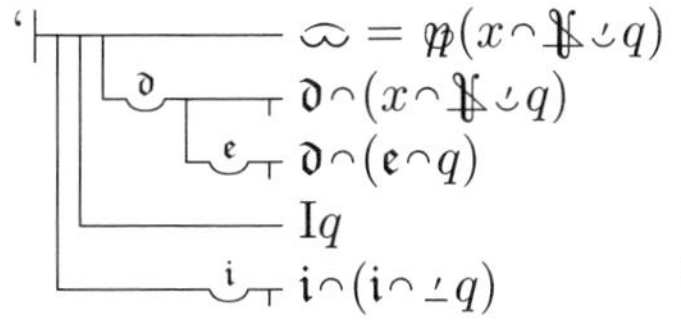 , (α

We use proposition (32) for this and need to prove that there is a relation which maps the cardinal number series into the q-series starting with x and whose converse maps the latter into the former. The obvious strategy is to correlate 0 with x, 1 with the member immediately following x in the q-series and, in this manner, each immediately following cardinal number to the immediately following member of the q-series. We always pair one member of the cardinal number series with one member of the q-series and form a series out of these pairs. The series-forming relation is determined thus: one pair stands in it to a second if the first member of the first pair stands in the $\mathfrak{f}$-relation to the first member of the second pair and the second member of the first pair stands in the q-relation to the second member of the second pair. If the pair $(n; y)$ belongs to the series that starts with the pair $(0; x)$, then n stands to y in the mapping relation that is to be exhibited. We now define the pair in this way:

$$\Vdash \grave{\varepsilon}(o \frown (a \frown \varepsilon)) = o; a \qquad (\Xi$$

The semicolon is here a two-sided function-sign. The expression

$$‘\Pi \frown (\Gamma; \Delta)’$$

is thus co-referential with

$$‘\Gamma \frown (\Delta \frown \Pi)’$$

provided ‘Γ’, ‘Δ’ and ‘Π’ refer to objects. For the extension of the relation given in the manner described above by, as I say, *coupling* the p-relation and the q-relation, I will introduce a simple sign by defining:

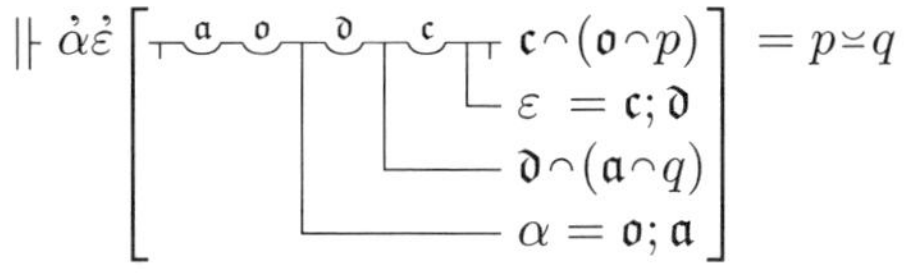

(O

Accordingly,

$$‘0; x \frown (\xi; \zeta \frown \smile(\mathfrak{f} \veebar q))’$$

indicates our mapping relation and

$$‘\grave{\alpha}\grave{\varepsilon}(0; x \frown (\varepsilon; \alpha \frown \smile(\mathfrak{f} \veebar q)))’$$

its extension. We now define

$$\Vdash \grave{\alpha}\grave{\varepsilon}(A \frown (\varepsilon; \alpha \frown \smile t)) = A \nwarrow t \qquad (\Pi$$

The propositions

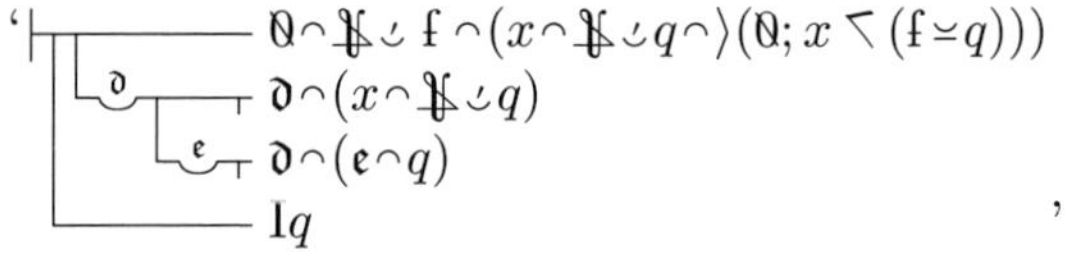

(β

and

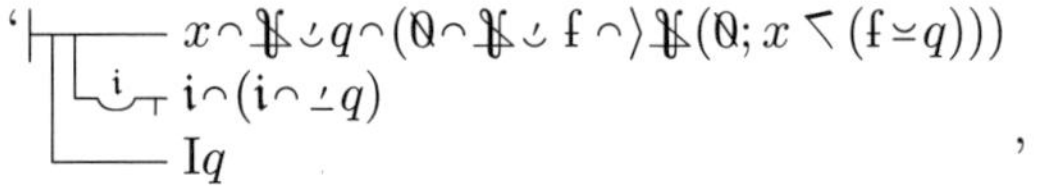

(γ

then have to be proven. Instead of (β) we first prove the somewhat more general proposition

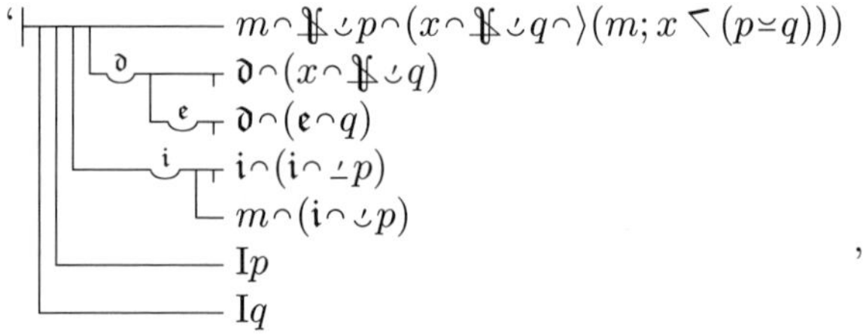

(δ

which we can then also use for the proof of (γ). We use (11) and thus have to deduce the proposition

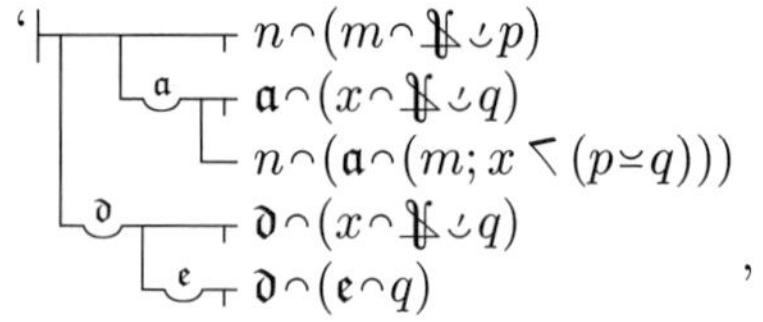

(ε

The proposition

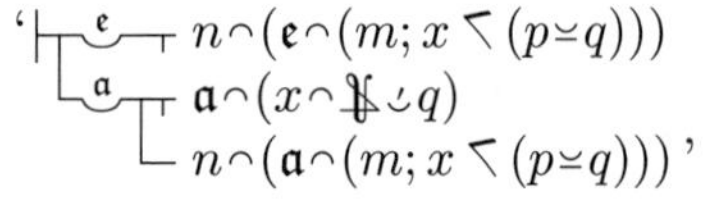

(ζ

is easily deduced from the proposition

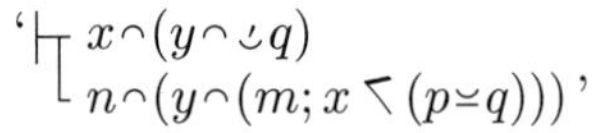

(η

and with the latter we can reduce (ε) to

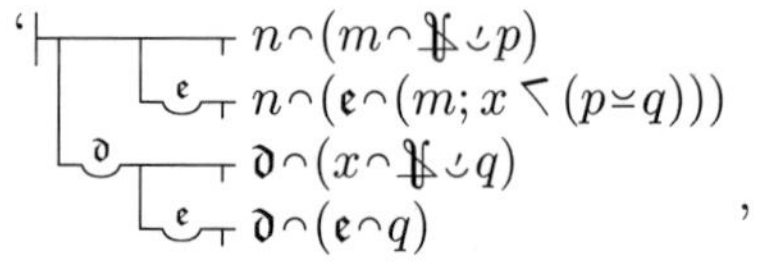

(ϑ

We prove this proposition using (144), by putting the function-marker

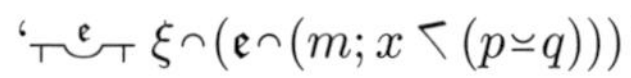

for '$F(\xi)$'. We then require the proposition

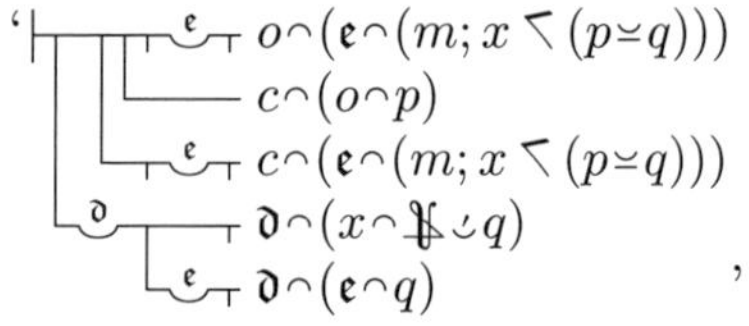

(ι

In order to aid comprehension, I give the following diagram of the *p*-series and the *q*-series placed next to each other:

p-series	*q*-series
m	*x*
⋮	⋮
c	*d*
o	*a*

It is to be shown:

"If the *q*-series starting with *x* proceeds without end, and if there is an object (d) which together with c forms a pair which belongs to the $p\smile q$-series starting with the pair $m;x$, then there is also an object (a) which forms such a pair with o provided c stands to o in the p-relation."

We first prove the proposition

$$\vdash\!\!\!\begin{array}{l} o\frown(a\frown(A\zigzag(p\smile q))) \\ d\frown(a\frown q) \\ c\frown(o\frown p) \\ c\frown(d\frown(A\zigzag(p\smile q)))\end{array}' \quad (\kappa$$ [1]

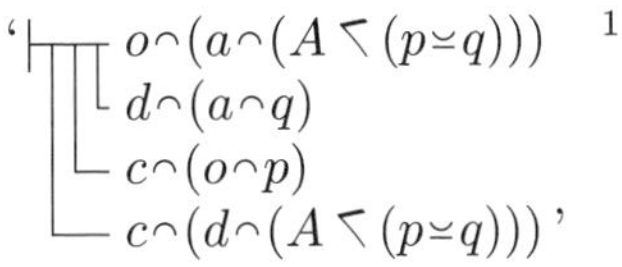

for which, because of definition (Π), we can write

$$\vdash\!\!\!\begin{array}{l} A\frown(o;a\frown\smile(p\smile q)) \\ d\frown(a\frown q) \\ c\frown(o\frown p) \\ A\frown(c;d\frown\smile(p\smile q))\end{array}' \quad (\lambda$$

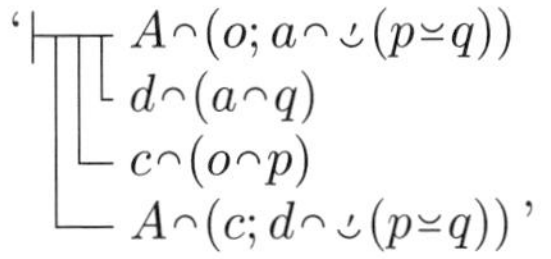

This can be proven straightforwardly by means of the proposition

$$\vdash\!\!\!\begin{array}{l} c;d\frown(o;a\frown(p\smile q)) \\ d\frown(a\frown q) \\ c\frown(o\frown p)\end{array}, \quad (\mu$$

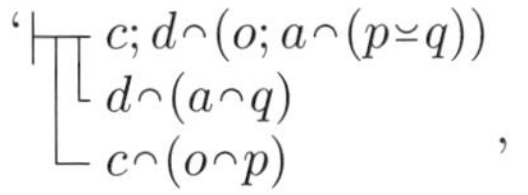

which follows from definition (O).

§145. *Construction*

$$O\vdash\dot{\alpha}\dot{\varepsilon}\begin{bmatrix}\begin{array}{l}{\vdash}\!\!\!\underset{a\quad o\quad \mathfrak{d}\quad c}{\frown\!\frown\!\frown}\!\!\!\underset{}{{\vdash}}\!\!\!\begin{array}{l}\mathfrak{c}\frown(\mathfrak{o}\frown\mathfrak{p}) \\ \varepsilon=\mathfrak{c};\mathfrak{d} \\ \mathfrak{d}\frown(\mathfrak{a}\frown q) \\ \alpha=\mathfrak{o};\mathfrak{a}\end{array}\end{array}\end{bmatrix}=p\smile q$$

(100): ─────────────────────

$$\vdash\!\!\!\underset{a\quad o\quad \mathfrak{d}\quad c}{\frown\!\frown\!\frown}\!\!\!\begin{array}{l}\mathfrak{c}\frown(\mathfrak{o}\frown\mathfrak{p}) \\ c;d=\mathfrak{c};\mathfrak{d} \\ \mathfrak{d}\frown(\mathfrak{a}\frown q) \\ o;a=\mathfrak{o};\mathfrak{a} \\ c;d\frown(o;a\frown(p\smile q))\end{array} \quad (\alpha$$

(IIa): ─ ─ ─ ─ ─ ─ ─ ─ ─ ─ ─ ─ ─ ─ ─

$$\vdash\!\!\!\underset{o\quad \mathfrak{d}\quad c}{\frown\!\frown}\!\!\!\begin{array}{l}\mathfrak{c}\frown(\mathfrak{o}\frown\mathfrak{p}) \\ c;d=\mathfrak{c};\mathfrak{d} \\ \mathfrak{d}\frown(\mathfrak{a}\frown q) \\ o;a=\mathfrak{o};a \\ c;d\frown(o;a\frown(p\smile q))\end{array} \quad (\beta$$

(IIa): ─ ─ ─ ─ ─ ─ ─ ─ ─ ─ ─ ─ ─ ─

$$\vdash\!\!\!\underset{\mathfrak{d}\quad c}{\frown}\!\!\!\begin{array}{l}\mathfrak{c}\frown(\mathfrak{o}\frown\mathfrak{p}) \\ c;d=\mathfrak{c};\mathfrak{d} \\ \mathfrak{d}\frown(\mathfrak{a}\frown q) \\ o;a=\mathfrak{o};a \\ c;d\frown(o;a\frown(p\smile q))\end{array} \quad (\gamma$$

(IIIe):: ─────────────────────

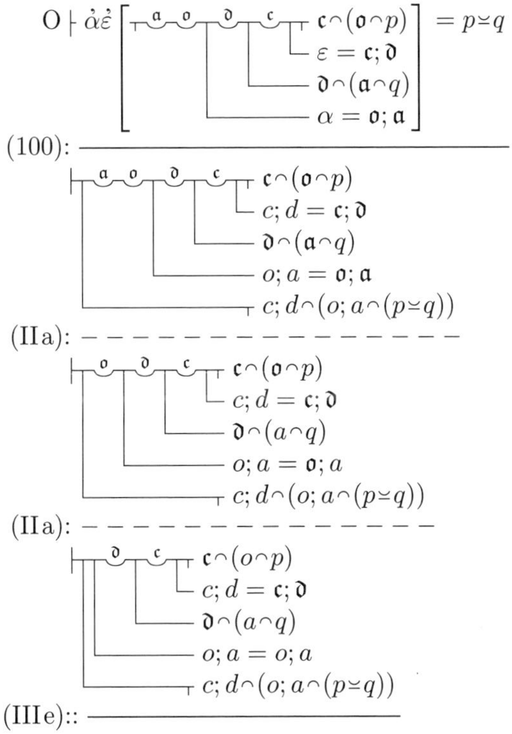

[1] Here "A" is written for "$m;x$".

$\vdash \mathfrak{d}\ \mathfrak{c}$ $\mathfrak{c} \frown (o \frown p)$
$c ; d = \mathfrak{c} ; \mathfrak{d}$
$\mathfrak{d} \frown (a \frown q)$
$c ; d \frown (o ; a \frown (p \veeeq q))$ (δ

(IIa): – – – – – – – – – – – – –

$\vdash \mathfrak{c}$ $\mathfrak{c} \frown (o \frown p)$
$c ; d = \mathfrak{c} ; d$
$d \frown (a \frown q)$
$c ; d \frown (o ; a \frown (p \veeeq q))$ (ε

(IIa): – – – – – – – – – – – –

$\vdash c \frown (o \frown p)$
$c ; d = c ; d$
$d \frown (a \frown q)$
$c ; d \frown (o ; a \frown (p \veeeq q))$ (ζ

(IIIe):: ————————————

$\vdash c \frown (o \frown p)$
$d \frown (a \frown q)$
$c ; d \frown (o ; a \frown (p \veeeq q))$ (η

×

$\vdash c ; d \frown (o ; a \frown (p \veeeq q))$
$d \frown (a \frown q)$
$c \frown (o \frown p)$ (208

(137): – – – – – – – – – –

$\vdash A \frown (o ; a \frown \smile(p \veeeq q))$
$d \frown (a \frown q)$
$c \frown (o \frown p)$
$A \frown (c ; d \frown \smile(p \veeeq q))$ (209

——— • ———

$\Pi \vdash \grave{\alpha}\grave{\varepsilon}(A \frown (\varepsilon ; \alpha \frown \smile t)) = A \prec t$

(10): ————————————

$\vdash F(o \frown (a \frown (A \prec t)))$
$F(A \frown (o ; a \frown \smile t))$ (210

——— • ———

210 $\vdash o \frown (a \frown (A \prec (p \veeeq q)))$
$A \frown (o ; a \frown \smile(p \veeeq q))$

(209):: – – – – – – – – – –

$\vdash o \frown (a \frown (A \prec (p \veeeq q)))$
$d \frown (a \frown q)$
$c \frown (o \frown p)$
$A \frown (c ; d \frown \smile(p \veeeq q))$ (α

(210): ————————————

$\vdash o \frown (a \frown (A \prec (p \veeeq q)))$
$d \frown (a \frown q)$
$c \frown (o \frown p)$
$c \frown (d \frown (A \prec (p \veeeq q)))$ (211

×

$\vdash d \frown (a \frown q)$
$o \frown (a \frown (A \prec (p \veeeq q)))$
$c \frown (o \frown p)$
$c \frown (d \frown (A \prec (p \veeeq q)))$ (α

(IIa):: – – – – – – – – – – – –

$\vdash d \frown (a \frown q)$
$\mathfrak{e}$ $o \frown (\mathfrak{e} \frown (A \prec (p \veeeq q)))$
$c \frown (o \frown p)$
$c \frown (d \frown (A \prec (p \veeeq q)))$ (β

$\vdash \mathfrak{e}$ $d \frown (\mathfrak{e} \frown q)$
$\mathfrak{e}$ $o \frown (\mathfrak{e} \frown (A \prec (p \veeeq q)))$
$c \frown (o \frown p)$
$c \frown (d \frown (A \prec (p \veeeq q)))$ (γ

(IIa): – – – – – – – – – – – – –

$\vdash d \frown (x \frown \nparallel \smile q)$
$\mathfrak{e}$ $o \frown (\mathfrak{e} \frown (A \prec (p \veeeq q)))$
$c \frown (o \frown p)$
$c \frown (d \frown (A \prec (p \veeeq q)))$
$\mathfrak{d}$ $\mathfrak{d} \frown (x \frown \nparallel \smile q)$
$\mathfrak{e}$ $\mathfrak{d} \frown (\mathfrak{e} \frown q)$ (δ

(23): ————————————

$\vdash x \frown (d \frown \smile q)$
$\mathfrak{e}$ $o \frown (\mathfrak{e} \frown (A \prec (p \veeeq q)))$
$c \frown (o \frown p)$
$c \frown (d \frown (A \prec (p \veeeq q)))$
$\mathfrak{d}$ $\mathfrak{d} \frown (x \frown \nparallel \smile q)$
$\mathfrak{e}$ $\mathfrak{d} \frown (\mathfrak{e} \frown q)$ (ε

×

$\vdash c \frown (d \frown (A \prec (p \veeeq q)))$
$\mathfrak{e}$ $o \frown (\mathfrak{e} \frown (A \prec (p \veeeq q)))$
$c \frown (o \frown p)$
$x \frown (d \frown \smile q)$
$\mathfrak{d}$ $\mathfrak{d} \frown (x \frown \nparallel \smile q)$
$\mathfrak{e}$ $\mathfrak{d} \frown (\mathfrak{e} \frown q)$ (212

§146. *Analysis*

We now have to eliminate the subcomponent in (212)

‘— $x^\frown(d^\frown \smile q)$’

For this, we employ the proposition (η) of §144, which follows from

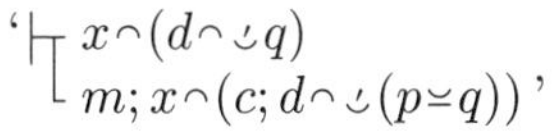
(α

We will first prove the proposition

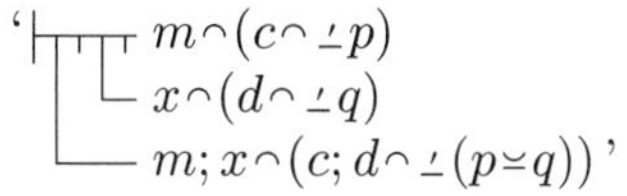
(β

which we will also need on other occasions. For this we need the proposition

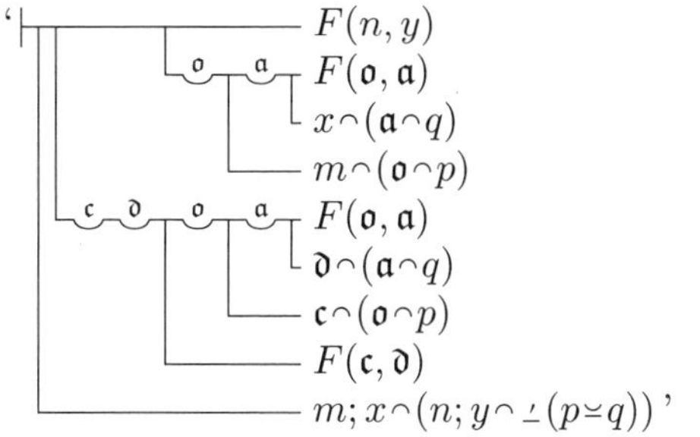
(γ

which is similar to proposition (123) and can be proven by means of it. To this end we write (123) in the form

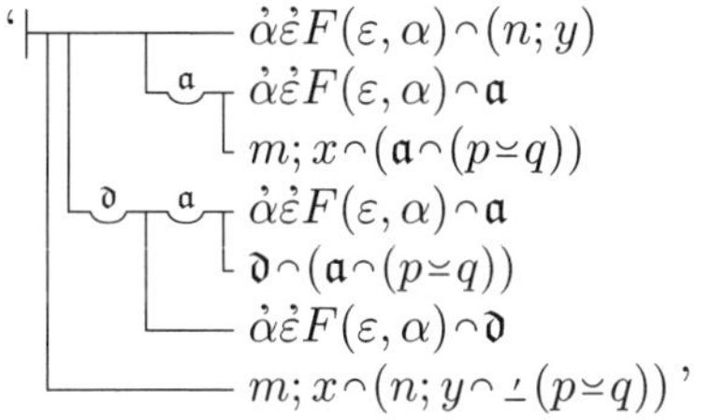

First, we have to prove the proposition

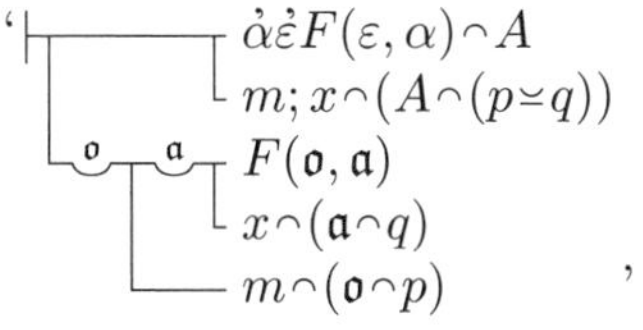
(δ

from which further the proposition

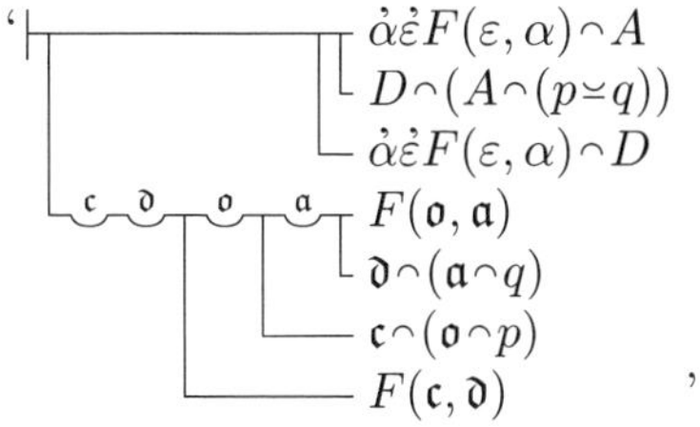
(ε

which we also need, follows. From the propositions

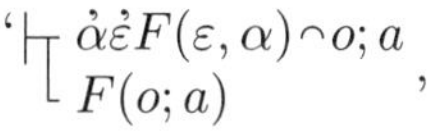
(ζ

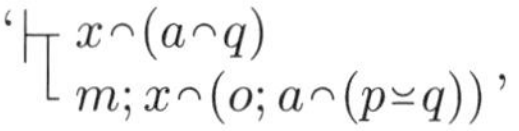
(η

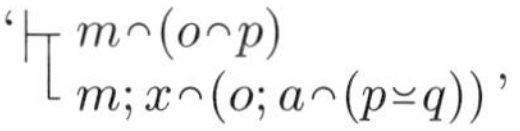
(ϑ

we can easily arrive at a proposition which differs from (δ) only in that ‘$o; a$’ stands in place of ‘A’. The supercomponent which consists of the first two lines may then be replaced by

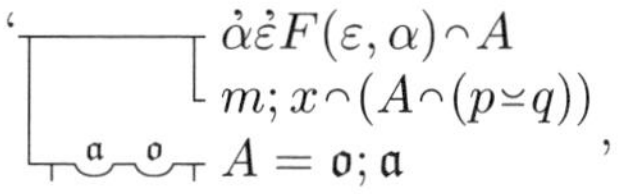

In order to eliminate the subcomponent ‘$\smile^{\mathfrak{a}}\smile^{\mathfrak{o}}\, A = \mathfrak{o};\mathfrak{a}$’, we use the proposition

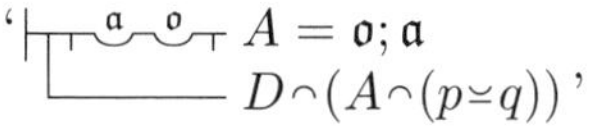
(ι

which follows from (O).

§147. *Construction*

$$\mathrm{O} \vdash \grave{\alpha}\grave{\varepsilon}\left[\begin{array}{l} \mathfrak{a}\ \mathfrak{o}\ \mathfrak{d}\ \mathfrak{c}\quad \mathfrak{c}\frown(\mathfrak{o}\frown p) \\ \quad \varepsilon = \mathfrak{c};\mathfrak{d} \\ \quad \mathfrak{d}\frown(\mathfrak{a}\frown q) \\ \quad \alpha = \mathfrak{o};\mathfrak{a} \end{array}\right] = p \succeq q$$

(14): ———————————

$$\vdash D\frown(A\frown(p\succeq q))$$
$$\mathfrak{a}\ \mathfrak{o}\ \mathfrak{d}\ \mathfrak{c}\quad \mathfrak{c}\frown(\mathfrak{o}\frown p)$$
$$D = \mathfrak{c};\mathfrak{d}$$
$$\mathfrak{d}\frown(\mathfrak{a}\frown q)$$
$$A = \mathfrak{o};\mathfrak{a} \qquad (213$$

—— • ——

$$\mathrm{IIa} \vdash \ \mathfrak{o}\quad A = \mathfrak{o};a$$
$$\mathfrak{a}\ \mathfrak{o}\quad A = \mathfrak{o};\mathfrak{a}$$

(IIa): - - - - - - - - -

$$\vdash A = o;a$$
$$\mathfrak{a}\ \mathfrak{o}\quad A = \mathfrak{o};\mathfrak{a} \qquad (\alpha$$

(Ia): - - - - - - - - -

$$\vdash c\frown(o\frown p)$$
$$D = c;d$$
$$d\frown(a\frown q)$$
$$A = o;a$$
$$\mathfrak{a}\ \mathfrak{o}\quad A = \mathfrak{o};\mathfrak{a} \qquad (\beta$$

$$\vdash \mathfrak{a}\ \mathfrak{o}\ \mathfrak{d}\ \mathfrak{c}\quad \mathfrak{c}\frown(\mathfrak{o}\frown p)$$
$$D = \mathfrak{c};\mathfrak{d}$$
$$\mathfrak{d}\frown(\mathfrak{a}\frown q)$$
$$A = \mathfrak{o};\mathfrak{a}$$
$$\mathfrak{a}\ \mathfrak{o}\quad A = \mathfrak{o};\mathfrak{a} \qquad (\gamma$$

(213): - - - - - - - - - - - -

$$\vdash D\frown(A\frown(p\succeq q))$$
$$\mathfrak{a}\ \mathfrak{o}\quad A = \mathfrak{o};\mathfrak{a} \qquad (214$$

—— • ——

$$\Xi \vdash \grave{\varepsilon}(o\frown(a\frown\varepsilon)) = o;a$$

(IIIh): ———————————

$$\vdash q\frown\grave{\varepsilon}(o\frown(a\frown\varepsilon)) = q\frown(o;a) \qquad (\alpha$$

(82): ———————————

$$\vdash o\frown(a\frown q) = q\frown(o;a) \qquad (215$$

—— • ——

$$215 \vdash o\frown(a\frown\grave{\alpha}\grave{\varepsilon}F(\varepsilon,\alpha)) = \grave{\alpha}\grave{\varepsilon}F(\varepsilon,\alpha)\frown(o;a)$$

(33): ———————————

$$\vdash F(o,a) = \grave{\alpha}\grave{\varepsilon}F(\varepsilon,\alpha)\frown(o;a) \qquad (216$$

(IIIc): ———————————

$$\vdash G(\grave{\alpha}\grave{\varepsilon}F(\varepsilon,\alpha)\frown(o;a))$$
$$G(F(o,a)) \qquad (217$$

—— • ——

§148. *Analysis*

We will use (213) in order to prove the propositions (η) and (ϑ) of §146, requiring additionally the proposition

$$\text{‘}\vdash c\frown(e\frown p)$$
$$m;x = c;d$$
$$d\frown(i\frown q)$$
$$o;a = e;i$$
$$m\frown(o\frown p)$$
$$x\frown(a\frown q)\text{’} \qquad (\alpha$$

that we can prove using the proposition

$$\text{‘}\vdash m = c$$
$$x = d$$
$$m;x = c;d\text{’}$$

The latter follows from (Ξ).

§149. *Construction*

$$\Xi \vdash \grave{\varepsilon}(m\frown(x\frown\varepsilon)) = m;x$$

(IIIc): ———————————

$$\vdash \grave{\varepsilon}(m\frown(x\frown\varepsilon)) = c;d$$
$$m;x = c;d \qquad (\alpha$$

(IIIa): ———————————

$$\vdash \begin{array}{l} \dot{\varepsilon}(m\frown(x\frown\varepsilon)) = \dot{\varepsilon}(c\frown(d\frown\varepsilon)) \\ m;x = c;d \\ \dot{\varepsilon}(c\frown(d\frown\varepsilon)) = c;d \end{array} \qquad (\beta$$

(Ξ):: ——————————

$$\vdash \begin{array}{l} \dot{\varepsilon}(m\frown(x\frown\varepsilon)) = \dot{\varepsilon}(c\frown(d\frown\varepsilon)) \\ m;x = c;d \end{array} \qquad (\gamma$$

(Vb): − − − − − − − − − − − −

$$\vdash \begin{array}{l} m\frown\left[x\frown\dot{\alpha}\dot{\varepsilon}\left(\begin{array}{l}\varepsilon = c\\ \alpha = d\end{array}\right)\right] = c\frown\left[d\frown\dot{\alpha}\dot{\varepsilon}\left(\begin{array}{l}\varepsilon = c\\ \alpha = d\end{array}\right)\right] \\ m;x = c;d \end{array} \qquad (\delta$$

(33): ——————————

$$\vdash \begin{array}{l} \left(\begin{array}{l}m = c\\ x = d\end{array}\right) = c\frown\left[d\frown\dot{\alpha}\dot{\varepsilon}\left(\begin{array}{l}\varepsilon = c\\ \alpha = d\end{array}\right)\right] \\ m;x = c;d \end{array} \qquad (\varepsilon$$

(33): ——————————

$$\vdash \begin{array}{l} \left(\begin{array}{l}m = c\\ x = d\end{array}\right) = \left(\begin{array}{l}c = c\\ d = d\end{array}\right) \\ m;x = c;d \end{array} \qquad (\zeta$$

(IIIa): − − − − − − − − − − − −

$$\vdash \begin{array}{l} m = c \\ x = d \\ c = c \\ d = d \\ m;x = c;d \end{array} \qquad (\eta$$

(Ie):: − − − − − − −

$$\vdash \begin{array}{l} m = c \\ x = d \\ c = c \\ d = d \\ m;x = c;d \end{array} \qquad (\vartheta$$

(IIIe, IIIe):: ══════════

$$\vdash \begin{array}{l} m = c \\ x = d \\ m;x = c;d \end{array} \qquad (218$$

(Id): − − − − − − −

$$\vdash \begin{array}{l} x = d \\ m;x = c;d \end{array} \qquad (219$$

——— • ———

$$218 \vdash \begin{array}{l} m = c \\ x = d \\ m;x = c;d \end{array}$$

(Ib): − − − − − − −

$$\vdash \begin{array}{l} m = c \\ m;x = c;d \end{array} \qquad (220$$

(IIIh): − − − − − −

$$\vdash \begin{array}{l} f(m,x) = f(c,x) \\ m;x = c;d \end{array} \qquad (\alpha$$

(IIIc): ——————————

$$\vdash \begin{array}{l} f(m,x) = f(c,d) \\ m;x = c;d \\ x = d \end{array} \qquad (\beta$$

(219):: − − − − − − − − − −

$$\vdash \begin{array}{l} f(m,x) = f(c,d) \\ m;x = c;d \end{array} \qquad (221$$

(IIIc): − − − − − − − − −

$$\vdash \begin{array}{l} f(c,d) \\ f(m,x) \\ m;x = c;d \end{array} \qquad (222$$

——— • ———

$$222 \vdash \begin{array}{l} c\frown(o\frown p) \\ m;x = c;d \\ d\frown(a\frown q) \\ m\frown(o\frown p) \\ x\frown(a\frown q) \end{array}$$

(222): ——————————

$$\vdash \begin{array}{l} c\frown(e\frown p) \\ m;x = c;d \\ d\frown(i\frown q) \\ o;a = e;i \\ m\frown(o\frown p) \\ x\frown(a\frown q) \end{array} \qquad (\alpha$$

$\vdash$ [$\mathfrak{a}$ $\mathfrak{o}$ $\mathfrak{d}$ $\mathfrak{c}$] $\mathfrak{c}\frown(\mathfrak{o}\frown p)$
$m;x = \mathfrak{c};\mathfrak{d}$
$\mathfrak{d}\frown(\mathfrak{a}\frown q)$
$o;a = \mathfrak{o};\mathfrak{a}$
$m\frown(o\frown p)$
$x\frown(a\frown q)$ (β

(213): - - - - - - - - - - - - -

$\vdash$ $m;x\frown(o;a\frown(p\stackrel{\smile}{=}q))$
$m\frown(o\frown p)$
$x\frown(a\frown q)$ (γ

×

$\vdash$ $m\frown(o\frown p)$
$x\frown(a\frown q)$
$m;x\frown(o;a\frown(p\stackrel{\smile}{=}q))$ (223

(Id): - - - - - - - - - - -

$\vdash$ $x\frown(a\frown q)$
$m;x\frown(o;a\frown(p\stackrel{\smile}{=}q))$ (224

——— • ———

223 $\vdash$ $m\frown(o\frown p)$
$x\frown(a\frown q)$
$m;x\frown(o;a\frown(p\stackrel{\smile}{=}q))$

(Ib): - - - - - - - - - - -

$\vdash$ $m\frown(o\frown p)$
$m;x\frown(o;a\frown(p\stackrel{\smile}{=}q))$ (225

(IIa): - - - - - - - - - -

$\vdash$ [$\mathfrak{a}$] $F(o,\mathfrak{a})$
$x\frown(\mathfrak{a}\frown q)$
$m;x\frown(o;a\frown(p\stackrel{\smile}{=}q))$
[$\mathfrak{o}$ $\mathfrak{a}$] $F(\mathfrak{o},\mathfrak{a})$
$x\frown(\mathfrak{a}\frown q)$
$m\frown(\mathfrak{o}\frown p)$ (α

(IIa): - - - - - - - - - - - - - - -

$\vdash$ $F(o,a)$
$x\frown(a\frown q)$
$m;x\frown(o;a\frown(p\stackrel{\smile}{=}q))$
[$\mathfrak{o}$ $\mathfrak{a}$] $F(\mathfrak{o},\mathfrak{a})$
$x\frown(\mathfrak{a}\frown q)$
$m\frown(\mathfrak{o}\frown p)$ (β

(224):: - - - - - - - - - - - - - -

$\vdash$ $F(o,a)$
$m;x\frown(o;a\frown(p\stackrel{\smile}{=}q))$
[$\mathfrak{o}$ $\mathfrak{a}$] $F(\mathfrak{o},\mathfrak{a})$
$x\frown(\mathfrak{a}\frown q)$
$m\frown(\mathfrak{o}\frown p)$ (γ

(217): ————————————————

$\vdash$ $\dot{\alpha}\dot{\varepsilon}F(\varepsilon,\alpha)\frown(o;a)$
$m;x\frown(o;a\frown(p\stackrel{\smile}{=}q))$
[$\mathfrak{o}$ $\mathfrak{a}$] $F(\mathfrak{o},\mathfrak{a})$
$x\frown(\mathfrak{a}\frown q)$
$m\frown(\mathfrak{o}\frown p)$ (δ

(IIIb): - - - - - - - - - - - - - -

$\vdash$ $A = o;a$
[$\mathfrak{o}$ $\mathfrak{a}$] $F(\mathfrak{o},\mathfrak{a})$
$x\frown(\mathfrak{a}\frown q)$
$m\frown(\mathfrak{o}\frown p)$
$\dot{\alpha}\dot{\varepsilon}F(\varepsilon,\alpha)\frown A$
$m;x\frown(A\frown(p\stackrel{\smile}{=}q))$ (ε

$\vdash$ [$\mathfrak{a}$ $\mathfrak{o}$] $A = \mathfrak{o};\mathfrak{a}$
[$\mathfrak{o}$ $\mathfrak{a}$] $F(\mathfrak{o},\mathfrak{a})$
$x\frown(\mathfrak{a}\frown q)$
$m\frown(\mathfrak{o}\frown p)$
$\dot{\alpha}\dot{\varepsilon}F(\varepsilon,\alpha)\frown A$
$m;x\frown(A\frown(p\stackrel{\smile}{=}q))$ (ζ

(214): - - - - - - - - - - - - -

$\vdash$ $m;x\frown(A\frown(p\stackrel{\smile}{=}q))$
[$\mathfrak{o}$ $\mathfrak{a}$] $F(\mathfrak{o},\mathfrak{a})$
$x\frown(\mathfrak{a}\frown q)$
$m\frown(\mathfrak{o}\frown p)$
$\dot{\alpha}\dot{\varepsilon}F(\varepsilon,\alpha)\frown A$
$m;x\frown(A\frown(p\stackrel{\smile}{=}q))$ (η

×

$\vdash$ $\dot{\alpha}\dot{\varepsilon}F(\varepsilon,\alpha)\frown A$
$m;x\frown(A\frown(p\stackrel{\smile}{=}q))$
[$\mathfrak{o}$ $\mathfrak{a}$] $F(\mathfrak{o},\mathfrak{a})$
$x\frown(\mathfrak{a}\frown q)$
$m\frown(\mathfrak{o}\frown p)$ (226

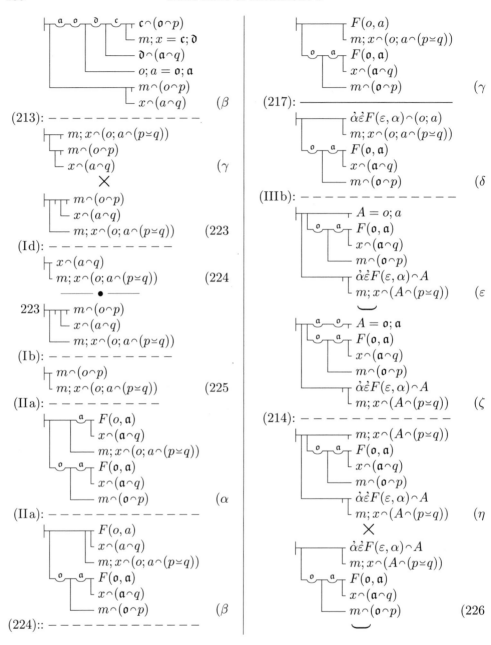

⊢ (𝔞) $\dot{\alpha}\dot{\varepsilon}F(\varepsilon,\alpha)\frown\mathfrak{a}$
$m;x\frown(\mathfrak{a}\frown(p\underline{\smile}q))$
(𝔬)(𝔞) $F(\mathfrak{o},\mathfrak{a})$
$x\frown(\mathfrak{a}\frown q)$
$m\frown(\mathfrak{o}\frown p)$ (227

———— • ————

IIa ⊢ (𝔠) $D=\mathfrak{c};d$
(𝔡)(𝔠) $D=\mathfrak{c};\mathfrak{d}$

(IIa): - - - - - - - -

⊢ $D=c;d$
(𝔡)(𝔠) $D=\mathfrak{c};\mathfrak{d}$ (α

(Ia): - - - - - - - -

⊢ $c\frown(o\frown p)$
$D=c;d$
$d\frown(a\frown q)$
$A=o;a$
(𝔡)(𝔠) $D=\mathfrak{c};\mathfrak{d}$ (β

⊢ (𝔞)(𝔬)(𝔡)(𝔠) $\mathfrak{c}\frown(\mathfrak{o}\frown p)$
$D=\mathfrak{c};\mathfrak{d}$
$\mathfrak{d}\frown(\mathfrak{a}\frown q)$
$A=\mathfrak{o};\mathfrak{a}$
(𝔡)(𝔠) $D=\mathfrak{c};\mathfrak{d}$ (γ

(213): - - - - - - - - - - - - -

⊢ $D\frown(A\frown(p\underline{\smile}q))$
(𝔡)(𝔠) $D=\mathfrak{c};\mathfrak{d}$ (228

———— • ————

216 ⊢ $F(n,y)=\dot{\alpha}\dot{\varepsilon}F(\varepsilon,\alpha)\frown(n;y)$

(IIIa): ————————————

⊢ $G(F(n,y))$
$G(\dot{\alpha}\dot{\varepsilon}F(\varepsilon,\alpha)\frown(n;y))$ (229

———— • ————

IIa ⊢ (𝔡)(𝔬)(𝔞) $F(\mathfrak{o},\mathfrak{a})$
$\mathfrak{d}\frown(\mathfrak{a}\frown q)$
$c\frown(\mathfrak{o}\frown p)$
$F(c,\mathfrak{d})$
(𝔠)(𝔡)(𝔬)(𝔞) $F(\mathfrak{o},\mathfrak{a})$
$\mathfrak{d}\frown(\mathfrak{a}\frown q)$
$\mathfrak{c}\frown(\mathfrak{o}\frown p)$
$F(\mathfrak{c},\mathfrak{d})$

(IIa): - - - - - - - - - - - - -

⊢ (𝔬)(𝔞) $F(\mathfrak{o},\mathfrak{a})$
$d\frown(\mathfrak{a}\frown q)$
$c\frown(\mathfrak{o}\frown p)$
$F(c,d)$
(𝔠)(𝔡)(𝔬)(𝔞) $F(\mathfrak{o},\mathfrak{a})$
$\mathfrak{d}\frown(\mathfrak{a}\frown q)$
$\mathfrak{c}\frown(\mathfrak{o}\frown p)$
$F(\mathfrak{c},\mathfrak{d})$ (α

(226): - - - - - - - - - - - -

⊢ $\dot{\alpha}\dot{\varepsilon}F(\varepsilon,\alpha)\frown A$
$c;d\frown(A\frown(p\underline{\smile}q))$
$F(c,d)$
(𝔠)(𝔡)(𝔬)(𝔞) $F(\mathfrak{o},\mathfrak{a})$
$\mathfrak{d}\frown(\mathfrak{a}\frown q)$
$\mathfrak{c}\frown(\mathfrak{o}\frown p)$
$F(\mathfrak{c},\mathfrak{d})$ (β

(217): ————————————

⊢ $\dot{\alpha}\dot{\varepsilon}F(\varepsilon,\alpha)\frown A$
$c;d\frown(A\frown(p\underline{\smile}q))$
$\dot{\alpha}\dot{\varepsilon}F(\varepsilon,\alpha)\frown(c;d)$
(𝔠)(𝔡)(𝔬)(𝔞) $F(\mathfrak{o},\mathfrak{a})$
$\mathfrak{d}\frown(\mathfrak{a}\frown q)$
$\mathfrak{c}\frown(\mathfrak{o}\frown p)$
$F(\mathfrak{c},\mathfrak{d})$ (γ

(IIIb): - - - - - - - - - - - - - - -

⊢ $D=c;d$
(𝔠)(𝔡)(𝔬)(𝔞) $F(\mathfrak{o},\mathfrak{a})$
$\mathfrak{d}\frown(\mathfrak{a}\frown q)$
$\mathfrak{c}\frown(\mathfrak{o}\frown p)$
$F(\mathfrak{c},\mathfrak{d})$
$\dot{\alpha}\dot{\varepsilon}F(\varepsilon,\alpha)\frown A$
$D\frown(A\frown(p\underline{\smile}q))$
$\dot{\alpha}\dot{\varepsilon}F(\varepsilon,\alpha)\frown D$ (δ

⊢ (𝔡)(𝔠) $D=\mathfrak{c};\mathfrak{d}$
(𝔠)(𝔡)(𝔬)(𝔞) $F(\mathfrak{o},\mathfrak{a})$
$\mathfrak{d}\frown(\mathfrak{a}\frown q)$
$\mathfrak{c}\frown(\mathfrak{o}\frown p)$
$F(\mathfrak{c},\mathfrak{d})$
$\dot{\alpha}\dot{\varepsilon}F(\varepsilon,\alpha)\frown A$
$D\frown(A\frown(p\underline{\smile}q))$
$\dot{\alpha}\dot{\varepsilon}F(\varepsilon,\alpha)\frown D$ (ε

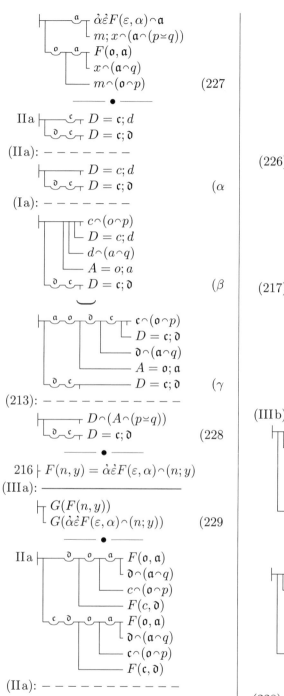

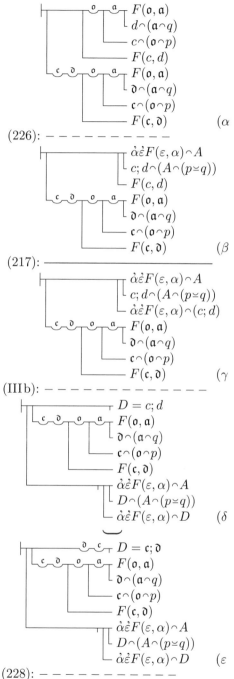

(228): - - - - - - - - - - - -

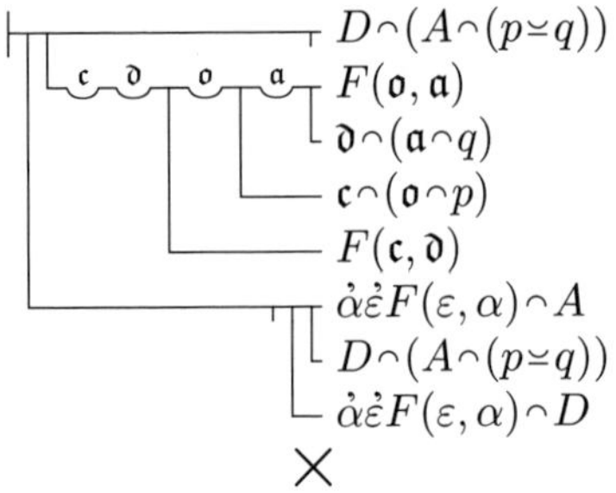

(ζ

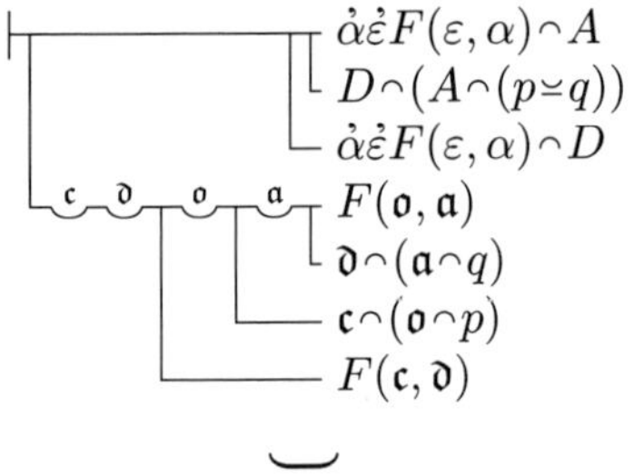

(η

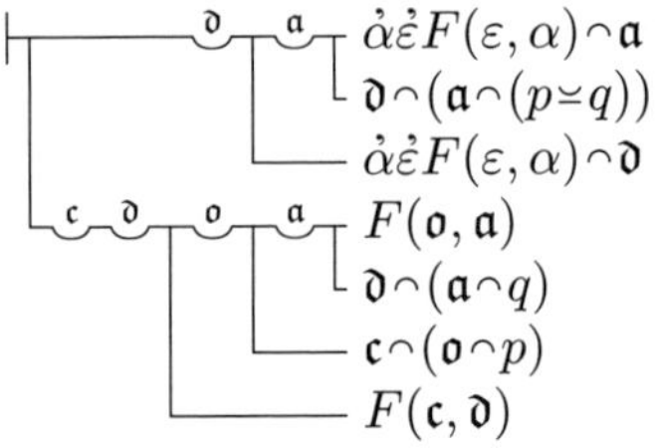

(230

(123): - - - - - - - - - - - - - -

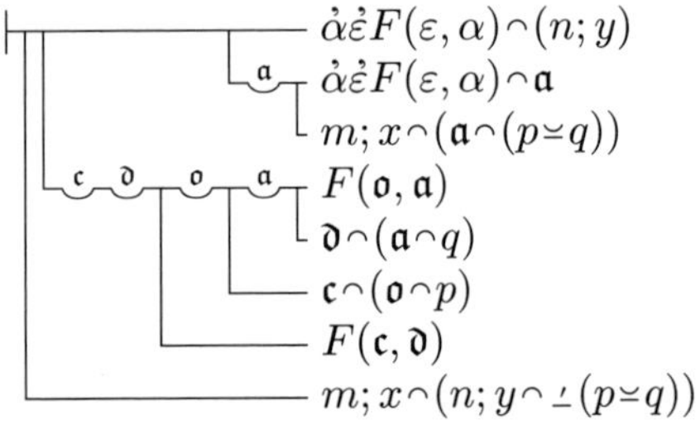

(α

(229): - - - - - - - - - - - - - - - - - -

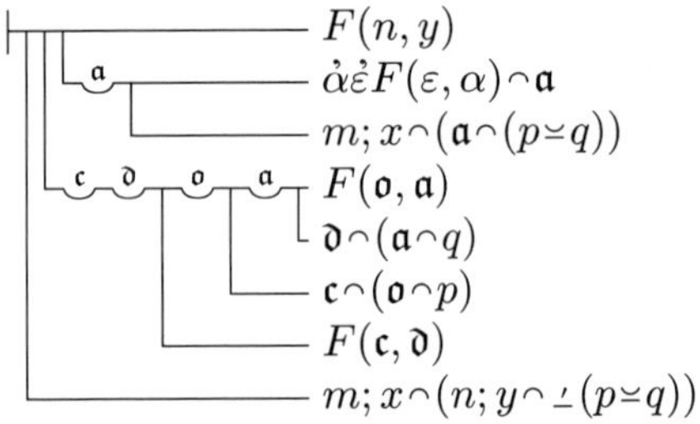

(β

(227):: -

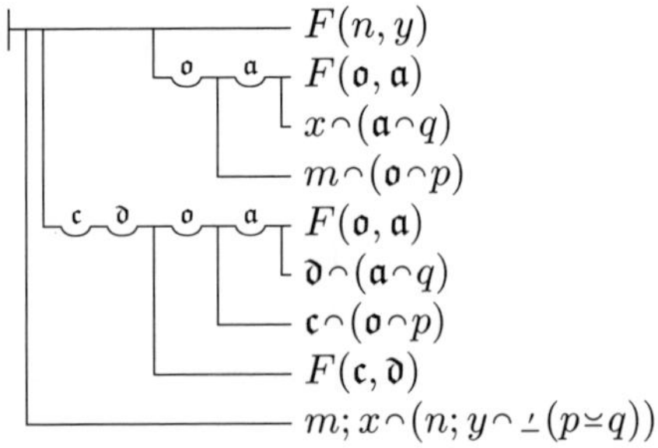

(231

§150. *Analysis*

In order now to derive the proposition (β) of §146, we replace the function-marker '$F(\xi,\zeta)$' in (231) by

$$`\begin{array}{l} \top\!\!\top\; m\frown(\xi\frown\underline{\prime}p) \\ \quad\llcorner\; x\frown(\zeta\frown\underline{\prime}q) \end{array}'$$

The propositions that are needed for this are straightforwardly proven from (133) and (131).

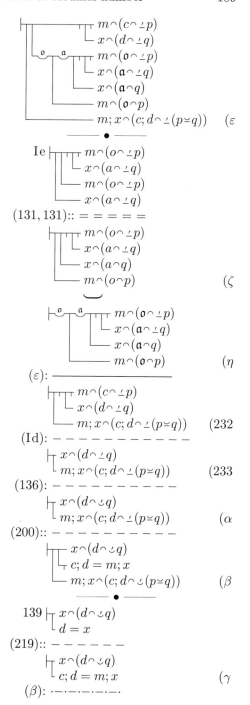

§151. *Construction*

$$133 \vdash \begin{array}{l} m\frown(o\frown\underline{\prime}p) \\ c\frown(o\frown p) \\ m\frown(c\frown\underline{\prime}p) \end{array}$$

(Ie): — — — — — —

$$\vdash \begin{array}{l} m\frown(o\frown\underline{\prime}p) \\ x\frown(a\frown\underline{\prime}q) \\ c\frown(o\frown p) \\ m\frown(c\frown\underline{\prime}p) \\ x\frown(a\frown\underline{\prime}q) \end{array} \qquad (\alpha$$

(133):: — — — — — — — —

$$\vdash \begin{array}{l} m\frown(o\frown\underline{\prime}p) \\ x\frown(a\frown\underline{\prime}q) \\ c\frown(o\frown p) \\ m\frown(c\frown\underline{\prime}p) \\ d\frown(a\frown q) \\ x\frown(d\frown\underline{\prime}q) \end{array} \qquad (\beta$$

(Ib, Id):: = = = = = = =

$$\vdash \begin{array}{l} m\frown(o\frown\underline{\prime}p) \\ x\frown(a\frown\underline{\prime}q) \\ d\frown(a\frown q) \\ c\frown(o\frown p) \\ m\frown(c\frown\underline{\prime}p) \\ x\frown(d\frown\underline{\prime}q) \end{array} \qquad (\gamma$$

$$\vdash\!\smile^{\mathfrak{c}}\!\smile^{\mathfrak{d}}\!\smile^{\mathfrak{o}}\!\smile^{\mathfrak{a}} \begin{array}{l} m\frown(\mathfrak{o}\frown\underline{\prime}p) \\ x\frown(\mathfrak{a}\frown\underline{\prime}q) \\ \mathfrak{d}\frown(\mathfrak{a}\frown q) \\ \mathfrak{c}\frown(\mathfrak{o}\frown p) \\ m\frown(\mathfrak{c}\frown\underline{\prime}p) \\ x\frown(\mathfrak{d}\frown\underline{\prime}q) \end{array} \qquad (\delta$$

(231): ————————————

$$\vdash \begin{array}{l} m\frown(c\frown\underline{\prime}p) \\ x\frown(d\frown\underline{\prime}q) \\ \smile^{\mathfrak{o}}\!\smile^{\mathfrak{a}}\; m\frown(\mathfrak{o}\frown\underline{\prime}p) \\ \quad x\frown(\mathfrak{a}\frown\underline{\prime}q) \\ \quad x\frown(\mathfrak{a}\frown q) \\ \quad m\frown(\mathfrak{o}\frown p) \\ m;x\frown(c;d\frown\underline{\prime}(p\simeq q)) \end{array} \qquad (\varepsilon$$

——— • ———

$$\text{Ie} \vdash \begin{array}{l} m\frown(o\frown\underline{\prime}p) \\ x\frown(a\frown\underline{\prime}q) \\ m\frown(o\frown\underline{\prime}p) \\ x\frown(a\frown\underline{\prime}q) \end{array}$$

(131, 131):: = = = = =

$$\vdash \begin{array}{l} m\frown(o\frown\underline{\prime}p) \\ x\frown(a\frown\underline{\prime}q) \\ x\frown(a\frown q) \\ m\frown(o\frown p) \end{array} \qquad (\zeta$$

$$\vdash\!\smile^{\mathfrak{o}}\!\smile^{\mathfrak{a}} \begin{array}{l} m\frown(\mathfrak{o}\frown\underline{\prime}p) \\ x\frown(\mathfrak{a}\frown\underline{\prime}q) \\ x\frown(\mathfrak{a}\frown q) \\ m\frown(\mathfrak{o}\frown p) \end{array} \qquad (\eta$$

(ε): ————————

$$\vdash \begin{array}{l} m\frown(c\frown\underline{\prime}p) \\ x\frown(d\frown\underline{\prime}q) \\ m;x\frown(c;d\frown\underline{\prime}(p\simeq q)) \end{array} \qquad (232$$

(Id): — — — — — — — — — — —

$$\vdash \begin{array}{l} x\frown(d\frown\underline{\prime}q) \\ m;x\frown(c;d\frown\underline{\prime}(p\simeq q)) \end{array} \qquad (233$$

(136): — — — — — — — — — — —

$$\vdash \begin{array}{l} x\frown(d\frown\overset{\prime}{\smile}q) \\ m;x\frown(c;d\frown\underline{\prime}(p\simeq q)) \end{array} \qquad (\alpha$$

(200):: — — — — — — — — — — —

$$\vdash \begin{array}{l} x\frown(d\frown\overset{\prime}{\smile}q) \\ c;d = m;x \\ m;x\frown(c;d\frown\overset{\prime}{\smile}(p\simeq q)) \end{array} \qquad (\beta$$

——— • ———

$$139 \vdash \begin{array}{l} x\frown(d\frown\overset{\prime}{\smile}q) \\ d = x \end{array}$$

(219):: — — — — — —

$$\vdash \begin{array}{l} x\frown(d\frown\overset{\prime}{\smile}q) \\ c;d = m;x \end{array} \qquad (\gamma$$

(β): -·-·-·-·-·

$$\vdash\; x\frown(d\frown\smile q) \;\; m;x\frown(c;d\frown\smile(p\simeq q)) \qquad (234$$

(210): ———

$$\vdash\; x\frown(d\frown\smile q) \;\; c\frown(d\frown(m;x\lessdot(p\simeq q))) \qquad (235$$

×

$$\vdash\; c\frown(d\frown(m;x\lessdot(p\simeq q))) \;\; x\frown(d\frown\smile q) \qquad (236$$

(212): -·-·-·-·-·-·-·-·-·-

$$\vdash\; c\frown(d\frown(m;x\lessdot(p\simeq q))) \;\; \overset{\mathfrak{e}}{\smile}\, o\frown(\mathfrak{e}\frown(m;x\lessdot(p\simeq q))) \;\; c\frown(o\frown p) \;\; \overset{\mathfrak{d}}{\smile}\,[\mathfrak{d}\frown(x\frown\nparallel\smile q) \;\; \overset{\mathfrak{e}}{\smile}\,\mathfrak{d}\frown(\mathfrak{e}\frown q)] \qquad (\alpha$$

$$\vdash\; \overset{\mathfrak{e}}{\smile}\, c\frown(\mathfrak{e}\frown(m;x\lessdot(p\simeq q))) \;\; \overset{\mathfrak{e}}{\smile}\, o\frown(\mathfrak{e}\frown(m;x\lessdot(p\simeq q))) \;\; c\frown(o\frown p) \;\; \overset{\mathfrak{d}}{\smile}\,[\mathfrak{d}\frown(x\frown\nparallel\smile q) \;\; \overset{\mathfrak{e}}{\smile}\,\mathfrak{d}\frown(\mathfrak{e}\frown q)] \qquad (\beta$$

×

$$\vdash\; \overset{\mathfrak{e}}{\smile}\, o\frown(\mathfrak{e}\frown(m;x\lessdot(p\simeq q))) \;\; c\frown(o\frown p) \;\; \overset{\mathfrak{e}}{\smile}\, c\frown(\mathfrak{e}\frown(m;x\lessdot(p\simeq q))) \;\; \overset{\mathfrak{d}}{\smile}\,[\mathfrak{d}\frown(x\frown\nparallel\smile q) \;\; \overset{\mathfrak{e}}{\smile}\,\mathfrak{d}\frown(\mathfrak{e}\frown q)] \qquad (\gamma$$

$$\vdash\; \overset{\mathfrak{d}}{\smile}\overset{\mathfrak{a}}{\smile}\,[\overset{\mathfrak{e}}{\smile}\,\mathfrak{a}\frown(\mathfrak{e}\frown(m;x\lessdot(p\simeq q))) \;\; \mathfrak{d}\frown(\mathfrak{a}\frown p)] \;\; \overset{\mathfrak{e}}{\smile}\,\mathfrak{d}\frown(\mathfrak{e}\frown(m;x\lessdot(p\simeq q))) \;\; \overset{\mathfrak{d}}{\smile}\,[\mathfrak{d}\frown(x\frown\nparallel\smile q) \;\; \overset{\mathfrak{e}}{\smile}\,\mathfrak{d}\frown(\mathfrak{e}\frown q)] \qquad (\delta$$

(144): - - - - - - - - - - - - - - -

$$\vdash\; \overset{\mathfrak{e}}{\smile}\, n\frown(\mathfrak{e}\frown(m;x\lessdot(p\simeq q))) \;\; \overset{\mathfrak{e}}{\smile}\, m\frown(\mathfrak{e}\frown(m;x\lessdot(p\simeq q))) \;\; \overset{\mathfrak{d}}{\smile}\,[\mathfrak{d}\frown(x\frown\nparallel\smile q) \;\; \overset{\mathfrak{e}}{\smile}\,\mathfrak{d}\frown(\mathfrak{e}\frown q)] \;\; m\frown(n\frown\smile p) \qquad (237$$

——— • ———

$$140 \vdash m;x\frown(m;x\frown\smile t)$$

(210): ———

$$\vdash m\frown(x\frown(m;x\lessdot t)) \qquad (238$$

——— • ———

$$\text{IIa} \vdash\; m\frown(x\frown(m;x\lessdot t)) \;\; \overset{\mathfrak{e}}{\smile}\, m\frown(\mathfrak{e}\frown(m;x\lessdot t))$$

×

$$\vdash\; \overset{\mathfrak{e}}{\smile}\, m\frown(\mathfrak{e}\frown(m;x\lessdot t)) \;\; m\frown(x\frown(m;x\lessdot t)) \qquad (\alpha$$

(238):: ———

$$\vdash \overset{\mathfrak{e}}{\smile}\, m\frown(\mathfrak{e}\frown(m;x\lessdot t)) \qquad (239$$

——— • ———

$$239 \vdash \overset{\mathfrak{e}}{\smile}\, m\frown(\mathfrak{e}\frown(m;x\lessdot(p\simeq q)))$$

(237): ———

$$\vdash\; \overset{\mathfrak{e}}{\smile}\, n\frown(\mathfrak{e}\frown(m;x\lessdot(p\simeq q))) \;\; \overset{\mathfrak{d}}{\smile}\,[\mathfrak{d}\frown(x\frown\nparallel\smile q) \;\; \overset{\mathfrak{e}}{\smile}\,\mathfrak{d}\frown(\mathfrak{e}\frown q)] \;\; m\frown(n\frown\smile p) \qquad (\alpha$$

(22): ———

$$\vdash\; \overset{\mathfrak{e}}{\smile}\, n\frown(\mathfrak{e}\frown(m;x\lessdot(p\simeq q))) \;\; \overset{\mathfrak{d}}{\smile}\,[\mathfrak{d}\frown(x\frown\nparallel\smile q) \;\; \overset{\mathfrak{e}}{\smile}\,\mathfrak{d}\frown(\mathfrak{e}\frown q)] \;\; n\frown(m\frown\nparallel\smile p) \qquad (\beta$$

×

$$\vdash\; n\frown(m\frown\nparallel\smile p) \;\; \overset{\mathfrak{e}}{\smile}\, n\frown(\mathfrak{e}\frown(m;x\lessdot(p\simeq q))) \;\; \overset{\mathfrak{d}}{\smile}\,[\mathfrak{d}\frown(x\frown\nparallel\smile q) \;\; \overset{\mathfrak{e}}{\smile}\,\mathfrak{d}\frown(\mathfrak{e}\frown q)] \qquad (240$$

——— • ———

$$236 \vdash\; n\frown(y\frown(m;x\lessdot(p\simeq q))) \;\; x\frown(y\frown\smile q)$$

(22): ———

$$\vdash\; n\frown(y\frown(m;x\lessdot(p\simeq q))) \;\; y\frown(x\frown\nparallel\smile q) \qquad (\alpha$$

(IIa):: - - - - - - - - - - -

$$\vdash\; n\frown(y\frown(m;x\lessdot(p\simeq q))) \;\; n\frown(y\frown(m;x\lessdot(p\simeq q))) \;\; \overset{\mathfrak{a}}{\smile}\,[\mathfrak{a}\frown(x\frown\nparallel\smile q) \;\; n\frown(\mathfrak{a}\frown(m;x\lessdot(p\simeq q)))] \qquad (\beta$$

(Ig): - - - - - - - - - - -

$$\vdash\; n\frown(y\frown(m;x\lessdot(p\simeq q))) \;\; \overset{\mathfrak{a}}{\smile}\,[\mathfrak{a}\frown(x\frown\nparallel\smile q) \;\; n\frown(\mathfrak{a}\frown(m;x\lessdot(p\simeq q)))] \qquad (\gamma$$

$$\vdash\; \overset{\mathfrak{e}}{\smile}\, n\frown(\mathfrak{e}\frown(m;x\lessdot(p\simeq q))) \;\; \overset{\mathfrak{a}}{\smile}\,[\mathfrak{a}\frown(x\frown\nparallel\smile q) \;\; n\frown(\mathfrak{a}\frown(m;x\lessdot(p\simeq q)))] \qquad (\delta$$

(240): - - - - - - - - - - -

⊢ n⌢(m⌢∤⌣p)
  𝔞: 𝔞⌢(x⌢∤⌣q)
    n⌢(𝔞⌢(m; x ⌒ (p≚q)))
  𝔡: 𝔡⌢(x⌢∤⌣q)
    𝔢: 𝔡⌢(𝔢⌢q) (ε

⊢ 𝔡: 𝔡⌢(m⌢∤⌣p)
  𝔞: 𝔞⌢(x⌢∤⌣q)
    𝔡⌢(𝔞⌢(m; x ⌒ (p≚q)))
  𝔡: 𝔡⌢(x⌢∤⌣q)
    𝔢: 𝔡⌢(𝔢⌢q) (ζ

(11): – – – – – – – – – – – – – – –

⊢ m⌢∤⌣p⌢(x⌢∤⌣q⌢)(m; x ⌒ (p≚q)))
  𝔡: 𝔡⌢(x⌢∤⌣q)
    𝔢: 𝔡⌢(𝔢⌢q)
  I(m; x ⌒ (p≚q)) (241

§152. *Analysis*

In (241) we have to replace the subcomponent

'— I(m; x ⌒ (p≚q))'

by others, in order to obtain our proposition (δ) of §144. The train of thought here is as follows. If the pairs $(b; d)$ and $(b; a)$ belong to the $(p \simeq q)$-series starting with $(m; x)$, then either $(b; d)$ must belong to the $(p \simeq q)$-series starting with $(b; a)$, or [$(b; a)$ must][e] follow after $(b; d)$ in this series, insofar the $(p \simeq q)$-relation is single-valued. If either $(b; a)$ follows after $(b; d)$ or $(b; d)$ follows after $(b; a)$ in this series, then b has to follow after itself in the p-series, which would contradict our subcomponent

'𝔦: 𝔦⌢(𝔦⌢⌣̲p)
  m⌢(𝔦⌢⌣p)'

The only remaining possibility is that $(b; d)$ coincides with $(b; a)$. In that case d also coincides with a.

We therefore need the proposition

'⊢ r⌢(n⌢⌣p)
  n⌢(r⌢⌣̲p)
  m⌢(n⌢⌣p)
  Ip
  m⌢(r⌢⌣p)' (α

in words: "If a first and a second objects belong to the p-series starting with a third, then the first precedes the second or belongs to the series starting with the second, provided the series-forming relation is single-valued."

We prove the proposition by means of (144), replacing the function-marker '$F(\xi)$' by '— ξ⌢(n⌢⌣p) / n⌢(ξ⌢⌣̲p)'. We then have to prove the proposition

'⊢ a⌢(n⌢⌣p)
  n⌢(a⌢⌣̲p)
  d⌢(a⌢p)
  d⌢(n⌢⌣p)
  n⌢(d⌢⌣̲p)
  Ip' (β

For this we need the proposition

'⊢ a⌢(n⌢⌣p)
  d⌢(a⌢p)
  Ip
  d⌢(n⌢⌣̲p)' (γ

which we deduce from the propositions

'⊢ a⌢(n⌢⌣p)
  𝔞: a⌢(𝔞⌢⌣p)
    d⌢(𝔞⌢p)
  d⌢(n⌢⌣̲p)' (δ

and

'⊢ $\mathfrak{a}$ — $a \frown (\mathfrak{a} \frown \breve{\ } p)$
　$d \frown (\mathfrak{a} \frown p)$
　$d \frown (a \frown p)$
　$\text{I}p$ ', (ε

§153. *Construction*

137 ⊢ $a \frown (m \frown \breve{\ } p)$
　$e \frown (m \frown p)$
　$a \frown (e \frown \breve{\ } p)$

⊢ $\mathfrak{d}$ $\mathfrak{a}$ — $a \frown (\mathfrak{a} \frown \breve{\ } p)$
　$\mathfrak{d} \frown (\mathfrak{a} \frown p)$
　$a \frown (\mathfrak{d} \frown \breve{\ } p)$ (α

(123): ——————

⊢ $a \frown (n \frown \breve{\ } p)$
　$\mathfrak{a}$ — $a \frown (\mathfrak{a} \frown \breve{\ } p)$
　$d \frown (\mathfrak{a} \frown p)$
　$d \frown (n \frown \underline{\ } p)$ (β

——— • ———

13 ⊢ $b = a$
　$d \frown (a \frown p)$
　$d \frown (b \frown p)$
　$\text{I}p$

(139): – – – – – –

⊢ $a \frown (b \frown \breve{\ } p)$
　$d \frown (b \frown p)$
　$d \frown (a \frown p)$
　$\text{I}p$ (γ

⊢ $\mathfrak{a}$ — $a \frown (\mathfrak{a} \frown \breve{\ } p)$
　$d \frown (\mathfrak{a} \frown p)$
　$d \frown (a \frown p)$
　$\text{I}p$ (δ

(β): – – – – – – – –

⊢ $a \frown (n \frown \breve{\ } p)$
　$d \frown (a \frown p)$
　$\text{I}p$
　$d \frown (n \frown \underline{\ } p)$ (242

——— • ———

131 ⊢ $n \frown (a \frown \underline{\ } p)$
　$n \frown (a \frown p)$

(IIIc): ——————

⊢ $n \frown (a \frown \underline{\ } p)$
　$d \frown (a \frown p)$
　$n = d$ (α

(130):: – – – – – –

⊢ $n \frown (a \frown \underline{\ } p)$
　$d \frown (a \frown p)$
　$d \frown (n \frown \underline{\ } p)$
　$d \frown (n \frown \breve{\ } p)$ (β

(Ia): – – – – – – – –

⊢ $a \frown (n \frown \breve{\ } p)$
　$n \frown (a \frown \underline{\ } p)$
　$d \frown (a \frown p)$
　$d \frown (n \frown \underline{\ } p)$
　$d \frown (n \frown \breve{\ } p)$ (γ

(242): ·–·–·–·–·–·

⊢ $a \frown (n \frown \breve{\ } p)$
　$n \frown (a \frown \underline{\ } p)$
　$d \frown (a \frown p)$
　$d \frown (n \frown \breve{\ } p)$
　$\text{I}p$ (δ

(I):: – – – – – – –

⊢ $a \frown (n \frown \breve{\ } p)$
　$n \frown (a \frown \underline{\ } p)$
　$d \frown (a \frown p)$
　$n \frown (d \frown \underline{\ } p)$
　$d \frown (n \frown \breve{\ } p)$
　$n \frown (d \frown \underline{\ } p)$
　$\text{I}p$ (ε

——— • ———

133 ⊢ $n \frown (a \frown \underline{\ } p)$
　$d \frown (a \frown p)$
　$n \frown (d \frown \underline{\ } p)$

×

⊢ $n \frown (d \frown \underline{\ } p)$
　$d \frown (a \frown p)$
　$n \frown (a \frown \underline{\ } p)$ (ζ

(ε): – – – – – – –

⊢ $a \frown (n \frown \smile p)$
$n \frown (a \frown \perp p)$
$d \frown (a \frown p)$
$d \frown (n \frown \smile p)$
$n \frown (d \frown \perp p)$
$\mathrm{I}p$ (η

⊢ $\mathfrak{d}$ $\mathfrak{a}$ $\mathfrak{a} \frown (n \frown \smile p)$
$n \frown (\mathfrak{a} \frown \perp p)$
$\mathfrak{d} \frown (\mathfrak{a} \frown p)$
$\mathfrak{d} \frown (n \frown \smile p)$
$n \frown (\mathfrak{d} \frown \perp p)$
$\mathrm{I}p$ (ϑ

(144): – – – – – – – – – – –

⊢ $r \frown (n \frown \smile p)$
$n \frown (r \frown \perp p)$
$m \frown (n \frown \smile p)$
$n \frown (m \frown \perp p)$
$\mathrm{I}p$
$m \frown (r \frown \smile p)$ (ι

(I):: – – – – – – – – –

⊢ $r \frown (n \frown \smile p)$
$n \frown (r \frown \perp p)$
$m \frown (n \frown \smile p)$
$\mathrm{I}p$
$m \frown (r \frown \smile p)$ (243

———— • ————

232 ⊢ $m \frown (b \frown \perp p)$
$x \frown (d \frown \perp q)$
$m; x \frown (b; d \frown \perp (p \asymp q))$

(Ib): – – – – – – – – – – – –

⊢ $m \frown (b \frown \perp p)$
$m; x \frown (b; d \frown \perp (p \asymp q))$ (244

×

⊢ $m; x \frown (b; d \frown \perp (p \asymp q))$
$m \frown (b \frown \perp p)$ (245

———— • ————

136 ⊢ $m \frown (b \frown \smile p)$
$m \frown (b \frown \perp p)$

(244):: – – – – – –

⊢ $m \frown (b \frown \smile p)$
$m; x \frown (b; d \frown \perp (p \asymp q))$ (α

(200):: – – – – – – – – – – –

⊢ $m \frown (b \frown \smile p)$
$b; d = m; x$
$m; x \frown (b; d \frown \smile (p \asymp q))$ (β

———— • ————

139 ⊢ $m \frown (b \frown \smile p)$
$b = m$

(220):: – – – – – –

⊢ $m \frown (b \frown \smile p)$
$b; d = m; x$ (γ

(β): ·–·–·–·–·–·

⊢ $m \frown (b \frown \smile p)$
$m; x \frown (b; d \frown \smile (p \asymp q))$ (246

———— • ————

II ⊢ $\dot{\alpha}\dot{\varepsilon}(A \frown (\varepsilon; \alpha \frown \smile t)) = A \nwarrow t$

(6): ————————————————

⊢ $F(A \frown (b; d \frown \smile t))$
$F(b \frown (d \frown (A \nwarrow t)))$ (247

———— • ————

243 ⊢ $b; a \frown (b; d \frown \smile (p \asymp q))$
$b; d \frown (b; a \frown \perp (p \asymp q))$
$m; x \frown (b; d \frown \smile (p \asymp q))$
$\mathrm{I}(p \asymp q)$
$m; x \frown (b; a \frown \smile (p \asymp q))$

(130): – – – – – – – – – – – – –

⊢ $b; d = b; a$
$b; a \frown (b; d \frown \perp (p \asymp q))$
$b; d \frown (b; a \frown \perp (p \asymp q))$
$m; x \frown (b; d \frown \smile (p \asymp q))$
$\mathrm{I}(p \asymp q)$
$m; x \frown (b; a \frown \smile (p \asymp q))$ (α

(245, 245):: = = = = = = = = = =

⊢ $b; d = b; a$
$b \frown (b \frown \perp p)$
$m; x \frown (b; d \frown \smile (p \asymp q))$
$\mathrm{I}(p \asymp q)$
$m; x \frown (b; a \frown \smile (p \asymp q))$ (β

(219): – – – – – – – – – – – – –

⊢ $d = a$
$b \frown (b \frown \perp p)$
$m; x \frown (b; d \frown \smile (p \asymp q))$
$\mathrm{I}(p \asymp q)$
$m; x \frown (b; a \frown \smile (p \asymp q))$ (γ

(IIa):: – – – – – – – – – – – – –

⊢ d = a
m⌢(b⌢⌣p)
i: i⌢(i⌢⌐p)
m⌢(i⌢⌣p)
m; x⌢(b; d⌢⌣(p≚q))
I(p≚q)
m; x⌢(b; a⌢⌣(p≚q)) (δ

(246):: – – – – – – – – – – – – – –

⊢ d = a
i: i⌢(i⌢⌐p)
m⌢(i⌢⌣p)
m; x⌢(b; d⌢⌣(p≚q))
I(p≚q)
m; x⌢(b; a⌢⌣(p≚q)) (ε

(247, 247):: = = = = = = = = = = = =

⊢ d = a
b⌢(a⌢(m; x ⋖ (p≚q)))
b⌢(d⌢(m; x ⋖ (p≚q)))
i: i⌢(i⌢⌐p)
m⌢(i⌢⌣p)
I(p≚q) (ζ

⊢ 𝔢 𝔡 𝔞: 𝔡 = 𝔞
𝔢⌢(𝔞⌢(m; x ⋖ (p≚q)))
𝔢⌢(𝔡⌢(m; x ⋖ (p≚q)))
i: i⌢(i⌢⌐p)
m⌢(i⌢⌣p)
I(p≚q) (η

(16): – – – – – – – – – – – – – –

⊢ I(m; x ⋖ (p≚q))
i: i⌢(i⌢⌐p)
m⌢(i⌢⌣p)
I(p≚q) (248

§154. *Analysis*

We now further prove the proposition

'⊢ I(p≚q)
Ip
Iq ', (α

For this we need the proposition

'⊢ o; a = e; i
o = e
a = i ', (β

which is to be derived from (Ξ). From this with (13) we gain the proposition

'⊢ o; a = e; i
c⌢(e⌢p)
c⌢(o⌢p)
Ip
d⌢(i⌢q)
d⌢(a⌢q)
Iq ', (γ

We introduce 'D' for '$o; a$' and 'A' for '$e; i$' and, after introducing German letters, apply (213).

§155. *Construction*

Ξ ⊢ ἐ(o⌢(a⌢ε)) = o; a

(IIIc): ————————

⊢ F(o; a)
F(ἐ(o⌢(a⌢ε))) (249

—— • ——

221 ⊢ f(m, x) = f(c, d)
m; x = c; d

(IIIa): – – – – – – – – – –

⊢ f(m, x)
f(c, d)
m; x = c; d (250

—— • ——

IIIh ⊢ o⌢(i⌢t) = e⌢(i⌢t)
o = e

(IIIa): ————————

⊢ o⌢(a⌢t) = e⌢(i⌢t)
o = e
a = i (α

$\vdash \mathfrak{a}\ o\frown(a\frown\mathfrak{a}) = e\frown(i\frown\mathfrak{a})$
$o = e$
$a = i$ (β

(V a): - - - - - - - - - - - -

$\vdash \dot{\varepsilon}(o\frown(a\frown\varepsilon)) = \dot{\varepsilon}(e\frown(i\frown\varepsilon))$
$o = e$
$a = i$ (γ

(249): ————————

$\vdash o; a = \dot{\varepsilon}(e\frown(i\frown\varepsilon))$
$o = e$
$a = i$ (δ

(249): ————————

$\vdash o; a = e; i$
$o = e$
$a = i$ (251

(13, 13):: = = = = =

$\vdash o; a = e; i$
$c\frown(e\frown p)$
$c\frown(o\frown p)$
$\mathrm{I}p$
$d\frown(i\frown q)$
$d\frown(a\frown q)$
$\mathrm{I}q$ (α

(III a): ————————

$\vdash D = e; i$
$c\frown(e\frown p)$
$c\frown(o\frown p)$
$\mathrm{I}p$
$d\frown(i\frown q)$
$d\frown(a\frown q)$
$\mathrm{I}q$
$D = o; a$ (β

×

$\vdash c\frown(o\frown p)$
$c\frown(e\frown p)$
$D = e; i$
$\mathrm{I}p$
$d\frown(i\frown q)$
$d\frown(a\frown q)$
$\mathrm{I}q$
$D = o; a$ (γ

(250): ————————

$\vdash m\frown(o\frown p)$
$c\frown(e\frown p)$
$D = e; i$
$\mathrm{I}p$
$d\frown(i\frown q)$
$x\frown(a\frown q)$
$\mathrm{I}q$
$D = o; a$
$m; x = c; d$ (δ

(III a): ————————

$\vdash m\frown(o\frown p)$
$E = m; x$
$x\frown(a\frown q)$
$D = o; a$
$c\frown(e\frown p)$
$D = e; i$
$\mathrm{I}p$
$d\frown(i\frown q)$
$\mathrm{I}q$
$E = c; d$ (ε

$\smile$

$\vdash \mathfrak{a}\ \mathfrak{o}\ \mathfrak{d}\ \mathfrak{c}\ \mathfrak{c}\frown(\mathfrak{o}\frown p)$
$E = \mathfrak{c}; \mathfrak{d}$
$\mathfrak{d}\frown(\mathfrak{a}\frown q)$
$D = \mathfrak{o}; \mathfrak{a}$
$c\frown(e\frown p)$
$D = e; i$
$\mathrm{I}p$
$d\frown(i\frown q)$
$\mathrm{I}q$
$E = c; d$ (ζ

(213): - - - - - - - - - - - -

$\vdash E\frown(D\frown(p \mathbin{\bar{\smile}} q))$
$c\frown(e\frown p)$
$D = e; i$
$\mathrm{I}p$
$d\frown(i\frown q)$
$\mathrm{I}q$
$E = c; d$ (η

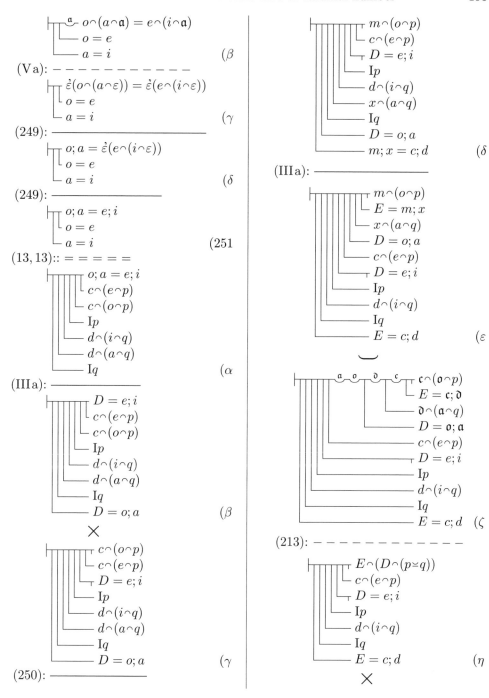

×

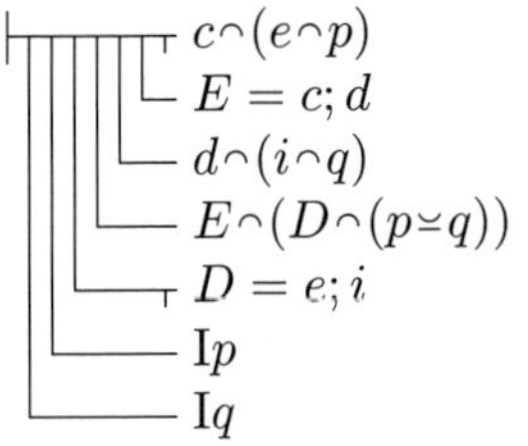

(ϑ

(III a): ————————

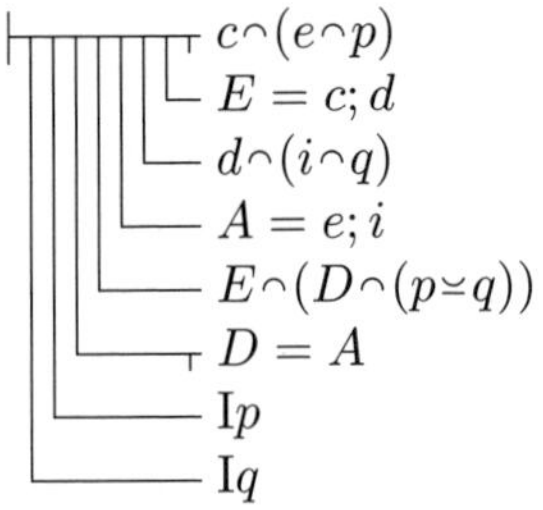

(ι

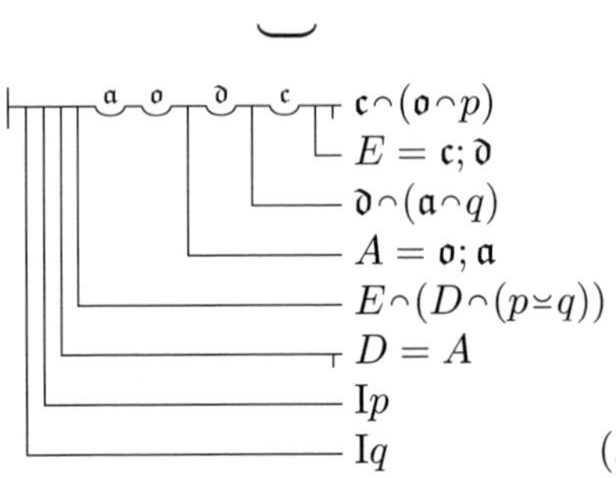

(κ

(213): — — — — — — — — — — — —

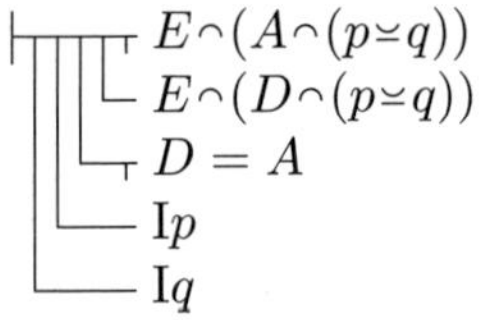

(λ

×

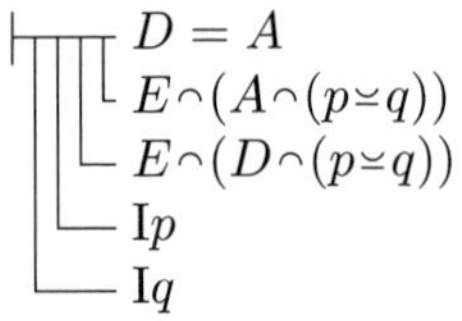

(μ

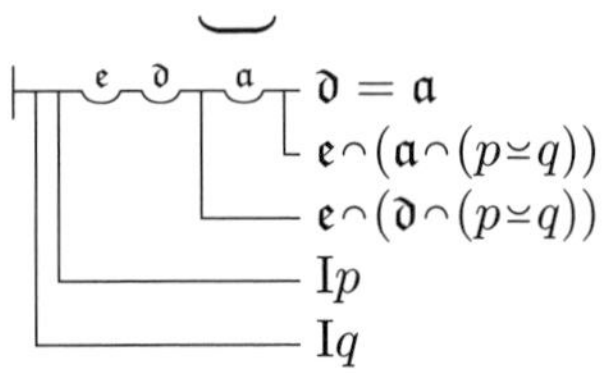

(ν

(16): — — — — — — — — — — —

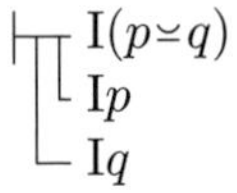

(252

(248): — — — — —

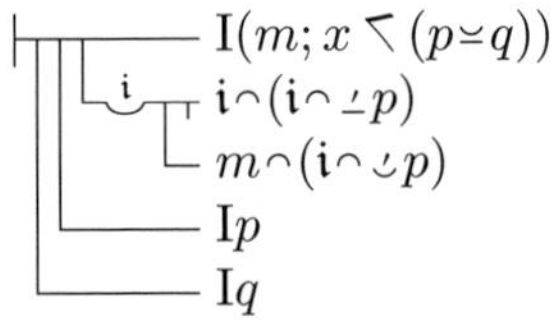

(253

(241): — — — — — — — — — — —

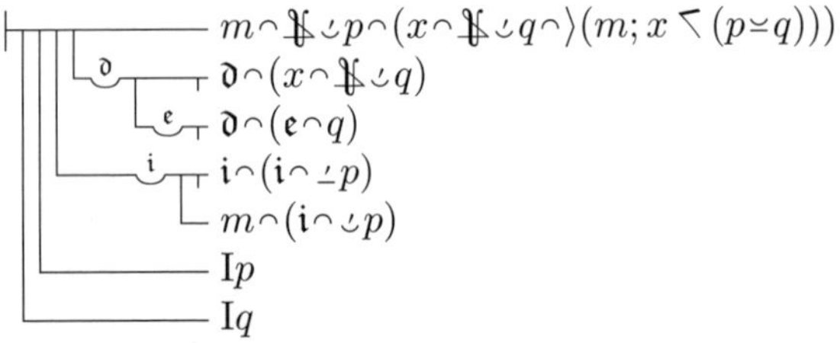

(254

——— • ———

145 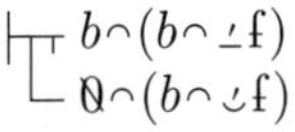

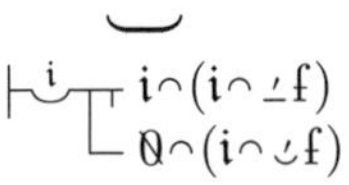

(255

(254): ————————

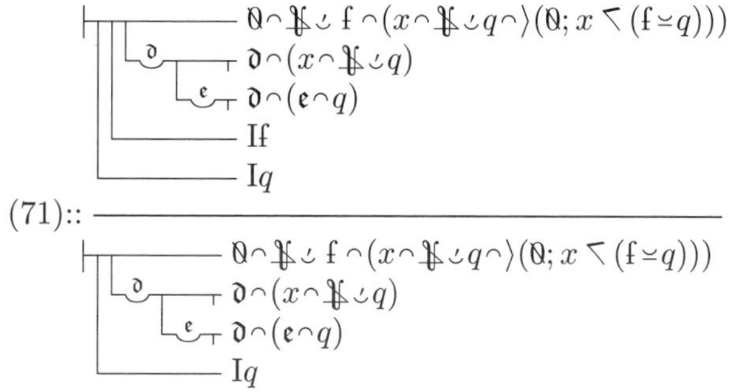

(α

(256

§156. *Analysis*

In (256) we have proposition (β) of §144. (γ) still remains to be proven. We will use (254) for this, by taking 'q' for 'p', '𝔒' for 'x', 'x' for 'm' and 'f' for 'q'. The resulting subcomponent

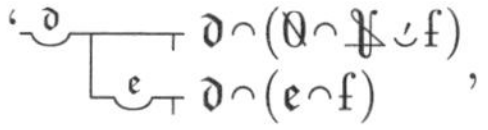

may be eliminated by using (156). It remains to derive the proposition

$$\text{`}\vdash x; m \nwarrow (q \simeq p) = \maltese(m; x \nwarrow (p \simeq q))\text{'} \qquad (\alpha$$

where, for greater generality, 'x' occurs instead of '𝔒' and 'p' instead of 'f'. This proposition may be reduced to

$$\text{`}\vdash x; m \frown (y; n \frown \smile (q \simeq p)) = m; x \frown (n; y \frown \smile (p \simeq q))\text{'} \qquad (\beta$$

We derive (β) from

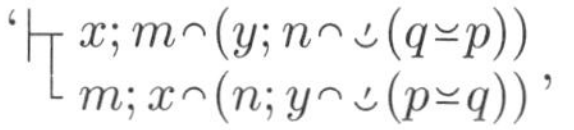

(γ

which we prove by means of the proposition

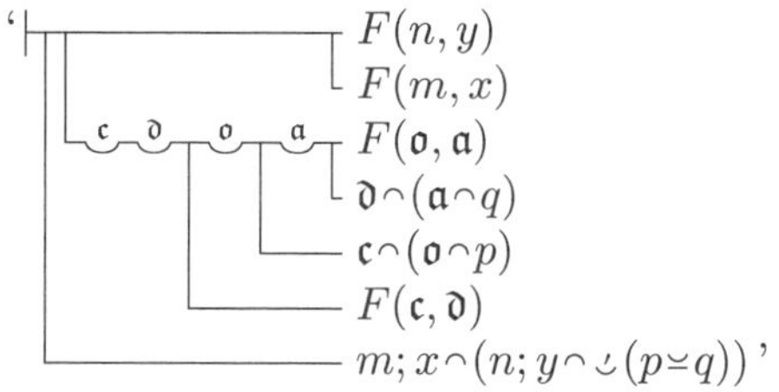

(δ

(δ) follows from (230) and (144).

§157. *Construction*

144 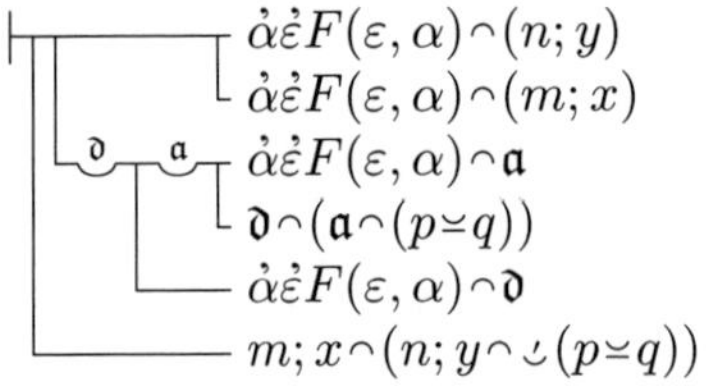

(230):: - - - - - - - - - - - - - -

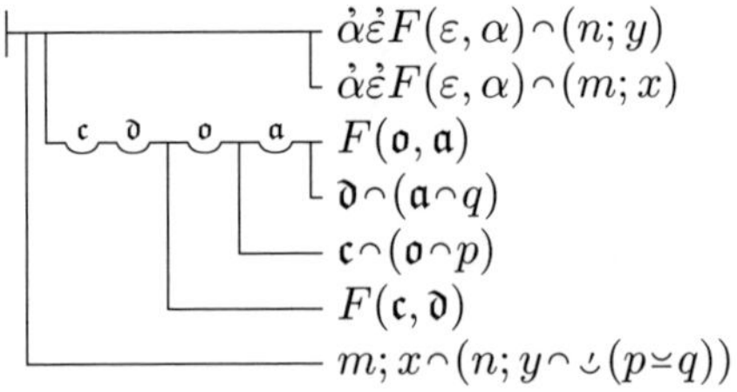 (α

(229): ————————————

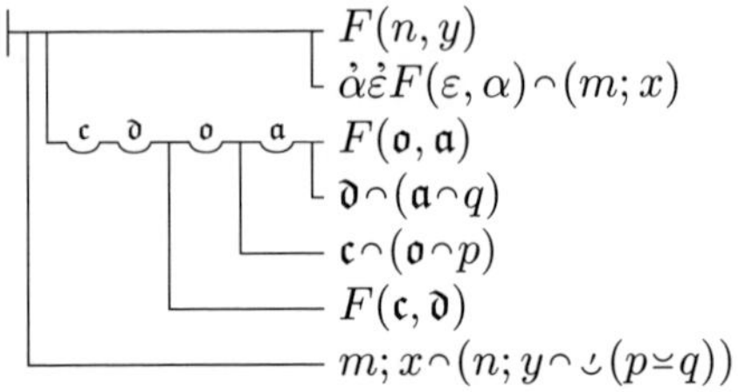 (β

(229): ————————————

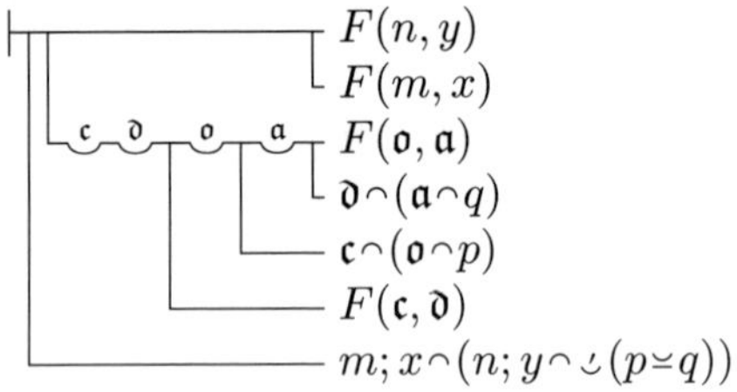 (257

———— • ————

209 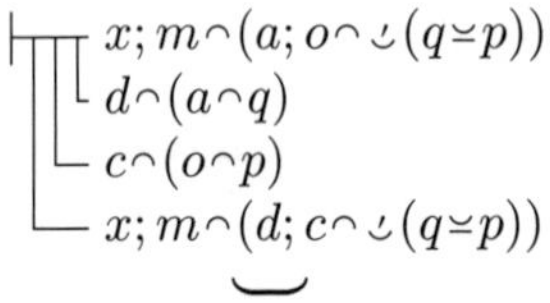

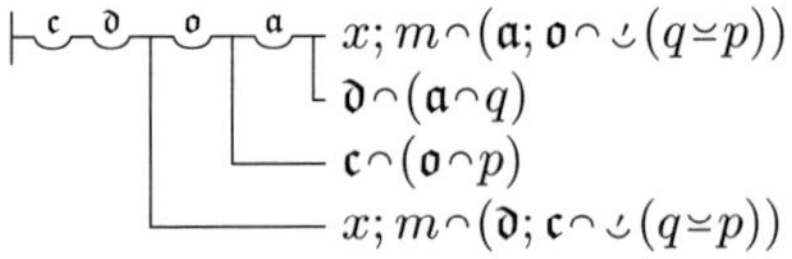 (α

(257): ————————————

$\vdash\!\!\top\; x; m \frown (y; n \frown \dot{\smile}(q \underline{\smile} p))$
$\quad\llcorner\; x; m \frown (x; m \frown \dot{\smile}(q \underline{\smile} p))$
$\quad\llcorner\; m; x \frown (n; y \frown \dot{\smile}(p \underline{\smile} q))$ (β

(140):: ―――――――――――――――――――

$\vdash\!\!\top\; x; m \frown (y; n \frown \dot{\smile}(q \underline{\smile} p))$
$\quad\llcorner\; m; x \frown (n; y \frown \dot{\smile}(p \underline{\smile} q))$ (258

(IVa): ―――――――――――――――――――

$\vdash\!\!\top\; (\text{—}\; x; m \frown (y; n \frown \dot{\smile}(q \underline{\smile} p))) = (\text{—}\; m; x \frown (n; y \frown \dot{\smile}(p \underline{\smile} q)))$
$\quad\llcorner\; m; x \frown (n; y \frown \dot{\smile}(p \underline{\smile} q))$
$\quad\llcorner\; x; m \frown (y; n \frown \dot{\smile}(q \underline{\smile} p))$ (α

(258):: ―――――――――――――――――――

$\vdash (\text{—}\; x; m \frown (y; n \frown \dot{\smile}(q \underline{\smile} p))) = (\text{—}\; m; x \frown (n; y \frown \dot{\smile}(p \underline{\smile} q)))$ (β

(203): ―――――――――――――――――――

$\vdash x; m \frown (y; n \frown \dot{\smile}(q \underline{\smile} p)) = (\text{—}\; m; x \frown (n; y \frown \dot{\smile}(p \underline{\smile} q)))$ (γ

(203): ―――――――――――――――――――

$\vdash x; m \frown (y; n \frown \dot{\smile}(q \underline{\smile} p)) = m; x \frown (n; y \frown \dot{\smile}(p \underline{\smile} q))$ (δ

(210): ―――――――――――――――――――

$\vdash x; m \frown (y; n \frown \dot{\smile}(q \underline{\smile} p)) = n \frown (y \frown (m; x \nwarrow (p \underline{\smile} q)))$ (ε

$\vdash\!\overset{\mathfrak{a}}{\smile}\; x; m \frown (\mathfrak{a}; n \frown \dot{\smile}(q \underline{\smile} p)) = n \frown (\mathfrak{a} \frown (m; x \nwarrow (p \underline{\smile} q)))$ (ζ

(Va): ―――――――――――――――――――

$\vdash \dot{\varepsilon}(x; m \frown (\varepsilon; n \frown \dot{\smile}(q \underline{\smile} p))) = \dot{\varepsilon}[n \frown (\varepsilon \frown (m; x \nwarrow (p \underline{\smile} q)))]$ (η

$\vdash\!\overset{\mathfrak{a}}{\smile}\; \dot{\varepsilon}(x; m \frown (\varepsilon; \mathfrak{a} \frown \dot{\smile}(q \underline{\smile} p))) = \dot{\varepsilon}[\mathfrak{a} \frown (\varepsilon \frown (m; x \nwarrow (p \underline{\smile} q)))]$ (ϑ

(Va): ―――――――――――――――――――

$\vdash \dot{\alpha}\dot{\varepsilon}(x; m \frown (\varepsilon; \alpha \frown \dot{\smile}(q \underline{\smile} p))) = \dot{\alpha}\dot{\varepsilon}[\alpha \frown (\varepsilon \frown (m; x \nwarrow (p \underline{\smile} q)))]$ (ι

(IIIc): ―――――――――――――――――――

$\vdash\!\!\top\; \dot{\alpha}\dot{\varepsilon}(x; m \frown (\varepsilon; \alpha \frown \dot{\smile}(q \underline{\smile} p))) = \natural(m; x \nwarrow (p \underline{\smile} q))$
$\quad\llcorner\; \dot{\alpha}\dot{\varepsilon}[\alpha \frown (\varepsilon \frown (m; x \nwarrow (p \underline{\smile} q)))] = \natural(m; x \nwarrow (p \underline{\smile} q))$ (κ

(E):: ―――――――――――――――――――

$\vdash \dot{\alpha}\dot{\varepsilon}(x; m \frown (\varepsilon; \alpha \frown \dot{\smile}(q \underline{\smile} p))) = \natural(m; x \nwarrow (p \underline{\smile} q))$ (λ

(IIIc): ―――――――――――――――――――

$\vdash\!\!\top\; x; m \nwarrow (q \underline{\smile} p) = \natural(m; x \nwarrow (p \underline{\smile} q))$
$\quad\llcorner\; \dot{\alpha}\dot{\varepsilon}(x; m \frown (\varepsilon; \alpha \frown \dot{\smile}(q \underline{\smile} p))) = x; m \nwarrow (q \underline{\smile} p)$ (μ

(II):: ―――――――――――――――――――

$\vdash x; m \nwarrow (q \underline{\smile} p) = \natural(m; x \nwarrow (p \underline{\smile} q))$ (259

(IIIc): ―――――――――――――――――――

$\vdash\!\!\top\; F(\natural(m; x \nwarrow (p \underline{\smile} q)))$
$\quad\llcorner\; F(x; m \nwarrow (q \underline{\smile} p))$ (260

——— • ———

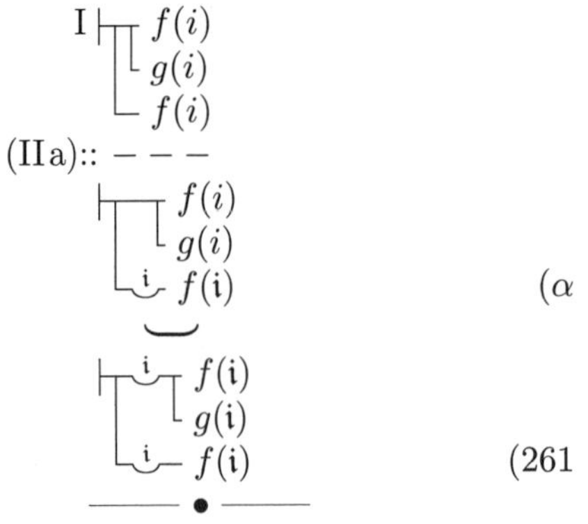
I
(IIa)::
(α
(261

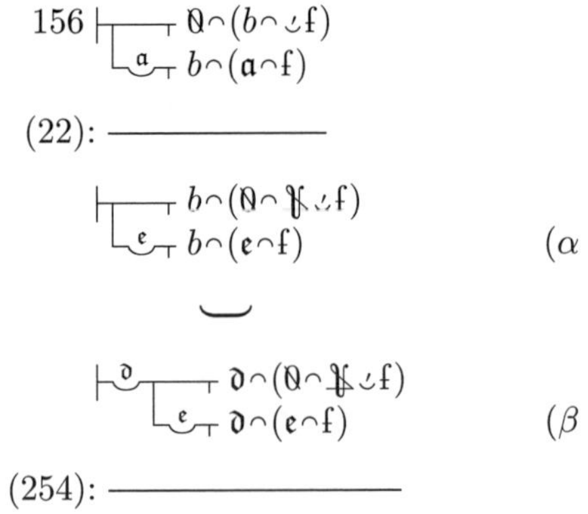
156
(22):
(α
(β
(254):

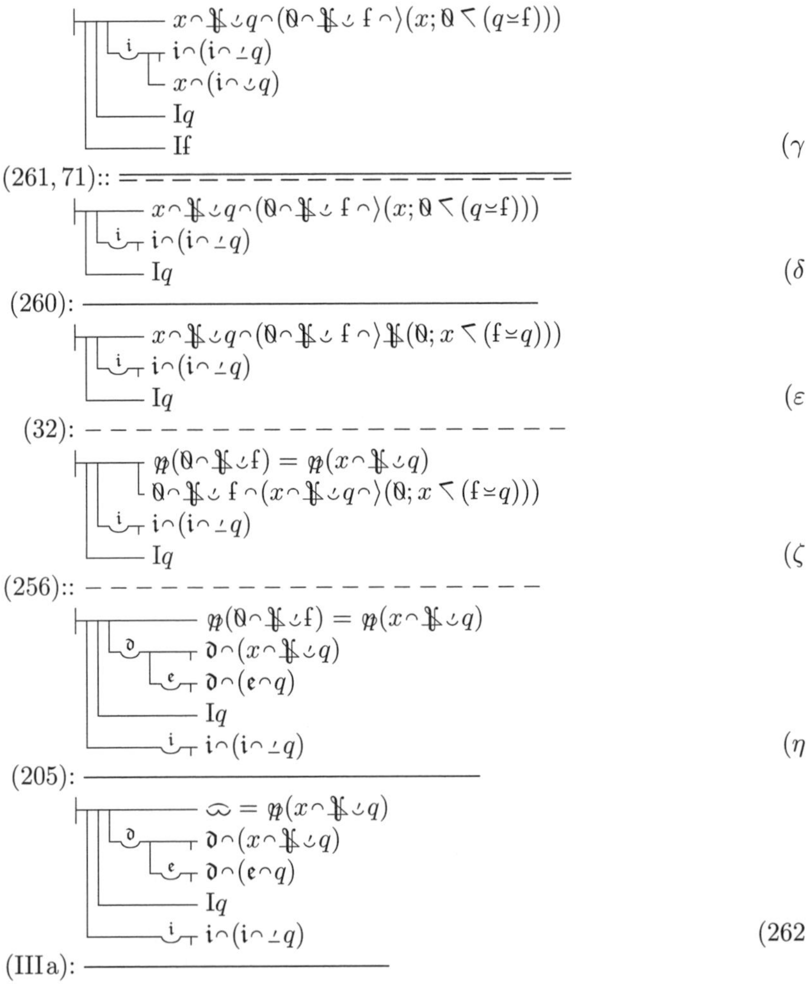
(γ
(261, 71)::
(δ
(260):
(ε
(32):
(ζ
(256)::
(η
(205):
(262
(IIIa):

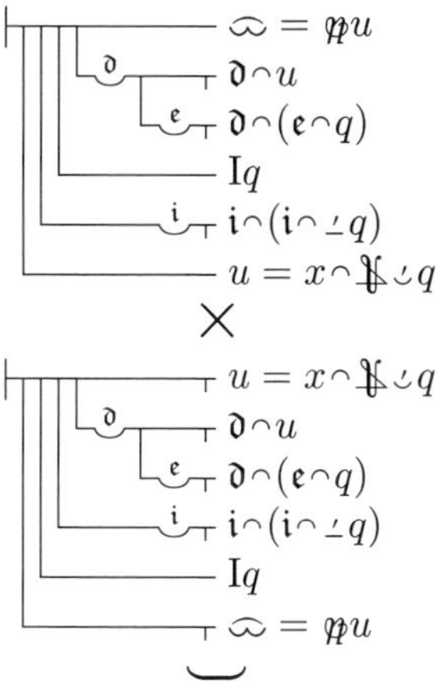

(α)   (β)   (γ)   (263

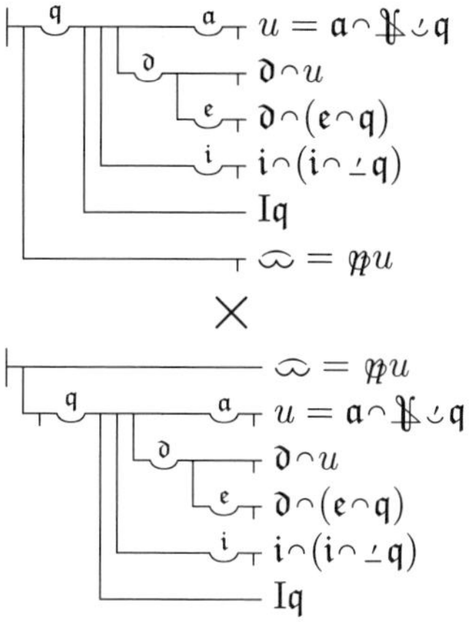

K. Proof of the proposition

$$\vdash \begin{array}{l} \emptyset \frown (\mathfrak{p}u \smile \mathfrak{f}) \\ \mathfrak{A} \stackrel{q}{\frown} u = \mathfrak{A} \underline{\mathfrak{s}} q \end{array},$$

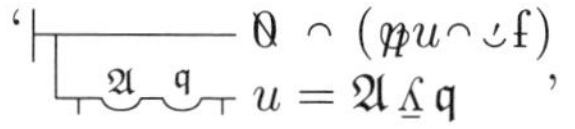

a) Proof of the proposition

$$\vdash \begin{array}{l} \mathfrak{p}(x; y \underline{\mathfrak{s}} q) = \mathfrak{p}(\mathfrak{1}; n \underline{\mathfrak{s}} \mathfrak{f}) \\ y \frown (y \frown \underline{} q) \\ Iq \\ x; \mathfrak{1} \frown (y; n \frown \smile (q \underline{\smile} \mathfrak{f})) \end{array},$$

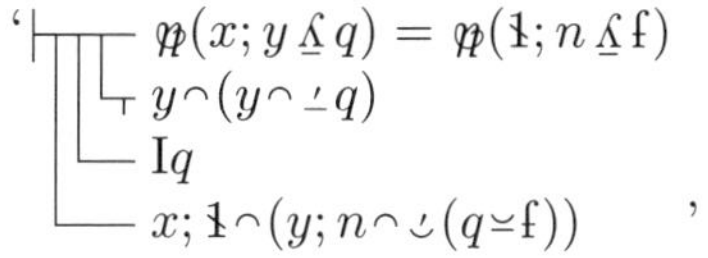

§158. *Analysis*

For finite cardinal numbers we can prove a proposition similar to the last, namely that the cardinal number of a concept is finite if the objects falling under it can be ordered into a *simple* (non-branching, not looping back into itself) series starting with a certain object and ending with a certain object. For this, we require an abbreviation which we will introduce as follows:

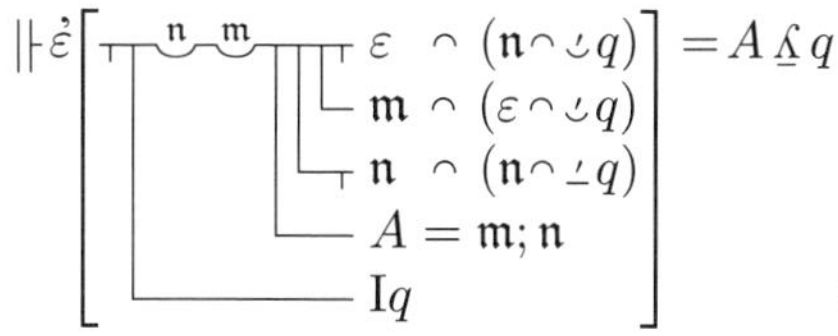

$$\Vdash \hat{\varepsilon} \left[\begin{array}{l} \underset{n}{\frown} \underset{m}{\frown} \left[\begin{array}{l} \varepsilon \frown (n \frown \smile q) \\ \mathfrak{m} \frown (\varepsilon \frown \smile q) \\ n \frown (n \frown \underline{} q) \\ A = \mathfrak{m}; n \end{array} \right] = A \underline{\mathfrak{s}} q \\ Iq \end{array} \right] \quad (P$$

If Γ, Δ, Θ are objects and Υ the extension of a relation, then

$$`\Gamma \frown (\Delta; \Theta \underline{\mathfrak{s}} \Upsilon)`$$

says that Γ belongs to the Υ-series starting with Δ and ending with Θ, where the Υ-relation is single-valued and Θ does not follow itself in the Υ-series. We express this in words, in short, as follows: "Γ *belongs to the Υ-series running from Δ to Θ*". With the notation thus explained our proposition assumes the form displayed in the main heading. For

$$\Delta; \Theta \underline{\mathfrak{s}} \Upsilon$$

is the extension of the concept *belonging*

to the Υ-series running from Δ to Θ.

We first prove the proposition

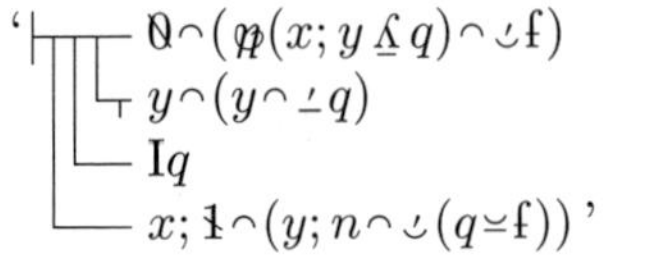

$$\vdash \begin{array}{l} \Theta \frown (\mathfrak{p}(x; y \,\underline{\measuredangle}\, q) \frown \smile \mathfrak{f}) \\ \quad y \frown (y \frown \underline{\smile}\, q) \\ \mathrm{I}q \\ x; \mathbf{1} \frown (y; n \frown \smile (q \veebar \mathfrak{f})) \end{array} \qquad (\alpha$$

from which the subcomponents then are to be removed. We derive (α) from (234) in the form

$$\vdash \begin{array}{l} \mathbf{1} \frown (n \frown \smile \mathfrak{f}) \\ x; \mathbf{1} \frown (y; n \frown \smile (q \veebar \mathfrak{f})) \end{array}$$

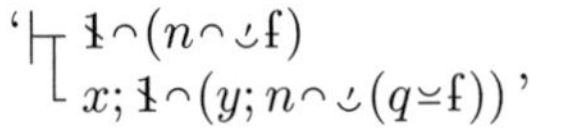

and the proposition

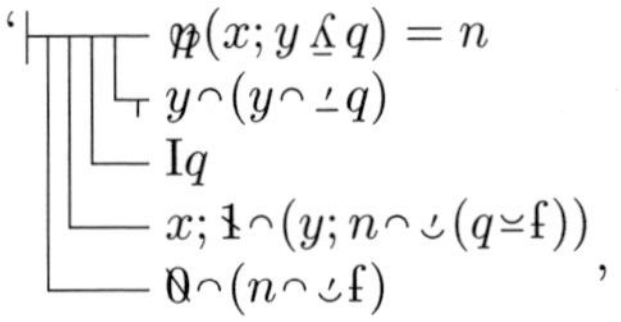

$$\vdash \begin{array}{l} \mathfrak{p}(x; y \,\underline{\measuredangle}\, q) = n \\ \quad y \frown (y \frown \underline{\smile}\, q) \\ \mathrm{I}q \\ x; \mathbf{1} \frown (y; n \frown \smile (q \veebar \mathfrak{f})) \\ \Theta \frown (n \frown \smile \mathfrak{f}) \end{array} \qquad (\beta$$

which we prove using the propositions

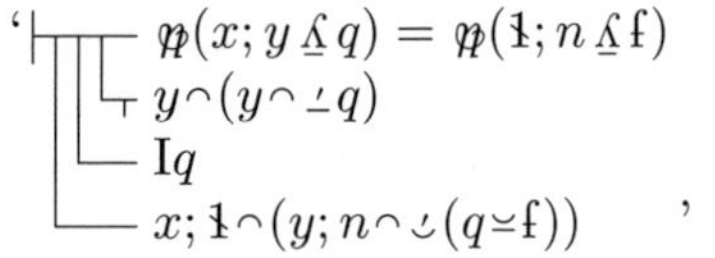

$$\vdash \begin{array}{l} \mathfrak{p}(x; y \,\underline{\measuredangle}\, q) = \mathfrak{p}(\mathbf{1}; n \,\underline{\measuredangle}\, \mathfrak{f}) \\ \quad y \frown (y \frown \underline{\smile}\, q) \\ \mathrm{I}q \\ x; \mathbf{1} \frown (y; n \frown \smile (q \veebar \mathfrak{f})) \end{array} \qquad (\gamma$$

$$\vdash \begin{array}{l} n = \mathfrak{p}(\mathbf{1}; n \,\underline{\measuredangle}\, \mathfrak{f}) \\ \Theta \frown (n \frown \smile \mathfrak{f}) \end{array} \qquad (\delta$$

(γ) is to be reduced to the more general proposition

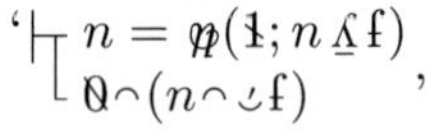

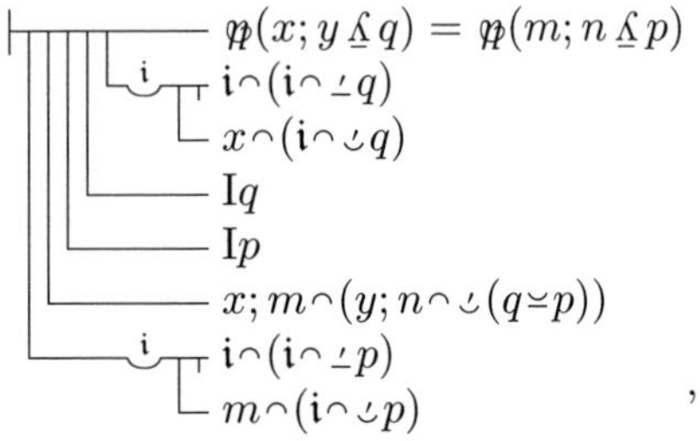

$$\vdash \begin{array}{l} \mathfrak{p}(x; y \,\underline{\measuredangle}\, q) = \mathfrak{p}(m; n \,\underline{\measuredangle}\, p) \\ \quad \overset{\mathfrak{i}}{\smile} \; \mathfrak{i} \frown (\mathfrak{i} \frown \underline{\smile}\, q) \\ \qquad x \frown (\mathfrak{i} \frown \smile q) \\ \mathrm{I}q \\ \mathrm{I}p \\ x; m \frown (y; n \frown \smile (q \veebar p)) \\ \quad \overset{\mathfrak{i}}{\smile} \; \mathfrak{i} \frown (\mathfrak{i} \frown \underline{\smile}\, p) \\ \qquad m \frown (\mathfrak{i} \frown \smile p) \end{array} \qquad (\varepsilon$$

whose proof requires the proposition

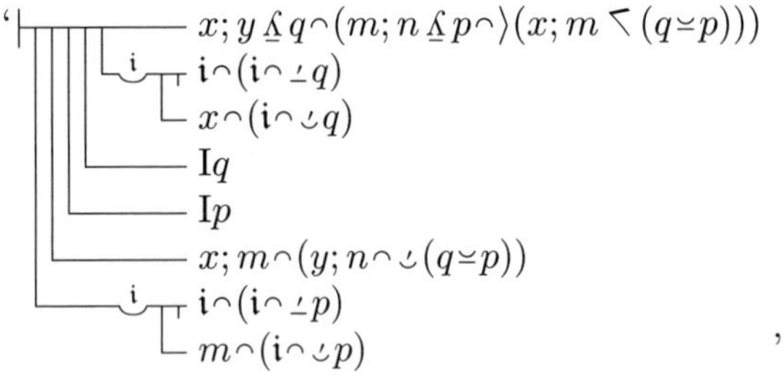

$$\vdash \begin{array}{l} x; y \,\underline{\measuredangle}\, q \frown (m; n \,\underline{\measuredangle}\, p \frown \rangle (x; m \,\nwarrow (q \veebar p))) \\ \quad \overset{\mathfrak{i}}{\smile} \; \mathfrak{i} \frown (\mathfrak{i} \frown \underline{\smile}\, q) \\ \qquad x \frown (\mathfrak{i} \frown \smile q) \\ \mathrm{I}q \\ \mathrm{I}p \\ x; m \frown (y; n \frown \smile (q \veebar p)) \\ \quad \overset{\mathfrak{i}}{\smile} \; \mathfrak{i} \frown (\mathfrak{i} \frown \underline{\smile}\, p) \\ \qquad m \frown (\mathfrak{i} \frown \smile p) \end{array} \qquad (\zeta$$

According to (11), this resolves into the propositions (253) and

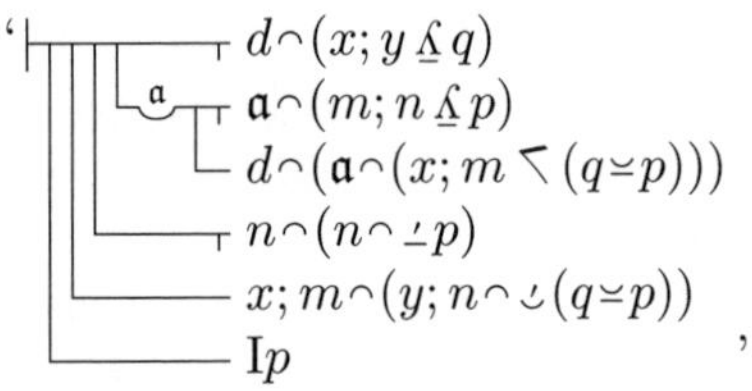

$$\vdash \begin{array}{l} d \frown (x; y \,\underline{\measuredangle}\, q) \\ \quad \overset{\mathfrak{a}}{\smile} \; \mathfrak{a} \frown (m; n \,\underline{\measuredangle}\, p) \\ \qquad d \frown (\mathfrak{a} \frown (x; m \,\nwarrow (q \veebar p))) \\ n \frown (n \frown \underline{\smile}\, p) \\ x; m \frown (y; n \frown \smile (q \veebar p)) \\ \mathrm{I}p \end{array} \qquad (\eta$$

We derive (η) from

$$\vdash \begin{array}{l} d \frown (x; y \,\underline{\measuredangle}\, q) \\ \quad \overset{\mathfrak{a}}{\smile} \; \mathfrak{a} \frown (m; n \,\underline{\measuredangle}\, p) \\ \qquad d \frown (\mathfrak{a} \frown (x; m \,\nwarrow (q \veebar p))) \\ \quad \overset{\mathfrak{e}}{\smile} \; x; m \frown (d; \mathfrak{e} \frown \smile (q \veebar p)) \\ n \frown (n \frown \underline{\smile}\, p) \\ x; m \frown (y; n \frown \smile (q \veebar p)) \\ \mathrm{I}p \end{array} \qquad (\vartheta$$

by removing the subcomponent

$$\overset{\mathfrak{e}}{\smile} \; x; m \frown (d; \mathfrak{e} \frown \smile (q \veebar p))$$

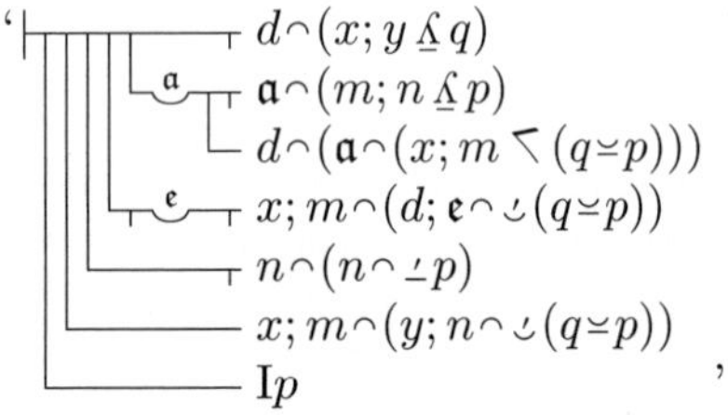

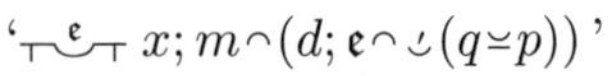

We prove (ϑ) by means of the proposition

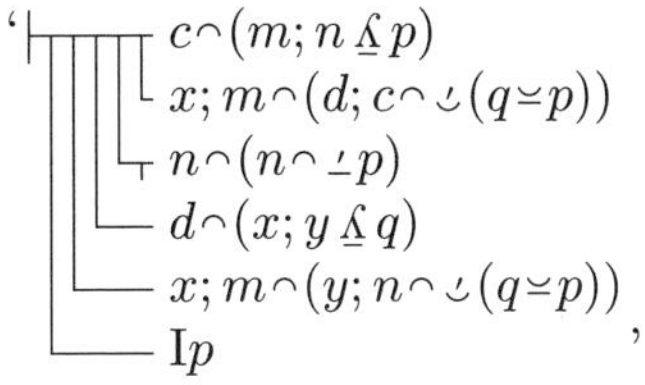

(ι

which leads back to the propositions (234) in the form

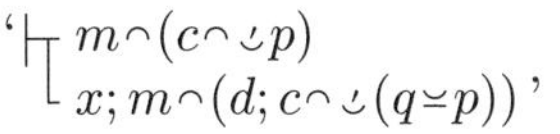

and

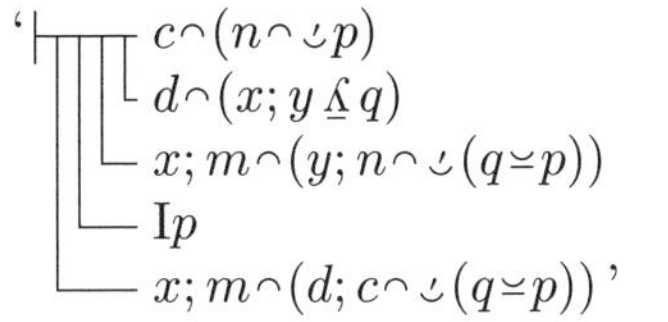

(κ

According to (243) we have

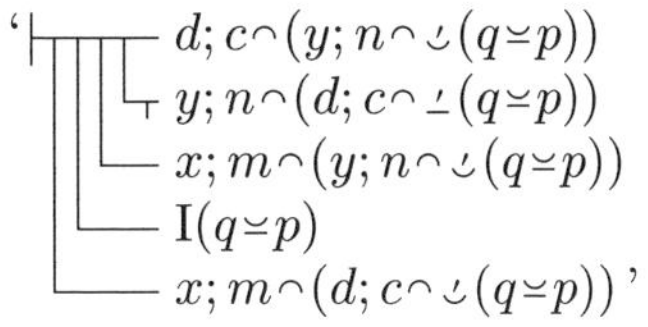

By means of (244) we prove

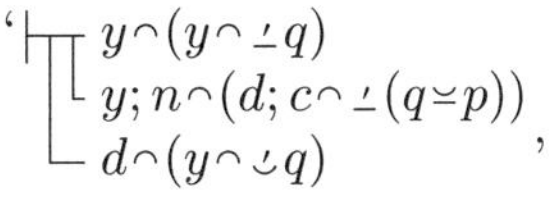

(λ

with which we can remove the subcomponent

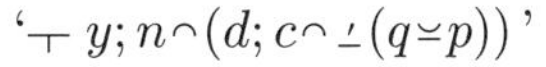

To begin with, we will draw the immediate consequences from our definition (P).

§159. *Construction*

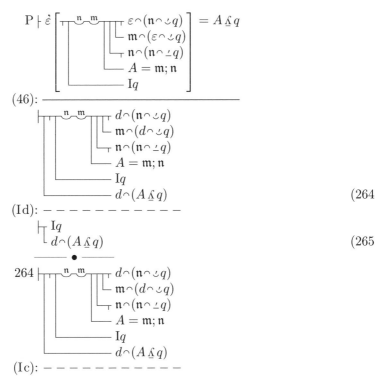

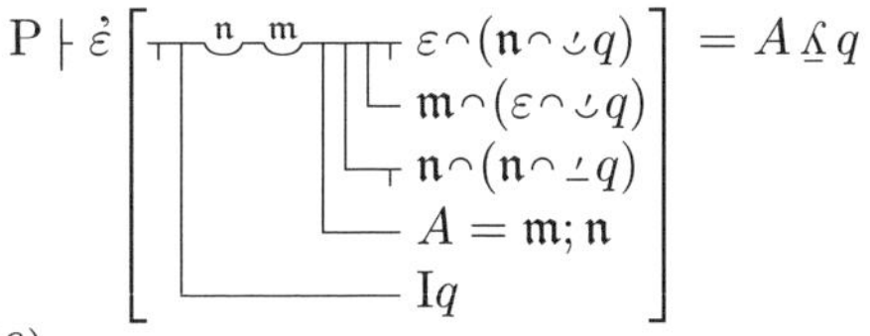

(46):

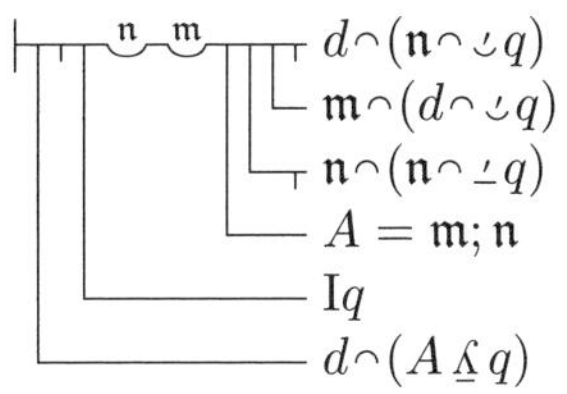

(264

(Id): – – – – – – – – – – – –

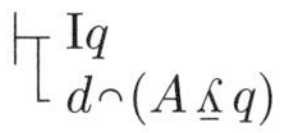

(265

264

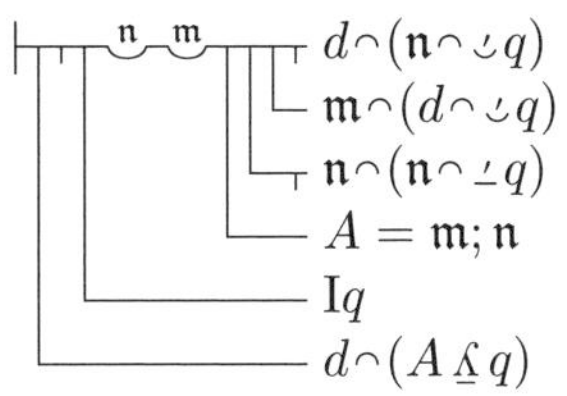

(Ic): – – – – – – – – – – – –

$\vdash$ $\mathfrak{n}$ $\mathfrak{m}$
$d \frown (\mathfrak{n} \frown \smile q)$
$\mathfrak{m} \frown (d \frown \smile q)$
$\mathfrak{n} \frown (\mathfrak{n} \frown \underline{\smile} q)$
$A = \mathfrak{m}; \mathfrak{n}$
$d \frown (A \,\underline{\measuredangle}\, q)$ (α

×

$\vdash$ $d \frown (A \,\underline{\measuredangle}\, q)$
$\mathfrak{n}$ $\mathfrak{m}$
$d \frown (\mathfrak{n} \frown \smile q)$
$\mathfrak{m} \frown (d \frown \smile q)$
$\mathfrak{n} \frown (\mathfrak{n} \frown \underline{\smile} q)$
$A = \mathfrak{m}; \mathfrak{n}$ (266

——— • ———

222 $\vdash$ $d \frown (n \frown \smile q)$
$m \frown (d \frown \smile q)$
$n \frown (n \frown \underline{\smile} q)$
$x; y = m; n$
$d \frown (y \frown \smile q)$
$x \frown (d \frown \smile q)$
$y \frown (y \frown \underline{\smile} q)$

‿

$\vdash$ $\mathfrak{n}$ $\mathfrak{m}$
$d \frown (\mathfrak{n} \frown \smile q)$
$\mathfrak{m} \frown (d \frown \smile q)$
$\mathfrak{n} \frown (\mathfrak{n} \frown \underline{\smile} q)$
$x; y = \mathfrak{m}; \mathfrak{n}$
$d \frown (y \frown \smile q)$
$x \frown (d \frown \smile q)$
$y \frown (y \frown \underline{\smile} q)$ (α

(266): - - - - - - - - - - -

$\vdash$ $d \frown (x; y \,\underline{\measuredangle}\, q)$
$d \frown (y \frown \smile q)$
$x \frown (d \frown \smile q)$
$y \frown (y \frown \underline{\smile} q)$ (β

×

$\vdash$ $d \frown (y \frown \smile q)$
$x \frown (d \frown \smile q)$
$y \frown (y \frown \underline{\smile} q)$
$d \frown (x; y \,\underline{\measuredangle}\, q)$ (267

(Ic): - - - - - - - - -

$\vdash$ $d \frown (y \frown \smile q)$
$x \frown (d \frown \smile q)$
$d \frown (x; y \,\underline{\measuredangle}\, q)$ (268

(Ib): - - - - - - - -

$\vdash$ $d \frown (y \frown \smile q)$
$d \frown (x; y \,\underline{\measuredangle}\, q)$ (269

——— • ———

268 $\vdash$ $d \frown (y \frown \smile q)$
$x \frown (d \frown \smile q)$
$d \frown (x; y \,\underline{\measuredangle}\, q)$

(Id): - - - - - - - -

$\vdash$ $x \frown (d \frown \smile q)$
$d \frown (x; y \,\underline{\measuredangle}\, q)$ (270

——— • ———

267 $\vdash$ $d \frown (y \frown \smile q)$
$x \frown (d \frown \smile q)$
$y \frown (y \frown \underline{\smile} q)$
$d \frown (x; y \,\underline{\measuredangle}\, q)$

(Id): - - - - - - - - -

$\vdash$ $y \frown (y \frown \underline{\smile} q)$
$d \frown (x; y \,\underline{\measuredangle}\, q)$ (271

×

$\vdash$ $d \frown (x; y \,\underline{\measuredangle}\, q)$
$y \frown (y \frown \underline{\smile} q)$ (272

——— • ———

P $\vdash$ $\grave{\varepsilon}$ [$\mathfrak{n}$ $\mathfrak{m}$
$\varepsilon \frown (\mathfrak{n} \frown \smile p)$
$\mathfrak{m} \frown (\varepsilon \frown \smile p)$
$\mathfrak{n} \frown (\mathfrak{n} \frown \underline{\smile} p)$
$m; n = \mathfrak{m}; \mathfrak{n}$
Ip] $= m; n \,\underline{\measuredangle}\, p$

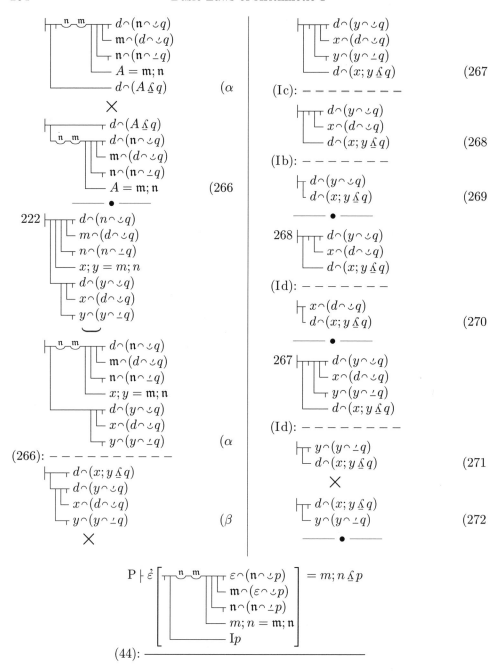

(44): ———————————

$c\frown(m; n \mathbin{\underline{\measuredangle}} p)$
$\mathfrak{n}\ \mathfrak{m}$ $c\frown(\mathfrak{n}\frown\smile p)$
$\mathfrak{m}\frown(c\frown\smile p)$
$\mathfrak{n}\frown(\mathfrak{n}\frown\perp p)$
$m; n = \mathfrak{m}; \mathfrak{n}$
$\mathrm{I}p$ (α

(If):: - - - - - - - - - - - -

$c\frown(m; n \mathbin{\underline{\measuredangle}} p)$
$\mathfrak{n}\ \mathfrak{m}$ $c\frown(\mathfrak{n}\frown\smile p)$
$\mathfrak{m}\frown(c\frown\smile p)$
$\mathfrak{n}\frown(\mathfrak{n}\frown\perp p)$
$m; n = \mathfrak{m}; \mathfrak{n}$
$\mathrm{I}p$ (β

×

$\mathfrak{n}\ \mathfrak{m}$ $c\frown(\mathfrak{n}\frown\smile p)$
$\mathfrak{m}\frown(c\frown\smile p)$
$\mathfrak{n}\frown(\mathfrak{n}\frown\perp p)$
$m; n = \mathfrak{m}; \mathfrak{n}$
$c\frown(m; n \mathbin{\underline{\measuredangle}} p)$
$\mathrm{I}p$ (γ

(IIa): - - - - - - - - - - - -

$\mathfrak{m}$ $c\frown(n\frown\smile p)$
$\mathfrak{m}\frown(c\frown\smile p)$
$n\frown(n\frown\perp p)$
$m; n = \mathfrak{m}; n$
$c\frown(m; n \mathbin{\underline{\measuredangle}} p)$
$\mathrm{I}p$ (δ

(IIa): - - - - - - - - - - -

$c\frown(n\frown\smile p)$
$m\frown(c\frown\smile p)$
$n\frown(n\frown\perp p)$
$m; n = m; n$
$c\frown(m; n \mathbin{\underline{\measuredangle}} p)$
$\mathrm{I}p$ (ε

(IIIe):: ———————————

$c\frown(n\frown\smile p)$
$m\frown(c\frown\smile p)$
$n\frown(n\frown\perp p)$
$c\frown(m; n \mathbin{\underline{\measuredangle}} p)$
$\mathrm{I}p$ (273

×

$c\frown(m; n \mathbin{\underline{\measuredangle}} p)$
$m\frown(c\frown\smile p)$
$n\frown(n\frown\perp p)$
$c\frown(n\frown\smile p)$
$\mathrm{I}p$ (274

——— • ———

133 $x\frown(a\frown\perp q)$
$d\frown(a\frown q)$
$x\frown(d\frown\perp q)$

$\mathfrak{d}\ \mathfrak{a}$ $x\frown(\mathfrak{a}\frown\perp q)$
$\mathfrak{d}\frown(\mathfrak{a}\frown q)$
$x\frown(\mathfrak{d}\frown\perp q)$ (α

(128): ———————————

$x\frown(y\frown\perp q)$
$x\frown(d\frown\perp q)$
$d\frown(y\frown\perp q)$ (275

(200):: - - - - - - -

$x\frown(y\frown\perp q)$
$x\frown(d\frown\perp q)$
$y = d$
$d\frown(y\frown\smile q)$ (α

(IIIa): -·-·-·-·-·-·-

$x\frown(y\frown\perp q)$
$x\frown(d\frown\perp q)$
$d\frown(y\frown\smile q)$ (276

——— • ———

244 $y\frown(d\frown\perp q)$
$y; n\frown(d; c\frown\perp(q \smile p))$

(276): - - - - - - - - - -

$y\frown(y\frown\perp q)$
$y; n\frown(d; c\frown\perp(q \smile p))$
$d\frown(y\frown\smile q)$ (α

×

$y; n\frown(d; c\frown\perp(q \smile p))$
$y\frown(y\frown\perp q)$
$d\frown(y\frown\smile q)$ (β

(271, 269):: = = = = = = = =

$y; n\frown(d; c\frown\perp(q \smile p))$
$d\frown(x; y \mathbin{\underline{\measuredangle}} q)$ (γ

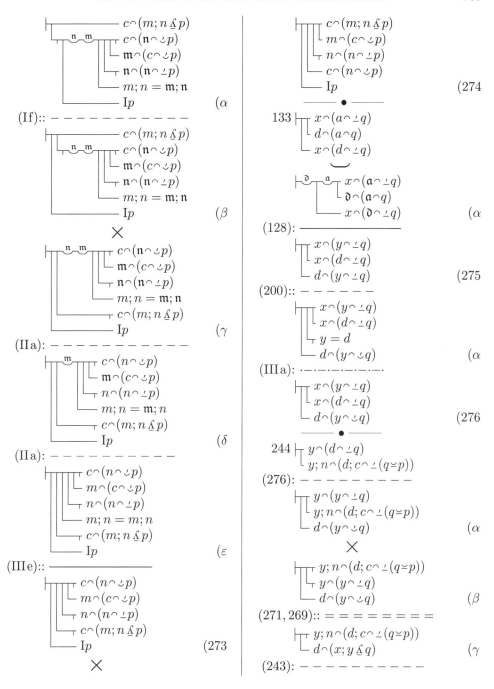

(243): - - - - - - - - - - -

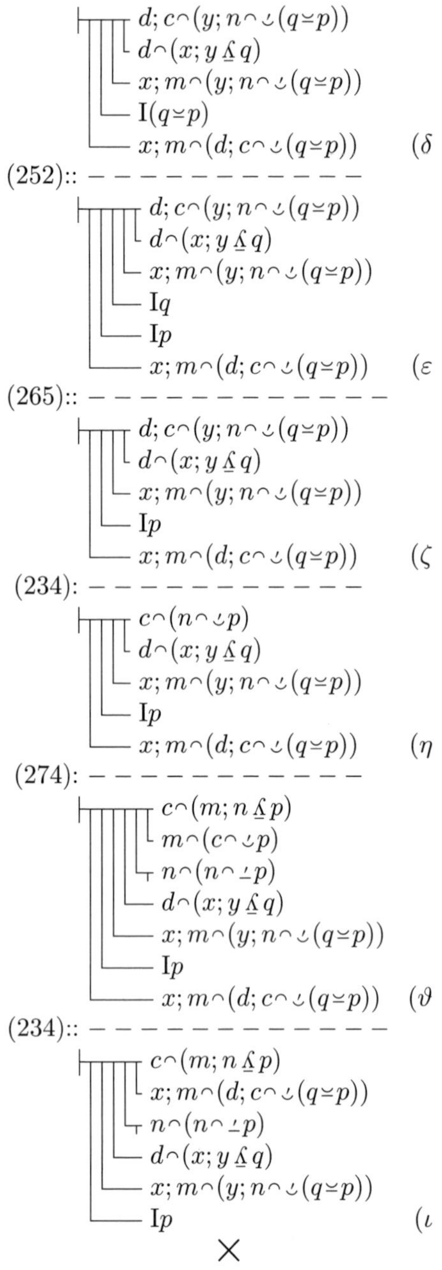
d; c⌒(y; n⌒⌣(q≚p))
d⌒(x; y ≰ q)
x; m⌒(y; n⌒⌣(q≚p))
I(q≚p)
x; m⌒(d; c⌒⌣(q≚p))
(δ
(252)::
Iq
Ip
(ε
(265)::
(ζ
(234):
c⌒(n⌒⌣p)
(η
(274):
c⌒(m; n ≰ p)
m⌒(c⌒⌣p)
n⌒(n⌒⊥p)
(ϑ
(234)::
(ι
×

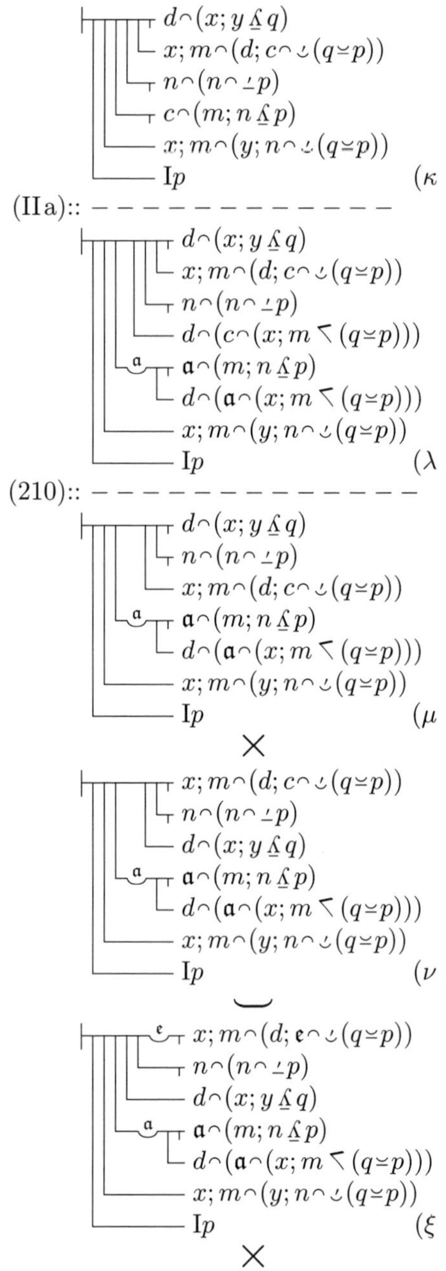
d⌒(x; y ≰ q)
x; m⌒(d; c⌒⌣(q≚p))
n⌒(n⌒⊥p)
c⌒(m; n ≰ p)
x; m⌒(y; n⌒⌣(q≚p))
Ip
(κ
(IIa)::
d⌒(c⌒(x; m ⊼ (q≚p)))
𝔞⌒(m; n ≰ p)
d⌒(𝔞⌒(x; m ⊼ (q≚p)))
(λ
(210)::
(μ
×
(ν
x; m⌒(d; 𝔢⌒⌣(q≚p))
(ξ
×

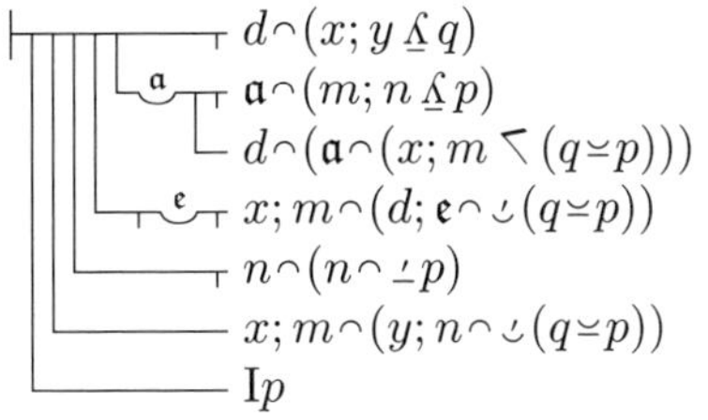

(277

§160. *Analysis*

Now the subcomponent

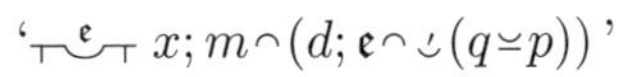

has to be eliminated (compare §158). This is accomplished by means of the proposition

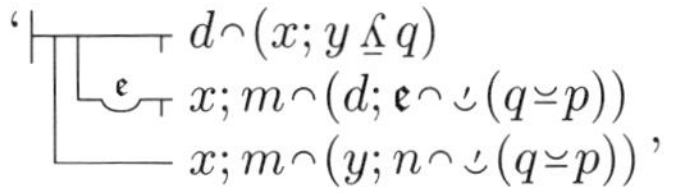

(α

which we prove by means of (257), replacing the function-marker '$F(\xi, \zeta)$' by

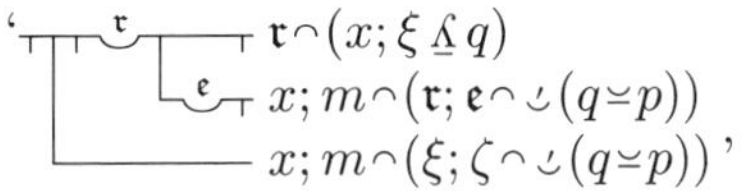

For this we require the proposition

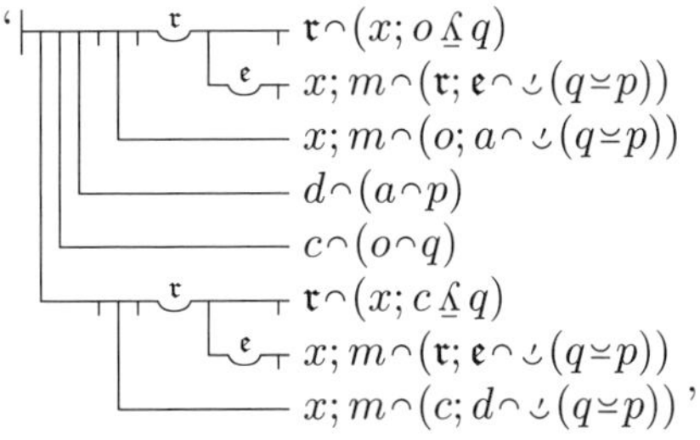

(β

This proposition is to be reduced to the proposition

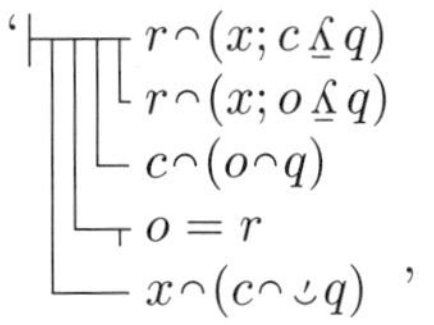

(γ

which follows from

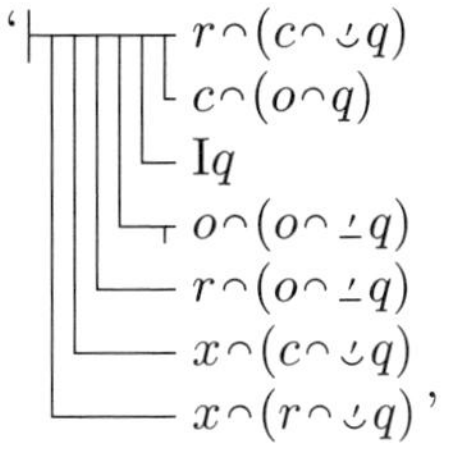

(δ

In order to prove (δ) we use proposition (243) in the form

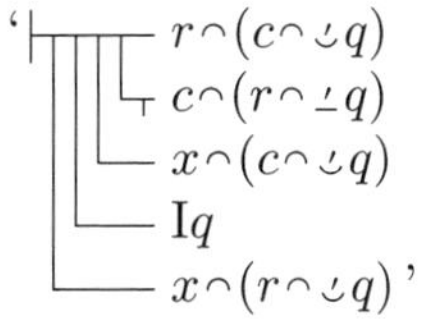

and show that, given our conditions, r cannot follow c in the q-series, since in that case, according to (242), r would belong to the q-series starting with o and therefore o would follow itself in the q-series.

§161. *Construction*

134

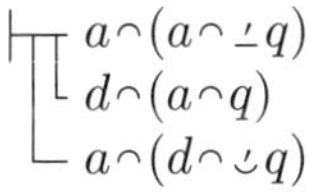

(242):: – – – – – –

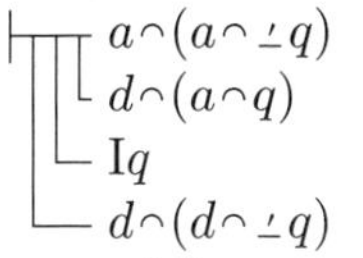

(278

×

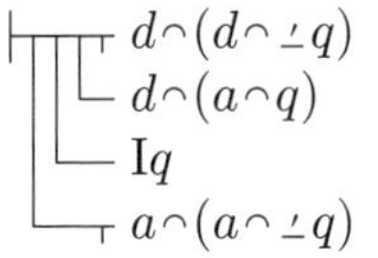

(279

•

275

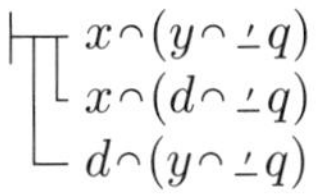

(200):: – – – – – –

$\vdash$ $x \frown (y \frown \dot{-} q)$
$\quad$ $\top\; d = x$
$\quad$ $x \frown (d \frown \dot{\smile} q)$
$\quad$ $d \frown (y \frown \dot{-} q)$ (α

(IIIc): -·-·-·-·-·-·

$\vdash$ $x \frown (y \frown \dot{-} q)$
$\quad$ $x \frown (d \frown \dot{\smile} q)$
$\quad$ $d \frown (y \frown \dot{-} q)$ (280

•

280 $\vdash$ $o \frown (o \frown \dot{-} q)$
$\quad$ $o \frown (r \frown \dot{\smile} q)$
$\quad$ $r \frown (o \frown \dot{-} q)$

(242):: - - - - - -

$\vdash$ $o \frown (o \frown \dot{-} q)$
$\quad$ $c \frown (o \frown q)$
$\quad$ $\mathrm{I}q$
$\quad$ $c \frown (r \frown \dot{-} q)$
$\quad$ $r \frown (o \frown \dot{-} q)$ (α

×

$\vdash$ $\top\; c \frown (r \frown \dot{-} q)$
$\quad$ $c \frown (o \frown q)$
$\quad$ $\mathrm{I}q$
$\quad$ $\top\; o \frown (o \frown \dot{-} q)$
$\quad$ $r \frown (o \frown \dot{-} q)$ (β

(243): - - - - - - -

$\vdash$ $\top\; r \frown (c \frown \dot{\smile} q)$
$\quad$ $c \frown (o \frown q)$
$\quad$ $\mathrm{I}q$
$\quad$ $\top\; o \frown (o \frown \dot{-} q)$
$\quad$ $r \frown (o \frown \dot{-} q)$
$\quad$ $x \frown (c \frown \dot{\smile} q)$
$\quad$ $x \frown (r \frown \dot{\smile} q)$ (γ

(274): - - - - - - - -

$\vdash$ $\top\; r \frown (x; c \,\underline{\wedge}\, q)$
$\quad$ $x \frown (r \frown \dot{\smile} q)$
$\quad$ $\top\; c \frown (c \frown \dot{-} q)$
$\quad$ $c \frown (o \frown q)$
$\quad$ $\mathrm{I}q$
$\quad$ $\top\; o \frown (o \frown \dot{-} q)$
$\quad$ $r \frown (o \frown \dot{-} q)$
$\quad$ $x \frown (c \frown \dot{\smile} q)$ (δ

(279, 200):: = = = = = =

$\vdash$ $r \frown (x; c \,\underline{\wedge}\, q)$
$\quad$ $x \frown (r \frown \dot{\smile} q)$
$\quad$ $c \frown (o \frown q)$
$\quad$ $\mathrm{I}q$
$\quad$ $\top\; o \frown (o \frown \dot{-} q)$
$\quad$ $\top\; o = r$
$\quad$ $r \frown (o \frown \dot{\smile} q)$
$\quad$ $x \frown (c \frown \dot{\smile} q)$ (ε

(269, 270):: = = = = = =

$\vdash$ $r \frown (x; c \,\underline{\wedge}\, q)$
$\quad$ $r \frown (x; o \,\underline{\wedge}\, q)$
$\quad$ $c \frown (o \frown q)$
$\quad$ $\mathrm{I}q$
$\quad$ $\top\; o \frown (o \frown \dot{-} q)$
$\quad$ $\top\; o = r$
$\quad$ $x \frown (c \frown \dot{\smile} q)$ (ζ

(271, 265):: = = = = =

$\vdash$ $r \frown (x; c \,\underline{\wedge}\, q)$
$\quad$ $r \frown (x; o \,\underline{\wedge}\, q)$
$\quad$ $c \frown (o \frown q)$
$\quad$ $\top\; o = r$
$\quad$ $x \frown (c \frown \dot{\smile} q)$ (η

×

$\vdash$ $r \frown (x; o \,\underline{\wedge}\, q)$
$\quad$ $\top\; r \frown (x; c \,\underline{\wedge}\, q)$
$\quad$ $c \frown (o \frown q)$
$\quad$ $\top\; o = r$
$\quad$ $x \frown (c \frown \dot{\smile} q)$ (ϑ

•

IIa $\vdash$ $x; m \frown (r; a \frown \dot{\smile} (q \simeq p))$
$\quad$ $\overset{\mathfrak{e}}{\smile}\top\; x; m \frown (r; \mathfrak{e} \frown \dot{\smile} (q \simeq p))$

(IIIa): ————————

$\vdash$ $x; m \frown (o; a \frown \dot{\smile} (q \simeq p))$
$\quad$ $\overset{\mathfrak{e}}{\smile}\top\; x; m \frown (r; \mathfrak{e} \frown \dot{\smile} (q \simeq p))$
$\quad$ $o = r$ (ι

(Ia): - - - - - - - - - - - -

$\vdash$ $\top\; r \frown (x; o \,\underline{\wedge}\, q)$
$\quad$ $x; m \frown (o; a \frown \dot{\smile} (q \simeq p))$
$\quad$ $\overset{\mathfrak{e}}{\smile}\top\; x; m \frown (r; \mathfrak{e} \frown \dot{\smile} (q \simeq p))$
$\quad$ $o = r$ (κ

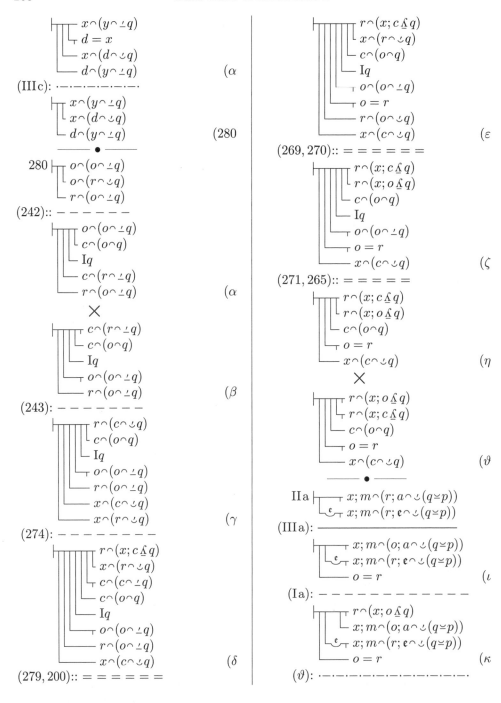

(ϑ): -·-·-·-·-·-·-·-·-·-·-

$$\vdash r \frown (x; o \underline{\curlywedge} q)$$
$$x; m \frown (o; a \frown \acute{\smile}(q \simeq p))$$
$$\smile^{\mathfrak{e}}\; x; m \frown (r; \mathfrak{e} \frown \acute{\smile}(q \simeq p))$$
$$r \frown (x; c \underline{\curlywedge} q)$$
$$c \frown (o \frown q)$$
$$x \frown (c \frown \acute{\smile} q) \qquad (\lambda$$

(IIa):: – – – – – – – – – – – – – –

$$\vdash r \frown (x; o \underline{\curlywedge} q)$$
$$\smile^{\mathfrak{e}}\; x; m \frown (r; \mathfrak{e} \frown \acute{\smile}(q \simeq p))$$
$$x; m \frown (o; a \frown \acute{\smile}(q \simeq p))$$
$$\smile^{\mathfrak{r}}\; \mathfrak{r} \frown (x; c \underline{\curlywedge} q)$$
$$\smile^{\mathfrak{e}}\; x; m \frown (\mathfrak{r}; \mathfrak{e} \frown \acute{\smile}(q \simeq p))$$
$$c \frown (o \frown q)$$
$$x \frown (c \frown \acute{\smile} q) \qquad (\mu$$

$$\vdash \smile^{\mathfrak{r}}\; \mathfrak{r} \frown (x; o \underline{\curlywedge} q)$$
$$\smile^{\mathfrak{e}}\; x; m \frown (\mathfrak{r}; \mathfrak{e} \frown \acute{\smile}(q \simeq p))$$
$$x; m \frown (o; a \frown \acute{\smile}(q \simeq p))$$
$$\smile^{\mathfrak{r}}\; \mathfrak{r} \frown (x; c \underline{\curlywedge} q)$$
$$\smile^{\mathfrak{e}}\; x; m \frown (\mathfrak{r}; \mathfrak{e} \frown \acute{\smile}(q \simeq p))$$
$$c \frown (o \frown q)$$
$$x \frown (c \frown \acute{\smile} q) \qquad (\nu$$

(Ie): – – – – – – – – – – – – – – –

$$\vdash \smile^{\mathfrak{r}}\; \mathfrak{r} \frown (x; o \underline{\curlywedge} q)$$
$$\smile^{\mathfrak{e}}\; x; m \frown (\mathfrak{r}; \mathfrak{e} \frown \acute{\smile}(q \simeq p))$$
$$x; m \frown (o; a \frown \acute{\smile}(q \simeq p))$$
$$x; m \frown (o; a \frown \acute{\smile}(q \simeq p))$$
$$\smile^{\mathfrak{r}}\; \mathfrak{r} \frown (x; c \underline{\curlywedge} q)$$
$$\smile^{\mathfrak{e}}\; x; m \frown (\mathfrak{r}; \mathfrak{e} \frown \acute{\smile}(q \simeq p))$$
$$c \frown (o \frown q)$$
$$x \frown (c \frown \acute{\smile} q) \qquad (\xi$$

(209, 246):: = = = = = = = = = = = = = =

$$\vdash \smile^{\mathfrak{r}}\; \mathfrak{r} \frown (x; o \underline{\curlywedge} q)$$
$$\smile^{\mathfrak{e}}\; x; m \frown (\mathfrak{r}; \mathfrak{e} \frown \acute{\smile}(q \simeq p))$$
$$x; m \frown (o; a \frown \acute{\smile}(q \simeq p))$$
$$d \frown (a \frown p)$$
$$c \frown (o \frown q)$$
$$x; m \frown (c; d \frown \acute{\smile}(q \simeq p))$$
$$\smile^{\mathfrak{r}}\; \mathfrak{r} \frown (x; c \underline{\curlywedge} q)$$
$$\smile^{\mathfrak{e}}\; x; m \frown (\mathfrak{r}; \mathfrak{e} \frown \acute{\smile}(q \simeq p)) \qquad (o$$

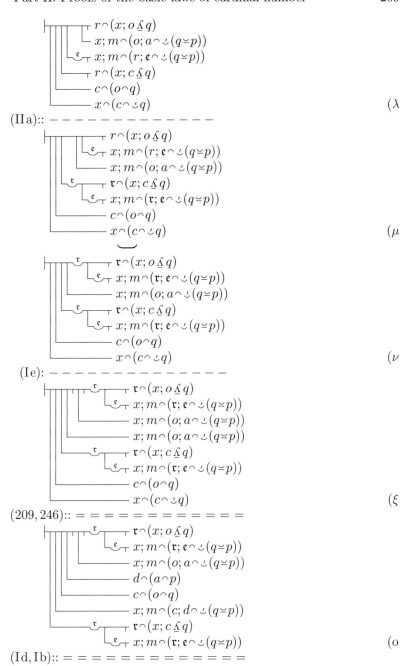

(Id, Ib):: = = = = = = = = = = = = = =

$$\mathfrak{r}\frown(x; o \mathrel{\underline{\wedge}} q)$$
$$x; m\frown(\mathfrak{r}; \mathfrak{e}\frown\smile(q\asymp p))$$
$$x; m\frown(o; a\frown\smile(q\asymp p))$$
$$d\frown(a\frown p)$$
$$c\frown(o\frown q)$$
$$\mathfrak{r}\frown(x; c \mathrel{\underline{\wedge}} q)$$
$$x; m\frown(\mathfrak{r}; \mathfrak{e}\frown\smile(q\asymp p))$$
$$x; m\frown(c; d\frown\smile(q\asymp p))$$ (π

$$\mathfrak{r}\frown(x; \mathfrak{o} \mathrel{\underline{\wedge}} q)$$
$$x; m\frown(\mathfrak{r}; \mathfrak{e}\frown\smile(q\asymp p))$$
$$x; m\frown(\mathfrak{o}; \mathfrak{a}\frown\smile(q\asymp p))$$
$$\mathfrak{d}\frown(\mathfrak{a}\frown p)$$
$$\mathfrak{c}\frown(\mathfrak{o}\frown q)$$
$$\mathfrak{r}\frown(x; \mathfrak{c} \mathrel{\underline{\wedge}} q)$$
$$x; m\frown(\mathfrak{r}; \mathfrak{e}\frown\smile(q\asymp p))$$
$$x; m\frown(\mathfrak{c}; \mathfrak{d}\frown\smile(q\asymp p))$$ (ρ

(257): ————————

$$\mathfrak{r}\frown(x; y \mathrel{\underline{\wedge}} q)$$
$$x; m\frown(\mathfrak{r}; \mathfrak{e}\frown\smile(q\asymp p))$$
$$x; m\frown(y; n\frown\smile(q\asymp p))$$
$$\mathfrak{r}\frown(x; x \mathrel{\underline{\wedge}} q)$$
$$x; m\frown(\mathfrak{r}; \mathfrak{e}\frown\smile(q\asymp p))$$
$$x; m\frown(x; m\frown\smile(q\asymp p))$$
$$x; m\frown(y; n\frown\smile(q\asymp p))$$ (σ

(Ib): – – – – – – – – – – – – – –

$$\mathfrak{r}\frown(x; y \mathrel{\underline{\wedge}} q)$$
$$x; m\frown(\mathfrak{r}; \mathfrak{e}\frown\smile(q\asymp p))$$
$$\mathfrak{r}\frown(x; x \mathrel{\underline{\wedge}} q)$$
$$x; m\frown(\mathfrak{r}; \mathfrak{e}\frown\smile(q\asymp p))$$
$$x; m\frown(x; m\frown\smile(q\asymp p))$$
$$x; m\frown(y; n\frown\smile(q\asymp p))$$ (τ

(Ie):: – – – – – – – – – – – – – –

$$\mathfrak{r}\frown(x; y \mathrel{\underline{\wedge}} q)$$
$$x; m\frown(\mathfrak{r}; \mathfrak{e}\frown\smile(q\asymp p))$$
$$\mathfrak{r}\frown(x; x \mathrel{\underline{\wedge}} q)$$
$$x; m\frown(\mathfrak{r}; \mathfrak{e}\frown\smile(q\asymp p))$$
$$x; m\frown(x; m\frown\smile(q\asymp p))$$
$$x; m\frown(y; n\frown\smile(q\asymp p))$$ (υ

(140):: ————————

$$\vdash \mathfrak{r} \frown (x; y \,\underline{\measuredangle}\, q) \;\; x; m \frown (\mathfrak{r}; \mathfrak{e} \frown \dot{\smile}(q \stackrel{\smile}{=} p)) \;\; \mathfrak{r} \frown (x; x \,\underline{\measuredangle}\, q) \;\; x; m \frown (\mathfrak{r}; \mathfrak{e} \frown \dot{\smile}(q \stackrel{\smile}{=} p)) \;\; x; m \frown (y; n \frown \dot{\smile}(q \stackrel{\smile}{=} p))$$ (281

———•———

270 $\vdash x \frown (r \frown \dot{\smile} q) \;\; r \frown (x; x \,\underline{\measuredangle}\, q)$

(200): — — — — — —

$$\vdash x \frown (r \frown \underline{\acute{\ }} q) \;\; r = x \;\; r \frown (x; x \,\underline{\measuredangle}\, q)$$ (α

(276): — — — — — — —

$$\vdash x \frown (x \frown \underline{\acute{\ }} q) \;\; r = x \;\; r \frown (x; x \,\underline{\measuredangle}\, q) \;\; r \frown (x \frown \dot{\smile} q)$$ (β

(269):: — — — — — — —

$$\vdash x \frown (x \frown \underline{\acute{\ }} q) \;\; r = x \;\; r \frown (x; x \,\underline{\measuredangle}\, q)$$ (γ

×

$$\vdash r = x \;\; x \frown (x \frown \underline{\acute{\ }} q) \;\; r \frown (x; x \,\underline{\measuredangle}\, q)$$ (δ

(271):: — — — — — — —

$$\vdash r = x \;\; r \frown (x; x \,\underline{\measuredangle}\, q)$$ (282

(IIIa): — — — — — —

$$\vdash x; m \frown (r; m \frown \dot{\smile}(q \stackrel{\smile}{=} p)) \;\; x; m \frown (x; m \frown \dot{\smile}(q \stackrel{\smile}{=} p)) \;\; r \frown (x; x \,\underline{\measuredangle}\, q)$$ (α

(140):: ————————————

$$\vdash x; m \frown (r; m \frown \dot{\smile}(q \stackrel{\smile}{=} p)) \;\; r \frown (x; x \,\underline{\measuredangle}\, q)$$ (β

×

$$\vdash r \frown (x; x \,\underline{\measuredangle}\, q) \;\; x; m \frown (r; m \frown \dot{\smile}(q \stackrel{\smile}{=} p))$$ (γ

(IIa):: — — — — — — — — — — — —

$$\vdash r \frown (x; x \,\underline{\measuredangle}\, q) \;\; x; m \frown (r; \mathfrak{e} \frown \dot{\smile}(q \stackrel{\smile}{=} p))$$ (δ

$\smile$

$$\vdash \mathfrak{r} \frown (x; x \,\underline{\measuredangle}\, q) \;\; x; m \frown (\mathfrak{r}; \mathfrak{e} \frown \dot{\smile}(q \stackrel{\smile}{=} p))$$ (ε

(281): ————————————

$$\vdash \mathfrak{r} \frown (x; y \,\underline{\measuredangle}\, q) \;\; x; m \frown (\mathfrak{r}; \mathfrak{e} \frown \dot{\smile}(q \stackrel{\smile}{=} p)) \;\; x; m \frown (y; n \frown \dot{\smile}(q \stackrel{\smile}{=} p))$$ (ζ

(IIa): — — — — — — — — — — — — —

$$\vdash d \frown (x; y \,\underline{\measuredangle}\, q) \;\; x; m \frown (d; \mathfrak{e} \frown \dot{\smile}(q \stackrel{\smile}{=} p)) \;\; x; m \frown (y; n \frown \dot{\smile}(q \stackrel{\smile}{=} p))$$ (η

(277): ·—·—·—·—·—·—·—·—·—·—·

$$\vdash d \frown (x; y \,\underline{\measuredangle}\, q) \;\; \mathfrak{a} \frown (m; n \,\underline{\measuredangle}\, p) \;\; d \frown (\mathfrak{a} \frown (x; m \,\overline{<}\, (q \stackrel{\smile}{=} p))) \;\; n \frown (n \frown \underline{\acute{\ }} p) \;\; x; m \frown (y; n \frown \dot{\smile}(q \stackrel{\smile}{=} p)) \;\; \mathrm{I} p$$ (ϑ

$\smile$

$$\vdash \mathfrak{d} \frown (x; y \,\underline{\measuredangle}\, q) \;\; \mathfrak{a} \frown (m; n \,\underline{\measuredangle}\, p) \;\; \mathfrak{d} \frown (\mathfrak{a} \frown (x; m \,\overline{<}\, (q \stackrel{\smile}{=} p))) \;\; n \frown (n \frown \underline{\acute{\ }} p) \;\; x; m \frown (y; n \frown \dot{\smile}(q \stackrel{\smile}{=} p)) \;\; \mathrm{I} p$$ (ι

(11): — — — — — — — — — — — — — —

$$\vdash x; y \,\underline{\measuredangle}\, q \frown (m; n \,\underline{\measuredangle}\, p \frown \rangle (x; m \,\overline{<}\, (q \stackrel{\smile}{=} p))) \;\; \mathrm{I}(x; m \,\overline{<}\, (q \stackrel{\smile}{=} p)) \;\; n \frown (n \frown \underline{\acute{\ }} p) \;\; x; m \frown (y; n \frown \dot{\smile}(q \stackrel{\smile}{=} p)) \;\; \mathrm{I} p$$ (κ

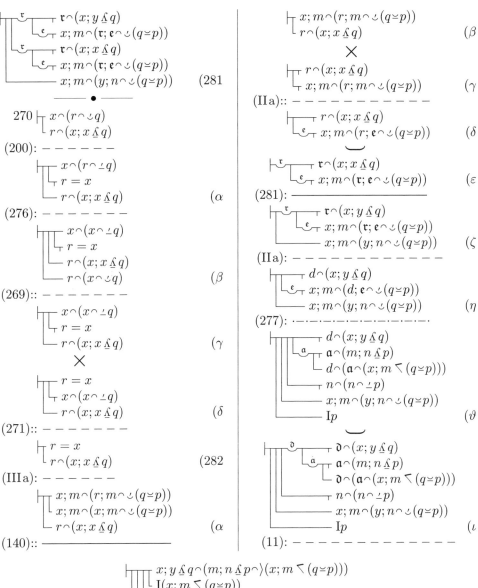

(253):: — — — — — — — — — — — — — — — — — —

$\vdash x;y\,\underline{\triangle}\,q\frown(m;n\,\underline{\triangle}\,p\frown\rangle(x;m\nwarrow(q\asymp p)))$
$\quad \overset{\mathfrak{i}}{\smile}\ \mathfrak{i}\frown(\mathfrak{i}\frown\perp q)$
$\quad\quad x\frown(\mathfrak{i}\frown\smile q)$
$\quad \mathrm{I}q$
$\quad \mathrm{I}p$
$\quad n\frown(n\frown\perp p)$
$\quad x;m\frown(y;n\frown\smile(q\asymp p))$ (λ

——— • ———

234 $\vdash m\frown(n\frown\smile p)$
$\quad x;m\frown(y;n\frown\smile(q\asymp p))$

(IIa): - - - - - - - - - - -

$\vdash n\frown(n\frown\perp p)$
$\quad x;m\frown(y;n\frown\smile(q\asymp p))$
$\quad \overset{\mathfrak{i}}{\smile}\ \mathfrak{i}\frown(\mathfrak{i}\frown\perp p)$
$\quad\quad m\frown(\mathfrak{i}\frown\smile p)$ (μ

(λ): - - - - - - - - - - - - -

$\vdash x;y\,\underline{\triangle}\,q\frown(m;n\,\underline{\triangle}\,p\frown\rangle(x;m\nwarrow(q\asymp p)))$
$\quad \overset{\mathfrak{i}}{\smile}\ \mathfrak{i}\frown(\mathfrak{i}\frown\perp q)$
$\quad\quad x\frown(\mathfrak{i}\frown\smile q)$
$\quad \mathrm{I}q$
$\quad \mathrm{I}p$
$\quad x;m\frown(y;n\frown\smile(q\asymp p))$
$\quad \overset{\mathfrak{i}}{\smile}\ \mathfrak{i}\frown(\mathfrak{i}\frown\perp p)$
$\quad\quad m\frown(\mathfrak{i}\frown\smile p)$ (283

——— • ———

260 $\vdash m;n\,\underline{\triangle}\,p\frown(x;y\,\underline{\triangle}\,q\frown\rangle\not\Downarrow(x;m\nwarrow(q\asymp p)))$
$\quad m;n\,\underline{\triangle}\,p\frown(x;y\,\underline{\triangle}\,q\frown\rangle(m;x\nwarrow(p\asymp q)))$

(32): - - - - - - - - - - - - - - - - - - -

$\vdash \mathfrak{P}(x;y\,\underline{\triangle}\,q)=\mathfrak{P}(m;n\,\underline{\triangle}\,p)$
$\quad x;y\,\underline{\triangle}\,q\frown(m;n\,\underline{\triangle}\,p\frown\rangle(x;m\nwarrow(q\asymp p)))$
$\quad m;n\,\underline{\triangle}\,p\frown(x;y\,\underline{\triangle}\,q\frown\rangle(m;x\nwarrow(p\asymp q)))$ (α

(283, 283):: = = = = = = = = = = = = = =

$\vdash \mathfrak{P}(x;y\,\underline{\triangle}\,q)=\mathfrak{P}(m;n\,\underline{\triangle}\,p)$
$\quad \overset{\mathfrak{i}}{\smile}\ \mathfrak{i}\frown(\mathfrak{i}\frown\perp q)$
$\quad\quad x\frown(\mathfrak{i}\frown\smile q)$
$\quad \mathrm{I}q$
$\quad \mathrm{I}p$
$\quad x;m\frown(y;n\frown\smile(q\asymp p))$
$\quad \overset{\mathfrak{i}}{\smile}\ \mathfrak{i}\frown(\mathfrak{i}\frown\perp p)$
$\quad\quad m\frown(\mathfrak{i}\frown\smile p)$
$\quad m;x\frown(n;y\frown\smile(p\asymp q))$ (β

(258):: - - - - - - - - - - - - - - - -

$$\vdash \mathfrak{p}(x; y \mathbin{\underset{\frown}{\wedge}} q) = \mathfrak{p}(m; n \mathbin{\underset{\frown}{\wedge}} p)$$
$$\stackrel{\mathfrak{i}}{\smile} \mathfrak{i} \frown (\mathfrak{i} \frown \underline{\ } q)$$
$$x \frown (\mathfrak{i} \frown \smile q)$$
$$\mathrm{I} q$$
$$\mathrm{I} p$$
$$x; m \frown (y; n \frown \smile (q \simeq p))$$
$$\stackrel{\mathfrak{i}}{\smile} \mathfrak{i} \frown (\mathfrak{i} \frown \underline{\ } p)$$
$$m \frown (\mathfrak{i} \frown \smile p) \qquad (284$$

———•———

136 $\vdash d \frown (y \frown \smile q)$
$d \frown (y \frown \underline{\ } q)$

(132):: – – – – – –

$$\vdash d \frown (y \frown \smile q)$$
$$d \frown (a \frown q)$$
$$a \frown (y \frown \smile q) \qquad (285$$

———•———

145 $\vdash b \frown (b \frown \underline{\ } \mathfrak{f})$
$\mathfrak{0} \frown (b \frown \smile \mathfrak{f})$

(285):: – – – – – –

$$\vdash b \frown (b \frown \underline{\ } \mathfrak{f})$$
$$\mathfrak{0} \frown (\mathfrak{1} \frown \mathfrak{f})$$
$$\mathfrak{1} \frown (b \frown \smile \mathfrak{f}) \qquad (\alpha$$

(110):: ————————

$$\vdash b \frown (b \frown \underline{\ } \mathfrak{f})$$
$$\mathfrak{1} \frown (b \frown \smile \mathfrak{f}) \qquad (286$$

$$\vdash \stackrel{\mathfrak{i}}{\smile} \mathfrak{i} \frown (\mathfrak{i} \frown \underline{\ } \mathfrak{f})$$
$$\mathfrak{1} \frown (\mathfrak{i} \frown \smile \mathfrak{f}) \qquad (287$$

(284): ————————

$$\vdash \mathfrak{p}(x; y \mathbin{\underset{\frown}{\wedge}} q) = \mathfrak{p}(\mathfrak{1}; n \mathbin{\underset{\frown}{\wedge}} \mathfrak{f})$$
$$\stackrel{\mathfrak{i}}{\smile} \mathfrak{i} \frown (\mathfrak{i} \frown \underline{\ } q)$$
$$x \frown (\mathfrak{i} \frown \smile q)$$
$$\mathrm{I} q$$
$$\mathrm{I} \mathfrak{f}$$
$$x; \mathfrak{1} \frown (y; n \frown \smile (q \simeq \mathfrak{f})) \qquad (\alpha$$

(71):: ————————————————

$$\vdash \mathfrak{p}(x; y \mathbin{\underset{\frown}{\wedge}} q) = \mathfrak{p}(\mathfrak{1}; n \mathbin{\underset{\frown}{\wedge}} \mathfrak{f})$$
$$\stackrel{\mathfrak{i}}{\smile} \mathfrak{i} \frown (\mathfrak{i} \frown \underline{\ } q)$$
$$x \frown (\mathfrak{i} \frown \smile q)$$
$$\mathrm{I} q$$
$$x; \mathfrak{1} \frown (y; n \frown \smile (q \simeq \mathfrak{f})) \qquad (\beta$$

(261):: – – – – – – – – – – – – –

$$\vdash \mathfrak{p}(x; y \mathbin{\underset{\frown}{\wedge}} q) = \mathfrak{p}(\mathfrak{1}; n \mathbin{\underset{\frown}{\wedge}} \mathfrak{f})$$
$$\stackrel{\mathfrak{i}}{\smile} \mathfrak{i} \frown (\mathfrak{i} \frown \underline{\ } q)$$
$$\mathrm{I} q$$
$$x; \mathfrak{1} \frown (y; n \frown \smile (q \simeq \mathfrak{f})) \qquad (288$$

§162. *Analysis*

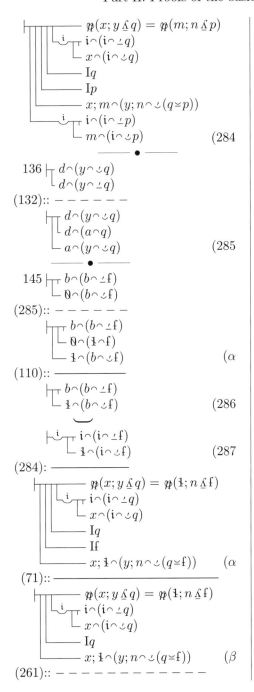

In order to replace in (288) the subcomponent

$$‘\stackrel{\mathfrak{i}}{\smile}\!\top \mathfrak{i} \frown (\mathfrak{i} \frown \underline{\ } q)’$$

by ‘$\top y \frown (y \frown \underline{\ } q)$’, we exchange in (288) ‘q’ with ‘$y \frown \smile q \frown q$’. For the proposition

$$‘\vdash \mathfrak{i} \frown (\mathfrak{i} \frown \underline{\ } (y \frown \smile q \frown q))$$
$$\top y \frown (y \frown \underline{\ } q)$$
$$\mathrm{I} q \qquad ’, \qquad (\alpha$$

is easily proven. In order to bring ‘q’ back into the proposition in place of ‘$y \frown \smile q \frown q$’, we apply the propositions (189),

$$‘\vdash \mathfrak{p}(x; y \mathbin{\underset{\frown}{\wedge}} q) = \mathfrak{p}(x; y \mathbin{\underset{\frown}{\wedge}} (y \frown \smile q \frown q))$$
$$\top y \frown (y \frown \underline{\ } q)$$
$$\mathrm{I} q \qquad ’ \quad (\beta$$

and

$$‘\vdash x; \mathfrak{1} \frown (y; n \frown \smile (y \frown \smile q \frown q \simeq \mathfrak{f}))$$
$$x; \mathfrak{1} \frown (y; n \frown \smile (q \simeq \mathfrak{f})) \qquad ’, \qquad (\gamma$$

By means of (257), we prove the more general proposition resulting from (γ) by the replacement of ‘$\mathfrak{1}$’ by ‘m’ and ‘$\mathfrak{f}$’ by ‘p’. For this we need the proposition

$$‘\vdash x; m \frown (a; o \frown \smile (y \frown \smile q \frown q \simeq p))$$
$$a \frown (y \frown \smile q)$$
$$c \frown (o \frown p)$$
$$d \frown (a \frown q)$$
$$x; m \frown (d; c \frown \smile (y \frown \smile q \frown q \simeq p))$$
$$d \frown (y \frown \smile q) \qquad ’, \qquad (\delta$$

which follows from (209) and (197).

§163. *Construction*

$$197 \vdash d \frown (a \frown (y \frown \smile q \,\hat{=}\, q))$$
$$d \frown (a \frown q)$$
$$a \frown (y \frown \smile q)$$

(209): – – – – – – – – – –

$$\vdash x; m \frown (a; o \frown \smile (y \frown \smile q \,\hat{=}\, q \,\check{=}\, p))$$
$$c \frown (o \frown p)$$
$$d \frown (a \frown q)$$
$$a \frown (y \frown \smile q)$$
$$x; m \frown (d; c \frown \smile (y \frown \smile q \,\hat{=}\, q \,\check{=}\, p)) \quad (\alpha$$

(I):: – – – – – – – – – – – – – – –

$$\vdash x; m \frown (a; o \frown \smile (y \frown \smile q \,\hat{=}\, q \,\check{=}\, p))$$
$$c \frown (o \frown p)$$
$$d \frown (a \frown q)$$
$$a \frown (y \frown \smile q)$$
$$d \frown (y \frown \smile q)$$
$$x; m \frown (d; c \frown \smile (y \frown \smile q \,\hat{=}\, q \,\check{=}\, p))$$
$$d \frown (y \frown \smile q) \quad (\beta$$

(285):: – – – – – – – – – – – – – – – – –

$$\vdash x; m \frown (a; o \frown \smile (y \frown \smile q \,\hat{=}\, q \,\check{=}\, p))$$
$$a \frown (y \frown \smile q)$$
$$c \frown (o \frown p)$$
$$d \frown (a \frown q)$$
$$x; m \frown (d; c \frown \smile (y \frown \smile q \,\hat{=}\, q \,\check{=}\, p))$$
$$d \frown (y \frown \smile q) \quad (\gamma$$

$$\vdash^{\mathfrak{c}}\smile^{\mathfrak{d}}\smile^{\mathfrak{o}}\smile^{\mathfrak{a}}\smile\; x; m \frown (\mathfrak{o}; \mathfrak{a} \frown \smile (y \frown \smile q \,\hat{=}\, q \,\check{=}\, p))$$
$$\mathfrak{o} \frown (y \frown \smile q)$$
$$\mathfrak{d} \frown (\mathfrak{a} \frown p)$$
$$\mathfrak{c} \frown (\mathfrak{o} \frown q)$$
$$x; m \frown (\mathfrak{c}; \mathfrak{d} \frown \smile (y \frown \smile q \,\hat{=}\, q \,\check{=}\, p))$$
$$\mathfrak{c} \frown (y \frown \smile q) \quad (\delta$$

(257): ————————————————————

$$\vdash x; m \frown (a; o \frown \smile (y \frown \smile q \,\hat{=}\, q \,\check{=}\, p))$$
$$a \frown (y \frown \smile q)$$
$$x; m \frown (x; m \frown \smile (y \frown \smile q \,\hat{=}\, q \,\check{=}\, p))$$
$$x \frown (y \frown \smile q)$$
$$x; m \frown (a; o \frown \smile (q \,\check{=}\, p)) \quad (\varepsilon$$

(I):: – – – – – – – – – – – – – – –

$$\vdash x; m \frown (a; o \frown \smile (y \frown \smile q \,\hat{=}\, q \,\check{=}\, p))$$
$$a \frown (y \frown \smile q)$$
$$x; m \frown (x; m \frown \smile (y \frown \smile q \,\hat{=}\, q \,\check{=}\, p))$$
$$x; m \frown (a; o \frown \smile (q \,\check{=}\, p)) \quad (\zeta$$

(140):: ——————————————

$$\vdash\!\!\top\!\top\; x;m\frown(a;o\frown\smile(y\frown\smile q\doteqdot q\asymp p))$$
$$\quad\llcorner\; a\frown(y\frown\smile q)$$
$$\llcorner\; x;m\frown(a;o\frown\smile(q\asymp p)) \qquad (289$$

——— • ———

$$140 \vdash y\frown(y\frown\smile q)$$

(289): ————————

$$\vdash\!\!\top\; x;m\frown(y;n\frown\smile(y\frown\smile q\doteqdot q\asymp p))$$
$$\llcorner\; x;m\frown(y;n\frown\smile(q\asymp p)) \qquad (290$$

§164. *Analysis*

In order to prove proposition (β) of §162 we need the propositions

$$\text{‘}\vdash\!\!\top\!\top\; x\frown(d\frown\smile(y\frown\smile q\doteqdot q))$$
$$\quad\llcorner\; d\frown(y\frown\smile q)$$
$$\llcorner\; x\frown(d\frown\smile q)\text{’} \qquad (\alpha$$

(194) and (189). We prove proposition (α) using (144).

§165. *Construction*

$$194 \vdash\!\!\top\; x\frown(y\frown\underline{\smile} q)$$
$$\llcorner\; x\frown(y\frown\underline{\smile}(u\doteqdot q))$$

$\times$

$$\vdash\!\!\top\; x\frown(y\frown\underline{\smile}(u\doteqdot q))$$
$$\llcorner\!\top\; x\frown(y\frown\underline{\smile} q) \qquad (291$$

——— • ———

$$137 \vdash\!\!\top\!\top\; x\frown(a\frown\smile(y\frown\smile q\doteqdot q))$$
$$\quad\llcorner\; d\frown(a\frown(y\frown\smile q\doteqdot q))$$
$$\llcorner\; x\frown(d\frown\smile(y\frown\smile q\doteqdot q))$$

(197):: — — — — — — — — — — —

$$\vdash\!\!\top\!\top\!\top\; x\frown(a\frown\smile(y\frown\smile q\doteqdot q))$$
$$\quad\llcorner\; d\frown(a\frown q)$$
$$\quad\llcorner\; a\frown(y\frown\smile q)$$
$$\llcorner\; x\frown(d\frown\smile(y\frown\smile q\doteqdot q)) \qquad (\alpha$$

(I):: — — — — — — — — — — —

$$\vdash\!\!\top\!\top\!\top\!\top\; x\frown(a\frown\smile(y\frown\smile q\doteqdot q))$$
$$\quad\llcorner\; d\frown(a\frown q)$$
$$\quad\llcorner\; a\frown(y\frown\smile q)$$
$$\quad\llcorner\; d\frown(y\frown\smile q)$$
$$\llcorner\!\top\; x\frown(d\frown\smile(y\frown\smile q\doteqdot q))$$
$$\llcorner\; d\frown(y\frown\smile q) \qquad (\beta$$

(285):: — — — — — — — — — — — —

$$\vdash\!\!\top\!\top\!\top\; x\frown(a\frown\smile(y\frown\smile q\doteqdot q))$$
$$\quad\llcorner\; a\frown(y\frown\smile q)$$
$$\quad\llcorner\; d\frown(a\frown q)$$
$$\llcorner\!\top\; x\frown(d\frown\smile(y\frown\smile q\doteqdot q))$$
$$\llcorner\; d\frown(y\frown\smile q) \qquad (\gamma$$

$\smile$

$$\vdash\!\overset{\mathfrak{d}}{\smile}\!\!\top\!\overset{\mathfrak{a}}{\smile}\!\!\top\!\top\; x\frown(\mathfrak{a}\frown\smile(y\frown\smile q\doteqdot q))$$
$$\quad\llcorner\; \mathfrak{a}\frown(y\frown\smile q)$$
$$\quad\llcorner\; \mathfrak{d}\frown(\mathfrak{a}\frown q)$$
$$\llcorner\!\top\; x\frown(\mathfrak{d}\frown\smile(y\frown\smile q\doteqdot q))$$
$$\llcorner\; \mathfrak{d}\frown(y\frown\smile q) \qquad (\delta$$

(144): ————————————————

$$\vdash\!\!\top\!\top\!\top\; x\frown(d\frown\smile(y\frown\smile q\doteqdot q))$$
$$\quad\llcorner\; d\frown(y\frown\smile q)$$
$$\quad\llcorner\!\top\; x\frown(x\frown\smile(y\frown\smile q\doteqdot q))$$
$$\quad\llcorner\; x\frown(y\frown\smile q)$$
$$\llcorner\; x\frown(d\frown\smile q) \qquad (\varepsilon$$

(I):: — — — — — — — — — — —

$$\vdash\!\!\top\!\top\!\top\; x\frown(d\frown\smile(y\frown\smile q\doteqdot q))$$
$$\quad\llcorner\; d\frown(y\frown\smile q)$$
$$\quad\llcorner\; x\frown(x\frown\smile(y\frown\smile q\doteqdot q))$$
$$\llcorner\; x\frown(d\frown\smile q) \qquad (\zeta$$

(140):: ————————————————

$$\vdash\!\!\top\!\top\; x\frown(d\frown\smile(y\frown\smile q\doteqdot q))$$
$$\quad\llcorner\; d\frown(y\frown\smile q)$$
$$\llcorner\; x\frown(d\frown\smile q) \qquad (292$$

——— • ———

$$140 \vdash y\frown(y\frown\smile q)$$

(292): ————————

$$\vdash\!\!\top\; d\frown(y\frown\smile(y\frown\smile q\doteqdot q))$$
$$\llcorner\; d\frown(y\frown\smile q) \qquad (293$$

(274): — — — — — — — — — —

$$\vdash\!\!\top\!\top\!\top\!\top\; d\frown(x;y\,\underline{\wedge}\,(y\frown\smile q\doteqdot q))$$
$$\quad\llcorner\; x\frown(d\frown\smile(y\frown\smile q\doteqdot q))$$
$$\quad\llcorner\!\top\; y\frown(y\frown\underline{\smile}(y\frown\smile q\doteqdot q))$$
$$\quad\llcorner\; d\frown(y\frown\smile q)$$
$$\llcorner\; \mathrm{I}(y\frown\smile q\doteqdot q) \qquad (\alpha$$

(291, 292):: = = = = = = = = = = =

$\vdash d \frown (x; y \,\underline{\measuredangle}\, (y \frown \dot{\smile} q \frown\!\!= q))$
$\quad d \frown (y \frown \dot{\smile} q)$
$\quad x \frown (d \frown \dot{\smile} q)$
$\quad y \frown (y \frown \dot{-} q)$
$\quad \mathrm{I}(y \frown \dot{\smile} q \frown\!\!= q)$ (β

(269, 270):: = = = = = = = = = =

$\vdash d \frown (x; y \,\underline{\measuredangle}\, (y \frown \dot{\smile} q \frown\!\!= q))$
$\quad d \frown (x; y \,\underline{\measuredangle}\, q)$
$\quad y \frown (y \frown \dot{-} q)$
$\quad \mathrm{I}(y \frown \dot{\smile} q \frown\!\!= q)$ (γ

(189):: – – – – – – – – – – –

$\vdash d \frown (x; y \,\underline{\measuredangle}\, (y \frown \dot{\smile} q \frown\!\!= q))$
$\quad d \frown (x; y \,\underline{\measuredangle}\, q)$
$\quad y \frown (y \frown \dot{-} q)$
$\quad \mathrm{I}q$ (δ

(271, 265):: = = = = = = = = =

$\vdash d \frown (x; y \,\underline{\measuredangle}\, (y \frown \dot{\smile} q \frown\!\!= q))$
$\quad d \frown (x; y \,\underline{\measuredangle}\, q)$ (294

———•———

274 $\vdash d \frown (x; y \,\underline{\measuredangle}\, q)$
$\quad x \frown (d \frown \dot{\smile} q)$
$\quad y \frown (y \frown \dot{-} q)$
$\quad d \frown (y \frown \dot{\smile} q)$
$\quad \mathrm{I}q$

(201, 201):: = = = = =

$\vdash d \frown (x; y \,\underline{\measuredangle}\, q)$
$\quad x \frown (d \frown \dot{\smile} (y \frown \dot{\smile} q \frown\!\!= q))$
$\quad y \frown (y \frown \dot{-} q)$
$\quad d \frown (y \frown \dot{\smile} (y \frown \dot{\smile} q \frown\!\!= q))$
$\quad \mathrm{I}q$ (α

(269, 270):: = = = = = = = = =

$\vdash d \frown (x; y \,\underline{\measuredangle}\, q)$
$\quad d \frown (x; y \,\underline{\measuredangle}\, (y \frown \dot{\smile} q \frown\!\!= q))$
$\quad y \frown (y \frown \dot{-} q)$
$\quad \mathrm{I}q$ (β

(IV a): – – – – – – – – – – – –

$\vdash (\text{—}\, d \frown (x; y \,\underline{\measuredangle}\, q)) = (\text{—}\, d \frown (x; y \,\underline{\measuredangle}\, (y \frown \dot{\smile} q \frown\!\!= q)))$
$\quad d \frown (x; y \,\underline{\measuredangle}\, (y \frown \dot{\smile} q \frown\!\!= q))$
$\quad d \frown (x; y \,\underline{\measuredangle}\, q)$
$\quad y \frown (y \frown \dot{-} q)$
$\quad \mathrm{I}q$ (γ

(294):: ————————————————

$\vdash (\text{—}\, d \frown (x; y \,\underline{\measuredangle}\, q)) = (\text{—}\, d \frown (x; y \,\underline{\measuredangle}\, (y \frown \dot{\smile} q \frown\!\!= q)))$
$\quad y \frown (y \frown \dot{-} q)$
$\quad \mathrm{I}q$ (δ

$\smile$

$\vdash \overset{\mathfrak{a}}{\smile} (\text{—}\, \mathfrak{a} \frown (x; y \,\underline{\measuredangle}\, q)) = (\text{—}\, \mathfrak{a} \frown (x; y \,\underline{\measuredangle}\, (y \frown \dot{\smile} q \frown\!\!= q)))$
$\quad y \frown (y \frown \dot{-} q)$
$\quad \mathrm{I}q$ (ε

(96): –

$\vdash \mathfrak{p}(x; y \,\underline{\measuredangle}\, q) = \mathfrak{p}(x; y \,\underline{\measuredangle}\, (y \frown \dot{\smile} q \frown\!\!= q))$
$\quad y \frown (y \frown \dot{-} q)$
$\quad \mathrm{I}q$ (295

———•———

278 $\vdash a \frown (a \frown \dot{-} q)$
$\quad d \frown (a \frown q)$
$\quad d \frown (d \frown \dot{-} q)$
$\quad \mathrm{I}q$

$\smile$

$\vdash \overset{\mathfrak{d}}{\smile} \overset{\mathfrak{a}}{\smile} \mathfrak{a} \frown (\mathfrak{a} \frown \dot{-} q)$
$\quad \mathfrak{d} \frown (\mathfrak{a} \frown q)$
$\quad \mathfrak{d} \frown (\mathfrak{d} \frown \dot{-} q)$
$\quad \mathrm{I}q$ (α

(144): – – – – – – – – – –

$\vdash y \frown (y \frown \bot q)$
$\quad i \frown (i \frown \bot q)$
$\quad \mathrm{I}q$
$\quad i \frown (y \frown \cup q)$ (296

(194):: – – – – – –

$\vdash y \frown (y \frown \bot q)$
$\quad i \frown (i \frown \bot (y \frown \cup q \asymp q))$
$\quad \mathrm{I}q$
$\quad i \frown (y \frown \cup q)$ (α

×

$\vdash i \frown (i \frown \bot (y \frown \cup q \asymp q))$
$\quad y \frown (y \frown \bot q)$
$\quad \mathrm{I}q$
$\quad i \frown (y \frown \cup q)$ (β

(193): –·–·–·–·–·–·–·–·–·

$\vdash i \frown (i \frown \bot (y \frown \cup q \asymp q))$
$\quad y \frown (y \frown \bot q)$
$\quad \mathrm{I}q$ (γ

$\smile$

$\vdash^{\mathfrak{i}} \mathfrak{i} \frown (\mathfrak{i} \frown \bot (y \frown \cup q \asymp q))$
$\quad y \frown (y \frown \bot q)$
$\quad \mathrm{I}q$ (297

(288): – – – – – – – – – – – –

$\vdash \mathfrak{p}(x; y \underline{\wedge} (y \frown \cup q \asymp q)) = \mathfrak{p}(\mathfrak{1}; n \underline{\wedge} \mathfrak{f})$
$\quad y \frown (y \frown \bot q)$
$\quad \mathrm{I}q$
$\quad \mathrm{I}(y \frown \cup q \asymp q)$
$\quad x; \mathfrak{1} \frown (y; n \frown \cup (y \frown \cup q \asymp q \preceq \mathfrak{f}))$ (α

(290, 189):: = = = = = = = = = = =

$\vdash \mathfrak{p}(x; y \underline{\wedge} (y \frown \cup q \asymp q)) = \mathfrak{p}(\mathfrak{1}; n \underline{\wedge} \mathfrak{f})$
$\quad y \frown (y \frown \bot q)$
$\quad \mathrm{I}q$
$\quad x; \mathfrak{1} \frown (y; n \frown \cup (q \preceq \mathfrak{f}))$ (β

(III a): ————————————

$\vdash \mathfrak{p}(x; y \underline{\wedge} q) = \mathfrak{p}(\mathfrak{1}; n \underline{\wedge} \mathfrak{f})$
$\quad y \frown (y \frown \bot q)$
$\quad \mathrm{I}q$
$\quad x; \mathfrak{1} \frown (y; n \frown \cup (q \preceq \mathfrak{f}))$
$\quad \mathfrak{p}(x; y \underline{\wedge} q) = \mathfrak{p}(x; y \underline{\wedge} (y \frown \cup q \asymp q))$ (γ

(295):: – – – – – – – – – – – – – –

$\vdash \mathfrak{p}(x; y \underline{\wedge} q) = \mathfrak{p}(\mathfrak{1}; n \underline{\wedge} \mathfrak{f})$
$\quad y \frown (y \frown \bot q)$
$\quad \mathrm{I}q$
$\quad x; \mathfrak{1} \frown (y; n \frown \cup (q \preceq \mathfrak{f}))$ (298

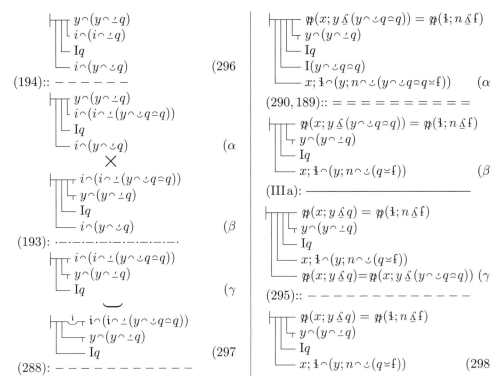

b) Proof of the proposition

'$\vdash n = \mathfrak{p}(\mathfrak{1}; n \underline{\wedge} \mathfrak{f})$
$\quad \mathfrak{0} \frown (n \frown \cup \mathfrak{f})$'

and end of section K

§166. *Analysis*

We prove proposition (δ) of §158 by means of (160). For this we require the proposition

'$\vdash \mathfrak{p}(\mathfrak{1}; n \underline{\wedge} \mathfrak{f}) = \mathfrak{p}\dot{\varepsilon}\left(\begin{array}{l} \varepsilon = \mathfrak{0} \\ \varepsilon \frown (n \frown \cup \mathfrak{f}) \end{array} \right)$
$\quad \mathfrak{0} \frown (n \frown \cup \mathfrak{f})$' (α

which follows using (IV a) and (96) from the propositions

'$\vdash a \frown (\mathfrak{1}; n \underline{\wedge} \mathfrak{f})$
$\quad a = \mathfrak{0}$
$\quad \mathfrak{0} \frown (n \frown \cup \mathfrak{f})$
$\quad a \frown (n \frown \cup \mathfrak{f})$' (β

'$\vdash a = \mathfrak{0}$
$\quad a \frown (n \frown \cup \mathfrak{f})$
$\quad a \frown (\mathfrak{1}; n \underline{\wedge} \mathfrak{f})$' (γ

(β) is to be reduced to the proposition

'$\vdash \mathfrak{1} \frown (a \frown \cup \mathfrak{f})$
$\quad a = \mathfrak{0}$
$\quad \mathfrak{0} \frown (n \frown \cup \mathfrak{f})$
$\quad a \frown (n \frown \cup \mathfrak{f})$' (δ

which can be derived from (242). We first obtain

'$\vdash \mathfrak{1} \frown (a \frown \cup \mathfrak{f})$
$\quad \mathfrak{0} \frown (a \frown \bot \mathfrak{f})$' (ε

It now turns on proving the proposition

$$\text{‘}\vdash \begin{array}{l} \mathfrak{q}\frown(a\frown\dot{\smile}\mathfrak{f}) \\ \mathfrak{q}\frown(n\frown\dot{\smile}\mathfrak{f}) \\ a\frown(n\frown\dot{\smile}\mathfrak{f})\text{’} \end{array} \qquad (\zeta$$

This follows from the proposition

$$\text{‘}\vdash \begin{array}{l} m\frown(n\frown\dot{\smile}p) \\ \neg\, n\frown(m\frown\dot{-}p) \\ m\frown(r\frown\dot{\smile}p) \\ n\frown(r\frown\dot{\smile}p) \\ \mathrm{I}\nparallel p \quad \text{’} \end{array} \qquad (\eta$$

which is similar to (243) and can be derived from it. Writing (243) thus

$$\text{‘}\vdash \begin{array}{l} n\frown(m\frown\dot{\smile}\nparallel p) \\ \neg\, m\frown(n\frown\dot{-}\nparallel p) \\ r\frown(m\frown\dot{\smile}\nparallel p) \\ r\frown(n\frown\dot{\smile}\nparallel p) \\ \mathrm{I}\nparallel p \quad \text{’} \end{array}$$

it then remains to prove the propositions

$$\text{‘}\vdash \begin{array}{l} m\frown(n\frown\dot{-}p) \\ n\frown(m\frown\dot{-}\nparallel p)\text{’} \end{array} \qquad (\vartheta$$

$$\text{‘}\vdash \begin{array}{l} n\frown(m\frown\dot{-}\nparallel p) \\ m\frown(n\frown\dot{-}p) \quad \text{’} \end{array} \qquad (\iota$$

and others similar to them, which can be accomplished by means of (123).

§167. *Construction*

$$23 \vdash \begin{array}{l} a\frown(d\frown p) \\ d\frown(a\frown\nparallel p) \end{array}$$

(129): − − − − − −

$$\vdash \begin{array}{l} a\frown(n\frown\dot{-}p) \\ d\frown(a\frown\nparallel p) \\ d\frown(n\frown\dot{-}p) \end{array} \qquad (\alpha$$

$$\vdash \forall\mathfrak{d}\,\forall\mathfrak{a} \begin{array}{l} \mathfrak{a}\frown(n\frown\dot{-}p) \\ \mathfrak{d}\frown(\mathfrak{a}\frown\nparallel p) \\ \mathfrak{d}\frown(n\frown\dot{-}p) \end{array} \qquad (\beta$$

(123): ———————

$$\vdash \begin{array}{l} m\frown(n\frown\dot{-}p) \\ \forall\mathfrak{a}\begin{array}{l} \mathfrak{a}\frown(n\frown\dot{-}p) \\ n\frown(\mathfrak{a}\frown\nparallel p) \end{array} \\ n\frown(m\frown\dot{-}\nparallel p) \end{array} \qquad (\gamma$$

——— • ———

$$131 \vdash \begin{array}{l} a\frown(n\frown\dot{-}p) \\ a\frown(n\frown p) \end{array}$$

(23):: − − − − − −

$$\vdash \begin{array}{l} a\frown(n\frown\dot{-}p) \\ n\frown(a\frown\nparallel p) \end{array} \qquad (\delta$$

$$\vdash \forall\mathfrak{a} \begin{array}{l} \mathfrak{a}\frown(n\frown\dot{-}p) \\ n\frown(\mathfrak{a}\frown\nparallel p) \end{array} \qquad (\varepsilon$$

(γ): ———————

$$\vdash \begin{array}{l} m\frown(n\frown\dot{-}p) \\ n\frown(m\frown\dot{-}\nparallel p) \end{array} \qquad (299$$

×

$$\vdash \begin{array}{l} \neg\, n\frown(m\frown\dot{-}\nparallel p) \\ \neg\, m\frown(n\frown\dot{-}p) \end{array} \qquad (300$$

——— • ———

$$\mathrm{IIIf} \vdash \begin{array}{l} n = m \\ m = n \end{array}$$

(139): − − − −

$$\vdash \begin{array}{l} m\frown(n\frown\dot{\smile}p) \\ m = n \end{array} \qquad (301$$

——— • ———

$$22 \vdash \begin{array}{l} a\frown(d\frown\nparallel p) \\ d\frown(a\frown p) \end{array}$$

(129): − − − − − −

$$\vdash \begin{array}{l} a\frown(m\frown\dot{-}\nparallel p) \\ d\frown(a\frown p) \\ d\frown(m\frown\dot{-}\nparallel p) \end{array} \qquad (\alpha$$

$$\vdash \forall\mathfrak{d}\,\forall\mathfrak{a} \begin{array}{l} \mathfrak{a}\frown(m\frown\dot{-}\nparallel p) \\ \mathfrak{d}\frown(\mathfrak{a}\frown p) \\ \mathfrak{d}\frown(m\frown\dot{-}\nparallel p) \end{array} \qquad (\beta$$

(123): ———————

$$\vdash \begin{array}{l} n\frown(m\frown\dot{-}\nparallel p) \\ \forall\mathfrak{a}\begin{array}{l} \mathfrak{a}\frown(m\frown\dot{-}\nparallel p) \\ m\frown(\mathfrak{a}\frown p) \end{array} \\ m\frown(n\frown\dot{-}p) \end{array} \qquad (\gamma$$

——— • ———

131 $\vdash a\frown(m\frown\dot{\bot}\between p)$
$\quad a\frown(m\frown\between p)$

(22):: - - - - - - - -

$\vdash a\frown(m\frown\dot{\bot}\between p)$
$\quad m\frown(a\frown p)$ (δ

$\vdash^{\mathfrak{a}} \mathfrak{a}\frown(m\frown\dot{\bot}\between p)$
$\quad m\frown(\mathfrak{a}\frown p)$ (ε

(γ): ———————

$\vdash n\frown(m\frown\dot{\bot}\between p)$
$\quad m\frown(n\frown\dot{\bot} p)$ (302

(136): - - - - - - - -

$\vdash n\frown(m\frown\dot{\smile}\between p)$
$\quad m\frown(n\frown\dot{\bot} p)$ (α

(200):: - - - - - - - -

$\vdash n\frown(m\frown\dot{\smile}\between p)$
$\quad n = m$
$\quad m\frown(n\frown\dot{\smile} p)$ (β

(301): ·—·—·—·—·—·—·—·

$\vdash n\frown(m\frown\dot{\smile}\between p)$
$\quad m\frown(n\frown\dot{\smile} p)$ (303

——— • ———

299 $\vdash m\frown(n\frown\dot{\bot} p)$
$\quad n\frown(m\frown\dot{\bot}\between p)$

(136): - - - - - - - -

$\vdash m\frown(n\frown\dot{\smile} p)$
$\quad n\frown(m\frown\dot{\bot}\between p)$ (α

(200):: - - - - - - - -

$\vdash m\frown(n\frown\dot{\smile} p)$
$\quad m = n$
$\quad n\frown(m\frown\dot{\smile}\between p)$ (β

(301): ·—·—·—·—·—·—·—·

$\vdash m\frown(n\frown\dot{\smile} p)$
$\quad n\frown(m\frown\dot{\smile}\between p)$ (304

(243):: - - - - - - - -

$\vdash m\frown(n\frown\dot{\smile} p)$
$\quad m\frown(n\frown\dot{\bot}\between p)$
$\quad r\frown(m\frown\dot{\smile}\between p)$
$\quad r\frown(n\frown\dot{\smile}\between p)$
$\quad \mathrm{I}\between p$ (α

(303, 303):: = = = = = =

$\vdash m\frown(n\frown\dot{\smile} p)$
$\quad m\frown(n\frown\dot{\bot}\between p)$
$\quad m\frown(r\frown\dot{\smile} p)$
$\quad n\frown(r\frown\dot{\smile} p)$
$\quad \mathrm{I}\between p$ (β

(300):: - - - - - - - - - -

$\vdash m\frown(n\frown\dot{\smile} p)$
$\quad n\frown(m\frown\dot{\bot} p)$
$\quad m\frown(r\frown\dot{\smile} p)$
$\quad n\frown(r\frown\dot{\smile} p)$
$\quad \mathrm{I}\between p$ (305

——— • ———

242 $\vdash \mathbb{1}\frown(a\frown\dot{\smile}\mathfrak{f})$
$\quad \mathbb{0}\frown(\mathbb{1}\frown\mathfrak{f})$
$\quad \mathrm{I}\mathfrak{f}$
$\quad \mathbb{0}\frown(a\frown\dot{\bot}\mathfrak{f})$

(110, 71):: ═══════

$\vdash \mathbb{1}\frown(a\frown\dot{\smile}\mathfrak{f})$
$\quad \mathbb{0}\frown(a\frown\dot{\bot}\mathfrak{f})$ (306

——— • ———

305 $\vdash \mathbb{0}\frown(a\frown\dot{\smile}\mathfrak{f})$
$\quad a\frown(\mathbb{0}\frown\dot{\bot}\mathfrak{f})$
$\quad \mathbb{0}\frown(n\frown\dot{\smile}\mathfrak{f})$
$\quad a\frown(n\frown\dot{\smile}\mathfrak{f})$
$\quad \mathrm{I}\between\mathfrak{f}$

(126, 89):: ═══════

$\vdash \mathbb{0}\frown(a\frown\dot{\smile}\mathfrak{f})$
$\quad \mathbb{0}\frown(n\frown\dot{\smile}\mathfrak{f})$
$\quad a\frown(n\frown\dot{\smile}\mathfrak{f})$ (307

(200): - - - - - -

$\vdash \mathbb{0}\frown(a\frown\dot{\bot}\mathfrak{f})$
$\quad a = \mathbb{0}$
$\quad \mathbb{0}\frown(n\frown\dot{\smile}\mathfrak{f})$
$\quad a\frown(n\frown\dot{\smile}\mathfrak{f})$ (α

(306): - - - - - - -

$\vdash \mathbb{1}\frown(a\frown\dot{\smile}\mathfrak{f})$
$\quad a = \mathbb{0}$
$\quad \mathbb{0}\frown(n\frown\dot{\smile}\mathfrak{f})$
$\quad a\frown(n\frown\dot{\smile}\mathfrak{f})$ (308

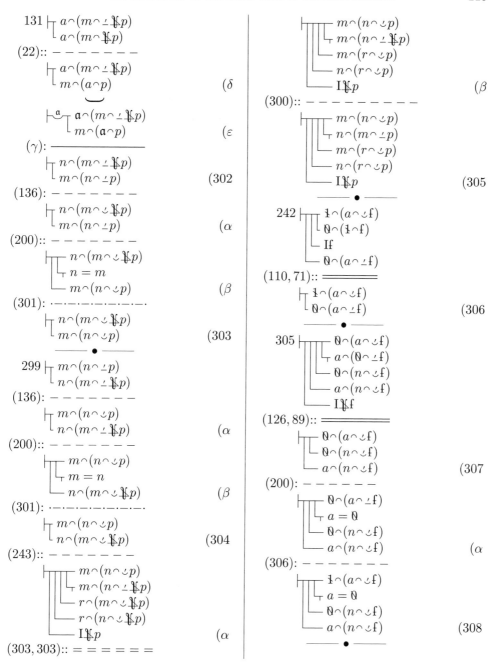

——— • ———

274 $\vdash$ $a \frown (\mathfrak{1}; n \,\underline{\curlywedge}\, \mathfrak{f})$
$\mathfrak{1} \frown (a \frown \smile\mathfrak{f})$
$n \frown (n \frown \perp\mathfrak{f})$
$a \frown (n \frown \smile\mathfrak{f})$
$\mathrm{I}\mathfrak{f}$

(145, 71):: $=====$

$\vdash$ $a \frown (\mathfrak{1}; n \,\underline{\curlywedge}\, \mathfrak{f})$
$\mathfrak{1} \frown (a \frown \smile\mathfrak{f})$
$\mathfrak{0} \frown (n \frown \smile\mathfrak{f})$
$a \frown (n \frown \smile\mathfrak{f})$ (309

(308):: - - - - - - -

$\vdash$ $a \frown (\mathfrak{1}; n \,\underline{\curlywedge}\, \mathfrak{f})$
$a = \mathfrak{0}$
$\mathfrak{0} \frown (n \frown \smile\mathfrak{f})$
$a \frown (n \frown \smile\mathfrak{f})$ (α

(Ic, Id):: = = = = = =

$\vdash$ $a \frown (\mathfrak{1}; n \,\underline{\curlywedge}\, \mathfrak{f})$
$a = \mathfrak{0}$
$a \frown (n \frown \smile\mathfrak{f})$
$\mathfrak{0} \frown (n \frown \smile\mathfrak{f})$ (310

———•———

270 $\vdash$ $\mathfrak{1} \frown (a \frown \smile\mathfrak{f})$
$a \frown (\mathfrak{1}; n \,\underline{\curlywedge}\, \mathfrak{f})$

(132): - - - - - -

$\vdash$ $\mathfrak{0} \frown (a \frown \perp\mathfrak{f})$
$\mathfrak{0} \frown (\mathfrak{1} \frown \mathfrak{f})$
$a \frown (\mathfrak{1}; n \,\underline{\curlywedge}\, \mathfrak{f})$ (α

(110):: ————

$\vdash$ $\mathfrak{0} \frown (a \frown \perp\mathfrak{f})$
$a \frown (\mathfrak{1}; n \,\underline{\curlywedge}\, \mathfrak{f})$ (311

(IIId): - - - - - -

$\vdash$ $a = \mathfrak{0}$
$a \frown (\mathfrak{1}; n \,\underline{\curlywedge}\, \mathfrak{f})$
$\mathfrak{0} \frown (\mathfrak{0} \frown \perp\mathfrak{f})$ (α

(126):: ————

$\vdash$ $a = \mathfrak{0}$
$a \frown (\mathfrak{1}; n \,\underline{\curlywedge}\, \mathfrak{f})$ (312

(If): - - - - - -

$\vdash$ $a = \mathfrak{0}$
$a \frown (n \frown \smile\mathfrak{f})$
$a \frown (\mathfrak{1}; n \,\underline{\curlywedge}\, \mathfrak{f})$
$a \frown (n \frown \smile\mathfrak{f})$ (α

(269):: - - - - - - -

$\vdash$ $a = \mathfrak{0}$
$a \frown (n \frown \smile\mathfrak{f})$
$a \frown (\mathfrak{1}; n \,\underline{\curlywedge}\, \mathfrak{f})$ (β

(IVa): ————

$\vdash$ $(\text{—}\, a \frown (\mathfrak{1}; n \,\underline{\curlywedge}\, \mathfrak{f})) = \left(\text{—} \begin{matrix} a = \mathfrak{0} \\ a \frown (n \frown \smile\mathfrak{f}) \end{matrix}\right)$
$a \frown (\mathfrak{1}; n \,\underline{\curlywedge}\, \mathfrak{f})$
$a = \mathfrak{0}$
$a \frown (n \frown \smile\mathfrak{f})$ (γ

(310):: -

$\vdash$ $(\text{—}\, a \frown (\mathfrak{1}; n \,\underline{\curlywedge}\, \mathfrak{f})) = \left(\text{—} \begin{matrix} a = \mathfrak{0} \\ a \frown (n \frown \smile\mathfrak{f}) \end{matrix}\right)$
$\mathfrak{0} \frown (n \frown \smile\mathfrak{f})$ (δ

(77): ————————

$\vdash$ $(\text{—}\, a \frown (\mathfrak{1}; n \,\underline{\curlywedge}\, \mathfrak{f})) = \left[\text{—}\, a \frown \dot{\varepsilon}\left(\begin{matrix} \varepsilon = \mathfrak{0} \\ \varepsilon \frown (n \frown \smile\mathfrak{f}) \end{matrix}\right)\right]$
$\mathfrak{0} \frown (n \frown \smile\mathfrak{f})$ (ε

$\vdash$ $\overset{\mathfrak{a}}{\smile}$ $(\text{—}\, \mathfrak{a} \frown (\mathfrak{1}; n \,\underline{\curlywedge}\, \mathfrak{f})) = \left[\text{—}\, \mathfrak{a} \frown \dot{\varepsilon}\left(\begin{matrix} \varepsilon = \mathfrak{0} \\ \varepsilon \frown (n \frown \smile\mathfrak{f}) \end{matrix}\right)\right]$
$\mathfrak{0} \frown (n \frown \smile\mathfrak{f})$ (ζ

(96): -

⊢ 𝔭(1; n ⩥ f) = 𝔭ε̇(ε = 0 ; ε ⌢ (n ⌢ ∪f))
  0 ⌢ (n ⌢ ∪f) (313

(IIIa): - - - - - - - - - - - - - -

⊢ n = 𝔭(1; n ⩥ f)
  n = 𝔭ε̇(ε = 0 ; ε ⌢ (n ⌢ ∪f))
  0 ⌢ (n ⌢ ∪f) (α

(160):: - - - - - - - - - - - - - -

⊢ n = 𝔭(1; n ⩥ f)
  0 ⌢ (n ⌢ ∪f) (314

(IIIa): - - - - - - - -

⊢ 𝔭(x; y ⩥ q) = n
  𝔭(x; y ⩥ q) = 𝔭(1; n ⩥ f)
  0 ⌢ (n ⌢ ∪f) (α

(298):: - - - - - - - - - - - - - -

⊢ 𝔭(x; y ⩥ q) = n
  y ⌢ (y ⌢ ∸q)
  Iq
  x; 1 ⌢ (y; n ⌢ ∪(q ⊻ f))
  0 ⌢ (n ⌢ ∪f) (β

(IIIa): - - - - - - - - - - - - -

⊢ 0 ⌢ (𝔭(x; y ⩥ q) ⌢ ∪f)
  y ⌢ (y ⌢ ∸q)
  Iq
  x; 1 ⌢ (y; n ⌢ ∪(q ⊻ f))
  0 ⌢ (n ⌢ ∪f) (315

———— • ————

110 ⊢ 0 ⌢ (1 ⌢ f)

(285): ————

⊢ 0 ⌢ (n ⌢ ∪f)
  1 ⌢ (n ⌢ ∪f) (316

(234):: - - - - - -

⊢ 0 ⌢ (n ⌢ ∪f)
  x; 1 ⌢ (y; n ⌢ ∪(q ⊻ f)) (317

(315): - - - - - - - - - - -

⊢ 0 ⌢ (𝔭(x; y ⩥ q) ⌢ ∪f)
  y ⌢ (y ⌢ ∸q)
  Iq
  x; 1 ⌢ (y; n ⌢ ∪(q ⊻ f)) (α

×

⊢ x; 1 ⌢ (y; n ⌢ ∪(q ⊻ f))
  y ⌢ (y ⌢ ∸q)
  Iq
  0 ⌢ (𝔭(x; y ⩥ q) ⌢ ∪f) (β

⊢ 𝔢 x; 1 ⌢ (y; 𝔢 ⌢ ∪(q ⊻ f))
  y ⌢ (y ⌢ ∸q)
  Iq
  0 ⌢ (𝔭(x; y ⩥ q) ⌢ ∪f) (318

§168. *Analysis*

The last two transitions serve to remove ‘*n*’. We now prove the proposition

‘⊢ 𝔢 x; 1 ⌢ (y; 𝔢 ⌢ ∪(q ⊻ f))
  x ⌢ (y ⌢ ∪q) ’ (α

by means of (144), requiring for this the proposition

‘⊢ 𝔢 x; 1 ⌢ (a; 𝔢 ⌢ ∪(q ⊻ f))
  d ⌢ (a ⌢ q)
  𝔢 x; 1 ⌢ (d; 𝔢 ⌢ ∪(q ⊻ f)) ’ (β

which we derive from (209).

§169. *Construction*

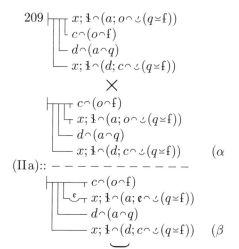

209 ⊢ x; 1 ⌢ (a; o ⌢ ∪(q ⊻ f))
  c ⌢ (o ⌢ f)
  d ⌢ (a ⌢ q)
  x; 1 ⌢ (d; c ⌢ ∪(q ⊻ f))

×

⊢ c ⌢ (o ⌢ f)
  x; 1 ⌢ (a; o ⌢ ∪(q ⊻ f))
  d ⌢ (a ⌢ q)
  x; 1 ⌢ (d; c ⌢ ∪(q ⊻ f)) (α

(IIa):: - - - - - - - - - - - -

⊢ c ⌢ (o ⌢ f)
  𝔢 x; 1 ⌢ (a; 𝔢 ⌢ ∪(q ⊻ f))
  d ⌢ (a ⌢ q)
  x; 1 ⌢ (d; c ⌢ ∪(q ⊻ f)) (β

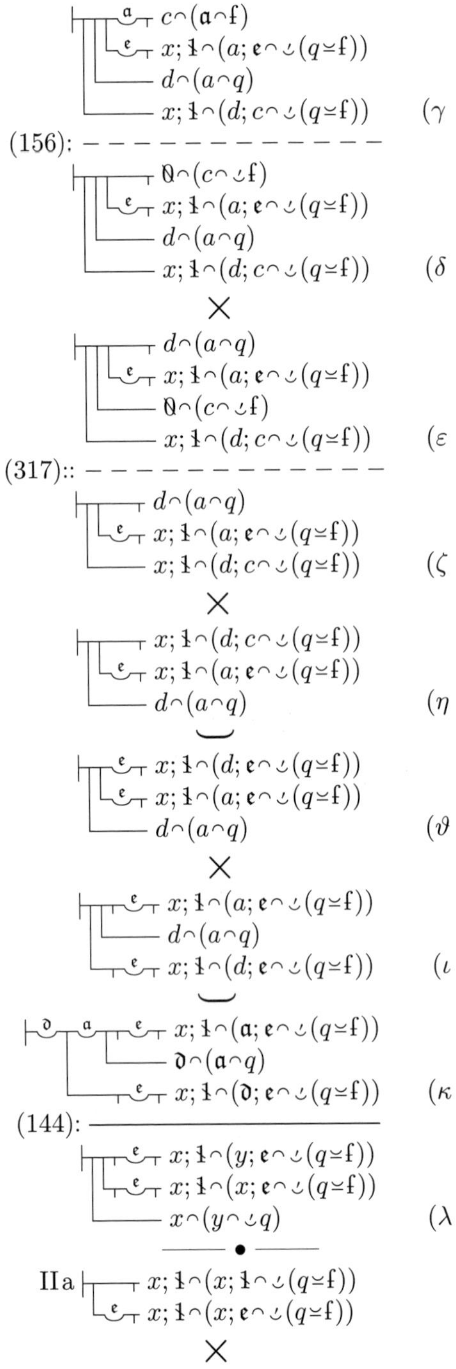

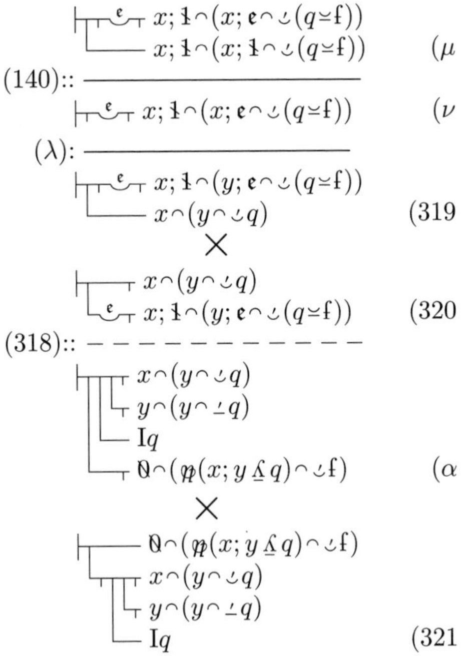

§170. *Analysis*

The subcomponent in (321) will now be eliminated. To this end, we prove the proposition

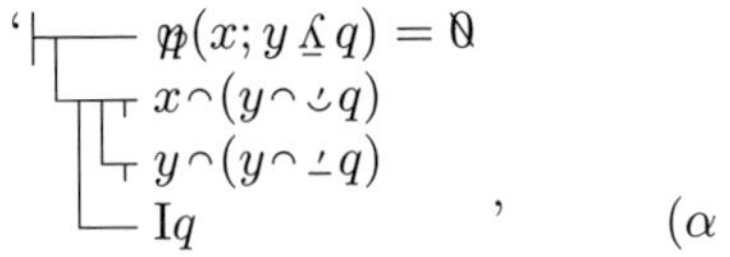

using (97), (271), (265) and the proposition

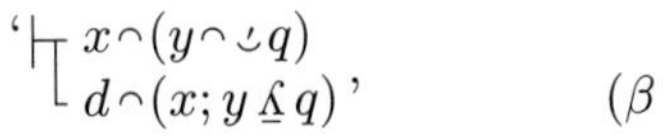

which follows from (269) and (270) using the proposition

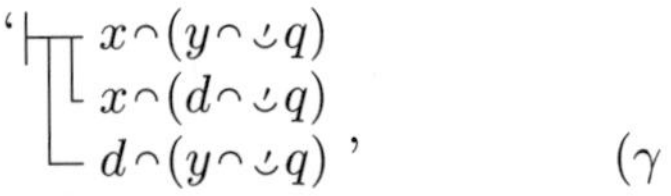

which is proven using (144).

§171. *Construction*

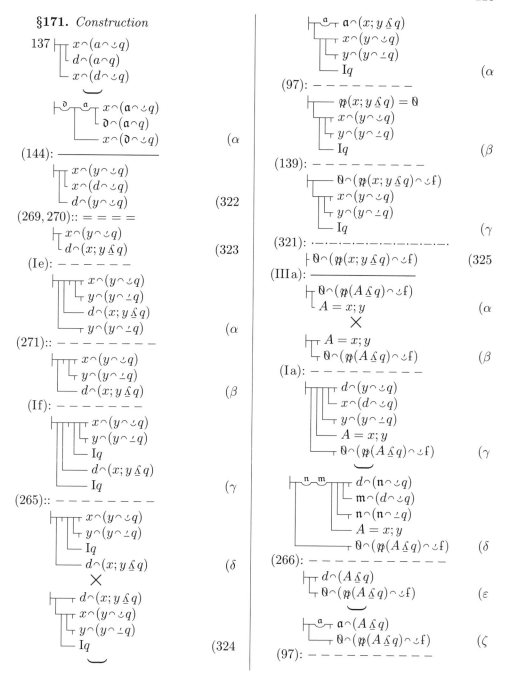

137 $\vdash$ $x \frown (a \frown \smile q)$
$d \frown (a \frown q)$
$x \frown (d \frown \smile q)$

$\vdash^{\mathfrak{d}}\ ^{\mathfrak{a}}$ $x \frown (\mathfrak{a} \frown \smile q)$
$\mathfrak{d} \frown (\mathfrak{a} \frown q)$
$x \frown (\mathfrak{d} \frown \smile q)$ (α

(144): ————————

$\vdash$ $x \frown (y \frown \smile q)$
$x \frown (d \frown \smile q)$
$d \frown (y \frown \smile q)$ (322

(269, 270):: = = = =

$\vdash$ $x \frown (y \frown \smile q)$
$d \frown (x; y \,\underline{\measuredangle}\, q)$ (323

(Ie): – – – – – –

$\vdash$ $x \frown (y \frown \smile q)$
$y \frown (y \frown \underline{\smile} q)$
$d \frown (x; y \,\underline{\measuredangle}\, q)$
$y \frown (y \frown \underline{\smile} q)$ (α

(271):: – – – – – – – – –

$\vdash$ $x \frown (y \frown \smile q)$
$y \frown (y \frown \underline{\smile} q)$
$d \frown (x; y \,\underline{\measuredangle}\, q)$ (β

(If): – – – – – – –

$\vdash$ $x \frown (y \frown \smile q)$
$y \frown (y \frown \underline{\smile} q)$
Iq
$d \frown (x; y \,\underline{\measuredangle}\, q)$
Iq (γ

(265):: – – – – – – – – –

$\vdash$ $x \frown (y \frown \smile q)$
$y \frown (y \frown \underline{\smile} q)$
Iq
$d \frown (x; y \,\underline{\measuredangle}\, q)$ (δ

×

$\vdash$ $d \frown (x; y \,\underline{\measuredangle}\, q)$
$x \frown (y \frown \smile q)$
$y \frown (y \frown \underline{\smile} q)$
Iq (324

$\vdash^{\mathfrak{a}}$ $\mathfrak{a} \frown (x; y \,\underline{\measuredangle}\, q)$
$x \frown (y \frown \smile q)$
$y \frown (y \frown \underline{\smile} q)$
Iq (α

(97): – – – – – – – –

$\vdash$ $\mathfrak{p}(x; y \,\underline{\measuredangle}\, q) = \mathfrak{Q}$
$x \frown (y \frown \smile q)$
$y \frown (y \frown \underline{\smile} q)$
Iq (β

(139): – – – – – – – – –

$\vdash$ $\mathfrak{Q} \frown (\mathfrak{p}(x; y \,\underline{\measuredangle}\, q) \frown \smile \mathfrak{f})$
$x \frown (y \frown \smile q)$
$y \frown (y \frown \underline{\smile} q)$
Iq (γ

(321): ·—·—·—·—·—·—·—·—·—·

$\vdash$ $\mathfrak{Q} \frown (\mathfrak{p}(x; y \,\underline{\measuredangle}\, q) \frown \smile \mathfrak{f})$ (325

(IIIa): ————————

$\vdash$ $\mathfrak{Q} \frown (\mathfrak{p}(A \,\underline{\measuredangle}\, q) \frown \smile \mathfrak{f})$
$A = x; y$ (α

×

$\vdash$ $A = x; y$
$\mathfrak{Q} \frown (\mathfrak{p}(A \,\underline{\measuredangle}\, q) \frown \smile \mathfrak{f})$ (β

(Ia): – – – – – – – – –

$\vdash$ $d \frown (y \frown \smile q)$
$x \frown (d \frown \smile q)$
$y \frown (y \frown \underline{\smile} q)$
$A = x; y$
$\mathfrak{Q} \frown (\mathfrak{p}(A \,\underline{\measuredangle}\, q) \frown \smile \mathfrak{f})$ (γ

$\vdash^{\mathfrak{n}}\ ^{\mathfrak{m}}$ $d \frown (\mathfrak{n} \frown \smile q)$
$\mathfrak{m} \frown (d \frown \smile q)$
$\mathfrak{n} \frown (\mathfrak{n} \frown \underline{\smile} q)$
$A = x; y$
$\mathfrak{Q} \frown (\mathfrak{p}(A \,\underline{\measuredangle}\, q) \frown \smile \mathfrak{f})$ (δ

(266): – – – – – – – – – – –

$\vdash$ $d \frown (A \,\underline{\measuredangle}\, q)$
$\mathfrak{Q} \frown (\mathfrak{p}(A \,\underline{\measuredangle}\, q) \frown \smile \mathfrak{f})$ (ε

$\vdash^{\mathfrak{a}}$ $\mathfrak{a} \frown (A \,\underline{\measuredangle}\, q)$
$\mathfrak{Q} \frown (\mathfrak{p}(A \,\underline{\measuredangle}\, q) \frown \smile \mathfrak{f})$ (ζ

(97): – – – – – – – – – –

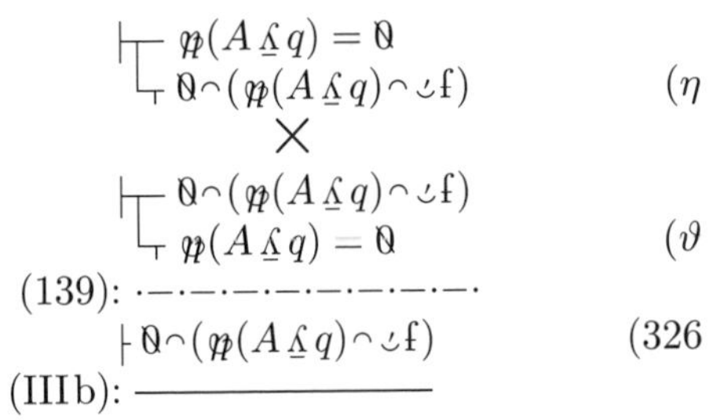

$$\begin{array}{ll}\vdash\!\!\!\top\; \mathfrak{p}(A \mathbin{\underline{\measuredangle}} q) = \mathfrak{O} & \\ \quad\llcorner\; \mathfrak{O} \frown (\mathfrak{p}(A \mathbin{\underline{\measuredangle}} q) \frown \smile\mathfrak{f}) & (\eta \\ \quad\times & \\ \vdash\!\!\!\top\; \mathfrak{O} \frown (\mathfrak{p}(A \mathbin{\underline{\measuredangle}} q) \frown \smile\mathfrak{f}) & \\ \quad\llcorner\; \mathfrak{p}(A \mathbin{\underline{\measuredangle}} q) = \mathfrak{O} & (\vartheta \\ (139): \cdot - \cdot - \cdot - \cdot - \cdot - \cdot - \cdot & \\ \vdash \mathfrak{O} \frown (\mathfrak{p}(A \mathbin{\underline{\measuredangle}} q) \frown \smile\mathfrak{f}) & (326 \\ (\mathrm{III\,b}): \overline{\qquad\qquad\qquad} & \end{array}$$

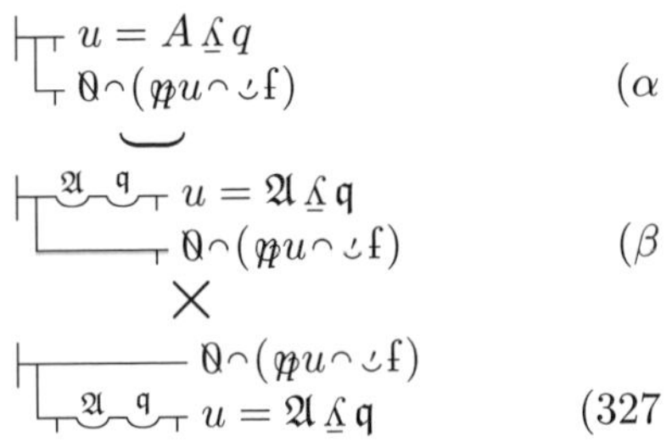

$$\begin{array}{ll}\vdash\!\!\!\top\; u = A \mathbin{\underline{\measuredangle}} q & \\ \quad\llcorner\; \mathfrak{O} \frown (\mathfrak{p}u \frown \smile\mathfrak{f}) & (\alpha \\ \quad\smile & \\ \vdash\!\!\!\top\!\overset{\mathfrak{A}}{\smile}\!\overset{\mathfrak{q}}{\smile}\; u = \mathfrak{A} \mathbin{\underline{\measuredangle}} \mathfrak{q} & \\ \quad\llcorner\!\!-\!\!-\; \mathfrak{O} \frown (\mathfrak{p}u \frown \smile\mathfrak{f}) & (\beta \\ \quad\times & \\ \vdash\!\!\!\top\!-\!-\; \mathfrak{O} \frown (\mathfrak{p}u \frown \smile\mathfrak{f}) & \\ \quad\llcorner\!\overset{\mathfrak{A}}{\smile}\!\overset{\mathfrak{q}}{\smile}\; u = \mathfrak{A} \mathbin{\underline{\measuredangle}} \mathfrak{q} & (327 \end{array}$$

Λ. Proof of the proposition

$$\text{‘}\begin{array}{l}\vdash\!\!\!\top\!\overset{\mathfrak{A}}{\smile}\!\overset{\mathfrak{q}}{\smile}\; \dot{\varepsilon}(-\, \varepsilon \frown u) = \mathfrak{A} \mathbin{\underline{\measuredangle}} \mathfrak{q} \\ \quad\llcorner\!\!-\!\!-\; \mathfrak{O} \frown (\mathfrak{p}u \frown \smile\mathfrak{f}) \end{array}\text{’}$$

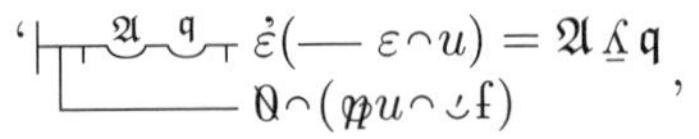

§172. *Analysis*

We attempt to express in words the proposition to be proven like this:

> "If the cardinal number of a concept is finite, then the objects that fall under it can be ordered into a simple series running from a specific object to a specific object."

This expression is not perfect insofar as according to it the proposition seems not to apply to the cardinal number Zero. However, we may take a series-forming relation such that no object belongs to its series running from Δ to Θ by preventing from ever occurring what is demanded by definition (P) in order for an object to belong to this series running from Δ to Θ.

By (314) we have

$$\text{‘}\begin{array}{l}\vdash\!\!\!\top\; \mathfrak{p}u = \mathfrak{p}(\mathfrak{1}; \mathfrak{p}u \mathbin{\underline{\measuredangle}} \mathfrak{f}) \\ \quad\llcorner\; \mathfrak{O} \frown (\mathfrak{p}u \frown \smile\mathfrak{f}) \end{array}\text{’}$$

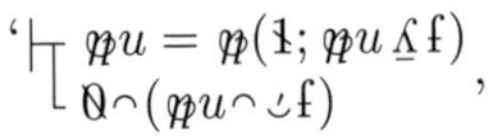

According to this, there is a relation which maps the $(\mathfrak{1}; \mathfrak{p}u \mathbin{\underline{\measuredangle}} \mathfrak{f})$-concept into the u-concept while its converse maps the latter into the former. This will be our p-relation. We now show that we can take the $(\mathfrak{P}p \smile \mathfrak{f} \smile p)$-relation as series-forming. It is first to be shown that every object falling under the u-concept belongs to the $(\mathfrak{P}p \smile \mathfrak{f} \smile p)$-series going from x to y, where $\mathfrak{1}$ stands to x, and $\mathfrak{p}u$ stands to y in the p-relation. More generally we write 'm' instead of '$\mathfrak{1}$', 'n' instead of '$\mathfrak{p}u$' and 'q' instead of '$\mathfrak{f}$' and prove the proposition

$$\text{‘}\begin{array}{ll}\vdash\!\!\top\!\!\top\!\!\top\!\!\top\!\!\top\; c \frown (x; y \mathbin{\underline{\measuredangle}} (\mathfrak{P}p \smile q \smile p)) & \\ \quad\llcorner\; c \frown u & \\ \quad\llcorner\; m \frown (x \frown p) & \\ \quad\llcorner\; m; n \mathbin{\underline{\measuredangle}} q \frown (u \frown \rangle p) & \\ \quad\llcorner\; u \frown (m; n \mathbin{\underline{\measuredangle}} q \frown \rangle \mathfrak{P}p) & \\ \quad\llcorner\; n \frown (y \frown p) & \text{’} \quad (\alpha \end{array}$$

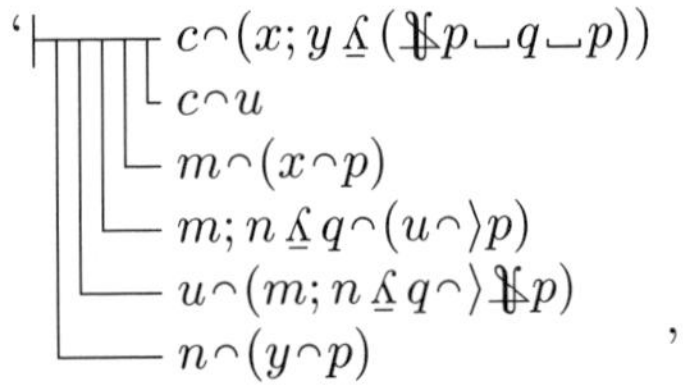

which, using (8), follows from

$$\text{‘}\begin{array}{ll}\vdash\!\!\top\!\!\top\!\!\top\!\!\top\!\!\top\!\!\top\; c \frown (x; y \mathbin{\underline{\measuredangle}} (\mathfrak{P}p \smile q \smile p)) & \\ \quad\llcorner\; c \frown (a \frown \mathfrak{P}p) & \\ \quad\llcorner\; a \frown (m; n \mathbin{\underline{\measuredangle}} q) & \\ \quad\llcorner\; m \frown (x \frown p) & \\ \quad\llcorner\; m; n \mathbin{\underline{\measuredangle}} q \frown (u \frown \rangle p) & \\ \quad\llcorner\; \mathrm{I}\mathfrak{P}p & \\ \quad\llcorner\; n \frown (y \frown p) & \text{’} \quad (\beta \end{array}$$

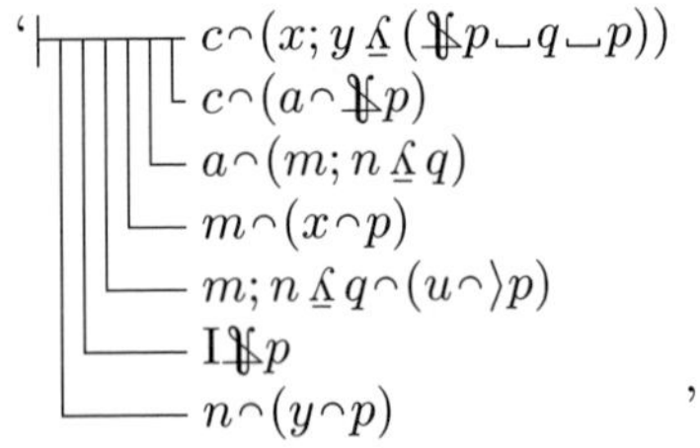

The following diagram will assist the understanding

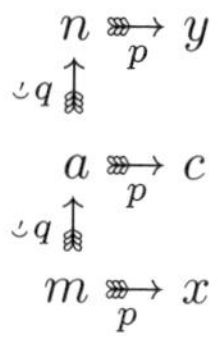

In order to prove (β), we require amongst others the following propositions

$$\text{`}\vdash\begin{array}{l} x\frown(c\frown\smile(\not\S p\smile q\smile p)) \\ a\frown(c\frown p) \\ a\frown(m;n\,\underline{\measuredangle}\,q) \\ m\frown(x\frown p) \\ m;n\,\underline{\measuredangle}\,q\frown(u\frown\rangle p) \end{array}\text{'} \qquad (\gamma$$

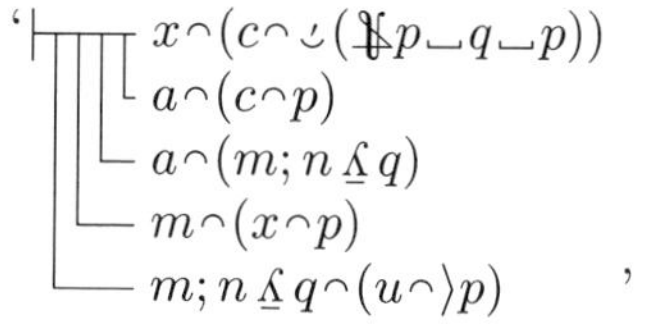

and

$$\text{`}\vdash\begin{array}{l} c\frown(y\frown\smile(\not\S p\smile q\smile p)) \\ n\frown(y\frown p) \\ a\frown(c\frown p) \\ a\frown(m;n\,\underline{\measuredangle}\,q) \\ m;n\,\underline{\measuredangle}\,q\frown(u\frown\rangle p) \end{array}\text{'} \qquad (\delta$$

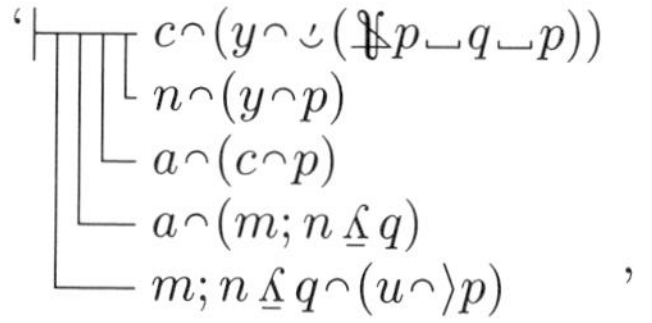

which we derive from

$$\text{`}\vdash\begin{array}{l} r\frown(c\frown\smile(\not\S p\smile q\smile p)) \\ a\frown(c\frown p) \\ a\frown(n\frown\smile q) \\ s\frown(r\frown p) \\ m\frown(s\frown\smile q) \\ b\frown(m;n\,\underline{\measuredangle}\,q) \\ m;n\,\underline{\measuredangle}\,q\frown(u\frown\rangle p) \\ s\frown(a\frown\smile q) \end{array}\text{'} \qquad (\varepsilon$$

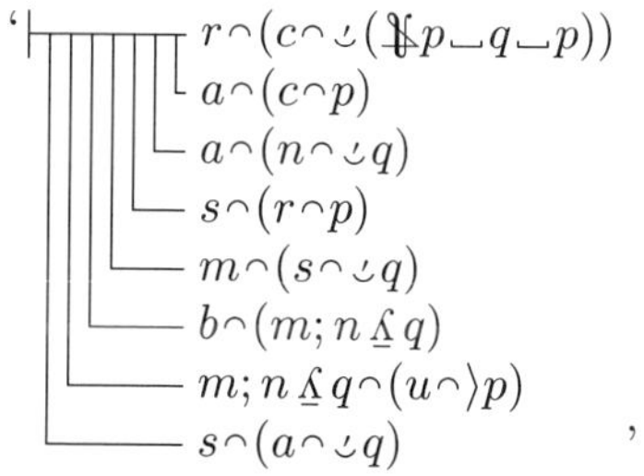

by first letting r coincide with x, s with m and b with a, and then c with y, a with n and b with s, and finally writing 'a' for 's' and 'c' for 'r'. For this, compare the following diagram

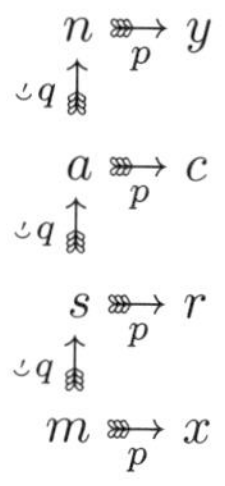

To derive (ε), we employ (152), replacing the function-marker '$F(\xi)$' by

$$\text{`}\smile^{\mathfrak{e}}\begin{array}{l} r\frown(\mathfrak{e}\frown\smile(\not\S p\smile q\smile p)) \\ \xi\frown(\mathfrak{e}\frown p) \\ \xi\frown(n\frown\smile q) \end{array}\text{'}$$

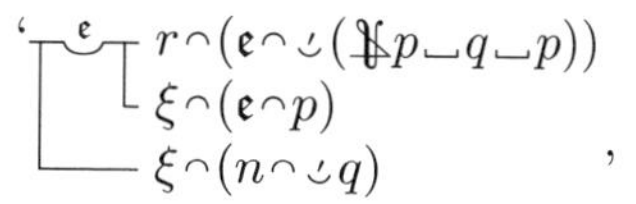

In doing so, as in the proof of (186), the propositions (183) and (185) are to be used, thus introducing the subcomponent

$$\text{`}— m;n\,\underline{\measuredangle}\,q\frown(u\frown\rangle p)\text{'}$$

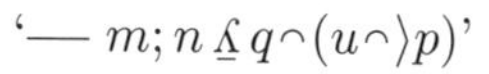

§173. *Construction*

$$183 \vdash\begin{array}{l} d\frown(m;n\,\underline{\measuredangle}\,q) \\ \smile^{\mathfrak{a}}\, d\frown(\mathfrak{a}\frown p) \\ m;n\,\underline{\measuredangle}\,q\frown(u\frown\rangle p) \end{array}$$

(273): – – – – – – – – – – –

$$\vdash\begin{array}{l} d\frown(n\frown\smile q) \\ m\frown(d\frown\smile q) \\ n\frown(n\frown\underline{\smile} q) \\ \smile^{\mathfrak{a}}\, d\frown(\mathfrak{a}\frown p) \\ m;n\,\underline{\measuredangle}\,q\frown(u\frown\rangle p) \\ \mathrm{I}q \end{array} \qquad (\alpha$$

(271, 265):: = = = = = = = = =

$$\vdash\begin{array}{l} d\frown(n\frown\smile q) \\ m\frown(d\frown\smile q) \\ b\frown(m;n\,\underline{\measuredangle}\,q) \\ \smile^{\mathfrak{a}}\, d\frown(\mathfrak{a}\frown p) \\ m;n\,\underline{\measuredangle}\,q\frown(u\frown\rangle p) \end{array} \qquad (\beta$$

(185):: – – – – – – – – – – –

$$\vdash\begin{array}{l} d\frown(n\frown\smile q) \\ m\frown(d\frown\smile q) \\ b\frown(m;n\,\underline{\measuredangle}\,q) \\ a\frown(c\frown p) \\ d\frown(a\frown q) \\ r\frown(c\frown\smile(\not\S p\smile q\smile p)) \\ \smile^{\mathfrak{e}}\, r\frown(\mathfrak{e}\frown\smile(\not\S p\smile q\smile p)) \\ \quad d\frown(\mathfrak{e}\frown p) \\ m;n\,\underline{\measuredangle}\,q\frown(u\frown\rangle p) \end{array} \qquad (\gamma$$

×

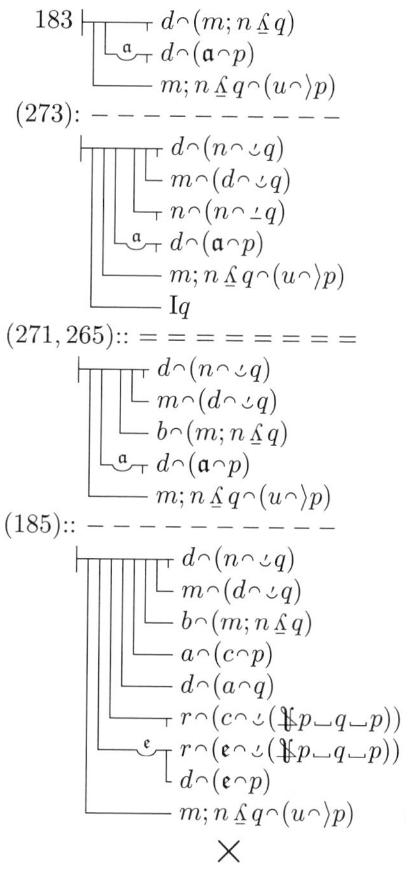

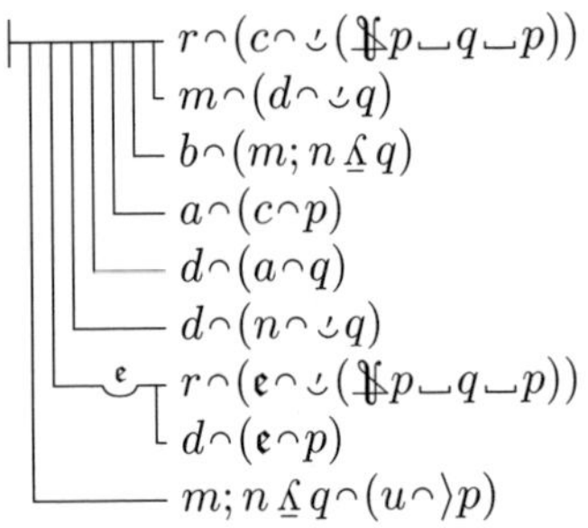

(δ

(I):: - - - - - - - - - - - - - -

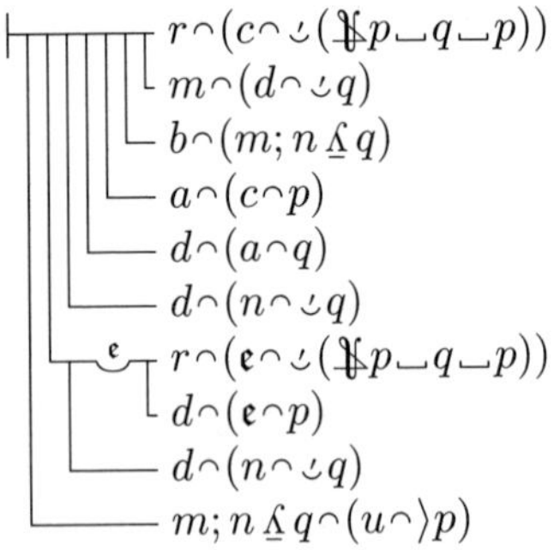

(ε

(285):: - - - - - - - - - - - - - -

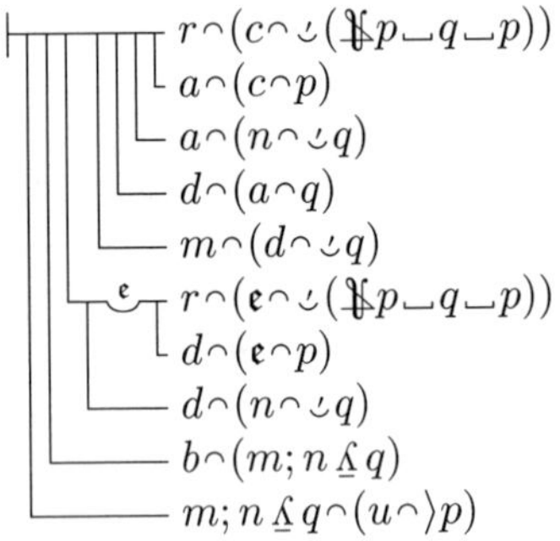

(ζ

(322):: - - - - - - - - - - - - - -

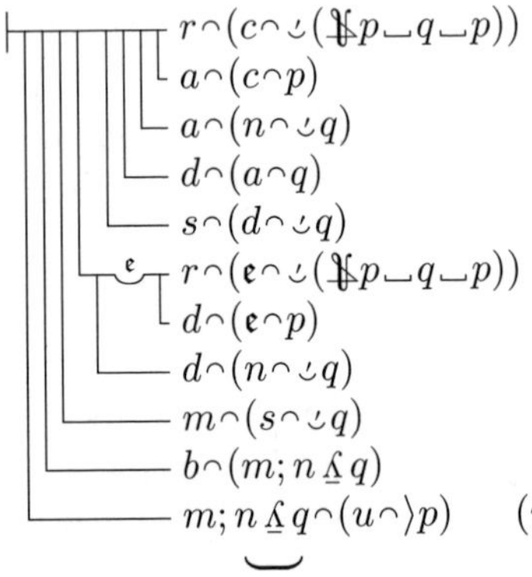

(η

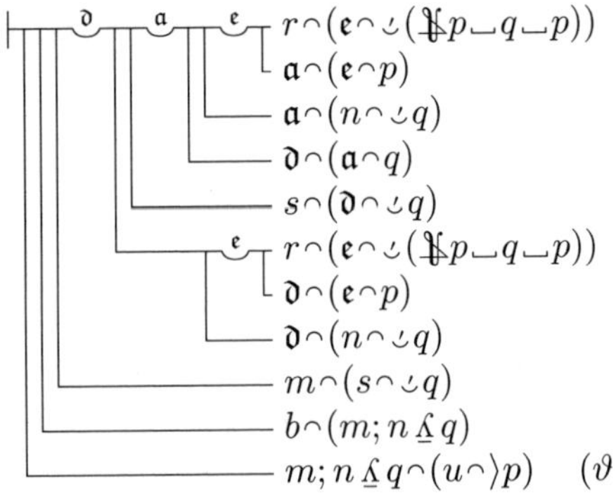

(ϑ

(152):: - - - - - - - - - - - - - -

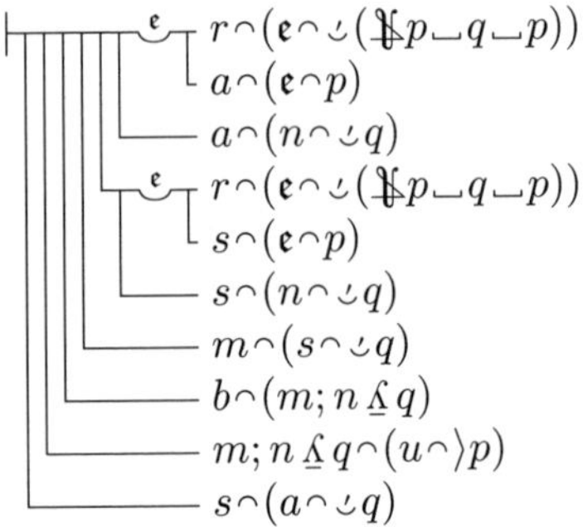

(ι

(I):: - - - - - - - - - - - - - -

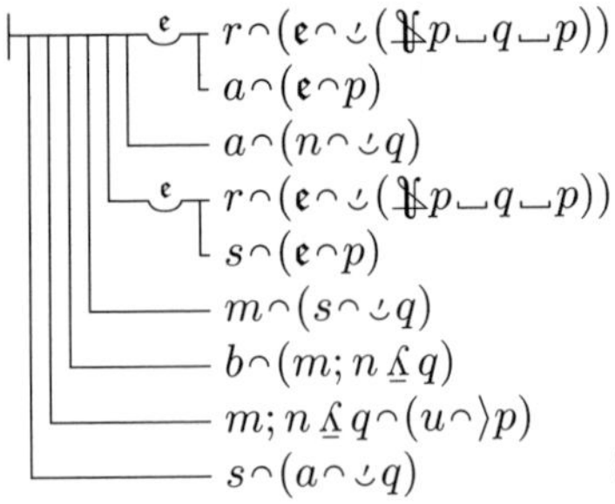

(κ

(182):: - - - - - - - - - - - - - -

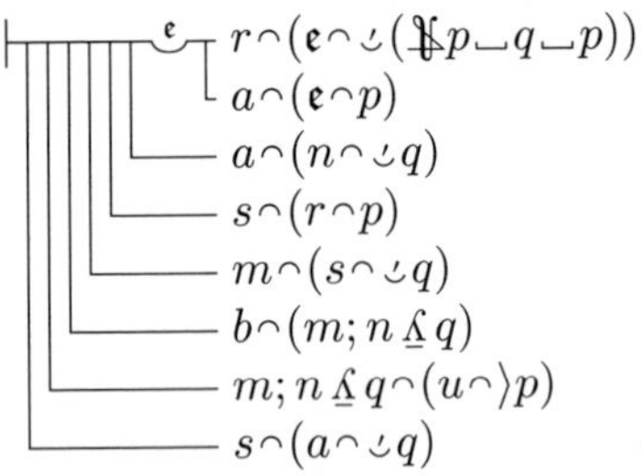

(λ

(IIa): - - - - - - - - - - - - - -

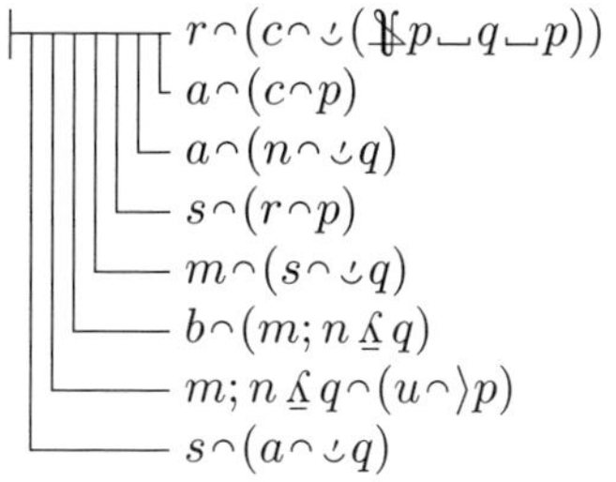

(328

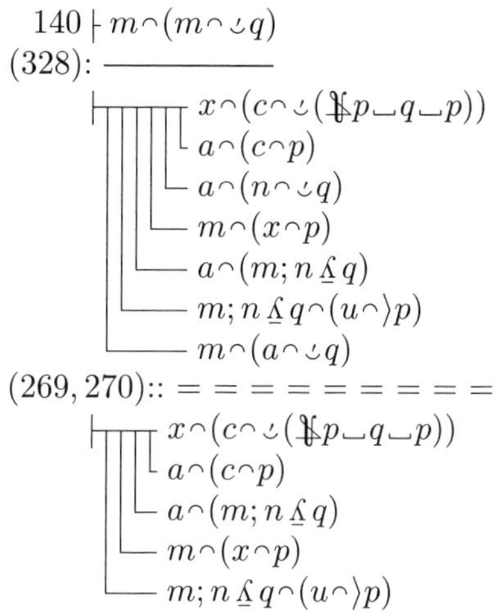

(α

(329

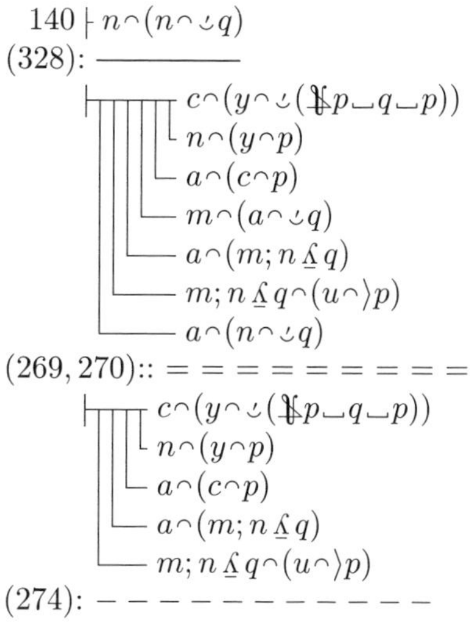

(α

(β

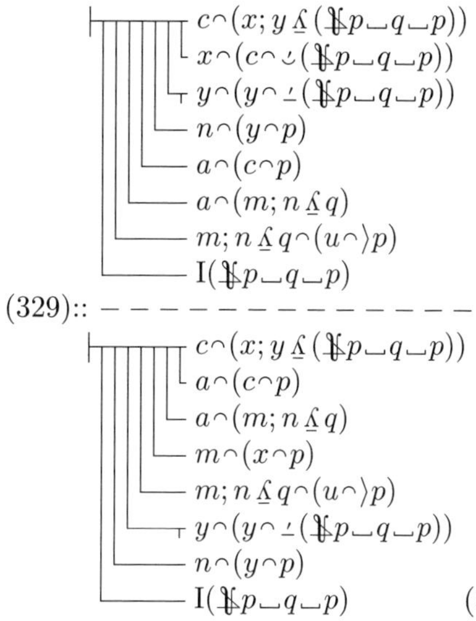

(γ

(330

§174. *Analysis*

In order to eliminate the subcomponent

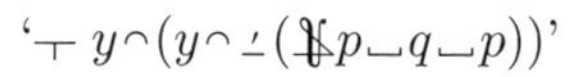

we prove the proposition

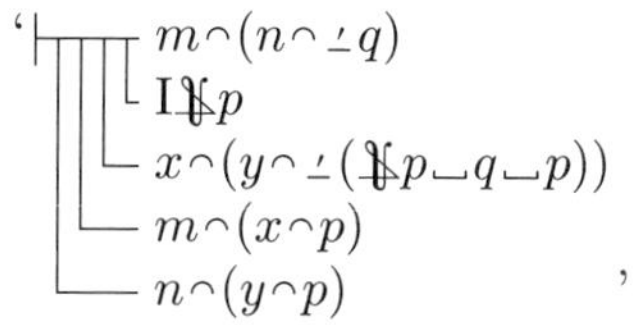

(α

which, using (177), we reduce to

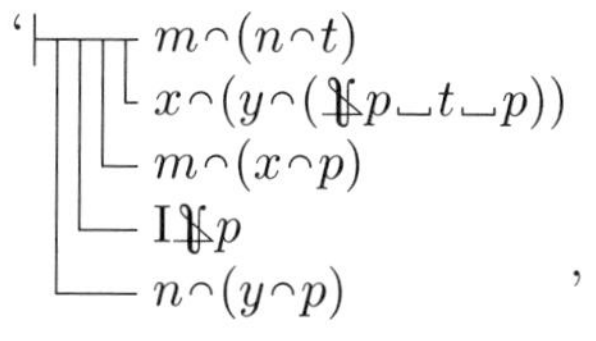

(β

We show that there are objects s and a such that s stands to x and a to y in the p-relation, and that a follows after s in the q-series. From the single-valuedness of the converse of the p-relation it then follows that s coincides with m and a

with n, thus that n follows after m in the q-series. The subcomponent

‘— I($\between p \smile q \smile p$)’

is also to be removed. This is easily done by means of (17).

§175. *Construction*

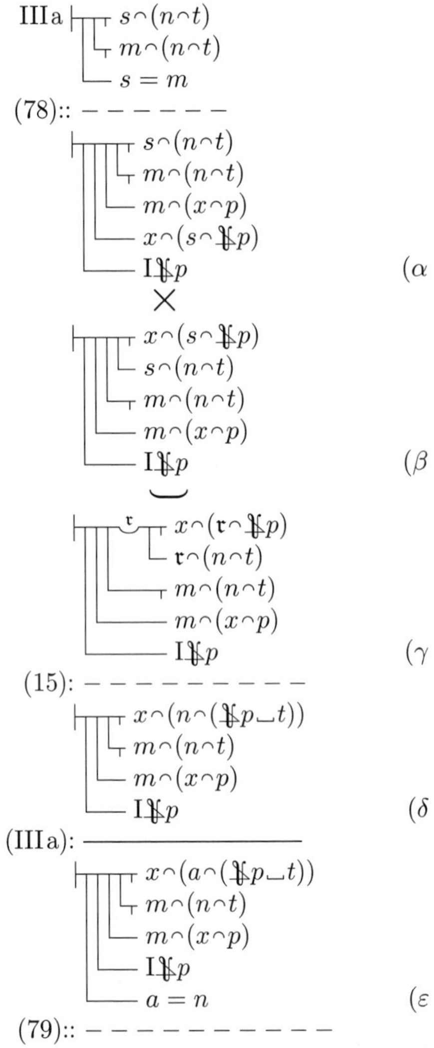

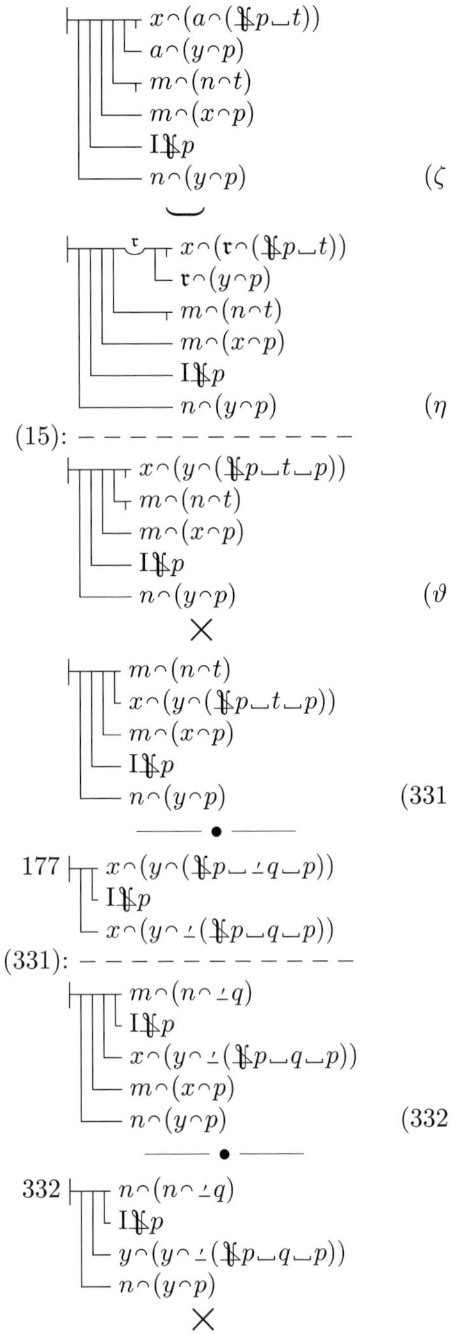

$\vdash y \frown (y \frown \dot{-}(\nparallel p \smile q \smile p))$
$I \nparallel p$
$n \frown (n \frown \dot{-} q)$
$n \frown (y \frown p)$ (α

(271):: – – – – – – – – – – – –

$\vdash y \frown (y \frown \dot{-}(\nparallel p \smile q \smile p))$
$I \nparallel p$
$a \frown (m; n \precsim q)$
$n \frown (y \frown p)$ (333

———— • ————

17 $\vdash I(\nparallel p \smile q)$
Iq
$I \nparallel p$

(17): – – – – – –

$\vdash I(\nparallel p \smile q \smile p)$
Ip
Iq
$I \nparallel p$ (α

(265, 18):: = = = = = =

$\vdash I(\nparallel p \smile q \smile p)$
$m; n \precsim q \frown (u \frown \rangle p)$
$a \frown (m; n \precsim q)$
$I \nparallel p$ (β

(330): – – – – – – – – – –

$\vdash c \frown (x; y \precsim (\nparallel p \smile q \smile p))$
$a \frown (c \frown p)$
$a \frown (m; n \precsim q)$
$m \frown (x \frown p)$
$m; n \precsim q \frown (u \frown \rangle p)$
$y \frown (y \frown \dot{-}(\nparallel p \smile q \smile p))$
$n \frown (y \frown p)$
$I \nparallel p$ (γ

(23, 333):: = = = = = = = = = = = =

$\vdash c \frown (x; y \precsim (\nparallel p \smile q \smile p))$
$c \frown (a \frown \nparallel p)$
$a \frown (m; n \precsim q)$
$m \frown (x \frown p)$
$m; n \precsim q \frown (u \frown \rangle p)$
$I \nparallel p$
$n \frown (y \frown p)$ (δ

×

$\vdash a \frown (m; n \precsim q)$
$c \frown (a \frown \nparallel p)$
$c \frown (x; y \precsim (\nparallel p \smile q \smile p))$
$m \frown (x \frown p)$
$m; n \precsim q \frown (u \frown \rangle p)$
$I \nparallel p$
$n \frown (y \frown p)$ (ε

$\smile$

$\vdash \overset{\mathfrak{a}}{\smile}\ \mathfrak{a} \frown (m; n \precsim q)$
$c \frown (\mathfrak{a} \frown \nparallel p)$
$c \frown (x; y \precsim (\nparallel p \smile q \smile p))$
$m \frown (x \frown p)$
$m; n \precsim q \frown (u \frown \rangle p)$
$I \nparallel p$
$n \frown (y \frown p)$ (ζ

(8): – – – – – – – – – – – – – – –

$\vdash c \frown u$
$c \frown (x; y \precsim (\nparallel p \smile q \smile p))$
$m \frown (x \frown p)$
$m; n \precsim q \frown (u \frown \rangle p)$
$I \nparallel p$
$n \frown (y \frown p)$
$u \frown (m; n \precsim q \frown \rangle \nparallel p)$ (η

(18):: – – – – – – – – – – – – – –

$\vdash c \frown u$
$c \frown (x; y \precsim (\nparallel p \smile q \smile p))$
$m \frown (x \frown p)$
$m; n \precsim q \frown (u \frown \rangle p)$
$u \frown (m; n \precsim q \frown \rangle \nparallel p)$
$n \frown (y \frown p)$ (ϑ

×

$\vdash c \frown (x; y \precsim (\nparallel p \smile q \smile p))$
$c \frown u$
$m \frown (x \frown p)$
$m; n \precsim q \frown (u \frown \rangle p)$
$u \frown (m; n \precsim q \frown \rangle \nparallel p)$
$n \frown (y \frown p)$ (334

§176. *Analysis*

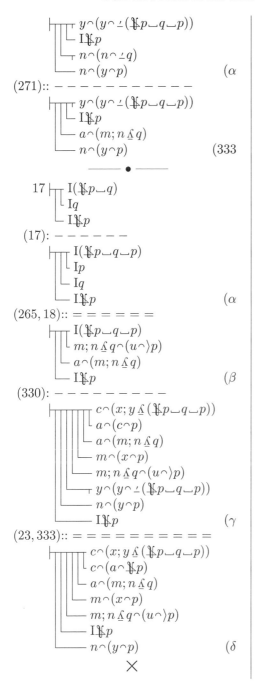

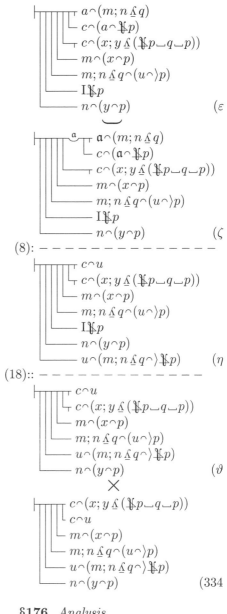

We now have proven proposition (α) of §172. What remains to be derived is the proposition

‘⊢ $c \frown u$
$c \frown (x; y \,\underline{\measuredangle}\, (\rtimes p \llcorner q \llcorner p))$
$m \frown (x \frown p)$
$u \frown (m; n \,\underline{\measuredangle}\, q \frown \rangle \rtimes p)$
$n \frown (y \frown p)$
$n \frown (n \frown \dot{-} q)$
$m; n \,\underline{\measuredangle}\, q \frown (u \frown \rangle p)$
$\mathrm{I}q$ ’ (α

It is to be derived using (179) from the propositions

‘⊢ $x \frown (c \frown \smile(\rtimes p \llcorner q \llcorner p))$
$m \frown (x \frown p)$
$\mathrm{I}\rtimes p$
$\mathfrak{r}$: $m \frown (\mathfrak{r} \frown \smile q)$
$\mathfrak{r} \frown (c \frown p)$ ’ (β

and

‘⊢ $c \frown (y \frown \smile(\rtimes p \llcorner q \llcorner p))$
$n \frown (y \frown p)$
$\mathfrak{r}$: $\mathfrak{r} \frown (n \frown \smile q)$
$\mathfrak{r} \frown (c \frown p)$
$\mathrm{I}\rtimes p$ ’ (γ

by inferring from the single-valuedness of the $\rtimes p$-relation that the same object which stands to c in the p-relation also belongs both to the q-series starting with m and to the one ending with n. Instead of (β) and (γ) we first prove the propositions that, while having the same subcomponents, have as supercomponents

‘⊤ $x \frown (c \frown (\rtimes p \llcorner \smile q \llcorner p))$’

and

‘⊤ $c \frown (y \frown (\rtimes p \llcorner \smile q \llcorner p))$’

Using proposition (180) we can move to (β). In order to arrive at (γ) we need the similar proposition

‘⊢ $c \frown (y \frown (\rtimes p \llcorner \smile q \llcorner p))$
$n \frown (y \frown p)$
$c \frown (y \frown \smile(\rtimes p \llcorner q \llcorner p))$
$\mathrm{I}\rtimes p$ ’, (δ

which we derive from (177).

§177. *Construction*

136 ⊢ $a \frown (n \frown \smile q)$
$a \frown (n \frown \dot{-} q)$

(174): — — — — — —

⊢ $c \frown (y \frown (\rtimes p \llcorner \smile q \llcorner p))$
$n \frown (y \frown p)$
$a \frown (n \frown \dot{-} q)$
$c \frown (a \frown \rtimes p)$ (α

×

⊢ $c \frown (a \frown \rtimes p)$
$a \frown (n \frown \dot{-} q)$
$n \frown (y \frown p)$
$c \frown (y \frown (\rtimes p \llcorner \smile q \llcorner p))$ (β

⌣

⊢ $\mathfrak{r}$: $c \frown (\mathfrak{r} \frown \rtimes p)$
$\mathfrak{r} \frown (n \frown \dot{-} q)$
$n \frown (y \frown p)$
$c \frown (y \frown (\rtimes p \llcorner \smile q \llcorner p))$ (γ

(15): — — — — — — — — — — — — —

⊢ $c \frown (n \frown (\rtimes p \llcorner \dot{-} q))$
$n \frown (y \frown p)$
$c \frown (y \frown (\rtimes p \llcorner \smile q \llcorner p))$ (δ

⌣

⊢ $\mathfrak{r}$: $c \frown (\mathfrak{r} \frown (\rtimes p \llcorner \dot{-} q))$
$\mathfrak{r} \frown (y \frown p)$
$c \frown (y \frown (\rtimes p \llcorner \smile q \llcorner p))$ (ε

(15): — — — — — — — — — — — — —

⊢ $c \frown (y \frown (\rtimes p \llcorner \dot{-} q \llcorner p))$
$c \frown (y \frown (\rtimes p \llcorner \smile q \llcorner p))$ (ζ

×

⊢ $c \frown (y \frown (\rtimes p \llcorner \smile q \llcorner p))$
$c \frown (y \frown (\rtimes p \llcorner \dot{-} q \llcorner p))$ (η

(177):: — — — — — — — — — — —

⊢ $c \frown (y \frown (\rtimes p \llcorner \smile q \llcorner p))$
$\mathrm{I}\rtimes p$
$c \frown (y \frown \dot{-}(\rtimes p \llcorner q \llcorner p))$ (ϑ

—— • ——

140 ⊢ $n \frown (n \frown \smile q)$

(184): ————————

⊢ $y \frown (y \frown (\rtimes p \llcorner \smile q \llcorner p))$
$n \frown (y \frown p)$ (ι

(IIIc): ————————————

⊢ c⌢(y⌢(∦p⌣ᵕq⌣p))
n⌢(y⌢p)
y = c (κ

(130):: – – – – – – – – – – –

⊢ c⌢(y⌢(∦p⌣ᵕq⌣p))
n⌢(y⌢p)
c⌢(y⌢∸(∦p⌣q⌣p))
c⌢(y⌢ᵕ(∦p⌣q⌣p)) (λ

(ϑ): –·–·–·–·–·–·–·–·–·–·–

⊢ c⌢(y⌢(∦p⌣ᵕq⌣p))
n⌢(y⌢p)
c⌢(y⌢ᵕ(∦p⌣q⌣p))
I∦p (335

×

⊢ c⌢(y⌢ᵕ(∦p⌣q⌣p))
n⌢(y⌢p)
c⌢(y⌢(∦p⌣ᵕq⌣p))
I∦p (336

——— • ———

23 ⊢ a⌢(c⌢p)
c⌢(a⌢∦p)

(IIa): – – – – – –

⊢ a⌢(n⌢ᵕq)
c⌢(a⌢∦p)
𝔯 𝔯⌢(n⌢ᵕq)
𝔯⌢(c⌢p) (α

×

⊢ c⌢(a⌢∦p)
a⌢(n⌢ᵕq)
𝔯 𝔯⌢(n⌢ᵕq)
𝔯⌢(c⌢p) (β

⏝

⊢ 𝔯 c⌢(𝔯⌢∦p)
𝔯⌢(n⌢ᵕq)
𝔯 𝔯⌢(n⌢ᵕq)
𝔯⌢(c⌢p) (γ

(15): – – – – – – – –

⊢ c⌢(n⌢(∦p⌣ᵕq))
𝔯 𝔯⌢(n⌢ᵕq)
𝔯⌢(c⌢p) (δ

(IIIa): ————————————

⊢ c⌢(e⌢(∦p⌣ᵕq))
𝔯 𝔯⌢(n⌢ᵕq)
𝔯⌢(c⌢p)
e = n (ε

(79):: – – – – – – – – – – – –

⊢ c⌢(e⌢(∦p⌣ᵕq))
e⌢(y⌢p)
𝔯 𝔯⌢(n⌢ᵕq)
𝔯⌢(c⌢p)
n⌢(y⌢p)
I∦p (ζ

⏝

⊢ 𝔯 c⌢(𝔯⌢(∦p⌣ᵕq))
𝔯⌢(y⌢p)
𝔯 𝔯⌢(n⌢ᵕq)
𝔯⌢(c⌢p)
n⌢(y⌢p)
I∦p (η

(15): – – – – – – – – – – – –

⊢ c⌢(y⌢(∦p⌣ᵕq⌣p))
𝔯 𝔯⌢(n⌢ᵕq)
𝔯⌢(c⌢p)
n⌢(y⌢p)
I∦p (ϑ

(336): – – – – – – – – – – – – – –

⊢ c⌢(y⌢ᵕ(∦p⌣q⌣p))
n⌢(y⌢p)
𝔯 𝔯⌢(n⌢ᵕq)
𝔯⌢(c⌢p)
I∦p (337

——— • ———

IIIa ⊢ s⌢(a⌢ᵕq)
m⌢(a⌢ᵕq)
s = m

(IIa):: – – – – – – –

⊢ s⌢(a⌢ᵕq)
a⌢(c⌢p)
𝔯 m⌢(𝔯⌢ᵕq)
𝔯⌢(c⌢p)
s = m (α

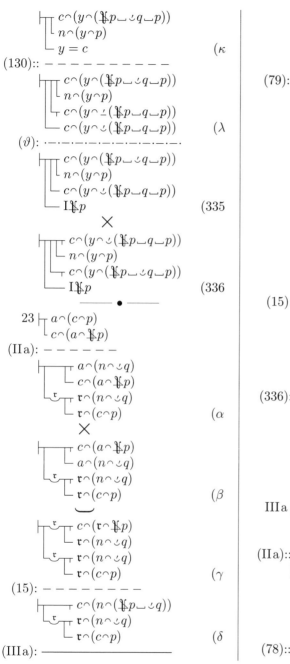

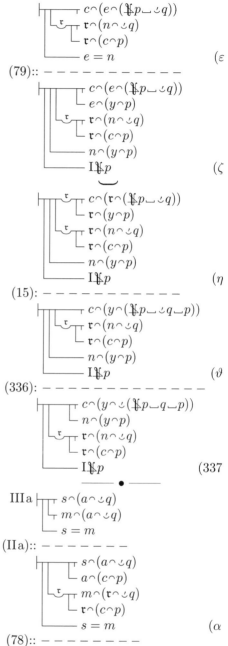

(78):: – – – – – – – –

$s \frown (a \frown \smile q)$
$a \frown (c \frown p)$
$\mathfrak{r}$: $m \frown (\mathfrak{r} \frown \smile q)$
$\mathfrak{r} \frown (c \frown p)$
$m \frown (x \frown p)$
$x \frown (s \frown \nparallel p)$
$\mathrm{I} \nparallel p$ (β

×

$x \frown (s \frown \nparallel p)$
$s \frown (a \frown \smile q)$
$a \frown (c \frown p)$
$\mathfrak{r}$: $m \frown (\mathfrak{r} \frown \smile q)$
$\mathfrak{r} \frown (c \frown p)$
$m \frown (x \frown p)$
$\mathrm{I} \nparallel p$ (γ

$\mathfrak{r}$: $x \frown (\mathfrak{r} \frown \nparallel p)$
$\mathfrak{r} \frown (a \frown \smile q)$
$a \frown (c \frown p)$
$\mathfrak{r}$: $m \frown (\mathfrak{r} \frown \smile q)$
$\mathfrak{r} \frown (c \frown p)$
$m \frown (x \frown p)$
$\mathrm{I} \nparallel p$ (δ

(15): - - - - - - - - - -

$x \frown (a \frown (\nparallel p \smile \smile q))$
$a \frown (c \frown p)$
$\mathfrak{r}$: $m \frown (\mathfrak{r} \frown \smile q)$
$\mathfrak{r} \frown (c \frown p)$
$m \frown (x \frown p)$
$\mathrm{I} \nparallel p$ (ε

$\mathfrak{r}$: $x \frown (\mathfrak{r} \frown (\nparallel p \smile \smile q))$
$\mathfrak{r} \frown (c \frown p)$
$\mathfrak{r}$: $m \frown (\mathfrak{r} \frown \smile q)$
$\mathfrak{r} \frown (c \frown p)$
$m \frown (x \frown p)$
$\mathrm{I} \nparallel p$ (ζ

(15): - - - - - - - - - - - -

$x \frown (c \frown (\nparallel p \smile \smile q \smile p))$
$\mathfrak{r}$: $m \frown (\mathfrak{r} \frown \smile q)$
$\mathfrak{r} \frown (c \frown p)$
$m \frown (x \frown p)$
$\mathrm{I} \nparallel p$ (η

———— • ————

180 $x \frown (c \frown (\nparallel p \smile \smile q \smile p))$
$m \frown (x \frown p)$
$\mathrm{I} \nparallel p$
$x \frown (c \frown \smile (\nparallel p \smile q \smile p))$

×

$x \frown (c \frown \smile (\nparallel p \smile q \smile p))$
$m \frown (x \frown p)$
$\mathrm{I} \nparallel p$
$x \frown (c \frown (\nparallel p \smile \smile q \smile p))$ (ϑ

(η):: - - - - - - - - - - - -

$x \frown (c \frown \smile (\nparallel p \smile q \smile p))$
$m \frown (x \frown p)$
$\mathrm{I} \nparallel p$
$\mathfrak{r}$: $m \frown (\mathfrak{r} \frown \smile q)$
$\mathfrak{r} \frown (c \frown p)$ (338

———— • ————

IIa $g(a) = (— a \frown u)$
$\mathfrak{a}$: $g(\mathfrak{a}) = (— \mathfrak{a} \frown u)$

(46): - - - - - - - - - - -

$g(a) = (— f(a))$
$\mathfrak{a}$: $g(\mathfrak{a}) = (— \mathfrak{a} \frown u)$
$\dot{\varepsilon}(— f(\varepsilon)) = u$ (α

$\mathfrak{a}$: $g(\mathfrak{a}) = (— f(\mathfrak{a}))$
$\mathfrak{a}$: $g(\mathfrak{a}) = (— \mathfrak{a} \frown u)$
$\dot{\varepsilon}(— f(\varepsilon)) = u$ (β

(Va): - - - - - - - - - - -

$\dot{\varepsilon} g(\varepsilon) = \dot{\varepsilon}(— f(\varepsilon))$
$\mathfrak{a}$: $g(\mathfrak{a}) = (— \mathfrak{a} \frown u)$
$\dot{\varepsilon}(— f(\varepsilon)) = u$ (γ

(IIIc): ————————————

$\dot{\varepsilon} g(\varepsilon) = u$
$\mathfrak{a}$: $g(\mathfrak{a}) = (— \mathfrak{a} \frown u)$
$\dot{\varepsilon}(— f(\varepsilon)) = u$ (339

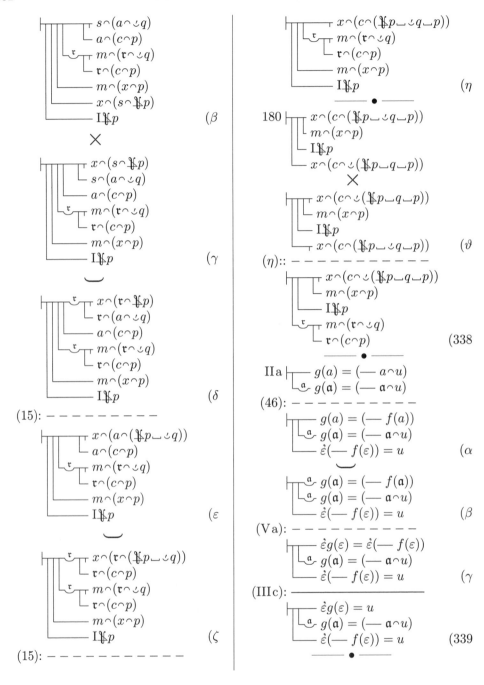

———— • ————

$\mathrm{P} \vdash \grave{\varepsilon}\Big[$ (𝔫 𝔪) $\varepsilon \frown (\mathfrak{n} \frown \smile q)$
$\mathfrak{m} \frown (\varepsilon \frown \smile q)$
$\mathfrak{n} \frown (\mathfrak{n} \frown \underline{\ \ } q)$
$A = \mathfrak{m}; \mathfrak{n}$
$\mathrm{I}q \Big] = A \,\underline{\measuredangle}\, q$

(339): ————————————

$\vdash \grave{\varepsilon} g(\varepsilon) = A \,\underline{\measuredangle}\, q$
(𝔞) $g(\mathfrak{a}) = (—\ \mathfrak{a} \frown (A \,\underline{\measuredangle}\, q))$ (340

———— • ————

179 $\vdash a \frown (m; n \,\underline{\measuredangle}\, q)$
$a \frown (c \frown p)$
$c \frown u$
$m; n \,\underline{\measuredangle}\, q \frown (u \frown \rangle p)$

(273): - - - - - - - - - -

$\vdash a \frown (n \frown \smile q)$
$m \frown (a \frown \smile q)$
$n \frown (n \frown \underline{\ \ } q)$
$a \frown (c \frown p)$
$c \frown u$
$m; n \,\underline{\measuredangle}\, q \frown (u \frown \rangle p)$
$\mathrm{I}q$ (α

(III a): ————————————

$\vdash s \frown (n \frown \smile q)$
$m \frown (a \frown \smile q)$
$n \frown (n \frown \underline{\ \ } q)$
$s \frown (c \frown p)$
$c \frown u$
$m; n \,\underline{\measuredangle}\, q \frown (u \frown \rangle p)$
$\mathrm{I}q$
$s = a$ (β

(79):: - - - - - - - - - - - -

$\vdash s \frown (n \frown \smile q)$
$s \frown (c \frown p)$
$m \frown (a \frown \smile q)$
$n \frown (n \frown \underline{\ \ } q)$
$c \frown u$
$m; n \,\underline{\measuredangle}\, q \frown (u \frown \rangle p)$
$\mathrm{I}q$
$a \frown (c \frown p)$
$\mathrm{I}\rangle\!\!\!\langle p$ (γ

$\vdash$ (𝔯) $\mathfrak{r} \frown (n \frown \smile q)$
$\mathfrak{r} \frown (c \frown p)$
$m \frown (a \frown \smile q)$
$n \frown (n \frown \underline{\ \ } q)$
$c \frown u$
$m; n \,\underline{\measuredangle}\, q \frown (u \frown \rangle p)$
$\mathrm{I}q$
$a \frown (c \frown p)$
$\mathrm{I}\rangle\!\!\!\langle p$ (δ

(337): - - - - - - - - - - - - - -

$\vdash c \frown (y \frown \smile (\rangle\!\!\!\langle p \smile q \smile p))$
$n \frown (y \frown p)$
$m \frown (a \frown \smile q)$
$n \frown (n \frown \underline{\ \ } q)$
$c \frown u$
$m; n \,\underline{\measuredangle}\, q \frown (u \frown \rangle p)$
$\mathrm{I}q$
$a \frown (c \frown p)$
$\mathrm{I}\rangle\!\!\!\langle p$ (ε

×

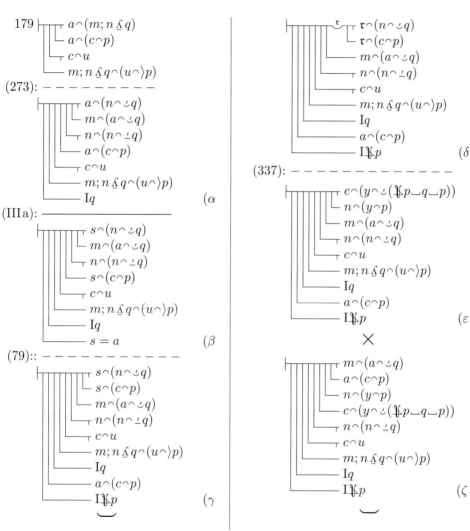

$\vdash m \frown (a \frown \smile q)$
$a \frown (c \frown p)$
$n \frown (y \frown p)$
$c \frown (y \frown \smile (\rangle\!\!\!\langle p \smile q \smile p))$
$n \frown (n \frown \underline{\ \ } q)$
$c \frown u$
$m; n \,\underline{\measuredangle}\, q \frown (u \frown \rangle p)$
$\mathrm{I}q$
$\mathrm{I}\rangle\!\!\!\langle p$ (ζ

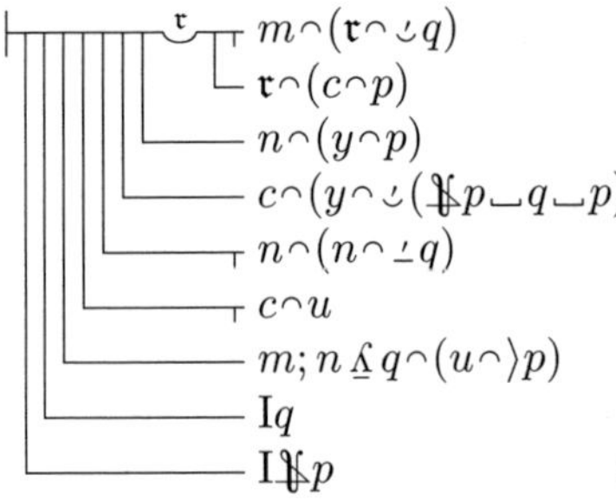

(η

(338): – – – – – – – – – – – – – – – –

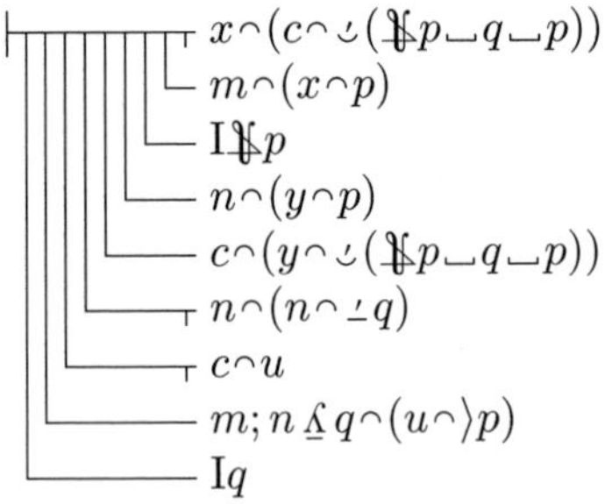

(ϑ

×

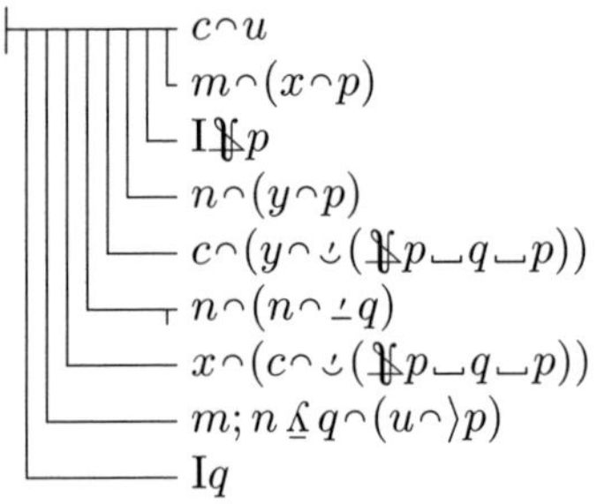

(ι

(269, 270):: = = = = = = = = = = =

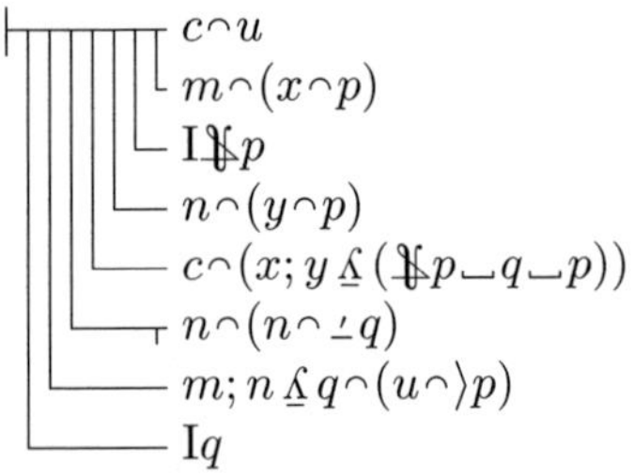

(κ

(18):: – – – – – – – – – – – – – –

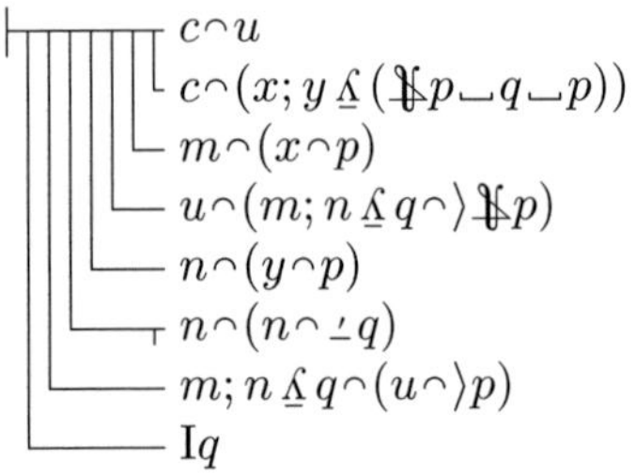

(λ

(IV a): – – – – – – – – – – – – – – –

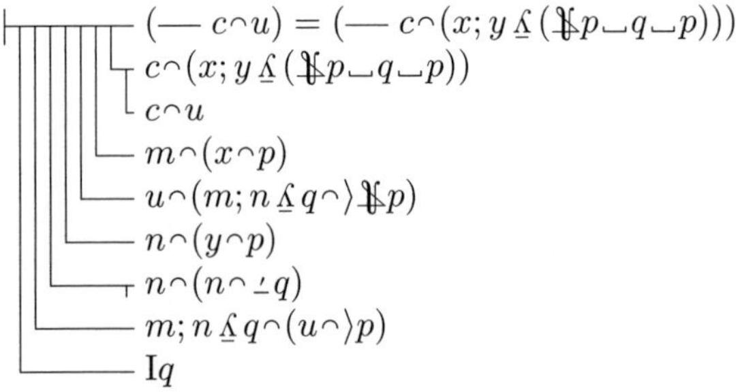

(μ

(334):: –

$$\begin{array}{l}\vdash (-\, c\frown u) = (-\, c\frown(x;y\,\underline{\lessdot}\,(\between p\smile q\smile p))) \\ m\frown(x\frown p) \\ u\frown(m;n\,\underline{\lessdot}\,q\frown\rangle\between p) \\ n\frown(y\frown p) \\ \neg\, n\frown(n\frown\underline{\perp}q) \\ m;n\,\underline{\lessdot}\,q\frown(u\frown\rangle p) \\ \mathrm{I}q\end{array} \qquad (\nu$$

$$\smile$$

$$\begin{array}{l}\vdash \overset{\mathfrak{a}}{\smile}\, (-\, \mathfrak{a}\frown u) = (-\, \mathfrak{a}\frown(x;y\,\underline{\lessdot}\,(\between p\smile q\smile p))) \\ m\frown(x\frown p) \\ u\frown(m;n\,\underline{\lessdot}\,q\frown\rangle\between p) \\ n\frown(y\frown p) \\ \neg\, n\frown(n\frown\underline{\perp}q) \\ m;n\,\underline{\lessdot}\,q\frown(u\frown\rangle p) \\ \mathrm{I}q\end{array} \qquad (\xi$$

(340): –

$$\begin{array}{l}\vdash \dot{\varepsilon}(-\, \varepsilon\frown u) = x;y\,\underline{\lessdot}\,(\between p\smile q\smile p) \\ m\frown(x\frown p) \\ u\frown(m;n\,\underline{\lessdot}\,q\frown\rangle\between p) \\ n\frown(y\frown p) \\ \neg\, n\frown(n\frown\underline{\perp}q) \\ m;n\,\underline{\lessdot}\,q\frown(u\frown\rangle p) \\ \mathrm{I}q\end{array} \qquad (o$$

$$\times$$

$$\begin{array}{l}\vdash \neg\, m\frown(x\frown p) \\ \neg\, \dot{\varepsilon}(-\, \varepsilon\frown u) = x;y\,\underline{\lessdot}\,(\between p\smile q\smile p) \\ u\frown(m;n\,\underline{\lessdot}\,q\frown\rangle\between p) \\ n\frown(y\frown p) \\ \neg\, n\frown(n\frown\underline{\perp}q) \\ m;n\,\underline{\lessdot}\,q\frown(u\frown\rangle p) \\ \mathrm{I}q\end{array} \qquad (\pi$$

(IIa):: – – – – – – – – – – – – – – –

$$\begin{array}{l}\vdash \neg\, m\frown(x\frown p) \\ \overset{\mathfrak{q}}{\smile}\neg\, \dot{\varepsilon}(-\, \varepsilon\frown u) = x;y\,\underline{\lessdot}\,\mathfrak{q} \\ u\frown(m;n\,\underline{\lessdot}\,q\frown\rangle\between p) \\ n\frown(y\frown p) \\ \neg\, n\frown(n\frown\underline{\perp}q) \\ m;n\,\underline{\lessdot}\,q\frown(u\frown\rangle p) \\ \mathrm{I}q\end{array} \qquad (\rho$$

(IIa):: – – – – – – – – – – – –

$$\begin{array}{l}\vdash \neg\, m\frown(x\frown p) \\ \overset{\mathfrak{A}}{\smile}\overset{\mathfrak{q}}{\smile}\neg\, \dot{\varepsilon}(-\, \varepsilon\frown u) = \mathfrak{A}\,\underline{\lessdot}\,\mathfrak{q} \\ u\frown(m;n\,\underline{\lessdot}\,q\frown\rangle\between p) \\ n\frown(y\frown p) \\ \neg\, n\frown(n\frown\underline{\perp}q) \\ m;n\,\underline{\lessdot}\,q\frown(u\frown\rangle p) \\ \mathrm{I}q\end{array} \qquad (\sigma$$

(Ia): – – – – – – – – – – – – – – – –

$$\begin{array}{l}\vdash \neg\, x\frown u \\ m\frown(x\frown p) \\ \overset{\mathfrak{A}}{\smile}\overset{\mathfrak{q}}{\smile}\neg\, \dot{\varepsilon}(-\, \varepsilon\frown u) = \mathfrak{A}\,\underline{\lessdot}\,\mathfrak{q} \\ u\frown(m;n\,\underline{\lessdot}\,q\frown\rangle\between p) \\ n\frown(y\frown p) \\ \neg\, n\frown(n\frown\underline{\perp}q) \\ m;n\,\underline{\lessdot}\,q\frown(u\frown\rangle p) \\ \mathrm{I}q\end{array} \qquad (\tau$$

$$\smile$$

$$\begin{array}{l}\vdash \overset{\mathfrak{a}}{\smile}\neg\, \mathfrak{a}\frown u \\ m\frown(\mathfrak{a}\frown p) \\ \overset{\mathfrak{A}}{\smile}\overset{\mathfrak{q}}{\smile}\neg\, \dot{\varepsilon}(-\, \varepsilon\frown u) = \mathfrak{A}\,\underline{\lessdot}\,\mathfrak{q} \\ u\frown(m;n\,\underline{\lessdot}\,q\frown\rangle\between p) \\ n\frown(y\frown p) \\ \neg\, n\frown(n\frown\underline{\perp}q) \\ m;n\,\underline{\lessdot}\,q\frown(u\frown\rangle p) \\ \mathrm{I}q\end{array} \qquad (341$$

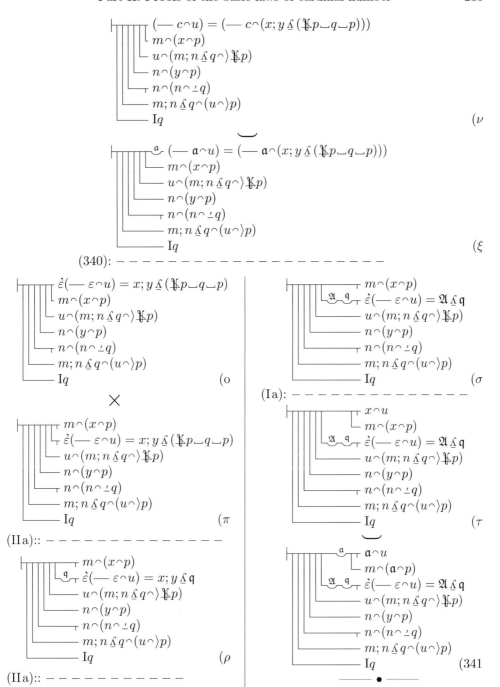

——— • ———

140 ⊦ $m \frown (m \frown \smile q)$

(273): ————————.

$m \frown (n \frown \smile q)$
$n \frown (n \frown \perp q)$
$m \frown (m; n \underline{\measuredangle} q)$
$\mathrm{I}q$ (342

(8):: - - - - - - - - -

$m \frown (n \frown \smile q)$
$n \frown (n \frown \perp q)$
$\mathfrak{a}$ $\mathfrak{a} \frown u$
$m \frown (\mathfrak{a} \frown p)$
$m; n \underline{\measuredangle} q \frown (u \frown \rangle p)$
$\mathrm{I}q$ (α

(341):: - - - - - - - - - - - -

$m \frown (n \frown \smile q)$
$n \frown (n \frown \perp q)$
$\mathfrak{A}$ $\mathfrak{q}$ $\dot{\varepsilon}(— \varepsilon \frown u) = \mathfrak{A} \underline{\measuredangle} \mathfrak{q}$
$u \frown (m; n \underline{\measuredangle} q \frown \rangle \nparallel p)$
$n \frown (y \frown p)$
$m; n \underline{\measuredangle} q \frown (u \frown \rangle p)$
$\mathrm{I}q$ (β

(Ia): - - - - - - - - - - - - - - - -

$y \frown u$
$n \frown (y \frown p)$
$m \frown (n \frown \smile q)$
$n \frown (n \frown \perp q)$
$\mathfrak{A}$ $\mathfrak{q}$ $\dot{\varepsilon}(— \varepsilon \frown u) = \mathfrak{A} \underline{\measuredangle} \mathfrak{q}$
$u \frown (m; n \underline{\measuredangle} q \frown \rangle \nparallel p)$
$m; n \underline{\measuredangle} q \frown (u \frown \rangle p)$
$\mathrm{I}q$ (γ

$\mathfrak{a}$ $\mathfrak{a} \frown u$
$n \frown (\mathfrak{a} \frown p)$
$m \frown (n \frown \smile q)$
$n \frown (n \frown \perp q)$
$\mathfrak{A}$ $\mathfrak{q}$ $\dot{\varepsilon}(— \varepsilon \frown u) = \mathfrak{A} \underline{\measuredangle} \mathfrak{q}$
$u \frown (m; n \underline{\measuredangle} q \frown \rangle \nparallel p)$
$m; n \underline{\measuredangle} q \frown (u \frown \rangle p)$
$\mathrm{I}q$ (343

——— • ———

140 ⊦ $n \frown (n \frown \smile q)$

(274): ————————

$n \frown (m; n \underline{\measuredangle} q)$
$m \frown (n \frown \smile q)$
$n \frown (n \frown \perp q)$
$\mathrm{I}q$ (344

×

$m \frown (n \frown \smile q)$
$n \frown (m; n \underline{\measuredangle} q)$
$n \frown (n \frown \perp q)$
$\mathrm{I}q$ (α

(8):: - - - - - - - - -

$m \frown (n \frown \smile q)$
$\mathfrak{a}$ $\mathfrak{a} \frown u$
$n \frown (\mathfrak{a} \frown p)$
$m; n \underline{\measuredangle} q \frown (u \frown \rangle p)$
$n \frown (n \frown \perp q)$
$\mathrm{I}q$ (β

(343):: - - - - - - - - - - - -

$m \frown (n \frown \smile q)$
$m \frown (n \frown \smile q)$
$n \frown (n \frown \perp q)$
$\mathfrak{A}$ $\mathfrak{q}$ $\dot{\varepsilon}(— \varepsilon \frown u) = \mathfrak{A} \underline{\measuredangle} \mathfrak{q}$
$u \frown (m; n \underline{\measuredangle} q \frown \rangle \nparallel p)$
$m; n \underline{\measuredangle} q \frown (u \frown \rangle p)$
$\mathrm{I}q$ (γ

(Ig): - - - - - - - - - - - - - - -

$m \frown (n \frown \smile q)$
$n \frown (n \frown \perp q)$
$\mathrm{I}q$
$\mathfrak{A}$ $\mathfrak{q}$ $\dot{\varepsilon}(— \varepsilon \frown u) = \mathfrak{A} \underline{\measuredangle} \mathfrak{q}$
$u \frown (m; n \underline{\measuredangle} q \frown \rangle \nparallel p)$
$m; n \underline{\measuredangle} q \frown (u \frown \rangle p)$ (δ

(324): - - - - - - - - - - - - - - -

$a \frown (m; n \underline{\measuredangle} q)$
$\mathfrak{A}$ $\mathfrak{q}$ $\dot{\varepsilon}(— \varepsilon \frown u) = \mathfrak{A} \underline{\measuredangle} \mathfrak{q}$
$u \frown (m; n \underline{\measuredangle} q \frown \rangle \nparallel p)$
$m; n \underline{\measuredangle} q \frown (u \frown \rangle p)$ (ε

(I): ————————————————

$a \frown (m; n \underline{\measuredangle} q)$
$c \frown (a \frown \nparallel p)$
$\mathfrak{A}$ $\mathfrak{q}$ $\dot{\varepsilon}(— \varepsilon \frown u) = \mathfrak{A} \underline{\measuredangle} \mathfrak{q}$
$u \frown (m; n \underline{\measuredangle} q \frown \rangle \nparallel p)$
$m; n \underline{\measuredangle} q \frown (u \frown \rangle p)$ (ζ

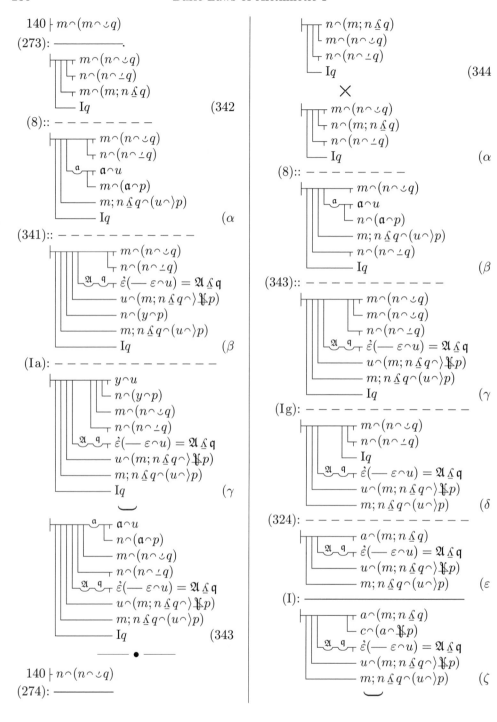

$$\vdash \overset{\mathfrak{a}}{\smile}\ \mathfrak{a}\frown(m;n\mathbin{\underline{\measuredangle}}q)$$
$$c\frown(\mathfrak{a}\frown\nmid p)$$
$$\overset{\mathfrak{A}\ \mathfrak{q}}{\smile\smile}\ \dot{\varepsilon}(\text{—}\ \varepsilon\frown u)=\mathfrak{A}\mathbin{\underline{\measuredangle}}\mathfrak{q}$$
$$u\frown(m;n\mathbin{\underline{\measuredangle}}q\frown\rangle\nmid p)$$
$$m;n\mathbin{\underline{\measuredangle}}q\frown(u\frown\rangle p) \quad (\eta$$

(8): - - - - - - - - - - - - - -

$$\vdash c\frown u$$
$$\overset{\mathfrak{A}\ \mathfrak{q}}{\smile\smile}\ \dot{\varepsilon}(\text{—}\ \varepsilon\frown u)=\mathfrak{A}\mathbin{\underline{\measuredangle}}\mathfrak{q}$$
$$u\frown(m;n\mathbin{\underline{\measuredangle}}q\frown\rangle\nmid p)$$
$$m;n\mathbin{\underline{\measuredangle}}q\frown(u\frown\rangle p) \quad (\vartheta$$

×

$$\vdash m;n\mathbin{\underline{\measuredangle}}q\frown(u\frown\rangle p)$$
$$u\frown(m;n\mathbin{\underline{\measuredangle}}q\frown\rangle\nmid p)$$
$$\overset{\mathfrak{A}\ \mathfrak{q}}{\smile\smile}\ \dot{\varepsilon}(\text{—}\ \varepsilon\frown u)=\mathfrak{A}\mathbin{\underline{\measuredangle}}\mathfrak{q}$$
$$c\frown u \quad (\iota$$

$\smile$

$$\vdash \overset{\mathfrak{q}}{\smile}\ m;n\mathbin{\underline{\measuredangle}}q\frown(u\frown\rangle\mathfrak{q})$$
$$u\frown(m;n\mathbin{\underline{\measuredangle}}q\frown\rangle\nmid\mathfrak{q})$$
$$\overset{\mathfrak{A}\ \mathfrak{q}}{\smile\smile}\ \dot{\varepsilon}(\text{—}\ \varepsilon\frown u)=\mathfrak{A}\mathbin{\underline{\measuredangle}}\mathfrak{q}$$
$$c\frown u \quad (\kappa$$

(49): - - - - - - - - - - - - - -

$$\vdash \mathfrak{P}(m;n\mathbin{\underline{\measuredangle}}q)=\mathfrak{P}u$$
$$\overset{\mathfrak{A}\ \mathfrak{q}}{\smile\smile}\ \dot{\varepsilon}(\text{—}\ \varepsilon\frown u)=\mathfrak{A}\mathbin{\underline{\measuredangle}}\mathfrak{q}$$
$$c\frown u \quad (\lambda$$

×

$$\vdash c\frown u$$
$$\overset{\mathfrak{A}\ \mathfrak{q}}{\smile\smile}\ \dot{\varepsilon}(\text{—}\ \varepsilon\frown u)=\mathfrak{A}\mathbin{\underline{\measuredangle}}\mathfrak{q}$$
$$\mathfrak{P}(m;n\mathbin{\underline{\measuredangle}}q)=\mathfrak{P}u \quad (\mu$$

$\smile$

$$\vdash \overset{\mathfrak{a}}{\smile}\ \mathfrak{a}\frown u$$
$$\overset{\mathfrak{A}\ \mathfrak{q}}{\smile\smile}\ \dot{\varepsilon}(\text{—}\ \varepsilon\frown u)=\mathfrak{A}\mathbin{\underline{\measuredangle}}\mathfrak{q}$$
$$\mathfrak{P}(m;n\mathbin{\underline{\measuredangle}}q)=\mathfrak{P}u \quad (\nu$$

×

$$\vdash \overset{\mathfrak{A}\ \mathfrak{q}}{\smile\smile}\ \dot{\varepsilon}(\text{—}\ \varepsilon\frown u)=\mathfrak{A}\mathbin{\underline{\measuredangle}}\mathfrak{q}$$
$$\overset{\mathfrak{a}}{\smile}\ \mathfrak{a}\frown u$$
$$\mathfrak{P}(m;n\mathbin{\underline{\measuredangle}}q)=\mathfrak{P}u \quad (345$$

§178. *Analysis*

We remove the subcomponent

$$\text{‘}\overset{\mathfrak{a}}{\smile}\ \mathfrak{a}\frown u\text{’}$$

by supplying a series-forming relation such that no object belongs to its series running from an object to an object, as was stated in §172. Such a relation is equality since every object in the series of this relation follows itself.

§179. *Construction*

$$37 \vdash y\frown(y\frown\dot{\alpha}\dot{\varepsilon}(\varepsilon=\alpha))$$

(131): ————————

$$\vdash y\frown(y\frown\!\perp\dot{\alpha}\dot{\varepsilon}(\varepsilon=\alpha)) \quad (\alpha$$

(272): ————————

$$\vdash c\frown(x;y\mathbin{\underline{\measuredangle}}\dot{\alpha}\dot{\varepsilon}(\varepsilon=\alpha)) \quad (\beta$$

(Ia): ————————

$$\vdash c\frown u$$
$$c\frown(x;y\mathbin{\underline{\measuredangle}}\dot{\alpha}\dot{\varepsilon}(\varepsilon=\alpha)) \quad (\gamma$$

(IVa): ————————

$$\vdash (\text{—}\ c\frown u)=(\text{—}\ c\frown(x;y\mathbin{\underline{\measuredangle}}\dot{\alpha}\dot{\varepsilon}(\varepsilon=\alpha)))$$
$$c\frown(x;y\mathbin{\underline{\measuredangle}}\dot{\alpha}\dot{\varepsilon}(\varepsilon=\alpha))$$
$$c\frown u \quad (\delta$$

——— • ———

$$\text{IIa} \vdash c\frown u$$
$$\overset{\mathfrak{a}}{\smile}\ \mathfrak{a}\frown u$$

(Ia): - - - - -

$$\vdash c\frown(x;y\mathbin{\underline{\measuredangle}}\dot{\alpha}\dot{\varepsilon}(\varepsilon=\alpha))$$
$$c\frown u$$
$$\overset{\mathfrak{a}}{\smile}\ \mathfrak{a}\frown u \quad (\varepsilon$$

(δ): - - - - - - - - - - - -

$$\vdash (\text{—}\ c\frown u)=(\text{—}\ c\frown(x;y\mathbin{\underline{\measuredangle}}\dot{\alpha}\dot{\varepsilon}(\varepsilon=\alpha)))$$
$$\overset{\mathfrak{a}}{\smile}\ \mathfrak{a}\frown u \quad (\zeta$$

$\smile$

$$\vdash \overset{\mathfrak{a}}{\smile}\ (\text{—}\ \mathfrak{a}\frown u)=(\text{—}\ \mathfrak{a}\frown(x;y\mathbin{\underline{\measuredangle}}\dot{\alpha}\dot{\varepsilon}(\varepsilon=\alpha)))$$
$$\overset{\mathfrak{a}}{\smile}\ \mathfrak{a}\frown u \quad (\eta$$

(340): -

$$\vdash \dot{\varepsilon}(-\, \varepsilon \frown u) = x ; y \mathbin{\underline{\wedge}} \dot{\alpha}\dot{\varepsilon}(\varepsilon = \alpha)$$
$$\mathfrak{a} \quad \mathfrak{a} \frown u \qquad (\vartheta$$

×

$$\vdash \mathfrak{a} \quad \mathfrak{a} \frown u$$
$$\dot{\varepsilon}(-\, \varepsilon \frown u) = x ; y \mathbin{\underline{\wedge}} \dot{\alpha}\dot{\varepsilon}(\varepsilon = \alpha) \qquad (\iota$$

(IIa):: - - - - - - - - - - - - - -

$$\vdash \mathfrak{a} \quad \mathfrak{a} \frown u$$
$$\mathfrak{q} \quad \dot{\varepsilon}(-\, \varepsilon \frown u) = x ; y \mathbin{\underline{\wedge}} \mathfrak{q} \qquad (\kappa$$

(IIa):: - - - - - - - - - - - - - -

$$\vdash \mathfrak{a} \quad \mathfrak{a} \frown u$$
$$\mathfrak{A} \quad \mathfrak{q} \quad \dot{\varepsilon}(-\, \varepsilon \frown u) = \mathfrak{A} \mathbin{\underline{\wedge}} \mathfrak{q} \qquad (\lambda$$

×

$$\vdash \mathfrak{A} \quad \mathfrak{q} \quad \dot{\varepsilon}(-\, \varepsilon \frown u) = \mathfrak{A} \mathbin{\underline{\wedge}} \mathfrak{q}$$
$$\mathfrak{a} \quad \mathfrak{a} \frown u \qquad (\mu$$

(345): ·—·—·—·—·—·—·—·—·—·—·

$$\vdash \mathfrak{A} \quad \mathfrak{q} \quad \dot{\varepsilon}(-\, \varepsilon \frown u) = \mathfrak{A} \mathbin{\underline{\wedge}} \mathfrak{q}$$
$$\wp(m ; n \mathbin{\underline{\wedge}} q) = \wp u \qquad (346$$

(IIIf):: - - - - - - - - - - - - -

$$\vdash \mathfrak{A} \quad \mathfrak{q} \quad \dot{\varepsilon}(-\, \varepsilon \frown u) = \mathfrak{A} \mathbin{\underline{\wedge}} \mathfrak{q}$$
$$\wp u = \wp(m ; n \mathbin{\underline{\wedge}} q) \qquad (347$$

——— • ———

$$347 \vdash \mathfrak{A} \quad \mathfrak{q} \quad \dot{\varepsilon}(-\, \varepsilon \frown u) = \mathfrak{A} \mathbin{\underline{\wedge}} \mathfrak{q}$$
$$\wp u = \wp(1 ; \wp u \mathbin{\underline{\wedge}} \mathfrak{f})$$

(314):: - - - - - - - - - - - - -

$$\vdash \mathfrak{A} \quad \mathfrak{q} \quad \dot{\varepsilon}(-\, \varepsilon \frown u) = \mathfrak{A} \mathbin{\underline{\wedge}} \mathfrak{q}$$
$$0 \frown (\wp u \frown \smile \mathfrak{f}) \qquad (348$$

Appendices

1. Table of the basic laws

and propositions immediately following from them

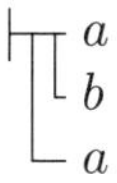

(I (§18))

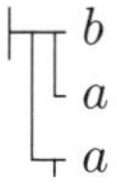

(Ia (§49))

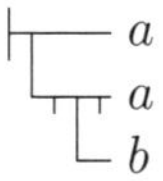

(Ib (§49))

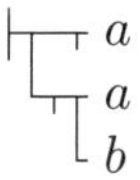

(Ic (§49))

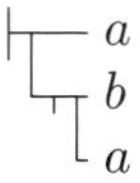

(Id (§49))

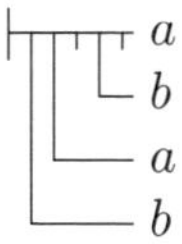

(Ie (§49))

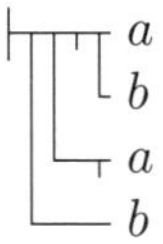

(If (§49))

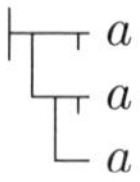

(Ig (§49))

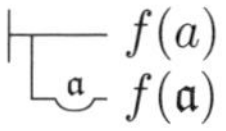

(IIa (§20))

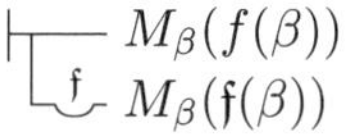

(IIb (§25))

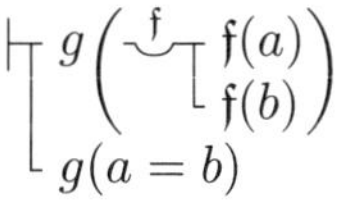

(III (§20))

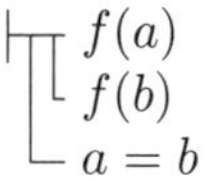

(IIIa (§50))

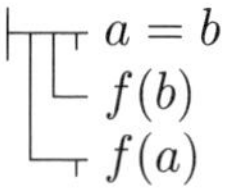

(IIIb (§50))

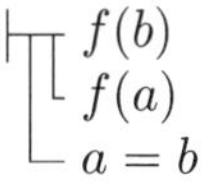

(IIIc (§50))

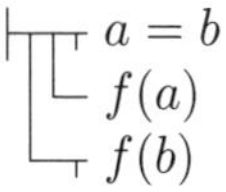

(IIId (§50))

$$\vdash a = a \qquad \text{(IIIe (§50)}$$

•

$$\begin{array}{l}\vdash\!\top\, b = a\\ \quad\llcorner\, a = b\end{array} \qquad \text{(IIIf (§50)}$$

•

$$\vdash\!\top (\text{—}\, a) = (\neg\, a) \qquad \text{(IIIg (§50)}$$

•

$$\begin{array}{l}\vdash\!\top\, f(a) = f(b)\\ \quad\llcorner\, a = b\end{array} \qquad \text{(IIIh (§50)}$$

•

$$\vdash (\text{—}\, a = b) = (a = b) \qquad \text{(IIIi (§50)}$$

•

$$\begin{array}{l}\vdash\!\top (\text{—}\, a) = (\text{—}\, b)\\ \quad\llcorner\!\neg (\text{—}\, a) = (\neg\, b)\end{array} \qquad \text{(IV (§18)}$$

•

$$\begin{array}{l}\vdash\!\top\!\top (\text{—}\, a) = (\text{—}\, b)\\ \quad\;\;\llcorner\!\top\, b\\ \quad\;\;\;\;\llcorner\, a\\ \quad\llcorner\!\top\, a\\ \quad\;\;\llcorner\, b\end{array} \qquad \text{(IVa (§51)}$$

•

$$\vdash (\text{—}\, a) = (\neg\neg\, a) \qquad \text{(IVb (§51)}$$

•

$$\begin{array}{l}\vdash\!\top\, f(\text{—}\, a)\\ \quad\llcorner\, f(\neg\neg\, a)\end{array} \qquad \text{(IVc (§51)}$$

•

$$\begin{array}{l}\vdash\!\top\, f(\neg\neg\, a)\\ \quad\llcorner\, f(\text{—}\, a)\end{array} \qquad \text{(IVd (§51)}$$

•

$$\vdash (a = b) = (b = a) \qquad \text{(IVe (§51)}$$

•

$$\vdash (\grave{\varepsilon} f(\varepsilon) = \grave{\alpha} g(\alpha)) = (\forall\mathfrak{a}\; f(\mathfrak{a}) = g(\mathfrak{a})) \qquad \text{(V (§20)}$$

•

$$\begin{array}{l}\vdash\!\top\, F(\grave{\varepsilon} f(\varepsilon)) = F(\grave{\alpha} g(\alpha))\\ \quad\llcorner\!\forall\mathfrak{a}\; f(\mathfrak{a}) = g(\mathfrak{a})\end{array} \qquad \text{(Va (§52)}$$

•

$$\begin{array}{l}\vdash\!\top\, f(a) = g(a)\\ \quad\llcorner\, \grave{\varepsilon} f(\varepsilon) = \grave{\alpha} g(\alpha)\end{array} \qquad \text{(Vb (§52)}$$

•

$$\vdash a = \backslash\grave{\varepsilon}(a = \varepsilon) \qquad \text{(VI (§18)}$$

•

$$\begin{array}{l}\vdash\!\top\, a = \backslash\grave{\varepsilon} f(\varepsilon)\\ \quad\llcorner\!\forall\mathfrak{a}\; f(\mathfrak{a}) = (a = \mathfrak{a})\end{array} \qquad \text{(VIa (§52)}$$

2. Table of definitions

$$\Vdash \backslash\grave{\alpha}\left(\neg\forall\mathfrak{g}\!\top\!\begin{array}{l}\top\, \mathfrak{g}(a) = \alpha\\ \llcorner\, u = \grave{\varepsilon}\mathfrak{g}(\varepsilon)\end{array}\right) = a \frown u \qquad \text{(A}$$

(Relation of an object falling within the extension of a concept. §34, p. 53.)[1]

$$\Vdash \grave{\alpha}\grave{\varepsilon}\left(\neg\forall\mathfrak{r}\!\top\!\begin{array}{l}\top\, \varepsilon \frown (\mathfrak{r} \frown p)\\ \llcorner\, \mathfrak{r} \frown (\alpha \frown q)\end{array}\right) = p \sqcup q \qquad \text{(B}$$

(Composite relation. §54, p. 72.)

$$\Vdash \left(\forall\mathfrak{e}\,\forall\mathfrak{d}\,\forall\mathfrak{a}\!\top\!\begin{array}{l}\top\, \mathfrak{d} = \mathfrak{a}\\ \;\llcorner\, \mathfrak{e} \frown (\mathfrak{a} \frown p)\\ \llcorner\, \mathfrak{e} \frown (\mathfrak{d} \frown p)\end{array}\right) = \mathrm{I}p \qquad (\Gamma$$

(Single-valuedness of a relation. §37, p. 55.)

[1] These short hints in words which I add to the concept-script definitions are not exhaustive and make no claim to be of the strictest precision.

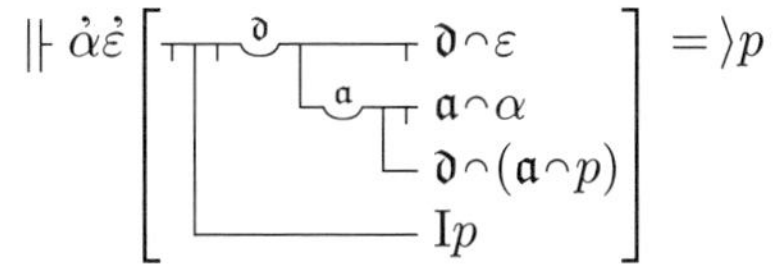

(Δ

(Mapping-into by a relation. §38, p. 56.)

$$\Vdash \dot{\alpha}\dot{\varepsilon}(\alpha\frown(\varepsilon\frown p)) = \rangle\!\!\!\setminus p \qquad \text{(E}$$

(Converse of a relation. §39, p. 57.)

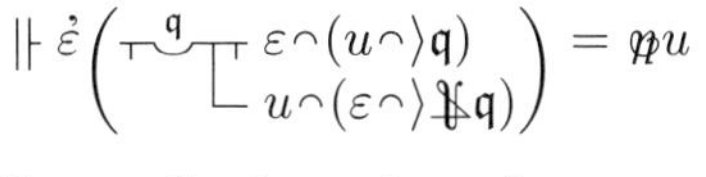

(Z

(The cardinal number of a concept; i.e., the cardinal number of objects falling under a concept. §40, p. 57.)

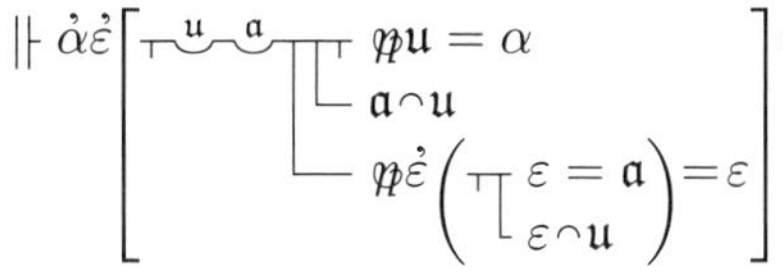

(H

(Relation of a cardinal number to the one immediately following. §43, p. 58.)

$$\Vdash \mathfrak{P}\dot{\varepsilon}(\top\, \varepsilon = \varepsilon) = 0 \qquad (\Theta$$

(The cardinal number Zero. §41, p. 58.)

$$\Vdash \mathfrak{P}\dot{\varepsilon}(\varepsilon = 0) = 1 \qquad \text{(I}$$

(The cardinal number One. §42, p. 58.)

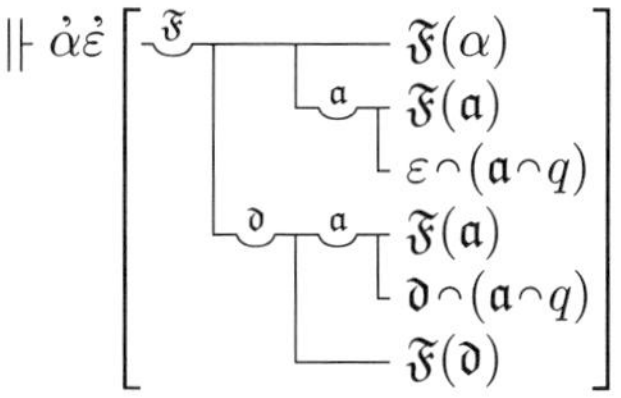

(K

(The following of an object after an object in the series of a relation. §45, p. 60.)

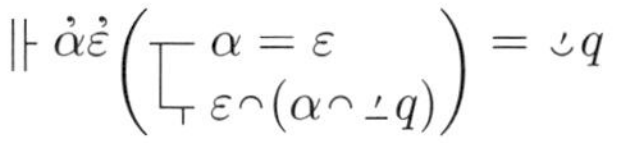

(Λ

(The relation of an object belonging to the series of a relation starting with an object. §46, p. 60.)

$$\Vdash \mathfrak{P}(0\frown \rangle\!\!\!\setminus \smile \mathfrak{f}) = \infty \qquad \text{(M}$$

(The cardinal number Endlos. §122, p. 150.)

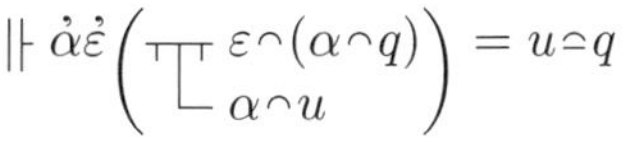

(N

(§138, p. 171.)

$$\Vdash \dot{\varepsilon}(o\frown(a\frown\varepsilon)) = o;a \qquad (\Xi$$

(The pair. §144, p. 179.)

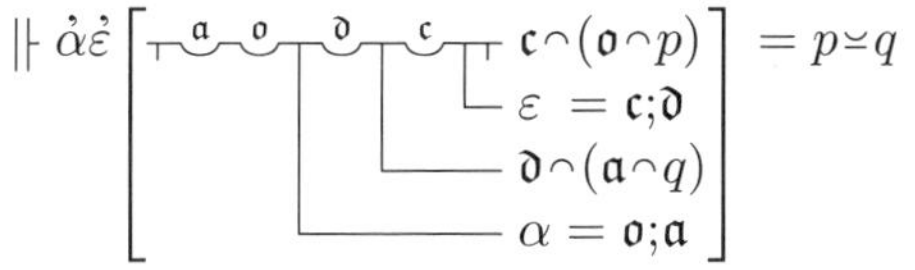

(O

(Coupling of a relation with a relation. §144, p. 179.)

$$\Vdash \dot{\alpha}\dot{\varepsilon}(A\frown(\varepsilon;\alpha\frown\smile t)) = A \lessdot t \qquad (\Pi$$

(§144, p. 179.)

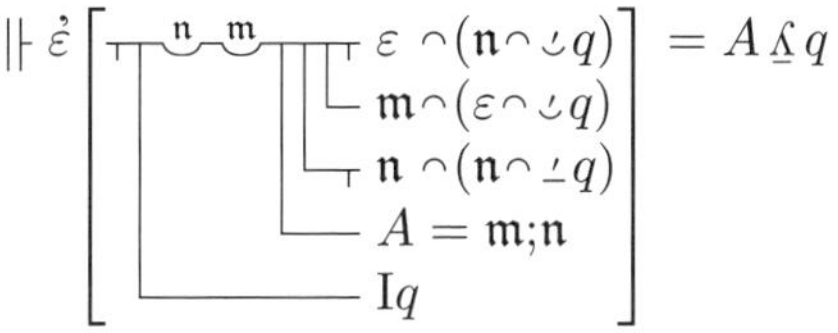

(P

(The circumstance that an object belongs to a series running from an object to an object. §158, p. 201.)

3. Table of the important theorems

$$\vdash f(a) = a \frown \grave{\varepsilon} f(\varepsilon) \qquad (1$$

•

$$\begin{array}{l} \vdash F(a \frown \grave{\varepsilon} f(\varepsilon)) \\ \quad F(f(a)) \end{array} \qquad (77$$

•

$$\begin{array}{l} \vdash F(f(a)) \\ \quad F(a \frown \grave{\varepsilon} f(\varepsilon)) \end{array} \qquad (82$$

•

$$\vdash f(a,b) = a \frown (b \frown \grave{\alpha}\grave{\varepsilon} f(\varepsilon, \alpha)) \qquad (2$$

•

$$\begin{array}{l} \vdash F(f(a,b)) \\ \quad F(a \frown (b \frown q)) \\ \quad \grave{\alpha}\grave{\varepsilon} f(\varepsilon, \alpha) = q \end{array} \qquad (6$$

•

$$\begin{array}{l} \vdash F(a \frown (b \frown q)) \\ \quad F(f(a,b)) \\ \quad \grave{\alpha}\grave{\varepsilon} f(\varepsilon, \alpha) = q \end{array} \qquad (10$$

•

$$\begin{array}{l} \vdash F(f(a,b)) \\ \quad F(a \frown (b \frown \grave{\alpha}\grave{\varepsilon} f(\varepsilon, \alpha))) \end{array} \qquad (33$$

•

$$\begin{array}{l} \vdash F(a \frown (b \frown \grave{\alpha}\grave{\varepsilon} f(\varepsilon, \alpha))) \\ \quad F(f(a,b)) \end{array} \qquad (36$$

•

$$\begin{array}{l} \vdash F(\top a \frown \grave{\varepsilon}(\top f(\varepsilon))) \\ \quad F(\text{—}\ f(a)) \end{array} \qquad (58$$

•

$$\begin{array}{l} \vdash d \frown (m \frown (p \smile q)) \\ \quad e \frown (m \frown q) \\ \quad d \frown (e \frown p) \end{array} \qquad (5$$

If an object (d) stands in a (p-) relation to a second (e) and if the second object (e) stands in a (q-)relation to a third (m), then the first object stands to the third in the relation composed of the first and the second.[1]

$$\begin{array}{l} \vdash d \frown (e \frown (\rtimes p \smile q \smile p)) \\ \quad c \frown (e \frown p) \\ \quad b \frown (c \frown q) \\ \quad d \frown (b \frown \rtimes p) \end{array} \qquad (174$$

•

$$\begin{array}{l} \vdash e \frown (d \frown (p \smile q)) \\ \quad \smile^{\mathfrak{r}} \; e \frown (\mathfrak{r} \frown p) \\ \qquad \mathfrak{r} \frown (d \frown q) \end{array} \qquad (15$$

•

$$\begin{array}{l} \vdash \mathrm{I}q \\ \quad \smile^{\mathfrak{e}} \smile^{\mathfrak{d}} \smile^{\mathfrak{a}} \; \mathfrak{d} = \mathfrak{a} \\ \qquad \mathfrak{e} \frown (\mathfrak{a} \frown q) \\ \qquad \mathfrak{e} \frown (\mathfrak{d} \frown q) \end{array} \qquad (16$$

•

$$\begin{array}{l} \vdash d = a \\ \quad b \frown (a \frown q) \\ \quad b \frown (d \frown q) \\ \quad \mathrm{I}q \end{array} \qquad (13$$

If a relation is single-valued and an object (b) stands to a second (d) and a third (a) in this relation, then the second (d) and the third (a) will coincide.

$$\begin{array}{l} \vdash \mathrm{I}(p \smile q) \\ \quad \mathrm{I}q \\ \quad \mathrm{I}p \end{array} \qquad (17$$

A relation composed of two relations is single-valued if these two relations are.

[1]The translations that I append to the concept-script propositions reflect their principal content, but do not always exhaust the whole content.

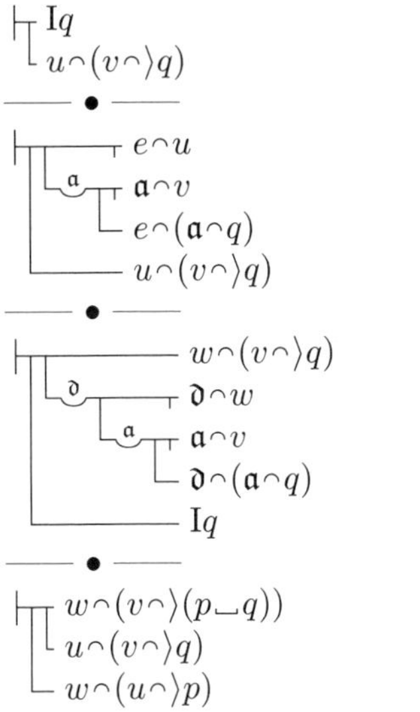

(18 (8 (11 (19

If a relation maps one concept into a second, and this second concept is mapped into a third by a second (q-) relation, then the relation composed of the first and second relation maps the first concept into the third.

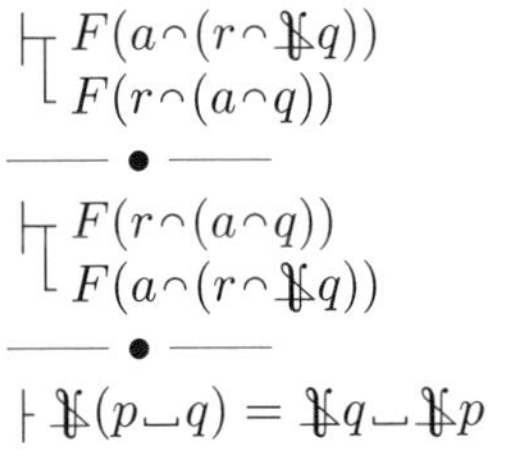

(22 (23 (24

The converse of a relation which is composed of a first and second, is composed of the converse of the second and the converse of the first.

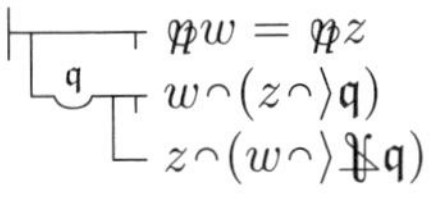

(49

The cardinal number of objects falling under a first (w-)concept does not coincide with the cardinal number of objects falling under a second (z-)concept if there is no relation that maps the first into the second and whose converse also maps the second into the first.

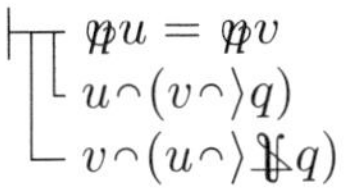

(32

The cardinal number of the objects falling under the (u-)concept coincides with the cardinal number of those falling under a second (v-)concept if a relation maps the first into the second concept whose converse maps the second into the first.

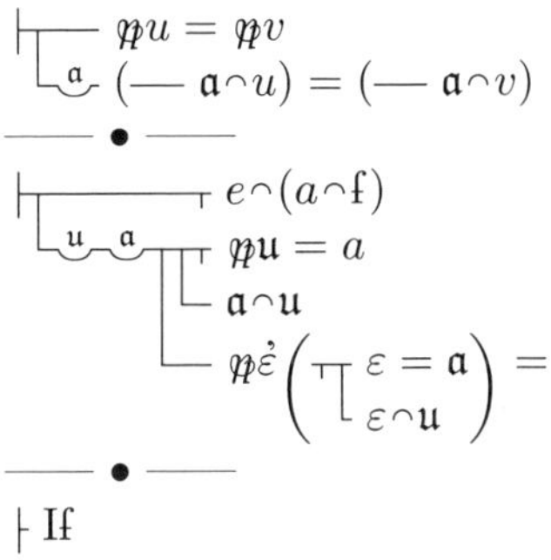

(96 (68 (71

The relation of a cardinal number to that immediately following it in the cardinal number series is single-valued.

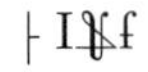

(89

The relation of a cardinal number to that immediately preceding it in the cardinal number series is single-valued.

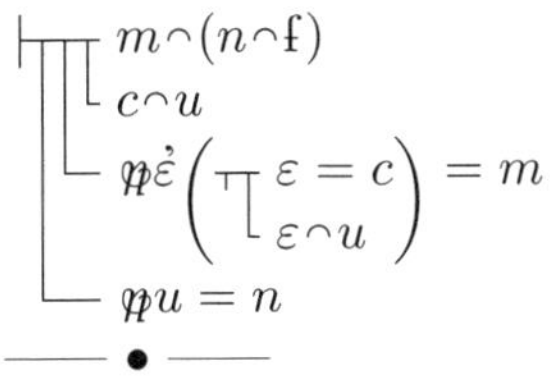

(101

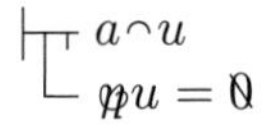

(94

If Zero is the cardinal number of objects that fall under a concept, then no object falls under the concept.

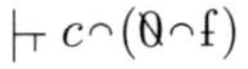

(108

In the cardinal number series nothing immediately precedes Zero.

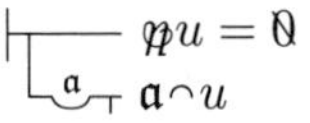

(97

If no object falls under a concept, then Zero is the cardinal number of the objects falling under that concept.

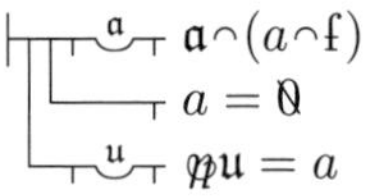

(107

For every cardinal number distinct from Zero, there is one immediately preceding it in the cardinal number series.

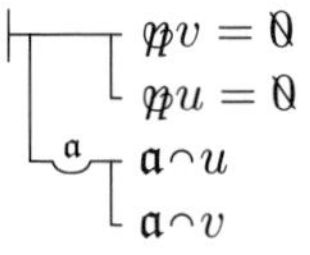

(99

If Zero is the cardinal number of the objects falling under a first concept, then Zero is also the cardinal number of the objects that fall under a concept subordinated to the first.

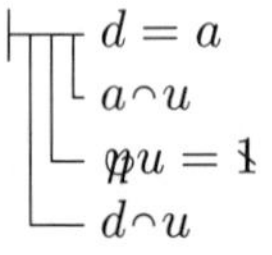

(117

If One is the cardinal number of objects falling under a concept and if a first object falls under this concept and likewise a second, then these objects coincide.

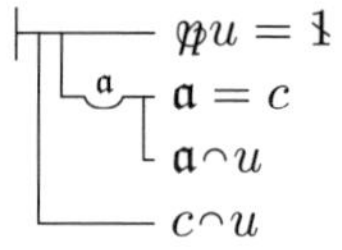

(121

If an object falls under a concept and if every object that falls under this concept coincides with the former, then One is the cardinal number of the objects falling under the concept.

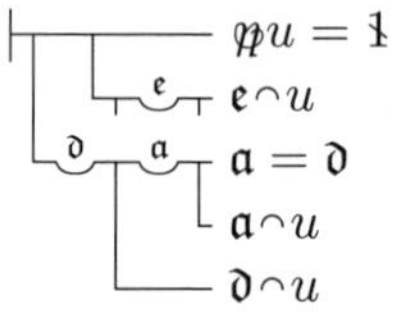

(122

One is the cardinal number of the objects falling under a concept if there is an object that falls under it and every object that falls under the concept coincides with any object that falls under it.

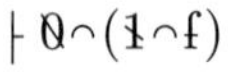

(110

The cardinal number One immediately follows the cardinal number Zero in the cardinal number series.

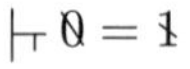

(111

The cardinal number Zero is distinct from the cardinal number One.

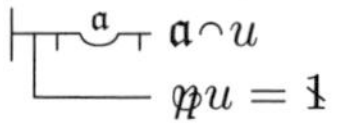

(113

If One is the cardinal number of objects falling under a concept, then there is an object falling under the concept.

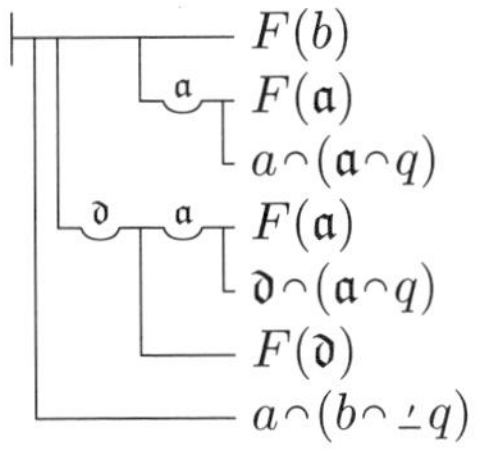

(123

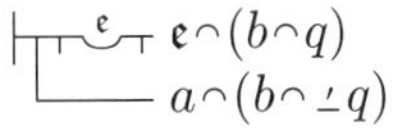

(124

If an object follows an object in a series, then there is an object which stands to the former in the series-forming relation.

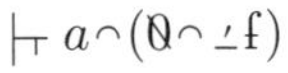

(126

Nothing precedes the cardinal number Zero in the cardinal number series.

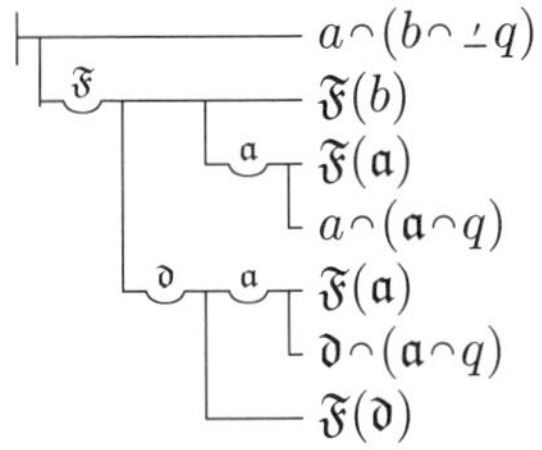

(127

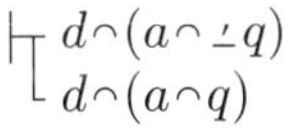

(131

A first object precedes a second in a series if it stands to it in the series-forming relation.

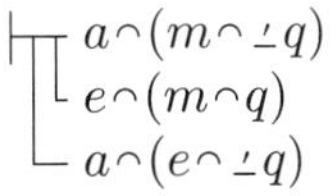

(133

If an object follows a second in a series and stands to a third in the series-forming relation, then the third follows the second in this series.

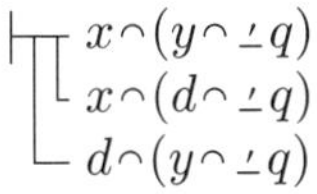

(275

If an object follows a second in a series and precedes a third in it, then the third also follows the second in this series.

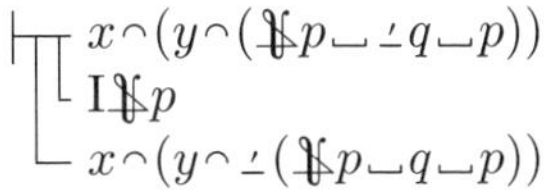

(177

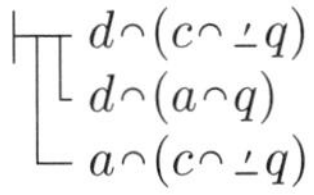

(129

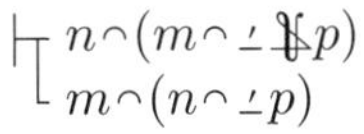

(302

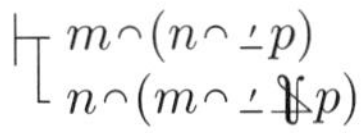

(299

An object follows a second in the series of a relation if the second follows the first in the series of the converse of the relation.

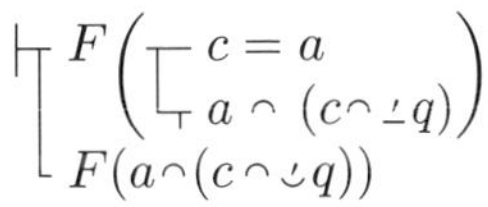

(130

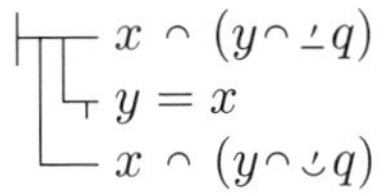

(200

If an object belongs to a series starting with a second, then it either coincides with it or follows it in this series.

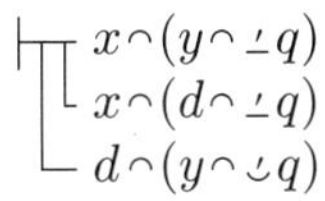

(276

$$\vdash \begin{array}{l} x \frown (y \frown \underline{\smile} q) \\ x \frown (d \frown \dot{\smile} q) \\ d \frown (y \frown \underline{\smile} q) \end{array} \quad (280$$

—— • ——

$$\vdash \begin{array}{l} F(b) \\ F(a) \\ \smile^{\mathfrak{d}}\smile^{\mathfrak{a}}\ F(\mathfrak{a}) \\ \quad \mathfrak{d} \frown (\mathfrak{a} \frown q) \\ \quad F(\mathfrak{d}) \\ a \frown (b \frown \dot{\smile} q) \end{array} \quad (144$$

—— • ——

$$\vdash \begin{array}{l} a \frown (m \frown \underline{\smile} q) \\ e \frown (m \frown q) \\ a \frown (e \frown \dot{\smile} q) \end{array} \quad (134$$

If an object belongs to a series starting with a second and stands to a third in the series-forming relation, then the third follows the second in this series.

$$\vdash \begin{array}{l} d \frown (c \frown \underline{\smile} q) \\ d \frown (a \frown q) \\ a \frown (c \frown \dot{\smile} q) \end{array} \quad (132$$

If an object belongs to a series ending with a second and if a third stands to it in the series-forming relation, then the second follows the third in this series.

$$\vdash \begin{array}{l} a \frown (m \frown \dot{\smile} q) \\ a \frown (m \frown \underline{\smile} q) \end{array} \quad (136$$

—— • ——

$$\vdash \begin{array}{l} a \frown (m \frown \dot{\smile} q) \\ e \frown (m \frown q) \\ a \frown (e \frown \dot{\smile} q) \end{array} \quad (137$$

—— • ——

$$\vdash \begin{array}{l} x \frown (y \frown \dot{\smile} q) \\ x \frown (d \frown \dot{\smile} q) \\ d \frown (y \frown \dot{\smile} q) \end{array} \quad (322$$

If an object (d) belongs both to a series ending with a second (y) and to a series of the same relation starting with a third (x), then the second also belongs to a series starting with the third.

$$\vdash \begin{array}{l} d \frown (y \frown \dot{\smile} q) \\ d \frown (a \frown q) \\ a \frown (y \frown \dot{\smile} q) \end{array} \quad (285$$

—— • ——

$$\vdash \begin{array}{l} a \frown (b \frown \dot{\smile} q) \\ b = a \end{array} \quad (139$$

—— • ——

$$\vdash a \frown (a \frown \dot{\smile} q) \quad (140$$

Every object belongs to the series starting with the object itself of any relation.

$$\vdash \begin{array}{l} c \frown (y \frown (\not\Join p \smile \dot{\smile} q \smile p)) \\ n \frown (y \frown p) \\ c \frown (y \frown \dot{\smile} (\not\Join p \smile q \smile p)) \\ \mathrm{I} \not\Join p \end{array} \quad (335$$

—— • ——

$$\vdash \begin{array}{l} x \frown (y \frown (\not\Join p \smile \dot{\smile} q \smile p)) \\ m \frown (x \frown p) \\ \mathrm{I} \not\Join p \\ x \frown (y \frown \dot{\smile} (\not\Join p \smile q \smile p)) \end{array} \quad (180$$

—— • ——

$$\vdash \begin{array}{l} F(b) \\ F(a) \\ \smile^{\mathfrak{d}}\smile^{\mathfrak{a}}\ F(\mathfrak{a}) \\ \quad \mathfrak{d} \frown (\mathfrak{a} \frown q) \\ \quad a \frown (\mathfrak{d} \frown \dot{\smile} q) \\ \quad F(\mathfrak{d}) \\ a \frown (b \frown \dot{\smile} q) \end{array} \quad (152$$

—— • ——

$$\vdash \begin{array}{l} \smile^{\mathfrak{e}}\ a \frown (\mathfrak{e} \frown \dot{\smile} q) \\ \quad \mathfrak{e} \frown (b \frown q) \\ a \frown (b \frown \underline{\smile} q) \end{array} \quad (141$$

If an object (b) follows a second object (a) in a series, then there is an object which stands to the first (b) in the series-forming relation and which belongs to the series of this relation starting with the second (a) (p. 143).

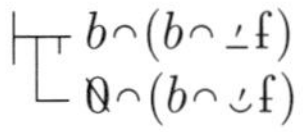

(145

No finite cardinal number follows itself in the cardinal number series (pp. 137 and 144).

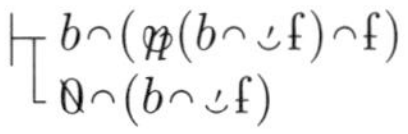

(155

The cardinal number of the members of the cardinal number series ending with a finite cardinal number (b) follows immediately after this cardinal number (b) in the cardinal number series.

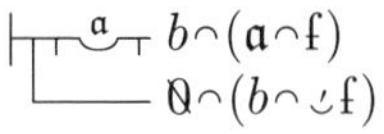

(157

For every finite cardinal number there is a member of the cardinal number series immediately following it.

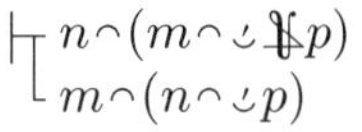

(303

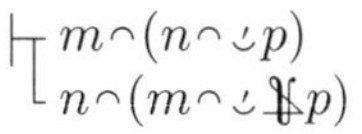

(304

An object (n) belongs to the series of a (p-)relation starting with a second (m) if the second (m) belongs to the series of the converse relation starting with (n).

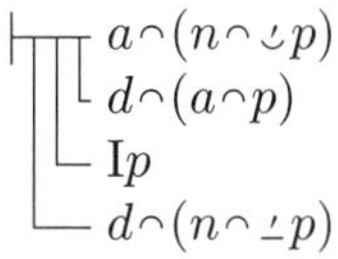

(242

If an object (d) precedes a second (n) in a series whose series-forming relation is single-valued, and if it stands to a third (a) in this relation, then the second (n) belongs to the series of that relation starting with the third (a).

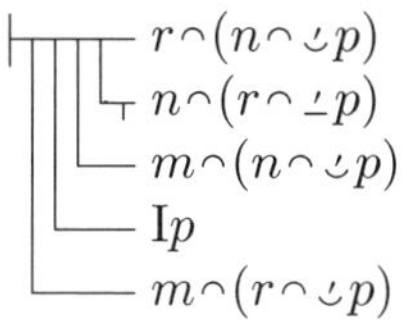

(243

If an object (r) belongs to a series starting with a second (m), whose series-forming relation is single-valued, and if a third object (n) belongs to the same series, then the latter (n) belongs to the series of this relation starting with the first (r) or precedes it in the series.

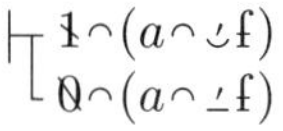

(306

If an object follows Zero in the cardinal number series, then it belongs to the cardinal number series starting with One.

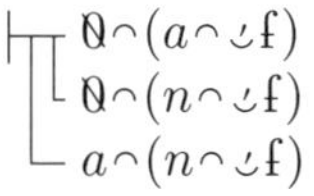

(307

If an object belongs to the cardinal number series ending with a finite cardinal number, then it is itself a finite cardinal number.

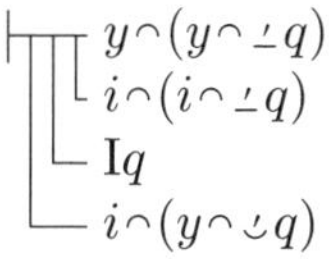

(296

If an object (y) belongs to a series starting with a second (i) whose series-forming relation is single-valued and if the second object (i) follows itself in the series of this relation, then the first (y) also follows itself.

$$\vdash\!\!\top F(\infty)$$
$$\quad\llcorner F(\mathfrak{p}(0 \frown \mathfrak{f} \smile \mathfrak{f})) \qquad (205$$

—— • ——

$$\vdash \infty \frown (\infty \frown \mathfrak{f}) \qquad (165$$

Endlos immediately follows itself in the cardinal number series.

$$\vdash\!\!\top \infty = \mathfrak{p}\dot{\varepsilon}\left(\begin{array}{l}\top \varepsilon \frown v \\ \llcorner \varepsilon \frown u\end{array}\right)$$
$$\quad\llcorner \infty = \mathfrak{p}u$$
$$\quad\llcorner 0 \frown (\mathfrak{p}v \frown \smile \mathfrak{f}) \qquad (172$$

If Endlos is the cardinal number of a concept and if the cardinal number of another concept is finite, then Endlos is the cardinal number of the concept *falling under the first or under the second concept* (p. 154).

$$\vdash\!\!\top 0 \frown (\infty \frown \smile \mathfrak{f}) \qquad (167$$

Endlos is not a finite cardinal number.

$$\vdash\!\!\top e \frown (a \frown q)$$
$$\quad\llcorner e \frown (a \frown (u \,\underline{\frown}\, q)) \qquad (188$$

—— • ——

$$\vdash\!\!\top \mathrm{I}(u \,\underline{\frown}\, q)$$
$$\quad\llcorner \mathrm{I}q \qquad (189$$

—— • ——

$$\vdash\!\!\top x \frown (y \frown \underline{\angle}\, q)$$
$$\quad\llcorner x \frown (y \frown \underline{\angle}\,(u \,\underline{\frown}\, q)) \qquad (194$$

—— • ——

$$\vdash\!\!\top x \frown (y \frown \smile q)$$
$$\quad\llcorner x \frown (y \frown \smile (u \,\underline{\frown}\, q)) \qquad (201$$

—— • ——

$$\vdash\!\!\top y \frown u$$
$$\quad\llcorner d \frown (y \frown (u \,\underline{\frown}\, q)) \qquad (191$$

—— • ——

$$\vdash\!\!\top d \frown (a \frown (u \,\underline{\frown}\, q))$$
$$\quad\llcorner d \frown (a \frown q)$$
$$\quad\llcorner a \frown u \qquad (197$$

—— • ——

$$\vdash\!\!\top\!\!\smile^{\mathfrak{q}}\!\top \;\smile^{\mathfrak{a}}\!\top\; \dot{\varepsilon}(-\!\!-\, \varepsilon \frown u) = \mathfrak{a} \frown \mathfrak{f} \smile \mathfrak{q}$$
$$\quad \llcorner\smile^{\mathfrak{d}}\!\top\; \mathfrak{d} \frown u$$
$$\qquad \llcorner\smile^{\mathfrak{e}}\!\top\; \mathfrak{d} \frown (\mathfrak{e} \frown \mathfrak{q})$$
$$\quad \llcorner\smile^{\mathfrak{i}}\!\top\; \mathfrak{i} \frown (\mathfrak{i} \frown \underline{\angle}\, \mathfrak{q})$$
$$\quad \llcorner \mathrm{I}\mathfrak{q}$$
$$\llcorner \infty = \mathfrak{p}u \qquad (207$$

If Endlos is the cardinal number of objects falling under a concept, then these can be ordered into a non-branching series which starts with a specific object and, without looping back into itself, proceeds endlessly (p. 160).

$$\vdash\!\!\top F(o; a)$$
$$\quad\llcorner F(\dot{\varepsilon}(o \frown (a \frown \varepsilon))) \qquad (249$$

—— • ——

$$\vdash\!\!\top o; a = e; i$$
$$\quad\llcorner o = e$$
$$\quad\llcorner a = i \qquad (251$$

If an object coincides with a second and a third coincides with a fourth, then the pair consisting of the first and the third coincides with that consisting of the second and the fourth.

$$\vdash o \frown (a \frown q) = q \frown (o; a) \qquad (215$$

—— • ——

$$\cdot \vdash\!\!\top x = d$$
$$\quad\llcorner m; x = c; d \qquad (219$$

If one pair coincides with another, then the second member of the first coincides with the second member of the second.

$$\vdash\!\!\top m = c$$
$$\quad\llcorner m; x = c; d \qquad (220$$

—— • ——

$$\vdash\!\!\top f(m, x) = f(c, d)$$
$$\quad\llcorner m; x = c; d \qquad (221$$

—— • ——

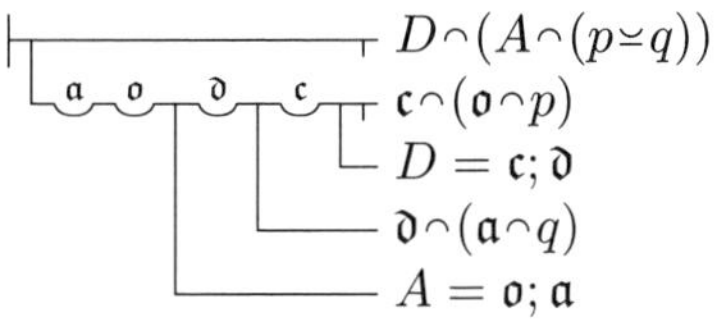
D⌢(A⌢(p⋍q))
𝔠⌢(𝔬⌢p)
D = 𝔠; 𝔡
𝔡⌢(𝔞⌢q)
A = 𝔬; 𝔞

(213

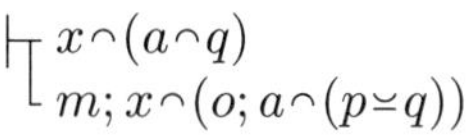
x⌢(a⌢q)
m; x⌢(o; a⌢(p⋍q))

(224

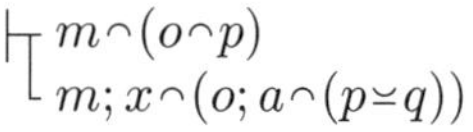
m⌢(o⌢p)
m; x⌢(o; a⌢(p⋍q))

(225

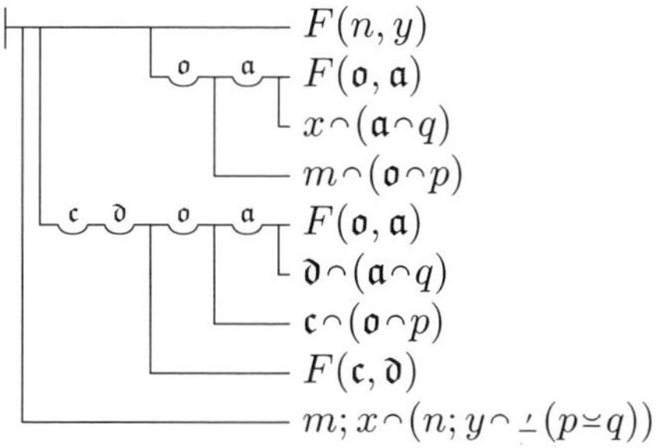
F(n, y)
F(𝔬, 𝔞)
x⌢(𝔞⌢q)
m⌢(𝔬⌢p)
F(𝔬, 𝔞)
𝔡⌢(𝔞⌢q)
𝔠⌢(𝔬⌢p)
F(𝔠, 𝔡)
m; x⌢(n; y⌢∸(p⋍q))

(231

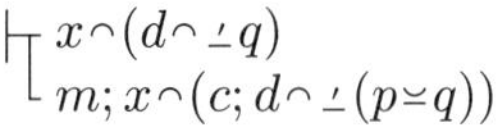
x⌢(d⌢∸q)
m; x⌢(c; d⌢∸(p⋍q))

(233

If a pair follows a second one in the series of a coupled relation, then the second member of the first pair (d) follows the second member of the second pair (x) in a series whose series-forming relation is the second member of the coupled relation.

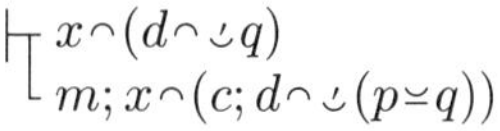
x⌢(d⌢⌣̇q)
m; x⌢(c; d⌢⌣̇(p⋍q))

(234

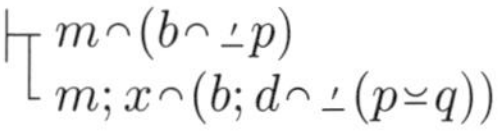
m⌢(b⌢∸p)
m; x⌢(b; d⌢∸(p⋍q))

(244

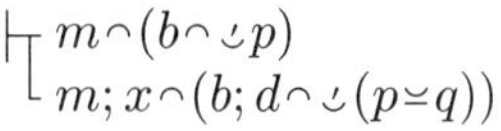
m⌢(b⌢⌣̇p)
m; x⌢(b; d⌢⌣̇(p⋍q))

(246

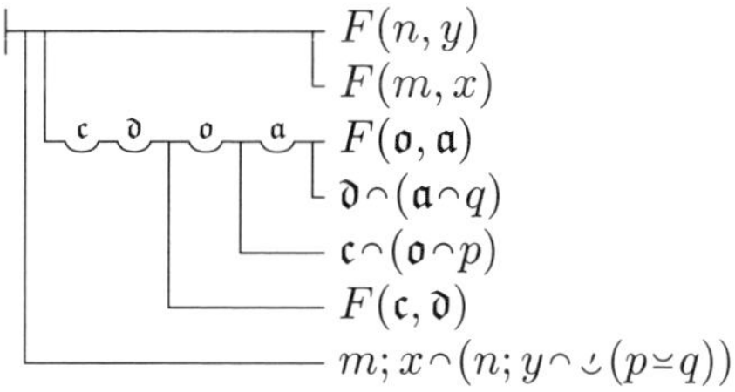
F(n, y)
F(m, x)
F(𝔬, 𝔞)
𝔡⌢(𝔞⌢q)
𝔠⌢(𝔬⌢p)
F(𝔠, 𝔡)
m; x⌢(n; y⌢⌣̇(p⋍q))

(257

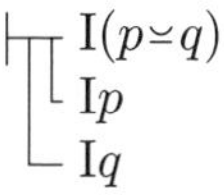
I(p⋍q)
Ip
Iq

(252

If a relation is single-valued and so is a second, then the relation formed by coupling the first and the second is also single-valued.

$$\vdash \begin{array}{l} c;d\frown(o;a\frown(p\underline{\smile}q)) \\ d\frown(a\frown q) \\ c\frown(o\frown p) \end{array} \qquad (208$$

If an object (c) stands to a second (o) in a (p-)relation and if a third object (d) stands to a fourth (a) in a second (q-) relation, then the pair consisting of the first and third object ($c; d$) stands to the pair consisting of the second and fourth ($o; a$) in the relation formed by coupling the first and the second relation.

$$\vdash \begin{array}{l} A\frown(o;a\frown\dot{\smile}(p\underline{\smile}q)) \\ d\frown(a\frown q) \\ c\frown(o\frown p) \\ A\frown(c;d\frown\dot{\smile}(p\underline{\smile}q)) \end{array} \qquad (209$$

$$\vdash \begin{array}{l} x;m\frown(y;n\frown\dot{\smile}(q\underline{\smile}p)) \\ m;x\frown(n;y\frown\dot{\smile}(p\underline{\smile}q)) \end{array} \qquad (258$$

$$\vdash \begin{array}{l} F(A\frown(b;d\frown\dot{\smile}t)) \\ F(b\frown(d\frown(A\nwarrow t))) \end{array} \qquad (247$$

$$\vdash \begin{array}{l} \mathrm{I}(m;x\nwarrow(p\underline{\smile}q)) \\ \smile^{\mathfrak{i}}\ \mathfrak{i}\frown(\mathfrak{i}\frown\underline{\prime}\,p) \\ m\frown(\mathfrak{i}\frown\dot{\smile}p) \\ \mathrm{I}p \\ \mathrm{I}q \end{array} \qquad (253$$

$$\vdash \begin{array}{l} F(o\frown(a\frown(A\nwarrow t))) \\ F(A\frown(o;a\frown\dot{\smile}t)) \end{array} \qquad (210$$

$$\vdash m\frown(x\frown(m;x\nwarrow t)) \qquad (238$$

$$\vdash \begin{array}{l} x\frown(d\frown\dot{\smile}q) \\ c\frown(d\frown(m;x\nwarrow(p\underline{\smile}q))) \end{array} \qquad (235$$

$$\vdash \begin{array}{l} o\frown(a\frown(A\nwarrow(p\underline{\smile}q))) \\ d\frown(a\frown q) \\ c\frown(o\frown p) \\ c\frown(d\frown(A\nwarrow(p\underline{\smile}q))) \end{array} \qquad (211$$

$$\vdash x;m\nwarrow(q\underline{\smile}p) = \mathfrak{U}(m;x\nwarrow(p\underline{\smile}q)) \qquad (259$$

$$\vdash \begin{array}{l} \infty = \mathfrak{p}u \\ \smile^{\mathfrak{q}}\ \smile^{\mathfrak{a}}\ u = \mathfrak{a}\frown\mathfrak{U}\dot{\smile}\mathfrak{q} \\ \smile^{\mathfrak{d}}\ \mathfrak{d}\frown u \\ \smile^{\mathfrak{e}}\ \mathfrak{d}\frown(\mathfrak{e}\frown\mathfrak{q}) \\ \smile^{\mathfrak{i}}\ \mathfrak{i}\frown(\mathfrak{i}\frown\underline{\prime}\,\mathfrak{q}) \\ \mathrm{I}\mathfrak{q} \end{array} \qquad (263$$

Endlos is the cardinal number of the objects falling under a concept if they can be ordered into a series which starts with a certain object and proceeds endlessly, without branching and without looping back into itself (p. 179).

$$\vdash \begin{array}{l} c\frown(m;n\,\underline{\measuredangle}\,p) \\ m\frown(c\frown\dot{\smile}p) \\ n\frown(n\frown\underline{\prime}\,p) \\ c\frown(n\frown\dot{\smile}p) \\ \mathrm{I}p \end{array} \qquad (274$$

An object (c) belongs to the series of a relation running from a second (m) to a third (n) if this relation is single-valued, if the third object (n) does not follow itself in the series of this relation, and if finally the first object (c) belongs to the series of this relation starting with the second (m) as well as the one ending with the third (n).

$$\vdash \begin{array}{l} n\frown(m;n\,\underline{\measuredangle}\,q) \\ m\frown(n\frown\dot{\smile}q) \\ n\frown(n\frown\underline{\prime}\,q) \\ \mathrm{I}q \end{array} \qquad (344$$

$$\vdash \begin{array}{l} \mathrm{I}q \\ d\frown(A\,\underline{\measuredangle}\,q) \end{array} \qquad (265$$

$$\vdash\; d\frown(y\frown\smile q) \quad \lnot\; d\frown(x;y\,\measuredangle\,q) \qquad (269$$

———•———

$$\vdash\; x\frown(d\frown\smile q) \quad \lnot\; d\frown(x;y\,\measuredangle\,q) \qquad (270$$

If an object belongs to a series running from a second to a third, then it belongs to the series of the same relation starting with the second.

$$\vdash\; a = \mathfrak{0} \quad \lnot\; a\frown(\mathfrak{1};n\,\measuredangle\,\mathfrak{f}) \qquad (312$$

If an object belongs to the cardinal number series running from One to a second object, then it is distinct from Zero.

$$\vdash\; n = \wp(\mathfrak{1};n\,\measuredangle\,\mathfrak{f}) \quad \mathfrak{0}\frown(n\frown\smile\mathfrak{f}) \qquad (314$$

Any finite cardinal number is the cardinal number of members of the cardinal number series running from One to itself.

$$\vdash\; x\frown(y\frown\smile q) \quad d\frown(x;y\,\measuredangle\,q) \qquad (323$$

———•———

$$\vdash\; y\frown(y\frown\perp q) \quad \lnot\; d\frown(x;y\,\measuredangle\,q) \qquad (271$$

———•———

$$\vdash\; r = x \quad r\frown(x;x\,\measuredangle\,q) \qquad (282$$

———•———

$$\vdash\; \wp(x;y\,\underline{\measuredangle}\,q) = \wp(\mathfrak{1};n\,\measuredangle\,\mathfrak{f}) \quad y\frown(y\frown\perp q) \quad \mathrm{I}q \quad x;\mathfrak{1}\frown(y;n\frown\smile(q\succeq\mathfrak{f})) \qquad (298$$

(p. 202.)

$$\vdash\; \mathfrak{0}\frown(\wp(x;y\,\underline{\measuredangle}\,q)\frown\smile\mathfrak{f}) \qquad (325$$

The cardinal number of members of a series running from an object to an object is finite.

$$\vdash\; \mathfrak{0}\frown(\wp u\frown\smile\mathfrak{f}) \quad \overset{\mathfrak{A}\;\;\mathfrak{q}}{\smile\smile}\; u = \mathfrak{A}\,\underline{\measuredangle}\,\mathfrak{q} \qquad (327$$

If the objects falling under a concept can be ordered into a series that runs from a specific object to a specific object, then their cardinal number is finite.

$$\vdash\; \grave{\varepsilon}g(\varepsilon) = A\,\measuredangle\,q \quad \overset{\mathfrak{a}}{\smile}\; g(\mathfrak{a}) = (\text{—}\;\mathfrak{a}\frown(A\,\measuredangle\,q)) \qquad (340$$

———•———

$$\vdash\; \overset{\mathfrak{A}\;\;\mathfrak{q}}{\smile\smile}\; \grave{\varepsilon}(\text{—}\;\varepsilon\frown u) = \mathfrak{A}\,\underline{\measuredangle}\,\mathfrak{q} \quad \mathfrak{0}\frown(\wp u\frown\smile\mathfrak{f}) \qquad (348$$

If the cardinal number of objects falling under a concept is finite, then they can be ordered into series that runs from a specific object to a specific object (p. 224).

Index

The numerals specify the pages.

all 24.
and 21.
argument 6, 37, 40.
argument places 6, 8, 13, 15.

belongs to 60, 201.
bracket 10, 11, 35, 64.

cardinal number 57, 58.
cardinal number series 58, its infinity 144.
characteristic mark 3.
composite relation 72.
concavity 13.
concept 3, 8, 38, 57, *u*-concept 71.
concept-script proposition 9, 44.
conditional-stroke 20.
consequences 25 ff.
content-stroke 9.
contraposition 27.
converse of a relation 57.
correctly formed 45.
corresponding 12, 13, 15, 41.
coupled relation 179.
couple 179.

definition 44.
designate 7.
double-stroke of definition 44.
double value-range 55.

ending 60.
Endlos 150.
equation 44.
every 24.
express 7, 50.
extension of a concept 8.
extension of a relation 55.

falls under 8.
False, the 7.
finite 60, 137.
fitting 41.
follows 59, immediately in the cardinal number series 58.
function 5, 6, 8, 11, 13, 37, 41.
function-letter 34, 42.
function-marker 33, 44.
function-name 44, two-sided 64.
fuse 21.
fusion 10, 14, 20, 29.

generality 11, 12, 31, 34.

horizontal 9, 10, 14, 20.

if 24.
immediately follows in the cardinal number series 58.
in need of completion 5, 8.
indicate 31, 32.
inferences 25 ff.
infinity of the cardinal number series 144.

judgement 9.
judgement-stroke 9.

kind of argument and argument places 40, 43.

label 25, 66.
letters, German 13, capital Greek 9, small Greek vowel 15, small Greek as label 66, Roman 31.
level 37, 38, 41.

map 57, 71.
marker 33, 44.
minus-sign 9.

name 7, 32, 43, 44.
negation-stroke 10.
neither–nor 21.
no 24.

object 3, 7, 37.
object-letter 34, 42.
object-marker 33, 44.
One 58.
one-sided 64.
or 22.

pair 179.
particular 24.
permutability 22.
precede 60.
proper name 7, 43.
property 3.
proposition 9, 44.

refer 7, 31, 46.
reference 7, 46.
referential 46.
regular 49.
related 8.
relation 8, 57, *p*-relation 71.
reverse 27.
Roman 31, 33, 42, 44.
run 201.

saturated 37.
scope 13, 15, 31, 35.
sense 7, 51.
series 59, 60, simple 201.
sign 43.
simple series 201.
single-valuedness 39, 55
smooth breathing 15.
some 24.
stand in 8.
starting with 60.
subcomponent 22.
subordinate 24.
subordination 24.
supercomponent 22.
superordinate 24.

there is 12.
thought 7, 9, 50.
transition-signs 44.
Truth, the 7.
truth-value 7.
two-sided 64.

unequal 39.
universal 24.
unsaturated 5, 6, 37.

value 6.
value-range 7, 18.

Zero 58.

GRUNDGESETZE DER ARITHMETIK.

Begriffsschriftlich abgeleitet

von

Dr. G. FREGE
PROFESSOR AN DER UNIVERSITÄT JENA.

II. Band.

JENA
Verlag von Hermann Pohle
1903.

BASIC LAWS

OF ARITHMETIC.

Derived using concept-script

by

DR. G. FREGE
PROFESSOR AT THE UNIVERSITY OF JENA

Volume II

JENA
Verlag von Hermann Pohle
1903

Table of contents

Preliminary remark .... page 1

M. *Proof of the proposition*

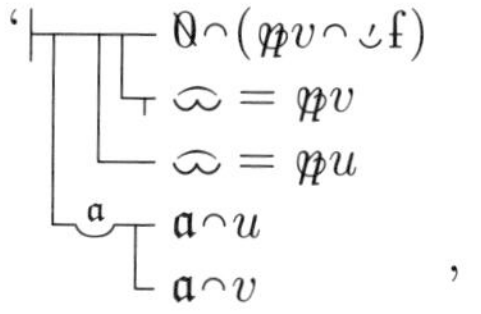

a) Proof of the proposition

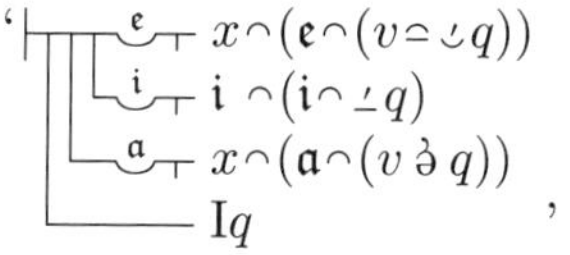

§1 to §6. Definition of the function $\xi \grave{\text{ə}} \zeta$. Propositions (349) to (359) ...... page 1

b) Proof of the proposition

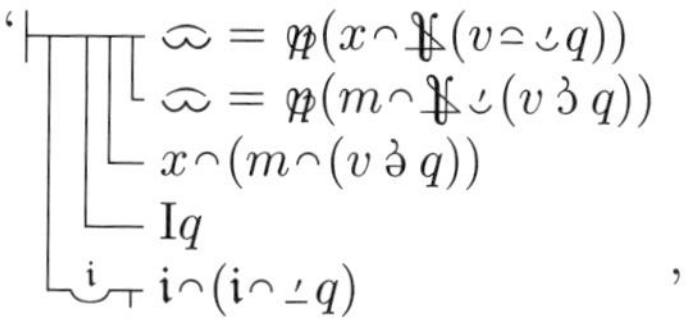

§7 to §20. Definition of the function $\xi \grave{\text{ə}} \zeta$. Propositions up to (408) .... page 7

c) Proof of the proposition

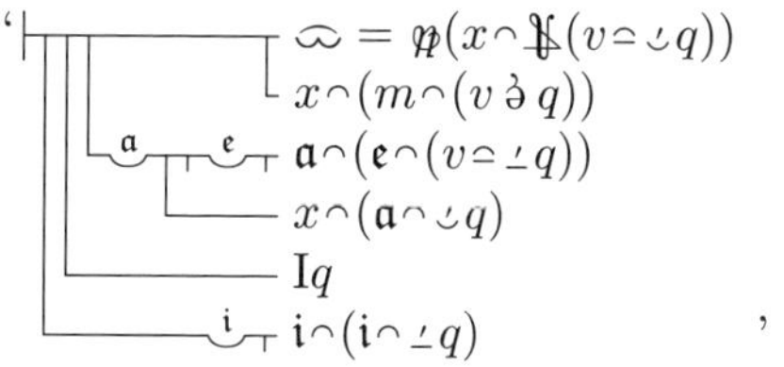

§21 to §24. Propositions up to (416) .... page 25

d) Proof of the proposition

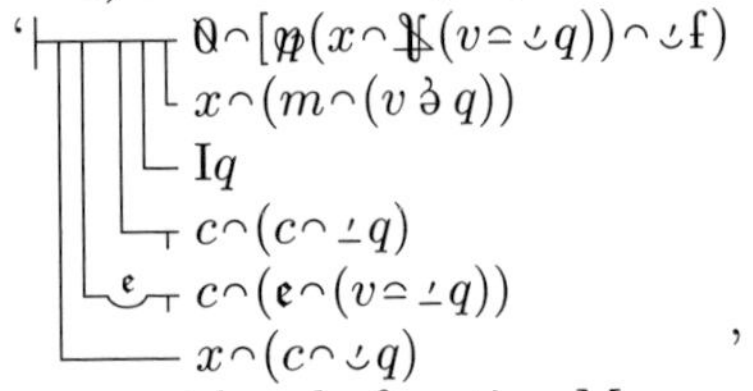

and end of section M

§25 to §28. Propositions up to (428) .... page 30

N. *Proof of the proposition*

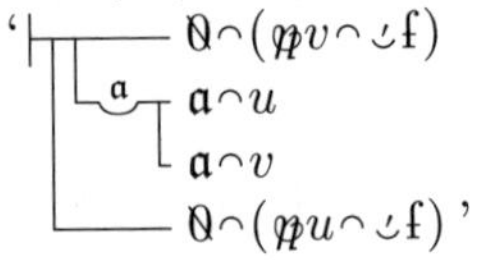

§29 to §32. Propositions up to (443) .... page 37

Ξ. *Proof of the proposition*

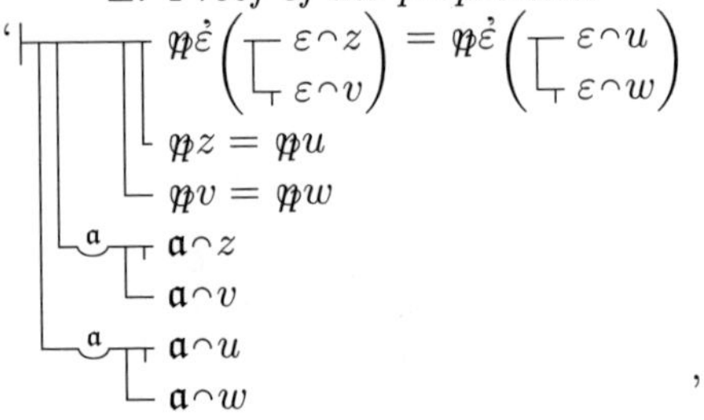

a) Proof of the proposition

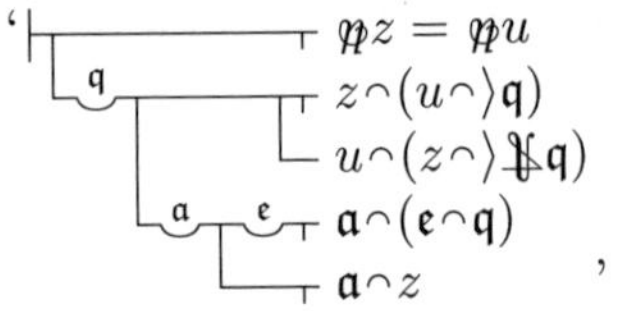

§33 to §36. Propositions up to (453) .... page 44

b) Proof of the proposition

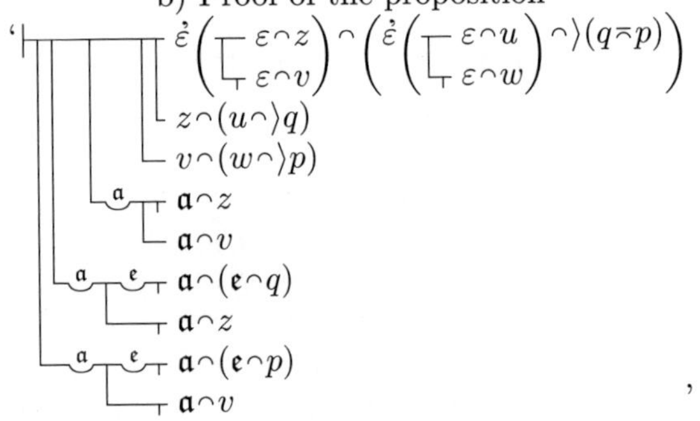

§37 to §40. Definition of the function $\xi \smallfrown \zeta$. Propositions up to (463) ...... page 48

c) Proof of the proposition

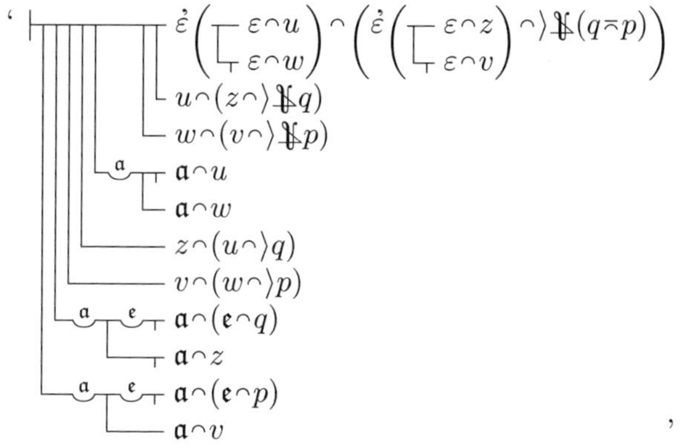

and end of section Ξ

§41 to §44. Propositions up to (469) .... page 52

O. *Corollaries*

a) Proof of the proposition

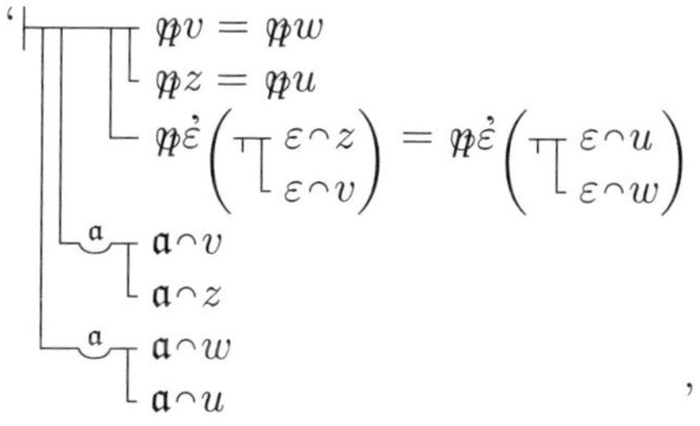

§45 and §46. Propositions up to (472) .... page 58

b) Proof of the proposition

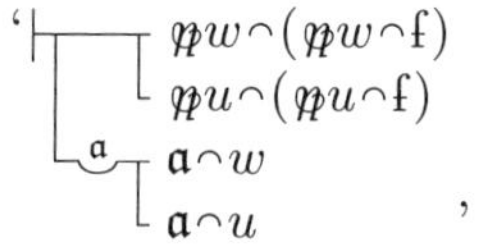

§47 to §50. Propositions up to (476) .... page 61

c) Proof of the propositions

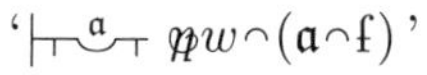 and 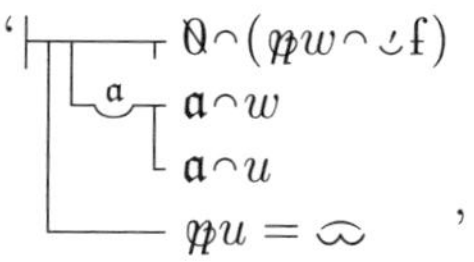

§51 to §54. Propositions up to (484) .... page 66

III. The real numbers

1. Critique of theories concerning irrational numbers

a) Principles of definition

§55. Preliminary remark .... page 69

1. Principle of completeness

§56. A concept must have sharp boundaries .... page 69
§57. Inadmissibility of piecemeal definition .... " 70
§58. Apology for defining in a piecemeal fashion owing to the development of the science .... " 70
§59. Untenability of Heine's definition of the equality of number-signs .... " 72
§60. Heine's presumed defense and its refutation .... " 73
§61. Without final definitions we do not have final theorems .... " 74
§62. The analogous holds of relations. The relations of being greater than and equality .... " 74
§63. Consequence for first-level functions with one argument .... " 75
§64. The analogous demand for first-level functions with two arguments .... " 75
§65. The same also holds for arithmetical signs which are function-names .... " 77

2. Principle of the simplicity of the explained expression

§66. Elucidation and justification of this principle .... page 79
§67. Violation of both principles of definition at once .... " 80

b) Cantor's theory of the irrational numbers

§68. G. Cantor assigns to each of his fundamental series a number *b*. Doubts about the sense. Assumption that the number *b* is a sign and intended to designate the fundamental series .... page 80
§69. Cantor's explanation of the expressions "equal to Zero", "greater than Zero", and "less than Zero" violate both of our principles of definition " 81
§70. Objection by Illigens against Cantor's theory .... " 81
§71. Further objections by Illigens. Concerns against this .... " 82
§72. Pringsheim's proposition that the rational numbers may represent well determined quantities but do not have to .... " 83
§73. The real number is a magnitude-ratio .... " 84
§74. Cantor's response to Illigens's objections run together sign and what is designated. In his own explanation there is no talk of the here mentioned abstract objects of thought .... " 85
§75. The quantitative determination of concrete magnitudes by means of abstract ones according to Cantor. We have in the expression explained by Cantor nothing that is capable of explanation .... " 86
§76. Cantor's numerical magnitudes are superfluous. Enlisting geometry gives us what matters most, the ratio. Therefore it is crucial .... " 88
§77. Attempt to understand Cantor's explanation in such a way that numerical magnitudes are not signs but perhaps abstract objects of thought.

This fails because an assignment is only possible once one knows that what is to be assigned .... page 88
§78. Cantor assign in his apparent explanation of the expression "equal to Zero" the number Zero to certain fundamental series and in the apparent explanations of "positive" and "negative" merely makes gestures at further assignments. Nothing is hereby defined .... " 90
§79. Cantor's definitions of sum, difference and product are flawed in multiple respects .... " 90
§80. Cantor's stipulation that to a fundamental series all of whose members are the rational number *a* is assigned this number *a* contains a contradiction and does not bring us closer to our goal .... " 91
§81. Cantor's definition of being equal to, greater than and less than. Double understanding of the equality sign. In the case of the first, we cannot get beyond the rational numbers, in the case of the second, we have a violation against our first principle of definition. The iridescing between being known and being unknown .... " 91
§82. The definition of sum, being greater than, etc., appear to be intended to create the irrational numbers, a violation against our second principle . " 92
§83. The illusion is dispelled if one uses, instead of the old, already explained words and signs, entirely new ones .... " 93
§84. Summarising review of the results of the examination of the Cantorian theory .... " 94
§85. An earlier account given by Cantor is equally flawed .... " 95

c) The theories of the irrationals of E. Heine and J. Thomae

§86. Preliminary characterisation of these theories .... page 96
§87. Heine's fundamental statement .... " 96
§88. Thomae's fundamental statement .... " 97
§89. Reason for preferring formal to contentual arithmetic .... " 98
§90. Formal arithmetic and concept-script as games .... " 99
§91. In formal arithmetic the equations and inequations do not express thoughts; on this standpoint applications are thus also impossible .... " 100
§92. Formal arithmetic relieves itself at the expense of the sciences in which applications are made .... " 101
§93. In the calculating game there are neither theorems, nor proofs, nor definitions, but they occur in the theory of the game. The possibility of a theory of the calculating game is out into doubt .... " 101
§94. Nothing at all of the numbers proper is required in the calculating game .... " 102
§95. Thomae's expression that a content is attributed to the number-signs according to their behaviour under the game rules .... " 103
§96. Thomae's expressions that certain properties are attributed to the chess pieces and that they are external signs for their behaviour which is constrained by these properties .... " 103

§ 97. Thomae's admissions that number-figures are sometimes also used as number-signs ..... page 104
§ 98. What are signs? ..... " 105
§ 99. Equal-shaped signs are not the same sign ..... " 106
§100. Differently in formal arithmetic. Here we have figures. Equal-shaped figures. Differences between formal and contentual arithmetic ..... " 107
§101. Calculating in formal arithmetic ..... " 108
§102. What is adding, multiplying, subtracting in the calculating game? ..... " 109
§103. *Zero*, the *negative* and *rational numbers* in Thomae ..... " 110
§104. The operations of calculation according to Heine ..... " 110
§105. The introduction of the negative in Heine ..... " 112
§106. The calculating rules of Thomae's ..... " 113
§107. Sense of Thomae's formulae as game rules ..... " 114
§108. Actions in the calculating game. Double role of equations ..... " 115
§109. Double system of rules in formal arithmetic. In order to avoid this, the rules must be expressed in words. Replacement of formulae by means of a rule ..... " 116
§110. Rules of formal arithmetic, laws of contentual arithmetic and moral laws ..... " 117
§111. Incompleteness of Thomae's inventory of rules ..... " 117
§112. Formal subtraction as the converse of formal addition ..... " 118
§113. The special position of Zero, points to a prohibitive rule ..... " 119
§114. Attempt to express this prohibitive rule ..... " 120
§115. Uncertainty in the application of this rule ..... " 121
§116. Further inadequacy of Thomae's inventory of rules and attempted solution ..... " 122
§117. Thomae's statement that division cannot always be performed in a way free from contradiction ..... " 122
§118. Thomae's statement regarding the freedom from contradiction of the formations ..... " 123
§119. Does formal arithmetic admit of a foundation entirely free from contradiction? ..... " 123
§120. Not every contentual arithmetic is based on sense perception ..... " 125
§121. The ordering in a series and being greater than in Thomae ..... " 125
§122. Positive and negative, rational number-figure in Thomae ..... " 126
§123. The infinite in Thomae ..... " 127
§124. Introduction of the irrationals. Heine's number-series ..... " 128
§125. Thomae's definition of an infinite sequence ..... " 129
§126. Thomae's Zero-sequences. There are no number-figures that are not written down ..... " 130
§127. Can all terms of a sequence not be written down? ..... " 131
§128. The formal arithmetician breaks faith with his plan ..... " 131
§129. Attempt better to understand Thomae's opinion. Obstacles ..... " 133

§130. Neither the circumstance that a proposition follows from the instruction for the continuation of a series nor this proposition itself can be of service for the definition of the Zero-instruction .... page 133
§131. Because of the finitude of the collection of number-figures formal arithmetic is incapable of defining the irrationals. Veiling of this state of affairs .... " 134
§132. The group '(0 0 0 . . 0 . .)' is no number-sequence .... " 135
§133. The group '(a_1 a_2 a_3 . . a_n . .)' is to be conceived like a single letter without paying heed to its composition .... " 135
§134. A sign is assigned to a sequence by means of the equality-sign. How is the resulting equation to be understood? .... " 136
§135. The rational number as assigned sign. Polysemy .... " 137
§136. Failure of formal arithmetic .... " 138
§137. Review of formal arithmetic .... " 138

d) The creation of new objects according to R. Dedekind, H. Hankel, O. Stolz

§138. The three main advantages of Dedekind's theory. Pronounced opposition to formal arithmetic .... page 140
§139. The cut. The creation of an irrational number. Is it possible? The power of creation is limited .... " 141
§140. A limitless creation would be too convenient to be admissible ...... " 141
§141. The alternating numbers according to H. Hankel. The proofs conducted using them founder. Their properties remain obscure. By means of the stipulation not even a class is determined to which the alternating units belong, to an even lesser extent itself .... " 142
§142. A greater formal rigour of proof would reveal the error. Proofs that are conducted with the imaginary unit are often equally deficient .. " 144
§143. O. Stolz's creative definitions. Highly consequential restriction of the power of creation .... " 144
§144. From the non-obviousness of a contradiction one cannot infer its absence .... " 145
§145. Is Stolz's theory formal in the sense of Thomae? Great difference between the conceptions of Stolz and Thomae. Dedekind. G. Cantor " 146
§146. Our introduction of value-ranges is different from the creation of numbers by the mathematicians .... " 147
§147. Our procedure is not really new, is performed in full awareness of its logical admissibility. Without it, a scientific justification of arithmetic would be impossible .... " 148

e) Weierstrass's theory

§148. Difficulties that oppose the exact capturing and assessment of this theory .... page 149
§149. The ginger-biscuit viewpoint concerning cardinal numbers .... " 149
§150. The two errors of the Weierstrassian theory of the cardinal numbers, smuggling in of the number proper .... " 150

§151. Struggle of the Weierstrassian theory against the nature of the subject matter. Singular and plural in the word "unit". Proper name or concept word? Is the equality-sign the identity-sign? The value of an aggregate .... page 151
§152. Varying references of the plus-sign. The miracle of objects that occur repeatedly .... " 151
§153. The number as aggregate of abstract units or of the repeatedly occurring One. Three conceptions of number in Weierstrass .... " 152
§154. The higher numbers in Weierstrass. Senseless equation, senseless proposition. The explanations are insufficient or are lacking completely " 153
.
§155. The recognition of higher numbers is based, according to Kossak, on their being defined. A creation whose justification remains doubtful " 153

f) Review and Prospects

§156. Formal and contentual arithmetic. Neither way has led to the goal so far .... page 154
§157. The real numbers as magnitude-ratios. The domain of the cardinal numbers cannot be extended to that of the real numbers .... " 155
§158. Number-signs do not refer to line segments. Breaking away from geometry. Logical nature of arithmetic. "Formal" in a different sense " 156
§159. Middle way between the geometrical foundation and the approaches attempted in recent times. Separation from any specific kind of magnitude, without neglecting measurement. Ways of application. Concern .... " 156

g) Magnitude

§160. Failed attempts to explain the word "magnitude" .... page 157
§161. Reason for the failures is asking the wrong question. Class. What properties must a class have in order to be a domain of magnitudes? " 158
§162. Remark by Gauss. Relation. The domain of magnitudes as a class of Relations .... " 159
§163. Example: distance Relations .... " 160
§164. Preliminary rebuttal of our concern of §159 .... " 160

2. The theory of magnitude

A. *Propositions concerning the composition and converse of Relations in general*

§165. The associative principle is to be proven for all compositions of Relations .... page 163
§166. Proof of the associative principle. Propositions (485) to (491) .... " 163
§167 to §170. The commutative principle in certain series. Definition of the function $_*\xi$. Propositions up to (502) .... " 166

Propositions demonstrating the similarity between the converse of

Relations and reversing the sign

§171 and §172. Propositions up to (508) .... page 168

B. *The positival class*

a) Definitions of the functions ðξ and ʒξ and consequences

§173 and §174. Definitions of the function ðξ and consequences. Propositions up to (517) .... page 168

§175 and §176. Positival class. Definition of the function ʒξ and consequences. Propositions up to (544) .... " 170

b) Proof of the proposition

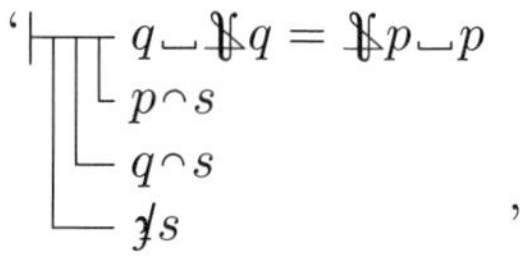

§177 and §178. Propositions up to (559) .... page 176

c) Proof of the proposition

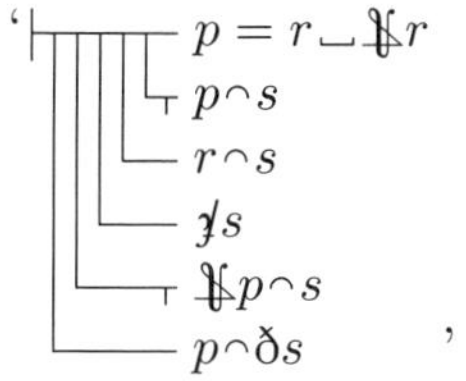

§179 and §180. Propositions up to (561) .... page 180

d) Proof of the propositions

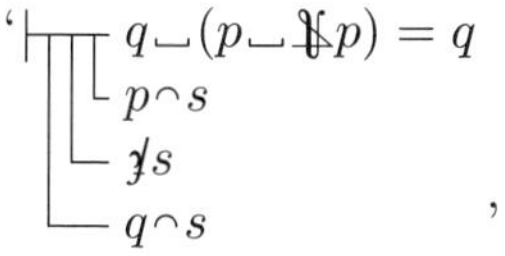

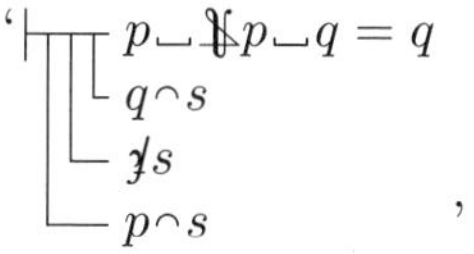

and consequences

§181 to §186. Propositions up to (585) .... page 181

e) Propositions concerning the greater and the lesser in a positival class

§187 to §192. Propositions up to (589) .... page 185

Γ. *The limit*

Definitions of the functions ξ ⅃ ζ and $\xi \wr \zeta$

§193 to §196. There is only one s-limit of u. Propositions up to (602) ... page 187

Δ. *The positive class*

a) Definitions of the function $\mathfrak{p}\xi$ and consequences

§197 and §198. Propositions up to (607) .... page 189

b) Proof of the proposition

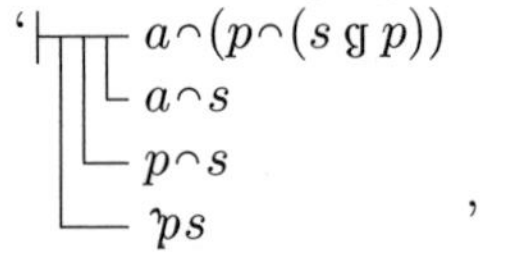

(Archimedean Axiom)

§199 to §214. Definition of the function $\xi \mathfrak{g} \zeta$. Propositions up to (636) .. page 191

E. *Proof of the proposition*

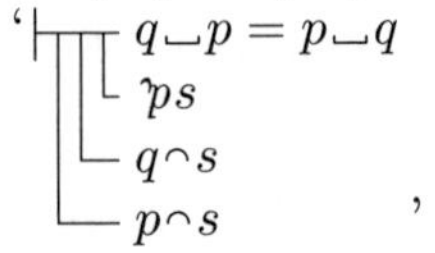

(Commutative Principle in a positive class)

a) Proof of the proposition

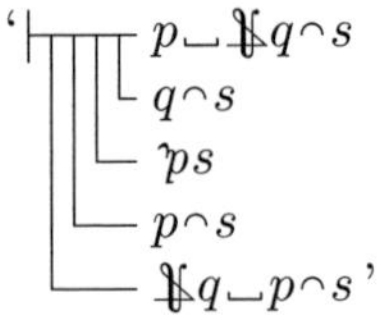

§215 and §216. Propositions up to (638) .... page 204

b) Proof of the proposition

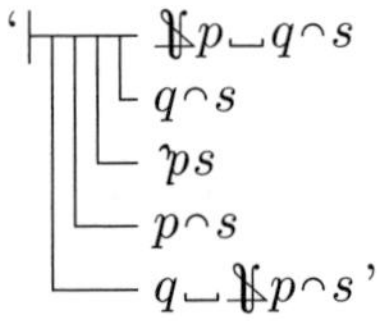

§217 and §218. Propositions up to (641) .... page 207

c) Proof of the proposition

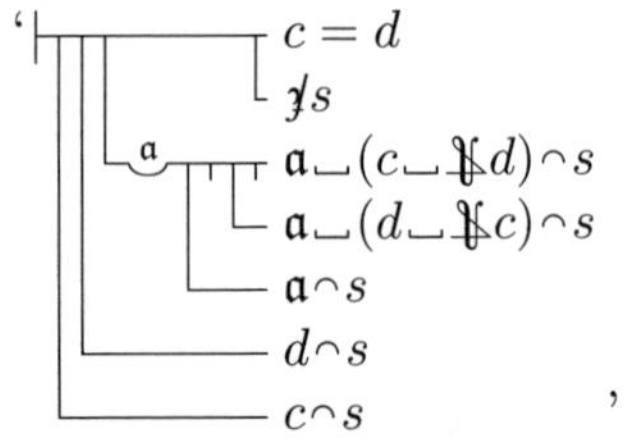

§219 and §220. Propositions up to (644) .... page 209

d) Proof of the proposition

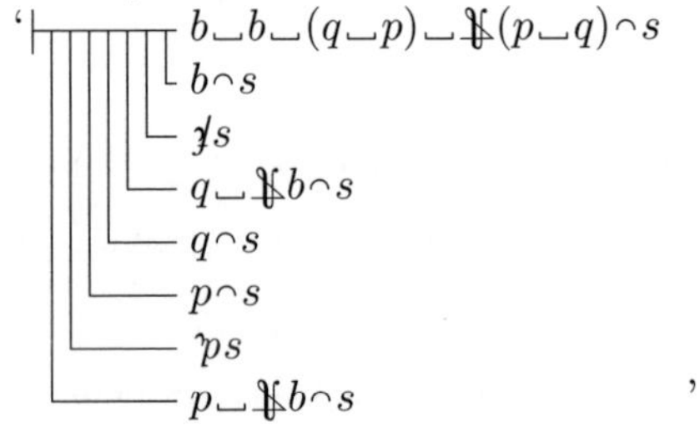

§221 to §230. Propositions up to (666) .... page 211

e) Proof of the proposition

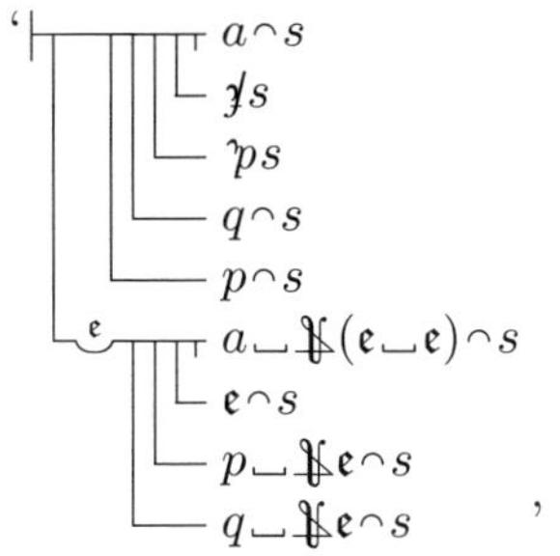

and end of the section E

§231 to §238. Propositions up to (678) .... page 230

Z. Proof of the proposition

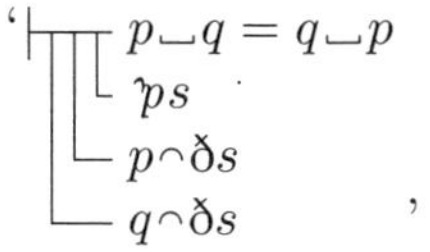

§239 to §244. Propositions up to (689) .... page 239
§245. The next task .... " 243

Appendices

1. Table of definitions .... page 244
2. Table of important theorems .... " 245

Afterword .... " 253

Index .... " 266

Preliminary remark

The two propositions that received the indices δ and η in the derivation of (281) on p. 208 of the first volume are here listed again, in order to assign them different indices for use in what follows.

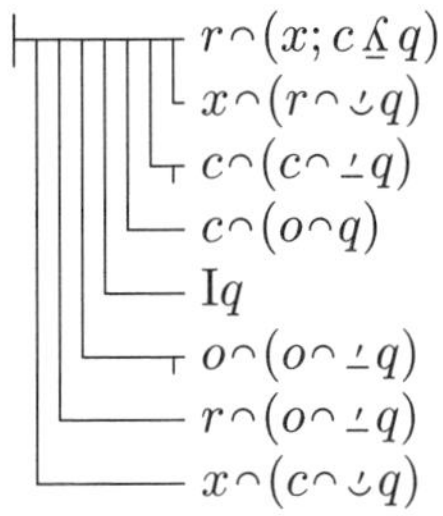

(280a

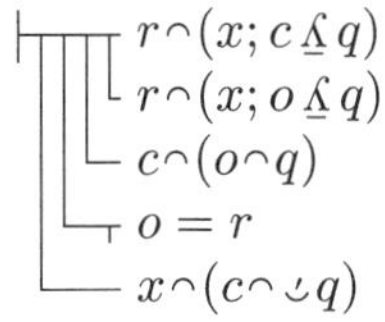

(280b

M. Proof of the proposition

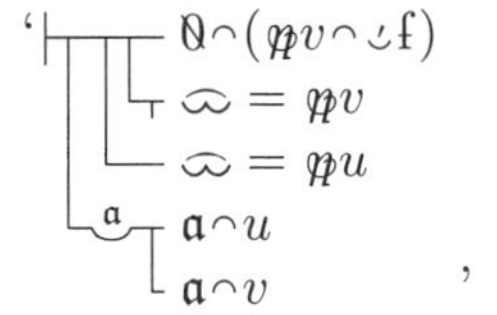

a) Proof of the proposition

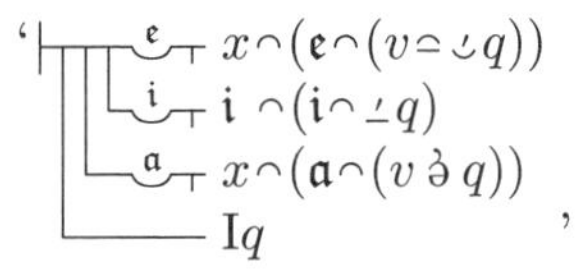

§1. *Analysis*

The proposition mentioned in the main heading can be expressed in words like this:

"If a concept is superordinated to a second and if Endlos is the cardinal number of the first, then the cardinal number of the second is also Endlos or it is finite."

To begin with, we know, according to (207), that the objects falling under the u-concept can be ordered into a simple series that starts with a specific object and, without looping back into itself, proceeds endlessly. If we now select from this series those objects that fall under the v-concept, then it is to be shown that these objects—if there are any—can likewise be ordered into a series that either also proceeds endlessly or else ends with some object. With propositions (263) and (327) we then reach the goal. This ordering is most easily accomplished by starting with the first object of the original series that falls under the v-concept, and then proceeding to the next object of the original series that falls under the v-concept. We first have to show that there is an object in the q-series starting with x that is the first to fall under the v-concept, if there is any object in this series at all that falls under the v-concept. Now, what does it mean to say that "the object y is the first to fall under the v-concept in the q-series starting with x"? y must fall under the v-concept and belong to the q-series starting with x; but no object that precedes it in this series must fall under the v-concept. In our symbolism

$$\tau \begin{array}{l} \Gamma \frown (\Delta \frown (\Theta \fallingdotseq \smile \Upsilon \smile \_ \Upsilon)) \\ \Gamma \frown (\Delta \frown (\Theta \fallingdotseq \smile \Upsilon)) \end{array}$$

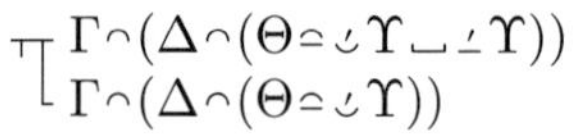

is the truth-value of: that Δ belongs to the Υ-series starting with Γ and falls under the Θ-concept, but that there is no object which belongs to the Υ-series starting with Γ that falls under the Θ-concept and precedes Δ in the Υ-series. We now give this definition:

$$\vdash \dot{\alpha}\dot{\varepsilon}\left(\tau \begin{array}{l} \varepsilon \frown (\mathfrak{a} \frown (v \fallingdotseq \smile q \smile \_ q)) \\ \varepsilon \frown (\mathfrak{a} \frown (v \fallingdotseq \smile q)) \end{array}\right) = v \eth q \quad (\Sigma$$

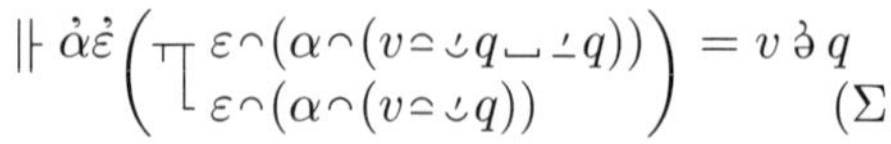

Accordingly, $\Gamma \frown (\Delta \frown (\Theta \eth \Upsilon))$ is the truth-value of: that Δ is the first in the Υ-series starting with Γ to fall under the Θ-concept. In the second heading we thus essentially have our proposition. We will prove it using the proposition

$$`\tau \begin{array}{l} \tau \begin{array}{l} \mathfrak{e} \frown (x; y \mathbin{\underline{\mathfrak{s}}} q) \\ \mathfrak{e} \frown v \end{array} \\ {}_{\mathfrak{a}}\tau \begin{array}{l} x \frown (\mathfrak{a} \frown (v \eth q)) \\ x \frown (y \frown \smile q) \end{array} \end{array} \quad , \quad (\alpha$$

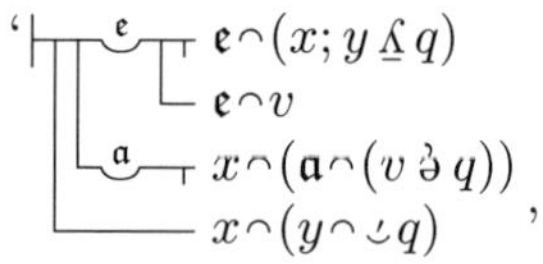

which we express in words like this: "There is no object that belongs to the q-series running from x to y and that falls under the v-concept, if there is no object in the q-series starting with x which is the first to fall under the v-concept, and if y belongs to the q-series starting with x."

Thus, if there is a member of the q-series starting with x that falls under the v-concept, then we can take it as the y in our proposition (α) and so reach our goal of section a. We derive proposition (α) by means of (152) for which we require the proposition

$$`\tau \begin{array}{l} \tau \begin{array}{l} \mathfrak{e} \frown (x; a \mathbin{\underline{\mathfrak{s}}} q) \\ \mathfrak{e} \frown v \\ d \frown (a \frown q) \end{array} \\ \mathfrak{e}\tau \begin{array}{l} \mathfrak{e} \frown (x; d \mathbin{\underline{\mathfrak{s}}} q) \\ \mathfrak{e} \frown v \end{array} \\ {}_{\mathfrak{a}}\tau \begin{array}{l} x \frown (d \frown \smile q) \\ x \frown (\mathfrak{a} \frown (v \eth q)) \end{array} \end{array}' \quad (\beta$$

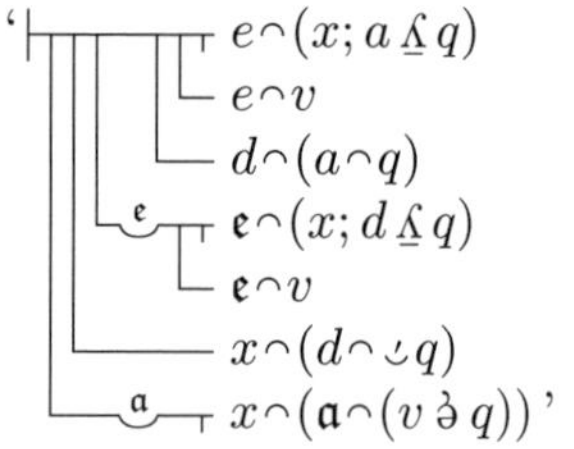

We here distinguish the cases where a falls under the v-concept and where it does not. In the second case, we show by means of proposition (280b)[1] that no member of the q-series running from x to a then falls under the v-concept either.

---

[1] Compare the preliminary remark.

§2. *Construction*

280b
$$\vdash \begin{array}{l} e \frown (x; d \,\underline{\measuredangle}\, q) \\ e \frown (x; a \,\underline{\measuredangle}\, q) \\ d \frown (a \frown q) \\ a = e \\ x \frown (d \frown \smile q) \end{array}$$

$$\times$$

$$\vdash \begin{array}{l} a = e \\ e \frown (x; a \,\underline{\measuredangle}\, q) \\ d \frown (a \frown q) \\ e \frown (x; d \,\underline{\measuredangle}\, q) \\ x \frown (d \frown \smile q) \end{array} \qquad (\alpha$$

(III a): - - - - - - - - -

$$\vdash \begin{array}{l} a \frown v \\ e \frown v \\ e \frown (x; a \,\underline{\measuredangle}\, q) \\ d \frown (a \frown q) \\ e \frown (x; d \,\underline{\measuredangle}\, q) \\ x \frown (d \frown \smile q) \end{array} \qquad (\beta$$

(II a):: - - - - - - - - -

$$\vdash \begin{array}{l} a \frown v \\ e \frown v \\ e \frown (x; a \,\underline{\measuredangle}\, q) \\ d \frown (a \frown q) \\ \overset{\mathfrak{e}}{\smile} \; \mathfrak{e} \frown (x; d \,\underline{\measuredangle}\, q) \\ \quad \mathfrak{e} \frown v \\ x \frown (d \frown \smile q) \end{array} \qquad (349$$

§3. *Analysis*

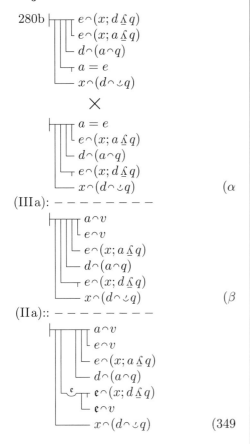

In order to prove proposition (β) of §1, we come to the case where a falls under the v-concept. Then a itself is the first member of the q-series starting with x to fall under the v-concept; there is thus such a member. We have to prove the proposition

$$`\vdash \begin{array}{l} x \frown (a \frown (v \grave{\partial} q)) \\ \overset{\mathfrak{e}}{\smile} \; \mathfrak{e} \frown (x; d \,\underline{\measuredangle}\, q) \\ \quad \mathfrak{e} \frown v \\ d \frown (a \frown q) \\ e \frown (x; a \,\underline{\measuredangle}\, q) \\ x \frown (d \frown \smile q) \\ a \frown v \end{array} \text{'} \qquad (\alpha$$

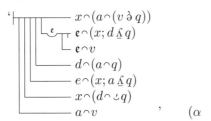

For this we need the proposition

$$`\vdash \begin{array}{l} x \frown (a \frown (v \grave{\partial} q)) \\ \overset{\mathfrak{r}}{\smile} \; x \frown (\mathfrak{r} \frown (v \simeq \smile q)) \\ \quad \mathfrak{r} \frown (a \frown \bot q) \\ d \frown (a \frown q) \\ x \frown (d \frown \smile q) \\ a \frown v \end{array} \text{'} \qquad (\beta$$

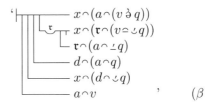

which straightforwardly follows from (Σ).

§4. *Construction*

$$\Sigma \vdash \dot{\alpha}\dot{\varepsilon}\left(\begin{array}{l} \varepsilon \frown (\alpha \frown (v \simeq \smile q \,\llcorner\, \bot q)) \\ \varepsilon \frown (\alpha \frown (v \simeq \smile q)) \end{array} \right) = v \grave{\partial} q$$

(10): ————————————————

$$\vdash \begin{array}{l} x \frown (a \frown (v \grave{\partial} q)) \\ x \frown (a \frown (v \simeq \smile q \,\llcorner\, \bot q)) \\ x \frown (a \frown (v \simeq \smile q)) \end{array} \qquad (\alpha$$

(I f):: - - - - - - - - - - -

$$\vdash \begin{array}{l} x \frown (a \frown (v \grave{\partial} q)) \\ x \frown (a \frown (v \simeq \smile q \,\llcorner\, \bot q)) \\ x \frown (a \frown (v \simeq \smile q)) \end{array} \qquad (\beta$$

(15, 197):: = = = = = = = = =

$$\vdash \begin{array}{l} x \frown (a \frown (v \grave{\partial} q)) \\ \overset{\mathfrak{r}}{\smile} \; x \frown (\mathfrak{r} \frown (v \simeq \smile q)) \\ \quad \mathfrak{r} \frown (a \frown \bot q) \\ x \frown (a \frown \smile q) \\ a \frown v \end{array} \qquad (350$$

(137):: - - - - - - - - - - -

$$\vdash \begin{array}{l} x \frown (a \frown (v \grave{\partial} q)) \\ \overset{\mathfrak{r}}{\smile} \; x \frown (\mathfrak{r} \frown (v \simeq \smile q)) \\ \quad \mathfrak{r} \frown (a \frown \bot q) \\ d \frown (a \frown q) \\ x \frown (d \frown \smile q) \\ a \frown v \end{array} \qquad (351$$

———— • ————

279
$$\vdash \begin{array}{l} d \frown (d \frown \bot q) \\ d \frown (a \frown q) \\ \mathrm{I}q \\ a \frown (a \frown \bot q) \end{array}$$

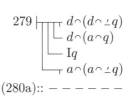

(280a):: - - - - - -

$$\begin{array}{l} c\frown(x;d\,\underline{\lesssim}\,q) \\ x\frown(c\frown\smile q) \\ d\frown(a\frown q) \\ \mathrm{I}q \\ \neg\, a\frown(a\frown\underline{\perp} q) \\ c\frown(a\frown\underline{\perp} q) \\ x\frown(d\frown\smile q) \end{array} \qquad (352$$

(265, 271):: = = = = =

$$\begin{array}{l} c\frown(x;d\,\underline{\lesssim}\,q) \\ x\frown(c\frown\smile q) \\ d\frown(a\frown q) \\ e\frown(x;a\,\underline{\lesssim}\,q) \\ c\frown(a\frown\underline{\perp} q) \\ x\frown(d\frown\smile q) \end{array} \qquad (\alpha$$

(188):: - - - - - - - - -

$$\begin{array}{l} c\frown(x;d\,\underline{\lesssim}\,q) \\ x\frown(c\frown(v\bumpeq\smile q)) \\ d\frown(a\frown q) \\ e\frown(x;a\,\underline{\lesssim}\,q) \\ c\frown(a\frown\underline{\perp} q) \\ x\frown(d\frown\smile q) \end{array} \qquad (\beta$$

×

$$\begin{array}{l} x\frown(c\frown(v\bumpeq\smile q)) \\ \neg\, c\frown(x;d\,\underline{\lesssim}\,q) \\ d\frown(a\frown q) \\ e\frown(x;a\,\underline{\lesssim}\,q) \\ c\frown(a\frown\underline{\perp} q) \\ x\frown(d\frown\smile q) \end{array} \qquad (\gamma$$

(IIa):: - - - - - - - - -

$$\begin{array}{l} x\frown(c\frown(v\bumpeq\smile q)) \\ c\frown v \\ \forall\mathfrak{e}\,[\mathfrak{e}\frown(x;d\,\underline{\lesssim}\,q) \ \text{/}\ \mathfrak{e}\frown v] \\ d\frown(a\frown q) \\ e\frown(x;a\,\underline{\lesssim}\,q) \\ c\frown(a\frown\underline{\perp} q) \\ x\frown(d\frown\smile q) \end{array} \qquad (\delta$$

(192): ·—·—·—·—·—·—·—·—·—·

$$\begin{array}{l} x\frown(c\frown(v\bumpeq\smile q)) \\ c\frown(a\frown\underline{\perp} q) \\ \forall\mathfrak{e}\,[\mathfrak{e}\frown(x;d\,\underline{\lesssim}\,q) \ \text{/}\ \mathfrak{e}\frown v] \\ d\frown(a\frown q) \\ e\frown(x;a\,\underline{\lesssim}\,q) \\ x\frown(d\frown\smile q) \end{array} \qquad (\varepsilon$$

$\smile$

$$\begin{array}{l} \forall\mathfrak{r}\,[x\frown(\mathfrak{r}\frown(v\bumpeq\smile q)) \ \text{/}\ \mathfrak{r}\frown(a\frown\underline{\perp} q)] \\ \forall\mathfrak{e}\,[\mathfrak{e}\frown(x;d\,\underline{\lesssim}\,q) \ \text{/}\ \mathfrak{e}\frown v] \\ d\frown(a\frown q) \\ e\frown(x;a\,\underline{\lesssim}\,q) \\ x\frown(d\frown\smile q) \end{array} \qquad (\zeta$$

(351): - - - - - - - - - - -

$$\begin{array}{l} x\frown(a\frown(v\,\grave{\partial}\,q)) \\ \forall\mathfrak{e}\,[\mathfrak{e}\frown(x;d\,\underline{\lesssim}\,q) \ \text{/}\ \mathfrak{e}\frown v] \\ d\frown(a\frown q) \\ e\frown(x;a\,\underline{\lesssim}\,q) \\ x\frown(d\frown\smile q) \\ a\frown v \end{array} \qquad (\eta$$

(349):: - - - - - - - - - - -

$$\begin{array}{l} x\frown(a\frown(v\,\grave{\partial}\,q)) \\ \forall\mathfrak{e}\,[\mathfrak{e}\frown(x;d\,\underline{\lesssim}\,q) \ \text{/}\ \mathfrak{e}\frown v] \\ d\frown(a\frown q) \\ e\frown(x;a\,\underline{\lesssim}\,q) \\ x\frown(d\frown\smile q) \\ e\frown v \end{array} \qquad (\vartheta$$

×

$$\begin{array}{l} e\frown(x;a\,\underline{\lesssim}\,q) \\ e\frown v \\ d\frown(a\frown q) \\ \forall\mathfrak{e}\,[\mathfrak{e}\frown(x;d\,\underline{\lesssim}\,q) \ \text{/}\ \mathfrak{e}\frown v] \\ x\frown(d\frown\smile q) \\ \neg\, x\frown(a\frown(v\,\grave{\partial}\,q)) \end{array} \qquad (\iota$$

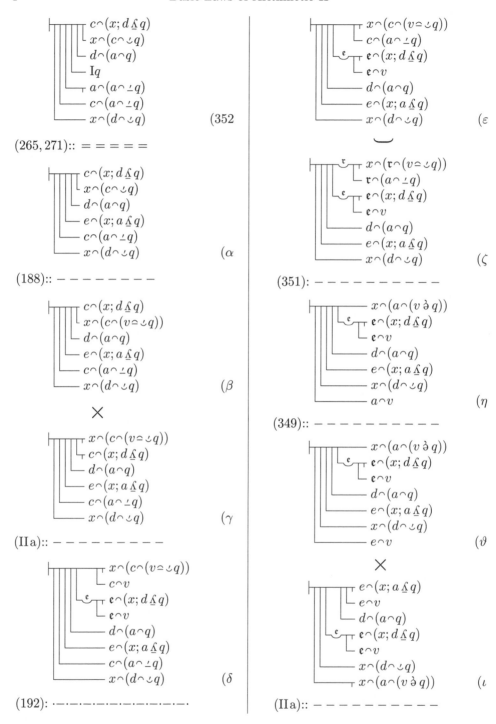

(IIa):: - - - - - - - - - - -

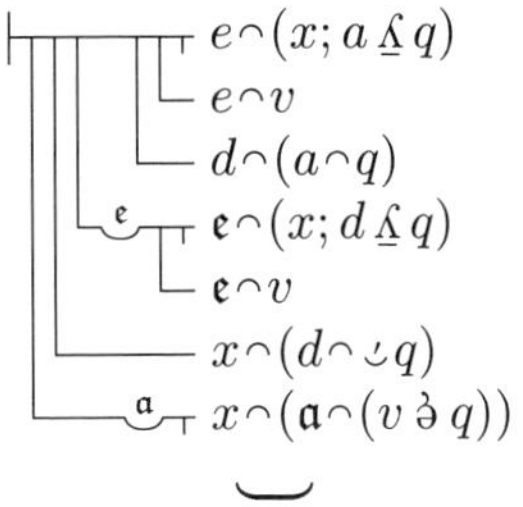

(κ

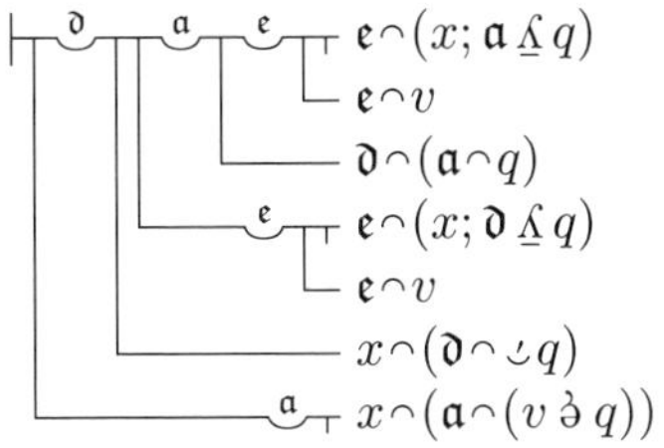

(λ

(152): - - - - - - - - - - - - -

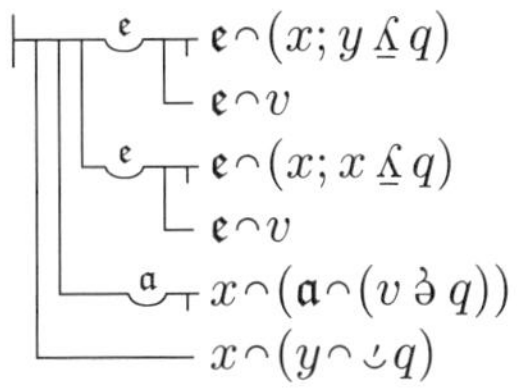

(353

§5. *Analysis*

We have to eliminate the subcomponent

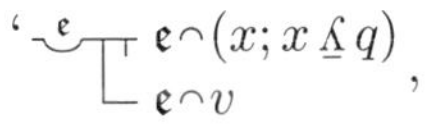

by showing that x itself would fall under the v-concept if there were a member of the q-series running from x to x which fell under the v-concept. In that case, x would be the member of the q-series starting with x that was the first to fall under the v-concept. Using (282), we prove the proposition

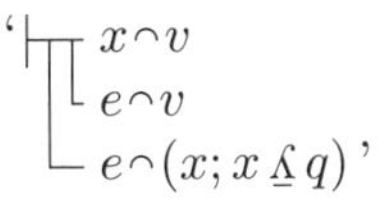

(α

In order to apply (350), we then still have to derive the proposition

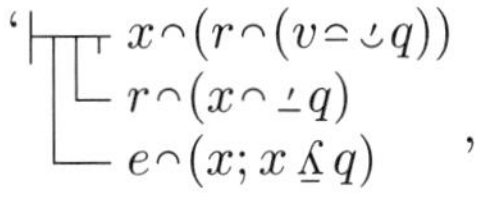

(β

§6. *Construction*

188

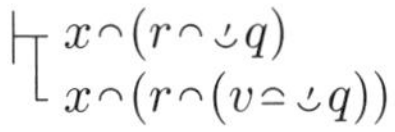

(280): - - - - - - - -

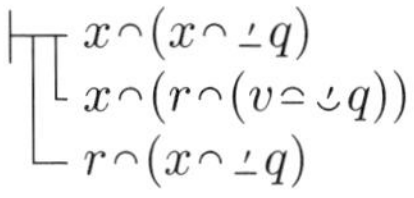

(α

×

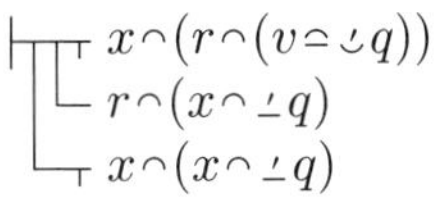

(β

(271):: - - - - - - - - -

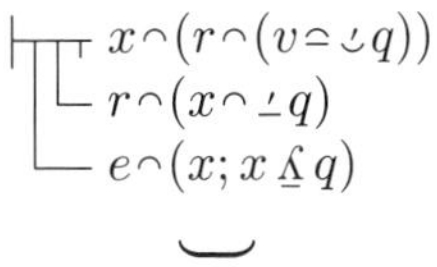

(γ

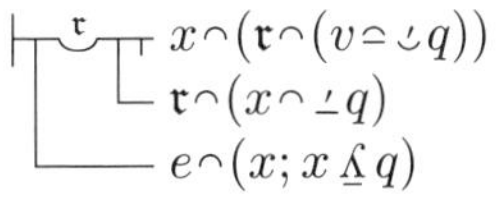

(δ

(350): - - - - - - - - - - -

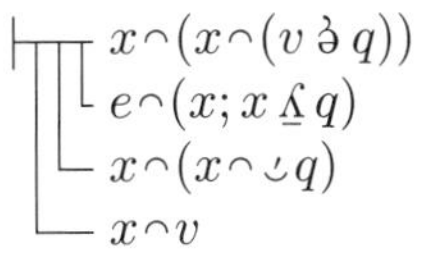

(ε

(140):: - - - - - - - - -

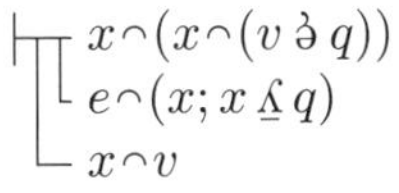

(ζ

282

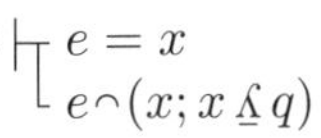

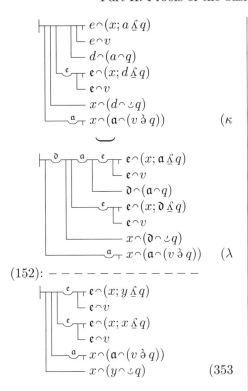

(IIIc): - - - - - - -

$x \frown v$
$e \frown v$
$e \frown (x; x \underline{\measuredangle} q)$ (η

(ζ): - - - - - -

$x \frown (x \frown (v \eth q))$
$e \frown (x; x \underline{\measuredangle} q)$
$e \frown v$ (ϑ

×

$e \frown (x; x \underline{\measuredangle} q)$
$e \frown v$
$x \frown (x \frown (v \eth q))$ (ι

(IIa):: - - - - - - - - -

$e \frown (x; x \underline{\measuredangle} q)$
$e \frown v$
$\mathfrak{a}$ $x \frown (\mathfrak{a} \frown (v \eth q))$ (κ

$\mathfrak{e}$ $\mathfrak{e} \frown (x; x \underline{\measuredangle} q)$
$\mathfrak{e} \frown v$
$\mathfrak{a}$ $x \frown (\mathfrak{a} \frown (v \eth q))$ (λ

(353): - - - - - - - - - -

$\mathfrak{e}$ $\mathfrak{e} \frown (x; y \underline{\measuredangle} q)$
$\mathfrak{e} \frown v$
$\mathfrak{a}$ $x \frown (\mathfrak{a} \frown (v \eth q))$
$x \frown (y \frown \breve{} q)$ (354

×

$x \frown (y \frown \breve{} q)$
$\mathfrak{a}$ $x \frown (\mathfrak{a} \frown (v \eth q))$
$\mathfrak{e}$ $\mathfrak{e} \frown (x; y \underline{\measuredangle} q)$
$\mathfrak{e} \frown v$ (355

——— • ———

187 $x \frown (y \frown p)$
$y \frown v$
$x \frown (y \frown (v \doteqdot p))$

×

$x \frown (y \frown (v \doteqdot p))$
$x \frown (y \frown p)$
$y \frown v$ (356

——— • ———

344 $y \frown (x; y \underline{\measuredangle} q)$
$x \frown (y \frown \breve{} q)$
$y \frown (y \frown \perp q)$
$\mathrm{I}q$

×

$x \frown (y \frown \breve{} q)$
$y \frown (x; y \underline{\measuredangle} q)$
$y \frown (y \frown \perp q)$
$\mathrm{I}q$ (α

(IIa):: - - - - - - -

$x \frown (y \frown \breve{} q)$
$y \frown v$
$\mathfrak{e}$ $\mathfrak{e} \frown (x; y \underline{\measuredangle} q)$
$\mathfrak{e} \frown v$
$y \frown (y \frown \perp q)$
$\mathrm{I}q$ (β

(355): ·–·–·–·–·–·–·–·

$x \frown (y \frown \breve{} q)$
$y \frown v$
$y \frown (y \frown \perp q)$
$\mathrm{I}q$
$\mathfrak{a}$ $x \frown (\mathfrak{a} \frown (v \eth q))$ (357

(356): - - - - - - - - - -

$x \frown (y \frown (v \doteqdot \breve{} q))$
$y \frown (y \frown \perp q)$
$\mathrm{I}q$
$\mathfrak{a}$ $x \frown (\mathfrak{a} \frown (v \eth q))$ (358

(IIa):: - - - - - - - - - -

$x \frown (y \frown (v \doteqdot \breve{} q))$
$\mathfrak{i}$ $\mathfrak{i} \frown (\mathfrak{i} \frown \perp q)$
$\mathfrak{a}$ $x \frown (\mathfrak{a} \frown (v \eth q))$
$\mathrm{I}q$ (358a

$\mathfrak{e}$ $x \frown (\mathfrak{e} \frown (v \doteqdot \breve{} q))$
$\mathfrak{i}$ $\mathfrak{i} \frown (\mathfrak{i} \frown \perp q)$
$\mathfrak{a}$ $x \frown (\mathfrak{a} \frown (v \eth q))$
$\mathrm{I}q$ (359

b) Proof of the proposition

$$\text{‘}\vdash\!\!\begin{array}{l} \infty = \mathfrak{p}(x \frown \mathfrak{f}(v \asymp \smile q)) \\ \infty = \mathfrak{p}(m \frown \mathfrak{f}\smile(v \,\check{\jmath}\, q)) \\ x \frown (m \frown (v \,\grave{\jmath}\, q)) \\ \mathrm{I}q \\ \smile^{\mathfrak{i}}\!\top\, \mathfrak{i} \frown (\mathfrak{i} \frown \angle q) \end{array} \text{’}$$

§7. *Analysis*

We now tie in with §1 again, by defining a relation by means of which we can order the members of the q-series starting with x that fall under the v-concept, as was stated there:

$$\Vdash \dot{\alpha}\dot{\varepsilon}\left(\begin{array}{l} \varepsilon \frown (\alpha \frown (v \asymp \angle q \smile \angle q)) \\ \varepsilon \frown (\alpha \frown (v \asymp \angle q)) \end{array}\right) = v \,\check{\jmath}\, q \qquad \text{(T}$$

By ‘m’ we indicate the member of the q-series starting with x that is the first to fall under the v-concept; we have to prove the two propositions

$$\text{‘}\vdash\!\!\begin{array}{l} x \frown (c \frown (v \asymp \smile q)) \\ m \frown (c \frown \smile(v \,\check{\jmath}\, q)) \\ x \frown (m \frown (v \,\grave{\jmath}\, q)) \end{array} \text{’} \qquad (\alpha$$

$$\text{‘}\vdash\!\!\begin{array}{l} m \frown (c \frown \smile(v \,\check{\jmath}\, q)) \\ x \frown (c \frown (v \asymp \smile q)) \\ x \frown (m \frown (v \,\grave{\jmath}\, q)) \\ \mathrm{I}q \\ \smile^{\mathfrak{i}}\!\top\, \mathfrak{i} \frown (\mathfrak{i} \frown \angle q) \end{array} \text{’} \qquad (\beta$$

from which it will follow that the $(v \,\check{\jmath}\, q)$-relation in fact provides the required ordering, and so that Endlos is the cardinal number of the members of the q-series starting with x that fall under the v-concept if Endlos is the cardinal number of the members of the $(v \,\check{\jmath}\, q)$-series starting with m, m is the first member of the q-series starting with x to fall under the v-concept, the q-relation is single-valued, and no object follows after itself in the q-series. This is the proposition mentioned in the heading. We prove (α) using (144).

§8. *Construction*

$$\mathrm{T} \vdash \dot{\alpha}\dot{\varepsilon}\left(\begin{array}{l} \varepsilon \frown (\alpha \frown (v \asymp \angle q \smile \angle q)) \\ \varepsilon \frown (\alpha \frown (v \asymp \angle q)) \end{array}\right) = v \,\check{\jmath}\, q$$

(14): ———————————————

$$\vdash\!\!\begin{array}{l} d \frown (a \frown (v \,\check{\jmath}\, q)) \\ d \frown (a \frown (v \asymp \angle q \smile \angle q)) \\ d \frown (a \frown (v \asymp \angle q)) \end{array} \qquad (360$$

(Ia):: — — — — — — — — — — —

$$\vdash\!\!\begin{array}{l} d \frown (a \frown (v \,\check{\jmath}\, q)) \\ d \frown (a \frown (v \asymp \angle q)) \end{array} \qquad (\alpha$$

×

$$\vdash\!\!\begin{array}{l} d \frown (a \frown (v \asymp \angle q)) \\ d \frown (a \frown (v \,\check{\jmath}\, q)) \end{array} \qquad (361$$

(188): — — — — — — —

$$\vdash\!\!\begin{array}{l} d \frown (a \frown \angle q) \\ d \frown (a \frown (v \,\check{\jmath}\, q)) \end{array} \qquad (362$$

(280): — — — — — — —

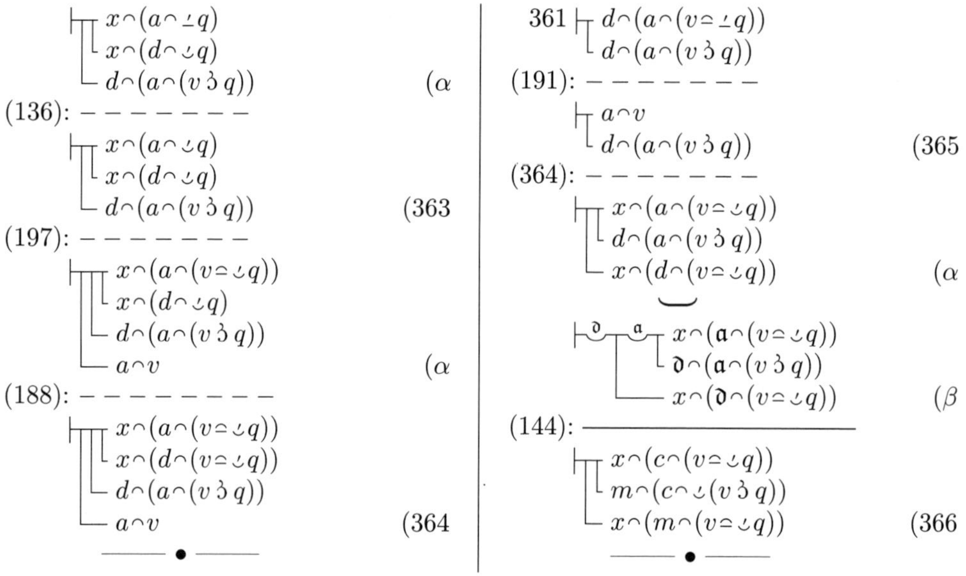

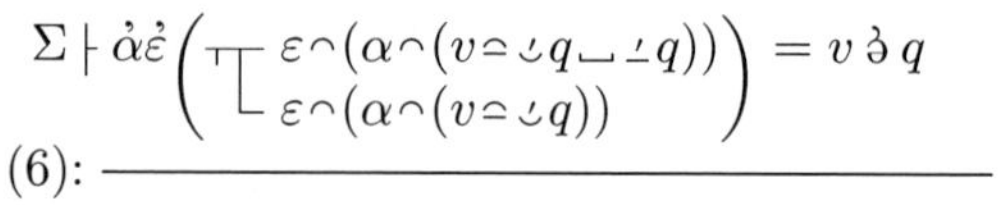

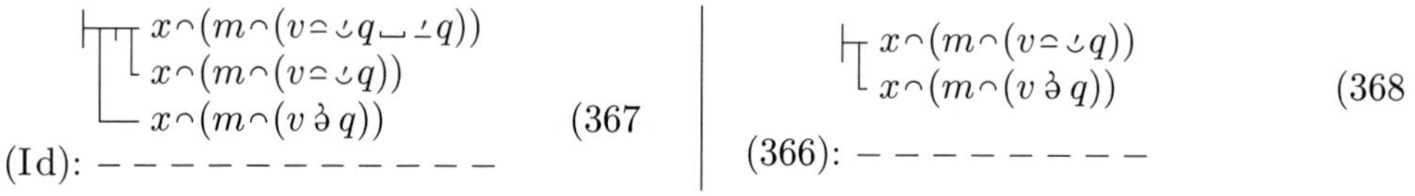

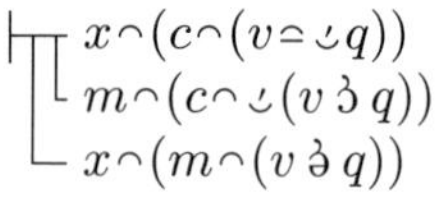

(369

§9. *Analysis*

In order to prove proposition (β) of §7 we first derive the proposition

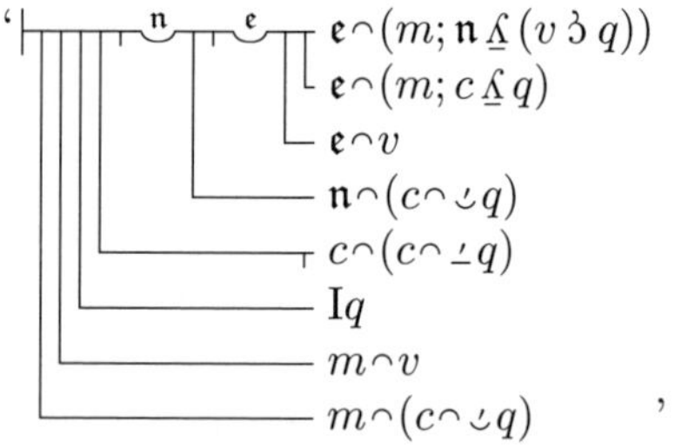

(α

i.e., we show that, under certain conditions, the q-series ending with c contains a member such that every member of the q-series running from m to c that falls under the v-concept belongs to the $(v \mathbin{\smallfrown\!\!\!\!\!\,\raisebox{0pt}{}} q)$-series running from m to that member. From this proposition we then conclude that if c falls under the v-concept and belongs to the q-series starting with m, it also belongs to the $(v \mathbin{\smallfrown\!\!\!\!\!\,\raisebox{0pt}{}} q)$-series starting with m. The condition that c belongs to the q-series starting with m can be replaced by the condition that it belongs to the q-series starting with x if m is the first member of the q-series starting with x which falls under the v-concept. We now prove the proposition (α) by means of (152) and for this require the proposition

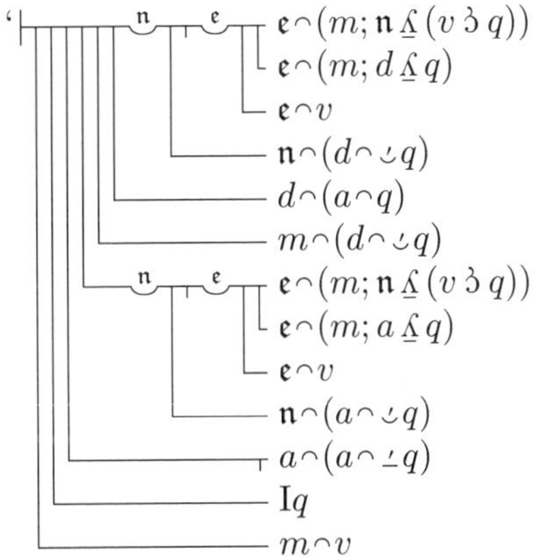

’ $(\beta$

In order to derive this, we distinguish the cases where a falls under the v-concept and where it does not. In the latter we have the proposition

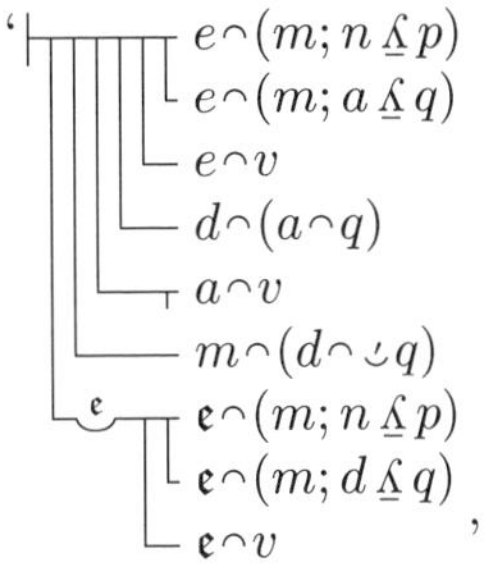

$(\gamma$

in which ‘p’ is written instead of ‘$v \mathbin{\smallfrown\!\!\!\!\!\,\raisebox{0pt}{}} q$’. This follows straightforwardly from (280b).

§10. *Construction*

IIIb 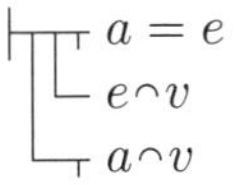

(280b): – – – – –

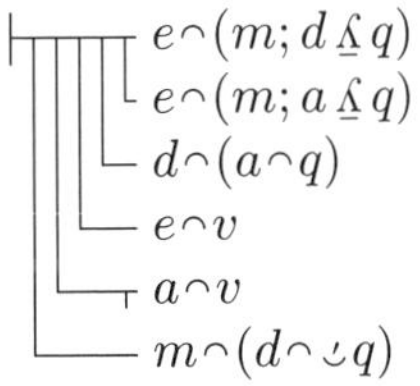

$(\alpha$

(IIa): – – – – – – – –

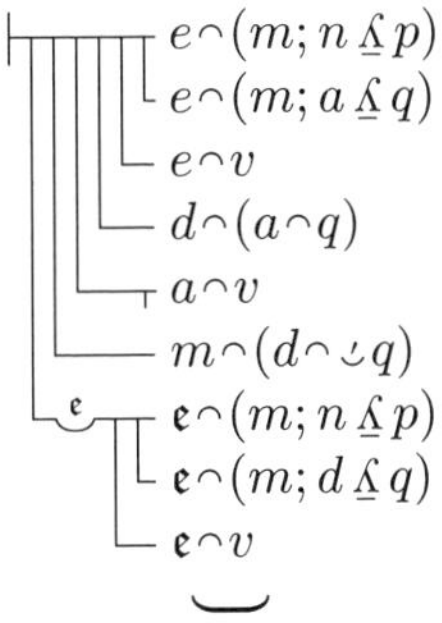

$(\beta$

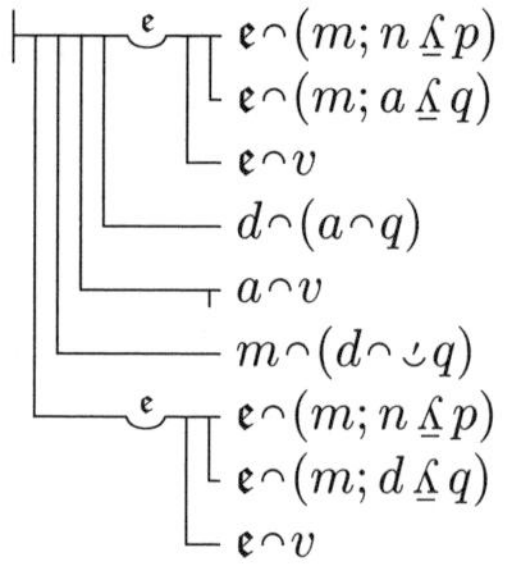
(γ

×

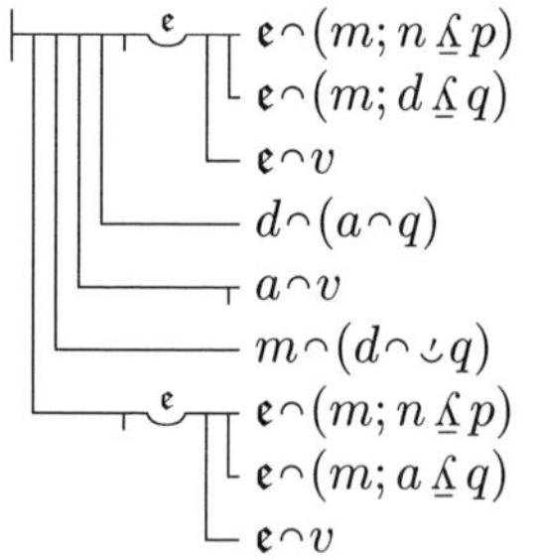
(δ

(IIa):: – – – – – – – – – – –

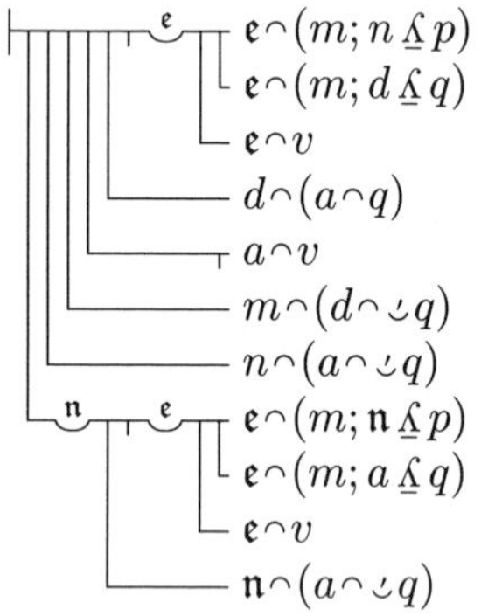
(ε

(137):: – – – – – – – – – – –

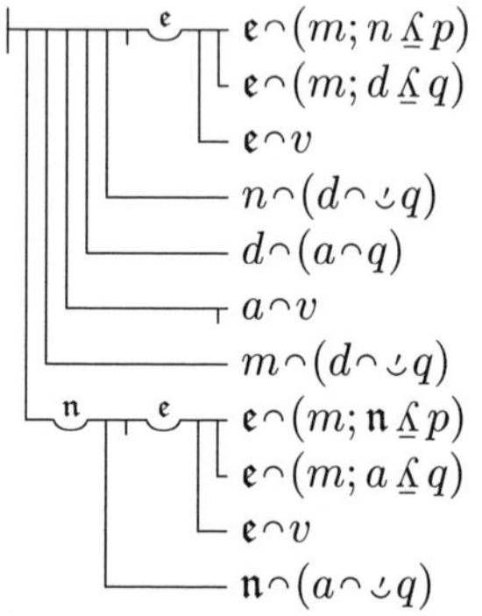
(370

§11. *Analysis*

In order to prove proposition (β) of §9 we now focus on the case where a falls under the v-concept and derive the proposition

‘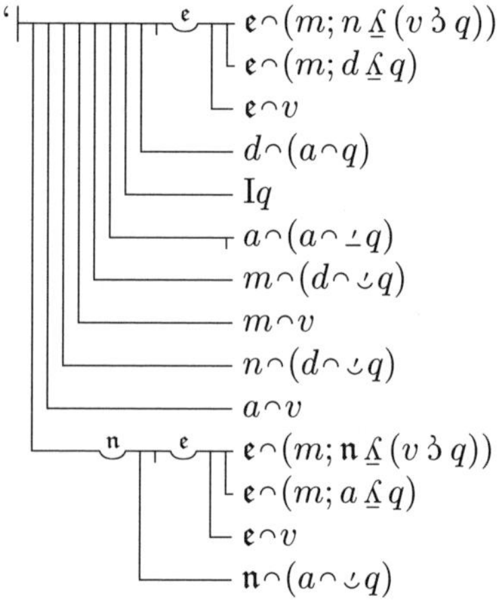’ (α

Distinguishing the case where n stands to a in the $(v \mathfrak{Z}\, q)$-relation from its opposite, we prove the two propositions

‘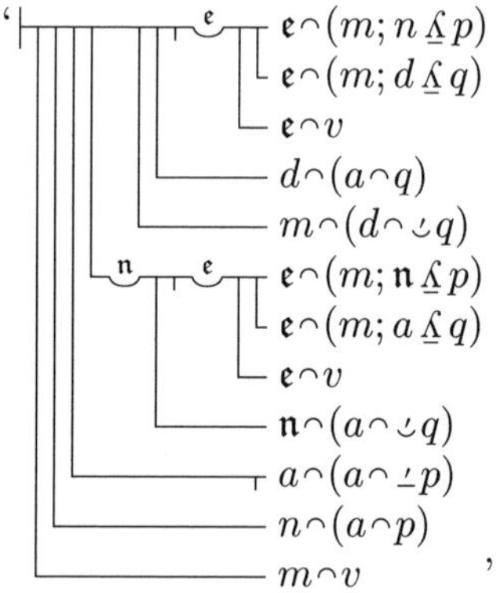’ (β

where ‘p’ takes the place of ‘$(v \mathfrak{Z}\, q)$’, and

$\mathfrak{e}\frown(m; n \,\underline{\wedge}\, (v \,\eth\, q))$
$\mathfrak{e}\frown(m; d \,\underline{\wedge}\, q)$
$\mathfrak{e}\frown v$
$d\frown(a\frown q)$
$\mathrm{I}q$
$a\frown(a\frown \underline{\cdot}\, q)$
$m\frown(d\frown \smile q)$
$m\frown v$
$n\frown(a\frown(v \,\eth\, q))$
$n\frown(d\frown \smile q)$
$a\frown v$ ' (γ

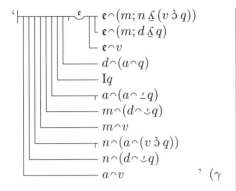

For the proof of (β) we again distinguish the case where e coincides with a from its opposite. We therefore have to prove the propositions

$a\frown(m; a \,\underline{\wedge}\, p)$
$n\frown(a\frown p)$
$d\frown(a\frown q)$
$e\frown(m; a \,\underline{\wedge}\, q)$
$m\frown(d\frown \smile q)$
$m\frown v$
$\mathfrak{e}\frown(m; n \,\underline{\wedge}\, p)$
$\mathfrak{e}\frown(m; d \,\underline{\wedge}\, q)$
$\mathfrak{e}\frown v$
$a\frown(a\frown \underline{\cdot}\, p)$ ' (δ

$e\frown(m; a \,\underline{\wedge}\, p)$
$e\frown(m; a \,\underline{\wedge}\, q)$
$d\frown(a\frown q)$
$a = e$
$m\frown(d\frown \smile q)$
$e\frown v$
$\mathfrak{e}\frown(m; n \,\underline{\wedge}\, p)$
$\mathfrak{e}\frown(m; d \,\underline{\wedge}\, q)$
$\mathfrak{e}\frown v$
$a\frown(a\frown \underline{\cdot}\, p)$
$n\frown(a\frown p)$ ' (ε

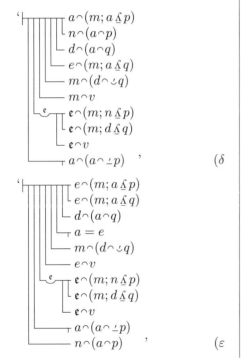

The former is to be reduced to (344) and for this purpose we have to show that, given our assumptions, a belongs to the p-series starting with m. For this we show that m belongs to the p-series running from m to n, from which it follows that n belongs to the p-series starting with m (269).

§12. *Construction*

342 $m\frown(d\frown \smile q)$
$d\frown(d\frown \underline{\cdot}\, q)$
$m\frown(m; d \,\underline{\wedge}\, q)$
$\mathrm{I}q$

×

$m\frown(m; d \,\underline{\wedge}\, q)$
$d\frown(d\frown \underline{\cdot}\, q)$
$m\frown(d\frown \smile q)$
$\mathrm{I}q$ (371

(279):: - - - - - - - -

$m\frown(m; d \,\underline{\wedge}\, q)$
$d\frown(a\frown q)$
$\mathrm{I}q$
$a\frown(a\frown \underline{\cdot}\, q)$
$m\frown(d\frown \smile q)$ (372

(271, 265):: = = = = =

$m\frown(m; d \,\underline{\wedge}\, q)$
$d\frown(a\frown q)$
$e\frown(m; a \,\underline{\wedge}\, q)$
$m\frown(d\frown \smile q)$ (373

•

269 $m\frown(n\frown \smile p)$
$m\frown(m; n \,\underline{\wedge}\, p)$

(137): - - - - - - - -

$m\frown(a\frown \smile p)$
$n\frown(a\frown p)$
$m\frown(m; n \,\underline{\wedge}\, p)$ (α

(344): - - - - - - -

$a\frown(m; a \,\underline{\wedge}\, p)$
$n\frown(a\frown p)$
$m\frown(m; n \,\underline{\wedge}\, p)$
$a\frown(a\frown \underline{\cdot}\, p)$
$\mathrm{I}p$ (β

(265):: - - - - - - - -

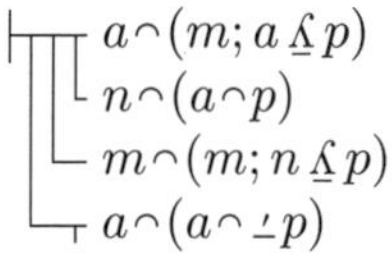

$$\begin{array}{l} \vdash a\frown(m;a\,\underline{\measuredangle}\,p) \\ \quad n\frown(a\frown p) \\ \quad m\frown(m;n\,\underline{\measuredangle}\,p) \\ \quad \top\, a\frown(a\frown\dot{-}p) \end{array}$$ (γ

(IIa):: - - - - - - - - -

$$\begin{array}{l} \vdash a\frown(m;a\,\underline{\measuredangle}\,p) \\ \quad n\frown(a\frown p) \\ \quad m\frown(m;d\,\underline{\measuredangle}\,q) \\ \quad m\frown v \\ \quad \overset{\mathfrak{e}}{\smile}\top\, \mathfrak{e}\frown(m;n\,\underline{\measuredangle}\,p) \\ \qquad \mathfrak{e}\frown(m;d\,\underline{\measuredangle}\,q) \\ \qquad \mathfrak{e}\frown v \\ \quad \top\, a\frown(a\frown\dot{-}p) \end{array}$$ (δ

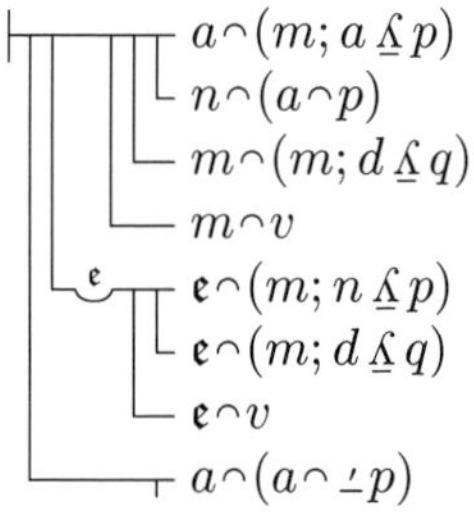

(373):: - - - - - - - - - -

$$\begin{array}{l} \vdash a\frown(m;a\,\underline{\measuredangle}\,p) \\ \quad n\frown(a\frown p) \\ \quad d\frown(a\frown q) \\ \quad e\frown(m;a\,\underline{\measuredangle}\,q) \\ \quad m\frown(d\frown\dot{\smile}q) \\ \quad m\frown v \\ \quad \overset{\mathfrak{e}}{\smile}\top\, \mathfrak{e}\frown(m;n\,\underline{\measuredangle}\,p) \\ \qquad \mathfrak{e}\frown(m;d\,\underline{\measuredangle}\,q) \\ \qquad \mathfrak{e}\frown v \\ \quad \top\, a\frown(a\frown\dot{-}p) \end{array}$$ (ε

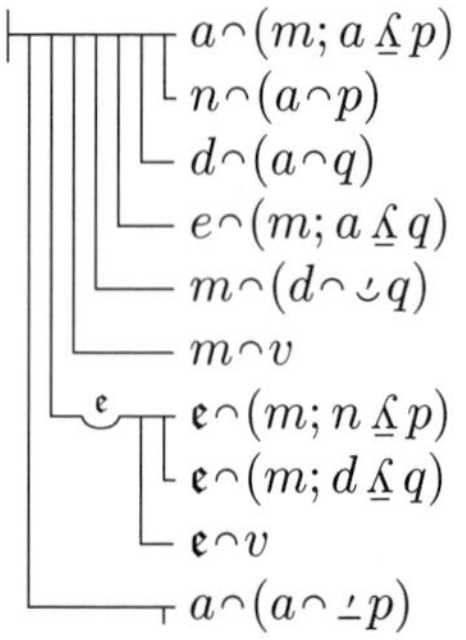

(IIIc): ————————————

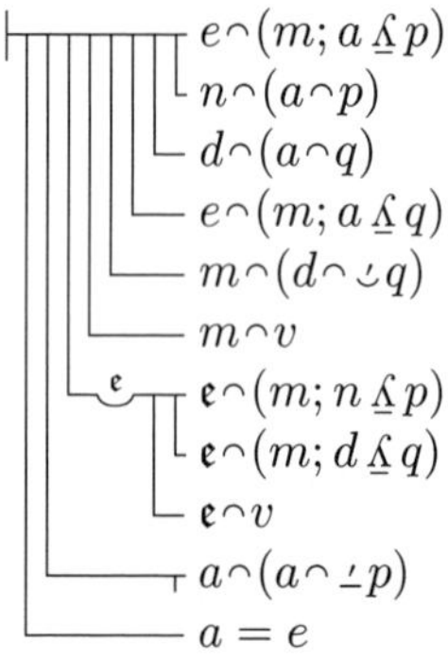

$$\begin{array}{l} \vdash e\frown(m;a\,\underline{\measuredangle}\,p) \\ \quad n\frown(a\frown p) \\ \quad d\frown(a\frown q) \\ \quad e\frown(m;a\,\underline{\measuredangle}\,q) \\ \quad m\frown(d\frown\dot{\smile}q) \\ \quad m\frown v \\ \quad \overset{\mathfrak{e}}{\smile}\top\, \mathfrak{e}\frown(m;n\,\underline{\measuredangle}\,p) \\ \qquad \mathfrak{e}\frown(m;d\,\underline{\measuredangle}\,q) \\ \qquad \mathfrak{e}\frown v \\ \quad \top\, a\frown(a\frown\dot{-}p) \\ \quad a = e \end{array}$$ (374

§13. *Analysis*

We now prove proposition (ε) of §11. Proposition (280b) serves this purpose for us together with the proposition

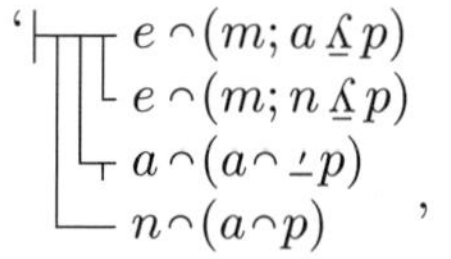

$$\begin{array}{l} \text{‘}\vdash e\frown(m;a\,\underline{\measuredangle}\,p) \\ \quad e\frown(m;n\,\underline{\measuredangle}\,p) \\ \quad \top\, a\frown(a\frown\dot{-}p) \\ \quad n\frown(a\frown p)\text{’}, \end{array}$$ (α

which can straightforwardly be deduced using (137).

§14. *Construction*

$$\begin{array}{l} 137 \vdash e\frown(a\frown\dot{\smile}p) \\ \quad n\frown(a\frown p) \\ \quad e\frown(n\frown\dot{\smile}p) \end{array}$$

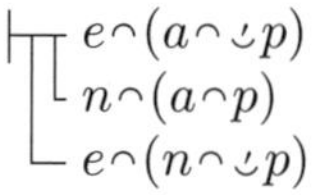

(269):: - - - - - -

$$\begin{array}{l} \vdash e\frown(a\frown\dot{\smile}p) \\ \quad n\frown(a\frown p) \\ \quad e\frown(m;n\,\underline{\measuredangle}\,p) \end{array}$$ (α

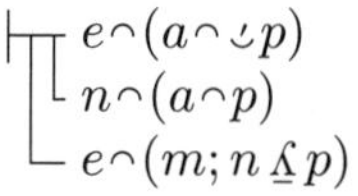

(274): - - - - - - -

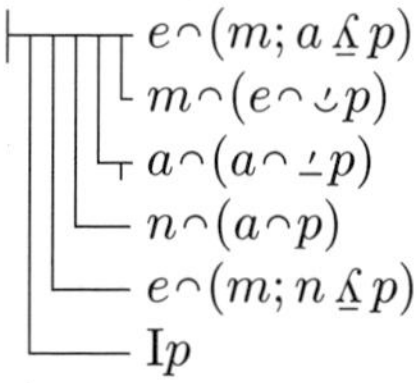

$$\begin{array}{l} \vdash e\frown(m;a\,\underline{\measuredangle}\,p) \\ \quad m\frown(e\frown\dot{\smile}p) \\ \quad \top\, a\frown(a\frown\dot{-}p) \\ \quad n\frown(a\frown p) \\ \quad e\frown(m;n\,\underline{\measuredangle}\,p) \\ \quad \mathrm{I}p \end{array}$$ (β

(270, 265):: = = = = = =

$$\begin{array}{l} \vdash e\frown(m;a\,\underline{\measuredangle}\,p) \\ \quad e\frown(m;n\,\underline{\measuredangle}\,p) \\ \quad \top\, a\frown(a\frown\dot{-}p) \\ \quad n\frown(a\frown p) \end{array}$$ (375

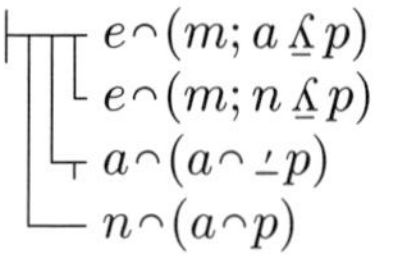

(IIa):: - - - - - - -

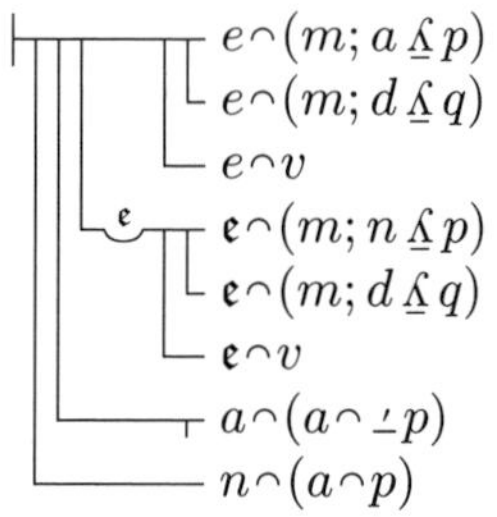

$$\begin{array}{l} \vdash e\frown(m;a\,\underline{\measuredangle}\,p) \\ \quad e\frown(m;d\,\underline{\measuredangle}\,q) \\ \quad e\frown v \\ \quad \overset{\mathfrak{e}}{\smile}\top\, \mathfrak{e}\frown(m;n\,\underline{\measuredangle}\,p) \\ \qquad \mathfrak{e}\frown(m;d\,\underline{\measuredangle}\,q) \\ \qquad \mathfrak{e}\frown v \\ \quad \top\, a\frown(a\frown\dot{-}p) \\ \quad n\frown(a\frown p) \end{array}$$ (α

(280b):: - - - - - - - - - -

$e \frown (m; a \,\underline{\measuredangle}\, p)$
$e \frown (m; a \,\underline{\measuredangle}\, q)$
$d \frown (a \frown q)$
$a = e$
$m \frown (d \frown \smile q)$
$e \frown v$
$\mathfrak{e}$: $\mathfrak{e} \frown (m; n \,\underline{\measuredangle}\, p)$
$\mathfrak{e} \frown (m; d \,\underline{\measuredangle}\, q)$
$\mathfrak{e} \frown v$
$a \frown (a \frown \perp p)$
$n \frown (a \frown p)$ (β

(374): ·—·—·—·—·—·—·—·

$e \frown (m; a \,\underline{\measuredangle}\, p)$
$e \frown (m; a \,\underline{\measuredangle}\, q)$
$e \frown v$
$d \frown (a \frown q)$
$m \frown (d \frown \smile q)$
$\mathfrak{e}$: $\mathfrak{e} \frown (m; n \,\underline{\measuredangle}\, p)$
$\mathfrak{e} \frown (m; d \,\underline{\measuredangle}\, q)$
$\mathfrak{e} \frown v$
$a \frown (a \frown \perp p)$
$n \frown (a \frown p)$
$m \frown v$ (γ

$\smile$

$\mathfrak{e}$: $\mathfrak{e} \frown (m; a \,\underline{\measuredangle}\, p)$
$\mathfrak{e} \frown (m; a \,\underline{\measuredangle}\, q)$
$\mathfrak{e} \frown v$
$d \frown (a \frown q)$
$m \frown (d \frown \smile q)$
$\mathfrak{e}$: $\mathfrak{e} \frown (m; n \,\underline{\measuredangle}\, p)$
$\mathfrak{e} \frown (m; d \,\underline{\measuredangle}\, q)$
$\mathfrak{e} \frown v$
$a \frown (a \frown \perp p)$
$n \frown (a \frown p)$
$m \frown v$ (δ

×

$\mathfrak{e}$: $\mathfrak{e} \frown (m; n \,\underline{\measuredangle}\, p)$
$\mathfrak{e} \frown (m; d \,\underline{\measuredangle}\, q)$
$\mathfrak{e} \frown v$
$d \frown (a \frown q)$
$m \frown (d \frown \smile q)$
$\mathfrak{e}$: $\mathfrak{e} \frown (m; a \,\underline{\measuredangle}\, p)$
$\mathfrak{e} \frown (m; a \,\underline{\measuredangle}\, q)$
$\mathfrak{e} \frown v$
$a \frown (a \frown \perp p)$
$n \frown (a \frown p)$
$m \frown v$ (ε

(IIa):: — — — — — — — — — — — —

$\mathfrak{e}$: $\mathfrak{e} \frown (m; n \,\underline{\measuredangle}\, p)$
$\mathfrak{e} \frown (m; d \,\underline{\measuredangle}\, q)$
$\mathfrak{e} \frown v$
$d \frown (a \frown q)$
$m \frown (d \frown \smile q)$
$a \frown (a \frown \smile q)$
$\mathfrak{n}$, $\mathfrak{e}$: $\mathfrak{e} \frown (m; \mathfrak{n} \,\underline{\measuredangle}\, p)$
$\mathfrak{e} \frown (m; a \,\underline{\measuredangle}\, q)$
$\mathfrak{e} \frown v$
$\mathfrak{n} \frown (a \frown \smile q)$
$a \frown (a \frown \perp p)$
$n \frown (a \frown p)$
$m \frown v$ (ζ

(140):: ————————————————

$\mathfrak{e}$: $\mathfrak{e} \frown (m; n \,\underline{\measuredangle}\, p)$
$\mathfrak{e} \frown (m; d \,\underline{\measuredangle}\, q)$
$\mathfrak{e} \frown v$
$d \frown (a \frown q)$
$m \frown (d \frown \smile q)$
$\mathfrak{n}$, $\mathfrak{e}$: $\mathfrak{e} \frown (m; \mathfrak{n} \,\underline{\measuredangle}\, p)$
$\mathfrak{e} \frown (m; a \,\underline{\measuredangle}\, q)$
$\mathfrak{e} \frown v$
$\mathfrak{n} \frown (a \frown \smile q)$
$a \frown (a \frown \perp p)$
$n \frown (a \frown p)$
$m \frown v$ (376

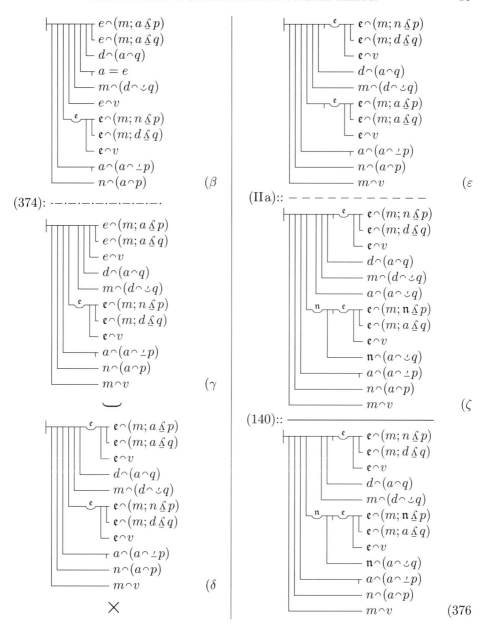

§15. *Analysis*

We now have to prove proposition (γ) of §11. For this we require the proposition

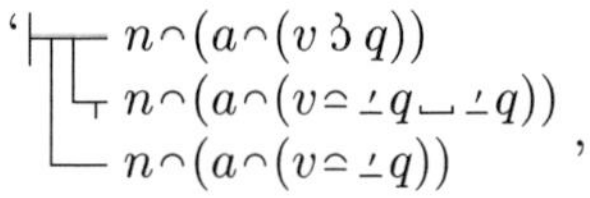

(α

which can straightforwardly be derived from (T). At this point, the subcomponent

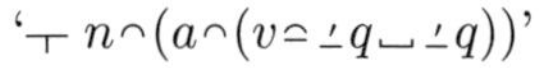

remains to be eliminated. It thus must be shown that there can be no member that follows n in the q-series and precedes a in the q-series and that falls under the v-concept. If r were such a member, then on our assumptions it would have to belong to the q-series running from m to d:

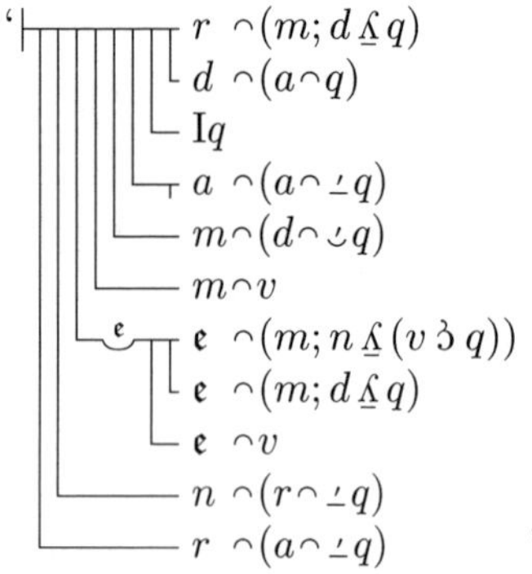

(β

Then, however, r would also belong to the ($v \mathbin{\text{ɔ}} q$)-series running from m to n and thus also to the q-series ending with n, according to the proposition to be proven:

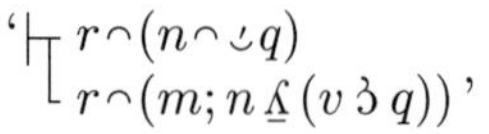

(γ

Since, according to our assumption, n precedes r in the q-series, n would have to follow itself in the q-series, a case which we excluded. In order to prove (γ), we derive the proposition

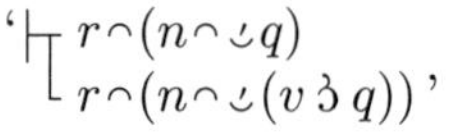

(δ

using (144) and (363).

§16. *Construction*

136 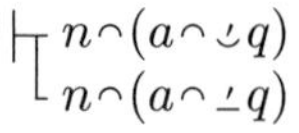

(275):: — — — — — —

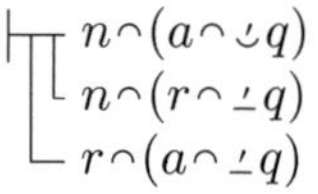

(377

(296): — — — — — —

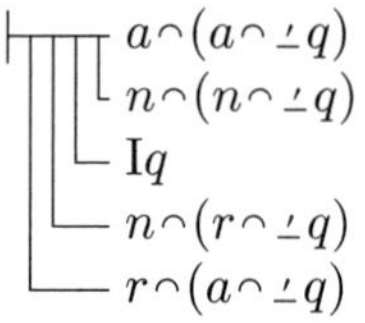

(378

363 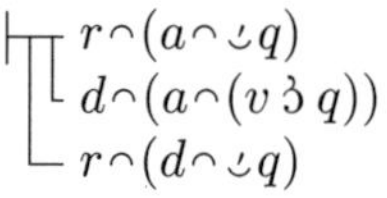

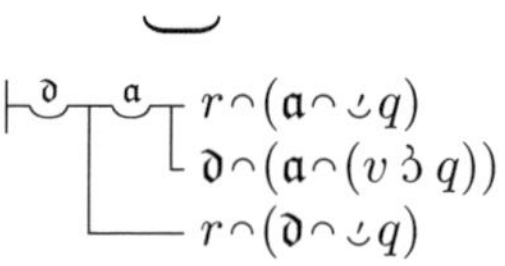

(α

(144): ————————————

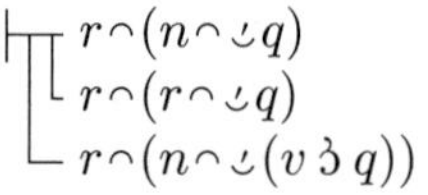

(β

(140):: ————————————

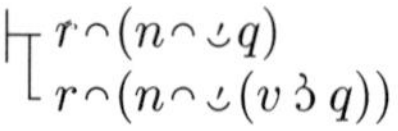

(379

(269):: — — — — — — — —

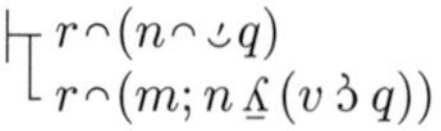

(380

(276): — — — — — — — — —

n⌢(n⌢⊥q)
n⌢(r⌢⊥q)
r⌢(m; n ∡ (v ɜ q)) (α

(378): – – – – – – – – – –

a⌢(a⌢⊥q)
n⌢(r⌢⊥q)
r⌢(m; n ∡ (v ɜ q))
Iq
r⌢(a⌢⊥q) (β

(IIa):: – – – – – – – – – – –

a⌢(a⌢⊥q)
n⌢(r⌢⊥q)
r⌢(m; d ∡ q)
r⌢v
ε 𝔢⌢(m; n ∡ (v ɜ q))
𝔢⌢(m; d ∡ q)
𝔢⌢v
Iq
r⌢(a⌢⊥q) (381

•

280 m⌢(r⌢⊥q)
m⌢(n⌢⌣q)
n⌢(r⌢⊥q)

(136): – – – – – –

m⌢(r⌢⌣q)
m⌢(n⌢⌣q)
n⌢(r⌢⊥q) (382

(380):: – – – – – –

m⌢(r⌢⌣q)
m⌢(m; n ∡ (v ɜ q))
n⌢(r⌢⊥q) (α

(IIa):: – – – – – – – – – –

m⌢(r⌢⌣q)
m⌢(m; d ∡ q)
m⌢v
ε 𝔢⌢(m; n ∡ (v ɜ q))
𝔢⌢(m; d ∡ q)
𝔢⌢v
n⌢(r⌢⊥q) (β

(372):: – – – – – – – – – – – –

m⌢(r⌢⌣q)
d⌢(a⌢q)
Iq
a⌢(a⌢⊥q)
m⌢(d⌢⌣q)
m⌢v
ε 𝔢⌢(m; n ∡ (v ɜ q))
𝔢⌢(m; d ∡ q)
𝔢⌢v
n⌢(r⌢⊥q) (γ

(352): – – – – – – – – – – – –

r⌢(m; d ∡ q)
d⌢(a⌢q)
Iq
a⌢(a⌢⊥q)
m⌢(d⌢⌣q)
m⌢v
ε 𝔢⌢(m; n ∡ (v ɜ q))
𝔢⌢(m; d ∡ q)
𝔢⌢v
n⌢(r⌢⊥q)
r⌢(a⌢⊥q) (δ

(381): – – – – – – – – – – – –

a⌢(a⌢⊥q)
n⌢(r⌢⊥q)
d⌢(a⌢q)
Iq
a⌢(a⌢⊥q)
m⌢(d⌢⌣q)
m⌢v
ε 𝔢⌢(m; n ∡ (v ɜ q))
𝔢⌢(m; d ∡ q)
𝔢⌢v
r⌢(a⌢⊥q)
r⌢v (ε

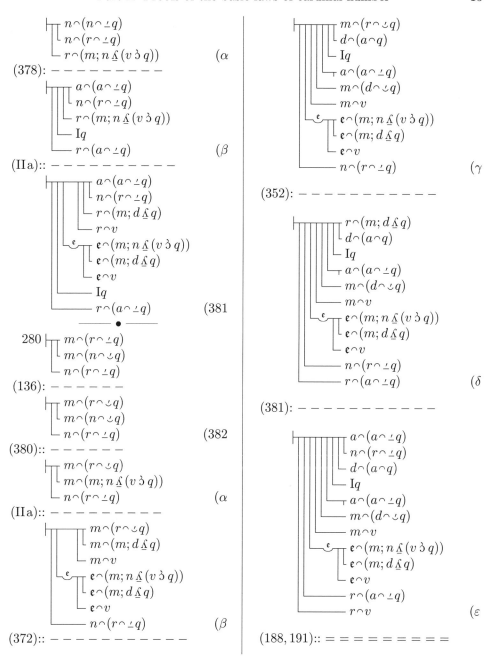

(188, 191):: = = = = = = = = = =

$a \frown (a \frown \underline{\ } q)$
$n \frown (r \frown (v \simeq \underline{\ } q))$
$d \frown (a \frown q)$
$\mathrm{I}q$
$a \frown (a \frown \underline{\ } q)$
$m \frown (d \frown \smile q)$
$m \frown v$
$\mathfrak{e} \frown (m; n \, \underline{\measuredangle} \, (v \, \eth \, q))$
$\mathfrak{e} \frown (m; d \, \underline{\measuredangle} \, q)$
$\mathfrak{e} \frown v$
$r \frown (a \frown \underline{\ } q)$ (ζ
×

$n \frown (r \frown (v \simeq \underline{\ } q))$
$r \frown (a \frown \underline{\ } q)$
$d \frown (a \frown q)$
$\mathrm{I}q$
$a \frown (a \frown \underline{\ } q)$
$m \frown (d \frown \smile q)$
$m \frown v$
$\mathfrak{e} \frown (m; n \, \underline{\measuredangle} \, (v \, \eth \, q))$
$\mathfrak{e} \frown (m; d \, \underline{\measuredangle} \, q)$
$\mathfrak{e} \frown v$ (η

$n \frown (\mathfrak{r} \frown (v \simeq \underline{\ } q))$
$\mathfrak{r} \frown (a \frown \underline{\ } q)$
$d \frown (a \frown q)$
$\mathrm{I}q$
$a \frown (a \frown \underline{\ } q)$
$m \frown (d \frown \smile q)$
$m \frown v$
$\mathfrak{e} \frown (m; n \, \underline{\measuredangle} \, (v \, \eth \, q))$
$\mathfrak{e} \frown (m; d \, \underline{\measuredangle} \, q)$
$\mathfrak{e} \frown v$ (383

———•———

$$T \vdash \dot{\alpha}\dot{\varepsilon}\left(\begin{matrix}\varepsilon \frown (\alpha \frown (v \simeq \underline{\ } q \llcorner \underline{\ } q)) \\ \varepsilon \frown (\alpha \frown (v \simeq \underline{\ } q))\end{matrix}\right) = v \, \eth \, q$$

(10): ————————————

$n \frown (a \frown (v \, \eth \, q))$
$n \frown (a \frown (v \simeq \underline{\ } q \llcorner \underline{\ } q))$
$n \frown (a \frown (v \simeq \underline{\ } q))$ (α

(If):: — — — — — — — — — — —

$n \frown (a \frown (v \, \eth \, q))$
$n \frown (a \frown (v \simeq \underline{\ } q \llcorner \underline{\ } q))$
$n \frown (a \frown (v \simeq \underline{\ } q))$ (384

(197):: — — — — — — — — — — —

$n \frown (a \frown (v \, \eth \, q))$
$n \frown (a \frown (v \simeq \underline{\ } q \llcorner \underline{\ } q))$
$n \frown (a \frown \underline{\ } q)$
$a \frown v$ (α

(15):: — — — — — — — — — — —

$n \frown (a \frown (v \, \eth \, q))$
$n \frown (\mathfrak{r} \frown (v \simeq \underline{\ } q))$
$\mathfrak{r} \frown (a \frown \underline{\ } q)$
$n \frown (a \frown \underline{\ } q)$
$a \frown v$ (385

(134):: — — — — — — — — — — —

$n \frown (a \frown (v \, \eth \, q))$
$n \frown (\mathfrak{r} \frown (v \simeq \underline{\ } q))$
$\mathfrak{r} \frown (a \frown \underline{\ } q)$
$d \frown (a \frown q)$
$n \frown (d \frown \smile q)$
$a \frown v$ (α

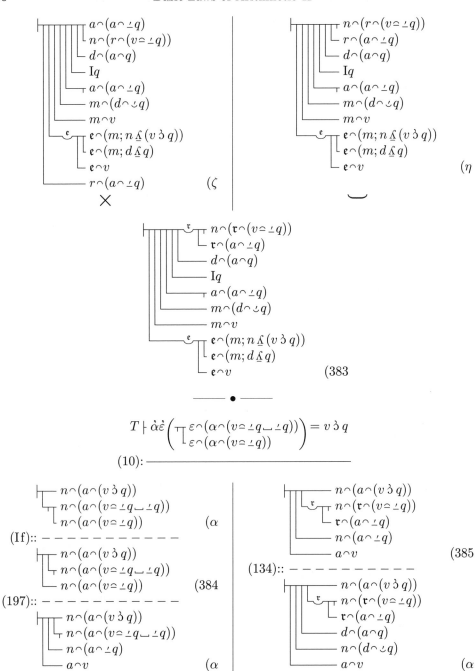

(383):: — — — — — — — — — — —

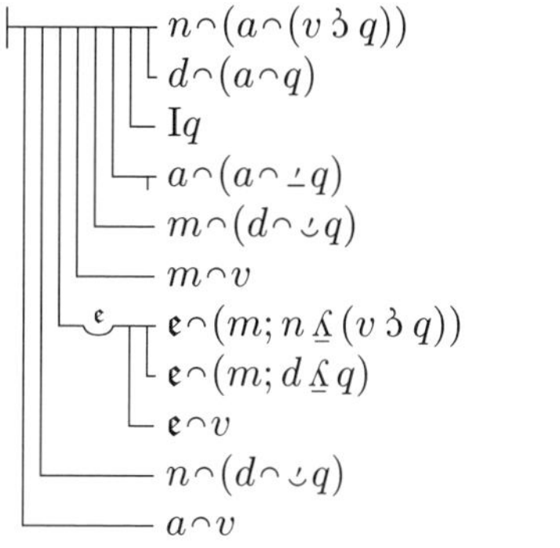

$$
\begin{array}{l}
\vdash n \frown (a \frown (v \mathbin{ↄ} q)) \\
\quad d \frown (a \frown q) \\
\quad \mathrm{I}q \\
\quad a \frown (a \frown \dot{\bot} q) \\
\quad m \frown (d \frown \dot{\smile} q) \\
\quad m \frown v \\
\quad \overset{\mathfrak{e}}{\smile}\ \mathfrak{e} \frown (m; n \underline{\measuredangle} (v \mathbin{ↄ} q)) \\
\qquad \mathfrak{e} \frown (m; d \underline{\measuredangle} q) \\
\qquad \mathfrak{e} \frown v \\
\quad n \frown (d \frown \dot{\smile} q) \\
\quad a \frown v
\end{array}
\qquad (\beta
$$

$\times$

$$
\begin{array}{l}
\vdash \overset{\mathfrak{e}}{\smile}\ \mathfrak{e} \frown (m; n \underline{\measuredangle} (v \mathbin{ↄ} q)) \\
\qquad \mathfrak{e} \frown (m; d \underline{\measuredangle} q) \\
\qquad \mathfrak{e} \frown v \\
\quad d \frown (a \frown q) \\
\quad \mathrm{I}q \\
\quad a \frown (a \frown \dot{\bot} q) \\
\quad m \frown (d \frown \dot{\smile} q) \\
\quad m \frown v \\
\quad n \frown (a \frown (v \mathbin{ↄ} q)) \\
\quad n \frown (d \frown \dot{\smile} q) \\
\quad a \frown v
\end{array}
\qquad (\gamma
$$

(376): ·–·–·–·–·–·–·–·–·–·–·–·

$$
\begin{array}{l}
\vdash \overset{\mathfrak{e}}{\smile}\ \mathfrak{e} \frown (m; n \underline{\measuredangle} (v \mathbin{ↄ} q)) \\
\qquad \mathfrak{e} \frown (m; d \underline{\measuredangle} q) \\
\qquad \mathfrak{e} \frown v \\
\quad d \frown (a \frown q) \\
\quad \mathrm{I}q \\
\quad a \frown (a \frown \dot{\bot} q) \\
\quad m \frown (d \frown \dot{\smile} q) \\
\quad m \frown v \\
\quad n \frown (d \frown \dot{\smile} q) \\
\quad a \frown v \\
\quad \overset{\mathfrak{n}}{\smile}\ \overset{\mathfrak{e}}{\smile}\ \mathfrak{e} \frown (m; \mathfrak{n} \underline{\measuredangle} (v \mathbin{ↄ} q)) \\
\qquad \mathfrak{e} \frown (m; a \underline{\measuredangle} q) \\
\qquad \mathfrak{e} \frown v \\
\qquad \mathfrak{n} \frown (a \frown \dot{\smile} q) \\
\quad a \frown (a \frown \dot{\bot} (v \mathbin{ↄ} q))
\end{array}
\qquad (386
$$

§15. *Analysis*[f]

We eliminate the subcomponent

$$‘\top a \frown (a \frown \dot{\bot} (v \mathbin{ↄ} q))’$$

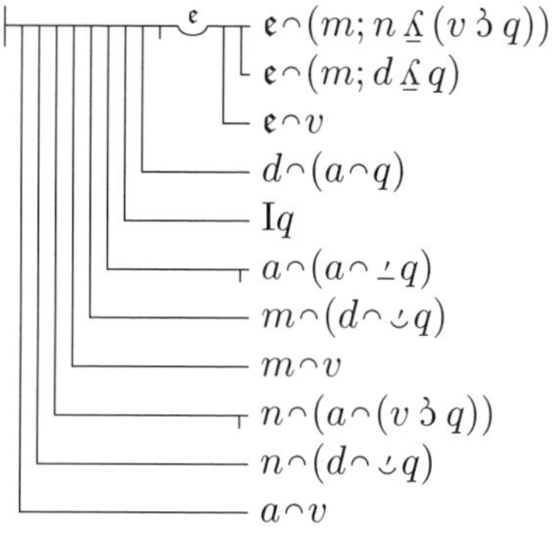

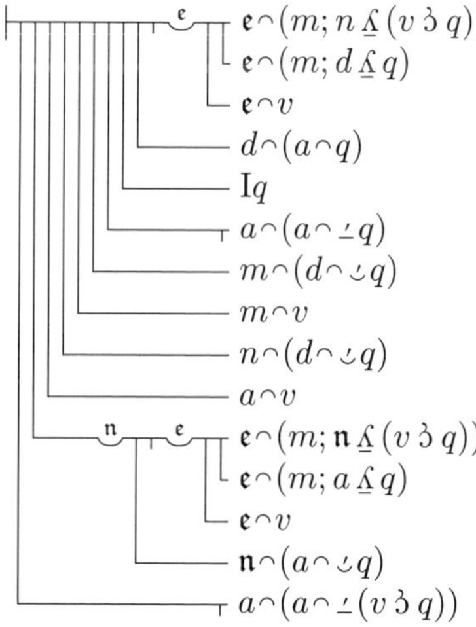

from (386) by applying the proposition

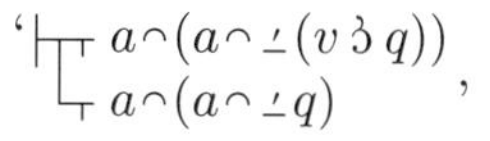

$$
‘\begin{array}{l}
\vdash\!\top a \frown (a \frown \dot{\bot} (v \mathbin{ↄ} q)) \\
\quad \top a \frown (a \frown \dot{\bot} q)
\end{array}’ \qquad (\alpha
$$

which is to be reduced to the proposition

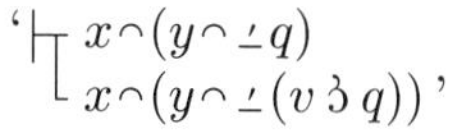

$$
‘\begin{array}{l}
\vdash x \frown (y \frown \dot{\bot} q) \\
\quad x \frown (y \frown \dot{\bot} (v \mathbin{ↄ} q))
\end{array}’ \qquad (\beta
$$

This follows by means of (123). However, we do better to prove first the somewhat more contentful proposition

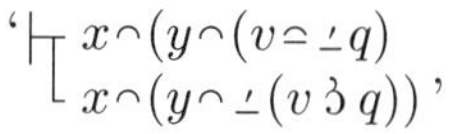

$$
‘\begin{array}{l}
\vdash x \frown (y \frown (v \Bumpeq \dot{\bot} q)) \\
\quad x \frown (y \frown \dot{\bot} (v \mathbin{ↄ} q))
\end{array}’ \qquad (\gamma
$$

which we will make use of later. The derivation is similar to that of (366).

§16. *Construction*

$$
\begin{array}{l}
361 \vdash x \frown (a \frown (v \Bumpeq \dot{\bot} q)) \\
\qquad x \frown (a \frown (v \mathbin{ↄ} q))
\end{array}
$$

$$\smile$$

$$
\begin{array}{l}
\vdash\!\overset{\mathfrak{a}}{\smile}\ x \frown (\mathfrak{a} \frown (v \Bumpeq \dot{\bot} q)) \\
\qquad x \frown (\mathfrak{a} \frown (v \mathbin{ↄ} q))
\end{array}
\qquad (\alpha
$$

———— • ————

$$
\begin{array}{l}
362 \vdash d \frown (a \frown \dot{\bot} q) \\
\qquad d \frown (a \frown (v \mathbin{ↄ} q))
\end{array}
$$

(275): – – – – – – –

$$
\begin{array}{l}
\vdash x \frown (a \frown \dot{\bot} q) \\
\quad d \frown (a \frown (v \mathbin{ↄ} q)) \\
\quad x \frown (d \frown \dot{\bot} q)
\end{array}
\qquad (\beta
$$

(197): – – – – – – –

$$
\begin{array}{l}
\vdash x \frown (a \frown (v \Bumpeq \dot{\bot} q)) \\
\quad d \frown (a \frown (v \mathbin{ↄ} q)) \\
\quad x \frown (d \frown \dot{\bot} q) \\
\quad a \frown v
\end{array}
\qquad (\gamma
$$

(365, 188):: = = = = =

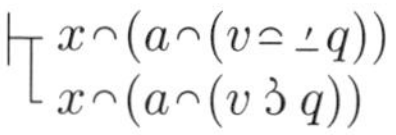

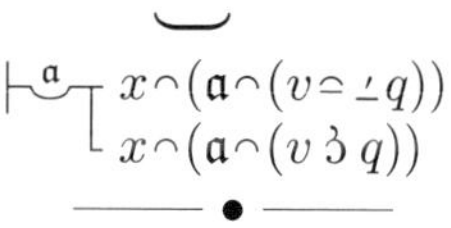

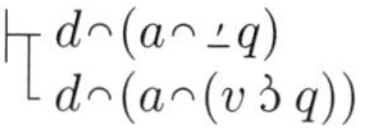

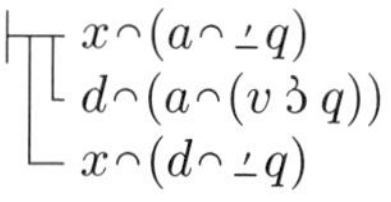

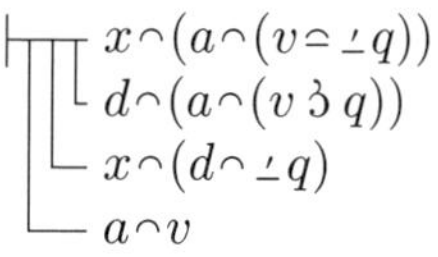

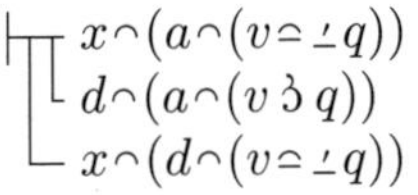
(δ

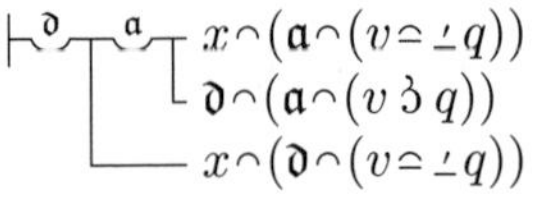
(ε

(123): ———

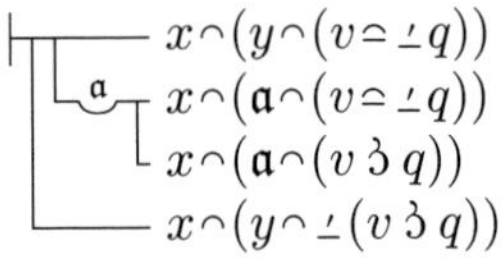
(ζ

(α):: ———

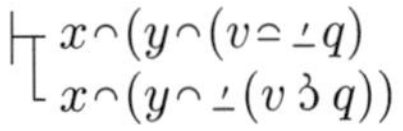
(387

(188): – – – – – – – –

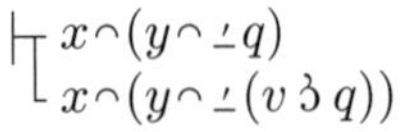
(388

———•———

388 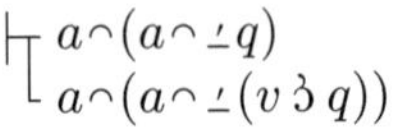

×

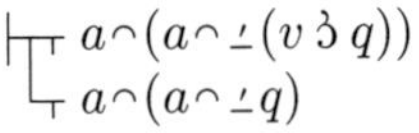
(389

(386): – – – – – – – –

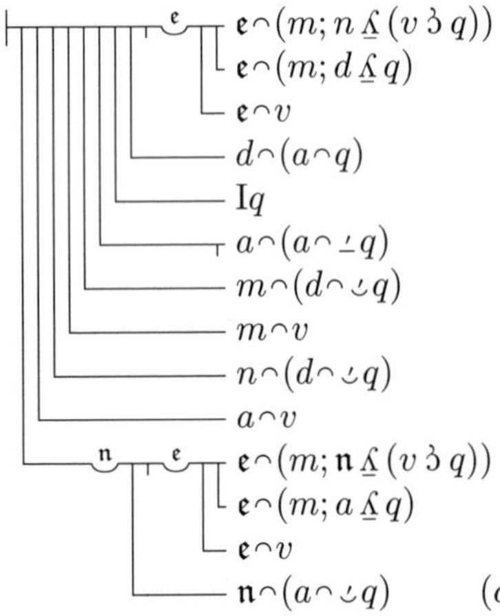
(α

(370): ·—·—·—·—·—·—·—·—·—·—·

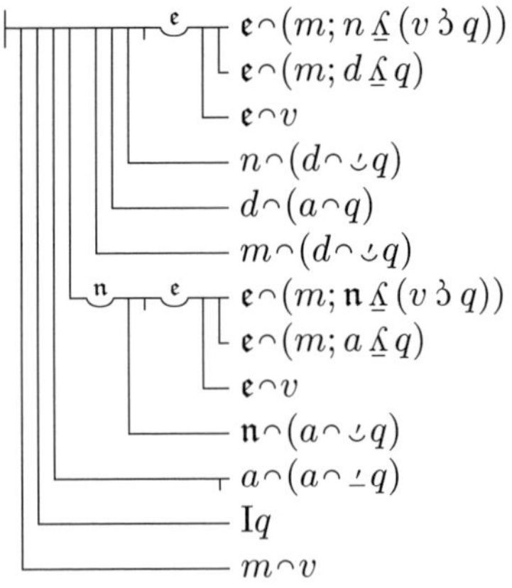
(β

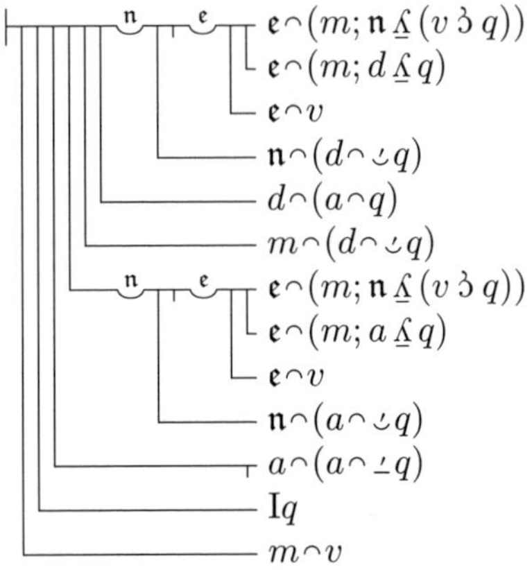
(γ

×

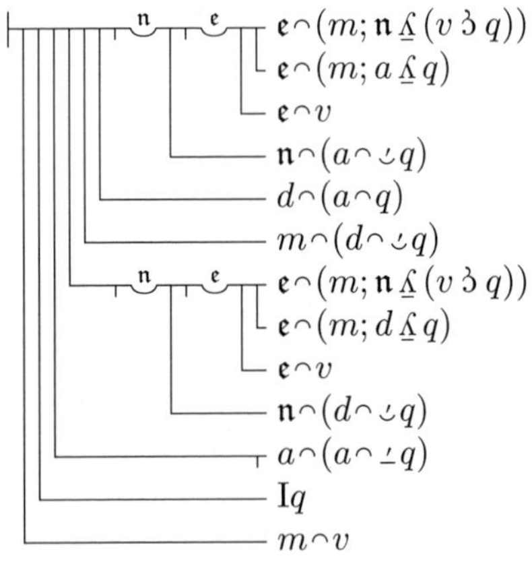
(δ

(I):: – – – – – – – – – – – – – – – –

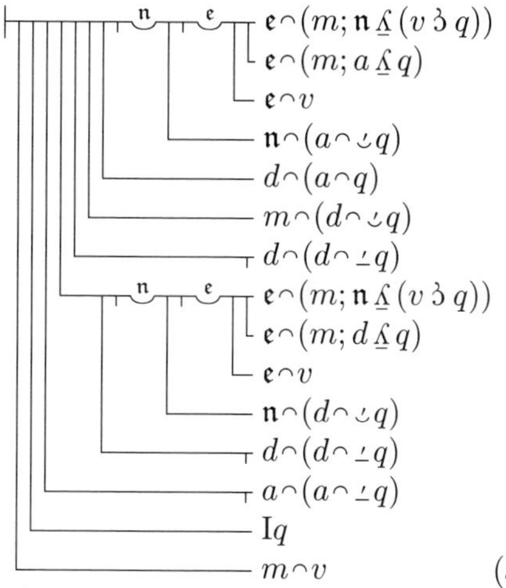

(ε

(279):: — — — — — — — — — — — — — —

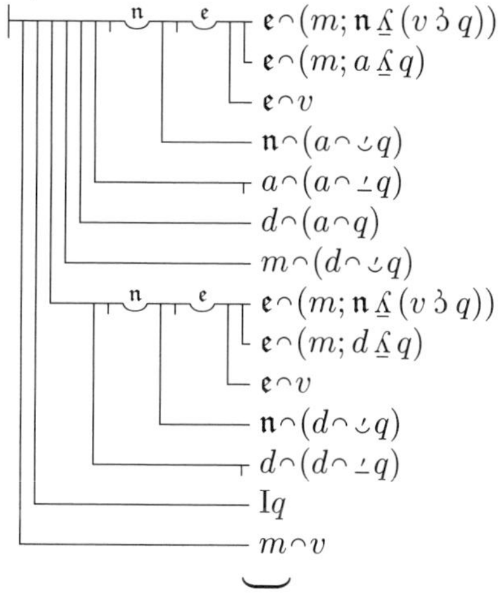

(ζ

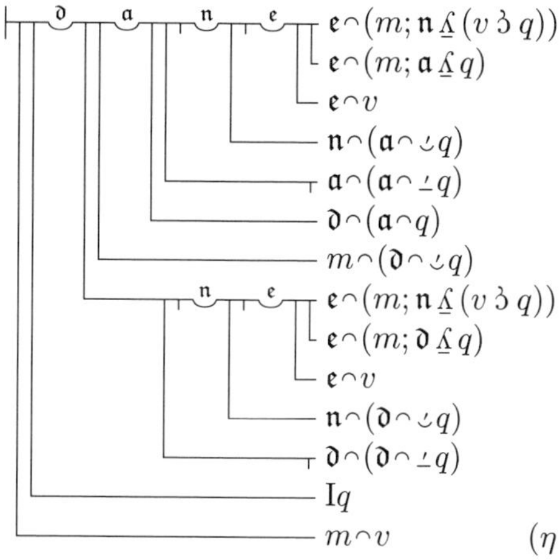

(η

(152): — — — — — — — — — — — — — —

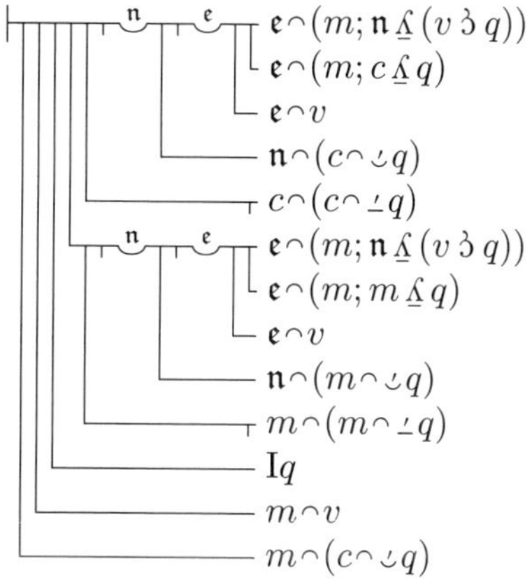

(390

§17. *Analysis*

We will now prove the proposition

‘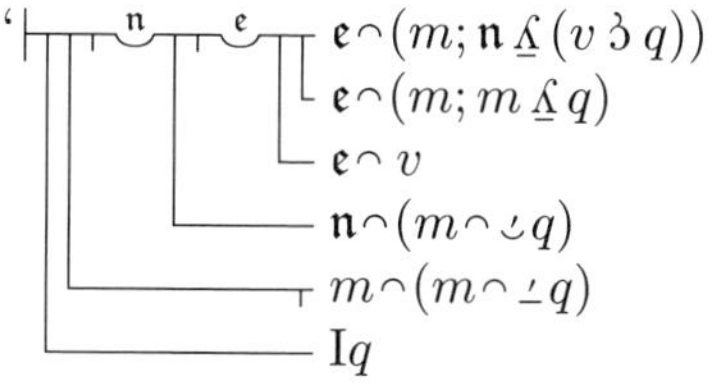’,

(α

which is to be reduced to the proposition

'⊢ $e \frown (m; m \,\underline{\measuredangle}\, (v \between q))$
$m \frown (m \frown \underline{\ \ } q)$
$\mathrm{I}q$
$e \frown (m; m \,\underline{\measuredangle}\, q)$ ' (β

which we derive from (344) and (282). In addition, we also require the proposition

'⊢ $\mathrm{I}(v \between q)$
$\mathrm{I}q$ ' (γ

which follows from (243) and the proposition

'⊢ $d \frown (a \frown \underline{\ \ } q)$
$e \frown (a \frown (v \between q))$
$e \frown (d \frown (v \between q))$ ' (δ

§18. *Construction*

140 ⊢ $m \frown (m \frown \smile p)$

(344): ————————

⊢ $m \frown (m; m \,\underline{\measuredangle}\, p)$
$m \frown (m \frown \underline{\ \ } p)$
$\mathrm{I}q$ (391

—— • ——

136 ⊢ $e \frown (d \frown \smile q)$
$e \frown (d \frown \underline{\ \ } q)$

(362):: – – – – –

⊢ $e \frown (d \frown \smile q)$
$e \frown (d \frown (v \between q))$ (392

—— • ——

360 ⊢ $e \frown (a \frown (v \between q))$
$e \frown (a \frown (v \cong \underline{\ \ } q \smile \underline{\ \ } q))$
$e \frown (a \frown (v \cong \underline{\ \ } q))$

(I):: – – – – – – – – – – –

⊢ $e \frown (a \frown (v \between q))$
$e \frown (a \frown (v \cong \underline{\ \ } q \smile \underline{\ \ } q))$ (393

(5):: – – – – – – – – – – –

⊢ $e \frown (a \frown (v \between q))$
$d \frown (a \frown \underline{\ \ } q)$
$e \frown (d \frown (v \cong \underline{\ \ } q))$ (394

×

⊢ $d \frown (a \frown \underline{\ \ } q)$
$e \frown (a \frown (v \between q))$
$e \frown (d \frown (v \cong \underline{\ \ } q))$ (395

(361):: – – – – – – – – –

⊢ $d \frown (a \frown \underline{\ \ } q)$
$e \frown (a \frown (v \between q))$
$e \frown (d \frown (v \between q))$ (396

(243): – – – – – – – – –

⊢ $a \frown (d \frown \smile q)$
$e \frown (a \frown (v \between q))$
$e \frown (d \frown (v \between q))$
$e \frown (d \frown \smile q)$
$\mathrm{I}q$
$e \frown (a \frown \smile q)$ (α

(392, 392):: = = = = = =

⊢ $a \frown (d \frown \smile q)$
$e \frown (a \frown (v \between q))$
$e \frown (d \frown (v \between q))$
$\mathrm{I}q$ (β

(130): ————————

⊢ $d = a$
$a \frown (d \frown \underline{\ \ } q)$
$e \frown (a \frown (v \between q))$
$e \frown (d \frown (v \between q))$
$\mathrm{I}q$ (γ

(396):: – – – – – – – – – –

⊢ $d = a$
$e \frown (a \frown (v \between q))$
$e \frown (d \frown (v \between q))$
$\mathrm{I}q$ (δ

⊢ $\mathfrak{e}\ \mathfrak{d}\ \mathfrak{a}$ — $\mathfrak{d} = \mathfrak{a}$
$\mathfrak{e} \frown (\mathfrak{a} \frown (v \between q))$
$\mathfrak{e} \frown (\mathfrak{d} \frown (v \between q))$
$\mathrm{I}q$ (ε

(16): – – – – – – – – – – – –

⊢ $\mathrm{I}(v \between q)$
$\mathrm{I}q$ (397

(391): – – – –

m⌢(m; m ⨽̲ (v ꝫ q))
m⌢(m⌢⨼(v ꝫ q))
Iq (α

(389):: - - - - - - - - - - -

m⌢(m; m ⨽̲ (v ꝫ q))
m⌢(m⌢⨼q)
Iq (β

(IIIa): ————————

e⌢(m; m ⨽̲ (v ꝫ q))
m⌢(m⌢⨼q)
Iq
e = m (γ

(282):: - - - - - - - - - - -

e⌢(m; m ⨽̲ (v ꝫ q))
m⌢(m⌢⨼q)
Iq
e⌢(m; m ⨽̲ q) (δ

(I): - - - - - - - - - - -

e⌢(m; m ⨽̲ (v ꝫ q))
e⌢(m; m ⨽̲ q)
e⌢v
m⌢(m⌢⨼q)
Iq (ε

‿

𝔢 𝔢⌢(m; m ⨽̲ (v ꝫ q))
𝔢⌢(m; m ⨽̲ q)
𝔢⌢v
m⌢(m⌢⨼q)
Iq (398

——— • ———

I ⊢ F(n)
G(n)
F(n)

(IIa):: - - - -

F(n)
G(n)
𝔫 F(𝔫) (α

‿

𝔫 F(𝔫)
G(𝔫)
𝔫 F(𝔫) (β

×

𝔫 F(𝔫)
𝔫 F(𝔫)
G(𝔫) (399

——— • ———

140 ⊢ m⌢(m⌢⌣q)

(IIa):: ————————

𝔢 𝔢⌢(m; m ⨽̲ (v ꝫ q))
𝔢⌢(m; m ⨽̲ q)
𝔢⌢v
𝔫 𝔢 𝔢⌢(m; 𝔫 ⨽̲ (v ꝫ q))
𝔢⌢(m; m ⨽̲ q)
𝔢⌢v
𝔫⌢(m⌢⌣q) (α

×

𝔫 𝔢 𝔢⌢(m; 𝔫 ⨽̲ (v ꝫ q))
𝔢⌢(m; m ⨽̲ q)
𝔢⌢v
𝔫⌢(m⌢⌣q)
𝔢 𝔢⌢(m; m ⨽̲ (v ꝫ q))
𝔢⌢(m; m ⨽̲ q)
𝔢⌢v (β

(398):: - - - - - - - - - - - -

𝔫 𝔢 𝔢⌢(m; 𝔫 ⨽̲ (v ꝫ q))
𝔢⌢(m; m ⨽̲ q)
𝔢⌢v
𝔫⌢(m⌢⌣q)
m⌢(m⌢⨼q)
Iq (γ

(390): - - - - - - - - - - - - - -

𝔫 𝔢 𝔢⌢(m; 𝔫 ⨽̲ (v ꝫ q))
𝔢⌢(m; c ⨽̲ q)
𝔢⌢v
𝔫⌢(c⌢⌣q)
c⌢(c⌢⨼q)
Iq
m⌢v
m⌢(c⌢⌣q) (δ

(399): - - - - - - - - - - - - - -

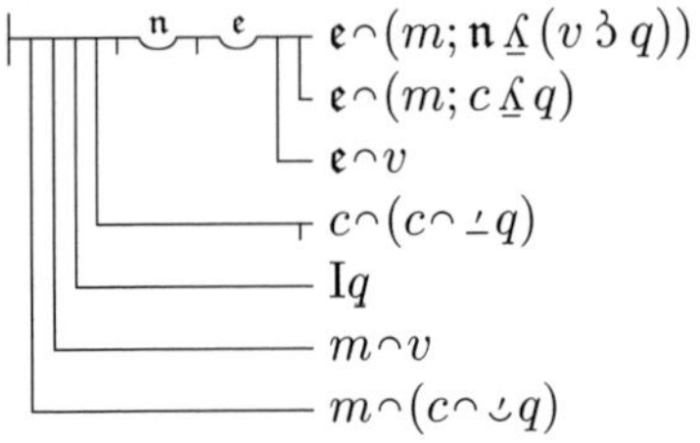
(400

368 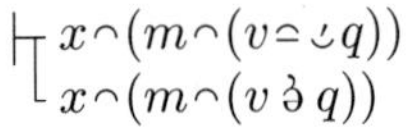

(188): – – – – – – – – –

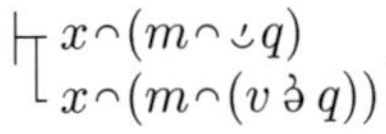
(401

368 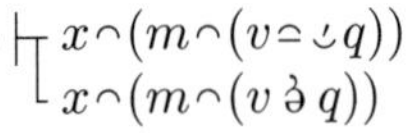

(191): – – – – – – – – –

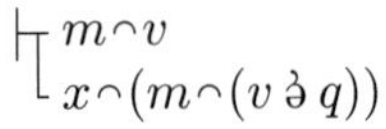
(402

(400): – – – – – – – – –

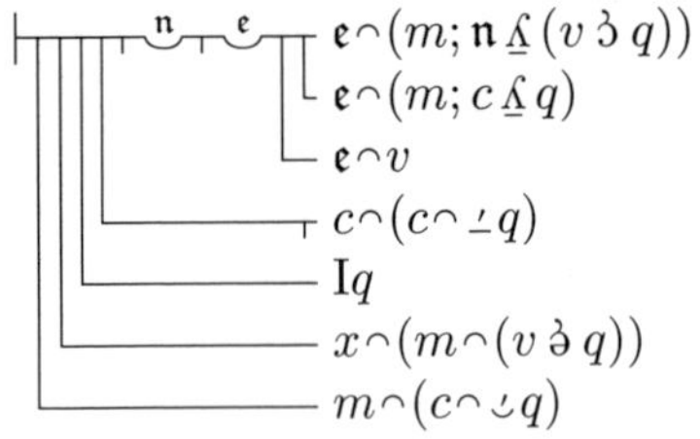
(403

§19. *Analysis*

In order to prove proposition (β) of §7 using (403), we require the proposition

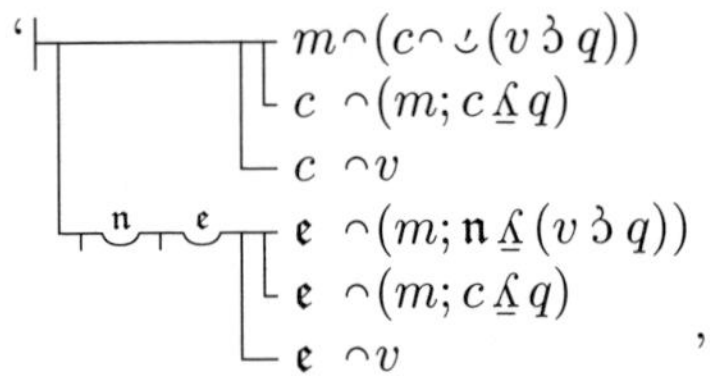
(α

To eliminate the subcomponent

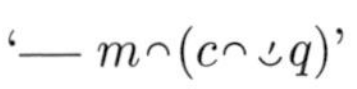

we then also require the proposition

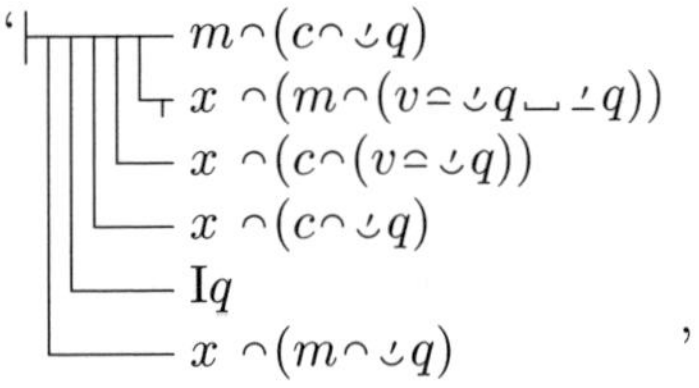
(β

which follows from (243). The subcomponents

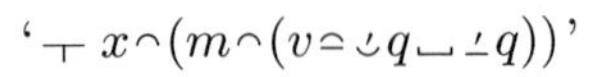

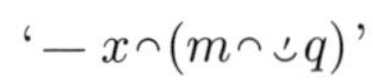

are to be replaced by '— x⌒(m⌒(v ∂ q))'.

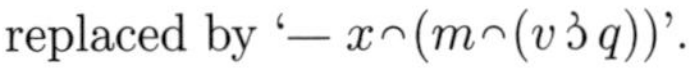

§20. *Construction*

367 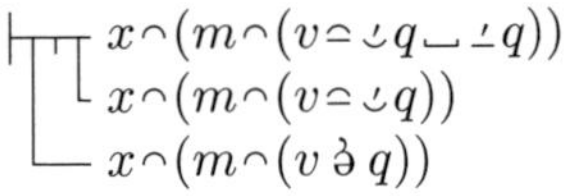

(Ic):: – – – – – – – – – – – –

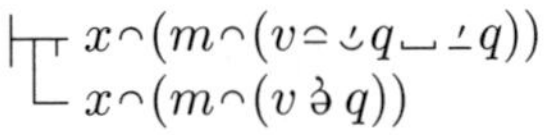
(404

270 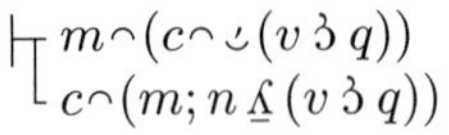

(IIa):: – – – – – – – – – –

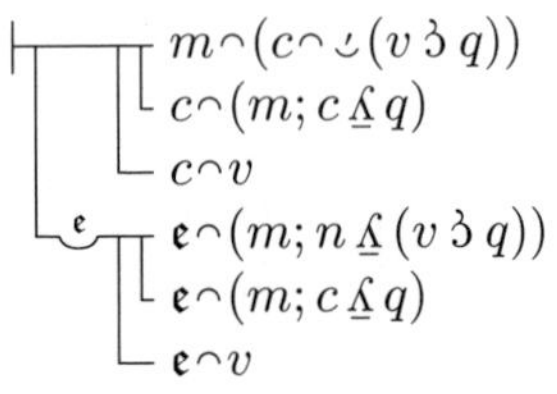
(α

×

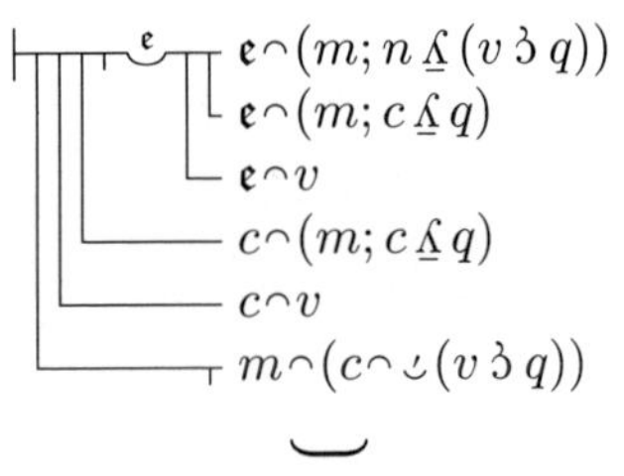
(β

$$\vdash\ \mathfrak{n}\ \mathfrak{e}\ \ \mathfrak{e}\frown(m;\mathfrak{n}\,\underline{\measuredangle}\,(v\,\breve{\jmath}\,q))$$
$$\mathfrak{e}\frown(m;c\,\underline{\measuredangle}\,q)$$
$$\mathfrak{e}\frown v$$
$$c\frown(m;c\,\underline{\measuredangle}\,q)$$
$$c\frown v$$
$$m\frown(c\frown\acute{\smile}(v\,\breve{\jmath}\,q))\quad(\gamma$$

×

$$\vdash\ m\frown(c\frown\acute{\smile}(v\,\breve{\jmath}\,q))$$
$$c\frown(m;c\,\underline{\measuredangle}\,q)$$
$$c\frown v$$
$$\mathfrak{n}\ \mathfrak{e}\ \ \mathfrak{e}\frown(m;\mathfrak{n}\,\underline{\measuredangle}\,(v\,\breve{\jmath}\,q))$$
$$\mathfrak{e}\frown(m;c\,\underline{\measuredangle}\,q)$$
$$\mathfrak{e}\frown v\quad(\delta$$

(403, 344):: = = = = = = = = = =

$$\vdash\ m\frown(c\frown\acute{\smile}(v\,\breve{\jmath}\,q))$$
$$m\frown(c\frown\smile q)$$
$$c\frown(c\frown\perp q)$$
$$\mathrm{I}q$$
$$c\frown v$$
$$x\frown(m\frown(v\,\grave{\partial}\,q))\quad(405$$

(IIa):: – – – – – – – – – – –

$$\vdash\ m\frown(c\frown\acute{\smile}(v\,\breve{\jmath}\,q))$$
$$m\frown(c\frown\smile q)$$
$$\mathfrak{i}\ \ \mathfrak{i}\frown(\mathfrak{i}\frown\perp q)$$
$$\mathrm{I}q$$
$$c\frown v$$
$$x\frown(m\frown(v\,\grave{\partial}\,q))\quad(406$$

•

$$5\ \vdash\ x\frown(m\frown(v\simeq\smile q\,\llcorner\!\lrcorner\,\perp q))$$
$$c\frown(m\frown\perp q)$$
$$x\frown(c\frown(v\simeq\smile q))$$

×

$$\vdash\ c\frown(m\frown\perp q)$$
$$x\frown(m\frown(v\simeq\smile q\,\llcorner\!\lrcorner\,\perp q))$$
$$x\frown(c\frown(v\simeq\smile q))\quad(\alpha$$

(243): – – – – – – – – – – –

$$\vdash\ m\frown(c\frown\smile q)$$
$$x\frown(m\frown(v\simeq\smile q\,\llcorner\!\lrcorner\,\perp q))$$
$$x\frown(c\frown(v\simeq\smile q))$$
$$x\frown(c\frown\smile q)$$
$$\mathrm{I}q$$
$$x\frown(m\frown\smile q)\quad(\beta$$

(401, 404):: = = = = = = = = = =

$$\vdash\ m\frown(c\frown\smile q)$$
$$x\frown(m\frown(v\,\grave{\partial}\,q))$$
$$x\frown(c\frown(v\simeq\smile q))$$
$$x\frown(c\frown\smile q)$$
$$\mathrm{I}q\quad(\gamma$$

(188):: – – – – – – – – – –

$$\vdash\ m\frown(c\frown\smile q)$$
$$x\frown(m\frown(v\,\grave{\partial}\,q))$$
$$x\frown(c\frown(v\simeq\smile q))$$
$$\mathrm{I}q\quad(407$$

(406): – – – – – – – – –

$$\vdash\ m\frown(c\frown\acute{\smile}(v\,\breve{\jmath}\,q))$$
$$x\frown(m\frown(v\,\grave{\partial}\,q))$$
$$x\frown(c\frown(v\simeq\smile q))$$
$$\mathrm{I}q$$
$$\mathfrak{i}\ \ \mathfrak{i}\frown(\mathfrak{i}\frown\perp q)$$
$$c\frown v\quad(\alpha$$

(191):: – – – – – – – – – – –

$$\vdash\ m\frown(c\frown\acute{\smile}(v\,\breve{\jmath}\,q))$$
$$x\frown(c\frown(v\simeq\smile q))$$
$$x\frown(m\frown(v\,\grave{\partial}\,q))$$
$$\mathrm{I}q$$
$$\mathfrak{i}\ \ \mathfrak{i}\frown(\mathfrak{i}\frown\perp q)\quad(\beta$$

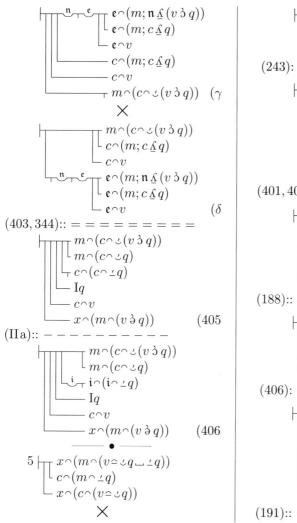

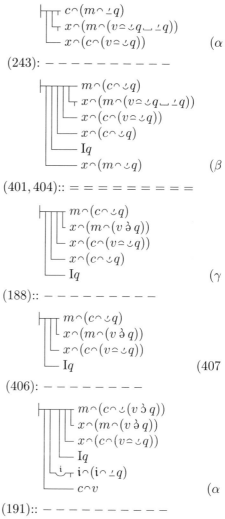

(IVa): – – – – – – – – – –

$$\vdash\ [-\ x \frown (c \frown (v \doteq \smile q))] = [-\ m \frown (c \frown \smile (v \,\mathfrak{z}\, q))]$$
$$x \frown (m \frown (v \,\dot{\eth}\, q))$$
$$\mathrm{I}q$$
$$\overset{\mathfrak{i}}{\smile}\ \mathfrak{i} \frown (\mathfrak{i} \frown \underline{\cdot}\, q)$$
$$x \frown (c \frown (v \doteq \smile q))$$
$$m \frown (c \frown \smile (v \,\mathfrak{z}\, q)) \qquad (\gamma$$

(369):: -

$$\vdash\ [-\ x \frown (c \frown (v \doteq \smile q))] = [-\ m \frown (c \frown \smile (v \,\mathfrak{z}\, q))]$$
$$x \frown (m \frown (v \,\dot{\eth}\, q))$$
$$\mathrm{I}q$$
$$\overset{\mathfrak{i}}{\smile}\ \mathfrak{i} \frown (\mathfrak{i} \frown \underline{\cdot}\, q) \qquad (\delta$$

(22): ————————————————————————

$$\vdash\ [-\ c \frown (x \frown \not\S (v \doteq \smile q))] = [-\ m \frown (c \frown \smile (v \,\mathfrak{z}\, q))]$$
$$x \frown (m \frown (v \,\dot{\eth}\, q))$$
$$\mathrm{I}q$$
$$\overset{\mathfrak{i}}{\smile}\ \mathfrak{i} \frown (\mathfrak{i} \frown \underline{\cdot}\, q) \qquad (\varepsilon$$

(22): ————————————————————————

$$\vdash\ [-\ c \frown (x \frown \not\S (v \doteq \smile q))] = [-\ c \frown (m \frown \not\S \smile (v \,\mathfrak{z}\, q))]$$
$$x \frown (m \frown (v \,\dot{\eth}\, q))$$
$$\mathrm{I}q$$
$$\overset{\mathfrak{i}}{\smile}\ \mathfrak{i} \frown (\mathfrak{i} \frown \underline{\cdot}\, q) \qquad (\zeta$$

$$\smile$$

$$\vdash\ \overset{\mathfrak{a}}{\smile}\ [-\ \mathfrak{a} \frown (x \frown \not\S (v \doteq \smile q))] = [-\ \mathfrak{a} \frown (m \frown \not\S \smile (v \,\mathfrak{z}\, q))]$$
$$x \frown (m \frown (v \,\dot{\eth}\, q))$$
$$\mathrm{I}q$$
$$\overset{\mathfrak{i}}{\smile}\ \mathfrak{i} \frown (\mathfrak{i} \frown \underline{\cdot}\, q) \qquad (\eta$$

(96): -

$$\vdash\ \not{\mathfrak{p}}(x \frown \not\S (v \doteq \smile q)) = \not{\mathfrak{p}}(m \frown \not\S \smile (v \,\mathfrak{z}\, q))$$
$$x \frown (m \frown (v \,\dot{\eth}\, q))$$
$$\mathrm{I}q$$
$$\overset{\mathfrak{i}}{\smile}\ \mathfrak{i} \frown (\mathfrak{i} \frown \underline{\cdot}\, q) \qquad (\vartheta$$

(IIIa): - - - - - - - - - - - - - - - - - - -

$$\vdash\ \varpi = \not{\mathfrak{p}}(x \frown \not\S (v \doteq \smile q))$$
$$\varpi = \not{\mathfrak{p}}(m \frown \not\S \smile (v \,\mathfrak{z}\, q))$$
$$x \frown (m \frown (v \,\dot{\eth}\, q))$$
$$\mathrm{I}q$$
$$\overset{\mathfrak{i}}{\smile}\ \mathfrak{i} \frown (\mathfrak{i} \frown \underline{\cdot}\, q) \qquad (408$$

c) Proof of the proposition

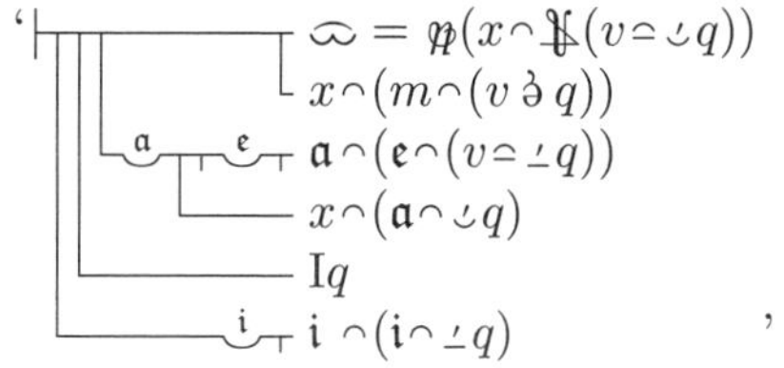

§21. *Analysis*

In order to arrive from (408) at the proposition that, given our assumptions about q and m, Endlos is the cardinal number of the members of the q-series starting with x that fall under the v-concept, if for each member of the q-series starting with x there is one following it that falls under the v-concept, we must use (262). Before we do this, we will reshape this proposition to suit our purposes. Specifically, we have to write 'v ɔ̀ q' for 'q' in (262). Then, in the resulting subcomponents, '— I(v ɔ̀ q)' and '$\smile^{\mathfrak{i}}\!\!\top$ $\mathfrak{i}$⌒($\mathfrak{i}$⌒(vɔ̀q))', 'vɔ̀q' is to be replaced by 'q'. Then we have to introduce in place of the subcomponent

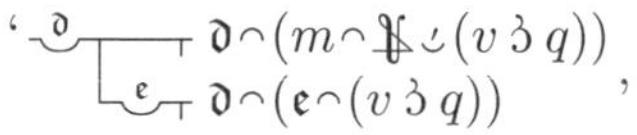

the subcomponent

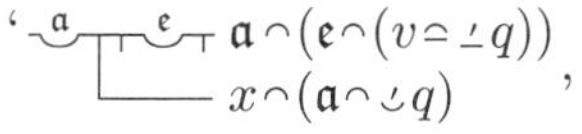

The proposition required for this,

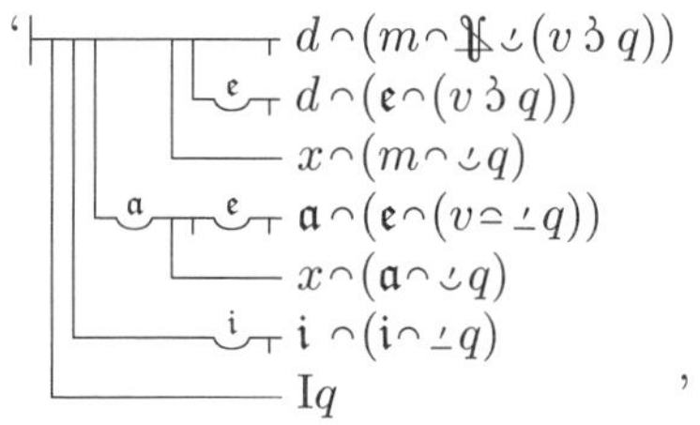 (α

is proven by means of the propositions

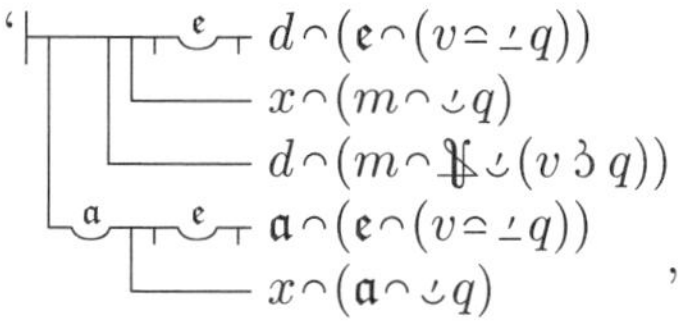 (β

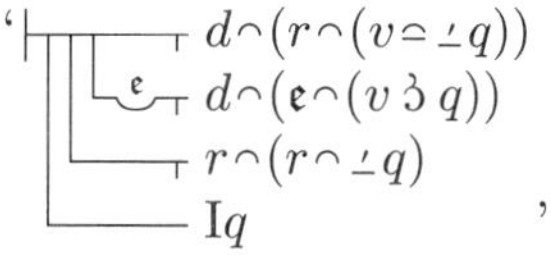 (γ

In order to prove (β) using IIa, we require the proposition

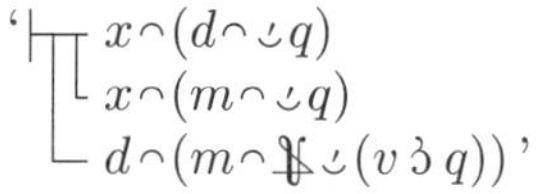 (δ

which follows by means of (322) from

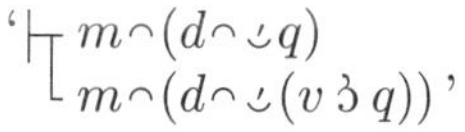 (ε

§22. *Construction*

388 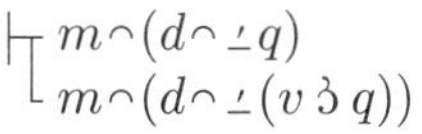

(136): — — — — — — — —

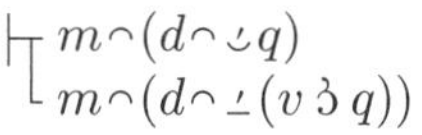 (α

(200):: — — — — — — — —

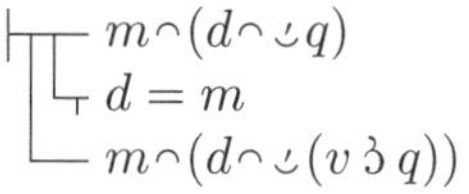 (β

(139): ·—·—·—·—·—·—·—·—·

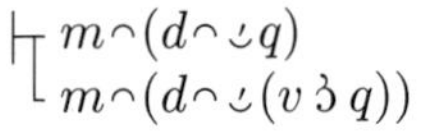

$$\vdash \begin{array}{l} m \frown (d \frown \smile q) \\ m \frown (d \frown \smile (v \eth q)) \end{array}$$ (409

(23):: — — — — — — — — —

$$\vdash \begin{array}{l} m \frown (d \frown \smile q) \\ d \frown (m \frown \natural \smile (v \eth q)) \end{array}$$ (α

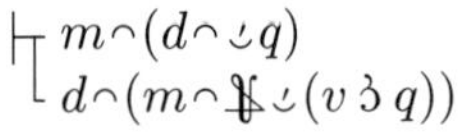

(322): — — — — — — — — — —

$$\vdash \begin{array}{l} x \frown (d \frown \smile q) \\ x \frown (m \frown \smile q) \\ d \frown (m \frown \natural \smile (v \eth q)) \end{array}$$ (β

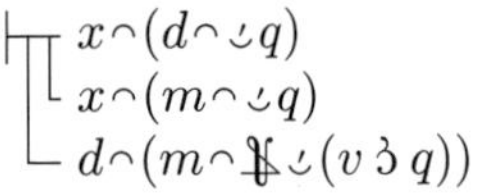

(IIa): — — — — — — — — — —

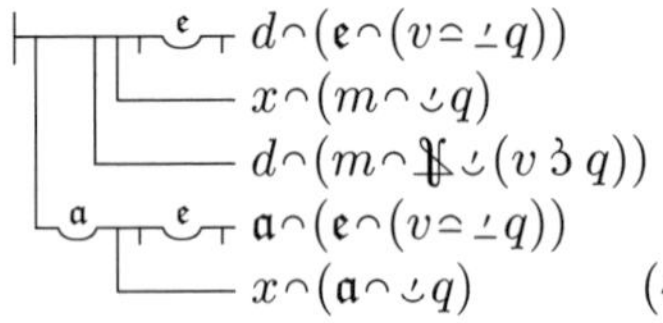

$$\vdash \begin{array}{l} \overset{\mathfrak{e}}{\smile}\; d \frown (\mathfrak{e} \frown (v \simeq \perp q)) \\ x \frown (m \frown \smile q) \\ d \frown (m \frown \natural \smile (v \eth q)) \\ \overset{\mathfrak{a}}{\smile}\;\overset{\mathfrak{e}}{\smile}\; \mathfrak{a} \frown (\mathfrak{e} \frown (v \simeq \perp q)) \\ x \frown (\mathfrak{a} \frown \smile q) \end{array}$$ (410

§23. *Analysis*

We now prove proposition (γ) of §21. This proposition says: if there is no object that is the first one following d in the q-series to fall under the v-concept, then there is no object at all that falls under the v-concept and follows d in the q-series, provided the q-relation conforms to our assumptions. It is similar to proposition (358) and can be derived from it. If d stands to no object in the q-relation, then there is no object which follows d in the q-series and thus also no such object that falls under the v-concept. If, however, d stands to an object a in the q-relation, then we can show that an object n is not the first in the q-series starting with a to fall under the v-concept if n is not the first among the objects following d in the q-series to fall under the v-concept:

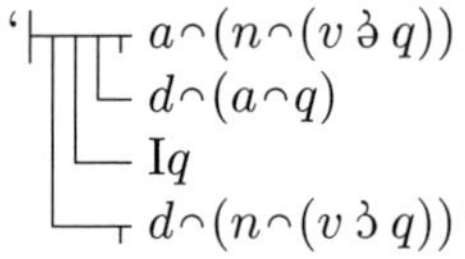

$$\text{‘}\vdash \begin{array}{l} a \frown (n \frown (v \eth q)) \\ d \frown (a \frown q) \\ \mathrm{I}q \\ d \frown (n \frown (v \eth q))\text{’} \end{array}$$ (α

We also have to derive the proposition

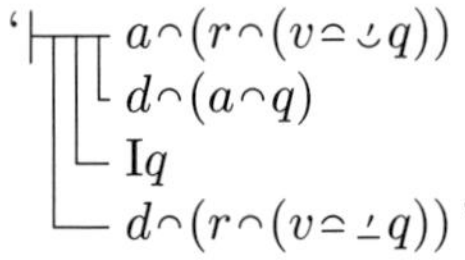

$$\text{‘}\vdash \begin{array}{l} a \frown (r \frown (v \simeq \smile q)) \\ d \frown (a \frown q) \\ \mathrm{I}q \\ d \frown (r \frown (v \simeq \perp q))\text{’} \end{array}$$ (β

which can easily be accomplished by means of (242). In order to prove (α) we start with the proposition (384):

$$\vdash \begin{array}{l} d \frown (n \frown (v \eth q)) \\ d \frown (n \frown (v \simeq \perp q \smile \perp q)) \\ d \frown (n \frown (v \simeq \perp q)) \end{array}$$

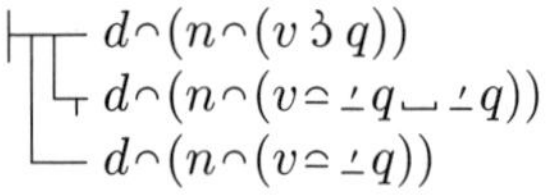

and then have to prove the propositions

$$\text{‘}\vdash \begin{array}{l} d \frown (n \frown (v \simeq \perp q)) \\ d \frown (a \frown q) \\ a \frown (n \frown (v \simeq \smile q))\text{’} \end{array}$$ (γ

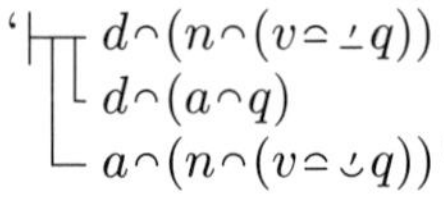

and

$$\text{‘}\vdash \begin{array}{l} d \frown (n \frown (v \simeq \perp q \smile \perp q)) \\ d \frown (a \frown q) \\ \mathrm{I}q \\ a \frown (n \frown (v \simeq \smile q \smile \perp q))\text{’} \end{array}$$ (δ

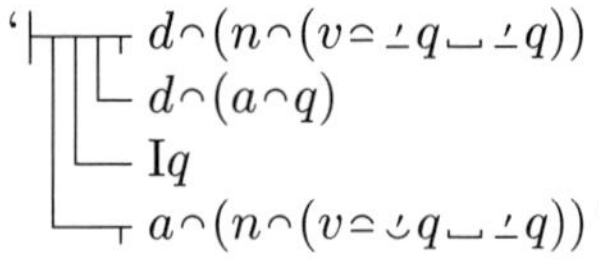

(γ) straightforwardly follows from (132). (δ) can be derived by means of (15):

$$\vdash \begin{array}{l} d \frown (n \frown (v \simeq \perp q \smile \perp q)) \\ \overset{\mathfrak{r}}{\smile}\; d \frown (\mathfrak{r} \frown (v \simeq \perp q)) \\ \mathfrak{r} \frown (n \frown \perp q) \end{array}$$

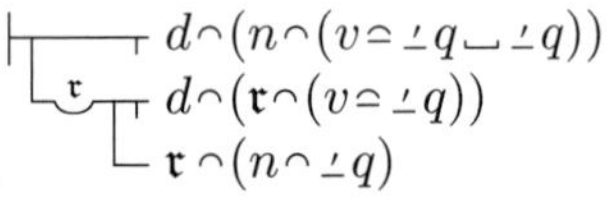

This is accomplished using (5):

$$\vdash \begin{array}{l} a \frown (n \frown (v \simeq \smile q \smile \perp q)) \\ r \frown (n \frown \perp q) \\ a \frown (r \frown (v \simeq \smile q)) \end{array}$$

and (β).

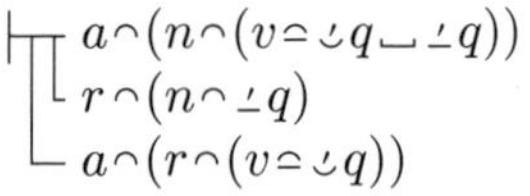

§24. *Construction*

242 $\vdash a\frown(r\frown\smile q)$
$d\frown(a\frown q)$
$\mathrm{I}q$
$d\frown(r\frown\perp q)$

(197): – – – – – –

$\vdash a\frown(r\frown(v\bumpeq\smile q))$
$d\frown(a\frown q)$
$\mathrm{I}q$
$d\frown(r\frown\perp q)$
$r\frown v$ (α

(188, 191):: = = = = = =

$\vdash a\frown(r\frown(v\bumpeq\smile q))$
$d\frown(a\frown q)$
$\mathrm{I}q$
$d\frown(r\frown(v\bumpeq\perp q))$ (411

(5): – – – – – – – –

$\vdash a\frown(n\frown(v\bumpeq\smile q\sqcup\perp q))$
$r\frown(n\frown\perp q)$
$d\frown(a\frown q)$
$\mathrm{I}q$
$d\frown(r\frown(v\bumpeq\perp q))$ (α

×

$\vdash d\frown(r\frown(v\bumpeq\perp q))$
$r\frown(n\frown\perp q)$
$d\frown(a\frown q)$
$\mathrm{I}q$
$a\frown(n\frown(v\bumpeq\smile q\sqcup\perp q))$ (β

$\vdash \overset{\mathfrak{r}}{\smile}\ d\frown(\mathfrak{r}\frown(v\bumpeq\perp q))$
$\mathfrak{r}\frown(n\frown\perp q)$
$d\frown(a\frown q)$
$\mathrm{I}q$
$a\frown(n\frown(v\bumpeq\smile q\sqcup\perp q))$ (γ

(15): – – – – – – – – – – – – – –

$\vdash d\frown(n\frown(v\bumpeq\perp q\sqcup\perp q))$
$d\frown(a\frown q)$
$\mathrm{I}q$
$a\frown(n\frown(v\bumpeq\smile q\sqcup\perp q))$ (δ

(384): – – – – – – – – – – – –

$\vdash d\frown(n\frown(v\,\mathfrak{Z}\,q))$
$d\frown(a\frown q)$
$\mathrm{I}q$
$a\frown(n\frown(v\bumpeq\smile q\sqcup\perp q))$
$d\frown(n\frown(v\bumpeq\perp q))$ (ε

———— • ————

132 $\vdash d\frown(n\frown\perp q)$
$d\frown(a\frown q)$
$a\frown(n\frown\smile q)$

(197): – – – – – –

$\vdash d\frown(n\frown(v\bumpeq\perp q))$
$d\frown(a\frown q)$
$a\frown(n\frown\smile q)$
$n\frown v$ (ζ

(188, 191):: = = = = = =

$\vdash d\frown(n\frown(v\bumpeq\perp q))$
$d\frown(a\frown q)$
$a\frown(n\frown(v\bumpeq\smile q))$ (η

(ε): – – – – – – – – –

$\vdash d\frown(n\frown(v\,\mathfrak{Z}\,q))$
$d\frown(a\frown q)$
$\mathrm{I}q$
$a\frown(n\frown(v\bumpeq\smile q\sqcup\perp q))$
$a\frown(n\frown(v\bumpeq\smile q))$ (ϑ

(368, 404):: = = = = = = = = =

$\vdash d\frown(n\frown(v\,\mathfrak{Z}\,q))$
$d\frown(a\frown q)$
$\mathrm{I}q$
$a\frown(n\frown(v\,\mathfrak{z}\,q))$ (ι

×

$\vdash a\frown(n\frown(v\,\mathfrak{z}\,q))$
$d\frown(a\frown q)$
$\mathrm{I}q$
$d\frown(n\frown(v\,\mathfrak{Z}\,q))$ (κ

(IIa):: – – – – – – – –

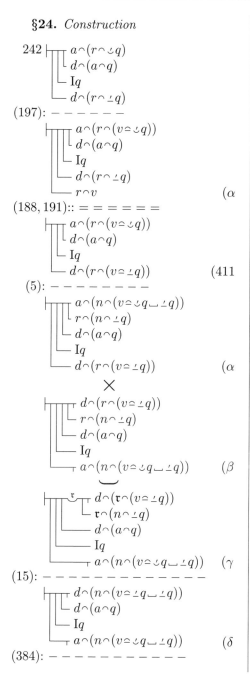

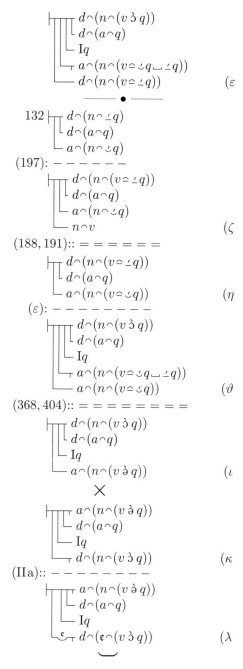

$\vdash a\frown(n\frown(v\,\mathfrak{z}\,q))$
$d\frown(a\frown q)$
$\mathrm{I}q$
$\overset{\mathfrak{e}}{\smile}\ d\frown(\mathfrak{e}\frown(v\,\mathfrak{Z}\,q))$ (λ

```
├─𝔞─ a⌢(𝔞⌢(v Ɛ q))
  ── d⌢(a⌢q)
  ── Iq
  ─𝔢─ d⌢(𝔢⌢(v Ɛ q))                    (μ
(358): - - - - - - - - - -
├── a⌢(r⌢(v⌢⌣q))
  ── r⌢(r⌢⊥q)
  ── d⌢(a⌢q)
  ── Iq
  ─𝔢─ d⌢(𝔢⌢(v Ɛ q))                    (ν
×
├─𝔢─ d⌢(𝔢⌢(v Ɛ q))
  ── r⌢(r⌢⊥q)
  ── d⌢(a⌢q)
  ── Iq
  ── a⌢(r⌢(v⌢⌣q))                      (ξ
(411):: - - - - - - - - - - - -
├─𝔢─ d⌢(𝔢⌢(v Ɛ q))
  ── r⌢(r⌢⊥q)
  ── d⌢(a⌢q)
  ── Iq
  ── d⌢(r⌢(v⌢⊥q))                      (o
×
├── d⌢(a⌢q)
  ── r⌢(r⌢⊥q)
  ─𝔢─ d⌢(𝔢⌢(v Ɛ q))
  ── Iq
  ── d⌢(r⌢(v⌢⊥q))                      (π
‿
├─𝔢─ d⌢(𝔢⌢q)
  ─𝔢─ d⌢(𝔢⌢(v Ɛ q))
  ── Iq
  ── r⌢(r⌢⊥q)
  ── d⌢(r⌢(v⌢⊥q))                      (412
——— • ———
IIa ├── d⌢(a⌢q)
    ─𝔢─ d⌢(𝔢⌢q)
(22): ————————
├── a⌢(d⌢⋠q)
  ─𝔢─ d⌢(𝔢⌢q)                          (α
‿
├─𝔢─ 𝔢⌢(d⌢⋠q)
  ─𝔢─ d⌢(𝔢⌢q)                          (β
(125):: - - - - - - - -
├── r⌢(d⌢⊥⋠q)
  ─𝔢─ d⌢(𝔢⌢q)                          (γ
×
├─𝔢─ d⌢(𝔢⌢q)
  ── r⌢(d⌢⊥⋠q)                         (δ
(302):: - - - - - - - - -
├─𝔢─ d⌢(𝔢⌢q)
  ── d⌢(r⌢⊥q)                          (413
(188):: - - - - - - - -
├─𝔢─ d⌢(𝔢⌢q)
  ── d⌢(r⌢(v⌢⊥q))                      (α
×
├── d⌢(r⌢(v⌢⊥q)
  ─𝔢─ d⌢(𝔢⌢q)                          (β
(412):: - - - - - - - - - -
├── d⌢(r⌢(v⌢⊥q))
  ── d⌢(r⌢(v⌢⊥q))
  ─𝔢─ d⌢(𝔢⌢(v Ɛ q))
  ── r⌢(r⌢⊥q)
  ── Iq                                (γ
(Ig): - - - - - - - - - - -
├── d⌢(r⌢(v⌢⊥q))
  ─𝔢─ d⌢(𝔢⌢(v Ɛ q))
  ── r⌢(r⌢⊥q)
  ── Iq                                (δ
(IIa):: - - - - - - - - - - -
├── d⌢(r⌢(v⌢⊥q))
  ─𝔢─ d⌢(𝔢⌢(v Ɛ q))
  ─𝔦─ 𝔦⌢(𝔦⌢⊥q)
  ── Iq                                (ε
‿
├─𝔢─ d⌢(𝔢⌢(v⌢⊥q))
  ─𝔢─ d⌢(𝔢⌢(v Ɛ q))
  ─𝔦─ 𝔦⌢(𝔦⌢⊥q)
  ── Iq                                (ζ
×
```

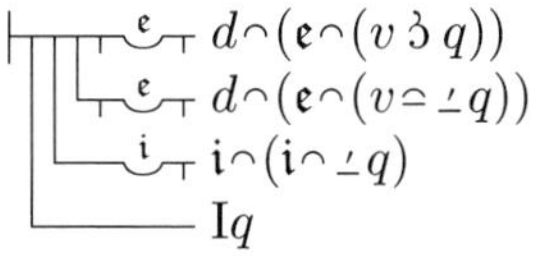
(η

(410):: – – – – – – – – – – –

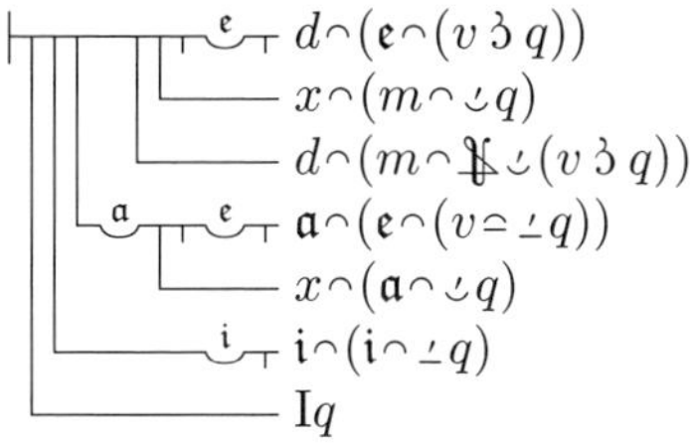
(ϑ

×

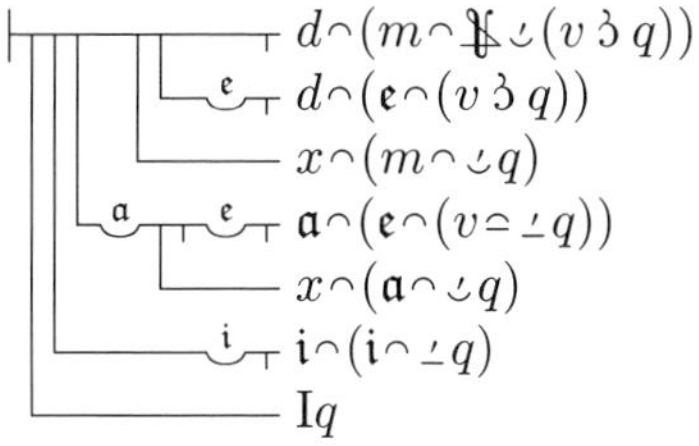
(ι

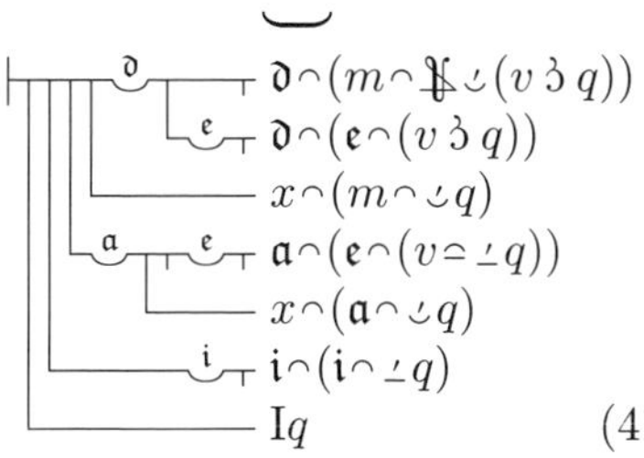
(414

———— • ————

389

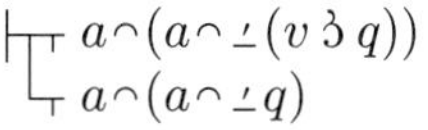

(IIa):: – – – – – – – – –

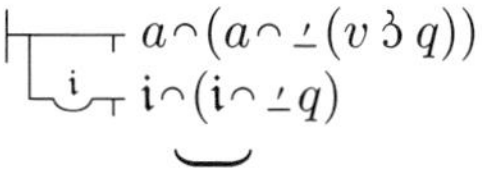
(α

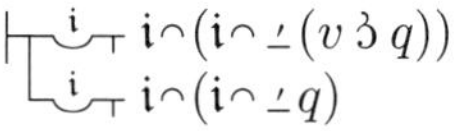
(415

(262): – – – – – – – – – –

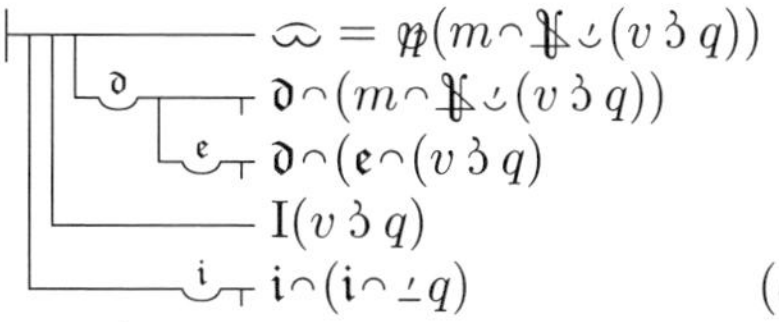
(α

(414, 397):: = = = = = = = = = = =

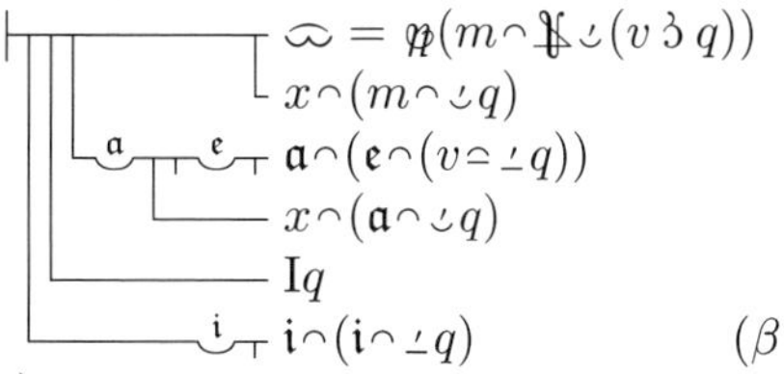
(β

(401):: – – – – – – – – – – – – – – –

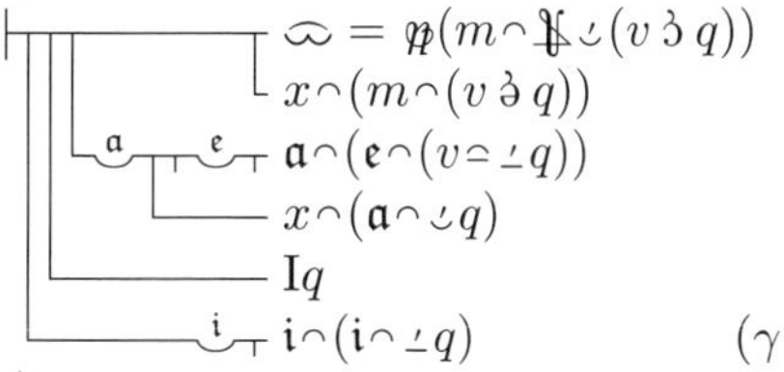
(γ

(408): – – – – – – – – – – – – – – –

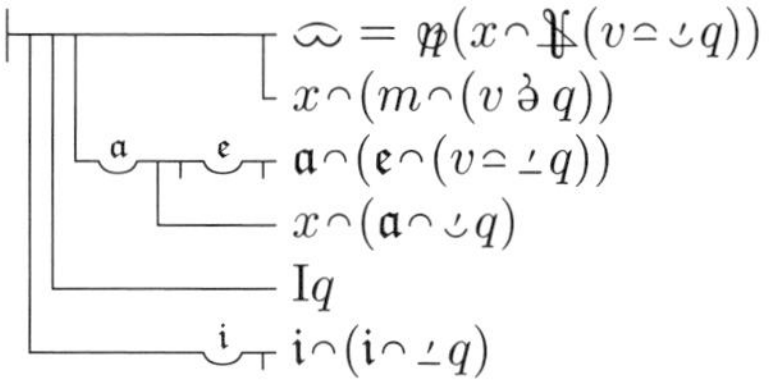
(416

d) Proof of the proposition

$$\begin{array}{l} \vdash\!\!\top\!\!\top\!\!\top\!\!\top\; \mathfrak{Q} \frown [\not{\mathfrak{p}}(x \frown \not{\mathfrak{X}}(v \,\underline{\frown}\, \smile q)) \frown \smile \mathfrak{f}] \\ \quad x \frown (m \frown (v \,\grave{\vartheta}\, q)) \\ \quad \mathrm{I}q \\ \quad c \frown (c \frown \perp q) \\ \quad \overset{\mathfrak{e}}{\smile}\!\top\; c \frown (\mathfrak{e} \frown (v \,\underline{\frown}\, \perp q)) \\ \quad x \frown (c \frown \smile q) \end{array} ,$$

and end of section M

§25. *Analysis*

Finally, it remains to be shown that the cardinal number of the members of the q-series starting with x that fall under the v-concept is finite if some member (c) of this series is succeeded by no member falling under the v-concept, provided our assumptions about q and m are fulfilled. This will be accomplished by means of proposition (326) in the form

$$\text{‘}\vdash \mathfrak{Q} \frown [\not{\mathfrak{p}}(m; n \,\underline{\measuredangle}\, (v \,\grave{\vartheta}\, q)) \frown \smile \mathfrak{f}]\text{’}$$

We have to show that, on our assumptions, the $(v \grave{\vartheta} q)$-series running from m to n contains all and only those members of the q-series starting with x that fall under the v-concept, if n is such a member that all members of the q-series running from m to c that fall under the v-concept belong to the $(v \grave{\vartheta} q)$-series running from m to n:

$$\begin{array}{l} \vdash\!\!\top\!\!\top\!\!\top\!\!\top\!\!\top\; [\!-\!\!-\, \mathfrak{a} \frown (x \frown \not{\mathfrak{X}}(v \,\underline{\frown}\, \smile q))] = [\!-\!\!-\, \mathfrak{a} \frown (m; n \,\underline{\measuredangle}\, (v \,\grave{\vartheta}\, q))] \\ \quad x \frown (m \frown (v \,\grave{\vartheta}\, q)) \\ \quad \mathrm{I}q \\ \quad c \frown (c \frown \perp q) \\ \quad \overset{\mathfrak{e}}{\smile}\!\top\; c \frown (\mathfrak{e} \frown (v \,\underline{\frown}\, \perp q)) \\ \quad x \frown (c \frown \smile q) \\ \quad \overset{\mathfrak{e}}{\smile}\!\top\!\!\top\; \mathfrak{e} \frown (m; n \,\underline{\measuredangle}\, (v \,\grave{\vartheta}\, q)) \\ \qquad \mathfrak{e} \frown (m; c \,\underline{\measuredangle}\, q) \\ \qquad \mathfrak{e} \frown v \end{array} , \qquad (\alpha$$

From (369) we straightforwardly derive the proposition

$$\begin{array}{l} \vdash\!\!\top\!\!\top\; x \frown (\mathfrak{a} \frown (v \,\underline{\frown}\, \smile q)) \\ \quad \mathfrak{a} \frown (m; n \,\underline{\measuredangle}\, (v \,\grave{\vartheta}\, q)) \\ \quad x \frown (m \frown (v \,\grave{\vartheta}\, q)) \end{array} , \qquad (\beta$$

In order to obtain in addition a proposition with

$$\begin{array}{l} \text{‘}\top\; \mathfrak{a} \frown (m; n \,\underline{\measuredangle}\, (v \,\grave{\vartheta}\, q)) \\ \quad x \frown (\mathfrak{a} \frown (v \,\underline{\frown}\, \smile q)) \end{array} ,$$

as supercomponent, we write (IIa) in the form

$$\begin{array}{l} \vdash\!\!\top\!\!\top\; \mathfrak{a} \frown (m; n \,\underline{\measuredangle}\, (v \,\grave{\vartheta}\, q)) \\ \quad \mathfrak{a} \frown (m; c \,\underline{\measuredangle}\, q) \\ \quad \mathfrak{a} \frown v \\ \quad \overset{\mathfrak{e}}{\smile}\!\top\!\!\top\; \mathfrak{e} \frown (m; n \,\underline{\measuredangle}\, (v \,\grave{\vartheta}\, q)) \\ \qquad \mathfrak{e} \frown (m; c \,\underline{\measuredangle}\, q) \\ \qquad \mathfrak{e} \frown v \end{array} ,$$

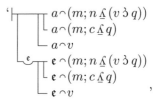

and apply to it the proposition

‘$\vdash$ $a \frown (m; c \,\underline{\curlyvee}\, q)$
$x \frown (m \frown (v \,\grave{\partial}\, q))$
$x \frown (a \frown (v \,\hat{=}\, \smile q))$
$\mathrm{I}q$
$c \frown (c \frown \perp q)$
$\mathfrak{e}$: $c \frown (\mathfrak{e} \frown (v \,\hat{=}\, \perp q))$
$x \frown (c \frown \smile q)$ ’, (γ

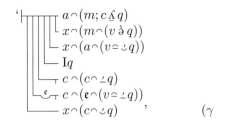

which we derive from (407) and

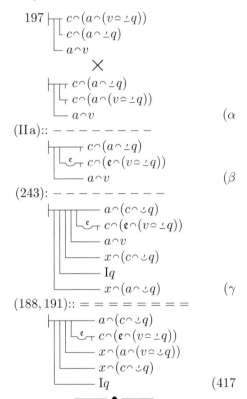

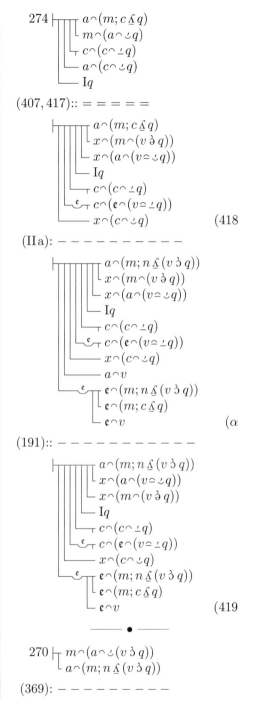

‘$\vdash$ $a \frown (c \frown \smile q)$
$\mathfrak{e}$: $c \frown (\mathfrak{e} \frown (v \,\hat{=}\, \perp q))$
$x \frown (a \frown (v \,\hat{=}\, \smile q))$
$x \frown (c \frown \smile q)$
$\mathrm{I}q$ ’, (δ

We prove (δ) using (243).

§26. *Construction*

197 $\vdash$ $c \frown (a \frown (v \,\hat{=}\, \perp q))$
$c \frown (a \frown \perp q)$
$a \frown v$

×

$\vdash$ $c \frown (a \frown \perp q)$
$c \frown (a \frown (v \,\hat{=}\, \perp q))$
$a \frown v$ (α

(IIa):: – – – – – – – – –

$\vdash$ $c \frown (a \frown \perp q)$
$\mathfrak{e}$: $c \frown (\mathfrak{e} \frown (v \,\hat{=}\, \perp q))$
$a \frown v$ (β

(243): – – – – – – – – – –

$\vdash$ $a \frown (c \frown \smile q)$
$\mathfrak{e}$: $c \frown (\mathfrak{e} \frown (v \,\hat{=}\, \perp q))$
$a \frown v$
$x \frown (c \frown \smile q)$
$\mathrm{I}q$
$x \frown (a \frown \smile q)$ (γ

(188, 191):: = = = = = = = =

$\vdash$ $a \frown (c \frown \smile q)$
$\mathfrak{e}$: $c \frown (\mathfrak{e} \frown (v \,\hat{=}\, \perp q))$
$x \frown (a \frown (v \,\hat{=}\, \smile q))$
$x \frown (c \frown \smile q)$
$\mathrm{I}q$ (417

——— • ———

274 $\vdash$ $a \frown (m; c \,\underline{\curlyvee}\, q)$
$m \frown (a \frown \smile q)$
$c \frown (c \frown \perp q)$
$a \frown (c \frown \smile q)$
$\mathrm{I}q$

(407, 417):: = = = = =

$\vdash$ $a \frown (m; c \,\underline{\curlyvee}\, q)$
$x \frown (m \frown (v \,\grave{\partial}\, q))$
$x \frown (a \frown (v \,\hat{=}\, \smile q))$
$\mathrm{I}q$
$c \frown (c \frown \perp q)$
$\mathfrak{e}$: $c \frown (\mathfrak{e} \frown (v \,\hat{=}\, \perp q))$
$x \frown (c \frown \smile q)$ (418

(IIa): – – – – – – – – – – –

$\vdash$ $a \frown (m; n \,\underline{\curlyvee}\, (v \,\grave{\partial}\, q))$
$x \frown (m \frown (v \,\grave{\partial}\, q))$
$x \frown (a \frown (v \,\hat{=}\, \smile q))$
$\mathrm{I}q$
$c \frown (c \frown \perp q)$
$\mathfrak{e}$: $c \frown (\mathfrak{e} \frown (v \,\hat{=}\, \perp q))$
$x \frown (c \frown \smile q)$
$a \frown v$
$\mathfrak{e}$: $\mathfrak{e} \frown (m; n \,\underline{\curlyvee}\, (v \,\grave{\partial}\, q))$
$\mathfrak{e} \frown (m; c \,\underline{\curlyvee}\, q)$
$\mathfrak{e} \frown v$ (α

(191):: – – – – – – – – – – – –

$\vdash$ $a \frown (m; n \,\underline{\curlyvee}\, (v \,\grave{\partial}\, q))$
$x \frown (a \frown (v \,\hat{=}\, \smile q))$
$x \frown (m \frown (v \,\grave{\partial}\, q))$
$\mathrm{I}q$
$c \frown (c \frown \perp q)$
$\mathfrak{e}$: $c \frown (\mathfrak{e} \frown (v \,\hat{=}\, \perp q))$
$x \frown (c \frown \smile q)$
$\mathfrak{e}$: $\mathfrak{e} \frown (m; n \,\underline{\curlyvee}\, (v \,\grave{\partial}\, q))$
$\mathfrak{e} \frown (m; c \,\underline{\curlyvee}\, q)$
$\mathfrak{e} \frown v$ (419

——— • ———

270 $\vdash$ $m \frown (a \frown \smile (v \,\grave{\partial}\, q))$
$a \frown (m; n \,\underline{\curlyvee}\, (v \,\grave{\partial}\, q))$

(369): – – – – – – – – – –

$x \frown (a \frown (v \supseteq \cup q))$
$a \frown (m; n \underline{\S} (v \Im q))$
$x \frown (m \frown (v \ni q))$ (α

(IV a): – – – – – – – – – –

$[— x \frown (a \frown (v \supseteq \cup q))] = [— a \frown (m; n \underline{\S} (v \Im q))]$
$a \frown (m; n \underline{\S} (v \Im q))$
$x \frown (a \frown (v \supseteq \cup q))$
$x \frown (m \frown (v \ni q))$ (β

(22): ——————————————

$[— a \frown (x \frown \nmid\!\S (v \supseteq \cup q))] = [— a \frown (m; n \underline{\S} (v \Im q))]$
$a \frown (m; n \underline{\S} (v \Im q))$
$x \frown (a \frown (v \supseteq \cup q))$
$x \frown (m \frown (v \ni q))$ (γ

(419):: –

$[— a \frown (x \frown \nmid\!\S (v \supseteq \cup q))] = [— a \frown (m; n \underline{\S} (v \Im q))]$
$x \frown (m \frown (v \ni q))$
$\mathrm{I}q$
$c \frown (c \frown \perp q)$
$\mathfrak{e}$: $c \frown (\mathfrak{e} \frown (v \supseteq \perp q))$
$x \frown (c \frown \cup q)$
$\mathfrak{e}$: $\mathfrak{e} \frown (m; n \underline{\S} (v \Im q))$
$\mathfrak{e} \frown (m; c \underline{\S} q)$
$\mathfrak{e} \frown v$ (δ

⌣

$\mathfrak{a}$: $[— \mathfrak{a} \frown (x \frown \nmid\!\S (v \supseteq \cup q))] = [— \mathfrak{a} \frown (m; n \underline{\S} (v \Im q))]$
$x \frown (m \frown (v \ni q))$
$\mathrm{I}q$
$c \frown (c \frown \perp q)$
$\mathfrak{e}$: $c \frown (\mathfrak{e} \frown (v \supseteq \perp q))$
$x \frown (c \frown \cup q)$
$\mathfrak{e}$: $\mathfrak{e} \frown (m; n \underline{\S} (v \Im q))$
$\mathfrak{e} \frown (m; c \underline{\S} q)$
$\mathfrak{e} \frown v$ (420

——— • ———

326 ⊢ $\mathfrak{Q} \frown [\mathfrak{p}(m; n \underline{\S} (v \Im q)) \frown \cup \mathfrak{f}]$

(III a): ——————————————

$\mathfrak{Q} \frown [\mathfrak{p}(x \frown \nmid\!\S (v \supseteq \cup q)) \frown \cup \mathfrak{f}]$
$\mathfrak{p}(x \frown \nmid\!\S (v \supseteq \cup q)) = \mathfrak{p}(m; n \underline{\S} (v \Im q))$ (α

(96):: – – – – – – – – – – – – – – – – – –

$\mathfrak{Q} \frown [\mathfrak{p}(x \frown \nmid\!\S (v \supseteq \cup q)) \frown \cup \mathfrak{f}]$
$\mathfrak{a}$: $[— \mathfrak{a} \frown (x \frown \nmid\!\S (v \supseteq \cup q))] = [— \mathfrak{a} \frown (m; n \underline{\S} (v \Im q))]$ (β

(420):: –

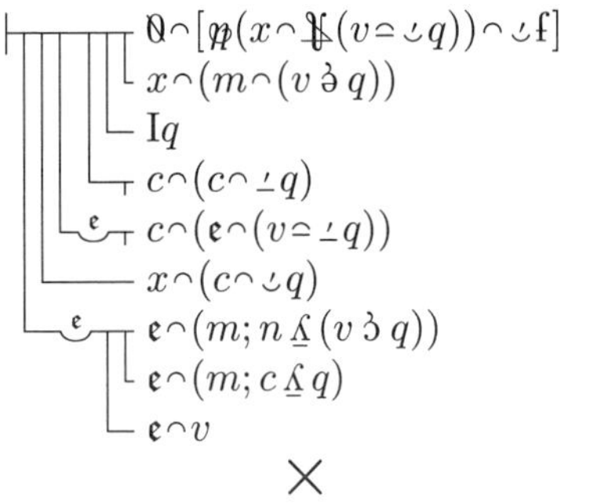

(γ

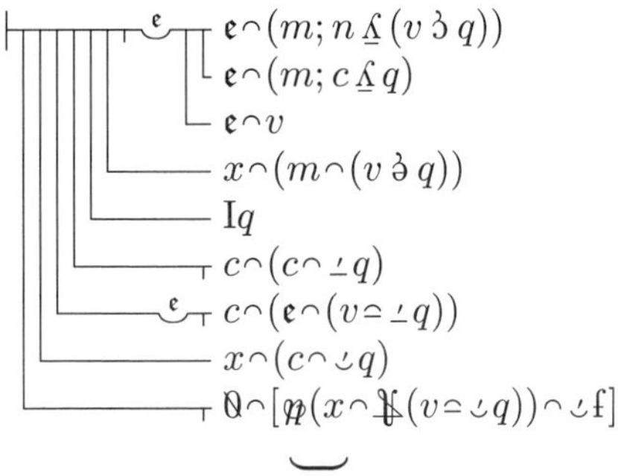

(δ

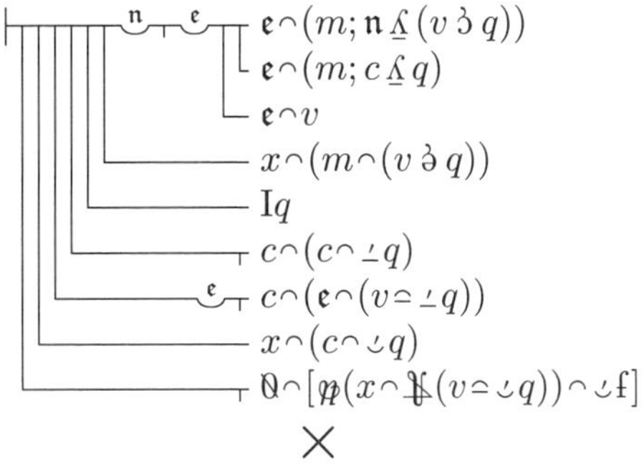

(ε

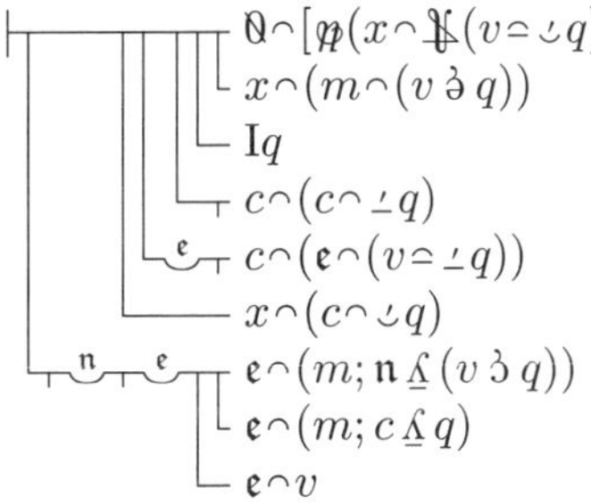

(ζ

(403):: — — — — — — — — — — — — — — — —

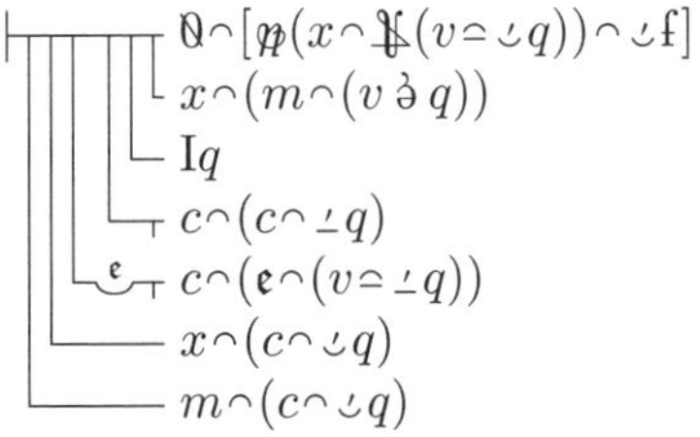

(421

§27. *Analysis*

We need to eliminate the subcomponent '— $m \frown (c \frown \smile q)$' from the last proposition. This is accomplished by means of the proposition

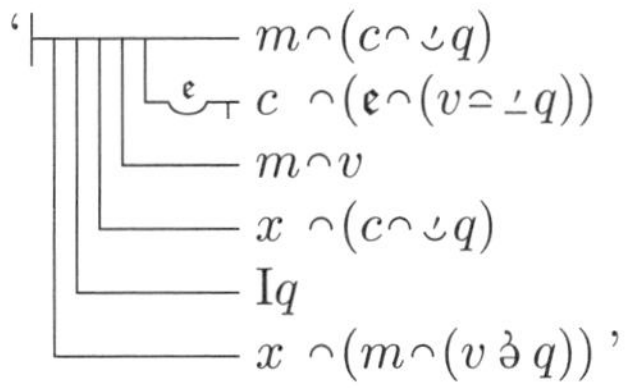

which we prove using (243), by showing that c cannot precede m in the q-series since then there would be a member following c, namely m, that fell under the v-concept.

§28. *Construction*

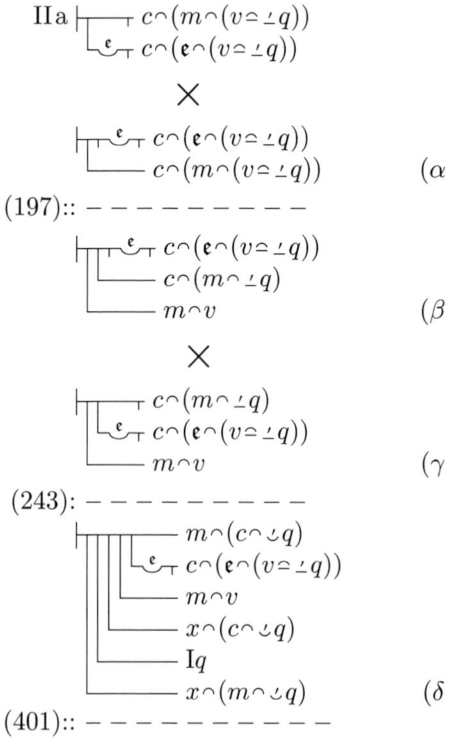

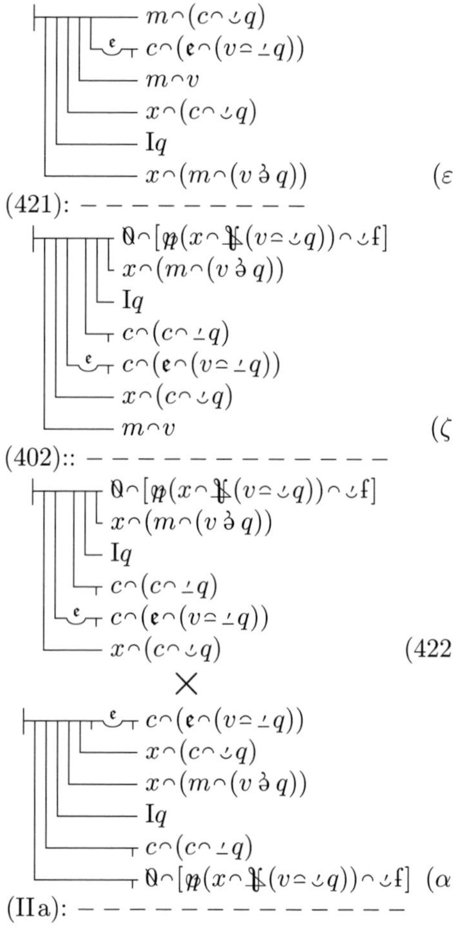

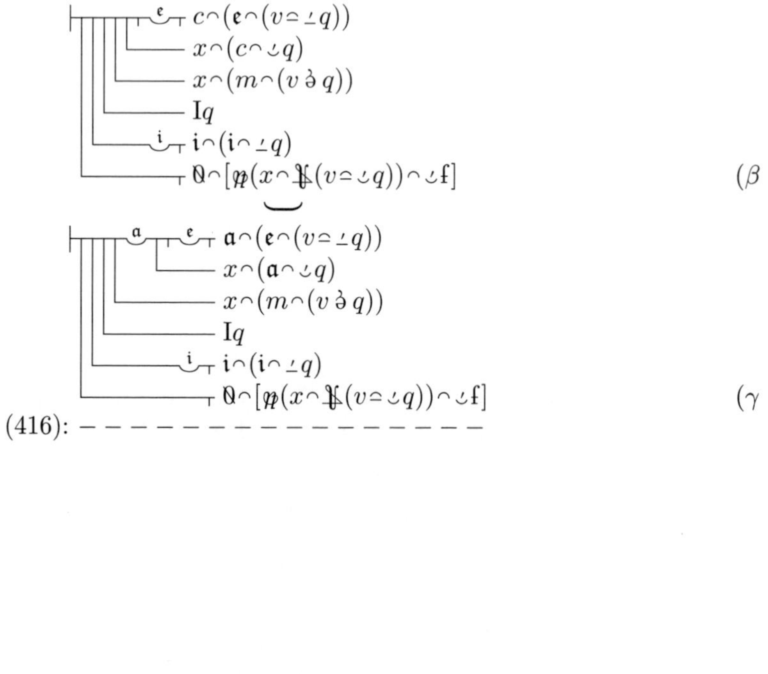

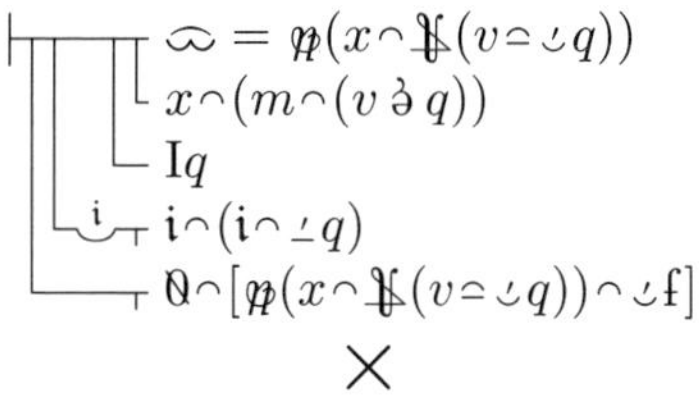

(δ

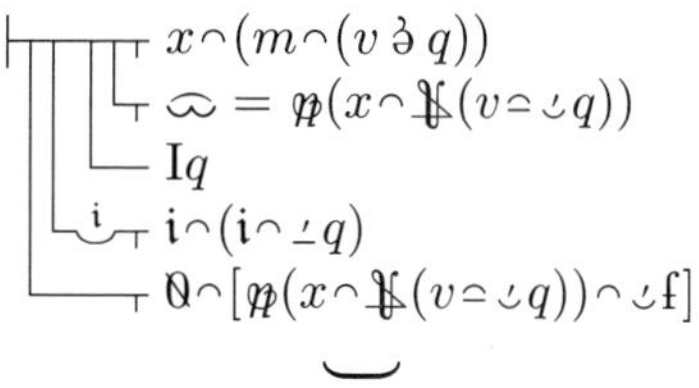

(ε

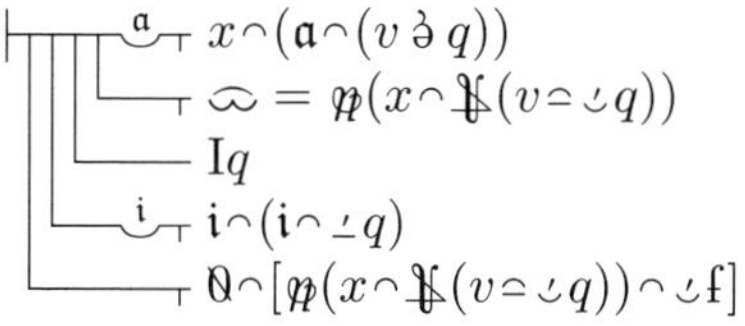

(ζ

(359):

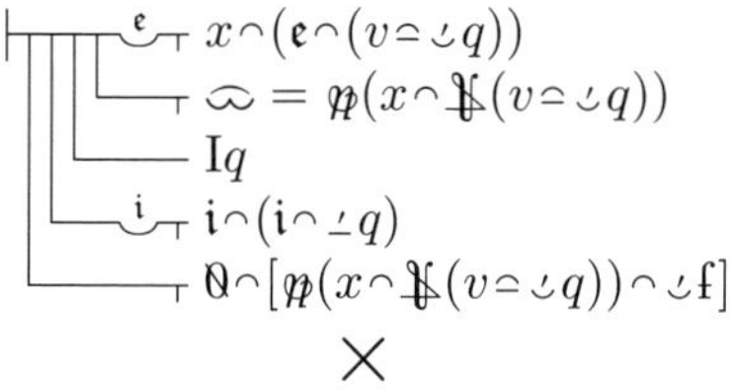

(η

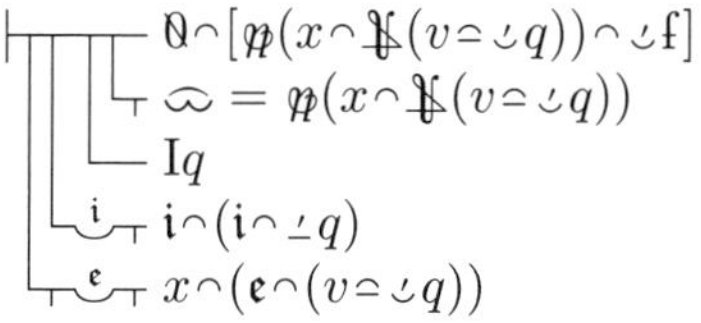

(423

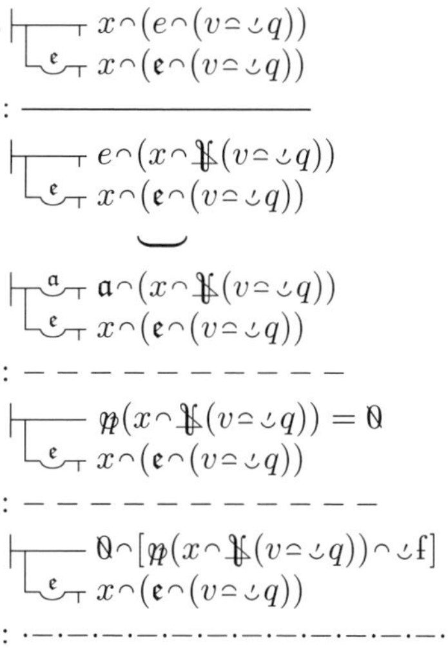

IIa

(22):

(α

(β

(97):

(γ

(139):

(δ

(423):

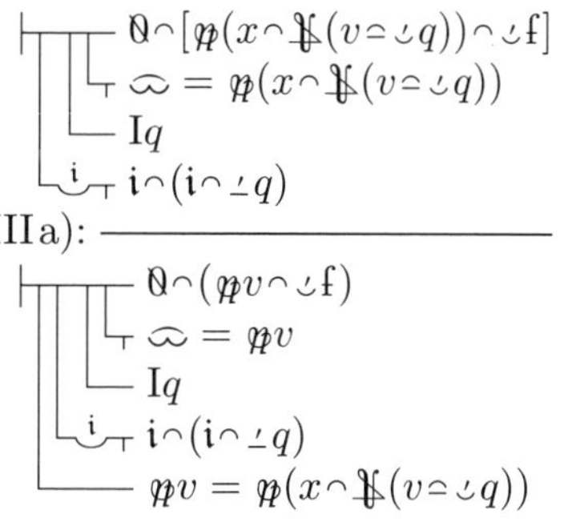

(424

(IIIa):

(425

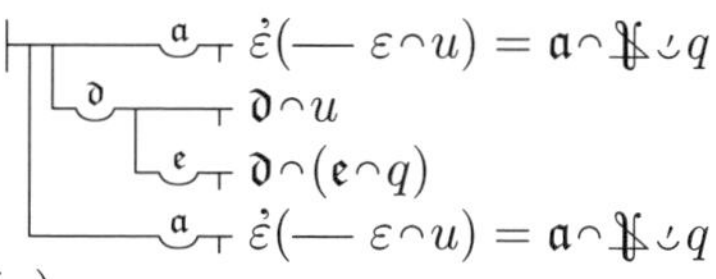

I

(IIa)::

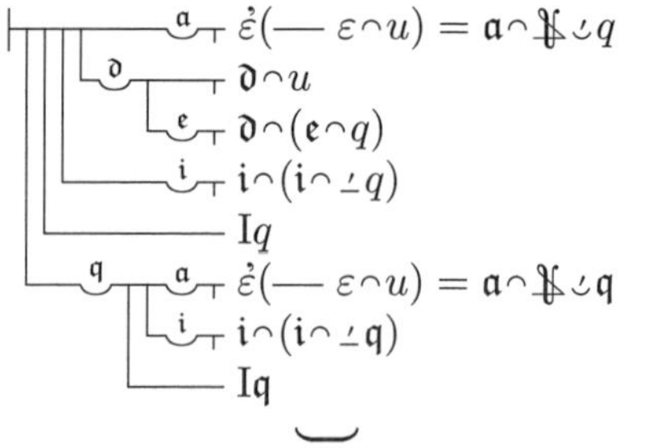

(α

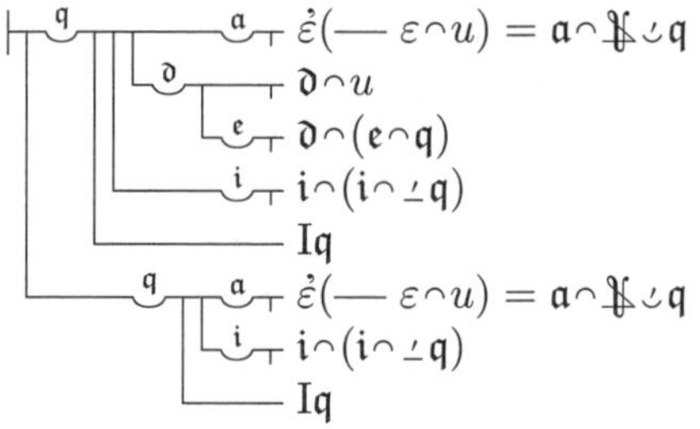

(β

(206): — — — — — — — — — — — — — — — — — —

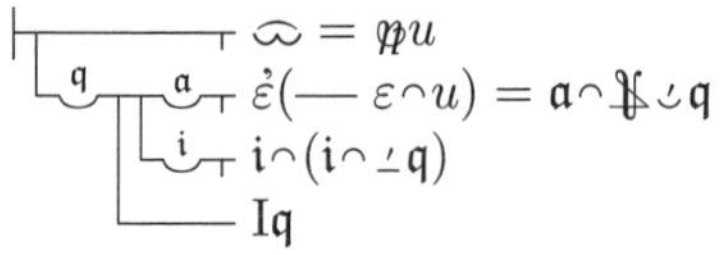

(426

——— • ———

191 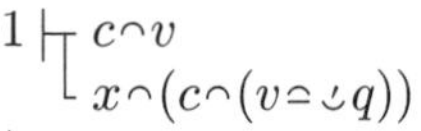

(IVa): ————————

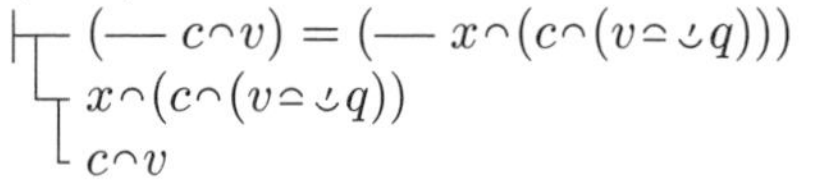

(α

(22): ————————————————

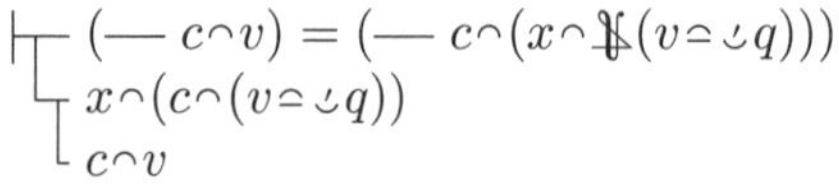

(β

——— • ———

23 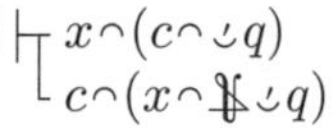

(IIIa): ————————

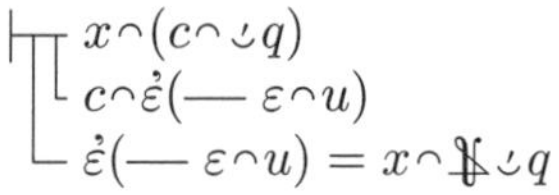

(γ

(77):: — — — — — — — — — — — —

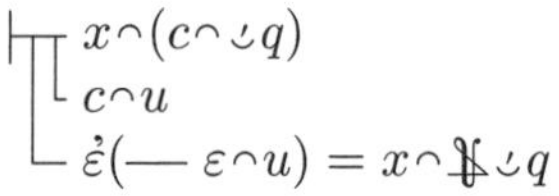

(δ

(197): — — — — — — — — — — — —

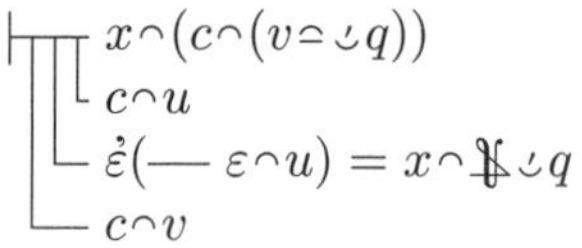

(ε

(IIa):: — — — — — — — — — — — —

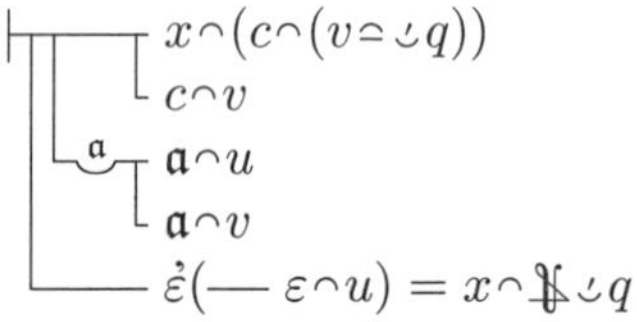

(ζ

(β): — — — — — — — — — — — — —

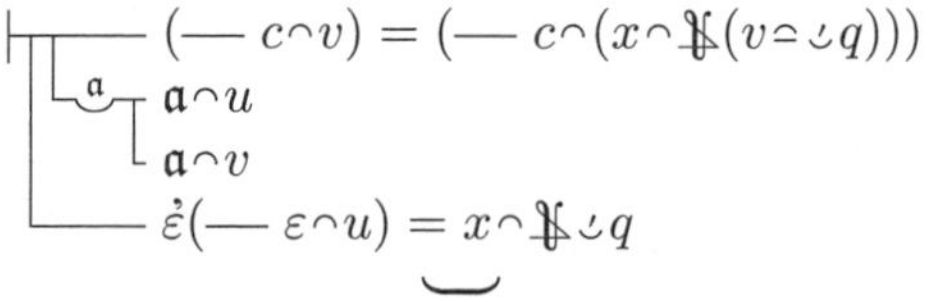

(η

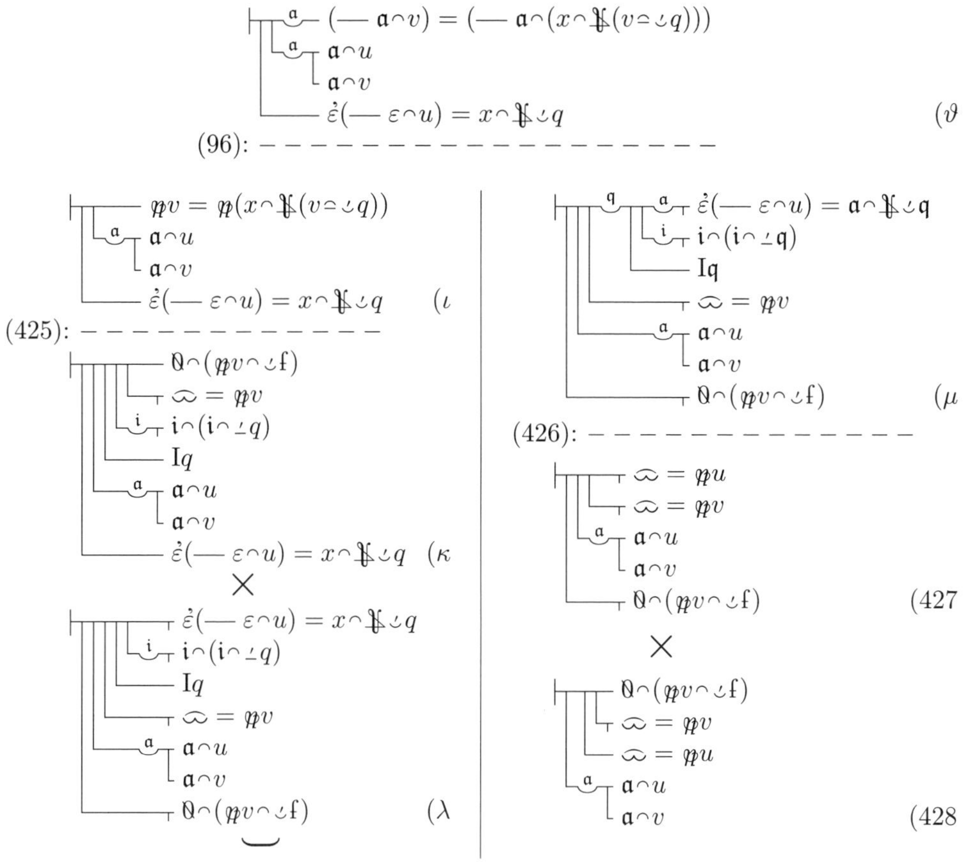

N. Proof of the proposition

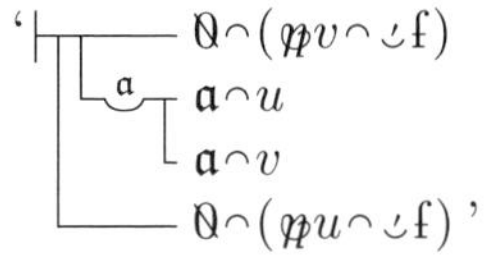

§29. *Analysis*

We now prove the proposition that a finite cardinal number belongs to a concept if a finite cardinal number belongs to a concept superordinate to it.

Let the u-concept be the superordinate and the v-concept be the subordinate. We have seen at the end of volume I that the objects falling under the u-concept can be ordered into a series that runs from a specific object to a specific object whenever the cardinal number of the u-concept is finite. Instead of the u-concept, we thus first take the q-series running from x to c and focus on the case where there are indeed objects that fall under the v-concept. Thus, let m be the first member in our series to fall under the v-concept and n be the last. We now apply our proposition (422), first eliminating 'm' by means of (357). We replace the subcomponents

'$— x \frown (c \frown \smile q)$', '$— x \frown (a \frown \smile q)$', and '$\overset{\mathfrak{e}}{\smile}\!\top c \frown (\mathfrak{e} \frown (v \simeq \,\underline{\cdot}\, q))$'

by the subcomponents

'$— a \frown v$' and '$\overset{\mathfrak{a}}{\smile}\!\top \mathfrak{a} \frown (x; c \underline{\measuredangle} q)$ / $\mathfrak{a} \frown v$'

We thus obtain the proposition

$$\vdash \mathfrak{Q} \frown [\not p(x \frown \not\!\!\!\!\mathfrak{X}(v \simeq \smile q)) \frown \smile \mathfrak{f}]$$
$$a \frown v$$
$$\overset{\mathfrak{a}}{\smile}\!\top \mathfrak{a} \frown (x; c \underline{\measuredangle} q)$$
$$\mathfrak{a} \frown v \quad , \qquad (\alpha$$

In place of '$\not p(x \frown \not\!\!\!\!\mathfrak{X}(v \simeq \smile q))$', we then introduce '$\not p v$' by means of the proposition

$$\vdash \not p v = \not p(x \frown \not\!\!\!\!\mathfrak{X}(v \simeq \smile q))$$
$$\overset{\mathfrak{a}}{\smile}\!\top \mathfrak{a} \frown (x; c \underline{\measuredangle} q)$$
$$\mathfrak{a} \frown v \quad , \qquad (\beta$$

which follows by means of (IVa) from (191) and the proposition

$$\vdash x \frown (a \frown (v \simeq \smile q))$$
$$a \frown v$$
$$\overset{\mathfrak{a}}{\smile}\!\top \mathfrak{a} \frown (x; c \underline{\measuredangle} q)$$
$$\mathfrak{a} \frown v \quad , \qquad (\gamma$$

§30. *Construction*

296 $\vdash c \frown (c \frown \,\underline{\cdot}\, q)$
$a \frown (a \frown \,\underline{\cdot}\, q)$
$\mathrm{I}q$
$a \frown (c \frown \smile q)$

×

$\vdash a \frown (a \frown \,\underline{\cdot}\, q)$
$c \frown (c \frown \,\underline{\cdot}\, q)$
$\mathrm{I}q$
$a \frown (c \frown \smile q)$ (429

—— • ——

269 $\vdash a \frown (c \frown \smile q)$
$a \frown (x; c \underline{\measuredangle} q)$

(276): — — — — — —

$\vdash c \frown (c \frown \,\underline{\cdot}\, q)$
$c \frown (a \frown \,\underline{\cdot}\, q)$
$a \frown (x; c \underline{\measuredangle} q)$ (α

(272): — — — — — —

$\vdash a \frown (x; c \underline{\measuredangle} q)$
$a \frown (x; c \underline{\measuredangle} q)$
$c \frown (a \frown \,\underline{\cdot}\, q)$ (β

(Ig): — — — — — — —

$\vdash a \frown (x; c \underline{\measuredangle} q)$
$c \frown (a \frown \,\underline{\cdot}\, q)$ (430

(188):: — — — — — —

$\vdash a \frown (x; c \underline{\measuredangle} q)$
$c \frown (a \frown (v \simeq \,\underline{\cdot}\, q))$ (α

×

$\vdash c \frown (a \frown (v \simeq \,\underline{\cdot}\, q))$
$a \frown (x; c \underline{\measuredangle} q)$ (β

(IIa):: — — — — — — — —

$\vdash c \frown (a \frown (v \simeq \,\underline{\cdot}\, q))$
$a \frown v$
$\overset{\mathfrak{a}}{\smile}\!\top \mathfrak{a} \frown (x; c \underline{\measuredangle} q)$
$\mathfrak{a} \frown v$ (γ

(192): -·-·-·-·-·-·-·-

$\vdash c \frown (a \frown (v \simeq \,\underline{\cdot}\, q))$
$\overset{\mathfrak{a}}{\smile}\!\top \mathfrak{a} \frown (x; c \underline{\measuredangle} q)$
$\mathfrak{a} \frown v$ (431

$\smile$

$\vdash \overset{\mathfrak{e}}{\smile}\!\top c \frown (\mathfrak{e} \frown (v \simeq \,\underline{\cdot}\, q))$
$\overset{\mathfrak{a}}{\smile}\!\top \mathfrak{a} \frown (x; c \underline{\measuredangle} q)$
$\mathfrak{a} \frown v$ (432

—— • ——

```
422 ├─ Ꝺ⌢[℘(x⌢ℵ(v⌢⌣q))⌢⌣𝔣]
     x⌢(m⌢(v ∂ q))
     Iq
     c⌢(c⌢⊥q)
     𝔢 c⌢(𝔢⌢(v⌢⊥q))
     x⌢(c⌢⌣q)
```

×

```
├─ x⌢(m⌢(v ∂ q))
   Ꝺ⌢[℘(x⌢ℵ(v⌢⌣q))⌢⌣𝔣]
   Iq
   c⌢(c⌢⊥q)
   𝔢 c⌢(𝔢⌢(v⌢⊥q))
   x⌢(c⌢⌣q)                                        (α
```

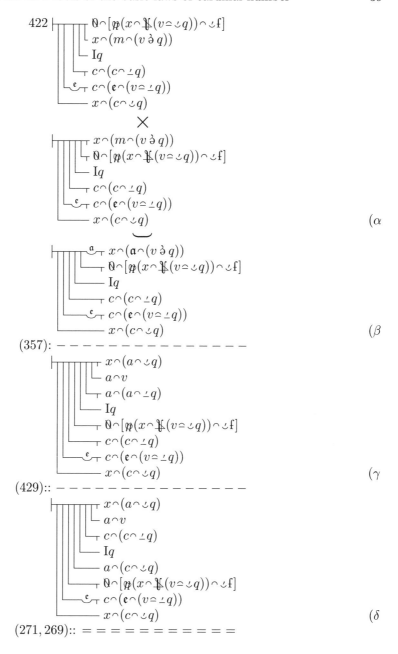

```
├─ 𝔞 x⌢(𝔞⌢(v ∂ q))
   Ꝺ⌢[℘(x⌢ℵ(v⌢⌣q))⌢⌣𝔣]
   Iq
   c⌢(c⌢⊥q)
   𝔢 c⌢(𝔢⌢(v⌢⊥q))
   x⌢(c⌢⌣q)                                        (β
(357): - - - - - - - - - - - - - - - -
├─ x⌢(a⌢⌣q)
   a⌢v
   a⌢(a⌢⊥q)
   Iq
   Ꝺ⌢[℘(x⌢ℵ(v⌢⌣q))⌢⌣𝔣]
   c⌢(c⌢⊥q)
   𝔢 c⌢(𝔢⌢(v⌢⊥q))
   x⌢(c⌢⌣q)                                        (γ
(429):: - - - - - - - - - - - - - - - -
├─ x⌢(a⌢⌣q)
   a⌢v
   c⌢(c⌢⊥q)
   Iq
   a⌢(c⌢⌣q)
   Ꝺ⌢[℘(x⌢ℵ(v⌢⌣q))⌢⌣𝔣]
   𝔢 c⌢(𝔢⌢(v⌢⊥q))
   x⌢(c⌢⌣q)                                        (δ
(271, 269):: = = = = = = = = = = = =
```

$x\frown(a\frown\grave{}q)$
$a\frown v$
$a\frown(x;c\,\underline{\measuredangle}\,q)$
$\mathrm{I}q$
$\mathfrak{d}\frown[\not{p}(x\frown\not{\mathfrak{t}}(v\asymp\grave{}q))\frown\grave{}\mathfrak{f}]$
$\mathfrak{e}$ $c\frown(\mathfrak{e}\frown(v\asymp\grave{}q))$
$x\frown(c\frown\grave{}q)$ (ε

(265, 323):: = = = = = = = = = = = = = = =

$x\frown(a\frown\grave{}q)$
$a\frown v$
$a\frown(x;c\,\underline{\measuredangle}\,q)$
$\mathfrak{d}\frown[\not{p}(x\frown\not{\mathfrak{t}}(v\asymp\grave{}q))\frown\grave{}\mathfrak{f}]$
$\mathfrak{e}$ $c\frown(\mathfrak{e}\frown(v\asymp\grave{}q))$ (ζ

×

$\mathfrak{d}\frown[\not{p}(x\frown\not{\mathfrak{t}}(v\asymp\grave{}q))\frown\grave{}\mathfrak{f}]$
$a\frown v$
$a\frown(x;c\,\underline{\measuredangle}\,q)$
$x\frown(a\frown\grave{}q)$
$\mathfrak{e}$ $c\frown(\mathfrak{e}\frown(v\asymp\grave{}q))$ (η

(432, 270):: = = = = = = = = = = = = =

$\mathfrak{d}\frown[\not{p}(x\frown\not{\mathfrak{t}}(v\asymp\grave{}q))\frown\grave{}\mathfrak{f}]$
$a\frown v$
$a\frown(x;c\,\underline{\measuredangle}\,q)$
$\mathfrak{a}$ $\mathfrak{a}\frown(x;c\,\underline{\measuredangle}\,q)$
$\mathfrak{a}\frown v$ (ϑ

(IIa):: – – – – – – – – – – – – –

$\mathfrak{d}\frown[\not{p}(x\frown\not{\mathfrak{t}}(v\asymp\grave{}q))\frown\grave{}\mathfrak{f}]$
$a\frown v$
$\mathfrak{a}$ $\mathfrak{a}\frown(x;c\,\underline{\measuredangle}\,q)$
$\mathfrak{a}\frown v$ (ι

(IIIa):: ————————————

$\mathfrak{d}\frown[\not{p}v\frown\grave{}\mathfrak{f}]$
$a\frown v$
$\mathfrak{a}$ $\mathfrak{a}\frown(x;c\,\underline{\measuredangle}\,q)$
$\mathfrak{a}\frown v$
$\not{p}v = \not{p}(x\frown\not{\mathfrak{t}}(v\asymp\grave{}q))$ (433

——— • ———

270 $x\frown(a\frown\grave{}q)$
$a\frown(x;c\,\underline{\measuredangle}\,q)$

(197): – – – – – –

$x\frown(a\frown(v\asymp\grave{}q))$
$a\frown v$
$a\frown(x;c\,\underline{\measuredangle}\,q)$ (α

(IIa):: – – – – – – – – –

$x\frown(a\frown(v\asymp\grave{}q))$
$a\frown v$
$\mathfrak{a}$ $\mathfrak{a}\frown(x;c\,\underline{\measuredangle}\,q)$
$\mathfrak{a}\frown v$ (β

——— • ———

191 $a\frown v$
$x\frown(a\frown(v\asymp\grave{}q))$

(IVa): ————————

$(\text{—}\, a\frown v) = (\text{—}\, x\frown(a\frown(v\asymp\grave{}q)))$
$x\frown(a\frown(v\asymp\grave{}q))$
$a\frown v$ (γ

(β):: – – – – – – – – – – – – – – – –

$$\vdash (— a \frown v) = (— x \frown (a \frown (v \bumpeq \smile q)))$$
$$\smile^{\mathfrak{a}}\top\ \mathfrak{a} \frown (x; c \,\underline{\measuredangle}\, q)$$
$$\mathfrak{a} \frown v \qquad (\delta$$

(22): ——————————

$$\vdash (— a \frown v) = (— a \frown (x \frown \S(v \bumpeq \smile q)))$$
$$\smile^{\mathfrak{a}}\top\ \mathfrak{a} \frown (x; c \,\underline{\measuredangle}\, q)$$
$$\mathfrak{a} \frown v \qquad (\varepsilon$$

$$\vdash\smile^{\mathfrak{a}}— (— \mathfrak{a} \frown v) = (— \mathfrak{a} \frown (x \frown \S(v \bumpeq \smile q)))$$
$$\smile^{\mathfrak{a}}\top\ \mathfrak{a} \frown (x; c \,\underline{\measuredangle}\, q)$$
$$\mathfrak{a} \frown v \qquad (\zeta$$

(96): - - - - - - - - - - - - - - - - - -

$$\vdash \mathfrak{p}v = \mathfrak{p}(x \frown \S(v \bumpeq \smile q))$$
$$\smile^{\mathfrak{a}}\top\ \mathfrak{a} \frown (x; c \,\underline{\measuredangle}\, q)$$
$$\mathfrak{a} \frown v \qquad (434$$

(433): - - - - - - - - - - - - -

$$\vdash \mathfrak{D} \frown (\mathfrak{p}v \frown \smile \mathfrak{f})$$
$$a \frown v$$
$$\smile^{\mathfrak{a}}\top\ \mathfrak{a} \frown (x; c \,\underline{\measuredangle}\, q)$$
$$\mathfrak{a} \frown v \qquad (435$$

§31. *Analysis*

We replace the lower subcomponent in (435) by

$$‘\smile^{\mathfrak{a}}\top\ \mathfrak{a} \frown u$$
$$\mathfrak{a} \frown v’ \quad \text{and}$$
$$‘— \dot{\varepsilon}(— \varepsilon \frown u) = x; c \,\underline{\measuredangle}\, q’$$

For this we require the proposition

$$‘\vdash a \frown (x; c \,\underline{\measuredangle}\, q)$$
$$a \frown v$$
$$\smile^{\mathfrak{a}}\top\ \mathfrak{a} \frown u$$
$$\mathfrak{a} \frown v$$
$$\dot{\varepsilon}(— \varepsilon \frown u) = x; c \,\underline{\measuredangle}\, q’ \qquad (\alpha$$

which follows from (44).

Then, ‘x’, ‘c’ and ‘q’ are to be eliminated. Since (345) is not entirely suitable for this, we derive the proposition

$$‘\vdash\smile^{\mathfrak{d}}\smile^{\mathfrak{a}}\smile^{\mathfrak{q}}\top\ \dot{\varepsilon}(— \varepsilon \frown u) = \mathfrak{d}; \mathfrak{a} \,\underline{\measuredangle}\, \mathfrak{q}$$
$$\smile^{\mathfrak{a}}\top\ \mathfrak{a} \frown u$$
$$\mathfrak{D} \frown (\mathfrak{p}u \frown \smile \mathfrak{f})’ \qquad (\beta$$

from (345). Finally, in order to eliminate ‘a’, we use the proposition

$$‘\vdash \mathfrak{D} \frown (\mathfrak{p}v \frown \smile \mathfrak{f})$$
$$\smile^{\mathfrak{a}}\top\ \mathfrak{a} \frown v’, \qquad (\gamma$$

which is derivable from (97).

§32. *Construction*

$$44 \vdash a \frown (x; c \,\underline{\measuredangle}\, q)$$
$$a \frown u$$
$$\dot{\varepsilon}(— \varepsilon \frown u) = x; c \,\underline{\measuredangle}\, q$$

(IIa):: - - - - - - - - - - -

$$\vdash a \frown (x; c \,\underline{\measuredangle}\, q)$$
$$a \frown v$$
$$\smile^{\mathfrak{a}}\top\ \mathfrak{a} \frown u$$
$$\mathfrak{a} \frown v$$
$$\dot{\varepsilon}(— \varepsilon \frown u) = x; c \,\underline{\measuredangle}\, q \qquad (436$$

$$\vdash\smile^{\mathfrak{a}}\top\ \mathfrak{a} \frown (x; c \,\underline{\measuredangle}\, q)$$
$$\mathfrak{a} \frown v$$
$$\smile^{\mathfrak{a}}\top\ \mathfrak{a} \frown u$$
$$\mathfrak{a} \frown v$$
$$\dot{\varepsilon}(— \varepsilon \frown u) = x; c \,\underline{\measuredangle}\, q \qquad (\alpha$$

(435): - - - - - - - - - - - - - -

$$\vdash \mathfrak{D} \frown (\mathfrak{p}v \frown \smile \mathfrak{f})$$
$$a \frown v$$
$$\smile^{\mathfrak{a}}\top\ \mathfrak{a} \frown u$$
$$\mathfrak{a} \frown v$$
$$\dot{\varepsilon}(— \varepsilon \frown u) = x; c \,\underline{\measuredangle}\, q \qquad (\beta$$

$$\times$$

$\vdash \dot{\varepsilon}(-\varepsilon \frown u) = x; c \,\underline{\measuredangle}\, q$
$a \frown v$
$[\mathfrak{a}]\ \mathfrak{a} \frown u$
$\mathfrak{a} \frown v$
$\not{0} \frown (\not{\wp} v \frown \dot{\smile}\mathfrak{f})$ (γ

$\vdash [\mathfrak{d}\ \mathfrak{a}\ \mathfrak{q}]\ \dot{\varepsilon}(-\varepsilon \frown u) = \mathfrak{d}; \mathfrak{a} \,\underline{\measuredangle}\, \mathfrak{q}$
$a \frown v$
$[\mathfrak{a}]\ \mathfrak{a} \frown u$
$\mathfrak{a} \frown v$
$\not{0} \frown (\not{\wp} v \frown \dot{\smile}\mathfrak{f})$ (δ

×

$\vdash a \frown v$
$[\mathfrak{d}\ \mathfrak{a}\ \mathfrak{q}]\ \dot{\varepsilon}(-\varepsilon \frown u) = \mathfrak{d}; \mathfrak{a} \,\underline{\measuredangle}\, \mathfrak{q}$
$[\mathfrak{a}]\ \mathfrak{a} \frown u$
$\mathfrak{a} \frown v$
$\not{0} \frown (\not{\wp} v \frown \dot{\smile}\mathfrak{f})$ (437

——— • ———

314 $\vdash n = \not{\wp}(\not{1}; n \,\underline{\measuredangle}\, \mathfrak{f})$
$\not{0} \frown (n \frown \dot{\smile}\mathfrak{f})$

(IIIf): - - - - - - -

$\vdash \not{\wp}(\not{1}; n \,\underline{\measuredangle}\, \mathfrak{f}) = n$
$\not{0} \frown (n \frown \dot{\smile}\mathfrak{f})$ (438

——— • ———

IIa $\vdash a \frown u$
$a \frown v$
$[\mathfrak{a}]\ \mathfrak{a} \frown u$
$\mathfrak{a} \frown v$

×

$\vdash a \frown v$
$a \frown u$
$[\mathfrak{a}]\ \mathfrak{a} \frown u$
$\mathfrak{a} \frown v$ (α

(IIa):: - - - - -

$\vdash a \frown v$
$[\mathfrak{a}]\ \mathfrak{a} \frown u$
$[\mathfrak{a}]\ \mathfrak{a} \frown u$
$\mathfrak{a} \frown v$ (439

——— • ———

IIa $\vdash \dot{\varepsilon}(-\varepsilon \frown u) = m; n \,\underline{\measuredangle}\, q$
$[\mathfrak{q}]\ \dot{\varepsilon}(-\varepsilon \frown u) = m; n \,\underline{\measuredangle}\, \mathfrak{q}$

(IIa):: - - - - - - - - - - - -

$\vdash \dot{\varepsilon}(-\varepsilon \frown u) = m; n \,\underline{\measuredangle}\, q$
$[\mathfrak{a}\ \mathfrak{q}]\ \dot{\varepsilon}(-\varepsilon \frown u) = m; \mathfrak{a} \,\underline{\measuredangle}\, \mathfrak{q}$ (α

(IIa):: - - - - - - - - - - - - -

$\vdash \dot{\varepsilon}(-\varepsilon \frown u) = m; n \,\underline{\measuredangle}\, q$
$[\mathfrak{d}\ \mathfrak{a}\ \mathfrak{q}]\ \dot{\varepsilon}(-\varepsilon \frown u) = \mathfrak{d}; \mathfrak{a} \,\underline{\measuredangle}\, \mathfrak{q}$ (β

(IIId): - - - - - - - - - - - - -

$\vdash A = m; n$
$\dot{\varepsilon}(-\varepsilon \frown u) = A \,\underline{\measuredangle}\, q$
$[\mathfrak{d}\ \mathfrak{a}\ \mathfrak{q}]\ \dot{\varepsilon}(-\varepsilon \frown u) = \mathfrak{d}; \mathfrak{a} \,\underline{\measuredangle}\, \mathfrak{q}$ (γ

(Ia): - - - - - - - - - - - - -

$\vdash a \frown (n \frown \dot{\smile} q)$
$m \frown (a \frown \dot{\smile} q)$
$n \frown (n \frown \underline{\smile} q)$
$A = m; n$
$\dot{\varepsilon}(-\varepsilon \frown u) = A \,\underline{\measuredangle}\, q$
$[\mathfrak{d}\ \mathfrak{a}\ \mathfrak{q}]\ \dot{\varepsilon}(-\varepsilon \frown u) = \mathfrak{d}; \mathfrak{a} \,\underline{\measuredangle}\, \mathfrak{q}$ (δ

$\vdash [\mathfrak{n}\ \mathfrak{m}]\ a \frown (\mathfrak{n} \frown \dot{\smile} q)$
$\mathfrak{m} \frown (a \frown \dot{\smile} q)$
$\mathfrak{n} \frown (\mathfrak{n} \frown \underline{\smile} q)$
$A = \mathfrak{m}; \mathfrak{n}$
$\dot{\varepsilon}(-\varepsilon \frown u) = A \,\underline{\measuredangle}\, q$
$[\mathfrak{d}\ \mathfrak{a}\ \mathfrak{q}]\ \dot{\varepsilon}(-\varepsilon \frown u) = \mathfrak{d}; \mathfrak{a} \,\underline{\measuredangle}\, \mathfrak{q}$ (ε

(266): - - - - - - - - - - - - -

$\vdash a \frown (A \,\underline{\measuredangle}\, q)$
$\dot{\varepsilon}(-\varepsilon \frown u) = A \,\underline{\measuredangle}\, q$
$[\mathfrak{d}\ \mathfrak{a}\ \mathfrak{q}]\ \dot{\varepsilon}(-\varepsilon \frown u) = \mathfrak{d}; \mathfrak{a} \,\underline{\measuredangle}\, \mathfrak{q}$ (ζ

(46): - - - - - - - - - - - - -

$\vdash a \frown u$
$\dot{\varepsilon}(-\varepsilon \frown u) = A \,\underline{\measuredangle}\, q$
$[\mathfrak{d}\ \mathfrak{a}\ \mathfrak{q}]\ \dot{\varepsilon}(-\varepsilon \frown u) = \mathfrak{d}; \mathfrak{a} \,\underline{\measuredangle}\, \mathfrak{q}$ (η

$\vdash [\mathfrak{a}]\ \mathfrak{a} \frown u$
$\dot{\varepsilon}(-\varepsilon \frown u) = A \,\underline{\measuredangle}\, q$
$[\mathfrak{d}\ \mathfrak{a}\ \mathfrak{q}]\ \dot{\varepsilon}(-\varepsilon \frown u) = \mathfrak{d}; \mathfrak{a} \,\underline{\measuredangle}\, \mathfrak{q}$ (ϑ

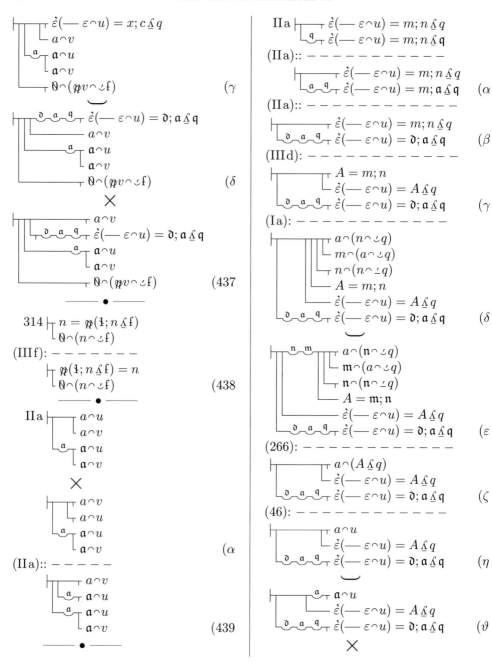

×

$\vdash \dot{\varepsilon}(-\ \varepsilon \frown u) = A \mathbin{\underline{\measuredangle}} q$
$\mathfrak{a}$ [concavity] $\mathfrak{a} \frown u$
$\mathfrak{d}\ \mathfrak{a}\ \mathfrak{q}$ [concavities] $\dot{\varepsilon}(-\ \varepsilon \frown u) = \mathfrak{d};\mathfrak{a} \mathbin{\underline{\measuredangle}} \mathfrak{q}$ (ι

$\vdash$ $\mathfrak{A}\ \mathfrak{q}$ [concavities] $\dot{\varepsilon}(-\ \varepsilon \frown u) = \mathfrak{A} \mathbin{\underline{\measuredangle}} \mathfrak{q}$
$\mathfrak{a}$ [concavity] $\mathfrak{a} \frown u$
$\mathfrak{d}\ \mathfrak{a}\ \mathfrak{q}$ [concavities] $\dot{\varepsilon}(-\ \varepsilon \frown u) = \mathfrak{d};\mathfrak{a} \mathbin{\underline{\measuredangle}} \mathfrak{q}$ (κ

×

$\vdash$ $\mathfrak{d}\ \mathfrak{a}\ \mathfrak{q}$ [concavities] $\dot{\varepsilon}(-\ \varepsilon \frown u) = \mathfrak{d};\mathfrak{a} \mathbin{\underline{\measuredangle}} \mathfrak{q}$
$\mathfrak{A}\ \mathfrak{q}$ [concavities] $\dot{\varepsilon}(-\ \varepsilon \frown u) = \mathfrak{A} \mathbin{\underline{\measuredangle}} \mathfrak{q}$
$\mathfrak{a}$ [concavity] $\mathfrak{a} \frown u$ (λ

(345):: - - - - - - - - - - - - - -

$\vdash$ $\mathfrak{d}\ \mathfrak{a}\ \mathfrak{q}$ [concavities] $\dot{\varepsilon}(-\ \varepsilon \frown u) = \mathfrak{d};\mathfrak{a} \mathbin{\underline{\measuredangle}} \mathfrak{q}$
$\mathfrak{a}$ [concavity] $\mathfrak{a} \frown u$
$\wp(\mathfrak{1}; \wp u \mathbin{\underline{\measuredangle}} \mathfrak{f}) = \wp u$ (μ

(438):: - - - - - - - - - - - - - -

$\vdash$ $\mathfrak{d}\ \mathfrak{a}\ \mathfrak{q}$ [concavities] $\dot{\varepsilon}(-\ \varepsilon \frown u) = \mathfrak{d};\mathfrak{a} \mathbin{\underline{\measuredangle}} \mathfrak{q}$
$\mathfrak{a}$ [concavity] $\mathfrak{a} \frown u$
$\mathfrak{0} \frown (\wp u \frown \smile \mathfrak{f})$ (440

(437): - - - - - - - - - - - - - -

$\vdash a \frown v$
$\mathfrak{a}$ [concavity] $\mathfrak{a} \frown u$
$\mathfrak{0} \frown (\wp u \frown \smile \mathfrak{f})$
$\mathfrak{a}$ [concavity] $\mathfrak{a} \frown u$
$\mathfrak{a} \frown v$
$\mathfrak{0} \frown (\wp v \frown \smile \mathfrak{f})$ (α

(439):·—·—·—·—·—·—·

$\vdash a \frown v$
$\mathfrak{a}$ [concavity] $\mathfrak{a} \frown u$
$\mathfrak{a} \frown v$
$\mathfrak{0} \frown (\wp u \frown \smile \mathfrak{f})$
$\mathfrak{0} \frown (\wp v \frown \smile \mathfrak{f})$ (β

$\vdash$ $\mathfrak{a}$ [concavity] $\mathfrak{a} \frown v$
$\mathfrak{a}$ [concavity] $\mathfrak{a} \frown u$
$\mathfrak{a} \frown v$
$\mathfrak{0} \frown (\wp u \frown \smile \mathfrak{f})$
$\mathfrak{0} \frown (\wp v \frown \smile \mathfrak{f})$ (γ

×

$\vdash \mathfrak{0} \frown (\wp v \frown \smile \mathfrak{f})$
$\mathfrak{a}$ [concavity] $\mathfrak{a} \frown u$
$\mathfrak{a} \frown v$
$\mathfrak{0} \frown (\wp u \frown \smile \mathfrak{f})$
$\mathfrak{a}$ [concavity] $\mathfrak{a} \frown v$ (441

—— • ——

140 $\vdash \mathfrak{0} \frown (\mathfrak{0} \frown \smile q)$

(IIIa): ————

$\vdash \mathfrak{0} \frown (\wp v \frown \smile q)$
$\wp v = \mathfrak{0}$ (α

(97):: - - - - - -

$\vdash \mathfrak{0} \frown (\wp v \frown \smile q)$
$\mathfrak{a}$ [concavity] $\mathfrak{a} \frown v$ (442

(441): ·—·—·—·—·—·—·

$\vdash \mathfrak{0} \frown (\wp v \frown \smile \mathfrak{f})$
$\mathfrak{a}$ [concavity] $\mathfrak{a} \frown u$
$\mathfrak{a} \frown v$
$\mathfrak{0} \frown (\wp u \frown \smile \mathfrak{f})$ (443

Ξ. Proof of the proposition

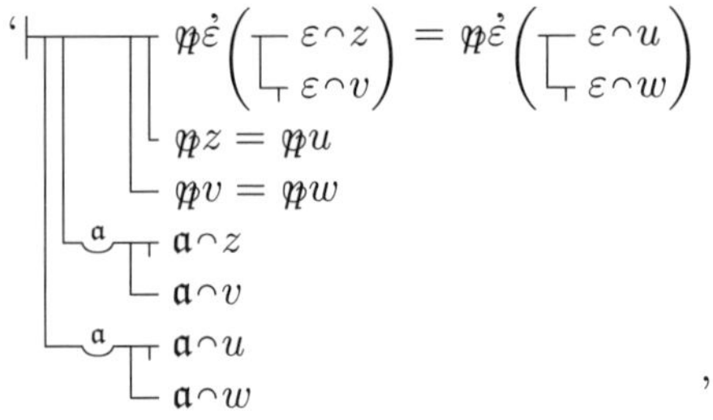

,

a) Proof of the proposition

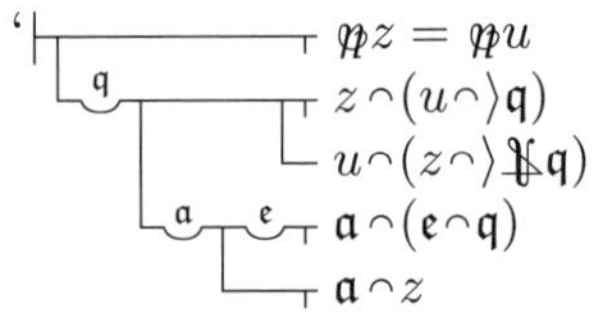

,

§33. *Analysis*

"The sum of two cardinal numbers is determined by them"; in this expression, the thought of the proposition in our main heading is most easily recognised, and it may thus be mentioned although the definite article of the subject in fact anticipates the predication of determinateness and although the use of the the word "sum" here is different from our later use for the numbers. For we here call $\mathfrak{P}\dot{\varepsilon}\left(\begin{array}{l}\varepsilon\frown z\\ \varepsilon\frown v\end{array}\right)$ sum of $\mathfrak{P}z$ and $\mathfrak{P}v$ provided no object falls both under the z- and the v-concept. Also infinite cardinal numbers are here to be considered. If we wanted to prove the proposition merely for finite cardinal numbers, a different approach would be more suitable. We first prove the proposition of the subsidiary heading, which adds somewhat to (49). For a proposition is obtained from it by contraposition which may be expressed thus: "If the cardinal number of the z-concept coincides with that of the u-concept, then there is a relation which maps the z-concept into the u-concept and whose converse maps the latter into the former, and in which any object which does not fall under the z-concept stands to no object". Let the q-relation be thus, and let the p-relation be constituted analogously with respect to the v-concept and the w-concept; then the $\dot{\alpha}\dot{\varepsilon}\left(\begin{array}{l}\varepsilon\frown(\alpha\frown q)\\ \varepsilon\frown(\alpha\frown p)\end{array}\right)$-relation maps the $\dot{\varepsilon}\left(\begin{array}{l}\varepsilon\frown z\\ \varepsilon\frown v\end{array}\right)$-concept into the $\dot{\varepsilon}\left(\begin{array}{l}\varepsilon\frown u\\ \varepsilon\frown w\end{array}\right)$, and its converse maps the latter into the former. In order for this relation and its converse to be single-valued, we have to assume for the q-relation that any object that does not fall under the z-concept stands to no object in the q-relation, and that the corresponding must hold of the p-relation.

In order to derive the proposition of the subsidiary heading from (49), we show that the $\between(z \frown \between q)$-relation has the required properties if the q-relation maps the z-concept into the u-concept and if its converse maps the latter into the former. We then prove the propositions

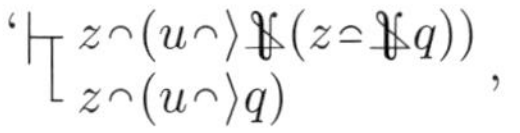
(α

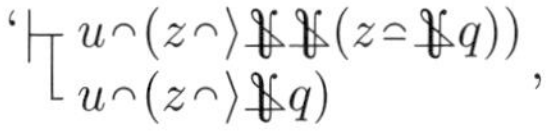
(β

In order to derive (α) we need, according to (11), the proposition

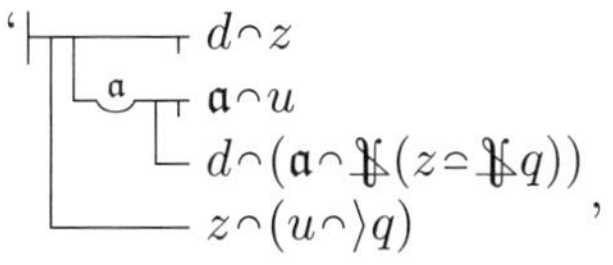
(γ

and for this in turn the proposition

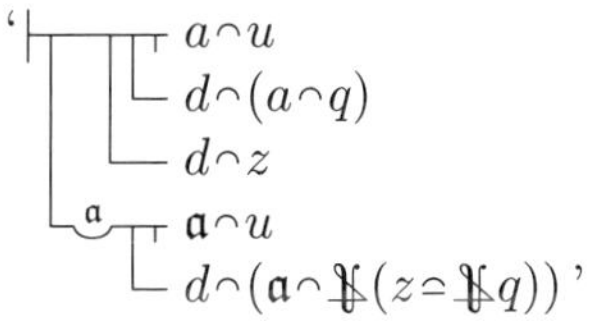
(δ

which follows from (197).

§34. *Construction*

22

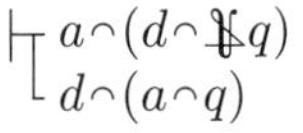

(197): – – – – – –

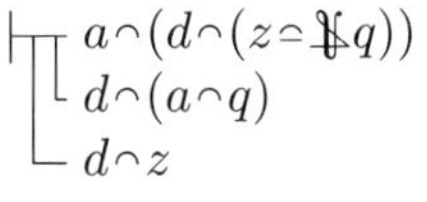
(α

(22): ————————

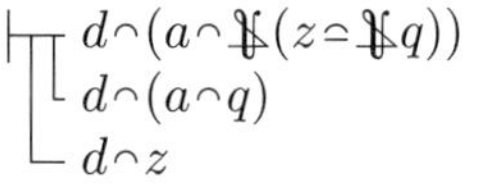
(444

(IIa): – – – – – – – – – –

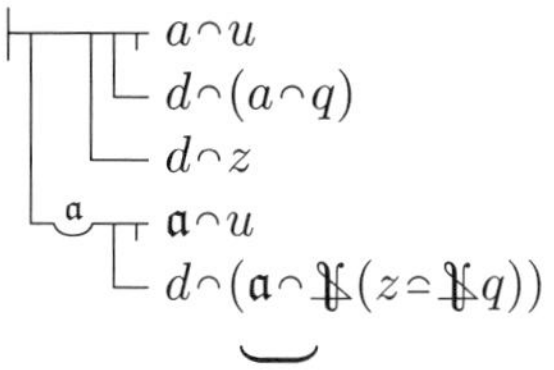
(α

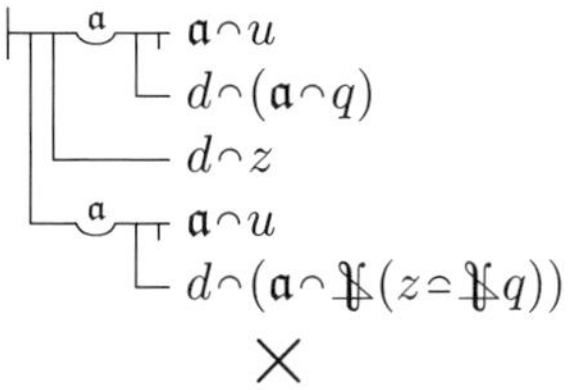
(β

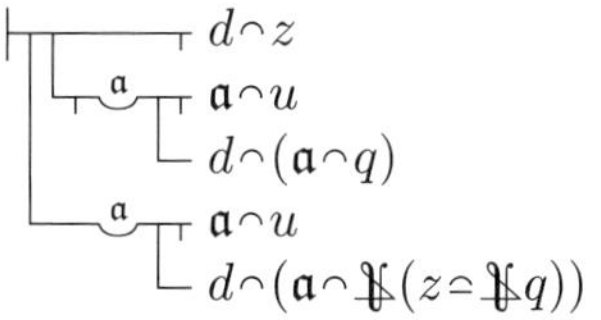
(γ

(8): -·-·-·-·-·-·-·-·-·-·-

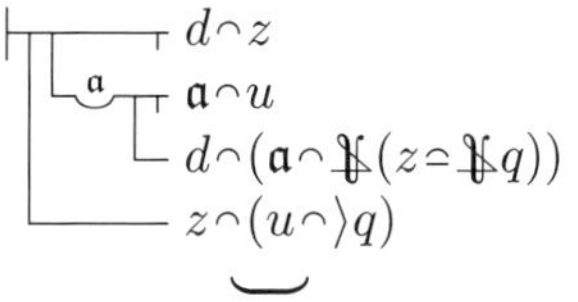
(δ

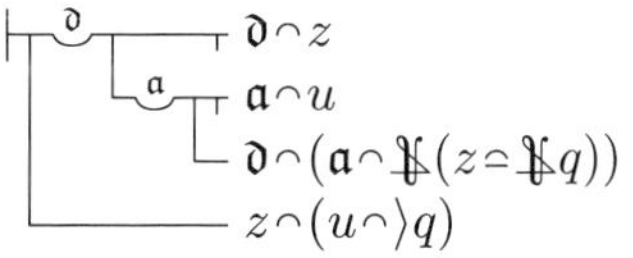
(ε

(11): – – – – – – – – – – – – – –

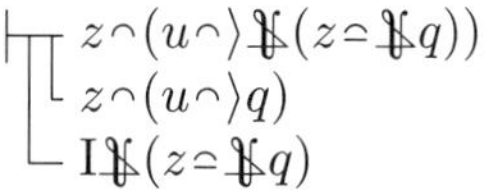
(445

———— • ————

23

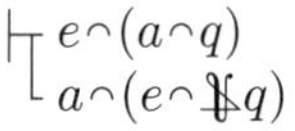

(188): – – – – – –

⊢ $e \frown (a \frown q)$
  $a \frown (e \frown (z \frown \mathfrak{\\$}q))$ (α

(14): ————————

⊢ $e \frown (a \frown q)$
  $e \frown (a \frown \mathfrak{\\$}(z \frown \mathfrak{\\$}q))$ (446

(13): - - - - - - - - - -

⊢ $d = a$
  $e \frown (a \frown \mathfrak{\\$}(z \frown \mathfrak{\\$}q))$
  $e \frown (d \frown q)$
  $\mathrm{I}q$ (α

(446):: - - - - - - - - - -

⊢ $d = a$
  $e \frown (a \frown \mathfrak{\\$}(z \frown \mathfrak{\\$}q))$
  $e \frown (d \frown \mathfrak{\\$}(z \frown \mathfrak{\\$}q))$
  $\mathrm{I}q$ (β

‿

⊢ 𝔢 𝔡 𝔞 $\mathfrak{d} = \mathfrak{a}$
  $\mathfrak{e} \frown (\mathfrak{a} \frown \mathfrak{\\$}(z \frown \mathfrak{\\$}q))$
  $\mathfrak{e} \frown (\mathfrak{d} \frown \mathfrak{\\$}(z \frown \mathfrak{\\$}q))$
  $\mathrm{I}q$ (γ

(16): - - - - - - - - - - - - - -

⊢ $\mathrm{I}\mathfrak{\\$}(z \frown \mathfrak{\\$}q)$
  $\mathrm{I}q$ (447

(18):: - - - - - -

⊢ $\mathrm{I}\mathfrak{\\$}(z \frown \mathfrak{\\$}q)$
  $z \frown (u \frown \rangle q)$ (α

(445): - - - - - -

⊢ $z \frown (u \frown \rangle \mathfrak{\\$}(z \frown \mathfrak{\\$}q))$
  $z \frown (u \frown \rangle q)$ (448

×

⊢ $z \frown (u \frown \rangle q)$
  $z \frown (u \frown \rangle \mathfrak{\\$}(z \frown \mathfrak{\\$}q))$ (449

§35. *Analysis*

In order to prove proposition (β) of §33 we require the proposition

'⊢ $d \frown u$
  𝔞 $\mathfrak{a} \frown z$
  $d \frown (\mathfrak{a} \frown \mathfrak{\\$}\mathfrak{\\$}(z \frown \mathfrak{\\$}q))$
  $u \frown (z \frown \rangle \mathfrak{\\$}q)$ ', (α

which, in turn, is to be reduced to

'⊢ $a \frown z$
  $d \frown (a \frown \mathfrak{\\$}q)$
  𝔞 $\mathfrak{a} \frown z$
  $d \frown (\mathfrak{a} \frown \mathfrak{\\$}\mathfrak{\\$}(z \frown \mathfrak{\\$}q))$ ' (β

§36. *Construction*

26 ⊢ $d \frown (a \frown \mathfrak{\\$}\mathfrak{\\$}(z \frown \mathfrak{\\$}q))$
  $d \frown (a \frown (z \frown \mathfrak{\\$}q))$

(197):: - - - - - - - - - - -

⊢ $d \frown (a \frown \mathfrak{\\$}\mathfrak{\\$}(z \frown \mathfrak{\\$}q))$
  $d \frown (a \frown \mathfrak{\\$}q)$
  $a \frown z$ (450

×

⊢ $a \frown z$
  $d \frown (a \frown \mathfrak{\\$}q)$
  $d \frown (a \frown \mathfrak{\\$}\mathfrak{\\$}(z \frown \mathfrak{\\$}q))$ (α

(IIa): -·-·-·-·-·-·-·-·-

⊢ $a \frown z$
  $d \frown (a \frown \mathfrak{\\$}q)$
  𝔞 $\mathfrak{a} \frown z$
  $d \frown (\mathfrak{a} \frown \mathfrak{\\$}\mathfrak{\\$}(z \frown \mathfrak{\\$}q))$ (β

‿

⊢ 𝔞 $\mathfrak{a} \frown z$
  $d \frown (\mathfrak{a} \frown \mathfrak{\\$}q)$
  𝔞 $\mathfrak{a} \frown z$
  $d \frown (\mathfrak{a} \frown \mathfrak{\\$}\mathfrak{\\$}(z \frown \mathfrak{\\$}q))$ (γ

(8): - - - - - - - - - - - - -

⊢ $d \frown u$
  𝔞 $\mathfrak{a} \frown z$
  $d \frown (\mathfrak{a} \frown \mathfrak{\\$}\mathfrak{\\$}(z \frown \mathfrak{\\$}q))$
  $u \frown (z \frown \rangle \mathfrak{\\$}q)$ (δ

‿

⊢ 𝔡 $\mathfrak{d} \frown u$
  𝔞 $\mathfrak{a} \frown z$
  $\mathfrak{d} \frown (\mathfrak{a} \frown \mathfrak{\\$}\mathfrak{\\$}(z \frown \mathfrak{\\$}q))$
  $u \frown (z \frown \rangle \mathfrak{\\$}q)$ (ε

(11): - - - - - - - - - - - - - -

$\vdash$ $u \frown (z \frown) 𝔖𝔖(z \asymp 𝔖q))$
$u \frown (z \frown) 𝔖q)$
$\mathrm{I}𝔖𝔖(z \asymp 𝔖q)$ (ζ

(29):: - - - - - - - - - - -

$\vdash$ $u \frown (z \frown) 𝔖𝔖(z \asymp 𝔖q))$
$u \frown (z \frown) 𝔖q)$
$\mathrm{I}(z \asymp 𝔖q)$ (η

(189):: - - - - - - - - - - -

$\vdash$ $u \frown (z \frown) 𝔖𝔖(z \asymp 𝔖q))$
$u \frown (z \frown) 𝔖q)$
$\mathrm{I}𝔖q$ (ϑ

(18):: - - - - - - - - - - -

$\vdash$ $u \frown (z \frown) 𝔖𝔖(z \asymp 𝔖q))$
$u \frown (z \frown) 𝔖q)$ (451

——— • ———

191 $\vdash$ $a \frown z$
$e \frown (a \frown (z \asymp 𝔖q))$

(22): ————————

$\vdash$ $a \frown z$
$a \frown (e \frown 𝔖(z \asymp 𝔖q))$ (452

×

$\vdash$ $a \frown (e \frown 𝔖(z \asymp 𝔖q))$
$a \frown z$ (α

$\vdash$ 𝔞 𝔢 $\mathfrak{a} \frown (\mathfrak{e} \frown 𝔖(z \asymp 𝔖q))$
$\mathfrak{a} \frown z$ (β

(IIa): ————————————

$\vdash$ $z \frown (u \frown) 𝔖(z \asymp 𝔖q))$
$u \frown (z \frown) 𝔖𝔖(z \asymp 𝔖q))$
𝔮 $z \frown (u \frown) \mathfrak{q})$
$u \frown (z \frown) 𝔖\mathfrak{q})$
𝔞 𝔢 $\mathfrak{a} \frown (\mathfrak{e} \frown \mathfrak{q})$
$\mathfrak{a} \frown z$ (γ

(451):: - - - - - - - - - - - - - -

$\vdash$ $z \frown (u \frown) 𝔖(z \asymp 𝔖q))$
$u \frown (z \frown) 𝔖q)$
𝔮 $z \frown (u \frown) \mathfrak{q})$
$u \frown (z \frown) 𝔖\mathfrak{q})$
𝔞 𝔢 $\mathfrak{a} \frown (\mathfrak{e} \frown \mathfrak{q})$
$\mathfrak{a} \frown z$ (δ

(449): - - - - - - - - - - - - - -

$\vdash$ $z \frown (u \frown) q)$
$u \frown (z \frown) 𝔖q)$
𝔮 $z \frown (u \frown) \mathfrak{q})$
$u \frown (z \frown) 𝔖\mathfrak{q})$
𝔞 𝔢 $\mathfrak{a} \frown (\mathfrak{e} \frown \mathfrak{q})$
$\mathfrak{a} \frown z$ (ε

$\vdash$ 𝔮 $z \frown (u \frown) 𝔖\mathfrak{q})$
$u \frown (z \frown) 𝔖\mathfrak{q})$
𝔮 $z \frown (u \frown) \mathfrak{q})$
$u \frown (z \frown) 𝔖\mathfrak{q})$
𝔞 𝔢 $\mathfrak{a} \frown (\mathfrak{e} \frown \mathfrak{q})$
$\mathfrak{a} \frown z$ (ζ

(49): - - - - - - - - - - -

$\vdash$ $\natural z = \natural u$
𝔮 $z \frown (u \frown) \mathfrak{q})$
$u \frown (z \frown) 𝔖\mathfrak{q})$
𝔞 𝔢 $\mathfrak{a} \frown (\mathfrak{e} \frown \mathfrak{q})$
$\mathfrak{a} \frown z$ (453

b) Proof of the proposition

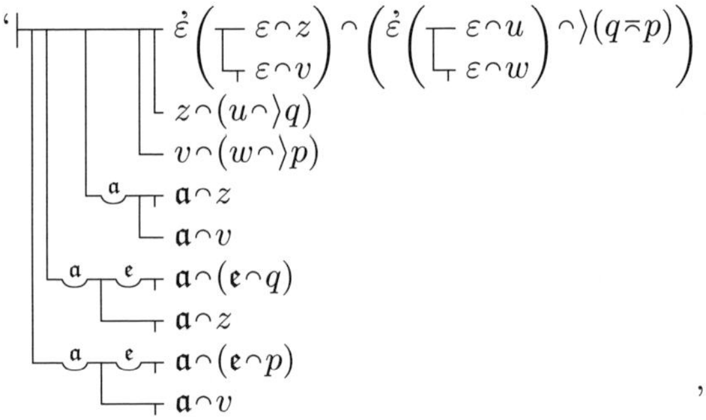

§37. *Analysis*

We now define:

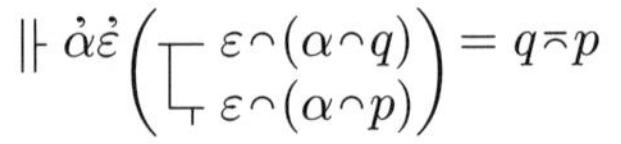

(Υ

and prove the proposition

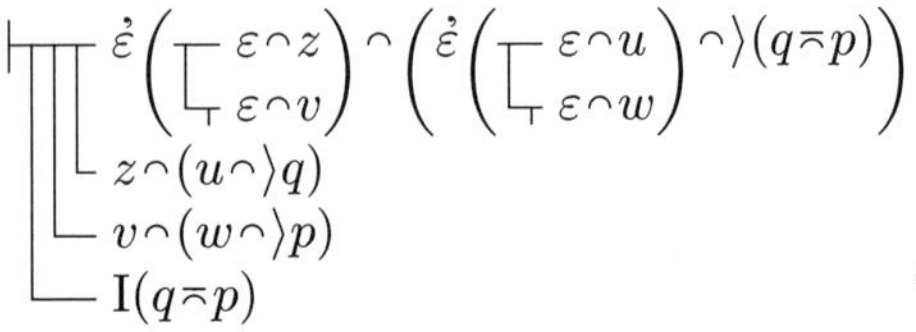

by means of (8) and (11).

§38. *Construction*

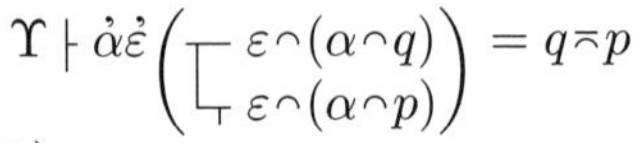

(10): ————————

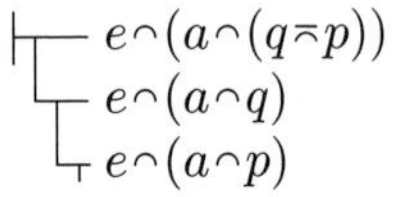

(454

(I):: - - - - - - - -

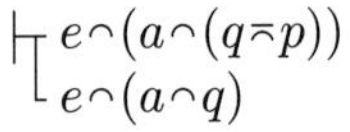

(455

——— • ———

454

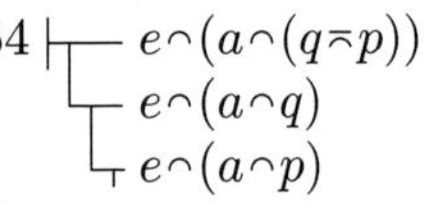

(Ia):: - - - - - - - -

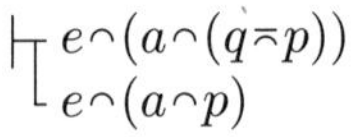

(456

——— • ———

If

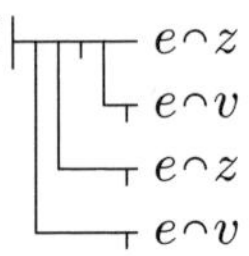

(77): ————

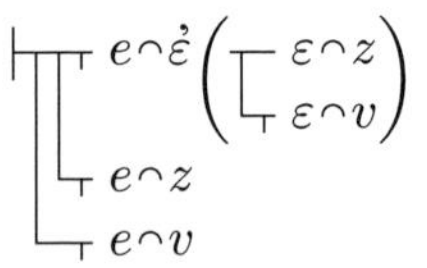

(α

(8):: - - - - - - - -

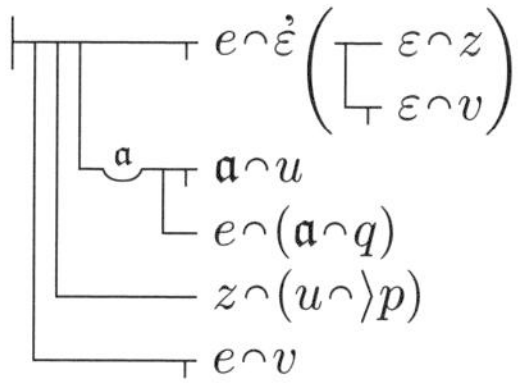

(457

Ic 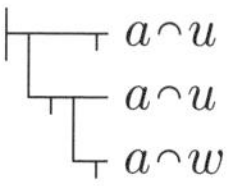

(77): ————

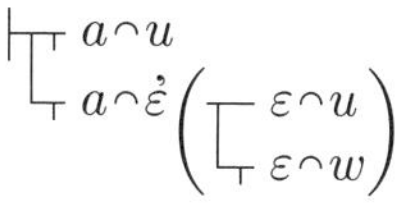

(α

(IIa):: - - - - - - - - -

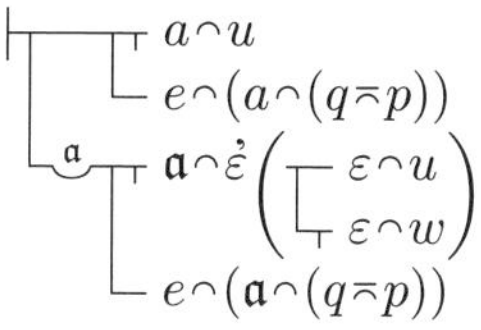

(β

(455):: - - - - - - - - - - -

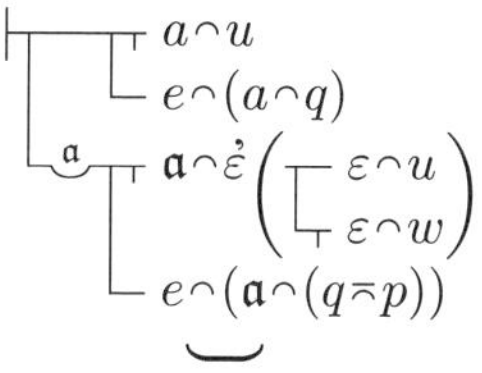

(γ

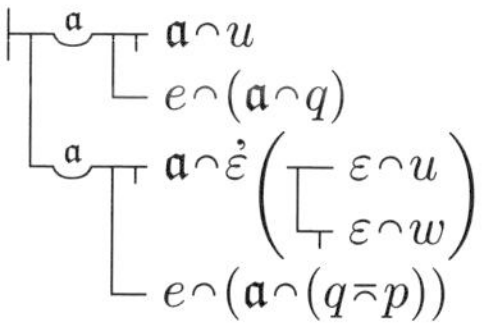

(δ

(457): - - - - - - - - - - -

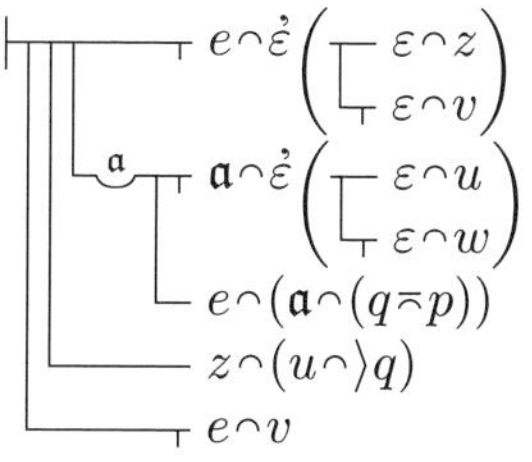

(458

Id 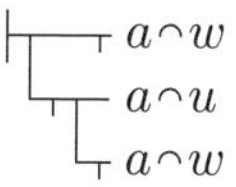

(77): ————

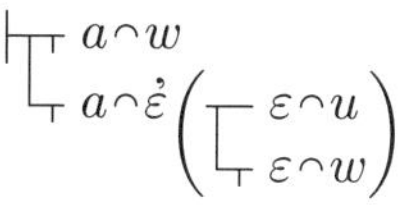

(α

(IIa):: - - - - - - - - -

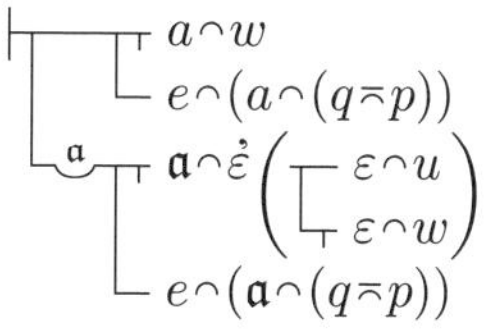

(β

(456):: - - - - - - - - - - -

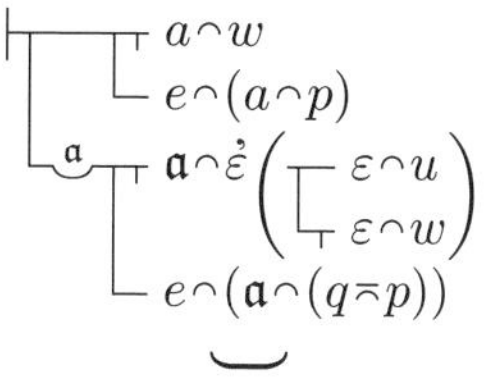

(γ

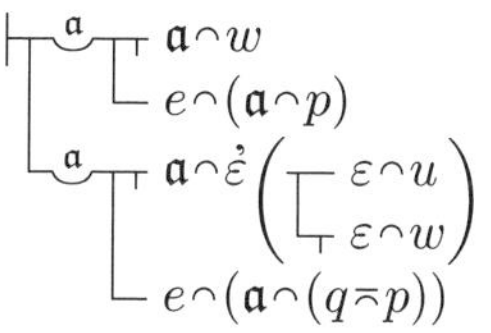

(δ

(8): - - - - - - - - - - -

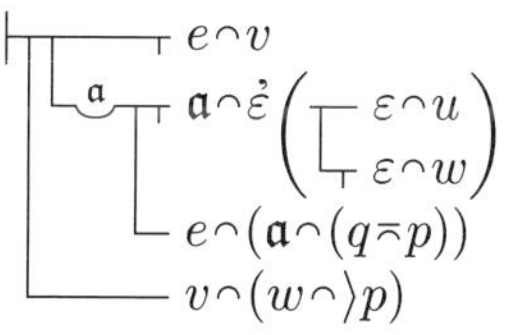

(ε

(458): - - - - - - - - - - -

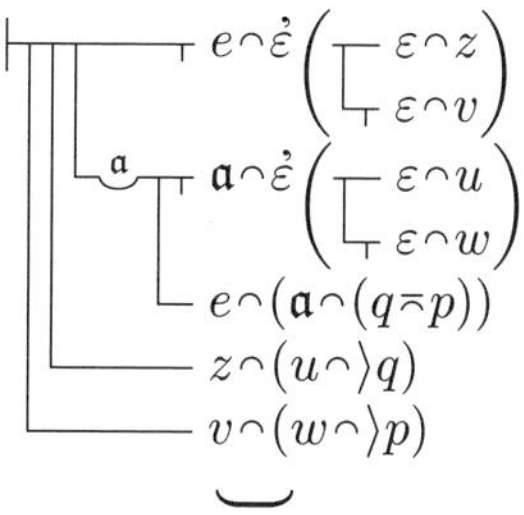

(ζ

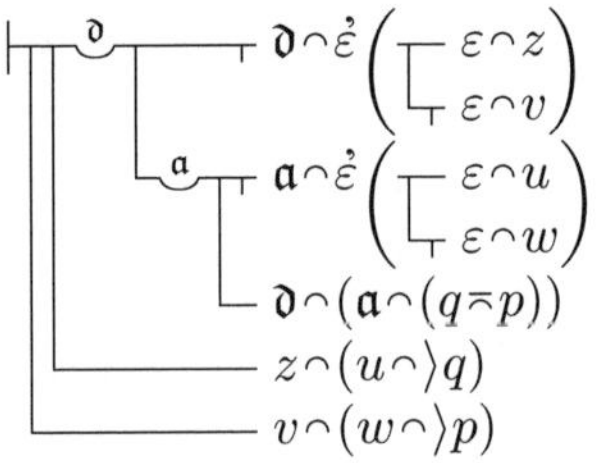

$$\vdash\ \begin{array}{l}\partial\frown\dot{\varepsilon}\left(\begin{array}{l}\varepsilon\frown z\\ \tau\ \varepsilon\frown v\end{array}\right)\\ {}^{\mathfrak{a}}\mathfrak{a}\frown\dot{\varepsilon}\left(\begin{array}{l}\varepsilon\frown u\\ \tau\ \varepsilon\frown w\end{array}\right)\\ \partial\frown(\mathfrak{a}\frown(q\bar{\frown}p))\\ z\frown(u\frown)q)\\ v\frown(w\frown)p)\end{array}$$ (η

(11): – – – – – – – – – – – –

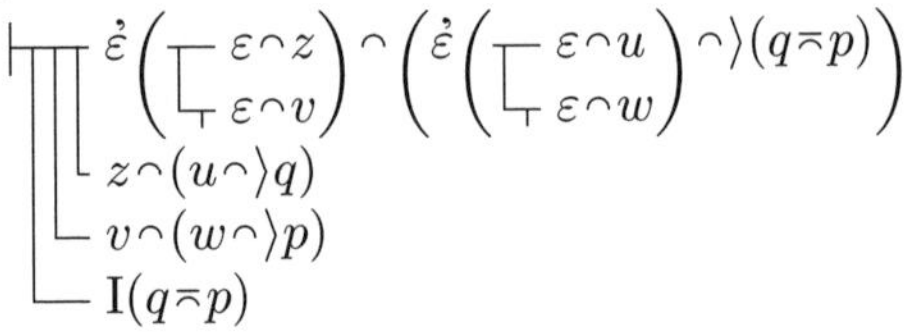

$$\vdash\ \begin{array}{l}\dot{\varepsilon}\left(\begin{array}{l}\varepsilon\frown z\\ \tau\ \varepsilon\frown v\end{array}\right)\frown\left(\dot{\varepsilon}\left(\begin{array}{l}\varepsilon\frown u\\ \tau\ \varepsilon\frown w\end{array}\right)\frown\rangle(q\bar{\frown}p)\right)\\ z\frown(u\frown)q)\\ v\frown(w\frown)p)\\ \mathrm{I}(q\bar{\frown}p)\end{array}$$ (459

§39. *Analysis*

We eliminate the subcomponent

'$\mathrm{I}(q\bar{\frown}p)$'

by means of the proposition

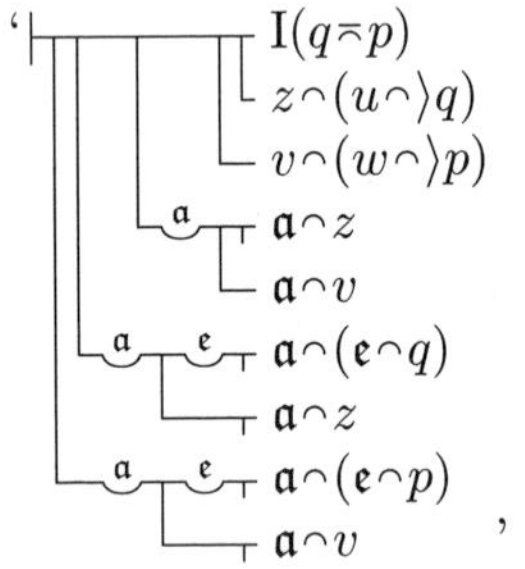

'$\vdash\ \begin{array}{l}\mathrm{I}(q\bar{\frown}p)\\ z\frown(u\frown)q)\\ v\frown(w\frown)p)\\ {}^{\mathfrak{a}}\tau\ \mathfrak{a}\frown z\\ \mathfrak{a}\frown v\\ {}^{\mathfrak{a}}{}^{\mathfrak{e}}\tau\ \mathfrak{a}\frown(\mathfrak{e}\frown q)\\ \tau\ \mathfrak{a}\frown z\\ {}^{\mathfrak{a}}{}^{\mathfrak{e}}\tau\ \mathfrak{a}\frown(\mathfrak{e}\frown p)\\ \tau\ \mathfrak{a}\frown v\end{array}$,   (α

for whose proof we require a proposition with the supercomponent

'$\vdash\ \begin{array}{l}d=a\\ e\frown(\mathfrak{a}\frown(q\bar{\frown}p))\\ e\frown(d\frown(q\bar{\frown}p))\end{array}$'

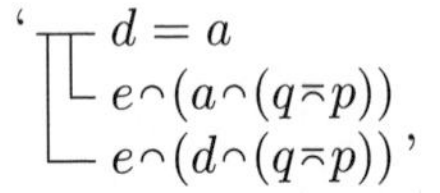

Here, we distinguish the case where e falls under the v-concept from its opposite.

§40. *Construction*

$T \vdash \dot{\alpha}\dot{\varepsilon}\left(\begin{array}{l}\varepsilon\frown(\alpha\frown q)\\ \tau\ \varepsilon\frown(\alpha\frown p)\end{array}\right) = q\bar{\frown}p$

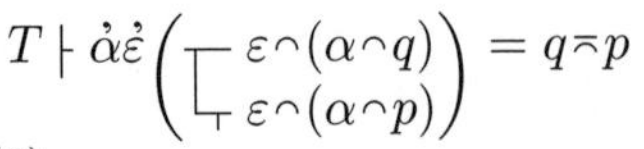

(6): ─────────────

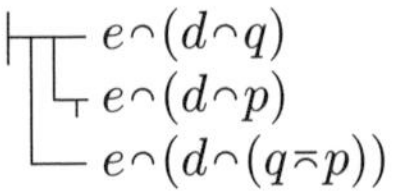

$\vdash\ \begin{array}{l}e\frown(d\frown q)\\ \tau\ e\frown(d\frown p)\\ e\frown(d\frown(q\bar{\frown}p))\end{array}$ (460

×

$\vdash\ \begin{array}{l}e\frown(d\frown p)\\ \tau\ e\frown(d\frown q)\\ e\frown(d\frown(q\bar{\frown}p))\end{array}$ (461

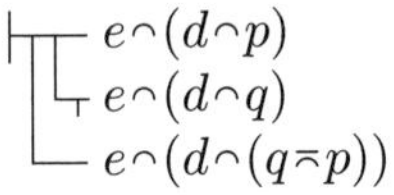

(IIa):: – – – – – – – –

$\vdash\ \begin{array}{l}e\frown(d\frown p)\\ {}^{\mathfrak{e}}\tau\ e\frown(\mathfrak{e}\frown q)\\ e\frown(d\frown(q\bar{\frown}p))\end{array}$ (α

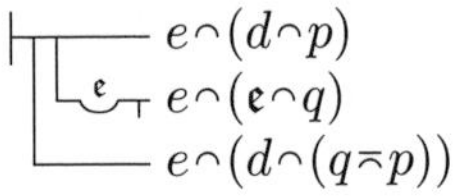

(IIa):: – – – – – – – –

$\vdash\ \begin{array}{l}e\frown(d\frown p)\\ \tau\ e\frown z\\ {}^{\mathfrak{a}}{}^{\mathfrak{e}}\tau\ \mathfrak{a}\frown(\mathfrak{e}\frown q)\\ \tau\ \mathfrak{a}\frown z\\ e\frown(d\frown(q\bar{\frown}p))\end{array}$ (β

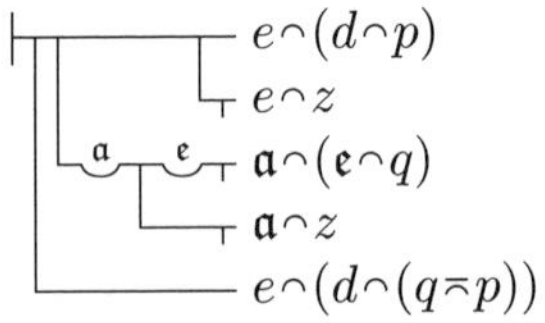

(IIa):: – – – – – – – – – –

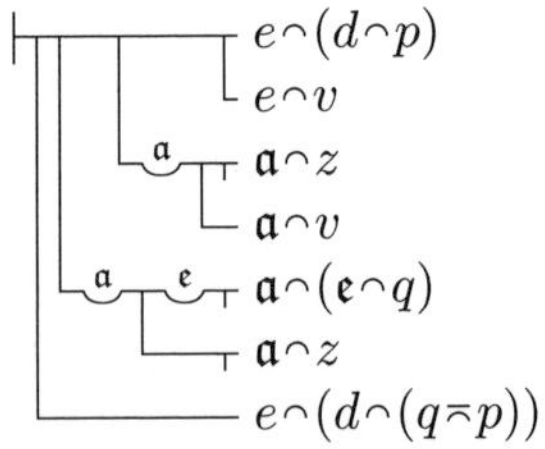

$\vdash\ \begin{array}{l}e\frown(d\frown p)\\ e\frown v\\ {}^{\mathfrak{a}}\tau\ \mathfrak{a}\frown z\\ \mathfrak{a}\frown v\\ {}^{\mathfrak{a}}{}^{\mathfrak{e}}\tau\ \mathfrak{a}\frown(\mathfrak{e}\frown q)\\ \tau\ \mathfrak{a}\frown z\\ e\frown(d\frown(q\bar{\frown}p))\end{array}$ (γ

(13): – – – – – – – – – –

$d = a$
$e \frown (a \frown p)$
$e \frown v$
$\mathfrak{a} \frown z$
$\mathfrak{a} \frown v$
$\mathfrak{a} \frown (\mathfrak{e} \frown q)$
$\mathfrak{a} \frown z$
$e \frown (d \frown (q \bar{\frown} p))$
$\mathrm{I}p$ (δ

(γ):: – – – – – – – – – – – –

$d = a$
$e \frown (a \frown (q \bar{\frown} p))$
$e \frown v$
$\mathfrak{a} \frown z$
$\mathfrak{a} \frown v$
$\mathfrak{a} \frown (\mathfrak{e} \frown q)$
$\mathfrak{a} \frown z$
$e \frown (d \frown (q \bar{\frown} p))$
$\mathrm{I}p$ (ε

———— • ————

IIa $e \frown (d \frown q)$
$e \frown (\mathfrak{e} \frown p)$

(460): – – – – – –

$e \frown (d \frown q)$
$e \frown (\mathfrak{e} \frown p)$
$e \frown (d \frown (q \bar{\frown} p))$ (ζ

(IIa):: – – – – – – – – – –

$e \frown (d \frown q)$
$e \frown v$
$\mathfrak{a} \frown (\mathfrak{e} \frown p)$
$\mathfrak{a} \frown v$
$e \frown (d \frown (q \bar{\frown} p))$ (η

(13): – – – – – – – – – – –

$d = a$
$e \frown (a \frown q)$
$e \frown v$
$\mathfrak{a} \frown (\mathfrak{e} \frown p)$
$\mathfrak{a} \frown v$
$e \frown (d \frown (q \bar{\frown} p))$
$\mathrm{I}q$ (ϑ

(η):: – – – – – – – – – – – –

$d = a$
$e \frown (a \frown (q \bar{\frown} p))$
$e \frown v$
$\mathfrak{a} \frown (\mathfrak{e} \frown p)$
$\mathfrak{a} \frown v$
$e \frown (d \frown (q \bar{\frown} p))$
$\mathrm{I}q$ (ι

(ε): ·–·–·–·–·–·–·–·–·–·–·

$d = a$
$e \frown (a \frown (q \bar{\frown} p))$
$e \frown (d \frown (q \bar{\frown} p))$
$\mathrm{I}q$
$\mathrm{I}p$
$\mathfrak{a} \frown z$
$\mathfrak{a} \frown v$
$\mathfrak{a} \frown (\mathfrak{e} \frown q)$
$\mathfrak{a} \frown z$
$\mathfrak{a} \frown (\mathfrak{e} \frown p)$
$\mathfrak{a} \frown v$ (κ

$\mathfrak{d} = \mathfrak{a}$
$\mathfrak{e} \frown (\mathfrak{a} \frown (q \bar{\frown} p))$
$\mathfrak{e} \frown (\mathfrak{d} \frown (q \bar{\frown} p))$
$\mathrm{I}q$
$\mathrm{I}p$
$\mathfrak{a} \frown z$
$\mathfrak{a} \frown v$
$\mathfrak{a} \frown (\mathfrak{e} \frown q)$
$\mathfrak{a} \frown z$
$\mathfrak{a} \frown (\mathfrak{e} \frown p)$
$\mathfrak{a} \frown v$ (λ

(16): – – – – – – – – – – – – –

$\mathrm{I}(q \bar{\frown} p)$
$\mathrm{I}q$
$\mathrm{I}p$
$\mathfrak{a} \frown z$
$\mathfrak{a} \frown v$
$\mathfrak{a} \frown (\mathfrak{e} \frown q)$
$\mathfrak{a} \frown z$
$\mathfrak{a} \frown (\mathfrak{e} \frown p)$
$\mathfrak{a} \frown v$ (μ

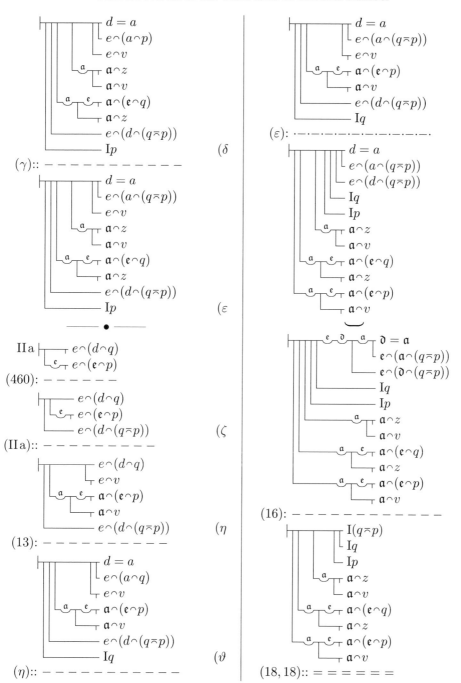

(18, 18):: = = = = = =

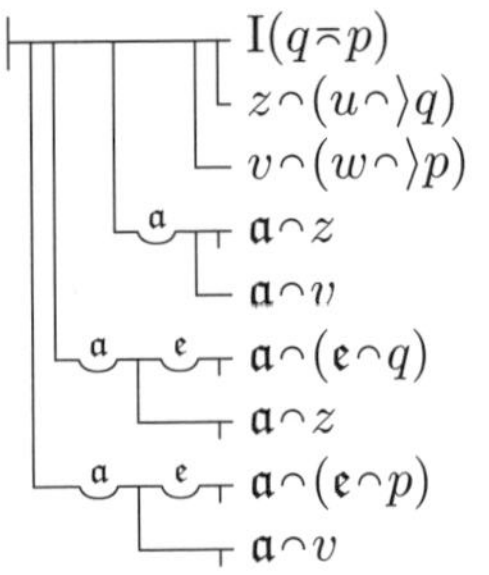

(462

(459): – – – – – – – – – – –

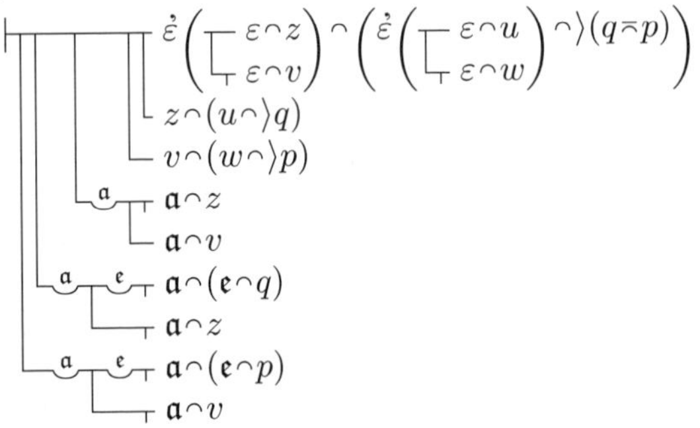

(463

c) Proof of the proposition

‘

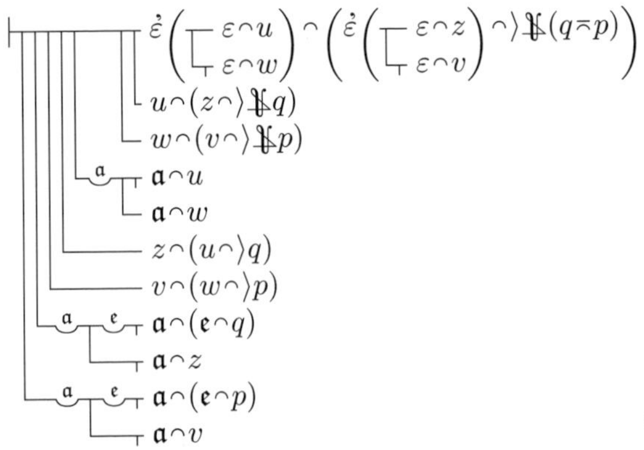

’

and end of section Ξ

§41. *Analysis*

In order to obtain a proposition with the supercomponent

‘ 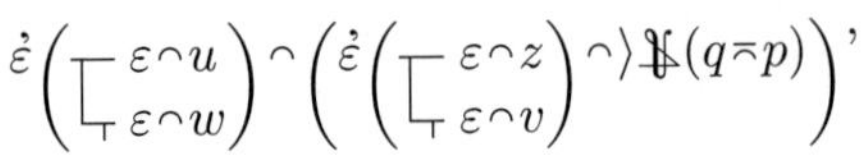 ’

we put ‘$\mathfrak{U}q$’ for ‘q’, ‘$\mathfrak{U}p$’ for ‘p’ in (463), and interchange ‘z’ with ‘u’, and ‘v’ with ‘w’. We then require the proposition

$$\text{‘}\vdash \mathfrak{U}(q \,\bar{\frown}\, p) = \mathfrak{U}q \,\bar{\frown}\, \mathfrak{U}p\text{’} \qquad (\alpha$$

§42. *Construction*

$$21 \vdash d\frown(e\frown q) = e\frown(d\frown\natural q)$$

(IIIh): ———

$$\vdash \left(\begin{array}{l} d\frown(e\frown q) \\ \llcorner\, d\frown(e\frown p) \end{array}\right) = \left(\begin{array}{l} e\frown(d\frown\natural q) \\ \llcorner\, d\frown(e\frown p) \end{array}\right) \qquad (\alpha$$

(IIIc): ———

$$\vdash \begin{array}{l} \left(\begin{array}{l} d\frown(e\frown q) \\ \llcorner\, d\frown(e\frown p) \end{array}\right) = \left(\begin{array}{l} e\frown(d\frown\natural q) \\ \llcorner\, e\frown(d\frown\natural p) \end{array}\right) \\ \llcorner\, d\frown(e\frown p) = e\frown(d\frown\natural p) \end{array} \qquad (\beta$$

(21):: ———

$$\vdash \left(\begin{array}{l} d\frown(e\frown q) \\ \llcorner\, d\frown(e\frown p) \end{array}\right) = \left(\begin{array}{l} e\frown(d\frown\natural q) \\ \llcorner\, e\frown(d\frown\natural p) \end{array}\right) \qquad (\gamma$$

$$\vdash\!\overset{\mathfrak{d}}{\smile}\!\overset{\mathfrak{a}}{\smile}\; \left(\begin{array}{l} \mathfrak{d}\frown(\mathfrak{a}\frown q) \\ \llcorner\, \mathfrak{d}\frown(\mathfrak{a}\frown p) \end{array}\right) = \left(\begin{array}{l} \mathfrak{a}\frown(\mathfrak{d}\frown\natural q) \\ \llcorner\, \mathfrak{a}\frown(\mathfrak{d}\frown\natural p) \end{array}\right) \qquad (\delta$$

(20): ———

$$\vdash \dot{\alpha}\dot{\varepsilon}\left(\begin{array}{l} \alpha\frown(\varepsilon\frown q) \\ \llcorner\, \alpha\frown(\varepsilon\frown p) \end{array}\right) = \dot{\alpha}\dot{\varepsilon}\left(\begin{array}{l} \varepsilon\frown(\alpha\frown\natural q) \\ \llcorner\, \varepsilon\frown(\alpha\frown\natural p) \end{array}\right) \qquad (\varepsilon$$

(IIIc): ———

$$\vdash \begin{array}{l} \dot{\alpha}\dot{\varepsilon}\left(\begin{array}{l} \alpha\frown(\varepsilon\frown q) \\ \llcorner\, \alpha\frown(\varepsilon\frown p) \end{array}\right) = \natural q\,\bar{\frown}\,\natural p \\ \llcorner\, \dot{\alpha}\dot{\varepsilon}\left(\begin{array}{l} \varepsilon\frown(\alpha\frown\natural q) \\ \llcorner\, \varepsilon\frown(\alpha\frown\natural p) \end{array}\right) = \natural q\,\bar{\frown}\,\natural p \end{array} \qquad (\zeta$$

(Υ):: ———

$$\vdash \dot{\alpha}\dot{\varepsilon}\left(\begin{array}{l} \alpha\frown(\varepsilon\frown q) \\ \llcorner\, \alpha\frown(\varepsilon\frown p) \end{array}\right) = \natural q\,\bar{\frown}\,\natural p \qquad (\eta$$

(IIIc): ———

$$\vdash \begin{array}{l} \natural\dot{\alpha}\dot{\varepsilon}\left(\begin{array}{l} \varepsilon\frown(\alpha\frown q) \\ \llcorner\, \varepsilon\frown(\alpha\frown p) \end{array}\right) = \natural q\,\bar{\frown}\,\natural p \\ \llcorner\, \dot{\alpha}\dot{\varepsilon}\left(\begin{array}{l} \alpha\frown(\varepsilon\frown q) \\ \llcorner\, \alpha\frown(\varepsilon\frown p) \end{array}\right) = \natural\dot{\alpha}\dot{\varepsilon}\left(\begin{array}{l} \varepsilon\frown(\alpha\frown q) \\ \llcorner\, \varepsilon\frown(\alpha\frown p) \end{array}\right) \end{array} \qquad (\vartheta$$

(40):: ———

$$\vdash \natural\dot{\alpha}\dot{\varepsilon}\left(\begin{array}{l} \varepsilon\frown(\alpha\frown q) \\ \llcorner\, \varepsilon\frown(\alpha\frown p) \end{array}\right) = \natural q\,\bar{\frown}\,\natural p \qquad (\iota$$

(IIIc): ———

$$\vdash \begin{array}{l} \natural(q\,\bar{\frown}\,p) = \natural q\,\bar{\frown}\,\natural p \\ \llcorner\, \dot{\alpha}\dot{\varepsilon}\left(\begin{array}{l} \varepsilon\frown(\alpha\frown q) \\ \llcorner\, \varepsilon\frown(\alpha\frown p) \end{array}\right) = q\,\bar{\frown}\,p \end{array} \qquad (\kappa$$

(Υ):: ———

$$\vdash \natural(q\,\bar{\frown}\,p) = \natural q\,\bar{\frown}\,\natural p \qquad (464$$

§43. *Analysis*

By making the changes in (463) that were described in §41 we obtain the subcomponent

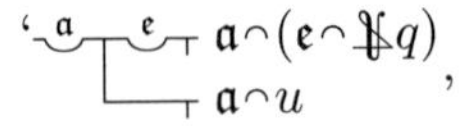

which we eliminate using the proposition

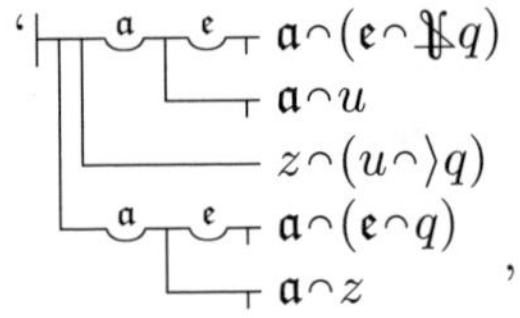
(α

For this we require the proposition

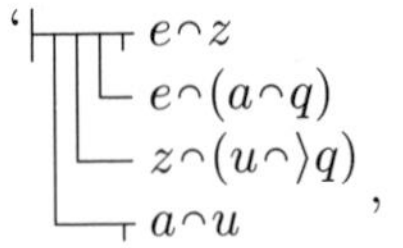
(β

which we prove by means of (13).

§44. *Construction*

IIIc 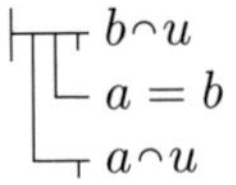

(13):: – – – – –

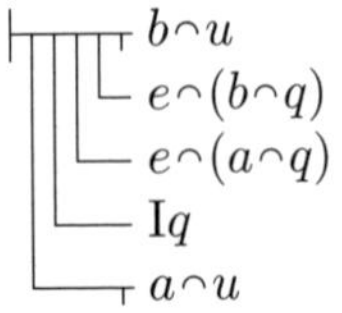
(α

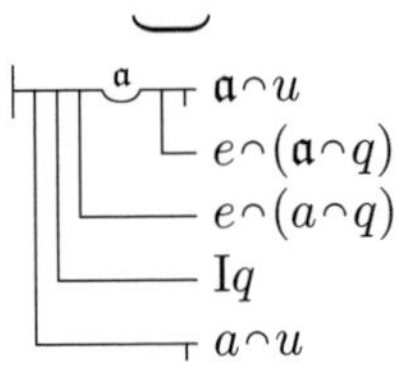
(β

(8): – – – – – – – – –

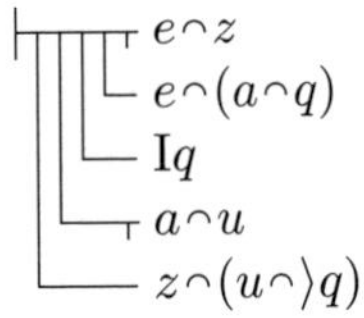
(γ

(18):: – – – – – – – –

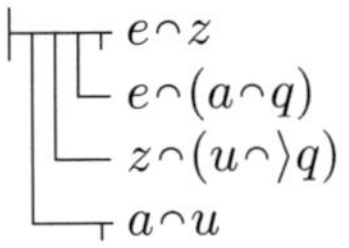
(465

×

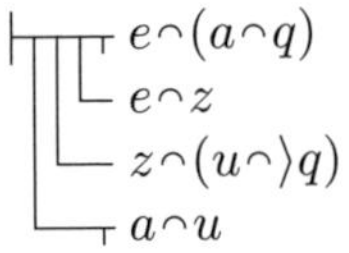
(466

IIa 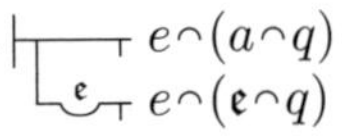

(IIa):: – – – – – – – –

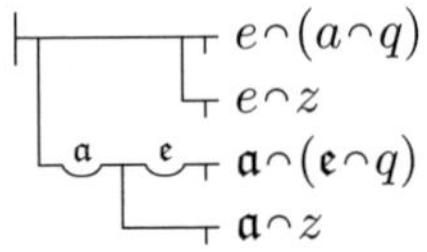
(α

(466): ·—·—·—·—·—·—·

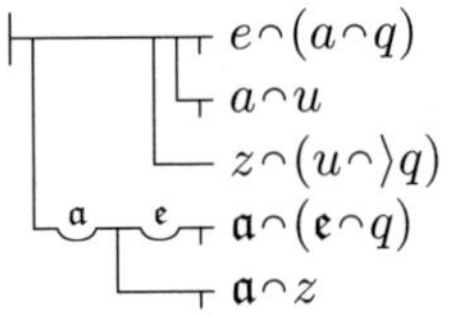
(β

(22): ————————

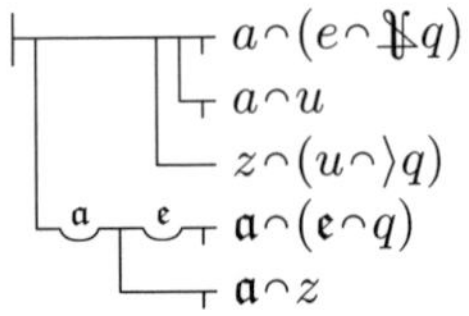
(γ

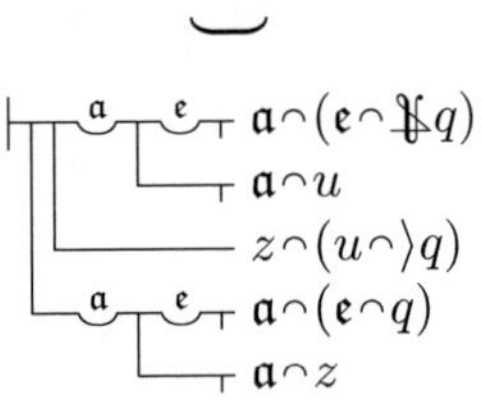
(467

464 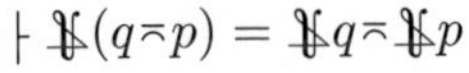

(IIIa): ————————

$$\vdash \dot{\varepsilon}\left(\begin{smallmatrix}\varepsilon\frown u\\ \varepsilon\frown w\end{smallmatrix}\right)\frown\left(\dot{\varepsilon}\left(\begin{smallmatrix}\varepsilon\frown z\\ \varepsilon\frown v\end{smallmatrix}\right)\frown\rangle\not\mathfrak{U}(q\bar{\frown}p)\right)$$
$$\dot{\varepsilon}\left(\begin{smallmatrix}\varepsilon\frown u\\ \varepsilon\frown w\end{smallmatrix}\right)\frown\left(\dot{\varepsilon}\left(\begin{smallmatrix}\varepsilon\frown z\\ \varepsilon\frown v\end{smallmatrix}\right)\frown\rangle(\not\mathfrak{U}q\bar{\frown}\not\mathfrak{U}p)\right) \qquad (\alpha$$

(463):: –

$$\vdash \dot{\varepsilon}\left(\begin{smallmatrix}\varepsilon\frown u\\ \varepsilon\frown w\end{smallmatrix}\right)\frown\left(\dot{\varepsilon}\left(\begin{smallmatrix}\varepsilon\frown z\\ \varepsilon\frown v\end{smallmatrix}\right)\frown\rangle\not\mathfrak{U}(q\bar{\frown}p)\right)$$
$$u\frown(z\frown\rangle\not\mathfrak{U}q)$$
$$w\frown(v\frown\rangle\not\mathfrak{U}p)$$
$$\underset{\smile}{\mathfrak{a}}\;\mathfrak{a}\frown u \;\;\; \mathfrak{a}\frown w$$
$$\underset{\smile}{\mathfrak{a}}\;\underset{\smile}{\mathfrak{e}}\;\mathfrak{a}\frown(\mathfrak{e}\frown\not\mathfrak{U}q) \;\;\; \mathfrak{a}\frown u$$
$$\underset{\smile}{\mathfrak{a}}\;\underset{\smile}{\mathfrak{e}}\;\mathfrak{a}\frown(\mathfrak{e}\frown\not\mathfrak{U}p) \;\;\; \mathfrak{a}\frown w \qquad (\beta$$

(467, 467):: =

$$\vdash \dot{\varepsilon}\left(\begin{smallmatrix}\varepsilon\frown u\\ \varepsilon\frown w\end{smallmatrix}\right)\frown\left(\dot{\varepsilon}\left(\begin{smallmatrix}\varepsilon\frown z\\ \varepsilon\frown v\end{smallmatrix}\right)\frown\rangle\not\mathfrak{U}(q\bar{\frown}p)\right)$$
$$u\frown(z\frown\rangle\not\mathfrak{U}q)$$
$$w\frown(v\frown\rangle\not\mathfrak{U}p)$$
$$\underset{\smile}{\mathfrak{a}}\;\mathfrak{a}\frown u \;\;\; \mathfrak{a}\frown w$$
$$z\frown(u\frown\rangle q)$$
$$v\frown(w\frown\rangle p)$$
$$\underset{\smile}{\mathfrak{a}}\;\underset{\smile}{\mathfrak{e}}\;\mathfrak{a}\frown(\mathfrak{e}\frown q) \;\;\; \mathfrak{a}\frown z$$
$$\underset{\smile}{\mathfrak{a}}\;\underset{\smile}{\mathfrak{e}}\;\mathfrak{a}\frown(\mathfrak{e}\frown p) \;\;\; \mathfrak{a}\frown v \qquad (468$$

(32): –

$$\vdash \not{p}\dot{\varepsilon}\left(\begin{smallmatrix}\varepsilon\frown z\\ \varepsilon\frown v\end{smallmatrix}\right) = \not{p}\dot{\varepsilon}\left(\begin{smallmatrix}\varepsilon\frown u\\ \varepsilon\frown w\end{smallmatrix}\right)$$
$$\dot{\varepsilon}\left(\begin{smallmatrix}\varepsilon\frown z\\ \varepsilon\frown v\end{smallmatrix}\right)\frown\left(\dot{\varepsilon}\left(\begin{smallmatrix}\varepsilon\frown u\\ \varepsilon\frown w\end{smallmatrix}\right)\frown\rangle(q\bar{\frown}p)\right)$$
$$u\frown(z\frown\rangle\not\mathfrak{U}q)$$
$$w\frown(v\frown\rangle\not\mathfrak{U}p)$$
$$\underset{\smile}{\mathfrak{a}}\;\mathfrak{a}\frown u \;\;\; \mathfrak{a}\frown w$$
$$z\frown(u\frown\rangle q)$$
$$v\frown(w\frown\rangle p)$$
$$\underset{\smile}{\mathfrak{a}}\;\underset{\smile}{\mathfrak{e}}\;\mathfrak{a}\frown(\mathfrak{e}\frown q) \;\;\; \mathfrak{a}\frown z$$
$$\underset{\smile}{\mathfrak{a}}\;\underset{\smile}{\mathfrak{e}}\;\mathfrak{a}\frown(\mathfrak{e}\frown p) \;\;\; \mathfrak{a}\frown v \qquad (\alpha$$

(463):: –

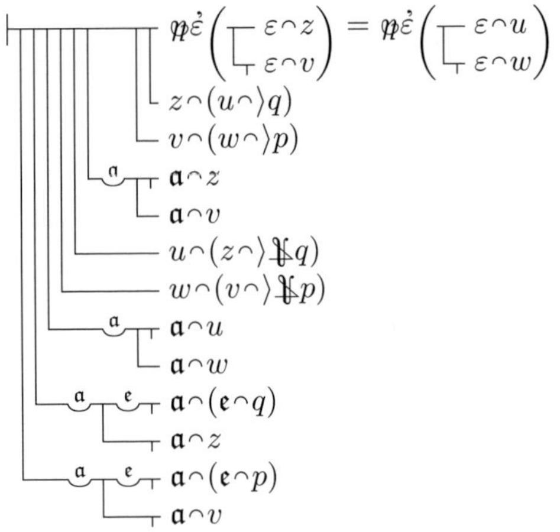

(β

×

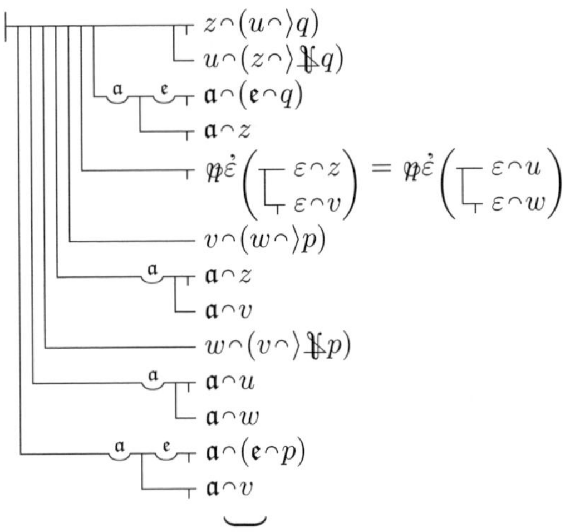

(γ

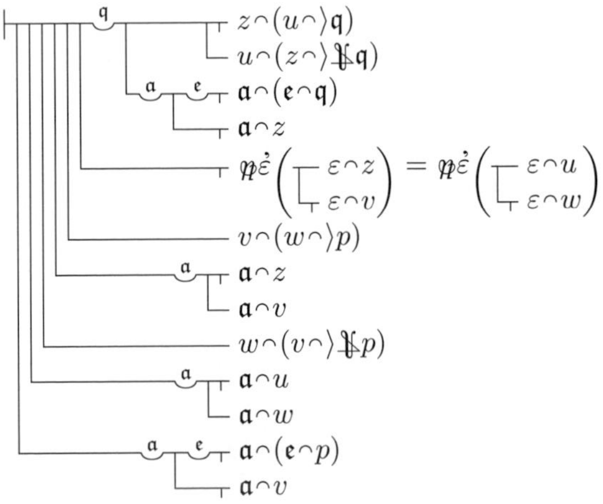

(δ

(453): —

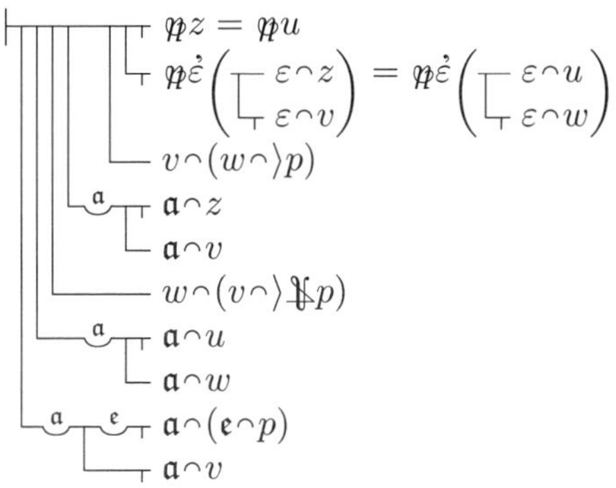
𝔭z = 𝔭u
𝔭ε̇(ε⌢z, ε⌢v) = 𝔭ε̇(ε⌢u, ε⌢w)
v⌢(w⌢⟩p)
𝔞⌢z
𝔞⌢v
w⌢(v⌢⟩p)
𝔞⌢u
𝔞⌢w
𝔞⌢(𝔢⌢p)
𝔞⌢v

(ε

×

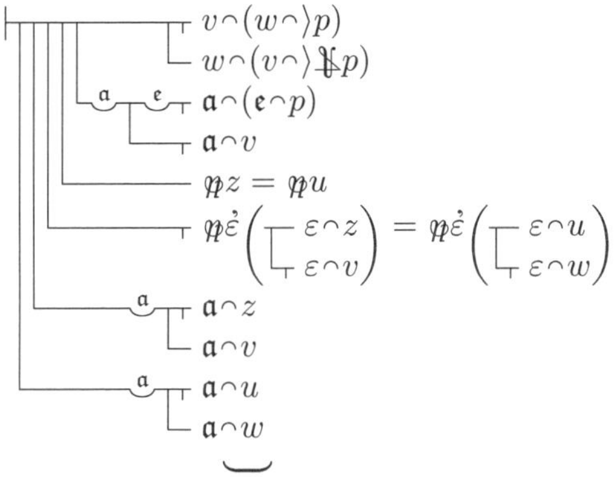
v⌢(w⌢⟩p)
w⌢(v⌢⟩p)
𝔞⌢(𝔢⌢p)
𝔞⌢v
𝔭z = 𝔭u
𝔭ε̇(ε⌢z, ε⌢v) = 𝔭ε̇(ε⌢u, ε⌢w)
𝔞⌢z
𝔞⌢v
𝔞⌢u
𝔞⌢w

(ζ

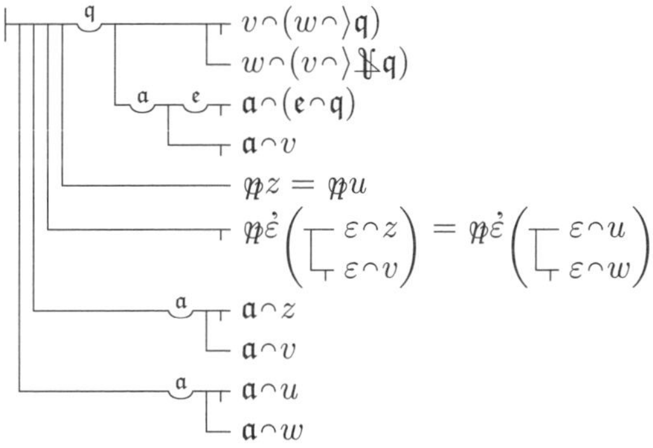
v⌢(w⌢⟩𝔮)
w⌢(v⌢⟩𝔮)
𝔞⌢(𝔢⌢𝔮)
𝔞⌢v
𝔭z = 𝔭u
𝔭ε̇(ε⌢z, ε⌢v) = 𝔭ε̇(ε⌢u, ε⌢w)
𝔞⌢z
𝔞⌢v
𝔞⌢u
𝔞⌢w

(η

(453): -

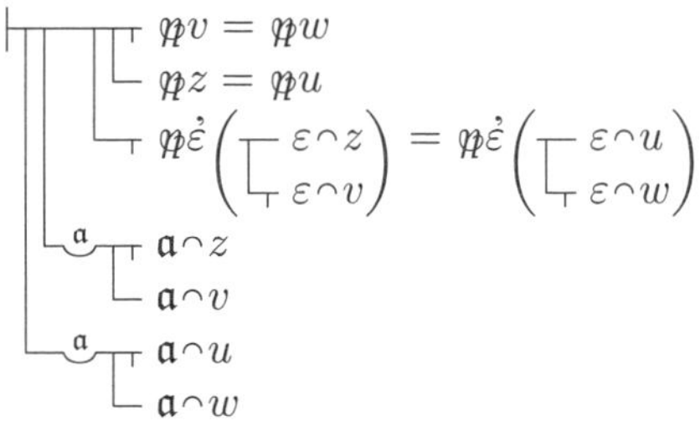
𝔭v = 𝔭w
𝔭z = 𝔭u
𝔭ε̇(ε⌢z, ε⌢v) = 𝔭ε̇(ε⌢u, ε⌢w)
𝔞⌢z
𝔞⌢v
𝔞⌢u
𝔞⌢w

(ϑ

×

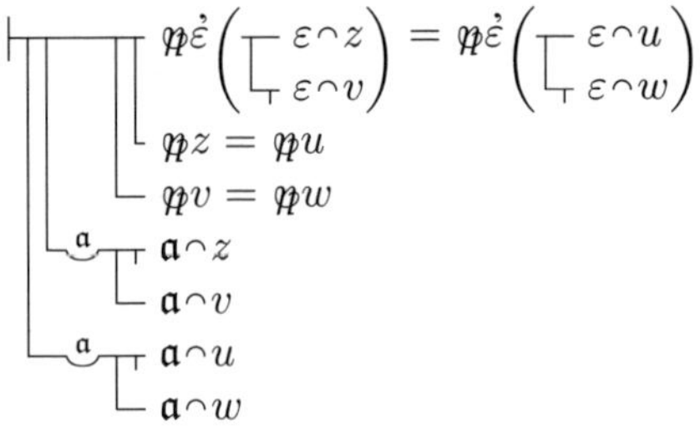

$$\begin{array}{l}
\vdash \prod \mathfrak{p}\grave{\varepsilon}\left(\underset{\varepsilon\frown v}{\vdash \varepsilon\frown z}\right) = \mathfrak{p}\grave{\varepsilon}\left(\underset{\varepsilon\frown w}{\vdash \varepsilon\frown u}\right) \\
\quad \vdash \mathfrak{p}z = \mathfrak{p}u \\
\quad \vdash \mathfrak{p}v = \mathfrak{p}w \\
\quad \underset{a}{\vdash} \underset{\mathfrak{a}\frown v}{\vdash \mathfrak{a}\frown z} \\
\quad \underset{a}{\vdash} \underset{\mathfrak{a}\frown w}{\vdash \mathfrak{a}\frown u}
\end{array} \qquad (469$$

O. Corollaries

a) Proof of the proposition

$$`\begin{array}{l}
\vdash \mathfrak{p}v = \mathfrak{p}w \\
\quad \vdash \mathfrak{p}z = \mathfrak{p}u \\
\quad \vdash \mathfrak{p}\grave{\varepsilon}\left(\prod \underset{\varepsilon\frown v}{\varepsilon\frown z}\right) = \mathfrak{p}\grave{\varepsilon}\left(\prod \underset{\varepsilon\frown w}{\varepsilon\frown u}\right) \\
\quad \underset{a}{\vdash} \underset{\mathfrak{a}\frown z}{\vdash \mathfrak{a}\frown v} \\
\quad \underset{a}{\vdash} \underset{\mathfrak{a}\frown u}{\vdash \mathfrak{a}\frown w}
\end{array}\text{'},$$

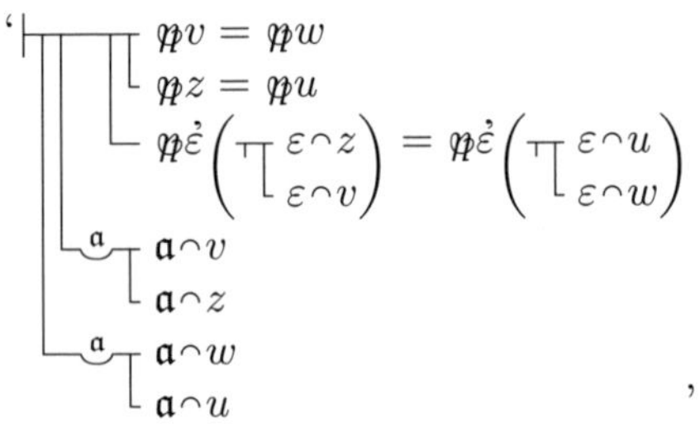

§45. *Analysis*

By leaving 'z' and 'u' unchanged in (469) but inserting '$\grave{\varepsilon}\left(\prod \underset{\varepsilon\frown v}{\varepsilon\frown z}\right)$' for '$v$' and '$\grave{\varepsilon}\left(\prod \underset{\varepsilon\frown w}{\varepsilon\frown u}\right)$' for '$w$', and adding the subcomponents: '$\underset{a}{\vdash} \underset{\mathfrak{a}\frown z}{\vdash \mathfrak{a}\frown v}$' and '$\underset{a}{\vdash} \underset{\mathfrak{a}\frown u}{\vdash \mathfrak{a}\frown w}$' we straightforwardly arrive at the proposition in our heading.

If we then insert '$\grave{\varepsilon}\left(\prod \underset{\varepsilon\frown u}{\varepsilon = c}\right)$' for '$z$' and '$\grave{\varepsilon}\left(\prod \underset{\varepsilon\frown w}{\varepsilon = c}\right)$' for '$v$' and add the condition that c falls under the u-concept, we obtain the proposition

$$`\begin{array}{l}
\vdash \mathfrak{p}w\frown(\mathfrak{p}w\frown \mathfrak{f}) \\
\quad \vdash \mathfrak{p}u\frown(\mathfrak{p}u\frown \mathfrak{f}) \\
\quad \underset{a}{\vdash} \underset{\mathfrak{a}\frown u}{\vdash \mathfrak{a}\frown w}
\end{array}\text{'},\qquad (\alpha$$

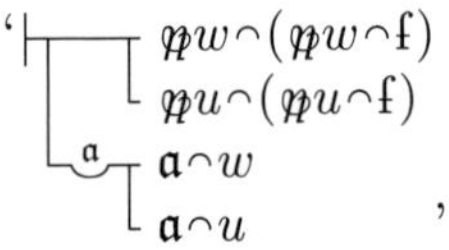

We can make use of this in order to prove that every cardinal number stands in the f-relation to a cardinal number, something which has only been proven for finite cardinal numbers in volume I. Assume that a cardinal number n belongs to the w-concept. Then either there is an object c that does not fall under the latter, in which case $\mathfrak{p}w$ stands in the f-relation to $\mathfrak{p}\grave{\varepsilon}\left(\vdash \underset{\varepsilon = c}{\varepsilon \frown w}\right)$; or there is no such object, in which case the $(\mathbf{0}\frown \mathbin{\reflectbox{\textit{X}}}\cup \mathfrak{f})$-concept is subordinate to the w-concept and we find that according to proposition (α) $\mathfrak{p}w$ stands in the f-relation to itself.

In order to prove the proposition in our heading, we have to show that $\mathfrak{p}\grave{\varepsilon}\left(\vdash \underset{\varepsilon\frown \grave{\varepsilon}\left(\prod \underset{\varepsilon\frown v}{\varepsilon\frown z}\right)}{\varepsilon\frown z}\right)$ coincides with $\mathfrak{p}v$ if the z-concept is subordinate to the v-concept.

§46. *Construction*

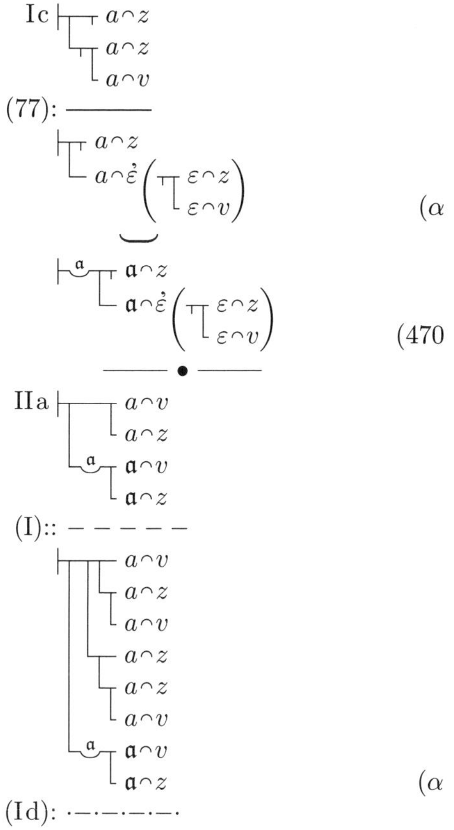

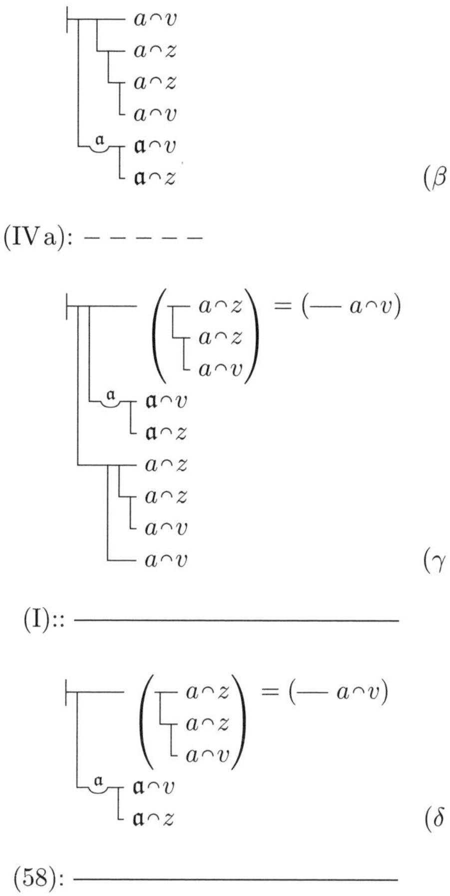

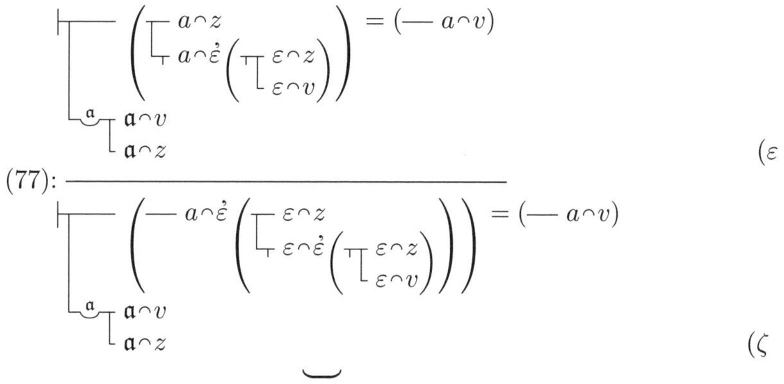

$$\vdash\!\begin{array}{|l} \overset{\mathfrak{a}}{\smile}\left(-\,\mathfrak{a}\frown\grave{\varepsilon}\left(\begin{array}{l}\varepsilon\frown z\\ \neg\ \varepsilon\frown\grave{\varepsilon}\left(\neg\begin{array}{l}\varepsilon\frown z\\ \varepsilon\frown v\end{array}\right)\end{array}\right)\right)=(-\,\mathfrak{a}\frown v)\\ \overset{\mathfrak{a}}{\smile}\begin{array}{l}\mathfrak{a}\frown v\\ \mathfrak{a}\frown z\end{array}\end{array} \qquad (\eta$$

(96): -

$$\vdash\!\begin{array}{|l} \mathfrak{p}\grave{\varepsilon}\left(\begin{array}{l}\varepsilon\frown z\\ \neg\ \varepsilon\frown\grave{\varepsilon}\left(\neg\begin{array}{l}\varepsilon\frown z\\ \varepsilon\frown v\end{array}\right)\end{array}\right)=\mathfrak{p}v\\ \overset{\mathfrak{a}}{\smile}\begin{array}{l}\mathfrak{a}\frown v\\ \mathfrak{a}\frown z\end{array}\end{array} \qquad (471$$

(IIIc): - - - - - - - - - - - - - - - - - - -

$$\vdash\!\begin{array}{|l} \mathfrak{p}v=\mathfrak{p}w\\ \mathfrak{p}\grave{\varepsilon}\left(\begin{array}{l}\varepsilon\frown z\\ \neg\ \varepsilon\frown\grave{\varepsilon}\left(\neg\begin{array}{l}\varepsilon\frown z\\ \varepsilon\frown v\end{array}\right)\end{array}\right)=\mathfrak{p}w\\ \overset{\mathfrak{a}}{\smile}\begin{array}{l}\mathfrak{a}\frown v\\ \mathfrak{a}\frown z\end{array}\end{array} \qquad (\alpha$$

(IIIa): ―――――――――――――――――――

$$\vdash\!\begin{array}{|l} \mathfrak{p}v=\mathfrak{p}w\\ \mathfrak{p}\grave{\varepsilon}\left(\begin{array}{l}\varepsilon\frown z\\ \neg\ \varepsilon\frown\grave{\varepsilon}\left(\neg\begin{array}{l}\varepsilon\frown z\\ \varepsilon\frown v\end{array}\right)\end{array}\right)=\mathfrak{p}\grave{\varepsilon}\left(\begin{array}{l}\varepsilon\frown u\\ \neg\ \varepsilon\frown\grave{\varepsilon}\left(\neg\begin{array}{l}\varepsilon\frown u\\ \varepsilon\frown w\end{array}\right)\end{array}\right)\\ \overset{\mathfrak{a}}{\smile}\begin{array}{l}\mathfrak{a}\frown v\\ \mathfrak{a}\frown z\end{array}\\ \mathfrak{p}\grave{\varepsilon}\left(\begin{array}{l}\varepsilon\frown u\\ \neg\ \varepsilon\frown\grave{\varepsilon}\left(\neg\begin{array}{l}\varepsilon\frown u\\ \varepsilon\frown w\end{array}\right)\end{array}\right)=\mathfrak{p}w\end{array} \qquad (\beta$$

(471):: -

$$\vdash\!\begin{array}{|l} \mathfrak{p}v=\mathfrak{p}w\\ \mathfrak{p}\grave{\varepsilon}\left(\begin{array}{l}\varepsilon\frown z\\ \neg\ \varepsilon\frown\grave{\varepsilon}\left(\neg\begin{array}{l}\varepsilon\frown z\\ \varepsilon\frown v\end{array}\right)\end{array}\right)=\mathfrak{p}\grave{\varepsilon}\left(\begin{array}{l}\varepsilon\frown u\\ \neg\ \varepsilon\frown\grave{\varepsilon}\left(\neg\begin{array}{l}\varepsilon\frown u\\ \varepsilon\frown w\end{array}\right)\end{array}\right)\\ \overset{\mathfrak{a}}{\smile}\begin{array}{l}\mathfrak{a}\frown v\\ \mathfrak{a}\frown z\end{array}\\ \overset{\mathfrak{a}}{\smile}\begin{array}{l}\mathfrak{a}\frown w\\ \mathfrak{a}\frown u\end{array}\end{array} \qquad (\gamma$$

(469):: -

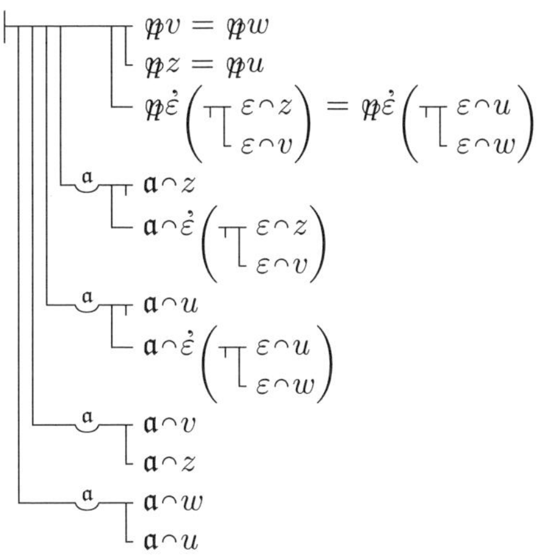

(δ

(470, 470):: ═══════════════

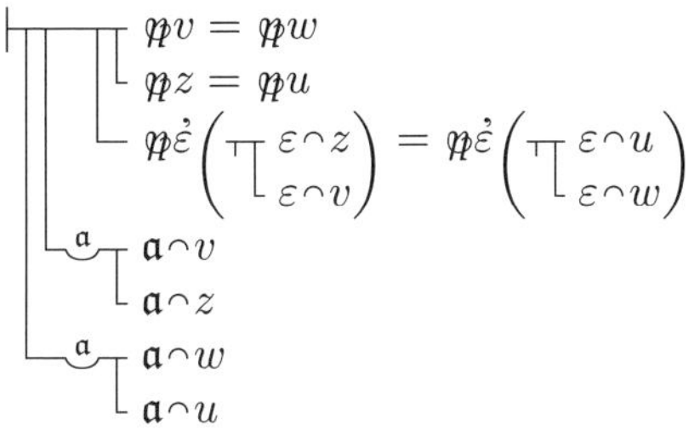

(472

b) Proof of the proposition

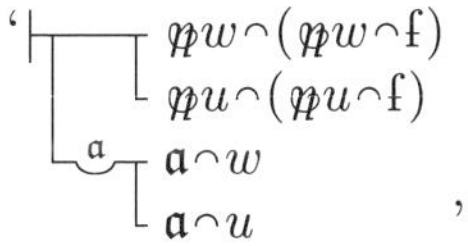

§47. *Analysis*

In order to arrive at our proposition (α) of §45, we have to prove that

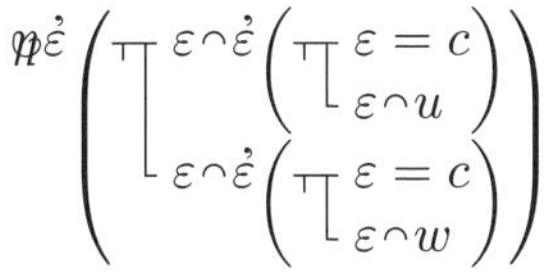

coincides with 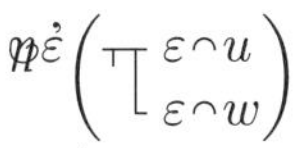 if c falls under the u-concept.

§48. *Construction*

If 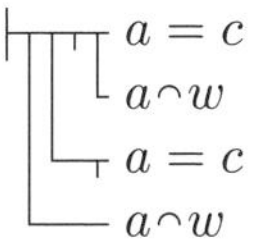

(IIa):: – – – – –

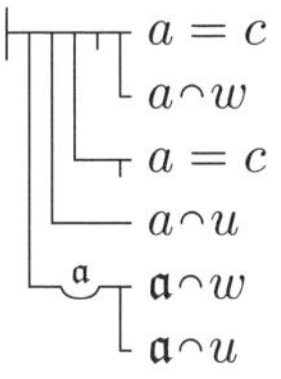

(α

(Ic, Id):: = = = =

$a = c$
$a \frown w$
$a = c$
$a \frown u$
$\mathfrak{a} \frown w$
$\mathfrak{a} \frown u$ (β

(77): ———

$a \frown \grave{\varepsilon}\left(\begin{matrix}\varepsilon = c\\ \varepsilon \frown w\end{matrix}\right)$
$a = c$
$a \frown u$
$\mathfrak{a} \frown w$
$\mathfrak{a} \frown u$ (γ

(77): ———

$a \frown \grave{\varepsilon}\left(\begin{matrix}\varepsilon = c\\ \varepsilon \frown w\end{matrix}\right)$
$a \frown \grave{\varepsilon}\left(\begin{matrix}\varepsilon = c\\ \varepsilon \frown u\end{matrix}\right)$
$\mathfrak{a} \frown w$
$\mathfrak{a} \frown u$ (δ

$\mathfrak{a} \frown \grave{\varepsilon}\left(\begin{matrix}\varepsilon = c\\ \varepsilon \frown w\end{matrix}\right)$
$\mathfrak{a} \frown \grave{\varepsilon}\left(\begin{matrix}\varepsilon = c\\ \varepsilon \frown u\end{matrix}\right)$
$\mathfrak{a} \frown w$
$\mathfrak{a} \frown u$ (473

——— • ———

Ia
$a = c$
$a \frown u$
$a \frown u$

(Ic):: – – – –

$a = c$
$a \frown u$
$a \frown u$
$a \frown w$ (α

(Ie): – – – –

$a = c$
$a \frown u$
$a = c$
$a \frown w$
$a \frown u$
$a \frown w$
$a = c$
$a \frown w$ (β

——— • ———

IIIb
$a = c$
$c \frown u$
$a \frown u$

(If): – – – – –

$a = c$
$a \frown w$
$c \frown u$
$a \frown u$
$a \frown w$ (γ

(Ic, Id):: = = = = =

$a = c$
$a \frown w$
$a \frown u$
$a \frown w$
$c \frown u$ (δ

(β): – – – – –

$a = c$
$a \frown u$
$a = c$
$a \frown w$
$a \frown u$
$a \frown w$
$c \frown u$ (ε

——— • ———

I
$a = c$
$a \frown u$
$a = c$
$a \frown u$

×

$a \frown u$
$a = c$
$a = c$
$a \frown u$ (ζ

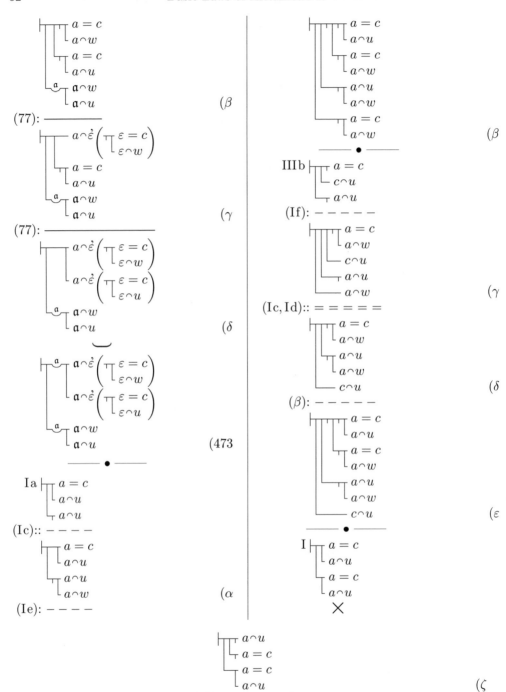

(If): – – – – –

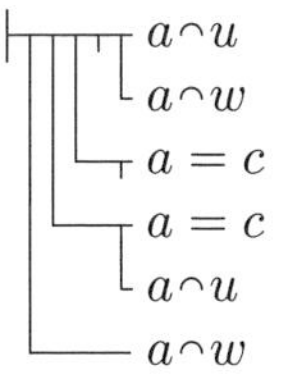

(η

(Ic, Id):: = = = = =

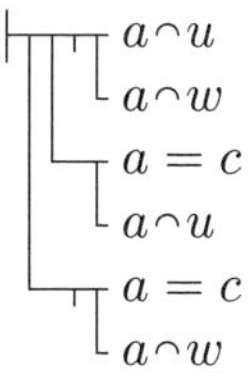

(ϑ

(Ib, Id):: = = = = =

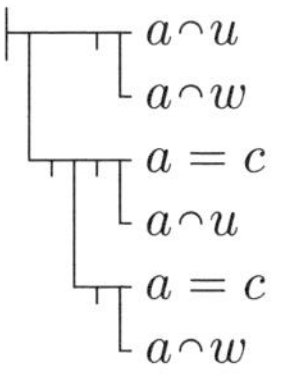

(ι

(IVa): ———

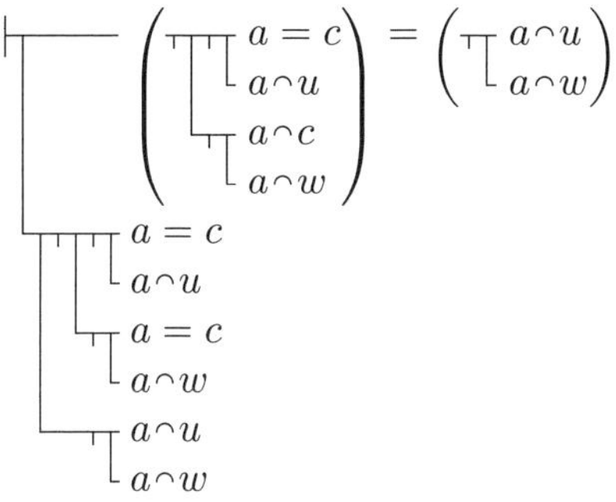

(κ

(ε):: - - - - - - - - - - - - - - -

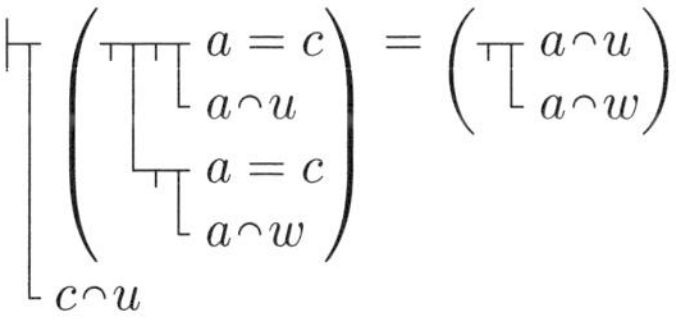

(λ

(77): ———

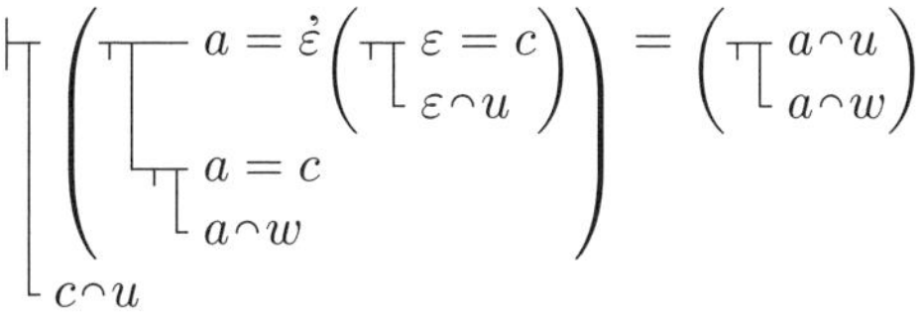

(μ

(77): ———

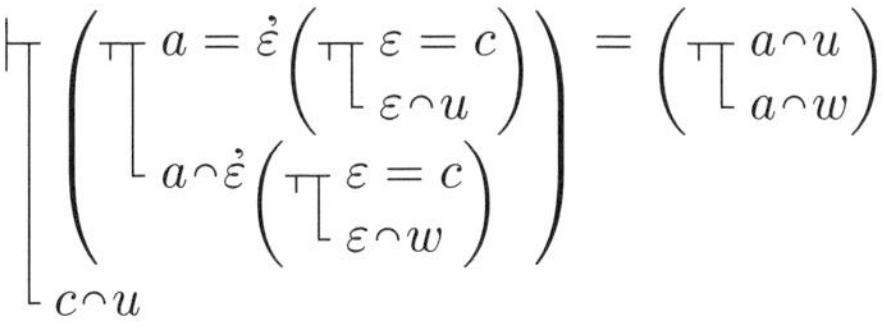

(ν

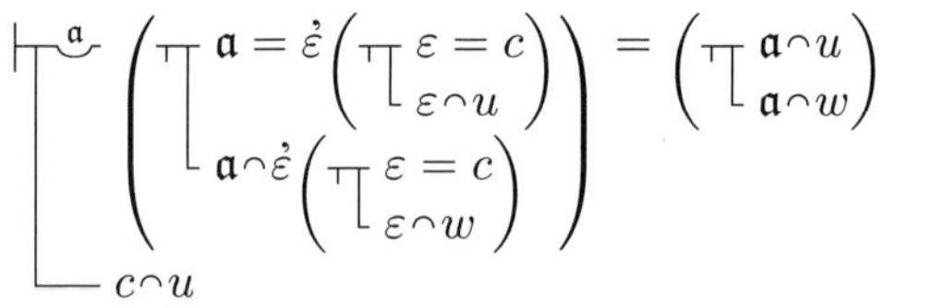

(ξ

(Va): –

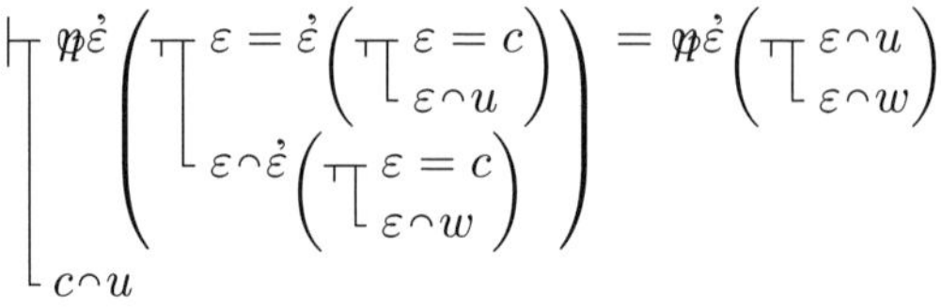

(o

(472): –

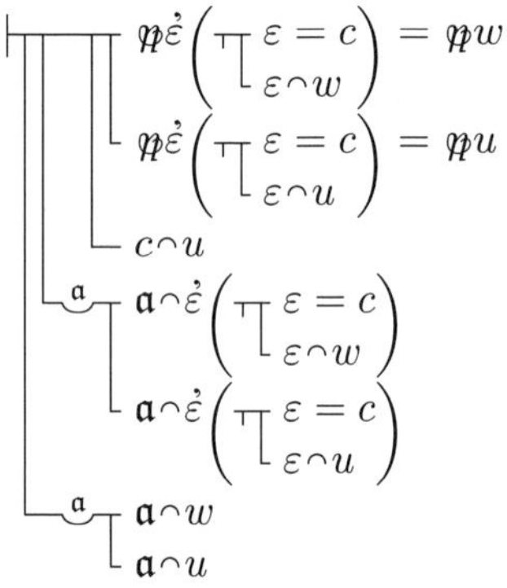

(π

(473):: – – – – – – – – – – – – –

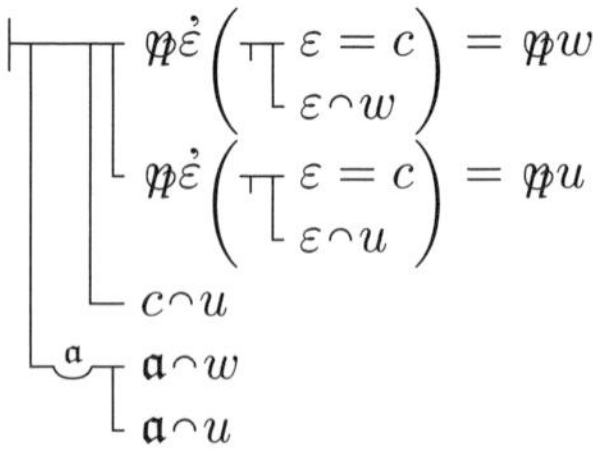

(ρ

(102): – – – – – – – – – – – – –

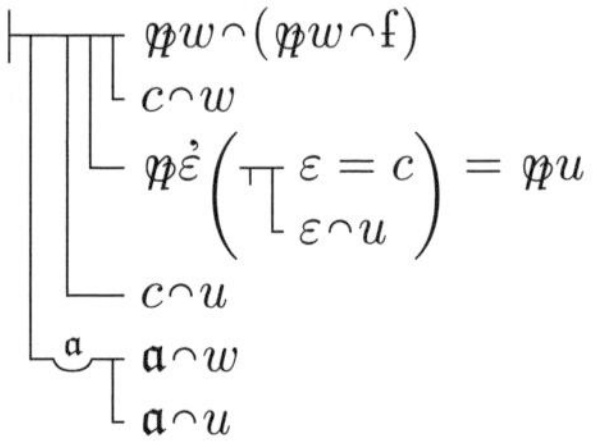

(σ

(IIa):: – – – – – – – – – – – – –

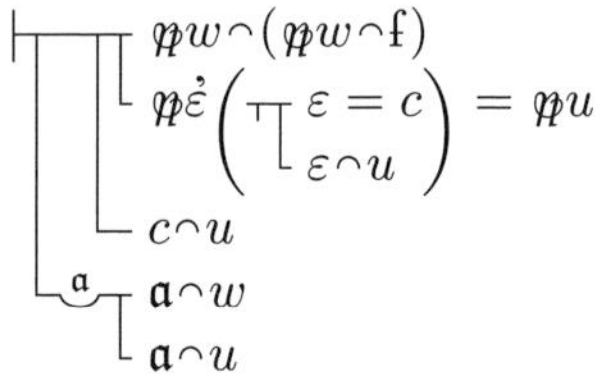

(τ

×

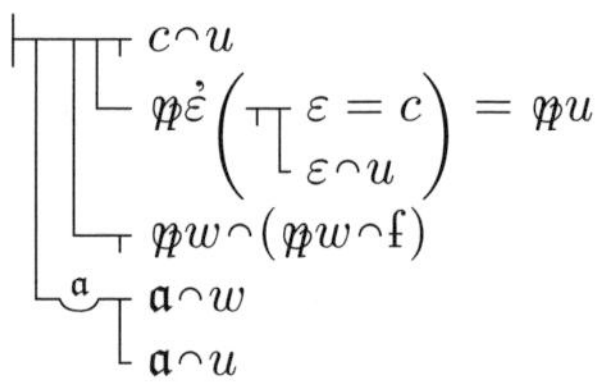

(υ

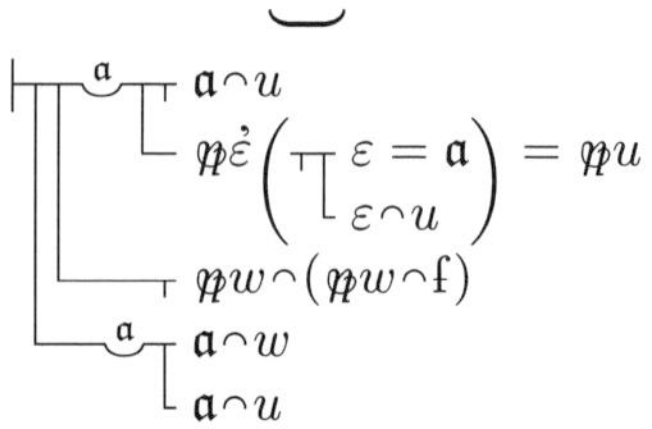

(474

§49. *Analysis*

In order now to obtain a proposition with the supercomponent ‘$\top\ \mathfrak{p}u\frown(\mathfrak{p}u\frown\mathfrak{f})$’, we cannot make use of (68) here but instead require the proposition

'⊢ $e \frown (\wp u \frown \mathfrak{f})$
 ⌣𝔞 $\mathfrak{a} \frown u$
 $\wp\dot{\varepsilon}\left(\varepsilon = \mathfrak{a},\ \varepsilon \frown u\right) = e$'

which we derive first.

§50. *Construction*

89 ⊢ I$\mathfrak{f}$

(79): ——

⊢ $\dot{\varepsilon}\left(\varepsilon = a,\ \varepsilon \frown u\right) = e$
 $e \frown (\wp u \frown \mathfrak{f})$
 $\wp\dot{\varepsilon}\left(\varepsilon = a,\ \varepsilon \frown u\right) \frown (\wp u \frown \mathfrak{f})$ (α

(108):: - - - - - - - - - - - - -

⊢ $\dot{\varepsilon}\left(\varepsilon = a,\ \varepsilon \frown u\right) = e$
 $e \frown (\wp u \frown \mathfrak{f})$
 $a \frown u$ (β

(IIa): - - - - - - - - -

⊢ $a \frown u$
 $a \frown u$
 $e \frown (\wp u \frown \mathfrak{f})$
 ⌣𝔞 $\mathfrak{a} \frown u$
 $\wp\dot{\varepsilon}\left(\varepsilon = \mathfrak{a},\ \varepsilon \frown u\right) = e$ (γ

(Ig): - - - - - - - - - - - -

⊢ $a \frown u$
 $e \frown (\wp u \frown \mathfrak{f})$
 ⌣𝔞 $\mathfrak{a} \frown u$
 $\wp\dot{\varepsilon}\left(\varepsilon = \mathfrak{a},\ \varepsilon \frown u\right) = e$ (δ

⌣

⊢ ⌣𝔞 $\mathfrak{a} \frown u$
 $e \frown (\wp u \frown \mathfrak{f})$
 ⌣𝔞 $\mathfrak{a} \frown u$
 $\wp\dot{\varepsilon}\left(\varepsilon = \mathfrak{a},\ \varepsilon \frown u\right) = e$ (ε

(97): - - - - - - - - - - - -

⊢ $\wp u = \emptyset$
 $e \frown (\wp u \frown \mathfrak{f})$
 ⌣𝔞 $\mathfrak{a} \frown u$
 $\wp\dot{\varepsilon}\left(\varepsilon = \mathfrak{a},\ \varepsilon \frown u\right) = e$ (ζ

(IIIc): - - - - - - - - - - - -

⊢ $e \frown (\emptyset \frown \mathfrak{f})$
 $e \frown (\wp u \frown \mathfrak{f})$
 ⌣𝔞 $\mathfrak{a} \frown u$
 $\wp\dot{\varepsilon}\left(\varepsilon = \mathfrak{a},\ \varepsilon \frown u\right) = e$ (η

×

⊢ $e \frown (\wp u \frown \mathfrak{f})$
 $e \frown (\emptyset \frown \mathfrak{f})$
 ⌣𝔞 $\mathfrak{a} \frown u$
 $\wp\dot{\varepsilon}\left(\varepsilon = \mathfrak{a},\ \varepsilon \frown u\right) = e$ (ϑ

(108):: ——————————

⊢ $e \frown (\wp u \frown \mathfrak{f})$
 ⌣𝔞 $\mathfrak{a} \frown u$
 $\wp\dot{\varepsilon}\left(\varepsilon = \mathfrak{a},\ \varepsilon \frown u\right) = e$ (475

—— • ——

475 ⊢ $\wp u \frown (\wp u \frown \mathfrak{f})$
 ⌣𝔞 $\mathfrak{a} \frown u$
 $\wp\dot{\varepsilon}\left(\varepsilon = \mathfrak{a},\ \varepsilon \frown u\right) = \wp u$

(474):: - - - - - - - - - - - - -

⊢ $\wp u \frown (\wp u \frown \mathfrak{f})$
 $\wp w \frown (\wp w \frown \mathfrak{f})$
 ⌣𝔞 $\mathfrak{a} \frown w$
 $\mathfrak{a} \frown u$ (α

×

⊢ $\wp w \frown (\wp w \frown \mathfrak{f})$
 $\wp u \frown (\wp u \frown \mathfrak{f})$
 ⌣𝔞 $\mathfrak{a} \frown w$
 $\mathfrak{a} \frown u$ (476

c) Proof of the propositions

'$\vdash\!\!\overset{\mathfrak{a}}{\smile}\!\!\!\top\; \wp w \frown (\mathfrak{a} \frown \mathfrak{f})$'

and

'$\vdash$ $\eth \frown (\wp w \frown \acute{\iota}\mathfrak{f})$
$\quad \overset{\mathfrak{a}}{\smile}\!\top\; \mathfrak{a} \frown w$
$\qquad \mathfrak{a} \frown u$
$\quad \wp u = \infty$ ',

§51. *Analysis*

In order to use (476) in the way announced in §45, we require the proposition

'$\vdash \wp w \frown \left(\wp\grave{\varepsilon}\left(\begin{array}{l} \varepsilon \frown w \\ \top\, \varepsilon = c \end{array}\right) \frown \mathfrak{f}\right)$
$\quad \top\, c \frown w$ ', (α

which is to be derived from (103). For this, the proposition

'$\vdash \wp w = \wp\grave{\varepsilon}\left(\begin{array}{l} \top\, \varepsilon = c \\ \varepsilon \frown \grave{\varepsilon}\left(\begin{array}{l} \varepsilon \frown w \\ \top\, \varepsilon = c \end{array}\right) \end{array}\right)$
$\quad \top\, c \frown w$ ', (β

is needed.

§52. *Construction*

IIIe $\vdash c = c$

(Ia): ———

$\vdash$ $c \frown w$
$\quad \top\, c = c$ (α

(77): ———

$\vdash c \frown \grave{\varepsilon}\left(\begin{array}{l} \varepsilon \frown w \\ \top\, \varepsilon = c \end{array}\right)$ (477

——— • ———

IIId $\vdash$ $a = c$
$\quad a \frown w$
$\quad \top\, c \frown w$

(If): - - - - -

$\vdash$ $a = c$
$\quad a \frown w$
$\quad \top\, a = c$
$\quad a \frown w$
$\quad \top\, c \frown w$
$\quad a \frown w$
$\quad \top\, a = c$ (α

(I):: - - - - - -

$\vdash$ $a = c$
$\quad a \frown w$
$\quad \top\, a = c$
$\quad a \frown w$
$\quad \top\, c \frown w$ (β

——— • ———

I $\vdash$ $a \frown w$
$\quad \top\, a = c$
$\quad a \frown w$
$\quad \top\, a = c$

(Ic, Id):: = = =

$\vdash$ $a \frown w$
$\quad a = c$
$\quad a \frown w$
$\quad \top\, a = c$ (γ

(IVa): ———

$\vdash (\!-\!\!-\, a \frown w) = \left(\begin{array}{l} \top\, a = c \\ a \frown w \\ \top\, a = c \end{array}\right)$
$\quad a = c$
$\quad a \frown w$
$\quad \top\, a = c$
$\quad a \frown w$ (δ

(β):: - - - - - - - - - - - - -

$$\vdash \left(-\, a \frown w\right) = \left(\begin{array}{l} a = c \\ a \frown w \\ a = c \end{array}\right)$$
$$c \frown w$$

(ε

(77): ————————————————

$$\vdash \left(-\, a \frown w\right) = \left(\begin{array}{l} a = c \\ a \frown \grave{\varepsilon}\left(\begin{array}{l} \varepsilon \frown w \\ \varepsilon = c \end{array}\right) \end{array}\right)$$
$$c \frown w$$

(ζ

(77): ————————————————

$$\vdash \left(-\, a \frown w\right) = \left(-\, a \frown \grave{\varepsilon}\left(\begin{array}{l} \varepsilon = c \\ \varepsilon \frown \grave{\varepsilon}\left(\begin{array}{l} \varepsilon \frown w \\ \varepsilon = c \end{array}\right) \end{array}\right)\right)$$
$$c \frown w$$

(η

$$\vdash \overset{\mathfrak{a}}{\smile} \left(-\, \mathfrak{a} \frown w\right) = \left(-\, \mathfrak{a} \frown \grave{\varepsilon}\left(\begin{array}{l} \varepsilon = c \\ \varepsilon \frown \grave{\varepsilon}\left(\begin{array}{l} \varepsilon \frown w \\ \varepsilon = c \end{array}\right) \end{array}\right)\right)$$
$$c \frown w$$

(ϑ

(96): –

$$\vdash \mathfrak{p}w = \mathfrak{p}\grave{\varepsilon}\left(\begin{array}{l} \varepsilon = c \\ \varepsilon \frown \grave{\varepsilon}\left(\begin{array}{l} \varepsilon \frown w \\ \varepsilon = c \end{array}\right) \end{array}\right)$$
$$c \frown w$$

(478

(III a): – – – – – – – – – – – – – – –

$$\vdash \mathfrak{p}w \frown \left(\mathfrak{p}\grave{\varepsilon}\left(\begin{array}{l} \varepsilon \frown w \\ \varepsilon = c \end{array}\right) \frown \mathfrak{f}\right)$$
$$\mathfrak{p}\grave{\varepsilon}\left(\begin{array}{l} \varepsilon = c \\ \varepsilon \frown \grave{\varepsilon}\left(\begin{array}{l} \varepsilon \frown w \\ \varepsilon = c \end{array}\right) \end{array}\right) \frown \left(\mathfrak{p}\grave{\varepsilon}\left(\begin{array}{l} \varepsilon \frown w \\ \varepsilon = c \end{array}\right) \frown \mathfrak{f}\right)$$
$$c \frown w$$

(α

(103):: –

$$\vdash \mathfrak{p}w \frown \left(\mathfrak{p}\grave{\varepsilon}\left(\begin{array}{l} \varepsilon \frown w \\ \varepsilon = c \end{array}\right) \frown \mathfrak{f}\right)$$
$$c \frown \grave{\varepsilon}\left(\begin{array}{l} \varepsilon \frown w \\ \varepsilon = c \end{array}\right)$$
$$c \frown w$$

(β

(477):: ————————————————

$$\vdash \begin{array}{l} \wp w \frown \left(\wp \dot{\varepsilon} \left(\begin{array}{l} \varepsilon \frown w \\ \varepsilon = c \end{array} \right) \frown \mathfrak{f} \right) \\ c \frown w \end{array}$$ (479

×

$$\vdash \begin{array}{l} c \frown w \\ \wp w \frown \left(\wp \dot{\varepsilon} \left(\begin{array}{l} \varepsilon \frown w \\ \varepsilon = c \end{array} \right) \frown \mathfrak{f} \right) \end{array}$$ (α

(IIa):: - - - - - - - - - - - -

$$\vdash \begin{array}{l} c \frown w \\ \smile^{\mathfrak{a}}\ \wp w \frown (\mathfrak{a} \frown \mathfrak{f}) \end{array}$$ (β

(I): - - - - - - - -

$$\vdash \begin{array}{l} c \frown w \\ c \frown (0 \frown \natural \smile \mathfrak{f}) \\ \smile^{\mathfrak{a}}\ \wp w \frown (\mathfrak{a} \frown \mathfrak{f}) \end{array}$$ (γ

$\smile$

$$\vdash \begin{array}{l} \smile^{\mathfrak{a}} \begin{array}{l} \mathfrak{a} \frown w \\ \mathfrak{a} \frown (0 \frown \natural \smile \mathfrak{f}) \end{array} \\ \smile^{\mathfrak{a}}\ \wp w \frown (\mathfrak{a} \frown \mathfrak{f}) \end{array}$$ (δ

(476): - - - - - - - - -

$$\vdash \begin{array}{l} \wp w \frown (\wp w \frown \mathfrak{f}) \\ \wp(0 \frown \natural \smile \mathfrak{f}) \frown (\wp(0 \frown \natural \smile \mathfrak{f}) \frown \mathfrak{f}) \\ \smile^{\mathfrak{a}}\ \wp w \frown (\mathfrak{a} \frown \mathfrak{f}) \end{array}$$ (ε

(IIIc): ————————————

$$\vdash \begin{array}{l} \wp w \frown (\wp w \frown \mathfrak{f}) \\ \infty \frown (\infty \frown \mathfrak{f}) \\ \smile^{\mathfrak{a}}\ \wp w \frown (\mathfrak{a} \frown \mathfrak{f}) \\ \wp(0 \frown \natural \smile \mathfrak{f}) = \infty \end{array}$$ (ζ

(M, 165):: ═══════════

$$\vdash \begin{array}{l} \wp w \frown (\wp w \frown \mathfrak{f}) \\ \smile^{\mathfrak{a}}\ \wp w \frown (\mathfrak{a} \frown \mathfrak{f}) \end{array}$$ (η

×

$$\vdash \begin{array}{l} \smile^{\mathfrak{a}}\ \wp w \frown (\mathfrak{a} \frown \mathfrak{f}) \\ \wp w \frown (\wp w \frown \mathfrak{f}) \end{array}$$ (ϑ

(IIa):: - - - - - - - - -

$$\vdash \begin{array}{l} \smile^{\mathfrak{a}}\ \wp w \frown (\mathfrak{a} \frown \mathfrak{f}) \\ \smile^{\mathfrak{a}}\ \wp w \frown (\mathfrak{a} \frown \mathfrak{f}) \end{array}$$ (ι

(Ig): ————————

$$\vdash \smile^{\mathfrak{a}}\ \wp w \frown (\mathfrak{a} \frown \mathfrak{f})$$ (480

(IIIc): ————————

$$\vdash \begin{array}{l} \smile^{\mathfrak{a}}\ n \frown (\mathfrak{a} \frown \mathfrak{f}) \\ \wp w = n \end{array}$$ (481

×

$$\vdash \begin{array}{l} \wp w = n \\ \smile^{\mathfrak{a}}\ n \frown (\mathfrak{a} \frown \mathfrak{f}) \end{array}$$ (α

$\smile$

$$\vdash \begin{array}{l} \smile^{\mathfrak{u}}\ \wp \mathfrak{u} = n \\ \smile^{\mathfrak{a}}\ n \frown (\mathfrak{a} \frown \mathfrak{f}) \end{array}$$ (482

×

$$\vdash \begin{array}{l} \smile^{\mathfrak{a}}\ n \frown (\mathfrak{a} \frown \mathfrak{f}) \\ \smile^{\mathfrak{u}}\ \wp \mathfrak{u} = n \end{array}$$ (483

§53. *Analysis*

Using (145) and (165), we draw from proposition (476) the further corollary that no finite cardinal number belongs to a concept if the cardinal number Endlos belongs to one of its subordinate concepts.

§54. *Construction*

$$145 \vdash \begin{array}{l} \wp w \frown (\wp w \frown \bar{\smile} \mathfrak{f}) \\ 0 \frown (\wp w \frown \smile \mathfrak{f}) \end{array}$$

×

$$\vdash \begin{array}{l} 0 \frown (\wp w \frown \smile \mathfrak{f}) \\ \wp w \frown (\wp w \frown \bar{\smile} \mathfrak{f}) \end{array}$$ (α

(131):: - - - - - - - -

$$\vdash \begin{array}{l} 0 \frown (\wp w \frown \smile \mathfrak{f}) \\ \wp w \frown (\wp w \frown \mathfrak{f}) \end{array}$$ (β

(476):: - - - - - - - -

$$\vdash \begin{array}{l} 0 \frown (\wp w \frown \smile \mathfrak{f}) \\ \wp u \frown (\wp u \frown \mathfrak{f}) \\ \smile^{\mathfrak{a}} \begin{array}{l} \mathfrak{a} \frown w \\ \mathfrak{a} \frown u \end{array} \end{array}$$ (γ

(IIIc): ————————

$$\vdash \begin{array}{l} 0 \frown (\wp w \frown \smile \mathfrak{f}) \\ \infty \frown (\infty \frown \mathfrak{f}) \\ \smile^{\mathfrak{a}} \begin{array}{l} \mathfrak{a} \frown w \\ \mathfrak{a} \frown u \end{array} \\ \wp u = \infty \end{array}$$ (δ

(165):: ————————————

$$\vdash \begin{array}{l} 0 \frown (\wp w \frown \smile \mathfrak{f}) \\ \smile^{\mathfrak{a}} \begin{array}{l} \mathfrak{a} \frown w \\ \mathfrak{a} \frown u \end{array} \\ \wp u = \infty \end{array}$$ (484

III. The real numbers

1. Critique of theories concerning irrational numbers

a) Principles of definition

§55. Before examining what foremost mathematicians have taught us about numbers, in particular about the irrational numbers, it will be beneficial to lay down and justify in advance some principles of definition that are disregarded by nearly all authors in this area, so that we need not discuss the issue in detail on every occasion. In volume I, we already laid down such principles for concept-script; here, we will mainly be concerned with definitions in ordinary language. To be sure, only those differences will emerge between the two expositions that are rooted in the different nature of these means of expression.

1. Principle of completeness

§56. A definition of a concept (a possible predicate) must be complete; it has to determine unambiguously for every object whether it falls under the concept or not (whether the predicate can be applied to it truly). Thus, there must be no object for which, after the definition, it remains doubtful whether it falls under the concept, even though it may not always be possible, for us humans, with our deficient knowledge, to decide the question. Figuratively, we can also express it like this: a concept must have sharp boundaries. If one pictures a concept with respect to its extension as a region in a plane, then this is, of course, merely an analogy and must be treated with care, though it can be of service here. A concept without sharp boundaries would correspond to a region that would not have a sharp borderline everywhere but would, in places, be completely blurred, merging with its surroundings. This would not really be a region at all; and, correspondingly, a concept without sharp definition is wrongly called a concept. Logic cannot recognise such concept-like constructions as concepts; it is impossible to formulate exact laws concerning them. The law of excluded middle is in fact just the requirement, in another form, that concepts have sharp boundaries. Any object Δ either falls under the concept Φ or it does not fall under it: *tertium non datur*. Would, for example, the proposition, "Every square root of 9 is odd", have any graspable sense if *square root of 9* were a concept without sharp boundaries? Does the question, "Are we still Christians?", indeed have a sense if it is not determined to whom the predicate *Christian* can be truly applied and from whom it must be withheld?

§57. From this now follows the inadmissibility of piecemeal definition, which is so popular in mathematics. This consists in providing a definition for a special case—for example, for the positive whole numbers—and putting it to use and then, after various theorems, following it up with a second explanation for a different case—for example, for the negative whole numbers and for Zero—at which point, all too often, the mistake is committed of once again making determinations for the case already dealt with. While one may in fact avoid contradictions, they are not ruled out in principle by this method. In most cases one does not attain completion either, with cases remaining for which no determination is made; and some are naïve enough to use the word or sign in these cases too, as if they had attributed a reference to it. Such piecemeal definition is to be compared with the procedure of drawing the boundary of a region in segments, perhaps without ever going on to connect them. The cardinal error, however, is to use the sign (word) before it has been completely explained: in theorems and often for the further development of the explanation itself. In the exact sciences, a word or sign may not be used so long as its reference has not been completely explained or is not otherwise known—least of all in order to develop its own explanation.

§58. To be sure, we have to grant that the development of the science which occurred in the conquest of ever wider domains of numbers, almost inevitably demands such a practice; and this demand could be used as an apology.[1] Indeed, it would have

[1] G. Peano says in part VI of the *Revue de mathématique*, pp. 60–61: "Mr Frege wishes every sign to have a unique definition. And that is my opinion too if we are dealing with a sign that does not contain variable letters (F_2 §1 P7). But if what is being defined does contain variable letters, i.e., is a function of these letters, then I see, in general, the need to give conditional definitions of that expression, or definitions with hypotheses (*ibid.* P7′), and to give as many definitions as there are kinds of entity to which we apply that operation. Thus the formula a + b is first defined when *a* and *b* are integers, then a second time when they are fractions, then again when they are irrationals or complex numbers. The same sign, +, is found between infinite and between transfinite numbers (F_1 VI), and then needs to be given a new definition. It is found between two vectors, and will be defined anew; and so on. And as the science progresses, the reference of the same formula is extended further and further. The various references of the notation a + b have properties in common; but these are insufficient to specify all the values that this expression can have.

The same happens with the formula a = b; in some cases its reference can be assumed as a primitive idea, in others it is defined, and in arithmetic in particular, given equality between integers, one defines equality between rationals, between irrationals, between imaginary numbers, etc. In geometry one is used to defining the equality of two areas, of two volumes, the equality of two vectors, etc. As the science progresses, the need to extend the reference of the formula a = b is felt more and more. The various references have properties in common; but I do not see how they suffice to specify all the been possible to replace the old signs and notations by new ones, and actually, this is what logic requires; but this is a decision that is hard to make. And this reluctance to introduce new signs or words is the cause of many unclarities in mathematics. One might have revoked the old explanations as invalid and started the science afresh with new ones, but a further reason why such a clean separation never occurred was because it was supposed that the old definitions could not be dispensed with for the beginnings of the science. Here, pedagogic needs may have played a role. And so people got used to piecemeal definition; and what was at first a dubious makeshift became customary and was accepted amongst the correct methods of science, with the result that hardly anyone complains anymore if a sign is first explained for a restricted domain and then used in a further explanation of the same sign for a broader domain; for general custom has a legitimising power, just as fashion can bestow a cachet of beauty upon the ugliest of costumes. So the point has to be emphasised: concept-like constructions that are still in flux and have not yet received final and sharp boundaries, logic cannot recognise as concepts; and therefore it must reject all piecemeal definition. After all, if

possible references of equality.

Moreover, the opinions of the various authors concerning the concept of equality are very diverse; and a study of this issue would be very useful, especially if carried out with the help of symbols, rather than words."[a]

Peano here appeals to a practical need. However, he does not thereby undermine the grounds given in my letter to him. What logic requires of definition may be difficult to comply with unfailingly, but possible it must be.

At best, one may regard several conditional definitions of the same sign as valid if their form clearly shows that together they cover all possible cases and that for no case multiple determinations are made, and if none of these partial explanations is used before all have been given, hence not in another partial explanation either. These explanations can then be combined into the form of a single explanation. Even so, it will be better to avoid this kind of definition wherever possible.

With respect to the equality-sign, we will do well to abide by our stipulation, according to which equality is complete coincidence, identity. To be sure, bodies of equal volume are not identical; however, they have the same volume. In such a case, the signs on each side of the equality-sign are not to be taken as signs for bodies but for their volumes, or alternatively for the measuring numbers resulting by measurement in terms of the same unit of volume. We will not be speaking of vectors as equal but of a particular specification—let us call it "directed length"—for these vectors, which may be the same for different vectors. On this account, the progress of science will not require extending the reference of the formula 'a = b'; rather, it is merely new determinations (*modi*) of the objects that will be subjected to investigation.

In the last proposition, Peano calmly speaks a momentous word. If mathematicians' opinions about equality diverge, then this means nothing less than that mathematicians disagree with respect to the content of their science; and if one regards the essence of the science as being thoughts, rather than words or signs, then this means that there is no one united mathematical science, that mathematicians do not, in fact, understand each other. For the sense of nearly all arithmetical propositions and of many geometrical propositions depends, directly or indirectly, on the sense of the word "equal".

the first definition is already complete and has drawn sharp boundaries, then the second definition either draws the same boundaries and is then to be rejected since its content should be proven as a theorem, or it draws different boundaries and thereby contradicts the first. For instance, one can define a conic section as the intersection of a plane with a conical surface of rotation. But once one has done so, one must not redefine it as, for example, a curve whose equation in Cartesian coordinates is of second degree; for this is something that must now be proven. Nor, on the other hand, can a conic section be defined as plane-figure whose equation in linear coordinates is of second degree; for this would also cover a pair of points that cannot be regarded as the intersection of a plane with a conic surface. Here, therefore, the boundary of the concept is not the same, and it would be a mistake to use the same term, "conic section". Thus, if a second definition is not rendered inadmissible by the first in either of these ways, then this is possible only because of the incompleteness of the first which has left the concept unfinished, that is, in a state in which it must not be employed at all, in particular, not in definitions.

§59. It will not be futile to provide an example acting as a counterbalance to the abstractness of these considerations. E. Heine lays down the following definition:[1]

> "Number-signs are called equal or are interchangeable when they belong to equal number-series, and not equal or not interchangeable when they belong to unequal number-series (§1 Def. 3)."

What would one say to the following definition:

> "Signs are called white when they belong to white objects"?

After all, I am permitted to take a circular black blot as a sign for the white paper lying in front of me, provided I have not used it as a sign for anything else. And, according to the definition, such a blot would be white. Against this one should say: in the expression, "when they belong to white objects", the definition presupposes the reference of the word "white" to be known; for if it were not, then it would not have been determined which signs belong to white objects. Very well! If the word "white" is already known, then one cannot want to explain it again. It should strike one as entirely self-evident that a word should not be explained by means of itself, for otherwise it is treated in one breath as both known and unknown. If it is known, then explanation is at best superfluous; but if it is not known, then it cannot serve in an explanation. This is so obvious, and yet it is so often ignored. The situation

[1] *Die Elemente der Functionenlehre*, Crelle, vol. 74, §2, Def. 2. From my here raising only one objection against this definition, it should be not concluded that I regard it as flawless otherwise.

is the same regarding Heine's definition. With the words, "when they belong to equal number-series", the reference of the word "equal" is presupposed as known, yet it is the same word that is intended to be explained.

§60. Heine would probably reply by remarking that the reference of the word "equal" is not presupposed as known for all cases, but is provided in his Def. 3 in §1 only for unbracketed number-series, whereas he here speaks of bracketed number-series and other signs. In addition to the considerations mentioned above, it should be objected that a double explanation of a word or sign is badly amiss since in this case there is still doubt whether the definitions contradict each other. At least a proof that there is no contradiction should be required; yet this obligation is usually neglected, and indeed no trace of such a proof can be found in Heine. In general, we have to reject any method of definition that makes the legitimacy of a definition depend upon proofs that have to be carried out beforehand; otherwise ascertaining that a proof has been conducted with the necessary rigour becomes extraordinarily complicated, since an investigation is needed whether any propositions are first to be proven before any definition is laid down, an investigation which is almost always missing. Indeed, such a gap is usually never felt and it thus poses a special danger to rigour. In arithmetic, it simply will not do to make some assertion without proof, or with a pseudo-proof, and then to wait and see whether anyone succeeds in refuting it; on the contrary, we must demand that any assertion that is not completely self-evident must actually be proven, and this includes that the expressions or signs used are correctly introduced, so long as they cannot be regarded as generally known.

It is, moreover, very easy to avoid multiple explanations of the same sign. Instead of first explaining it for a restricted domain and then using it to explain itself for a wider domain, that is, instead of employing the same sign twice over, one need only choose different signs and to confine the reference of the first to the restricted domain once and for all, so that the first definition is now also complete and draws sharp boundaries. Then the logical relation between the references of the two signs is not prejudged in any way, and one may investigate it without it being possible that the result of this investigation should call the legitimacy of the definitions into question.

It really is worth the travails to invent a new sign if we can thereby forestall not insignificant logical concerns and safeguard the rigour of proof. Yet some mathematicians' sense for logical cleanliness and precision seems so underdeveloped that they use one word with three or four different references, rather than make the monstrous decision to invent a new word.

§61. Piecemeal definition also renders the status of theorems uncertain. If, for example, the words 'square root of 9' have been explained restricting the domain to the positive whole numbers, then one may, for example, prove the proposition that there is only one square root of 9, which will be immediately overturned when one extends one's considerations to the negative numbers and supplements the definition accordingly. But who knows whether we have arrived at a final proposition? Who can know whether we may not come under pressure to accept four square roots of 9? How indeed can we know that there are no more than two square roots of -1? So long as there are no final and complete definitions, it is impossible. It might be objected that some propositions would then no longer be valid. But the same reason would speak against the admission of a second square root of 9. Thus we never have entirely firm ground beneath our feet. Without final definitions there will be no final theorems. We never escape from provisionality and instability.

§62. In the case of relations the issue is very similar to that for concepts: logic can acknowledge them only provided it is determined of every first and second object whether or not the former bears the relation to the latter. Here too we have a *tertium non datur*: the case of its being undecided is excluded. If this requirement on a relation were not met, then the concepts which can be arrived at by its partial saturation (vol. I, §30) would likewise not all have sharp boundaries; so they would not, strictly speaking, be concepts but inadmissible pseudo-concepts. For instance, if the relation of being greater than is not completely defined, then neither can one be sure whether a concept-like construction obtained from a partial saturation of it, such as being *greater than Zero*, or *positive*, is a proper concept. That would require, e.g., that it also be determined whether the Moon is greater than Zero. One could now stipulate that only numbers can stand in this relation, and then conclude from it that the Moon, since it is not a number, is not greater than Zero. To accomplish this, however, would require a complete explanation of the word "number", which is usually lacking.

In particular in the case of the greater-than relation, piecemeal, that is, incomplete, definition is, as it were, good form in mathematics. One first explains the expression, "greater than", just for the case of the positive whole numbers, thus incompletely. The pseudo-relation obtained in this way, which ought not to be used at all, is then employed anyway in order to supplement the first definition, although one surely cannot always tell at what point the definition of the relation of being greater than is to be considered final. The case of the relation of equality is similar: here too piecemeal definition is good form indeed.[1] Nonetheless, we have to stand firm: without complete and final definitions one never has solid ground under foot, one is not certain of the validity of one's theorems, one cannot be confident in applying logical laws, which after all presuppose sharp boundaries for concepts and hence also for relations.

§63. There is a straightforward consequence here concerning those functions that are neither concepts nor relations. Take, for example, the expression, "the half of something", which presents itself as the name of such a function. The word "something" marks the place for the argument-name, corresponding to the letter 'ξ' in '$\frac{1}{2}\xi$'. Such an expression can then become part of a concept-name, for instance "something whose half is less than One".

Now, if the latter is really taken to refer to a concept with sharp boundaries, then it must also be determined for, say, the Moon whether its half is less than One. For this to be the case, the expression "the half of the Moon" must have a reference; i.e., there must be one and only one object that is thereby designated. Yet according to common linguistic usage, this is not the case, since no one knows which half of the Moon is intended. Thus, a more precise stipulation needs to be made here, so that for every object it is determined which object the half of it is; otherwise it is not permissible to use the expression, "the half of x", with the definite article. The result is that a first-level function with one argument always has to be so constituted that an object results as its value, whatever object may be taken as its argument—by whatever object the function may be saturated.[2]

§64. We must make the analogous demand regarding functions with two arguments. The expression

[1] In the practice of giving proofs, probably all mathematicians treat equality as identity, although in the theory most do not want to accept it. Yet no one will say, e.g., that the equation '$4 . x - 3 = 3$' has both $\frac{6}{4}$ and $\frac{3}{2}$ as roots, since, although $\frac{6}{4} = \frac{3}{2}$, $\frac{6}{4}$ and $\frac{3}{2}$ do not coincide. If they do not coincide, then they are different, and our equation has at least two different roots. It is remarkable to see what a glaring conflict there is for many mathematicians between their stated theories and their tacitly performed practice. If, however, in mathematics equality is identity, then its multiple definition is an utterly nonsensical procedure.

[2] Compare with what has been said in the first volume about functions; also the author's work *Function und Begriff* (Jena, Pohle, 1891).

"the sum of a first and second object"

presents itself as a name of such a function. Hence in this case too it has to be determined for every first and every second object which object is the sum of the first and the second, and there must always be such an object. If that is not the case, then it is not determined either which object added to itself results in One. Thus, in this case, the words "something that added to itself results in One" do not refer to a concept with sharp boundaries, and therefore to nothing logically usable at all. And the question, how many objects there are that added to themselves result in One, cannot be answered.

But might one not stipulate that the expression "the sum of a first and a second object" is to have reference only when both objects are numbers? Then one would surely suppose that *something that added to itself results in One* is a concept with sharp boundaries; for one now knows that no object falls under it that is not a number. For example, the Moon does not fall under it, since the sum of the Moon and of the Moon is not One. This is false; for the proposition "the sum of the Moon and the Moon is One" is now neither true nor false; for in both cases the words "the sum of the Moon and the Moon" would have to refer to something, which is exactly what the suggested stipulation denies. Our proposition is rather to be compared with the proposition "Scylla had six dragon gullets". This proposition too is neither true nor false but fiction, since the proper name "Scylla" designates nothing. Such propositions can be the object of a scientific examination, e.g., one concerned with mythology, but no scientific investigation can be carried out using them. If our proposition, "the sum of the Moon and the Moon is not One", were scientific, then it would say that the reference of the words "the sum of the Moon and the Moon" does not coincide with the reference of the word "One"; yet, according to the stipulation suggested above, the former reference would not be there. Consequently, one could truthfully say neither that it coincides with the reference of "One" nor that it does not coincide with it. Thus the question whether the sum of the Moon and the Moon is One, or whether the Moon falls under the concept *something that added to itself results in One* would have no answer; in other words: what we just labelled a concept would not be a proper concept since sharp boundaries are wanting. However, once the expression "a added to b results in c" has been introduced, then the formation of the concept-name "something that added to itself results in One" cannot be prevented. Now, if one actually attempted properly to prevent the formation of such inadmissible, yet linguistically possible, concept-names by laying down a law, then it would soon be abandoned as excessively difficult and probably unrealisable. The only feasible course, if one wants to use these words at all, is to explain, "sum", "adding", and others in such a way that concept-names, formed from them in a linguistically correct manner, refer to concepts with sharp boundaries and are thus admissible.

Thus, the demand here imposed that every first-level function with two arguments have an object as value for every first object as first argument and every second object as second argument, is also a consequence of the fact that concepts must have sharp boundaries, that no expression can be allowed which according to its construction appears to refer to a concept but in fact only pretends to do so, just as proper names are inadmissible that do not actually designate an object.

§65. What has been said about expressions in words also holds good for arithmetical signs. If the addition-sign is completely explained, then we have in

$$`\xi + \xi = \zeta'$$

the name of a relation, that of single to double. If this is not the case, then we cannot say whether the equation

$$`x + x = 1'$$

has a single or multiple solutions. Now, someone will reply: I prohibit that anything but numbers be considered at all. We dealt with a similar objection above; here we will shed further light on the matter from other angles. Anyone who wants to rule out of consideration all objects save numbers must first say what he understands by "number", after which any further extension is not permissible. Such a restriction would have to be incorporated into the explanation, so that it would take, for example, the form: "if a and b are numbers, then $a + b$ refers to", etc. We would have a conditional explanation.[1] However, the addition-sign has been explained only if the reference of every possible combination of signs of the form '$a + b$' has been determined, whatever referential proper names might be inserted for 'a' and 'b'. Yet if one explains such combinations of signs just for the case, for example, where signs of whole real numbers are to be taken for 'a' and 'b', then one has indeed only explained those combinations, rather than the addition-sign, thereby violating a second basic principle of definition which remains to be discussed. And still one unwittingly imagines that the reference of the addition-sign is known and treats it accordingly also in those cases for which no explanation has been given.

As soon as one aims for generality in one's propositions, one will also employ in

[1] Compare the author's letter to Mr G. Peano, *Revue de mathématique*, vol. VI, pp. 53ff.

arithmetical formulae beyond signs for specific objects—the proper names, e.g., '2'—letters that merely indicate,[1] rather than designate, and thus be led without noticing beyond the region for which the signs have been explained. One might attempt to address the ensuing dangers by not allowing these letters to indicate objects in general, as we did, but only those within a region with strict boundaries. Let us suppose that the concept *number* is sharply defined and that it is stipulated that the Roman letters indicate only numbers and that the addition-sign is explained only for numbers. Then to the proposition

'$a + b = b + a$'

we have to add in thought the conditions that a and b are numbers; and these are easily forgotten since they are not stated.[2] But let us for once resolve not to forget them! According to a well-known law of logic, we may transform the proposition

"if a is a number and b is a number, then $a + b = b + a$"

into the proposition

"if $a + b$ is not equal to $b + a$ and a is a number, then b is not a number";

and here it is impossible to maintain the restriction to the domain of numbers. The pressures of the subject matter ineluctably work towards breach of such barriers. But then the conditional clause

"if $a + b$ is not equal to $b + a$"

has no sense, owing to the incomplete explanation of the addition-sign.

Here, once again, we see that the laws of logic presuppose concepts with sharp boundaries and also, therefore, complete explanations of function-names—e.g., of the plus-sign.[3] In the first volume, we expressed this point thus: every function-name must have a reference. Accordingly, all conditional definitions, and all piecemeal defining, are to be rejected. Every sign must be completely explained in one stroke, so that, in our terms, it receives a reference.

All these points are intimately connected, and may be regarded as flowing from the principle of completeness for definitions.

---

[1] Compare, vol. I, pp. 31 and 32.

[2] When expanding the domain of numbers, e.g., does one in fact always think that the sense of those conditions is thereby being changed, that all general propositions proven hitherto acquire a different thought-content, that also the proofs become obsolete?

[3] It goes without saying that certain functions cannot be defined owing to their logical simplicity; nevertheless these also have to have values for every argument.

2. Principle of the simplicity of the explained expression[1]

§66. That the reference of an expression together with that of one of its parts does not always determine the reference of the remaining part is obvious. One must not, therefore, explain a sign or word by explaining an expression in which it occurs and whose remaining parts are known. For otherwise an investigation will first be required whether a solution for the unknown—to make use of a readily understood algebraic image—is possible, and whether the unknown is uniquely determined. Yet it is, as remarked above, infeasible to make the legitimacy of a definition depend on the results of such an investigation, one which, moreover, might not even be practicable. Rather, a definition must have the character of an equation, solved for the unknown, on whose other side nothing unknown occurs.

Even less will it do to explain two things by means of a single definition; every definition must, on the contrary, contain a single sign whose reference it stipulates. After all, one cannot determine two unknowns by means of a single equation either.

It sometimes occurs that a whole system of definitions is laid down, each containing several words that are to be explained, such that each of these words occurs in several of these definitions. This is comparable to a system of equations with several unknowns, where it is once again entirely left open to question whether there is a solution, and whether it is uniquely determined.

To be sure, one may regard any sign, any word, as consisting of parts; however we deny it simplicity only if the reference of the whole would follow from the references of the parts according to the general rules of grammar or of the notation, and if these parts also occur in other combinations and are treated as autonomous signs with their own reference. In this sense, therefore, we may say: the explained expression—the explained sign—must be simple. Otherwise, it might occur that the parts were also explained separately and that these explanations contradicted that of the whole.

Names of functions cannot, of course, appear on their own on one side of a definitional equation, owing to their distinctive unsaturatedness; rather, their argument-places have always somehow to be filled. This, as we have seen,[2] is accomplished in concept-script by Roman letters, which must then also feature on the other side. In ordinary language, indeterminately indicating pronouns and particles ("something",

[1] Vol. I, §33, 3.
[2] Vol. I, §33, 5.

"what", "it") fulfil this role. This constitutes no violation of our principle, since these letters, pronouns, particles refer to nothing, but merely indicate.

§67. Often, both of our principles of definition are flouted at once, for example, by explaining the equality-sign together with what stands on its right and left. In such a case, the equality-sign is already explained beforehand, though incompletely. Thus, a peculiar twilight is generated by treating the equality-sign half as known and then again half as unknown. On the one hand, it seems one is supposed to recall the earlier definition and elicit from it something to determine what now occurs to the right and left. But on the other, this earlier explanation does not suffice for the case at hand. Something similar also happens with other signs. This twilight is required by some mathematicians for the performance of their logical conjuring tricks. The results that are to be gained in this way may be obtained in an irreproachable manner by our conversion of the generality of an equality into an equality of value-ranges in accordance with basic law V (vol. I, §3, §9, §20).

Without claiming to have provided hereby a complete overview of everything that should be observed when defining, I will content myself with this exposition of these two principles, which are most often flouted by the mathematicians.

b) Cantor's theory of the irrational numbers

§68. G. Cantor first defines[1] his fundamental series:

> "Every set[2] (a_ν) which can[b] be characterised by the condition:
>
> $$\operatorname*{Lim}_{\nu=\infty}(a_{\nu+\mu} - a_\nu) = 0 \qquad \text{(for arbitrary } \mu)$$
>
> I call a fundamental series,[c] and assign to it a number b to be defined by it."

It is an atrocious state of affairs that the use of the word "number" by mathematicians fluctuates: sometimes it is the number-signs that are called numbers, sometimes it is their reference. Every mathematical author who uses the word "number" should properly state how he intends to use it.[3] In the present case too there is doubt about the sense of Cantor's proposition. If the word "number" is intended to have the reference of "number-sign", then the proposition has to be understood in the following way: I give to any fundamental series a specific sign, 'b', by means of which the series is designated. In that case different fundamental series must receive different signs.

[1] *Math. Annalen* XXI, p. 567.
[2] Set of rational numbers.
[3] We have chosen the latter.

However, fundamental series are distinct whenever there is a number that pertains to one of them, but not the other. Factually, this bestowal of signs is completely immaterial; for I must already have designated the series in some way in order to be able to say: to the fundamental series thus specified I give the sign 'b'. Whether, when I want to speak of the series, I now use the sign 'b' or the original notation cannot make any difference to the facts; at most, the new notation may be more convenient owing to its greater simplicity.

§69. Cantor now says further:

> "Such a fundamental series[c] presents ... three cases: either its members (a_ν), for sufficiently large value ν, are less in their absolute value than an arbitrarily given number; or, from a certain ν on, they are greater than some specifiable rational[d] number, ϱ; or, from a definite ν on, they are less than some specifiable negative rational magnitude, $-\varrho$. In the first case I say that b equals Zero; in the second I say that b is larger than Zero, or positive; in the third that b is less than Zero, or negative."

If our conjecture about the sense of the first quoted proposition of Cantor's is correct, then without recourse to 'b' we can express the above like this: "Of a fundamental series which, for sufficiently large ν, has members, a_ν, that are less in their absolute value than an arbitrarily given number, I say that it is equal to Zero; of a fundamental series whose members are, from a certain ν onwards, greater than some specifiable rational number, ϱ, I say that it is greater than Zero, or positive; of a fundamental series whose members are, from a specific ν onwards, less than some specifiable negative rational number, $-\varrho$, I say that it is less than Zero, or negative." Whether we now take this wording, or Cantor's original, we have in both cases three explanations of the expressions "equal to Zero", "greater than Zero", "less than Zero". These definitions are flawed since the expressions being explained are not simple but contain the words "greater" and "less", which must be presupposed as known since they contribute to the explanation—violation of both of our principles of definition. Moreover, the words "Zero" and "equal" also have to be presupposed as known; and then the expressions "equal to Zero", "greater than Zero", "less than Zero" are completely known, and must not be explained again. If they were not, the earlier definitions would have been incomplete—violation against our first principle of definition.

§70. In his article "Zur Weierstrass-Cantorschen Theorie der Irrationalzahlen",[1] Illigens understands Cantor's theory in such a way that the number b is to be

[1] *Math. Annalen* XXXIII, pp. 155ff.

understood as a sign for the series' being given by some law. Accordingly, it may at first seem that the number b is intended to represent a proposition; however, the subsequent frequent use of the expression, "number-series sign", makes it plain that Illigens also understands Cantor's theory as we have here attempted to do. He objects that such a number-series sign does not, in contrast to the rational numbers, designate a quantity, that the words "greater" and "less" here have a completely different sense than in the case of the rational numbers; this, in turn, entails the same for the word "limit"; the mere use of the word "greater" cannot turn b into a quantity sign, and the rational numbers belong to the new numbers by being used as number-series signs, but not insofar as they designate quantities; the goal of making the rational numbers appear as a specific kind of number-series sign is thus missed. We could surely add, in Illigens's spirit, that in that case the rational numbers—our author apparently takes them to be number-signs—would be ambiguous, in one case designating quantities, in the other number-series (fundamental series). Illigens adds: "The number-series signs introduced, despite all the names given to them by various definitions, are not able to become concepts of quantity."

Certainly! A sign can never become a concept. Yet perhaps one need not assume that Cantor confused sign and what is designated in this way. Nevertheless there is something correct about this objection provided one construes it so that despite all the names given to them by various definitions, the number-series do not become quantities.

§71. Illigens further contends that if $\sqrt{2}$ were taken as a mere number-series sign (for the series 1.4, 1.41, 1.414, ...), then on the left side of the equation $(\sqrt{2})^2 = 2$, we would have a mere sign for the number-series 1.4^2, 1.41^2, 1.414^2, ... It would hardly occur to anyone, however, to equate such a sign, for which any arbitrary thing could be chosen, with the quantity designated by the number 2. The objection is only justified if Cantor mistakes the sign for what is designated, which is what should first be shown. When the equality-sign is written between other signs, the sign on the left, '$(\sqrt{2})^2$', is not thereby equated with the reference of that on the right; rather, the reference of that on the left is equated with the reference of the sign standing on the right. Here accordingly a number-series is equated with the reference of the number-sign '2', i.e., with the number 2. The justification for this remains to be examined.

One may indeed object to Cantor that he has neglected this examination, and that he is aiming to stipulate by means of a pseudo-definition what in fact ought to be proven.

In Cantor, the fundamental series 1.4^2, 1.41^2, 1.414^2, ... must be presupposed as known, along with the number 2 and the reference of the word "equal". Whether, therefore, this fundamental series is equal to the number 2 cannot be the subject of an arbitrary stipulation but must come out as a result. To be sure, this objection only holds provided that the sense of Cantor's theory assumed here is correct. Later, we will have to explore yet another construal.

Illigens further contends that in Cantor's theory one cannot say what a line of length $\sqrt{2}$ metres is.

There is surely much that is true in these objections; but this would be more salient if Illigens had not used the word "number" with the reference number-sign and in general had distinguished more sharply between number and number-sign; for when he speaks of the rational numbers and calls one number greater than another, this does not fit with the reference number-sign. The danger of such inexactness is always present when one understands by "number" not the reference of a number-sign but instead the sign itself; for the view is well entrenched that the objects of arithmetic are numbers. It then always suggests itself to turn the number-signs from mere auxiliary tools of arithmetical research into its objects, and thereby to create worrying confusions.[1]

Secondly, it strikes me as worrying, at least as inviting misunderstandings, that Illigens calls rational numbers quantity-signs and states that they designate quantities. According to common linguistic usage, lengths, surfaces, angles, periods of time, masses, forces may all indeed be called quantities. But is it then right to say that the number $\frac{2}{3}$ or the number-sign '$\frac{2}{3}$' designates a specific length, or a specific angle, or indeed both?

§72. A. Pringsheim contends[2] that the rational numbers feature as signs that *may* represent well determined quantities but do not *have* to. Clearly, this author too understands the rational numbers as the kind of figures that are artificially produced by a writing instrument on a writing surface or by a printing press. Now: either these figures are, for arithmetical purposes, nothing more than figures, in which case it is irrelevant to this science what ideas may be connected to them, provided that, while engaged in arithmetic, one sets aside these ideas. This is irrelevant to arithmetic in

[1] Admittedly, there is also the view according to which numbers are neither signs that refer to something, nor the non-sensory reference of such signs, but figures that are treated according to certain rules, similar to pieces in chess.[f] In that case, numbers are neither auxiliary tools of research nor objects of study, but objects of manipulation. This will have to be examined later.

[2] *Encyklopädie der math. Wissenschaften* I A 3, p. 55.

the same way as someone's associating the idea of an elephant with the figure of a triangle is irrelevant to geometry, provided he entirely refrains from it while doing geometry. Now, someone might want to say that "a triangle may still represent an elephant, but it does not have to"; this, however, would scarcely further one's insight into the nature of geometry or of the triangle. Or, the figures termed rational numbers by Pringsheim are signs for something, serving for the expression of arithmetical thoughts, just as the sign '♃' is used by astronomers to designate Jupiter. In this case, how peculiar it would be to say that this sign may designate Jupiter, but does not have to! The astronomer will simply say: "I designate Jupiter by the sign '♃'", and thereby the issue will be settled; any more chatter about the sign is then superfluous. So: either it is essential for arithmetic that the number-signs refer to something; in that case, what they refer to is the main issue, the object of investigation, and they themselves are mere auxiliary tools on which few words need to be spent; or the number-signs are themselves the objects of arithmetic; then it is irrelevant whether someone or other associates another reference with them, and there is no need to talk about it in doing arithmetic. In the first case, the number-signs simply designate something, namely numbers; in the second case they designate nothing, at least in arithmetic. In neither case, however, can we find anything decisive or enlightening in the proposition that numbers may represent quantities but do not have to.

§73. What then is the substance of the assertion that number-signs designate quantities? Let us look at the applications of arithmetical laws in geometry, astronomy, physics. Here, indeed, numbers occur in relation to magnitudes, such as length, mass, light intensity, electric charge; and a superficial consideration might suggest that the same number-sign refers sometimes to a length, sometimes to a mass, then to a light intensity, and this would, in turn, seem to support Pringsheim's claim that there is a certain relation, albeit a relaxed one, between the number-signs and the quantities. Let us look at this more carefully. Indeed, what is it that we are applying when we make use of an arithmetical proposition? The sound of words? Groups of curious figures that consist of printer's ink? Or do we apply a thought-content that we associate with these words, with these signs? What is it that we prove when we prove an arithmetical proposition? That sound? Those figures? Or this thought-content? Unquestionably the latter! Well then, in that case, however, there must also be a determinate thought in the proposition, which we would not have if the occurring number-signs and number-words referred now to this, now to that.

If we look more closely, we realise that a number-sign cannot on its own designate a length, force, and so on, but only in connection with the designation of a measure, a unit, such as metre, gramme, and so on. What in that case does a number-sign refer to on its own? Clearly a magnitude-ratio. And this suggests itself so strongly that we must not be surprised to encounter this insight[e] early.[1] If we now understand by "number" the reference of a number-sign, then *real number* is the same as magnitude-ratio. What, then, has been accomplished by defining *real number* as magnitude-ratio? Initially, it seems that one expression has merely been replaced by another. Nevertheless it is a step forward. For, first, no-one will confuse a magnitude-ratio with a written or printed sign; and by this the source of innumerable misunderstandings and errors is blocked. Second, by means of the expression "magnitude-ratio" or "ratio of a magnitude to a magnitude", the manner in which real numbers are connected to magnitudes is alluded to. The main thing is, no doubt, still to be done. So far, we only have words that point roughly in the direction where a solution is to be found. It remains to be established more precisely what these words refer to. Already, however, we will no longer say that a number or a number-sign sometimes designates a length, sometimes a mass, sometimes a light intensity; rather we will say that a length can have the same ratio to another length, as a mass to another mass, as a light intensity to another light intensity;[2] and this same ratio is the same number and can be designated by the same number-sign.

If Illigens understands by his term "quantities" magnitude-ratios or, what on our present view would be to refer to the same, real numbers and is of the opinion that, according to Cantor, a number-series sign does not designate a magnitude-ratio, then he is correct. Only the fundamental series and the number b are mentioned in Cantor's definition, and the latter is the number-series sign. There is here no mention of a magnitude-ratio. The number-series sign simply designates the fundamental series and for that reason alone must not also designate a magnitude-ratio; otherwise it would be ambiguous.

§74. Unfortunately, Cantor's response[3] to Illigens's objections turns out to be short and unclear. We do not gather with certainty whether Illigens is correct to assume that the numbers b, b', etc., are signs, or whether they designate fundamental series, knowledge of which would surely be most necessary to clarify the matter. Instead, signs and what is designated are thrown together in a jumble. Cantor writes:

[1] Compare Baumann, *Die Lehren von Zeit, Raum und Mathematik* I, p. 475.
[2] I propose to ban the expression, "named number", as it merely causes confusion.
[3] *Math. Annalen* XXXIII, p. 476.

> "But it has never been asserted by me or by others, that the signs b, b', b'', ... are *concrete* magnitudes in the proper sense of the word. As *abstract objects of thought* they are magnitudes merely in an improper or figurative sense of the word."

Here, b, b', etc., are indeed called signs, but at the same time also abstract objects of thought. To take signs that are, for instance, written with chalk on a blackboard or with ink on paper, that can be seen with the bodily eye, as abstract objects of thought demands a strong faith indeed, the kind of faith which moves mountains, and creates the irrational numbers.

Probably, Cantor's view is that the signs b, b', etc., are to designate abstract objects of thought. We can distinguish physical from logical objects, by which of course no exhaustive classification is intended to be given. The former are in the proper sense actual; the latter not so, though no less objective because of that. While they cannot act on our senses, nonetheless they are graspable by our logical faculties. Such logical objects include our cardinal numbers; and it is probable that the remaining numbers also belong here. Accordingly, if Cantor understands by the expression, "abstract object of thought", what we term a logical object, then there seems to be ample agreement between us concerning the subject matter. What a pity that these abstract objects do not occur at all in Cantor's explanations! We have fundamental series and signs b, b', etc. The latter cannot by any stretch of the imagination be regarded as abstract objects of thought, nor can fundamental series be thereby intended. So if abstract objects of thought are the most important thing, then the most important thing is simply missing from Cantor's definitions. Which abstract objects of thought, for example, is supposed to be designated by the sign '$\sqrt{2}$'? We are not told. We grope at inessential superficialities and let the core slip away.

§75. Cantor takes it to be crucial here that by means of these abstract magnitudes, b, b', b''..., one can effect a precise quantitative determination of concrete magnitudes proper, for example, a geometric line segment. Thus, far from being merely a pleasing bonus, the application within geometry is crucial. However, if it is crucial, then this goes against Cantor's theory, since what is crucial does not occur in the definition of numerical magnitude. It is only after b, b', b'' ... are introduced that the determination of distances by means of numerical magnitudes is provided.[1] The earlier introduction of numerical magnitudes is purely arithmetical, but does not contain what is crucial; the latter statement, how distances are determined by

[1] *Math. Annalen* V, p. 127.

numerical magnitudes, contains what is crucial, but it is not purely arithmetical. And thus the goal that Cantor set himself is surely missed. In the former definition one has the fundamental series on the one hand and the signs b, b', b'' ... on the other, nothing more. The issue would be different if we had a purely arithmetical or logical definition of *ratio*[1] enabling us to infer that there are ratios and that they include irrational ones. Then what is crucial would be provided by this definition, and the determination of a distance by means of a unit and a ratio (real number) would merely have the status of an elucidatory example, which could be omitted.

Let us, however, look more closely at the way in which, according to Cantor, these quantitative determinations of concrete magnitudes by means of abstract magnitudes come about. It is presupposed as known how a distance can be determined by a rational number. Each member of a fundamental series thus corresponds to a determinate distance, and so to a determinate point which has this distance from a given starting point on a given straight line to a given side. As the fundamental series progresses these points converge to a point, which is thereby determined. Then one reads:

> "This is expressed by saying: *the distance of the point to be determined from the point o is equal to b*, where b is the numerical magnitude corresponding to the series (1).[2]"

Here the mistake should first be noted that there is no mention at all of the unit in the explained expression, although it is necessary for the determination. From this, the deceptive impression can arise that b, b', b''... are distances, although only ratios are in question and these could equally well occur with strength of electrical current, mechanical work, etc. Still, this might surely be improved. However, which expression is actually being explained? It has to be presupposed as known what the distance from a point to another point is; the so-called numerical magnitude (b) was just introduced; and the word, "equal", is surely already known too. Thus, everything in the explained expression is known, and if everything were in order the sense of the proposition, "The distance of the point to be determined from the point o is equal to b", would be known too, so that the explanation would be superfluous at best and faulty for that reason.

If it is unknown what b is, which perhaps accords with the truth of the matter if not with Cantor's intention, then we actually have something capable of explanation; however the explanation is not purely arithmetical. The proposition would have been more acceptable if in place of "distance" Cantor had said "ratio of the distance to the unit of length". What a ratio of distances might be is here, of course, to be

[1] The useless arithmetical ratios are of course not intended.

[2] the fundamental series.

considered as unknown, therefore capable of explanation—and exactly there lies the core of the issue—but this too would encounter difficulties.

§76. Let us, however, proceed to the main issue. Manifestly, Cantor's numerical magnitudes b, b' ... are completely superfluous for the determination of distances, whether they now be signs, abstract objects of thought, or both at the same time. Indeed, their intrusion makes the issue even more tangled without proving of any benefit. One can leave out numerical magnitudes entirely, and one will find that the fundamental series itself completely suffices for the purpose, regardless of whether or not a sign b is assigned to it.

Without the stated determination of distances, we have only fundamental series on the one hand and signs on the other, but the most important thing is missing. Since the signs b, b' ... are inessential, we in fact have only the fundamental series. These series may serve for a determination of ratios but only after we have learned what a magnitude-ratio is, and exactly that is what is missing.

Suppose we know some line spectra but no metals. Would it be of any use to assign the sign "*K*" to one of the spectra, "*Na*" to another, and "*Fe*" to a third? Very little! Would it be of any use to call these signs metals? It might perhaps be of use in order to deceive ourselves about our ignorance, but we would not be any better acquainted with the metals proper. First we have to know the metals; then we can discover how to determine the metals by means of spectra. In that case we have the objects that are of concern to us, and the previous assignment of signs contributes nothing to this. Before we have the metals themselves, the signs are merely useless, empty shells and any hope that they might, on their own, fill themselves with new content is in vain.

So it is here, too. First we must know the magnitude-ratios, the real numbers; then we can discover how to determine ratios by means of fundamental series. It is odd to credit the assignment of the signs b, b', b'' ... with creative force.[1] Enlisting geometry is crucial since it is in this way that one appropriates the content at which all endeavours are directed. The crux is then to be found in geometry, however, and Cantor's theory is in no way purely arithmetical.

§77. However, is the view that the so-called numbers that Cantor assigns to his fundamental series are signs actually correct? Although Cantor contents himself with

[1] The fundamental importance attributed to this act is distinctive of the peculiar view of many recent mathematicians concerning signs and what they are supposed to achieve; it should be enshrined in special ceremonies!

it, the serious problems that accompany it strongly suggest preferring something different. I surmise roughly the following: associated with every fundamental series there is a specific number, which need not be rational. So some of these numbers are new, hitherto unconsidered, and ought thus to be determined by the fundamental series with which they are associated. The sign '*b*' then does not designate the fundamental series but rather the number associated with it. The latter, thus, is then itself no sign but rather what Cantor calls an abstract object of thought.[1]

One now surely recognises the cliff at which this attempt must fail. We do not know anything about the kind of association which is supposed to obtain between the number and the fundamental series. Yet something unknown cannot be determined by means of a completely unknown relation. The error is to draw together into a single act both the assignment to a fundamental series and the definition of the new numbers. Though we can assign a defined number to a fundamental series, we cannot assign a number that is still to be defined and which we thus do not have yet at all.[2] Let us resume our analogy and again adopt a standpoint in which we are acquainted with no metals but only line spectra. Now we say: to this spectrum we assign a metal, sodium, thereby defined. But how do we obtain this metal? Not, in any case, by means of this assignment which cannot supply anything that we do not already have. All we know of are some spectra, but we do not yet know anything about spectral analysis, have not yet any idea of the special relationship in which spectra stand to metals which are still unknown to us. The name "sodium" still hangs in the air. And so too Cantor's sign '*b*' at first hangs in the air; we do not yet have anything that we can designate with it. Thus when in the place cited above we find it written that

> "in the first case I say that *b* equals Zero; in the second I say that *b* is larger than Zero, or positive; in the third that *b* is less than Zero, or negative",

the sign '*b*' is used without further ado as if it designated something, although it is not at all established that anything might be found which could be assigned to the fundamental series as intended. For so far an intention has merely been pronounced; it still remains to be seen whether it can be realised.

---

[1] In this case, the expression of Cantor's quoted above, according to which the sign '*b*' is itself an abstract object of thought, would be imprecise and would have to be so understood that '*b*' designates such an object of thought. Such imprecisions enjoy great popularity, but are no more acceptable for that.

[2] One might just as well say: "They are hanging the thief who is still to be captured."

§78. We now consider the propositions quoted above, in which it is stated when a number assigned (or to be assigned?) to a number-series is equal to Zero, greater than Zero or less than Zero. Our question is: Is something thereby defined? Or is something assigned? Or what is the purpose of these propositions otherwise? Cantor writes:

> "In the first case I say that b equals Zero."

We first ask: is "equals" here to be understood in the sense of "coincides with"? This is likely, for otherwise it would have to be assumed that there are numbers which, although all equal to Zero, nevertheless would be distinct from each other. In that case it would not be determined which of these numbers was supposed to be assigned to the fundamental series in this first case. If, however, "a is equal to b" does say "a coincides with b", then there is only one number equal to Zero, namely the number Zero itself, and the sense thus is: in the first case, I assign the number Zero to the fundamental series. Here, accordingly, we clearly have no definition, but rather an assignment of a number already known. Against this there is no objection to be made; however, we thereby gain no new number either.

Now, Cantor proceeds to say that in the second case, b is greater than Zero, or positive. Here, in fact, the greater-than relation, along with Zero and also what it is to be positive, must be presupposed as completely defined and known. Although it is improbable that Cantor had such definitions, which could be applied to the new numbers without further ado, we will nevertheless assume this to begin with. If now b also had already been defined, then nothing further would remain to be stipulated; but one would simply have to investigate whether b was greater than Zero. Clearly, b is thus not yet defined here; we do not know it yet. Then this at most can be a gesture at an assignment: in the second case, one may assign to a fundamental series some number that is greater than Zero. Which of all the positive numbers should be taken remains, of course, completely undecided. And when Cantor says, in the third case, that b is less than Zero, or negative, then this too will be merely a gesture at an assignment.

In sum, so far we have assigned an already known number to certain fundamental series, and have made gestures at further assignment; but nothing has yet been defined.

§79. Cantor continues:

> "Now to the elementary operations. Let (a_ν) and (a'_ν) be *fundamental series* by which the numbers b and b' are determined; then it follows that $(a_\nu \pm a'_\nu)$ and $(a_\nu \,.\, a'_\nu)$ are also fundamental series, determining accordingly three new numbers which will serve me as definitions of the sum and the difference, $b \pm b'$, and of the product, $b \,.\, b'$."

This proposition is badly flawed. So far, it is, at best, of Zero alone that one can say that it is determined by specific fundamental series. We thus have to take Zero for b as well as for b'. Then, however, it is a theorem that $(a_\nu + a'_\nu)$ determines Zero, consequently not a definition. The proposition can be proven, hence everything in it must be known. Accordingly, this is no place for a definition. So it is very fortunate that what the apparent definition states coincides with something that can be proven. However, given the principles of definition followed by Cantor, or rather the lack of any healthy principle of definition that is coming to light here, it would be equally possible that something is determined by this definition whose falsity could be proven.

If it is here presupposed that every fundamental series determines a number, then the intention is taken for the deed. Excepting the case in which Zero is assigned, one has so far merely announced an intention to assign, and made some gestures at how it may be accomplished; nothing more.

Moreover, the words "sum", "difference", "product" are explained by means of themselves; thus so far they have been only incompletely explained, a violation of our first principle.

Thus, in reality nothing is hereby achieved, but rather something has been falsely presented as a definition which ought to have been proven as a theorem.

§80. Passing over what is said about the quotient, I continue with the examination of Cantor's exposition. It is stated there:

> "The elementary operations on a number b given by a fundamental series (a_ν) and a directly given rational number a are included in those just stipulated, by letting $a'_\nu = a$, $b' = a$."

Herein lies a contradiction. In this context, the expression "a number b given by a fundamental series (a_ν)" presupposes that a number is given by every fundamental series. If that were the case, then a number would also be given by the fundamental series whose members are all a, and there would be no place left for the stipulation that this number be a; rather, one would now have to investigate what number was given by this fundamental series, an investigation which would surely be fruitless since, indeed, merely the intention to assign a number to every fundamental series, by obeying certain rules, has been announced. This plan is of course taken further by means of the last stipulation, but without any new numbers thereby being brought to light. At bottom, we have not come closer to our goal at all.

§81. We take a further step. Cantor writes:

> "Now finally, we come to the definitions of being equal to, greater than,

and less than for two numbers b and b' (where b' can also be $= a$); namely, we say that $b = b'$ or $b > b'$ or $b < b'$ depending on whether $b - b'$ is equal to Zero, or greater or less than Zero."

As regards the equality-sign, we have assumed above that it should refer to complete coincidence. If we also make this assumption here, which, granted, is not completely in accord with the above wording, then we are here not given a definition of equality; rather, it is known. In that case, the point can only be to specify the numbers that are to be assigned to the fundamental series somewhat more closely. For the intention is then announced to assign the same number to all those series for which $\lim\limits_{\nu=\infty} (a_\nu - a'_\nu) = 0$. Now, whenever a number is already assigned to one of the two series, then the number to be assigned to the other is thereby determined too. Since, however, it is only rational numbers that are assigned so far, we will not advance any further in this way.

However, our conjecture that the equality-sign is used by Cantor as the identity-sign does seem to founder here. The explicit wording is "definitions of being equal to, greater than, and less than", from which it may be gathered that the reference of the signs '=', '>', '<' is not here presupposed as completely known. They are surely to be regarded as partly known, so that a violation of our first principle of definition may well be present. This mistake causes a peculiar illusion. Our mental field of vision is in a similar state to that of our bodily one when colours compete: at one moment the words "equal", "greater", "less", "sum", "product" appear as known, then immediately after as unknown, and then again as known. When, for example, the words "greater than" are explained by means of themselves, they appear in the same proposition partly as known—at the point when they are used to provide an explanation—partly as unknown—at the point when they are being explained.

§82. These definitions of sum, difference, product, being equal, greater and less than now seem properly to create the new numbers themselves for the first time, to give the signs 'b' a content. One involuntarily concludes from the ostensibly known reference of the word "greater" that what is greater is a magnitude,[g] be it abstract or concrete. Since the words "sum", "difference", "equal", "greater", etc. have already been explained, or at least are presupposed as explained, one gathers the impression that one already knows what a sum or a difference is, what equal to, or greater than another is; and by means of this prior knowledge, a content may then be conferred, no doubt in an unclear manner, on the signs 'b', 'b'', etc. by employing them in propositions such as '$b > b'$', '$b + b' < b''$'. What initially presents itself as an explanation of the signs '+', '>', etc. at the very next moment claims to determine further that which, according to Cantor, is to be assigned to a fundamental series. This illusion, however, is only possible because the former signs are now once again regarded as known. So, these definitions iridesce in two colours, sometimes defining sum, product, greater than, etc., sometimes determining the new numbers. But this is inconsistent.

An analogy may facilitate understanding. If one gives the definition "A judgement is hard if it attributes the property *hard* to a thing", then one makes the mistake of explaining the word "hard" by means of itself, of regarding it, with respect to its reference, in a single breath to be both known and unknown. Now, one could infer, on the basis of the ostensibly known reference of "hard", that everything that is hard is a physical body and then conclude further: therefore the judgements in which the property *hard* is attributed to an object are physical bodies. The matter is very similar here; here too the words "greater", "sum", etc. are explained by means of themselves, here too one concludes from the ostensibly known reference of these words that the new objects are numbers like the rational numbers, and that they can serve the same purpose as the latter. The difference is only that we do not yet have these new numbers, that they should really hereby be created for the first time, whereas we do by contrast already have the judgements.

One cannot define two things by means of the definition: both being greater than and the irrational numbers—violation of our second principle.

§83. Piecemeal definition here creates the twilight necessary to effect the illusion. It is dispelled immediately, if we put in the place of words and signs that are treated as unknown, newly created words and signs with which neither a sense nor the appearance of a sense is already associated. So, if we replace

"positive"	by	"albig",
"negative"	"	"bebig",
"equal"	"	"azig",
"greater than"	"	"bezig",
"less than"	"	"zezig",
"Zero"	"	"poll",
"sum"	"	"arung",
"difference"	"	"berung",
"product"	"	"asal",
the symbols '>'	"	'ᔓ',
'<'	"	'ᘔ',
'='	"	'ω',
'+'	"	'ш',
'−'	"	'ϖ',
'·'	"	'∪',

then Cantor's explanations are turned into:

> "Such a fundamental series presents ... three cases: either its members (a_ν), for sufficiently large value ν, are less in their absolute value than an arbitrarily given number; or, from a certain ν on, they are greater than some specifiable rational number, ϱ; or, from a specific ν on, they are less than some specifiable negative rational magnitude, $-\varrho$. In the first case I say that b is azig poll; in the second I say that b is bezig poll or albig; in the third that b is zezig poll, or bebig."[1]

> "Let (a_ν) and (a'_ν) be fundamental series by which the numbers b and b' are determined; then it follows that $(a_\nu + a'_\nu)$, $(a_\nu - a'_\nu)$, $(a_\nu \,.\, a'_\nu)$ are also fundamental series, determining accordingly three new numbers which will serve me as definitions of the arung b ⩃ b' and the berung b ⩂ b' and the asal b ∪ b'."

> "Now finally, we come to the definitions of being azig, bezig and zezig, between two numbers, b and b';[h] namely, we say that b ∞ b' or b ∾ b' or b ∿ b' depending on whether b ⩂ b' is azig poll or bezig poll or zezig poll."[2]

These definitions can fix the references of the new words and signs with at least the same right as the earlier incomplete references can be completed by Cantor's definitions. As one can see, however, the sense thus conferred on these new words and signs does not suffice for the purpose at hand. And so any appearance that these definitions have determined the new numbers any further vanishes. In the case of Cantor's definitions, this appearance is generated only by sinning against our first principle, so that the words "equal", "greater", etc., are put into constant oscillation between being known and being unknown. The original explanations seem to inject something into the new numbers, although they are not specified for this extended usage. As soon as the oscillation is stilled, the illusion disappears. If this is what Illigens intends by the words "The number-series signs introduced, despite all names given to them by various definitions, are not able to become concepts of quantity" then we concur, but must find fault with his imprecise mode of expression.

§84. Let us review summarily the results of our examination of the Cantorian theory. We distinguished two interpretations: according to the first, the numbers that

---

[1] The expressions "azig poll", "bezig poll" and "zezig poll" must be regarded as inseparable wholes, in accordance with our second principle.

[2] As one can see, we really should have created a third tier of expression for "equal", "greater" and "less".

Cantor aims to assign to his fundamental series are signs; according to the second, they are abstract objects of thought.

In the first case, we have to regard the series themselves as the reference of the so-called signs. The assignment of these numbers is inessential; basically, all we have are the fundamental series, while the main thing, namely the numbers themselves, the magnitude-ratios, is missing. Cantor tries to address this deficiency by showing how his numbers can provide for a quantitative determination of distances. First, however, his numbers are completely superfluous here, and second, the crux is thereby shifted into geometry, and with this the theory ceases to be purely arithmetical.

In the second case, there is no more than the mere intention to assign new numbers to the fundamental series. A grasp of the abstract ideas is not achieved, and we cannot assign them before we have them. Cantor does sometimes claim that his fundamental series determine numbers, but he himself contradicts this. All that is achieved is to assign rational numbers to some fundamental series; we do not even manage securely to assign to the fundamental series

$$\frac{1}{2}, \frac{2}{3}, \frac{3}{4}, \ldots \frac{\nu}{\nu+1}, \ldots$$

the number One; rather all one recognises is the intention to assign to it a number equal to One, and "equal" here is likely not used in the sense of "coinciding with". Which of all the numbers equal to One is to be assigned to the series remains uncertain. Since *sum, difference, product, equal, greater, less* are defined piecemeal, so in contravention of our first principle, the false impression arises that the number-signs have been given a reference. In sum, Cantor's theory does not in any way fulfil its aim.

§85. I append here a consideration of a somewhat different account that Cantor has given earlier.[1] Here too he starts with the series which he later calls fundamental series, and says:

> "This constitution of the series (1), I express in words: 'The series (1) has a definite limit *b*'."

This definition is flawed since it contains the letter '*b*', which is replaced by other signs, e.g., '*b*′', in other series of the same constitution. Thereby, a difference enters into the explained expressions which corresponds to nothing in the explaining expression; rather, we always have the same constitution. If Cantor had said: "This constitution of the series (1), I express in words: 'The series (1) is a fundamental series'", this objection would lapse; but at the same time, everything that was tied to this

---

[1] *Math. Annalen* V, p. 123.

particular *b* would then be lost. The subsequent determinations of equality, being greater than, sum, product of such *b* would drop away, and with it everything that is important. This is, presumably, the reason why Cantor later preferred a different account.

c) The theories of the irrationals of E. Heine and J. Thomae

§86. At first glance the theories of E. Heine[1] and J. Thomae[2] seem almost to coincide with Cantor's. Here we have number-series, number-sequences similar to Cantor's fundamental series. Here too signs are assigned to these series and this deed is considered to be of particular importance. In general, the manner of proceeding, viewed superficially, is very similar to the Cantorian. But in fact there is far less similarity than one is initially inclined to suppose, so a particularly detailed examination of these theories appears to be called for. They will not differ from each other essentially. Heine's basic thought is merely worked out more precisely by Thomae. And this basic thought will depart significantly from Cantor's view. For, as it seems to me, the latter author does not view the number-signs as empty, although the way he writes does not perhaps rule out all doubt and he himself has likely not brought the matter explicitly to mind. What is essential for him is, no doubt, that which is designated by the signs, which of course he wants to capture by means of the signs, rather than the signs themselves.

Now, what is peculiar to Heine's theory, and what has been pronounced by Thomae yet more explicitly, is precisely that the signs are everything for them. Both authors also agree in not carrying this basic thought through to the end, but eventually allow their signs to designate something after all, namely those number-series or number-sequences that correspond to Cantor's fundamental series. However, while we may surely assume that the latter fundamental series consist of *abstract objects of thought*, to use Cantor's own expression, the former number-series or number-sequences must be thought of as composed from written or printed, visible, material figures. These series will thus presumably be groups of such figures presenting themselves to the eye as series through their spatial configuration. We have here, then, the peculiar state of affairs that certain signs designate series or sequences whose members again designate such series, and so on, into the infinite.

§87. I now continue with Heine's and Thomae's statements concerning this matter

[1] Crelles J., vol. 74, p. 172.

[2] *Elementare Theorie der analytischen Functionen einer complexen Veränderlichen*, 2nd ed. Halle a. S. 1898, §§1 to 11.

and will subsequently enquire after the motive for proposing these theories.

Heine writes:

> "When I do not want to stop at the positive rational numbers, I do not answer the question, what is a number, by defining number conceptually, still less by introducing the irrationals as limits whose *existence* would be a presupposition. Regarding the definition, I take a purely formal point of view *by calling certain tangible signs numbers*, so that the existence of these numbers is not in question."

Here, Heine emphasises existence twice, and rightly so; for we have seen how unsatisfactory Cantor's answer is to the question of existence in particular. So this is why Heine calls certain signs numbers, to safeguard the existence of these numbers, albeit in an empirical rather than purely logical or arithmetical way. Now, the real goal that all these theories of the irrationals are meant to serve is of course to display arithmetic as free from all foreign additions, including geometrical ones; to ground it on logic alone. This goal is certainly a proper one; but here it is missed. If one has no qualms about relying on the apprehensibility of signs, then one might as well help oneself to spatial intuition and determine the irrational numbers as ratios of lengths.

We see that number-signs here have an entirely different importance from that attributed to them before the rise of formal theories. They are no longer external auxiliaries, like blackboard and chalk; rather, they comprise the essential components of the theory itself.

Here we cannot but ask the question: do these signs, just by being called numbers, have the properties of the true numbers which we have provisionally characterised as magnitude-ratios?

§88. Thomae writes:

> "The formal conception of the numbers works within more modest limits than the logical. It asks not, what are numbers and what do they demand,[a] but rather it asks, what do we require of the numbers in arithmetic. Now, for the formal conception, arithmetic is a game with signs which one may well call empty, thereby conveying that (in the calculating game) they do not have any content except that which is attributed to them with respect to their behaviour under certain combinatorial rules (game rules). A chess player makes use of his pieces in a similar fashion: he attributes certain properties to them that constrain their behaviour in the game, and the pieces are only external signs for this behaviour. There is, of course, a significant difference between chess and arithmetic. The rules of chess are arbitrary; the system of the rules of arithmetic is such that, by means of

> simple axioms, the numbers can be related to intuitive manifolds and, consequently, can perform an essential service for us in the knowledge[e] of nature."

In other words:

In arithmetic, the only rules that come into consideration are those according to which the arithmetical signs are to be handled, but not what these signs refer to. One might note that a difference from Heine's viewpoint is that Thomae rejects the question about the essence of the numbers as immaterial, while Heine answers that numbers are signs. However, since both agree that arithmetic has to concern itself with signs, this difference is inessential. Heine calls these signs numbers; Thomae on the other hand seems to understand by "number" something whose essence does not come into consideration in arithmetic which is therefore not a sign but perhaps the reference of a sign. On the other hand, since he speaks also of the reference of the numbers, he stamps them as signs after all, so that a consistent manner of speaking is certainly wanting here. It is these references of the number-signs, which are indeed assumed by Thomae but viewed as lying outside the scope of arithmetic, that we have always called numbers. Thus we find that these numbers proper or magnitude-ratios are, according to this mathematician, to be excluded from arithmetic. So we have here a peculiar arithmetic, wholly different from that which views the numbers proper as its objects and which, in contrast to formal arithmetic, we want to call contentual. Accordingly, we may surely assume that Cantor takes his stand on the ground of contentual arithmetic, Heine and Thomae, in contrast, on that of formal arithmetic. The difference cuts deep. To be sure, a future historiographer may perhaps discover that both sides lack in consistent execution, and thus the contrast again loses some of its sharpness.

§89. Now, what might be the reason to prefer the formal to the contentual? Thomae replies:

> "The formal point of view elevates us above all metaphysical difficulties; this is the gain that it offers to us."

The difficulties spoken of here will likely be those that we encountered in the examination of Cantor's theories, namely to apprehend the numbers proper and to demonstrate that there are these numbers. In formal arithmetic, we do not need to justify the rules of the game; we simply lay them down. We do not need to prove that there are numbers with certain properties; we simply introduce the figures for whose deployment we give rules. We then view these rules as properties of the figures and so can—so it at least appears—create things at will with the desired properties. That we thereby, at least initially, economise on intellectual effort is obvious. Admittedly, Thomae contrasts the arbitrariness of the rules of chess with the rules of arithmetic, which enable the numbers to be of considerable service in the knowledge[e] of nature; but this contrast only comes to light when we are dealing with applications and are leaving the ground of formal arithmetic behind. If we do not look beyond its boundaries, then its rules seem as arbitrary to us as those of chess. The applicability cannot, however, be coincidence; but in formal arithmetic we spare ourselves any account of why we lay down the rules in exactly this way and not in any other.

§90. Let us try to make the essence of formal arithmetic even clearer to ourselves. The obvious question is of course: how is it distinguished from a mere game? Thomae answers by pointing to the service it can provide in the explanation of nature. This can rest only on the number-signs referring to something; the chess pieces in contrast refer to nothing. If one accords arithmetic a higher honour than chess, it can only be founded in this. According to Thomae, however, what makes the difference lies outside arithmetic, so that it is, in and of itself, of the same rank as chess and better termed an art or a game than a science. Although the number-signs designate something, we may in Thomae's view ignore this and regard them merely as figures[b] to be manipulated according to certain rules. If one wanted to go back to their reference, then a justification of the rules would be found in just these references; however, this happens behind the scenes, so to speak; on the stage of formal arithmetic none of this is to be noticed.

Now it is quite right to say that we too could have introduced our rules of inference and the other laws of the concept-script as arbitrary stipulations, without ever speaking of the reference and the sense of the signs. The signs would then indeed be treated as figures. What we took to be an external representation of an inference would then be comparable to a move in chess, the mere transition from one configuration of pieces to another, without any transition from one thought to another corresponding to it. One could give our formulae I to VI and definitions A to H of the first volume to someone as a starting point—comparable to the initial position of the pieces in chess—tell him the rules according to which he is allowed to carry out transformations, and now set him the task of reaching our proposition (71) of the first volume from these starting points; all this without him having any inkling of these signs having sense and reference or of the thoughts that the formulae express. It would even be conceivable that he might accomplish the task in just the same manner as we did. That it would take mental effort is evident, just as in the similar task in chess of arriving at a given end position from the initial set up according to the rules of the game, where there would be no talk of thoughts expressed by different positions, and no move could be interpreted as an inference. Thus although mental effort would be involved, the train of thought which accompanied the matter for us and which, in fact, bestowed interest upon it, would be completely missing. Possible it may be, but hardly of advantage; for the task would have been rendered not easier but significantly harder by banishing the accompaniment of thought.

§91. While equations and inequations in contentual arithmetic are propositions which express thoughts, in formal arithmetic they are to be compared to the positions of chess pieces which are modified according to certain rules without regard for a sense. For if a sense had to be observed, the rules could not be laid down arbitrarily; rather, they would have to be fashioned in such a way that from formulae expressing true thoughts one could only ever derive formulae which also expressed true thoughts. With this, the standpoint of formal arithmetic would be abandoned. On that standpoint the rules for manipulating the signs are, on the contrary, laid down altogether arbitrarily. It is only in retrospect that one can ask whether the signs may be given a sense compatible with the rules previously laid down; but this issue already lies beyond formal arithmetic and will only come into question when applications are to be made. Then, however, it will indeed have to be considered; for without a thought-content, no application will be possible. Why can no application be made of a configuration of chess pieces? Because, obviously, it does not express a thought. If it did, and if a move in chess in accordance with the rules corresponded to a transition from one thought to another, then applications of chess would be conceivable as well. Why can one make applications of arithmetical equations? Solely because they express thoughts. How could an equation which expressed nothing, which was nothing but a group of figures that could be transformed according to certain rules into other groups of figures, be applied? Now, it is applicability alone which elevates arithmetic above a game to the rank of a science. Applicability thus necessarily belongs to it. Is it appropriate, then, to exclude from arithmetic what it needs to be a science?

§92. What is really gained by this? To be sure: arithmetic is relieved of a labour; but is the task thereby removed altogether? The formal arithmetician tries to shift it onto the shoulders of his colleagues, the geometer, the physicist, the astronomer; but they gratefully refuse to take it on themselves; and thus it falls through the cracks between the sciences into the void. The clean separation of the various areas of knowledge may be good; but not if it happens in such a way that an area is left for which no one is willing to take responsibility. We know that the same magnitude-ratio (the same number) can occur with lengths, periods of time, masses, moments of inertia, and so on, and it thereby becomes likely that the problem of making use of arithmetic is at least partially to be solved independently of those sciences in which the application is to occur. It is therefore appropriate to ask this work of the arithmetician, at least to the extent that he can carry it out without trespassing into those special areas of knowledge. Part of the task is, above all, that he connects a sense with his formulae; and this sense will then be so general that with the assistance of geometric axioms, and physical and astronomical observations and hypotheses, multiple applications of it can be made in these sciences.

This, it seems to me, can be demanded of arithmetic. For otherwise it could happen that this science should treat its formulae merely as groups of figures without sense; and that then, however, a physicist, wanting to make an application of it, might straightaway presuppose, quite without justification, that a thought had been proven to be true. At best an illusion of knowledge[e] would be created. The divide between arithmetical formulae and their applications would not be bridged. For that, it is necessary that the formulae express a sense and that the rules find their grounding in the reference of the signs. Knowledge[e] must stand as the goal, and everything that occurs must be determined thereby.

§93. Formal arithmetic withdraws itself from this aim. As a game with figures, there are no more theorems and proofs to be found in it than there are in chess. Admittedly, there can be theorems in a theory of chess, but not in chess itself. Formal arithmetic knows only rules. A theory of formal arithmetic, however, is similarly conceivable; and in it there will be theorems which say, e.g., that from a certain group of figures one can arrive at another group of figures in accordance with the rules of the game.

Are definitions possible in formal arithmetic? At least not those which stipulate reference for arithmetical signs; for reference is to be left out of consideration completely.

In place of definitions there will here be introductions of new figures, with the addition of rules for manipulating them. So if the expression "formal definition" occurs in Thomae's writings, no more than that, surely, is to be understood by it. In the theory of formal arithmetic, proper definitions are also possible; but these do not give the figures a reference, which, after all, is to be left out of consideration; rather, they explain expressions by means of which the theorems of this theory can be abbreviated.

The distinction between the game itself and its theory is not made by Thomae, but it is essential for a correct understanding of the matter. When we encounter theorems in Thomae's exposition, then it is to be assumed that they belong to the theory of the game. These propositions only seem to say something about the figures, for the properties of the latter are almost completely irrelevant and come into consideration only in so far as they allow us to distinguish them. On the contrary, it is the rules of the game whose properties are brought to light by means of these propositions. In the same way, the theory of chess does not in fact examine the chess pieces; rather, what matters are the rules and what follows from them.

This formal arithmetic is, to be sure, different from chess in that it is always possible to introduce new figures with new rules, while in chess everything stands fixed. And thereby the possibility of a theory of the calculating game is once again put in doubt. For one might suspect that there are no final propositions here at all. By introducing new figures one may both render possible something that was formerly impossible and also, conversely, render impossible something that was formerly possible. In chess, at least, the presence of new pieces could prevent some moves. That something similar cannot occur in arithmetic must first be proven, before the possibility of a theory of this calculating game can be taken as assured.

§94. According to Thomae, the question, "what does arithmetic require of the numbers", is likely to be answered like this: in arithmetic we require of the numbers only their signs, which are however not treated as such but rather as figures; and we require rules according to which these figures are manipulated. Here, we do not extract these rules from the reference of the signs, rather we lay them down on our own absolute authority, reserving complete freedom in principle and acknowledging no necessity to justify these rules, while admittedly we exercise this freedom with an eye to possible applications, for without them arithmetic would be a game and nothing more.

Accordingly, Thomae's question may also be answered in this way: in the calculating game we require of the numbers absolutely nothing; for the number-signs are here completely divorced from their reference, the numbers proper, and could be arbitrarily replaced by other figures.

§95. Some of Thomae's words might perhaps seem to contradict the interpretation of his view attempted here, according to which the number-signs of formal arithmetic are treated as if they designate nothing. When Thomae says, e.g., "Now, for the formal conception, arithmetic is a game with signs which one may well call empty, thereby conveying that (in the calculating game) they do not have any content except that which is attributed to them with respect to their behaviour under certain combinatorial rules (game rules)" then it appears as if the signs are to be treated not as entirely empty after all, but rather that a certain content is ascribed to them which is also taken into consideration in arithmetic. Yet, this appearance arises merely from the somewhat inappropriate manner of expression, which is no doubt prompted by a recoil from the emptiness of the signs. Can one say that a content is attributed to the chess pieces with respect to their behaviour under the rules of chess? I acknowledge that the chess pieces are there, and also that rules have been laid down for their manipulation; but I know nothing of any content. It cannot simply be said that the black king designates something as a consequence of these rules, like the name "Sirius" designates a certain fixed star. On the contrary, the correct manner of expression is surely that the rules of chess treat of the black king.

Additionally, it seems unhappy to me that there is talk of the behaviour of the signs in relation to the rules. It does not suffice for my behaving in some way in relation to the state laws that I am subject to them, but rather that I obey or disobey them. So since neither the chess pieces nor the number-figures[c] have wills of their own, it will be the player, or calculator, who by observing or not observing the rules manifests behaviour in relation to them, not the pieces or figures.[d] All that survives then is the utterly simple point that certain rules treat of the arithmetical figures.

§96. When it is now added that "A chess player makes use of his pieces in a similar fashion: he attributes certain properties to them that constrain their behaviour in the game, and the pieces are only external signs for this behaviour", then this too cannot really be precise; for basically, chess pieces acquire no new properties as a result of the rules being laid down; they can still be moved in the most diverse manner; it is merely that some of these movements accord with the game rules, others do not. Even this accord does not properly have its source in the laying down of the rules; it is just that we can only judge it after the rules are known to us. Moreover, I cannot find that a pawn in chess is an external sign of its behaviour; rather, I keep coming back to the simple expression: the rules of chess treat of the manipulation of the chess pieces. Would it not be a bizarre mode of expression if, instead of "the Prussian constitution accords certain rights and duties to the King", one proposed to say, "the King of Prussia is an external sign of his constitutional behaviour"? I must maintain that the use of expressions like "being an external sign for something", "behaving in relation to the rules" merely obscures the utterly simple matter without adding anything to what is said by the proposition, "the rules of chess treat of the manipulation of the chess pieces".

Purely on the grounds that a rule not seldom treats of several pieces,[1] and that several rules may concern the same piece, the relation of a piece to a rule is not at all comparable to that of a sign to a sense or a reference. In any case it is not brought about by the game rules that a configuration of chess pieces expresses a thought, and the analogous point holds for the formulae of the arithmetical game.

§97. A little further on Thomae writes:

> "However, there are cases in arithmetic too where the numbers[2] do not just receive a formal reference,[g] e.g., in the proposition, 'this equation is of degree 3', etc."

This seems to say that the number-signs are awarded a formal reference in addition to their proper reference, where the latter only exceptionally comes into consideration in arithmetic, and if that were true, it would be alarming because of the ambiguity; but the offense here is probably merely that of an unhappily chosen expression. It is likely that what is intended is only that in some cases the number-signs cannot be treated just as figures, but that sometimes we must draw on their reference. To be sure, it is striking that in formal arithmetic, or in its theory, something besides the game rules can enter into consideration. How could it conceivably be that the reference of the chess pieces, which is irrelevant to the game itself, could be important for the theory of chess?

---

[1] The squares of the chess board must here really also be counted amongst the pieces.

[2] The word "numbers" here apparently stands instead of "number-signs"; for only then can one speak of a reference, while above where the question "what are numbers and what do they demand?"[f] was pushed to the side, it was apparently the reference of the number-signs that was intended. In what follows, Thomae regularly uses the word "number" to refer to *number-signs*, or better to *number-figures*. When it is otherwise, special mention will be made.

The admission that even in formal arithmetic the number-signs are not always used merely as figures is worrying for Thomae's theory; it is thereby conceded that one will not always be able to maintain the formal standpoint. It is clear that in the calculating game itself any reference of the signs cannot come into consideration. One can ask after reference only where signs are components of propositions expressing thoughts. Now, such propositions might occur in the theory of the game; but it can only cause confusion in the highest degree if one allows the figures of the game to serve simultaneously in the exposition of its theory as signs that have a reference. For the handling of these signs will in this case conform to their reference, while rules laid down wholly arbitrarily hold in the game itself. That both ways of handling them are in agreement cannot simply be assumed. To avoid such confusion arising from the number-signs' playing this double role, one ought to give the number-signs used in the exposition of the theory of the game, in so far as reference is attributed to them, a shape different from that of the mere number-figures.

§98. It may be useful at this point to treat of the signs in more detail, since the claims made by Heine and others that numbers are signs have stamped them as objects of mathematics, and they have therefore gained an importance beyond what they would have as mere auxiliary means for thinking and communicating. The fluctuating use of language here so easily gives rise to misunderstandings that we cannot proceed too cautiously and must not be squeamish about stating even what is self-evident in order to ensure that we have a common starting point.

What are signs? I will restrict attention to formations that are created by means of writing or printing on the surface of a physical object (blackboard, paper); for clearly it is only these that are intended when the numbers are termed signs. But we will not term every such formation a sign—we will, e.g., not in general consider a blot worthy of this honour—but only those that serve us in designating, expressing or asserting something. We do not thereby intend to say something of the sign when we use it; rather, as a rule, it is its reference that usually matters most to us. The astronomer, for example, means the planet Jupiter when he uses the sign '♃'; towards the latter he is indeed indifferent, it is just an arbitrarily chosen means for expressing thoughts, and remains completely out of consideration. The utility of signs lies in this representation. As an exception, it does of course occur that one wants to talk about the sign itself, and this case will crop up in examination of formal arithmeticians. In order to forestall any doubt these cases too must be distinguished externally. In such cases what is most useful is to put the signs in quotation marks. For greater clarity one can also prefix the words, "the sign". This may seem pedantic but it is by no means superfluous. Had this distinction always been kept sharply in focus, then perhaps a presentation like Heine's, in which exactly this iridescence in two colours seems essential, would never have been possible. Amongst mathematicians certain modes of expression have become customary, and because of them one has gotten so used to this iridescence that one no longer notices it. We thus find expressions like

"Let a designate the smallest root of equation (1)",

and if in what follows the letter 'a' occurs, one means by it the smallest root of that equation. We here have the iridescence. In the first proposition one means the sign, later its reference. One should thus write either

"Let 'a' designate the smallest root of equation (1)"

or

"Let a be the smallest root of equation (1)".

If one so wished, one could perhaps fill volumes with such examples from the writings of recent mathematicians.[1] This looks like an insignificant trifle, yet such carelessness has, it certainly seems, become a source of great confusion. And if it should transpire that the formal theories of arithmetic have drawn nourishment from exactly this, then the matter is certainly not to be taken lightly.

§99. Signs would have little use if they could not serve to designate the same thing repeatedly and in different contexts, while making it easily recognisable that the same thing is intended. This is achieved by using signs that are as similar as possible on the different occasions. No doubt, one will hardly ever succeed in reproducing the exact same shape; and even if one once succeeded, our eyes would not be sufficiently sharp to recognise this with certainty. But that is not necessary at all. For if the sign has the sole purpose of serving mutual communication between people, as well as that of people with themselves—reflection—, then the writer needs only the intention to produce a sign similar to the one made earlier, and only needs to succeed in this to the extent that the reader recognises the intention correctly. In what follows,

[1] The following just catches my eye:
"On the number of different values that a function of given *letters* can acquire through their interchange", *Math. Annalen* XXXIII, p. 584.[h]

we propose to understand by "equal-shaped signs" those that, according to the intention of the writer, are meant to be equal-shaped in order to designate the same thing. Now, it is common to express oneself loosely and to speak of equal-shaped signs as one and the same sign, although in fact I create another object every time I write down an equality-sign. These formations differ in their position, time of creation and probably also in shape. Maybe it will be said here that one may abstract from these differences and therefore regard these figures as the same sign. How much abstraction is supposed to make possible! No abstraction can make what is distinct coincide, and if one regards them as the same nevertheless, one simply makes an error. If, abstracting from the distinction between my house and that of my neighbour, I were to regard them as the same, and consequently act in the other house as in my own, then the erroneousness of my abstraction would soon be made clear to me. What may perhaps be achieved by way of abstraction is a concept, and if we want to call the extension of a concept "class" for short, then we can reckon all equal-shaped signs to the same class. Yet this class is not the sign; I cannot create it by writing, but can only ever create single objects that belong to it. If one nonetheless speaks of these as if they were the same sign, then the coincidence of the reference is transferred to the sign.

§100. All this holds of the common and correct use of signs. In formal arithmetic their role is different: they are not supposed to designate something else; rather they are themselves the objects of concern. Occasionally, as in Heine, they emphasise their own constitution—their tangibility—and throw it into the scales as evidence. Accordingly, we had better term them figures here, since the aim of designating something does not come into consideration at all. Figures are then to be termed equal-shaped whose perhaps recognisable differences in shape are taken to be irrelevant to the rules for their manipulation. In their regular use, equal-shaped signs are supposed to have the same reference and will therefore be treated in many respects as if they were the same, since they come into consideration only as signs for their reference. This ground drops away in the case of equal-shaped figures. We are not allowed to treat the white pawn pieces on the chess board as a single piece.[i] In laying down the rules of chess, and in the theory, we may not use the word "pawn" with the definite article in the singular as a proper name; for there are several. Whereas in contentual arithmetic expressions like "the number One" or "the One" or also simply "One" are admissible as proper names, these must be rejected in the theory of formal arithmetic; for there are very many One-figures. New ones are always formed and old ones destroyed. Here one may, to be sure, say "a One-figure", "some One-figures", "all One-figures", but not "the One-figure", except where such specification is added to pick out a single One-figure unambiguously.

Further, the following difference between formal and contentual arithmetic may be noted. In the latter, the word, "One", and the sign, '1', have the same reference, namely the proper, non-sensory number One, while in the theory of formal arithmetic, the words, "One-figure", or "One", when erroneously used in its place, refer to the concept under which the sign, '1', and all that are of equal shape to it, fall.

§101. Let us now consider Thomae's theory in more detail. We read in §2:

> "Once the concept of the whole number and of counting is acquired, then two kinds of calculating can be introduced as special kinds of counting in a simple and natural way,
>
> addition and multiplication.
>
> It is in their nature that these can always be performed within the realm of the whole numbers. If one, however, proceeds to invert these kinds of calculating, introducing the new kinds of calculating,
>
> subtraction and division,
>
> then these cannot always be performed within the realm of the whole numbers."

Here, as it seems, we pick up the thread from what is known about the whole numbers, from connections that are grounded in their nature. We thus seem to have abandoned the formal standpoint for the time being in order to acquaint ourselves with kinds of calculating which we then want to transfer into formal arithmetic. Accordingly, it is the reference of the number-signs that is intended by the whole numbers here, and not the number-signs themselves. On the formal standpoint, naturally, we have no use for adding and multiplying as something pertaining to the numbers themselves; but, as soon as we let the number-signs designate the numbers, what concerns the numbers themselves is reflected in the signs, and we obtain techniques within the realm of the signs which serve to solve problems that are posed within the realm of the numbers themselves. Such manipulation of the signs here passes as calculating. The rules of this calculating find their justification in the nature of the numbers themselves and their relations to each other. But now we can completely ignore the reference of the number-signs, treat them as figures, and regard the rules, without asking for their justification, as arbitrarily laid down game-rules. And we can now also calculate in accordance with them using figures of which we do not know whether they indeed are signs, nor whether the calculating rules may be in any way connected with the reference of these signs.

§102. All this issues so immediately from the project of a formal arithmetic in Thomae's sense that a doubt concerning the correctness of this elucidation is hardly possible. However, something now seems to me to be missing, namely a statement of what, in the arithmetical game, adding, multiplying, subtracting and dividing are. In chess, we first have to acquaint ourselves with the pieces in order to be able to understand the rules. We expect something similar here. On which figures are these actions performed? What is the state of affairs like before the addition and what is it like afterwards? Likewise, we need to know the analogue concerning subtraction; only then can we judge in which circumstances subtracting is possible. For we must constantly bear in mind that subtracting is here not an operation in thought but an external action, a handling of figures.

Now, if subtracting one figure from a second consists in writing the latter—or one of equal shape—to the left and the former to the right of the same minus-sign, then nothing prevents me from subtracting a Three-figure from a Two-figure. Moreover, I can just as well subtract a calendar sign of the Moon from a calendar sign of the Sun by treating these signs as mere figures. The introduction of new figures to make the subtraction possible is then not required.

We just have to keep it quite clearly before our eyes that here in the arithmetical game nothing at all hinges on reference. So: it cannot be judged when a subtraction is possible before we know which figures are allowed to come into consideration, and what is to be done with them. This must be described to us as precisely as castling in chess.

How, for example, subtraction is to be understood as the converse of addition will be our focus later, after we have acquainted ourselves with some rules of formal arithmetic.

Initially, we do not know at all what in the calculating game addition and subtraction are. In any case, the addition of number-signs is entirely different from the addition of numbers. If a conqueror burns a city, he does not thereby burn the name of the city; what happens to a thing need not thereby happen to its name or sign. Now one may suspect that adding two number-signs is supposed to mean writing down a third number-sign in such a way that the number which, in contentual arithmetic, is the reference of this sign is the sum of the numbers designated by the first two number-signs. However, contentual arithmetic would then be presupposed for all numbers, while here it is assumed as known only for positive whole numbers. Otherwise formal arithmetic would surely be superfluous. Accordingly, one would not know what addition was for two number-figures that do not both designate positive whole numbers in contentual arithmetic. Alternatively, one might think of calling addition a procedure of moving back and forth in a series of number-figures; but this too would not be applicable in sufficient generality. So perhaps only this conjecture remains: one adds two number-figures by writing two that are of equal shape to them next to each other separated by the addition-cross. An analogue can be said for multiplication, by letting the multiplication-dot stand in place of the addition-cross. According to these explanations, then, all inscribable figures can be added and multiplied, whether they have a reference in some context or not.

§103. Thomae continues:

> "If one demands, however, that these operations always be performable, then one arrives at new number-formations, at Zero, at the negative numbers, and fractions. These can be understood as formations that are purely formal, i.e., as concepts whose content is exhausted by their behaviour with respect to the operations of calculation."

Comprehension is here impeded by the mode of expression. The word "concept" is blatantly not used here in our sense, nor in the sense of the logicians; it is here not mentioned at all, for example, that objects might fall under these concepts. What is it that behaves, to use Thomae's expression, with respect to the operations of calculation? Or, of what is it that the calculating rules treat, as we prefer to say? Of figures, which may, for example, be written with chalk on a blackboard. However, these are no more concepts than are chess pieces, but rather belong to the realm of physical bodies. We thus arrive at the view that these new number-formations are intended to be put forward as figures which are generated by writing or printing, which refer to nothing, or whose reference is at least of no concern to us; but for which rules are laid down regarding their manipulation. There can, hence, surely be no doubt that the *Zero*, the *negative numbers* and the *fractions* of which Thomae speaks are not really intended as numbers in our sense, but as number-figures.

As we have already seen, the introduction of these new number-figures is, from the standpoint of formal arithmetic, not needed at all in order to make subtraction and division always performable; yet all the same it will be possible.

§104. First, let us now have a look at the way in which Heine treats the numbers. He writes:

> "A main emphasis is to be placed on the operations of calculation, and a number-sign must be chosen, or equipped with an apparatus, in such a way that some guidance is provided for defining these operations."

A puzzling announcement! If I chose instead of the Three sign a capital Roman 'U', might this provide less guidance concerning the definition of the operations? And how can one recognise that a sign provides such guidance? Finally, what are we to understand by the apparatus with which a number-sign is allegedly equipped? Where, for example, does the sign '3' have its apparatus? One should have supposed that it be visible, or even tangible, since the number itself is tangible according to Heine.

We here notice a difference between Heine's and Thomae's conception. For the latter, specifically, the operations of calculation are already there, and the behaviour of the new figures is, as it were, subject to them, while in Heine the figures have to be formed first, and it is only later, as it were, that the operations of calculation are defined. The latter seems to be more appropriate; for how might I lay down rules without mentioning the figures of which they treat?

Heine continues:

> "Operations of calculation are what we call rules according to which two numbers which are connected by operation-signs can be exchanged for a single one."

This has obviously gone awry. One might as well say: "The knitting of socks is what we call a rule according to which one produces a sock from a thread by means of knitting needles." What Heine means to say is:

> "Operations of calculation are exchanges, performed in accordance with certain rules, of a group, consisting of two numbers and an operation-sign separating them, by a single number."

One may add that the operation-sign here marks the rule that is to be applied. Heine is thinking, for example, of the case where the group '$3 + 5$' is exchanged for the sign '8'. If the group '$3 + 5$' is used in a proposition of contentual arithmetic, then we can put the sign '8' in its place without jeopardising truth, since both designate the same object, the same number proper, and everything that holds good of the object designated by '$3 + 5$' must therefore also hold good of the object designated by '8'. Further, progress in knowledge[e] will be made in many instances of such replacement, in so far as the senses of the co-referential signs, and thereby also the thoughts expressed by the two propositions, will be different. The goal of knowledge[e] thus determines the rule that the group '$3+5$' can be replaced by the sign '8'. This goal requires the rules to be so constituted that, if a new proposition is derived from true propositions[1] in accordance with them, it too is true. Whether the rules are of this kind can, naturally, be judged only after the signs have been given reference; for beforehand we cannot use them to form propositions that express a true thought. This holds for contentual arithmetic; in formal arithmetic, the rules are independent of a sense. The goal of knowledge[j] does not determine their content; rather they are laid down arbitrarily.

§105. Heine proceeds:

> "Initially, these rules are stipulated in such a way that they yield the result of ordinary calculation when the only numbers introduced are 0, 1, 2, 3, … etc."

This is actually imprecise; for the result of ordinary calculation—i.e., surely, calculations in contentual arithmetic—is a number proper, not a number-sign, and thus no number according to Heine's use of the term. Heine here borrows from a foreign theory. He adds:

> "The impossibility, in many cases, of subtraction prompts the introduction of new signs or numbers: for each already given sign, a, one introduces a further sign, $neg(a)$, extending the definition of the operations in a suitable matter, so that they also deliver a result for the new numbers, but, applied to the previous numbers, the same result as before."

At this point, various questions arise; first, how is "the impossibility of subtraction" to be understood? What is probably meant is that the rule is not universally applicable, the rule, that is, according to which that number-sign is to be taken as the result of subtraction which, in contentual arithmetic restricted to the non-negative whole numbers, designates the difference. This rule says nothing about which number '$3 - 5$' may be exchanged for. Yet that does not in fact need to motivate the introduction of new signs. For one could stipulate that '$3 - 5$', like '$5 - 3$', may be exchanged for '2'. Since the aim of these rules here lies beyond our horizon, from a formal standpoint, any rule is as good as any other as long as conflict within the rules is avoided.

Now, further it is mentioned that the definitions of the operations are to be extended. What is probably meant is that the rules are to be supplemented.

Heine proceeds:

> "Accordingly, an appropriate definition of subtraction shows us that $neg(a) = 0 - a$ must hold."

In place of "appropriate definition of subtraction", this probably ought to say "appro-

[1] More precisely: from propositions which express true thoughts.

priate laying down of rules for the exchange of figure-groups of the form '$a - b$' by other figures". Yet we cannot judge at all what is appropriate since the purpose is not known to us. Indeed, we do not even know whether such an exchange is needed at all, or whether we might not rest content with '$3 - 5$'. Thus, we have not been told how the supplemented rule might best be formulated; and so a central point remains completely obscure. How can we come to know the game, how can we understand its theory, if we are not so much as advised of its complete rules! We are, it seems, tacitly referred to our knowledge of contentual arithmetic. But if we have that, we do not need formal arithmetic.

Next comes the assertion that it must be the case that $neg(a) = 0 - a$. This is unintelligible. We have here a group of signs with which we associate no thought; consequently, nothing can be asserted either. '$neg(a)$' is a mere figure for us, as are the equality-sign, the minus-sign, and the Zero-sign. Since Heine makes an assertion here, he thinks he expresses a thought, but he probably does not know which one himself. Such is the nemesis of formal arithmetic, that it cannot avoid pronouncing propositions which are intended to express thoughts, but of which no-one can give themselves any exact account.

Of course we ourselves use the equality-sign to say that the reference of the figure-group standing to its left coincides with the reference of the one standing to the right. This is here inapplicable since we do not have any reference. But what the equality-sign might otherwise be intended to express we do not know. Whatever it might be, what stands to the right and to the left will have to refer to something, and one must presume that we are concerned with a proposition of the theory of the game, as of course neither any reference of the figures nor any assertion can feature in the game itself.

§106. We find something similar in Thomae. There we read:

"These rules are contained in the formulae:

$$a + a' = a' + a \qquad a + (a' + a'') = (a + a') + a'' = a + a' + a'' \qquad (a' - a) + a = a'$$
$$aa' = a'a \qquad a(a'a'') = (aa')a'' = aa'a'' \qquad (a' : a)a = a'$$
$$a(a' + a'') = aa' + aa''."$$

This is a surprise. What would someone say who asked for the rules of chess and instead of any answer was shown a group of chess pieces on the chess board? Probably that he could discern no rule in it since he attached no sense at all to those pieces or to their arrangement. It only seems that the case is different here because we already know the plus-sign, the equality-sign, and the use of the letters from contentual arithmetic; whereas here we want to practice formal arithmetic, whence the question arises whether those signs are here to be treated as signs at all or merely as figures. In the latter case, there would be no foreseeing how a rule could be given thereby. If, however, they are to be treated as signs, they cannot in any case have the same reference as in contentual arithmetic; for then we would have a proposition of contentual arithmetic, and not a rule of formal arithmetic.

§107. Although the account leaves us stranded at this point, we can still try to identify the senses which these formulae are supposed to carry here by asking what follows for the management of the signs from the senses that these signs have in contentual arithmetic. Let us initially disregard the letters that are clearly supposed to lend generality to the rules, and consider the formula,

$$`2 + \tfrac{1}{2} = \tfrac{1}{2} + 2'.$$

In contentual arithmetic, this says that the sum of 2 and $\frac{1}{2}$ coincides with the sum of $\frac{1}{2}$ and 2. What follows from this for the signs? Evidently, that a sign-group shaped like '$2 + \frac{1}{2}$' may be replaced everywhere by one shaped like '$\frac{1}{2} + 2$', and conversely. Here it is evidently presupposed that equal-shaped signs or sign-groups refer to the same. We have thereby laid down the rule of formal arithmetic which corresponds to our proposition of contentual arithmetic and it is this, we may surmise, that Thomae would express by the formula,

$$`2 + \tfrac{1}{2} = \tfrac{1}{2} + 2'.$$

For Thomae, accordingly, the formula,

$$`a + a' = a' + a',$$

will say: a figure-group consisting of two number-figures to the left and right respectively of a plus-figure may be replaced by a figure-group of the same kind in which the number-figures have exchanged their positions with respect to the plus-figure. Prior to that, it would need to have been stated which figures are the number-figures.

Let us now recall that the theory of the game is to be distinguished from the game itself. The actions of the game are indeed performed according to the rules; the rules, however, are not items in the game, but the basis of the theory of the game. The moves of chess are indeed performed according to rules; but these rules are expressed neither by any configuration of chess pieces nor by any move; for it is no part of the role of the pieces in chess to express anything; they are, rather, moved subject to rules. So, if formal arithmetic is viewed as a game, then the formula '$a + a' = a' + a$', as an expression of a rule of this game, is part of the basis of its theory, upon which inferences can be constructed within the theory; but it is not something with which changes are made in the game, not an item in the game, not comparable to a configuration of chess pieces, but rather to an expression in words of a rule of chess.

§108. Let us now ask what corresponds to a move in chess, what actions are governed by the rules of formal arithmetic. If we identify the senses of Thomae's formulae cited above in the same way we did with the first formula, then it transpires that each of them allows for replacing one figure-group by another, or by a single figure. We can best understand this by thinking of the figures as written with chalk on a blackboard. Thus, for example, we erase a figure-group shaped like '$2 + \frac{1}{2}$' and write down a group like '$\frac{1}{2} + 2$' in its stead. This action corresponds to a move in chess and occurs according to a rule of formal arithmetic. From the standpoint of contentual arithmetic it may seem silly even to mention the chalk, the erasure, in short, the whole external activity; but let us not forget that the calculating game precisely consists in such external activities.

The erased figure-group will have been part of a larger group; and, recalling contentual arithmetic, one might surmise that the latter figure-group will be what we call an equality or inequality. Thus, from a group like, say, '$(2 + \frac{1}{2}) \,.\, 5 = 12 + \frac{1}{2}$' will arise one shaped like '$(\frac{1}{2} + 2) \,.\, 5 = 12 + \frac{1}{2}$'. Each of these will correspond to a configuration of chess pieces. Now, though, we have already acquainted ourselves with equations as expressions of the rules. We thus observe that equations are here playing a double role: first, in the game itself, where they no more express anything than configurations of chess pieces do, and second, in the theory of the game, where they initially express rules, but then also, so we may expect, consequences of the rules. So now think through the analogy with chess! Here, the game-rules would be expressed by groups of chess pieces which could also occur in the game itself. So general propositions would have to be laid down stating how one should understand groups of chess pieces as rules or as theorems of the theory. In other words: a language would need to have been given whose expressive resources would be the chess pieces and their configurations on the chess board. So a figure-group might need to be regarded in two ways: first, as occurring in the game itself, when it would not express anything but would simply have originated from another group by a move and, similarly, could then be transformed by a further move into another one; second, as occurring in the theory of the game, when it would be a theorem and thus would have a sense. An inference would then be represented as a transition to a new configuration, and the rules according to which such transitions must occur would result from the logical laws and by the manner in which the chess pieces by means of their configuration express a sense. Therefore these rules could not be laid down arbitrarily, and it is not to be expected that they would be in agreement with the rules of chess. Because of this double role of the pieces, and the resulting duplicity of the rules in the game itself and in its theory, a clear insight into the matter would be obstructed to such an extent that one might suppose the double role was made up for the specific purpose of engendering maximal confusion.

§109. Such a double role of the figures we now find in formal arithmetic. Here too we must first have rules for the manipulation of the figures in the game itself, and these can be laid down completely arbitrarily without regard for any sense. Second, we must then have rules for how we are to manage these same figures as signs in the theory of the game; and these cannot be arbitrary but must be guided by the senses which these signs express, by means of their arrangement, in the theory of the game. Now it is a major error that no distinction at all is made between these two systems of rules, rather that their agreement is presupposed without any attempt at proof. On the contrary, it should initially be supposed that the game-rules cease to be valid as soon as the equations are no longer understood as senseless arrangements of figures, but instead as theorems of the theory of the game. The only effective way of clearing this thicket is this: that one not use number-figures and calculation-figures (like '+' and '=') in two ways but confine them to the game itself, and instead state the rules and the theorems of the theory of the game in words of ordinary language. In the investigation of formal arithmetic to follow, an equation is thus supposed to say nothing, to have no sense, but is for us merely a group of figures which is to be treated according to the game-rules. What is termed the equality-sign in contentual arithmetic is now just a figure which does not refer to a relation.

We have seen that the rules in Thomae's inventory allow the replacement of one figure or figure-group by another. One could express this in words as we did for the first rule. But it seems more appropriate to let Thomae's equations stand as figure-groups from which the game starts, similar to the initial position of the chess pieces. In that case they express no rules, have no sense at all. One then lays down the following as a rule: given an equation and a formula containing a component of equal shape to what is on one side of it, it is permissible to replace this by a component of equal shape to what is on the other side of the equation.

§110. This rule too issues a permission as do the rules of chess, where nothing may happen which is not allowed by a rule. One can add further rules concerning the replaceability of letters by others or by number-figures. These rules too will allow something. In truth, nobody can thereby be given a freedom which he did not already have; the rules are not laid down in the name of reason or of nature; it is merely that by their means some actions are acknowledged as admissible in the calculating game. Truth is not the issue here as it is in contentual arithmetic; what the arithmetical legislator wishes to accept as lawful is up to him, and in this he is not restricted by any concern for the references of the figures, for officially there are none for him. As guidelines for action the rules of formal arithmetic are more closely related to moral laws than to the laws of contentual arithmetic, which can be misapprehended but not transgressed.

§111. If we compare Thomae's rules, or the rule we laid down instead, to those of chess, then we notice that they treat of all number-figures indiscriminately while in chess different rules hold for pawns and knights. If there were no other rules in formal arithmetic, then there would be no point in using number-figures of different shapes. If all the rules that treat of the Two-figures likewise held of the Three-figures and *vice versa*, then the distinction between these shapes would be pointless. If in chess all figures were to be manipulated like pawns, then they might as well all be shaped like pawns.

If we now look back at contentual arithmetic, we notice that while there are indeed laws which hold of all numbers, in no way does everything that holds of one number also hold of any other. Quite the opposite: every number is essentially different from every other and must therefore also have its own peculiar sign. If, then, formal arithmetic is not to lose all connection with contentual arithmetic, if the various shapes of the number-figures are not to be a burdensome surplus, then the rules listed by Thomae must be supplemented with others that are not supposed to hold of all number-figures, so that to every difference in shape there also corresponds some difference in the rules. Such rules will be, e.g., that a number-figure shaped like '$1+1$' may be replaced by a Two-figure, a number-figure shaped like '$2+1$' by a Three-figure, and further that a figure-group of the shape '$1-1$' may be replaced by a Zero-figure, a figure-group like '$\frac{1}{2}+\frac{1}{2}$' by a One-figure, and so on. Now, since there is no fixed limit to the extent of the number-figures, the inventory of rules will also, it seems, never be finalised once and for all. In any case, we may note a severe incompleteness in Thomae.

§112. Above, in connection with the question of the nature of formal subtraction, we resolved to try to construe this operation as the converse of addition at some point. Here is the place to do it. Subtracting a first number-figure from a second will then mean writing down a third figure, or group of figures, such that each figure-group that results by adding the latter to the first according to the rules of formal arithmetic may be replaced by a figure of equal shape to the second. Here addition is the action described above. If, e.g., a Two-figure is to be subtracted from a Three-figure, then we can use Thomae's rule according to which a figure-group shaped like '(3 − 2) + 2' is replaceable by a Three-figure. We see from this that every figure-group like '3 − 2' is a solution to the problem. But solutions of this shape are not the only acceptable ones; there are also solutions of shapes like '2 − 1' and '1' and very many more. Granted, this does not follow from Thomae's rules; but we have already seen that they are incomplete. We can presume that there will be rules according to which figure-groups like '(2 − 1) + 2' and '1 + 2' are replaceable by a Three-figure. Consequently, formal subtraction will be, as mathematicians would say, a many-valued operation, since a subtraction problem will allow of various solutions. This will constitute an essential difference from contentual arithmetic, where subtraction[1] is single-valued. If, in contentual arithmetic, one presents 3 − 2, or 2 − 1, or 1, as solution to a subtraction problem, then it is the reference that is intended, which is the same. If One-figures, or groups that are shaped like '2 − 1', are written in different places on the blackboard, then all of them have the same reference and, despite the difference in location and shape of the signs written down for the solution, there is still only a single solution in contentual arithmetic. Here, in the calculating game, the figures and figure-groups are themselves solutions, and since they differ in location and shape, we have many solutions.

One could try to avoid this consequence by appealing to the formal reference that the number-figures have in Thomae. One might for example say: since all the figures and groups that were presented as solutions are to be manipulated according to the same rules, they have the same formal reference, and the latter is the proper solution. Against this it is to be objected:

[1] Evidently, formal subtraction and subtraction in contentual arithmetic really agree in name only; and this situation would be more clearly recognisable if this apparent agreement were also avoided by the choice of another word.

first, that this formal reference, as outlined above, is not to be accepted at all;

second, that this formal reference, were it admissible, would be the same for all number-figures, provided one assumes only Thomae's rules, which are supposed to hold equally for all number-figures;

third, that no further calculation with the result of the subtraction would be possible, since the calculating rules treat of number-figures, not of their dubious formal reference. In chess, after all, one makes moves with the pieces themselves, not with a certain something supposedly common to, say, all black pawns.

These points about subtraction transfer in essentials to division.

§113. Now, at the end of §2 Thomae says concerning lower arithmetic:[k]

> "It proves the single-valuedness (non-contradictoriness) of the four fundamental operations for all numbers with the exception of Zero, for which only addition and subtraction and multiplication are single-valued, while division is not, and so cannot be performed in a way free from contradiction. A quotient whose denominator is 0 has no reference,[m] and Zero occupies a singular position amongst the numbers."

The lower arithmetic that proves this can only be a contentual one, and so nothing is gained for formal arithmetic; for adding, multiplying, subtracting, dividing in contentual arithmetic are indeed entirely different from the operations with the same name in the calculating game. Single-valuedness does not feature here. Nor, moreover, should we acknowledge any reason why special provision ought to be made for the Zero-figures. In any case, there is no special mention of the Zero-figures in Thomae's inventory of rules, where all number-figures are treated completely equally. It defies understanding why the figure-groups '2 : 0' or '$\frac{2}{0}$' cannot be regarded as solutions to a division problem just as, say, '2 : 3' or '$\frac{2}{3}$' can.

In the proposition,

> "A quotient whose denominator is Zero has no reference"

figure-groups like those cited above are apparently called quotients.[1] Now, their lack of reference is no reason for formal arithmetic not to use such groups; for it does not ask after reference at all, and its having no need for it is, indeed, the real reason why it is preferred to contentual arithmetic. For formal arithmetic, '$\frac{2}{3}$' no more has

[1] If the quotient itself were a reference, one could hardly speak of its reference.

a reference than '$\frac{2}{0}$', and both groups may be handled according to the rules equally well.[1]

What is it, then, that Thomae really wants to say by denying any reference to a quotient with denominator Zero? Surely, that a figure-group in which a Zero-figure stands to the right of a colon or beneath a fraction bar is impermissible. Now, is this to say that the rules of the calculating game fail to issue permission for it? Or that they downright forbid it? In the first case, we would have a proposition of the theory of the game, comparable, for example, to the proposition in chess that a bishop which stands on a white square cannot reach a black one. There is no prohibition against this, but the permission provided for its movements is insufficient to make it possible. In the present case, however, one of Thomae's rules allows the occurrence of such a figure-group which is above presented as impermissible. For according to this rule, the replacement of a Two-figure by the group '$(2 : 0) . 0$' is allowed. So if writing down a Two-figure is allowed at all, it would accordingly also be permitted to write down a group shaped like '$2 : 0$'. If, on the contrary, this is supposed to be impermissible, then that it is so does not follow from the rules previously listed, and what is required for this is, rather, a prohibition; and our proposition,

"A quotient whose denominator is Zero has no reference"

will accordingly have to be regarded as a prohibitive rule, or as a proposition of the theory of the game. If it is the latter, then it is a consequence of the game-rules. Amongst those there will then have to be at least one prohibitive rule, since no prohibition capable of limiting a permissive rule can follow solely from permissive rules. In any case, there must be at least one prohibitive rule alongside the permissive rules. On the standpoint taken by the formal theory, the only reason for such a prohibition that can be recognised is, of course, just as for all other rules, the will of the legislator.

§114. Another question arises here: are figure-groups like '$2 : (1 - 1)$', '$2 : (2 - 2)$', '$2 : (6 - 2 . 3)$' admissible? Obviously, Thomae would deny this. But—I hear the objection—this is of course evident; for $1 - 1$ is Zero, likewise $2 - 2$ and $6 - 2 . 3$. Certainly, in contentual arithmetic '$1 - 1$', '$2 - 2$' and '$6 - 2 . 3$' have the same reference as the sign '0'; but let us not forget that here we find ourselves in formal arithmetic! Here we are concerned with these figure-groups themselves, not with what they refer to; and in that case there is no denying that they differ from each other and from the

[1] Formal arithmetic seems to break character here. Moreover, it sometimes seems as if formal arithmetic is indeed contentual arithmetic which, simply in order to escape inconvenient questions, tries to wear a formal character; however, it does not properly succeed in doing so.

Zero-figure not merely in the manner of two pawns of the same colour in chess but rather as pieces of a different kind, like knight and bishop. Now, figure-groups which are shaped like, say, '1 − 1', '2 − 2', etc., will be replaceable by Zero-figures. This, of course, does not appear as one of Thomae's rules; but these are, as we have seen, incomplete. So that is not yet a reason to consider figure-groups like '2 : (1 − 1)' as impermissible. In that case, however, one could arrive at a contradiction in the rules by obtaining a prohibited group if one chose to replace the group shaped like '1 − 1' by a Zero-figure. One might here assume some restriction on the rule allowing replacements by means of a prohibitive rule. But obviously Thomae will also prohibit figure-groups formed like '2 : (1 − 1)'.[1]

The sense of Thomae's proposition, "A quotient whose denominator is Zero has no reference", understood as a prohibitive rule, will be better expressed like this:

> "It is prohibited to form figure-groups in which either a Zero-figure, or a figure-group which, in formal arithmetic, is replaceable according to some rule by a Zero-figure, stands on the right-hand side of the division-colon."

§115. In applying this rule (or this theorem of the theory of the calculating game) there results an uncertainty; for one has to know which figure-groups are, directly or indirectly, replaceable by a Zero-figure; and this question cannot be answered with certainty before all rules of the game are laid down. And so this now is the question, whether a complete inventory of the rules can be given at all. Before this can be done, surely all classes of figures which are permissible at all must have been introduced. If a replacement by a Zero-figure does not appear to be possible according to an incomplete inventory of rules, then a rule added later can make this possible and thereby render inadmissible a figure-group which seemed admissible before; and, on the other hand, if some rules allow the replacement, a prohibition following later can rescind this permission again.

In order to make the matter certain, one would have to lay down the principle that all prohibitive rules are to have greater force than the permissive ones, and one would have to give a complete list of at least all rules that prohibit replacement of a figure, or a figure-group, by a Zero, so that each such replacement which is not explicitly forbidden would be allowed; or conversely, one would have to give a complete list of all rules that permit such a replacement. Either, however, would be difficult to carry out

[1] The Zero-figures, therefore, have no singular position in formal arithmetic.

because of the unsurveyable variety of figures and figure-groups. Until this is done, our rule is incomplete and consequently inapplicable.

§116. The inadequacy of Thomae's inventory of rules attracts further attention in other respects. In particular, nowhere is it stated how one can replace a figure-group which consists of two number-figures separated by a minus-bar with something else. As a result, these rules allow for the replacement of the figure-group '$(3 + 2) - 2$' by a Three-figure neither directly, nor indirectly. Let us recall how the corresponding proposition of contentual arithmetic is proven. There we find for example:

> By definition, $(3+2)-2$ is the number which, increased by 2, yields $(3+2)$. This number is 3; therefore $(3 + 2) - 2$ and 3 coincide.

The single-valuedness of subtraction is essential here, and this, as we have seen, does not hold in formal arithmetic. It is also essential that the sign-group '$(3 + 2) - 2$' has a reference to which one can point with the definite article ("*the* number which") and the demonstrative pronoun ("*This* number"). This, too, fails in the calculating game. One cannot speak of a definition here at all; for the minus-bar here is no subtraction-sign, it does not refer at all; instead of a definition, there are here only rules for the manipulation of this bar; but what is missing is precisely a rule that is usable for our purpose. Hence, the possibility is also missing, as far as Thomae's rules are concerned, of transforming the figure-group '$((3 + 2) - 2) - 3$' into one shaped like '$3 - 3$'. As a result, we are therefore free to view the group '$2 : [((3 + 2) - 2) - 3]$' as admissible.

According to Thomae's rules we cannot replace the figure-group '$(3 \,.\, 2) : 2$' by a Three-figure either; for a rule according to which a figure-group consisting of two number-figures separated by a colon may be replaced by something else is nowhere to be found.

We will try to remedy this deficiency not by laying down a new rule, but by using the rule given at the end of §109 and adding to Thomae's formulae the following two: '$(a + a') - a' = a$' and '$(a \,.\, a') : a' = a$'. These too are now figure-groups from which the game starts. Here, as also elsewhere, a rule is presupposed which states how one may replace letters by number-figures.

§117. If we ignore here the prohibition laid down above (§114), we could produce from the figure-groups '$3 \,.\, 0 = 0$' and '$(3 \,.\, 0) : 0 = 3$', gained from our second new formula by replacing the letters with number-figures, the group '$0 : 0 = 3$', and likewise the group '$0 : 0 = 4$'. From those two, we could then further gain the group '$3 = 4$'. And in this, perhaps, lies the reason for Thomae's pronouncement that division cannot always be performed in such a way as to yield a single value, and thus not in a way free from contradiction.[1] But here, in formal arithmetic, there is initially no contradiction at all. Why should a group like '$3 = 4$' not be allowed? In contentual arithmetic, of course, it cannot occur with any claim to validity, since what matters there are the references of the number-signs, which are distinct. This reason lapses here. Writing down a figure-group like '$3 = 4$' has not been prohibited, at least this far. It is only when such a prohibition is decreed that a contradiction arises, or better a conflict within the rules, which are partly prohibitive, partly permissive.

§118. Now, Thomae says in §2 immediately after his inventory of rules:

> "Subtraction and division become addition and multiplication through the introduction of the new numbers. Since none but these *four*, or, if you will, *two* operations of calculation, are recognised throughout arithmetic, new formations are free from contradiction within arithmetic as a whole if they are free from contradiction with respect to the four (or two) basic operations."

This pronouncement is difficult to understand. First, one can doubt whether no operations other than the ones mentioned are recognised throughout arithmetic. The taking of limits, at least, does not appear to be reducible to these operations. Incidentally, in order to be able to make such a pronouncement, one would really have to know the whole of arithmetic, which is impossible since this science is not completed and probably never will be.

Furthermore, it is conspicuous that freedom from contradiction is predicated of a figure. It would sound peculiar were the suspicion to be voiced that a chess piece might contain a contradiction. But let us recall Thomae's manner of expression whereby the circumstance that rules, amongst other things, treat of a certain figure is presented as a matter of the behaviour of the figure with respect to the rules, and whereby this behaviour, in turn, is designated as the content of the figure; then we see that any conflict which obtains between the rules of chess, for example, seems to be transferred into the interior of a chess piece. So, in order to obtain any understanding, we will have to transfer the contradiction back into the rules again.

§119. Now we have to enquire further what should be understood by a contradiction with respect to the basic operations. Here, most likely, we should draw on

[1] Here, the "thus" is noteworthy, since the extraction of square roots cannot in general be performed in such a way as to yield a single value, without thereby giving rise to a contradiction.

Thomae's proposition[1] adverted to above that division by Zero cannot be carried out without contradiction. It is therefore to be surmised that according to Thomae a figure contains a contradiction with respect to an operation when this operation cannot be carried out with the figure without contradiction. Thus, Zero-figures would not be free from contradiction with respect to division, and therefore not free from contradiction in arithmetic generally.

Such a contradiction can only arise when the specific rules that hold of a class of figures conflict with the general rules that hold of all number-figures. So since, according to Thomae's inventory of rules, the latter are all permissive, a contradiction is to be feared only if the rules treating of a specific class of figures also include prohibitive ones. This, indeed, does happen with the Zero-figures. But it also occurs in other cases. For when introducing a new class of figures, it will always be necessary to determine that these figures may not constitute the left-hand side of an equation whose right-hand side is a Zero-figure. Let us, for instance, introduce the class of figures shaped like '$\sqrt{2}$'. Now, if we did not know whether an equation like '$0 = \sqrt{2}$' was allowed, then neither would we know whether we may build a group like '$2 : \sqrt{2}$'; so we do not know either whether we may replace a Two-figure by a group like '$(2 : \sqrt{2}) \,.\, \sqrt{2}$'. Moreover, it would not help us should we have failed in many attempts to generate an equation like '$0 = \sqrt{2}$' for figures shaped like '$\sqrt{2}$', according to the general and the specific rules; for many failed attempts are not yet a proof of impossibility, and in particular not in the case when one cannot base the attempts on a completed inventory of rules. Figure-groups like '$0 = 1 - \sqrt{2}$' and countless others would likewise have to be prohibited.

Now, whether all these prohibitive rules might be combined into a single one, or into a few, will not concern us here any further. In any case, amongst the specific rules holding of figures shaped like '$\sqrt{2}$' prohibitive ones also occur, and that they cannot come into conflict with the general ones cannot be inferred from these general rules' being in harmony with each other. Each newly introduced class of figures will demand specific rules, amongst which there will also be prohibitive ones. For if the same rules were supposed to hold for the new figures as for those of a previous class, then there would be no reason to choose a special shape. If, e.g., exactly the same rules are supposed to hold for figures shaped like 'i' as for One-figures, then there would be no need for the former figures. On the other hand, if the rules are in part different, then the mutual agreement of the rules treating of One-figures does not vouchsafe the same concerning figures shaped like 'i'.

[1] *Op. cit.*, end of §2.

Proof of the proposition that formal arithmetic admits of a foundation entirely free from contradiction is accordingly lacking, and quite to the contrary, that it is true is subject to serious doubts. Thomae's opinion to the contrary rests on the mistaken belief that the rules listed in his §2 constitute a complete inventory, and in particular on his complete oversight of the prohibitive rules which each new class of figures necessarily brings with it.

§120. Thomae continues:

> "Since in the further development of the concept of number the formal conception, which is free from connections to sensory objects, has to enter at a certain point, we already choose it for the negative numbers and the fractions."

Here it is presupposed that each conception of numbers that makes no connection between them and the sensory objects is formal in Thomae's sense, or expressed conversely, that each contentual arithmetic places the numbers in some connection with sensory objects. This is a mistake. In our first volume, it is clear that nothing was more foreign to us than to treat number-signs as figures and to call these figures numbers; but, equally, nothing was more foreign to us than to base arithmetic on sense perception and to term heaps of sensory objects numbers. By the cardinal number Zero we do not understand a certain roundish figure; rather, for us the latter is only a sign of what we intend and what we recognise as being there even though it is neither a physical body nor a property of such. Thus, however much we agree that arithmetic has to guard itself against any mention of sensory things, and accordingly that the number-signs refer to nothing sensory, to the same extent we emphasise, on the other hand, that these signs do not fail to refer on that account, and reject calling the signs themselves numbers.

§121. Let us now turn to §3 of Thomae's exposition. There we read:

> "The rational numbers[n] may be ordered in series or they may be subsumed under the concept of magnitude. It is the case that $3 > 2$ and $3 > -4$ and $9:10 > 8:9$, since in $9:10 = 81:90$, the numerator is greater than the numerator 80 in $80:90 = 8:9$, while the denominators are equal."

It is striking here that ordering in series and subsuming under the concept magnitude are treated as the same. Does it make a book a magnitude that it can be ordered in a series together with other books? Is it then quite irrelevant how they are ordered? In that case everything would really be magnitude, e.g., the strokes of a clock, the letters of the word "magnitude". We can also order chess pieces in series; are they therefore magnitudes? It might be thought, rather, that the recognition of things as magnitudes preceded the order and delivered the principle for it.

Usually, when things are ordered in series, each thing has at least one neighbour. This does not appear to be the case here. The entire passage can be understood only if contentual arithmetic is known; and then formal arithmetic is superfluous. If we adopt Thomae's formal standpoint, then we lack all instruction how to order number-figures. No doubt, the sign '>' is supposed to serve this purpose; but we do not know its reference. According to the idea of formal arithmetic we would have to understand it as a figure which—at least in the calculating game—would have no reference, but for whose manipulation rules would be provided. In this case, however, such rules are completely missing; and such an understanding is ruled out by '$3 > 2$' occurring as a component in the declarative sentence "It is the case that $3 > 2$" from which one gathers that it is to be connected to a thought. '$3 > 2$' is, accordingly, not something with which changes are made in the course of the calculating game—comparable to a configuration of chess pieces; rather, it will be a proposition of the theory of the game. For the reasons developed earlier (§109), we here discard the use of number-figures and attempt to express the content in words, like this:

> "Every Three-figure is greater than any Two-figure"

where of course in choosing the words "every" and "any" we trust to luck. Here, the reference of the words "greater than" remains to be determined. In any case, they are to be taken neither in the geometrical sense nor in that of contentual arithmetic; for in the latter we have a relation between the numbers themselves, the references of the number-signs, references which are not present in the calculating game. What the sense of the proposition

> "The number-figure a is greater than the number-figure b"

is supposed to be, we are not told. We can merely surmise that we are dealing with some relation by which the number-figures are ordered in series. The nature of this relation, and its connection to the game-rules, remains obscure to us.

§122. Thomae's proposition that Zero provides the limit for a positive number's becoming smaller and smaller is, from the standpoint of formal arithmetic, incomprehensible. Does a number-figure, then, change during the game? All by itself? In one of the properties essential to the way it is treated in the game? Also incomprehensible are the propositions that amongst the positive numbers there is no smallest, that a rational number which is not negative but less than every positive number that may be given, is necessarily Zero. Which Zero? There are many Zero-figures. Is '$1 - 1$' a Zero-figure? Immediately afterwards it is said:

> "In this important proposition a number, the number Zero, is recognised by a negative criterion."

The number Zero? Which? "Zero" is here treated as a proper name. This is correct in contentual arithmetic, but false in formal arithmetic. What, in the latter, is a rational number-figure? What is a positive one? What is a negative number-figure? The propositions cited should belong to the theory of the game and follow from the game-rules. How and from which remains obscure. The words, "rational", "positive" and "negative" should designate properties of number-figures which come into consideration in the application of the rules, like for example "black" and "white" in chess. Moreover, small accidental differences in the shape or colour of the figures of which the rules do not treat cannot come into consideration within the theory of game. Now we ask: what are the rules that take account of the just-mentioned properties of the figures? Which rule makes a distinction between positive and negative number-figures, between rational and non-rational?[o] No answer! Indeed, we see the formal arithmetician breaks character yet again. The formal conception is a shield that is held up as long as questions concerning the reference of the signs threaten. When this danger passes, it is dropped; for it is in the end a burden, after all.

§123. Thomae believes that it is possible to advance from a number to ever greater and greater numbers since nothing places a boundary on new formation by addition. Yet certainly a boundary is set to the new formation, just as in the case of the growth of a city. We have neither an infinite blackboard, nor infinitely much chalk at our disposal. Number-figures are simply formations created by writing. An immaterial, non-spatial number-figure may be compared to a castle in the air; but even castles in the air have their boundaries. Fantasy lends wings; but even those will tire in the end.

It may be doubted whether a new formation of number-figures by means of addition occurs. Groups of number-figures and addition-crosses are so created. Whether such groups are to be viewed as number-figures, however, remains doubtful, since nowhere is it stated what a number-figure is.

It is notable that Thomae, as in the case of Zero, also senses a danger of contradiction in the infinite. The rules he himself lists are all permissive, thus cannot come into conflict with each other, no matter which figure one may treat in accordance with them, and whether an Eight-figure lies sideways or stands up can make no difference unless further specific rules hold for the sideways Eight-figures. None of these, however, are listed here yet.

The infinite is explicitly called a concept here, without any statement of a reason. Why is it not simply a figure? The actual infinite, rightly advocated by G. Cantor, is indeed no figure and ought to have no place at all in formal arithmetic.

§124. How, now, are the irrationals introduced into formal arithmetic? At first glance, exactly as in Cantor, whose fundamental series correspond to Heine's number-series and Thomae's number-sequences. However, the members of Cantor's fundamental series are not visible, tangible figures but, it appears, figures of a non-sensory kind, whereas Heine's number-series and Thomae's number-sequences are apparently supposed to consist of sensory figures. If this reading is right, then this similarity of the theories is merely superficial and inessential. Although clear pronouncements by these authors on this matter seem to be missing, it may be surmised that the series-forming relation in Heine and Thomae is a spatial relation. We assume that Heine's number-series, as well as Thomae's sequences of numbers, are groups of number-figures written adjacently and not too far apart from left to right, and that each of these figures is called a term of the sequence. We will have to add, further, that between single terms only empty space on the blackboard is to be visible.

Now, we gather from Heine's exposition that such a series is to proceed into the infinite. In order to produce it, however, we need an infinitely long blackboard, an infinite amount of chalk and an infinitely long time. One might here protest that it is egregiously atrocious to try to bring down such a lofty flight of the mind by such a pedestrian objection; but that does not refute the objection. If one turns the numbers into tangible figures and leans on this tangibility for an assurance of their existence, well, then one must also accept all the conditions of such material being. We here recognise a peculiar nemesis: in Heine the existence of the number-series, and with it that of the irrational numbers, is destroyed by the very tangibility of the numbers that was supposed to safeguard them.

Heine now introduces the requirement that a sign be allocated to each number-series and says:

> "One introduces the series itself as a sign, placed in square parentheses in such a way that, e.g., the sign belonging to the series a, b, c, etc. is $[a, b, c$, etc.]."

In order to be able to put this into practice, one would first have to conjure the art of putting an infinitely proceeding series into parentheses.

Heine gives the further definition:

> "We call the sign belonging to a number-series a more general number or number-sign."

Accordingly, a number-series put in square parentheses, if there were such a thing, would be a more general number. From this we gather that number-series are to form the proper object of Heine's study not in their original nakedness but clothed, rather, in square brackets.[1]

§125. Thomae attempts to sidestep the difficulty presented for formal arithmetic by the proceeding of a series into the infinite by presenting a definition of an infinite sequence. He states in §5:

> "A sequence of (in the first instance, rational) numbers ($a_1\ a_2\ a_3\ .\ .\ a_n\ .\ .$) is called an infinite sequence if no term in it is last but rather, if according to an instruction to be given, more and more new terms can always be formed."

If we did not know what Thomae was aiming at, we might think of an ordering of number-figures looping back into itself. Since this is clearly not what is intended, a sequence of number-figures will always have to have two ends, and some term will always be the last. Because of the latter proposition starting with "but rather", we may assume though that "last" here is not to be taken in its usual sense. Below I write a number-sequence:[2]

$$2\quad 3\quad 5$$

and ask: is it infinite in Thomae's view? If the Two-figure is called first term, then according to linguistic custom the Five-figure should be called last term. Accordingly we would have no infinite sequence. But in Thomae's explanation the emphasis is placed on possibility. On Thomae's account the Five-figure is not a last term if, according to an instruction to be given, more and more new terms can always be formed. However it is not required that more and always more new terms are actually formed; the possibility is enough; the sequence need never contain more than those three terms, and it will still be infinite provided that there is the possibility. Is there? For an omnipotent god, yes; for a human being, no. We here run up against the difficult concept of the possible, but we see that for the answer to our question it is entirely irrelevant what terms our sequence consists of. Sequences are not hereby divided into finite and infinite but rather will either all be finite or all infinite, according to the sense that is attached to the word "can". We have a similar case in a row of houses[p] which, from a city, extends gradually into the countryside. We may give this definition: "A row of houses is called an infinite row if no house in it is last but rather, if, according to an instruction to be given, more and more new houses can always be built." If we here assume human ability and understand the word "ever"

[1] Incidentally, the definite article in front of "sign belonging to a number-series" is noteworthy, since one may of course assign different signs to the same number-series—a possibility which Heine indeed exploits.

[2] In what follows, it will be called the sequence F. It is to provide the underlying example for the considerations to follow.

in the strictest sense, then, accordingly, no row of houses is infinite. Under the same assumptions and for the same reason, no sequence of number-figures is infinite; for we foresee that the possibility of continuation will give out at some point.

Surely we recognise how pointless it is to try to deceive ourselves about the limits of our abilities by means of a definition.

§126. For what purpose however do we actually need infinite number sequences? The answer to this question might enable us to recognise what we ought to understand by this expression. Thomae writes:

> "A sequence $(\delta_1\ \delta_2\ \delta_3\ ..\ \delta_n\ ..)$ is called a Zero-sequence, and is assigned the number Zero by means of the equality-sign
>
> $$0 = (\delta_1\ \delta_2\ \delta_3\ ..\ \delta_n\ ..)$$
>
> when the numbers $\delta_1\ \delta_2\ \delta_3\ ..\ \delta_n\ ..$ become arbitrarily small as their indices increase, so that for each number σ, however small, an n can be found such that all terms $\delta_n\ \delta_{n+1}\ \delta_{n+2}\ ...$ taken absolutely are less than σ."

Here, the notation is confusing. We do not know yet what, e.g., an index within a number-sequence is. In the formation '$(\delta_1\ \delta_2\ \delta_3\ ..\ \delta_n\ ..)$' indices are admittedly to be seen; but this collection of letters, digits, dots and brackets is no number-sequence.

We will surely not go wrong if we understand by the index of a term in a sequence an ordinal number which states the position that the term has in the sequence. Accordingly, the n which is mentioned in Thomae's explanation will not be a number-figure but rather a number. Suppose that in one case we have found such an n to be $9^{(9^9)}$! May we now use the words, "the $9^{(9^9)}$th term of the sequence F"? Not before we know that the sequence contains $9^{(9^9)}$ terms. Otherwise, "the $9^{(9^9)}$th term of the sequence F" is a proper name without reference since a $9^{(9^9)}$th term presupposes a $(9^{(9^9)} - 1)$th one. Here it has to be observed that there are no terms unless they are written down; for the terms are number-figures, and number-figures are formations created by writing. Yet could we not speak of the $9^{(9^9)}$th term of the sequence F with the addition "if there were such a term"? Yes, in the same way as we might speak of the oldest man who lives at one hundred degrees latitude north, if there were such a man. One may spin interesting fables about such things; but that has no place in science.

So, if Thomae understands by 'δ_n' the nth term in a sequence, then he enters the realm of fiction once the number n is so great that the presence of so many terms can no longer be assumed with certainty.

Where there is mention of an arbitrarily small number σ, we have to remember that the reference of "number" here is the same as that of "number-figure", that there is only a finite collection of number-figures on the entire earth, and that we cannot change this by writing down new number-figures. It will not do to point to infinitely many possible number-figures; for only the actual ones are number-figures. Something merely possible is not a number-figure at all. Perhaps we have an idea of a number-figure and also think it is possible to write it; but then we only have an idea, not a number-figure. Moreover, it is exceedingly doubtful whether it is possible to construct infinitely many number-figures. So, there can be no talk here at all of an unbounded convergence toward the Zero-figures, even if we are prepared to admit a relation between number-figures designated by the words, "absolutely less than".

§127. Regarding the word "all", Thomae makes the following remark:

> "Since one cannot write down all terms, here, as in similar cases, one must understand by ""all"" however many terms one may construct, or, to speak negatively, from a certain index onwards no term is $>\sigma$."

Here, several things are to be remembered. The Zero-sequences are undoubtedly to be infinite; i.e., it is supposed to be possible to construct, i.e., to write down, more and always more new terms. Now we find it stated that all terms cannot be written down. From this we can conclude that a Zero-sequence according to Thomae consists of, first, terms that are written down, second, terms that are not written down but can be written down and, third, so it would seem, terms that cannot be written down. The analogue of this would be, say, an infinite row of houses consisting, first, of houses that are built, second, of houses that are not built but can be built and, third, of houses that are neither built nor can be built. Such a row of houses would thus start in the actual, and extend through the realm of the merely possible into the impossible. A remarkable row of houses!

§128. Might not the formal arithmetician here once again have broken faith with his plan? We know how likely this is since it is, after all, quite customary to regard the number-figures as number-signs, namely as proper names which designate something. For this reason the example of the houses is so liberating, since here the association with contentual arithmetic lapses, while there is no difference which would affect our question; for both houses and number-figures are products of intentional human activity, and this is all that matters. Indeed, one could use the houses themselves instead of the number-figures. Building would correspond to writing down, demolition to erasing. Writing down and erasing are the actions in the calculating game after all. Thus one could straightforwardly assimilate all game-rules to this case.

In contentual arithmetic, there is indeed nothing strange about saying that all terms of a sequence cannot be written down; for what is written down here are signs for the terms. The terms themselves are not created by this, and their being is not affected in any way by being written or not being written. It is utterly different in the calculating game! Here, the number-figures are themselves terms. No more are there unwritten number-figures than there are unbuilt houses. If only three terms are written down, then the sequence too consists of only three terms. In that case, how could one still say that all terms of the sequence cannot be written down! What can a number-sequence be in the calculating game other than a whole, a group, consisting of written figures? If such a group cannot be written down, then it cannot come into being. And because there have been no such number-sequences from eternity, there are therefore none at all, and there will be none. If a number-sequence is written down, on the other hand, then all its terms are written down; for only in them it has its being.

In his remark, Thomae uses the expression, "however many terms one may construct"; if here, instead of "may construct", it said, "has constructed and will construct", then there would be nothing to say against it. However, terms which are merely possible and are never written down are not terms in formal arithmetic.

It is no more to be feared that trees grow to the sky than that number-sequences may be continued without end. At some point each will have reached its greatest length. Let us consider our sequence F at this moment. Let the cardinal number of its terms then be n; let the $(n-1)$th term be a Two-figure, the nth a One-figure. We will then perhaps be able to say: all terms following the $(n-2)$th are less than a Three-figure; likewise also: all terms following the $(n-1)$th are less than a Two-figure, and finally: all terms following the nth are less than a One-figure; since, that is, no term either follows the nth or will follow. Negatively, we can also say: no term following the nth is greater than a One-figure. Accordingly, our sequence would be a Zero-sequence. With the same right, we could of course assert: no term following the nth is less than a Nine-figure. For since there is no term at all following the nth, there is also none which is less than a Nine-figure.

This is the manner in which Thomae's words have, strictly speaking, to be understood; but obviously, they are not intended to be understood in this manner.

§129. Let us consider the following case. Suppose an instruction V for the continuation of our sequence F is given to us. And now, suppose, that without knowing anything about the future length of the sequence we can deduce from the character of the instruction the proposition that all terms that there may be at some point which are written down according to our instruction and marked by an index greater than One Hundred are less than a One-figure. Then if such a proposition is deducible for any positive number-figure σ in place of the One-figure, where one has merely to choose another number instead of the number One Hundred, then Thomae will probably declare the sequence a Zero-sequence.

However, here one must always keep in mind that the collection of positive number-figures is finite, and always will be.

In addition, we note that the possibility of such deductions primarily depends on the instruction, but that the latter is not determined by the sequence F as it is displayed before our eyes. This would be more suitable to legitimize the definition of a Zero-instruction than that of a Zero-sequence.

§130. However not even a Zero-instruction could be defined in that way. That the proposition "all terms constructed according to the instruction V whose index is greater than One Hundred are less than a One-figure" followed from the nature of the instruction V cannot be used as a characteristic mark. It might work if it followed entirely formally, as, for example, the proposition "there is a B" follows from "A is a B", completely independently of the reference of 'A' and 'B', only presupposing that these signs refer to something. Obviously, however, that is not the case here. Although we do not know the reference of the words "less than" in the theory of the calculating game, it inevitably will come into consideration here and propositions will have to be drawn upon which illuminate the character of this relation, propositions which naturally we do not know at this point, but whose validity we have to assume. Perhaps it will be necessary to add further propositions which give more detailed information about the objects, concepts and relations which may be mentioned in the instruction. And these propositions may treat of other objects, concepts and relations which in turn demand that we draw on yet other propositions. Maybe the very proposition to be proven is amongst them.

Thus, since the instruction on its own is not sufficient, and since it cannot be stated precisely which additional propositions are to be drawn upon, there is really no sense in saying that such and such follows from the instruction. For the purposes of definition, we must therefore forgo any recourse to the circumstance that a proposition follows from the instruction V. All we can do, for that purpose, is to employ the proposition itself, for example the proposition that for every positive number-figure σ there is an index n such that a term of a sequence is less than σ if it is constructed according to the instruction V and if the index is greater than n.

A hypothetical thought, however, is always true when the condition is never fulfilled. Hence one can always assert truly: if a man lives without nourishment for a thousand years, he grows green hair. Accordingly, a statement of what would happen under circumstances which are not fulfilled and will never be, can in no way be of service for the definition of a Zero-sequence.

§131. We here recognise the irremediable mismatch between what the introduction of the irrationals demands and what formal arithmetic can offer. In order to introduce the irrationals we require infinitely many numbers, and formal arithmetic only has a finite collection of number-figures. No amount of definition, no amount of twisting and turning can make it otherwise. Indeed, Thomae too presupposes that his sequences have infinitely many terms by using signs like 'δ_n', without upper bound for n, to represent referential proper names, even though there are infinitely many cases where there is nothing that could be designated by means of such a sign.

If an infinite number-sequence consists of number-figures and nothing further, and if number-figures are formations created by writing, then such a number-sequence can likewise be written down. Let it be done! What will the result be? A series which starts with a figure and ends with a figure. Now, one can no doubt provide a definition according to which this written sequence is nevertheless infinite; but what is the point? The infinity required for the introduction of the irrationals is not achieved in this way. What use is the word "infinite" to us if the thing that matters is missing!

Since actuality is not sufficient, possibility or even impossibility is meant to step into the breach; as we have seen, in vain. If merely possible figures could serve as a substitute for actual ones, we would not need the actual figures.

This state of affairs is veiled by the inevitable use of contentual arithmetic as a supplement. It indeed shimmers everywhere so distinctly through the veil of formal arithmetic that sometimes we believe it is the sole thing we see. Here, however, one overlooks that much of what has good reason in contentual arithmetic is unjustified in formal arithmetic. Time and again, the profound differences are forgotten. Some reader may have condemned as entirely unmathematical the intrusion of time, when we assumed, for instance, that the length of a sequence changes in time. This rebuke would be completely justified on the standpoint of proper or contentual arithmetic; in formal arithmetic, however, time is introduced by the subject matter itself; for while the numbers proper are timeless, number-figures come into being and decay in time, and it is also in time that the actions of the game occur.

§132. Thomae writes:

> "The simplest Zero-sequence is naturally (0 0 0 . . 0 . .)."

We next ask: does this figure-group designate a number-sequence, or is it one? According to the usage of signs that is otherwise customary, we would assume the former; however, since the Zero-figures are not signs here, but rather figures, we probably have to plump for that figure-group's being a number-sequence, according to Thomae's intention. But is it really so? We have assumed that the number-sequence consist of number-figures and nothing further; but here we also see dots and brackets. We might be permitted to discount the latter as mere clothing. What, though, have the dots to do with the number-sequence? Are they intended to represent Zero-figures? Then why do the represented Zero-figures not stand there themselves? That would be simpler. We would then have a sequence consisting of eight Zero-figures. Of course, whether we would then have a Zero-sequence in Thomae's sense is doubtful. Thus, we do not have even a sequence of number-figures but merely a series which consists partly of number-figures, partly of dots. It is only the group of the first three Zero-figures that we can take here as a sequence of number-figures; even the fourth Zero-figure cannot be counted in, since it it is separated from them by dots. Are the dots intended to encourage us to imagine indefinitely many more Zero-figures? Images of Zero-figures are no Zero-figures, and an image of indefinitely many Zero-figures is in any case quite blurry. Then this whole figure-group would not stand alone like a group of chess pieces, but instead the images excited by it would be the most important thing. We would now once again have something like a sign after all. Would the image associated with it then be a Zero-sequence? Would it consist of number-figures? Hardly! Twist and turn as we may, we do not reach a conception of our figure-group which is compatible with the fundamental thought of formal arithmetic.

§133. The issue becomes yet more difficult if we have letters with indices instead of number-figures. Nothing is said concerning the use of letters in formal arithmetic, although it will diverge from the use in contentual arithmetic. How, e.g., is a group of letters with indices such as we have in the proposition,

> "A sequence of, in the first place, rational numbers $(a_1\ a_2\ a_3\ .\ .\ a_n\ .\ .)$ is called an infinite sequence"

to be understood? This is reminiscent of the turn of phrase, "a commander Caesar". Here, "Caesar" is a proper name and one could likewise understand '$(a_1\ a_2\ a_3\ .\ .\ a_n\ .\ .)$' as a proper name of a number-sequence. But obviously no specific sequence is supposed to be designated here. Nor is this group itself a number-sequence. One may next surmise that it merely indicates a sequence, as when one says, for example, "a prime number p". Here the letter is not a proper name, but rather stands in for one. Letters are written in place of proper names in order to confer generality upon the treatment. However, it is always possible to arrive at a specific instance by replacing the letters by proper names, e.g., by writing the proper name '7' in place of the letter 'p'. Let it be well remarked: the proper name, not the figure! Accordingly, if in our case 'a_1', 'a_2', etc., are taken as standing in for proper names, that is, for number-signs, then what would be obtained in the transition to a particular instance would be, for example, '(3 7 1 . . 2 . .)'; and in this each number-sign designates a number. Of course, what the whole is supposed to designate remains unclear since no explanation is to hand concerning what such a manner of combination of signs refers to. In any case, we now find ourselves in the territory of contentual arithmetic. But even if we wanted, erroneously, to view the number-signs as figures, the whole figure-group neither designates a number-sequence nor is it one, as we have seen above.

Thus we do not attain a satisfactory conception of '$(a_1\ a_2\ a_3\ .\ .\ a_n\ .\ .)$' if we dissolve it into its components. So we will have to accept it without paying heed to its composition. But in that case, a single letter will serve equally well, and we will be able to say "a number-sequence F" just as we say, for example, "a prime number p".

§134. Thomae now continues in §5:

> "We assign to such a sequence a sign and express this by the equality-sign,
>
> $$a = (a_1\ a_2\ a_3\ .\ .\ a_n\ .\ .)."$$

In G. Cantor we have already encountered this assignment of a sign as a particularly important act; it also to be found in Heine. Here, obviously, the letter on the left, 'a', stands in for a proper name. However, the right-hand side does the same. For it, we can write a single letter, 'F', as above,

$$\text{`}a = F\text{'}.$$

If we now want to consider a particular instance, then we must insert proper names both for 'F' and for 'a'. In that case, though, we already have a proper name of the sequence under consideration, namely the one inserted for 'F', and we do not need to assign another one to it.

Let us take a particular instance. We write with chalk on a blackboard, from left to right, following one another, a Two-figure, a Three-figure and a Five-figure. Let us assume that what is thereby created is an infinite number-sequence in Thomae's sense. In order to be able to predicate something of it, we give it a turned-over Roman A, 'Ɐ', as sign or proper name and can now for instance write: "The sequence Ɐ consists of a Two-figure, a Three-figure and a Five-figure". If for some purpose we want to give yet another sign to this sequence, nothing stands in the way and we can, for instance, write

‘ꓭ = Ɐ’

in which the equality-sign refers to coinciding, to identity, thus to what we call equality. This equation results from Thomae's

$$‘a = (a_1\ a_2\ a_3\ .\ .\ a_n\ .\ .)’$$

by introducing proper names on both left and right for the signs that merely indicate. Now it is, of course, unlikely that this captures Thomae's intent. It is more likely that he would tell us to write an equality-sign to the left of our sequence, and then our sign ‘Ɐ’ to the left of that again, so that what would result would be a sign- and figure-group of the form

‘Ɐ = 2 3 5’.

And it is thereby, he might say, that the sign ‘Ɐ’ is assigned to our sequence. This would, of course, be untenable. For, first, it would require that the left-hand side of the equation, ‘$a = (a_1\ a_2\ a_3\ .\ .\ a_n\ .\ .)$’, be treated in a wholly different way from the right-hand side. On the left, a proper name, ‘Ɐ’, would be inserted for ‘a’, but on the right there would be the object (the sequence) itself. That would contravene all principles concerning the use of letters in mathematics. And a strange mixture of signs and figures would thereby emerge. The equality-sign would be used neither as a mere figure, as in the game, nor in the way that Thomae applies it in the theory of the game, nor yet in the way that it is used in contentual arithmetic; rather, it would be taken to say that the sign on the left, ‘Ɐ’, is to designate the sequence on the right. The interchangeability of the left- and the right-hand sides of the equation would fail here; for when the sequence stands on the left, but ‘Ɐ’ stands on the right, then the sequence would thereby be presented as a sign for the figure ‘Ɐ’, which is something different entirely.

Whether we have now captured Thomae's intent or not, it is in any case not permissible to treat the equality-sign in so arbitrary a manner, as if it had never been used before.

§135. The following propositions appear to confirm that we have correctly captured Thomae's intent. They read:

> "For this sign a, we take under certain circumstances a rational number. If, namely, from a specific term onwards, the same number always reappears, so that
>
> $$a_{n+1} = a_{n+2} = a_{n+3} =\ .\ .\ = a$$
>
> then we choose the number a as sign for the sequence. But also when the terms in $(a_1\ a_2\ a_3\ .\ .\ a_n\ .\ .)$ differ from those of the sequence $(a\ a\ a\ .\ .\ a\ .\ .)$ only by the terms of a *Zero-sequence*, which we will define immediately,

we assign the number a as sign to the sequence $(a_1\ a_2\ .\ .\ a_n\ .\ .)$ by means of the equality-sign."[1]

However, we have to remember here that different things are given the same sign, which violates all principles of designation. This, of course, is veiled by the association with contentual arithmetic. Here, most likely, the thought shimmers through that all of these sequences determine one and the same number in our sense, one and the same magnitude-ratio, in a manner that is here not stateable, of course, since for this the question "what is a magnitude-ratio?" would have to be answered in advance. If Thomae now assigns to the sequence a sign a, then probably he wishes at bottom to assign to that magnitude-ratio a sign, without being fully conscious of this, and then the univocality of the sign is indeed preserved. This ceremonial assignment of a sign is to pass muster for what should really be accomplished, namely an explanation of magnitude-ratios and a proof that there are such things. Since the fruit is missing, empty husks are offered instead.

§136. The greatest surprise, however, is that the plan of formal arithmetic here fails completely, since number-figures are now used as signs after all. If the aim was to generate confusion, one could not do any better than this. For in formal arithmetic, after all, everything depended on the number-figures being mere figures, not signs. For figures, arbitrary rules can be laid down; for signs, the rules follow from the references.

The question arises now whether number-figures occurring as terms of a sequence are to be regarded as signs which in turn themselves refer to sequences. In that case the sequence whose terms they are would also have to refer to something; but what could that be? We would enter an infinite regress if we proposed always to regard the terms of a sequence as signs for other sequences in turn. Accordingly we might surmise that the number-figures, when they are terms, are not to be understood as signs. But it is doubtful whether such a distinction can be accomplished. In any case, such a double use of equal-shaped formations would be disconcerting.

§137. It will be unnecessary to study Thomae's exposition any further. This attempt at a formal arithmetic must be regarded as having failed, for the very reason that it cannot be carried through consistently. Number-figures are, in the end, used as signs again after all. The inventory of game rules laid down by Thomae himself is incomplete,

[1] What does the equality-sign refer to in

$$\text{`}a_{n+1} = a_{n+2} = a_{n+3} = \ .\ .\ = a\text{'?}$$

indeed, we had to surmise that such an inventory could never be completed at all, that in addition to permissive rules, one would have to lay down prohibitive rules as well, and that an uncertainty would thereby be generated concerning what was permissible, an uncertainty which probably could never be entirely dispelled. We tried to remove, as far as possible, the unclarity resulting from the lack of a distinction between the game itself and its theory. But it did not seem possible to provide a theory of the game before all its rules were present. We saw that notations and expressions were taken over indiscriminately and without further explanation from contentual arithmetic, for instance, 'greater' and 'less', whose role in the calculating game, although apparently most important, remained obscure. Formal arithmetic proved to be incapable of defining the irrationals, since only a finite collection of number-figures are available to it.

Many mathematicians are probably unclear how far the basic thought of formal arithmetic can carry. Formal arithmetic is apparently understood as contentual arithmetic, relieved of the obligation to state the reference of the signs. Indeed, the conception of the numbers as figures is only brought to bear at the start when this obligation is pressing. Later on one slips without noticing it back into contentual arithmetic. And yet this conception has further consequences which can become a burden; it causes so complete an alteration of arithmetic from the ground up that it seems scarcely acceptable to use the one name, "arithmetic", for both the formal and the contentual approaches. Formal arithmetic can keep itself alive only by betraying itself.[1] This illusion of life is made easier by the haste with which mathematicians, in order to get to the more important topics, pass over the foundations of their science, if indeed they concern themselves with them at all. Much is totally ignored, other things are touched on only fleetingly, nothing is worked out in detail. Thus a theory whose weakness would immediately be revealed in any serious attempt actually to work it out can assume an appearance of stability. And hereby the way to its refutation is shown. One merely has to follow the paths of thought on which one has placed foot and see where they lead. To take formal arithmetic seriously is to overcome it; and this is what we have done.[2]

[1] He who enjoys paradoxes might say: the correct understanding of the formal theory is to understand it incorrectly.

[2] H. von Helmholtz seems to adhere to a formal theory in his essay *Zählen und Messen erkenntnistheoretisch betrachtet* (*Philosophische Aufsätze*, dedicated to E. Zeller on the occasion of the 50th anniversary of the doctorate) when he for instance says: "I regard arithmetic, or the theory of the pure numbers, as a method grounded on purely psychological facts through which the consistent application of a system of signs (namely, of the numbers) of unlimited extension and unlimited possibility of

d) The creation of new objects according to R. Dedekind, H. Hankel, O. Stolz

§138. We now turn to the exposition given by R. Dedekind in his work *Stetigkeit und Irrationale Zahlen.*[1] There, he writes in §1, p. 6:

> "If one wishes to express that the signs a and b refer to one and the same rational number, then one sets both $a = b$ and $b = a$."

Here the sharpness of the distinction between the sign and that which it is supposed to refer to is pleasing and noteworthy, as is the conception of the equality-sign which coincides exactly with our own. Against this Thomae remarks:[2]

> "For if equality or the equality-sign = were supposed to refer[q] to mere identity, we would come to a stop at the trivial knowledge[e] or, if one prefers, the necessity of thought that a is a ($a = a$)."

This is an error. The knowledge[e] that the evening star is the same as the morning star is much more valuable than a mere application of the proposition '$a = a$', not the mere product of a necessity of thought. The explanation rests on the fact that the senses of signs or words (evening star, morning star) can be different while their reference is the same, and that it is precisely the sense of the proposition—in addition to its reference, the truth-value—that determines the value for our knowledge.[e]

We may conclude from Dedekind's proposition quoted above that, for him, numbers are not signs but rather what signs refer to.

These three points:

1) the sharp distinction between a sign and its reference;
2) the explanation of the equality-sign as the identity-sign;
3) the conception of numbers as reference of number-signs, not as the latter themselves;

are very closely related and let Dedekind's conception appear in the most pronounced opposition to any formal theory that considers the signs, or figures, as the proper objects

---

refinement is taught. Arithmetic investigates in particular which different ways of connecting these signs (operations of calculation) lead to the same final result."

Here too, since their reference has disappeared from view, the signs receive a magical power. The unclarity is only increased by the additional invocation of psychology and the empirical. Helmholtz wants to found arithmetic empirically, whether it bends or breaks. So he does not ask: how far can one get without invoking experiential facts? Rather he asks: what is the quickest way I can drag in some facts of sensory experience? Anyone with this aspiration achieves it very easily in the same way, by confounding the applications of the arithmetical propositions with the latter themselves. As if the questions of the truth of a thought, and of its applicability, were not entirely different! I may very well accept the truth of a proposition without knowing whether I will ever be able to apply it. But let's just mix everything nicely together! And never distinguish what is distinct! Clarity will then surely take care of itself. Hardly ever have I seen anything more unphilosophical than this philosophical essay, and hardly ever has the sense of the epistemological issue been more misunderstood than here.

[1] Braunschweig, Vieweg & Sohn, 1892.

[2] *Op. cit.*, p. 2.

of arithmetic. This makes all the more remarkable the approval that Dedekind accords to Heine's conception, when he says about the essay discussed above:

> "In essence, I completely agree with the content of this work, as indeed it cannot be otherwise."

In reality, there is no such agreement at all. On the contrary, Dedekind's views may be closer to Cantor's.

§139. After Dedekind has called a cut any division of the system of the rational numbers into two classes such that every number of the first class is less than every number of the second, and after he has then shown that every rational number generates a cut, or indeed two cuts, but that there are cuts that cannot be generated by any rational numbers, he says in §4, p. 14:

> "Whenever, then, we are presented with a cut (A_1, A_2) that is not generated by a rational number, we create for ourselves a new, irrational number a, which we regard as completely defined by this cut; we will say that the number a corresponds to this cut, or that it generates this cut."

This creating is the heart of the matter. To begin with it should be noted that it is quite different from what occurs in formal arithmetic when figures and specific rules for their manipulation are introduced. There the difficulty is how to recognise whether these new rules can conflict with previously established ones and, if so, how to settle such conflict. Here the question is whether creating is possible at all; whether, if it is possible, it is so without constraint; or whether certain laws have to be obeyed while creating. In the last case, before one could carry out an act of creation, one would first have to prove that, according to these laws, the creation is justified. These examinations are entirely missing here, and thus the main point is missing; what is missing is that on which the cogency of proofs conducted using irrational numbers depends.

In any case, the power of creation, if there is such a thing, cannot be limitless, as is seen from the obvious impossibility of creating an object which combines within itself contradictory properties.

§140. The following consideration leads to the same result. It is not unusual in mathematics that, in order to prove a proposition, an auxiliary object is required; this is an object which the proposition itself does not mention. In geometry one has auxiliary lines, auxiliary points. Likewise auxiliary numbers occur in arithmetic. For instance, a square root of -1 is required in order to prove propositions that only concern real numbers. When, in number theory, we prove using indices that, provided δ is the greatest common factor of n and $p-1$, the congruences '$x^n \equiv 1$' and '$x^\delta \equiv 1$' have the same square roots for the prime number modulus p, we require a primitive root, namely the basis of the indices, as an auxiliary number. In our proofs, too, auxiliary objects have already occurred; compare, e.g., vol. I, §94. There we have also seen how we get rid of such an object in turn; for no mention of it is to be made in the proposition to be proven, although we require some of its properties for the proof. In the number-theoretic proposition mentioned above, for example, we require the property of being a primitive root of the prime number p. In that case, we must initially introduce conditional clauses that express that some object has these properties. Once we know of such an object, we can eliminate these conditions. If we cannot supply such an object, as in the present instance, where we do not speak of this or that specific prime number but rather of prime numbers in general, we must at least prove that there always will be such an object—a primitive root for the prime number p. How much easier this would be if one could simply create the required object! If we do not know whether there is a number whose square is -1, then we simply create one. If we do not know whether some prime number has a primitive root, then we simply create one. If we do not know whether there is a straight line that passes through specific points, then we simply create one. This, unfortunately, is too convenient to be correct. Certain constraints on creating have to be acknowledged. For an arithmetician who accepts the possibility of creation in general, the most important thing will be a lucid development of the laws that are to govern it, in order then to prove of each individual act of creation that it is sanctioned by these laws. Otherwise, everything will be imprecise, and the proofs will descend to a mere illusion, to a gratifying self-deception.

§141. At the beginning of section 7 of his *Theorie der complexen Zahlensysteme*, *Hankel* says:

> "We consider in this section the numbers $\alpha, \beta, \ldots$, which are composed linearly of units $\iota_1, \ldots \iota_n$ whose rules of multiplication are stated in the relations
>
> $$\iota_1\iota_1 = 0, \quad \iota_2\iota_2 = 0, \quad .\,.\ \iota_n\iota_n = 0, \quad \iota_k\iota_m = -\iota_m\iota_k."$$

With these so-called units, he then proves, e.g., the proposition regarding the multiplication for determinants; or rather, he imagines himself to have proven it. In fact, it is no more than a perplexing sleight of hand; for nowhere is it proven that there are such units, nowhere it is proven that it is legitimate to create them. It is not even proven that the properties attributed to these units do not contradict each other. Indeed, it remains obscure what these properties really are, for nowhere is it stated what is to be understood as a product in this case. The propositions '$\iota_1\iota_1 = 0$', etc., mentioned above, should really be introduced as conditions, and the law of multiplication for determinants should be presented as depending on these conditions. To discharge them remains an unresolved task when the proof is conducted in this way. It would be possible if 'ι_1', 'ι_2', etc., were proper names of objects satisfying these conditions. For numbers of this kind, we do not know what a product or what a sum is. However, let us suppose we did know this, then we would know of ι_1 that it has the property that $\iota_1\iota_1 = 0$, a property it shares with ι_2, ι_3, etc. Further, we would know certain relations in which ι_1 ought to stand to the likewise unknown ι_2, ι_3, etc. It is clear that ι_1 is not specified thereby. We do not know how many such objects there are, nor whether there are any at all. Not even the class to which these objects could belong is determined. Let us assume that such a class contains the objects,

$$\iota_1, \iota_2, \ldots \iota_9.$$

Then the class that contains only the objects,

$$\iota_1\,\iota_2\,\iota_3,\;\; \iota_4\,\iota_5\,\iota_6,\;\; \iota_7\,\iota_8\,\iota_9$$

has the same general constitution; and so likewise does the class that contains only the objects,

$$\iota_1\,\iota_4\,\iota_7,\;\; \iota_2\,\iota_5\,\iota_8,\;\; \iota_3\,\iota_6\,\iota_9$$

and many more. Since, accordingly, not even the class to which these objects belong is determined, they are themselves even less so, and it is impossible to regard 'ι_1', 'ι_2', etc., as referential proper names, similar to '2' and '3'. All that remains is to regard them, like 'a', 'b', 'c', as indicating objects, rather than referring to or designating objects. In that case, however, all depends on whether there are such objects satisfying the conditions mentioned above. The conditions are not even complete, for the condition is missing that the product of an ordinary number and the product of some of the ιs is distinct from the product of a different ordinary number and the same product of ιs. Without it, we could not infer $a = b$ from

$$a \,.\, \iota_1\,\iota_2\,\iota_3 = b \,.\, \iota_1\,\iota_2\,\iota_3.$$

Now, the proof that there are such objects ι is lacking. Perhaps Hankel believed he could create them with the words mentioned above; but he still owes the proof that he was justified in so creating them.

§142. If we had attempted to conduct the proof Hankel gives of this proposition for determinants in our concept-script, we would so to speak have run headlong into this obstacle. That it is so easily overlooked in Hankel's conduct of the proof, results from a failure to formulate the assumptions in the manner of Euclid's, paying the closest attention to making no use of any other. Were one to do this, it would not be so easy to cause the assumptions to vanish by sleight of hand.

Incidentally, there are many proofs conducted with the imaginary unit that stand on no firmer a footing than that of Hankel's just mentioned. If the error in the latter is more striking, then this is due not to any essential logical difference, but rather to one's already being more accustomed to imaginary units than to alternating numbers.[r] One needs only to use a word or sign as a proper name often enough for the impression to arise that this proper name designates something, and this impression will gather strength over time so that, in the end, nobody has any doubts.

§143. Creative definitions are a first-class invention. Otto Stolz writes:[1]

> "6. *Definition.* When lim $(f : g)$ is a positive number or $+\infty$, a thing distinct from the moments is to exist, designated by $\mathfrak{u}(f) : \mathfrak{u}(g)$, which satisfies the equation
>
> $$\mathfrak{u}(g) \,.\, \{\mathfrak{u}(f) : \mathfrak{u}(g)\} = \mathfrak{u}(f)."$$

Let us compare this with the following:

> "*Definition.* If the points A, B, C, D, E, F are so positioned that the connecting lines AD, BE, CF go through the same point, then a thing is to exist which is a straight line passing through the respective points of intersection of the connecting lines AB and DE, BC and EF, CA and FD."

The cases will be regarded as entirely different, yet more precise examination will not disclose any essential logical difference. The latter definition is superfluous; in its place, one proposes a theorem, which is then proven. That it spares one a proof, however, is the inestimable advantage of creative definition. And this advantage is effortlessly achieved: one merely has to choose the word "definition", instead of the word

[1] *Vorlesungen über allgemeine Arithmetik,* First part, p. 211. Leipzig Teubner, 1885.

"theorem", as the heading. This, however, is urgently required, since the nature of the proposition might otherwise easily be misapprehended.

A different example of a creative definition is found on p. 34 of the work mentioned. There we read:

> '1. *Definition.* "If in case D_1) no magnitude of the system (I) fulfils the equation $b \circ x = a$, then it shall be satisfied by *one and only one new thing not occurring in* (I), which may be designated by $a \smile b$ since this symbol has not yet been used. Thus, one has
>
> $$b \circ (a \smile b) = (a \smile b) \circ b = a."^{1}$$
>
> Since the new objects possess no further properties, any may be attributed to them at will, provided they do not contradict each other.'

The creation, thus, takes place in distinct steps. After the first, the thing indeed is there, but it is, so to say, stark naked, lacking the most essential properties, which are attributed to it only by further acts of creation, whereupon it may be greeted as the lucky bearer of these properties. To be sure, this creative power is constrained by the addition that these properties must not contradict each other; a self-evident yet highly consequential restriction. How is it to be recognised that properties do not contradict each other? There seems to be no other criterion than to find the properties in question in one and the same object. In that case, however, the creative power that many mathematicians award themselves is as good as worthless. For before performing the act of creation, they now have to prove that the properties that they want to attribute to the object to be created, or which is already created, do not contradict each other; and this they can do, it seems, only by proving that there is an object which has all these properties. If they can do that, there is no need to create such an object in the first place.

§144. Or is there perhaps a different way to prove the freedom from contradiction? If there were, this would be of the highest significance for all mathematicians who ascribe the power of creation to themselves. And yet hardly anyone seems concerned to find such a method of proof. Why not? Probably because of the view that it is superfluous to prove freedom from contradiction since any contradiction would surely

[1] Of $\circ$ it is said on p. 26: "The connection $\circ$ is labelled thesis." One might infer from the definite article that the sign '$\circ$' has a determined reference. This, however, is not the case: it is supposed merely to indicate a connection. How, though, "connection" and "result of a connection" are to be understood is not said.

be noticed immediately. How nice if it were so! How easy all proofs would then be! The proof of the Pythagorean theorem would then go as follows:

> "Assume the square of the hypotenuse is not of equal area with the squares of the two other sides taken together; then there would be a contradiction between this assumption and the familiar axioms of geometry. Therefore, our assumption is false, and the square of the hypotenuse is of an area exactly equal to the squares of the two other sides taken together."

The law of reciprocity for quadratic residues could be proven just as easily:

> "Let p and q be prime numbers of which at least one is congruent to 1 modulo 4, and let p be a quadratic residue of q. If we now assume that q is not a quadratic residue of p, then that would obviously involve a contradiction with our premisses and the familiar basic laws of arithmetic—anyone who does not see it just does not count. Therefore, our assumption is false and q has to be the quadratic residue of p."

Absolutely any proof could be conducted following this pattern. Unfortunately, the method is too easy to be acceptable. Surely, we see that not every contradiction lies in plain view. Moreover, we lack a sure criterion for the cases where from the non-obviousness of a contradiction we may infer its absence. In these circumstances, the presumed creative power of the mathematicians has to be regarded as worthless, since, in exactly those cases where it would be worthwhile, its exercise is tied to conditions which, as it seems, cannot be fulfilled. Incidentally, how is it known that avoiding contradiction is the only constraint that creation has to obey?

§145. Stolz, like Thomae, terms his conception formal. So it may not be superflous to draw attention to the great difference between the two theories. Where Stolz creates a new—in any case non-sensory—thing, which he labels with a sign, Thomae introduces a new kind of figure together with associated rules. So, Stolz speaks of a thing designated by $\mathfrak{u}(f):\mathfrak{u}(g)$, and likewise of a thing that can be designated by $a \smile b$. We would have enclosed these signs in quotation marks to make clear that we are speaking just of the signs and not of their reference. Everywhere else Stolz distinguishes between the sign and what is designated as sharply as we do; and it does not even occur to him to put the signs themselves forward as the proper objects of arithmetic. Stolz's arithmetic is contentual, despite his use of the word "formal". Owing to the similarity in form, these differences in the matter are easily overlooked.

In fact, Thomae's theory of the arithmetical game is a completely different science from the arithmetic of Stolz. No proposition, even if it had exactly the same wording, has the same sense in Thomae and in Stolz; for the former, it concerns physical objects, namely the figures, and the rules arbitrarily laid down for their manipulation; for the latter, it is supposed to concern non-sensory objects. These, clearly, are fundamentally different matters, whether the numbers are figures for whose manipulation rules are laid down, or whether the numbers are references of number-signs and can be created. In both cases we encounter seemingly insurmountable difficulties. For Thomae, they consist in recognising whether the new rules can conflict with the old ones and of settling such conflict; for Stolz, they consist in proving that no contradiction obtains amongst the properties of the thing to be created, where the properties of the things that are already there will also come into consideration more often than not. In addition, there is the doubt whether creation is possible at all.

Dedekind's conception of creation agrees with that of Stolz; for him too the numbers are not signs but the reference of the number-signs. G. Cantor too is probably to be considered as belonging to this group, although his conception is less clearly articulated.[1]

§146. It has thus become plausible that creating proper is not available to the mathematician, or at least, that it is tied to conditions that make it worthless. Against this, it could be pointed out that in the first volume (§3, §9, §10) we ourselves created new objects, namely value-ranges. What did we in fact do there? Or to begin with: what did we not do? We did not list properties and then say: we create a thing that has these properties. Rather, we said: if one function (of first-level with one argument) and a second function are so constituted that both always have the same value for the same argument, then one may say instead: the value-range of the first function is the same as the value-range of the second. We then recognise something in common to both functions and this we call the value-range both of the first function and of the second function. That we have the right so to acknowledge what is common, and that, accordingly, we can convert the generality of an equality into an equality (identity), must be regarded as a basic law of logic. This conversion is not to be taken as a definition; neither the words "the same", nor the equality-sign, nor the expression

[1] It is hard to say what standpoint H. Hankel adopts in his *Theorie der complexen Zahlensysteme* (Leipzig, 1867), since he makes opposing claims. Probably he has not sharply distinguished between sign and what is designated.

"value-range", nor a combination of signs like '$\grave{\varepsilon}\Phi(\varepsilon)$', nor both at the same time, are thereby explained. For the proposition

"The value-range of the first function is the same as that of the second"

is composite and contains the expression "the same" as a component, which has to be considered as completely known. In the same way, the sign '$\grave{\varepsilon}\Phi(\varepsilon) = \grave{\alpha}\Psi(\alpha)$' is composite and contains the already known equality-sign as a component. Thus, if we wanted to regard our stipulation in I, §3, as definition, then this would, indeed, be an offense against our second principle of definition.[1]

§147. It is certainly clear that the mentioned possibility of conversion has, in fact, always been made use of; it is just that the coinciding is predicated of the functions themselves, rather than of the value-ranges. If a first function always has the same value for the same argument as a second, it is customary to say: "The first function is the same as the second" or "Both functions coincide". Even though this expression is different from our own, still, the generality of an equality is here converted into an equality (identity).[2]

When logicians have long spoken of the extension of a concept and mathematicians have spoken of sets, classes, and manifolds, then such a conversion forms the basis of this too; for, one may well take it that what mathematicians call a set, etc., is really nothing but the extension of a concept, even if they are not always clearly aware of this.

We are thus not really doing anything new by means of this conversion; but we do it in full awareness and by appealing to a basic law of logic. And what we do in this way is completely different from the arbitrary, lawless creation of numbers by many mathematicians.

---

[1] In general, we should not regard the stipulations about the primitive signs in the first volume as definitions. Only what is logically composite can be defined; what is simple can only be pointed to.

[2] Equally, only very few mathematicians would think twice about expressing by '$f = g$' the circumstance that $f(\xi)$ always has the same value for the same argument as the function $g(\xi)$. However, this embodies a mistake which springs from a faulty conception of the nature of a function. An isolated function-letter without an argument-place is surely a complete aberration, just as an isolated function-sign like '*sin*' is an aberration. For what characterises a function, in contrast to an object, is just this unsaturatedness, that it requires completion by an argument, and this must feature in the notation too. The impermissibility of a notation like '$f = g$' emerges from the fact that in particular cases it immediately falters. If we take, for instance, $\xi^2 - 1$ for $f(\xi)$, and $(\xi - 1) . (\xi + 1)$ for $g(\xi)$, then it is striking that nothing can be written to correspond to the equation '$f = g$'. When, however, the notation is correct, it must always be possible to make such a transition from the general to the particular in the symbolism. Even though the notation '$f = g$' thus cannot be accepted as correct, it is apparent nonetheless that mathematicians have made use of the possibility of our conversion.

If there are logical objects at all—and the objects of arithmetic are such—then there must also be a means to grasp them, to recognise them. The basic law of logic which permits the transformation of the generality of an equality into an equality serves for this purpose. Without such a means, a scientific foundation of arithmetic would be impossible. For us it serves the purposes that other mathematicians intend to achieve by the creation of new numbers. Our hope is thus that from the eight functions whose names are listed in I, §31, we can develop, as from one seed, the whole wealth of objects and functions that mathematics deals with. Can our procedure be called a creation? The discussion of this question can easily degenerate into a quarrel about words. In any case, our creation, if one wishes so to call it, is not unconstrained and arbitrary, but rather the way of proceeding, and its permissibility, is settled once and for all. And with this, all the difficulties and concerns that otherwise put into question the logical possibility of creation vanish; and by means of our value-ranges we may hope to achieve everything that these other approaches fall short of.

e) *Weierstrass's theory*

§148. It is to be expected that we would also subject the views of so outstanding a mathematician as Weierstrass to a more searching examination, especially since—in contrast to many others—he regards the foundation of arithmetic as worthy of particular attention and his statements on the topic are widely admired for their clarity. But this encounters peculiar difficulties. For the most part we have to rely on indirect sources[1] that deviate from each other in their manner of expression. Perhaps Weierstrass did not always have exactly the same opinion; perhaps some points have not been properly understood by his students. In these circumstances, we have to try to glean the genuine Weierstrassian theory from different expositions, wherein errors, of course, may easily occur.

§149. First of all, it is to be applauded that Weierstrass aims to provide a deeper foundation than most mathematicians. He begins, as we do, with the cardinal numbers.

---

[1] Kossak, *Die Elemente der Arithmetik* (*Programm des Friedrichs-Werderschen Gymnasiums*, Berlin, 1872);

Biermann, *Theorie der analytischen Functionen* (Leipzig 1887);

Handwritten *Collegienhefte*.[s]

What G. Cantor in *Math. Annalen* XXI, p. 565, portraits as Weierstrass's way of defining the irrational numbers seems heavily coloured by Cantor's hand and will likely be pressured by the same difficulties as Cantor's own theory. For example, here too there is talk of numbers that are *to be defined*.

But immediately we are astonished that he found nothing worthy of consideration in what others before him have thought about the subject, and that he has failed to see any of the cliffs that threaten here.

Comparison of the writings mentioned above leaves hardly any doubt that Weierstrass had adopted the childlike ginger-biscuits viewpoint, or at least that he thought he did; for it is foreseeable that he had to be forced out of it time and again by the nature of the subject matter itself. To the question of the nature of cardinal numbers, we are given answers such as "a series of kindred things", "an object consisting of elements of a single kind"; in brief: according to Weierstrass, a heap of ginger biscuits is a number.[1] If a man who had never thought about the matter was woken from his sleep with the question, "What is a number?", then, in his initial confusion, he would be likely to produce expressions similar to those produced by Weierstrass: "set", "heap", "series of things", "object consisting of kindred parts", etc.; and whether he were to add the word "kindred" or omit it would be inessential since nothing would be further specified thereby anyway.

§150. Here both of the possible principal errors have been committed. The first consists in the confusion of a number with its bearer or substrate, similar to the confusion of *color* with *pigmentum.* The second consists in taking as the bearer of a number neither a concept nor the extension of a concept, but what is designated by the words, "aggregate", "series of things", "object that consists of kindred parts". The difference is that an aggregate consists of objects which are held together by relations, and which can be called parts of the aggregate. With the destruction of the parts, the whole is also destroyed. In contrast, what constitutes a concept—or its extension—is not the objects that fall under it, but its characteristic marks; these are the properties that an object must have in order to fall under the concept. Empty concepts are possible, empty aggregates are an aberration. It is determined by a concept which objects fall under it; it is not determined by an aggregate what should count as its parts, whether, for example, in a regiment they are the individual soldiers, the companies, or the battalions; or whether, in a chair, they are the atoms, the molecules, or the artificially joined pieces of wood.

It is foreseeable that this theory has to founder immediately when it comes to multiplication, and that the numbers proper, since they are not what Weierstrass explains

[1] Kossak and Biermann take the issue into the psychological by using expressions such as "idea", "abstracting", "fixing"; this is not helpful to the theory and surely not in the spirit of Weierstrass; at least I find nothing of this kind in the *Collegienhefte* with which I am familiar.

as numerical magnitudes, will have to be smuggled in somehow; and this is then effected with complete naïvity by means of expressions like, "it depends on the set", "how often", "of equal cardinal number".[1] Expressions like "a-times" occur, which is completely nonsensical if one understands a Weierstrassian numerical magnitude by 'a'—a railway train, for instance.

§151. In general, a schism runs through the Weierstrassian theory, a struggle between the explanations expressly stated and what the nature of the subject matter demands. This schism is exhibited in the conception and in the use of the word "unity", the equality-sign, and the plus-sign. According to the explanations, the number has to consist of units, while the nature of the subject matter pushes for a single unit, at least when the number is not complex. So, we see a struggle between singular and plural in the word "unit", and a change in standpoint. Sometimes "unit" is used as concept word (*nomen appellativum*) co-referential with, say, "element", sometimes as proper name (*nomen proprium*) co-referential with "One". It is clear that arithmetic has no use for different Ones, but only for *the* number One. The linguistic abomination of the plural is here a suitable garb for its factual impossibility. According to Weierstrass equality is not identity, but it is identity that the nature of the subject matter demands.

The value or worth of an aggregate or of a number is distinguished from the aggregate itself, and is thus apparently intended as the number proper. This is also a way in which the latter is smuggled in; nowhere is it said what the value or the worth is. There are passages according to which the value of a numerical magnitude remains unaltered after certain changes, in such a way that we can infer that equal numbers (numerical magnitudes) have the same value. Now we ask: to what, then, does a number-sign, such as '2', actually refer according to Weierstrass? To a Weierstrassian numerical magnitude, e.g., the system consisting of Earth and Moon; or to the value, the worth of such an aggregate? In the latter case, the equality-sign in '$1 + 1 = 2$' would have to be viewed as the identity-sign, in contradiction with Weierstrass's explanation.

§152. Further, the plus-sign has to vary in reference, depending on whether it stands between signs of Weierstrassian numerical magnitudes (of railway trains, rows of books) or between signs for values of such numerical magnitudes. The explanation only fits the first case, which, however, for arithmetic cannot come under consideration. According to that explanation,

---

[1] "*Two numerical magnitudes of the new kind are equal* if they can be transformed in such a way that they both contain the same elements and each of the latter are equal in cardinal number." Biermann, *op. cit.*, §4, similarly Kossak, *op. cit.*, p. 21. So, *numerical magnitudes* contain elements that are equal in *cardinal number*. Why not equal in numerical magnitude?

'♁ + ☾'

should designate the system, or aggregate, consisting of Earth and Moon, while

'☾ + ☾'

if it had a reference at all, could only designate the Moon; and likewise '1 + 1' could only refer to 1. This is obviously useless for arithmetic.

This struggle between the requirements of arithmetic and Weierstrass's theory also generates the miracle of objects that occur repeatedly, which, incidentally, can be observed in other mathematical authors too. If only these gentlemen would, to begin with, attempt to occur repeatedly themselves! Has any one of them ever seen a repeatedly occurring grain of sand? Is this not perhaps merely a loose manner of expression? Not, in any case, a completely innocent one! One could attempt to replace it with a precise one, but one will thereby withdraw a mainstay of Weierstrass's theory.

§153. Alongside this—needless to say, never clearly stated—conception, whereby a number-sign designates the value of a numerical magnitude, another can be found. According to this, it designates a numerical magnitude itself, only of such a kind as to consist, not of concrete, but of abstract units, or indeed of a single such unit, namely the only abstract unit there is, or One. Just as a row of books consists of books, the number 3 consists, in this case, of abstract units, or better it consists of the—of course repeatedly occurring—One. Needless to say, we are not told what the latter is. Probably it is so abstract that, in order to think it, one should think nothing at all. Only a little less difficult is even the abstract cow, from which, doubtless, one may form a herd of cows.[1] As cow, it is of course not yet abstract enough to make '50' seem suitable as a sign for a herd consisting of it. For that, an even greater abstraction would be required.

By now we already recognise the sources of all these muddles in the two principal errors previously mentioned.

Three conceptions of number (cardinal number) are thus to be distinguished in Weierstrass:

1) a number is an aggregate of concrete things (ginger biscuits, railway wagons, books);
2) a number is a property (value, worth) of such an aggregate;
3) a number is an aggregate of abstract things or of one single repeatedly occurring abstract thing.

[1] The milk that we can obtain from it will leave nothing to be desired in terms of its abstractness. See the author's *Grundlagen der Arithmetik* (Breslau, Koebner, 1884) and *Die Zahlen des Herrn Schubert* (Jena, Pohle, 1899).

The students of Weierstrass are in the convenient positions of always being able to select from the three standpoints the one which best suits the demands of the moment. If there were only the second, then a confusion between a cardinal number and its bearer would be avoided; but it is exactly this conception that is the least clearly articulated, and the statements concerning it are clearly forced only by the pressures of the subject matter itself.

§154. This is the foundation on which Weierstrass aims to build his theory of the higher numbers. That it is shaky is obvious. And thus his theoretical edifice concerning Zero, and the negative, rational and irrational numbers, displays worrying cracks. Immediately when Zero is introduced, the equality-sign and the plus-sign are used in a case to which his explanations are not applicable, so that, strictly speaking, the equation '$(a - a) + b = b$' does not have a sense for Weierstrass. No more does the expression "a number which, added to a, results in Zero" have a sense for Weierstrass, since his explanation of addition does not apply to this case.

Transformations of a numerical magnitude are mentioned in which either many elements are replaced by a single one or, conversely, one element may be represented by many; but nowhere is it said how one is to recognise the possibility of such a replacement, or in what respect such a representation may occur. It is also said that an element is equivalent to many others; but how one is supposed to recognise the equivalence remains obscure.

§155. On what, then, for Weierstrass, is a recognition of the negative, rational and irrational numbers to be based? They are, as it seems, simply created, without any examination of the possibility of such creation. According to Kossak,[1] the legitimation of the negative numbers is based purely on their being defined. What this is probably meant to say is that one merely has to put forward a concept in order to be certain that something falls under it.[2] Now, we have long recognised that numbers cannot be physical objects, but only logical ones. But that there are these objects also has to be demonstrated, and to accomplish this it is not sufficient to supply a concept under which they fall. In this respect, there is no difference between physical and logical objects. Although Weierstrass gave examples in his lectures (income and expenditure, line segments), they serve, as it seems, merely to elucidate and not to justify.

---

[1] *Op. cit.*, p. 17.

[2] One may here compare Kant's critique of the ontological argument for the existence of God. *Kritik der reinen Vernunft*, ed. Hartenstein, p. 405.

Weierstrass regards the sum of infinitely many positive summands not as the limit of a sum, but as a sum. That there is this sum seems to him as certain as for the case of finitely many summands; and yet it does not agree with his explanation of addition. It is only that he regards it as necessary specifically to explain equality in this case (violation of our first principle of definition), and thereby arrives at finitude. This is connected to the fact that the word "sum" and the plus-sign as well as the equality-sign have no fixed references, as we have seen above.

A more thoroughgoing criticism of Weierstrass's foundation for the irrational numbers, now that we have shown its basis to be completely unsafe, is unnecessary.

f) Review and Prospects

§156. By way of summary, let us now review the attempts to introduce the higher numbers, and ask what benefit we can draw from this for our own efforts.

We distinguish two main directions in which these attempts have moved: the formal and the contentual. The former takes numbers to be certain figures produced by writing, which are treated in accordance with arbitrary rules. In a different context, these figures may indeed also be signs that refer to something; but the formal arithmetician completely abstains from this. For contentual arithmetic, these figures are mere signs, number-signs, external auxiliary tools, standing for the proper objects. It seems that it is the inevitable fate of formal arithmetic to flow time and again into the channel of contentual arithmetic and that it faces, amongst others, the following difficulty. When introducing figures of a new kind, it must, at the same time, lay down new rules for their manipulation, and indeed prohibitive as well as permissive rules. Next it would have to be shown that these new rules can come into conflict neither amongst themselves nor with the old ones. This task has apparently not even been tackled in any serious way, let alone been resolved.

For us, only a contentual arithmetic can come into consideration. But the accounts attempted on this basis have also remained unsuccessful, as we have seen, at least insofar as they aim to remain purely arithmetical.

Either one does not distinguish between concept and object and so believes that by presenting a concept, one thereby also obtains an object with the desired properties; or one acknowledges that a concept can also be empty, and demands of it that it be free from contradiction. As obvious as the necessity of this condition is, its sufficiency is no less questionable. That there is no contradiction—some may now claim—is immediately obvious since if one were present, it would have to be recognised straight away. This is easily refuted by mention of indirect proofs. That there is no contradiction present is thus something to be proven. Now, admittedly we have the proposition that a concept is free from contradiction when an object falls under it; but this proposition is of no use in our very case. So unless a completely new principle is found whereby we can prove freedom from contradiction, we cannot advance any further on this path.[1]

§157. Nevertheless, this consideration of the futile attempts has not been completely unfruitful. We have been reminded of our transformation of the generality of an equality into an equality of value-ranges that promises to accomplish what the creative definitions of other mathematicians are not capable of. We have understood the real numbers as magnitude-ratios and so excluded formal arithmetic in the sense described above. Thereby we have indicated magnitudes as the objects between which such a ratio obtains.[2]

Since the cardinal numbers are not ratios, we have to distinguish them from the positive whole numbers. Therefore it is not possible to extend the domain of the cardinal numbers to that of the real numbers; they are simply completely separate domains. The cardinal numbers answer the question: "How many objects of a certain kind are there?" while the real numbers can be considered as measuring numbers that state how large a magnitude is in comparison to a unit. Some readers might have wondered about

---

[1] Like us, L. Kronecker seems to have regarded the attempts made so far to provide an arithmetical foundation for the irrationals as misconceived; for him, irrational algebraic magnitudes are foreign to arithmetic proper; accordingly, he proposed to keep them separate and amongst the numbers recognises only the positive whole numbers as objects of arithmetic. Every attempt to provide a purely arithmetical foundation for the theory of the real numbers will ultimately, it seems, amount to tracing everything back to the cardinal numbers, so that every arithmetical proposition can be accredited to the theory of the cardinal numbers. In the same sense one may also assign every proposition of the theory of algebraic areas and curves to the theory of points, straight lines and planes. It makes a difference, however, whether one takes the expression, "curve of fourth order", as merely dispensable, or whether one proposes to ban it from geometry altogether on the ground that the concept thereby designated is foreign to geometry. And Kronecker seems to regard the irrational numbers not merely as dispensable but rather as downright non-arithmetical, so that any proofs that involve them rely on something that clouds the purity of arithmetic. We may well approve of the principle that arithmetic should not, if at all possible, make use in its proofs of any grounds borrowed from geometry or any other alien science. The question arises, however, whether it may not yet be possible to define the irrational numbers purely arithmetically; and we will attempt to open up such a way.

Incidentally, the theory of the cardinal numbers on which Kornecker makes everything rest is untenable. According to him, a cardinal number is a whole consisting of number-signs. That leads to the absurd consequence that a people who used different number-signs would also have different cardinal numbers, so that their arithmetic would deal with objects completely different from ours and so would be a completely different science. Moreover, there would be only finitely many cardinal numbers, since only finitely many number-signs have, obviously, so far been written down, and only finitely many wholes can be based on them. (Compare *Ueber den Zahlbegriff* by Leopold Kronecker, *Philosoph. Aufsätze*, dedicated to E. Zeller on the occasion of the 50th anniversary of the doctorate.)

[2] Here we are in agreement with Newton.

our formula '$a \frown (b \frown \mathfrak{f})$', expecting '$a + 1 = b$' in its place; however the $\mathfrak{f}$-relation of a cardinal number to the one immediately following it is distinct from the relation $\xi + 1 = \zeta$. The former applies only to cardinal numbers, the latter applies not just to the positive whole numbers but also to others, so that, following its converse, we can proceed from the positive whole numbers, past Zero, to the negative ones, while going backwards past the cardinal number $\mathbb{0}$ is not possible.[1] For this reason also, we distinguish between the cardinal numbers $\mathbb{0}$ and $\mathbb{1}$ and the numbers 0 and 1.

§158. At first it may seem that we will have to rely on geometry; yet by taking a real number as a magnitude-ratio, we have repudiated the idea that a real number is, for instance, a line segment, and that a number-sign refers to a line segment. One does not always clearly distinguish between a line segment and the measuring number which belongs to it in ratio to some unit line segment. Thus, one speaks, for example, of a line segment a and in what follows understands by 'a' the measuring number. The resulting confusion gave rise to the opinion that the number-sign refers—or could at least refer—to a line segment. To be sure, it refers to something, but not to anything geometrical. Rather, we have exactly the same magnitude-ratio occurring in the case of periods of times, masses, light intensities, etc., as in the case of line segments. Thus, the real number detaches itself from specific kinds of magnitude and, as it were, floats above them. And this is why it does not seem appropriate to focus our considerations too closely on geometric constructions. They may well be used to facilitate the understanding, but one has to beware of making anything rest on them as foundation. For if arithmetical propositions can be proven without recourse to geometrical axioms, then they must be. Otherwise one unnecessarily denies the autonomy of arithmetic and its logical nature.

Someone may, perhaps, describe this treatment of arithmetic too as formal, but in that case the word is not used in the sense presented above. It would then mark the purely logical nature of arithmetic, but without any claim that the number-signs are contentless figures that are treated according to arbitrary rules. Rather the rules here follow necessarily from the reference of the signs, and this reference is to the proper objects of arithmetic; what is arbitrary is only the notation.

§159. So the path to be taken here steers between the old approach, still preferred by H. Hankel, of a geometrical foundation for the theory of the irrational numbers and

---

[1] Thus, we could not have introduced the plus-sign at an earlier stage without explaining it incompletely and piecemeal, thereby violating our first principle of definition.

the approaches pursued in recent times. From the former we retain the conception of a real number as a magnitude-ratio, or measuring number, but separate it from geometry and indeed from all specific kinds of magnitudes, thereby coming closer to the more recent efforts. At the same time, however, we avoid the emerging problems of the latter approaches, that either measurement does not feature at all, or that it features without any internal connection grounded in the nature of the number itself, but is merely tacked on externally, from which it follows that we would, strictly speaking, have to state specifically for each kind of magnitude how it should be measured, and how a number is thereby obtained. Any general criteria for where the numbers can be used as measuring numbers and what shape their application will then take, are here entirely lacking.

So we can hope, on the one hand, not to let slip away from us the ways in which arithmetic is applied in specific areas of knowledge, without, on the other hand, contaminating arithmetic with the objects, concepts, relations borrowed from these sciences and endangering its special nature and autonomy. One may surely expect arithmetic to present the ways in which arithmetic is applied, even though the application itself is not its subject matter.

The attempt will have to show whether our plan is executable. Here the following concern may arise. If the positive square root of 2 is a magnitude-ratio, then in order to define it, it seems necessary to supply magnitudes which have this ratio to each other. But where should we obtain them from, if appeal to geometrical or physical magnitudes is forbidden? And yet we do indeed require such a magnitude-ratio, since otherwise even the sign '$\sqrt{2}$' may not be used.

Before we attempt to eradicate this concern, we have to clarify the reference of the word "magnitude".

g) Magnitude

§160. The word "magnitude" ("quantity") is, for some mathematicians, used to refer to more or less the same as "real number" or simply "number". This manner of use is probably connected to the fact that a measuring number is not always kept distinct from the magnitude which it determines with respect to a unit (a measure); it cannot come into consideration for us here.

What a magnitude is surely has never been stated satisfactorily. Going through the attempts to explain it, one often comes across the word "similar" or something like it. It is demanded of magnitudes that one can compare, add, substract amongst ones that are similar, or that one can analyse a magnitude into parts that are similar to it.[1] Clearly, nothing is said by the word "similar"; for things can be similar in one

[1] Otto Stolz, *Vorlesungen über allgemeine Arithmetik* (Leipzig, Teubner, 1885), introduction.

respect that are dissimilar in another. Thus the question, whether an object is similar to another, cannot be answered by "yes" or "no": what is missing is the first logical requirement, the sharp boundary.

Others define magnitude with the words "greater" and "less", or "enlarge" and "reduce", by which, however, nothing is gained; for it remains unexplained in what the relation of greater than, or the task of enlargement, consists. Likewise, it is of no avail to use the words "addition", "sum", "multiplication", unless they are sufficiently explained.

When H. Hankel says[1]

> "By the sum of two magnitudes, *a* and *b*, one understands a new magnitude, resulting from their synthesis",

then he uses a likewise unexplained word "synthesis", and it remains doubtful whether there might not be different syntheses for two given objects.

If one has explained words for a specific context, one must not then presume to connect a sense with them in other contexts. One goes, as it seems, in circles, repeatedly explaining a word by another which, in turn, requires explanation, without thereby getting any closer to the heart of the matter.

§161. The reason for these failures lies in asking the wrong question. There are many different kinds of magnitudes: lengths, angles, periods of time, masses, temperatures, etc., and it will scarcely be possible to say how objects that belong to these kinds of magnitude differ from other objects that do not belong to any kind of magnitude. Moreover, little would be gained thereby; for we still lack any way of recognising which of these magnitudes belong to the same domain of magnitudes.

Instead of asking which properties an object must have in order to be a magnitude, one needs to ask: how must a concept be constituted in order for its extension to be a domain of magnitudes?[2] Henceforth, for brevity, we will speak of

---

[1] *Theorie der complexen Zahlensysteme* (Leipzig, Voss, 1867) §14.

[2] An attempt in this direction is made by O. Stolz, writing in the work cited above: "... a magnitude concept will be a concept of such a kind that each two things falling under it are explained as either equal or unequal." Here, indeed, an attempt is made to state how the concept is constituted, as requested above; however, in the immediately following proposition,

> 'In other words: "we call magnitude each thing that is intended to be set equal or unequal to another."'

the attempt is abandoned once again. If being explained were what the issue depended on, then a concept would become a magnitude concept from a specific moment onwards and not before. In order to decide the question whether a concept was, at a certain time, a magnitude concept, some difficult historical research might have to be undertaken. And there is this additional point. Either the reference of the words "equal" and "unequal" is already known in advance; in which case it cannot then be a matter of explanation whether any things are equal or unequal; or these words are not previously "*class*" rather than "extension of a concept". The question can then also be framed like this: what properties must a class have in order to be a domain of magnitudes? A thing is a magnitude not in itself but only insofar it belongs, with other objects, to a class that is a domain of magnitudes.

§162. Since our initial concern is to obtain a foundation just for the theory of the real numbers, we will simplify the matter by leaving absolute magnitudes out of consideration, focusing only on those domains of magnitudes in which there is an opposition corresponding to that between the positives and negatives for the measuring numbers. Here a remark by Gauss (*Werke*, vol. II, p. 176) may prove helpful:

> "Positive and negative numbers can find application only where that which is counted has an opposite, so that thinking them in union amounts to annihilation. Properly speaking, this presupposition is met only when what is counted are Relations[t] between any two objects and not substances (objects thinkable in themselves). It is thereby postulated that these objects are ordered in a specific way in a series, e.g., $A, B, C, \ldots$, and that the Relation of A to B can be regarded as equal to the Relation of B to C, etc. Now, nothing more here pertains to the concept of opposing than the exchange of the members of the Relation in such a way that if the Relation (or the transition) from A to B counts as $+1$, then the Relation from B to A must be represented as -1. Thus, insofar as a series is unlimited on both sides, each real number represents the Relation between a member which is arbitrarily chosen as a beginning and a specific member of the series."[u]

In essence, we can agree with this thought, omitting however the restriction to whole numbers and replacing "what is counted" by "what is measured". It seems that Gauss thinks of a Relation as determined by the objects amongst which it holds and requires

known: in that case, one is free to take any randomly formed words, e.g., "azig", "bezig", in their place, and we have seen that no further specification can be achieved in this way. One might think a third case is possible, namely that up until the relevant time the words "equal" and "unequal" are neither completely known, nor entirely unknown, to the extent that certain properties of their reference are known. However, before one may talk at all about the references of these words, they have to be completely determined; only then may one ask whether these references have the desired properties. Yet such explanations cannot in any way be decisive for whether any concept is a magnitude concept. Either a concept is a magnitude concept, in which case that it is so is independent of any explanation of "equal" or "unequal"; or it is not a magnitude concept, in which case that too is independent of such an explanation. All it does is to establish a connection between these words and their intended reference; its effect does not extend beyond that. We need, simply, to be entirely clear that an explanation can never change anything in the subject matter itself, that it merely concerns our notation or designation.

a postulate about the equality of Relations. We however regard a relation as definable without appeal to specific objects standing in it, so that, in general, to acknowledge a relation is not yet to make a statement that there are objects standing in it. If, then, a relation is given in which A stands to B, then it is thereby simultaneously decided whether B stands to C, and C to D, in this same relation, and an ordering of objects in a series is thereby provided by itself. Of course, not every relation will produce a series proceeding into the infinite; not every relation, therefore, is suitable for our purpose. Further examination will have to show what restrictions are necessary.

Let us, for short, say "*Relation*"[v] instead of "extension of a relation", then we can say: the magnitudes considered by us are Relations. Accordingly, the ratios of magnitudes, or real numbers, will be regarded as Relations on Relations. Our domains of magnitudes are classes of Relations, namely extensions of concepts that are subordinate to the concept *Relation*. Reversal of the signs will be analogous to taking the converse of relations (K and $\mathfrak{X}K$). To the addition of measuring numbers will correspond the composition of Relations ($K \smile \Pi$). Thus the sign '$\mathfrak{X}$' will be comparable to the minus-sign and the sign '$\smile$' to the addition sign; the formula '$A \smile \mathfrak{X}B$' will correspond to '$a - b$', and '$A \smile \mathfrak{X}A$' to the sign for Zero.

§163. Let us, for example, take as objects the points on a straight line. Between them occur distance relations. Point B is, for example, separated from A by a certain line segment in a certain direction (say, to the right). If the point D is equally far from C in the same direction, then C stands to D in the same distance relation as A to B. Obviously, the distance Relation from B to A is the converse of that from A to B. Further, the converse of the distance Relation corresponds to the converse of the direction. It is moreover immediately obvious that the composition of such distance Relations matches the geometrical addition of the line segments.

§164. We can now approach the question raised earlier (§159) from where we obtain the magnitudes whose ratios are real numbers. They will be Relations; and they must not be empty, i.e., they must not be extensions of relations in which no objects stand to each other. For such relations have the same extension; there is only one empty Relation. We could not define any real number with it. If q is the empty Relation, then $\mathfrak{X}q$ is the same; likewise $q \smile \mathfrak{X}q$. Also the composition of Relations on our domain of magnitudes cannot result in the empty Relation; but that would happen if there were no object to which some object stood in the first Relation and which also stood to some object in the second Relation.

We thus require a class of objects that stand to each other in the Relations of our domain of magnitudes, and, in particular, this class has to contain infinitely many objects. Now, the concept *finite cardinal number* has an infinite cardinal number which we have called Endlos; this infinity, however, will not yet suffice. Call the extension of a concept which is subordinate to the concept *finite cardinal number* a *class of finite cardinal numbers*, then the concept *class of finite cardinal numbers* has an infinite cardinal number greater than Endlos; i.e., the concept *finite cardinal number* can be mapped into the concept *class of finite cardinal numbers*, but the latter, conversely, cannot be mapped into the former.

Now it would have be to shown that Relations hold between the classes of finite cardinal numbers which may be understood as members of a domain of magnitudes. The matter of course will turn out somewhat differently, as we will see next.

Let us for the moment assume knowledge of the irrational numbers. Every positive number a can be designated in this way:

$$`r + \sum_{k=1}^{k=\infty} \left\{ \frac{1}{2^{n_k}} \right\}',$$

where 'r' is a positive whole number or 0, and 'n_1', 'n_2', etc., are to be understood as positive whole numbers whose cardinal number we can take to be infinite. In this way, to each positive rational or irrational number a belongs a pair whose first member (r) is a positive whole number or 0 and whose second member is a class of positive whole numbers (class of n_k). In place of the whole numbers we can also take cardinal numbers, so that now to every positive real number belongs a pair whose first member is a cardinal number and whose second member is a class of cardinal numbers that does not contain $\mathfrak{0}$. Now, if a, b, and c are positive numbers and $a + b = c$, then for each b a relation obtains between the pairs that belong to a and to c. And this relation can be defined without drawing on the real numbers a, b, c, and thus without presupposing knowledge of the real numbers. So we have relations each of which is again characterised by a pair (that which belongs to b). To this the converses are to be added. A single-valued correspondence obtains between the extensions of these relations (these Relations) and the positive and negative real numbers. To the addition of the numbers b and b' corresponds the composition of the respective Relations. The class of these Relations is now a domain that suffices for our plan. These hints may suffice for the time being to dispel doubts regarding the feasibility of this plan. I do not mean to say that we will follow this path in every detail. Now I just wish to draw attention to two points in particular. First: neither the classes of finite cardinal numbers, nor the mentioned pairs, nor the Relations between these pairs, are irrational numbers. No more are they signs of irrational numbers; they are not formations produced by writing or print; rather, they are simply classes, pairs, Relations. Second: the mentioned relations between the pairs can be defined without presupposing as known the connection to the irrational numbers that we began with to assist understanding. Hence there must, at the outset, be no talk of the irrational numbers in the execution; rather we will start with classes of finite cardinal numbers, amongst which we will define certain relations without so much as mentioning any connection to addition of real numbers here. In this manner, we will succeed in defining the real number purely arithmetically or logically as a ratio of magnitudes that are demonstrably there, so that no doubt that there are irrational numbers can remain.

First however we must answer the question: what properties must a class of Relations have to be a domain of magnitudes?

2. The theory of magnitude

A. Propositions concerning the composition and converse of Relations in general

§165. *Analysis*

The demarcation of a domain of magnitudes results from the demand that the laws which are essential to addition, known under the names of the commutative and associative principles, must hold. The question is thus to be posed like this: what properties must a class of Relations have in order for the commutative and associative laws to hold for the composition of the Relations in it? Regarding the latter law, it turns out that it holds generally, so that no further specification is needed. In order to prove it, we require some propositions concerning the equality of Relations, which we derive first. Here we must note the following. The condition that p is a Relation can be expressed by means of the subcomponent

$$\text{‘}\dot{\alpha}\dot{\varepsilon}(\text{—}\ f(\varepsilon, \alpha)) = p\text{’}$$

or alternatively by the subcomponent

$$\text{‘}\dot{\alpha}\dot{\varepsilon}(\text{—}\ \varepsilon \frown (\alpha \frown p)) = p\text{’}$$

§166. *Construction*

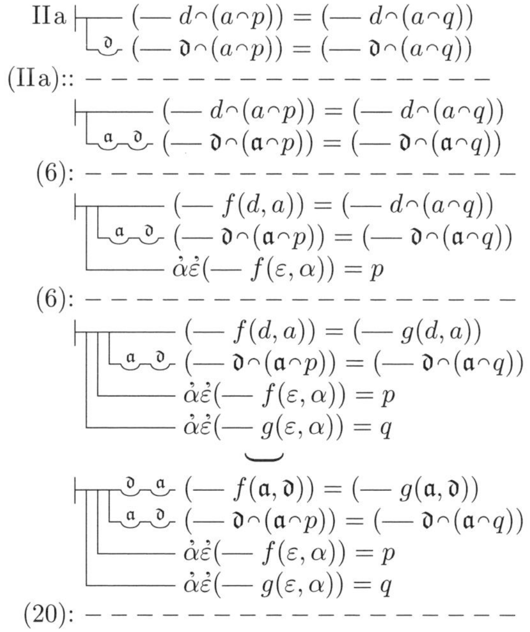

(α

(β

(γ

(δ

$$\vdash \dot{\alpha}\dot{\varepsilon}(-\, f(\varepsilon,\alpha)) = \dot{\alpha}\dot{\varepsilon}(-\, g(\varepsilon,\alpha))$$
$$\quad \mathfrak{a}\ \mathfrak{d}\ (-\, \mathfrak{d}\frown(\mathfrak{a}\frown p)) = (-\, \mathfrak{d}\frown(\mathfrak{a}\frown q))$$
$$\quad \dot{\alpha}\dot{\varepsilon}(-\, f(\varepsilon,\alpha)) = p$$
$$\quad \dot{\alpha}\dot{\varepsilon}(-\, g(\varepsilon,\alpha)) = q \qquad (\varepsilon$$

(IIIc): ———————————

$$\vdash p = \dot{\alpha}\dot{\varepsilon}(-\, g(\varepsilon,\alpha))$$
$$\quad \mathfrak{a}\ \mathfrak{d}\ (-\, \mathfrak{d}\frown(\mathfrak{a}\frown p)) = (-\, \mathfrak{d}\frown(\mathfrak{a}\frown q))$$
$$\quad \dot{\alpha}\dot{\varepsilon}(-\, f(\varepsilon,\alpha)) = p$$
$$\quad \dot{\alpha}\dot{\varepsilon}(-\, g(\varepsilon,\alpha)) = q \qquad (\zeta$$

(IIIc): ———————————

$$\vdash p = q$$
$$\quad \mathfrak{a}\ \mathfrak{d}\ (-\, \mathfrak{d}\frown(\mathfrak{a}\frown p)) = (-\, \mathfrak{d}\frown(\mathfrak{a}\frown q))$$
$$\quad \dot{\alpha}\dot{\varepsilon}(-\, f(\varepsilon,\alpha)) = p$$
$$\quad \dot{\alpha}\dot{\varepsilon}(-\, g(\varepsilon,\alpha)) = q \qquad (485$$

——— • ———

$$\mathrm{B} \vdash \dot{\alpha}\dot{\varepsilon}\left(\ \mathfrak{r}\ \ \varepsilon\frown(\mathfrak{r}\frown p) \ ;\ \mathfrak{r}\frown(\alpha\frown r)\right) = p \lrcorner r$$

(485): ———————————

$$\vdash p \lrcorner r = q$$
$$\quad \mathfrak{a}\ \mathfrak{d}\ (-\, \mathfrak{d}\frown(\mathfrak{a}\frown(p \lrcorner r))) = (-\, \mathfrak{d}\frown(\mathfrak{a}\frown q))$$
$$\quad \dot{\alpha}\dot{\varepsilon}(-\, g(\varepsilon,\alpha)) = q \qquad (486$$

——— • ———

$$\mathrm{B} \vdash \dot{\alpha}\dot{\varepsilon}\left(\ \mathfrak{r}\ \ \varepsilon\frown(\mathfrak{r}\frown s) \ ;\ \mathfrak{r}\frown(\alpha\frown t)\right) = s \lrcorner t$$

(486): ———————————

$$\vdash p \lrcorner r = s \lrcorner t$$
$$\quad \mathfrak{a}\ \mathfrak{d}\ (-\, \mathfrak{d}\frown(\mathfrak{a}\frown(p \lrcorner r))) = (-\, \mathfrak{d}\frown(\mathfrak{a}\frown(s \lrcorner t))) \qquad (487$$

——— • ———

$$5 \vdash e\frown(a\frown(q \lrcorner t))$$
$$\quad r\frown(a\frown t)$$
$$\quad e\frown(r\frown q)$$

(5): - - - - - - -

$$\vdash d\frown(a\frown(p \lrcorner (q \lrcorner t)))$$
$$\quad r\frown(a\frown t)$$
$$\quad e\frown(r\frown q)$$
$$\quad d\frown(e\frown p) \qquad (\alpha$$

×

$$\vdash d\frown(e\frown p)$$
$$\quad e\frown(r\frown q)$$
$$\quad r\frown(a\frown t)$$
$$\quad d\frown(a\frown(p \lrcorner (q \lrcorner t))) \qquad (\beta$$

$$\vdash \mathfrak{r}\ \ d\frown(\mathfrak{r}\frown p)$$
$$\quad \mathfrak{r}\frown(r\frown q)$$
$$\quad r\frown(a\frown t)$$
$$\quad d\frown(a\frown(p \lrcorner (q \lrcorner t))) \qquad (\gamma$$

(15): - - - - - - - - - - - - -

$$\vdash d\frown(r\frown(p\smile q))$$
$$r\frown(a\frown t)$$
$$d\frown(a\frown(p\smile(q\smile t))) \qquad (\delta$$

$$\vdash^{\mathfrak{r}} d\frown(\mathfrak{r}\frown(p\smile q))$$
$$\mathfrak{r}\frown(a\frown t)$$
$$d\frown(a\frown(p\smile(q\smile t))) \qquad (\varepsilon$$

(15): - - - - - - - - - - - -

$$\vdash d\frown(a\frown(p\smile q\smile t))$$
$$d\frown(a\frown(p\smile(q\smile t))) \qquad (\zeta$$

×

$$\vdash d\frown(a\frown(p\smile(q\smile t)))$$
$$d\frown(a\frown(p\smile q\smile t)) \qquad (488$$

——— • ———

$$5 \vdash d\frown(r\frown(p\smile q))$$
$$e\frown(r\frown q)$$
$$d\frown(e\frown p)$$

(5): - - - - - - - -

$$\vdash d\frown(a\frown(p\smile q\smile t))$$
$$r\frown(a\frown t)$$
$$e\frown(r\frown q)$$
$$d\frown(e\frown p) \qquad (\alpha$$

×

$$\vdash e\frown(r\frown q)$$
$$r\frown(a\frown t)$$
$$d\frown(a\frown(p\smile q\smile t))$$
$$d\frown(e\frown p) \qquad (\beta$$

$$\vdash^{\mathfrak{r}} e\frown(\mathfrak{r}\frown q)$$
$$\mathfrak{r}\frown(a\frown t)$$
$$d\frown(a\frown(p\smile q\smile t))$$
$$d\frown(e\frown p) \qquad (\gamma$$

(15): - - - - - - - - - - - -

$$\vdash e\frown(a\frown(q\smile t))$$
$$d\frown(a\frown(p\smile q\smile t))$$
$$d\frown(e\frown p) \qquad (\delta$$

×

$$\vdash d\frown(e\frown p)$$
$$e\frown(a\frown(q\smile t))$$
$$d\frown(a\frown(p\smile q\smile t)) \qquad (\varepsilon$$

$$\vdash^{\mathfrak{r}} d\frown(\mathfrak{r}\frown p)$$
$$\mathfrak{r}\frown(a\frown(q\smile t))$$
$$d\frown(a\frown(p\smile q\smile t)) \qquad (\zeta$$

(15): - - - - - - - - - - -

$$\vdash d\frown(a\frown(p\smile(q\smile t)))$$
$$d\frown(a\frown(p\smile q\smile t)) \qquad (\eta$$

×

$$\vdash d\frown(a\frown(p\smile q\smile t))$$
$$d\frown(a\frown(p\smile(q\smile t))) \qquad (\vartheta$$

(IVa): ————————————

$$\vdash [\text{—}\, d\frown(a\frown(p\smile(q\smile t)))] = [\text{—}\, d\frown(a\frown(p\smile q\smile t))]$$
$$d\frown(a\frown(p\smile(q\smile t)))$$
$$d\frown(a\frown(p\smile q\smile t)) \qquad (\iota$$

(488): ————————————————————————

$$\vdash [\text{—}\, d\frown(a\frown(p\smile(q\smile t)))] = [\text{—}\, d\frown(a\frown(p\smile q\smile t))] \qquad (\kappa$$

$$\vdash^{\mathfrak{a}\ \mathfrak{d}} [\text{—}\, \mathfrak{d}\frown(\mathfrak{a}\frown(p\smile(q\smile t)))] = [\text{—}\, \mathfrak{d}\frown(\mathfrak{a}\frown(p\smile q\smile t))] \qquad (\lambda$$

(487): ————————————————————————

$$\vdash p\smile(q\smile t) = p\smile q\smile t \qquad (489$$

(IIIa): ————————————

$$\vdash F(p\smile(q\smile t))$$
$$F(p\smile q\smile t) \qquad (490$$

——— • ———

$$489 \vdash p\smile(q\smile t) = p\smile q\smile t$$

(IIIc): ————————————

$$\vdash F(p\smile q\smile t)$$
$$F(p\smile(q\smile t)) \qquad (491$$

§167. *Analysis*

In the proposition just proven we have the associative law for the composition of Relations. The commutative law does not hold without restriction here. We first prove it for the members of a series such as

$$K,\ K \smile K,\ K \smile (K \smile K),\ K \smile (K \smile (K \smile K))\ \ldots$$

It will be desirable to have a more concise notation for the series-forming Relation associated with such a series. We thus define:

$$\Vdash \grave{\alpha}\grave{\varepsilon}(t \smile \varepsilon = \alpha) = {}_* t \qquad (\Phi$$

and we draw from it the following consequences.

§168. *Construction*

$\Phi \vdash \grave{\alpha}\grave{\varepsilon}(t \smile \varepsilon = \alpha) = {}_* t$

(6): ———————

$\vdash F(t \smile d = a)$
$\quad \llcorner F(d \frown (a \frown_* t))$ (492

———•———

$492 \vdash t \smile d = a$
$\quad \llcorner d \frown (a \frown_* t)$

(IIIc): – – – – –

$\vdash F(a)$
$\quad \llcorner F(t \smile d)$
$\quad \llcorner d \frown (a \frown_* t)$ (493

———•———

$492 \vdash t \smile d = a$
$\quad \llcorner d \frown (a \frown_* t)$

(IIIa): – – – – –

$\vdash F(t \smile d)$
$\quad \llcorner F(a)$
$\quad \llcorner d \frown (a \frown_* t)$ (494

———•———

$\Phi \vdash \grave{\alpha}\grave{\varepsilon}(t \smile \varepsilon = \alpha) = {}_* t$

(10): ———————

$\vdash F(d \frown (a \frown_* t))$
$\quad \llcorner F(t \smile d = a)$ (495

———•———

$\text{III e} \vdash t \smile d = t \smile d$

(495): ———————

$\vdash d \frown (t \smile d \frown_* t)$ (496

———•———

$\text{III c} \vdash d = a$
$\quad \llcorner t \smile e = a$
$\quad \llcorner t \smile e = d$

(492,492):: = = = =

$\vdash d = a$
$\quad \llcorner e \frown (a \frown_* t)$
$\quad \llcorner e \frown (d \frown_* t)$ (α

$\vdash^{\mathfrak{e}}\smile^{\mathfrak{d}}\smile^{\mathfrak{a}}\smile\ \mathfrak{d} = \mathfrak{a}$
$\quad \llcorner \mathfrak{e} \frown (\mathfrak{a} \frown_* t)$
$\quad \llcorner \mathfrak{e} \frown (\mathfrak{d} \frown_* t)$ (β

(16): ———————

$\vdash \mathrm{I}_* t$ (497

§169. *Analysis*

In order to prove the proposition

'$\vdash p \smile q = q \smile p$
$\quad \llcorner t \frown (p \frown \dot{\smile}_* t)$
$\quad \llcorner t \frown (q \frown \dot{\smile}_* t)$ ' (α

using (144), we need the proposition

'$\vdash p \smile a = a \smile p$
$\quad \llcorner d \frown (a \frown_* t)$
$\quad \llcorner p \smile d = d \smile p$
$\quad \llcorner t \frown (p \frown \dot{\smile}_* t)$ ' (β

or the proposition

'$\vdash p \smile (t \smile d) = t \smile d \smile p$
$\quad \llcorner p \smile d = d \smile p$
$\quad \llcorner t \frown (p \frown \dot{\smile}_* t)$ ',

which we prove using the proposition

'$\vdash p \smile t = t \smile p$
$\quad \llcorner t \frown (p \frown \dot{\smile}_* t)$ ' (δ

The latter is to be derived using (144).

§170. *Construction*

IIIh $$\vdash \begin{array}{l} t \sqcup (d \sqcup t) = t \sqcup (t \sqcup d) \\ d \sqcup t = t \sqcup d \end{array}$$

(491): ———————

$$\vdash \begin{array}{l} t \sqcup d \sqcup t = t \sqcup (t \sqcup d) \\ d \sqcup t = t \sqcup d \end{array} \qquad (\alpha$$

(493): - - - - - - - - - -

$$\vdash \begin{array}{l} a \sqcup t = t \sqcup a \\ d \frown (a \frown_* t) \\ d \sqcup t = t \sqcup d \end{array} \qquad (\beta$$

$$\vdash \overset{\mathfrak{d}}{\smile}\overset{\mathfrak{a}}{\smile} \begin{array}{l} \mathfrak{a} \sqcup t = t \sqcup \mathfrak{a} \\ \mathfrak{d} \frown (\mathfrak{a} \frown_* t) \\ \mathfrak{d} \sqcup t = t \sqcup \mathfrak{d} \end{array} \qquad (\gamma$$

(144): - - - - - - - - - -

$$\vdash \begin{array}{l} p \sqcup t = t \sqcup p \\ t \sqcup t = t \sqcup t \\ t \frown (p \frown \acute{\smile}_* t) \end{array} \qquad (\delta$$

(IIIe):: ———————

$$\vdash \begin{array}{l} p \sqcup t = t \sqcup p \\ t \frown (p \frown \acute{\smile}_* t) \end{array} \qquad (498$$

(IIIa):: - - - - - - - -

$$\vdash \begin{array}{l} F(p \sqcup t) \\ F(t \sqcup p) \\ t \frown (p \frown \acute{\smile}_* t) \end{array} \qquad (499$$

——— • ———

498 $$\vdash \begin{array}{l} p \sqcup t = t \sqcup p \\ t \frown (p \frown \acute{\smile}_* t) \end{array}$$

(IIIc): - - - - - -

$$\vdash \begin{array}{l} F(t \sqcup p) \\ F(p \sqcup t) \\ t \frown (p \frown \acute{\smile}_* t) \end{array} \qquad (500$$

——— • ———

IIIh $$\vdash \begin{array}{l} t \sqcup (p \sqcup d) = t \sqcup (d \sqcup p) \\ p \sqcup d = d \sqcup p \end{array}$$

(491): ———————

$$\vdash \begin{array}{l} t \sqcup p \sqcup d = t \sqcup (d \sqcup p) \\ p \sqcup d = d \sqcup p \end{array} \qquad (\alpha$$

(491): ———————

$$\vdash \begin{array}{l} t \sqcup p \sqcup d = t \sqcup d \sqcup p \\ p \sqcup d = d \sqcup p \end{array} \qquad (\beta$$

(499): ———————

$$\vdash \begin{array}{l} p \sqcup t \sqcup d = t \sqcup d \sqcup p \\ p \sqcup d = d \sqcup p \\ t \frown (p \frown \acute{\smile}_* t) \end{array} \qquad (\gamma$$

(490): ———————

$$\vdash \begin{array}{l} p \sqcup (t \sqcup d) = t \sqcup d \sqcup p \\ p \sqcup d = d \sqcup p \\ t \frown (p \frown \acute{\smile}_* t) \end{array} \qquad (\delta$$

(493): ———————

$$\vdash \begin{array}{l} p \sqcup a = a \sqcup p \\ d \frown (a \frown_* t) \\ p \sqcup d = d \sqcup p \\ t \frown (p \frown \acute{\smile}_* t) \end{array} \qquad (\varepsilon$$

$$\vdash \overset{\mathfrak{d}}{\smile}\overset{\mathfrak{a}}{\smile} \begin{array}{l} p \sqcup \mathfrak{a} = \mathfrak{a} \sqcup p \\ \mathfrak{d} \frown (\mathfrak{a} \frown_* t) \\ p \sqcup \mathfrak{d} = \mathfrak{d} \sqcup p \\ t \frown (p \frown \acute{\smile}_* t) \end{array} \qquad (\zeta$$

(144): - - - - - - - - - - -

$$\vdash \begin{array}{l} p \sqcup q = q \sqcup p \\ p \sqcup t = t \sqcup p \\ t \frown (p \frown \acute{\smile}_* t) \\ t \frown (q \frown \acute{\smile}_* t) \end{array} \qquad (\eta$$

(498):: - - - - - - - -

$$\vdash \begin{array}{l} p \sqcup q = q \sqcup p \\ t \frown (p \frown \acute{\smile}_* t) \\ t \frown (q \frown \acute{\smile}_* t) \end{array} \qquad (501$$

(IIIc): - - - - - - - -

$$\vdash \begin{array}{l} f(q \sqcup p) \\ f(p \sqcup q) \\ t \frown (p \frown \acute{\smile}_* t) \\ t \frown (q \frown \acute{\smile}_* t) \end{array} \qquad (502$$

Propositions demonstrating the similarity between the converse of Relations and reversing the sign

§171. *Analysis*

We first prove a proposition that says that the double converse leaves a double value-range—thereby also a Relation—unchanged. Then we appeal to (24) in order to derive some propositions that correspond to those arithmetical propositions that concern the elimination of a bracket that has a prefixed minus-sign.

§172. *Construction*

$$40 \vdash \dot{\alpha}\dot{\varepsilon}f(\alpha,\varepsilon) = \maltese\dot{\alpha}\dot{\varepsilon}f(\varepsilon,\alpha)$$

(IIIh): ———————————

$$\vdash \maltese\dot{\alpha}\dot{\varepsilon}f(\alpha,\varepsilon) = \maltese\maltese\dot{\alpha}\dot{\varepsilon}f(\varepsilon,\alpha) \qquad (\alpha$$

(IIIa): ———————————

$$\vdash\!\!\!\top\begin{array}[t]{l} \dot{\alpha}\dot{\varepsilon}f(\varepsilon,\alpha) = \maltese\maltese\dot{\alpha}\dot{\varepsilon}f(\varepsilon,\alpha) \\ \llcorner\ \dot{\alpha}\dot{\varepsilon}f(\varepsilon,\alpha) = \maltese\dot{\alpha}\dot{\varepsilon}f(\alpha,\varepsilon) \end{array} \qquad (\beta$$

(40):: ———————————

$$\vdash \dot{\alpha}\dot{\varepsilon}f(\varepsilon,\alpha) = \maltese\maltese\dot{\alpha}\dot{\varepsilon}f(\varepsilon,\alpha) \qquad (503$$

(IIIc): ———————————

$$\vdash\!\!\!\top\begin{array}[t]{l} q = \maltese\maltese q \\ \llcorner\ \dot{\alpha}\dot{\varepsilon}f(\varepsilon,\alpha) = q \end{array} \qquad (504$$

——— • ———

$$24 \vdash \maltese(p\smile q) = \maltese q\smile\maltese p$$

(IIIc): ———————————

$$\vdash\!\!\!\top\begin{array}[t]{l} f(\maltese q\smile\maltese p) \\ \llcorner\ f(\maltese(p\smile q)) \end{array} \qquad (505$$

——— • ———

$$505 \vdash\!\!\!\top\begin{array}[t]{l} F(t\smile(\maltese q\smile\maltese p)) \\ \llcorner\ F(t\smile\maltese(p\smile q)) \end{array}$$

(491): ———————————

$$\vdash\!\!\!\top\begin{array}[t]{l} F(t\smile\maltese q\smile\maltese p) \\ \llcorner\ F(t\smile\maltese(p\smile q)) \end{array} \qquad (506$$

——— • ———

$$24 \vdash \maltese(p\smile q) = \maltese q\smile\maltese p$$

(IIIa): ———————————

$$\vdash\!\!\!\top\begin{array}[t]{l} f(\maltese(p\smile q)) \\ \llcorner\ f(\maltese q\smile\maltese p) \end{array} \qquad (507$$

——— • ———

$$507 \vdash\!\!\!\top\begin{array}[t]{l} F(t\smile\maltese(p\smile q)) \\ \llcorner\ F(t\smile(\maltese q\smile\maltese p)) \end{array}$$

(491): ———————————

$$\vdash\!\!\!\top\begin{array}[t]{l} F(t\smile\maltese(p\smile q)) \\ \llcorner\ F(t\smile\maltese q\smile\maltese p) \end{array} \qquad (508$$

B. The positival class

a) Definitions of the functions $\mathfrak{d}\xi$ *and* $\mathfrak{j}\xi$ *and consequences*

§173. *Analysis*

Proposition (501) contains the commutative law for the composition of Relations that belong to a series like

$$K, K\smile K, K\smile(K\smile K)\ldots$$

Accordingly, we can regard the class of members of this series as a domain of magnitudes and define each positive rational number as the ratio of two magnitudes in such a domain of magnitudes. The negative may easily be introduced by extending the series backwards past the initial member. The main difficulty, however, concerns the irrationals. These we can only obtain as limits; and in order to define the limit, we require the relation of the lesser to the greater. Now, at this point such a relation offers itself, namely that of following in our series. This, however, is of no use for Relations that do not belong to that series. We will therefore need to impose certain conditions on a relation that have to be fulfilled in order for it to be regarded as one of the lesser to the greater, and following in our series has to emerge as a special case of this relation. Then we can use such a relation to demarcate the domain of magnitudes by saying: all Relations belong to the domain of magnitudes that stand to one and the same Relation in this relation. However, it serves our purpose better to reduce the relation of the lesser to the greater to the positive. For one can either explain the positive as that which is greater than the null magnitude, or one can say: a is said to be greater than b if the Relation composed of a and the converse of b is positive. We cannot simply talk of the class of the positive with the definite article, since there will be such a class in every domain of magnitudes. The expression "*positive class*" will for us rather be the name of a concept, and we will put the question this way: what properties must a class have in order to count as a positive class? Once we have a positive class, we can then demarcate the corresponding domain of magnitudes. Every Relation belongs to it that either belongs to the positive class, or is the converse of a Relation that belongs to the positive class, or is composed of a Relation that belongs to the positive class and its converse (null magnitude). If, accordingly, Σ is a positive class, then Π belongs to the corresponding domain of magnitudes if

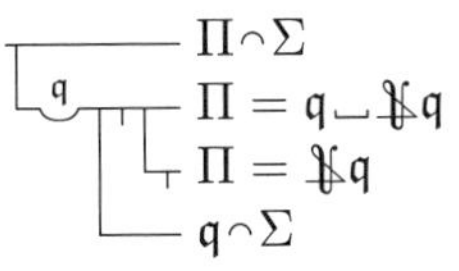

is the True. Thereby we arrive at the following definition:

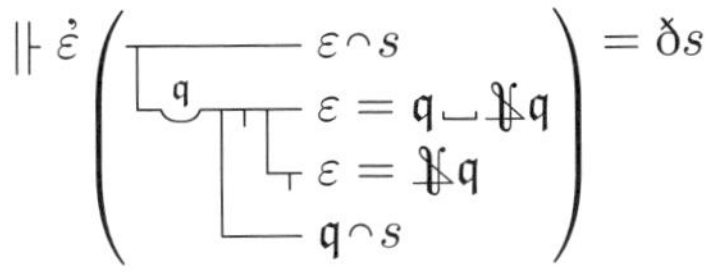

(X

So, if Σ is a positive class, then $\mathfrak{d}\Sigma$ is the corresponding domain of magnitudes, and $\Pi \frown \mathfrak{d}\Sigma$ is the truth-value of: that Π belongs to this domain of magnitudes. We read "$\Pi \frown \mathfrak{d}\Sigma$" more simply as: "$\Pi$ belongs to the Σ-domain". Let us first draw the most straightforward consequences from this definition.

§174. *Construction*

X 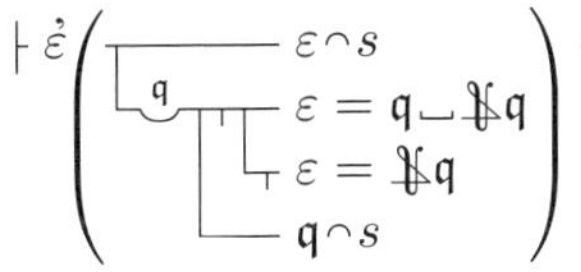

(44): ————————————

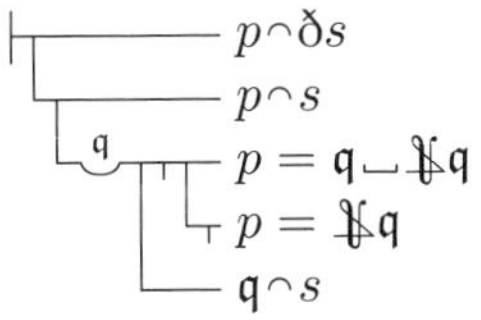

(509

(I):: — — — — — — — — — —

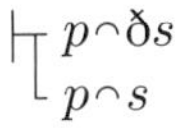

(510

——— • ———

509 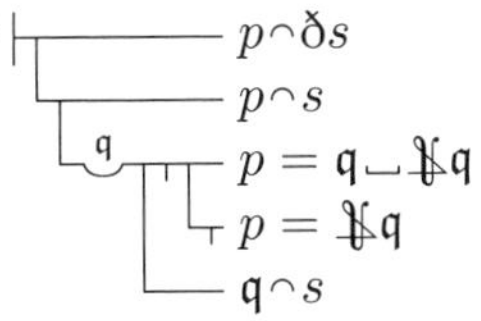

(Ia):: — — — — — — — — — —

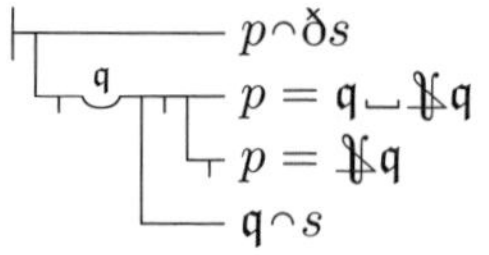
(α

×

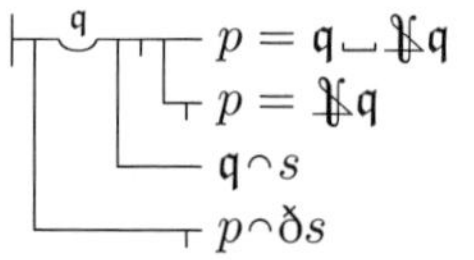
(β

(IIa): - - - - - - - - - -

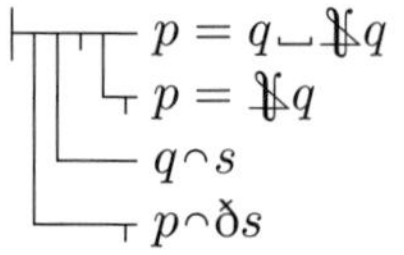
(511

(Ic): - - - - - - - -

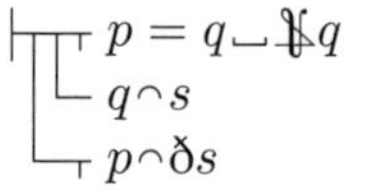
(α

×

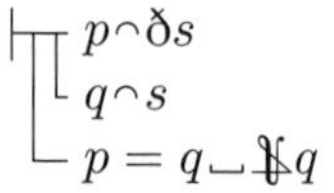
(512

——— • ———

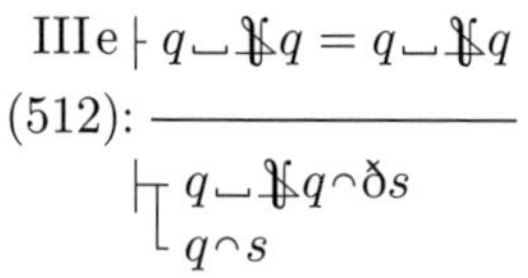

IIIe 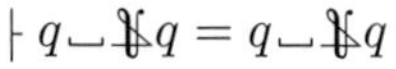

(512): ————————

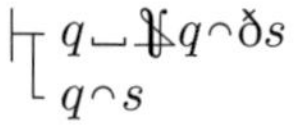
(513

——— • ———

511 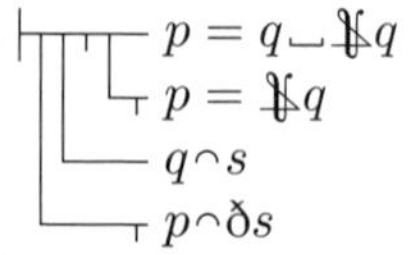

(Id): - - - - - - - - -

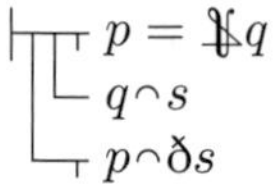
(α

×

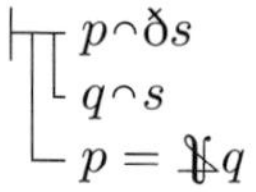
(514

——— • ———

IIIe 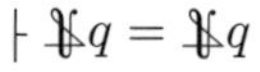

514: ————————

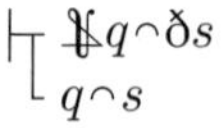
(515

——— • ———

X 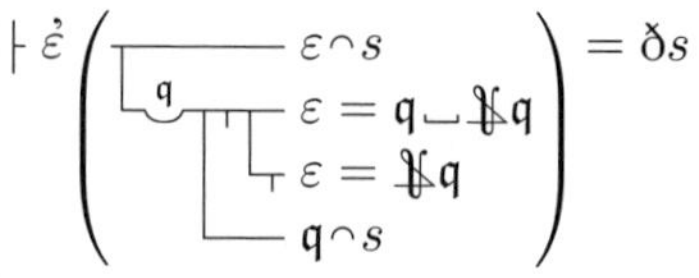

(46): ————————————————

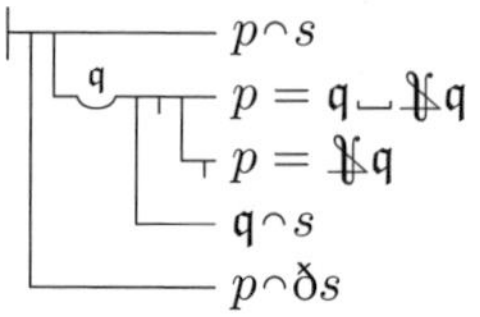
(516

×

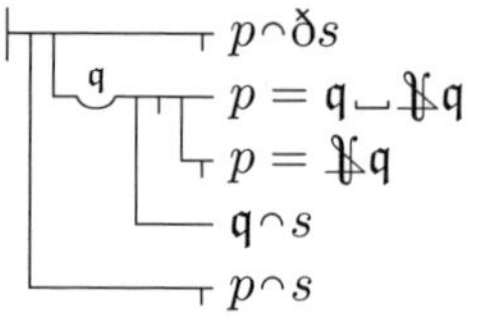
(517

§175. *Analysis*

Now, when is a class a positive class? If the corresponding domain of magnitudes is to be continuous, then the following must obtain. If a magnitude in this domain has a property Φ that also belongs to all lesser magnitudes, while there is also is a magnitude in this domain that does not have the property Φ, then there must be an upper bound in the domain of all magnitudes with the property Φ; i.e., there here must be a magnitude such that all those less than it have the property Φ, yet each magnitude greater than it is greater than at least one magnitude that does not have the property Φ. But this is merely preliminary! As one can see, we require for

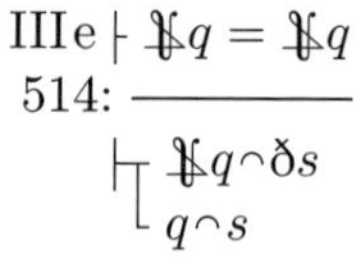

the definition of limit a relation of the lesser to the greater that we aimed to explain in terms of the positive class. We will thus not achieve our goal in one step. Before defining the concept, positive class, we propose a further concept—we will call it *positival class*—with which we can define the upper limit. Using this, we then arrive at the positive class.

The specifications that are of importance in the first instance are the following:

What belongs to a positival class must be a Relation such that both it and its converse are single-valued. The Relation composed of such a Relation and its converse, being a null magnitude, must not belong to the positival class. Further, if two Relations belong to the same positival class, then also the Relation composed of them (as their sum) must belong to this positival class, and the Relation composed of the former and the converse of the latter (as difference) must belong to the domain of magnitudes of the positival class. The same must hold of the Relation composed of the latter and the converse of the former.

If Γ stands to some object in the Relation Π, then we will say that Γ *occurs as first member of the Relation* Π; and if any object stands in the Relation Π to Δ, then we say Δ *occurs as second member of the Relation* Π. For every Relation, there is a first class of objects that can occur as first members of the Relation, and a second class of objects that can occur as second members of the Relation. We now require that the first class belonging to a Relation Π coincides with the second class belonging to a Relation K if Π and K belong to the same positival class. Thereby it is also stated that the first class belonging to Π coincides with the second. Accordingly, for each positival class Σ there is a single class of objects that can occur both as first and as second members of each Relation belonging to Σ.

This leads us to the following definition:

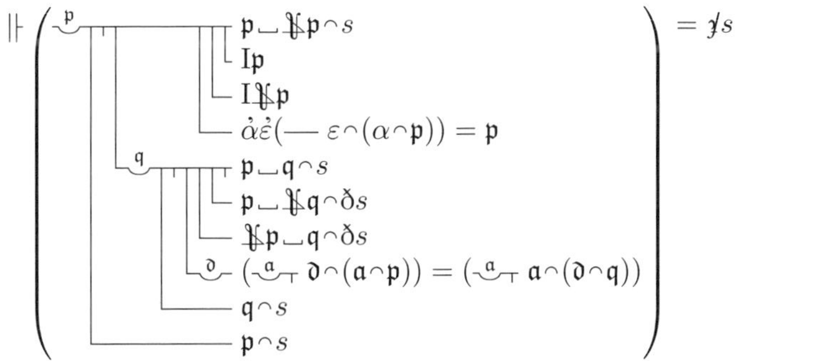

(Ψ

Accordingly, ⟊Σ is the truth-value of: that Σ is a *positival class*. In laying down this definition, I have tried to include only the necessary specifications and only those that are independent of each other. That this has been achieved cannot, of course, be proven,[a] but it becomes probable when multiple attempts to reduce some of these specifications to others fail. In particular, it does not seem possible to dispense with the line

“$\natural\mathfrak{p}\smile\mathfrak{q}\frown\mathfrak{d}s$”.

Should such an attempt nevertheless later succeed, then even if no logical error has been demonstrated in our definition, still a blemish would have been discovered. We now draw the following consequences from our definition.

§176. *Construction*

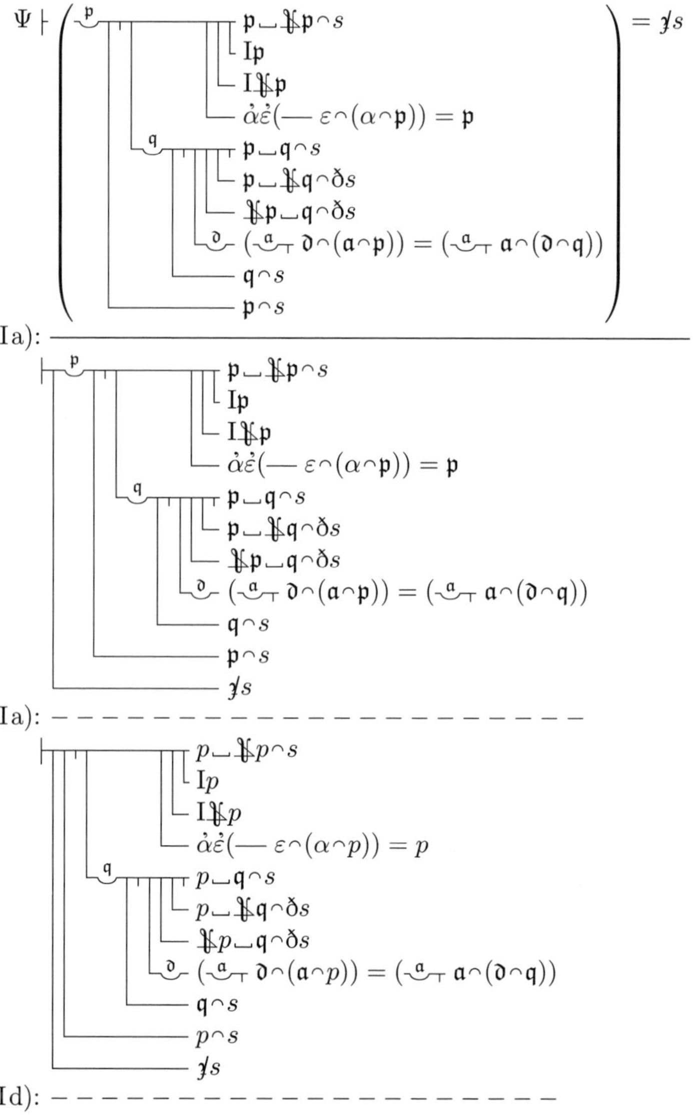

(IIIa):

(α

(IIa):

(518

(Id):

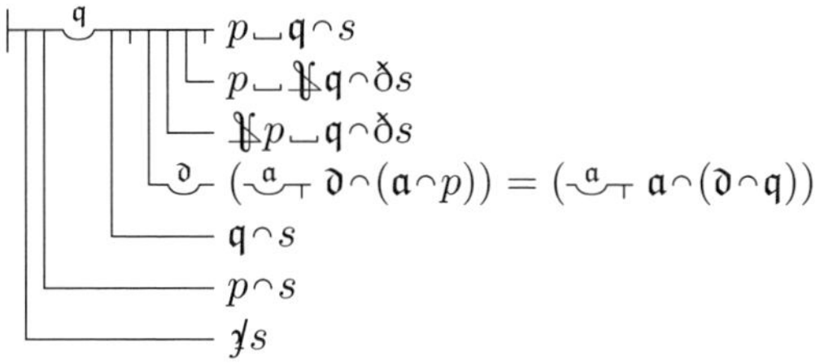

(α

(IIa): –

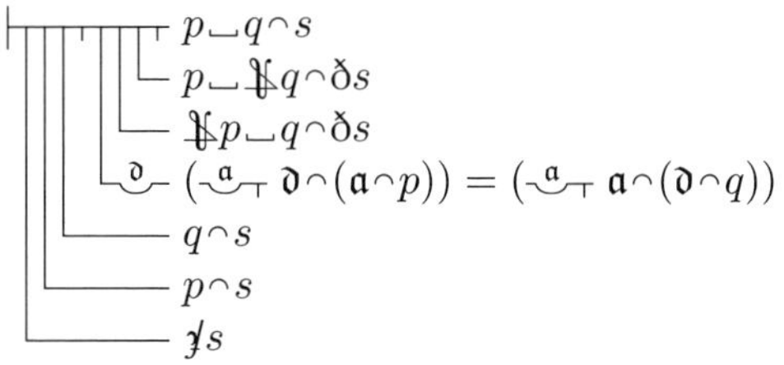

(519

(Id): –

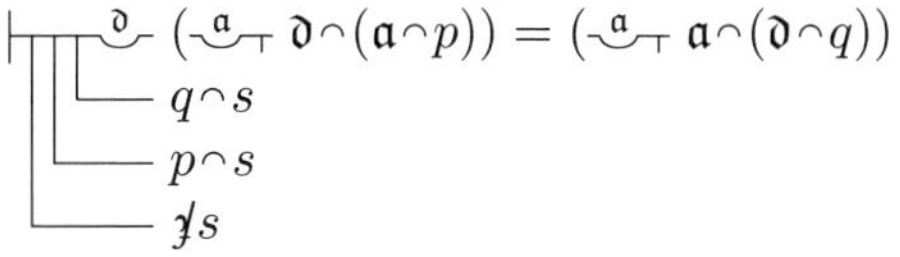

(α

(IIa): – – – – – – – – – – – – – – – – – –

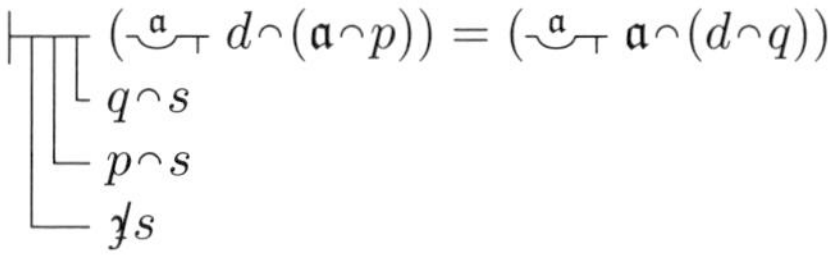

(520

(IIIa): – – – – – – – – – – – – – – – – – –

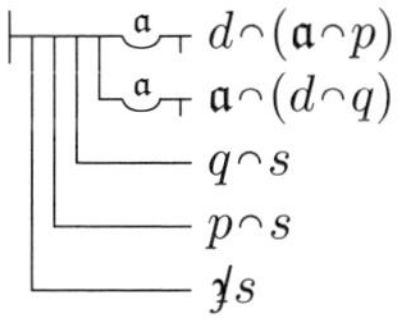

(521

(IIa): – – – – – – – – –

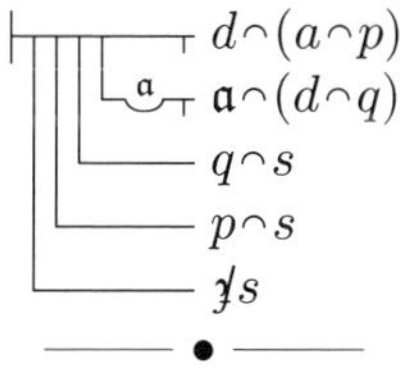

(522

IIIc 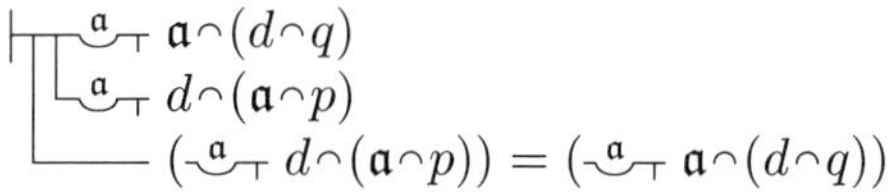

(520):: –

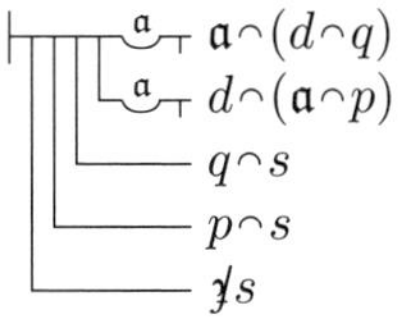

(523

(IIa): – – – – – – – – –

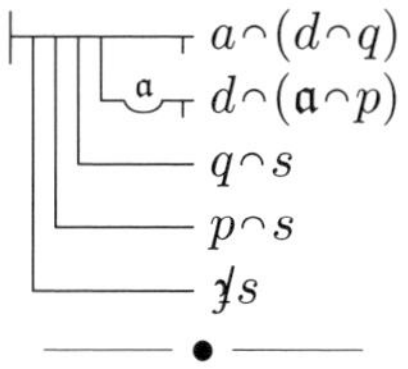

(524

519 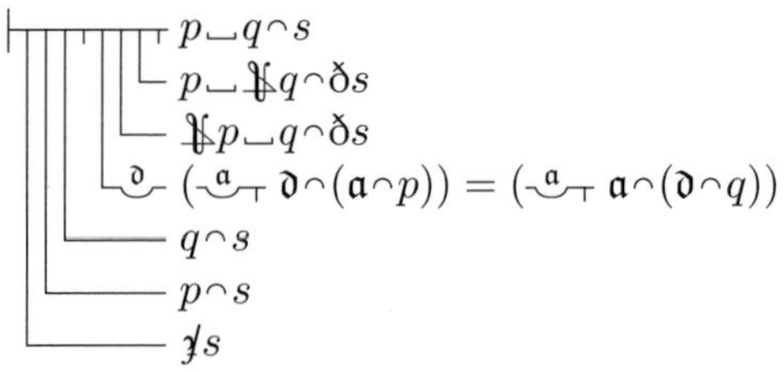

(Ic): –

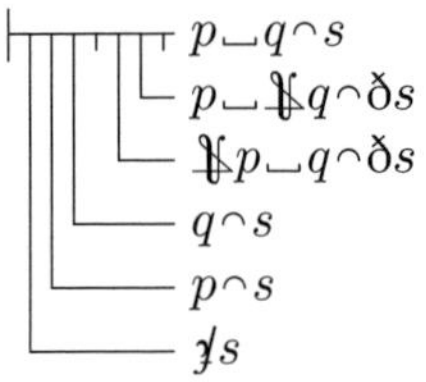

(525

(Id): – – – – – – – – –

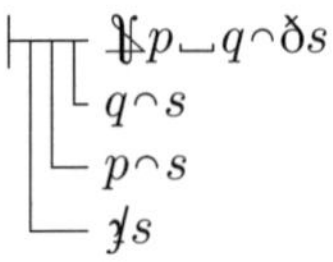

(526

———— • ————

525 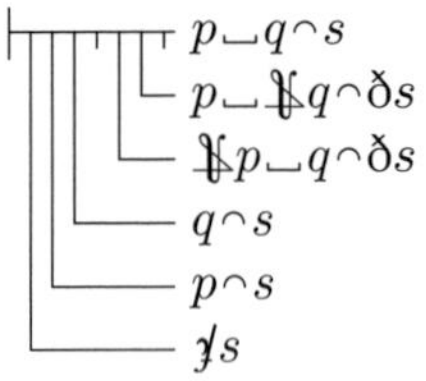

(Ic): – – – – – – – – –

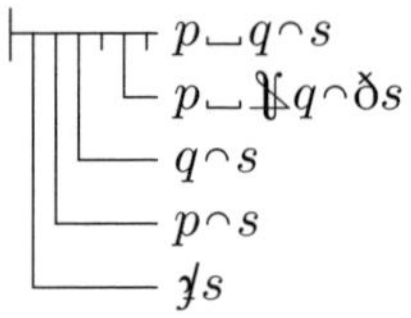

(527

(Id): – – – – – – – – –

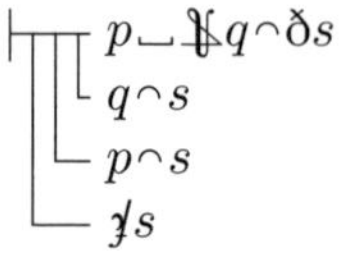

(528

———— • ————

527 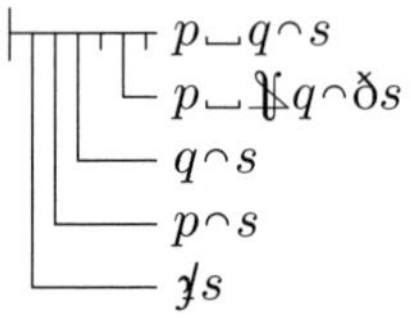

(Ib): – – – – – – – – –

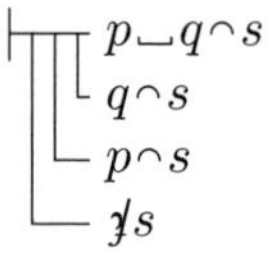

(529

———— • ————

518 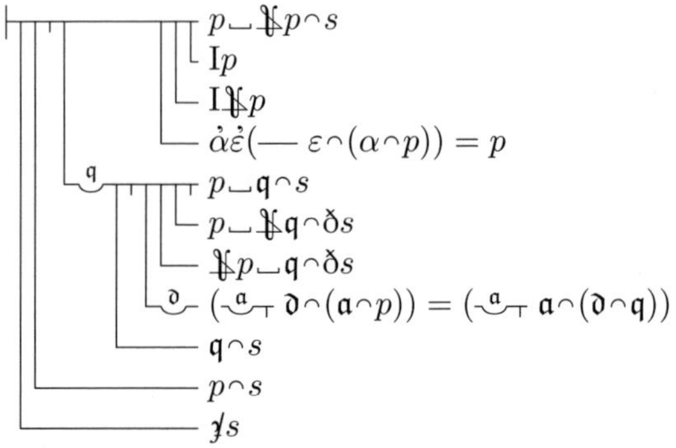

(Ic): –

$\vdash p \smile ∦p \frown s$
$\quad Ip$
$\quad I∦p$
$\quad \dot{\alpha}\dot{\varepsilon}(\text{—}\ \varepsilon \frown (\alpha \frown p)) = p$
$\quad p \frown s$
$\quad ∤s$ (530

(Id): - - - - - - - - - - - - - -

$\vdash \dot{\alpha}\dot{\varepsilon}(\text{—}\ \varepsilon \frown (\alpha \frown p)) = p$
$\quad p \frown s$
$\quad ∤s$ (531

——— • ———

530 $\vdash p \smile ∦p \frown s$
$\quad Ip$
$\quad I∦p$
$\quad \dot{\alpha}\dot{\varepsilon}(\text{—}\ \varepsilon \frown (\alpha \frown p)) = p$
$\quad p \frown s$
$\quad ∤s$

(Ic): - - - - - - - - - - - - - -

$\vdash p \smile ∦p \frown s$
$\quad Ip$
$\quad I∦p$
$\quad p \frown s$
$\quad ∤s$ (532

(Id): - - - - - - - -

$\vdash I∦p$
$\quad p \frown s$
$\quad ∤s$ (533

——— • ———

532 $\vdash p \smile ∦p \frown s$
$\quad Ip$
$\quad I∦p$
$\quad p \frown s$
$\quad ∤s$

(Ic): - - - - - - - -

$\vdash p \smile ∦p \frown s$
$\quad Ip$
$\quad p \frown s$
$\quad ∤s$ (534

(Id): - - - - - -

$\vdash Ip$
$\quad p \frown s$
$\quad ∤s$ (535

——— • ———

534 $\vdash p \smile ∦p \frown s$
$\quad Ip$
$\quad p \frown s$
$\quad ∤s$

(Ic): - - - - - -

$\vdash p \smile ∦p \frown s$
$\quad p \frown s$
$\quad ∤s$ (536

(IIId): - - - - - -

$\vdash p = p \smile ∦p$
$\quad p \frown s$
$\quad ∤s$ (537

——— • ———

529 $\vdash p \smile ∦p \frown s$
$\quad ∦p \frown s$
$\quad p \frown s$
$\quad ∤s$

×

$\vdash ∦p \frown s$
$\quad \top\ p \smile ∦p \frown s$
$\quad p \frown s$
$\quad ∤s$ (α

(536):: - - - - - -

$\vdash ∦p \frown s$
$\quad p \frown s$
$\quad ∤s$ (538

——— • ———

504 $\vdash q = ∦∦q$
$\quad \dot{\alpha}\dot{\varepsilon}(\text{—}\ \varepsilon \frown (\alpha \frown q)) = q$

(531):: - - - - - - - - - - -

$\vdash q = ∦∦q$
$\quad q \frown s$
$\quad ∤s$ (539

(IIIc): - - - - - -

$\vdash F(∦∦q)$
$\quad F(q)$
$\quad q \frown s$
$\quad ∤s$ (540

——— • ———

$$539 \vdash \left[\begin{array}{l} q = \natural\natural q \\ q \frown s \\ \jmath s \end{array}\right.$$

(III a): – – – – – –

$$\vdash \left[\begin{array}{l} F(q) \\ F(\natural\natural q) \\ q \frown s \\ \jmath s \end{array}\right. \qquad (541$$

———— • ————

$$505 \vdash \left[\begin{array}{l} f(\natural\natural q \smile \natural p) \\ f(\natural(p \smile \natural q)) \end{array}\right.$$

(541): – – – – – – – –

$$\vdash \left[\begin{array}{l} f(q \smile \natural p) \\ f(\natural(p \smile \natural q)) \\ q \frown s \\ \jmath s \end{array}\right. \qquad (542$$

———— • ————

$$507 \vdash \left[\begin{array}{l} f(\natural(p \smile \natural q)) \\ f(\natural\natural q \smile \natural p) \end{array}\right.$$

(540):: – – – – – – – –

$$\vdash \left[\begin{array}{l} f(\natural(p \smile \natural q)) \\ f(q \smile \natural p) \\ q \frown s \\ \jmath s \end{array}\right. \qquad (543$$

———— • ————

$$505 \vdash \left[\begin{array}{l} f(\natural q \smile \natural\natural p) \\ f(\natural(\natural p \smile q)) \end{array}\right.$$

(541): – – – – – – – –

$$\vdash \left[\begin{array}{l} f(\natural q \smile p) \\ f(\natural(\natural p \smile q)) \\ p \frown s \\ \jmath s \end{array}\right. \qquad (544$$

b) Proof of the proposition

$$‘\vdash \left[\begin{array}{l} q \smile \natural q = \natural p \smile p \\ p \frown s \\ q \frown s \\ \jmath s \end{array}\right.’$$

§177. *Analysis*

If the Relation Π belongs to the positival class Σ, then we call $\Pi \smile \natural\Pi$ a null Relation of the Σ-domain. We call $\natural\Pi \smile \Pi$ the same. We now prove the proposition mentioned in the heading by means of (487). From this it follows that there is always only one null Relation in the domain of a positival class. For the proof we require the propositions

$$‘\vdash \left[\begin{array}{l} d \frown (a \frown (\natural p \smile p)) \\ d \frown (a \frown (q \smile \natural q)) \\ p \frown s \\ q \frown s \\ \jmath s \end{array}\right.’, \qquad (\alpha$$

$$‘\vdash \left[\begin{array}{l} d \frown (a \frown (q \smile \natural q)) \\ d \frown (a \frown (\natural p \smile p)) \\ p \frown s \\ q \frown s \\ \jmath s \end{array}\right.’, \qquad (\beta$$

For the proof of (α) the propositions

$$‘\vdash \overset{\mathfrak{a}}{\smile} \left[\begin{array}{l} \mathfrak{a} \frown (d \frown p) \\ d \frown (d \frown (\natural p \smile p)) \end{array}\right.’ \qquad (\gamma$$

$$‘\vdash \left[\begin{array}{l} a = d \\ d \frown (a \frown (q \smile \natural q)) \\ q \frown s \\ \jmath s \end{array}\right.’, \qquad (\delta$$

are needed. For the proof of (β) we require the propositions

$$‘\vdash \overset{\mathfrak{a}}{\smile} \left[\begin{array}{l} a \frown (\mathfrak{a} \frown q) \\ a \frown (a \frown (q \smile \natural q)) \end{array}\right.’ \qquad (\varepsilon$$

$$\text{‘}\vdash \begin{array}{l} a = d \\ d \frown (a \frown (\nparallel p \smile p)) \\ p \frown s \\ \mathfrak{Z} s \end{array} \text{’} \qquad (\zeta$$

§178. *Construction*

$$78 \vdash \begin{array}{l} a = d \\ d \frown (e \frown q) \\ e \frown (a \frown \nparallel q) \\ \mathrm{I} \nparallel q \end{array}$$

$$\times$$

$$\vdash \begin{array}{l} \neg\, d \frown (e \frown q) \\ e \frown (a \frown \nparallel q) \\ \neg\, a = d \\ \mathrm{I} \nparallel q \end{array} \qquad (\alpha$$

$$\vdash \begin{array}{l} \overset{\mathfrak{r}}{\smile}\neg\, d \frown (\mathfrak{r} \frown q) \\ \quad \mathfrak{r} \frown (a \frown \nparallel q) \\ \neg\, a = d \\ \mathrm{I} \nparallel q \end{array} \qquad (\beta$$

(15): - - - - - - - - -

$$\vdash \begin{array}{l} \neg\, d \frown (a \frown (q \smile \nparallel q)) \\ \neg\, a = d \\ \mathrm{I} \nparallel q \end{array} \qquad (\gamma$$

$$\times$$

$$\vdash \begin{array}{l} a = d \\ d \frown (a \frown (q \smile \nparallel q)) \\ \mathrm{I} \nparallel q \end{array} \qquad (545$$

(533):: - - - - - - - - -

$$\vdash \begin{array}{l} a = d \\ d \frown (a \frown (q \smile \nparallel q)) \\ q \frown s \\ \mathfrak{Z} s \end{array} \qquad (546$$

(IIIa): - - - - - - - - - -

$$\vdash \begin{array}{l} f(a) \\ f(d) \\ d \frown (a \frown (q \smile \nparallel q)) \\ q \frown s \\ \mathfrak{Z} s \end{array} \qquad (547$$

——— • ———

$$22 \vdash \begin{array}{l} \neg\, d \frown (a \frown \nparallel p) \\ a \frown (d \frown p) \end{array}$$

(5): - - - - - -

$$\vdash \begin{array}{l} \neg\, d \frown (d \frown (\nparallel p \smile p)) \\ a \frown (d \frown p) \end{array} \qquad (\alpha$$

$$\times$$

$$\vdash \begin{array}{l} \neg\, a \frown (d \frown p) \\ \neg\, d \frown (d \frown (\nparallel p \smile p)) \end{array} \qquad (\beta$$

$$\vdash \begin{array}{l} \overset{\mathfrak{a}}{\smile}\neg\, \mathfrak{a} \frown (d \frown p) \\ \neg\, d \frown (d \frown (\nparallel p \smile p)) \end{array} \qquad (548$$

(521): - - - - - - - - - -

$$\vdash \begin{array}{l} \overset{\mathfrak{a}}{\smile}\neg\, d \frown (\mathfrak{a} \frown q) \\ \neg\, d \frown (d \frown (\nparallel p \smile p)) \\ p \frown s \\ q \frown s \\ \mathfrak{Z} s \end{array} \qquad (\alpha$$

——— • ———

$$\mathrm{IIa} \vdash \begin{array}{l} \neg\, d \frown (r \frown q) \\ \overset{\mathfrak{a}}{\smile}\neg\, d \frown (\mathfrak{a} \frown q) \end{array}$$

(I): ————————

$$\vdash \begin{array}{l} \neg\, d \frown (r \frown q) \\ r \frown (a \frown \nparallel q) \\ \overset{\mathfrak{a}}{\smile}\neg\, d \frown (\mathfrak{a} \frown q) \end{array} \qquad (\beta$$

$$\vdash \begin{array}{l} \overset{\mathfrak{r}}{\smile}\neg\, d \frown (\mathfrak{r} \frown q) \\ \quad \mathfrak{r} \frown (a \frown \nparallel q) \\ \overset{\mathfrak{a}}{\smile}\neg\, d \frown (\mathfrak{a} \frown q) \end{array} \qquad (\gamma$$

(15): - - - - - - - - -

$$\vdash \begin{array}{l} \neg\, d \frown (a \frown (q \smile \nparallel q)) \\ \overset{\mathfrak{a}}{\smile}\neg\, d \frown (\mathfrak{a} \frown q) \end{array} \qquad (\delta$$

(α):: - - - - - - - - - -

$$\vdash \begin{array}{l} \neg\, d \frown (a \frown (q \smile \nparallel q)) \\ \neg\, d \frown (d \frown (\nparallel p \smile p)) \\ p \frown s \\ q \frown s \\ \mathfrak{Z} s \end{array} \qquad (\varepsilon$$

$$\times$$

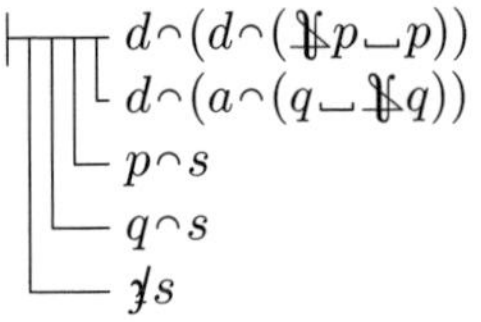

(ζ

(547): - - - - - - - - - -

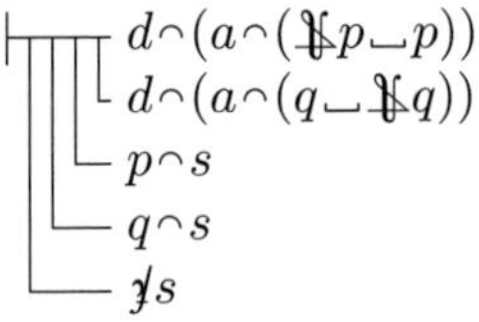

(549

(IVa): - - - - - - - - - -

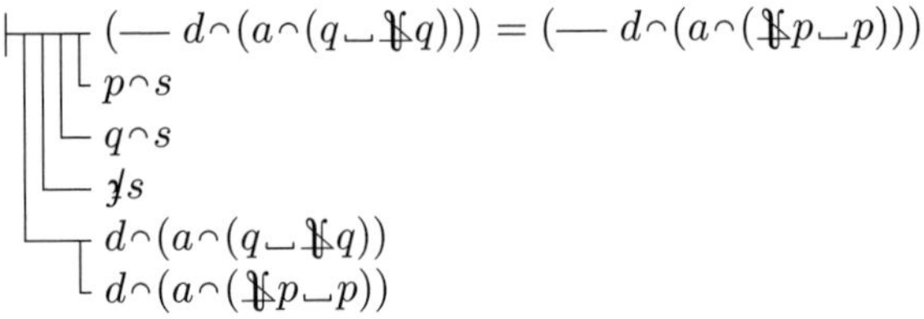

(550

---

23 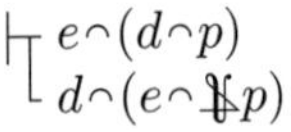

(13): - - - - - -

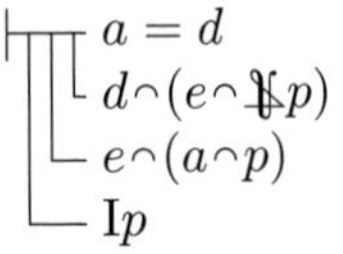

(α

×

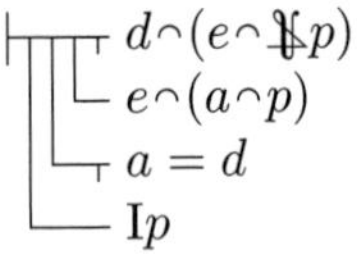

(β

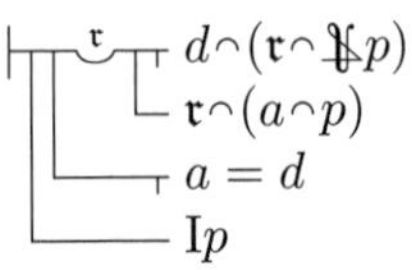

(γ

(15): - - - - - - -

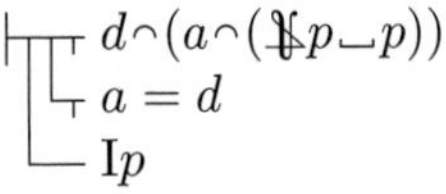

(δ

×

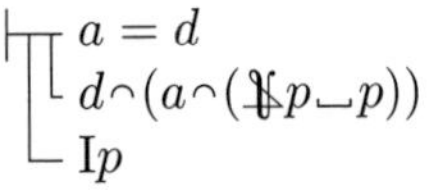

(ε

(535):: - - - - - - - -

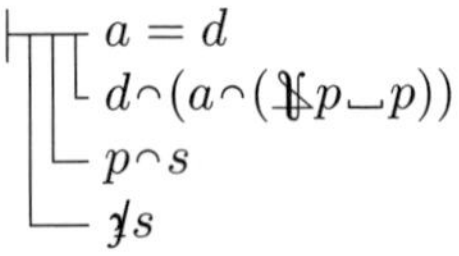

(551

(IIIc): - - - - - - - - - -

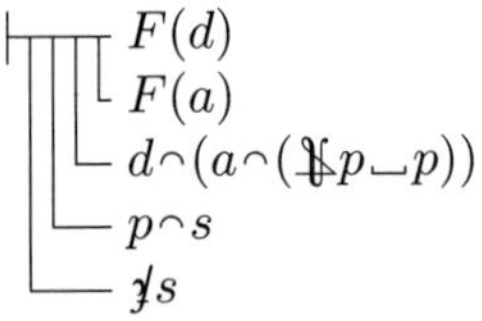

(552

---

22 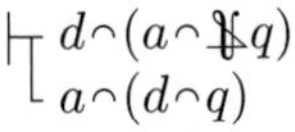

(5): - - - - - -

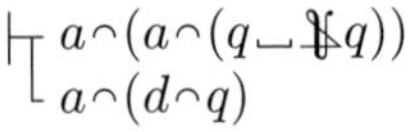

(553

×

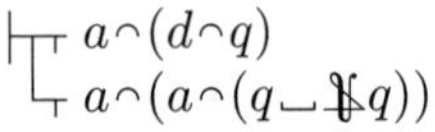

(554

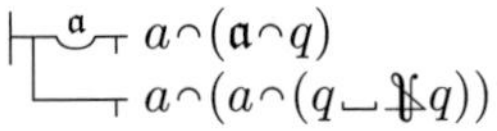

(555

(523): - - - - - - - - - -

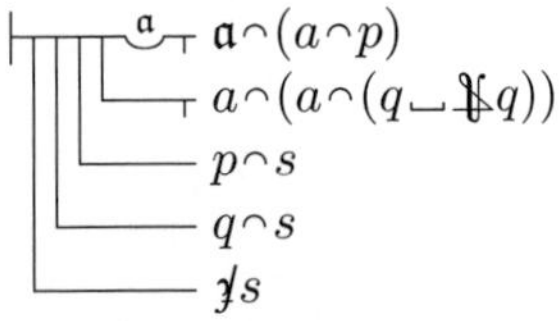

(α

IIa ⊢ $r \frown (a \frown p)$
$\quad \mathfrak{a}$ ⌣ $\mathfrak{a} \frown (a \frown p)$

(Ia): – – – – – –

⊢ $d \frown (r \frown \rangle\!\!\!\langle p)$
$\quad r \frown (a \frown p)$
$\quad \mathfrak{a}$ ⌣ $\mathfrak{a} \frown (a \frown p)$ (β

⌣

⊢ $\mathfrak{r}$ ⌣ $d \frown (\mathfrak{r} \frown \rangle\!\!\!\langle p)$
$\quad \mathfrak{r} \frown (a \frown p)$
$\quad \mathfrak{a}$ ⌣ $\mathfrak{a} \frown (a \frown p)$ (γ

(15): – – – – – – – –

⊢ $d \frown (a \frown (\rangle\!\!\!\langle p \smile p))$
$\quad \mathfrak{a}$ ⌣ $\mathfrak{a} \frown (a \frown p)$ (δ

(α):: – – – – – – – – – –

⊢ $d \frown (a \frown (\rangle\!\!\!\langle p \smile p))$
$\quad a \frown (a \frown (q \smile \rangle\!\!\!\langle q))$
$\quad p \frown s$
$\quad q \frown s$
$\quad \not\mathfrak{J} s$ (ε

×

⊢ $a \frown (a \frown (q \smile \rangle\!\!\!\langle q))$
$\quad d \frown (a \frown (\rangle\!\!\!\langle p \smile p))$
$\quad p \frown s$
$\quad q \frown s$
$\quad \not\mathfrak{J} s$ (ζ

(552): ———————————

⊢ $d \frown (a \frown (q \smile \rangle\!\!\!\langle q))$
$\quad d \frown (a \frown (\rangle\!\!\!\langle p \smile p))$
$\quad p \frown s$
$\quad q \frown s$
$\quad \not\mathfrak{J} s$ (η

(550): – – – – – – – – – –

⊢ $(— d \frown (a \frown (q \smile \rangle\!\!\!\langle q))) = (— d \frown (a \frown (\rangle\!\!\!\langle p \smile p)))$
$\quad p \frown s$
$\quad q \frown s$
$\quad \not\mathfrak{J} s$ (ϑ

⌣

⊢ $\mathfrak{a}$ ⌣ $\mathfrak{d}$ ⌣ $(— \mathfrak{d} \frown (\mathfrak{a} \frown (q \smile \rangle\!\!\!\langle q))) = (— \mathfrak{d} \frown (\mathfrak{a} \frown (\rangle\!\!\!\langle p \smile p)))$
$\quad p \frown s$
$\quad q \frown s$
$\quad \not\mathfrak{J} s$ (ι

(487): –

⊢ $q \smile \rangle\!\!\!\langle q = \rangle\!\!\!\langle p \smile p$
$\quad p \frown s$
$\quad q \frown s$
$\quad \not\mathfrak{J} s$ (556

(IIIa): – – – – – – – – – –

⊢ $f(q \smile \rangle\!\!\!\langle q)$
$\quad f(\rangle\!\!\!\langle p \smile p)$
$\quad p \frown s$
$\quad q \frown s$
$\quad \not\mathfrak{J} s$ (557

——— • ———

IIIc ⊢ $f(\rangle\!\!\!\langle p \smile p)$
$\quad f(q \smile \rangle\!\!\!\langle q)$
$\quad q \smile \rangle\!\!\!\langle q = \rangle\!\!\!\langle p \smile p$

(556):: – – – – – – – – –

⊢ $f(\rangle\!\!\!\langle p \smile p)$
$\quad f(q \smile \rangle\!\!\!\langle q)$
$\quad p \frown s$
$\quad q \frown s$
$\quad \not\mathfrak{J} s$ (558

(557): – – – – – – – –

⊢ $f(p \smile \rangle\!\!\!\langle p)$
$\quad f(q \smile \rangle\!\!\!\langle q)$
$\quad p \frown s$
$\quad q \frown s$
$\quad \not\mathfrak{J} s$ (559

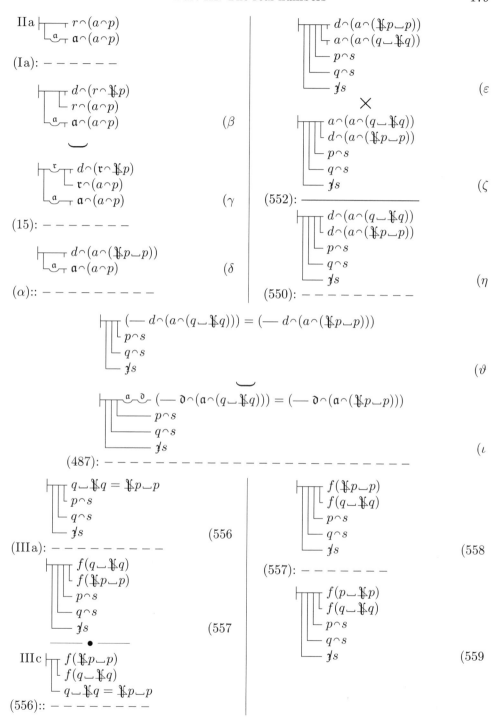

c) Proof of the proposition

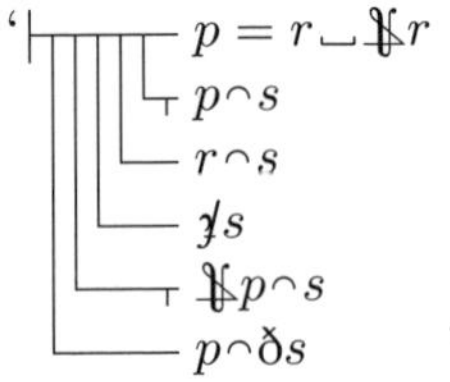

§179. *Analysis*

We now prove the proposition that a Relation belonging to the domain of a positival class is a null Relation when neither it nor its converse belongs to the positival class itself. For this we use (516).

§180. *Construction*

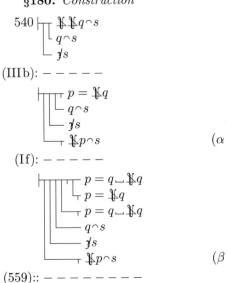

540 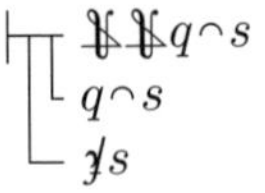

(IIIb): – – – – –

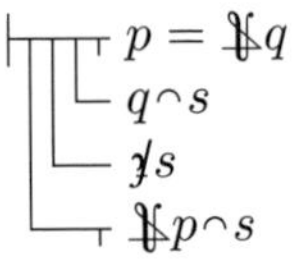
(α

(If): – – – – –

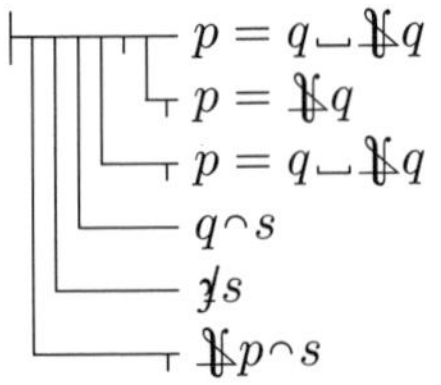
(β

(559):: – – – – – – – –

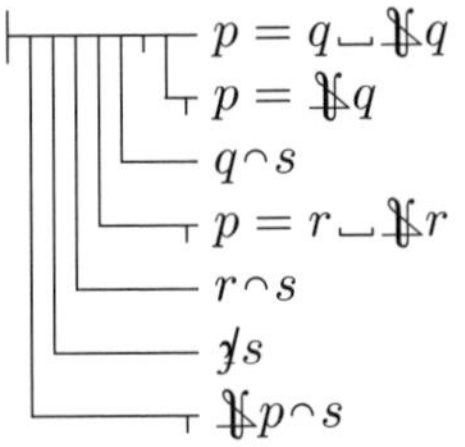
(γ

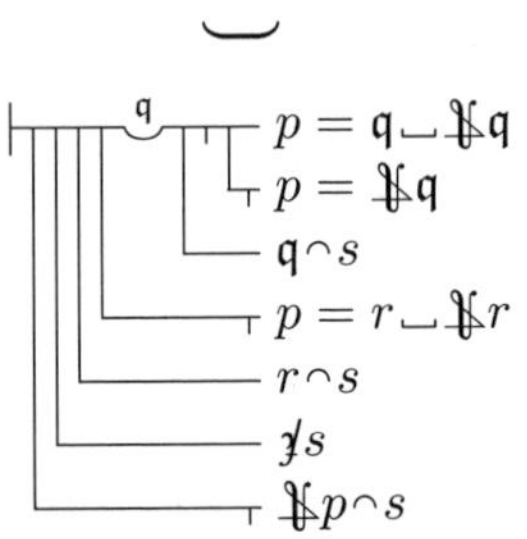
(δ

(516): – – – – – – – – – – –

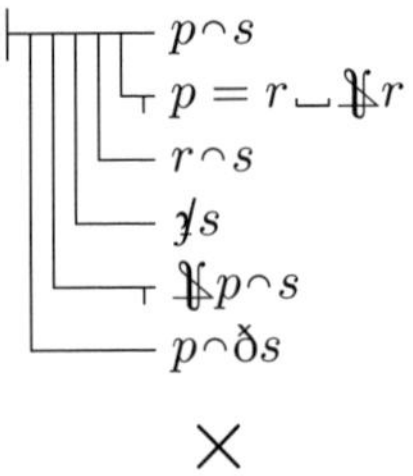
(560

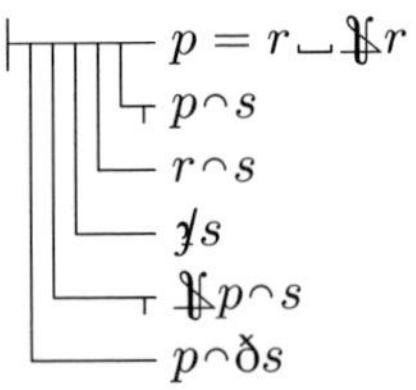
(561

d) Proof of the propositions

$$‘\vdash\begin{array}{l} q\smile(p\smile\nparallel p)=q \\ p\frown s \\ \not\mathfrak{z}\,s \\ q\frown s \end{array} \quad ’,$$

$$‘\vdash\begin{array}{l} p\smile\nparallel p\smile q=q \\ q\frown s \\ \not\mathfrak{z}\,s \\ p\frown s \end{array} \quad ’,$$

and consequences

§181. *Analysis*

The propositions mentioned in the heading say that a Relation that belongs to a positival class remains unchanged if it is composed with the null Relation of the domain of its positival class. In order to prove the first proposition, we make use of (486); and for this the propositions

$$‘\vdash\begin{array}{l} d\frown(a\frown(q\smile(p\smile\nparallel p))) \\ d\frown(a\frown q) \\ q\frown s \\ p\frown s \\ \not\mathfrak{z}\,s \end{array} \quad ’, \qquad (\alpha$$

$$‘\vdash\begin{array}{l} d\frown(a\frown q) \\ d\frown(a\frown(q\smile(p\smile\nparallel p))) \\ p\frown s \\ \not\mathfrak{z}\,s \end{array} \quad ’, \qquad (\beta$$

are required.

§182. *Construction*

$$555 \quad \vdash\!\overset{\mathfrak{a}}{\smile}\begin{array}{l} \top\, a\frown(\mathfrak{a}\frown p) \\ \top\, a\frown(a\frown(p\smile\nparallel p)) \end{array}$$

(524): – – – – – – – – – – –

$$\vdash\begin{array}{l} \top\, d\frown(a\frown q) \\ \top\, a\frown(a\frown(p\smile\nparallel p)) \\ q\frown s \\ p\frown s \\ \not\mathfrak{z}\,s \end{array} \qquad (\alpha$$

×

$$\vdash\begin{array}{l} a\frown(a\frown(p\smile\nparallel p)) \\ d\frown(a\frown q) \\ q\frown s \\ p\frown s \\ \not\mathfrak{z}\,s \end{array} \qquad (\beta$$

(5): – – – – – – – – – –

$$\vdash\begin{array}{l} d\frown(a\frown(q\smile(p\smile\nparallel p))) \\ d\frown(a\frown q) \\ q\frown s \\ p\frown s \\ \not\mathfrak{z}\,s \end{array} \qquad (\gamma$$

——— • ———

$$\text{IIIc} \quad \vdash\begin{array}{l} \top\, d\frown(e\frown q) \\ \top\, d\frown(a\frown q) \\ a=e \end{array}$$

(546):: – – – – – – –

$$\vdash\begin{array}{l} \top\, d\frown(e\frown q) \\ e\frown(a\frown(p\smile\nparallel p)) \\ p\frown s \\ \not\mathfrak{z}\,s \\ \top\, \underbrace{d\frown(a\frown q)} \end{array} \qquad (\delta$$

$$\vdash\!\overset{\mathfrak{r}}{\smile}\begin{array}{l} \top\, d\frown(\mathfrak{r}\frown q) \\ \mathfrak{r}\frown(a\frown(p\smile\nparallel p)) \\ p\frown s \\ \not\mathfrak{z}\,s \\ \top\, d\frown(a\frown q) \end{array} \qquad (\varepsilon$$

(15): – – – – – – – – – – – –

$$\vdash\begin{array}{l} \top\, d\frown(a\frown(q\smile(p\smile\nparallel p))) \\ p\frown s \\ \not\mathfrak{z}\,s \\ \top\, d\frown(a\frown q) \end{array} \qquad (\zeta$$

×

$$\vdash \begin{array}{l} d \frown (a \frown q) \\ d \frown (a \frown (q \smile (p \smile \not\mathcal{J}\!\!\!/ p))) \\ p \frown s \\ \not\mathcal{J} s \end{array} \qquad (\eta$$

(IV a): – – – – – – – – – – – –

$$\vdash \begin{array}{l} [\text{—}\, d \frown (a \frown (q \smile (p \smile \not\mathcal{J}\!\!\!/ p)))] = (\text{—}\, d \frown (a \frown q)) \\ p \frown s \\ \not\mathcal{J} s \\ d \frown (a \frown (q \smile (p \smile \not\mathcal{J}\!\!\!/ p))) \\ d \frown (a \frown q) \end{array} \qquad (\vartheta$$

(γ):: –

$$\vdash \begin{array}{l} [\text{—}\, d \frown (a \frown (q \smile (p \smile \not\mathcal{J}\!\!\!/ p)))] = (\text{—}\, d \frown (a \frown q)) \\ p \frown s \\ \not\mathcal{J} s \\ q \frown s \end{array} \qquad (\iota$$

$$\vdash \begin{array}{l} \overset{\mathfrak{a}}{\smile}\overset{\mathfrak{d}}{\smile} [\text{—}\, \mathfrak{d} \frown (\mathfrak{a} \frown (q \smile (p \smile \not\mathcal{J}\!\!\!/ p)))] = (\text{—}\, \mathfrak{d} \frown (\mathfrak{a} \frown q)) \\ p \frown s \\ \not\mathcal{J} s \\ q \frown s \end{array} \qquad (\kappa$$

(486): –

$$\vdash \begin{array}{l} q \smile (p \smile \not\mathcal{J}\!\!\!/ p) = q \\ p \frown s \\ \not\mathcal{J} s \\ q \frown s \\ \grave{\alpha}\grave{\varepsilon}(\text{—}\, \varepsilon \frown (\alpha \frown q)) = q \end{array} \qquad (\lambda$$

(531):: – – – – – – – – – – – –

$$\vdash \begin{array}{l} q \smile (p \smile \not\mathcal{J}\!\!\!/ p) = q \\ p \frown s \\ \not\mathcal{J} s \\ q \frown s \end{array} \qquad (562$$

(III a): – – – – – – – – – –

$$\vdash \begin{array}{l} f(q \smile (p \smile \not\mathcal{J}\!\!\!/ p)) \\ f(q) \\ p \frown s \\ \not\mathcal{J} s \\ q \frown s \end{array} \qquad (563$$

———— • ————

$$\text{III c} \vdash \begin{array}{l} f(q) \\ f(q \smile (p \smile \not\mathcal{J}\!\!\!/ p)) \\ q \smile (p \smile \not\mathcal{J}\!\!\!/ p) = q \end{array}$$

(562):: – – – – – – – – –

$$\vdash \begin{array}{l} f(q) \\ f(q \smile (p \smile \not\mathcal{J}\!\!\!/ p)) \\ p \frown s \\ \not\mathcal{J} s \\ q \frown s \end{array} \qquad (564$$

———— • ————

$$562 \vdash \begin{array}{l} q \smile (p \smile \not\mathcal{J}\!\!\!/ p) = q \\ p \frown s \\ \not\mathcal{J} s \\ q \frown s \end{array}$$

(558): – – – – – – – – – –

$$\vdash \begin{array}{l} q \smile (\not\mathcal{J}\!\!\!/ p \smile p) = q \\ p \frown s \\ \not\mathcal{J} s \\ q \frown s \end{array} \qquad (565$$

(491): – – – – – – – – – –

$$\vdash \begin{array}{l} q \smile \not\mathcal{J}\!\!\!/ p \smile p = q \\ p \frown s \\ \not\mathcal{J} s \\ q \frown s \end{array} \qquad (566$$

(III c): – – – – – – – –

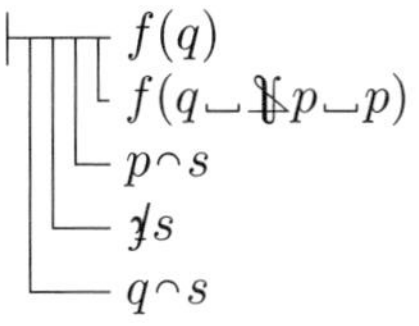

(567

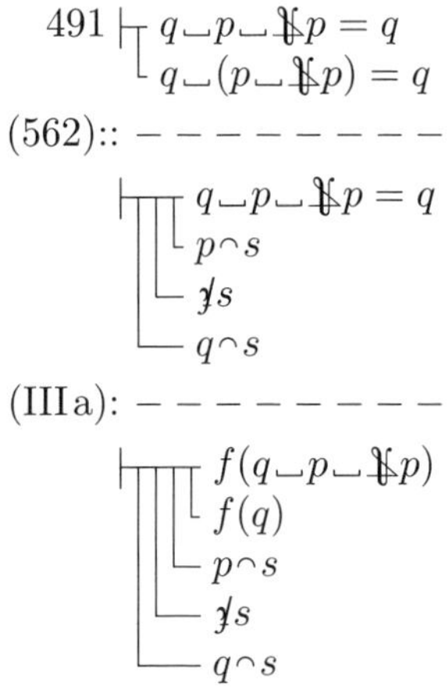

(568

(569

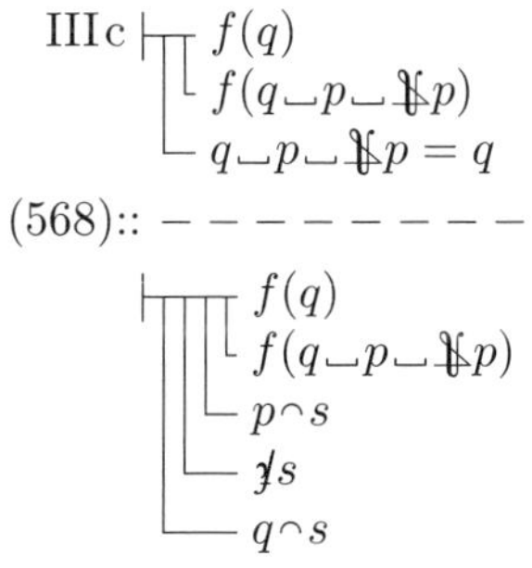

(570

§183. *Analysis*

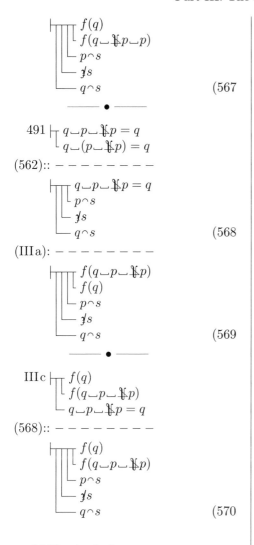

Using (559), we can now reduce the second proposition of our heading to (566), writing 'q' for 'p' in (566). From the proposition thus obtained, we can draw further consequences in a similar manner as with (562).

§184. *Construction*

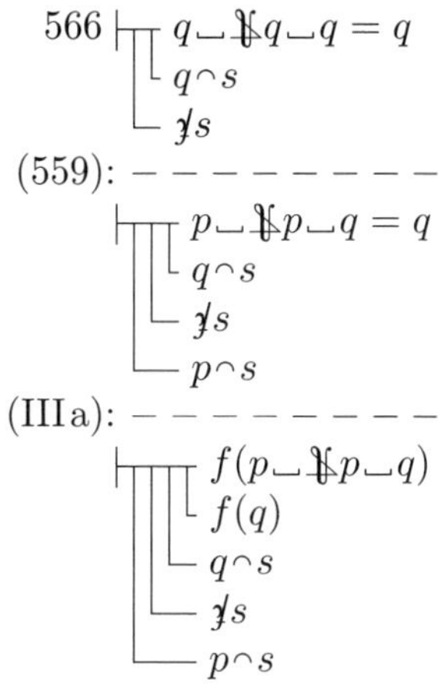

(571

(572

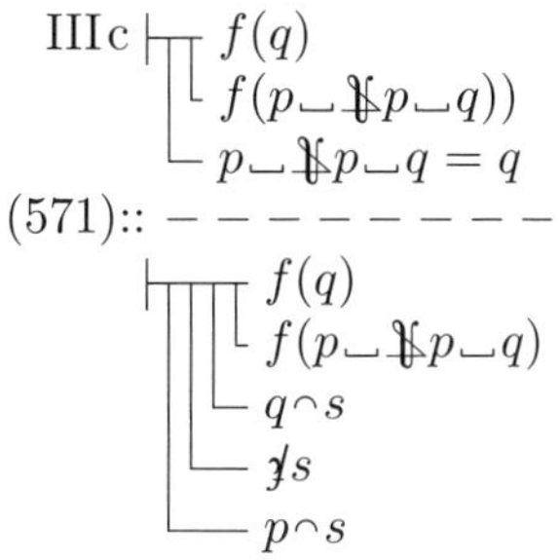

(573

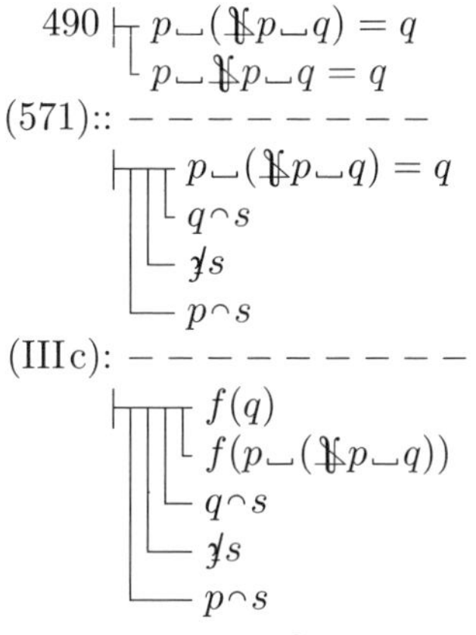

(574

(575

IIIh $\vdash\!\!\top\; \not\backslash(p\smile\not\backslash p\smile q) = \not\backslash q$
$\quad\llcorner\; p\smile\not\backslash p\smile q = q$

(505): ————————

$\vdash\!\!\top\; \not\backslash q\smile\not\backslash(p\smile\not\backslash p) = \not\backslash q$
$\quad\llcorner\; p\smile\not\backslash p\smile q = q$ (α

(542): ————————

$\vdash\!\!\top\; \not\backslash q\smile(p\smile\not\backslash p) = \not\backslash q$
$\quad\llcorner\; p\smile\not\backslash p\smile q = q$
$\quad\llcorner\; p\frown s$
$\quad\llcorner\; \not{\mathfrak{z}}s$ (β

(571):: – – – – – – – – – – – –

$\vdash\!\!\top\; \not\backslash q\smile(p\smile\not\backslash p) = \not\backslash q$
$\quad\llcorner\; q\frown s$
$\quad\llcorner\; \not{\mathfrak{z}}s$
$\quad\llcorner\; p\frown s$ (576

(IIIa): – – – – – – – – – – – –

$\vdash\!\!\top\; f(\not\backslash q\smile(p\smile\not\backslash p))$
$\quad\llcorner\; f(\not\backslash q)$
$\quad\llcorner\; q\frown s$
$\quad\llcorner\; \not{\mathfrak{z}}s$
$\quad\llcorner\; p\frown s$ (577

——— • ———

571 $\vdash\!\!\top\; p\smile\not\backslash p\smile q = q$
$\quad\llcorner\; q\frown s$
$\quad\llcorner\; \not{\mathfrak{z}}s$
$\quad\llcorner\; p\frown s$

(558): ————————

$\vdash\!\!\top\; \not\backslash p\smile p\smile q = q$
$\quad\llcorner\; q\frown s$
$\quad\llcorner\; \not{\mathfrak{z}}s$
$\quad\llcorner\; p\frown s$ (578

(490): – – – – – – – – –

$\vdash\!\!\top\; \not\backslash p\smile(p\smile q) = q$
$\quad\llcorner\; q\frown s$
$\quad\llcorner\; \not{\mathfrak{z}}s$
$\quad\llcorner\; p\frown s$ (579

(IIIc): – – – – – – – – – –

$\vdash\!\!\top\; f(q)$
$\quad\llcorner\; f(\not\backslash p\smile(p\smile q))$
$\quad\llcorner\; q\frown s$
$\quad\llcorner\; \not{\mathfrak{z}}s$
$\quad\llcorner\; p\frown s$ (580

——— • ———

578 $\vdash\!\!\top\; \not\backslash p\smile p\smile q = q$
$\quad\llcorner\; q\frown s$
$\quad\llcorner\; \not{\mathfrak{z}}s$
$\quad\llcorner\; p\frown s$

(IIIc): – – – – – – – – –

$\vdash\!\!\top\; f(q)$
$\quad\llcorner\; f(\not\backslash p\smile p\smile q)$
$\quad\llcorner\; q\frown s$
$\quad\llcorner\; \not{\mathfrak{z}}s$
$\quad\llcorner\; p\frown s$ (581

——— • ———

IIIh $\vdash\!\!\top\; \not\backslash(\not\backslash p\smile(p\smile q)) = \not\backslash q$
$\quad\llcorner\; \not\backslash p\smile(p\smile q) = q$

(544): ————————

$\vdash\!\!\top\; \not\backslash(p\smile q)\smile p = \not\backslash q$
$\quad\llcorner\; \not\backslash p\smile(p\smile q) = q$
$\quad\llcorner\; p\frown s$
$\quad\llcorner\; \not{\mathfrak{z}}s$ (α

(505): ————————

$\vdash\!\!\top\; \not\backslash q\smile\not\backslash p\smile p = \not\backslash q$
$\quad\llcorner\; \not\backslash p\smile(p\smile q) = q$
$\quad\llcorner\; p\frown s$
$\quad\llcorner\; \not{\mathfrak{z}}s$ (β

(579):: – – – – – – – – – – –

$\vdash\!\!\top\; \not\backslash q\smile\not\backslash p\smile p = \not\backslash q$
$\quad\llcorner\; q\frown s$
$\quad\llcorner\; \not{\mathfrak{z}}s$
$\quad\llcorner\; p\frown s$ (582

(IIIc): – – – – – – – – – – –

$\vdash\!\!\top\; f(\not\backslash q)$
$\quad\llcorner\; f(\not\backslash q\smile\not\backslash p\smile p)$
$\quad\llcorner\; q\frown s$
$\quad\llcorner\; \not{\mathfrak{z}}s$
$\quad\llcorner\; p\frown s$ (583

§185. *Analysis*

From (580) we can further deduce the proposition: "If a Relation that is composed of a second and a third, coincides with the second, then the third does not belong to the same positival class as the second."

For this purpose we show that, if it belonged to the same positival class as the second, the third Relation would be a null Relation; according to (537), however, this is impossible.

§186. *Construction*

IIIh ⊢ $Ɉc \smile (c \smile p) = Ɉc \smile c$
  $c \smile p = c$

(580): ————————

⊢ $p = Ɉc \smile c$
  $c \smile p = c$
  $p \frown s$
  $ʒ̸s$
  $c \frown s$ (α

(557): ————————

⊢ $p = p \smile Ɉp$
  $c \smile p = c$
  $p \frown s$
  $ʒ̸s$
  $c \frown s$ (β

×

⊢ $c \smile p = c$
  $p = p \smile Ɉp$
  $p \frown s$
  $ʒ̸s$
  $c \frown s$ (γ

(537):: − − − − − − − − −

⊢ $c \smile p = c$
  $p \frown s$
  $ʒ̸s$
  $c \frown s$ (584

×

⊢ $p \frown s$
  $c \smile p = c$
  $ʒ̸s$
  $c \frown s$ (585

e) Propositions concerning the greater and the lesser in a positival class

§187. *Analysis*

If Σ is a positival class and Π and K belong to its domain, then we may read

'$\Pi \smile ɈK \frown \Sigma$'

as: "Π is greater than K in the Σ-domain".

We now prove the proposition:

"If of two Relations that belong to the same positival class the first (q) is greater than the second (r), then the Relation composed of the first and a third (t) belongs to the positival class (s) if the Relation composed of the second (r) and third (t) belongs to it."

If the third Relation is regarded as the converse of a Relation (p), then we obtain as a special case the proposition:

"If a Relation (q) is greater than a second (r) which is itself greater than a third (p), then the first (q) is greater than the third (p) if the Relations belong to the same positival class."

§188. *Construction*

491 ⊢ $q \smile Ɉr \smile r \smile t \frown s$
  $q \smile Ɉr \smile (r \smile t) \frown s$

(567): ————————

$\vdash q \smile t \frown s$
$q \smile \mathfrak{F} r \smile (r \smile t) \frown s$
$r \frown s$
$\mathfrak{J} s$
$q \frown s$ (α

(529):: - - - - - - - - - - -

$\vdash q \smile t \frown s$
$r \smile t \frown s$
$q \smile \mathfrak{F} r \frown s$
$r \frown s$
$\mathfrak{J} s$
$q \frown s$ (586

§189. *Analysis*

We now prove the proposition that a Relation coincides with a second when the first is neither greater nor less than the second, while both belong to the same positival class. We reduce this proposition to (561).

§190. *Construction*

IIIh $\vdash a \smile \mathfrak{F} c \smile c = c \smile \mathfrak{F} c \smile c$
$a \smile \mathfrak{F} c = c \smile \mathfrak{F} c$

(567): ————————————

$\vdash a = c \smile \mathfrak{F} c \smile c$
$a \smile \mathfrak{F} c = c \smile \mathfrak{F} c$
$c \frown s$
$\mathfrak{J} s$
$a \frown s$ (α

(567): ————————————

$\vdash a = c$
$a \smile \mathfrak{F} c = c \smile \mathfrak{F} c$
$c \frown s$
$\mathfrak{J} s$
$a \frown s$ (β

(561):: - - - - - - - - - -

$\vdash a = c$
$a \smile \mathfrak{F} c \frown s$
$c \frown s$
$\mathfrak{J} s$
$\mathfrak{F}(a \smile \mathfrak{F} c) \frown s$
$a \smile \mathfrak{F} c \frown \mathfrak{d} s$
$a \frown s$ (γ

(543, 528):: = = = = = =

$\vdash a = c$
$a \smile \mathfrak{F} c \frown s$
$c \frown s$
$\mathfrak{J} s$
$c \smile \mathfrak{F} a \frown s$
$a \frown s$ (587

×

$\vdash c \smile \mathfrak{F} a \frown s$
$a \smile \mathfrak{F} c \frown s$
$c \frown s$
$\mathfrak{J} s$
$a = c$
$a \frown s$ (588

§191. *Analysis*

We now further prove that a Relation which belongs to a positival class is not greater than a second Relation if the latter is greater than the former. This follows from (538).

§192. *Construction*

538 $\vdash \mathfrak{F}(q \smile \mathfrak{F} p) \frown s$
$q \smile \mathfrak{F} p \frown s$
$\mathfrak{J} s$

(542): ————————————

$\vdash p \smile \mathfrak{F} q \frown s$
$q \smile \mathfrak{F} p \frown s$
$p \frown s$
$\mathfrak{J} s$ (589

Γ. The limit

Definitions of the functions ξ ⅃ ζ and ξ ł ζ

§193. *Analysis*

As announced in §175, we will now define the upper limit in a positival class. Instead of saying "upper limit of those Relations in a positival class Σ which belong to a class Φ", we will say for short:

"Σ-*limit of* Φ".

When, then, do we say that Δ is a Σ-limit of Φ? This requires

1. that Σ be a positival class;
2. that Δ belongs to the Σ-class;
3. that every Relation less than Δ that belongs to the class Σ belongs to the class Φ;
4. that all Relations in Σ that are greater than Δ are greater than at least one Relation in Σ that does not belong to the class Φ.

Before defining the limit we will stipulate for the sake of brevity:

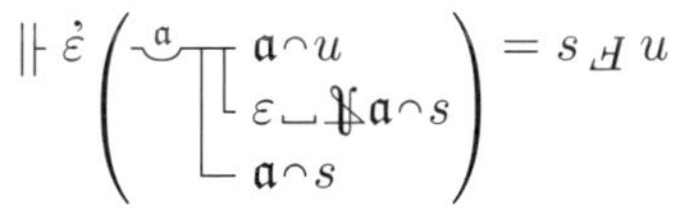

(Ω

We will use this notation for the definition of the limit:

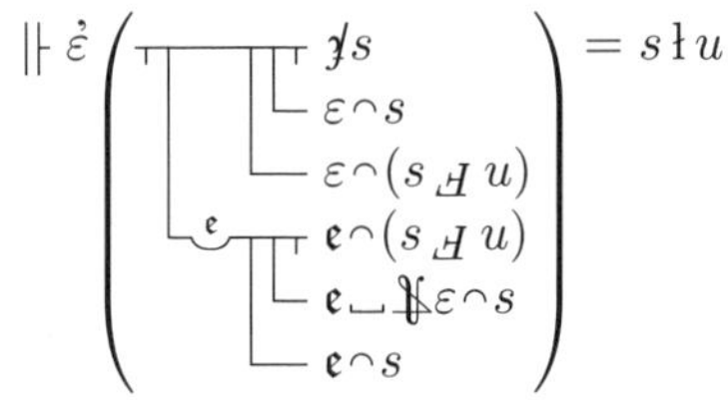

(AA

Accordingly, we read

"Δ⌢(Σ ł Φ)"

as

"Δ is a Σ-limit of Φ".

We then draw the following consequences from our definitions.

§194. *Construction*

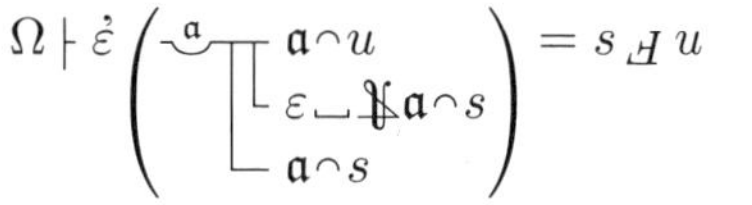

(44): ——————————

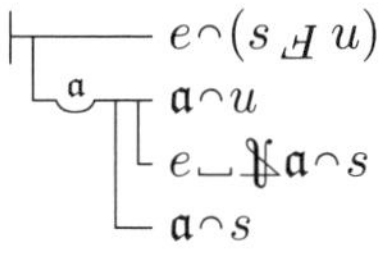

(590

—— • ——

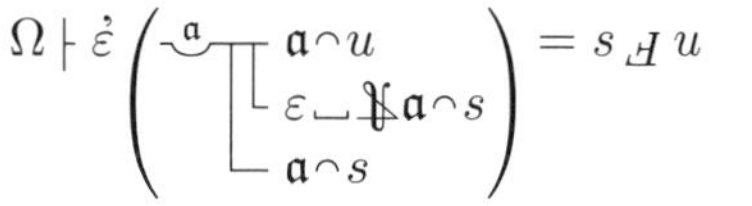

(46): ——————————

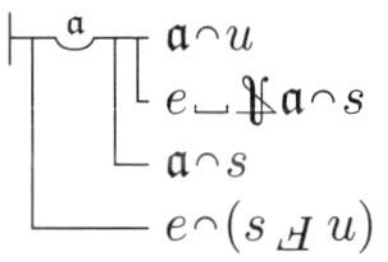

(α

(IIa): – – – – – – –

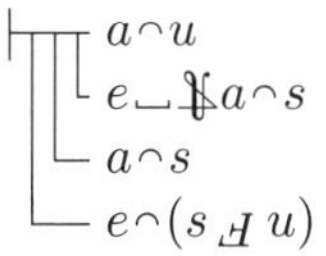

(591

—— • ——

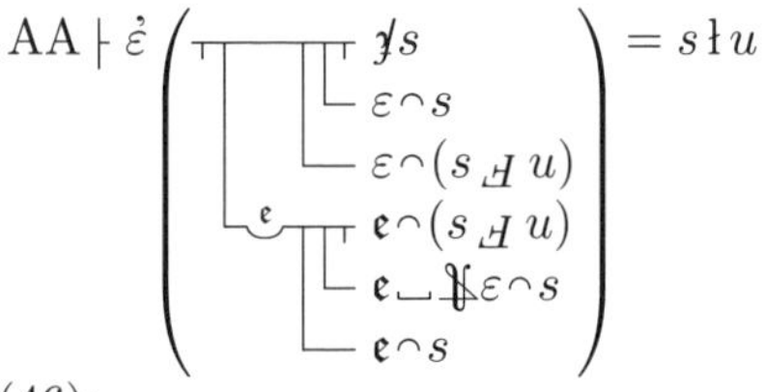

(46): ——————————

$$\vdash\; \mathfrak{J}s;\; p \frown s;\; p \frown (s \,\lrcorner\!\!\!\dashv u);\; \mathfrak{e}\!: \mathfrak{e} \frown (s \,\lrcorner\!\!\!\dashv u),\ \mathfrak{e} \smile \rangle\!\!\!\rangle p \frown s,\ \mathfrak{e} \frown s;\; p \frown (s \,\natural\, u) \quad (592$$

(Id): - - - - - - - -

$$\vdash\; \mathfrak{e}\!: \mathfrak{e} \frown (s \,\lrcorner\!\!\!\dashv u),\ \mathfrak{e} \smile \rangle\!\!\!\rangle p \frown s,\ \mathfrak{e} \frown s;\; p \frown (s \,\natural\, u) \quad (\alpha$$

(IIa): - - - - - - -

$$\vdash\; e \frown (s \,\lrcorner\!\!\!\dashv u);\; e \smile \rangle\!\!\!\rangle p \frown s;\; e \frown s;\; p \frown (s \,\natural\, u) \quad (593$$

×

$$\vdash\; p \frown (s \,\natural\, u);\; e \smile \rangle\!\!\!\rangle p \frown s;\; e \frown s;\; e \frown (s \,\lrcorner\!\!\!\dashv u) \quad (594$$

•

592 $$\vdash\; \mathfrak{J}s;\; p \frown s;\; p \frown (s \,\lrcorner\!\!\!\dashv u);\; \mathfrak{e}\!: \mathfrak{e} \frown (s \,\lrcorner\!\!\!\dashv u),\ \mathfrak{e} \smile \rangle\!\!\!\rangle p \frown s,\ \mathfrak{e} \frown s;\; p \frown (s \,\natural\, u)$$

(Ic): - - - - - - - - -

$$\vdash\; \mathfrak{J}s;\; p \frown s;\; p \frown (s \,\lrcorner\!\!\!\dashv u);\; p \frown (s \,\natural\, u) \quad (595$$

(Id): - - - - - - - -

$$\vdash\; p \frown (s \,\lrcorner\!\!\!\dashv u);\; p \frown (s \,\natural\, u) \quad (596$$

(591): - - - - -

$$\vdash\; a \frown u;\; p \smile \rangle\!\!\!\rangle a \frown s;\; a \frown s;\; p \frown (s \,\natural\, u) \quad (597$$

•

595 $$\vdash\; \mathfrak{J}s;\; p \frown s;\; p \frown (s \,\lrcorner\!\!\!\dashv u);\; p \frown (s \,\natural\, u)$$

(Ic): - - - - - - - -

$$\vdash\; \mathfrak{J}s;\; p \frown s;\; p \frown (s \,\natural\, u) \quad (598$$

(Id): - - - - - -

$$\vdash\; p \frown s;\; p \frown (s \,\natural\, u) \quad (599$$

•

598 $$\vdash\; \mathfrak{J}s;\; p \frown s;\; p \frown (s \,\natural\, u)$$

(Ib): - - - - - -

$$\vdash\; \mathfrak{J}s;\; p \frown (s \,\natural\, u) \quad (600$$

•

AA $$\vdash\; \dot{\varepsilon}\left(\mathfrak{J}s;\; \varepsilon \frown s;\; \varepsilon \frown (s \,\lrcorner\!\!\!\dashv u);\; \mathfrak{e}\!: \mathfrak{e} \frown (s \,\lrcorner\!\!\!\dashv u),\ \mathfrak{e} \smile \rangle\!\!\!\rangle \varepsilon \frown s,\ \mathfrak{e} \frown s \right) = s \,\natural\, u$$

(44): ————————————

$$\vdash\; p \frown (s \,\natural\, u);\; \Big[\mathfrak{J}s;\; p \frown s;\; p \frown (s \,\lrcorner\!\!\!\dashv u);\; \mathfrak{e}\!: \mathfrak{e} \frown (s \,\lrcorner\!\!\!\dashv u),\ \mathfrak{e} \smile \rangle\!\!\!\rangle p \frown s,\ \mathfrak{e} \frown s\Big] \quad (\alpha$$

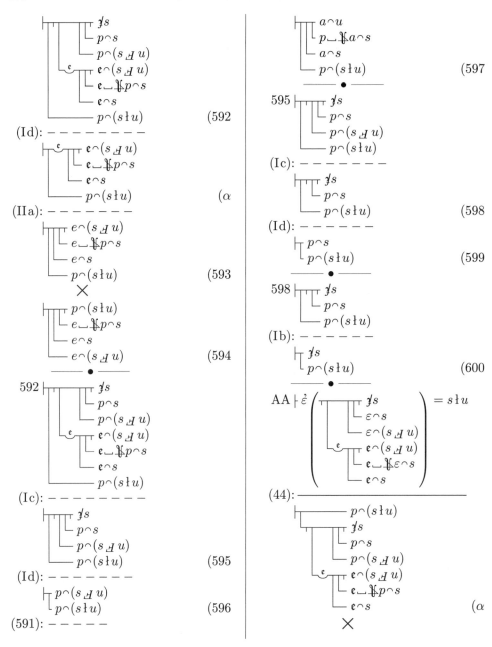

×

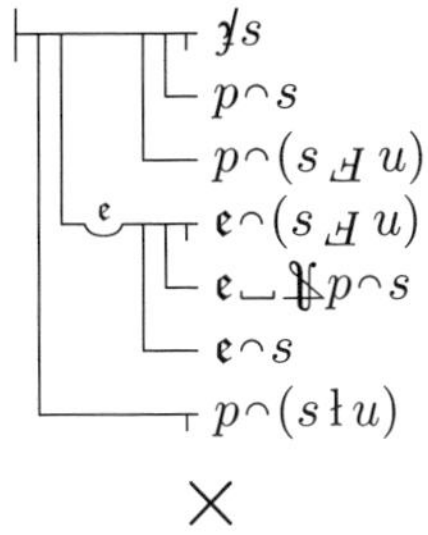

(β

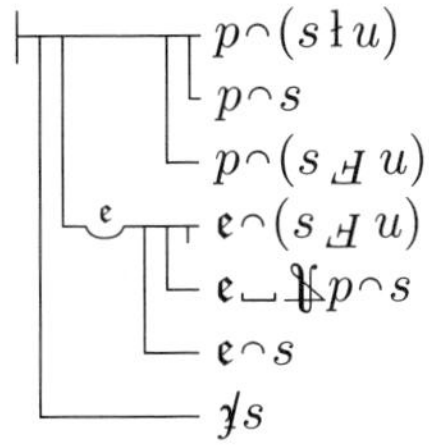

(601

§195. *Analysis*

We now prove that there is no more than one Σ-limit of a class; in signs:

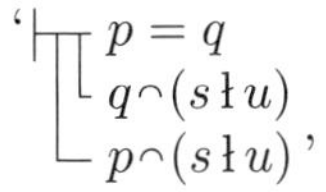

For this we use (587), showing of two such limits that the first is neither greater nor less than the second.

§196. *Construction*

594 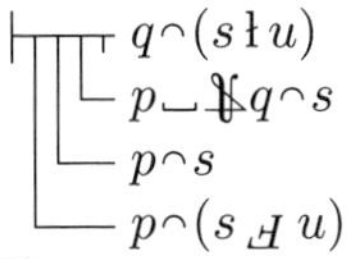

(599, 596):: = = = = =

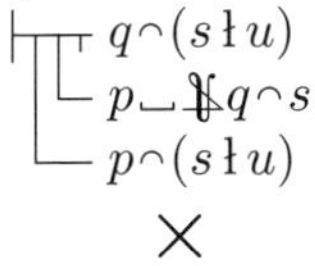

(α

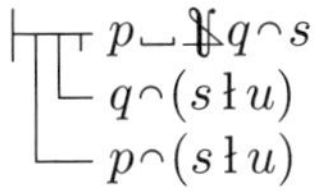

(β

(587): – – – – – –

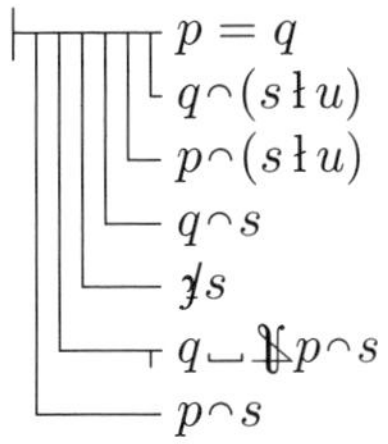

(γ

(β, 600):: = = = = = = = =

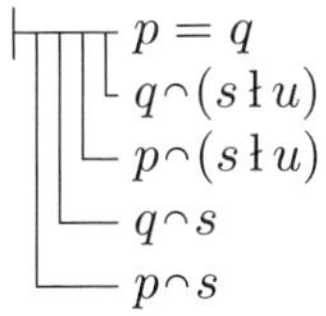

(δ

(599, 599):: = = = = =

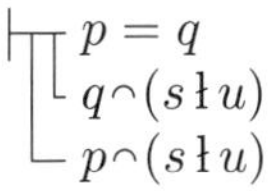

(602

Δ. The positive class

a) Definitions of the function ·p*ξ and consequences*

§197. *Analysis*

In §175 we asked the question: when is a class a positive class? This question has not yet been answered. At that point, we saw the need first to define the wider concept of the positival class. Having used it to defined the Σ-limit of a class, we can now answer the original question. For a class Σ to be a *positive class*, it must have the following properties:

1. it must be a positival class;
2. for every Relation belonging to it there must be another that also belongs to it and is less;
3. if there is a Relation in the class Σ such that all Relations in Σ that are less belong to a class Φ while there is a Relation in Σ that does not belong to the class Φ, then there must be a Σ-limit of Φ.

We now lay down the following definition:

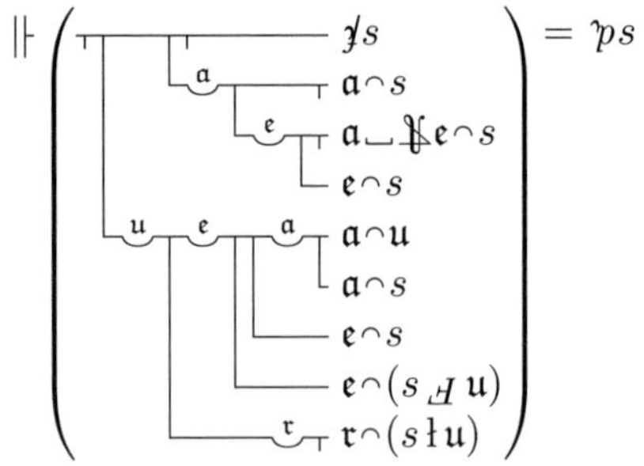

(AB

We read '$\mathfrak{p}\Sigma$': 'Σ is a positive class'. Let us now draw from (AB) the following consequences.

§198. *Construction*

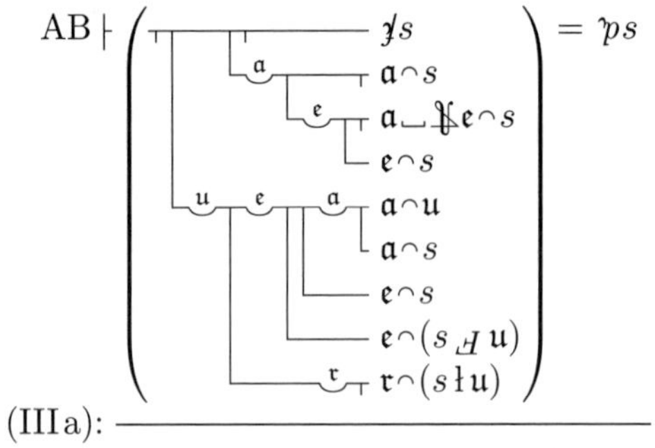

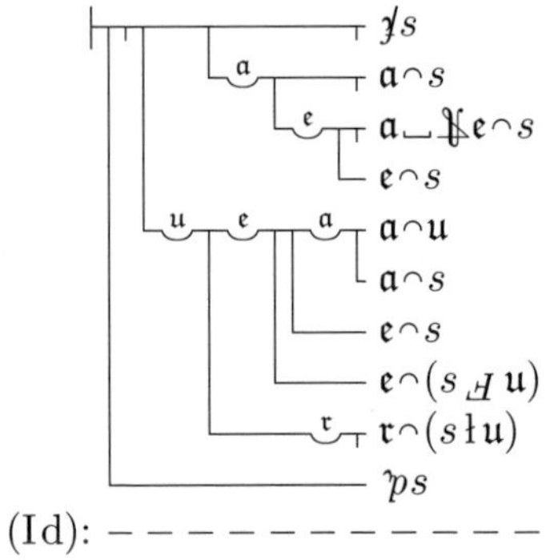

(603

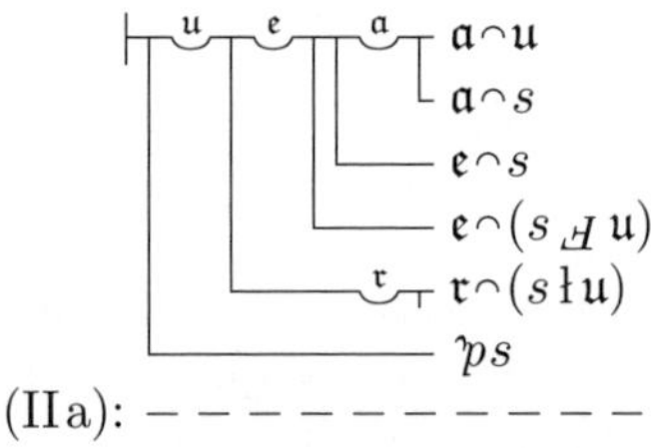

(α

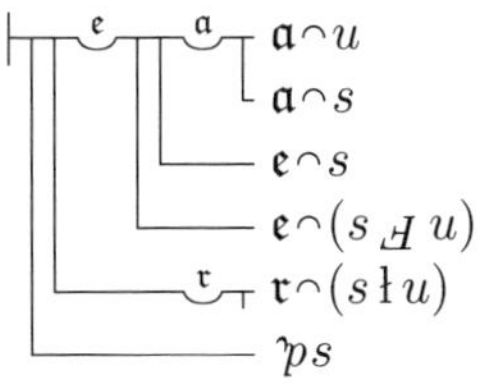

(β

(IIa): – – – – – – – – – –

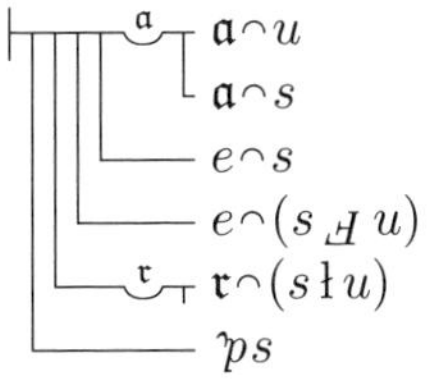

(γ

(IIa): – – – – – – – – –

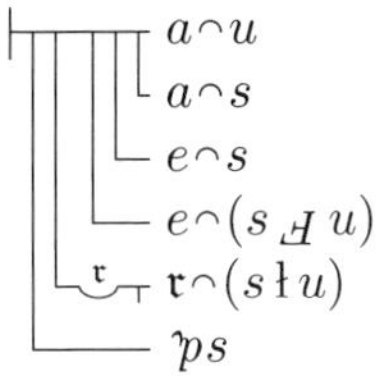

(604

•

603

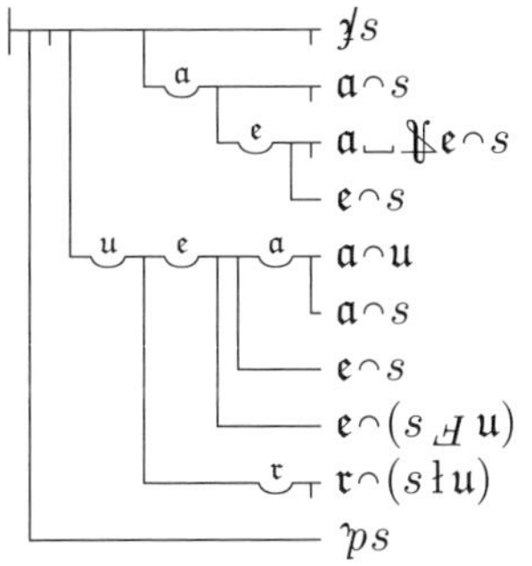

(Ic): – – – – – – – – – – – –

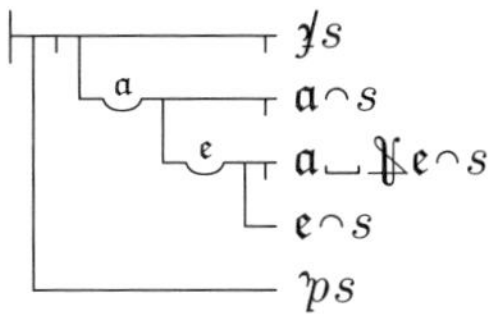

(605

(Id): – – – – – – – – – –

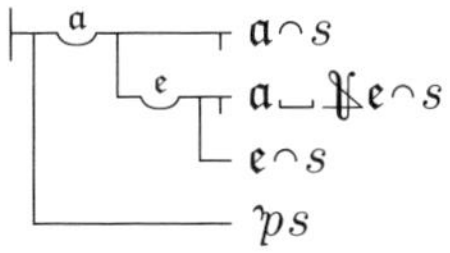

(IIa): – – – – – – – – – –

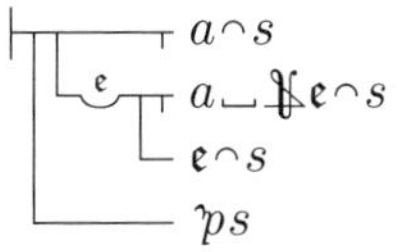

(606

•

605

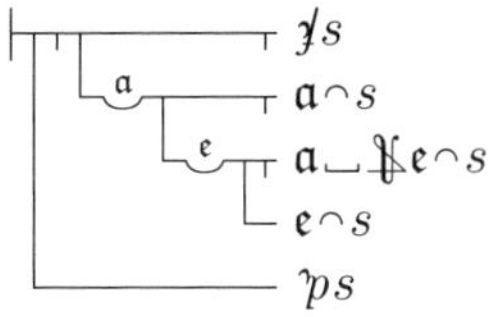

(Ib): – – – – – – – – – –

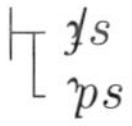

(607

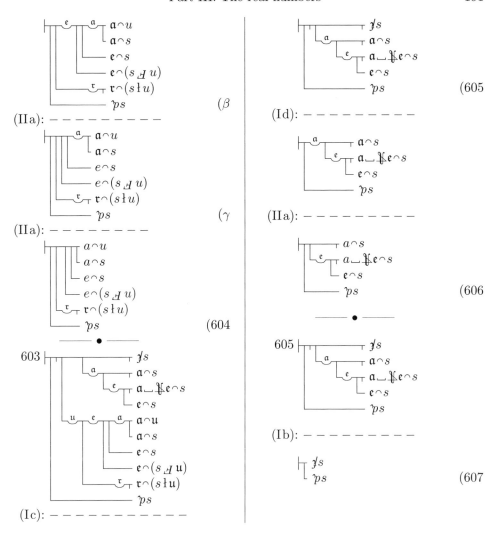

b) Proof of the proposition

‘

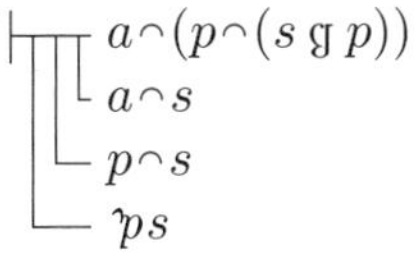

’

§199. *Analysis*

We now prove the proposition: “If two Relations belong to the same positive class then there is a multiple of the one that is not less than the other.”

(Archimedian Axiom)

The thought that there is a multiple of Π that is not less than A may be expressed thus:

$$\neg \underset{t}{\vdash} \begin{array}{l} A \smile \mathfrak{X} t \frown \Sigma \\ \Pi \frown (t \frown \smile_* \Pi) \end{array}$$

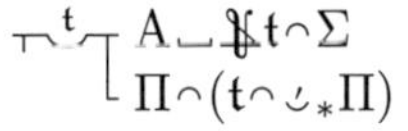

provided that Σ is the positive class. For the sake of brevity, we introduce a specific notation by defining:

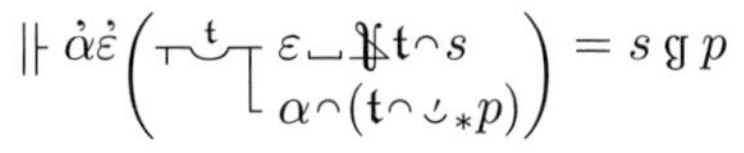

$$\Vdash \dot{\alpha}\dot{\varepsilon} \left(\neg \underset{t}{\vdash} \begin{array}{l} \varepsilon \smile \mathfrak{X} t \frown s \\ \alpha \frown (t \frown \smile_* p) \end{array} \right) = s \mathfrak{g} p \qquad (A\Gamma$$

Accordingly, if Σ is a positive class and A and Π belong to it, we read:

'$A \frown (\Pi \frown (\Sigma \mathfrak{g} \Pi))$'

as "there is a multiple of Π that is not less than A".

With this notation, we have in the heading the proposition mentioned in words above.

Let us first draw some consequences from $(A\Gamma)$.

§200. Construction

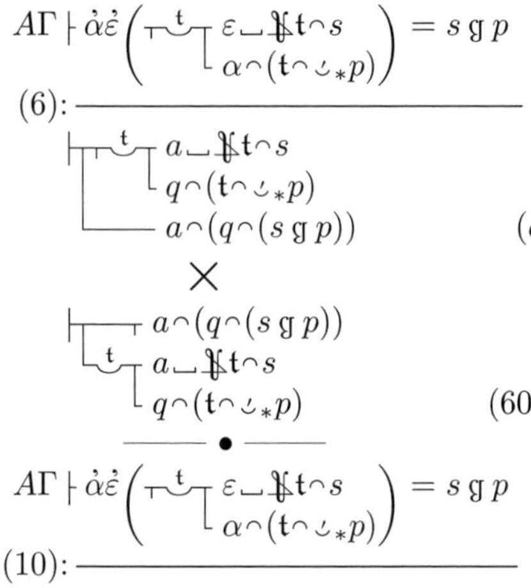

$$A\Gamma \vdash \dot{\alpha}\dot{\varepsilon} \left(\neg \underset{t}{\vdash} \begin{array}{l} \varepsilon \smile \mathfrak{X} t \frown s \\ \alpha \frown (t \frown \smile_* p) \end{array} \right) = s \mathfrak{g} p$$

(6): ─────────────

$$\vdash \underset{t}{\vdash} \begin{array}{l} a \smile \mathfrak{X} t \frown s \\ q \frown (t \frown \smile_* p) \\ a \frown (q \frown (s \mathfrak{g} p)) \end{array} \qquad (\alpha$$

$\times$

$$\vdash \underset{t}{\vdash} \begin{array}{l} a \frown (q \frown (s \mathfrak{g} p)) \\ a \smile \mathfrak{X} t \frown s \\ q \frown (t \frown \smile_* p) \end{array} \qquad (608$$

$$A\Gamma \vdash \dot{\alpha}\dot{\varepsilon} \left(\neg \underset{t}{\vdash} \begin{array}{l} \varepsilon \smile \mathfrak{X} t \frown s \\ \alpha \frown (t \frown \smile_* p) \end{array} \right) = s \mathfrak{g} p$$

(10): ─────────────

$$\vdash \begin{array}{l} a \frown (q \frown (s \mathfrak{g} p)) \\ a \smile \mathfrak{X} t \frown s \\ q \frown (t \frown \smile_* p) \end{array} \qquad (\alpha$$

$\times$

$$\vdash \underset{t}{\vdash} \begin{array}{l} a \smile \mathfrak{X} t \frown s \\ q \frown (t \frown \smile_* p) \\ a \frown (q \frown (s \mathfrak{g} p)) \end{array} \qquad (\beta$$

(IIa): ─ ─ ─ ─ ─ ─ ─ ─ ─

$$\vdash \begin{array}{l} a \smile \mathfrak{X} t \frown s \\ q \frown (t \frown \smile_* p) \\ \dot{a} \frown (q \frown (s \mathfrak{g} p)) \end{array} \qquad (609$$

$\times$

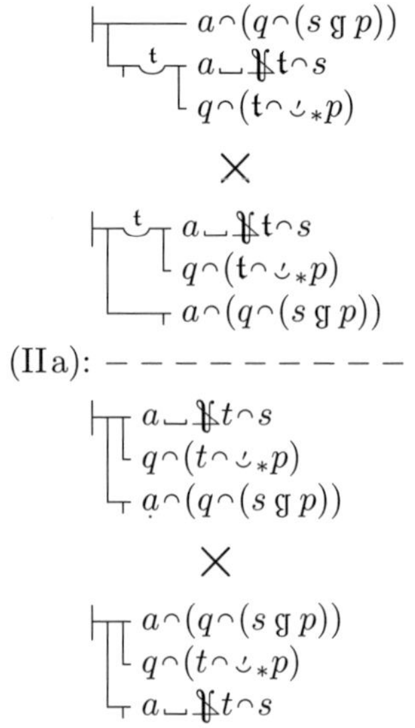

$$\vdash \begin{array}{l} a \frown (q \frown (s \mathfrak{g} p)) \\ q \frown (t \frown \smile_* p) \\ a \smile \mathfrak{X} t \frown s \end{array} \qquad (610$$

§201. Analysis

In order to prove the proposition of §199, we make the following observation. We assume that p and q belong to the positive class s. If now every member of the $_*p$-series starting with q were less than a, then according to (604) there would be an s-limit of $q \frown (s \mathfrak{g} p)$; for in the class s there are also Relations—for instance, q—that are reached or surpassed by members of that series, and this then also holds of all that are less. This s-limit m of $q \frown (s \mathfrak{g} p)$ is obviously dependent on q. We will show that $c \smile p$ is the s-limit of $p \smile p \frown (s \mathfrak{g} p)$ if c is the s-limit of $p \frown (s \mathfrak{g} p)$. Now, obviously, the two s-limits coincide, so that $c \smile p$ would have to coincide with c. In that case, according to (585), p could not belong to the class s. From this results the falsity of the assumption that every member of the p-series starting with p—i.e., each multiple of p—is less than a.

Our first task will be to prove the proposition

$$`\vdash \begin{array}{l} c\smile p\frown(s\,\natural\,(p\smile p\frown(s\,\mathfrak{g}\,p))) \\ p\frown s \\ c\frown(s\,\natural\,(p\frown(s\,\mathfrak{g}\,p))) \end{array} \text{'} \qquad (\alpha$$

According to (601), we require for this the proposition

$$`\vdash \begin{array}{l} c\smile p\frown(s\,\not\natural\,(p\smile p\frown(s\,\mathfrak{g}\,p))) \\ p\frown s \\ \not\mathfrak{f}s \\ c\frown(s\,\natural\,(p\frown(s\,\mathfrak{g}\,p))) \end{array} \text{'} \qquad (\beta$$

According to (597) we have

$$\vdash \begin{array}{l} r\frown(q\frown(s\,\mathfrak{g}\,p)) \\ c\smile\not\natural r\frown s \\ r\frown s \\ c\frown(s\,\natural\,(q\frown(s\,\mathfrak{g}\,p))) \end{array}$$

in which 'r' is then to be replaced by '$a\smile\not\natural p$', after '$r\frown\mathfrak{d}s$' is introduced for '$r\frown s$'. The transition to (β) is mediated by the proposition

$$`\vdash \begin{array}{l} a\frown(q\smile b\frown(s\,\mathfrak{g}\,p)) \\ a\smile\not\natural b\frown(q\frown(s\,\mathfrak{g}\,p)) \end{array} \text{'} \qquad (\gamma$$

which is derived from

$$`\vdash \begin{array}{l} q\smile b\frown(t\smile b\frown\smile_{*}p) \\ q\frown(t\frown\smile_{*}p) \end{array} \text{'} \qquad (\delta$$

The latter we prove using (144).

§202. *Construction*

$$490 \vdash \begin{array}{l} p\smile(d\smile b) = a\smile b \\ p\smile d\smile b = a\smile b \end{array}$$

(IIIh):: – – – – – – – – – –

$$\vdash \begin{array}{l} p\smile(d\smile b) = a\smile b \\ p\smile d = a \end{array} \qquad (\alpha$$

(492):: – – – – – – – – – –

$$\vdash \begin{array}{l} p\smile(d\smile b) = a\smile b \\ d\frown(a\frown_{*}p) \end{array} \qquad (\beta$$

(495): ————————————

$$\vdash \begin{array}{l} d\smile b\frown(a\smile b\frown_{*}p) \\ d\frown(a\frown_{*}p) \end{array} \qquad (611$$

(137): – – – – – – – –

$$\vdash \begin{array}{l} r\frown(a\smile b\frown\smile_{*}p) \\ d\frown(a\frown_{*}p) \\ r\frown(d\smile b\frown\smile_{*}p) \end{array} \qquad (\alpha$$

$$\vdash^{\mathfrak{d}}\!\smile\!^{\mathfrak{a}}\!\smile \begin{array}{l} r\frown(\mathfrak{a}\smile b\frown\smile_{*}p) \\ \mathfrak{d}\frown(\mathfrak{a}\frown_{*}p) \\ r\frown(\mathfrak{d}\smile b\frown\smile_{*}p) \end{array} \qquad (\beta$$

(144): ————————————————

$$\vdash \begin{array}{l} r\frown(t\smile b\frown\smile_{*}p) \\ r\frown(q\smile b\frown\smile_{*}p) \\ q\frown(t\frown\smile_{*}p) \end{array} \qquad (612$$

——— • ———

$$140 \vdash q\smile b\frown(q\smile b\frown\smile_{*}p)$$

(612): ————————————

$$\vdash \begin{array}{l} q\smile b\frown(t\smile b\frown\smile_{*}p) \\ q\frown(t\frown\smile_{*}p) \end{array} \qquad (613$$

——— • ———

$$506 \vdash \begin{array}{l} a\smile\not\natural b\smile\not\natural t\frown s \\ a\smile\not\natural(t\smile b)\frown s \end{array}$$

(609):: – – – – – – –

$$\vdash \begin{array}{l} a\smile\not\natural b\smile\not\natural t\frown s \\ q\smile b\frown(t\smile b\frown\smile_{*}p) \\ a\frown(q\smile b\frown(s\,\mathfrak{g}\,p)) \end{array} \qquad (\alpha$$

(613):: – – – – – – – – –

$$\vdash \begin{array}{l} a\smile\not\natural b\smile\not\natural t\frown s \\ q\frown(t\frown\smile_{*}p) \\ a\frown(q\smile b\frown(s\,\mathfrak{g}\,p)) \end{array} \qquad (\beta$$

$$\vdash^{\mathfrak{t}}\!\smile \begin{array}{l} a\smile\not\natural b\smile\not\natural\mathfrak{t}\frown s \\ q\frown(\mathfrak{t}\frown\smile_{*}p) \\ a\frown(q\smile b\frown(s\,\mathfrak{g}\,p)) \end{array} \qquad (\gamma$$

(608): – – – – – – – – – –

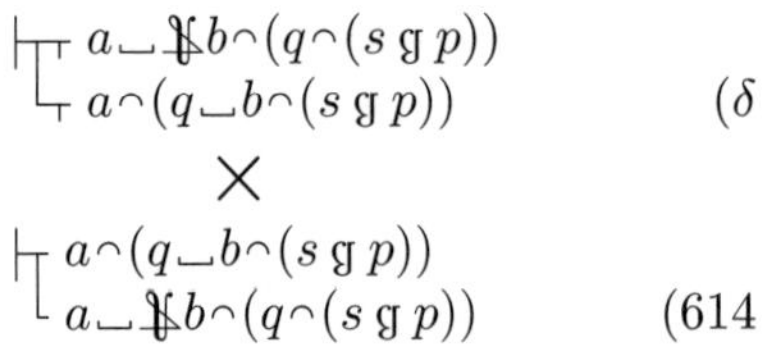

§203. *Analysis*

In order to replace '$r \frown s$' by '$r \frown \eth s$', as indicated in §201, we prove the proposition

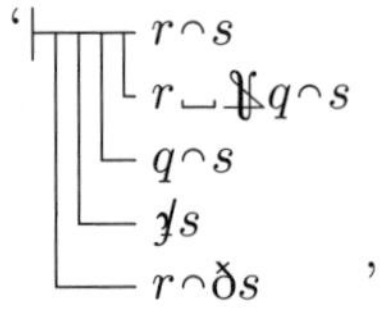

i.e., "a magnitude is positive if it is greater than a positive magnitude in its domain."

§204. *Construction*

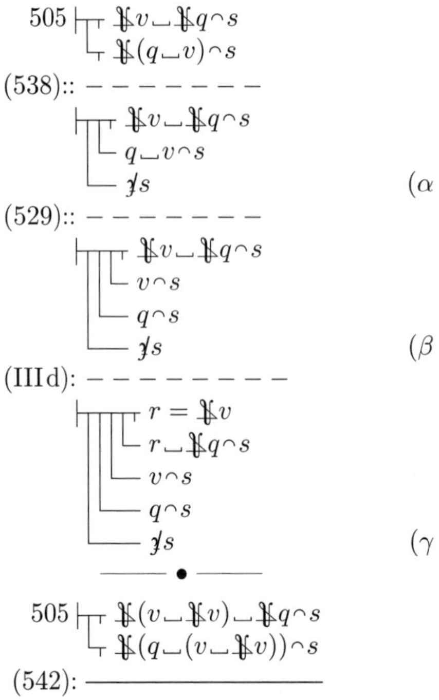

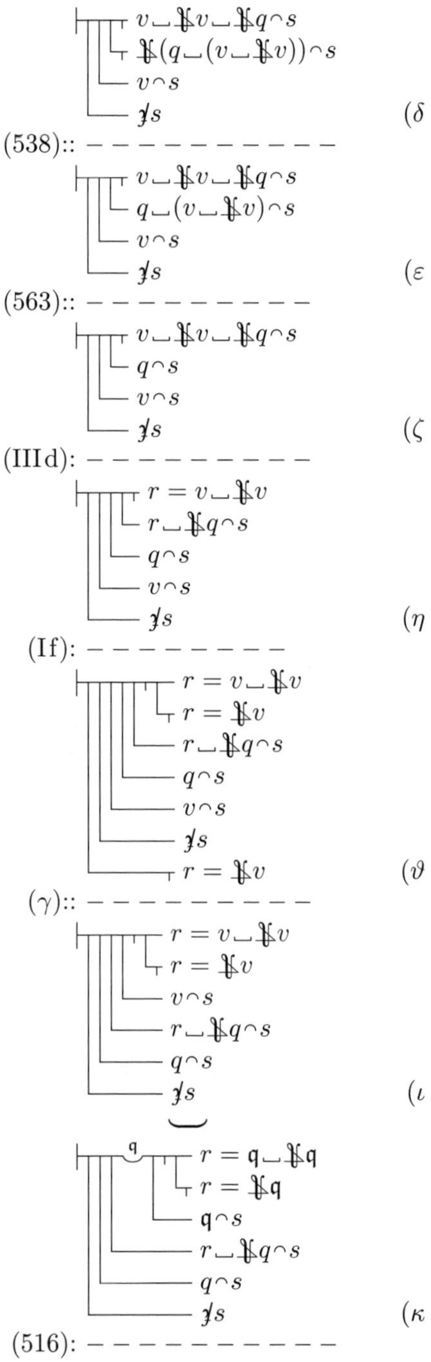

⊢ r⌢s
  r⌣Ꝁq⌢s
  q⌢s
  ʒ̸s
  r⌢ðs (615

(609):: – – – – – – – –

⊢ r⌢s
  q⌢(q⌢⌣̇∗p)
  r⌢(q⌢(s 𝔤 p))
  q⌢s
  ʒ̸s
  r⌢ðs (α

(140):: ————————————

⊢ r⌢s
  r⌢(q⌢(s 𝔤 p))
  q⌢s
  ʒ̸s
  r⌢ðs (β

×

⊢ r⌢(q⌢(s 𝔤 p))
  r⌢s
  q⌢s
  ʒ̸s
  r⌢ðs (γ

(597): ·–·–·–·–·–·–·–·

⊢ r⌢(q⌢(s 𝔤 p))
  q⌢s
  ʒ̸s
  r⌢ðs
  c⌣Ꝁr⌢s
  c⌢(s ł (q⌢(s 𝔤 p))) (616

——— • ———

490 ⊢ F(c⌣(p⌣Ꝁa))
  F(c⌣p⌣Ꝁa)

(543): ————————————

⊢ F(c⌣Ꝁ(a⌣Ꝁp))
  F(c⌣p⌣Ꝁa)
  p⌢s
  ʒ̸s (617

(607):: – – – – – – – – – –

⊢ F(c⌣Ꝁ(a⌣Ꝁp))
  F(c⌣p⌣Ꝁa)
  p⌢s
  ʒ̸s (618

——— • ———

614 ⊢ a⌢(p⌣p⌢(s 𝔤 p))
  a⌣Ꝁp⌢(p⌢(s 𝔤 p))

(616):: – – – – – – – – – –

⊢ a⌢(p⌣p⌢(s 𝔤 p))
  p⌢s
  ʒ̸s
  a⌣Ꝁp⌢ðs
  c⌣Ꝁ(a⌣Ꝁp)⌢s
  c⌢(s ł (p⌢(s 𝔤 p))) (α

(617, 528):: = = = = = = = = =

⊢ a⌢(p⌣p⌢(s 𝔤 p))
  c⌣p⌣Ꝁa⌢s
  a⌢s
  p⌢s
  ʒ̸s
  c⌢(s ł (p⌢(s 𝔤 p))) (β

⊢ 𝔞 ⌣ 𝔞⌢(p⌣p⌢(s 𝔤 p))
  c⌣p⌣Ꝁ𝔞⌢s
  𝔞⌢s
  p⌢s
  ʒ̸s
  c⌢(s ł (p⌢(s 𝔤 p))) (γ

(590): – – – – – – – – – – – – – –

⊢ c⌣p⌢(s ꟻ (p⌣p⌢(s 𝔤 p)))
  p⌢s
  ʒ̸s
  c⌢(s ł (p⌢(s 𝔤 p))) (619

§205. *Analysis*

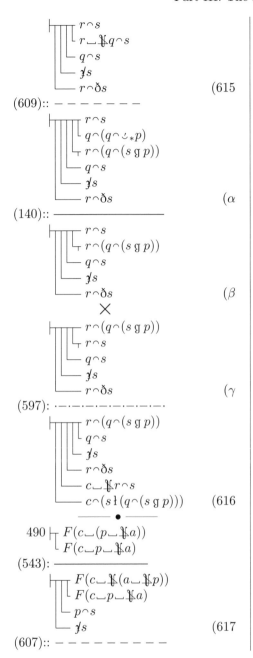

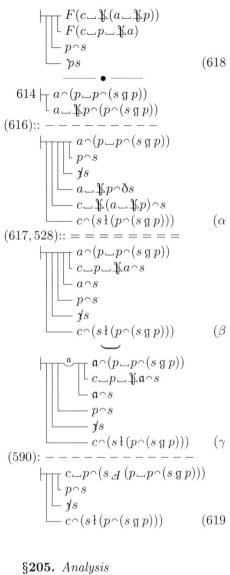

In order to prove proposition (α) of §201, we require, by (601), in addition to the proposition just proven, the further proposition

'⊢ e⌢(s ⅎ (p⌣p⌢(s g p)))
 e⌣ℵ(c⌣p)⌢s
 e⌢s
 p⌢s
 c⌣p⌢s
 ⅎs
 c⌢(s ł (p⌢(s g p))) ' (α

which says that, under our assumptions, there is for every Relation e greater than $c⌣p$, a Relation that is less than e and positive and not reached by any member of the p-series starting with $p⌣p$, provided that c is an s-limit of $(p⌢(s\,g\,p))$. We show that $c⌣p$ itself is not reached. From c being an s-limit of $p⌢(s\,g\,p)$ it follows that c is not surpassed by members of the $_*p$-series starting with p and hence is not reached. From this it then follows that $c⌣p$ is not reached by members of the $_*p$-series starting with $p⌣p$.

We first prove the proposition

'⊢ c⌣p⌢(p⌣p⌢(s g p))
 p⌢s
 ⅎs
 c⌢(p⌢(s g p)) ' (β

which is to be reduced to the proposition

'⊢ p⌢(t⌣ℵp⌢⌣*p)
 p⌢s
 ⅎs
 p⌣p⌢(t⌢⌣*p) ' (γ

The latter follows from (612).

§206. *Construction*

491 ⊢ F(c⌣b⌣ℵt)
 F(c⌣(b⌣ℵt))

(542):: - - - - - - - -

⊢ F(c⌣b⌣ℵt)
 F(c⌣ℵ(t⌣ℵb))
 b⌢s
 ⅎs (620

———•———

612 ⊢ p⌢(t⌣ℵp⌢⌣*p)
 p⌢(p⌣p⌣ℵp⌢⌣*p)
 p⌣p⌢(t⌢⌣*p)

(569):: - - - - - - - - - - -

⊢ p⌢(t⌣ℵp⌢⌣*p)
 p⌢(p⌢⌣*p)
 p⌢s
 ⅎs
 p⌣p⌢(t⌢⌣*p) (α

(140):: ————————

⊢ p⌢(t⌣ℵp⌢⌣*p)
 p⌢s
 ⅎs
 p⌣p⌢(t⌢⌣*p) (621

(609): - - - - - - - - - -

⊢ c⌣ℵ(t⌣ℵp)⌢s
 p⌢s
 ⅎs
 p⌣p⌢(t⌢⌣*p)
 c⌢(p⌢(s g p)) (α

(620): ————————

⊢ c⌣p⌣ℵt⌢s
 p⌣p⌢(t⌢⌣*p)
 p⌢s
 ⅎs
 c⌢(p⌢(s g p)) (β

⊢ ⌣t c⌣p⌣ℵt⌢s
 p⌣p⌢(t⌢⌣*p)
 p⌢s
 ⅎs
 c⌢(p⌢(s g p)) (γ

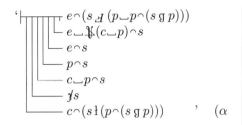

(608): - - - - - - - - - - -

$$\begin{array}{l} \vdash\!\!\top\!\!\top\!\!\top\; c \smile p \frown (p \smile p \frown (s\, \mathfrak{g}\, p)) \\ \quad\;\; p \frown s \\ \quad\;\; \not\mathfrak{I} s \\ \quad\;\; c \frown (p \frown (s\, \mathfrak{g}\, p)) \end{array} \qquad (622$$

§207. *Analysis*

To execute the plan of §205 we prove that, under our assumptions, if c is an s-limit of $p \frown (s\, \mathfrak{g}\, p)$, c cannot be reached by the members of the ${}_*p$-series starting with p, that is, the proposition

$$\text{‘}\begin{array}{l} \vdash\!\!\top\!\!\top\!\!\top\; c \frown (p \frown (s\, \mathfrak{g}\, p)) \\ \quad\;\; c \frown (s\, \mathfrak{l}\, (p \frown (s\, \mathfrak{g}\, p))) \\ \quad\;\; \not\mathfrak{I} s \\ \quad\;\; p \frown s \end{array}\text{’}, \qquad (\alpha$$

If c were reached exactly by the member t of the ${}_*p$-series starting with p, so that $t = c$, then c would be surpassed by the next member of this series, $p \smile t$. So in every case where c is reached by a member of our series, there is also one by which it is surpassed. If, however, t is a member of our series, and if a is less than t, then a is surpassed by a member of the series, which, if t is greater than c, contradicts that c is to be an s-limit of $p \frown (s\, \mathfrak{g}\, p)$ by (594). In this manner we prove the proposition

$$\text{‘}\begin{array}{l} \vdash\!\!\top\!\!\top\!\!\top\; t \smile \not\mathfrak{h}\, c \frown s \\ \quad\;\; c \frown (s\, \mathfrak{l}\, (p \frown (s\, \mathfrak{g}\, p))) \\ \quad\;\; t \frown s \\ \quad\;\; p \frown (t \frown \acute{\smile}\, {}_*p) \\ \quad\;\; \not\mathfrak{I} s \end{array}\text{’}, \qquad (\beta$$

From this we deduce using (588) that, under our assumptions, t is less than c, since the case where $t = c$ is, as seen above, also excluded. Next the subcomponent ‘$t \frown s$’ has to be eliminated. This is done by means of the proposition to be proven next:

$$\text{‘}\begin{array}{l} \vdash\!\!\top\!\!\top\!\!\top\; t \frown s \\ \quad\;\; q \frown s \\ \quad\;\; p \frown s \\ \quad\;\; \not\mathfrak{I} s \\ \quad\;\; q \frown (t \frown \acute{\smile}\, {}_*p) \end{array}\text{’} \qquad (\gamma$$

§208. *Construction*

$$529 \begin{array}{l} \vdash\!\!\top\!\!\top\; p \smile d \frown s \\ \quad\;\; d \frown s \\ \quad\;\; p \frown s \\ \quad\;\; \not\mathfrak{I} s \end{array}$$

(493): ———————

$$\begin{array}{l} \vdash\!\!\top\!\!\top\!\!\top\; a \frown s \\ \quad\;\; d \frown (a \frown {}_*p) \\ \quad\;\; d \frown s \\ \quad\;\; p \frown s \\ \quad\;\; \not\mathfrak{I} s \end{array} \qquad (\alpha$$

$$\begin{array}{l} \vdash\!\!\top\!\!\overset{\mathfrak{d}}{\smile}\!\!\top\!\!\overset{\mathfrak{a}}{\smile}\!\!\top\; \mathfrak{a} \frown s \\ \quad\;\; \mathfrak{d} \frown (\mathfrak{a} \frown {}_*p) \\ \quad\;\; \mathfrak{d} \frown s \\ \quad\;\; p \frown s \\ \quad\;\; \not\mathfrak{I} s \end{array} \qquad (\beta$$

(144): - - - - - - - - - -

$$\begin{array}{l} \vdash\!\!\top\!\!\top\!\!\top\; t \frown s \\ \quad\;\; q \frown s \\ \quad\;\; p \frown s \\ \quad\;\; \not\mathfrak{I} s \\ \quad\;\; q \frown (t \frown \acute{\smile}\, {}_*p) \end{array} \qquad (623$$

(607):: - - - - - - - -

$$\begin{array}{l} \vdash\!\!\top\!\!\top\!\!\top\; t \frown s \\ \quad\;\; q \frown s \\ \quad\;\; p \frown s \\ \quad\;\; \mathfrak{p} s \\ \quad\;\; q \frown (t \frown \acute{\smile}\, {}_*p) \end{array} \qquad (624$$

———— • ————

496 ⊢ $t \frown (p \smile t \frown_{*} p)$

(137): ————————

⊢ $q \frown (p \smile t \frown \dot{\smile}_{*} p)$
  $q \frown (t \frown \dot{\smile}_{*} p)$ (625

——— • ———

529 ⊢ $c \smile p \frown s$
  $p \frown s$
  $c \frown s$
  $\not{\jmath} s$

(600, 599):: = = =

⊢ $c \smile p \frown s$
  $p \frown s$
  $c \frown (s \,\text{ł}\, u)$ (626

——— • ———

610 ⊢ $a \frown (p \frown (s \,\mathfrak{g}\, p))$
  $p \frown (t \frown \dot{\smile}_{*} p)$
  $a \smile \not{\S} t \frown s$

(589):: – – – – – – – –

⊢ $a \frown (p \frown (s \,\mathfrak{g}\, p))$
  $t \smile \not{\S} a \frown s$
  $a \frown s$
  $p \frown (t \frown \dot{\smile}_{*} p)$
  $\not{\jmath} s$ (α

⊢ $\overset{\mathfrak{a}}{\smile}$ $\mathfrak{a} \frown (p \frown (s \,\mathfrak{g}\, p))$
  $t \smile \not{\S} \mathfrak{a} \frown s$
  $\mathfrak{a} \frown s$
  $p \frown (t \frown \dot{\smile}_{*} p)$
  $\not{\jmath} s$ (β

(590): – – – – – – – – – –

⊢ $t \frown (s \,\text{Ⅎ}\, (p \frown (s \,\mathfrak{g}\, p)))$
  $p \frown (t \frown \dot{\smile}_{*} p)$
  $\not{\jmath} s$ (γ

(594): – – – – – – – – – –

⊢ $c \frown (s \,\text{ł}\, (p \frown (s \,\mathfrak{g}\, p)))$
  $t \smile \not{\S} c \frown s$
  $t \frown s$
  $p \frown (t \frown \dot{\smile}_{*} p)$
  $\not{\jmath} s$ (δ

×

⊢ $t \smile \not{\S} c \frown s$
  $c \frown (s \,\text{ł}\, (p \frown (s \,\mathfrak{g}\, p)))$
  $t \frown s$
  $p \frown (t \frown \dot{\smile}_{*} p)$
  $\not{\jmath} s$ (ε

——— • ———

IIId ⊢ $t = c$
  $p \smile t \smile \not{\S} t \frown s$
  $p \smile t \smile \not{\S} c \frown s$

(569):: – – – – – – – –

⊢ $t = c$
  $p \frown s$
  $t \frown s$
  $\not{\jmath} s$
  $p \smile t \smile \not{\S} c \frown s$ (ζ

(588): – – – – – – – – –

⊢ $c \smile \not{\S} t \frown s$
  $t \smile \not{\S} c \frown s$
  $c \frown s$
  $\not{\jmath} s$
  $p \frown s$
  $t \frown s$
  $p \smile t \smile \not{\S} c \frown s$ (η

(ε, ε):: = = = = = = = = =

⊢ $c \smile \not{\S} t \frown s$
  $c \frown (s \,\text{ł}\, (p \frown (s \,\mathfrak{g}\, p)))$
  $t \frown s$
  $p \frown (t \frown \dot{\smile}_{*} p)$
  $\not{\jmath} s$
  $c \frown s$
  $p \frown s$
  $p \smile t \frown s$
  $p \frown (p \smile t \frown \dot{\smile}_{*} p)$ (ϑ

(529, 625):: = = = = = = = = =

⊢ $c \smile \not{\S} t \frown s$
  $c \frown (s \,\text{ł}\, (p \frown (s \,\mathfrak{g}\, p)))$
  $t \frown s$
  $p \frown (t \frown \dot{\smile}_{*} p)$
  $\not{\jmath} s$
  $c \frown s$
  $p \frown s$ (ι

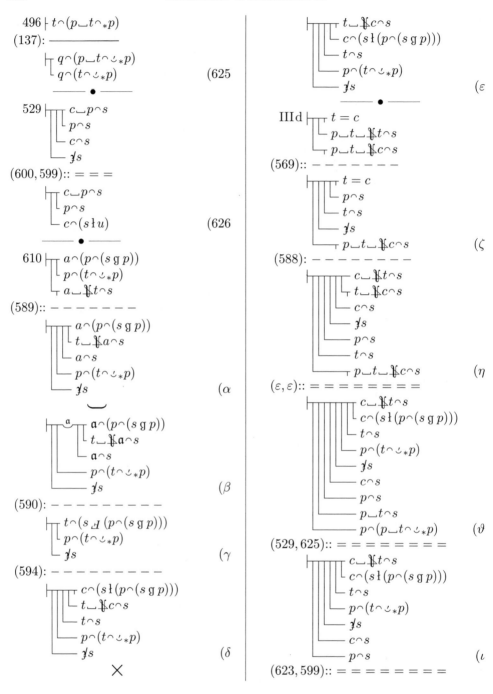

(623, 599):: = = = = = = = = =

```
⊢ c⌣∦t⌢s
  p⌢(t⌢⌣̇*p)
  c⌢(s ł (p⌢(s ɡ p)))
  ɟs
  p⌢s                                   (κ
```

```
⊢ ‿t‿ c⌣∦t⌢s
      p⌢(t⌢⌣̇*p)
  c⌢(s ł (p⌢(s ɡ p)))
  ɟs
  p⌢s                                   (λ
```

(608): - - - - - - - - - - - -

```
⊢ c⌢(p⌢(s ɡ p))
  c⌢(s ł (p⌢(s ɡ p)))
  ɟs
  p⌢s                                   (μ
```

(622): - - - - - - - - - - -

```
⊢ c⌣p⌢(p⌣p⌢(s ɡ p))
  p⌢s
  ɟs
  c⌢(s ł (p⌢(s ɡ p)))                   (ν
```

×

```
⊢ p⌢s
  c⌣p⌢(p⌣p⌢(s ɡ p))
  ɟs
  c⌢(s ł (p⌢(s ɡ p)))                   (ξ
```

(591):: - - - - - - - - - - - -

```
⊢ p⌢s
  e⌣∦(c⌣p)⌢s
  c⌣p⌢s
  e⌢(s Ⅎ (p⌣p⌢(s ɡ p)))
  ɟs
  c⌢(s ł (p⌢(s ɡ p)))                   (o
```

×

```
⊢ e⌢(s Ⅎ (p⌣p⌢(s ɡ p)))
  e⌣∦(c⌣p)⌢s
  p⌢s
  c⌣p⌢s
  ɟs
  c⌢(s ł (p⌢(s ɡ p)))                   (π
```

(I): ————————————————

```
⊢ e⌢(s Ⅎ (p⌣p⌢(s ɡ p)))
  e⌣∦(c⌣p)⌢s
  e⌢s
  p⌢s
  c⌣p⌢s
  ɟs
  c⌢(s ł (p⌢(s ɡ p)))                   (ρ
```

```
⊢ ‿e‿ e⌢(s Ⅎ (p⌣p⌢(s ɡ p)))
      e⌣∦(c⌣p)⌢s
      e⌢s
  p⌢s
  c⌣p⌢s
  ɟs
  c⌢(s ł (p⌢(s ɡ p)))                   (σ
```

(601): - - - - - - - - - - - - - - -

```
⊢ c⌣p⌢(s ł (p⌣p⌢(s ɡ p)))
  c⌣p⌢s
  c⌣p⌢(s Ⅎ (p⌣p⌢(s ɡ p)))
  p⌢s
  ɟs
  c⌢(s ł (p⌢(s ɡ p)))                   (τ
```

(619, 626):: = = = = = = = = = = =

```
⊢ c⌣p⌢(s ł (p⌣p⌢(s ɡ p)))
  p⌢s
  c⌢(s ł (p⌢(s ɡ p)))
  ɟs                                    (υ
```

(600):: - - - - - - - - - - - - -

```
⊢ c⌣p⌢(s ł (p⌣p⌢(s ɡ p)))
  p⌢s
  c⌢(s ł (p⌢(s ɡ p)))                   (627
```

§209. *Analysis*

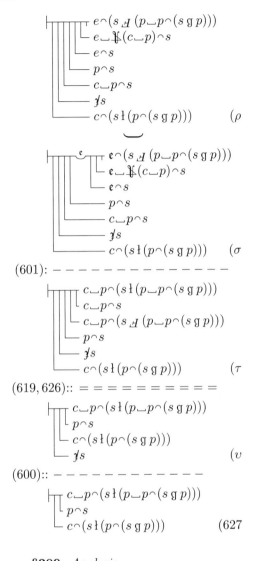

To execute the plan of §201 we now prove that the s-limits of $p⌣p⌢(s\,ɡ\,p)$ and $p⌢(s\,ɡ\,p)$ coincide. For this, we will need to use (602). We still have to show that every s-limit of $p⌣p⌢(s\,ɡ\,p)$ is also an s-limit of $p⌢(s\,ɡ\,p)$:

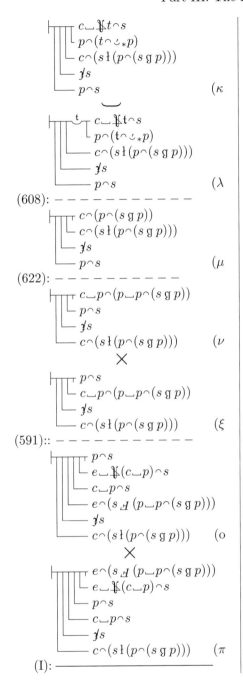

$$`\vdash \begin{array}{l} d\frown(s\,\mathfrak{l}\,(p\frown(s\,\mathfrak{g}\,p))) \\ d\frown(s\,\mathfrak{l}\,(p\smile p\frown(s\,\mathfrak{g}\,p))) \\ p\frown s \end{array}\text{'} \qquad (\alpha$$

For this, by (601), the proposition

$$`\vdash \begin{array}{l} d\frown(s\mathrel{H}(q\frown(s\,\mathfrak{g}\,p))) \\ d\frown(s\,\mathfrak{l}\,(p\smile p\frown(s\,\mathfrak{g}\,p))) \end{array}\text{'} \qquad (\beta$$

is required. We will prove it using the proposition

$$`\vdash \begin{array}{l} q\frown(t\frown\smile_*p) \\ p\smile q\frown(t\frown\smile_*p) \end{array}\text{'} \qquad (\gamma$$

which straightforwardly follows from (285).

§210. *Construction*

$$496 \vdash q\frown(p\smile q\frown_*p)$$

(285): ——————

$$\vdash \begin{array}{l} q\frown(t\frown\smile_*p) \\ p\smile q\frown(t\frown\smile_*p) \end{array} \qquad (628$$

(609): - - - - - - - -

$$\vdash \begin{array}{l} a\smile\not\mathfrak{F}t\frown s \\ p\smile q\frown(t\frown\smile_*p) \\ \neg\, a\frown(q\frown(s\,\mathfrak{g}\,p)) \end{array} \qquad (\alpha$$

$$\vdash \overset{\mathfrak{t}}{\smile} \begin{array}{l} a\smile\not\mathfrak{F}\mathfrak{t}\frown s \\ p\smile q\frown(\mathfrak{t}\frown\smile_*p) \\ \neg\, a\frown(q\frown(s\,\mathfrak{g}\,p)) \end{array} \qquad (\beta$$

(608): - - - - - - - - - -

$$\vdash \begin{array}{l} a\frown(p\smile q\frown(s\,\mathfrak{g}\,p)) \\ \neg\, a\frown(q\frown(s\,\mathfrak{g}\,p)) \end{array} \qquad (\gamma$$

$$\times$$

$$\vdash \begin{array}{l} a\frown(q\frown(s\,\mathfrak{g}\,p)) \\ a\frown(p\smile q\frown(s\,\mathfrak{g}\,p)) \end{array} \qquad (629$$

(597):: - - - - - - - - - -

$$\vdash \begin{array}{l} a\frown(q\frown(s\,\mathfrak{g}\,p)) \\ d\smile\not\mathfrak{F}a\frown s \\ a\frown s \\ d\frown(s\,\mathfrak{l}\,(p\smile q\frown(s\,\mathfrak{g}\,p))) \end{array} \qquad (\alpha$$

$$\vdash \overset{\mathfrak{a}}{\smile} \begin{array}{l} \mathfrak{a}\frown(q\frown(s\,\mathfrak{g}\,p)) \\ d\smile\not\mathfrak{F}\mathfrak{a}\frown s \\ \mathfrak{a}\frown s \\ d\frown(s\,\mathfrak{l}\,(p\smile q\frown(s\,\mathfrak{g}\,p))) \end{array} \qquad (\beta$$

(590): - - - - - - - - - - - - - -

$$\vdash \begin{array}{l} d\frown(s\mathrel{H}(q\frown(s\,\mathfrak{g}\,p))) \\ d\frown(s\,\mathfrak{l}\,(p\smile q\frown(s\,\mathfrak{g}\,p))) \end{array} \qquad (630$$

§211. *Analysis*

In order to prove proposition (α) of §209, we require, by (601), in addition to the proposition just proven, the further proposition

$$`\vdash \begin{array}{l} \neg\, e\frown(s\mathrel{H}(p\frown(s\,\mathfrak{g}\,p))) \\ e\smile\not\mathfrak{F}d\frown s \\ e\frown s \\ p\frown s \\ d\frown(s\,\mathfrak{l}\,(p\smile p\frown(s\,\mathfrak{g}\,p))) \\ \not{\jmath}s \end{array}\text{'} \qquad (\alpha$$

In order to prove it, we need, by (594), the proposition

$$`\vdash \begin{array}{l} e\frown(s\mathrel{H}(p\smile p\frown(s\,\mathfrak{g}\,p))) \\ p\frown s \\ e\frown(s\mathrel{H}(p\frown(s\,\mathfrak{g}\,p))) \\ \not{\jmath}s \end{array}\text{'} \qquad (\beta$$

which, using (590) and (591), is to be reduced to

$$`\vdash \begin{array}{l} a\frown(p\smile p\frown(s\,\mathfrak{g}\,p)) \\ p\frown s \\ a\frown(p\frown(s\,\mathfrak{g}\,p)) \\ \not{\jmath}s \end{array}\text{'} \qquad (\gamma$$

The latter is to be derived using (608) from

$$`\vdash \begin{array}{l} a\smile\not\mathfrak{F}t\frown s \\ p\frown(t\frown\smile_*p) \\ p\frown s \\ \neg\, a\frown(p\smile p\frown(s\,\mathfrak{g}\,p)) \\ \not{\jmath}s \end{array}\text{'} \qquad (\delta$$

To prove (δ) we need the propositions

‘ $\vdash p \smile q \frown (p \smile t \frown \dot{\smile}_{*} p)$
  $q \frown (t \frown \dot{\smile}_{*} p)$ ’, (ε

and

‘ $\vdash a \smile \not{\mathfrak{P}} t \frown s$
  $p \frown s$
  $a \smile \not{\mathfrak{P}}(p \smile t) \frown s$
  $\not{\jmath} s$
  $t \frown s$ ’, (ζ

We prove (ε) using (144).

§212. *Construction*

IIIh $\vdash p \smile (p \smile d) = p \smile a$
  $p \smile d = a$

(495): ———————

$\vdash p \smile (p \smile d) = p \smile a$
  $d \frown (a \frown_{*} p)$ (α

(495): ———————

$\vdash p \smile d \frown (p \smile a \frown_{*} p)$
  $d \frown (a \frown_{*} p)$ (β

(137): – – – – – – – – –

$\vdash p \smile q \frown (p \smile a \frown \dot{\smile}_{*} p)$
  $d \frown (a \frown_{*} p)$
  $p \smile q \frown (p \smile d \frown \dot{\smile}_{*} p)$ (γ

$\vdash^{\mathfrak{d}}\smile^{\mathfrak{a}}\smile\ p \smile q \frown (p \smile \mathfrak{a} \frown \dot{\smile}_{*} p)$
  $\mathfrak{d} \frown (\mathfrak{a} \frown_{*} p)$
  $p \smile q \frown (p \smile \mathfrak{d} \frown \dot{\smile}_{*} p)$ (δ

(144): ———————

$\vdash p \smile q \frown (p \smile t \frown \dot{\smile}_{*} p)$
  $p \smile q \frown (p \smile q \frown \dot{\smile}_{*} p)$
  $q \frown (t \frown \dot{\smile}_{*} p)$ (ε

(140):: ———————

$\vdash p \smile q \frown (p \smile t \frown \dot{\smile}_{*} p)$
  $q \frown (t \frown \dot{\smile}_{*} p)$ (631

———— • ————

505 $\vdash a \smile (\not{\mathfrak{P}} t \smile \not{\mathfrak{P}} p) \frown s$
  $a \smile \not{\mathfrak{P}}(p \smile t) \frown s$

(529): – – – – – – – – –

$\vdash a \smile (\not{\mathfrak{P}} t \smile \not{\mathfrak{P}} p) \smile p \frown s$
  $p \frown s$
  $a \smile \not{\mathfrak{P}}(p \smile t) \frown s$
  $\not{\jmath} s$ (α

(490): ———————

$\vdash a \smile (\not{\mathfrak{P}} t \smile \not{\mathfrak{P}} p \smile p) \frown s$
  $p \frown s$
  $a \smile \not{\mathfrak{P}}(p \smile t) \frown s$
  $\not{\jmath} s$ (β

(583): ———————

$\vdash a \smile \not{\mathfrak{P}} t \frown s$
  $p \frown s$
  $a \smile \not{\mathfrak{P}}(p \smile t) \frown s$
  $\not{\jmath} s$
  $t \frown s$ (632

(609):: – – – – – – – – – –

$\vdash a \smile \not{\mathfrak{P}} t \frown s$
  $p \frown s$
  $p \smile p \frown (p \smile t \frown \dot{\smile}_{*} p)$
  $\top a \frown (p \smile p \frown (s \mathfrak{g} p))$
  $\not{\jmath} s$
  $t \frown s$ (α

(623, 631):: = = = = = = = =

$\vdash a \smile \not{\mathfrak{P}} t \frown s$
  $p \frown (t \frown \dot{\smile}_{*} p)$
  $p \frown s$
  $\top a \frown (p \smile p \frown (s \mathfrak{g} p))$
  $\not{\jmath} s$ (β

$\vdash^{\mathfrak{t}}\smile\ a \smile \not{\mathfrak{P}} \mathfrak{t} \frown s$
  $p \frown (\mathfrak{t} \frown \dot{\smile}_{*} p)$
  $p \frown s$
  $\top a \frown (p \smile p \frown (s \mathfrak{g} p))$
  $\not{\jmath} s$ (γ

(608): – – – – – – – – – – –

$\vdash a \frown (p \frown (s \mathfrak{g} p))$
  $p \frown s$
  $\top a \frown (p \smile p \frown (s \mathfrak{g} p))$
  $\not{\jmath} s$ (δ

×

⊢ a⌢(p‿p⌢(s 𝔤 p))
  p⌢s
  a⌢(p⌢(s 𝔤 p))
  ∤s (ε

(591):: – – – – – – – – – –

⊢ a⌢(p‿p⌢(s 𝔤 p))
  e‿∦a⌢s
  a⌢s
  p⌢s
  e⌢(s ⅎ (p⌢(s 𝔤 p)))
  ∤s (ζ

‿

⊢ 𝔞 ⌣ 𝔞⌢(p‿p⌢(s 𝔤 p))
  e‿∦𝔞⌢s
  𝔞⌢s
  p⌢s
  e⌢(s ⅎ (p⌢(s 𝔤 p)))
  ∤s (η

(590): – – – – – – – – – – – – –

⊢ e⌢(s ⅎ (p‿p⌢(s 𝔤 p)))
  p⌢s
  e⌢(s ⅎ (p⌢(s 𝔤 p)))
  ∤s (ϑ

(594): – – – – – – – – – – – – –

⊢ d⌢(s ł (p‿p⌢(s 𝔤 p)))
  e‿∦d⌢s
  e⌢s
  p⌢s
  e⌢(s ⅎ (p⌢(s 𝔤 p)))
  ∤s (ι

×

⊢ e⌢(s ⅎ (p⌢(s 𝔤 p)))
  e‿∦d⌢s
  e⌢s
  p⌢s
  d⌢(s ł (p‿p⌢(s 𝔤 p)))
  ∤s (κ

‿

⊢ 𝔢 ⌣ 𝔢⌢(s ⅎ (p⌢(s 𝔤 p)))
  𝔢‿∦d⌢s
  𝔢⌢s
  p⌢s
  d⌢(s ł (p‿p⌢(s 𝔤 p)))
  ∤s (λ

(601): – – – – – – – – – – – – – –

⊢ d⌢(s ł (p⌢(s 𝔤 p)))
  d⌢s
  d⌢(s ⅎ (p⌢(s 𝔤 p)))
  p⌢s
  d⌢(s ł (p‿p⌢(s 𝔤 p)))
  ∤s (μ

(600, 630):: = = = = = = = = = =

⊢ d⌢(s ł (p⌢(s 𝔤 p)))
  d⌢s
  d⌢(s ł (p‿p⌢(s 𝔤 p)))
  p⌢s (ν

(599):: – – – – – – – – – – – –

⊢ d⌢(s ł (p⌢(s 𝔤 p)))
  d⌢(s ł (p‿p⌢(s 𝔤 p)))
  p⌢s (633

§213. *Analysis*

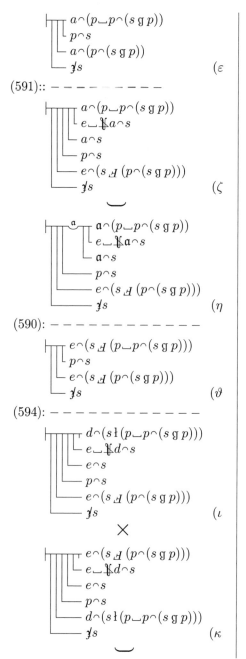

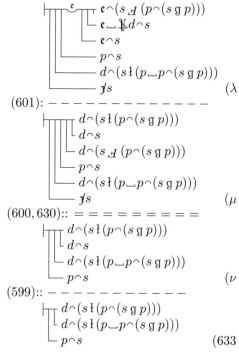

Using propositions (627) and (633) we can now bring the plan of §201 to a close and prove the proposition cited in the heading of section *b*.

§214. *Construction*

585 ⊢ p⌢s
  c‿p = c
  ∤s
  c⌢s

(599, 602):: = = =

⊢ p⌢s
  c⌢(s ł (p⌢(s 𝔤 p)))
  c‿p⌢(s ł (p⌢(s 𝔤 p)))
  ∤s (α

(633, 600):: = = = = = = = = =

$\vdash p \frown s$; $c \frown (s \nmid (p \frown (s \mathfrak{g}\, p)))$; $c \smile p \frown (s \nmid (p \smile p \frown (s \mathfrak{g}\, p)))$ (β

(627):: - - - - - - - - - - - -

$\vdash p \frown s$; $c \frown (s \nmid (p \frown (s \mathfrak{g}\, p)))$; $p \frown s$ (γ

×

$\vdash c \frown (s \nmid (p \frown (s \mathfrak{g}\, p)))$; $p \frown s$ (δ

$\vdash \mathfrak{r}$: $\mathfrak{r} \frown (s \nmid (p \frown (s \mathfrak{g}\, p)))$; $p \frown s$ (ε

(604): - - - - - - - - - - -

$\vdash a \frown (p \frown (s \mathfrak{g}\, p))$; $a \frown s$; $p \frown s$; $p \frown (s \lrcorner (p \frown (s \mathfrak{g}\, p)))$; $\urcorner p s$ (634

———— • ————

140 $\vdash p \frown (p \frown \smile_* p)$

(609): ————————

$\vdash a \smile \not\mathfrak{F} p \frown s$; $a \frown (p \frown (s \mathfrak{g}\, p))$ (α

(589): - - - - - - - -

$\vdash p \smile \not\mathfrak{F} a \frown s$; $a \frown (p \frown (s \mathfrak{g}\, p))$; $p \frown s$; $\nmid s$ (β

×

$\vdash a \frown (p \frown (s \mathfrak{g}\, p))$; $p \smile \not\mathfrak{F} a \frown s$; $p \frown s$; $\nmid s$ (γ

(I): ————————

$\vdash a \frown (p \frown (s \mathfrak{g}\, p))$; $p \smile \not\mathfrak{F} a \frown s$; $a \frown s$; $p \frown s$; $\nmid s$ (δ

$\vdash \mathfrak{a}$: $\mathfrak{a} \frown (p \frown (s \mathfrak{g}\, p))$; $p \smile \not\mathfrak{F} \mathfrak{a} \frown s$; $\mathfrak{a} \frown s$; $p \frown s$; $\nmid s$ (ε

(590): - - - - - - - - - - -

$\vdash p \frown (s \lrcorner (p \frown (s \mathfrak{g}\, p)))$; $p \frown s$; $\nmid s$ (ζ

(634): - - - - - - - - - - -

$\vdash a \frown (p \frown (s \mathfrak{g}\, p))$; $a \frown s$; $p \frown s$; $\nmid s$; $\urcorner p s$ (η

(607):: - - - - - - - -

$\vdash a \frown (p \frown (s \mathfrak{g}\, p))$; $a \frown s$; $p \frown s$; $\urcorner p s$ (635

×

$\vdash a \frown s$; $a \frown (p \frown (s \mathfrak{g}\, p))$; $p \frown s$; $\urcorner p s$ (636

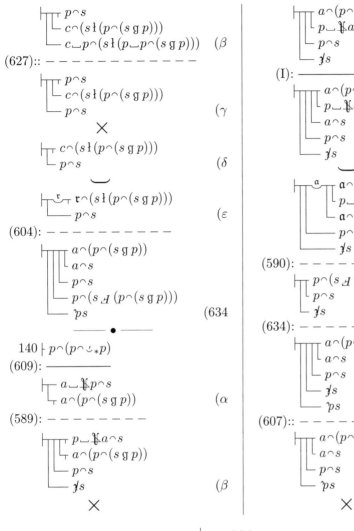

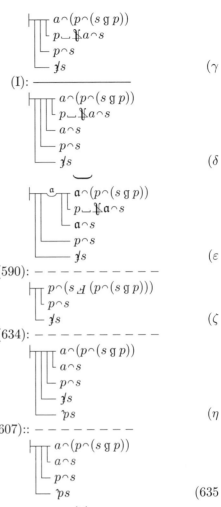

E. Proof of the proposition

$$\text{‘}\vdash \begin{array}{l} q \smile p = p \smile q \\ \wp s \\ q \frown s \\ p \frown s \end{array} \text{’}$$

a) Proof of the proposition

$$\text{‘}\vdash \begin{array}{l} p \smile \mathfrak{I} q \frown s \\ q \frown s \\ \wp s \\ p \frown s \\ \mathfrak{I} q \smile p \frown s \end{array} \text{’}$$

§215. *Analysis*

We now take as our goal the commutative principle, initially within a positive class. In preparation for it, we prove the proposition mentioned in the heading for this section *a*. For this we need the proposition

$$\text{‘}\vdash \begin{array}{l} q \smile p \smile \mathfrak{I} q \frown s \\ q \frown s \\ \wp s \\ p \frown s \end{array} \text{’} \qquad (\alpha$$

in which we then replace ‘p’ by ‘$\mathfrak{I} q \smile p$’. For the proof we distinguish the cases $p \smile \mathfrak{I} q \frown s$, $p = q$, and $q \smile \mathfrak{I} p \frown s$, the first two of which offer no difficulties. In the last case, we make use of (635) by first proving the proposition

$$\text{‘}\vdash \begin{array}{l} q \smile \mathfrak{I} t \frown s \\ p \frown (t \frown \smile_* p) \\ q \smile \mathfrak{I} p \frown s \\ \neg\, q \smile \mathfrak{I}(q \smile \mathfrak{I} p) \frown s \\ \mathfrak{f} s \\ q \frown s \end{array} \text{’} \qquad (\beta$$

using (152). For this we need the proposition

$$\text{‘}\vdash \begin{array}{l} q \smile \mathfrak{I} a \frown s \\ d \frown (a \frown_* p) \\ p \frown (d \frown \smile_* p) \\ q \smile \mathfrak{I} d \frown s \\ \neg\, q \smile \mathfrak{I}(q \smile \mathfrak{I} p) \frown s \\ q \smile \mathfrak{I} p \frown s \\ \mathfrak{f} s \\ q \frown s \end{array} \text{’} \qquad (\gamma$$

which follows from

$$\text{‘}\vdash \begin{array}{l} q \smile \mathfrak{I} p \smile \mathfrak{I} d \frown s \\ q \smile \mathfrak{I} d \frown s \\ \neg\, q \smile \mathfrak{I}(q \smile \mathfrak{I} p) \frown s \\ q \smile \mathfrak{I} p \frown s \\ \mathfrak{f} s \\ q \frown s \end{array} \text{’} \qquad (\delta$$

with the help of (500). (δ) in turn follows from

$$\text{‘}\vdash \begin{array}{l} q \smile \mathfrak{I} p \smile \mathfrak{I} d \frown s \\ q \smile \mathfrak{I} d \frown s \\ q \smile \mathfrak{I} p \smile \mathfrak{I} q \frown s \\ \mathfrak{f} s \\ q \frown s \\ q \smile \mathfrak{I} p \frown s \end{array} \text{’} \qquad (\varepsilon$$

§216. *Construction*

$$508 \vdash \begin{array}{l} f(q \smile \mathfrak{I}(d \smile p)) \\ f(q \smile \mathfrak{I} p \smile \mathfrak{I} d) \end{array}$$

(500): ———————————

$\vdash f(q \smile \between(p \smile d))$
$\quad f(q \smile \between p \smile \between d)$
$\quad p \frown (d \frown \smile_* p)$ (α

(493): ———

$\vdash f(q \smile \between a)$
$\quad f(q \smile \between p \smile \between d)$
$\quad p \frown (d \frown \smile_* p)$
$\quad d \frown (a \frown_* p)$ (β

•

491 $\vdash q \smile \between p \smile \between q \smile q \smile \between d \frown s$
$\quad q \smile \between p \smile \between q \smile (q \smile \between d) \frown s$

(567): ———

$\vdash q \smile \between p \smile \between d \frown s$
$\quad q \smile \between p \smile \between q \smile (q \smile \between d) \frown s$
$\quad q \frown s$
$\quad \nmid s$
$\quad q \smile \between p \frown s$ (γ

(529):: - - - - - - - - - - - -

$\vdash q \smile \between p \smile \between d \frown s$
$\quad q \smile \between d \frown s$
$\quad q \smile \between p \smile \between q \frown s$
$\quad \nmid s$
$\quad q \frown s$
$\quad q \smile \between p \frown s$ (δ

(588):: - - - - - - - - - -

$\vdash q \smile \between p \smile \between d \frown s$
$\quad q \smile \between d \frown s$
$\quad \neg\, q \smile \between (q \smile \between p) \frown s$
$\quad q \smile \between p \frown s$
$\quad \nmid s$
$\quad \neg\, q = q \smile \between p$
$\quad q \frown s$ (ε

(IIIc): ·-·-·-·-·-·-·-·-·-·

$\vdash q \smile \between p \smile \between d \frown s$
$\quad q \smile \between d \frown s$
$\quad \neg\, q \smile \between (q \smile \between p) \frown s$
$\quad q \smile \between p \frown s$
$\quad \nmid s$
$\quad q \frown s$ (ζ

(β): ———

$\vdash q \smile \between a \frown s$
$\quad d \frown (a \frown_* p)$
$\quad p \frown (d \frown \smile_* p)$
$\quad q \smile \between d \frown s$
$\quad \neg\, q \smile \between (q \smile \between p) \frown s$
$\quad q \smile \between p \frown s$
$\quad \nmid s$
$\quad q \frown s$ (η

$\smile$

$\vdash^{\mathfrak{d}}\ ^{\mathfrak{a}}\ q \smile \between \mathfrak{a} \frown s$
$\quad \mathfrak{d} \frown (\mathfrak{a} \frown_* p)$
$\quad p \frown (\mathfrak{d} \frown \smile_* p)$
$\quad q \smile \between \mathfrak{d} \frown s$
$\quad \neg\, q \smile \between (q \smile \between p) \frown s$
$\quad q \smile \between p \frown s$
$\quad \nmid s$
$\quad q \frown s$ (ϑ

(152): - - - - - - - - - - - - - -

$\vdash q \smile \between t \frown s$
$\quad p \frown (t \frown \smile_* p)$
$\quad q \smile \between p \frown s$
$\quad \neg\, q \smile \between (q \smile \between p) \frown s$
$\quad \nmid s$
$\quad q \frown s$ (ι

$\smile$

$\vdash^{\mathfrak{t}}\ q \smile \between \mathfrak{t} \frown s$
$\quad p \frown (\mathfrak{t} \frown \smile_* p)$
$\quad q \smile \between p \frown s$
$\quad \neg\, q \smile \between (q \smile \between p) \frown s$
$\quad \nmid s$
$\quad q \frown s$ (κ

(608): - - - - - - - - - - - -

$\vdash \neg\, q \frown (p \frown (s \,\mathfrak{g}\, p))$
$\quad q \smile \between p \frown s$
$\quad \neg\, q \smile \between (q \smile \between p) \frown s$
$\quad \nmid s$
$\quad q \frown s$ (λ

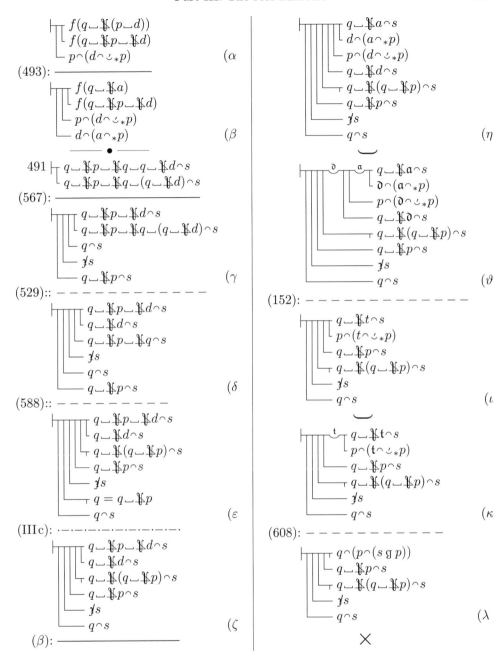

×

$$\begin{array}{l} \vdash q \smile \nparallel(q \smile \nparallel p) \frown s \\ \quad \neg\; q \smile \nparallel p \frown s \\ \quad q \frown (p \frown (s \mathfrak{g} p)) \\ \quad \not{\jmath} s \\ \quad q \frown s \end{array} \qquad (\mu$$

(635):: - - - - - - - - - - -

$$\begin{array}{l} \vdash q \smile \nparallel(q \smile \nparallel p) \frown s \\ \quad \neg\; q \smile \nparallel p \frown s \\ \quad q \frown s \\ \quad p \frown s \\ \quad \gamma p s \\ \quad \not{\jmath} s \end{array} \qquad (\nu$$

(542): ————————————

$$\begin{array}{l} \vdash q \smile (p \smile \nparallel q) \frown s \\ \quad \neg\; q \smile \nparallel p \frown s \\ \quad q \frown s \\ \quad p \frown s \\ \quad \gamma p s \\ \quad \not{\jmath} s \end{array} \qquad (\xi$$

(588):: - - - - - - - - - - -

$$\begin{array}{l} \vdash q \smile (p \smile \nparallel q) \frown s \\ \quad \neg\; p \smile \nparallel q \frown s \\ \quad q \frown s \\ \quad \not{\jmath} s \\ \quad \neg\; p = q \\ \quad p \frown s \\ \quad \gamma p s \end{array} \qquad (o$$

——— • ———

$$\begin{array}{l} \text{III a} \vdash q \smile (p \smile \nparallel q) \frown s \\ \quad \neg\; q \smile (q \smile \nparallel q) \frown s \\ \quad p = q \end{array}$$

(564): ————————

$$\begin{array}{l} \vdash q \smile (p \smile \nparallel q) \frown s \\ \quad \neg\; q \frown s \\ \quad p = q \\ \quad \not{\jmath} s \end{array} \qquad (\pi$$

(o): ·—·—·—·—·—·—·—·

$$\begin{array}{l} \vdash q \smile (p \smile \nparallel q) \frown s \\ \quad \neg\; p \smile \nparallel q \frown s \\ \quad q \frown s \\ \quad \not{\jmath} s \\ \quad p \frown s \\ \quad \gamma p s \end{array} \qquad (\rho$$

(529): ·—·—·—·—·—·—·—·—·

$$\begin{array}{l} \vdash q \smile (p \smile \nparallel q) \frown s \\ \quad \neg\; q \frown s \\ \quad \not{\jmath} s \\ \quad p \frown s \\ \quad \gamma p s \end{array} \qquad (\sigma$$

(607):: - - - - - - - - -

$$\begin{array}{l} \vdash q \smile (p \smile \nparallel q) \frown s \\ \quad \neg\; q \frown s \\ \quad \gamma p s \\ \quad p \frown s \end{array} \qquad (\tau$$

(491): ————————

$$\begin{array}{l} \vdash q \smile p \smile \nparallel q \frown s \\ \quad \neg\; q \frown s \\ \quad \gamma p s \\ \quad p \frown s \end{array} \qquad (637$$

——— • ———

$$\begin{array}{l} 575 \vdash p \smile \nparallel q \frown s \\ \quad \neg\; q \smile (\nparallel q \smile p) \smile \nparallel q \frown s \\ \quad p \frown s \\ \quad \not{\jmath} s \\ \quad q \frown s \end{array}$$

(637, 607):: = = = = = = = =

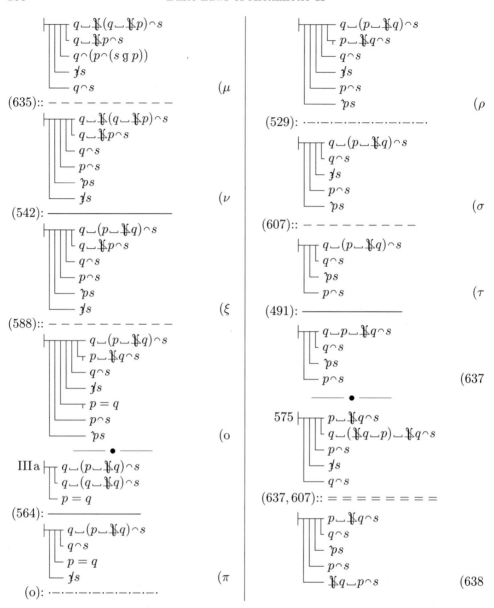

$$\begin{array}{l} \vdash p \smile \nparallel q \frown s \\ \quad \neg\; q \frown s \\ \quad \gamma p s \\ \quad p \frown s \\ \quad \nparallel q \smile p \frown s \end{array} \qquad (638$$

b) Proof of the proposition

$$
\text{`}\vdash\begin{array}{l}
\not\mathfrak{h}p\smile q\frown s \\
q\frown s \\
\wp s \\
p\frown s \\
q\smile\not\mathfrak{h}p\frown s\text{'}
\end{array}
$$

§217. *Analysis*

As a further preparation for establishing the commutative principle, we prove the proposition of our heading. We could easily do so using a proposition similar to proposition (637):

$$
\text{`}\vdash\begin{array}{l}
\not\mathfrak{h}q\smile p\smile q\frown s \\
q\frown s \\
\wp s \\
p\frown s
\end{array}\text{'},
$$

If we wanted to prove this in a manner similar to (637), we would also need to make a transition similar to that from (δ) to (ε) (§216), and for this we would require a proposition such as

$$
\text{`}\vdash\begin{array}{l}
\not\mathfrak{h}p\smile q\frown s \\
\top\ \not\mathfrak{h}q\smile p\frown s \\
q\frown s \\
\mathfrak{y}s \\
\top\ p=q \\
p\frown s
\end{array}\text{'},\qquad(\alpha
$$

which would have to be derived by (526), as (588) is derived by (528). So far, we have made no use of (526) in order to explore whether one might get by without it. Should this be feasible, we could strike out the seventh line on the left-hand side of the definitional equation (Ψ). But it is here that its indispensability shows itself. At any rate, all attempts to do without the proposition (526) have invariably faltered at the same obstruction, namely that we cannot arrive at '$\not\mathfrak{h}\mathrm{B}\smile\mathrm{A}\frown\Sigma$' from '$\top\ \not\mathfrak{h}\mathrm{A}\smile\mathrm{B}\frown\Sigma$' if the coinciding of A and B is excluded.

Once we have settled on making use of (526), we can make our proof shorter by using (638), showing that neither '$\not\mathfrak{h}q\smile p\frown s$' nor '$p=q$' is compatible with '$q\smile\not\mathfrak{h}p\frown s$'. From this, using ($\alpha$), our proposition follows. The incompatibility of '$p=q$' with '$q\smile\not\mathfrak{h}p\frown s$' follows from (536). The incompatibility of '$\not\mathfrak{h}q\smile p\frown s$' with '$q\smile\not\mathfrak{h}p\frown s$' follows from (589) and (638). We first prove (α) in a manner similar to (588).

§218. *Construction*

$$
\text{IIIh}\vdash\begin{array}{l}
q\smile(\not\mathfrak{h}q\smile p)=q\smile(q\smile\not\mathfrak{h}q) \\
\not\mathfrak{h}q\smile p=q\smile\not\mathfrak{h}q
\end{array}
$$

(575): ———————————

$$
\vdash\begin{array}{l}
p=q\smile(q\smile\not\mathfrak{h}q) \\
\not\mathfrak{h}q\smile p=q\smile\not\mathfrak{h}q \\
p\frown s \\
\mathfrak{y}s \\
q\frown s
\end{array}\qquad(\alpha
$$

(564): ———————————

$$
\vdash\begin{array}{l}
p=q \\
\not\mathfrak{h}q\smile p=q\smile\not\mathfrak{h}q \\
p\frown s \\
\mathfrak{y}s \\
q\frown s
\end{array}\qquad(\beta
$$

(561):: - - - - - - - - - -

⊢ p = q
  ⅌q⌣p⌒s
  q⌒s
  ɉs
  ⅌(⅌q⌣p)⌒s
  ⅌q⌣p⌒ðs
  p⌒s (γ

(526):: - - - - - - - - - -

⊢ p = q
  ⅌q⌣p⌒s
  q⌒s
  ɉs
  ⅌(⅌q⌣p)⌒s
  p⌒s (δ

(544): ———————————

⊢ p = q
  ⅌q⌣p⌒s
  q⌒s
  ɉs
  ⅌p⌣q⌒s
  p⌒s (639

×

⊢ ⅌p⌣q⌒s
  ⅌q⌣p⌒s
  q⌒s
  ɉs
  p = q
  p⌒s (640

——— • ———

IIIc ⊢ q⌣⅌p⌒s
  p⌣⅌p⌒s
  p = q

(536):: - - - - - -

⊢ q⌣⅌p⌒s
  p⌒s
  ɉs
  p = q (α

(Ia): - - - - - -

⊢ ⅌p⌣q⌒s
  q⌣⅌p⌒s
  p⌒s
  ɉs
  p = q (β

——— • ———

589 ⊢ q⌣⅌p⌒s
  p⌣⅌q⌒s
  q⌒s
  ɉs

(638):: - - - - - - -

⊢ q⌣⅌p⌒s
  q⌒s
  ʾps
  p⌒s
  ⅌q⌣p⌒s
  ɉs (γ

×

⊢ ⅌q⌣p⌒s
  q⌒s
  ʾps
  p⌒s
  q⌣⅌p⌒s
  ɉs (δ

(640): - - - - - - - - -

⊢ ⅌p⌣q⌒s
  q⌒s
  ʾps
  p⌒s
  q⌣⅌p⌒s
  ɉs
  p = q (ε

(β): ·—·—·—·—·—·—·

⊢ ⅌p⌣q⌒s
  q⌒s
  ʾps
  p⌒s
  q⌣⅌p⌒s
  ɉs (ζ

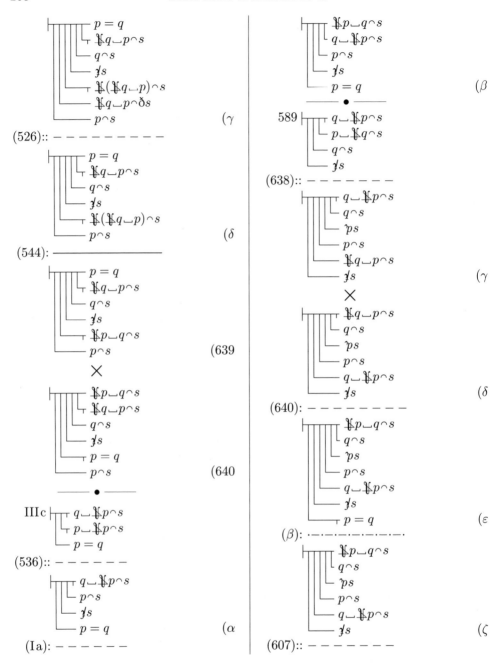

(607):: - - - - - - -

$$\begin{array}{l} \vdash\!\top\!\top\!\top\; \between p\smile q\frown s \\ \quad\llcorner\; q\frown s \\ \quad\llcorner\; \not{p}\, s \\ \quad\llcorner\; p\frown s \\ \quad\llcorner\; q\smile \between p\frown s \end{array} \qquad (641$$

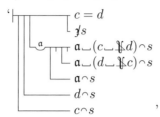

c) Proof of the proposition

$$\begin{array}{l} `\vdash\!\top\!\top\!\top\; c = d \\ \quad\llcorner\; \not{\jmath} s \\ \quad\llcorner\; \overset{\mathfrak{a}}{\smile}\!\top\!\top\; \mathfrak{a}\smile(c\smile\between d)\frown s \\ \qquad\llcorner\; \mathfrak{a}\smile(d\smile\between c)\frown s \\ \qquad\llcorner\; \mathfrak{a}\frown s \\ \quad\llcorner\; d\frown s \\ \quad\llcorner\; c\frown s \end{array} \quad ,$$

§219. *Analysis*

For the proof of the commutative principle we need the proposition mentioned in the heading, whose precise expression in words will be cumbersome. To assist understanding the following free translation may be allowed:

"If in a domain of magnitudes, the two differences $(d\smile\between c)$ and $(c\smile\between d)$ of the positive magnitudes c and d are less than any positive magnitude, then the magnitudes c and d coincide."

We reduce this proposition to the following:

$$\begin{array}{l} `\vdash\!\top\!\top\!\top\; b = p\smile\between p \\ \quad\llcorner\; \not{\jmath} s \\ \quad\llcorner\; \overset{\mathfrak{a}}{\smile}\!\top\!\top\; \mathfrak{a}\smile b\frown s \\ \qquad\llcorner\; \mathfrak{a}\smile\between b\frown s \\ \qquad\llcorner\; \mathfrak{a}\frown s \\ \quad\llcorner\; p\frown s \\ \quad\llcorner\; b\frown \delta s \end{array} \quad ,$$

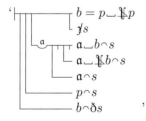

which likewise can be expressed in words freely like this:

"A magnitude b is a null magnitude if both it and its converse $\between b$ results in a positive magnitude when composed with any positive magnitude in the same domain."

For short, I here call a *positive magnitude* any Relation which belongs to the positival class (s). The proof is to be conducted using (516). It is shown that b is neither positive nor can it be the converse of any positive Relation (q). In the first case $b\smile\between b$ is not positive; in the second $q\smile b$ is not positive.

§220. *Construction*

$$\begin{array}{l} 536\; \vdash\!\top\!\top\; q\smile\between q\frown s \\ \quad\llcorner\; q\frown s \\ \quad\llcorner\; \not{\jmath} s \end{array}$$

(IIId): – – – – – –

$$\begin{array}{l} \vdash\!\top\!\top\!\top\; b = \between q \\ \quad\llcorner\; q\smile b\frown s \\ \quad\llcorner\; q\frown s \\ \quad\llcorner\; \not{\jmath} s \end{array} \qquad (\alpha$$

(If): – – – – – –

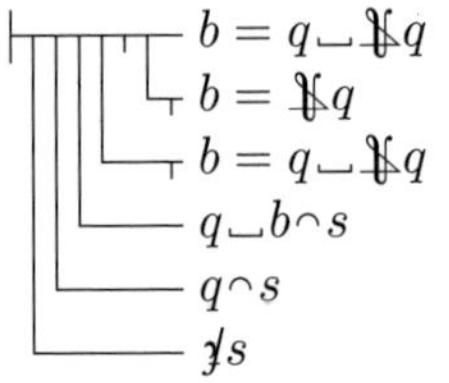

(β

(559): ———————

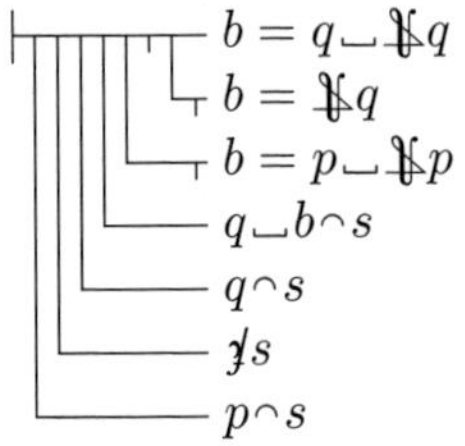

(γ

(Ib):: – – – – – – – – – –

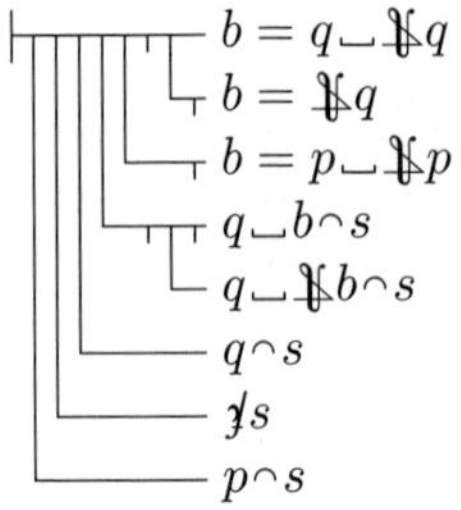

(δ

(IIa):: – – – – – – – – – –

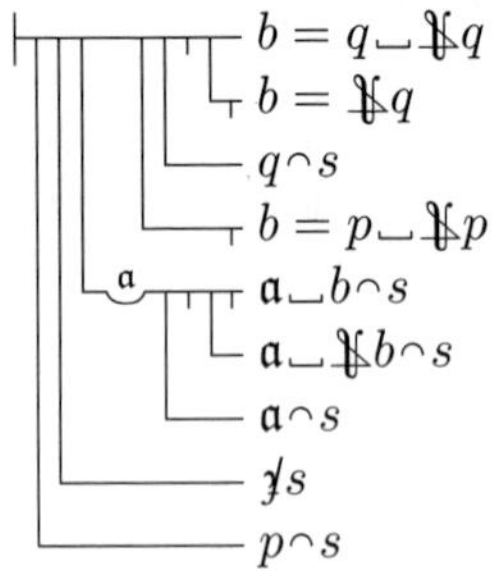

(ε

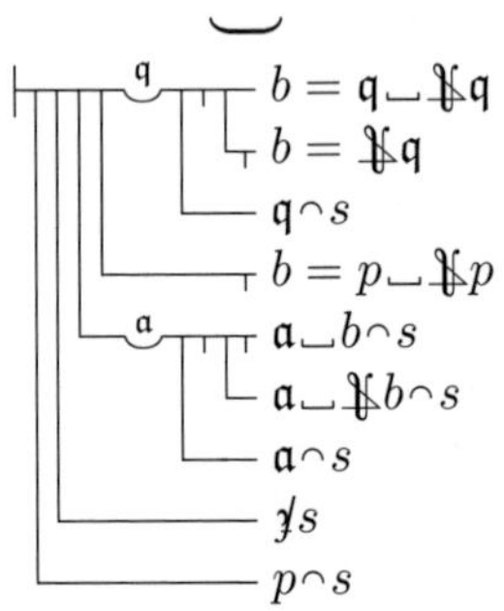

(ζ

(516): – – – – – – – – – – – –

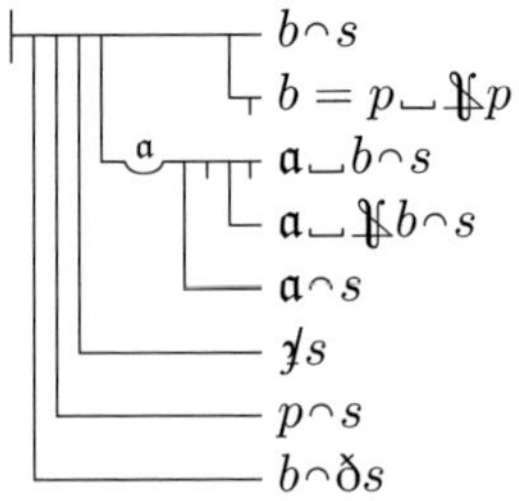

(η

×

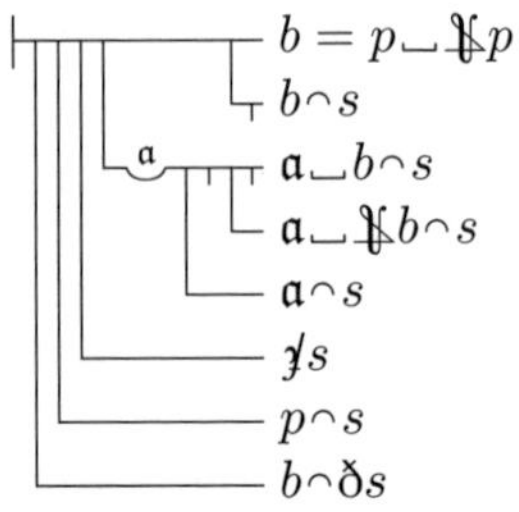

(ϑ

536 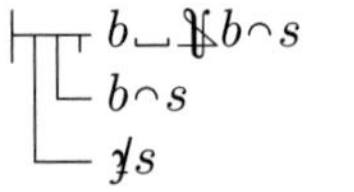

×

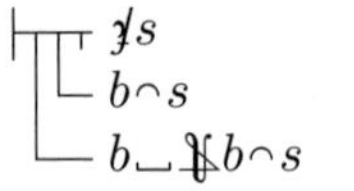

(ι

(Id):: – – – – – – – –

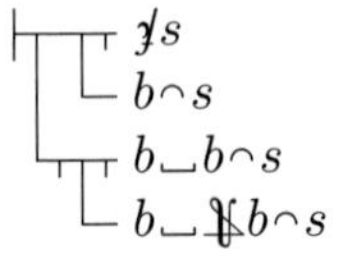

(κ

(IIa):: – – – – – – – –

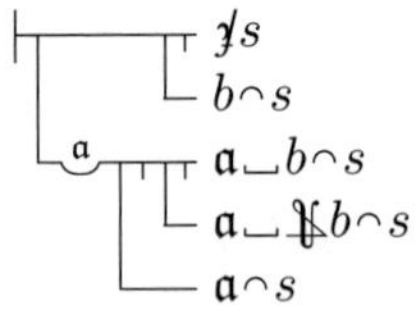

(λ

×

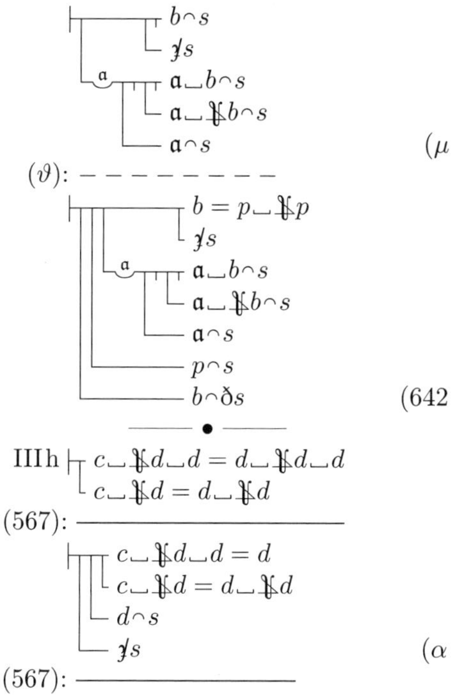

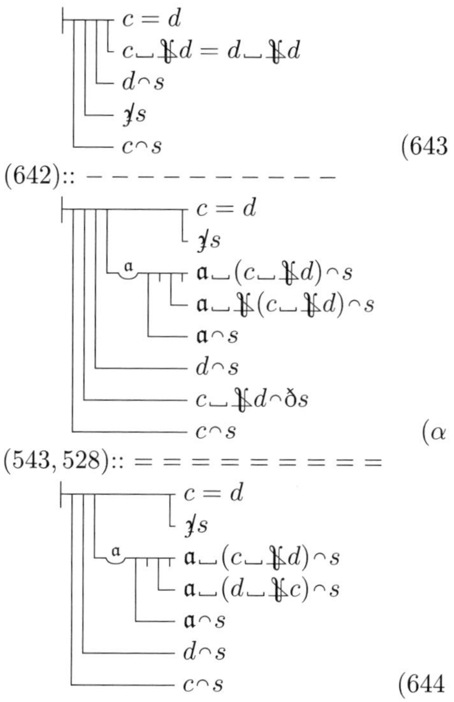

d) Proof of the proposition

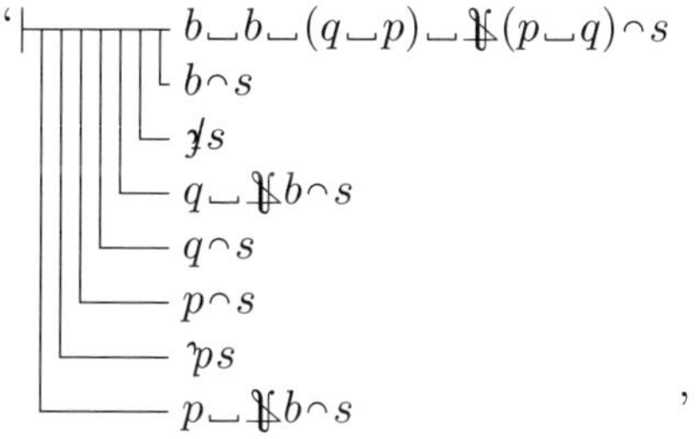

§221. *Analysis*

Using (644), we now prove the commutative law by replacing '*d*' by '*p*⌣*q*' and '*c*' by '*q*⌣*p*'. We enclose *p*⌣*q* and *q*⌣*p* between the same limits, which we then let approach each other arbitrarily closely.

For this purpose, we take a positive magnitude *b* (compare §219) which is less than *p* and less than *q*. According to (635), there is then a multiple of *b* that is not less than *p*. Consequently, according to (357), there is also an *o* that, first, is not less than *p* and is such that *c*, immediately preceding it, is less than *p*. In this manner one can enclose *p* between two magnitudes *c* and *o*, that differ by *b*. Likewise, one can enclose *q* between the magnitudes d and a that differ by b. So c will be less than p and o will not be less than p. Likewise, d will be less and a not less than q. Here $b \smile c$ coincides with o and $b \smile d$ with a. For this we need the proposition

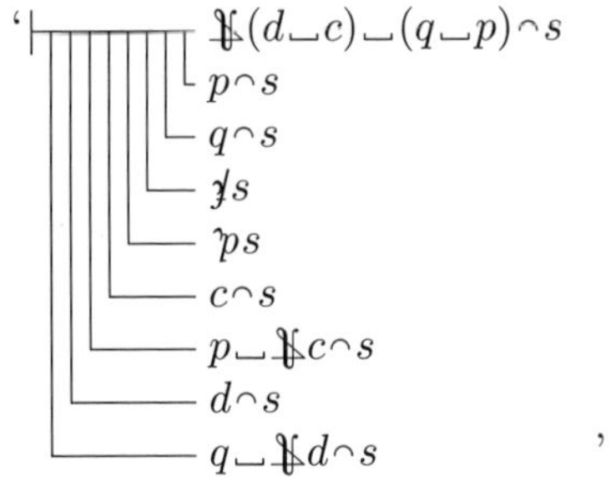
(α

The same content can be written with different lettering, so that the supercomponent becomes

$$\text{‘}— \ddagger(p \smile q) \smile (o \smile a) \frown s\text{’}$$

From both forms we obtain, according to (529), a proposition with the supercomponent

$$\text{‘}— \ddagger(p \smile q) \smile (o \smile a) \smile (\ddagger c \smile \ddagger d \smile (q \smile p)) \frown s\text{’}$$

This, using the already proven commutative law (502) for multiples of the same magnitude, is then transformed into

$$\text{‘}— \ddagger(p \smile q) \smile (b \smile b \smile (q \smile p)) \frown s\text{’}$$

which, by (638), can then in turn be converted into

$$\text{‘}— b \smile b \smile (q \smile p) \smile \ddagger(p \smile q) \frown s\text{’}$$

In the proposition so obtained, we interchange ‘p’ and ‘q’. It then remains to be shown that b can be so chosen that $b \smile b$ is not greater than an arbitrarily small positive magnitude. Then we can apply (644).

We first deduce (α) from the two propositions

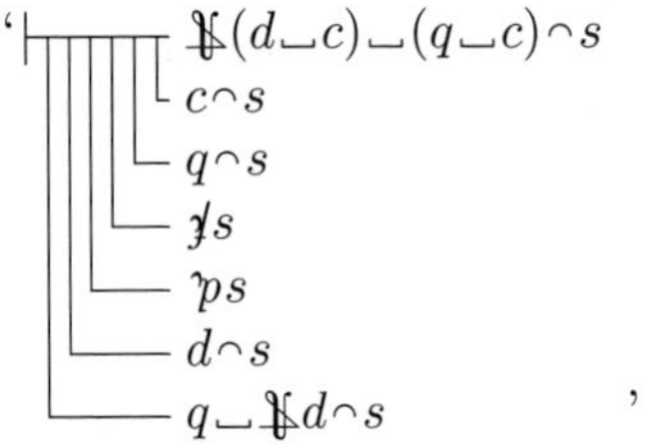
(β

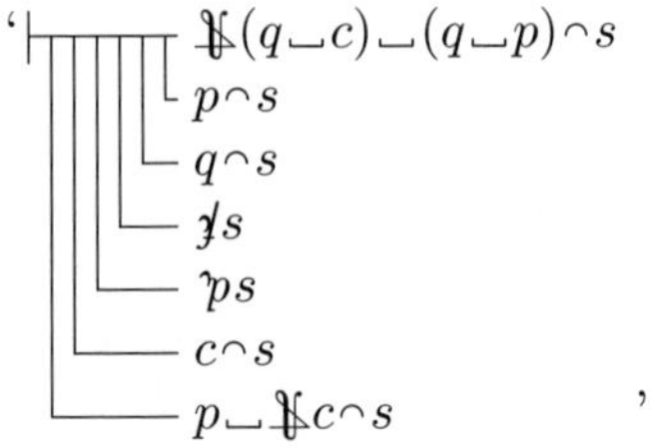
(γ

which are to be proven using (641).

§222. *Construction*

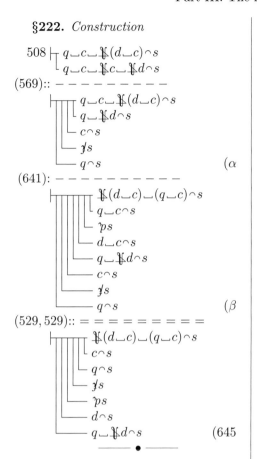

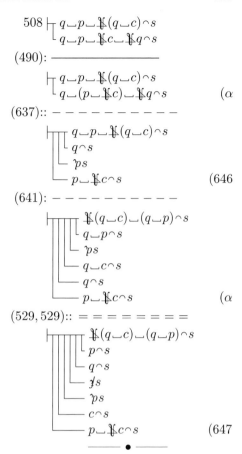

$508 \vdash q \smile c \smile \mathfrak{V}(d \smile c) \frown s$
$\quad q \smile c \smile \mathfrak{V}c \smile \mathfrak{V}d \frown s$

(569):: - - - - - - - - - -

$\vdash q \smile c \smile \mathfrak{V}(d \smile c) \frown s$
$\quad q \smile \mathfrak{V}d \frown s$
$\quad c \frown s$
$\quad \not\jmath s$
$\quad q \frown s$ (α

(641): - - - - - - - - - - -

$\vdash \mathfrak{V}(d \smile c) \smile (q \smile c) \frown s$
$\quad q \smile c \frown s$
$\quad \daleth p s$
$\quad d \smile c \frown s$
$\quad q \smile \mathfrak{V}d \frown s$
$\quad c \frown s$
$\quad \not\jmath s$
$\quad q \frown s$ (β

(529, 529):: = = = = = = = = = =

$\vdash \mathfrak{V}(d \smile c) \smile (q \smile c) \frown s$
$\quad c \frown s$
$\quad q \frown s$
$\quad \not\jmath s$
$\quad \daleth p s$
$\quad d \frown s$
$\quad q \smile \mathfrak{V}d \frown s$ (645

———— • ————

$508 \vdash q \smile p \smile \mathfrak{V}(q \smile c) \frown s$
$\quad q \smile p \smile \mathfrak{V}c \smile \mathfrak{V}q \frown s$

(490): ————————————

$\vdash q \smile p \smile \mathfrak{V}(q \smile c) \frown s$
$\quad q \smile (p \smile \mathfrak{V}c) \smile \mathfrak{V}q \frown s$ (α

(637):: - - - - - - - - - - -

$\vdash q \smile p \smile \mathfrak{V}(q \smile c) \frown s$
$\quad q \frown s$
$\quad \daleth p s$
$\quad p \smile \mathfrak{V}c \frown s$ (646

(641): - - - - - - - - - - -

$\vdash \mathfrak{V}(q \smile c) \smile (q \smile p) \frown s$
$\quad q \smile p \frown s$
$\quad \daleth p s$
$\quad q \smile c \frown s$
$\quad q \frown s$
$\quad p \smile \mathfrak{V}c \frown s$ (α

(529, 529):: = = = = = = = = =

$\vdash \mathfrak{V}(q \smile c) \smile (q \smile p) \frown s$
$\quad p \frown s$
$\quad q \frown s$
$\quad \not\jmath s$
$\quad \daleth p s$
$\quad c \frown s$
$\quad p \smile \mathfrak{V}c \frown s$ (647

———— • ————

$490 \vdash \mathfrak{V}(d \smile c) \smile [q \smile c \smile (\mathfrak{V}(q \smile c) \smile (q \smile p))] \frown s$
$\quad \mathfrak{V}(d \smile c) \smile (q \smile c) \smile (\mathfrak{V}(q \smile c) \smile (q \smile p)) \frown s$

(575): ————————————————————

$\vdash \mathfrak{V}(d \smile c) \smile (q \smile p) \frown s$
$\quad \mathfrak{V}(d \smile c) \smile (q \smile c) \smile (\mathfrak{V}(q \smile c) \smile (q \smile p)) \frown s$
$\quad q \smile p \frown s$
$\quad \not\jmath s$
$\quad q \smile c \frown s$ (α

(529, 529): = = = = = = = = = = = = = = = = =

$$\begin{array}{l} \vdash \mathfrak{F}(d\smile c)\smile(q\smile p)\frown s \\ \quad \mathfrak{F}(d\smile c)\smile(q\smile c)\smile(\mathfrak{F}(q\smile c)\smile(q\smile p))\frown s \\ \quad p\frown s \\ \quad q\frown s \\ \quad \mathfrak{y}s \\ \quad c\frown s \end{array} \qquad (\beta$$

(529):: -

$$\begin{array}{l} \vdash \mathfrak{F}(d\smile c)\smile(q\smile p)\frown s \\ \quad \mathfrak{F}(q\smile c)\smile(q\smile p)\frown s \\ \quad \mathfrak{F}(d\smile c)\smile(q\smile c)\frown s \\ \quad \mathfrak{y}s \\ \quad p\frown s \\ \quad q\frown s \\ \quad c\frown s \end{array} \qquad (\gamma$$

(645, 647):: = = = = = = = =

$$\begin{array}{l} \vdash \mathfrak{F}(d\smile c)\smile(q\smile p)\frown s \\ \quad p\frown s \\ \quad q\frown s \\ \quad \mathfrak{y}s \\ \quad \mathfrak{p}s \\ \quad c\frown s \\ \quad p\smile\mathfrak{F}c\frown s \\ \quad d\frown s \\ \quad q\smile\mathfrak{F}d\frown s \end{array} \qquad (648$$

(505): ————————————

$$\begin{array}{l} \vdash \mathfrak{F}c\smile\mathfrak{F}d\smile(q\smile p)\frown s \\ \quad p\frown s \\ \quad q\frown s \\ \quad \mathfrak{y}s \\ \quad \mathfrak{p}s \\ \quad c\frown s \\ \quad p\smile\mathfrak{F}c\frown s \\ \quad d\frown s \\ \quad q\smile\mathfrak{F}d\frown s \end{array} \qquad (649$$

§223. *Analysis*

As we planned in §221, we now insert 'p' for 'd', 'q' for 'c', 'o' for 'q', and 'a' for 'p' in (648). We then obtain therein the subcomponents '— $o\smile\mathfrak{F}p\frown s$' and '— $a\smile\mathfrak{F}q\frown s$'. In place of these, '$\top\, p\smile\mathfrak{F}o\frown s$', '$\top\, p = o$', '$\top\, q\smile\mathfrak{F}a\frown s$', '$\top\, q = a$' are to be introduced, since o is supposed to be the first multiple of b that is not less than p and a the first that is not less than q. Instead of '$\top\, p = o$' and '$\top\, q = a$', one can bring in the subcomponent, '$\begin{array}{l}\top\, p = o \\ \quad q = a\end{array}$'.

By (529) we have

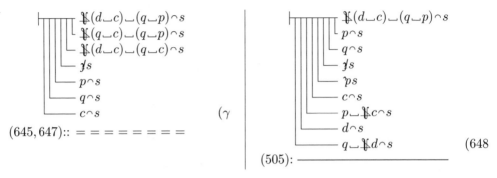

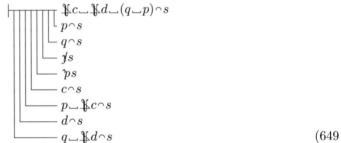

$$\begin{array}{l} \vdash \mathfrak{F}(p\smile q)\smile(o\smile a)\smile(\mathfrak{F}c\smile\mathfrak{F}d\smile(q\smile p))\frown s \\ \quad \mathfrak{F}c\smile\mathfrak{F}d\smile(q\smile p)\frown s \\ \quad \mathfrak{F}(p\smile q)\smile(o\smile a)\frown s \\ \quad \mathfrak{y}s \end{array}$$

In this,

'$\vdash$ ⋔$(p \smile q) \smile (o \smile a) \frown s$
$\quad\top p = o$
$\quad q = a$',

will have to be introduced for the penultimate subcomponent. In order, as stated above, to bring in the subcomponent '$\top p = o$ / $q = a$' for '$\top p = o$' and '$\top q = a$' in the proposition with the supercomponent

'— ⋔$(p \smile q) \smile (o \smile a) \frown s$'

we require propositions with this supercomponent, one having the subcomponents '$\top p = o$' and '— $q = a$', and another having the subcomponents '$\top q = a$' and '— $p = o$', the first of which follows from (645), the second from (647). For the transformation of (529) mentioned above, we require the proposition

'$\vdash$ ⋔$(p \smile q) \smile (o \smile a) \smile ($⋔$c \smile$ ⋔$d \smile (q \smile p)) \frown s$
⋔$c \smile$ ⋔$d \smile (q \smile p) \frown s$
$\top p = o$
$q = a$
ʒs
$o \smile a \frown s$', (α

which is to be proven using (581).

§224. *Construction*

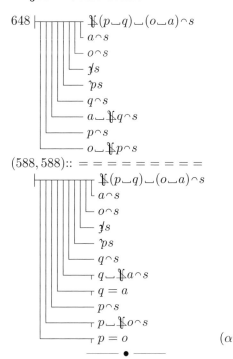

648 $\vdash$ ⋔$(p \smile q) \smile (o \smile a) \frown s$
$a \frown s$
$o \frown s$
ʒs
ƿs
$q \frown s$
$a \smile$ ⋔$q \frown s$
$p \frown s$
$o \smile$ ⋔$p \frown s$

(588, 588):: = = = = = = = = = =

$\vdash$ ⋔$(p \smile q) \smile (o \smile a) \frown s$
$a \frown s$
$o \frown s$
ʒs
ƿs
$q \frown s$
$\top q \smile$ ⋔$a \frown s$
$\top q = a$
$p \frown s$
$\top p \smile$ ⋔$o \frown s$
$\top p = o$ (α

———— • ————

IIIa $\vdash$ ⋔$(p \smile q) \smile (o \smile a) \frown s$
⋔$(p \smile a) \smile (o \smile a) \frown s$
$q = a$

(645):: - - - - - - - - - - -

$\vdash$ ⋔$(p \smile q) \smile (o \smile a) \frown s$
$a \frown s$
$o \frown s$
ʒs
ƿs
$p \frown s$
$o \smile$ ⋔$p \frown s$
$q = a$ (β

(588):: - - - - - - - - - - - - -

$\vdash$ ⋔$(p \smile q) \smile (o \smile a) \frown s$
$a \frown s$
$o \frown s$
ʒs
ƿs
$p \frown s$
$\top p \smile$ ⋔$o \frown s$
$\top p = o$
$q = a$ (γ

(I):: - - - - - - - - - - - - -

$\vdash \natural(p \smallsmile q) \smallsmile (o \smallsmile a) \frown s$
$a \frown s$
$o \frown s$
$\not\mathfrak{z}s$
$\mathfrak{p}s$
$p \frown s$
$p \smallsmile \natural o \frown s$
$q = a$
$p = o$
$q = a$ (δ

———— • ————

IIIa $\vdash \natural(p \smallsmile q) \smallsmile (o \smallsmile a) \frown s$
$\natural(o \smallsmile q) \smallsmile (o \smallsmile a) \frown s$
$p = o$

(647):: – – – – – – – – – – –

$\vdash \natural(p \smallsmile q) \smallsmile (o \smallsmile a) \frown s$
$a \frown s$
$o \frown s$
$\not\mathfrak{z}s$
$\mathfrak{p}s$
$q \frown s$
$a \smallsmile \natural q \frown s$
$p = o$ (ε

(588):: – – – – – – – – – – – – –

$\vdash \natural(p \smallsmile q) \smallsmile (o \smallsmile a) \frown s$
$a \frown s$
$o \frown s$
$\not\mathfrak{z}s$
$\mathfrak{p}s$
$q \frown s$
$q \smallsmile \natural a \frown s$
$q = a$
$p = o$ (ζ

(α): ·–·–·–·–·–·–·–·–·–·–·–·

$\vdash \natural(p \smallsmile q) \smallsmile (o \smallsmile a) \frown s$
$a \frown s$
$o \frown s$
$\not\mathfrak{z}s$
$\mathfrak{p}s$
$q \frown s$
$q \smallsmile \natural a \frown s$
$q = a$
$p \frown s$
$p \smallsmile \natural o \frown s$ (η

(δ): ·–·–·–·–·–·–·–·–·–·–·–·

$\vdash \natural(p \smallsmile q) \smallsmile (o \smallsmile a) \frown s$
$p = o$
$q = a$
$a \frown s$
$o \frown s$
$\not\mathfrak{z}s$
$\mathfrak{p}s$
$q \frown s$
$q \smallsmile \natural a \frown s$
$p \frown s$
$p \smallsmile \natural o \frown s$ (650

———— • ————

529 $\vdash \natural(p \smallsmile q) \smallsmile (o \smallsmile a) \smallsmile (\natural c \smallsmile \natural d \smallsmile (q \smallsmile p)) \frown s$
$\natural c \smallsmile \natural d \smallsmile (q \smallsmile p) \frown s$
$\natural(p \smallsmile q) \smallsmile (o \smallsmile a) \frown s$
$\not\mathfrak{z}s$

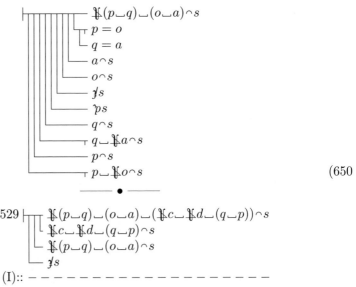

(I):: –

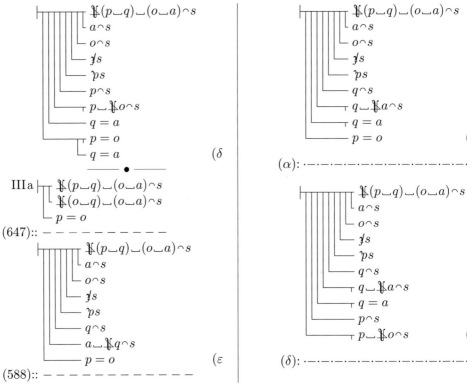

$$\vdash \between(p\smile q)\smile(o\smile a)\smile(\between c\smile\between d\smile(q\smile p))\frown s$$
$$\between c\smile\between d\smile(q\smile p)\frown s$$
$$p = o$$
$$q = a$$
$$\between(p\smile q)\smile(o\smile a)\frown s$$
$$p = o$$
$$q = a$$
$$\not\jmath s \qquad (\alpha$$

———•———

$$\text{IIIa}\ \vdash \between(p\smile a)\smile(o\smile a)\smile(\between c\smile\between d\smile(q\smile p))\frown s$$
$$\between(o\smile a)\smile(o\smile a)\smile(\between c\smile\between d\smile(q\smile p))\frown s$$
$$p = o$$

(IIIa): ————————————————

$$\vdash \between(p\smile q)\smile(o\smile a)\smile(\between c\smile\between d\smile(q\smile p))\frown s$$
$$\between(o\smile a)\smile(o\smile a)\smile(\between c\smile\between d\smile(q\smile p))\frown s$$
$$p = o$$
$$q = a \qquad (\beta$$

(Ib, Id):: = = = = = = = = = = = = = = = = = =

$$\vdash \between(p\smile q)\smile(o\smile a)\smile(\between c\smile\between d\smile(q\smile p))\frown s$$
$$\between(o\smile a)\smile(o\smile a)\smile(\between c\smile\between d\smile(q\smile p))\frown s$$
$$p = o$$
$$q = a \qquad (\gamma$$

(581): ————————————————

$$\vdash \between(p\smile q)\smile(o\smile a)\smile(\between c\smile\between d\smile(q\smile p))\frown s$$
$$\between c\smile\between d\smile(q\smile p)\frown s$$
$$p = o$$
$$q = a$$
$$\not\jmath s$$
$$o\smile a\frown s \qquad (\delta$$

(529):: –

$$\vdash \between(p\smile q)\smile(o\smile a)\smile(\between c\smile\between d\smile(q\smile p))\frown s$$
$$\between c\smile\between d\smile(q\smile p)\frown s$$
$$p = o$$
$$q = a$$
$$\not\jmath s$$
$$a\frown s$$
$$o\frown s \qquad (\varepsilon$$

(α): –·–·–·–·–·–·–·–·–·–·–·–·–·–·–·–·–·–·–

$$
\begin{array}{l}
\vdash \nparallel(p \smile q) \smile (o \smile a) \smile (\nparallel c \smile \nparallel d \smile (q \smile p)) \frown s \\
\quad \nparallel c \smile \nparallel d \smile (q \smile p) \frown s \\
\quad \not\jmath s \\
\quad a \frown s \\
\quad o \frown s \\
\quad \nparallel(p \smile q) \smile (o \smile a) \frown s \\
\quad p = o \\
\quad q = a
\end{array}
\qquad (\zeta
$$

(649, 650):: =

$$
\begin{array}{l}
\vdash \nparallel(p \smile q) \smile (o \smile a) \smile (\nparallel c \smile \nparallel d \smile (q \smile p)) \frown s \\
\quad p \frown s \\
\quad q \frown s \\
\quad \not\jmath s \\
\quad \daleth s \\
\quad c \frown s \\
\quad p \smile \nparallel c \frown s \\
\quad d \frown s \\
\quad q \smile \nparallel d \frown s \\
\quad a \frown s \\
\quad o \frown s \\
\quad q \smile \nparallel a \frown s \\
\quad p \smile \nparallel o \frown s
\end{array}
\qquad (\eta
$$

(623, 623):: =

$$
\begin{array}{l}
\vdash \nparallel(p \smile q) \smile (o \smile a) \smile (\nparallel c \smile \nparallel d \smile (q \smile p)) \frown s \\
\quad p \frown s \\
\quad q \frown s \\
\quad \not\jmath s \\
\quad \daleth s \\
\quad c \frown s \\
\quad p \smile \nparallel c \frown s \\
\quad d \frown s \\
\quad q \smile \nparallel d \frown s \\
\quad b \frown s \\
\quad b \frown (a \frown \smile_{*} b) \\
\quad b \frown (o \frown \smile_{*} b) \\
\quad q \smile \nparallel a \frown s \\
\quad p \smile \nparallel o \frown s
\end{array}
\qquad (651
$$

§225. *Analysis*

According to our plan of §221, we now prove the proposition

$$
\text{‘}\vdash\ \begin{array}{l} b \smile b = o \smile a \smile (\not{\mathfrak{I}}c \smile \not{\mathfrak{I}}d) \\ c \frown (o \frown_{*} b) \\ c \frown s \\ \text{ɟ}s \\ b \frown s \\ b \frown (a \frown \dot{\smile}_{*} b) \\ d \frown (a \frown_{*} b) \\ d \frown s \\ b \frown (o \frown \dot{\smile}_{*} b) \end{array}\text{’} \qquad (\alpha
$$

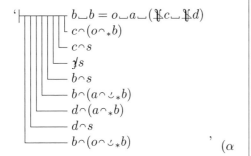

which we derive from the proposition

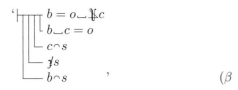

$$
\text{‘}\vdash\ \begin{array}{l} b = o \smile \not{\mathfrak{I}}c \\ b \smile c = o \\ c \frown s \\ \text{ɟ}s \\ b \frown s \end{array}\text{’} \qquad (\beta
$$

and (502). (β) follows from (570).

§226. *Construction*

$$
\text{IIIh} \vdash\ \begin{array}{l} b \smile c \smile \not{\mathfrak{I}}c = o \smile \not{\mathfrak{I}}c \\ b \smile c = o \end{array}
$$

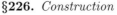

(570): ———————

$$
\vdash\ \begin{array}{l} b = o \smile \not{\mathfrak{I}}c \\ b \smile c = o \\ c \frown s \\ \text{ɟ}s \\ b \frown s \end{array} \qquad (652
$$

(492):: - - - - - - -

$$
\vdash\ \begin{array}{l} b = o \smile \not{\mathfrak{I}}c \\ c \frown (o \frown_{*} b) \\ c \frown s \\ \text{ɟ}s \\ b \frown s \end{array} \qquad (653
$$

(IIIa): - - - - - - -

$$
\vdash\ \begin{array}{l} f(b) \\ f(o \smile \not{\mathfrak{I}}c) \\ c \frown (o \frown_{*} b) \\ c \frown s \\ \text{ɟ}s \\ b \frown s \end{array} \qquad (654
$$

———— • ————

$$
489 \vdash a \smile (o \smile \not{\mathfrak{I}}c) = a \smile o \smile \not{\mathfrak{I}}c
$$

(IIIh): ———————

$$
\vdash a \smile (o \smile \not{\mathfrak{I}}c) \smile \not{\mathfrak{I}}d = a \smile o \smile \not{\mathfrak{I}}c \smile \not{\mathfrak{I}}d \qquad (\alpha
$$

(490): ———————

$$
\vdash a \smile (o \smile \not{\mathfrak{I}}c) \smile \not{\mathfrak{I}}d = a \smile o \smile (\not{\mathfrak{I}}c \smile \not{\mathfrak{I}}d) \qquad (\beta
$$

(654): ———————

$$
\vdash\ \begin{array}{l} a \smile b \smile \not{\mathfrak{I}}d = a \smile o \smile (\not{\mathfrak{I}}c \smile \not{\mathfrak{I}}d) \\ c \frown (o \frown_{*} b) \\ c \frown s \\ \text{ɟ}s \\ b \frown s \end{array} \qquad (\gamma
$$

(500): ———————

$$
\vdash\ \begin{array}{l} b \smile a \smile \not{\mathfrak{I}}d = a \smile o \smile (\not{\mathfrak{I}}c \smile \not{\mathfrak{I}}d) \\ c \frown (o \frown_{*} b) \\ c \frown s \\ \text{ɟ}s \\ b \frown s \\ b \frown (a \frown \dot{\smile}_{*} b) \end{array} \qquad (\delta
$$

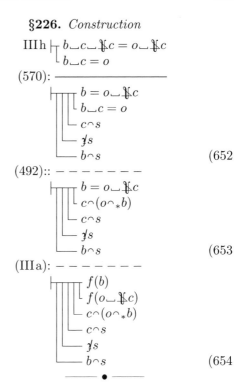

(490): ———————

$$\vdash \; b \smile (a \smile \between d) = a \smile o \smile (\between c \smile \between d)$$
$$c \frown (o \frown_{*} b)$$
$$c \frown s$$
$$\not\jmath s$$
$$b \frown s$$
$$b \frown (a \frown \dot{\smile}_{*} b) \qquad (\varepsilon$$

(654): ————————————————

$$\vdash \; b \smile b = a \smile o \smile (\between c \smile \between d)$$
$$c \frown (o \frown_{*} b)$$
$$c \frown s$$
$$\not\jmath s$$
$$b \frown s$$
$$b \frown (a \frown \dot{\smile}_{*} b)$$
$$d \frown (a \frown_{*} b)$$
$$d \frown s \qquad (\zeta$$

(502):: ————————————————

$$\vdash \; b \smile b = o \smile a \smile (\between c \smile \between d)$$
$$c \frown (o \frown_{*} b)$$
$$c \frown s$$
$$\not\jmath s$$
$$b \frown s$$
$$b \frown (a \frown \dot{\smile}_{*} b)$$
$$d \frown (a \frown_{*} b)$$
$$d \frown s$$
$$b \frown (o \frown \dot{\smile}_{*} b) \qquad (\eta$$

(IIIa): – – – – – – – – – – – – – –

$$\vdash \; \between (p \smile q) \smile (b \smile b \smile (q \smile p)) \frown s$$
$$\between (p \smile q) \smile (o \smile a \smile (\between c \smile \between d) \smile (q \smile p)) \frown s$$
$$c \frown (o \frown_{*} b)$$
$$c \frown s$$
$$\not\jmath s$$
$$b \frown s$$
$$b \frown (a \frown \dot{\smile}_{*} b)$$
$$d \frown (a \frown_{*} b)$$
$$d \frown s$$
$$b \frown (o \frown \dot{\smile}_{*} b) \qquad (\vartheta$$

——— • ———

490 $\vdash \; \between (p \smile q) \smile [o \smile a \smile (\between c \smile \between d \smile (q \smile p))] \frown s$
$\between (p \smile q) \smile (o \smile a) \smile (\between c \smile \between d \smile (q \smile p)) \frown s$

(491): ————————————————

$$\vdash \; \between (p \smile q) \smile [o \smile a \smile (\between c \smile \between d) \smile (q \smile p)] \frown s$$
$$\between (p \smile q) \smile (o \smile a) \smile (\between c \smile \between d \smile (q \smile p)) \frown s \qquad (\iota$$

(ϑ): – – – – – – – – – – – – – – – – – – –

$\vdash \!\!\!\not\mathfrak{P}(p \smile q) \smile (b \smile b \smile (q \smile p)) \frown s$
$\not\mathfrak{P}(p \smile q) \smile (o \smile a) \smile (\not\mathfrak{P}c \smile \not\mathfrak{P}d \smile (q \smile p)) \frown s$
$c \frown (o \frown_{*} b)$
$c \frown s$
$\daleth s$
$b \frown s$
$b \frown (a \frown \smile_{*} b)$
$d \frown (a \frown_{*} b)$
$d \frown s$
$b \frown (o \frown \smile_{*} b)$ (κ

(651):: -

$\vdash \!\!\!\not\mathfrak{P}(p \smile q) \smile (b \smile b \smile (q \smile p)) \frown s$
$p \frown s$
$q \frown s$
$\daleth s$
$\gimel p s$
$c \frown s$
$p \smile \not\mathfrak{P}c \frown s$
$d \frown s$
$q \smile \not\mathfrak{P}d \frown s$
$b \frown s$
$b \frown (a \frown \smile_{*} b)$
$b \frown (o \frown \smile_{*} b)$
$q \smile \not\mathfrak{P}a \frown s$
$p \smile \not\mathfrak{P}o \frown s$
$c \frown (o \frown_{*} b)$
$d \frown (a \frown_{*} b)$ (λ

(623, 623):: = = = = = = = = = = = = = = = = = =

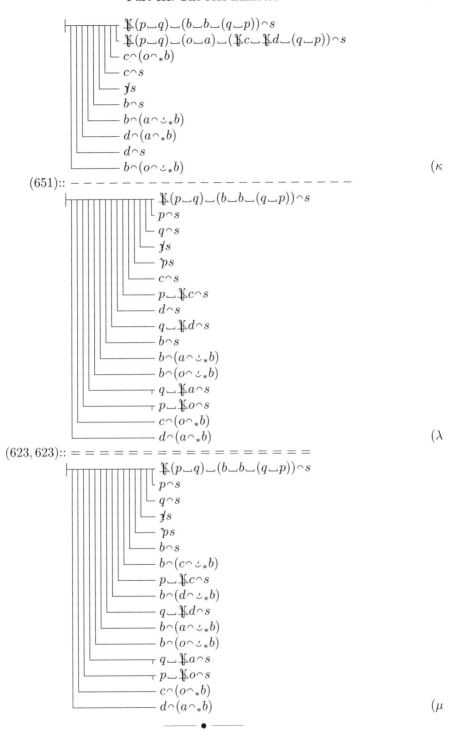

$\vdash \!\!\!\not\mathfrak{P}(p \smile q) \smile (b \smile b \smile (q \smile p)) \frown s$
$p \frown s$
$q \frown s$
$\daleth s$
$\gimel p s$
$b \frown s$
$b \frown (c \frown \smile_{*} b)$
$p \smile \not\mathfrak{P}c \frown s$
$b \frown (d \frown \smile_{*} b)$
$q \smile \not\mathfrak{P}d \frown s$
$b \frown (a \frown \smile_{*} b)$
$b \frown (o \frown \smile_{*} b)$
$q \smile \not\mathfrak{P}a \frown s$
$p \smile \not\mathfrak{P}o \frown s$
$c \frown (o \frown_{*} b)$
$d \frown (a \frown_{*} b)$ (μ

529 ⊢ b⌣b⌣(q⌣p)⌢s
q⌣p⌢s
b⌣b⌢s
ɟs

(529, 529):: = = = = = =

⊢ b⌣b⌣(q⌣p)⌢s
p⌢s
q⌢s
ɟs
b⌢s (ν

(638): − − − − − − − − −

⊢ b⌣b⌣(q⌣p)⌣∤(p⌣q)⌢s
p⌣q⌢s
ʹps
p⌢s
q⌢s
ɟs
b⌢s
∤(p⌣q)⌣(b⌣b⌣(q⌣p))⌢s (ξ

(μ, 529):: = = = = = = = = = = = =

⊢ b⌣b⌣(q⌣p)⌣∤(p⌣q)⌢s
q⌢s
p⌢s
ɟs
ʹps
b⌢s
b⌢(c⌢⌣*b)
p⌣∤c⌢s
b⌢(d⌢⌣*b)
q⌣∤d⌢s
b⌢(a⌢⌣*b)
b⌢(o⌢⌣*b)
q⌣∤a⌢s
p⌣∤o⌢s
c⌢(o⌢*b)
d⌢(a⌢*b) (o

×

⊢ p⌣∤c⌢s
b⌢(c⌢⌣*b)
c⌢(o⌢*b)
q⌢s
p⌢s
ɟs
ʹps
b⌢s
b⌣b⌣(q⌣p)⌣∤(p⌣q)⌢s
b⌢(d⌢⌣*b)
q⌣∤d⌢s
b⌢(a⌢⌣*b)
b⌢(o⌢⌣*b)
q⌣∤a⌢s
p⌣∤o⌢s
d⌢(a⌢*b) (π

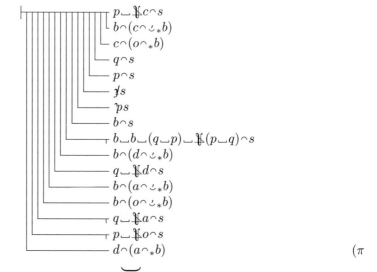

⏟

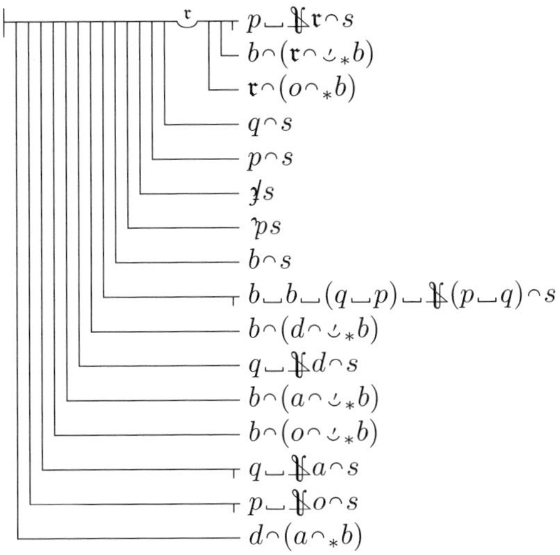

(655

§227. *Analysis*

The last transformations serve to eliminate ‘*c*’. Likewise, ‘*d*’ has to be eliminated. For this we need the proposition

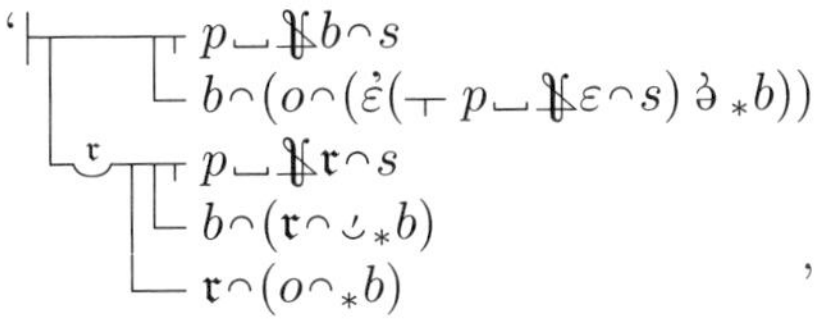

’ (α

which is to be reduced to the proposition

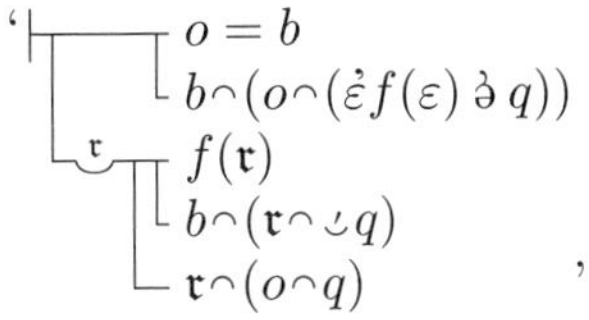

’ (β

For the proof of (β) we require the proposition

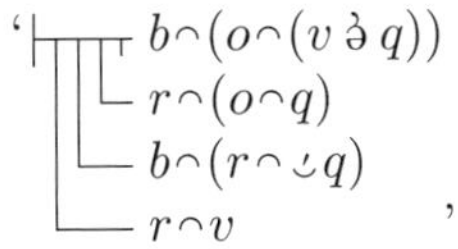

’ (γ

which follows from (404) using (5) and (197). We then replace ‘*v*’ by ‘$\dot{\varepsilon} f(\varepsilon)$’ and apply (142).

§228. *Construction*

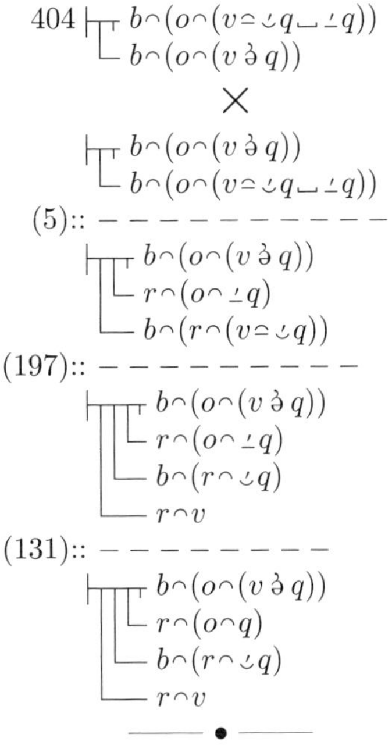

(α

(β

(656

(657

77 $\vdash r \frown \dot{\varepsilon} f(\varepsilon)$
$f(r)$

(657): — — — — —

$\vdash b \frown (o \frown (\dot{\varepsilon} f(\varepsilon) \grave{\partial}\, q))$
$r \frown (o \frown q)$
$b \frown (r \frown \smile q)$
$f(r)$ (658

(IIa):: — — — — — — — — — — —

$\vdash b \frown (o \frown (\dot{\varepsilon} f(\varepsilon) \grave{\partial}\, q))$
$r \frown (o \frown q)$
$b \frown (r \frown \smile q)$
$\smile^{\mathfrak{r}}$ $f(\mathfrak{r})$
$b \frown (\mathfrak{r} \frown \smile q)$
$\mathfrak{r} \frown (o \frown q)$ (α

$\times$

$\vdash b \frown (r \frown \smile q)$
$r \frown (o \frown q)$
$b \frown (o \frown (\dot{\varepsilon} f(\varepsilon) \grave{\partial}\, q))$
$\smile^{\mathfrak{r}}$ $f(\mathfrak{r})$
$b \frown (\mathfrak{r} \frown \smile q)$
$\mathfrak{r} \frown (o \frown q)$ (β

$\smile$

$\vdash \smile^{\mathfrak{e}}$ $b \frown (\mathfrak{e} \frown \smile q)$
$\mathfrak{e} \frown (o \frown q)$
$b \frown (o \frown (\dot{\varepsilon} f(\varepsilon) \grave{\partial}\, q))$
$\smile^{\mathfrak{r}}$ $f(\mathfrak{r})$
$b \frown (\mathfrak{r} \frown \smile q)$
$\mathfrak{r} \frown (o \frown q)$ (γ

(142): — — — — — — — — — — —

$\vdash b \frown (o \frown \perp q)$
$b \frown (o \frown (\dot{\varepsilon} f(\varepsilon) \grave{\partial}\, q))$
$\smile^{\mathfrak{r}}$ $f(\mathfrak{r})$
$b \frown (\mathfrak{r} \frown \smile q)$
$\mathfrak{r} \frown (o \frown q)$ (δ

(130): — — — — — — — — — — —

$\vdash o = b$
$b \frown (o \frown (\dot{\varepsilon} f(\varepsilon) \grave{\partial}\, q))$
$\smile^{\mathfrak{r}}$ $f(\mathfrak{r})$
$b \frown (\mathfrak{r} \frown \smile q)$
$\mathfrak{r} \frown (o \frown q)$
$b \frown (o \frown \smile q)$ (ε

(401):: — — — — — — — — — — —

$\vdash o = b$
$b \frown (o \frown (\dot{\varepsilon} f(\varepsilon) \grave{\partial}\, q))$
$\smile^{\mathfrak{r}}$ $f(\mathfrak{r})$
$b \frown (\mathfrak{r} \frown \smile q)$
$\mathfrak{r} \frown (o \frown q)$ (659

———— • ————

82 $\vdash f(o)$
$o \frown \dot{\varepsilon} f(\varepsilon)$

(402):: — — — — —

$\vdash f(o)$
$b \frown (o \frown (\dot{\varepsilon} f(\varepsilon) \grave{\partial}\, q))$ (660

———— • ————

IIIc $\vdash p \llcorner \nparallel b \frown s$
$p \llcorner \nparallel o \frown s$
$o = b$

(659, 660):: = = = =

$\vdash p \llcorner \nparallel b \frown s$
$b \frown (o \frown (\dot{\varepsilon}(\top p \llcorner \nparallel \varepsilon \frown s) \grave{\partial}\, {}_{*}b))$
$\smile^{\mathfrak{r}}$ $p \llcorner \nparallel \mathfrak{r} \frown s$
$b \frown (\mathfrak{r} \frown \smile_{*} b)$
$\mathfrak{r} \frown (o \frown_{*} b)$ (661

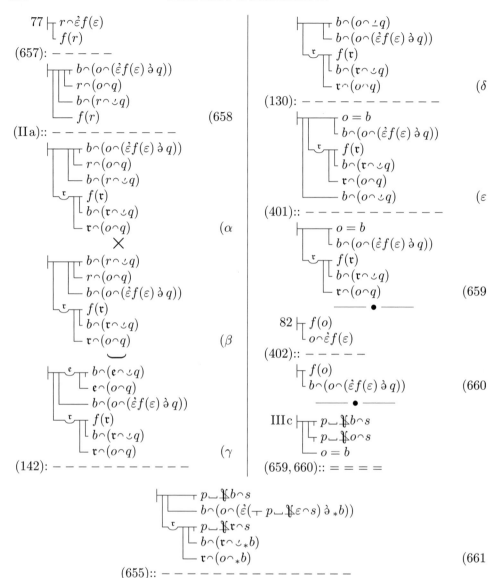

(655):: — — — — — — — — — — — — — — —

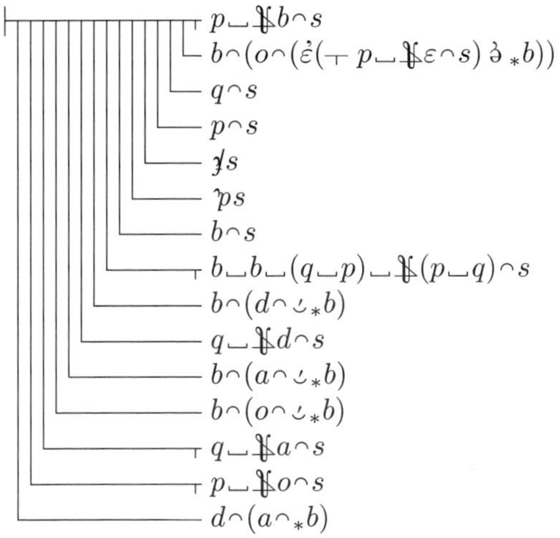

(α

(660, 401):: = = = = = = = = = = = = = = = = = =

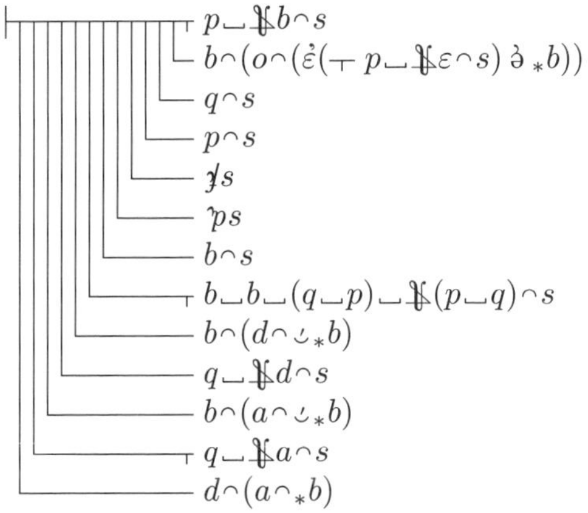

(β

$\times$

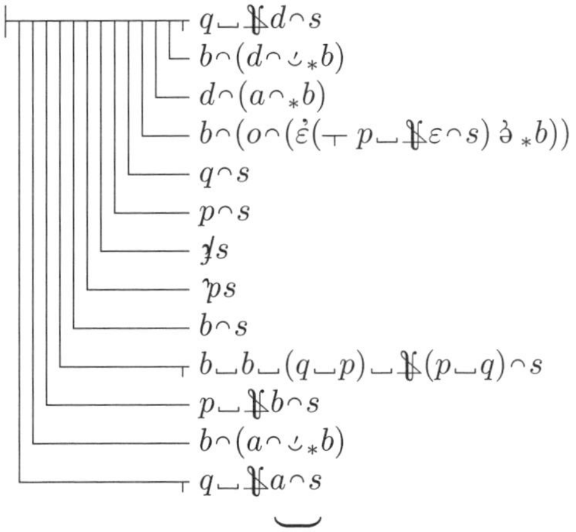

(γ

$$\vdash \overset{\mathfrak{r}}{\smile} q \smile \between \mathfrak{r} \frown s$$
$$b \frown (\mathfrak{r} \frown \grave{\smile}_* b)$$
$$\mathfrak{r} \frown (a \frown_* b)$$
$$b \frown (o \frown (\dot{\varepsilon}(\top p \smile \between \varepsilon \frown s) \eth_* b))$$
$$q \frown s$$
$$p \frown s$$
$$\not{\mathfrak{y}} s$$
$$\mathfrak{p} s$$
$$b \frown s$$
$$\top b \smile b \smile (q \smile p) \smile \between (p \smile q) \frown s$$
$$p \smile \between b \frown s$$
$$b \frown (a \frown \grave{\smile}_* b)$$
$$\top q \smile \between a \frown s \quad (\delta$$

(661): — — — — — — — — — — — — — — — — — —

$$\vdash \top q \smile \between b \frown s$$
$$b \frown (a \frown (\dot{\varepsilon}(\top q \smile \between \varepsilon \frown s) \eth_* b))$$
$$b \frown (o \frown (\dot{\varepsilon}(\top p \smile \between \varepsilon \frown s) \eth_* b))$$
$$q \frown s$$
$$p \frown s$$
$$\not{\mathfrak{y}} s$$
$$\mathfrak{p} s$$
$$b \frown s$$
$$\top b \smile b \smile (q \smile p) \smile \between (p \smile q) \frown s$$
$$p \smile \between b \frown s$$
$$b \frown (a \frown \grave{\smile}_* b)$$
$$\top q \smile \between a \frown s \quad (\varepsilon$$

(660, 401):: = = = = = = = = = = = = = = = = =

$$\vdash \top q \smile \between b \frown s$$
$$b \frown (a \frown (\dot{\varepsilon}(\top q \smile \between \varepsilon \frown s) \eth_* b))$$
$$b \frown (o \frown (\dot{\varepsilon}(\top p \smile \between \varepsilon \frown s) \eth_* b))$$
$$q \frown s$$
$$p \frown s$$
$$\not{\mathfrak{y}} s$$
$$\mathfrak{p} s$$
$$b \frown s$$
$$\top b \smile b \smile (q \smile p) \smile \between (p \smile q) \frown s$$
$$p \smile \between b \frown s \quad (\zeta$$

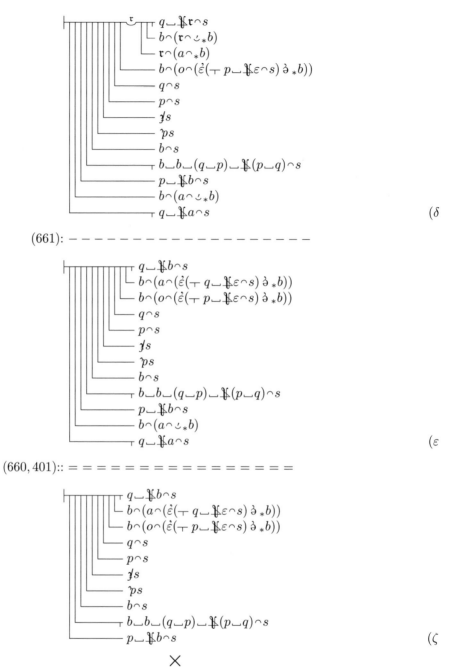

×

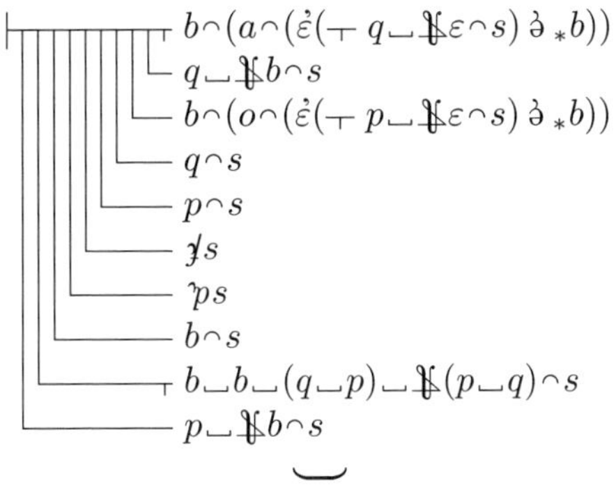

(η

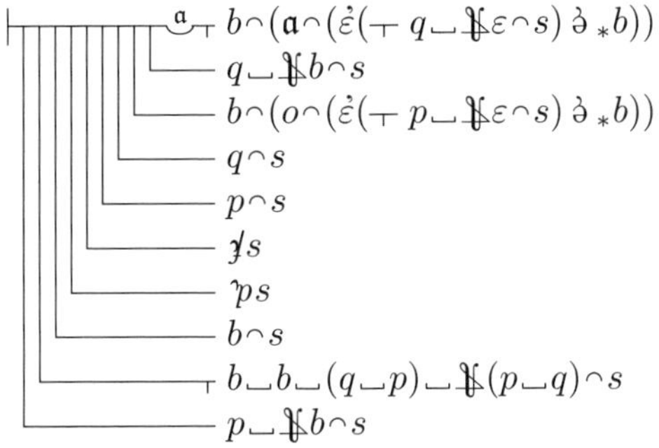

(662

§229. *Analysis*

We now replace the supercomponent in (662) by ‘$\top q \frown (b \frown (s \mathfrak{g}\, b))$’, which is then itself to be eliminated using (636). For this we require the proposition

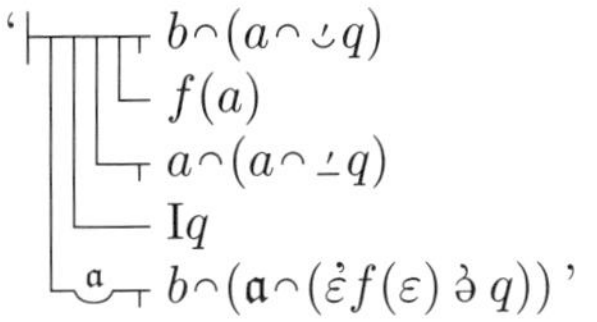

(α

which we prove using (357). Applying it to our case, where ‘$_*b$’ is to replace ‘q’, we then require, in addition to (497), the further proposition

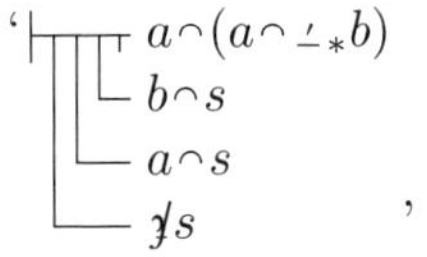

(β

which we prove using (123).

§230. *Construction*

77

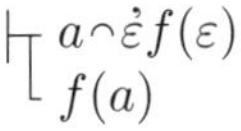

(357): – – – – –

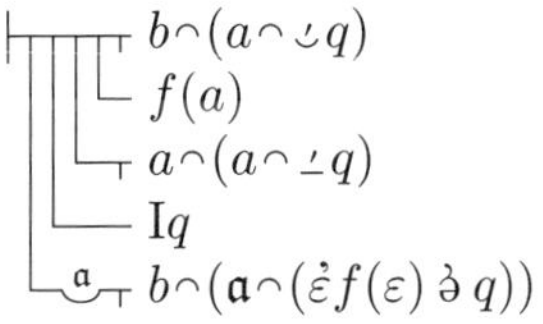

(663

491

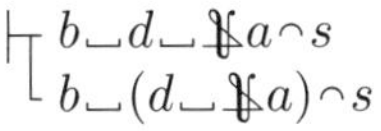

(529):: – – – – – – – –

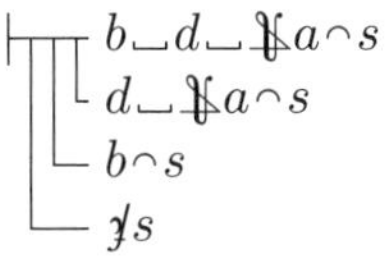

(α

(493): ——————

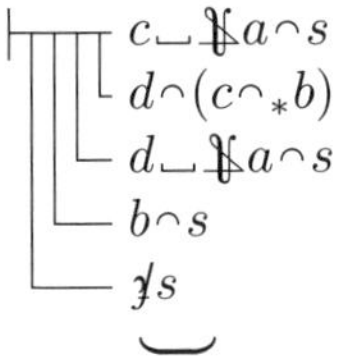

(β

$\vdash \smile^{\mathfrak{d}} \smile^{\mathfrak{a}}$ $\mathfrak{a} \smile \mathfrak{f}a \frown s$
$\mathfrak{d} \frown (\mathfrak{a} \frown_* b)$
$\mathfrak{d} \smile \mathfrak{f}a \frown s$
$b \frown s$
$\not\mathfrak{z}s$ (γ

(123): - - - - - - - - - -

$\vdash$ $a \smile \mathfrak{f}a \frown s$
$\smile^{\mathfrak{a}}$ $\mathfrak{a} \smile \mathfrak{f}a \frown s$
$a \frown (\mathfrak{a} \frown_* b)$
$b \frown s$
$\not\mathfrak{z}s$
$a \frown (a \frown \dot{\lrcorner}_* b)$ (δ

———— • ————

493 $\vdash$ $c \smile \mathfrak{f}a \frown s$
$b \smile a \smile \mathfrak{f}a \frown s$
$a \frown (c \frown_* b)$

(570): ————————

$\vdash$ $c \smile \mathfrak{f}a \frown s$
$a \frown (c \frown_* b)$
$b \frown s$
$a \frown s$
$\not\mathfrak{z}s$ (ε

$\vdash \smile^{\mathfrak{a}}$ $\mathfrak{a} \smile \mathfrak{f}a \frown s$
$a \frown (\mathfrak{a} \frown_* b)$
$b \frown s$
$a \frown s$
$\not\mathfrak{z}s$ (ζ

(δ): - - - - - - - - -

$\vdash$ $a \smile \mathfrak{f}a \frown s$
$b \frown s$
$a \frown s$
$\not\mathfrak{z}s$
$a \frown (a \frown \dot{\lrcorner}_* b)$ (η

×

$\vdash$ $a \frown (a \frown \dot{\lrcorner}_* b)$
$b \frown s$
$a \frown s$
$\not\mathfrak{z}s$
$\top$ $a \smile \mathfrak{f}a \frown s$ (ϑ

(536): - - - - - - - - -

$\vdash$ $a \frown (a \frown \dot{\lrcorner}_* b)$
$b \frown s$
$a \frown s$
$\not\mathfrak{z}s$ (664

(663): - - - - - - - - -

$\vdash$ $b \frown (a \frown \dot{\smile}_* b)$
$\top$ $q \smile \mathfrak{f}a \frown s$
$b \frown s$
$a \frown s$
$\not\mathfrak{z}s$
$\mathrm{I}_* b$
$\smile^{\mathfrak{a}}\top$ $b \frown (\mathfrak{a} \frown (\dot{\varepsilon}(\top q \smile \mathfrak{f}\varepsilon \frown s) \grave{\partial}_* b))$ (α

×

$\vdash$ $q \smile \mathfrak{f}a \frown s$
$b \frown (a \frown \dot{\smile}_* b)$
$b \frown s$
$a \frown s$
$\not\mathfrak{z}s$
$\mathrm{I}_* b$
$\smile^{\mathfrak{a}}\top$ $b \frown (\mathfrak{a} \frown (\dot{\varepsilon}(\top q \smile \mathfrak{f}\varepsilon \frown s) \grave{\partial}_* b))$ (β

(623, 497):: ══════════════

$\vdash$ $q \smile \mathfrak{f}a \frown s$
$b \frown (a \frown \dot{\smile}_* b)$
$b \frown s$
$\not\mathfrak{z}s$
$\smile^{\mathfrak{a}}\top$ $b \frown (\mathfrak{a} \frown (\dot{\varepsilon}(\top q \smile \mathfrak{f}\varepsilon \frown s) \grave{\partial}_* b))$ (γ

$\vdash \smile^{\mathfrak{t}}$ $q \smile \mathfrak{f}\mathfrak{t} \frown s$
$b \frown (\mathfrak{t} \frown \dot{\smile}_* b)$
$b \frown s$
$\not\mathfrak{z}s$
$\smile^{\mathfrak{a}}\top$ $b \frown (\mathfrak{a} \frown (\dot{\varepsilon}(\top q \smile \mathfrak{f}\varepsilon \frown s) \grave{\partial}_* b))$ (δ

(608): - - - - - - - - - - - - - -

$\vdash$ $q \frown (b \frown (s\ \mathfrak{g}\ b))$
$b \frown s$
$\not\mathfrak{z}s$
$\smile^{\mathfrak{a}}\top$ $b \frown (\mathfrak{a} \frown (\dot{\varepsilon}(\top q \smile \mathfrak{f}\varepsilon \frown s) \grave{\partial}_* b))$ (ε

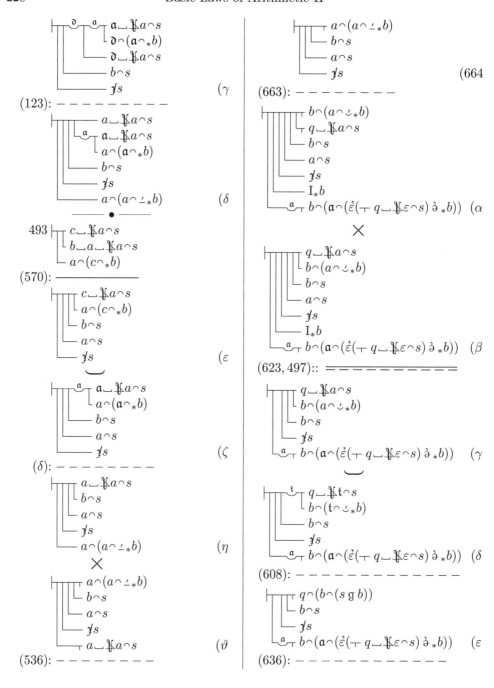

(636): - - - - - - - - - - - - -

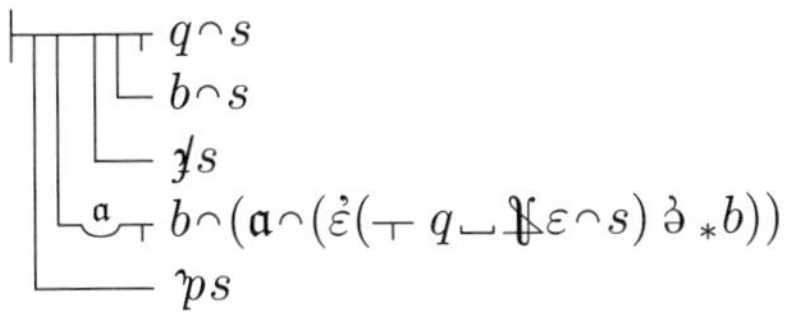

(665

(662):: - - - - - - - - - - - - - - - -

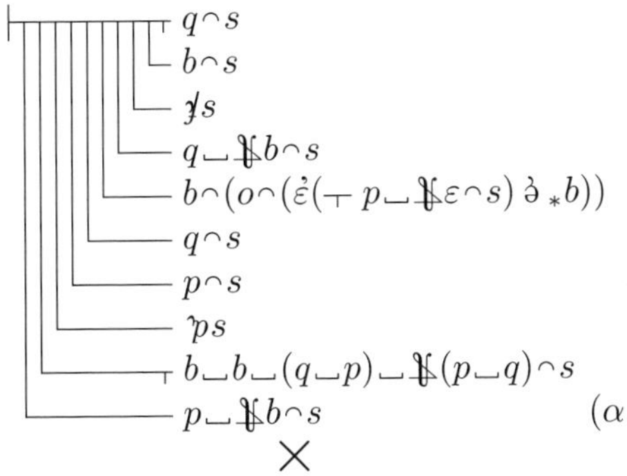

(α

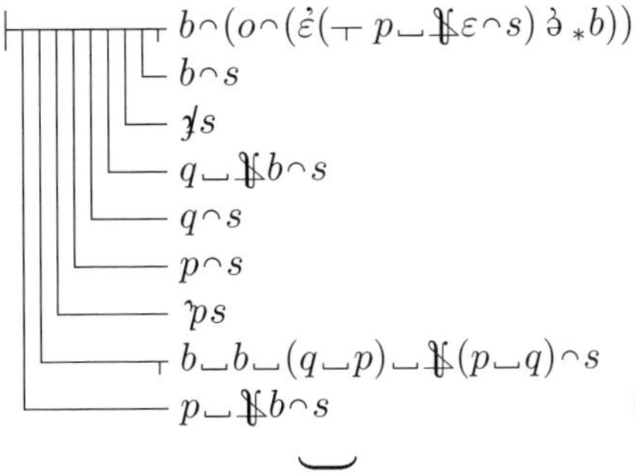

(β

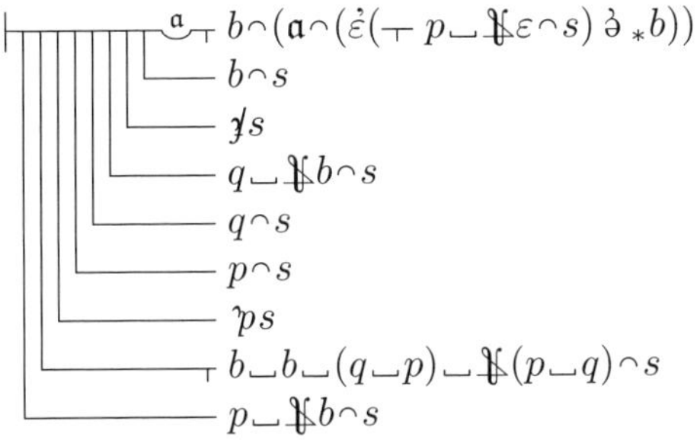

(γ

(665): - - - - - - - - - - - - - - - - - - -

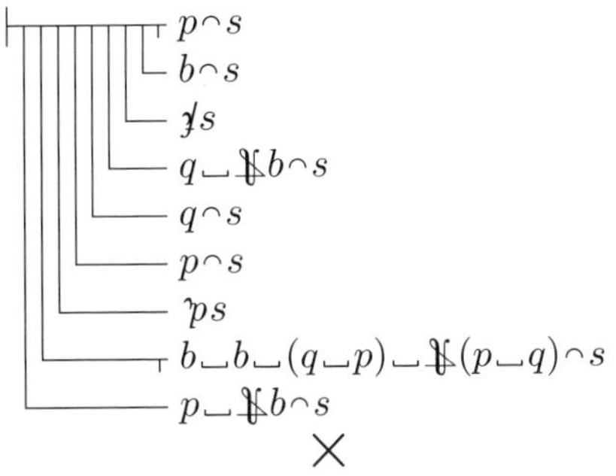

(δ

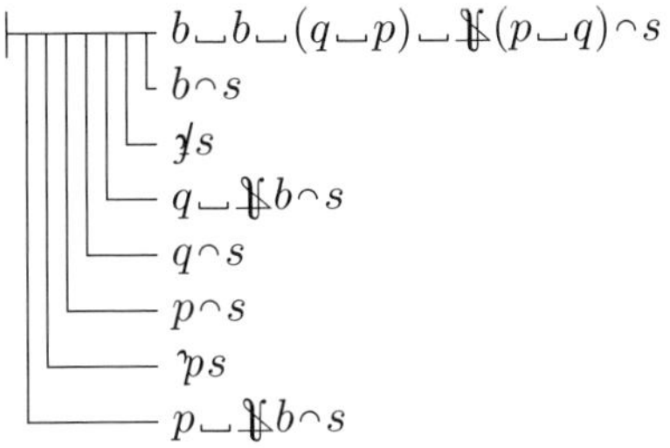

(666

e) Proof of the proposition

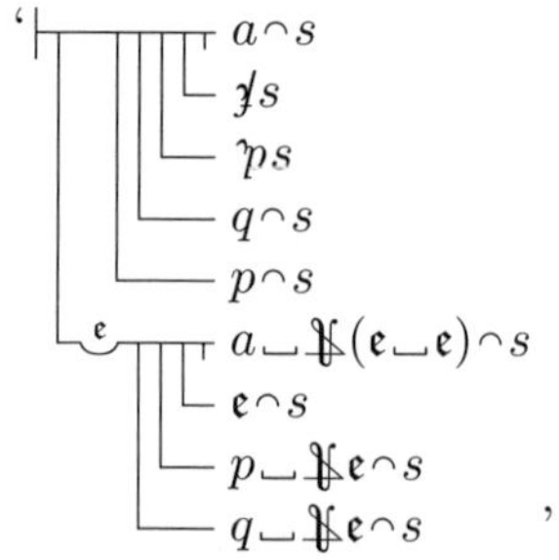

and end of the section E

§231. *Analysis*

The task now remains of proving that in the positive class s there is always a b such that $b \smile b$ is less than a, if a belongs to the positive class s. Then it will also have to be proven that there is such a b that, at the same time, is less than both p and q. Accordingly, we first derive the proposition

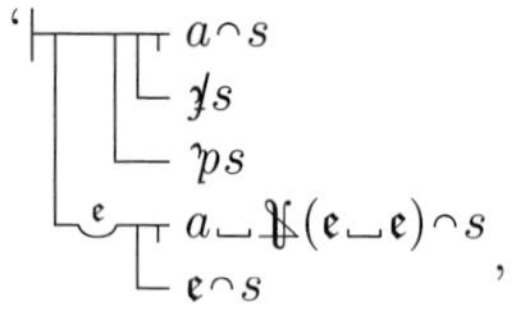

For this purpose we use proposition (606), which says that in a positive class there is always a Relation b that is less than a given Relation in that positive class. We distinguish the cases:

1. b is less than $a \smile \mathfrak{U}b$,
2. $a \smile \mathfrak{U}b$ is less than b,
3. $a \smile \mathfrak{U}b$ coincides with b.

In the first case, b itself is of the required kind, in the second case, this is $a \smile \mathfrak{U}b$, in the last case, any Relation in our positive class that is less than b satisfies our requirement, and by (606) there always is such a Relation.

§232. *Construction*

506 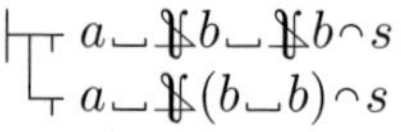

(IIa):: – – – – – – – – –

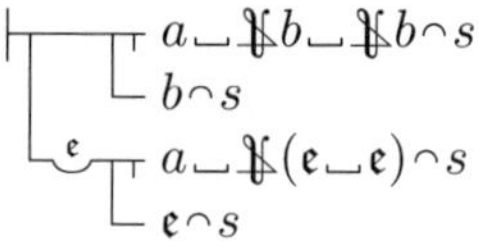 (α

(588): – – – – – – – – – –

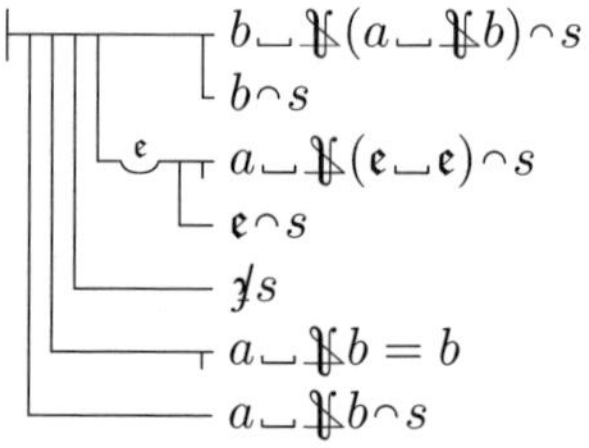 (β

(646): – – – – – – – – – – – – –

$$\vdash a \smile \rtimes b \smile b \smile \rtimes(a \smile \rtimes b \smile (a \smile \rtimes b)) \frown s$$
$$a \smile \rtimes b \frown s$$
$$\mathfrak{p} s$$
$$b \frown s$$
$$\mathfrak{e}:\ a \smile \rtimes(\mathfrak{e} \smile \mathfrak{e}) \frown s;\ \mathfrak{e} \frown s$$
$$\mathfrak{z} s$$
$$a \smile \rtimes b = b$$
(γ

(567): ————————————

$$\vdash a \smile \rtimes(a \smile \rtimes b \smile (a \smile \rtimes b)) \frown s$$
$$a \smile \rtimes b \frown s$$
$$\mathfrak{p} s$$
$$b \frown s$$
$$\mathfrak{e}:\ a \smile \rtimes(\mathfrak{e} \smile \mathfrak{e}) \frown s;\ \mathfrak{e} \frown s$$
$$\mathfrak{z} s$$
$$a \smile \rtimes b = b$$
$$a \frown s$$
(δ

×

$$\vdash a \smile \rtimes b = b$$
$$a \smile \rtimes b \frown s$$
$$\mathfrak{p} s$$
$$b \frown s$$
$$\mathfrak{e}:\ a \smile \rtimes(\mathfrak{e} \smile \mathfrak{e}) \frown s;\ \mathfrak{e} \frown s$$
$$\mathfrak{z} s$$
$$a \smile \rtimes(a \smile \rtimes b \smile (a \smile \rtimes b)) \frown s$$
$$a \frown s$$
(ε

(IIa):: – – – – – – – – – – – – – – – –

$$\vdash a \smile \rtimes b = b$$
$$a \smile \rtimes b \frown s$$
$$\mathfrak{p} s$$
$$b \frown s$$
$$\mathfrak{e}:\ a \smile \rtimes(\mathfrak{e} \smile \mathfrak{e}) \frown s;\ \mathfrak{e} \frown s$$
$$\mathfrak{z} s$$
$$a \frown s$$
(667

——— • ———

IIIh $\vdash a \smile \rtimes b \smile b = b \smile b$; $a \smile \rtimes b = b$

(567): ————————

$$\vdash a = b \smile b$$
$$a \smile \rtimes b = b$$
$$b \frown s$$
$$\mathfrak{z} s$$
$$a \frown s$$
(668

——— • ———

638 $\vdash q \smile p \smile \rtimes(d \smile c) \frown s$; $d \smile c \frown s$; $\mathfrak{p} s$; $q \smile p \frown s$; $\rtimes(d \smile c) \smile (q \smile p) \frown s$

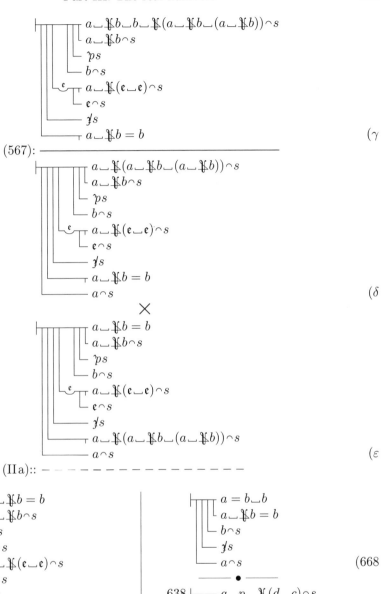

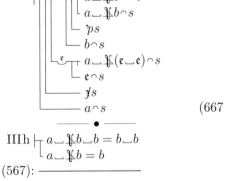

(529, 529):: = = = = = = = =

⊢ q⌣p⌣∦(d⌣c)⌢s
c⌢s
d⌢s
ʒs
ᵽs
p⌢s
q⌢s
∦(d⌣c)⌣(q⌣p)⌢s (α

(648):: - - - - - - - - - - - - - - - -

⊢ q⌣p⌣∦(d⌣c)⌢s
c⌢s
d⌢s
ʒs
ᵽs
p⌢s
q⌢s
p⌣∦c⌢s
q⌣∦d⌢s (669

—— • ——

IIIa ⊢ a⌣∦(c⌣c)⌢s
b⌣b⌣∦(c⌣c)⌢s
a = b⌣b

(668, 669):: = = = = = =

⊢ a⌣∦(c⌣c)⌢s
c⌢s
ʒs
ᵽs
b⌢s
b⌣∦c⌢s
a⌣∦b = b
a⌢s (α

×

⊢ b⌣∦c⌢s
c⌢s
ʒs
ᵽs
b⌢s
a⌣∦(c⌣c)⌢s
a⌣∦b = b
a⌢s (β

(667, IIa):: = = = = = = = =

⊢ b⌣∦c⌢s
c⌢s
ʒs
ᵽs
b⌢s
𝔢 a⌣∦(𝔢⌣𝔢)⌢s
𝔢⌢s
a⌣∦b⌢s
a⌢s (γ

‿

⊢ 𝔢 b⌣∦𝔢⌢s
𝔢⌢s
ʒs
ᵽs
b⌢s
𝔢 a⌣∦(𝔢⌣𝔢)⌢s
𝔢⌢s
a⌣∦b⌢s
a⌢s (δ

(606): - - - - - - - - - - - - -

⊢ b⌢s
ʒs
ᵽs
b⌢s
𝔢 a⌣∦(𝔢⌣𝔢)⌢s
𝔢⌢s
a⌣∦b⌢s
a⌢s (ε

×

⊢ a⌣∦b⌢s
b⌢s
ʒs
ᵽs
𝔢 a⌣∦(𝔢⌣𝔢)⌢s
𝔢⌢s
a⌢s (ζ

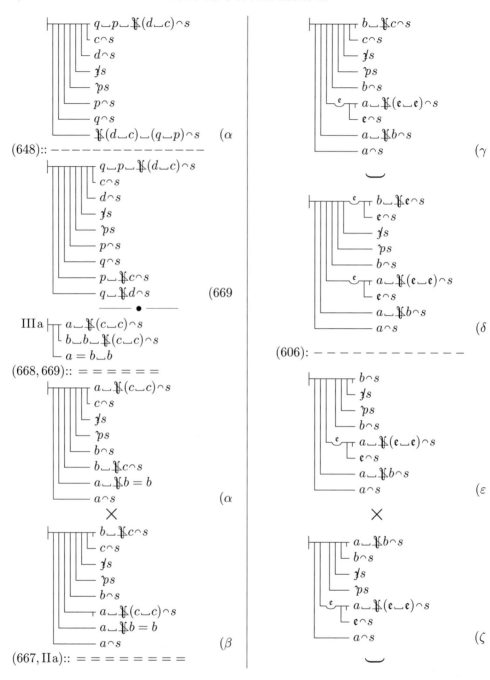

‿

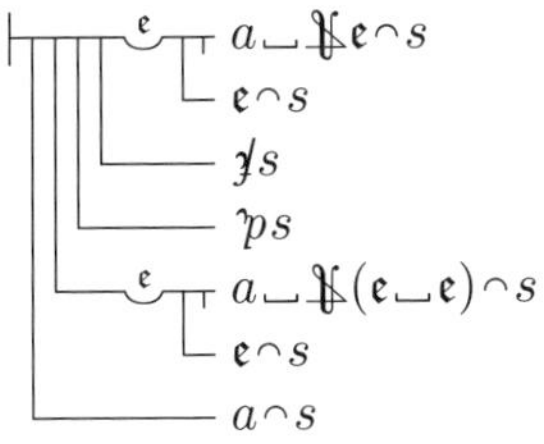

(η

(606): - - - - - - - - - - - -

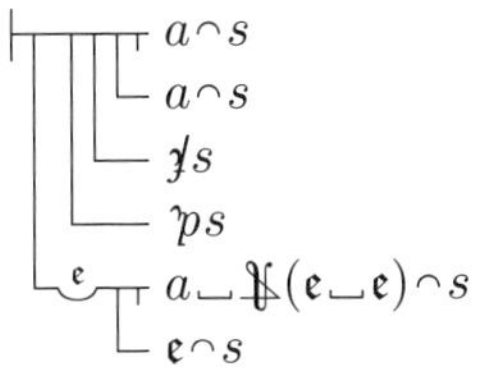

(ϑ

(Ig): - - - - - - - - - - -

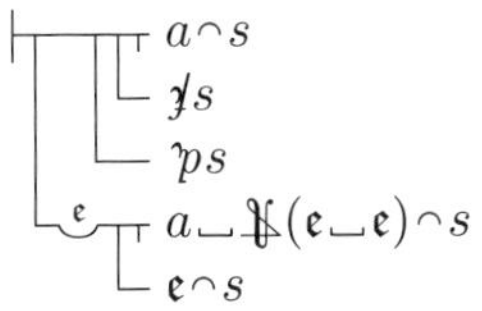

(670

§233. *Analysis*

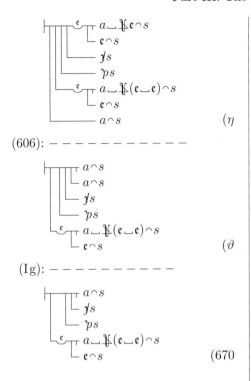

Using (670), we now prove the proposition

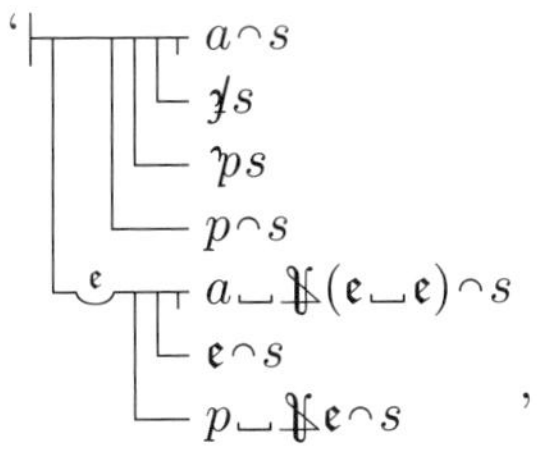

which in essentials—after a contraposition—says that in a positive class (s) there is a Relation (e) that is less than a given Relation (p) and whose double ($e \smile e$) is less than a given Relation (a), provided p and a belong to the same positive class (s). For this we assume that in our positive class a Relation e is known such that $e \smile e$ is less than a. Now two cases have to be distinguished:

1. e is less than p,
2. e is not less than p.

In the first case, we already have a Relation of the required kind. In the second case, there is in our positive class a Relation (b) less than p, which then is also less than e, of which it therefore holds that $b \smile b$ is less than $e \smile e$ and thus also less than a.

§234. *Construction*

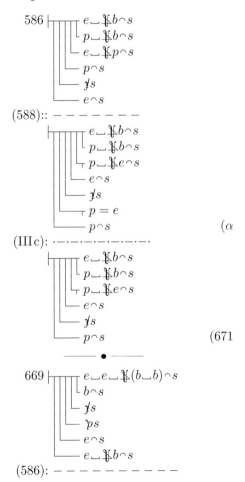

586 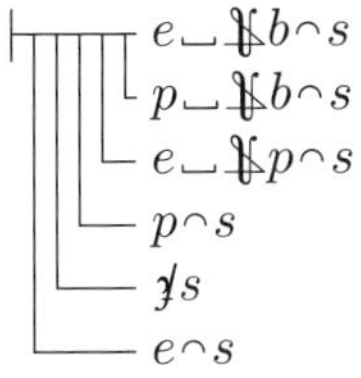

(588):: - - - - - - - -

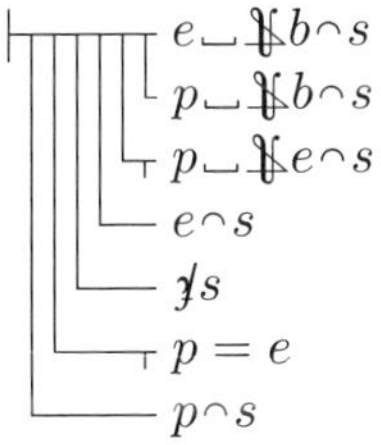

(α

(IIIc): -·-·-·-·-·-·-·-

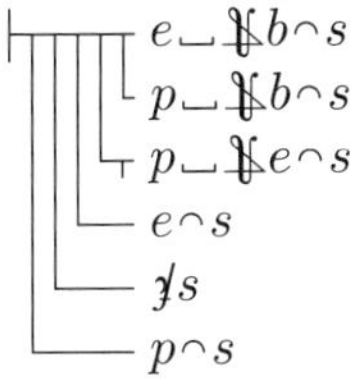

(671

———— • ————

669 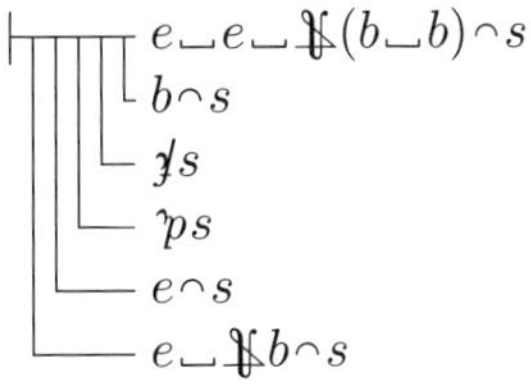

(586): - - - - - - - - - - -

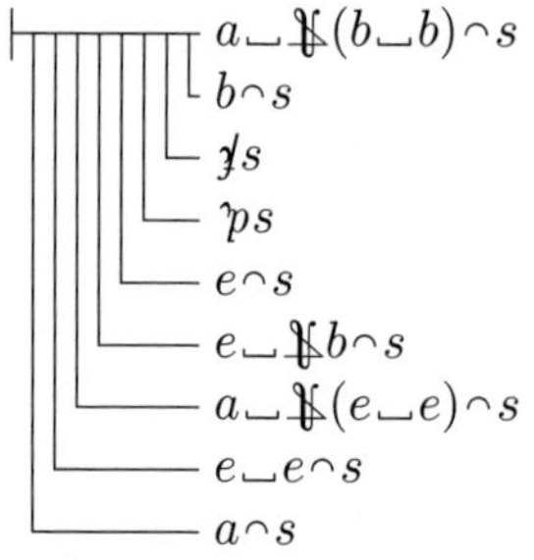

(α

(671, 529):: = = = = = = =

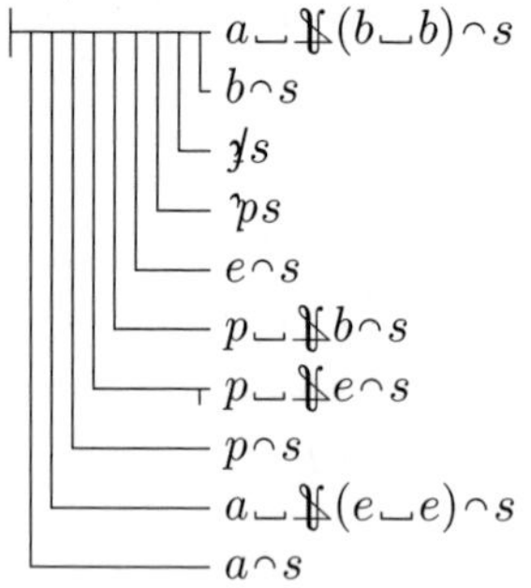

(β

×

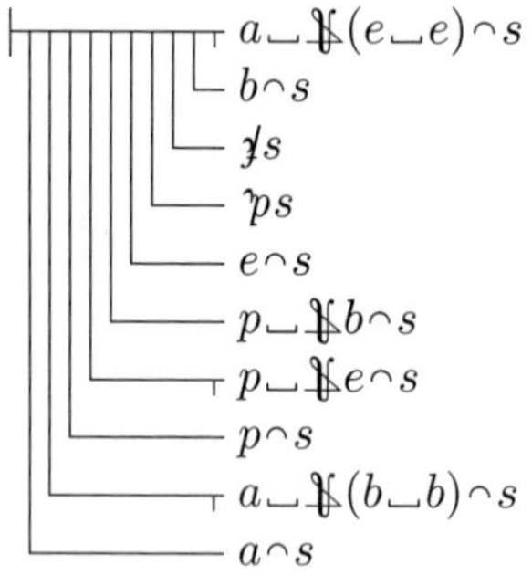

(γ

(IIa):: – – – – – – – – – – – –

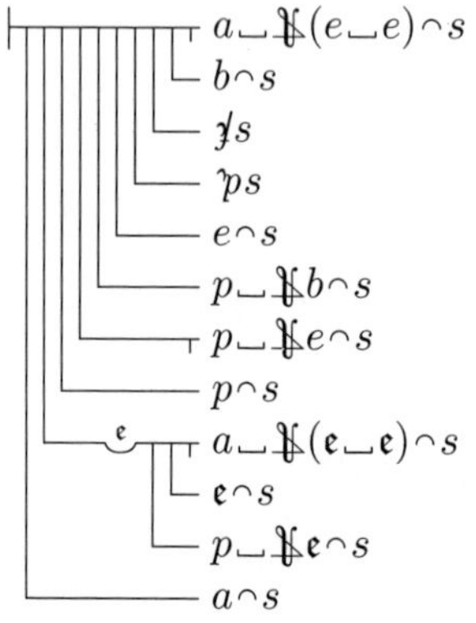

(δ

(IIa): ·–·–·–·–·–·–·–·–·–·–·–·

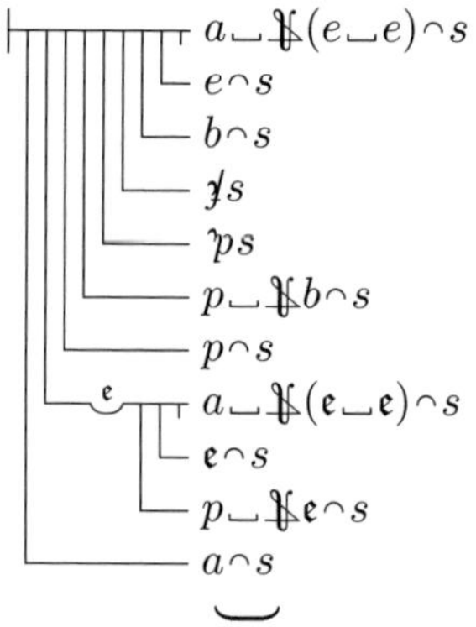

(ε

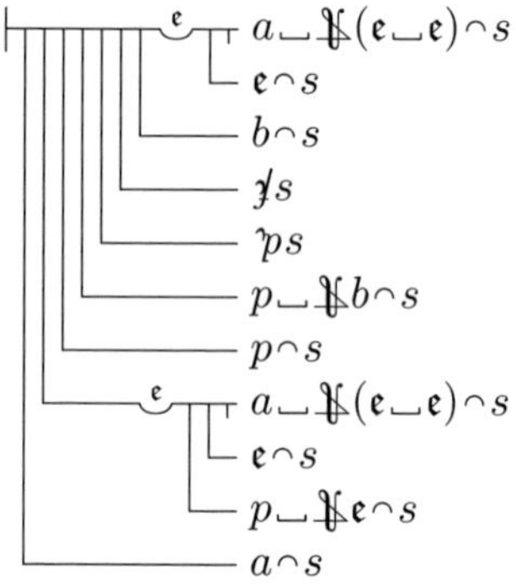

(ζ

(670): – – – – – – – – – – – – –

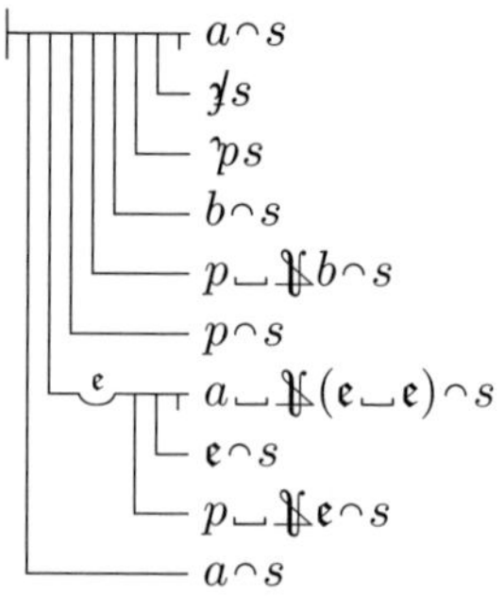

(η

×

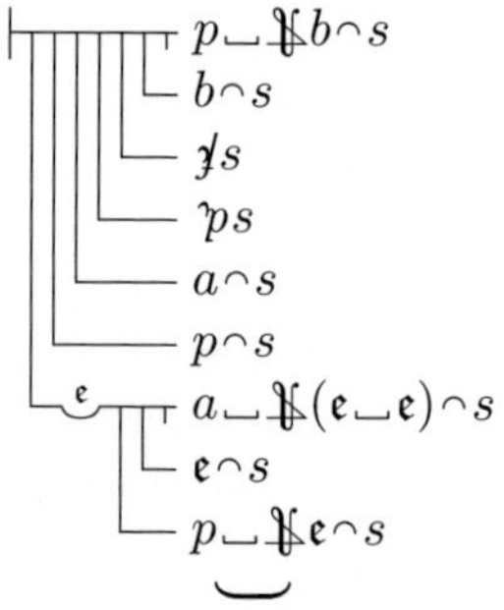

(ϑ

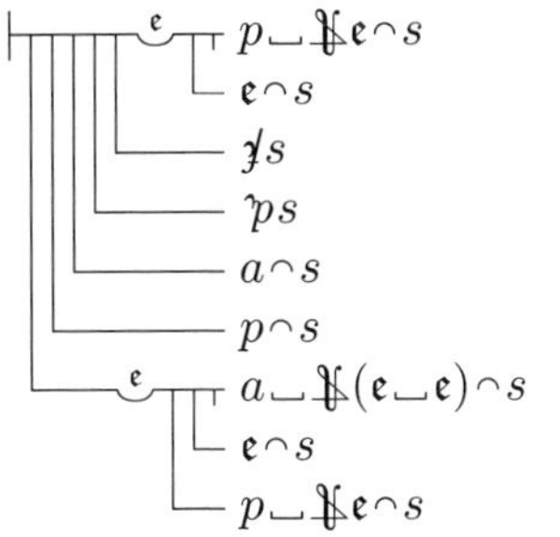

(ι

(606): - - - - - - - - - - - - -

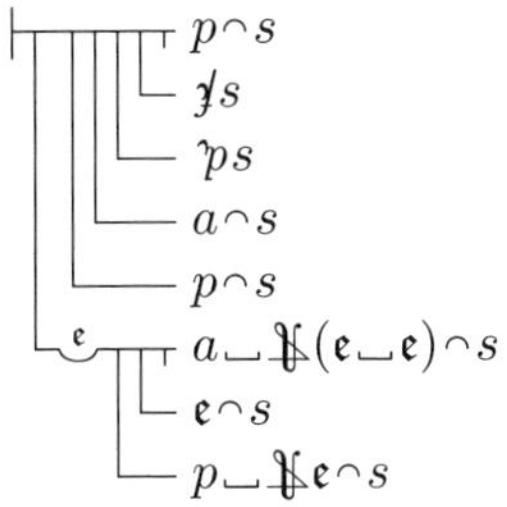

(κ

×

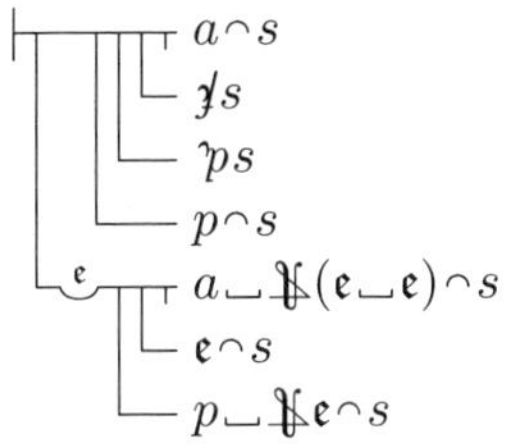

(672

§235. *Analysis*

In order, finally, to prove the proposition

‘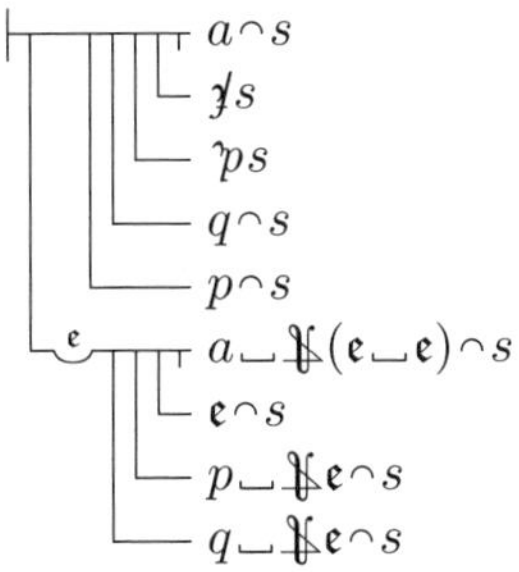’,

we will proceed essentially as above, distinguishing the cases where p is less than q and where p is not less than q.

§236. *Construction*

586 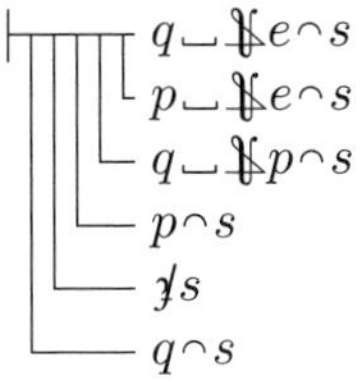

(IIa): - - - - - - - -

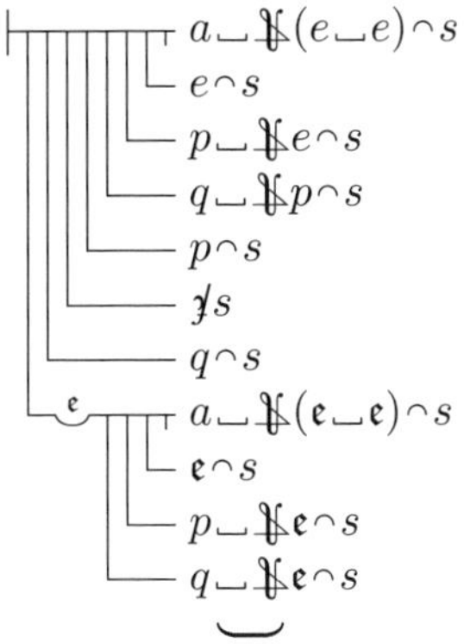

(α

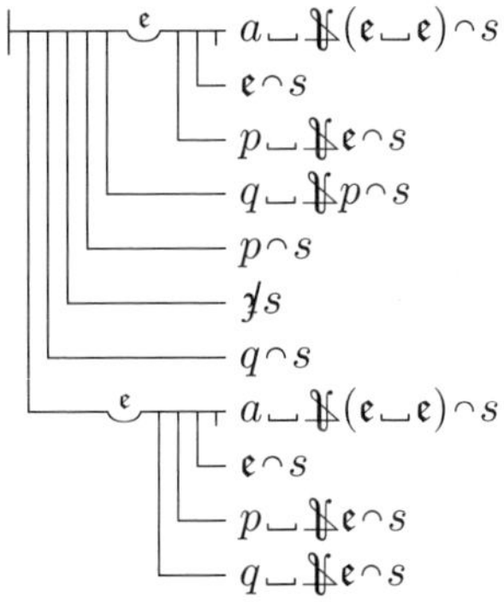

(β

(672): - - - - - - - - - - - - -

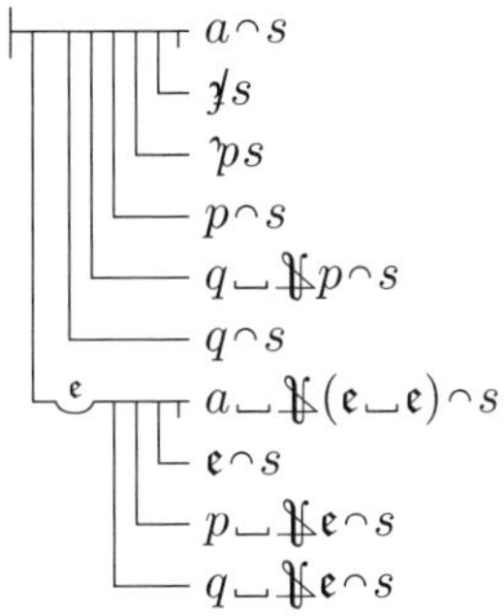

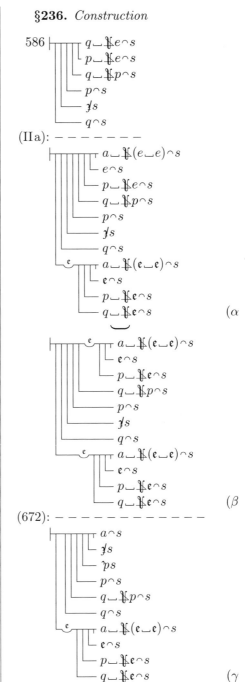

(γ

671 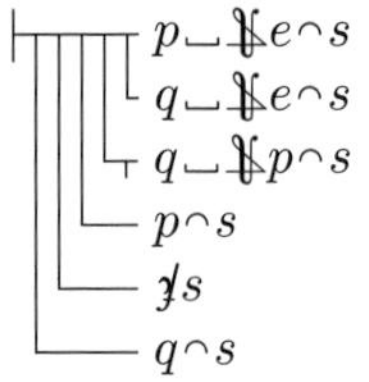

(II a): - - - - - - - -

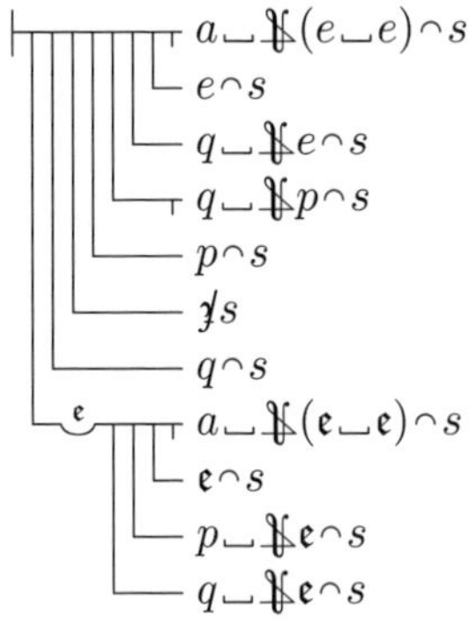 (δ

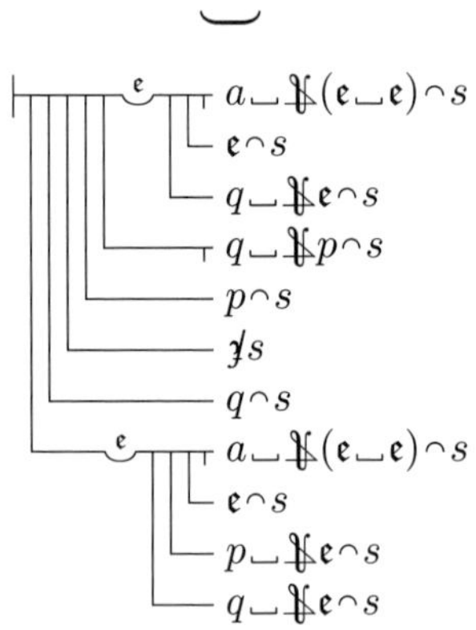 (ε

(672): - - - - - - - - - - - - - -

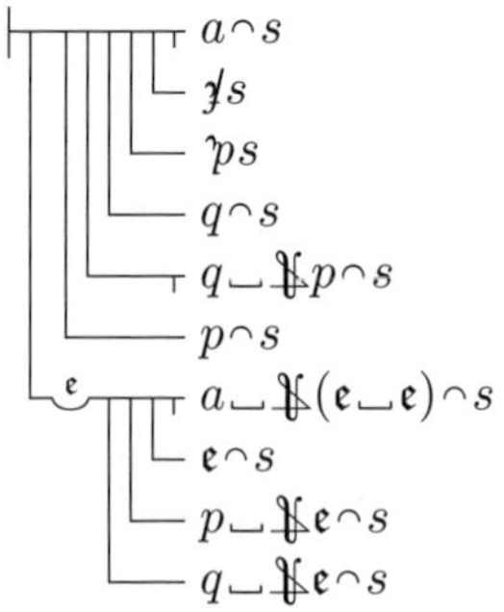 (ζ

(γ): -·-·-·-·-·-·-·-·-·-

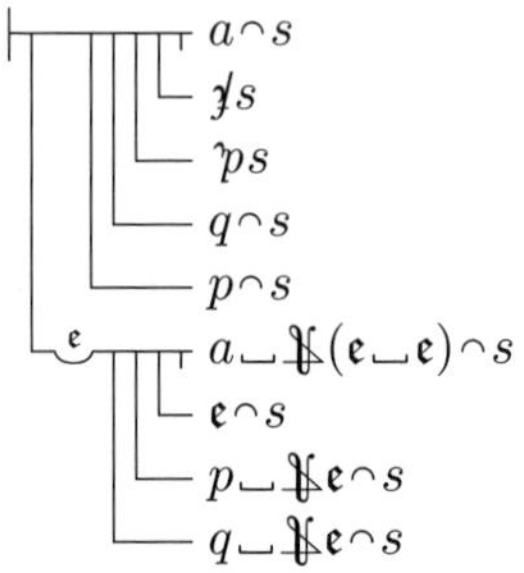 (673

§237. *Analysis*

We apply (673) to (666) to obtain the proposition

‘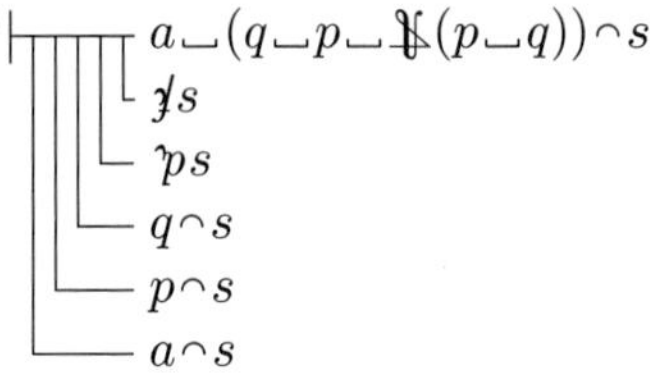 ’,

With the help of (644), we then reach the goal of our section.

§238. *Construction*

586 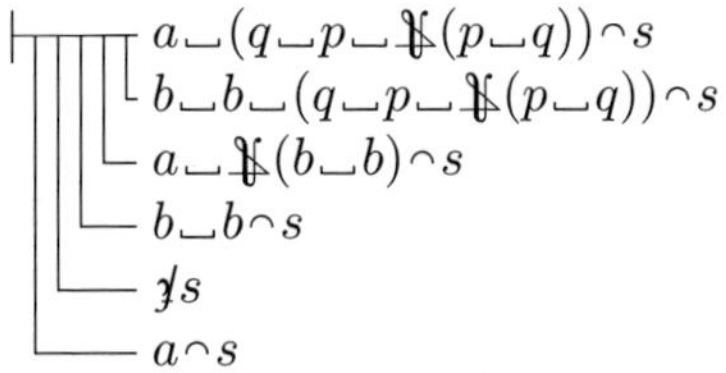

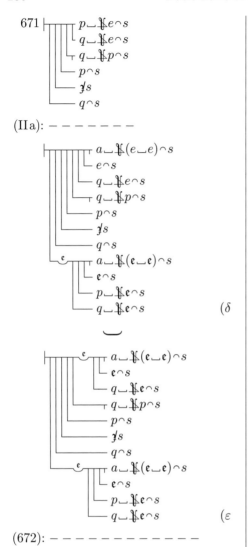

(666, 529):: = = = = = = = = = = = = = =

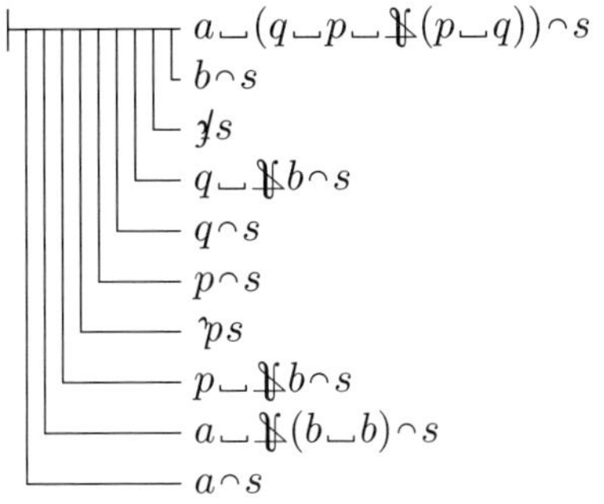

(α

×

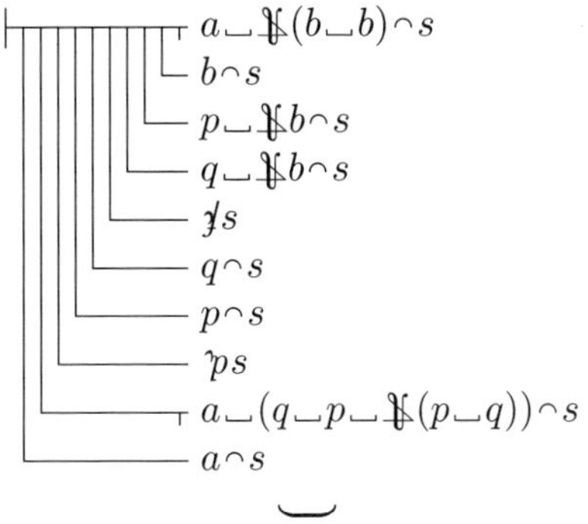

(β

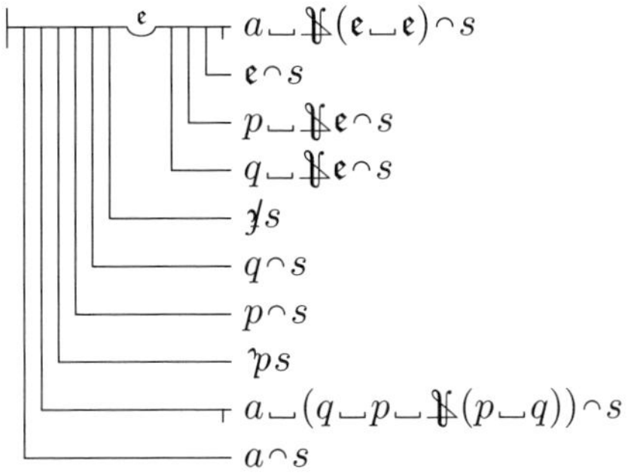

(γ

(673): - - - - - - - - - - - - - - - - -

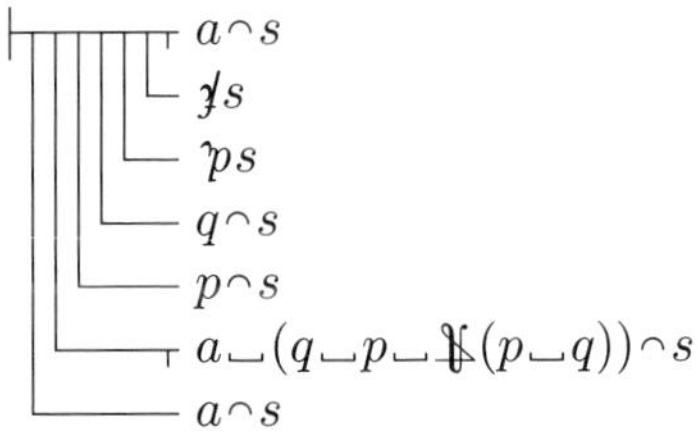

(δ

×

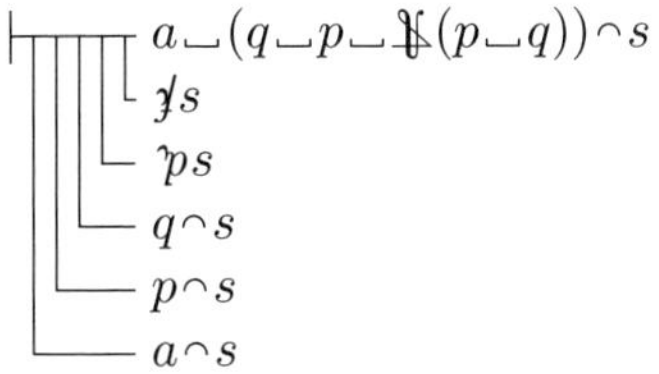

(ε

(Ie): - - - - - - - - - - - - - -

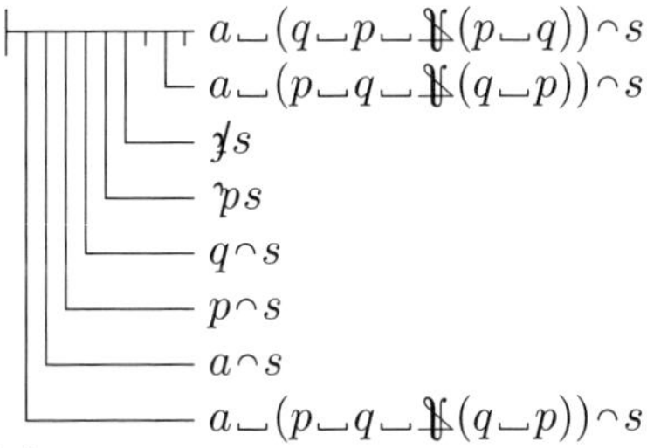

(ζ

(ε):: - - - - - - - - - - - - - -

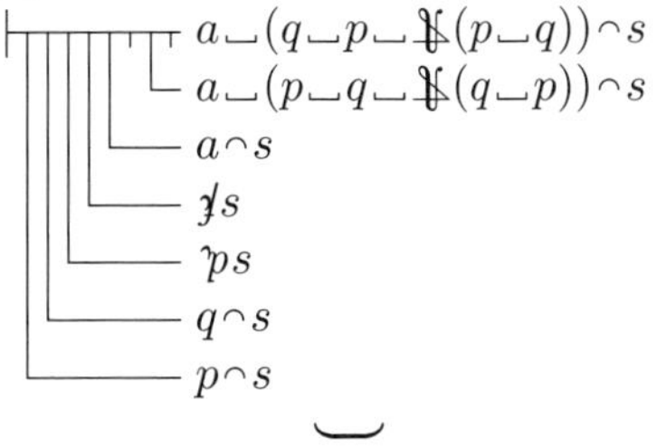

(η

$\vdash^{\mathfrak{a}}\ \mathfrak{a} \smile (q \smile p \smile \maltese(p \smile q)) \frown s$
$\quad \mathfrak{a} \smile (p \smile q \smile \maltese(q \smile p)) \frown s$
$\quad \mathfrak{a} \frown s$
$\quad \nmid s$
$\quad \wp s$
$\quad q \frown s$
$\quad p \frown s$ (ϑ

(644): – – – – – – – – – – – – – –

$\vdash q \smile p = p \smile q$
$\quad \nmid s$
$\quad \wp s$
$\quad q \frown s$
$\quad p \frown s$
$\quad p \smile q \frown s$
$\quad q \smile p \frown s$ (ι

(529, 529):: = = = = = =

$\vdash q \smile p = p \smile q$
$\quad \nmid s$
$\quad \wp s$
$\quad q \frown s$
$\quad p \frown s$ (κ

(607):: – – – – – – – – –

$\vdash q \smile p = p \smile q$
$\quad \wp s$
$\quad q \frown s$
$\quad p \frown s$ (674

(IIIh): – – – – – – – –

$\vdash q \smile p \smile \maltese q = p \smile q \smile \maltese q$
$\quad \wp s$
$\quad q \frown s$
$\quad p \frown s$ (α

(570): ———————————

$\vdash q \smile p \smile \maltese q = p$
$\quad \wp s$
$\quad q \frown s$
$\quad p \frown s$
$\quad \nmid s$ (β

(IIIh): – – – – – – – – –

$\vdash \maltese q \smile (q \smile p \smile \maltese q) = \maltese q \smile p$
$\quad \wp s$
$\quad q \frown s$
$\quad p \frown s$
$\quad \nmid s$ (γ

(491): ———————————

$\vdash \maltese q \smile (q \smile p) \smile \maltese q = \maltese q \smile p$
$\quad \wp s$
$\quad q \frown s$
$\quad p \frown s$
$\quad \nmid s$ (δ

(580): ———————————

$\vdash p \smile \maltese q = \maltese q \smile p$
$\quad \wp s$
$\quad q \frown s$
$\quad p \frown s$
$\quad \nmid s$ (ε

(607):: – – – – – – – – – –

$\vdash p \smile \maltese q = \maltese q \smile p$
$\quad \wp s$
$\quad q \frown s$
$\quad p \frown s$ (675

(IIIa): – – – – – – – – – –

$\vdash f(p \smile \maltese q)$
$\quad f(\maltese q \smile p)$
$\quad \wp s$
$\quad q \frown s$
$\quad p \frown s$ (676

——— • ———

IIIc $\vdash f(\maltese q \smile p)$
$\quad f(p \smile \maltese q)$
$\quad p \smile \maltese q = \maltese q \smile p$

(675):: – – – – – – – – –

$\vdash f(\maltese q \smile p)$
$\quad f(p \smile \maltese q)$
$\quad \wp s$
$\quad q \frown s$
$\quad p \frown s$ (677

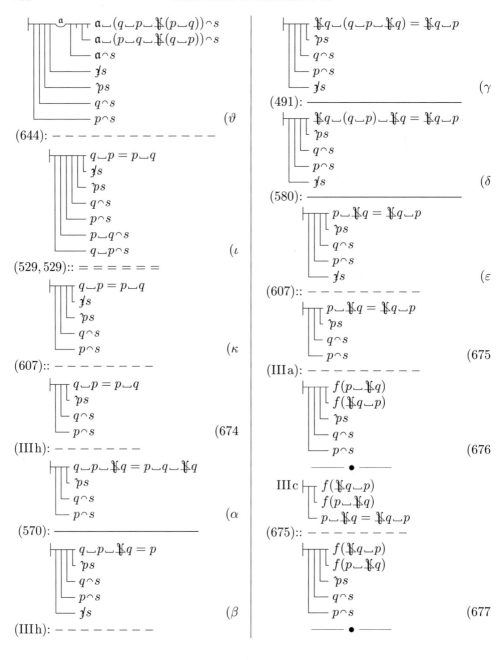

——— • ———

$$\text{IIIa}\vdash\begin{array}{l} f(q\smallsmile p) \\ f(p\smallsmile q) \\ q\smallsmile p = p\smallsmile q \end{array}$$

(674):: – – – – – – – –

$$\vdash\begin{array}{l} f(q\smallsmile p) \\ f(p\smallsmile q) \\ \mathcal{P}s \\ q\frown s \\ p\frown s \end{array} \qquad (678$$

Z. Proof of the proposition

$$‘\vdash\begin{array}{l} p\smallsmile q = q\smallsmile p \\ \mathcal{P}s \\ p\frown \eth s \\ q\frown \eth s \end{array} \quad ,$$

§239. *Analysis*

We now undertake the proof of the commutative law for the entire domain of magnitudes of a positive class.

The following propositions will serve as an intermediary stage

$$‘\vdash\begin{array}{l} p\smallsmile q = q\smallsmile p \\ \mathcal{P}s \\ q\frown s \\ p\frown \eth s \end{array} \quad , \qquad (\alpha$$

$$‘\vdash\begin{array}{l} p\smallsmile \mathfrak{I}b = \mathfrak{I}b\smallsmile p \\ \mathcal{P}s \\ b\frown s \\ p\frown \eth s \end{array} \quad , \qquad (\beta$$

$$‘\vdash\begin{array}{l} p\smallsmile (b\smallsmile \mathfrak{I}b) = b\smallsmile \mathfrak{I}b\smallsmile p \\ b\frown s \\ \mathcal{P}s \\ p\frown \eth s \end{array} \quad , \qquad (\gamma$$

of which we first prove (α).

§240. *Construction*

$$\text{IIIa}\vdash\begin{array}{l} a\smallsmile \mathfrak{I}a\smallsmile q = q\smallsmile (a\smallsmile \mathfrak{I}a) \\ a\smallsmile \mathfrak{I}a\smallsmile q = q \\ q\smallsmile (a\smallsmile \mathfrak{I}a) = q \end{array}$$

(562, 571):: = = = = = = = = =

$$\vdash\begin{array}{l} a\smallsmile \mathfrak{I}a\smallsmile q = q\smallsmile (a\smallsmile \mathfrak{I}a) \\ q\frown s \\ \not{y}s \\ a\frown s \end{array} \qquad (\alpha$$

(607):: – – – – – – – – – – – – – –

$$\vdash\begin{array}{l} a\smallsmile \mathfrak{I}a\smallsmile q = q\smallsmile (a\smallsmile \mathfrak{I}a) \\ q\frown s \\ \mathcal{P}s \\ a\frown s \end{array} \qquad (679$$

(IIIb): – – – – – – – – – – – – – –

$$\vdash\begin{array}{l} p = a\smallsmile \mathfrak{I}a \\ q\frown s \\ \mathcal{P}s \\ a\frown s \\ \top\; p\smallsmile q = q\smallsmile p \end{array} \qquad (\alpha$$

———— • ————

$$\text{IIIf}\vdash\begin{array}{l} \mathfrak{I}a\smallsmile q = q\smallsmile \mathfrak{I}a \\ q\smallsmile \mathfrak{I}a = \mathfrak{I}a\smallsmile q \end{array}$$

(675):: – – – – – – – – –

$$\vdash\begin{array}{l} \mathfrak{I}a\smallsmile q = q\smallsmile \mathfrak{I}a \\ \mathcal{P}s \\ a\frown s \\ q\frown s \end{array} \qquad (\beta$$

(IIIb): – – – – – – – – – –

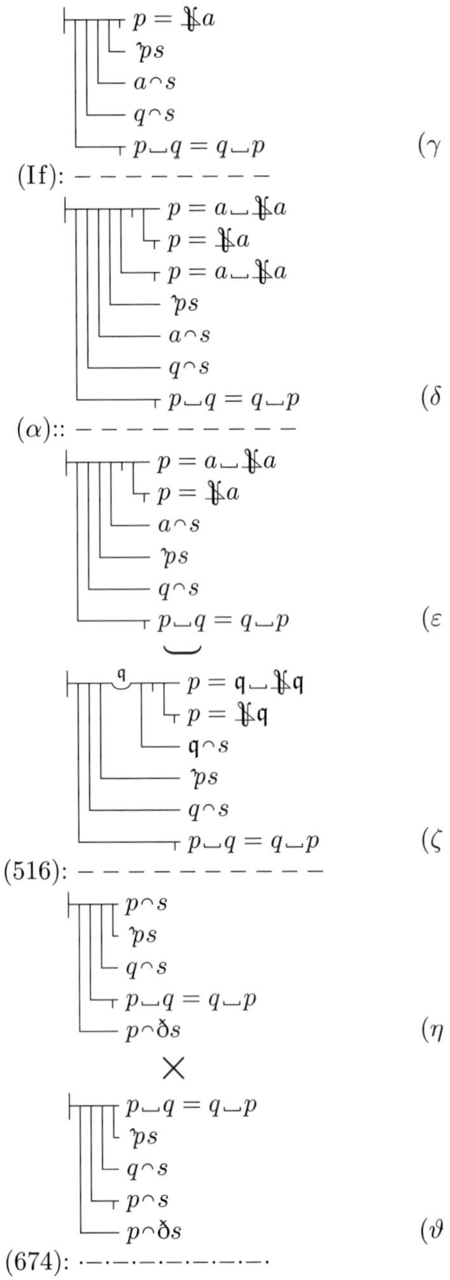

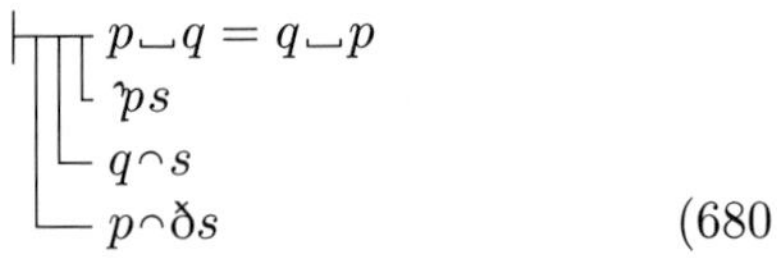

§241. *Analysis*

In a similar manner, we now prove proposition (β) of §239, taking for '*p*' first '$a \smile \not\,a$' and then '$\not\,a$', and finally applying (675).

§242. *Construction*

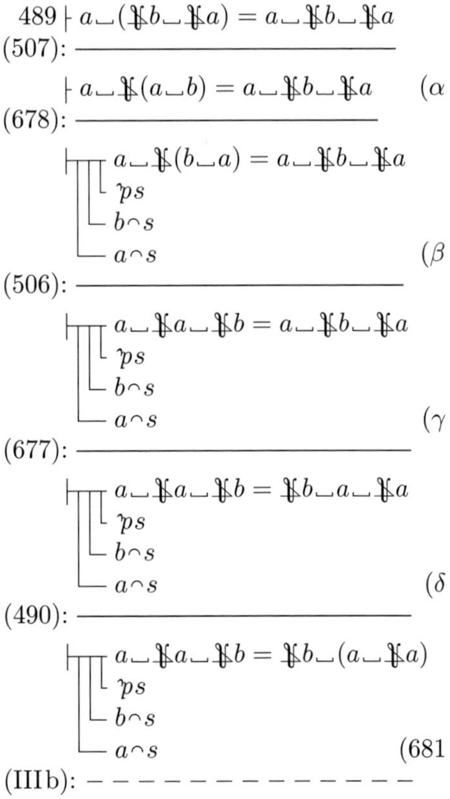

$\vdash p = a \smile \ddagger a$
$\text{ɿ}ps$
$b \frown s$
$a \frown s$
$p \smile \ddagger b = \ddagger b \smile p$ (α

•

IIIh $\vdash \ddagger(b \smile a) = \ddagger(a \smile b)$
$b \smile a = a \smile b$

(505): ———————

$\vdash \ddagger a \smile \ddagger b = \ddagger(a \smile b)$
$b \smile a = a \smile b$ (β

(505): ———————

$\vdash \ddagger a \smile \ddagger b = \ddagger b \smile \ddagger a$
$b \smile a = a \smile b$ (γ

(674): - - - - - - - - - - -

$\vdash \ddagger a \smile \ddagger b = \ddagger b \smile \ddagger a$
$\text{ɿ}ps$
$b \frown s$
$a \frown s$ (δ

(IIIb): - - - - - - - - - - -

$\vdash p = \ddagger a$
$\text{ɿ}ps$
$b \frown s$
$a \frown s$
$p \smile \ddagger b = \ddagger b \smile p$ (ε

(If): - - - - - - - - - - -

$\vdash p = a \smile \ddagger a$
$p = \ddagger a$
$p = a \smile \ddagger a$
$\text{ɿ}ps$
$b \frown s$
$a \frown s$
$p \smile \ddagger b = \ddagger b \smile p$ (ζ

(α):: - - - - - - - - - - -

$\vdash p = a \smile \ddagger a$
$p = \ddagger a$
$a \frown s$
$\text{ɿ}ps$
$b \frown s$
$p \smile \ddagger b = \ddagger b \smile p$ (η

$\smile$

$\vdash^{\mathfrak{q}} p = \mathfrak{q} \smile \ddagger \mathfrak{q}$
$p = \ddagger \mathfrak{q}$
$\mathfrak{q} \frown s$
$\text{ɿ}ps$
$b \frown s$
$p \smile \ddagger b = \ddagger b \smile p$ (ϑ

(516): - - - - - - - - - - - - -

$\vdash p \frown s$
$\text{ɿ}ps$
$b \frown s$
$p \smile \ddagger b = \ddagger b \smile p$
$p \frown \mathfrak{d}s$ (ι

$\times$

$\vdash p \smile \ddagger b = \ddagger b \smile p$
$\text{ɿ}ps$
$b \frown s$
$p \frown s$
$p \frown \mathfrak{d}s$ (κ

(675): ·—·—·—·—·—·—·—·

$\vdash p \smile \ddagger b = \ddagger b \smile p$
$\text{ɿ}ps$
$b \frown s$
$p \frown \mathfrak{d}s$ (682

(IIIb): - - - - - - - - - -

$\vdash q = \ddagger b$
$\text{ɿ}ps$
$b \frown s$
$p \frown \mathfrak{d}s$
$p \smile q = q \smile p$ (683

§243. *Analysis*

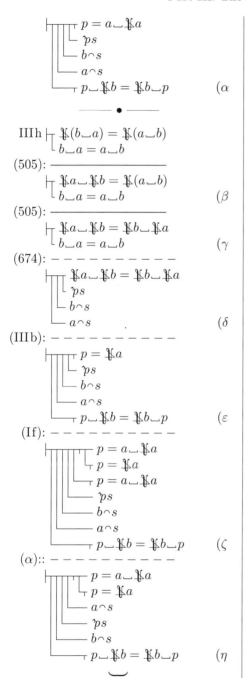

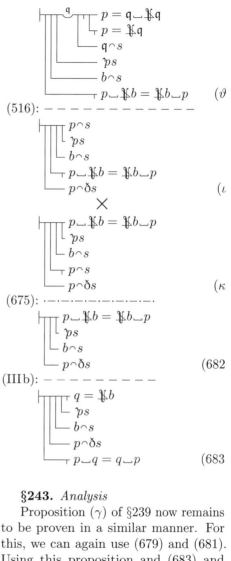

Proposition (γ) of §239 now remains to be proven in a similar manner. For this, we can again use (679) and (681). Using this proposition and (683) and (680), we arrive at the goal of section Z.

§244. *Construction*

IIIf $\vdash p \smile (b \smile \ddagger b) = b \smile \ddagger b \smile p$
$b \smile \ddagger b \smile p = p \smile (b \smile \ddagger b)$

(679):: - - - - - - - - - - - -

⊢ p⌣(b⌣∦b) = b⌣∦b⌣p
  p⌢s
  ℘s
  b⌢s (684

— • —

IIIf ⊢ ∦a⌣(b⌣∦b) = b⌣∦b⌣∦a
  b⌣∦b⌣∦a = ∦a⌣(b⌣∦b)

(681):: - - - - - - - - - - - - - -

⊢ ∦a⌣(b⌣∦b) = b⌣∦b⌣∦a
  ℘s
  a⌢s
  b⌢s (685

(IIIb): - - - - - - - - - - - - -

⊢ p = ∦a
  ℘s
  a⌢s
  b⌢s
  p⌣(b⌣∦b) = b⌣∦b⌣p (686

— • —

IIIe ⊢ a⌣∦a⌣(a⌣∦a) = a⌣∦a⌣(a⌣∦a)

(559): ————————————————

⊢ a⌣∦a⌣(b⌣∦b) = b⌣∦b⌣(a⌣∦a)
  b⌢s
  a⌢s
  ℊs (α

(607):: - - - - - - - - - - - - - - - - - -

⊢ a⌣∦a⌣(b⌣∦b) = b⌣∦b⌣(a⌣∦a)
  b⌢s
  a⌢s
  ℘s (687

(IIIb): - - - - - - - - - - - - - - - - - -

⊢ p = a⌣∦a
  b⌢s
  a⌢s
  ℘s
  p⌣(b⌣∦b) = b⌣∦b⌣p (α

(If): - - - - - - - - - - - - - -

⊢ p = a⌣∦a
  p = ∦a
  b⌢s
  a⌢s
  ℘s
  p⌣(b⌣∦b) = b⌣∦b⌣p
  p = ∦a (β

(686):: - - - - - - - - - - - - - -

⊢ p = a⌣∦a
  p = ∦a
  a⌢s
  b⌢s
  ℘s
  p⌣(b⌣∦b) = b⌣∦b⌣p (γ

⊢ 𝔮: p = 𝔮⌣∦𝔮
  p = ∦𝔮
  𝔮⌢s
  b⌢s
  ℘s
  p⌣(b⌣∦b) = b⌣∦b⌣p (δ

(516): - - - - - - - - - - - - - -

⊢ p⌢s
  b⌢s
  ℘s
  p⌣(b⌣∦b) = b⌣∦b⌣p
  p⌢𝔡s (ε

×

⊢ p⌣(b⌣∦b) = b⌣∦b⌣p
  b⌢s
  ℘s
  p⌢s
  p⌢𝔡s (ζ

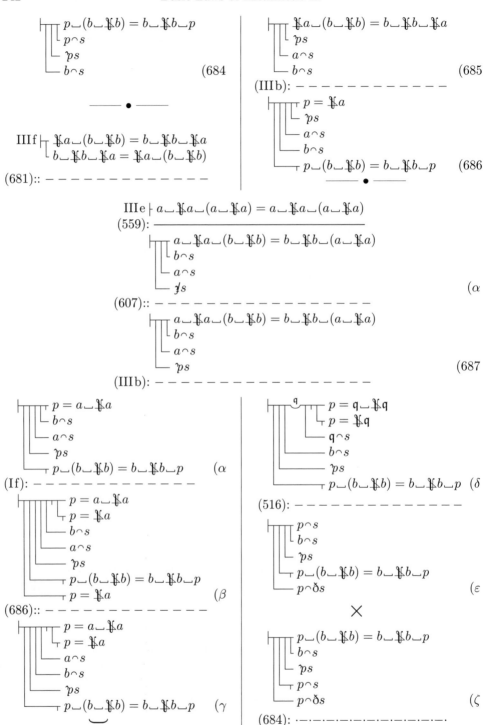

(684): ·–·–·–·–·–·–·–·–·–·–·

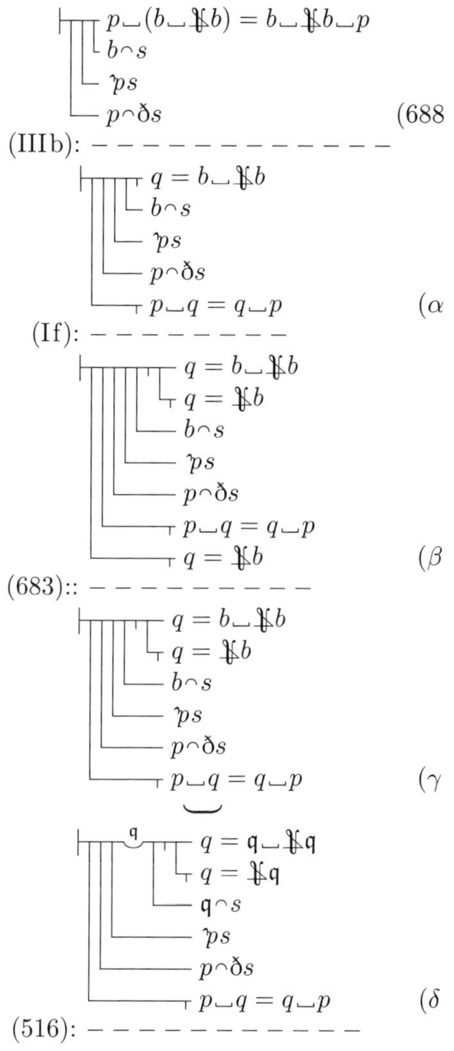

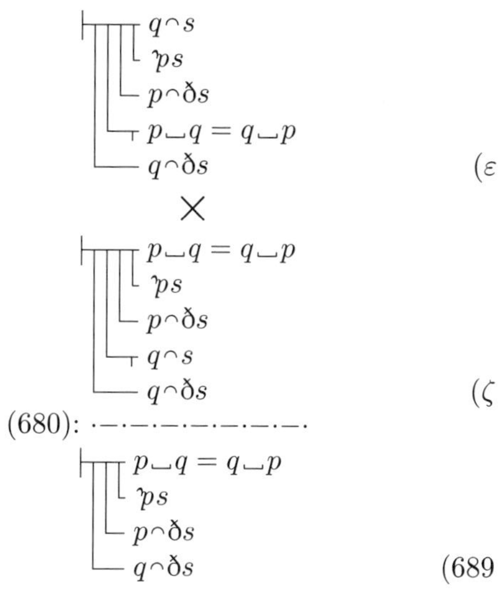

§245. The commutative law for the domain of a positive class is thus proven. The next task is now to show that there is a positive class, as indicated in §164. This opens the possibility of defining a real number as a ratio of magnitudes of a domain that belongs to a positive class. Moreover, we will then be able to prove that the real numbers themselves belong as magnitudes to the domain of a positive class.

Note *to §175, p. 172, first column*

That the presented specifications cannot be proven to be independent of each other should not be put forward unconditionally. It is of course conceivable that classes of Relations could be found for each of which all but one of the specifications applies, so that each of them does not apply to one of the examples. Whether it is possible, however, to give such examples at this stage of the enquiry, without presupposing geometry or the fractions, negative and irrational numbers, or empirical facts, is doubtful.

Appendices

1. Table of definitions

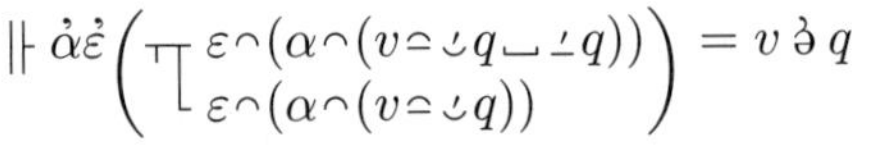

(Σ

(Relation: that an object, in a series starting with an object, is the first to belong to a class. §1, p. 2)

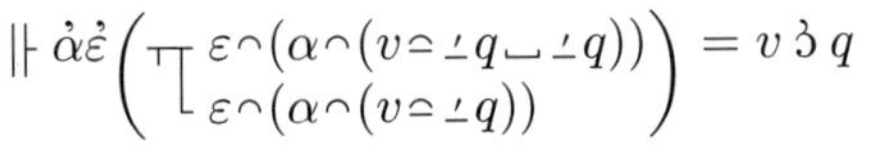

(T

(Relation: that an object in a series is the first after an object to belong to a class. §7, p. 7)

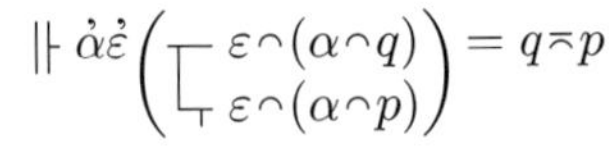

(Υ

(Combination of Relations. §37, p. 48)

$$\Vdash \dot{\alpha}\dot{\varepsilon}(t \smile \varepsilon = \alpha) = {}_{*}t$$

(Φ

(§167, p. 166)

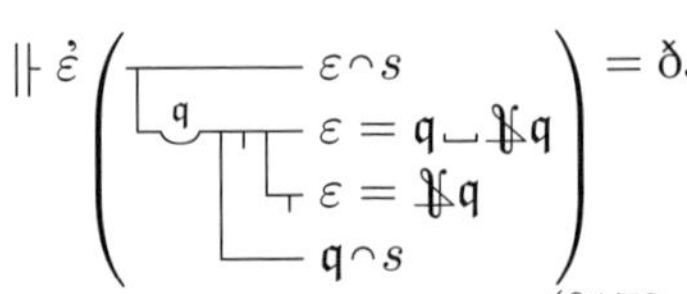

(X

(§173, p. 169)

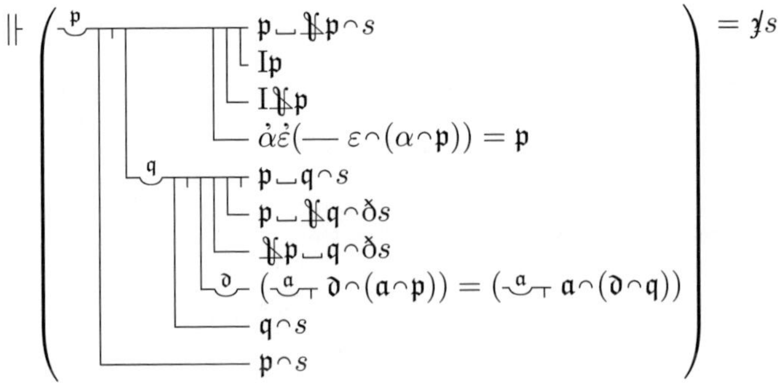

(Ψ

(Positival class §175, p. 171)

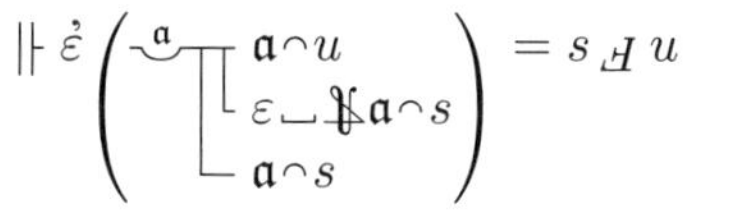

(Ω
(§193, p. 187)

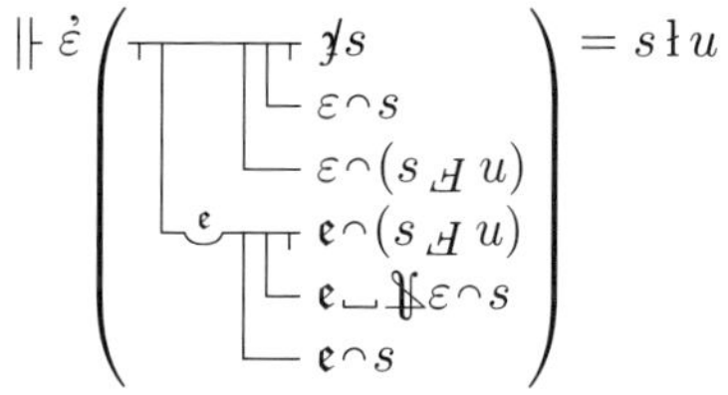

(AA

(Limit, §193, p. 187)

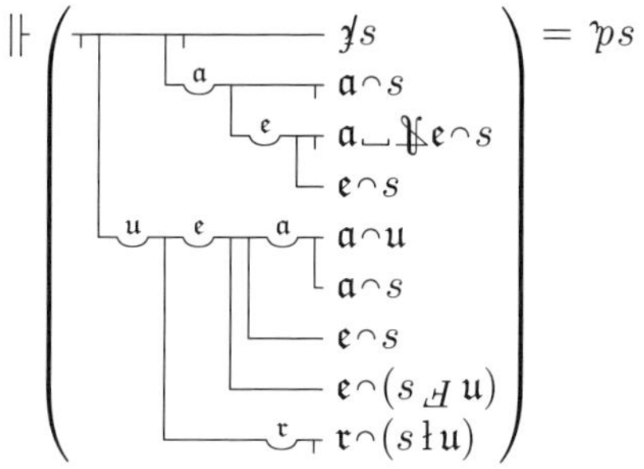

(AB

(Positive class, §197, p. 190)

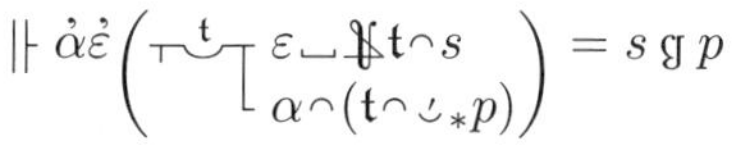

(AΓ
(§199, p. 192)

2. Table of important theorems

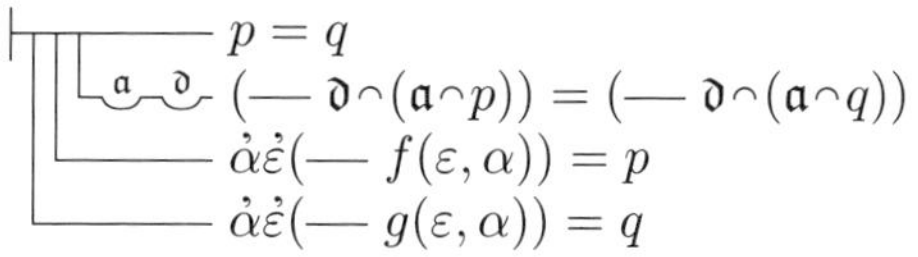

(485

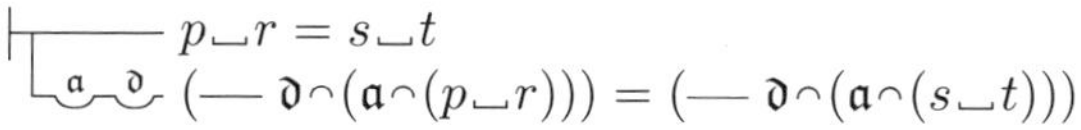

(487

$\vdash p \smile (q \smile t) = p \smile q \smile t$ (489

(The associative law for the composition of Relations.)

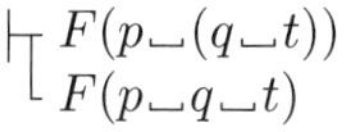

(490

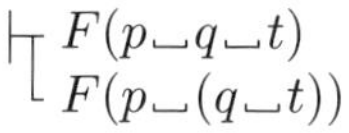

(491

$\vdash f(\not\Upsilon q \smile \not\Upsilon p)$
$\quad f(\not\Upsilon(p \smile q))$ (505

●

$\vdash F(t \smile \not\Upsilon q \smile \not\Upsilon p)$
$\quad F(t \smile \not\Upsilon(p \smile q))$ (506

●

$\vdash f(\not\Upsilon(p \smile q))$
$\quad f(\not\Upsilon q \smile \not\Upsilon p)$ (507

●

$\vdash F(t \smile \not\Upsilon(p \smile q))$
$\quad F(t \smile \not\Upsilon q \smile \not\Upsilon p)$ (508

●

$\vdash \overset{\mathfrak{e}}{\smile}\; d \frown (\mathfrak{e} \frown q)$
$\quad d \frown (r \frown \underline{\smile}\, q)$ (413

●

$\vdash n \frown (a \frown \dot\smile\, q)$
$\quad n \frown (r \frown \underline{\smile}\, q)$
$\quad r \frown (a \frown \underline{\smile}\, q)$ (377

●

$\vdash m \frown (r \frown \dot\smile\, q)$
$\quad m \frown (n \frown \dot\smile\, q)$
$\quad n \frown (r \frown \underline{\smile}\, q)$ (382

●

$\vdash d \frown (a \frown \not\Upsilon(z \bumpeq \not\Upsilon q))$
$\quad d \frown (a \frown q)$
$\quad d \frown z$ (444

●

$\vdash e \frown (a \frown q)$
$\quad e \frown (a \frown \not\Upsilon(z \bumpeq \not\Upsilon q))$ (446

●

$\vdash \mathrm{I}\not\Upsilon(z \bumpeq \not\Upsilon q)$
$\quad \mathrm{I}q$ (447

●

$\vdash m \frown (m; d \,\underline{\Delta}\, q)$
$\quad d \frown (d \frown \underline{\smile}\, q)$
$\quad m \frown (d \frown \dot\smile\, q)$
$\quad \mathrm{I}q$ (371

●

$\vdash e \frown (m; a \,\underline{\Delta}\, p)$
$\quad e \frown (m; n \,\underline{\Delta}\, p)$
$\quad a \frown (a \frown \underline{\smile}\, p)$
$\quad n \frown (a \frown p)$ (375

●

$\vdash m \frown (m; m \,\underline{\Delta}\, p)$
$\quad m \frown (m \frown \underline{\smile}\, p)$
$\quad \mathrm{I}q$ (391

●

$\vdash x \frown (a \frown (v \grave\partial\, q))$
$\quad \overset{\mathfrak{r}}{\smile}\; x \frown (\mathfrak{r} \frown (v \bumpeq \dot\smile\, q))$
$\qquad \mathfrak{r} \frown (a \frown \underline{\smile}\, q)$
$\quad x \frown (a \frown \dot\smile\, q)$
$\quad a \frown v$ (350

●

$\vdash \overset{\mathfrak{e}}{\smile}\; \mathfrak{e} \frown (x; y \,\underline{\Delta}\, q)$
$\qquad \mathfrak{e} \frown v$
$\quad \overset{\mathfrak{a}}{\smile}\; x \frown (\mathfrak{a} \frown (v \grave\partial\, q))$
$\quad x \frown (y \frown \dot\smile\, q)$ (354

●

$\vdash x \frown (y \frown \dot\smile\, q)$
$\quad y \frown v$
$\quad y \frown (y \frown \underline{\smile}\, q)$
$\quad \mathrm{I}q$
$\quad \overset{\mathfrak{a}}{\smile}\; x \frown (\mathfrak{a} \frown (v \grave\partial\, q))$ (357

●

$\vdash x \frown (m \frown (v \bumpeq \dot\smile\, q))$
$\quad x \frown (m \frown (v \grave\partial\, q))$ (368

●

$\vdash x \frown (m \frown \dot\smile\, q)$
$\quad x \frown (m \frown (v \grave\partial\, q))$ (401

●

$\vdash m \frown v$
$\quad x \frown (m \frown (v \grave\partial\, q))$ (402

●

$\vdash x \frown (m \frown (v \bumpeq \dot\smile\, q \smile \underline{\smile}\, q))$
$\quad x \frown (m \frown (v \grave\partial\, q))$ (404

●

$\vdash m \frown (c \frown \dot\smile\, q)$
$\quad x \frown (m \frown (v \grave\partial\, q))$
$\quad x \frown (c \frown (v \bumpeq \dot\smile\, q))$
$\quad \mathrm{I}q$ (407

●

$\vdash b \frown (o \frown (v \grave\partial\, q))$
$\quad r \frown (o \frown \underline{\smile}\, q)$
$\quad b \frown (r \frown \dot\smile\, q)$
$\quad r \frown v$ (656

●

$$\vdash \begin{array}{l} d \frown (a \frown (v \simeq \perp q)) \\ d \frown (a \frown (v \,ʒ\, q)) \end{array} \qquad (361$$

$$\vdash \begin{array}{l} d \frown (a \frown \perp q) \\ d \frown (a \frown (v \,ʒ\, q)) \end{array} \qquad (362$$

$$\vdash \begin{array}{l} a \frown v \\ d \frown (a \frown (v \,ʒ\, q)) \end{array} \qquad (365$$

$$\vdash \begin{array}{l} r \frown (n \frown \smile q) \\ r \frown (m; n \,\underline{\Delta}\, (v \,ʒ\, q)) \end{array} \qquad (380$$

$$\vdash \begin{array}{l} n \frown (a \frown (v \,ʒ\, q)) \\ n \frown (a \frown (v \simeq \perp q \smile \perp q)) \\ n \frown (a \frown (v \simeq \perp q)) \end{array} \qquad (384$$

$$\vdash \begin{array}{l} n \frown (a \frown (v \,ʒ\, q)) \\ \overset{\mathfrak{r}}{\smile}\; n \frown (\mathfrak{r} \frown (v \simeq \perp q)) \\ \quad \mathfrak{r} \frown (a \frown \perp q) \\ n \frown (a \frown \perp q) \\ a \frown v \end{array} \qquad (385$$

$$\vdash \begin{array}{l} x \frown (y \frown (v \simeq \perp q)) \\ x \frown (y \frown \perp (v \,ʒ\, q)) \end{array} \qquad (387$$

$$\vdash \begin{array}{l} x \frown (y \frown \perp q) \\ x \frown (y \frown \perp (v \,ʒ\, q)) \end{array} \qquad (388$$

$$\vdash \begin{array}{l} a \frown (a \frown \perp (v \,ʒ\, q)) \\ a \frown (a \frown \perp q) \end{array} \qquad (389$$

$$\vdash \begin{array}{l} d \frown (a \frown \perp q) \\ e \frown (a \frown (v \,ʒ\, q)) \\ e \frown (d \frown (v \simeq \perp q)) \end{array} \qquad (395$$

$$\vdash \begin{array}{l} \mathrm{I}(v \,ʒ\, q) \\ \mathrm{I}q \end{array} \qquad (397$$

$$\vdash \begin{array}{l} m \frown (d \frown \smile q) \\ m \frown (d \frown \smile (v \,ʒ\, q)) \end{array} \qquad (409$$

$$\vdash \begin{array}{l} \mathfrak{Q} \frown (\mathfrak{p} v \frown \smile \mathfrak{f}) \\ \frown = \mathfrak{p} v \\ \frown = \mathfrak{p} u \\ \overset{\mathfrak{a}}{\smile}\; \mathfrak{a} \frown u \\ \quad \mathfrak{a} \frown v \end{array} \qquad (428$$

$$\vdash \begin{array}{l} \mathfrak{Q} \frown (\mathfrak{p} v \frown \smile \mathfrak{f}) \\ \overset{\mathfrak{a}}{\smile}\; \mathfrak{a} \frown u \\ \quad \mathfrak{a} \frown v \\ \mathfrak{Q} \frown (\mathfrak{p} u \frown \smile \mathfrak{f}) \end{array} \qquad (443$$

$$\vdash \begin{array}{l} e \frown (a \frown (q \bar{\frown} p)) \\ e \frown (a \frown q) \end{array} \qquad (455$$

$$\vdash \begin{array}{l} e \frown (a \frown (q \bar{\frown} p)) \\ e \frown (a \frown p) \end{array} \qquad (456$$

$$\vdash \begin{array}{l} e \frown (d \frown q) \\ e \frown (d \frown p) \\ e \frown (d \frown (q \bar{\frown} p)) \end{array} \qquad (460$$

$$\vdash \not{\mathfrak{p}}(q \bar{\frown} p) = \not{\mathfrak{p}} q \bar{\frown} \not{\mathfrak{p}} p \qquad (464$$

$$\vdash \begin{array}{l} \mathfrak{p}\dot{\varepsilon}\left(\begin{array}{l} \varepsilon \frown z \\ \varepsilon \frown v \end{array}\right) = \mathfrak{p}\dot{\varepsilon}\left(\begin{array}{l} \varepsilon \frown u \\ \varepsilon \frown w \end{array}\right) \\ \mathfrak{p} z = \mathfrak{p} u \\ \mathfrak{p} v = \mathfrak{p} w \\ \overset{\mathfrak{a}}{\smile}\; \mathfrak{a} \frown z \\ \quad \mathfrak{a} \frown v \\ \overset{\mathfrak{a}}{\smile}\; \mathfrak{a} \frown u \\ \quad \mathfrak{a} \frown w \end{array} \qquad (469$$

$\vdash\smile^{\mathfrak{a}}\ \wp w\frown(\mathfrak{a}\frown\mathfrak{f})$ (480

—— • ——

$\vdash \mathfrak{O}\frown(\wp w\frown\dot{\smile}\mathfrak{f})$
$\quad \smile^{\mathfrak{a}}\ \mathfrak{a}\frown w$
$\qquad \mathfrak{a}\frown u$
$\quad \wp u = \frown$ (484

—— • ——

$\vdash F(t\smile d = a)$
$\quad F(d\frown(a\frown_{*}t))$ (492

—— • ——

$\vdash F(a)$
$\quad F(t\smile d)$
$\quad d\frown(a\frown_{*}t)$ (493

—— • ——

$\vdash F(d\frown(a\frown_{*}t))$
$\quad F(t\smile d = a)$ (495

—— • ——

$\vdash d\frown(t\smile d\frown_{*}t)$ (496

—— • ——

$\vdash \mathrm{I}_{*}t$ (497

—— • ——

$\vdash p\smile t = t\smile p$
$\quad t\frown(p\frown\dot{\smile}_{*}t)$ (498

—— • ——

$\vdash p\smile q = q\smile p$
$\quad t\frown(p\frown\dot{\smile}_{*}t)$
$\quad t\frown(q\frown\dot{\smile}_{*}t)$ (501

—— • ——

$\vdash d\smile b\frown(a\smile b\frown_{*}p)$
$\quad d\frown(a\frown_{*}p)$ (611

—— • ——

$\vdash p\frown s$
$\quad \smile^{\mathfrak{q}}\ p = \mathfrak{q}\smile\not{\mathfrak{X}}\mathfrak{q}$
$\qquad p = \not{\mathfrak{X}}\mathfrak{q}$
$\qquad \mathfrak{q}\frown s$
$\quad p\frown\eth s$ (516

—— • ——

$\vdash \smile^{\mathfrak{a}}\ d\frown(\mathfrak{a}\frown p)$
$\quad \smile^{\mathfrak{a}}\ \mathfrak{a}\frown(d\frown q)$
$\quad q\frown s$
$\quad p\frown s$
$\quad \not{\jmath}s$ (521

—— • ——

$\vdash \not{\mathfrak{X}}p\smile q\frown\eth s$
$\quad q\frown s$
$\quad p\frown s$
$\quad \not{\jmath}s$ (526

—— • ——

$\vdash p\smile\not{\mathfrak{X}}q\frown\eth s$
$\quad q\frown s$
$\quad p\frown s$
$\quad \not{\jmath}s$ (528

—— • ——

$\vdash p\smile q\frown s$
$\quad q\frown s$
$\quad p\frown s$
$\quad \not{\jmath}s$ (529

—— • ——

$\vdash \dot{\alpha}\dot{\varepsilon}(\text{—}\ \varepsilon\frown(\alpha\frown p)) = p$
$\quad p\frown s$
$\quad \not{\jmath}s$ (531

—— • ——

$\vdash \mathrm{I}\not{\mathfrak{X}}p$
$\quad p\frown s$
$\quad \not{\jmath}s$ (533

—— • ——

$\vdash \mathrm{I}p$
$\quad p\frown s$
$\quad \not{\jmath}s$ (535

—— • ——

$\vdash p\smile\not{\mathfrak{X}}p\frown s$
$\quad p\frown s$
$\quad \not{\jmath}s$ (536

—— • ——

$\vdash \not{\mathfrak{X}}p\frown s$
$\quad p\frown s$
$\quad \not{\jmath}s$ (538

—— • ——

$\vdash q = \not{\mathfrak{X}}\not{\mathfrak{X}}q$
$\quad q\frown s$
$\quad \not{\jmath}s$ (539

—— • ——

$\vdash f(q\smile\not{\mathfrak{X}}p)$
$\quad f(\not{\mathfrak{X}}(p\smile\not{\mathfrak{X}}q))$
$\quad q\frown s$
$\quad \not{\jmath}s$ (542

—— • ——

⊢ $f(⧹(p ⌣ ⧹q))$
$f(q ⌣ ⧹p)$
$q ⌢ s$
$ɏ s$ (543

———— • ————

⊢ $f(⧹q ⌣ p)$
$f(⧹(⧹p ⌣ q))$
$p ⌢ s$
$ɏ s$ (544

———— • ————

⊢ $a = d$
$d ⌢ (a ⌢ (q ⌣ ⧹q))$
$q ⌢ s$
$ɏ s$ (546

———— • ————

⊢ $a = d$
$d ⌢ (a ⌢ (⧹p ⌣ p))$
$p ⌢ s$
$ɏ s$ (551

———— • ————

⊢ $q ⌣ ⧹q = ⧹p ⌣ p$
$p ⌢ s$
$q ⌢ s$
$ɏ s$ (556

———— • ————

⊢ $f(p ⌣ ⧹p)$
$f(q ⌣ ⧹q)$
$p ⌢ s$
$q ⌢ s$
$ɏ s$ (559

———— • ————

⊢ $p = r ⌣ ⧹r$
$p ⌢ s$
$r ⌢ s$
$ɏ s$
$⧹p ⌢ s$
$p ⌢ 𝔡 s$ (561

———— • ————

⊢ $q ⌣ (p ⌣ ⧹p) = q$
$p ⌢ s$
$ɏ s$
$q ⌢ s$ (562

———— • ————

⊢ $q ⌣ ⧹p ⌣ p = q$
$p ⌢ s$
$ɏ s$
$q ⌢ s$ (566

———— • ————

⊢ $f(q)$
$f(q ⌣ ⧹p ⌣ p)$
$p ⌢ s$
$ɏ s$
$q ⌢ s$ (567

———— • ————

⊢ $q ⌣ p ⌣ ⧹p = q$
$p ⌢ s$
$ɏ s$
$q ⌢ s$ (568

———— • ————

⊢ $f(q ⌣ p ⌣ ⧹p)$
$f(q)$
$p ⌢ s$
$ɏ s$
$q ⌢ s$ (569

———— • ————

⊢ $f(q)$
$f(q ⌣ p ⌣ ⧹p)$
$p ⌢ s$
$ɏ s$
$q ⌢ s$ (570

———— • ————

⊢ $p ⌣ ⧹p ⌣ q = q$
$q ⌢ s$
$ɏ s$
$p ⌢ s$ (571

———— • ————

⊢ $p ⌣ (⧹p ⌣ q) = q$
$q ⌢ s$
$ɏ s$
$p ⌢ s$ (574

———— • ————

⊢ $f(q)$
$f(p ⌣ (⧹p ⌣ q))$
$q ⌢ s$
$ɏ s$
$p ⌢ s$ (575

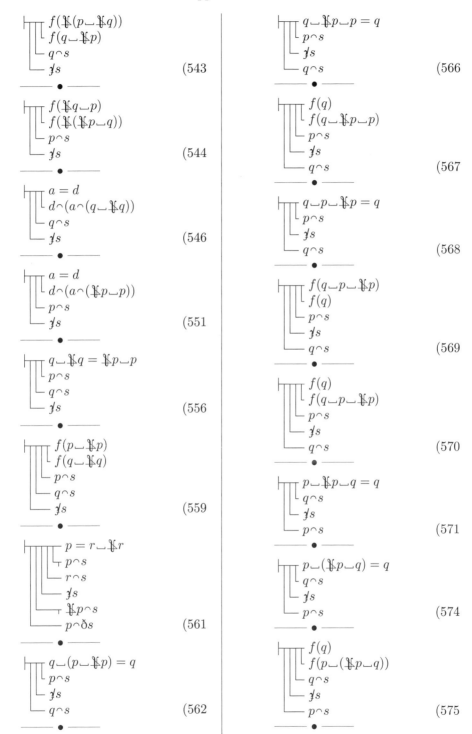

———— • ————

⊢ Ɥq⌣(p⌣Ɥp) = Ɥq
  q⌢s
  Ɉs
  p⌢s (576

———— • ————

⊢ Ɥp⌣p⌣q = q
  q⌢s
  Ɉs
  p⌢s (578

———— • ————

⊢ Ɥp⌣(p⌣q) = q
  q⌢s
  Ɉs
  p⌢s (579

———— • ————

⊢ Ɥq⌣Ɥp⌣p = Ɥq
  q⌢s
  Ɉs
  p⌢s (582

———— • ————

⊢ p⌢s
  c⌣p = c
  Ɉs
  c⌢s (585

———— • ————

⊢ q⌣t⌢s
  r⌣t⌢s
  q⌣Ɥr⌢s
  r⌢s
  Ɉs
  q⌢s (586

———— • ————

⊢ c⌣Ɥa⌢s
  ⊤ a⌣Ɥc⌢s
  c⌢s
  Ɉs
  ⊤ a = c
  a⌢s (588

———— • ————

⊢ p⌣Ɥq⌢s
  q⌣Ɥp⌢s
  p⌢s
  Ɉs (589

———— • ————

⊢ t⌢s
  q⌢s
  p⌢s
  Ɉs
  q⌢(t⌢⌣∗p) (623

———— • ————

⊢ p = q
  ⊤ Ɥq⌣p⌢s
  q⌢s
  Ɉs
  ⊤ Ɥp⌣q⌢s
  p⌢s (639

———— • ————

⊢ b = p⌣Ɥp
  Ɉs
  𝔞 ⊤ 𝔞⌣b⌢s
    𝔞⌣Ɥb⌢s
    𝔞⌢s
  p⌢s
  b⌢𝔡s (642

———— • ————

⊢ c = d
  Ɉs
  𝔞 ⊤ 𝔞⌣(c⌣Ɥd)⌢s
    𝔞⌣(d⌣Ɥc)⌢s
    𝔞⌢s
  d⌢s
  c⌢s (644

———— • ————

⊢ e⌣Ɥb⌢s
  p⌣Ɥb⌢s
  ⊤ p⌣Ɥe⌢s
  e⌢s
  Ɉs
  p⌢s (671

———— • ————

⊢ e⌢(s ꟻ u)
  𝔞 𝔞⌢u
    e⌣Ɥ𝔞⌢s
    𝔞⌢s (590

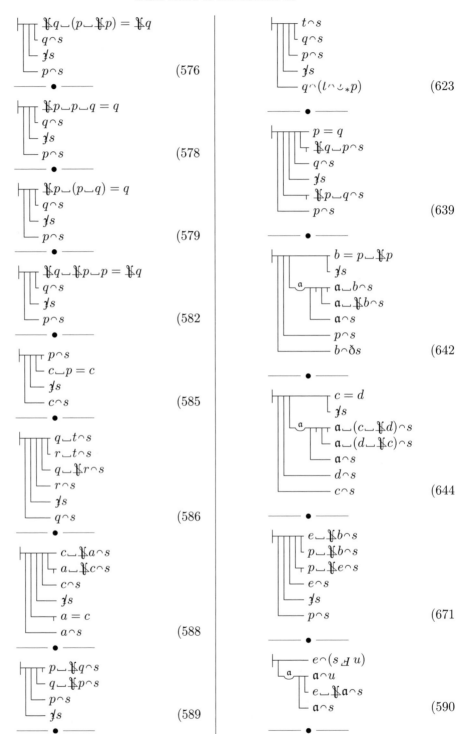

———— • ————

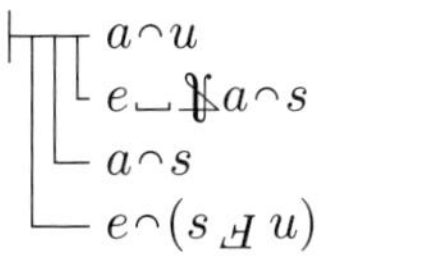

(591

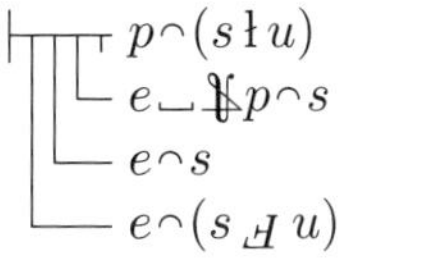

(594

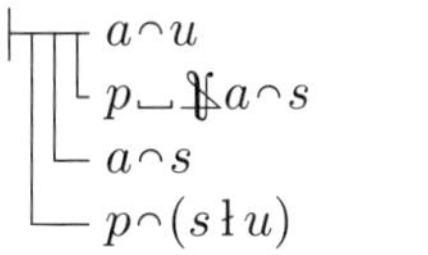

(597

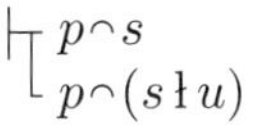

(599

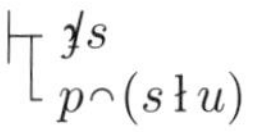

(600

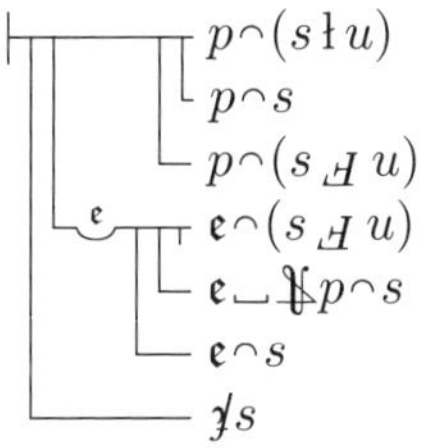

(601

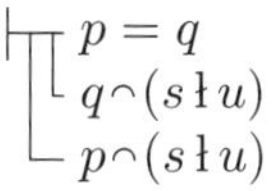

(602

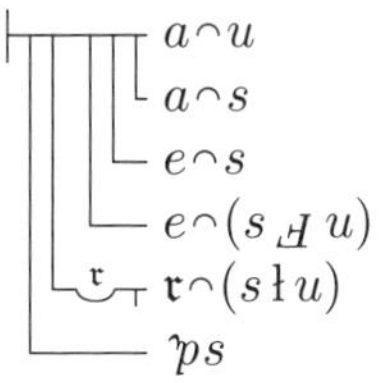

(604

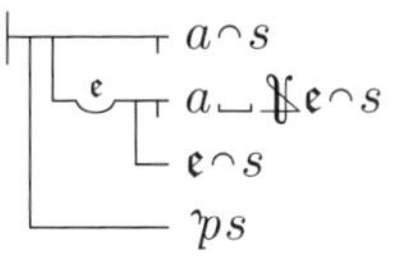

(606

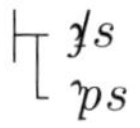

(607

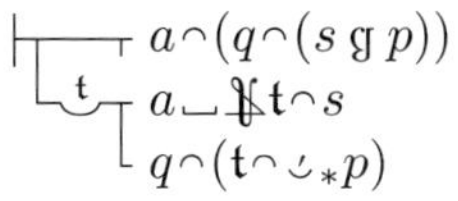

(608

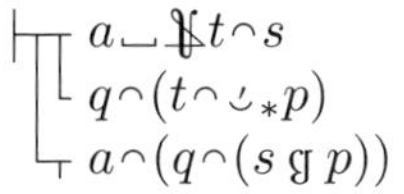

(609

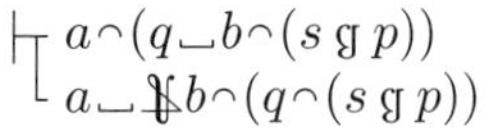

(614

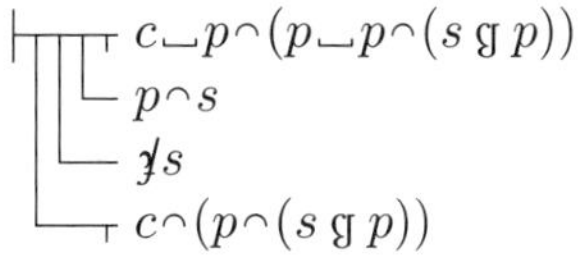

(622

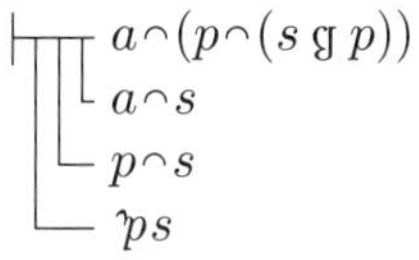

(635

(Archimedean Axiom)

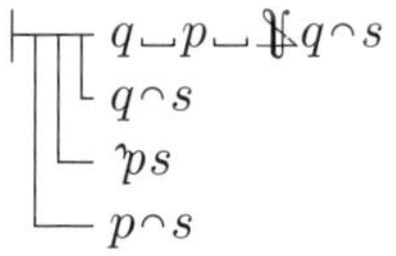

(637

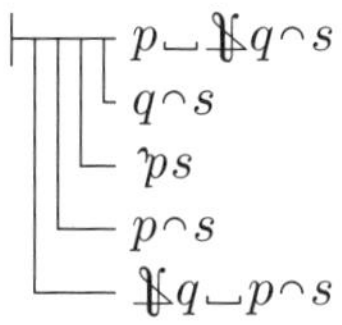

(638

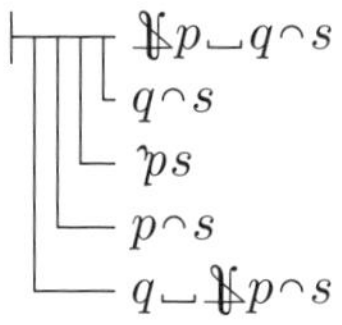

(641

$$\vdash q \smile p \smile \mathfrak{K}(q \smile c) \frown s$$
$$q \frown s$$
$$\mathfrak{p}s$$
$$p \smile \mathfrak{K}c \frown s \qquad (646$$

———•———

$$\vdash \mathfrak{K}(d \smile c) \smile (q \smile p) \frown s$$
$$p \frown s$$
$$q \frown s$$
$$\mathfrak{f}s$$
$$\mathfrak{p}s$$
$$c \frown s$$
$$p \smile \mathfrak{K}c \frown s$$
$$d \frown s$$
$$q \smile \mathfrak{K}d \frown s \qquad (648$$

———•———

$$\vdash (q \smile p) \smile \mathfrak{K}(d \smile c) \frown s$$
$$c \frown s$$
$$d \frown s$$
$$\mathfrak{f}s$$
$$\mathfrak{p}s$$
$$p \frown s$$
$$q \frown s$$
$$p \smile \mathfrak{K}c \frown s$$
$$q \smile \mathfrak{K}d \frown s \qquad (669$$

———•———

$$\vdash q \smile p = p \smile q$$
$$\mathfrak{p}s$$
$$q \frown s$$
$$p \frown s \qquad (674$$

———•———

$$\vdash p \smile \mathfrak{K}q = \mathfrak{K}q \smile p$$
$$\mathfrak{p}s$$
$$q \frown s$$
$$p \frown s \qquad (675$$

———•———

$$\vdash p \smile q = q \smile p$$
$$\mathfrak{p}s$$
$$p \frown \mathfrak{d}s$$
$$q \frown \mathfrak{d}s \qquad (689$$

(The commutative law in the domain of a positive class.)

Afterword

Hardly anything more unwelcome can befall a scientific writer than to have one of the foundations of his edifice shaken after the work is finished.

This is the position into which I was put by a letter from Mr Bertrand Russell as the printing of this volume was nearing completion. The matter concerns my basic law (V). I have never concealed from myself that it is not as obvious as the others nor as obvious as must properly be required of a logical law. Indeed, I pointed out this very weakness in the foreword to the first volume, p. VII. I would gladly have dispensed with this foundation if I had known of some substitute for it. Even now, I do not see how arithmetic can be founded scientifically, how the numbers can be apprehended as logical objects and brought under consideration, if it is not—at least conditionally—permissible to pass from a concept to its extension. May I always speak of the extension of a concept, of a class? And if not, how are the exceptions to be recognised? May one always infer from the extension of one concept's coinciding with that of a second that every object falling under the first concept also falls under the latter? These questions arise from Mr Russell's communication.

Solatium miseris, socios habuisse malorum.[a] This consolation, if it is one, is on my side also; for everyone who has made use of extensions of concepts, classes, sets in their proofs[1] is in the same position. What is at stake here is not my approach to a foundation in particular, but rather the very possibility of any logical foundation of arithmetic.

But to the matter itself! Mr Russell has discovered a contradiction which may now be presented.

Of the class of human beings no-one will want to claim that it is a human being. We have here a class that does not belong to itself. For I say that something belongs

[1] The systems of Mr R. Dedekind are also included here.

to a class when it falls under the concept whose extension is just that class. Let us now focus on the concept *class that does not belong to itself.* The extension of this concept, if it is permissible to speak of it, is accordingly the class of those classes that do not belong to themselves. We will call it the class K for short. Let us now ask whether this class K belongs to itself. Let us first assume that it does. If something belongs to a class, then it falls under the concept whose extension the class is. So if our class belongs to itself, then it is a class that does not belong to itself. Our first assumption thus leads to a contradiction with itself. But if, alternatively, we assume that our class K does not belong to itself, then it falls under the concept of which it is itself the extension, and thus does belong to itself. Thus again a contradiction!

What attitude should we take to this? Should we assume the law of excluded middle fails for classes? Or should we assume that there are cases where to an incontestable concept no class corresponds as its extension? In the first case we would find ourselves compelled to deny the full objecthood of classes. For if classes were proper objects, then the law of excluded middle would have to hold of them. They have, on the other hand, nothing unsaturated, predicative, about them whereby they would be characterised as, for example, functions, concepts, relations. What we customarily treat as a name of a class, e.g., "the class of prime numbers", has rather the nature of a proper name, cannot occur predicatively, but can occur as the grammatical subject of a singular proposition, e.g., "The class of prime numbers contains infinitely many objects". If we wanted to revoke the law of excluded middle for classes, we could consider taking classes—and presumably value-ranges in general—as improper objects. In that case, they would not be admissible as arguments for all first-level functions. There would, however, be some functions which could have both proper and improper objects as arguments. At least the relation of equality (identity) would be of this kind. One might try to avoid this by assuming a special kind of equality for improper objects. But that is surely ruled out. Identity is a relation given in so determinate a way that it is inconceivable that different kinds of it could occur. But now the result would be a great multitude of first-level functions, namely, first those that may have only proper objects as arguments, second those that can have both proper and improper objects as arguments, and finally also those that can have only improper objects as arguments. A further division would result from the values of the functions.

One would thus have to distinguish, first, functions that only had proper objects as values, second those that had both proper and improper objects as values, finally those that had only improper objects as values. Both distinctions amongst first-level functions would obtain simultaneously, so that nine kinds would result. To these in turn would correspond nine kinds of value-ranges, improper objects, which would have to be distinguished logically. The classes of proper objects would have to be distinguished from the classes of classes of proper objects, the Relations between proper objects from the classes of proper objects, from the classes of Relations between proper objects, and so on. An imponderable multiplicity of kinds would thus result; and objects which belonged to different kinds could never occur as arguments of the same function. It seems to be extraordinarily difficult, however, to devise complete legislation to decide in general which objects would be admissible as arguments for which functions. One may doubt, moreover, whether improper objects are legitimate.

If these difficulties scare us away from taking classes, and with them numbers, as improper objects, yet we do not want to recognise them as proper objects either, that is, as things that may occur as arguments of any first-level function, then presumably all that is left is to consider class-names as pseudo proper names which would thus in truth have no reference. They would then have to be viewed merely as parts of signs which would have a reference only as a whole.[1] It may of course be considered advantageous for certain purposes to form distinct signs that agree in some part without thereby rendering them composite. The simplicity of a sign indeed demands no more than that distinguishable parts within it do not autonomously have a reference. In this case, even what we are accustomed to regard as a number-sign would not in fact be a sign, but the non-autonomous part of a sign. An explanation of the sign '2' would be impossible; instead one would have to explain many signs which contain '2' as non-autonomous component, but are not to be thought of as composed logically out of '2' and another part. It would then be inadmissible to allow a letter to stand in for such a non-autonomous part; for as far as the content is concerned there would be no composition at all. The generality of arithmetical propositions would thus be lost. It would also not be comprehensible how one could talk of the cardinal number of a class, of the cardinal number of cardinal numbers.

I think that this suffices to render this way too apparently impassable. So presumably nothing remains but to recognise extensions of concepts or classes as objects in

[1] On this, compare vol. I, §29.

the full and proper sense of the word, but to concede at the same time that the erstwhile understanding of the words "extension of a concept" requires correction.

Before we consider that in more detail, it will be useful to track down the origin of the contradiction using our signs. That Δ is a class that does not belong to itself can be expressed thus:

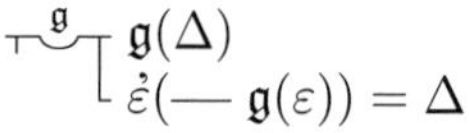

And the class of the classes not belonging to themselves may be designated like this:

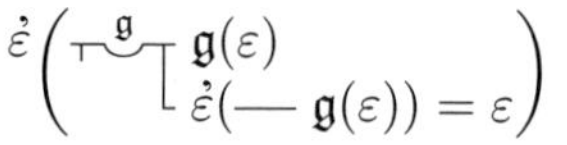 [1]

In the following derivation I will abbreviate this using the sign '∀' and omit the judgement-stroke because of the doubtful truth. Accordingly, I will express by

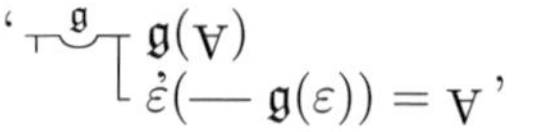

that the class ∀ does not belong to itself.[b]

According to (Vb) we now have

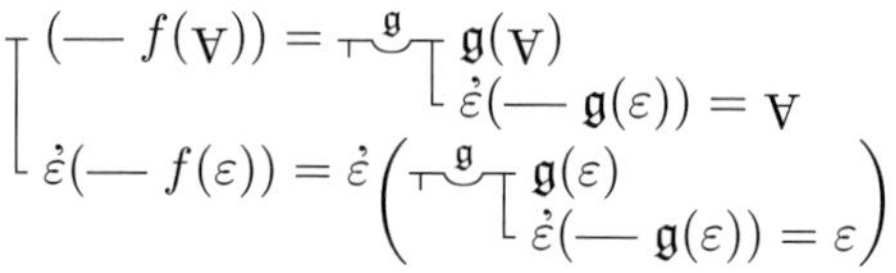

or, if we use the abbreviation and apply (IIIa)

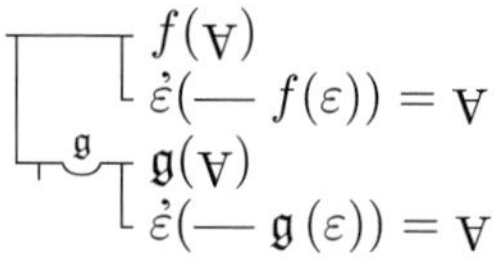

(α

Now we introduce for 'f' the German '𝔤':

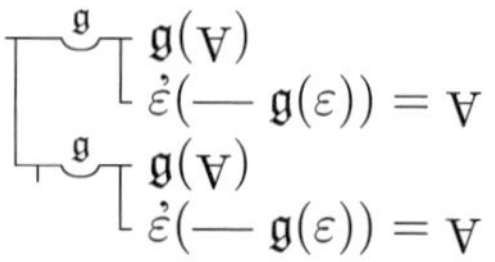

(β

i.e.: if ∀ does not belong to itself, it belongs to itself.[c] This is one direction.

For the other, we have by (IIb)

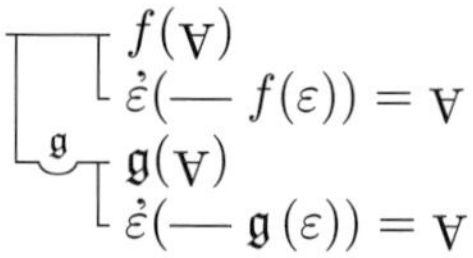

(γ

If we now take 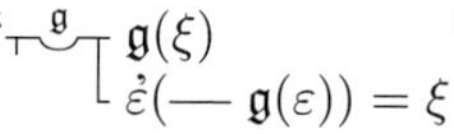 for $f(\xi)$:

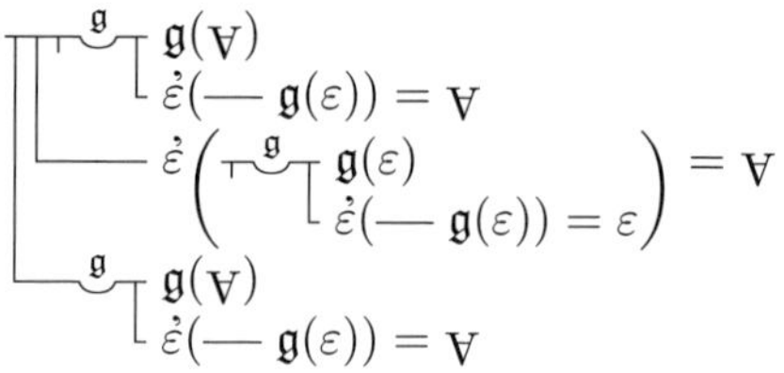

(δ

and taking into account our abbreviation:

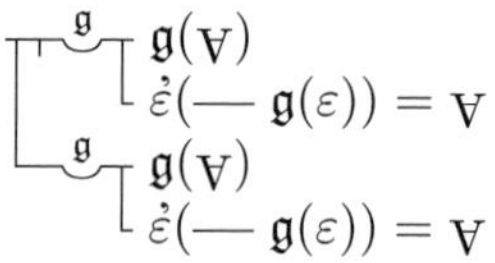

(ε

i.e.: if ∀ belongs to itself, it does not belong to itself.[c] From (ε) by (Ig)

[1] Regarding the use of Greek letters compare vol. I, §9.

$$\vdash\!\overset{\mathfrak{g}}{\smile}\!\top\; \mathfrak{g}(\forall) \quad \mathrm{if}\quad \dot{\varepsilon}(-\,\mathfrak{g}(\varepsilon)) = \forall \qquad (\zeta$$

follows, and from this using (β)

$$\overset{\mathfrak{g}}{\smile}\!\top\; \mathfrak{g}(\forall) \quad \mathrm{if}\quad \dot{\varepsilon}(-\,\mathfrak{g}(\varepsilon)) = \forall \qquad (\eta$$

The propositions (ζ) and (η) contradict each other. The error can only lie in our law (V b) which must therefore be false.

We now want to see how the matter stands if we use our sign '$\frown$'. '$\dot{\varepsilon}(\top\,\varepsilon\frown\varepsilon)$' will take the place of '$\forall$'. By taking '$\top\,\xi\frown\xi$' for '$f(\xi)$', '$-\,\xi$' for '$F(\xi)$', and '$\dot{\varepsilon}(\top\,\varepsilon\frown\varepsilon)$' for '$a$' in (82), we obtain

$$\vdash\!\top\; \dot{\varepsilon}(\top\,\varepsilon\frown\varepsilon)\frown\dot{\varepsilon}(\top\,\varepsilon\frown\varepsilon) \quad \mathrm{if}\quad \dot{\varepsilon}(\top\,\varepsilon\frown\varepsilon)\frown\dot{\varepsilon}(\top\,\varepsilon\frown\varepsilon) \qquad (\vartheta$$

from which

$$\vdash\!\top\; \dot{\varepsilon}(\top\,\varepsilon\frown\varepsilon)\frown\dot{\varepsilon}(\top\,\varepsilon\frown\varepsilon) \qquad (\iota$$

follows by (Ig). Using the same insertions, we obtain from (77):

$$\vdash\; \dot{\varepsilon}(\top\,\varepsilon\frown\varepsilon)\frown\dot{\varepsilon}(\top\,\varepsilon\frown\varepsilon) \quad \mathrm{if\ not}\quad \dot{\varepsilon}(\top\,\varepsilon\frown\varepsilon)\frown\dot{\varepsilon}(\top\,\varepsilon\frown\varepsilon) \qquad (\kappa$$

From that, using (ι),

$$\vdash\; \dot{\varepsilon}(\top\,\varepsilon\frown\varepsilon)\frown\dot{\varepsilon}(\top\,\varepsilon\frown\varepsilon) \qquad (\lambda$$

follows, which contradicts (ι). Thus at least one of the propositions (77) and (82) is false, and hence also (1) from which they follow. If the derivation of (1) in §55 of the first volume is reviewed, it will be found that there too use is made of (V b). Thus here too suspicion is cast on this proposition. With (V b), (V) itself also collapses, but not (V a). There is no obstacle to the transformation of the generality of an equality into an equality of value-range; only the converse transformation is shown to be not always allowed. With this it is of course recognised that my introduction of value-ranges in §3 of the first volume is not always permissible. We cannot in general use the words,

> "the function $\Phi(\xi)$ has the same value-range as the function $\Psi(\xi)$"

as co-referential with the words,

> "the functions $\Phi(\xi)$ and $\Psi(\xi)$ always have the same value for the same argument"

and must consider the possibility that there are concepts which—in the common sense of the word at least—have no extension. The legitimacy of our second-level function $\dot{\varepsilon}\,\varphi(\varepsilon)$ is thereby shattered. And yet such a function is indispensable for the foundation of arithmetic.

Let us now supplement our investigation by obtaining the falsity of (V b) as an end result instead of starting from (V b) and running into a contradiction. To avoid dependence in this on the already suspect value-range signs, we will carry out the derivation in full generality for a second-level function with one argument of the second kind,[1] using the notation of vol. I, §25. Our combination of signs

$$\text{`}\dot{\varepsilon}\left(\top\!\overset{\mathfrak{g}}{\smile}\!\top\; \mathfrak{g}(\varepsilon) \quad \mathrm{cond}\quad \dot{\varepsilon}(-\,\mathfrak{g}(\varepsilon)) = \varepsilon\right)\text{'}$$

[1] Vol. I, §23, p. 40.

is accordingly replaced by

$$`M_\beta\left(\begin{array}{l}\overset{\mathfrak{g}}{\smile}\!\top\ \mathfrak{g}(\beta)\\ \quad\llcorner\ M_\beta(\!-\!\!-\ \mathfrak{g}(\beta)) = \beta\end{array}\right)'$$

to which the specifications regarding value-range signs that we laid down in I, §9, for the scope of a Greek letter are to be applied analogously. Our formula contains an 'M' twice: the first in initial position, and the second within. At the argument-place of the former stands the function-marker

$$`\overset{\mathfrak{g}}{\smile}\!\top\ \begin{array}{l}\mathfrak{g}(\xi)\\ \llcorner\ M_\beta(\!-\!\!-\ \mathfrak{g}(\beta)) = \xi'\end{array}$$

while '— $\mathfrak{g}(\xi)$' stands at the argument-place of the latter. First we obtain the following:

$$\text{IIb}\ \vdash\ \begin{array}{l}\overset{\mathfrak{g}}{\smile}\!\top\ \mathfrak{g}(a)\\ \quad\llcorner\ M_\beta(\!-\!\!-\ \mathfrak{g}(\beta)) = a\\ \llcorner\ M_\beta\left(\begin{array}{l}\overset{\mathfrak{g}}{\smile}\!\top\ \mathfrak{g}(\beta)\\ \quad\llcorner\ M_\beta(\!-\!\!-\ \mathfrak{g}(\beta)) = \beta\end{array}\right) = a\\ \llcorner\ \overset{\mathfrak{g}}{\smile}\!\top\ \mathfrak{g}(a)\\ \qquad\llcorner\ M_\beta(\!-\!\!-\ \mathfrak{g}(\beta)) = a\end{array}$$

$$\times$$

$$\vdash\ \begin{array}{l}M_\beta\left(\begin{array}{l}\overset{\mathfrak{g}}{\smile}\!\top\ \mathfrak{g}(\beta)\\ \quad\llcorner\ M_\beta(\!-\!\!-\ \mathfrak{g}(\beta)) = \beta\end{array}\right) = a\\ \llcorner\ \overset{\mathfrak{g}}{\smile}\!\top\ \mathfrak{g}(a)\\ \qquad\llcorner\ M_\beta(\!-\!\!-\ \mathfrak{g}(\beta)) = a\end{array} \qquad (\mu$$

$$\times$$

$$\vdash\ \begin{array}{l}\overset{\mathfrak{g}}{\smile}\!\top\ \mathfrak{g}(a)\\ \quad\llcorner\ M_\beta(\!-\!\!-\ \mathfrak{g}(\beta)) = a\\ \llcorner\ M_\beta\left(\begin{array}{l}\overset{\mathfrak{g}}{\smile}\!\top\ \mathfrak{g}(\beta)\\ \quad\llcorner\ M_\beta(\!-\!\!-\ \mathfrak{g}(\beta)) = \beta\end{array}\right) = a\end{array} \qquad (\nu$$

If in this we introduce as an abbreviation

$$`\Phi(\xi)\text{' for '}\overset{\mathfrak{g}}{\smile}\!\top\ \begin{array}{l}\mathfrak{g}(\xi)\\ \llcorner\ M_\beta(\!-\!\!-\ \mathfrak{g}(\beta)) = \xi\end{array}\text{'}$$

and replace 'a' by '$M_\beta(\Phi(\beta))$', then we obtain from (ν)

$$\Phi(M_\beta(\Phi(\beta)))$$

i.e., the value of our second-level function for the concept $\Phi(\xi)$ falls under that very concept. But on the other hand we also obtain from (ν)

$$\overset{\mathfrak{g}}{\smile}\!\top\ \begin{array}{l}\mathfrak{g}(M_\beta(\Phi(\beta)))\\ \llcorner\ M_\beta(\!-\!\!-\ \mathfrak{g}(\beta)) = M_\beta(\Phi(\beta))\end{array}$$

i.e.: there is a concept for which, taken as argument, our second-level function receives the same value as for $\Phi(\xi)$ but under which this value does not fall. In other words: for every second-level function with one argument of the second kind, there are two concepts such that, when taken as arguments of this function, they result in the same value, and this value falls under the first of these concepts but not the second.

We may derive this in concept-script as follows:

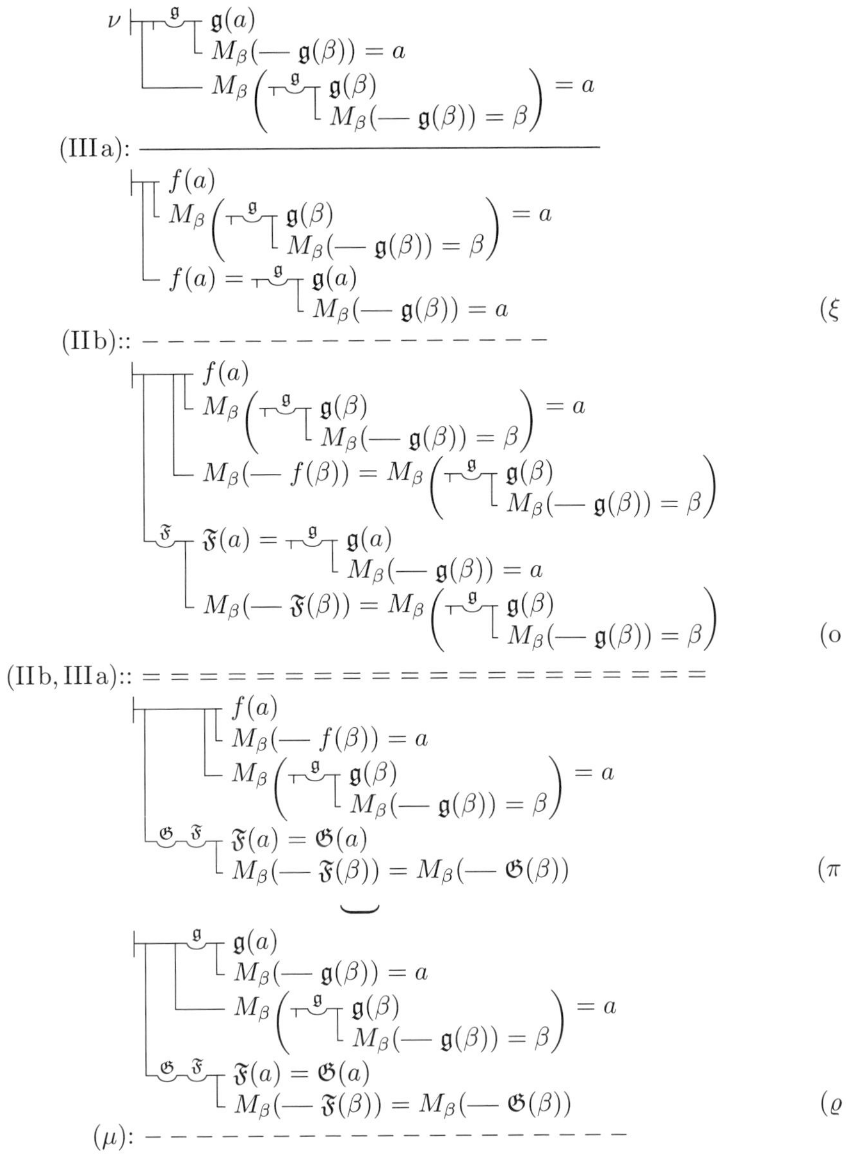

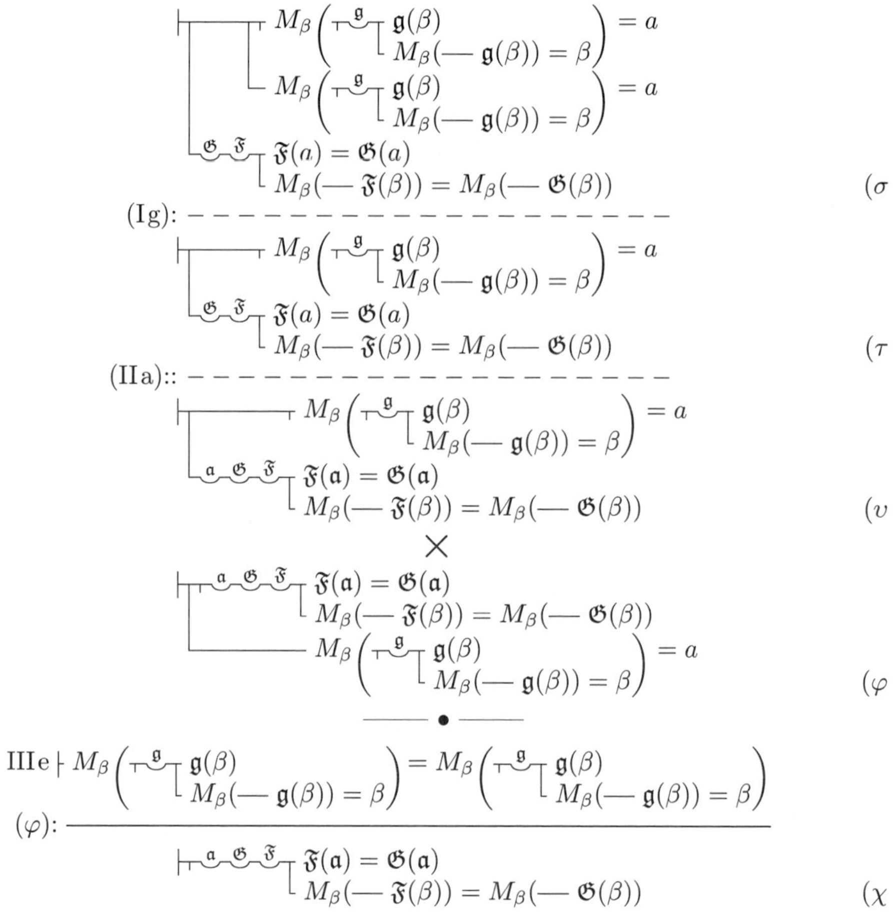

$$\begin{array}{l}
\vdash\!\!\!\!\!\top\!\!\!\!\top M_\beta\!\left(\!\!\top\!\!\!\!\top^{\mathfrak{g}}\begin{array}{l}\mathfrak{g}(\beta)\\ M_\beta(\!\!-\!\mathfrak{g}(\beta))=\beta\end{array}\!\!\right)=a\\
\quad\;\vdash M_\beta\!\left(\!\!\top\!\!\!\!\top^{\mathfrak{g}}\begin{array}{l}\mathfrak{g}(\beta)\\ M_\beta(\!\!-\!\mathfrak{g}(\beta))=\beta\end{array}\!\!\right)=a\\
\quad\;\;\;\vdash^{\mathfrak{G}\;\mathfrak{F}}\begin{array}{l}\mathfrak{F}(\mathfrak{a})=\mathfrak{G}(\mathfrak{a})\\ M_\beta(\!\!-\!\mathfrak{F}(\beta))=M_\beta(\!\!-\!\mathfrak{G}(\beta))\end{array}\qquad(\sigma
\end{array}$$

(Ig): -------------------------

$$\vdash M_\beta\!\left(\!\!\top\!\!\!\!\top^{\mathfrak{g}}\begin{array}{l}\mathfrak{g}(\beta)\\ M_\beta(\!\!-\!\mathfrak{g}(\beta))=\beta\end{array}\!\!\right)=a$$
$$\vdash^{\mathfrak{G}\;\mathfrak{F}}\begin{array}{l}\mathfrak{F}(\mathfrak{a})=\mathfrak{G}(\mathfrak{a})\\ M_\beta(\!\!-\!\mathfrak{F}(\beta))=M_\beta(\!\!-\!\mathfrak{G}(\beta))\end{array}\qquad(\tau$$

(IIa):: -------------------------

$$\vdash M_\beta\!\left(\!\!\top\!\!\!\!\top^{\mathfrak{g}}\begin{array}{l}\mathfrak{g}(\beta)\\ M_\beta(\!\!-\!\mathfrak{g}(\beta))=\beta\end{array}\!\!\right)=a$$
$$\vdash^{\mathfrak{a}\;\mathfrak{G}\;\mathfrak{F}}\begin{array}{l}\mathfrak{F}(\mathfrak{a})=\mathfrak{G}(\mathfrak{a})\\ M_\beta(\!\!-\!\mathfrak{F}(\beta))=M_\beta(\!\!-\!\mathfrak{G}(\beta))\end{array}\qquad(\upsilon$$

$$\times$$

$$\vdash^{\mathfrak{a}\;\mathfrak{G}\;\mathfrak{F}}\begin{array}{l}\mathfrak{F}(\mathfrak{a})=\mathfrak{G}(\mathfrak{a})\\ M_\beta(\!\!-\!\mathfrak{F}(\beta))=M_\beta(\!\!-\!\mathfrak{G}(\beta))\end{array}$$
$$M_\beta\!\left(\!\!\top\!\!\!\!\top^{\mathfrak{g}}\begin{array}{l}\mathfrak{g}(\beta)\\ M_\beta(\!\!-\!\mathfrak{g}(\beta))=\beta\end{array}\!\!\right)=a\qquad(\varphi$$

IIIe $\vdash M_\beta\!\left(\!\!\top\!\!\!\!\top^{\mathfrak{g}}\begin{array}{l}\mathfrak{g}(\beta)\\ M_\beta(\!\!-\!\mathfrak{g}(\beta))=\beta\end{array}\!\!\right)=M_\beta\!\left(\!\!\top\!\!\!\!\top^{\mathfrak{g}}\begin{array}{l}\mathfrak{g}(\beta)\\ M_\beta(\!\!-\!\mathfrak{g}(\beta))=\beta\end{array}\!\!\right)$

(φ): ———————————————————

$$\vdash^{\mathfrak{a}\;\mathfrak{G}\;\mathfrak{F}}\begin{array}{l}\mathfrak{F}(\mathfrak{a})=\mathfrak{G}(\mathfrak{a})\\ M_\beta(\!\!-\!\mathfrak{F}(\beta))=M_\beta(\!\!-\!\mathfrak{G}(\beta))\end{array}\qquad(\chi$$

i.e.: for every second-level function with one argument of the second kind there are concepts which, taken as its arguments, result in the same value, although not all objects that fall under one of these concepts also fall under the other.

Our proof was conducted without recourse to propositions or notations whose legitimacy would be in any way doubtful. Our proposition thus holds in particular for the second-level function $\dot{\varepsilon}\,\varphi(\varepsilon)$ if it is admissible, or in words:

If for every first-level concept one may in general speak of its extension, then the case arises of concepts' having the same extension although not all objects that fall under one of these concepts also fall under the other.

With this, however, the extension of a concept in the traditional sense of the phrase is in effect abolished. It must not be said that in general the expression "the extension of a first concept coincides with that of a second" is co-referential with the expression "all objects falling under first concept also fall under the second and *vice versa*". We see from the result of our derivation that it is not possible at all to assign to the words "the extension of the concept $\Phi(\xi)$" such a sense that in general one may infer from the equality of the extension of concepts that every object falling under one of them also falls under the other.

Our proposition can also be arrived at in another way, namely as follows:

IIb

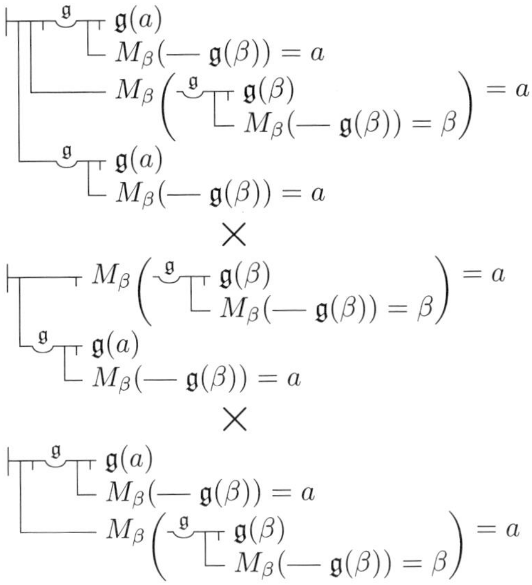

(ψ

(ω

If here we introduce as an abbreviation

'$\Psi(\xi)$' for '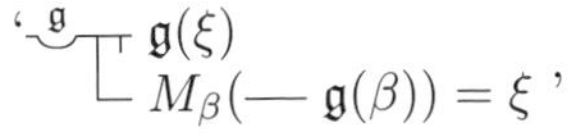 '

and replace 'a' by '$M_\beta(\Psi(\beta))$', then we obtain from (ω)

$$\top\, \Psi(M_\beta(\Psi(\beta)))$$

i.e., the value of our second-level function for the argument $\Psi(\xi)$ does not fall under the concept $\Psi(\xi)$. But on the other hand we also obtain from (ω)

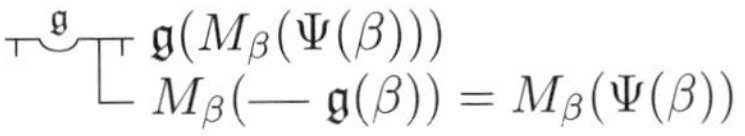

i.e.: there is a concept for which, as argument, our second-level function receives the same value as for $\Psi(\xi)$ and under which this value falls. Here again we thus have two concepts which, taken as arguments of the second-level function, result in the same value, which now falls under the second of these concepts but not the first. The derivation of the proposition (χ) from proposition

(ω) may proceed in a way similar to that from (ν).

Let us now try to take the function $\dot{\varepsilon}(-\, \varphi(\varepsilon))$ as the second-level function of our propositions. We then have in

$$\begin{array}{l} \top\!\!\smile^{\mathfrak{g}}\!\!\top\; \mathfrak{g}(\xi) \\ \qquad\;\llcorner\; \dot{\varepsilon}(-\, \mathfrak{g}(\varepsilon)) = \xi \end{array}$$

a concept under which its own extension falls. Then, however, there is by (ν) a concept whose extension coincides with that just mentioned but under which that extension does not fall. We would like to have an example of this. How might such a concept be found? It is not possible without a more precise specification of our function $\dot{\varepsilon}(-\, \varphi(\varepsilon))$ or of the extension of a concept; for our erstwhile criterion for the coinciding of extensions of concepts here leaves us in the lurch.

On the other hand, we have in

$$\begin{array}{l} -\!\!\smile^{\mathfrak{g}}\!\!\top\!\!\top\; \mathfrak{g}(\xi) \\ \qquad\quad\llcorner\; \dot{\varepsilon}(-\, \mathfrak{g}(\varepsilon)) = \xi \end{array}$$

a concept under which its own extension does not fall. However, by (ω) there is then a concept whose extension coincides with that just mentioned but under which that extension does fall. All this, naturally, is under the presupposition that the function-name '$\dot{\varepsilon}(-\, \varphi(\varepsilon))$' is logically legitimate.

In both cases, we see that it is the extension of the concept itself that brings about the exceptional case, by falling under only one of two concepts which have it as extension; and we see that the occurrence of this exception can in no way be avoided. It accordingly suggests itself to formulate the criterion of extensional equality like this: the extension of one concept coincides with that of another if every object, save the extension of the first concept, which falls under the first concept also falls under the second, and if conversely every object, save the extension of the second concept, which falls under the second concept also falls under the first.

Evidently, this cannot be regarded as a definition of, say, the extension of a concept, but rather merely as a statement of the characteristic constitution of that second-level function.

By transferring to value-ranges in general what we have already said of extensions of concepts, we arrive at the basic law

$$\begin{array}{ll} \vdash (\dot{\varepsilon}f(\varepsilon) = \dot{\alpha}g(\alpha)) = \smile^{\mathfrak{a}}\!\!\top & f(\mathfrak{a}) = g(\mathfrak{a}) \\ & \top\!\!\top\; \mathfrak{a} = \dot{\varepsilon}f(\varepsilon) \\ & \;\;\llcorner\!\!\top\; \mathfrak{a} = \dot{\alpha}\, g(\alpha) \end{array} \qquad \text{(V}'$$

which has to take the place of (V) (I, §20, p. 36). From this law, (V a) follows. In contrast, (V b) has to give way to the following propositions:

$$\begin{array}{l} \vdash\!\!\top\!\!\top\; f(a) = g(a) \\ \quad\;\; \llcorner\!\!\top\; a = \dot{\varepsilon}f(\varepsilon) \\ \quad\; \llcorner\; \dot{\varepsilon}f(\varepsilon) = \dot{\alpha}\, g(\alpha) \end{array} \qquad \text{(V}'\text{b}$$

or

$$\begin{array}{l} \vdash\!\!\top\!\!\top\; f(a) = g(a) \\ \quad\;\; \llcorner\!\!\top\; a = \dot{\alpha}\, g(\alpha) \\ \quad\; \llcorner\; \dot{\varepsilon}f(\varepsilon) = \dot{\alpha}\, g(\alpha) \end{array} \qquad \text{(V}'\text{c}$$

Let us convince ourselves that the contradiction arising earlier between the propositions (β) and (ε) is now avoided. We proceed as in the derivation of (β), using (V′c) instead of (Vb). Let ‘$\forall$’ again abbreviate

$$\text{‘}\dot{\varepsilon}\left(\smile^{\mathfrak{g}}\!\!\!\top\!\!\begin{array}{l}\mathfrak{g}(\varepsilon)\\ \llcorner\,\dot{\varepsilon}(-\,\mathfrak{g}(\varepsilon)) = \varepsilon\end{array}\right)\text{’}$$

We have by (V′c)

$$\begin{array}{l}
\vdash\!\top\!\!\top (-\, f(\forall)) = \smile^{\mathfrak{g}}\!\!\!\top\!\!\begin{array}{l}\mathfrak{g}(\forall)\\ \llcorner\,\dot{\varepsilon}(-\,\mathfrak{g}(\varepsilon)) = \forall\end{array}\\
\quad\llcorner\top\, \forall = \dot{\varepsilon}\left(\smile^{\mathfrak{g}}\!\!\!\top\!\!\begin{array}{l}\mathfrak{g}(\varepsilon)\\ \llcorner\,\dot{\varepsilon}(-\,\mathfrak{g}(\varepsilon)) = \varepsilon\end{array}\right)\\
\llcorner\, \dot{\varepsilon}(-\, f(\varepsilon)) = \dot{\varepsilon}\left(\smile^{\mathfrak{g}}\!\!\!\top\!\!\begin{array}{l}\mathfrak{g}(\varepsilon)\\ \llcorner\,\dot{\varepsilon}(-\,\mathfrak{g}(\varepsilon)) = \varepsilon\end{array}\right)
\end{array}$$

The use of the abbreviation yields

$$\begin{array}{l}
\vdash\!\top\!\!\top (-\, f(\forall)) = \smile^{\mathfrak{g}}\!\!\!\top\!\!\begin{array}{l}\mathfrak{g}(\forall)\\ \llcorner\,\dot{\varepsilon}(-\,\mathfrak{g}(\varepsilon)) = \forall\end{array}\\
\quad\llcorner\top\, \forall = \forall\\
\llcorner\, \dot{\varepsilon}(-\, f(\varepsilon)) = \dot{\varepsilon}\left(\smile^{\mathfrak{g}}\!\!\!\top\!\!\begin{array}{l}\mathfrak{g}(\varepsilon)\\ \llcorner\,\dot{\varepsilon}(-\,\mathfrak{g}(\varepsilon)) = \varepsilon\end{array}\right)
\end{array}$$

which is self-evident because of the subcomponent ‘$\top\, \forall = \forall$’ and precisely because of this can never lead to a contradiction.

We had (I, p. 17) stipulated that the extension of a concept under which only the True falls should be the True and that the extension of a concept under which only the False falls should be the False. These determinations suffer no alteration under the new conception of the extension of a concept.

Now what impact does this new conception have on the values of our function $\backslash\xi$ if we retain the determinations in I, §11? Let us assume $\Phi(\xi)$ is an empty concept. According to our erstwhile conception of the extension of a concept, $\backslash\dot{\varepsilon}\,\Phi(\varepsilon)$ coincided with $\dot{\varepsilon}\,\Phi(\varepsilon)$, since there was no object Δ such that $\dot{\varepsilon}(\Delta = \varepsilon)$ coincided with $\dot{\varepsilon}\,\Phi(\varepsilon)$. According to our new conception of the extension of a concept there is such an object, namely $\dot{\varepsilon}\,\Phi(\varepsilon)$ itself. The result, nevertheless, is the same, namely that $\backslash\dot{\varepsilon}\,\Phi(\varepsilon)$ coincides with $\dot{\varepsilon}\,\Phi(\varepsilon)$. The same will happen if $\dot{\varepsilon}\,\Phi(\varepsilon)$ is the unique object falling under the concept $\Phi(\xi)$. If we assume that Δ is the unique object falling under the concept $\Phi(\xi)$, then $\backslash\dot{\varepsilon}\,\Phi(\varepsilon)$ coincides with Δ. The same still holds if $\dot{\varepsilon}\,\Phi(\varepsilon)$ is the only thing, other than Δ, falling under the concept $\Phi(\xi)$; and here things are now different from before; for in this case $\backslash\dot{\varepsilon}\,\Phi(\varepsilon)$ would not have coincided with Δ before but with $\dot{\varepsilon}\,\Phi(\varepsilon)$. In all other cases there is no difference between the old and the new conception of the extension of a concept concerning the values of the function $\backslash\xi$, and our basic law (VI) holds now as it did before.

It remains to be asked how the new conception of the value-range influences the values of our function $\xi \frown \zeta$. In the case where Γ is a value-range, it is now no longer determined in every case what value a function whose value-range is Γ has for the argument Θ;[1] in particular, it is not determined in the case where Θ coincides with Γ. There can then be functions which have the same value-range Γ but have different values for the argument Γ. The extension of the concept

$$\smile^{\mathfrak{g}}\!\!\top\!\!\left[\begin{array}{l}\mathfrak{g}(\Gamma) = \xi \\ \Gamma = \grave{\varepsilon}\,\mathfrak{g}(\varepsilon)\end{array}\right.$$

can now no longer coincide with the extension of a concept $\Delta = \xi$, since Δ is the unique object falling under the latter, while all objects fall under the former. For if Γ is a value-range and E is an object, it will always be possible to supply a function $\mathrm{X}(\xi)$ such that $\grave{\varepsilon}\,\mathrm{X}(\varepsilon) = \Gamma$ and $\mathrm{X}(\Gamma) = \mathrm{E}$. According to the stipulation of I, §11,

$$\backslash\grave{\alpha}\left(\smile^{\mathfrak{g}}\!\!\top\!\!\left[\begin{array}{l}\mathfrak{g}(\Gamma) = \alpha \\ \Gamma = \grave{\varepsilon}\,\mathfrak{g}(\varepsilon)\end{array}\right.\right)$$

conincides with

$$\grave{\alpha}\left(\smile^{\mathfrak{g}}\!\!\top\!\!\left[\begin{array}{l}\mathfrak{g}(\Gamma) = \alpha \\ \Gamma = \grave{\varepsilon}\,\mathfrak{g}(\varepsilon)\end{array}\right.\right)$$

Accordingly if Γ is a value-range, then

$$\Gamma\frown\Gamma = \grave{\alpha}\left(\smile^{\mathfrak{g}}\!\!\top\!\!\left[\begin{array}{l}\mathfrak{g}(\Gamma) = \alpha \\ \Gamma = \grave{\varepsilon}\,\mathfrak{g}(\varepsilon)\end{array}\right.\right)$$

i.e., $\Gamma\frown\Gamma$ is the extension of an all-encompassing concept. If Γ is not a value-range, then $\Gamma\frown\Gamma$ is the extension of an empty concept. In the first case, $—\,\Gamma\frown\Gamma$ is the False:

$$\vdash\!\top\, \grave{\varepsilon}f(\varepsilon)\frown\grave{\varepsilon}f(\varepsilon) \qquad (\alpha'$$

This is important for the function $\wp\xi$. One might initially fear that according to our stipulations concepts with the same extension have to receive the same cardinal number, even though one more object falls under the one than under the other, namely the extension of the concept itself, so that in the end one would only obtain a single finite cardinal number. However, the concept $\Phi(\xi)$ does not come into consideration as far as $\wp\,\grave{\varepsilon}\,\Phi(\varepsilon)$ is concerned, but rather $—\,\xi\frown\grave{\varepsilon}\,\Phi(\varepsilon)$, and the extension $\grave{\varepsilon}\,\Phi(\varepsilon)$ does not fall under the latter even though it falls under the concept $\Phi(\xi)$.

If one repeats the derivation of (1) (I, §55) using (V′b) instead of (Vb), then instead of (1) one obtains the proposition (1′):

$$\vdash\!\!\left[\begin{array}{l}f(a) = a\frown\grave{\varepsilon}f(\varepsilon) \\ \top\, a = \grave{\varepsilon}f(\varepsilon)\end{array}\right. \qquad (1'$$

from which, instead of (77) and (82), the propositions (77′) and (82′) are to be derived:

$$\vdash\!\!\left[\begin{array}{l}F(a\frown\grave{\varepsilon}f(\varepsilon)) \\ F(f(a)) \\ \top\, a = \grave{\varepsilon}f(\varepsilon)\end{array}\right. \quad (77' \qquad\qquad \vdash\!\!\left[\begin{array}{l}F(f(a)) \\ F(a\frown\grave{\varepsilon}f(\varepsilon)) \\ \top\, a = \grave{\varepsilon}f(\varepsilon)\end{array}\right. \quad (82'$$

We now draw some further consequences.

$$\alpha' \vdash\!\top\, \grave{\varepsilon}f(\varepsilon)\frown\grave{\varepsilon}f(\varepsilon)$$

(IIIa): ———————

$$\vdash\!\!\left[\begin{array}{l}\top\, a\frown\grave{\varepsilon}f(\varepsilon) \\ a = \grave{\varepsilon}f(\varepsilon)\end{array}\right. \qquad (\beta'$$

(Ia): – – – – – –

$$\vdash\!\!\left[\begin{array}{l}f(a) \\ a\frown\grave{\varepsilon}f(\varepsilon) \\ a = \grave{\varepsilon}f(\varepsilon)\end{array}\right.$$

(82′): ·—·—·—·—·—·

[1] Compare I, p. 53.

$$\begin{array}{l} \vdash\!\!\top f(a) \\ \quad\llcorner\, a \frown \dot{\varepsilon} f(\varepsilon) \end{array} \qquad (82''$$

———— • ————

$$82'' \begin{array}{l} \vdash\!\!\top\!\top \dot{\varepsilon}(\top\, \varepsilon \frown \varepsilon) \frown \dot{\varepsilon}(\top\, \varepsilon \frown \varepsilon) \\ \quad\llcorner\, \dot{\varepsilon}(\top\, \varepsilon \frown \varepsilon) \frown \dot{\varepsilon}(\top\, \varepsilon \frown \varepsilon) \end{array}$$

$$\text{(Ig):} \overline{\qquad\qquad\qquad\qquad\qquad\qquad}$$

$$\vdash\!\!\top \dot{\varepsilon}(\top\, \varepsilon \frown \varepsilon) \frown \dot{\varepsilon}(\top\, \varepsilon \frown \varepsilon) \qquad (\gamma'$$

This follows in exactly the manner of (ι) above. No contradiction arises here, however, as we will shortly see. (γ') is merely a special case of (α').

$$77' \begin{array}{l} \vdash\!\!\top\!\top\!\!- \dot{\varepsilon}(\top\, \varepsilon \frown \varepsilon) \frown \dot{\varepsilon}(\top\, \varepsilon \frown \varepsilon) \\ \quad\llcorner\!\top \dot{\varepsilon}(\top\, \varepsilon \frown \varepsilon) \frown \dot{\varepsilon}(\top\, \varepsilon \frown \varepsilon) \\ \quad\llcorner\!\top \dot{\varepsilon}(\top\, \varepsilon \frown \varepsilon) = \dot{\varepsilon}(\top\, \varepsilon \frown \varepsilon) \end{array}$$

$$\times$$

$$\begin{array}{l} \vdash\!\!\top\!\!- \dot{\varepsilon}(\top\, \varepsilon \frown \varepsilon) = \dot{\varepsilon}(\top\, \varepsilon \frown \varepsilon) \\ \quad\llcorner\!\top \dot{\varepsilon}(\top\, \varepsilon \frown \varepsilon) \frown \dot{\varepsilon}(\top\, \varepsilon \frown \varepsilon) \end{array} \qquad (\delta'$$

$$(\gamma')\text{::} \overline{\qquad\qquad\qquad\qquad\qquad\qquad}$$

$$\vdash \dot{\varepsilon}(\top\, \varepsilon \frown \varepsilon) = \dot{\varepsilon}(\top\, \varepsilon \frown \varepsilon) \qquad (\varepsilon'$$

(ε') is a special case of (IIIe). A contradiction did not arise.

It would take us too far here to pursue the consequences of replacing (V) by (V′) any further. One of course cannot but acknowledge that subcomponents must be added to many propositions; but one surely need not worry that this will raise essential obstacles for the conduct of proof. In any case, an inspection of all hitherto established propositions will be required.

This question may be viewed as the fundamental problem of arithmetic: how are we to apprehend logical objects, in particular, the numbers? What justifies us to acknowledge numbers as objects? Even if this problem is not solved to the extent that I thought it was when composing this volume, I do not doubt that the path to the solution is found.

Jena, October 1902.

Index

The numerals specify the pages.

class 159.
class of finite cardinal numbers 161.
domain (Σ-domain) 169.
domain of magnitudes 158, 160.
first 2.
greater 185.
limit (Σ-limit of Φ) 187.
logical object 86, 149.
member (occurring as first, second member of a relation) 171.
null Relation 176.
positival class 171.
positive class 189.
Relation 160.

Translators' Notes

Notes to the Foreword

a *Anzahl* — See our Translators' Introduction.

b *anschauendes Erkennen.*

c *Erkenntnissthat.*

d "This is metaphysics, it will not be read!" — Frege alludes to the phrase, "*graeca sunt, non leguntur*": "This is Greek, it will not be read". The phrase was used to indicate that a certain passage was skipped because it was deemed too difficult to understand. It is echoed in the phrase, "It is all Greek to me".

e *festsetzen* — We diverged from our usual translation, "stipulate", since Frege uses "*festsetzen*" in a different sense here. See also our Translators' Introduction.

f *Vorstellender* — For the translations of "*Vorstellung*" ("idea"), and its cognates, "*Vorstellender*" ("bearer of ideas"), "*vorstellen*" ("ideate"), "*Vorstellen*" ("ideation"), etc., see our Translators' Introduction.

g *Vorstellens* — See our note f above.

h *Vorgestelltem* — See our note f above.

i Frege omits the Greek and Latin terms that Erdmann notes: "(*ὑποκείμενον*; *pars subjectiva, subdita; subjectum*)" after "subject", and "(*κατηγορούμενον*; *pars declarativa, praedicatum*)" after "predicate".

j The full passage in Erdmann's *Logik* (1892), p. 36, including the original emphases, is:

> "Das Vorstellen setzt sich zusammen aus den *Vorstellungen*, in denen uns Gegenstände gegeben werden, und den *Vorstellungsverläufen*, durch welche dieselben ihren Associationsbeziehungen gemäss erinnert, neu verknüpft oder prädikativ zerlegt werden."

English translation:

> "Ideation is composed of the *ideas* in which objects are given to us and the *passages of ideas* by which these are remembered, newly combined, or predicatively analysed according to their associative relations."

k Frege leaves out Erdmann's helpful example for a perception-mass (*Perceptionsmasse*): "*etwa das Geräusch eines vorrüberfahrenden Wagens*", i.e., "for example, the sound of a passing carriage".

l Frege leaves out Erdmann's additional examples:

> "... den wir als süss, braun, warm empfinden, als dreieckig, entfernt

> wahrnehmen, durch Vorstellungen gegeben ist, die selbst, süss, braun, warm, dreieckig, entfernt sind."

English translation:

> "... that we experience as sweet, brown, warm, perceive as triangular, distant, is presented by ideas which themselves are sweet, brown, warm, triangular, distant."

m The full sentence reads:

> "Das Transscendente soll dabei nicht als das Unerkennbare, nur in einer Grenzvorstellung ohne Wesensbestimmung Erreichbare (85) angenommen werden, sondern seine Transscendenz soll nur in der Unabhängigkeit vom Vorgestelltwerden bestehen."

English translation:

> "Here, the transcendent should not be regarded as the unknowable, reachable only in a limit-idea without determination of its essence (85), rather its transcendence is only to consist in its independence from being ideated."

n *Erkennen.*

o In the original, it says, "*die wir im Deutschen mit "es gibt" aussprechen*". We took the freedom to change "German" into "English".

p Frege uses the word "*Guckkasten*" here which we translate as "peep box", rather than "peepshow" which has been used in previous translations. Peep boxes are (often stereoscopic) devices into which pictures are inserted, which are then looked at through one or two holes in the front. These boxes were very popular around the turn of the twentieth century and are still to be found as children's toys today. "Peep box" was apparently the term most commonly used at the time in English, and it is devoid of the risqué connotations associated with "peepshow".

Notes to Part I

a Frege's use of words with the same stem, "*andeuten*" ("indicate") and "*bedeuten*" ("refer"), is unfortunately lost in translation. Compare also our Translators' Introduction.

b We added the missing italics, since Frege refers to the book here, rather than the system.

c Compare our note a above.

d Furth suggests to correct "proper name" into "function-name"; see Furth (1964), pp. 85–86 fn. It is not, in fact, a mistake; see Richard G. Heck, Jr (2012). *Reading Frege's* Grundgesetze. Oxford: Clarendon Press, p. 65, fn. 22.

e *beiderseits eindeutig zuordnet* — Austin renders this phrase as "correlates one to one" in his translation of *Grundlagen* (Frege, 1950). Compare our Translators' Introduction regarding our translation of "*eindeutig*" as "single-valued".

f For the remainder of this section, we have omitted periods and semicolons occurring in the original text adjacent to formulae and transition-signs.

g *Im Betreff der Wörter 'denselben' und 'dieselbe'* — The gender difference of the German words is lost in translation.

Notes to Part II

a Note that the German is "*Nummern*", not "*Zahlen*".

b The original uses semicircles, ${}_{\scriptscriptstyle(}p^{\scriptscriptstyle)}$, for quotation-marks *within* quotation marks here and below. This is not generally followed in the original.

c Regarding the quotation marks, see our note b above.

d We here follow Frege's suggestion noted in his *corrigenda* at the end of the first volume:

> "The proposition which has received the index 'κ' in the second column of p. 168 should have received as index a numeral—for example, '179a'—according to the principles adhered to here and should have been cited in the proof of proposition (181) with this index in the first column of p. 169." (Frege 1893, p. 254)

e The original misses out "$(b; a)$ *muss*" ("$(b; a)$ must") in this sentence. We assume that this is a slip.

f This non-standard way of counting sections is licensed by a simple application of Basic Law V. It is left as an exercise to the reader.

Notes to Part III.1, a–c (§§55–85)

a Frege quotes Peano from the Italian original, without German translation.

b Frege omits "*auch*" ("also") in this quotation from Cantor (Cantor (1883), p. 567; Cantor (1932), p. 186): the passage should read "which can also be characterised".

c Cantor's emphasis on "fundamental series" is neglected by Frege.

d Frege misses out the word "*positive*" in "*positive rationale Zahl*" ("positive rational number") in this quotation from Cantor (compare Cantor (1883), p. 567, or Cantor (1932), p. 186) and in the ensuing discussion. Frege's quotation of this passage is somewhat sloppy in other respects too: 'a_ν' should not be enclosed in parentheses, and "from a definite ν on" ("*von einem bestimmten* ν *an*") should be "from a certain ν on" ("*von einem gewissen* ν *an*") as immediately above; Frege also takes liberties regarding the punctuation.

e *Erkenntnis*.

f The German words are "*Figuren*" ("figures") and "*Schachfiguren*" ("chess pieces"). This analogy is unfortunately lost in translation. Below (§95ff), Frege also uses "*Zahlfiguren*" ("number-figures").

g *dass das Grössere eine Grösse sei* — Frege's play on the words "*das Grössere*" ("what is greater") and "*eine Grösse*" ("a magnitude") unfortunately cannot be captured in the English translation.

h Frege leaves out Cantor's parenthetical remark, "where b' can also be $= a$", which would here become "where b' can also be $\backsim\!\!\!\!\backsim a$". Compare the original quotation in §81 above.

Notes to Part III.1, d–g (§§86–164)

a *was sind und was wollen die Zahlen* — Thomae intends to allude to Dedekind's "Was sind und was *sollen* die Zahlen?" (our emphasis) but misquotes.

b *Figuren* — See our note f to Part III.1, a–c, above.

c *Schachfiguren / Zahlfiguren* — See our note f to Part III.1, a–c, above.

d There is only one word, "*Figuren*", in the original. Since this is ambiguous between *Schachfiguren* and *Zahlfiguren* (see our note f to Part III.1, a–c, above) we had to use the disjunction, "pieces or figures", in the translation.

e *Erkenntnis.*

f Compare our note a above.

g *formale Bedeutung* — See our Translators' Introduction.

h Frege is quoting the title of Alfred Bochert's article, "*Ueber die Zahl der verschiedenen Werthe, die eine Function gegebener Buchstaben durch Vertauschung derselben erlangen kann*" (Bochert 1889). The emphasis on "letters" ("*Buchstaben*") is Frege's.

i The impossibility to capture the ambiguity of the German "*Figuren*" in English ("figures" / "pieces") is particularly painful here. Compare our note f to Part III.1, a–c, above.

j *Erkenntniszweck.*

k *niedern Arithmetik* — For Frege, this is arithmetic of the natural numbers.

l [Left out to prevent confusion with '1' as a mark for Frege's footnote on p. 119.]

m *Bedeutung* — See our Translators' Introduction.

n "*Gemeine Zahlen*" is Thomae's expression for rational numbers (see Thomae (1898), §3, pp. 3–4), which however does not appear to have been widely used. The literal translation of "*gemein*" is "common".

o *gemeine und nicht gemeine* — See our note n above.

p *Häuserreihe* — The analogy between "*Häuserreihe*" ("row of houses") and "*Zahlenreihe*" ("number-series") is unfortunately lost in translation.

q *bedeuten* — See our Translators' Introduction.

r *alternirende Zahlen* — Alternating numbers were introduced by Hankel in his *Vorlesungen über complexe Zahlen* of 1867, section VII. Hankel (1867, p. 140) gives credit to Grassmann: "The system of alternating numbers and operations is already contained in Grassmann's *Ausdehnungslehre* of 1844, albeit in a presentation that is very abstract and difficult to understand."

s These handwritten notebooks are transcriptions of Weierstrass's lectures on the theory of functions (*"Functionentheorie"*), which appear to have been in wide circulation. They were treated as an authoritative source; compare, for instance, Benno Kerry (1886), "Ueber Anschauung und ihre psychische Verarbeitung", part 2, *Vierteljahrsschrift für wissenschaftliche Philosophie*, 10:419–467, p. 436, fn. 1: "I take the following lines from a notebook that was recommended and made available to me by Mr *Oberlehrer* Dr M. Simon (Strassburg), a former student of Weierstrass; given the wide circulation of the Weierstrassian notebooks, I believe it to be permissible to treat them like a printed work and use them for my purposes" (original in German, our translation). Kerry also quotes (*ibid.*) the passage from the *Collegienhefte* that Frege criticises below. Transcriptions of Weierstrass's lectures on the theory of functions are still in existence. For example, a notebook of 1876, attributed to the Mathematischer Verein Berlin, is now held at the University of Chicago and contains the exact passage quoted by Kerry. (Thanks to Sean Walsh and Dirk Schlimm for helping to locate the manuscript.)

t *Relation*, not: *Beziehung* — Compare our note v below.

u The passage Frege quotes from Gauss's *Werke* in fact starts in the last line of

p. 175 (Gauss 1863). In the last sentence of the quotation, it should read "... each real whole number represents a Relation ...". The "*ganze*" ("whole") in "*reelle ganze Zahl*" ("real whole number") that is present in Gauss's text is left out by Frege.

v *Relation* — We translate Frege's "*Relation*" as "Relation" (capitalised), and "*Beziehung*" as "relation" (not capitalised). Compare also our Translators' Introduction.

Notes to Part III.2

a See Frege's note at the end of the main text of this volume, p. 243.

Notes to the Afterword

a "It is comfort to the wretched to have companions in misery" — The Latin proverb traditionally starts with the synonymous "*Solamen*" instead.

b Frege misses out the negation in the German, thus expressing that the class *does* belong to itself. As Geach and Black note, this "is clearly a slip" (Geach & Black (1980), p. 217, fn. E).

c Another slip in Frege: he paraphrases the conditional the wrong way around (Geach & Black (1980), p. 218, fn. H).

Corrections

The following is a list of typos and misprints that are present in the original publication of *Grundgesetze* which we corrected in our translation. Where the typos were noticed by someone other than ourselves, the name of the finder is given in square brackets. ‘Frege 1’ refers to Frege’s own *corrigenda* from volume 1 of *Grundgesetze*, ‘Frege 2’ to those from volume 2; ‘Thiel’ marks Christian Thiel’s correction from the 1998 reprint of *Grundgesetze*; ‘Bhowmick’, ‘Siu’, and ‘Stirton’, indicate that the error in question was first brought to our attention by Nilanjan Bhowmick, Charlie Siu, and William Stirton, respectively: we are indebted to all of them for their careful reading. Compare also our section “General remarks on the translation” in the Translators’ Introduction.

Volume 1

XXX↓3	‘⟩’, in front of ‘ἀἐ’ at the end of the supercomponent missing in the original (compare proposition (84), p. 120)
11↑8	the original has ‘$x\,.\,(x-1) = x^2 - 1$’ instead of ‘$x\,.\,(x-1) = x^2 - x$’ [Thiel]
13↓15	the original has ‘$(\smile^{\mathfrak{a}}\, \xi + \mathfrak{a})$’ instead of ‘$(\smile^{\mathfrak{a}}\, \xi = \mathfrak{a})$’ [Thiel]
13↓16	the original has ‘$(\smile^{\mathfrak{a}}\, \mathfrak{a} + \xi)$’ instead of ‘$(\smile^{\mathfrak{a}}\, \mathfrak{a} = \xi)$’ [Thiel]
13↓17	the original has ‘$(\smile^{\mathfrak{a}}\, \mathfrak{a} + \mathfrak{a})$’ instead of ‘$(\smile^{\mathfrak{a}}\, \mathfrak{a} = \mathfrak{a})$’ [Thiel]
13↓20	the original has ‘$(\smile^{\mathfrak{a}}\, \mathfrak{a} + \mathfrak{a})$’ instead of ‘$(\smile^{\mathfrak{a}}\, \mathfrak{a} = \mathfrak{a})$’ [Thiel]
17↑12	the original has ‘$\Phi(\xi)$’ instead of ‘$\mathrm{X}(\xi)$’ [Thiel]
20↑2	the original has “*das Wahre*” (“the True”) instead of “*das Falsche*” (“the False”) [Frege 2]
23↓18	original has ‘$4 > 2$’ instead of ‘$3 > 2$’ [Thiel]
24↓15	original has ‘— 1’ (using the horizontal) instead of ‘-1’ (using the negation sign) [Thiel]
24↓16	original has ‘— 1’ (using the horizontal) instead of ‘-1’ (using the negation sign) [Thiel]
24↓17	original has ‘— 1’ (using the horizontal) instead of ‘-1’ (using the negation sign) [Thiel]
24↓18	original has ‘— 1’ (using the horizontal) instead of ‘-1’ (using the negation sign) [Thiel]

25↓17 original has '$2+3=4$' in the subcomponent instead of '$2+2=4$' [Thiel]

26b↑7 original has '$\vdash\!\top\Gamma$ / Θ' instead of '$\top\Gamma$ / Θ' [Thiel]

31b↑8, 11 the first premiss, '$\vdash\!\top x^8=1$ / $x^4=1$', and the conclusion, '$\vdash\!\top x^8=1$ / $x^2=1$', are swapped in Frege [Thiel suggests a different resolution]

53↓10 in the original, the quotation marks around 'u' are missing

53↑1 a misprint in the original makes '$\dot{\varepsilon}(\top\,\varepsilon=\varepsilon)$' look like '$\dot{\varepsilon}(\!-\!-\,\varepsilon=\varepsilon)$' [Thiel]

55b↑14 original has '$-\!-\,\xi\frown(\xi\frown\Delta)$' instead of '$-\!-\,\xi\frown(\zeta\frown\Delta)$' [Thiel]

55b↑4 original has '$\not p\varepsilon(\top\,\varepsilon=\varepsilon)$' instead of '$\not p\dot{\varepsilon}(\top\,\varepsilon=\varepsilon)$' [Thiel]

73a↑12 the Olms reprint, but not the original, misses out the opening quotation mark and the judgement-stroke of (ψ)

73b↓7 original has 'β' instead of 'β'' as index [Frege 2]; this is corrected in the reprints [Thiel]

77b↓15 original has '(Ie):' instead of '(Ie)::' [Frege 2]

86↓4 original has '$\not\mathfrak{f}p$' instead of '$\not\mathfrak{f}q$' in the lowest subcomponent of proposition (δ) [Frege 2]

86↓26–27 the original has the subcomponent '$\dot{\varepsilon}\left(\top\!\overset{\mathfrak{q}}{\smile}\!\top\,\varepsilon\frown(\upsilon\frown\rangle\mathfrak{q}) \;/\; \upsilon\frown(\varepsilon\frown\rangle\not\mathfrak{f}\mathfrak{q})\right)=\not p\upsilon$' in (ϑ) instead of '$\dot{\varepsilon}\left(\top\!\overset{\mathfrak{q}}{\smile}\!\top\!\top\,\varepsilon\frown(\upsilon\frown\rangle\mathfrak{q}) \;/\; \upsilon\frown(\varepsilon\frown\rangle\not\mathfrak{f}\mathfrak{q})\right)=\not p\upsilon$' [Frege 2]; this is corrected in the reprints [Thiel]

92↑9 original has '$\vdash f(a=a\frown\dot{\varepsilon}f(\varepsilon))$' instead of '$\vdash f(a)=a\frown\dot{\varepsilon}f(\varepsilon)$'; [Frege 2: superfluous parenthesis at the end, which is deleted in some reprints; Thiel: missing parenthesis after the first occurrence of 'a']

93↓8 original has '47::' instead of '(47)::' [Siu]

100↑1 original has the transition sign '$=\!=\!=\!=\!=\!=$' upside down

104a↓12 original has '$(\top\,\alpha\frown\dot{\varepsilon}(\top\,f(\varepsilon)))$' instead of '$(\top\,a\frown\dot{\varepsilon}(\top\,f(\varepsilon)))$' [Frege 2]

104b↓6 original has '$\top\!\top\,\varepsilon=c$ / $\varepsilon=\upsilon$' instead of '$\top\!\top\,\varepsilon=c$ / $\varepsilon\frown\upsilon$' in the supercomponent of (γ) [Frege 2]

104b↓18 original has '$\top\!\top\,\varepsilon=c$ / $\varepsilon=\upsilon$' instead of '$\top\!\top\,\varepsilon=c$ / $\varepsilon\frown\upsilon$' in the supercomponent [Frege 2]

109a↓16 original has '$\overset{\mathfrak{a}}{\smile}\,b\frown(\mathfrak{a}\frown q)$' instead of '$\overset{\mathfrak{a}}{\smile}\!\top\,b\frown(\mathfrak{a}\frown q)$' in lowest subcomponent of (63) [Frege 2]

111↓6–7 original has '$\dot{\varepsilon}\left(\top\!\top\,\varepsilon=c \;/\; \varepsilon=\upsilon\right)$' instead of '$\dot{\varepsilon}\left(\top\!\top\,\varepsilon=c \;/\; \varepsilon\frown\upsilon\right)$' in (μ) [Frege 2]

118b↓9 original has '8:' instead of '(8):' (Siu)

124b↑2 the reprint, but not the original, misses out the index (η) [Thiel]

126b↑9 original has '(IIIc)::' instead of '(IIIc):' [Frege 2]

129b↓4 original has '$(-\!-\,\mathfrak{a}\frown(\dot{\varepsilon}\top\,\varepsilon=\varepsilon))$' instead of '$(-\!-\,\mathfrak{a}\frown\dot{\varepsilon}(\top\,\varepsilon=\varepsilon))$' in the supercomponent [Bhowmick]

135a↓24 original lacks the negation-stroke in the subcomponent of (σ) [Frege 2]

142a↑16 original has '$a \frown (m \,\underline{\smile}\, q)$' instead of $a \frown (m \frown \underline{\smile}\, q)$' in the supercomponent of (α) [Frege 2]

147a↓5 original has '102):' instead of '(102):'

152a↑13–14 the subcomponent '— $\mathrm{I}\mathfrak{P}q$' is missing in the first formula of §124 in the original [Frege 1]

154a↓3 original has '$\mathfrak{p}\,\mathfrak{Q} \frown \mathfrak{P} \smile \mathfrak{f}) = \backsim$' instead of '$\mathfrak{p}(\mathfrak{Q} \frown \mathfrak{P} \smile \mathfrak{f}) = \backsim$' in the lowest supcomponent of (ε) [Thiel]

156b↑12 original has '$a \frown c$' instead of '$a = c$' in (ϑ) [Frege 2]

162b↑11–13 original has '$\begin{array}{l} \smile^{\mathfrak{d}}\smile^{\mathfrak{m}}\smile^{\mathfrak{a}}\; \mathfrak{m} \frown (\mathfrak{a} \frown t) \\ \qquad\quad \mathfrak{d} \frown (\mathfrak{a} \frown q) \\ \qquad\quad \mathfrak{m} \frown (\mathfrak{d} \frown t) \end{array}$' instead of '$\begin{array}{l} \smile^{\mathfrak{d}}\smile^{\mathfrak{m}}\smile^{\mathfrak{a}}\; \mathfrak{m} \frown (\mathfrak{a} \frown t) \\ \qquad\quad \mathfrak{d} \frown (\mathfrak{a} \frown q) \\ \qquad\quad \mathfrak{m} \frown (\mathfrak{d} \frown t) \end{array}$' as subcomponent in ($\delta$) [Stirton]

168b↑20 original has '(κ' as index instead of '(179a' [Frege 1]

169a↓12 original cites '(κ)' instead of '(179a)' [Frege 1]

189b↓7 original has '$m; x \frown (c; d \frown (p \,\underline{\smile}\, q))$' instead of '$m; x \frown (c; d \frown \underline{\smile}\, (p \,\underline{\smile}\, q))$' in the lowest subcomponent of (ε) [Stirton]

193a↑ original has '(200::' instead of '(200)::'

196↑5–6 original has '$\begin{array}{l} \vdash\!\top\; b \frown (b \frown \smile \mathfrak{f}) \\ \;\; \llcorner\; \mathfrak{Q} \frown (b \frown \smile \mathfrak{f}) \end{array}$' instead of '$\begin{array}{l} \vdash\!\top\; b \frown (b \frown \underline{\smile}\, \mathfrak{f}) \\ \;\; \llcorner\; \mathfrak{Q} \frown (b \frown \smile \mathfrak{f}) \end{array}$' [Frege 1]

196↑2–3 original has '$\begin{array}{l} \vdash\!\smile^{\mathfrak{i}}\!\top\; \mathfrak{i} \frown (\mathfrak{i} \frown \smile \mathfrak{f}) \\ \qquad \llcorner\; \mathfrak{Q} \frown (\mathfrak{i} \frown \underline{\smile}\, \mathfrak{f}) \end{array}$' instead of '$\begin{array}{l} \vdash\!\smile^{\mathfrak{i}}\!\top\; \mathfrak{i} \frown (\mathfrak{i} \frown \underline{\smile}\, \mathfrak{f}) \\ \qquad \llcorner\; \mathfrak{Q} \frown (\mathfrak{i} \frown \smile \mathfrak{f}) \end{array}$' [Frege 1]

216↑19 original has '$\mathfrak{p}(x; y \,\underline{\measuredangle}\, y) = \mathfrak{p}(x; y \,\underline{\measuredangle}\, (y \frown \smile q \,\underline{\frown}\, q))$' as a supercomponent of (295) instead of '$\mathfrak{p}(x; y \,\underline{\measuredangle}\, q) = \mathfrak{p}(x; y \,\underline{\measuredangle}\, (y \frown \smile q \,\underline{\frown}\, q))$' [Frege 1]

241↓5 original has '57' instead of '56' as a page-reference [Thiel]

241↓7 original has '56' instead of '57' as a page-reference [Thiel]

Volume 2

VIII↓12 original has "*ungültige*" ("invalid") instead of "*entgültige*" ("final") [Thiel]

11↑7 original has '$\mathfrak{e} \frown m; d \,\underline{\measuredangle}\, q)$' instead of '$\mathfrak{e} \frown (m; d \,\underline{\measuredangle}\, q)$' in proposition ($\varepsilon$)

24↓6 original has '$m \frown (c \frown (v \,\mathfrak{z}\, q))$' instead of '$m \frown (c \frown \smile (v \,\mathfrak{z}\, q))$' in the lowest subcomponent of (γ) [Stirton]

28↑14 original has '$d \frown (r \frown (v \frown \underline{\smile}\, q))$' instead of '$d \frown (r \frown (v \,\underline{\frown}\, \underline{\smile}\, q))$' in the lowest subcomponent of (π) [Stirton]

30↓9 original has '$\frown \underline{\smile}\, \mathfrak{f}$' instead of '$\frown \smile \mathfrak{f}$' at the end of the formula [Stirton]

38b↓19 original has '(188):' instead of '(188)::' [Stirton]

58a↑20 '$\grave{\varepsilon}$' missing in front of '$\left(\begin{array}{l} \top\; \varepsilon \frown u \\ \llcorner\; \varepsilon \frown w \end{array} \right)$' in the original

59↑4–6 original has large closing parenthesis missing in the supercomponent of proposition (ζ) [Stirton]

60↓1–3 original has large closing parenthesis missing in the supercomponent of proposition (η) [Stirton]

66↓16 original has '$\varepsilon \frown c$' instead of '$\varepsilon = c$' in the subcomponent of the most deeply embedded conditional of proposition (β) [Stirton]

113↓12 original has '*nega*' instead of '$neg(a)$'

150↑15 original has "color *oder* pigmentum" ("*color* or *pigmentum*") instead of "color *mit* pigmentum" ("*color* with *pigmentum*") [Frege 2]

169b↑11 original has '$q = s$' instead of '$q \frown s$' in the lowest subcomponent of proposition (509) [Stirton]

169b↑2 original has '$q = s$' instead of '$q \frown s$' in the lowest subcomponent [Stirton]

183a↓5 original has '$q \smile s$' instead of '$q \frown s$' in the lowest subcomponent of proposition (567) [Frege 2]

200b↑3 original has '$a \frown (p \sqcup p \frown (s \mathfrak{g}\, p)))$' instead of '$a \frown (p \sqcup p \frown (s \mathfrak{g}\, p))$' [Frege 2]; this is corrected in the reprints [Thiel]

215↓9 original has '$— \mathfrak{F}(p \sqcup q) \sqcup (o \sqcup a)$' instead of '$— \mathfrak{F}(p \sqcup q) \sqcup (o \sqcup a) \frown s$'

256↓17 original has '$\dot{\varepsilon}(— f(\varepsilon)$' instead of '$\dot{\varepsilon}(— f(\varepsilon))$' [Thiel]

Bibliography

The following is a bibliography of the works cited by Frege and, where available, their English translations. Also included are references to editions of *Grundgesetze* and previous partial translations. A complete bibliography of Frege's works and their English translations is available as an appendix to Edward N. Zalta (2012). Gottlob Frege. In Edward N. Zalta, editor, *The Stanford Encyclopedia of Philosophy* (Winter 2012 Edition). `http://plato.stanford.edu/archives/win2012/entries/frege/`.

Baumann, Johann Julius (1868). *Die Lehren von Raum, Zeit, und Mathematik in der neueren Philosophie: nach ihrem ganzen Einfluss dargestellt und beurteilt.* Vol. 1*: Suarez, Descartes, Spinoza, Hobbes, Locke, Newton.* Berlin: Georg Reimer.

Beaney, Michael, ed. (1997). *The Frege Reader.* Oxford: Blackwell.

Biermann, Otto (1887). *Theorie der analytischen Functionen.* Leipzig: B. G. Teubner.

Bochert, Alfred (1889). 'Ueber die Zahl der verschiedenen Werthe, die eine Function gegebener Buchstaben durch Vertauschung derselben erlangen kann', *Mathematische Annalen* XXXIII:584–590.

Cantor, Georg (1872). 'Ueber die Ausdehnung eines Satzes aus der Theorie der trigonometrischen Reihen', *Mathematische Annalen* V:123–132. Reprinted in Cantor (1932), pages 92–101.

——— (1883). 'Ueber unendliche, lineare Punktmannichfaltigkeiten. 5. Fortsetzung: Grundlagen einer allgemeinen Mannigfaltigkeitslehre', *Mathematische Annalen* XXI:545–590. Reprinted in Cantor (1932), pages 165–204; English translation by William Ewald in Ewald (1996), vol. II, pages 881–920.

——— (1889). 'Bemerkung mit Bezug auf den Aufsatz: Zur Weierstrass'-Cantor'schen Theorie der Irrationalzahlen in Math. Annalen. Bd. XXXIII, p. 154', *Mathematische Annalen* XXXIII:476. Reprinted in Cantor (1932), page 114.

——— (1932). *Gesammelte Abhandlungen mathematischen und philosophischen Inhalts.* Edited by Ernst Zermelo. Berlin: Springer.

Dedekind, Richard (1872). *Stetigkeit und Irrationale Zahlen.* Braunschweig: Friedrich Vieweg & Sohn. English translation by Wooster W. Beman as 'Continuity and Irrational Numbers' in Dedekind (1901); reprinted, with corrections by William Ewald, in Ewald (1996), vol. II, pages 765–779.

Dedekind, Richard (1888). *Was sind und was sollen die Zahlen?*, 1st edition. Braunschweig: Friedrich Vieweg & Sohn. English translation of the second edition (1893) by Wooster W. Beman as 'The Nature and Meaning of Numbers' in Dedekind (1901); reprinted, with corrections by William Ewald, in Ewald (1996), vol. II, pages 787–833.

——— (1901). *Essays on the Theory of Numbers.* Translated by Wooster W. Beman. Chicago: Open Court. Reprinted by Dover, New York, 1963.

Erdmann, Benno (1892). *Logik.* Halle a. d. Saale: Max Niemeyer.

Ewald, William Bragg, ed. (1996). *From Kant to Hilbert: A Source Book in the Foundations of Mathematics.* 2 vols. Oxford: Oxford University Press.

Feigl, Herbert and Wilfrid Sellars, eds. (1949). *Readings in Philosophical Analysis.* New York: Appleton–Century–Crofts.

Frege, Gottlob (1879). *Begriffsschrift: Eine der arithmetischen nachgebildete Formelsprache des reinen Denkens.* Halle a. d. Saale: Verlag L. Nebert. Reprinted in Frege (1964a). English translation by S. Bauer-Mengelberg in van Heijenoort (1967), pages 1–82; and by T. W. Bynum in Frege (1972).

——— (1884). *Die Grundlagen der Arithmetik. Eine logisch mathematische Untersuchung über den Begriff der Zahl.* Breslau: Wilhelm Koebner. English translation: Frege (1950).

——— (1885). 'Ueber formale Theorien der Arithmetik', *Jenaische Zeitschrift für Naturwissenschaft* 19 Supplement II (1886) (= Sitzungberichte der Jenaischen Gesellschaft für Medizin und Naturwissenschaft für das Jahr 1885):94–104. Reprinted in Frege (1967), pages 103–111; English translation by E.-H. W. Kluge as 'On Formal Theories of Arithmetic' in Frege (1984), pages 112–121.

——— (1891). *Function und Begriff: Vortrag, gehalten in der Sitzung vom 9. Januar 1891 der Jenaischen Gesellschaft für Medizin und Naturwissenschaft.* Jena: Hermann Pohle. Reprinted in Frege (1967), pages 125–142; English translation as 'Function and Concept' in Frege (1984), pages 137–156, and Beaney (1997), pages 130–148.

——— (1892a). 'Über Sinn und Bedeutung', *Zeitschrift für Philosophie und philosophische Kritik* 100:25–50. Reprinted in Frege (1967), pages 143–162; English translation by Max Black as 'Sense and Reference' Frege (1948), revised as 'On Sense and Meaning' in Geach and Black (1980), pages 56–78, and in Frege (1984), pages 157–177, and with revisions by Michael Beaney as 'On *Sinn* and *Bedeutung*' in Beaney (1997), pages 151–171; also translated by Herbert Feigl as 'On Sense and Nominatum' in Feigl and Sellars (1949), pages 85–102.

——— (1892b). 'Ueber Begriff und Gegenstand', *Vierteljahresschrift für wissenschaftliche Philosophie* 16:192–205. Reprinted in Frege (1967), pages 167–178; English translation by Peter Geach as 'On Concept and Object' Frege (1951), also in Geach and Black (1980), pages 42–55, and Frege (1984), pages 182–194, and with revisions by Michael Beaney in Beaney (1997), pages 181–193.

——— (1893). *Grundgesetze der Arithmetik: Begriffsschriftlich abgeleitet.* I. Band. Jena: Hermann Pohle. Reprinted in Frege (1962) and Frege (1998); partial English translation in Frege (1964b) and Beaney (1997).

——— (1896–1899). 'Lettera del Sig. G. Frege all'Editore', *Rivista di Matematica* 6:53–59. Reprinted in Frege (1967), pages 234–239, and Gabriel et al. (1976), pages 181–186; English translation by H. Kaal in Gabriel et al. (1980), pages 112–118.

——— (1899). *Über die Zahlen des Herrn H. Schubert.* Jena: Hermann Pohle. Reprinted in Frege (1967), pages 240–261; English translation by H. Kraal as 'On Mr. Schubert's Numbers' in Frege (1984), pages 249–272.

——— (1902). 'Letter to Russell, June 22, 1902', in Gabriel et al. (1976), pages 212–213. English translation: Gabriel et al. (1980), pages 131–133; first published in English translation in van Heijenoort (1967), pages 126–128.

——— (1903). *Grundgesetze der Arithmetik: Begriffsschriftlich abgeleitet.* II. Band. Jena: Hermann Pohle. Reprinted in Frege (1962) and Frege (1998); partial English translation in Geach and Black (1980) and Beaney (1997).

——— (1948). 'Sense and Reference', translated by Max Black, *The Philosophical Review* 57:207–230.

——— (1950). *The Foundations of Arithmetic.* Translated by J. L. Austin. Oxford: Blackwell.

——— (1951). 'On Concept and Object', translated by Peter Geach, *Mind* 60:168–180.

——— (1962). *Grundgesetze der Arithmetik I/II.* Facsimile reprint. Hildesheim: Olms.

——— (1964a). *Begriffsschrift und andere Aufsätze: Mit E. Husserls und H. Scholz' Anmerkungen.* Edited by Ignacio Angelelli. Hildesheim: Olms. English translation by T. W. Bynum: Frege (1972).

——— (1964b). *The Basic Laws of Arithmetic: exposition of the system.* Edited and translated, with an introduction, by Montgomery Furth. Berkeley and Los Angeles: University of California Press.

——— (1967). *Kleine Schriften.* Edited by Ignacio Angelelli. Hildesheim: Olms. 2nd ed., Olms, Hildesheim, 1990.

——— (1972). *Conceptual Notation and Related Articles.* Translated and edited by T. W. Bynum. Oxford: Clarendon Press.

——— (1984). *Collected Papers on Mathematics, Logic, and Philosophy.* Edited by Brian McGuinness. Oxford: Basil Blackwell.

——— (1998). *Grundgesetze der Arithmetik I/II.* Facsimile reprint, with critical additions by Christian Thiel. Hildesheim: Olms.

Gabriel, Gottfried, H. Hermes, F. Kambartel, Christian Thiel, A. Veraart, and Brian McGuinness, eds. (1980). *Gottlob Frege: Philosophical and Mathematical Correspondence.* Translated by H. Kaal. Chicago: University of Chicago Press.

Gabriel, Gottfried, H. Hermes, F. Kambartel, Christian Thiel, and A. Veraart, eds. (1976). *Gottlob Frege: Wissenschaftlicher Briefwechsel.* Hamburg: Meiner. Partial English Translation: Gabriel et al. (1980).

Gauss, Carl Friedrich (1831). 'Anzeige der *Theoria residuorum biquadraticorum. Commentatio secunda*', *Göttingische gelehrte Anzeigen.* Reprinted in Gauss (1876), pages 169–178; English translation by William Ewald in Ewald (1996), vol. I, pages 306–313.

——— (1876). *Werke.* Vol. II. Göttingen: Königliche Gesellschaft der Wissenschaften.

Geach, Peter and Max Black, eds. (1980). *Translations from the Philosophical Writings of Gottlob Frege*, 3rd edition. Oxford: Blackwell.

Grassmann, Hermann (1844). *Die lineale Ausdehnungslehre: ein neuer Zweig der Mathematik.* Leipzig: Otto Wigand. English translation in Grassmann (1995).

——— (1995). *A New Branch of Mathematics: The* Ausdehnungslehre *of 1844 and Other Works.* Translated by Lloyd C. Kannenberg. Chicago and La Salle, Ill.: Open Court.

Hankel, Hermann (1867). *Vorlesungen über die complexen Zahlen und ihrer Functionen. Teil 1: Theorie der complexen Zahlensysteme.* Leipzig: Leopold Voss.

Heine, Eduard (1872). 'Die Elemente der Functionenlehre', *Journal für die reine und angewandte Mathematik* (Crelle's *Journal*) 74:172–188.

Illigens, Eberhard (1889). 'Zur Weierstrass'-Cantor'schen Theorie der Irrationalzahlen', *Mathematische Annalen* XXXIII:155–160.

Kant, Immanuel (1853). *Kritik der reinen Vernunft.* Edited by G. Hartenstein. Leipzig: Leopold Voss.

Kossak, Ernst (1872). *Die Elemente der Arithmetik.* Berlin: Nicolai'sche Verlagsbuchhandlung.

Kronecker, Leopold (1887). 'Ueber den Zahlbegriff', in *Philosophische Aufsätze, Eduard Zeller zu seinem fünfzigjährigen Doctorjubiläm gewidmet.* Leipzig: Fues, pages 261–274. English translation as 'On the Concept of Number' in Ewald (1996), vol. II, pages 947–955.

Peano, Guiseppe (1986). 'Risposta [Reply to Frege (1896–1899)]', *Rivista di Matematica* 6:60–61. Reprinted in Gabriel et al. (1976), pages 186–188; English translation by H. Kaal in Gabriel et al. (1980), pages 118–120.

Pringsheim, Alfred (1898). 'Irrationalzahlen und Konvergenz unendlicher Prozesse', in Wilhelm Franz Meyer (ed.), *Encyklopädie der mathematischen Wissenschaften: mit Einschluss Ihrer Anwendungen.* Vol. I: *Arithmetik und Algebra.* Leipzig: B. G. Teubner, pages 47–146.

Russell, Bertrand (1902). 'Letter to Frege, June 16, 1902', in Gabriel et al. (1976), pages 211–212. English translation: Gabriel et al. (1980), pages 130–131; first published in English translation in van Heijenoort (1967), pages 124–125.

Schröder, Ernst (1890–1905). *Vorlesungen über die Algebra der Logik.* 3 vols. Leipzig: B. G. Teubner. Partial English translation in Geraldine Brady (2000). *From Peirce to Skolem: A Neglected Chapter in the History of Logic.* Amsterdam: North-Holland.

Stolz, Otto (1885). *Vorlesungen über allgemeine Arithmetik. Erster Theil: Allgemeines und Arithmetik der reellen Zahlen.* Leipzig: B. G. Teubner.

Thomae, J. (1898). *Elementare Theorie der analytischen Functionen einer complexen Veränderlichen,* 2nd edition. Halle a. d. Saale: Nebert.

van Heijenoort, Jean, ed. (1967). *From Frege to Gödel: A Source Book in Mathematical Logic, 1879–1931.* Cambridge, Mass.: Harvard University Press.

von Helmholtz, Hermann (1887). 'Zählen und Messen, erkenntnistheoretisch betrachtet', in *Philosophische Aufsätze, Eduard Zeller zu seinem fünfzigjährigen Doctorjubiläm gewidmet.* Leipzig: Fues, pages 17–52. English translation as 'Numbering and Measuring from an Epistemological Viewpoint' in Ewald (1996), vol. II, pages 727–752.

Appendix: How to read *Grundgesetze*

Roy T. Cook

Introduction

This appendix is intended to assist the reader in becoming comfortable with the notations, rules, and definitions of Frege's *Grundgesetze*. The intention is *not*, however, to provide a guide to "translating" Frege's formulas and proofs into modern notation. On the contrary, although contemporary notation will be used occasionally to illustrate some of Frege's ideas (and, importantly, will be used to draw contrasts between the workings of Frege's system and the workings of our own twenty-first century constructions), any attempted translation from *Grundgesetze* to a contemporary formalism will, in the end, fail. Frege's system is not equivalent to any contemporary, "living" formal system currently studied within logic and its philosophy. Instead, the formal system of *Grundgesetze* is unique, with its own strengths, weaknesses, and quirks, and the reader who wishes to learn more about Frege's logic and philosophy should strive to understand this unique formal calculus on its own terms, rather than embedding it into contemporary (and hence anachronistic, if one's goal is understanding Frege) systems.

Thus, the purpose of the discussion below is to provide the reader with the basic tools required to gain fluency in the language of *Grundgesetze*. One of the myths regarding the formal system(s) presented in *Begriffsschrift* and *Grundgesetze* is that these systems were abandoned for more "linear" notations because of the unwieldiness of Frege's notation. Regardless of what the actual reasons are for Frege's notations not catching on, the generally accepted story regarding their lack of clarity, economy, readability, and efficiency is just false. While Frege's notations and proofs no doubt appear rather strange to the reader trained in contemporary predicate logic with linear notation, there are real advantages to Frege's system (including, but not limited to, the two-dimensional notation providing a "geometric" aspect to the discovery and analysis of proofs). As a result, although gaining "literacy" in the system of *Grundgesetze* is not easy, the rewards—both in terms of Frege scholarship and in terms of more general lessons regarding the nature of logic—more than repay the effort involved.

One final note: The formal system presented in *Begriffsschrift* and the one presented above (and discussed below) are different in a number of respects. These changes are not merely further developments and refinements of an existing system, but represent genuine changes in Frege's thought. Since the system of *Grundgesetze* (and not the earlier system presented in *Begriffsschrift*) represents Frege's mature views on logic, I shall for the most part restrict my attention to explaining how the later system works. Readers interested in the differences between the two systems should consult the growing secondary literature on this topic.

Terminology

Frege introduces a number of technical terms for various types of expression that can be constructed within the formal language, or *concept-script*, of *Grundgesetze* (as well as for expressions that occur only within the metalanguage in which the formal language of *Grundgesetze* is formulated).

First, a *name* is any concept-script expression that names an object or a function. Names are further subdivided into proper names, truth-value-names, and function-names. A name is a *proper name* if and only if it is the name of an object. A name is a *truth-value-name* if and only if it is the name of a truth-value, hence is either a name of the True or a name of the False. Thus, truth-value-names are a proper sub-collection of the proper names (I §2).[1] Importantly, what we would intuitively think of as sentences within the formal system of *Grundgesetze*—that is, expressions not prefixed with a judgement-stroke—are truth-value-names, although concept-script propositions—the result of prefixing the judgement-stroke to concept-script expressions—are not names. As a result, I will avoid the term "sentence" altogether in what follows, calling the former either "truth-value-names" or "concept-script expressions" depending on context, and reserving "concept-script proposition" (or simply "proposition") for Frege's technical use of the term: concept-script expressions (i.e., truth-value-names) prefixed with the judgement-stroke.

A name is a *function-name* if and only if it is the name of a function. Functions, for Frege, are mappings from any *Grundgesetze* "entities" to objects, although of absolutely fundamental importance here is Frege's insistence that functions must be defined for all arguments. Every primitive concept-script expression within the formal system of *Grundgesetze* other than the judgement-stroke (and with the possible exception of the Roman letter generality device, see below) is a function-name. Hence, the only way to obtain proper names (and hence truth-value-names) within *Grundgesetze* is by combining two or more function-names of appropriate types (I §26).

Two particularly important types of function are concepts and relations. A *concept* is a unary function such that, for any argument (of the appropriate type), the value of the function applied to that argument is a truth-value (thus, Frege's understanding of a concept is much like the contemporary understanding of the characteristic function of that concept). A *relation* is a function with two (or more) arguments such that, for any pair (or n-tuple) of arguments (again, of the appropriate type), the value of the function applied to that pair is a truth-value (I §4, see also I §22).

In addition to distinguishing functions in terms of the number of arguments that they take and the kind of output that they deliver, Frege also subdivides functions in

[1] Throughout, 'I' and 'II' in such contexts refers to the first and second volume of *Grundgesetze*, respectively.

terms of the *kinds* of argument that they take. Thus, a function is a *1st-level function* if and only if it takes an object or objects as argument(s), a function is a *2nd-level function* if and only if it takes a 1st-level function or functions as argument(s), and a function is a *3rd-level function* if and only if it takes a 2nd-level function or functions as argument(s). Along similar lines, a function is a function of the *1st kind* if and only if it takes an object or objects (i.e., zero-ary functions) as argument(s), a function is a function of the *2nd kind* if and only if it takes a unary function or functions as argument(s), and a function is a function of the *3rd kind* if and only if it takes a binary function or functions as argument. Note that the 1st-level functions are exactly the functions of the 1st kind, although it is not the case that, for $n > 1$, that n^{th}-level functions are the same as functions of the nth kind. Needless to say, Frege could have extended these definitions indefinitely, defining n^{th}-level functions and functions of the nth kind for any n. Frege never needs functions of higher than 3rd-level and the 3rd kind, however, since the value-range operator and other devices within *Grundgesetze* allow us to "reduce" functions of higher kinds to functions of the 3rd-level and 3rd kind or lower. Additionally, although there is nothing in *Grundgesetze* that excludes functions of "mixed" levels or "mixed" kinds, Frege never uses such "mixed" functions, and provides no specific terminology for them (I §§21–23, see also §26).

A note on "metatheoretic" variables: We will follow Frege's notation in general, and his use of different fonts for different logical "types", whenever possible. Thus, we use the upper-case Greek letters 'Δ', 'Γ' 'Θ', 'Λ', etc. as ranging over arbitrary proper names (including arbitrary truth-value-names, the Fregean analogue of propositional letters), the upper-case Greek letters 'Φ', 'Ψ', etc. as ranging over 1st-level function-names (including concept-names, relation-names, etc). As a result, and contrary to standard contemporary practice, if we want to represent an arbitrary (modern) conditional we will write '$\Delta \supset \Gamma$' rather than '$\Phi \supset \Psi$'. Also following Frege, the lower case Greek letters 'ξ' and 'ζ' will be used as parameters when forming function-names (I §§1 and 4). Thus, if we wish to speak of the result of applying a 1st-level function symbol 'Φ' to an arbitrary proper name 'Δ', we will write '$\Phi(\Delta)$' whereas, if we wish to speak of the function itself, we write '$\Phi(\xi)$'. Special care should be taken in distinguishing the different notations used for generality. German *Fraktur* letters are used for the *bound* variables in concavity constructions, while *Roman* letters are used for the *bound* variables in applications of the Roman letter generality device. As a result, care should be taken when parsing concept-script expressions in Frege's *Grundgesetze* since the same *letter* can occur more than once, playing quite different roles (e.g., Fraktur '$\mathfrak{f}$', Fraktur '$\mathfrak{F}$', Roman 'f', and Roman 'F').

Finally, it is often noted that Frege is sloppy regarding the use/mention distinction. This way of putting it is, of course, anachronistic and unfair, since it was only with the advent of precise formal calculi such as the formal system presented in *Grundgesetze* that the need for care regarding the use/mention distinction became apparent. Nevertheless, Frege does often 'slide' from talk of names of objects to talk of the objects themselves and vice versa (and analogously for higher-order notions) while using the same expressions (in particular, the meta-variables 'Δ', 'Γ', '$\Phi(\xi)$', '$\Psi(\xi)$', etc.) for both. In what follows, meta-variables will sometimes be understood as ranging over the entities recognised within *Grundgesetze* (i.e., objects, functions, etc.) and sometimes as ranging over names of such entities. In the latter sort of case, the meta-variable will be enclosed in quotation marks (Frege's own notational convention). Further, a single variable will never be used both ways in the same context or discussion (unlike Frege's actual practice). This does not completely eliminate the problem, but minimises it enough for our purposes (for further discussion of this issue, see the discussion of metavariables and auxiliary names in Heck (2012)).

The Language of *Grundgesetze*

In this section we will examine the primitive vocabulary of the concept-script as developed within *Grundgesetze*, examining the objects and functions that are denoted by these concept-script expressions, and the way that each such expression combines with other expressions to form complex constructions. The discussion will not follow the order in which Frege introduces these notions within *Grundgesetze*. Instead, the presentation is ordered in a way that will, it is hoped, feel more "natural" to students of modern propositional and predicate logics. Thus, we shall first introduce the judgement-stroke, a historical precursor to the modern day single turnstile representing deducibility within a formal system, and the related definition-stroke. We then turn to Frege's "propositional logic"—that is, those operators that correspond, at least roughly, to the operations of modern propositional logic. These notions include the horizontal, the negation-stroke, the conditional-stroke, and (counterintuitively, perhaps, to a modern reader) the equality-sign. Next, we shall introduce Frege's two devices for expressing generality within *Grundgesetze*, the concavity and the Roman letter generality device. Finally, we shall look at those portions of the logic of *Grundgesetze* that do not correspond to anything appearing within most modern treatments of *logic*—the value-range operator and the backslash operator.

The Judgement-stroke

A *concept-script proposition* is the result of prefixing any concept-script expression of *Grundgesetze* with the *judgement-stroke*. A concept-script proposition consisting of an expression prefixed with the judgement-stroke is meant to be understood as asserting that the expression in question is a name of the True. For example, given some proper name 'Δ', '$\Delta = \Delta$' is (as we shall see) not a concept-script proposition (and hence does not represent the content of a judgement), but is instead a name of a truth-value (in particular, in this case a name of the True). The result of prefixing this expression with the judgement-stroke:

$$\vdash (\Delta = \Delta)$$

however, is a concept-script proposition, and expresses something like the judgement:

The concept-script expression '$\Delta = \Delta$' names the True.

Thus, the application of the judgement-stroke to a concept-script expression lacking the judgement-stroke transforms a name into a judgement expressing a thought (I §5). In short, the judgement-stroke operates as a kind of truth predicate—one which holds of a concept-script expression if and only if that expression is a name of the True (this account will be complicated somewhat in our discussion of the Roman letter generality device below, however). Since calling applications of the judgement-stroke to truth-value-names *true* can introduce needless confusion into our discussion, in what follows the term *correct* will be reserved for those concept-script propositions of the form:

$$\vdash \Delta$$

where 'Δ' is, in fact, a name of the True.

Frege suggests that the judgement-stroke proper (I §5), as well as the negation-stroke (see I §6), the conditional-stroke (see I §12), and the concavity (see I §8), can (and *should*!) be understood as consisting merely of the actual vertical "stroke" or line involved in the formalization, with the attached horizontal portion of the notation understood as separate occurrence of the horizontal. This is, of course, how logical operators are understood in *Begriffsschrift*, but we shall see some reasons for not taking Frege quite at his word in this regard. For now it is enough to note that, even if we *were* to understand the judgement-stroke proper as consisting solely of the vertical bar (and thus understand the symbol '$\vdash$' to be complex), in actual practice within *Grundgesetze* the judgement-stroke never appears without the attached horizontal.

The Definition-stroke

Any theorem that Frege proves within *Grundgesetze* is prefixed with the judgement-stroke, since each such theorem (if correctly derived, and ignoring the problems that eventually arise with regard to Basic Law V) is a correct judgement—that is, the concept-script expression within the scope of the judgement-stroke is (or should be) a name of the True (again, this will be complicated slightly in our discussion of the Roman letter generality device below). A second, albeit similar device is used when Frege introduces definitions. Definitions, of course, are not derived from basic laws and theorems, but are instead stipulated. Frege indicates this difference in status by introducing a different notation for definitions, the *definition-stroke* '$\Vdash$' (I §27).[2] For example, the definition of the application operator '$\frown$', is prefixed not with the judgement-stroke but with the definition-stroke:

$$\Vdash \grave{\dot{\alpha}}\left(\overset{\mathfrak{g}}{\smile}\!\!-\!\!\begin{array}{l}\mathfrak{g}(a)=\alpha\\ u=\grave{\varepsilon}\mathfrak{g}(\varepsilon)\end{array}\right)=a\frown u \qquad \text{(A}$$

The equalities provided by definitions are not judgements, and do not express thoughts, since they are merely stipulations regarding how novel notation (i.e., abbreviations) are to be used. Definitions can be turned into (correct) judgements by replacing the definition-stroke by the judgement-stroke. We shall have much more to say about definitions below.

The Horizontal

The *horizontal* is a unary function symbol that attaches to names of objects, and it names a function that always outputs a truth-value, regardless of the kind of object input—in Frege's terminology it names a 1st-level concept. The purpose of the horizontal is to map any object onto a truth-value—that is, application of the horizontal transforms any proper name into a name of a truth-value. If the horizontal

[2] Frege calls the definition-stroke the "double-stroke of definition"—see I §27.

is prefixed to the name of a truth-value 'Δ', then the resulting complex name:

$$— \Delta$$

names the same truth-value as is named by 'Δ'. If, however, 'Δ' does not name a truth-value, then '— Δ' names the False (I §5). The logical properties of the horizontal function are thus summed up in the following table:

Δ	— Δ
T	*T*
F	*F*
other	*F*

It should be emphasised that the table above is not a truth-table in the modern sense: Frege's logical operators are not truth-functions, since they are not functions from n-tuples of truth-values to truth-values. Instead, Frege's horizonal is a function from *any* object to a truth-value (similar comments apply to the negation-stroke and conditional-stroke, discussed below). In short, the horizontal '— ξ' can be understood, loosely, as denoting the truth-value of:

ξ is (i.e., is identical to) the True.

Frege's treatment of the logical operators as total functions does have some odd consequences. For example, as Frege himself notes (I §5), '— 2' names the False. (We will not see how to formulate the complex name for the cardinal number Two until our discussion of Frege's definitions. Thus, we are following Frege's own treatment in using the numeral '2' at this point merely as an intuitive example.) This is, of course, odd if Frege's notations are read as completely analogous with a modern understanding of the propositional connectives, and if '— ξ' is identified with any modern truth-functional operator (e.g., classical double negation), but it makes perfect sense if read along the lines outlined above.

The Negation-stroke

The *negation-stroke* is a unary function symbol that attaches to names of objects—in Frege's terminology it names a 1st-level concept. Like the horizontal, the negation-stroke transforms any proper name into a truth-value-name. If the negation-stroke is prefixed to the name of a truth-value 'Δ', then:

$$\top \Delta$$

names the True if 'Δ' names the False, and names the False if 'Δ' names the True. If, however, 'Δ' does not name a truth-value, then '$\top$ Δ' names the True (I §6). In short, the negation-stroke '$\top$ ξ' can be understood, loosely, as denoting the truth-value of:

ξ fails to be (i.e., fails to be identical to) the True.

Hence, the logical properties of the negation-stroke are summed up in the following table:

α	$\top \alpha$
T	*F*
F	*T*
other	*T*

As was the case with the horizontal above, Frege's treatment of the negation-stroke as a total function has odd consequences. For example. '$\top$ 2' names the True. Furthermore, Frege's treatment of negation as a total function complicates the behavior of double negations. Given a proper name 'Δ', there is no guarantee that: '$\top\!\top\ \Delta$' names the same object as 'Δ'. For example, '$\top\!\top$ 2' is a name of the False, not a name of Two. Of course, if 'Δ' is a truth-value-name, then the double negation of 'Δ' names the same truth-value as does 'Δ'. Hence, the triple negation of any proper name 'Δ', that is, '$\top\!\top\!\top\ \Delta$' names the same object as the negation of that proper name '$\top\ \Delta$', and '— Δ' names the same object as '$\top\!\top\ \Delta$' (and, although we have not officially discussed the equality sign yet, it turns out that '$\Delta = \top\!\top\ \Delta$' is a name of the True if and only if 'Δ' is a name of a truth-value).

The Conditional-stroke

Frege's *conditional-stroke* is a binary function symbol that attaches to names of objects—in Frege's terms it is a 1st-level relation. The conditional-stroke is meant to capture something like the modern material conditional—that is, "if—then". As is the case with all of the logical operators in *Grundgesetze*, the conditional-stroke is a total function: Given any two proper names 'Δ' and 'Γ' connected by the conditional-stroke (in that order—in attempting to draw as tight connections between Frege's conditional-stroke notation and the modern material conditional, I am taking the lower argument to be the "first" argument):

$$\begin{array}{l}\ulcorner\, \Gamma \\ \llcorner\, \Delta\end{array}$$

is a name of the True if either 'Δ' fails to name the True (i.e., either names the False or does not name a truth-value) or 'Γ' names the True, and names the False otherwise (I §12). Frege calls the lower component (what modern readers would term the antecedent) the *subcomponent* of the conditional, and the upper component (what modern readers would call the consequent) the *supercomponent* (although see below!). The logical properties of the conditional-stroke are summed up in the following table:

$\begin{array}{l}\ulcorner\, \Gamma \\ \llcorner\, \Delta\end{array}$	*T*	*F*	*other*
T	*T*	*F*	*F*
F	*T*	*T*	*T*
other	*T*	*T*	*T*

(The left-most vertical column represents the value of 'Δ', the top-most horizontal row the value of 'Γ'.)

As was the case with the horizontal and the negation-stroke, Frege's treatment of the conditional-stroke as a total function has odd consequences. For example, for any proper name 'Δ', both of:

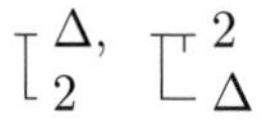

are names of the True. The counter-intuitiveness can be countered by reading the conditional-stroke '  ' as providing something like the truth-value of:

> It is not the case that both the subcomponent ξ is (i.e., is identical to) the True and the supercomponent ζ is (i.e., is identical to) something other than the True.

Conditional-stroke constructions can be parsed into supercomponent and subcomponents(s) in multiple ways. Consider:

This concept-script expression is analogous (but of course not equivalent, in any reasonable sense of equivalent) to '$\Delta \supset (\Gamma \supset \Theta)$' in modern notation. But Frege often treats concept-script expressions of this form as instead expressing something more akin to '$(\Delta \wedge \Gamma) \supset \Theta$'. Of course, these formulas are equivalent within classical logic. Frege takes advantage of this equivalence, introducing a systematic ambiguity into the usage of "subcomponent" and "supercomponent". When considering a complex concept-script expression involving multiple embedded conditional-strokes such as the example displayed above, we can, following the original, simple understanding, treat this as having a single subcomponent 'Δ' and single supercomponent:

But, we can also treat this concept-script expression as having two distinct subcomponents 'Δ' and 'Γ' and single supercomponent 'Θ'. Either reading is legitimate within *Grundgesetze*. Since many of Frege's rules of inference are formulated in terms of adding, eliminating, or repositioning supercomponents and subcomponents, the systematic ambiguity has profound implications for how proofs are constructed within *Grundgesetze*.

The multiple-subcomponent reading of complex, multiple-conditional-stroke concept-script expressions has a second immediate consequence. Frege notes that, on the reading of:

where Δ and Γ are the two subcomponents, each subcomponent plays exactly the same role as the other (I §12). As a result, the ordering of subcomponents does not matter, and as a result this concept-script expression names the same truth-value as:

Frege later (I §48) introduces a rule of inference that allows arbitrary re-ordering of subcomponents.

Frege goes on to note that these observations generalise to embedded conditionals (of the appropriate form) of any length. For example, given proper names Δ, Γ, Θ, Λ, and Ξ, we can parse the complex concept-script expression:

as having any of:

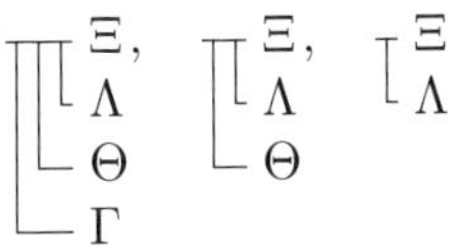

or Ξ as supercomponent (with one, two, three, or four subcomponents on each reading, respectively). This brings up a final observation regarding the differences between the conditional-stroke in *Grundgesetze* and a modern understanding of the conditional. Frege introduces the conditional-stroke as if it is a simple binary 1st-level function from pairs of objects to truth-values. And in one sense it is. In his manipulation of the conditional-stroke (and especially in the rules of inference for reasoning with the conditional-stroke), however, Frege treats the conditional-stroke much more like a kind of open-ended n-ary function-name that takes a single argument as supercomponent, but which can take any (finite) number of arguments as its subcomponents. Thus, Frege's discussion of, and later manipulation of, the conditional-stroke instead suggests that it should be read as a generalised connective along something like the following lines:

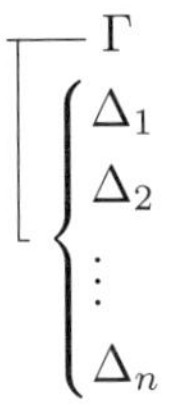

While this notation is extremely illuminating, since it emphasises both the two-dimensional aspect of Frege's notation and the equivalent status of each subcomponent, in the remainder of this appendix I shall use the slightly more efficient (typographically speaking):

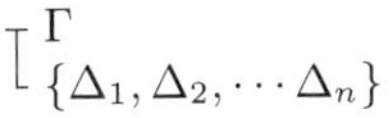

As we have already noted, complex concept-script expressions involving multiple conditional-strokes can often be analyzed along these lines in multiple ways.

Defined Propositional Connectives

Frege provides a brief translation manual for mimicking "or", "and", and "neither–nor" within *Grundgesetze*. Given two proper names 'Δ' and 'Γ', we can express the conjunction of 'Δ' and 'Γ' as:

The properties of this conjunction operator can be summed up as follows:

	*T*	*F*	*other*
T	*T*	*F*	*F*
F	*F*	*F*	*F*
other	*F*	*F*	*F*

Similarly, we can express the disjunction of proper names 'Δ' and 'Γ' as:

The properties of this disjunction operator can be summed up as follows:

	*T*	*F*	*other*
T	*T*	*T*	*T*
F	*T*	*F*	*F*
other	*T*	*F*	*F*

Finally, we can express "neither Δ nor Γ" as:

Its properties are summed up as follows:

	T	F	*other*
T	F	F	F
F	F	T	T
other	F	T	T

Note that, intuitively, this complex operation captures something along the lines of the truth-value of:

Neither ξ nor ζ names the True.

rather than the truth-value of:

Both ξ and ζ are names of the False.

although these are equivalent if both arguments are truth-value-names (I §12).

Although Frege does not point this out explicitly, the systematic ambiguity of the conditional-stroke allows for the formulation of elegant n-ary generalizations of conjunction, disjunction, and the "neither–nor" construction. Thus, given n proper names 'Δ_1', 'Δ_2'...'Δ_n', we can express:

Δ_1 and Δ_2 and ... and Δ_n

as:

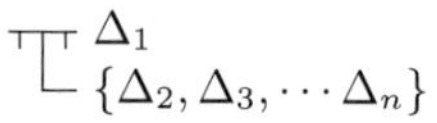

Likewise, we can express:

Δ_1 or Δ_2 or ... or Δ_n

as:

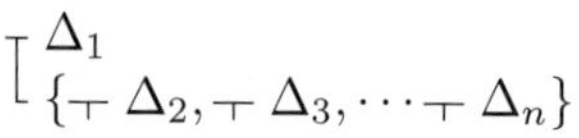

And we can express:

Neither Δ_1 nor Δ_2 nor ... nor Δ_n

as:

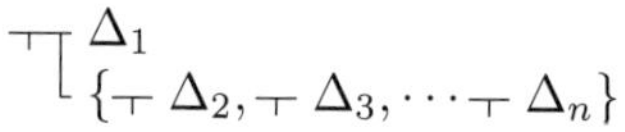

respectively. These constructions provide a nice guide for translating natural language constructions into the language of *Grundgesetze*, and vice versa, but care should be taken that they are not applied blindly. In particular, the systematic ambiguity in Frege's reading of the conditional-stroke discussed in the previous section trumps the informal translation manual given by Frege in (I §12). Thus, when Frege wishes to formalise something along the lines of (the truth-value of):

If both Δ and Γ then Θ.

he renders it as:

not as:

These are equivalent within *Grundgesetze*, but, since the deductive system of *Grundgesetze* takes advantage of the systematic ambiguity present in the conditional, the former formalization is preferable.

Fusion of Horizontals

As was noted above, Frege suggests that the negation-stroke, the conditional-stroke, and the judgement-stroke can be understood (and suggests, further, that they *should* be understood, see I §5, §6, and §12) as consisting merely of the actual vertical strokes or lines involved in their formalization, with the attached horizontal portions of their notation understood as separate occurrences of the horizontal. But, as already suggested in the earlier discussion, it is not clear that Frege's suggestion should be taken too literally. For example, it is not clear that the result of prefixing the vertical portion of the negation symbol '$\top$' (without its horizontal "parts") to '2'—that is, something like '$\shortmid$ 2'—is a well-formed concept-script expression in *Grundgesetze*. Further, the "metatheoretic" arguments for the claim that every concept-script expression in *Grundgesetze* has a reference, found in Volume I §§10 and 31, do not cover such cases, rendering these arguments incomplete (and obviously so) if Frege genuinely intended us to understand '$\top$' and '$\begin{array}{l}\top\\ \llcorner\end{array}$' as complex symbols.

Thus, it is perhaps better to read Frege here as suggesting that the judgement-stroke, the negation-stroke, and the conditional-stroke (as well as the concavity, discussed below) can be treated *as if* they contain occurrences of the horizontal.

Nevertheless, regardless of whether the negation-stroke, conditional-stroke, and judgement-stroke are complex, containing occurrences of the horizontal, or merely behave *as if* such horizontals are present, it follows that, for any name 'Δ', that all of:

$$\top \Delta, \quad \top(\!-\!\!\!- \Delta), \quad -\!\!\!-(\top \Delta), \quad \text{and} \quad -\!\!\!-(\top(\!-\!\!\!- \Delta))$$

must name the same truth-value (I §6). Along similar lines, it must be the case, for any names 'Δ' and 'Γ', that all of:

$$\begin{array}{l}\top \Gamma\\ \llcorner \Delta\end{array}, \quad -\!\!\!-\left(\begin{array}{l}\top \Gamma\\ \llcorner \Delta\end{array}\right), \quad \begin{array}{l}\top \Gamma\\ \llcorner (-\!\!\!- \Delta)\end{array}, \quad \begin{array}{l}\top (-\!\!\!- \Gamma)\\ \llcorner \Delta\end{array}$$

$$\begin{array}{l}\top (-\!\!\!- \Gamma)\\ \llcorner (-\!\!\!- \Delta)\end{array}, \quad -\!\!\!-\left(\begin{array}{l}\top (-\!\!\!- \Gamma)\\ \llcorner \Delta\end{array}\right), \quad -\!\!\!-\left(\begin{array}{l}\top \Gamma\\ \llcorner (-\!\!\!- \Delta)\end{array}\right), \quad \text{and} \quad -\!\!\!-\left(\begin{array}{l}\top (-\!\!\!- \Gamma)\\ \llcorner (-\!\!\!- \Delta)\end{array}\right)$$

name the same truth-value. Finally, horizontals must fuse with the judgement-stroke (and, presumably, with the definition-stroke, although Frege is not explicit about this), so that either both of:

$$\vdash \Delta, \quad \vdash (-\!\!\!- \Delta)$$

are correct concept-script propositions, or neither are. Frege's calls these equivalences, and the transformations that result from replacing one of the concept-script expressions listed above with another, equivalent formulation, the *fusion of horizontals*. Like the permutation of subcomponents, Frege introduces an explicit rule allowing us to fuse (and un-fuse) horizontals within derivations (I §48).

Finally, since all functions in *Grundgesetze* are total, Frege needs to define the value of the horizontal, the negation-stroke, and the conditional-stroke when these functions are applied to non-truth-values, and the sections above describe Frege's choices in this regard. Although Frege's choices are no doubt coherent (and, as we shall see, allow him to formulate rather intuitive rules of inference), the careful reader might wonder why Frege chose the particular functions he in fact chose. We are now in a position to provide an answer: Given that the horizontal must fuse with the judgement-stroke, the horizontal must only output the True when the input is the True. Further, this fact, plus the fact that the negation-stroke and the conditional-stroke must themselves fuse with horizontals, entails that the particular definitions of the negation-stroke and the conditional-stroke given by Frege are the only ones possible.

The Equality-sign

Frege's *equality sign* is defined as one would expect: If 'Δ' and 'Γ', are proper names, then:

$$\Gamma = \Delta$$

is a name of the True if 'Γ' names the same object as does 'Δ'; and is a name of the False if 'Γ' and 'Δ' are names of different objects (I §7). Of course, like Frege's other primitive operators, the equality-sign is not a relation symbol in the modern sense but is instead a binary 1st-level function.

While the definition of the equality-sign is relatively straightforward, Frege's use of it is somewhat different from how the equality symbol is used in modern predicate calculi. Frege uses the equality-sign when making everyday equality claims such as '2 + 2 = 4', but he also uses the same symbol flanked by truth-value-names in order to express the claim that the truth-value-names are names of the same truth-value—that is, that the concept-script expressions in question are truth-functionally equivalent. This explains an apparent oversight in Frege's discussion of defined propositional operators. Frege does not explicitly provide a definition of the material biconditional within *Grundgesetze*, although he could have easily defined the material biconditional along standard lines as the conjunction of two conditionals:

If 'Δ' and 'Γ' are both truth-value-names, then this defined notion and the function named by Frege's primitive equality-sign output the same value. If one or both of 'Δ' and 'Γ' are proper names but not truth-value-names, however, then the value of the complex material conditional applied to these arguments can differ from the value of the equality-sign applied to them. For example:

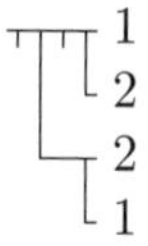

is a name of the True, while '1 = 2' is a name of the False. As a result, within the formalism of *Grundgesetze* the equality-sign plays (to modern eyes) two distinct roles. When attached to proper names generally, it provides the truth-value of the claim that those names are names of the same object. When attached to truth-value-names, however, it provides the truth-value of the claim that those truth-value-names denote the same truth-value—that is, it plays something analogous to the role of the material biconditional within modern logical calculi. Of course, on Frege's understanding, these are not really two separate tasks. Rather, the biconditional reading of the equality-sign is merely a special case of the more general "identity" reading.[3]

The Concavity and German Letters

In *Grundgesetze* Frege mobilises two distinct forms of universal quantification. The first of these is the concavity. The 1st-order version of the *concavity* names a 2nd-level function from 1st-level functions to objects. Given any 1st-level function-name 'Φ(ξ)':

$$\overset{\mathfrak{a}}{\smile}\ \Phi(\mathfrak{a})$$

[3] Here and below I have respected Frege's preference for the term "equality" rather than "identity". Frege's discusses the equality-versus-identity issue in II §§81, 134, 138, and 151, amongst other places.

names the True if and only if the function named by '$\Phi(\xi)$' outputs the True for every possible argument, and names the False otherwise (I §8). The crucial difference between the concavity and the Roman letter generality device, discussed below, is that the variable binding device in the concavity construction (i.e., the concavity itself, and the variable that occurs within it) is sensitive to scope in a manner in which the Roman letter generality device is not. In short, the concavity allows us to distinguish between, and formalise both, the wide-scope reading:

> For any $\mathfrak{a}$, it is not the case that $\mathfrak{a} = \mathfrak{a}$.

and the narrow-scope reading:

> It is not the case that, for any $\mathfrak{a}$, $\mathfrak{a} = \mathfrak{a}$.

by formalizing the former as '$\overset{\mathfrak{a}}{\smile}\!\top\ \mathfrak{a} = \mathfrak{a}$' and the latter as '$\top\!\overset{\mathfrak{a}}{\smile}\ \mathfrak{a} = \mathfrak{a}$'.

The concavity, like the negation-stroke and the conditional-stroke, fuses with horizontals. In other words, all of the following name the same truth-value (i.e., are equivalent):

$$\overset{\mathfrak{a}}{\smile}\ \Phi(\mathfrak{a}),\ \text{—}\ (\overset{\mathfrak{a}}{\smile}\ \Phi(\mathfrak{a})),\ \overset{\mathfrak{a}}{\smile}\ (\text{—}\ \Phi(\mathfrak{a})),\ \text{and}\ \text{—}\ (\overset{\mathfrak{a}}{\smile}\ (\text{—}\ \Phi(\mathfrak{a}))).$$

As a result, Frege also suggests that we can understand the concavity operator as consisting solely of the concavity, with the two horizontal "bits" flanking it being understood as independent occurrences of the horizontal. As already noted, however, this is probably best read as the suggestion that we can treat the concavity *as if* it involves two occurrences of the horizontal (I §8, see also the discussion of the fusion of horizontals above).

Frege introduces some terminology for statements constructed using the concavity. Given two concept-names '$\Phi(\xi)$' and '$\Psi(\xi)$':

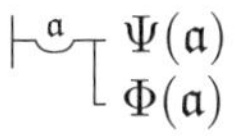

is a *universal affirmative proposition* (note that Frege only introduces terminology for propositions here), or the *subordination* of the concept named by '$\Phi(\xi)$' under the concept named by '$\Psi(\xi)$'. Hence, '$\Psi(\xi)$' names the *superordinate concept* and '$\Phi(\xi)$' names the *subordinate concept*. Frege also notes that we can formulate other familiar (Aristotelian) constructions, such as *universal negative propositions*:

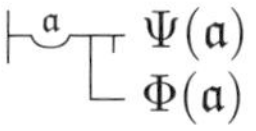

and *particular affirmative propositions*:

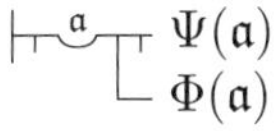

using combinations of the concavity, the negation-stroke, and the conditional-stroke (I §13).

Frege also allows higher-order quantification to be expressed via the concavity. Thus, if 'M_β' is a 2nd-level function-name (where the subscripted occurrence of 'β' binds occurrences of 'β' in the argument to which 'M_β' is applied, see I §25), then '$\overset{\mathfrak{f}}{\smile}\ M_\beta(\mathfrak{f}(\beta))$' is a name of the truth-value of the claim that, for every first-level function $\Phi(\xi)$, the result of applying the function named by 'M_β' to $\Phi(\xi)$ is the True (I §19).

Roman Letters

The second means for expressing generality within *Grundgesetze* is the *Roman letter generality device*. Restricting our attention for the moment to the simplest case, where the lower-case Roman letter 'x' *indicates* (in Frege's terminology) an object, and where '$\Phi(\xi)$' is any 1st-level function-name :

$$\vdash \Phi(x)$$

is a correct concept-script proposition if and only if the function named by '$\Phi(\xi)$'

outputs the True for every possible argument, and it is incorrect otherwise (I §17). Frege's claim that Roman letter 'x' indicates an object (and 'f' indicates a function, etc., see below) is akin to the modern idea that variables "range over" the domain of objects in question, but only roughly so.

The reader will have noticed that (following Frege) our explanation of the Roman letter generality device does not follow the general pattern utilised in our discussion of the horizontal, negation-stroke, conditional-stroke, or concavity. In short, we have not described the object or other entitity to which the concept-script expression '$\varphi(x)$' *refers* (where 'x' represents an occurrence of the Roman letter generality device), but instead have only explained when judgements involving the Roman letter generality device are *correct.*

Further, it should be noted that this sense of "correct" at work here is broader than that introduced in our discussion of the judgement-stroke, since we cannot understand:

$$\vdash \Phi(x)$$

as the judgement (or thought) that '$\Phi(x)$' is a name of the True. Unlike the other logical operations we have looked at, Frege does not stipulate that a concept-script expression containing an occurrence of the Roman letter generality device is to refer to a particular object (much less a truth-value). Rather, concept-script expressions containing instances of the Roman letter generality device are only provided with a "semantics" when such expressions are prefixed with an occurrence of the judgement-stroke. In fact, Frege *never* considers occurrences of the Roman letter generality device that are not prefixed with an occurrence of the judgement-stroke (apart from the Afterword to Volume II, which is decidedly more informal that the main body of *Grundgesetze*).

Thus, the Roman letter generality device cannot be understood as a simple function-name within the taxonomy of object and function-name types that Frege mobilises elsewhere in *Grundgesetze.* As a result, although Frege does not provide an account of what concept-script expressions containing occurrences of the Roman letter generality device denote when not flanked by the judgement-stroke (nor does he seem to think that such an explanation is required), he does introduce new terminology for such Roman letter constructions analogous to, but distinct from, the terminology introduced for object and function-names. Thus, a concept-script expression is a *Roman object-marker* if and only if it is an expression that (i) contains Roman letters and (ii) becomes a proper name when every Roman letter is replaced (uniformly) by a name. Similarly, a concept-script expression is a *Roman truth-value-marker* if and only if it is an expression that (i) contains Roman letters and (ii) becomes a truth-value-name when every Roman letter is replaced (uniformly) by a name. A concept-script expression is a *Roman function-marker* if and only it is an expression that (i) contains Roman letters and (ii) becomes a function-name when every Roman letter is replaced (uniformly) by a name. Analogous definitions of Roman concept-marker, Roman relation-marker, etc. proceed in terms of whether the result of uniformly replacing Roman letters by appropriate names results in a concept-name, relation-name, etc. (I §17).

Already implicit in the discussion above is the fact that, without a "quantifier" to "bind" the Roman letter 'x', the concept-script proposition '$\vdash\!\!\top x = x$' always takes the wide-scope reading analogous to:

for any x, it is not the case that $x = x$

rather than the narrow-scope reading analogous to:

it is not the case that, for any x, $x = x$

(and similarly for any concept-script proposition containing one or more occurrence of the Roman letter generality device). Frege explicitly opts for the wide-scope reading of the Roman letter generality device, writing that "the *scope* of a *Roman letter* should include everything that occurs in the proposition apart from the judgement-stroke" (I §17). Note Frege's assumption, in this quotation, that there always *will* be an occurrence of the judgement-stroke in constructions involving the Roman letter generality device and, further, that although he claims that the scope of a Roman letter occurring in a concept-script proposition includes *everything* in the proposition other than the judgement-stroke, he does not require that it include *only* what is contained in that proposition.

Every concept-script proposition involving the Roman letter generality device is, in a sense, "equivalent" to a concept-script proposition involving initial occurrence of the concavity. For example, if '$\Phi(\xi, \zeta)$' is a binary 1st-level function, then:

$$\vdash \Phi(x, y)$$

is a correct concept-script proposition if and only if:

$$\vdash\!\overset{\mathfrak{a}}{\smile}\!\overset{\mathfrak{e}}{\smile}\, \Phi(\mathfrak{a}, \mathfrak{e})$$

is a correct concept-script proposition. Of course, there is no substantial sense in which '$\Phi(x, y)$' and '$\overset{\mathfrak{a}}{\smile}\!\overset{\mathfrak{e}}{\smile}\, \Phi(\mathfrak{a}, \mathfrak{e})$' (without prefixed judgement-strokes) are equivalent, however, since the latter, but not the former, is a truth-value-name.

Given the fact that every concept-script proposition involving the Roman letter generality device is equivalent to one involving only occurrences of the concavity, the reader might wonder why Frege introduced the Roman letter generality device at all. The answer is simple: Judgements involving the Roman letter generality device play an important role in the deductive system of *Grundgesetze*. Frege notes that, given concept-script propositions:

$$\vdash\!\overset{\mathfrak{a}}{\smile}\!\!\begin{array}{l} \mathfrak{a}^4 = 1 \\ \llcorner\, \mathfrak{a}^2 = 1 \end{array}, \qquad \vdash\!\overset{\mathfrak{a}}{\smile}\!\!\begin{array}{l} \mathfrak{a}^8 = 1 \\ \llcorner\, \mathfrak{a}^4 = 1 \end{array}$$

we would like to be able to infer:

$$\vdash\!\overset{\mathfrak{a}}{\smile}\!\!\begin{array}{l} \mathfrak{a}^8 = 1 \\ \llcorner\, \mathfrak{a}^2 = 1 \end{array}$$

But we cannot apply hypothetical syllogism to these propositions, since they have the concavity, and not the conditional-stroke, as their "main" logical operator. If, however, we express the premises of this argument as:

$$\vdash\!\!\begin{array}{l} x^4 = 1 \\ \llcorner\, x^2 = 1 \end{array}, \qquad \vdash\!\!\begin{array}{l} x^8 = 1 \\ \llcorner\, x^4 = 1 \end{array}$$

however, then Frege claims that we *can* apply hypothetical syllogism, obtaining:

$$\vdash\!\!\begin{array}{l} x^8 = 1 \\ \llcorner\, x^2 = 1 \end{array}$$

Frege justifies this inference as follows:

> "Our stipulation regarding the *scope* of a *Roman letter* is only to demarcate its narrowest extent and not its widest. It thus remains permissible to let the scope extend to multiple [concept-script] propositions so that the Roman letters are suitable to serve in inferences in which the German letters, with their strict demarcation of scope, cannot serve. So, when our premisses are '$\vdash \begin{array}{l} x^8 = 1 \\ x^4 = 1 \end{array}$' and '$\vdash \begin{array}{l} x^4 = 1 \\ x^2 = 1 \end{array}$', then in order to make the inference to the conclusion '$\vdash \begin{array}{l} x^8 = 1 \\ x^2 = 1 \end{array}$', we temporarily expand the scope of 'x' to include both premises and conclusion, although each of these propositions holds even without this extension." (I §17)

Hence, when carrying out the inference above, Frege suggests that we temporarily extend the scope of the Roman letter 'x' so that it includes both premises and the conclusion. As a result, the Roman letter 'x' indicates the same object (whatever object this might be) uniformly throughout all three concept-script propositions, and we can apply hypothetical syllogism. Unsurprisingly, as we shall see in later sections, much of the deductive system of *Grundgesetze* is devoted to negotiating the subtle interactions between generalities involving the Roman letter generality device and generalities involving the concavity.

Frege also allows higher-order quantification to be expressed via the Roman letter generality device. Thus, if 'M_β' is a 2nd-level function-name, then:

$$\vdash M_\beta(f(\beta))$$

is a correct concept-script proposition if and only if, for every first-level function $\Phi(\xi)$, the result of applying the function named by 'M_β' to $\Phi(\xi)$ is the True. A particularly common, and initially non-obvious, instance of this sort of application is '$f(\Delta)$' where 'Δ' is a name. '$\vdash f(\Delta)$' is a correct concept-script proposition if and only if, for every first-level function $\Phi(\xi)$, the result of applying $\Phi(\xi)$ to the object named by 'Δ' is the True. Note, however, that this is really just an instance of the construction just given using 'M_β', where we understand the occurrence of 'Δ', not merely as a name of an object, but rather as the name of the 2nd-level function N_β such that, for any function $\Phi(\xi)$, the value of N_β applied to $\Phi(\xi)$ is the same as the value of $\Phi(\xi)$ applied to the object named by 'Δ' (I §19).

Needless to say, the sketch just given of the manner in which the Roman letter generality device operates within *Grundgesetze* is incomplete. Much more needs to be said regarding exactly how to understand "indication" and "scope" (amongst other things) in concept-script propositions involving Roman letters before one could claim to have presented a full explication of Frege's Roman letter generality device. The right way to work out the details of such an account is a matter of rather substantial controversy, however, and as a result the reader is encouraged to consult the growing secondary literature on these and related topics (e.g., Heck (2012), §3.2).

Value-ranges

Perhaps the most notorious primitive notion in *Grundgesetze* is Frege's value-range operator, given its central role in Russell's paradox and the collapse of Frege's *Grundgesetze*. The value-range symbol, or "smooth breathing", names a 2nd-level function from 1st-level functions to objects. Given any first-level function-name '$\Phi(\xi)$', the object named by the application of the unary 2nd-level 'smooth-breathing' operator to '$\Phi(\xi)$':

$$\dot{\varepsilon}(\Phi(\varepsilon))$$

is the value-range of the function named by '$\Phi(\xi)$'. The value-ranges corresponding to any two function-names '$\Phi(\xi)$' and '$\Psi(\xi)$' are equal if and only if the functions named by '$\Phi(\xi)$' and '$\Psi(\xi)$' output the same value for any argument (i.e., if and only if '$\Phi(\xi)$' and '$\Psi(\xi)$' name the same function, where functions are individuated extensionally). Hence, given any two 1st-level function-names '$\Phi(\xi)$' and '$\Psi(\xi)$', '$\dot{\varepsilon}(\Phi(\alpha)) = \dot{\alpha}(\Psi(\alpha))$' is a name of the True if and only if the functions named by '$\Phi(\xi)$' and '$\Psi(\xi)$' output the same value for every argument—that is, if and only if '$\smile^{\mathfrak{a}}\, \Phi(\mathfrak{a}) = \Psi(\mathfrak{a})$' is a name of the True (I §3, see also I §9).

Frege's use of the value-range operator is much broader than modern reformulations of the same or similar ideas, such as naïve set theories that admit a set corresponding to any well-formed predicate or to any property or concept. Frege's notion of value-range applies not only to concepts (i.e., 1st-level functions from objects to truth-values), but to any function whatsoever. Thus, the function named by '$\xi + 1$', which given any natural number as input, outputs the successor of that number, has a value-range—an object that "codes up" the argument–value pairs determined by the function. The value-range of the function named by '$\xi + 1$' is in some ways analogous to its graph, in the set-theoretic sense—that is, to:

$$\{< x, y >: y = x + 1\}$$

although it is important to note that Frege inverts the order of explanation here: value-ranges are not constructed from sets, but are primitive, and various set-like notions are constructed, within *Grundgesetze*, in terms of value-ranges.

Along these lines, one of the most important applications of the value-range operator is its application to 1st-level concepts, and the resulting objects, which Frege calls *extensions*, can be thought of, from a modern perspective and loosely speaking, as akin to the graphs of the characteristic functions of these concepts. Extensions do "behave" logically very similarly to (naïve) sets, and as we shall see below, Frege introduces an application operator '$\frown$' on value-ranges that behaves very much like the set membership relation '$\in$' when it is applied to extensions (although, like the value-range operator itself, it has much wider applicability). The sensitive reader should be wary of attributing too much of our own modern views about sets onto extensions, however.

Frege identifies another sub-class of objects that can be constructed using the value-range operator that do not correspond to anything widely used within modern mathematics: double value-ranges. Given any binary 1st-level function-name '$\Phi(\xi, \zeta)$', we form the double value-range of (the function named by)'$\Phi(\xi, \zeta)$' by applying the value-range operator to '$\Phi(\xi, \zeta)$' (binding the argument place marked by 'ξ'), obtaining the unary 1st-level function-name '$\dot{\varepsilon}(\Phi(\varepsilon, \zeta))$'. We now obtain the double value-range of '$\Phi(\xi, \zeta)$' by applying the value-range function a second time, to '$\dot{\varepsilon}(\Phi(\varepsilon, \zeta))$', to obtain '$\dot{\varepsilon}\dot{\alpha}(\Phi(\varepsilon, \alpha))$', which names the double value-range of the function named by '$\Phi(\xi, \zeta)$' (I §36).

Just as Frege insists that it must always be possible to transform the generality of an equality to the equality of (single) value-ranges, he also demands that a similar transformation is possible with double value-ranges. In other words, the truth-value named by ‘$\grave{\alpha}\grave{\varepsilon}(\Phi(\varepsilon,\alpha)) = \grave{\alpha}\grave{\varepsilon}(\Psi(\varepsilon,\alpha))$’ must be equal to the truth-value named by ‘$\smile^{\mathfrak{a}}\smile^{\mathfrak{e}}\ \Phi(\mathfrak{e},\mathfrak{a}) = \Psi(\mathfrak{e},\mathfrak{a})$’ (although, interestingly, this equality is never explicitly proven within *Grundgesetze*). Thus, just as thc value-range operator provides us with a means for introducing an object-level surrogate—a value-range—for each unary 1st-level function, it also provides us with a means for introducing a (more complex) surrogate for each binary 1st-level function—a double value-range. The technique can of course be generalised to so “reduce” any n-ary 1st-level relation to a corresponding “n-fold” value-range, although Frege, within *Grundgesetze*, only uses single and double value-ranges.

As we have seen, any binary function has a double value-range, but Frege introduces a special term—“Relation” (with a capital ‘R’)—to refer to the double value-range of a relation (i.e., the double value-range of a binary function that maps each pair of objects to a truth-value). Thus, the double value-range of any relation is a Relation (II §162, see also II §165). This terminology will be helpful in our discussion of Frege’s definitions below.[4]

Frege only defines value-ranges for 1st-level functions. Of course, he could have extended the notion in order to obtain object-level analogues of 2nd- and higher-level functions directly. His reasons for not doing so are simple: the value-range operator allows us to “reduce” 2nd- and higher-level functions to 1st-level functions via repeated applications of the value-range operator on 1st-level functions. For example, given a 2nd-level concept-name ‘M_β’ mapping 1st-level functions to truth-values, we can construct an object-level analogue by first constructing the name of the concept that holds of an object if and only if that object is the value-range of a 1st-level function that the concept named by ‘M_β’ maps to the True:

$$\top\!\!\smile^{\mathfrak{f}}\!\!\top\begin{array}{l} M_\beta(\mathfrak{f}(\beta)) \\ \xi = \grave{\varepsilon}(\mathfrak{f}(\varepsilon)) \end{array}$$

The object-level analogue of the 2nd-level concept named by ‘M_β’ is then the value-range of this 1st-level concept:

$$\grave{\alpha}\left(\top\!\!\smile^{\mathfrak{f}}\!\!\top\begin{array}{l} M_\beta(\mathfrak{f}(\beta)) \\ \alpha = \grave{\varepsilon}(\mathfrak{f}(\varepsilon)) \end{array}\right)$$

This maneuver is quite general, and, as a result, any time one might, within *Grundgesetze*, desire an object-level analogue of a 2nd- or higher-level function, one can instead use this trick (and tricks like it for functions more generally) to construct such an object indirectly. As a result, there is no need for a primitive value-range operator attaching directly to such higher-level functions.

The Backslash

The *backslash* is a unary 1st-level function, mapping objects to objects. Given any proper name ‘Γ’:

[4] Frege’s distinction is not marked by capitalization in the original German. Frege’s term for a relation (i.e., a binary first-level function from objects to truth-values) is “*Beziehung*”, and his term for Relation (i.e., the double value-range of a relation) is “*Relation*”.

$$\backslash\Gamma$$

is the name of an object named by 'Δ' if and only if 'Γ' is a name of the value-range of the concept that holds of an object if and only if it is the object named by 'Δ', if there is such a value-range. Otherwise, '$\backslash\Gamma$' names the same object as 'Γ' names. This can be put a bit more clearly if we introduce a bit of terminology. Given any proper name 'Δ', let us call the object named by '$\dot{\varepsilon}(\Delta = \varepsilon)$' the *singleton-extension* of (the object named by)'Δ'. Then '$\backslash\Gamma$' is a name of the object named by 'Δ' if and only if 'Γ' names the singleton-extension of (the object named by) 'Δ', and is co-referential with 'Γ' otherwise (I §11).

Read literally, the backslash is a kind of "singleton-stripping" device. The function named by '$\backslash\xi$' maps any singleton-extension onto the sole object "contained" in that singleton, and maps any non-singleton onto itself. While this gets the backslash right from a formal perspective, taken at face value it is somewhat misleading. In adding this function to *Grundgesetze*, Frege is not particularly interested in the extensions-theoretic function from singleton-extensions to their unique "members". Instead, Frege utilises the backslash as a definite description operator. In modern treatments, a definite description operator 'ι' attaches to predicates and, given a predicate '$\Phi(x)$', '$\iota x(\Phi(x))$' denotes the unique object that satisfies the predicate '$\Phi(\xi)$' (if there is such). Frege, however, in keeping with his general strategy of reducing levels via successive applications of the value-range operator, defines his definite description operator as applying, not to concepts (or the function-names denoting them), but rather to the value-ranges. Thus, where '$\Phi(\xi)$' is a concept-name, Frege achieves the same purpose as is served by the definite description operator 'ι' by applying the backslash, not directly to '$\Phi(\xi)$', but rather to '$\dot{\varepsilon}(\Phi(\varepsilon))$'—hence, '$\backslash\dot{\varepsilon}(\Phi(\varepsilon))$' denotes the unique object that is mapped to the True by the concept named by '$\Phi(\xi)$', if there is such, and denotes the object named by '$\dot{\varepsilon}(\Phi(\varepsilon))$' otherwise.[5]

The Logic of *Grundgesetze*

Rules of Inference

We have already mentioned two rules of inference that Frege applies for the most part without comment or label in derivations. First, there is the fusion of horizontals, which allows the addition or deletion of horizontals to instances of the horizontal itself, the negation-stroke, the conditional-stroke, and the concavity, and also allows the addition or deletion of such horizontals to the arguments of these function-names (the fusion of horizontals is first given explicitly as a rule at I §48). Second, Frege allows the permutation of subcomponents, and the fusion of equal subcomponents, within a conditional-stroke construction, again without comment or specific label (these are first given as explicit rules in I §48 and I §15, respectively). In this section we shall examine the remaining rules of inference within the deductive system of *Grundgesetze*. After discussing each of the rules individually, Frege provides a nice summary of the rules of inference in I §48.

[5] The term "backslash" is not Frege's—he refers to this function symbol as the "replacement for the definite article" (I §11) and does not seem to have a special name for the symbol itself.

Generalised Modus Ponens

The first rule of inference is *generalised modus poncns.* Loosely put: if one has proven a conditional, and one has also proven a subcomponent of that conditional, then one may infer the result of deleting that subcomponent from the conditional. Thus, given a concept-script proposition of the form:

$$\begin{array}{l} \vdash\!\top\, \Gamma \\ \quad \llcorner\, \{\Delta_1, \cdots \Delta_n\} \end{array}$$

(this is the *major premise* of the inference, recalling that there might be more than one way to parse a proposition along these lines) and a concept-script proposition '$\vdash \Delta_m$' ($1 \leq m \leq n$) (the *minor premise*), we can infer:

$$\begin{array}{l} \vdash\!\top\, \Gamma \\ \quad \llcorner\, \{\Delta_1, \cdots \Delta_{m-1}, \Delta_{m+1}, \cdots \Delta_n\} \end{array}$$

There are two ways in which applications of generalised modus ponens are represented, depending on whether application of the rule occurs immediately after an occurrence of the major premise, or immediately after an occurrence of the minor premise. If the former, then the subcomponent to be eliminated occurs earlier, and will be labeled with a lower-case Greek letter:

$$\vdash \Delta_m \qquad\qquad (\beta$$

Then the application of the rule will occur immediately after the major premise, and appear (in the notation introduced ealier) as:

$$(\beta){::}\; \frac{\begin{array}{l} \vdash\!\top\, \Gamma \\ \quad \llcorner\, \{\Delta_1, \cdots \Delta_n\} \end{array}}{\begin{array}{l} \vdash\!\top\, \Gamma \\ \quad \llcorner\, \{\Delta, \cdots \Delta_{m-1}, \Delta_{m+1}, \cdots \Delta_n\} \end{array}}$$

If, however, it is the major premise that occurs earlier in the proof (and is labeled by 'γ', say), then the application of the rule will occur immediately after the subcomponent to be eliminated, and will appear as:

$$(\gamma){:}\; \frac{\vdash \Delta_m}{\begin{array}{l} \vdash\!\top\, \Gamma \\ \quad \llcorner\, \{\Delta, \cdots \Delta_{m-1}, \Delta_{m+1}, \cdots \Delta_n\} \end{array}}$$

Applications of generalised modus ponens are indicated by a solid horizontal line—this is the first instance of a *transition sign.* A double colon is used to indicate an application of the rule of the first sort, and a single colon is used to indicate an application of the second sort (I §14).

Frege also allows for multiple, simultaneous applications of the generalised form of modus ponens. Thus, if we have, earlier in a derivation, derived both '$\vdash \Delta_k$' (at line β, say) and '$\vdash \Delta_m$' (at line γ, say), then, given some concept-script proposition that contains both Δ_k and Δ_m as subcomponents ($1 \leq k < m \leq n$), we may eliminate both subcomponents simultaneously:

$$(\beta, \gamma){::}\; \begin{array}{l} \vdash\!\top\, \Gamma \\ \quad \llcorner\, \{\Delta_1, \cdots \Delta_n\} \\ \hline\hline \vdash\!\top\, \Gamma \\ \quad \llcorner\, \{\Delta_1, \cdots \Delta_{k-1}, \Delta_{k+1} \cdots \Delta_{m-1}, \Delta_{m+1}, \cdots \Delta_n\} \end{array}$$

Note that multiple applications of generalised modus ponens only occur in the form (marked by the double colon) where the inference occurs immediately after the major premise, and the fact that multiple minor premises are involved is indicated both by the fact that two labels are listed and by the double horizontal line (I §14).[6]

Generalised Hypothetical Syllogism

The next rule of inference is *generalised hypothetical syllogism.* This rule is a generalised version of the familiar rule *hypothetical syllogism* (or transitivity) which, in the notation of *Grundgesetze*, would allow us to pass from concept-script propositions of the form:

$$\vdash\!\begin{array}{l}\Gamma\\ \Delta\end{array},\quad \vdash\!\begin{array}{l}\Delta\\ \Theta\end{array}$$

to:

$$\vdash\!\begin{array}{l}\Gamma\\ \Theta\end{array}$$

The version of this rule mobilised with *Grundgesetze* is a more powerful version of this rule that takes advantage of the "equal" status of subcomponents: Given a concept-script proposition, and a second concept-script proposition whose supercomponent is a subcomponent of the first, we can infer the concept-script proposition that results from replacing the relevant subcomponent in the first proposition with the subcomponents of the second proposition. Further, any subcomponent that occurs in both premises need only be included once in the inferred proposition (I §15).

There are, again, two forms of this rule. In either case, if we have concept-script propositions of the form:

$$\vdash\!\begin{array}{l}\Gamma\\ \{\Delta_1,\cdots\Delta_n\}\end{array},\quad \vdash\!\begin{array}{l}\Delta_k\\ \{\Theta_1,\cdots\Theta_m\}\end{array}$$

$(1 \leq k \leq n)$ we can then infer:

$$\vdash\!\begin{array}{l}\Gamma\\ \{\Delta_1,\cdots\Delta_{k-1},\Theta_1,\cdots\Theta_m,\Delta_{k+1},\cdots\Delta_n\}\end{array}$$

[6] The term "generalised modus ponens" is not Frege's—he calls this rule "Inferring (a)" (I §48).

Applications of generalised hypothetical syllogism are indicated by new type of transition sign, a dashed horizontal line '– – – – –' (prefixed with the appropriate Greek letters labeling premises, and single or double colon depending on whether it is the major or minor premise that occurs immediately above the inference). Like generalised modus ponens, Frege allows multiple simultaneous applications of this generalised form of hypothetical syllogism. Multiple applications of generalised hypothetical syllogism only occur in the form (marked by the double colon) where the inference occurs immediately after the major premise, and the fact that multiple minor premises are involved is indicated both by the fact that two labels are listed and by the double dashed horizontal line '= = = = =' (I §15).

Before moving on, it is worth noting the sort of justification Frege provides for this particular rule. Frege writes that:

> "For $\begin{array}{l}\Gamma\\ \Theta\end{array}$ is only the False if Θ is the True and Γ is not the True. However if Θ is the True then also Δ must be the True because otherwise $\begin{array}{l}\Delta\\ \Theta\end{array}$ would be the False. If, however, Δ is the True and were Γ not the True, then $\begin{array}{l}\Gamma\\ \Delta\end{array}$ would be the False. The case in which $\begin{array}{l}\Gamma\\ \Theta\end{array}$ is the False cannot, therefore, occur, and $\begin{array}{l}\Gamma\\ \Theta\end{array}$ is the True." (I §15)

Frege is quite careful here to distinguish between a concept-script expression being a name of the False and merely failing to be a name of the True (and similarly, between being a name of the True, and merely failing to be a name of the False).

Thus, this justification for hypothetical syllogism not only shows that Frege was, in fact, interested in providing and studying the (informal) semantics of the formal system of *Grundgesetze*, but also shows that he was quite careful to attend to the fact that his propositional connectives are, in fact, functions from *any* object to truth-values. As a result, Frege realised the need to verify that the rules of inference adopted in the deductive system are valid even when the arguments to which the relevant conditional-stroke instances are applied are not names of truth-values.[7]

Generalised Contraposition

The third rule of inference is *generalised contraposition.*[8] Like generalised modus ponens and generalised hypothetical syllogism, generalised contraposition is a more powerful version of the standard rule of contraposition, which, in the notation of *Grundgesetze* would allow us to move from:

$$\vdash\!\!\begin{array}{l}\Gamma\\ \Delta\end{array}$$

to:

$$\vdash\!\!\begin{array}{l}\Delta\\ \top\,\Gamma\end{array}$$

Generalised contraposition allows us to "switch" the supercomponent of a concept-script proposition with *any* subcomponent of it, provided one "simultaneously *reverses* the truth-value of each" (I §15).

Applications of generalised contraposition are represented using the transition sign '$\times$'. Thus:

$$\vdash\!\!\begin{array}{l}\Gamma\\ \{\Delta_1,\cdots\Delta_n\}\end{array}$$

$$\times$$

$$\vdash\!\!\begin{array}{l}\top\,\Delta_m\\ \{\Delta_1,\cdots\Delta_{m-1},\top\,\Gamma,\Delta_{m+1},\cdots\Delta_n\}\end{array}$$

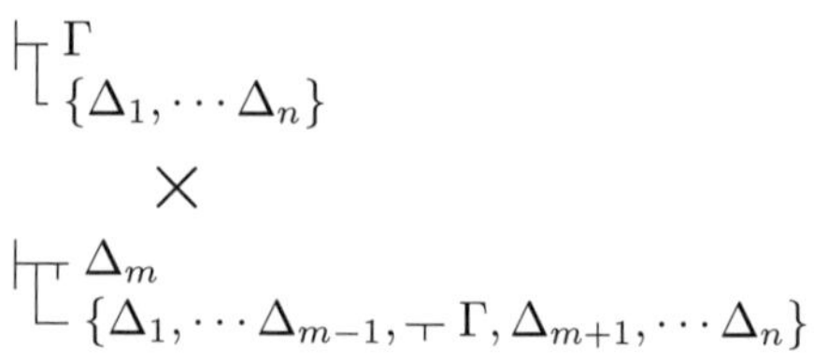

Note that the concept-script expressions that have been switched are prefixed with a negation-stroke. Frege's understanding of the phrase "*reverses* the truth-value" (I §15) is, however, more general than this. Hence, instead of adding a negation-stroke to each of the components that are switched when applying generalised contraposition, we can remove a negation if one is present. Thus, from:

$$\vdash\!\!\begin{array}{l}\top\,\Gamma\\ \Delta\end{array}$$

we can infer, via generalised contraposition, either of:

$$\vdash\!\!\begin{array}{l}\top\,\Delta\\ \top\top\,\Gamma\end{array},\qquad \vdash\!\!\begin{array}{l}\top\,\Delta\\ \Gamma\end{array}$$

Generalised Dilemma

The fourth rule of inference is *generalised dilemma.* This rule states that, given two concept-script propositions with the same supercomponent, where a subcomponent of one is the negation of a subcomponent of the other, we may infer the concept-script proposition whose supercomponent is the same as in the two premises, and whose subcomponents are exactly the subcomponents of the two premises, excepting the truth-functionally "opposed" subcomponents. Thus, if we have derived concept-script propositions of the form:

$$\vdash\!\!\begin{array}{l}\Delta\\ \{\Gamma_1,\cdots\Gamma_n,\Theta\}\end{array},\qquad \vdash\!\!\begin{array}{l}\Delta\\ \{\Omega_1,\cdots\Omega_m,\top\,\Theta\}\end{array}$$

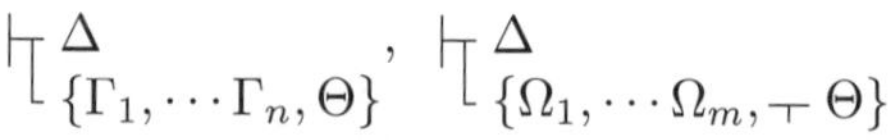

we can infer:

$$\vdash\!\!\begin{array}{l}\Delta\\ \{\Gamma_1,\cdots\Gamma_n,\Omega_1,\cdots\Omega_M\}\end{array}$$

[7] The term "generalised hypothetical syllogism" is not Frege's—he calls this rule "Inferring (b)" (I §48).

[8] Frege calls this rule "*Wendung*", which has been translated as "contraposition" (I §15).

Generalised dilemma is indicate by the dot-dashed line '—·—·—·—'. Note that the order of the subcomponents in the premises does not matter (the "opposed" subcomponents are written last in the example above merely for convenience), and that we only have the single-colon version of the rule, since in this case there is no privileging one premise as "major", the other as "minor" (I §16).⁹

Concavity Introduction

Frege's next rule of inference governs the interactions between the two devices for expressing generality within *Grundgesetze*: the Roman letter generality device and the concavity. *Concavity introduction* states that, given any concept-script proposition containing a Roman letter, we can infer any concept-script proposition that uniformly replaces the Roman letter with a German letter (of the appropriate type) and inserts a concavity containing the same German letter immediately in front of some supercomponent that contains all occurrences of the new German letter. Additionally, the particular German letter must be chosen so that it does not "conflict" with other German letters already present in the original proposition (see I, §8). Thus, if 'Δ' is any name not containing the Roman letter 'x', and '$\Phi(\xi)$' and '$\Psi(\xi)$' are two 1ˢᵗ-level function-names, then from:

$$\vdash \begin{array}{l} \Psi(x) \\ \Phi(x) \\ \Delta \end{array}$$

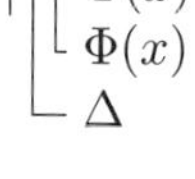

we can infer either of:

$$\vdash \underset{\mathfrak{a}}{\frown} \begin{array}{l} \Psi(\mathfrak{a}) \\ \Phi(\mathfrak{a}) \\ \Delta \end{array}, \quad \vdash \underset{\mathfrak{a}}{\frown} \begin{array}{l} \Psi(\mathfrak{a}) \\ \Phi(\mathfrak{a}) \\ \Delta \end{array}$$

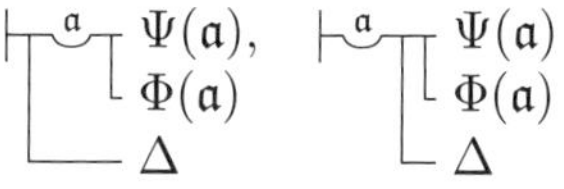

but not:

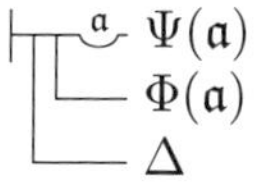

since in the final concept-script proposition the concavity does not bind all occurrences of the German letter '$\mathfrak{a}$' (I §17).

The transition symbol used to label applications of concavity introduction is '$\frown$'. This rule is, of course, not restricted to first-order instances of the Roman letter generality device. Thus, if 'Δ' is any name not containing the Roman letter 'f', and 'Γ' and 'Θ' are any two names, then from:

$$\vdash \begin{array}{l} f(\Theta) \\ f(\Gamma) \\ \Delta \end{array}$$

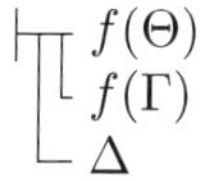

we can infer either of:

$$\vdash \underset{\mathfrak{f}}{\frown} \begin{array}{l} \mathfrak{f}(\Theta) \\ \mathfrak{f}(\Gamma) \\ \Delta \end{array}, \quad \vdash \underset{\mathfrak{f}}{\frown} \begin{array}{l} \mathfrak{f}(\Theta) \\ \mathfrak{f}(\Gamma) \\ \Delta \end{array}$$

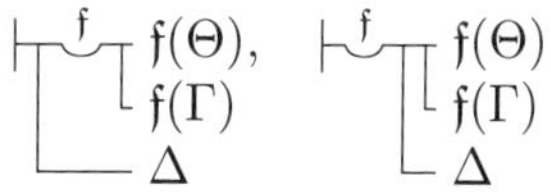

via an application of concavity introduction.¹⁰

Roman Letter Elimination

The reader will have noted that, while the concavity introduction rule discussed in the previous section allows us to move from one means of expressing generality to another, it does not allows us to move from a concept-script proposition expressing a generality (of either sort) to an instance. In short, we as of yet have no means for moving from a concept-script proposition of the form '$\vdash \Phi(x)$' to a proposition of the form '$\vdash \Phi(\Delta)$' where 'Δ' is a proper name. Frege rectifies this by introducing what we shall call the

---

⁹ The term "generalised dilemma" is not Frege's—he calls this rule "Inferring (c)" (I §48).

¹⁰ The term "concavity introduction" is not Frege's—he calls this rule "Transformation of a Roman letter into a German letter" (I §48).

Roman letter elimination rule, which he describes as follows:

> "we may effect a simple inference by uniformly replacing a Roman letter within the proposition by the same proper name or the same Roman object-marker." (I §9)

This rule allows for two sorts of application. The first, and simplest case, involves uniformly replacing a particular Roman letter with a proper name (the term "uniformly" captures what Frege intends by the somewhat more awkward "... by the same proper name..."). But the Roman letter elimination rule also allows us to replace a single Roman letter with a Roman object-marker. Hence, using this rule we can also obtain '$\vdash \Phi(\top\, y)$' from '$\vdash \Phi(x)$' by replacing the Roman letter 'x' with the Roman object-marker '$\top\, y$'. There is no transition symbol for this inference.

Frege also introduces a "higher-order" version of this rule, which allows us to replace any Roman function letter with either an appropriate function-name or a Roman function-marker of the appropriate type. Hence, given a concept-script proposition of the form '$\vdash f(\Delta)$' we can infer either '$\vdash\!\top \Delta$' (by replacing the Roman letter 'f' with the function-name '$\top$') or '$\vdash\!\top g(\top\, \Delta)$' (by replacing the Roman letter 'f' with the Roman function-marker '$\top\, g \top$'). One of the most visible and most important applications of the Roman letter elimination rule is when introducing "instances" of one of the basic laws. As we will see below, the laws of *Grundgesetze* are not schematic, but instead are formulated in terms of the Roman letter generality device. When using a basic law in a derivation, however, Frege does not require that we write the law explicitly. Instead, we can cite any instance of the law that results from applying the Roman letter elimination rule (one or more times) to the law itself.[11]

Combining Rules

Frege sometimes allows application of more than one rule in a single step. For example, if we are given three concept-script propositions:

$$\begin{array}{l}\vdash\!\top\, \Delta \\ \;\;\llcorner\, \{\Gamma_1, \cdots \Gamma_n\}\end{array}, \quad \begin{array}{l}\vdash\!\top\, \Gamma_k \\ \;\;\llcorner\, \{\Upsilon_1, \cdots \Upsilon_p\}\end{array}, \quad \vdash \Gamma_m$$

$(1 \leq k \leq m \leq n)$ then we can infer:

$$\begin{array}{l}\vdash\!\top\, \Delta \\ \;\;\llcorner\, \{\Gamma_1, \cdots \Gamma_{k-1}, \Upsilon_1, \cdots \Upsilon_p, \Gamma_{k+1} \cdots \Gamma_{m-1}, \Gamma_{m+1}, \cdots \Gamma_n\}\end{array}$$

via a simultaenous application of generalised modus ponens and generalised hypothetical syllogism. The fact that multiple rules are involved is indicated by a "double" transition sign, in this case the hybrid '=====' incorporating both the dashed horizontal line indicating generalised hypothetical syllogism and the solid line indicating generalised modus ponens.

Frege only allows multiple applications of inference rules in a single step in three kinds of case: two applications of generalised modus ponens (marked by a double solid line), two applications of generalised hypothetical syllogism (marked by a double dashed line), or one application of each as above (I §§14–15).

[11] The term "Roman letter elimination" is not Frege's—he calls this rule "Citing propositions: Replacement of Roman letters" (I §48).

Derivation Breaks

Frege introduces one last important piece of primitive notation, one that serves, in effect, as a sort of punctuation mark within long proofs, subdividing the proofs into what we might informally think of as lemmas. At certain points within a derivation we will want to introduce concept-script propositions that are not obtained from the immediately preceding propositions (for example, via the use of basic laws or definitions). Whenever such a new subderivation is introduced, Frege annotates this "shift of attention" by separating the two propositions with an occurrence of '——— • ———'. Note that this transition sign does not indicate the use of any particular rule, but is instead a sort of punctuation mark, indicating that the derivation has, in effect, gone off in a different direction from the concept-script propositions that immediately preceed it (I §50).

Basic Laws

In addition to the rules of inference just discussed, *Grundgesetze* contains six *basic laws* (some of which have multiple versions). Frege provides a summary of the basic laws in I §47.

Basic Law I

Frege's first basic law—Basic Law I (I §18)—is:

$$\vdash \begin{array}{c} a \\ b \\ a \end{array}$$

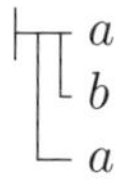

This is the *Grundgesetze* analogue of the following familiar classical tautology '$(\Delta \supset (\Gamma \supset \Delta))$' or, alternatively, of '$((\Delta \wedge \Gamma) \supset \Delta)$'. It is important to notice, however, that this law contains two instances of the Roman generality device. Thus, it is not, contrary to first appearances (and to modern eyes), a

schema, but is instead akin to a universally quantified formula. In short, this law expresses the fact that, given any two names (not just any two truth-value-names) 'Δ' and 'Γ':

$$\vdash \begin{array}{c} \Delta \\ \Gamma \\ \Delta \end{array}$$

is a name of the True. Given the discussion above, it should come as little surprise that this law is, in fact, correct even when the objects in question are not truth-values. Frege notes (I §18) that:

$$\vdash \begin{array}{c} a \\ a \end{array}$$

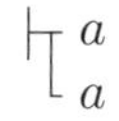

is a special instance of the formulation of Basic Law I above, obtained by replacing 'b' with 'a' and then fusing equal subcomponents. Given its obvious utility, Frege lists this as a second version of Basic Law I, one that we can use as a primitive law without explicitly deriving it.

Basic Law II

Basic Law II comes in two varieties, a "first-order" version and a "second-order" version. The first version, Basic Law IIa (I §20), is:

$$\vdash \begin{array}{c} f(a) \\ \text{\textfrak{a}}\, f(\text{\textfrak{a}}) \end{array}$$

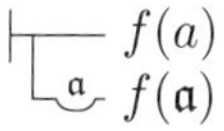

Again, care should be taken to remember that the 'a' appearing in the supercomponent (as well as the 'f' occurring in both subcomponent and supercomponent) is not schematic, but is an instance of the Roman letter generality device. The import of this law is nicely summed up by Frege himself, who described it as expressing the thought that "what holds of all objects, also holds of any" (I §20). This law, in the presence of generalised modus ponens, functions as a sort of converse to the concavity introduction rule, since it provides a means for inferring a Roman letter generality from a generality formulated using the concavity. Given a concept-script proposition of the form '$\vdash\!\!\overset{\mathfrak{a}}{\smile}\!\!-\, \Phi(\mathfrak{a})$', we can invoke an instance of Basic Law II:

$$\begin{array}{l}\vdash\!\!\!\top\!\!-\!\!-\; \Phi(a)\\ \;\,\llcorner\!\overset{\mathfrak{a}}{\smile}\!\!-\; \Phi(\mathfrak{a})\end{array}$$

and combine these, using Generalised modus ponens, to conclude '$\vdash \Phi(a)$'.

The second-order version of Basic Law II is called Basic Law IIb (I §25):

$$\begin{array}{l}\vdash\!\!\!\top\!\!-\!\!-\; M_{\beta}(f(\beta))\\ \;\,\llcorner\!\overset{\mathfrak{f}}{\smile}\!\!-\; M_{\beta}(\mathfrak{f}(\beta))\end{array}$$

This rule, in essence, allows us to replace a concavity binding a 1st-level function variable with an instance of the Roman letter generality device, and is used analogously to the first-order version.

Basic Law III

Basic Law III (I §20), the basic law governing the equality-sign, appears at first glance to be a variant of the indiscernibility of identicals:

$$\begin{array}{l}\vdash\!\!\!\top\; g\left(\overset{\mathfrak{f}}{\smile}\!\!\!\top\!\begin{array}{l}\mathfrak{f}(a)\\ \llcorner\,\mathfrak{f}(b)\end{array}\right)\\ \;\,\llcorner\; g(a=b)\end{array}$$

Basic Law III, loosely put, states that for any unary 1st-level function-name '$\Phi(\xi)$' and any proper names 'Δ' and 'Γ', it is not the case that the application of the function named by '$\Phi(\xi)$' to the truth-value named by '$\Delta = \Gamma$' is the True, while application of the function named by '$\Phi(\xi)$' to the truth-value named by:

$$\overset{\mathfrak{f}}{\smile}\!\!\!\top\!\begin{array}{l}\mathfrak{f}(\Delta)\\ \llcorner\,\mathfrak{f}(\Gamma)\end{array}$$

is something other than the True. In short, since for any two distinct objects there will be at least one function that maps the first to the True and the second to something other than the True, it follows that this law implies that the truth-value named by the equality is equal to the truth-value named by the generality. In practice, this law amounts to the claim that one can always replace an equality with the corresponding universally quantified concept-script expression in any concept-script proposition (i.e., as the argument of any function g).

Note that the presence of the Roman letter 'g' renders this much more general than the modern version of the indiscernibility of identicals. To see why, we need merely remember that the negation-stroke is one of the functions that can be substituted for 'g'. Thus, Basic Law III (plus an application of the Roman letter elimination rule) provides:

$$\begin{array}{l}\vdash\!\!\!\top\!\!\top\left(\overset{\mathfrak{f}}{\smile}\!\!\!\top\!\begin{array}{l}\mathfrak{f}(a)\\ \llcorner\,\mathfrak{f}(b)\end{array}\right)\\ \;\,\llcorner\!\top\; (a=b)\end{array}$$

which, by the generalised contraposition, gives us:

$$\begin{array}{l}\vdash\!\!\!\top\!\!-\!\!-\!\!-\; (a=b)\\ \;\,\llcorner\!\overset{\mathfrak{f}}{\smile}\!\!\!\top\!\begin{array}{l}\mathfrak{f}(a)\\ \llcorner\,\mathfrak{f}(b)\end{array}\end{array}$$

Thus, this law implies a *Grundgesetze* analogue of the identity of indiscernibles.

Basic Law IV

Basic Law IV (I §18):

$$\begin{array}{l}\vdash\!\!\!\top\; (-\, a) = (-\, b)\\ \;\,\llcorner\!\top\; (-\, a) = (\,\top\, b)\end{array}$$

appears, at first glance, to be nothing more than a *Grundgesetze* analogue of a familiar principle of classical propositional logic:

$$(\neg(\Delta \equiv \neg\Gamma)) \supset (\Delta \equiv \Gamma)$$

As usual, however, we should be careful

not to read this law as only applying to truth-value-names. Instead, instances of Basic Law IV name the True no matter what names are substituted in for 'a' and 'b'. Nevertheless, given that all occurrences of Roman letters within Basic Law IV are prefixed by the horizontal, the import of this principle is much the same as the classical analogue: Given any two truth-values, if the first is not equal to the negation of the second, then they are themselves equal. Thus, this principle expresses (amongst other things) Frege's commitment to classical logic.

Basic Law V

Given the central role that value-ranges and Basic Law V (I §20) have played in post-paradox discussions of Frege's logicism, it is somewhat surprising how little fanfare accompanies Frege's introduction of Basic Law V (just as value-ranges themselves are introduced earlier in *Grundgesetze* with equally little fanfare). Frege merely notes once again that:

> "a value-range equality can always be converted into the generality of an equality, and *vice versa*" (I §20)

and then states the law:

$$\vdash (\grave{\varepsilon}f(\varepsilon) = \grave{\alpha}g(\alpha)) = (\smile^{\mathfrak{a}} f(\mathfrak{a}) = g(\mathfrak{a}))$$

It is worth noting, however, that the *Grundgesetze* formulation of Basic Law V is a good bit more general than well-known modern representations of this law within higher-order logic such as:

$$(\forall X)(\forall Y)(\S(X) = \S(Y) \equiv (\forall z)(X(z) \equiv Y(z)))$$

The *Grundgesetze* formulation of Basic Law V entails not only that every concept has a corresponding extension, but in addition that any function whatever (concept or not) has a value-range. Thus, Basic Law V also captures something akin to:

$$(\forall f)(\forall g)(\S(f) = \S(g) \equiv (\forall z)(f(z) = g(z)))$$

Since within modern treatments of higher-order logic concepts are a different sort from functions, rather than being a sub-class of the class of functions as in *Grundgesetze*, Frege's formulation of Basic Law V is not equivalent to either of these modern formulations.

Basic Law VI

The final basic law of *Grundgesetze*, Basic Law VI (I §18), governs the behavior of the backslash:

$$\vdash a = \backslash\grave{\varepsilon}(a = \varepsilon)$$

This law makes explicit one part of the informal definition of the backslash operator discussed earlier. If 'Δ' is a name of the singleton-extension of the object named by Γ, then the result of applying the backslash to 'Δ' names the same object as does 'Γ'.

This basic law—or, more carefully, the lack of a second principle that we might expect to accompany it—leads to a striking insight into Frege's methodology. Basic

Law VI tells us exactly what results when we apply the backslash to the name of a singleton-extension. But it tells us nothing regarding what results when we apply the backslash to a name that is *not* the name of a singleton-extension. We might, quite fairly, wonder why Frege did not add a second law governing this latter case. It would not have been difficult to do. For example:

$$\vdash\!\!\begin{array}{l} \text{———} \backslash b = b \\ \llcorner\!\smile^{\mathfrak{a}}\!\top\; b = \acute{\varepsilon}(\mathfrak{a} = \varepsilon) \end{array}$$

would suffice. But Frege had no need for such a law in his derivation of arithmetic within the formal system of *Grundgesetze* (and we can presume he also had no need of it in his envisioned derivation of real and complex analysis). Thus, he did not need to add it to his stock of basic laws. Frege only needs to identity those logical principles that are required for his reconstructions of arithmetic and analysis—he did not need to identify *every* principle that might be a logical truth. In short, Frege's project did not require a logic that was proof-theoretically complete in the modern sense, and as a result we should not be surprised that the logic he did formulate would have appeared (to Frege, to his readers, and to us) to be rather obviously incomplete (had Basic Law V not rendered it inconsistent, and hence trivially complete).

The Definitions of *Grundgesetze*

Throughout *Grundgesetze* Frege introduces a number of definitions in order to streamline his presentation of ideas, derivations, and important theorems. These definitions divide rather naturally (although see below) into two categories: those introduced in Volume I and used within Frege's derivation of arithmetic and his study of the infinite, and those introduced in Volume II, which continues his examination of the infinite and also contains the beginnings of Frege's construction of the real numbers (the definitions from Volume I continue to play a central role in Volume II, however).

A terminological reminder: Recall that "Relation" (with capitalised "R") refers to double value-ranges of relations. Further, if 'Δ' is a name of the double value-range of a relation $\Phi(\xi, \zeta)$ (i.e., 'Δ' names the corresponding Relation), then we will say that the Relation named by 'Δ' *holds* of Γ and Θ, understanding this as shorthand for the claim that the relation $\Phi(\xi, \zeta)$ holds of Γ and Θ (i.e., $\Phi(\xi, \zeta)$ maps Γ and Θ to the True). As we shall see, this terminological shorthand will help to simplify the explanation of a number of Frege's definitions.

Understanding Frege's conception of definition within *Grundgesetze*, as well as the role played by the various definitions themselves, is a deep and interesting topic, worthy of a great deal of further scrutiny. Here I will merely give a rough explanation of the meaning and import of each of these definitions, and make a few further observations when appropriate. Further, I will in most cases only explain how each definition works when the notion in question is used in the intended sense. Thus, I will not explain what object is obtained when the composition operation (which is intended to map pairs of Relations to Relations) is applied to non-Relations. The reader should remember, however, that every function within *Grundgesetze*—hence, every defined function—is total. (A useful exercise for those wishing to gain a deeper understanding of the nuts-and-bolts of *Grundgesetze* is determining the values of various defined functions when they are applied to such "unintended" arguments!) Thus, the comments below should be taken in roughly the same spirit as Frege's own (much more brief) glosses of these notions given in Appendix 2 of Volume I:

> "These short hints in words which I add to the concept-script definitions are not exhaustive and make no claim to be of the strictest precision."

It is perhaps worth emphasizing, however, that in the discussion to follow it is exhaustiveness, rather than precision, that is lacking.

The Definitions of Volume I

The definitions given in Volume I of *Grundgesetze* consist, for the most part, of definitions of notions that are required for Frege's derivation of something akin to the second-order Peano axioms for arithmetic and his extension of these results to the infinite, as well as an early treatment of recursion. This treatment of arithmetic, recursion, and countable infinite numbers extends into Volume II, however. The more salient division, content-wise, is between Part II and Part III of *Grundgesetze.* The former, which bridges Volumes I and II, contains the treatment of arithmetic, broadly construed, while the latter contains the beginnings of Frege's construction of the real numbers. Frege organises his definitions in terms of the division between Volumes I and II, however (for example, in the tables of definitions that conclude each volume), rather than in terms of the Parts of *Grundgesetze*, and this structure is replicated here (in particular, the discussion below follows the order in which these definitions are presented in Appendix 2 of Volume I, rather than the order in which they are presented in the main text). Readers interested in further details regarding the role these defined notions play within Frege's investigations into cardinal numbers, countable infinities, and recursion should consult the excellent treatments of these issues by Richard Heck (2011, 2012), and Gregory Landini (2012).

Definition A: The Application Operator

The first, but arguably most important, definition in *Grundgesetze* is definition A (I §34)—the *application operation* '$\frown$':

$$\Vdash \backslash\grave{\alpha}\left(\neg^{\mathfrak{g}}\!\!\top\!\!\begin{array}{l} \mathfrak{g}(a) = \alpha \\ u = \grave{\varepsilon}\mathfrak{g}(\varepsilon)\end{array}\right) = a \frown u \qquad \text{(A}$$

Given two names 'Δ' and 'Γ', the reference of '$\Delta \frown \Gamma$' is determined as follows. If 'Γ' is a name of the value-range of some function $\Phi(\xi)$, then '$\Delta \frown \Gamma$' is a name of the object that results from applying $\Phi(\xi)$ to the object named by 'Δ'. If 'Γ' is not the name of a value-range, then '$\Delta \frown \Gamma$' names value-range of the function that maps every object to the False—that is, it is co-referential with '$\grave{\varepsilon}(\top\, \varepsilon = \varepsilon)$'. The application operator (along with the value-range operator) is one of Frege's central tools for "reducing" levels within *Grundgesetze*, since it allows him to replace talk of 1st-level functions with talk of their value-ranges (I §35).

Of particular interest is the case where 'Γ' names the extension of a concept $\Phi(\xi)$, where '$\Delta \frown \Gamma$' names the True if and only if $\Phi(\Delta)$ is the True. As a result, when applied to extensions, '$\frown$' is, in effect, a Fregean analogue of the set-theoretic membership relation '∈', and for the most part Frege uses '⌒' as a membership relation on extension of concepts.[12]

Definition B: Composition

Frege's second definition is definition B (I §54)—the definition of the *composition* of two Relations:

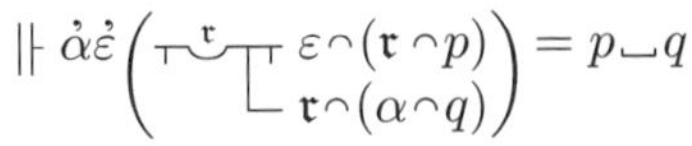

(B

If 'Π' and 'Ξ' are names of Relations , then 'Π⌣Ξ' is a name of the Relation that holds between two objects Δ and Γ if and only if there is a third object Θ such that the Relation named by 'Π' holds of Δ and Θ and the Relation named by 'Ξ' holds of Θ and Γ. While Frege' s definition captures something much like the familiar notion of composition of relations, it is important to note that, unlike modern treatments of composition (and other, similar operations on functions and relations), Frege's composition Relation does not map two relations onto the relation that is their composition. Instead, the composition Relation maps two Relations (i.e., the double value-ranges of two relations) onto the Relation (i.e., double value-range) that is their composition. In short, Frege's composition operator is a binary function from objects (i.e., double value-ranges) to objects, not from relations to relations (similar comments apply to many of the definitions below).

Definition Γ: Single-valuedness

The next definition is definition Γ (I §37, see also I §23)—the definition of *single-valuedness*:

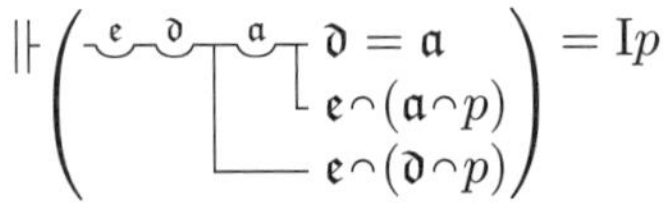

(Γ

'I(ξ)' is a concept-name—that is, it names a function that maps objects to truth-values. If 'Π' is the name of a Relation, then 'I(Π)' is a name of the True if and only if, for any objects Δ, Γ, and Θ, if the concept named by 'Π' holds of Δ and Γ, and the concept named by 'Π' holds of Δ and Θ, then Γ is equal to Θ. In short, a Relation is single-valued if and only if it is *functional.*

Definition Δ: Mapping Into

Frege's next definition is definition Δ (I §38)– the *mapping into* operation on Relations:

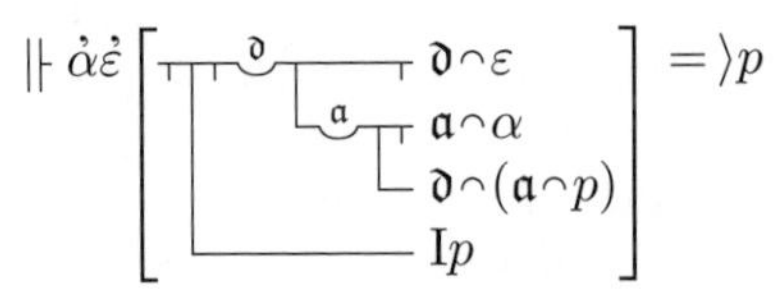

(Δ

[12] The term "application operator" is not Frege's—he calls this operation the "Relation of an object falling within the extension of a concept" (I Appendix 2) which only captures part of the meaning of the function named by 'ξ⌒ζ'.

If 'Π' names a single-valued (see Definition Γ) Relation, then '⟩Π' is the name of the Relation that holds between two value-ranges $\dot{\varepsilon}(\Phi(\varepsilon))$ and $\dot{\varepsilon}(\Psi(\varepsilon))$ if and only if, for any object Δ, if Φ(Δ) is the True, then there is an object Γ such that Ψ(Γ) is the True and the Relation named by 'Π' holds of Δ and Γ. In short, '⟩Π' is the Relation holding between two extensions if and only if the Relation named by 'Π' maps the "members" of the first extension one-to-one to "members" of the second extension.

Definition E: Converse

Next up is definition E (I §39)—the *converse* of a Relation:

$$\Vdash \dot{\alpha}\dot{\varepsilon}(\alpha\frown(\varepsilon\frown p)) = \mathfrak{U}p \qquad \text{(E}$$

The converse operator '$\mathfrak{U}(\xi)$' behaves exactly as advertised (with the standard caveat in place regarding the fact that it maps Relations to Relations, rather than relations to relations): If 'Π' is a name of Relation, then '$\mathfrak{U}(\Pi)$' is a name of the Relation such that, for any objects Δ and Γ, the Relation named by 'Π' holds of Δ and Γ (in that order) if and only if the Relation named by '$\mathfrak{U}(\Pi)$' holds of Γ and Δ (in that order).

Definition Z: Cardinal Number

Perhaps the most important definition in Volume I of *Grundgesetze* is definition Z (I §40)—the definition of *cardinal number*:

$$\Vdash \dot{\varepsilon}\left(\begin{array}{l} \top\!\smile^{\mathfrak{q}}\!\top\!\!\top\; \varepsilon\frown(u\frown\rangle\mathfrak{q}) \\ \qquad\quad\;\, \llcorner\; u\frown(\varepsilon\frown\rangle\mathfrak{U}\mathfrak{q}) \end{array}\right) = \mathfrak{p}u \qquad \text{(Z}$$

The cardinal number operator '$\mathfrak{p}(\xi)$' is the name of a function that maps each value-range onto the cardinal number of that value-range. In particular, if 'Δ' is the name of the value-range of a concept Φ(ξ), then '$\mathfrak{p}\Delta$' is a name of the extension of the concept that holds of the value-range of a concept Ψ(ξ) if and only if there is a Relation that maps Φ(ξ) one-to-one into Ψ(ξ) and maps Ψ(ξ) one-to-one into Φ(ξ) (i.e there is a Relation Π such that both ⟩Π and ⟩$\mathfrak{U}$Π hold of $\dot{\varepsilon}\Phi(\varepsilon)$ and $\dot{\varepsilon}\Psi(\varepsilon)$). Note that, if there is a Relation Π such that Π maps Φ(ξ) one-to-one into Ψ(ξ) and vice versa, then Π is, in effect, a bijection between the objects that Φ(ξ) maps to the True and the objects that Ψ(ξ) maps to the True. Hence, the cardinal number of a value-range is the extension of the concept that holds of all and only the value-ranges of those concepts that are equinumerous to it.

Definition H: Predecessor

Once we have cardinal numbers to hand, the obvious next step is to define the *predecessor Relation* holding of each cardinal number and the next. This is provided by definition H (I §43):

$$\Vdash \dot{\alpha}\dot{\varepsilon}\left[\begin{array}{l} \top\!\smile^{\mathfrak{u}}\!\smile^{\mathfrak{a}}\!\top\!\!\top\; \mathfrak{p}\mathfrak{u} = \alpha \\ \qquad\qquad\;\; \llcorner\; \mathfrak{a}\frown\mathfrak{u} \\ \qquad\qquad\;\; \llcorner\; \mathfrak{p}\dot{\varepsilon}\left(\begin{array}{l}\top\!\!\top\; \varepsilon = \mathfrak{a} \\ \;\;\llcorner\; \varepsilon\frown\mathfrak{u}\end{array}\right) = \varepsilon \end{array}\right] = \mathfrak{f} \qquad \text{(H}$$

'f' is a name of the Relation that holds between two objects Δ and Γ just in case there is a concept Φ(ξ) and object Ω such that Φ(Ω) is the True, Γ is the number of the value-range of Φ(ξ) (i.e., Γ = ℘ἐΦ(ε) is the True), and Δ is the number of the value-range of the concept that holds of all objects satisfying Φ(ξ) other than Ω, that is:

$$\Delta = \wp\dot{\varepsilon}\left(\begin{array}{l}\top\, \varepsilon = \Omega \\ \llcorner\, \varepsilon \frown \dot{\alpha}(\Phi(\alpha))\end{array}\right)$$

is the True. Note that the predecessor Relation is single-valued, that is, 'I(f)' is a name of the True. Further, note that if 'Δ' names the value-range of a concept that holds of (Dedekind) infinitely many objects, then the predecessor Relation holds between the object named by 'Δ' and itself.[13]

Definition Θ: Zero

Frege's identifies three numbers that are of particular interest in the derivation of arithmetic and study of the infinite that takes up the bulk of Volume I. The first of these is given by Definition Θ (I §41)—the definition of the cardinal number *Zero*:

$$\Vdash \wp\dot{\varepsilon}(\top\, \varepsilon = \varepsilon) = 0 \qquad (\Theta$$

'0' is a name of the cardinal number of the value-range of the concept that holds exactly those objects that are not self-identical—that is, it is the cardinal number of the value-range of the empty concept. Note that, contrary to what Frege's notation might suggest, '0' does not name the value-range of the empty concept, but instead is a name of the value-range of the concept holding solely of the value-range of the empty concept. In modern terminology, '0' is not a name of the empty set, but is instead a name of the *Grundgesetze* analogue of the singleton of the empty set.

Definition I: One

Also unsurprising is the fact that the second cardinal number identified explicitly by Frege is the one given by Definition I (I §42): The cardinal number *One*:

$$\Vdash \wp\dot{\varepsilon}(\varepsilon = 0) = 1 \qquad (I$$

'1' is a name of the cardinal number of the concept that holds of Zero (i.e., '0') and nothing else. In other words, '1' names the value-range of the concept that holds of exactly the value-ranges of those concepts that are equinumerous to the concept holding of exactly Zero. More succinctly, '1' names the *Grundgesetze* analogue of the set of all singletons.

Definition K: Strong Ancestral

Next up we have the first of two definitions that are central to the technical details of Frege's derivation of arithmetic within *Grundgesetze*, and have as a result become central to discussions of Frege's contributions to the philosophy of mathematics: the definitions of the (strong and weak) ancestral(s) of a Relation. First we have Definition K (I §45)– the definition of the *strong ancestral* of a Relation:

[13] The term "predecessor" is not Frege's—he calls this Relation the "Relation of a cardinal number to the one immediately following" (I Appendix 2).

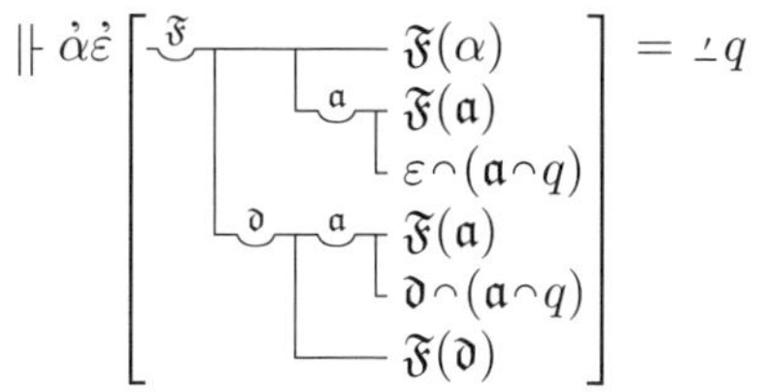

(K

Introducing the notion of a *hereditary concept* will be helpful here. A concept $\Phi(\xi)$ is *hereditary* with respect to a relation $\Psi(\xi,\zeta)$ if and only if, for any objects Δ and Γ, if $\Phi(\Delta)$ is the True and $\Psi(\Delta,\Gamma)$ is the True, then $\Phi(\Gamma)$ is also the True. Thus, a concept $\Phi(\xi)$ is hereditary with respect to a relation $\Psi(\xi,\zeta)$ if and only if:

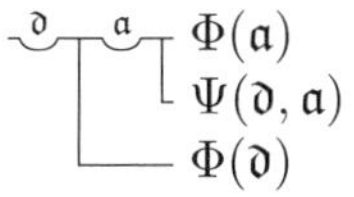

is a name of the True. Further, a concept $\Phi(\xi)$ is hereditary with respect to a Relation if and only if it is hereditary with respect to the corresponding relation.

If 'Π' names a Relation, then '$\underline{\smile}$Π' is a name of the Relation that holds between objects Δ and Γ if and only if, for any concept $\Xi(\xi)$, if both:

- $\Xi(\xi)$ is hereditary with respect to the Relation named by 'Π'
- for any object Θ, if the relation named by 'Π' holds of Δ and Θ then $\Xi(\Theta)$ is the True

then $\Xi(\Gamma)$ is the True. Simply put, the strong ancestral of a Relation named by 'Π' is the Relation that holds between two objects Δ and Γ if and only if either the Relation named by 'Π' holds of Δ and Γ, or there exists a sequence of objects $\Theta_1, \Theta_2, \ldots \Theta_m$ such that the Relation named by 'Π' holds of Δ and Θ_1, holds of Θ_1 and $\Theta_2, \ldots$ and holds of Θ_m and Γ. Note that, in general, the strong ancestral is not reflexive—that is, the strong ancestral of a Relation 'Π' need not hold between any object and itself.[14]

Definition Λ: Weak Ancestral

In Definition Λ (I §46) Frege provides a definition of the *weak ancestral* of a Relation:

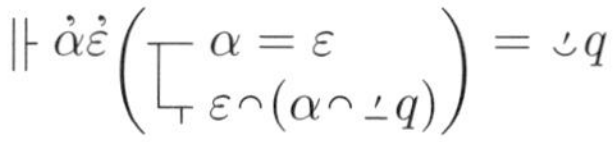

(Λ

Simply put, if 'Π' is a name of a Relation, then '$\underset{\smile}{=}$Π' is a name of the Relation that holds between two objects Δ and Γ if and only if either the strong ancestral of the Relation named by 'Π' holds of these objects (i.e., the Relation named by '$\underline{\smile}$Π)' holds of Δ and Γ), or they are equal (i.e., $\Delta = \Gamma$ is the True). In short, the weak ancestral is the reflexive version of the strong ancestral: If 'Π' names a Relation, then '$\underset{\smile}{=}$Π' names the Relation that holds between Δ and Γ if and only if either $\Delta = \Gamma$ is the True, or the Relation named by 'Π' holds of Δ and Γ, or there exists a sequence of objects $\Theta_1, \Theta_2, \ldots \Theta_m$ such that the Relation named by 'Π' holds of Δ and Θ_1, holds of Θ_1 and $\Theta_2, \ldots$ and holds of Θ_m and Γ.[15]

---

[14] The term "ancestral" is not Frege's—he calls this Relation "The following of an object after an object in the series of a relation" (I Appendix 2).

[15] Frege calls the weak ancestral Relation "The relation of an object belonging to the series of a relation starting with an object" (I Appendix 2).

Definition M: Endlos

The third, and final, cardinal number that Frege provides a special symbol for in *Grundgesetze* is introduced in Definition M (I §122)—the definition of the cardinal number *Endlos*:

$$\Vdash \mathfrak{p}(\mathfrak{0} \frown \mathfrak{p} \smile \mathfrak{f}) = \infty \qquad \text{(M}$$

Simply put, '∞' is a name of the cardinal number of the extension of the concept that holds of an object Δ if and only if the ancestral of the converse of the predecessor relation holds between Δ and Zero (in that order—note that this explains the presence of the converse relation symbol '$\mathfrak{p}(\xi)$', since we are, in effect, asking whether one can "reach" Zero starting at Δ).

Thus, the cardinal number Endlos, that is, the cardinal number named by '∞', is the cardinal number of the extension of the concept that holds of the finite cardinal numbers. Note, importantly, that the cardinal number named by '∞' need not be equal to either the number of cardinal numbers (including infinite cardinal numbers), or the number of all objects (the cardinal number sometimes called *Anti-Zero*), and intuitively there is no guarantee that any of these three numbers is equal to any of the others.

Definition N: Relation Restriction

Definition N (I §138) provides us with the *Relation restriction* operator on extensions and Relations:

$$\Vdash \grave{\alpha}\grave{\varepsilon}\left(\begin{array}{l} \top\!\!\top\, \varepsilon \frown (\alpha \frown q) \\ \llcorner\, \alpha \frown u \end{array} \right) = u \doteq q \qquad \text{(N}$$

If 'Δ' is a name of the extension of a concept $\Phi(\xi)$, and 'Π' is a name of a Relation, then '$\Delta \doteq \Pi$' is a name of the Relation that holds between two objects Γ and Θ if and only if the the Relation named by 'Π' holds of Γ and Θ, and $\Phi(\Theta)$ is the True. In short, '$\Delta \doteq \Pi$' names the result of restricting the Relation named by 'Π' to those pairs whose second element is a "member" of the extension named by 'Δ'. Note that the definition of '$\doteq$' does not place any restriction on the first argument of the Relation.[16]

Definition Ξ: Ordered Pair

In definition Ξ (I §144) Frege provides a very unfamiliar definition of a very familiar notion—the *ordered pair* of two objects:

$$\Vdash \grave{\varepsilon}(o \frown (a \frown \varepsilon)) = o; a \qquad \text{(}\Xi$$

If 'Δ' and 'Γ' are proper names, then '$\Delta;\Gamma$' is a name of the value-range of the concept that holds of all and only those Relations that hold of the objects named by 'Δ' and 'Γ'. Simply put, '$\Delta;\Gamma$' names something like the *Grundgesetze* analogue of the set of all Relations that relate the object named by 'Δ' to the object named by 'Γ'.

One of the most interesting things about this definition is that it reverses the order of explanation when compared to later treatments of ordered pairs. Frege's defines

[16] The term "relation restriction" is not Frege's—he provides no shorthand for this Relation in Appendix 2 of Volume II (but see the discussion of this notion in I §138).

ordered pairs in terms of the totality of Relations, rather than defining relations (or Relations) to be, or be akin to, sets or classes of ordered pairs.

Definition O: Coupling

Next up is Definition O (I §144), the *coupling* of two Relations:

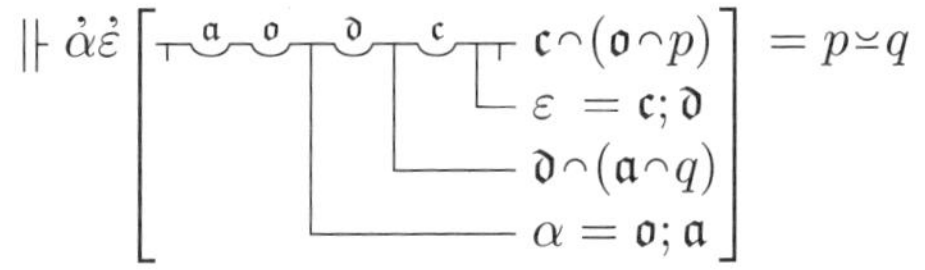

(O

The coupling operation takes two Relations as arguments, and provides a new Relation defined on ordered pairs of objects as output, where the first member of a pair in the range of the coupling of two Relations occurs in the range of the first Relation and the second member of the pair occurs in the range of the second Relation. In particular, if 'Π' and 'Ξ' are names of Relations , then '$\Pi \veebar \Xi$' names the Relation that holds between two ordered pairs $\Delta;\Gamma$ and $\Theta;\Omega$ just in case the Relation named by 'Π' holds of Δ and Θ and the Relation named by 'Ξ' holds of Γ and Ω.

Definition Π: Recursion

Frege next introduces an operation on objects and Relations that we shall call the *recursion* operation, because of its intimate role in Frege's treatment of recursive definitions. The recursion operation is given by Definition Π (I §144):

$$\Vdash \dot{\alpha}\dot{\varepsilon}(A \frown (\varepsilon; \alpha \frown \grave{\iota} t)) = A \lessdot t \qquad (\Pi$$

Frege typically uses upper-case 'A' as a variable (i.e., a Roman letter) ranging over ordered pairs. Thus, if 'Π' is a name of a Relation that takes ordered pairs as arguments, and 'Δ' and 'Γ' are proper names, then '$\Delta;\Gamma \lessdot \Pi$' names the Relation that holds of two objects Θ and Ω just in case the ordered pair named by '$\Delta;\Gamma$' is related to the ordered pair $\Theta;\Omega$ by the (weak) ancestral of the Relation named by 'Π'. The connection to recursion can be easily seen by considering the case where, for any natural number n, the Relation in question relates each pair of the form $n;\Gamma$ to a unique pair $n+1;\Omega$.[17]

Definition P: Occurring in a Bounded Series

Next up we have Definition P (I §158)—the definition of *occurring in a bounded series*:

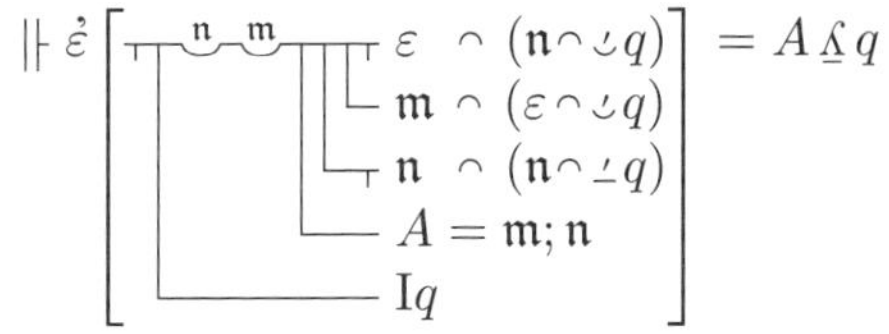

(P

Again, recalling that Frege uses upper-case 'A' to range over ordered pairs, we can understand this operation as follows: If 'Δ' and 'Γ' are proper names and 'Π' is a

[17] The term "recursion" is not Frege's—Frege provides no shorthand for this Relation in Appendix 2 of Volume II (but see the discussion of this notion in I §144).

name of a single-valued Relation, then '$\Delta; \Gamma \measuredangle \Pi$' is a name of the value-range of the concept that holds of an object Θ just in case:

- The object named by 'Δ' is related to Θ by the weak ancestral of the Relation named by 'Π' (i.e., the Relation named by '$\smile\!\Pi$' holds of the object named by 'Δ' and Θ).

- Θ is related to the object named by 'Γ' by the weak ancestral of the Relation named by 'Π' (i.e., the Relation named by '$\smile\!\Pi$' holds of Θ and the object named by 'Γ').

- The object named by 'Γ' is not related to itself by the strong ancestral of the Relation named by 'Π' (i.e., '$\top \Gamma \frown (\Gamma \frown \perp\!\Pi)$' is a name of the True).

In short, '$\Delta; \Gamma \measuredangle \Pi$' names the value-range of the concept that holds of an object Θ just in case Θ "lies between" the objects named by 'Δ' and 'Γ' in the sequence "determined" by the Relation named by 'Π'. Note that the requirement that the object named by 'Γ' not be related to itself by the strong ancestral rules out "loops" in this sequence—otherwise, Θ might occur "after" the object named by 'Δ' in the series, and "before" the object named by 'Γ' (i.e., Θ might satify the first two bullet points) by occurring in the loop rather than occurring between the objects named by 'Δ' and 'Γ'.[18]

The Definitions of Volume II

In Volume II of *Grundgesetze* Frege concludes his examination of arithmetic, infinite cardinals, and recursion (in the remainder of Part II), and then begins his reconstruction of the real numbers (in Part III). The basic idea underlying these constructions is that real numbers are ratios of magnitudes (or quantities). Thus, much of the extant portion of Frege's reconstruction of the real numbers found in Volume II of *Grundgesetze* deals with characterizing domains of magnitudes. Frege understands magnitudes to be double value-ranges of relations—that is, Relations—and hence domains of magnitudes are value-ranges of concepts holding of Relations. Further, the sum of two Relations named by 'Π' and 'Ξ' is the composition of those Relations (i.e., the Relation named by '$\Pi \llcorner\!\lrcorner \Xi$'), and the "negative" of a Relation named by 'Π' is its converse, (i.e., the Relation named by '$\mathfrak{V}\Pi$'). The remaining constructions will become clear in the discussion below. The reader who desires more details regarding how these operations are (or were intended to be) used in the actual reconstruction of the reals should consult the excellent Simons (1987).

Definition Σ: First Object in a Series with a Given Property

Frege's first definition in Volume II is Definition Σ (II §1):

$$\Vdash \dot{\alpha}\dot{\varepsilon}\left(\begin{array}{l} \top\!\!\!\top\, \varepsilon \frown (\alpha \frown (v \doteqdot \smile q \llcorner\!\lrcorner \perp q)) \\ \;\llcorner\, \varepsilon \frown (\alpha \frown (v \doteqdot \smile q)) \end{array} \right) = v \,\eth\, q \qquad (\Sigma$$

[18] Frege calls this Relation "the circumstance that an object belongs to a series running from an object to an object" (I Appendix 2).

If 'Δ' is a name of the value-range of a concept $\Phi(\xi)$, and 'Π' is a name of Relation, then '$\Delta \mathbin{\dot{\text{ȝ}}} \Pi$' is a name of the Relation that holds of two objects Γ and Θ if and only if Γ is related to Θ by the weak ancestral of the Relation named by 'Π' restricted to $\Phi(\xi)$, but Γ is *not* related to Θ by the composition of the weak ancestral of the Relation named by 'Π' restricted to $\Phi(\xi)$ and the strong ancestral of the Relation named by 'Π'. (Note that the supercomponent of this definition is parsed as '$\varepsilon\frown(\alpha\frown((\upsilon\bumpeq\acute{\smile}q)\sqcup\acute{\perp}q))$' and not as '$\varepsilon\frown(\alpha\frown(\upsilon\bumpeq(\acute{\smile}q\sqcup\acute{\perp}q)))$'.) To see the import of the second clause, imagine that Γ *were* related to Θ by the composition of the weak ancestral of the Relation named by 'Π' restricted to $\Phi(\xi)$ and the strong ancestral of the Relation named by 'Π'. If so, then there must be some Ω such that Γ is related to Ω by the weak ancestral of the Relation named by 'Π' restricted to $\Phi(\xi)$, and Ω is related to Θ by the strong ancestral of of the Relation named by 'Π'. But this entails that Θ is not the *first* object in the series beginning with Γ such that $\Phi(\xi)$ holds of that object, since the strong ancestral is not reflexive and $\Phi(\Omega)$.[19]

Definition T: Next Object in a Series with a Given Property

Definition T (II §7) introduces a stronger, irreflexive version of the notion introduced in the previous definition (Definition Σ):

$$\Vdash \dot{\alpha}\dot{\varepsilon}\left(\begin{array}{l}\top\!\!\!\ulcorner\ \varepsilon\frown(\alpha\frown(\upsilon\bumpeq\acute{\perp}q\sqcup\acute{\perp}q))\\ \llcorner\ \varepsilon\frown(\alpha\frown(\upsilon\bumpeq\acute{\perp}q))\end{array}\right) = \upsilon \mathbin{\text{ȝ}} q \qquad (\mathrm{T}$$

This definition is obtained by replacing the two occurrences of the weak ancestral relation symbol '$\acute{\smile}$' in Definition Σ with the strong ancestral relation symbol '$\acute{\perp}$'. As a result, for any proper names 'Δ' and 'Γ' and Relation-name 'Π', it cannot be the case that '$\Gamma\frown(\Gamma\frown(\Delta \mathbin{\text{ȝ}} \Pi))$' is a name of the True (however, '$\Gamma\frown(\Gamma\frown(\Delta \mathbin{\dot{\text{ȝ}}} \Pi))$' can name the True, as the reader is encouraged to verify).[20]

Definiton Υ: Combination of Relations

The next definition in Volume II is Definition Υ (II §37)—the combination of Relations:

$$\Vdash \dot{\alpha}\dot{\varepsilon}\left(\begin{array}{l}\ulcorner\ \varepsilon\frown(\alpha\frown q)\\ \llcorner\!\!\top\ \varepsilon\frown(\alpha\frown p)\end{array}\right) = q \mathbin{\overline{\frown}} p \qquad (\Upsilon$$

If 'Π'and 'Ξ' are names of Relations, then '$\Pi \mathbin{\overline{\frown}} \Xi$' is a name of the Relation that holds of objects Δ and Γ if and only if either the Relation named by 'Π' holds of Δ and Γ or the Relation named by 'Ξ' holds of Δ and Γ.

Definition Φ: Series of Composition

Definition Φ (II §167) provides the means to represent the result of repeatedly composing a Relation with itself:

---

[19] The term "first object in a series with a given property" is not Frege's—he calls this Relation "Relation: that an object, in a series starting with an object, is the first to belong to a class" (II Appendix 1).

[20] The term "next object in a series with a given property" is not Frege's—he calls this Relation "Relation: that an object in a series is the first after an object to belong to a class" (II Appendix 1).

$$\Vdash \dot{\alpha}\dot{\varepsilon}(t \smile \varepsilon = \alpha) = {}_*t \tag{Φ}$$

If 'Π' is a name of a Relation, then '$_*$Π' is a name of the Relation that holds between two objects Δ and Γ' if and only if Δ and Γ are themselves Relations, and Γ is equal to the composition of the Relation named by 'Π' and the Relation Δ.[21]

Definition X: Domain of Magnitudes

Although its import is not clear until one has digested the later definition of, and discussion of, positive classes and positival classes (see Definitions Ψ and AB below), Definition X (II §173) provides an account of the *domain of magnitudes* relative to a positival class:

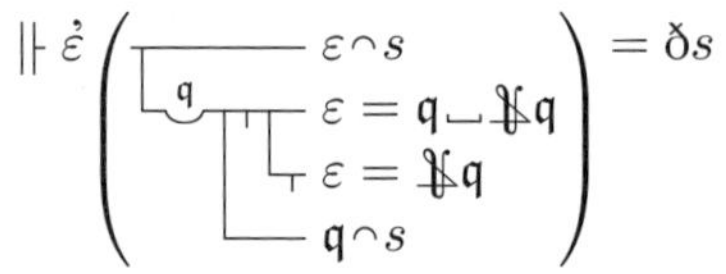

$$\Vdash \dot{\varepsilon}\left(\begin{array}{c} \varepsilon \smallfrown s \\ \varepsilon = q \smile \mathfrak{X}q \\ \varepsilon = \mathfrak{X}q \\ q \smallfrown s \end{array}\right) = \eth s \tag{X}$$

For our purposes here, we only need note that a positival class is the value-range of a concept that holds of Relations (i.e., it is *Grundgesetze* analogue of a set of Relations), and that these Relations are, intuitively speaking, the "positive" elements of a domain of magnitudes. Thus, if 'Σ' is the name of a positival class, then the domain of magnitudes correponding to this class (or simply the Σ-*domain*) named by 'ðΣ' is the value-range of the concept that holds of any Relation that is either in the class named by 'Σ', or is the converse of a Relation in the class named by 'Σ', or is the composition of some Relation in the class named by 'Σ' and the converse of that same Relation. In short, given a positival class, the corresponding domain of magnitudes is given by taking that class, plus the "negatives" of members of that class, plus the null element named by 'Π⌣𝔛Π' (for any Π).[22]

Definition Ψ: Positival Class

Next up is Definition Ψ (II §175)—the definition of a *positival class*:

$$\Vdash \left(\begin{array}{c} p \\ \quad p \smile \mathfrak{X}p \smallfrown s \\ \quad Ip \\ \quad I\mathfrak{X}p \\ \quad \dot{\alpha}\dot{\varepsilon}(— \varepsilon \smallfrown(\alpha \smallfrown p)) = p \\ q \\ \quad p \smile q \smallfrown s \\ \quad p \smile \mathfrak{X}q \smallfrown \eth s \\ \quad \mathfrak{X}p \smile q \smallfrown \eth s \\ \quad (\mathrel{\overset{a}{\smile}_\top} \eth \smallfrown (a \smallfrown p)) = (\mathrel{\overset{a}{\smile}_\top} a \smallfrown (\eth \smallfrown q)) \\ q \smallfrown s \\ p \smallfrown s \end{array}\right) = \jmath s \tag{Ψ}$$

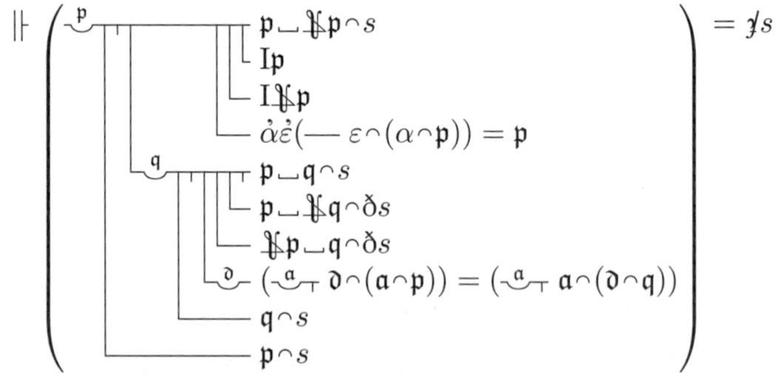

While this is no doubt Frege's most complex definition, it can be parsed as follows: A class named by 'Σ' is a positival class if and only if:

---

[21] The term "series of composition" is not Frege's—he does not give a shorthand for this Relation in Appendix 1 of Volume II (but see the discussion of this notion in II §167).

[22] Although Frege does not call this value-range a "domain of magnitudes" in Appendix 1 of Volume II, the term is used in this sense in II §173.

- Any object in the class named by 'Σ' must be a Relation (this is insured by the clause that reads '$\dot{\alpha}\dot{\varepsilon}(\text{—}\ \varepsilon\frown(\alpha\frown\mathfrak{p})) = \mathfrak{p}$'.
- If Π is any Relation in the class named by 'Σ', then both Π and the converse of Π must be single-valued (i.e., $\mathrm{I}(\Pi)$ and $\mathrm{I}(\text{⅄}\Pi)$ are the True).
- If Π and Ξ are Relations in the class named by 'Σ', then any object that can occur as the first argument in a true application of the Relation Π must also occur as the second argument in a true application of the Relation Ξ.
- If Π and Ξ are Relations in the class named by 'Σ', then $\text{⅄}\Pi\sqcup\Xi$ (the composition of the converse of Π and Ξ), and $\Pi\sqcup\text{⅄}\Xi$ (the composition of Π and the converse of Ξ) must be in the class named by '$\mathfrak{d}\Sigma$' (the domain of magnitudes relative to the class named by 'Σ'). Note that this does not imply that $\Pi\sqcup\text{⅄}\Xi$ or $\Xi\sqcup\text{⅄}\Pi$ themselves are in the class named by 'Σ'.
- If Π and Ξ are Relations in the class named by 'Σ', then $\Pi\sqcup\Xi$ (the composition of Π and Ξ) must be in the class named by 'Σ'.
- If Π is any Relation in the class named by 'Σ', then $\Pi\sqcup\text{⅄}\Pi$ (the composition of Π with its own converse) must not be in the class named by 'Σ'.

Intuitively, a class of Relations is a positival class if and only if (i) the Relations in the class, and the converses of the Relations in the class, are functional, (ii) the relations in the class all have the same domain and range, (iii) the class is closed under addition (where addition is $\xi\sqcup\zeta$), (iv) the class fails to contain the null element $\Pi\sqcup\text{⅄}\Pi$ (for any Π), and (v) the domain of magnitudes corresponding to the class (i.e., $\mathfrak{d}(\xi)$ applied to the class) is closed under subtraction of elements (where subtraction is the addition of negative elements, that is: $\xi\sqcup\text{⅄}\zeta$).

Definition Ω: Downwards Closure

The next definition of Volume II is definition Ω (II §193) —the definition of being *downwards closed* (relative to another class):

$$\Vdash \dot{\varepsilon}\left(\smile^{\mathfrak{a}}\!\!\begin{array}{l} \mathfrak{a}\frown u \\ \varepsilon\sqcup\text{⅄}\mathfrak{a}\frown s \\ \mathfrak{a}\frown s \end{array}\right) = s \,\underline{\text{Ⅎ}}\, u \qquad (\Omega$$

In applications of this definition, the first argument will be a positival class—that is, a class of "positive" magnitudes. Recall that, in the present context, '$\Delta\sqcup\text{⅄}\Omega\frown\Sigma$' can be read as expressing the claim that the Relation named by 'Δ' is "greater than" the Relation named by 'Ω' relative to the positival class named by 'Σ' (since this amounts to stating that the "sum" of the Relation named by 'Δ' and the "negative" of the Relation named by 'Ω' is in the positival class named by 'Σ' and is thus "positive").

If 'Σ' and 'Π' are names of two classes of Relations, then '$\Sigma \,\underline{\text{Ⅎ}}\, \Pi$' is a name of the value-range of the concept that holds of a Relation Δ if and only if, for any Relation Γ in the class named by 'Σ', if Γ is "less" than Δ (relative to the ordering imposed by Σ) then Γ is in the class named by 'Π'. Hence, '$\Sigma \,\underline{\text{Ⅎ}}\, \Pi$' is a name of the class of Relations such that the class of Relations named by 'Π' is "downwards closed" relative to any of those Relations (and relative to the ordering imposed by the class named by 'Σ').[23]

Definition AA: Limit

Frege's next definition is Definition AA (I §193) — the *limit* of a class (relative to another class).

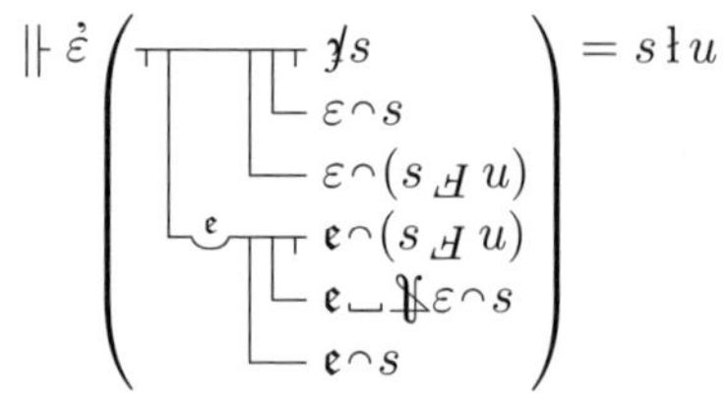

(AA

Like the definition of downward closure in Definition Ω above, the definition of limit is a definition of a class of Relations. If 'Σ' and 'Π' are both names of classes of Relations, then 'Σ⫮Π', is the value-range of the concept that holds of a Relation Δ if and only if:

- 'Σ' names a positival class.

- Δ is a member of the class named by 'Σ'.

- The class named by 'Π' is downwards closed relative to Δ (i.e., the class named by 'Π' contains all Relations in the class named by 'Σ' that are "less than" Δ relative to the ordering imposed by the class named by 'Σ').

- There is no Relation Γ such that Γ is "greater than" Δ (relative to the ordering imposed by the class named by 'Σ') and the class named by 'Π' is downwards closed relative to Γ (again, on the ordering imposed by the class named by 'Σ').

Hence, Δ is in the class named by 'Σ⫮Π' if and only if it is a "maximal" Relation relative to which the class named by 'Π' is downwards closed. Frege proves that the limit of the class named by 'Π' (relative to the ordering imposed by 'Σ'), if it exists, is unique (more carefully: that the extension defined has a single object as member—see II §195).

Definition AB: Positive Class

The penultimate definition of Volume II is Definition AB (II §197)—the definition of a *positive class*:

[23] The term "downward closure" is not Frege's—he does not provide a shorthand for this operation in Appendix 1 of Volume II (but see the discussion of this notion in II §193).

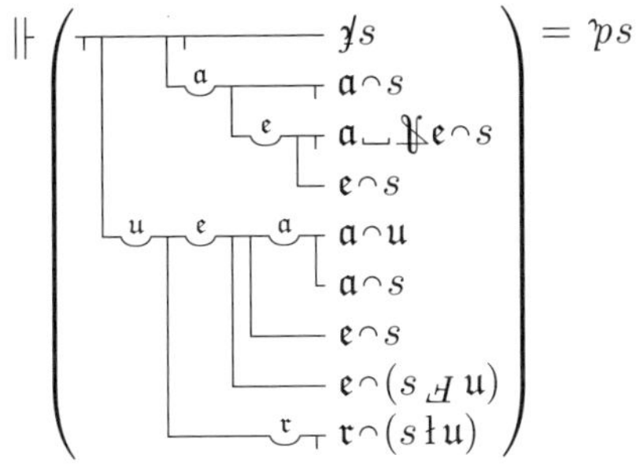

(AB

While this definition appears at first glance to be nearly as complex as the definiton of positival class discussed above, it can be straightforwardly parsed as a conjunction of three conditions. A class of Relations named by 'Σ' is a positive class if and only if:

- 'Σ' names a positival class.
- For any Relation Δ, if Δ is in the class named by 'Σ', then there is a Relation Γ such that Γ is in the class named by 'Σ' and Γ is "less than" Δ (relative to the ordering imposed by the class named by 'Σ').
- For any class of Relations Π, if there is a Relation Ω in the class named by 'Σ' such that Π is downwards closed relative to Ω, and there is a second Relation Ξ in the class named by 'Σ' such that Ξ is not in Π, then Π has a limit (in the sense defined in Definition AA).

In short, a class of Relations named by 'Σ' is a positive class if and only if the class is a positival class, is not bounded from below, and every proper sub-class of the class named by 'Σ' that is downwards closed relative to some Relation has a limit.

Definition AΓ: Archimedean Condition

The final definition of *Grundgesetze* is definition AΓ (II §199)– the definition of the *Archimedean condition*:

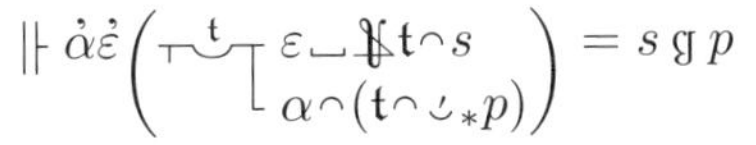

(AΓ

If 'Σ' is a name of a class of Relations, and 'Π' is a name of a Relation, then '$\Sigma\,\mathfrak{g}\,\Pi$' is a name of the Relation that holds between two Relations Δ and Γ if and only if there is a Relation Θ such that Γ is related to Θ by the weak ancestral of the series of composition of the Relation named by 'Π' (that is, abusing the use-mention distinction slightly: if Θ is one of Γ, $\Pi\smile\Gamma$, $\Pi\smile(\Pi\smile\Gamma)$, $\Pi\smile(\Pi\smile(\Pi\smile\Gamma))$,... etc.), and Δ is not greater than Θ (relative to the ordering imposed by the class named by 'Σ'). The special case of interest is:

$$\Delta\frown(\Pi\frown(\Sigma\,\mathfrak{g}\,\Pi))$$

which holds if and only if there is a Relation Θ such that the Relation named by 'Π' is related to Θ by the weak ancestral of the series of composition of the Relation named by 'Π' (that is, again abusing use-mention: if Θ is one of Π, $\Pi\smile\Pi$, $\Pi\smile(\Pi\smile\Pi)$, $\Pi\smile(\Pi\smile(\Pi\smile\Pi))$, etc.), and the Relation named by 'Δ' is not greater than Θ (relative to the ordering imposed by the class named by 'Σ'). Recalling that the composition operator '$\xi \smile \zeta$' corresponds to addition within Frege's reconstruction of the reals, this amounts to saying that we can "reach" the Relation named by 'Δ' by repeatly "adding" the Relation named by 'Π' to itself finitely many times (again, all of this relative to the ordering imposed by the class named by 'Σ'). Alternatively, paraphrasing Frege, we can just say that there is a multiple of the Relation named by 'Π' that is not less that the Relation named by 'Δ'.[24]

Conclusion

This concludes our brief tour of the notations, deductive system, and definitions of Frege's *Grundgesetze.* Of course, it is impossible to note every interesting, or odd, aspect of the formal system of *Grundgesetze* in a brief overview of this sort. Thus, there is much left to be explored and discovered. Nevertheless, the material presented in these few pages will, it is hoped, provide the reader with the necessary tools to engage directly with Frege's formal system, rather than translating his notations and derivations into modern formal systems—a practice that is likely to yield anachronistic and distorted results. Hence, I will conclude this essay with a call to arms: Now that you are prepared with the tools provided above, dive into *Grundgesetze* and see what new insights and quirks of Frege's formal treatment of logic and mathematics you can discover. I discovered much that was new to me in preparing this essay, but it is clear that much more work remains to be done![25]

References

Heck, Jr., Richard G. (2011). *Frege's Theorem.* Oxford: Clarendon Press.

——— (2012). *Reading Frege's* Grundgesetze. Oxford: Clarendon Press.

Landini, Gregory (2012). *Frege's Notations: What They Are and How They Mean.* New York: Palgrave Macmillan.

Simons, Peter (1987). 'Frege's Theory of Real Numbers', *History and Philosophy of Logic* 8:25–44.

[24] The term "Archimedean condition" is not Frege's—he does not provide a shorthand for this Relation in Appendix 1 of Volume II (but see the discussion of this notion in II §199).

[25] Thanks are due to everyone who has been involved in *Grundgesetze* translation project over the past eight years. Special thanks, however, are due to Philip Ebert, Richard Heck, Marcus Rossberg, and Crispin Wright for advice and guidance that proved invaluable during the writing of this appendix.

Index

actual **I** XVIII–XIX, XXII–XXV, **II** 86, 127, 131, 134.
addition **I** V, 39fn, **II** 108–110, 118–119, 123, 127, 153–154, 158, 160–163.
addition-sign **II** 77–78, 160.
aggregate **I** IX, **II** 150–152.
algebraic **I** XIX, **II** 79, 155fn.
all **I** 24.
and **I** 21–23, 25, 56.
arbiter **I** XIX.
Archimedean axiom **II** 191, 251.
argument **I** 6–8, 10–13, 16–21, 23–24, 32, 35–57, 60–61, 63–64, **II** 75, 77–78, 147–148, 257–258, 260–261, 264.
  first kind **I** 40–41.
  second kind **I** 40–42, 44, 46–48, 63, **II** 257–258, 260.
  third kind **I** 40–41, 44.
argument place **I** 6–8, 10, 13–15, 35–37, 39–52, 63–64, **II** 79, 148fn, 258.
  complete **I** 6, 8, 148fn.
  first kind **I** 40–41, 43, 47.
  proxy for **I** 37–38, 56.
  related **I** 8, 13, 15, 47, 52, 63.
  second kind **I** 40–41.
  third kind **I** 40–41, 43–44.
argument-name **II** 75.
arithmetic **I** V, VII–IX, XIII, 1, 3–4, 39fn, **II** 70fn, 73, 77–78, 83–88, 95, 97–104, 109, 117, 119, 123–125, 139, 141, 146–147, 149, 151–152, 154–157, 162, 168, 253, 255, 257, 265.
arithmetician **I** 142.
  formal **II** 101, 105, 127, 131, 154.
article, definite **I** IX, 19–20, **II** 44, 75, 107, 122, 129fn, 145fn, 169.
assign **I** 20, **II** 88–92, 95–96, 129–130, 136–138, 261.
associative law **II** 163, 166, 245.
associative principle **II** 163.
astronomy **I** XXIV, **II** 84.
axiom **I** VII, 1, 88, **II** 98, 101, 146, 156.

Barbara **I** 31.
basic law **I** VI–VII, XII, XVII, 60–61, 70, 239, **II** 80, 146, 253, 262–263.
basic law of logic **II** 147–149.
Baumann, Johannes Julius **II** 85.
Begriffsschrift **I** VI, VIII, IX–X, 5fn, 22, 26, 59.
being **I** XVIII, XX, XXII, XXIV, **II** 128, 132, 134.
belongs to a series **I** 60, 140–141, 149, 167, 169, 179, 181, 191, 201, 224, 230, 237, 241, 245–247, 250–251, **II** 2, 9, 11.
Biermann, Otto **II** 149–151.
blackboard **II** 86, 97, 105, 110, 115, 118, 127–128, 136.
Boole, George **I** 14.
bound, upper **II** 134, 170.
boundary (of a concept) **I** 9fn, **II** 69–78, 158.
bracket **I** 10–11, 34, 35, 64, **II** 129–130, 135, 168, 129–130, 135.

Cantor, Georg **II** 80–83, 85–92, 94–98, 127, 128, 136, 142, 147, 149.
cardinal number **I** V, VIII–X, 1, 3, 14, 57–60, 70, 87–88, 90, 102, 113–114, 127–129, 131–133, 137, 140, 144–145, 149–150, 154, 160, 162, 165, 179, 201, 224, 241, 243–245, 247–248, 250–251, **II** 1, 7, 25, 30, 37, 44, 58, 68, 86, 125, 132, 149–153, 155–156, 161–162, 255, 264.
cardinal number series **I** 58–60, 87, 113, 129, 132, 137, 140–141, 144–145, 149–150, 165, 167, 179, 243–245, 247–248, 250–251.
  its infinity **I** 144, 149.
Cartesian coordinates **II** 72.
chain of inferences **I** VI–VIII, X, 26.
chalk **II** 86, 97, 110, 115, 127–128, 136.
characteristic mark **I** XIV, XXIV, 3, 24, **II** 150.
Charlemagne **I** XXI.
chess **I** XIII, **II** 83fn, 97, 99–104, 107, 109–110, 113–117, 119–121, 123.
cite **I** 3, 25, 45, 62–67, 70, 74 **II** 89, 115, 119, 127, 202.
class **I** 2, 17, **II** 107, 121, 124–125, 141, 143, 148, 159–163, 168–171, 187, 189–190, 192, 243–244, 253–256.
  empty **I** 2.
class of finite cardinal numbers **II** 161–162.
co-referential, have the same reference **I** 7, 16–18, 45, 49, 51–52, 56, 179, **II** 108, 111, 114, 118, 120, 151, 257, 261.
cognition **I** VII, XXIV.
coincidence **I** IX, **II** 70fn, 92, 95, 107, 137, 148, 207, 253, 262.
collection **II** 130–131, 133–134, 139, *see also* set.
commutative law **II** 166, 168, 211–212, 239, 243, 252.
commutative principle **II** 204, 207, 209.
composed relation **I** 72, 77, 242–243.
composed Relation **II** 169, 171, 181, 185.
concavity **I** 13–15, 31–33, 35, 41, 45, 50, 61–63.
concept **I** V, VII, IX–X, XIfn, XII, XIV, XVIII, XXIV–XXV, 1–3, 5, 7–10, 14–15, 18–20, 24, 36, 38–40, 53, 56–60, 68–72, 74–75, 78, 82, 87–90, 94, 102, 110, 114, 117, 127–128, 131, 133, 137, 144, 149–150, 154, 157, 160, 162, 165, 167, 169, 171, 179, 201, 224, 240–241, 243–244, 248, 250–251, 253–255, 257–258, 260–262, 264, **II** 1–3, 5, 7, 9–10, 14, 25–26, 30, 34, 37–38, 44–45, 50, 58, 61, 68–70, 72, 74–78, 82, 107–108, 110, 125, 127, 148, 150, 153–155, 157–161, 169, 171, 189, 253–254, 256, 258, 260–264.
  empty **I** 3, 24fn, **II** 150, 154, 263–264.
  first-level **I** X, XXIV–XXV, 38, **II** 260.
  second-level **I** X, XXIV–XXV, 38, 40.
concept-name **II** 75–77.
concept-script **I** V, IX, XI–XII, 3, 5–6, 9, 12, 16, 25, 41–43, 89, 123, 240fn, 242, **II** 69, 79, 99, 144, 259.
concept-script proposition **I** XII, 9, 44, 50–51, 242fn, **II** 69, 76, 105, 112, *see also* proposition.
  correctly formed **I** 50.
conditional-stroke **I** 20, 23.
conduct of proof **I** VI, 1, 3, 25, 54, 60, 106, **II** 73, 141, 143–144, 146, 209, 260, 265.
congruence **II** 142.
conic section **I** 72.
conic surface **I** 72.
consequence **I** VI–VII, X, XII, XIX, 1, 25, 70, 129, 133, 141, 144–145, 203, **II** 75, 77, 103, 115, 118, 120, 139, 155fn, 166, 168–169,

172, 181, 183, 187, 189–190, 192, 264–265.
content **I** VIII, X, XVII, 9fn, 242fn, **II** 70fn, 72, 78fn, 84, 88, 92, 97, 100, 103, 110, 112, 123, 126, 212, 255.
content-stroke **I** X, 9fn.
contentual arithmetic **II** 98, 100, 107–120, 122–123, 125–127, 131–132, 134–139, 154.
contradiction **I** XV, 18fn, **II** 70, 73, 91, 121, 123–124, 127, 145–147, 151, 154–155, 253–254, 256–257, 262–263, 265.
 free from **II** 119, 123–125, 145, 154–155.
contraposition **I** 27–28, 30, 61, 71–73, 78, 87, 89, 94, 102–103, 108, 114, 139, 141, **II** 44, 233.
converse of a relation **I** 57, 70–72, 87, 89–90, 102, 114, 134, 137, 149, 160, 163, 179, 224, 227, 241, 243, 245, 247, **II** 44–45, 156, 160–161.
converse of a Relation **II** 160, 163, 168–169, 171, 180, 185, 209.
correlate **I** 56–57, 71, 88, 137, 179.
coupled relation **I** V, 179, 241, 249–50.
cow, abstract **II** 152.
creation **I** 88–89, **II** 107, 140–142, 145–149, 153.
criterion **I** 16–18, **II** 126, 145–146, 157, 262.
cube root **I** 25, 39.
curve **I** XIX, 5, **II** 72, 155fn.

declarative sentence **I** X, **II** 126.
Dedekind, Richard **I** VII–VIII, XI, 1–3, **II** 140, 141, 147, 253.
Deïanira **I** XXI.
definite article *see* article, definite.
definition **I** VI–VII, IX–X, XIII–XIV, 4, 7, 9fn, 11, 14, 18–19, 39–40, 43–45, 51–58, 63, 70, 72–73, 78, 81, 87, 92, 113, 128–129, 131, 141, 150, 162, 171, 179, 181, 203, 224, 240, **II** 2, 48, 69–70, 72–75, 77–82, 85–87, 89–95, 97, 99, 101–102, 111–112, 122, 128–130, 133–134, 137, 141, 144–145, 147–149, 153–155, 157–158, 160–162, 166, 168–169, 171–172, 187, 189–190, 207, 244, 262,
 creative **I** VI, XIII, **II** 140–142, 144–149, 153, 155.
 double stroke of **I** 44, **II** 144–145, 155.
 piecemeal **II** 70–71, 74–75, 78, 93, 95, 156fn.
definition-stroke **I** 45, 63.
derivation **I** V, XII, 65–69, 73, 89, **II** 1, 17, 256–257, 261, 263–264.
determine **I** IX, XX, 17, 1, 13, 15–18, 20, 50–51, 53, 59, 137, 179, **II** 44, 69–70, 72, 74–77, 79–80, 83, 86–92, 94–95, 97, 101, 111–112, 124, 126, 133, 138, 140, 143, 145fn, 150, 157–159, 264.
designation **I** IX, XIII, XVII, XIX–XXII, 1, 3, 5–9, 13, 16, 19–20, 52, 57, 60, 137, 150, **II** 75–78, 80–86, 89, 96, 99, 103, 105–112, 123, 127, 131, 134–138, 143–147, 150, 152, 155fn, 158fn, 161, 256.
distance **II** 87, 160.
 determinate **II** 86–88, 95.
division **II** 108, 110, 119, 123–124.
division-colon **II** 121.
domain (Σ-domain) **I** 7, 16, **II** 70–71, 73–74, 78, 155, 158–163, 168–171, 176, 180–181, 185, 194, 209, 239, 243, 252.
domain of magnitudes *see* magnitude, domain of.
dragon gullet **II** 76.

Eight-figure **II** 127.
element **I** 1–3, **II** 150–151, 153, *see also* member.
elucidation **I** XXIV, 5fn, 6fn, 53–54, **II** 87, 109, 153.

ending **I** 60, 144–145, 149, 167, 201, 230, 246–247, 250, **II** 8, 14.
Endlos **I** V, 150, 154, 160, 179, 241, 248, 250, **II** 1, 7, 25, 68, 161.
equality **I** 11, 14, 17, 36, 69, 87, 90, 237, **II** 70fn, 72, 75, 80, 92, 96, 115, 137, 140, 147–149, 151, 154–155, 158fn, 160, 163, 254, 257, 261–262.
equality-sign **I** IX, 11, 34, 45, 64, 66, **II** 70fn, 80, 82, 92, 107, 113, 116, 130, 136–138, 140, 147–148, 151, 153–154.
equation **I** IX, 5, 7, 9, 11–12, 18–19, 44–45, **II** 72, 75fn, 77, 79, 82, 100, 104, 106, 115–116, 124, 137, 144–145, 148fn, 153.
  definitional **I** 51–52, 73, **II** 79, 207.
equinumerous **I** 3, 56–57.
Erdmann, Benno **I** XIV–XVII, XIX–XXIII, XXV.
Euclid **I** VI, 88, **II** 144.
Euenus **I** XXI.
evening star **II** 140.
every **I** 24.
existence **I** XIII, XIIIfn, XXV, **II** 97, 128, 153fn.
experience **I** XXII, 1, 4, **II** 139fn.
explanation **I** X, XII, 8, 12, 15, 23, 25, 31fn, 36, 38, 41–48, 51–53, 58–59, 66, 201, **II** 70–74, 77–81, 86–88, 91–95, 99, 102, 110, 129–130, 136, 138–140, 148, 150–151, 153–154, 156fn, 157–158, 169, 171, 255.
express **I** XI, XIII, XVII, XXIII, XXV, 6–7, 12, 31–32, 34–35, 39, 42, 50–51, 56, 58–60, 72, 90, 128, 131, 160, 201, 224, **II** 1–2, 44, 69, 78, 81, 87, 95, 100–101, 104, 107, 111–116, 121, 125–126, 136, 140, 142, 163, 192, 209, 256.
expression **I** VIII, XI–XII, XXII–XXIII, 3, 5, 8–9, 11, 13–14, 24–25, 28, 31, 37, 42, 46, 53, 56, 94, 113, 179, 224, **II** 44, 70fn, 72–77, 79, 81–82, 85–87, 89fn, 91, 94–96, 102, 104, 106–107, 110, 114–115, 130, 132, 139, 147–148, 150–151, 153, 155fn, 169, 261.
extension of a concept **I** VII, X, 7–8, 14–15, 18, 57, 60, 71, 87, 150, 201, 240, **II** 69, 107, 148, 150, 158–161, 253–257, 260–264, *see also* class.
  of a relation **I** 55–60, 71–72, 179, 201, **II** 160–161, *see also* Relation.

faculty, logical **II** 86.
fall under **I** V, XIV, 3, 8, 19, 36, 38, 53, 56–59, 71–72, 87–90, 117, 127–128, 131, 133, 137, 149, 154, 160, 162, 165, 167, 169, 171, 179, 201, 224, 241, 243–244, 248, 250–251, **II** 1–3, 5, 7, 9–10, 14, 25–26, 30, 37–38, 44, 50, 58, 61, 69, 76, 110, 150, 153, 155, 158, 253–254, 258, 260–264.
False, the **I** X, 7, 9–12, 17–18, 20, 22–27, 30, 32, 34, 36, 38, 40–42, 48–49, 53, 55, **II** 263–264.
fiction **I** XXI, 2–3, **II** 76, 130.
figure **I** XIII, **II** 72, 83–84, 96, 98–105, 107–111, 113–147, 154, 156, *see also* number-figure.
  actual **II** 134.
  arithmetical **II** 103.
  contentless **II** 156.
  egg-shaped **I** XIII.
  equal-shaped **II** 107, 118.
  group of **II** 124.
  sensory **II** 128.
fill in **I** 6.
finite **I** V, 60, 137, 144–145, 149–150, 154, 160–162, 165, 201, 224, 247–248, 251, **II** 1, 30, 37, 44, 58, 68, 129, 131, 133–134, 139, 161–162, 264.
first (in a series) **II** 2–3, 5, 7, 9, 26, 38.
fitting **I** 41–42, 45, 48.

Five-figure **II** 129, 136–137.
fix **II** 94.
follows (in a series) **I** 59–60, 137, 140–141, 150, 162, 171, 227–228, 245–249, **II** 7, 14, 26.
  immediately in the cardinal number series **I** 58–60, 87, 132, 140, 144–145, 244, 247–248.
formal arithmetic **II** 98–105, 107–110, 112–123, 125–129, 132, 134–135, 138–139, 141, 154–155.
function **I** IX–X, 5–21, 23, 27, 32–44, 47, 49, 51–61, 63–66, **II** 70fn, 75–76, 78–79, 106fn, 148–149, 168, 187, 189, 254–255, 257, 262–264.
  corresponding **I** 12–13, 15–16, 35, 41, 63.
  first-level **I** 37–43, 45–55, 63–64, **II** 75, 147, 254–255.
  second-level **I** X, 37–42, 44, 46–49, 52, 54–55, **II** 257–258, 260–262.
  third-level **I** 41, 46–48.
  unequal-levelled **I** 39.
  with one argument **I** 8, 24, 35–51, 53–54, 63–64, **II** 75, 147, 257–258, 260.
  with two arguments **I** 8, 20, 35–37, 39–41, 43, 46–48, 52–56, 63–64, **II** 75, 77.
function-letter **I** 34–36, 41–42, 44–45, 62–63, 66–67, 127, **II** 148fn.
function-marker **I** 33, 44, 66–67, 74, 138, 141, 144, 147, 154, 180, 191, 207, 225, **II** 258.
function-name **I** 6, 9, 25, 32–33, 35–37, 43–44, 46–52, 54, 57, 63, 66–67, **II** 78, 262.
function-sign **I** 56, 64, 179, **II** 148fn.
Function und Begriff **I** X, 5, 36, **II** 75.
fundamental series **II** 80–83, 85–92, 94–96, 128.
fusion (of equal subcomponents) **I** VI, 29–31, 61, 85.
fusion (of horizontals) **I** 10, 14, 20, 23, 61, 64–67.

game **I** XIII–XIV, **II** 97–105, 108–109, 113–116, 120–122, 126–127, 132, 134, 137–139, 147.
  calculating **II** 97, 102–103, 105, 109, 115, 117–122, 126, 132–133, 139.
gaplessness **I** XII–VIII, X, XII.
Gauss, Carl Friedrich **II** 159.
general certainty **I** XV.
general validity **I** XV.
generality **I** 4, 11–12, 14, 18–19, 31, 34, 36, 42, 197, **II** 77, 80, 110, 114, 136, 147–149, 155, 255, 257.
genus, highest **I** XXI, XXIII.
geometry **I** 88–89, 123, **II** 70fn, 84, 86, 88, 95, 141, 146, 155–157, 243.
German letter **I** 13–14, 31–35, 45, 50, 62–63, 87, 114, 194.
ginger biscuit **II** 150, 152.
grammar **II** XV–XVI, 79, **II** 254.
greater **I** 21, 23–24, **II** 74–75, 81–83, 90–96, 125–126, 132–133, 139, 158, 161, 168–171, 185–187, 189, 194, 196–197, 212.
Greek letter **I** 9fn, 15, 43, 45, 51, 63, 66, **II** 256fn, 258.
Greek vowel **I** 15, 34–35, 45, 63.
Grundlagen der Arithmetik, Die **I** VIII, IX–XI, 1, 3, 14, 56–60, 72, 150, **II** 152.

Hankel, Hermann **II** 140, 142–144, 147, 156, 158.
Heine, Heinrich Eduard **I** XIII, **II** 72, 73, 96–98,105–107, 110–113, 128, 129, 136, 141.
Helmholtz, Hermann von **I** XI, **II** 139, 140.
horizontal **I** X, 9–11, 14, 20–21, 23, 32–33, 35, 44, 61, 64–67.
houses, row of **II** 129–131.
Husserl, Edmund **I** 3.

idea **I** XIV, XVIII–XXIII, XXIV, XXV, **II** 83–84, 131, 150fn.
idealism **I** XIX, XXII.
ideation **I** XIX–XXIV.

identity **I** XVII, **II** 70fn, 75, 92, 137, 140, 147–148, 151, 254.
if **I** 24.
Illigens, Eberhard **II** 81–83, 85, 94.
in need of completion **I** 5–6, 8, **II** 148fn.
indicate **I** 8, 31–32, 34–37, 42–44, 52, 63, 71–72, 93, 107, 127, 179, **II** 7, 78, 80, 136–137, 145fn.
  indeterminately **I** 5, 11.
indirect speech **I** X.
induction **I** XVI.
inferences **I** VI–VIII, X, XII, 1, 25–29, 31, 60, 62, 67, **II** 99–100, 114–115.
infinite **I** Vfn, 6, 88, 144, 150, **II** 44, 70fn, 96, 127–131, 134–136, 138, 154, 160–161, 254.
insight **I** VII–VIII, 1, 3, **II** 84–85, 116.
instantiated **I** XIV, 24fn.
iridesce **II** 93, 106.

judgement **I** XII, XVI–XVII, XX, XXII–XXIV, 1, 9, 50, **II** 93.
judgement-stroke **I** 9, 26, 31–33, 35, 44–45, 62, 64, 94, **II** 256.
Jupiter **II** 84, 105.

knowledge **II** 69, 85, 92, 98–99, 101, 111–113, 140, 157, 161, *see also* cognition *and* insight.
  intuitive **I** VII.
Kossak, Ernst **II** 149–151, 153.
Kronecker, Leopold **I** XI, **II** 155.

label **I** 25, 27, 34, 43, 62–63, 65–66.
law **I** VI, VII–VIII, XV–XVII, 1, 34–35, 42, 60, 63, 65, 70, **II** 69, 76, 82, 84, 99, 103, 117, 141–143, 146, 163, 166, 168, 211–212, 239, 243, 245, 252, 257, 262,
  logical **I** VII, XIV–XVII, 14, **II** 75, 78, 115, 253.
  of excluded middle **II** 69, 254.
  psychological **I** XV–XVI, XVIII.
law of thought **I** XV–XVI.
Leibniz, Gottfried Wilhelm **I** 14.
level **I** X, XXIV–XXV, 37–55, 63–64, 66, **II** 75, 77, 147, 254–255, 257–258, 260–262.
limit (Σ-limit of Φ) **I** 6, **II** 80, 82, 92, 95, 97, 117, 123, 126, 144, 154, 168, 171, 187, 189–190, 192, 196–197, 199, 211, 245.
  upper **II** 171, 187.
line **I** XIII, 54, 61, 88, **II** 83, 86–87, 142, 144, 153, 155–156, 160, 172.
  auxiliary **I** 88, 123, **II** 141.
logic **I** VII–VIII, XIV–XV, XIX, XXIII–XXVI, 1, 3, 7 14, 31, **II** 69, 70–71, 74, 78, 97, 147–149.
  psychological **I** XIV–XV, XXIII–XXV.
logical faculty **II** 86.
logical object **II** 86, 149, 153, 253, 265.
logician **I** VII, XIV, XXII, XXIV–XXV, 24fn, **II** 110, 148.
  psychological **I** XV, XVI–XVIII, XIX, XXV–XXVI.
looping **I** 179, 201, 248, 250, **II** 1, 129.

madness, hitherto unknown kind **I** XVI.
magnitude **II** 81, 84–88, 92, 94, 125–126, 145, 151–153, 155–163, 168, 170, 194, 209, 211–212, 239, 243.
  domain of **II** 158, 160.
  null **II** 169, 171, 209.
  numerical **II** 86–88, 151–153.
magnitude-ratio **II** 85, 88, 95, 97–98, 101, 138, 155–157.
manifold **II** 98, 148.
map **I** V, IX, 57, 70–72, 87–90, 102, 114, 117, 134, 137, 149, 160, 162, 165, 169, 179, 224, 241, 243, **II** 44–45, 161.
marker *see* function-marker *and* object-marker.
mathematician **I** VIII, XI–XII, XIVfn, XXV–XXVI, 1, 3, 88, **II** 69–70, 73, 75fn, 80, 88fn, 98, 106, 118, 139, 145–149, 155, 157.
mathematics **I** V–VII, XIX, 3, 5, **II** 70–71, 74–75, 80, 105, 137, 141, 149, 152.

measuring number *see* number, measuring.
member **I** IX, 58–59, 137, 167, 179, 247–249, 251, **II** 2–3, 5, 7, 9, 14, 25, 30, 34, 38, 81, 87, 91, 94, 96, 128, 159, 161, 166, 168, 192, 196–197, *see also* element.
  occurring as first, second member of a relation **I** 160, 179, 249, **II** 159, 171.
milk, abstract **II** 152.
Mill, John Stuart **I** XVIII.
mind **I** XIII, XV, XVIII, XX, XXIV, 2, 88, 128, **II** 92, 100.
minus-bar **II** 122.
minus-sign **I** 9, **II** 109, 113, 160, 168.
Moon **I** XVIII–XIX, XXI, **II** 74–76, 109, 151–152.
morning star **II** 140.
multiplication **I** V, 7, 11, 39fn, **II** 108–110, 119, 123, 142–143, 150, 158.

name **I** VI, XIII, XIV–XXV, XVII, XIX, XXI, 4, 6–7, 9, 11, 13–17, 19–21, 23, 32–33, 35–37, 39–41, 43–54, 56, 62–65, **II** 64, 75–79, 82, 85fn, 89, 94, 103, 107–109, 117–119, 127, 130–131, 134–137, 139, 143–144, 149, 151, 163, 169, 254–255, *see also* function-name *and* value-range-name.
  complex **I** 43, 47, 51.
  correctly formed **I** 39fn, 45, 48, 50–52.
  primitive **I** 48, 50–51.
  simple **I** 47–48, 51, 58, 60, 64.
negation **I** 11–12, 24–25, 31.
negation-stroke **I** 10, 23–25, 27, 30, 61–62.
neither–nor **I** 21.
Nessus **I** XXI.
Newton, Isaac **II** 155.
Nine-figure **II** 132.
no **I** 24.
notation **I** VI, 5, 9fn, 11, 15, 23, 34, 36, 42, 55, 201, **II** 70–71, 79, 81, 130, 139, 148fn, 156, 158fn, 166, 187, 192, 257, 260.
null Relation *see* Relation, null.
number **I** V, VIII, IX, XIfn, XIII–XIV, XVIII, XIX, XXV, 5–7, 10–11, 14, 24, 38–40, 54, 56, 58, **II** 44, 69–70, 74, 76–78, 80–86, 88–95, 97–99, 101–105, 107–112, 117, 119, 122–123, 125–128, 130–131, 133–134, 136–143, 146–157, 159–161, 253–255, 265.
  alternating **II** 144.
  auxiliary **II** 141–142.
  cardinal *see* cardinal number.
  complex **I** V, 6, **II** 70fn, 151.
  imaginary **II** 70fn.
  irrational **I** V, 14, 19, **II** 69–70, 80, 86, 93, 96–97, 128, 134, 139, 141, 149fn, 153–156, 161–162, 168, 243.
  measuring **II** 70fn, 155–157, 159–160.
  negative **I** V, 14, 19fn, **II** 70, 74, 81, 110, 112, 125–127, 153, 156, 159, 161, 243.
  positive **II** 70, 74, 90, 97, 110, 126–127, 133, 144, 154–156, 159, 161, 168.
  rational **I** V, **II** 80–84, 87, 89, 91–95, 97, 125–127, 129, 135, 137, 140–141, 153, 161, 168.
  real **I** 23, **II** 69, 77, 85, 87–88, 142, 155–157, 159–162, 243.
  statement of **I** IX, 3.
  whole **II** 70, 74, 77, 108, 110, 155–156, 159, 161.
number-figure **II** 103–105, 110, 114, 116–119, 122, 124, 126–135, 138–139, *see also* figure.
  actual **II** 131.
number-sequence **II** 96, 128–130, 132, 134–136.
number-series **I** XIX, 58–60, 87, 113, 129, 132, 137, 140–141, 144–145, 149–150, 165, 167, 179,

243–245, 247–248, 251, **II** 72–73, 82, 85, 90, 94, 96, 128–129.
number-sign **I** XIII, 4–6, 137, **II** 72, 80, 82–85, 95–99, 102–105, 108–109, 111–112, 123, 125–126, 128, 131, 136, 140, 147, 151–152, 154–156, 255.
numerical magnitude *see* magnitude, numerical.

object **I** V, IX–X, XIII–XV, XVII–XXV, 2–4, 7–11, 15–20, 23, 31–32, 34–41, 43–44, 48, 50, 52–60, 71–72, 87–90, 93, 107, 114, 117, 127–128, 131, 133, 137, 141, 149, 160, 162–163, 165, 167, 169, 171, 179, 181, 191, 201, 224, 227, 230, 237, 240–248, 250–251, **II** 1–2, 7, 26, 37–38, 44, 58, 69–70, 72, 74–78, 83–84, 88, 93, 98, 105, 107, 110–111, 125, 133, 137, 140–143, 145, 147–150, 152–161, 171, 244, 253–255, 260–265.
  abstract **II** 86.
  auxiliary **I** 123, **II** 141–142.
  logical **II** 86, 149, 153, 253, 265.
  non-sensory **II** 147.
  physical **II** 86, 105, 147, 153.
  sensory **II** 125.
object of thought, abstract **II** 86, 88–89, 95–96.
object-letter **I** 34–35, 41–42, 44, 51–52, 62–63.
object-marker **I** 33, 44–45, 52, 62.
objective **I** XIV, XVII–XVIII, XIX, XXII, 2, **II** 86.
  non-actual **I** XVIII, XIX.
One **I** XIV, XVIII, 58, 131–133, 137, 241, 244, 247, 251, **II** 75–76, 95, 107–108, 151–152.
One-figure **II** 108, 117–118, 124, 132–133.
one-sided **I** 64.
or **I** 21.
ordering **I** 26, **II** 2, 7, 129.
  in a series **I** 160, 179, 201, 224, 248, 250–251, **II** 1, 37, 125–126, 159–160.
ordinary language **I** 24–25, **II** 69, 79, 116.
ouch **I** XIX.

pair **I** 89, 179, 181, 191, 241, 248–250, **II** 72, 161–162.
particular **I** 24.
Peano, Guiseppe **II** 70, 71, 77.
peep box, psychological **I** XXV.
permutability **I** VI, 22, 26–28, 61, 65, 73.
physics **II** 84.
plane **I** 54, **II** 69, 72, 155fn.
plus-sign **I** 19, 38, **II** 78, 113, 151, 153–154, 156fn.
point **I** 88, **II** 72, 87, 142, 144, 155fn, 160.
  auxiliary **II** 141.
  determinate **II** 87.
positival class **II** 168, 171, 176, 180–181, 185–187, 189–190, 209, 244.
positive class **II** 169–171, 189–192, 204, 230, 233, 239, 243, 245, 252.
postulate **I** 88, **II** 159–160.
precede in a series **I** 59–60, 113, 129, 191, 244–245, 247, **II** 2, 14, 33.
predicate **I** XIX–XXIV, **II** 69–70.
predication **I** XX–XXII, XXIV, 3, 36, 66, **II** 44, 123, 136, 148.
Pringsheim, Alfred **II** 83, 84.
proceeds endlessly **I** 149, 160, 173, 179–180, 248, 250, **II** 1.
product **I** 5, 9fn, 62, **II** 90–93, 95–96, 143.
proof **I** V–VIII, XII, 1, 3, 25, 54, 60, 70, 73, 80–81, 86, 88, 94, 97, 101, 106, 109, 113–114, 121, 127–128, 131, 137, 139, 141, 144, 150, 154, 160, 162, 178, 180, 201–202, 217, 224–225, **II** 1, 7, 11, 25, 30, 37, 44, 48, 50, 52, 58, 61, 66, 73, 75fn, 78fn, 101, 116, 124–125, 138, 141–

146, 155, 176, 180–181, 191, 204, 207, 209, 211, 223, 230, 239, 253, 260, 265.
proper name **I** XIV, 7, 11, 19–21, 32–33, 35–37, 39–41, 43–53, 62, **II** 76–78, 107–108, 127, 130–131, 134–137, 143–144, 151, 254–255.
property **I** IX, XIII–XIV, XVIII–XIX, XXIV, 3, 16, 18fn, 38, 43, 89, 93, 162, 171, **II** 45, 70fn, 93, 97–99, 102–103, 125–127, 141–143, 145, 147, 150, 152, 154, 158–159, 162–163, 169–170, 189.
proposition **I** X, XVI–XVIII, XXI–XXII, 9, 21, 24–26, 29–33, 35, 42, 44–45, 50, 56, 61–63.
  concept-script **I** XI, 9, 25, 44, 50–51, 242fn, **II** 69, 76, 105, 112.
  general **II** 77–78, 115, 140, 147, 255.
  singular **II** 254.
pseudo-concept **II** 74.
pseudo-relation **II** 74.
psychology **I** XIV–XVI, XVIII, XIX–XXI, XXIII–XXV, 88, **II** 139fn, 150fn.

quantity **II** 82–85, 94, 157.
quotation mark **I** 4, 144fn, **II** 106, 146.
quotient **I** 6, 39, **II** 91, 119–121.

ratio **II** 85, 87–88, 97, 155–157, 160, 162, 168, 243.
  irrational **II** 87.
reference **I** IX–X, XII, XVI, XVIII, XXII, 4–5, 7, 9–12, 13–20, 22–23, 31–32, 34–35, 39fn, 41, 43, 45–54, 57, 179, **II** 70, 72–80, 82–85, 92–95, 98–110, 112–114, 116–123, 126–127, 130–131, 133, 136–140, 145–147, 151–152, 154, 156–158, 255.
  determinate **I** 11.
  formal **II** 104, 118–119.
  without **I** 9fn, 13, 19, 47, **II** 130.
referential **I** 46–52, 179, **II** 77, 134, 143.
regular *see* value-range name, regular.
related *see* argument place, related.
relation **I** V, IX–X, XII, XX, XXII, XXV, 3, 5, 8, 10, 20, 39–40, 55–60, 70–73, 77, 87–90, 93, 102, 114, 134, 137, 149, 160, 162–163, 165, 167, 169, 171, 179, 181, 201, 224, 227, 230, 237, 240–243, 245–248, 250–251, **II** 7, 44–45, 74–75, 77, 89–90, 116, 126, 131, 133, 142–143, 150, 156–158, 160–162, 168–169, 171, 254.
  equal-levelled **I** 39.
  series-forming **I** 141, 160, 179, 191, 224, 237, 245–247, 249, **II** 128.
  single-valued **I** V, IX, 3, 39–40, 55–57, 59, 71, 78, 89, 114, 144, 160, 163, 167, 191, 201, 227, 230, 240, 242–243, 247, 249–250, **II** 7, 119, 122.
  unequal-levelled **I** 39.
Relation **II** 159–163, 166, 168–169, 171, 176, 180–181, 185–187, 190–192, 196, 209, 230, 233, 243–245, 255, *see also* extension of a relation *and* value-range, double.
  null **II** 176, 180–181, 185.
  single-valued **II** 171.
replacement **I** 44, 47, 62–63, 213, **II** 111, 116, 120–122, 153.
retina **I** XXIII.
reverse **I** 27.
rigour **I** VII, XII, 88, **II** 73.
Roman letter **I** 31–35, 42, 44, 52–53, 62–63, 94, **II** 78–79.
rule **I** VI–VII, XIII, 13–15, 27, 29–30, 32–36, 41, 43, 61–67, 69–74, 76, 78, 84, 94–95, 101, 106, 124, **II** 79, 83fn, 91, 97–105, 107–127, 132, 138–139, 141–142, 146–147, 154, 156.
  permissive **II** 117, 120–121, 123–124, 127, 139, 154.

prohibitive **II** 120–125, 139, 154.
run in a series **I** 201–202, 224, 237, 241, 250–251, **II** 2, 5, 9, 11, 14, 30, 37–38.
Russell, Bertrand **II** 253.

saturated **I** 7–8, 37, **II** 74–75.
Schröder, Ernst **I** XIII, XXVII, 2–3.
scope **I** 11, 13–15, 31, 33–35, 45, 62–63, 95, **II** 258.
Scylla **II** 76.
sense **I** IXfn, X, XVI, XVIII, XXI–XXII, 7, 16fn, 20, 25fn, 45, 50–51, 59, 71, **II** 69–70, 78, 80–83, 87, 90, 93–95, 99–101, 104, 111–116, 121, 126, 129, 140, 153, 158, 261.
series **I** V, IX, 6, 59–60, 137, 140–141, 160, 162, 167, 169, 171, 173, 179–181, 191, 201–202, 207, 224, 227–228, 230, 237, 241, 245–251, **II** 1–3, 5, 7–9, 11, 14, 25–26, 30, 33, 37–38, 80–82, 87–88, 92, 95–96, 110, 125–126, 128–129, 134–135, 150, 159–160, 166, 168–169, 192–193, 196–197, 244.
  non-branching **I** 160, 179, 201, 248, 250.
  simple **I** 201, 224, **II** 1.
set **I** Vfn, 1, 3, **II** 80, 148, 150–151, 253, *see also* collection.
shape **II** 105–107, 114–118, 120–122, 124, 127, 138.
shimmer **II** 134, 138.
sign **I** V–VI, IX–XIII, 1, 4–11, 13–16, 23, 29, 32–33, 36–39, 41, 43–45, 50, 53, 60–62, 64, 67, 129, 141, 179, **II** 70–73, 77–86, 88–89, 92–109, 111–114, 116–118, 120, 125–129, 132–140, 144–148, 152, 154, 156–157, 160, 162, 168, 189, 255–258.
  complex **I** 35, 43.
  correctly formed **I** XII.
  equal-shaped **II** 107, 114, 138.
  primitive **I** IX–X, 5, 42, 45, 148fn.
  simple **I** 14, 35, 43, 51, 179, **II** 79.
single-valuedness *see* relation, single-valued.
smooth breathing **I** IX, 15, 35, 45, 63.
solipsism **I** XIX.
some **I** 24–25.
specify **I** 10–11, 64, **II** 70fn, 92.
square root **I** XXV, 7, 12, 15, 19–20, 24, 31, 39, **II** 69, 74, 123fn, 141–142, 157.
stand in **I** 45, 64, 183.
stand in for **II** 136, 255.
starting with **I** 60, 137, 140–141, 149, 167, 169, 179–181, 191, 201, 207, 230, 241, 245–247, 250–251, **II** 2–3, 5, 7, 9, 11, 25–26, 30, 129, 192–193, 196–197, 244.
statement of number **I** IX, 3.
stipulate **I** XII, 6fn, 9–11, 14, 17–20, 31, 49–50, 52–53, 64, **II** 70fn, 74–76, 78–79, 82–83, 90–91, 99, 101, 112, 148, 187, 263–264.
Stolz, Otto **I** XI, **II** 140, 144, 146, 147, 158.
subcomponent **I** VI, 22, 26–31, 33–35, 61–62, 65, 67, 73, 85, 91, 94–96, 107, 114, 117, 123–124, 144–145, 183, 191, 197, 202–203, 207, 213, 222, 225, 227–228, 230, 237, **II** 5, 14, 17, 22, 25, 33, 38, 41, 50, 54, 58, 163, 197, 214–215, 263, 265.
subjective **I** XIV, XVIII–XIX, XXIV, 2, 88.
subordinate **I** 2, 20, 24, 224, **II** 37, 58, 68, 160–161.
subtraction **I** 11, **II** 108–110, 112, 118–119, 122–123.
sum **I** 5, 9, 21, **II** 44, 76–77, 90–93, 95–96, 109, 114, 143, 154, 158, 171.
Sun **II** 109.
supercomponent **I** 22, 27–31, 33, 35, 61–62, 65, 91, 121, 145, 154, 183, 230, **II** 30, 50, 52, 64, 212, 215, 227.

superordinate **I** V, 24, **II** 1, 37.
system **I** VIII, IX, **II** 79, 97, 116, 139fn, 145, 151–152.
  Dedekind **I** 1–3, **II** 141, 253fn.

taking to be true **I** XV–XVII.
theorem **I** V, VII, XI, XIII, 242, **II** 70, 72, 74–75, 91, 101–102, 115–116, 121, 144–146, 245.
theory **I** V, VIII, **II** 69, 75fn, 80–83, 86, 88, 94–98, 101–102, 104–108, 112–116, 120–121, 126–128, 133, 137, 139–140, 142, 146–147, 149–153, 155–156, 159, 163.
there is **I** XXV, 12, 24, **II** 133.
thing **I** VIII, XIII, XIV, XIX, 1–3, 88, 93, 99, **II** 82, 93, 106–107, 109, 126, 128, 130, 138, 144–147, 150, 157–159, 255, 263.
  abstract **II** 152.
  concrete **II** 152.
  non-sensory **II** 146.
  sensory **II** 125.
Thomae, Carl Johannes **II** 96–99, 102–105, 108–111, 113–138, 140, 146, 147.
thought **I** X–XI, XIII, XVI–XVII, 6–7, 9, 50–51, **II** 44, 70fn, 78fn, 84, 99–101, 104–105, 111–113, 126, 133, 139fn.
  determinate **II** 84.
Three-figure **II** 109, 117–118, 122, 126, 132, 136–137.
tools, auxiliary **II** 83–84, 154.
transformation **I** 27, 62, **II** 99, 153, 215, 223, 257.
transformation of generality **I** 14, **II** 80, 147–149, 155, 257.
transition-sign **I** 44, 61–62.
True, the **I** X, 7–14, 16–28, 30, 32, 34–35, 38–43, 48–50, 53, 55, **II** 169, 263.
truth-value **I** X, 7–10, 12–13, 17–18, 21–22, 24, 28–30, 34–36, 38–39, 44–45, 48–51, 55, 57–60, 69, 107, **II** 2, 140, 169, 171.
Two-figure **II** 109, 117–118, 120, 124, 126, 129, 132, 136–137.
two-sided **I** 64, 179.

Ueber Begriff und Gegenstand **I** 3, 5, 8.
Über die Zahlen des Herrn H. Schubert **II** 152.
Ueber formale Theorien der Arithmetik **I** 3.
Über Sinn und Bedeutung **I** IX–X, 7, 9.
uniquely determined **II** 79.
unit **II** 70fn, 85, 87, 142–143, 151–152, 155–157.
  imaginary **II** 144.
universal **I** 24.
unsaturated **I** 5–6, 8, 37–38, **II** 79, 148fn, 254.

validity **I** VI, XV–XVI, **II** 70fn, 74–75, 116, 123, 133.
value **II** 70fn, 81, 94, 106fn, 123, 140, 151–152.
  of a function **I** 6–13, 16–21, 23–24, 27, 32, 35–43, 49, 52–61, **II** 75, 77–78, 81, 94, 106fn, 147–148, 151–152, 254–255, 257–258, 260–261, 263–264.
value-range **I** VII, IX–X, 7–8, 14–18, 20, 36, 42, 49–50, 52–55, **II** 80, 147–149, 155, 254–255, 257–258, 262–264.
  double **I** 54–56, 73, **II** 168, *see also* Relation.
  represented **I** 42, 52, 54–55.
value-range name **I** 49–50.
  regular **I** 49–50.

Weierstrass, Karl **II** 149–154.

Zero **I** XIV, 3, 14, 58, 127–129, 131–132, 137, 224, 241, 244–245, 247, 251, **II** 70, 74, 81, 89–93, 110, 113, 117, 119–121, 124–127, 130–135, 137, 153, 156, 160.
Zero-figure **II** 119–121, 124, 126, 131, 135.
Zero-sequence **II** 130–135, 137.